Molecular Genetics and Plant Breeding 2.0

Molecular Genetics and Plant Breeding 2.0

Editors

Andrés J. Cortés
Hai Du

MDPI

Basel • Beijing • Wuhan • Barcelona • Belgrade • Novi Sad • Cluj • Manchester

Editors
Andrés J. Cortés
Corporación Colombiana de Investigación Agropecuaria (AGROSAVIA)
Universidad Nacional de Colombia - Sede Medellín
Antioquia
Colombia

Hai Du
College of Agronomy and Biotechnology, Chongqing Engineering Research Center for Rapeseed
Southwest University
Chongqing
China

Editorial Office
MDPI AG
Grosspeteranlage 5
4052 Basel, Switzerland

This is a reprint of articles from the Special Issue published online in the open access journal *International Journal of Molecular Sciences* (ISSN 1422-0067) (available at: www.mdpi.com/journal/ijms/special_issues/gene_plant_breading_2nd).

For citation purposes, cite each article independently as indicated on the article page online and as indicated below:

Lastname, A.A.; Lastname, B.B. Article Title. *Journal Name* **Year**, *Volume Number*, Page Range.

ISBN 978-3-7258-1498-5 (Hbk)
ISBN 978-3-7258-1497-8 (PDF)
doi.org/10.3390/books978-3-7258-1497-8

Contents

About the Editors

Andrés J. Cortés

Dr. Cortés holds an associate research position as geneticist at the Colombian Agricultural Research Corporation (AGROSAVIA), CI La Selva. Dr. Cortés graduated as a plant geneticist (Ph.D.) from Uppsala University (Sweden) and as a biologist (B.Sc. Hons and M.Sc.) from Universidad de los Andes (Colombia). His research experiences include the International Center for Tropical Agriculture (CGIAR – CIAT), the University of Fribourg (Switzerland), the Swiss Federal Institute for Forest, Snow and Landscape Research (WSL – SLF), and the Swedish University of Agricultural Sciences (SLU). Dr. Cortés is interested in investigating the genetic adaptive potential in plants and trees of agro-ecological interest using genomic, evolutionary, and ecological tools. In particular, he has explored abiotic stress tolerance and genetic diversity in beans, rootstock-mediated inheritance in avocado trees, genetic adaptation to climate change in willow species, and adaptive diversification across various plant lineages in the Páramo—a rapidly evolving ecosystem that provides water resources for the Neotropics.

Hai Du

Dr. Hai Du, a professor at Southwest University and the Rapeseed Research Institute (China), explores the relationship between root development and nutrient use efficiency (NUE) and between stress tolerance and yield by identifying the key genes involved in root development (e.g., TFs and hormone signalling), nutrient utilization, and stress response (e.g., nutrient and drought stresses) processes using *Brassica napus* as a study system. As part of broader research interests, he (1) explores the molecular evolutionary mechanisms of key genes, gene families, and pathways and their relationship with the formation of important/novel traits in crops at genome-wide levels (e.g., the emergence of new gene families or novel functions); (2) analyses the relationship between polyploidy/genome-wide duplication events and the formation of important/novel traits in crops (e.g., the causal link between gene duplication and the emergence of major evolutionary innovations); and (3) assesses the relationships between genotypes and phenotypes (e.g., behavioural evolution and the evolution of novel traits). He is also interested in new sequencing skills (e.g., single-cell sequencing, DNA/RNA methylation sequencing, etc.), epigenomics, genome editing, genome breeding, and the creation and utilization of new germplasm/materials with a focus on rapeseed.

Preface

Classical breeding continues to form the foundation of contemporary agricultural science and the crop industry. After all, direct human-driven plant selection is as old as the dawn of agriculture itself; until very recently, it was the only reachable tool. Fortunately, the course of empirical testing and key innovations have enabled modern technologies that push our otherwise orthodox methodologies towards unimaginable frontiers in molecular screening, genome sequencing, 'omic'-enabled forecasting, computing power, and trans-genesis. All of the above have delivered the promise of expanding our capabilities to revolutionize the field of varietal customization and deployment. However, the successful translation of these approaches into the farmers' field have been fiercely debated ever since their inception. In particular, a major restraint is that these developments have failed to embrace novel transdisciplinary arenas where genomics, molecular biology, biotechnology, and bioinformatics effectively converge with traditional plant enhancement. So far, prevalent atomized views that ignore intrinsic trade-offs from the own biological nature of the system feedback a generalized public scepticism, while punishing the emergence of innovative varietal products and their eventual scalation into factual technological advancements. It is within this bottleneck that the current reprint aims to better illustrate how integrative molecular research could more efficiently open up new avenues for the exploration, utilization, and transformation of crop biodiversity in the quest for breeding crucial yield and sustainability targets by restocking the genetic potential of cultivated genepools.

This second compilation on Molecular Genetics and Plant Breeding spans 31 fascinating trans-disciplinary works. A main lesson that readers can grasp from the contributing chapters is that transgressive analytical approaches must be adopted as molecular and phenomic data continue to accumulate. A first step toward unifying the molecular biology and plant genetics paradigms is provided by predictive breeding through genomic-enabled selection and machine learning at the plant breeding triangle interface. This compilation also explores AI applications for crop breeding and advocates for a stronger integration of AI with "omics" techniques. Ecological modelling under current and future scenarios as part of climate vulnerability assessments, gap analyses of the available natural variation for pre-breeding, and signatures of genomic selection by historic Darwinian adaptation in crop–wild genepools may also add to the screening of underlying trait genetic determinants and potential gene editing. In order to provide training datasets that are sufficiently reliable for workable, indirect forecasting for selection, it is necessary to ensure both open access data and efficient crop mobilization networks.

Andrés J. Cortés and Hai Du
Editors

International Journal of
Molecular Sciences

MDPI

Editorial

Molecular Genetics Enhances Plant Breeding

Andrés J. Cortés [1,*,†] and Hai Du [2,3,4,*]

1 Corporación Colombiana de Investigación Agropecuaria AGROSAVIA, C.I. La Selva, Km 7 vía Rionegro—Las Palmas, Rionegro 054048, Colombia
2 College of Agronomy and Biotechnology, Chongqing Engineering Research Center for Rapeseed, Southwest University, Chongqing 400000, China
3 Integrative Science Center of Germplasm Creation in Western China (Chongqing) Science City and Southwest University, of Agronomy and Biotechnology, Southwest University, Chongqing 400000, China
4 Engineering Research Center, South Upland Agriculture, Ministry of Education, Chongqing 400000, China
* Correspondence: acortes@agrosavia.co (A.J.C.); haidu81@126.com (H.D.)
† Current address: Facultad de Ciencias Agrarias—Departamento de Ciencias Forestales, Universidad Nacional de Colombia—Sede Medellín, Medellín 050034, Colombia.

Human-driven plant selection, a practice as ancient as agriculture itself, has laid the foundations of plant breeding and contemporary farming [1]. The principles of classical breeding still comprise the nucleus of modern crop science and agricultural production [2]. Recent unthinkable methodological achievements in molecular screening, genome sequencing, 'omic' technologies, trans-genesis, and computational power have advanced disciplinary boundaries in order to revolutionize the field of varietal development and its broader connections with related disciplines. These advances have opened up new trans-disciplinary arenas in which classical plant improvement intersects with genomics, molecular biology, biotechnology, and bioinformatics [3]. Yet, the factual potential of these interplays is often disregarded by prevailing atomized views and skepticism, both of which penalize the dawn of emerging properties and their ultimate deployment into framers' fields [4]. Therefore, this second Special Issue of IJMS on "Molecular Genetics and Plant Breeding" exemplifies through 31 inter-disciplinary works (https://www.mdpi.com/journal/ijms/special_issues/gene_plant_breading_2nd, accessed on 6 June 2023) the ways in which disruptive molecular research opens up innovative pathways for the exploration, transformation, and utilization of crop biodiversity in order to improve essential breeding targets and replenish the genetic potential of the cultivated genepools in order to achieve greater yield and sustainability [5].

Citation: Cortés, A.J.; Du, H. Molecular Genetics Enhances Plant Breeding. *Int. J. Mol. Sci.* **2023**, *24*, 9977. https://doi.org/10.3390/ijms24129977

Received: 8 May 2023
Accepted: 29 May 2023
Published: 9 June 2023

1. Marker-Guided Pre-Breeding

Natural genetic variation continues to constitute a major source of crop innovation, and its exploration remains a key milestone in molecular pre-breeding. This Special Issue provides some insights in this regard by presenting a series of molecular characterizations of diverse germplasm (Table 1). For instance, López-Hernández and Cortés [6] assessed the scale and determinants of somaclonal coding diversity in mint (*Mentha* spp.), introduced to the northern Andes, using RNA-seq on 29 clonally propagated accessions. The authors found a single genetic cluster for *M.* × *piperita*, and three clusters for *M. spicata*, suggesting two independent introductions of the latter.

Meanwhile, marker-assisted pre-breeding is underdoing a transformation from the method of classically screening crop genepools to the assessment of their associated antagonistic biotic agents. For example, Choi et al. [7] elucidated the molecular phylogenetic origin of seven strains of soybean mosaic virus (SMV) that had been sampled from 150 *Glycine max* L. accessions. The use of RNA-seq and 143 SMV available genomes enabled concluding that recombination and plant hosts drive the genetic diversity of SMV.

Table 1. Compilation of 31 studies as part of this second Special Issue of IJMS on "Molecular Genetics and Plant Breeding". Four of the contributions were modern literature reviews (first rows).

Species	Goal	Sampling	Genotyping	Key Finding	Reference
			Reviews		
Cruciferous crops and *Plasmodiophora brassicae*	† Review clubroot control methods and breeding for resistant cultivars	Cruciferous crops	Resistance loci and R genes	Resistance loci offer feasible strategies for resistance breeding	Zhang et al. [8]
Various	† Review functionality of *MYB* genes in plant roots	Plant roots	*MYB* genes	*MYB* gene functionality spans responses from biotic to abiotic stresses	Chen et al. [9]
Various	† Review applications of miR393 for plant development and stresses	Various	miR393 targeting TIR1 and AFB auxin receptors	miR393 assists plant responses to biotic and abiotic stresses	Jiang et al. [10]
Various	† Review the applications of AI in crop breeding	Various	AI and "omics" tools	Integration of AI into "omics" tools needed for crop-improvement	Khan et al. [11]
			Germplasm molecular characterization		
Mint (*Mentha* spp.)	° Assess somaclonal coding diversity in mint at the northern Andes	A total of 29 clonally propagated mints from the northern Andes	RNA-seq	One and three clusters found for *M.* × *piperita* and *M. spicata*	López-Hernández and Cortés [6]
Soybean mosaic virus (SMV)	° Elucidate the molecular phylogenetic origin of SMV strains	A total of 7 SMV from 150 different soybean germplasm	RNA-seq and 143 SMV available genomes	Recombination and plant hosts drive the genetic diversity of SMV	Choi et al. [7]
Viromes of pepper *Capsicum annuum*	° Examine the viromes of 15 pepper cultivars through RNA-seq	A total of 15 pepper (*Capsicum annuum* L.) cultivars	RNA-seq of pepper viromes	First viromes in 15 major pepper cultivars through RNA-seq	Jo et al. [12]
			Genetic mapping		
Rice (*Oryza sativa*)	‡ Infer the genomic bases (QTLs) of germination under cold conditions	One hundred and twenty lines of the CNDH population	778 SSR markers	Four QTLs and 41 genes were recovered, and 25 were qRT-PCR tested	Kim et al. [13]
Sesame (*Sesamum indicum*)	‡ Disclose the genomic basis (GWAS) for lignan lignin biosynthesis	410 accessions	WGR: 5.38 M SNPs and 1.16 M InDels	*SiNST1* is a target gene for the molecular breeding of lignans content	Dossou et al. [14]
Soybean (*Glycine max*)	‡ Reveal the genomic architecture (QTLs) for drought tolerance	A total of 160 RILs drought-tolerant 'Jindou21' × control 'Zhongdou33'	WGR: 923,420 SNPs	Five QTLs may be useful for molecular marker-assisted selection	Ouyang et al. [15]
Soybean (*Glycine max*)	‡ Validate existing QTLs for MAS of pod-shattering tolerance	A total of 2 RIL families (154 and 153) + 102 varieties and elite lines	QTLs identified via WGR and 3 KASP markers	Recovered accuracy: 90.9% for RILs and 100% for varieties and lines	Seo et al. [16]
Wheat (*Triticum aestivum*)	‡ Reveal genomic basis (GWAS) for six traits in eight environments	In total, 509 European varieties (277) and breeding lines from the STH (232)	Total 13,499 DArTseq-derived SNP markers	A GWAS for heterosis revealed 1261 markers with significant effects	Mokrzycka et al. [17]
Wheat (*Triticum aestivum*)	‡ Reveal the genomic architecture (QTLs) of spike layer uniformity	A total of 300 RILs	Wheat 55 K SNP array: 53,063 SNPs and one KASP-marker	*QSlu.sicau-2B-2* is a target for MAS of spike-layer uniformity	Zhou et al. [18]

Table 1. *Cont.*

Species	Goal	Sampling	Genotyping	Key Finding	Reference
Gene functional validation with expression analysis, RNA-seq, and/or other "omic" techniques					
Arabidopsis thaliana	+ Validate the overexpression effects of *PIF4* for HSR	Columbia-0 (Col) ecotype	RNA-seq, qRT-PCR, and ChIP-qPCR	Overexpression of *PIF4* boosts basal thermotolerance	Yang et al. [19]
Arabidopsis thaliana	+ Characterize functionally *AtEAU1-AtEAU2* via RT-PCR	Columbia-0 (Col) ecotype	Two ABA-responsive EAR-motif-containing genes	*AtEAU1* and *AtEAU2* are novel repressors of ABA responses	Zhang et al. [20]
Cabbage (*Brassica oleracea*)	+ Explore the molecular mechanisms that enable ME via HS	Accession '01-88'	DNA methylation and miRNA	DNA methylation and miRNA interference regulate HS-induced ME	Kong et al. [21]
Chinese cabbage (*Brassica rapa*)	+ Map and validate *BrGOLDEN*, a dominant gene for carotenoid content	A total of 151 tri-crossed hybrids between '1900264' and '1900262'	Full-length *BrGOLDEN* sequence and qRT-PCR	*BrGOLDEN* gives insights into the regulation of carotenoid synthesis	Zhang et al. [22]
Corn (*Zea mays*) and rice (*Oryza sativa*)	+ Validate PHO-encoding genes phylogenetically derived	Mo17 inbred line of *Z. mays*	PHO1- and PHO2-encoding genes + qRT-PCR	ABA could up-regulate the expression of both PHO1 and PHO2	Yu et al. [23]
Rapeseed (*Brassica napus*)	+ Identify and qRT-PCR test nitrate transporter 2 (*NRT2*) genes	Cultivars 'Zhongshuang11' and 'Darmor-bzh'	A total of 31 and 19 *NRT2* genes respectively tested in 'Zhongshuang11' and 'Darmor-bzh'	Candidates provided for functional *BnaZSNRT2s* studies	Du et al. [24]
Rapeseed (*Brassica napus*)	+ Validate *ARI* gene family functionality for agronomic traits	Varietal 'ZS 11'	qRT-PCR of 39 *ARI* genes	Eight *BnARI* genes identified as candidates for key traits	Wahid et al. [25]
Rapeseed (*Brassica napus*)	+ Explore the functionality of Trihelix (*TH*) genes across Brassica species	Six Brassica species from the Brassica database (BRAD)	A total of 455 Trihelix (*TH*) genes available sequences and qRT-PCR	*BnaTH* genes are involved in response to drought, cold, and heat	Zhang et al. [26]
Ribes nigrum	+ Evaluate the molecular mechanisms related to BRV tolerance	Cultivar Aldoniai	RNA-seq	Novel transcripts for breeding BRV-tolerance are provided	Mažeikienė et al. [27]
Sacha Inchi (*Plukenetia volubilis*)	+ Recover transcriptomic and proteomic profiles of seeds at two stages	Three-year-old Peruvian trees introduced to South China	RNA-seq and iTRAQ	The study enables further research and utilization of RIPs	Liu et al. [28]
Sorghum (*Sorghum bicolor*)	+ Validate *DOF* genes functionality for starch biosynthesis	Cultivar 'BTx623'	qRT-PCR of 30 *DOF* genes	*SbDOF21* acts as a key master regulator for starch biosynthesis	Xiao et al. [29]
Wheat (*Triticum aestivum*)	+ Characterize a new Q allele (Q^{c5}) for compact spikes and good bread	Cultivar "Roblin"	Q gene and its new allele Q^{c5}	Q^{c5} boosts bread-making quality by repressing *SPR*	Guo et al. [30]
Trans-genesis targeting inter-specific standing variation for validation purposes					
Pitaya (*Hylocereus monacanthus*)	§ Map and qRT-PCR validate NAC candidate genes	Cultivar 'Hongguan No. 1'	A total of 64 NAC TFs (from *HuNAC1* to *HuNAC64* genes)	*HuNAC20* and *25* from Pitaya confer cold tolerance in *Arabidopsis*	Hu et al. [31]
Woody resurrection plant (*Myrothamnus flabellifolia*)	§ Overexpressed *MfWRKY40* in *Arabidopsis* for abiotic stress roles	*M. flabellifolia* and *A. thaliana* Columbia-0 (Col) ecotype	WRKY TFs (i.e., early dehydration-induced gene *MfWRKY40*)	*MfWRKY40* confers tolerance to drought and salinity in an *Arabidopsis*	Huang et al. [32]

Table 1. *Cont.*

Species	Goal	Sampling	Genotyping	Key Finding	Reference
Trans-genesis targeting intra-specific standing variation					
Rapeseed (*Brassica napus*)	$^{\lambda}$ Assess homozygous transcript overexpression lines for *γ-TMT*	Cultivar 'Zhongshuang11'	Gene γ-Tocopherol methyltransferase (*γ-TMT*)	Feasible to genetic engineer *γ-TMT* for salt tolerance breeding	Guo et al. [33]
Soybean (*Glycine max*)	$^{\lambda}$ Silence *GmBIR1* by BPMV-VIGS and explore the phenotypes	Cultivar 'Williams 82'	*BAK1*-interacting receptor-like kinase (*GmBIR1*)	*GmBIR1* is a negative regulator of immunity in soybean	Liu et al. [34]
Mutagenesis targeting intra-specific de novo mutations					
Wheat (*Triticum aestivum*)	$^{\Psi}$ Characterize mutant *Q* alleles for grain yield and grain protein	Wheat cultivar 'Shumai482' and its *S-Cp*1-1 mutant	Two new *Q* alleles (Q^{s1} and Q^{c1}-*N8*) obtained via mutagenesis	New *Q* alleles offer novel germplasm relevant to wheat breeding	Chen et al. [35]
Introgression breeding targeting inter-specific standing variation					
Rice (*Oryza sativa*)	$^{\theta}$ Breed for heat tolerance in rice via introgressed sorghum ancestry	Rice restorer line 'R21' and recipient restorer line Jin 'Hui 1'	WGR	The sorghum introgression in line 'R21' confers mobile heat tolerance	Zhang et al. [36]

Table is arranged top-down by research goals, species, and citations. Studies are clustered by goals as follows: † for reviews; ° for germplasm molecular characterization; ‡ for genetic mapping; + for gene functional validation with expression analysis, RNA-seq and/or other omic techniques; § for trans-genesis targeting inter-specific standing variation for validation purposes; λ for trans-genesis targeting intra-specific standing variation; Ψ for mutagenesis targeting intra-specific de novo mutations; and θ for introgression breeding targeting inter-specific standing variation. Abbreviations are as follows. ABA: abscisic acid; AFB: auxin signaling F-box; ARI: Ariadne proteins of ring-between-ring (RBR) finger protein subfamilies; BPMV-VIGS: bean-pod-mottle-virus-induced gene silencing; BRV: mite-transmitted blackcurrant reversion virus; CNDH: Cheongcheong Nagdong double haploid; DOF: C2-C2 zinc finger domain, EAR: ethylene-responsive element-binding-factor-associated amphiphilic repression; AI: artificial intelligence; GWAS: genome-wide association study; HS: heat shock; HSR: heat stress response; KASP: Kompetitive allele-specific PCR; MAS: marker-assisted selection; ME: microspore embryogenesis; PIF4: phytochrome interacting factor 4; PHO: starch phosphorylase; QTL: quantitative trait loci; RILs: recombinant inbreed lines; RIP: ribosome-inactivating protein; SNP: single-nucleotide polymorphism; SPR: storage protein repressor; SSR: simple sequence repeat; STH: Plant Breeding Strzelce, TIR1: transport inhibitor response1; TF: transcription factors; WGR: whole-genome re-sequencing.

Similarly, Jo et al. [12] systematically examined the viromes of 15 pepper (*Capsicum annuum* L.) cultivars through RNA-seq, enabling a high-throughput identification of the principal viromes present in commercially important pepper genotypes. As a promising perspective, these initial characterizations of natural variation may unleash a further potential when coupled with explicit estimates of their genomic bases, i.e., the genetic determinants of trait variation.

As such, genetic mapping arises as a key step in the pre-breeding pipeline that enables the genetic architecture of key traits to be disclosed, and ultimately provides candidate markers and genes for further marker-guided applications, such as parental screening, marker-assisted selection [37], and gene editing [38]. In this regard, and as part of this Special Issue, Kim et al. [13] genotyped 120 double haploid rice lines with 778 SSRs in order to reconstruct the genetic basis of germination under cold conditions. The authors managed to retrieve 4 QTLs and 41 genes, 25 of which were validated via qRT-PCR. Similarly, Dossou et al. [14] characterized 410 sesame (*S. indicum*) accessions using WGR, and performed GWAS for lignan–lignin biosynthesis. The team found that *SiNST1* is a major target gene for the molecular breeding of lignan content. In soybean, Ouyang et al. [15] and Seo et al. [16] performed WGR of RIL families with ca. 155 genotypes each. The teams, respectively, identified five QTLs for drought tolerance, and validated accuracies above 90% for existing QTLs for pod-shattering tolerance. Finally, Mokrzycka et al. [17] and Zhou et al. [18], respectively, genotyped 509 wheat accessions and 300 RILs with 13,499 DArT-SNPs and a 55 K SNP array. The authors discovered 1261 candidate markers for six agronomical traits, and the *QSlu.sicau-2B-2* MAS target for spike layer uniformity. A common denominator across all genetic mapping exercises is the need for a further validation of the associated markers [39] before any downstream plant improvement application. After all, they are prone to displaying inflated rates of false positives due to multiple comparison testing, population stratification, and intrinsic redundancy from linkage disequilibrium (LD) [40].

Marker validation often requires that a battery of downstream analyses be performed. The first step required is narrow mapping of the causal variants using target genotyping across extended panels. A second, more experimental approach involves examining the expression profiles of flanking candidate genes in terms of congruency and stability. Several teams pursued this goal as part of the current Special Issue. For example, Yang et al. [19] and Zhang et al. [20] used the Columbia-0 (Col) *A. thaliana* ecotype to validate the functional roles of *PIF4* and *AtEAU* (*1* and *2*) genes, respectively, through transcriptomic tools as boosters and repressors of basal thermotolerance and ABA response. Meanwhile, Kong et al. [21] and Zhang et al. [22], respectively, implemented epigenomics and qRT-PCR in cabbage (*B. oleracea*) and Chinese cabbage (*B. rapa*) to study the DNA methylation footprint during microspore embryogenesis induced by heat shock, and the functional role of *BrGOLDEN* in carotenoid biosynthesis. Interestingly, rapeseed (*B. napus*) was a widely considered study system by Du et al. [24], Wahid et al. [25], and Zhang et al. [26], who used qRT-PCR to corroborate the functionality of *BnaZSNRT2s*, *BnARI*, and *BnaTH* genes for nitrogen uptake, agronomical trait regulation, and abiotic stress tolerance, respectively. Other studies focused on key crops for food security. For example, while also considering the ABA pathway, Yu et al. [23] used qRT-PCR to demonstrate the up-regulation of PHO-encoding genes by ABA in maize and rice. On the other hand, Xiao et al. [29] validated *SbDOF21* as a key regulator for starch biosynthesis in sorghum, and Guo et al. [30] characterized a new *SPR*-repressor *Q* allele (Q^{c5}) for compact spikes and bread quality in wheat. The other studies which stand out are those by Mažeikienė et al. [27] and Liu et al. [28], who brought RNA-seq technology to the new promissory crops *Ribes nigrum* and Sacha Inchi (*Plukenetia volubilis*). Each team was able to provide transcriptomic resources to breed for tolerance to the BRV virus [27] and better comprehend the expression profiles of seeds at two different developmental stages [28].

Finally, the works by Hu et al. [31] and Huang et al. [32] utilized trans-genesis to move beyond the species boundary and validate in an *Arabidopsis* background candidate genes for cold (*HuNAC20* and *25*) and osmatic (*MfWRKY40*) stress tolerance, respectively, from two other exotic crops, namely, the cactus-fruit Pitaya (*Hylocereus monacanthus*) [31] and the woody resurrection plant (*Myrothamnus flabellifolia*) [32]. Despite these advances in germplasm characterization, genetic mapping, and functional validation, moving forward, elite cultivars are beginning to benefit from embracing genomic-enabled breeding.

2. Molecularly Enabled Breeding

Mobilizing germplasm's allelic novelty, leveraging marker genetic mapping, and functionally corroborating candidate genes to customize crops requires the modernization of classical plant breeding with phenomic and genomic tools [2]. Perhaps one of the promptest strategies for bridging long breeding cycles is trans-genesis. When it targets standing variation, segregating within the same species genepool, it simply provides a shortcut to the time-consuming alternative of recurrent backcrossing [37]. However, it also allows customizability for de novo variation. This Special Issue compiles an exquisite repertoire in this regard. For instance, Guo et al. [33] illustrated, once again in rapeseed, that it is feasible to genetically engineer *BnaC02.TMT.a*, a γ-*TMT* paralogue, for salt tolerance breeding via the use of homozygous transgenic lines. Similarly, Liu et al. [34] silenced *GmBIR1* with BPMV-VIGS to demonstrate that it is a negative regulator of immunity in soybean. Finally, Chen et al. [35] went one step further and explored mutagenesis as a reliable source of de novo allelic variation. The authors characterized mutant *Q* alleles (Q^{s1} and Q^{c1}-*N8*) for grain yield and grain protein in the wheat cultivar 'Shumai482' and its *S-Cp1-1* mutant [35], complementing the results achieved by Guo et al. [30] in terms of compact spikes and bread quality. In spite of these exciting results achieved for the development of trans-genesis, classical intercrossing has not lost validity as a breeding strategy, but rather it has been permeable to the latest genomic advances, such as marker-assisted backcrossing (MABC) [37], introgression breeding, and genomic-assisted recurrent selection [41].

Inter-specific introgression breeding, in particular, aims to break species boundaries to pyramid exotic variation from one species to elite commercial varieties from the other species [37]. However, inter-specific crossing conveys two main challenges, which are species incompatibility and polygenic trait variation. Therefore, coupling hybrid breeding with bridge genotypes and guiding molecular markers unleashes novel opportunities to improve the chances of success, pace, and precision of the target introgression. Such utility has been reinforced in rice by Zhang et al. [36], who bred for heat tolerance by introgressed sorghum ancestry. The authors utilized the rice restorer line 'R21' and the recipient restorer line Jin 'Hui 1', and were able to demonstrate with WGR that the sorghum introgression in line 'R21' confers mobile heat tolerance. Gene editing may speed up rice breeding for abiotic stresses [42], too. Bean breeders are also seeing a similar development by GBS-genotyping 87 advanced lines with inter-specific ancestries between common bean (*P. vulgaris*) and tepary bean (*P. acutifolius*) for heat and drought tolerance across four environments in coastal northern South America [43]. The authors found 47 associated loci and 90 flanking candidate genes for molecular-guided downstream selection. Meanwhile, the team also detected suitable allelic variation within the candidate genes of tepary bean (*P. acutifolius*) that transcends the adaptive genepool of common bean (*P. vulgaris*) [44]. These studies demonstrate that the integration of genomic- and gene-based strategies can leverage inter-specific adaptive variation via bridge genotypes in order to deliver candidate introgressed lines for heat tolerance [43]. In certain cases, grafting could also provide a fast track to harness such species diversity [45]. These success stories that intermingle modern genotyping technology with classical intercrossing exemplify, against any intellectual skepticism, the factual applicability of the molecular breeding paradigm. Still, there remains room for further developments.

3. Perspectives

As molecular and phenomic data continue to pile up, modern analytical techniques must be embraced. Predictive breeding, reached throughout genomic-enabled selection [41] and machine learning [11] at the interface of the plant breeding triangle [46], confers a primary way forward capable of bringing together molecular biology and plant genetic paradigms [47]. In this regard, Khan et al. [11] reviewed applications of crop breeding in this Special Issue, calling for a better integration of AI with "omics" tools. Still, effective crop mobilization networks and open access data must be assured in order to build sufficiently reliable training datasets without sampling bias or over-fitting prediction [48].

Meanwhile, biotic and abiotic stresses are becoming more common, jeopardizing global food production. In this sense, another review within this collection by Zhang et al. [8] was more concrete in envisioning gene targets for biotic pressures, specifically that clubroot control methods in cruciferous crops could be harnessed with allelic variation at the R genes, which are in turn susceptible to be gene-edited, eventually. Similarly, the reviews by Chen et al. [9] and Jiang et al. [10] prospected gene-enabled abiotic stress responses by looking at the *MYB* genes in plant roots and the miR393 during plant development, respectively.

The method of screening the genetic determinants of the abiotic susceptibility, and its potential gene editing, should be supplemented by explicit in situ eco-physiological indices targeting specific stresses, performing potential ecological niche modeling (ENM) under present and future scenarios as part of climate vulnerability assessments, undertaking gap analyses of the available variation for pre-breeding, exploring genomic selection signatures of historic adaptation, and unpicking the genome–environment associations (GEA) of current niche preferences [3]. After all, tackling the climate crisis and agrobiodiversity loss in the process of addressing food security [5] demands that we harness trans-disciplinary sustainable enterprises that promote integrative agendas among the otherwise disentangled fields of in situ and ex situ conversation, physiology, ecology, molecular genetics, plant breeding [2], conservation, seed delivery [4], food policy [49], and marketing.

Author Contributions: A.J.C. and H.D. conceived this Special Issue and jointly invited potential authors, handled manuscripts, recommended reviewers, and approved submissions. A.J.C. prepared a first draft, which was later on edited by H.D. All authors have read and agreed to the published version of the manuscript.

Funding: A.J.C. received support from Vetenskapsrådet (VR) and Kungliga Vetenskapsakademien (KVA) while conceiving and closing this Special Issue through grants 2016-00418/2022-04411 and BS2017-0036, respectively. A.J.C. also acknowledges Newton Fund's 527023146 grant, executed during the time this Special Issue was being processed, as well as Fulbright U.S. Specialist Program for supporting synergistic discussion with M.W. Blair on molecular breeding of tropical species in Rionegro (Antioquia, Colombia) during the summer of 2019.

Institutional Review Board Statement: Not applicable.

Informed Consent Statement: Not applicable.

Data Availability Statement: For original datasets, please refer to the published articles [6–36] within the second Special Issue of IJMS on "Molecular Genetics and Plant Breeding 2.0" (https://www.mdpi.com/journal/ijms/special_issues/gene_plant_breading_2nd, as accessed on 6 June 2023).

Acknowledgments: The Guest Editors acknowledge the time and effort contributed by all authors, reviewers, and editors who made possible this Special Issue on "Molecular Genetics and Plant Breeding 2.0". Support from M.J. Torres-Urrego is appreciated. MDPI's IJMS is thanked for inviting, encouraging, and assisting the Guest Editors to host this compilation. Section Managing Editor is also acknowledged for their continuous assistance.

Conflicts of Interest: The authors declare no conflict of interest.

References

1. Darwin, C. *The Variation of Animals and Plants under Domestication*; John Murray: London, UK, 1868.
2. Varshney, R.K.; Bohra, A.; Roorkiwal, M.; Barmukh, R.; Cowling, W.A.; Chitikineni, A.; Lam, H.-M.; Hickey, L.T.; Croser, J.S.; Bayer, P.E.; et al. Fast-forward breeding for a food-secure world. *Trends Genet.* **2021**, *37*, 1124–1136. [CrossRef] [PubMed]
3. Cortés, A.J.; López-Hernández, F.; Blair, M.W. Genome–Environment Associations, an Innovative Tool for Studying Heritable Evolutionary Adaptation in Orphan Crops and Wild Relatives. *Front. Genet.* **2022**, *13*, 910386. [CrossRef] [PubMed]
4. Peláez, D.; Aguilar, P.A.; Mercado, M.; López-Hernández, F.; Guzmán, M.; Burbano-Erazo, E.; Denning-James, K.; Medina, C.I.; Blair, M.W.; De Vega, J.J.; et al. Genotype Selection, and Seed Uniformity and Multiplication to Ensure Common Bean (*Phaseolus vulgaris* L.) var. Liborino. *Agronomy* **2022**, *12*, 2285. [CrossRef]
5. McCouch, S. Feeding the future. *Nature* **2013**, *499*, 23–24. [CrossRef] [PubMed]
6. López-Hernández, F.; Cortés, A.J. Whole Transcriptome Sequencing Unveils the Genomic Determinants of Putative Somaclonal Variation in Mint (*Mentha* L.). *Int. J. Mol. Sci.* **2022**, *23*, 5291. [CrossRef] [PubMed]
7. Choi, H.; Jo, Y.; Choi, S.Y.; Kim, S.-M.; Choi, Y.M.; Hong, J.-S.; Lee, B.C.; Cho, W.K. Evolution and Phylogeny of Soybean Mosaic Virus Based on 143 Complete Genomes. *Int. J. Mol. Sci.* **2022**, *24*, 22. [CrossRef] [PubMed]
8. Zhang, C.; Du, C.; Li, Y.; Wang, H.; Zhang, C.; Chen, P. Advances in Biological Control and Resistance Genes of Brassicaceae Clubroot Disease-The Study Case of China. *Int. J. Mol. Sci.* **2023**, *24*, 785. [CrossRef]
9. Chen, Z.; Wu, Z.; Dong, W.; Liu, S.; Tian, L.; Li, J.; Du, H. MYB Transcription Factors Becoming Mainstream in Plant Roots. *Int. J. Mol. Sci.* **2022**, *23*, 9262. [CrossRef]
10. Jiang, J.; Zhu, H.; Li, N.; Batley, J.; Wang, Y. The miR393-Target Module Regulates Plant Development and Responses to Biotic and Abiotic Stresses. *Int. J. Mol. Sci.* **2022**, *23*, 9477. [CrossRef]
11. Khan, M.H.U.; Wang, S.; Wang, J.; Ahmar, S.; Saeed, S.; Khan, S.U.; Xu, X.; Chen, H.; Bhat, J.A.; Feng, X. Applications of Artificial Intelligence in Climate-Resilient Smart-Crop Breeding. *Int. J. Mol. Sci.* **2022**, *23*, 11156. [CrossRef]
12. Jo, Y.; Choi, H.; Lee, J.H.; Moh, S.H.; Cho, W.K. Viromes of 15 Pepper (*Capsicum annuum* L.) Cultivars. *Int. J. Mol. Sci.* **2022**, *23*, 10507. [CrossRef] [PubMed]
13. Kim, N.; Jan, R.; Park, J.-R.; Asif, S.; Zhao, D.-D.; Kim, E.-G.; Jang, Y.-H.; Eom, G.-H.; Lee, G.-S.; Kim, K.-M. QTL Mapping and Candidate Gene Analysis for Seed Germination Response to Low Temperature in Rice. *Int. J. Mol. Sci.* **2022**, *23*, 7379. [CrossRef] [PubMed]

14. Dossou, S.S.K.; Song, S.; Liu, A.; Li, D.; Zhou, R.; Berhe, M.; Zhang, Y.; Sheng, C.; Wang, Z.; You, J.; et al. Resequencing of 410 Sesame Accessions Identifies SINST1 as the Major Underlying Gene for Lignans Variation. *Int. J. Mol. Sci.* **2023**, *24*, 1055. [CrossRef] [PubMed]
15. Ouyang, W.; Chen, L.; Ma, J.; Liu, X.; Chen, H.; Yang, H.; Guo, W.; Shan, Z.; Yang, Z.; Chen, S.; et al. Identification of Quantitative Trait Locus and Candidate Genes for Drought Tolerance in a Soybean Recombinant Inbred Line Population. *Int. J. Mol. Sci.* **2022**, *23*, 10828. [CrossRef]
16. Seo, J.-H.; Dhungana, S.K.; Kang, B.-K.; Baek, I.-Y.; Sung, J.-S.; Ko, J.-Y.; Jung, C.-S.; Kim, K.-S.; Jun, T.-H. Development and Validation of SNP and InDel Markers for Pod-Shattering Tolerance in Soybean. *Int. J. Mol. Sci.* **2022**, *23*, 2382. [CrossRef]
17. Mokrzycka, M.; Stojałowski, S.; Tyrka, M.; Matysik, P.; Żmijewska, B.; Marcinkowski, R.; Woźna-Pawlak, U.; Martofel, R.; Rokicki, M.; Rakoczy-Trojanowska, M.; et al. Genome-Wide Association Analysis for Hybrid Breeding in Wheat. *Int. J. Mol. Sci.* **2022**, *23*, 15321. [CrossRef]
18. Zhou, K.; Lin, Y.; Jiang, X.; Zhou, W.; Wu, F.; Li, C.; Wei, Y.; Liu, Y. Identification and Validation of Quantitative Trait Loci Mapping for Spike-Layer Uniformity in Wheat. *Int. J. Mol. Sci.* **2022**, *23*, 1052. [CrossRef]
19. Yang, J.; Qu, X.; Ji, L.; Li, G.; Wang, C.; Wang, C.; Zhang, Y.; Zheng, L.; Li, W.; Zheng, X. PIF4 Promotes Expression of HSFA2 to Enhance Basal Thermotolerance in *Arabidopsis*. *Int. J. Mol. Sci.* **2022**, *23*, 6017. [CrossRef]
20. Zhang, N.; Chen, S.; Adnan, A.; Wang, X.; Hussain, S.; Cheng, Y.; Li, Y.; Yuan, Y.; Wang, C.; Lin, R.; et al. AtEAU1 and AtEAU2, Two EAR Motif-Containing ABA Up-Regulated Novel Transcription Repressors Regulate ABA Response in *Arabidopsis*. *Int. J. Mol. Sci.* **2022**, *23*, 9053. [CrossRef]
21. Kong, C.; Su, H.; Deng, S.; Ji, J.; Wang, Y.; Zhang, Y.; Yang, L.; Fang, Z.; Lv, H. Global DNA Methylation and mRNA-miRNA Variations Activated by Heat Shock Boost Early Microspore Embryogenesis in Cabbage (*Brassica oleracea*). *Int. J. Mol. Sci.* **2022**, *23*, 5147. [CrossRef]
22. Zhang, L.; Zhang, S.; Dai, Y.; Wang, S.; Wang, C.; Li, F.; Zhang, H.; Chen, G.; Yuan, L.; Hou, J.; et al. Mapping and Validation of BrGOLDEN: A Dominant Gene Regulating Carotenoid Accumulation in *Brassica rapa*. *Int. J. Mol. Sci.* **2022**, *23*, 12442. [CrossRef] [PubMed]
23. Yu, G.; Shoaib, N.; Xie, Y.; Liu, L.; Mughal, N.; Li, Y.; Huang, H.; Zhang, N.; Zhang, J.; Liu, Y.; et al. Comparative Study of Starch Phosphorylase Genes and Encoded Proteins in Various Monocots and Dicots with Emphasis on Maize. *Int. J. Mol. Sci.* **2022**, *23*, 4518. [CrossRef] [PubMed]
24. Du, R.-J.; Wu, Z.-X.; Yu, Z.-X.; Li, P.-F.; Mu, J.-Y.; Zhou, J.; Li, J.-N.; Du, H. Genome-Wide Characterization of High-Affinity Nitrate Transporter 2 (NRT2) Gene Family in *Brassica napus*. *Int. J. Mol. Sci.* **2022**, *23*, 4965. [CrossRef] [PubMed]
25. Wahid, S.; Xie, M.; Sarfraz, S.; Liu, J.; Zhao, C.; Bai, Z.; Tong, C.; Cheng, X.; Gao, F.; Liu, S. Genome-Wide Identification and Analysis of Ariadne Gene Family Reveal Its Genetic Effects on Agronomic Traits of *Brassica napus*. *Int. J. Mol. Sci.* **2022**, *23*, 6265. [CrossRef]
26. Zhang, C.; Lu, L.; Gong, R.; Su, X.; Liu, F.; Zhang, R.; Hu, J. Conservation and Divergence of the Trihelix Genes in Brassica and Expression Profiles of BnaTH Genes in *Brassica napus* under Abiotic Stresses. *Int. J. Mol. Sci.* **2022**, *23*, 15766. [CrossRef]
27. Mažeikienė, I.; Juškytė, A.D.; Bendokas, V.; Stanys, V. De Novo Transcriptome Analysis of R. nigrum cv. Aldoniai in Response to Blackcurrant Reversion Virus Infection. *Int. J. Mol. Sci.* **2022**, *23*, 9560. [CrossRef]
28. Liu, G.; Wu, Z.; Peng, Y.; Shang, X.; Gao, L. Integrated Transcriptome and Proteome Analysis Provides Insight into the Ribosome Inactivating Proteins in *Plukenetia volubilis* Seeds. *Int. J. Mol. Sci.* **2022**, *23*, 9562. [CrossRef]
29. Xiao, Q.; Liu, T.; Ling, M.; Ma, Q.; Cao, W.; Xing, F.; Huang, T.; Zhang, Y.; Duan, H.; Liu, Z. Genome-Wide Identification of DOF Gene Family and the Mechanism Dissection of SbDof21 Regulating Starch Biosynthesis in Sorghum. *Int. J. Mol. Sci.* **2022**, *23*, 12152. [CrossRef]
30. Guo, Z.; Chen, Q.; Zhu, J.; Wang, Y.; Li, Y.; Li, Q.; Zhao, K.; Li, Y.; Tang, R.; Shi, X.; et al. The Qc5 Allele Increases Wheat Bread-Making Quality by Regulating SPA and SPR. *Int. J. Mol. Sci.* **2022**, *23*, 7581. [CrossRef]
31. Hu, X.; Xie, F.; Liang, W.; Liang, Y.; Zhang, Z.; Zhao, J.; Hu, G.; Qin, Y. *HuNAC20* and *HuNAC25*, Two Novel NAC Genes from Pitaya, Confer Cold Tolerance in Transgenic *Arabidopsis*. *Int. J. Mol. Sci.* **2022**, *23*, 2189. [CrossRef]
32. Huang, Z.; Wang, J.; Li, Y.; Song, L.; Chen, D.E.; Liu, L.; Jiang, C.-Z. A WRKY Protein, MfWRKY40, of Resurrection Plant *Myrothamnus flabellifolia* Plays a Positive Role in Regulating Tolerance to Drought and Salinity Stresses of *Arabidopsis*. *Int. J. Mol. Sci.* **2022**, *23*, 8145. [CrossRef] [PubMed]
33. Guo, Y.; Li, D.; Liu, T.; Liao, M.; Li, Y.; Zhang, W.; Liu, Z.; Chen, M. Effect of Overexpression of *γ-Tocopherol Methyltransferase* on α-Tocopherol and Fatty Acid Accumulation and Tolerance to Salt Stress during Seed Germination in *Brassica napus* L. *Int. J. Mol. Sci.* **2022**, *23*, 15933. [CrossRef] [PubMed]
34. Liu, D.-D.; Lan, H.-J.; Masoud, H.S.; Ye, M.-Y.; Dai, X.-Y.; Zhong, C.-L.; Tian, S.-N.; Liu, J.-Z. Silencing *GmBIR1* in Soybean Results in Activated Defense Responses. *Int. J. Mol. Sci.* **2022**, *23*, 7450. [CrossRef] [PubMed]
35. Chen, Q.; Guo, Z.; Shi, X.; Wei, M.; Fan, Y.; Zhu, J.; Zheng, T.; Wang, Y.; Kong, L.; Deng, M.; et al. Increasing the Grain Yield and Grain Protein Content of Common Wheat (*Triticum aestivum*) by Introducing Missense Mutations in the Q Gene. *Int. J. Mol. Sci.* **2022**, *23*, 10772. [CrossRef] [PubMed]
36. Zhang, T.; Li, X.; Zhao, Z.; Wu, R.; Yang, Z.; He, G. Sequencing and Genomic Analysis of Sorghum DNA Introgression Variant Line R21 and Recipient Rice Jin Hui 1 Revealed Repetitive Element Variation. *Int. J. Mol. Sci.* **2022**, *23*, 11864. [CrossRef]

37. Herzog, E.; Frisch, M. Selection strategies for marker-assisted backcrossing with high-throughput marker systems. *Theor. Appl. Genet.* **2011**, *123*, 251–260. [CrossRef]
38. Rodríguez-Leal, D.; Lemmon, Z.H.; Man, J.; Bartlett, M.E.; Lippman, Z.B. Engineering Quantitative Trait Variation for Crop Improvement by Genome Editing. *Cell* **2017**, *171*, 470–480.e478. [CrossRef]
39. Hirschhorn, J.N.; Daly, M.J. Genome-wide association studies for common diseases and complex traits. *Nat. Rev. Genet.* **2005**, *6*, 95–108. [CrossRef]
40. Wray, N.R.; Yang, J.; Hayes, B.J.; Price, A.L.; Goddard, M.E.; Visscher, P.M. Pitfalls of predicting complex traits from SNPs. *Nat. Rev. Genet.* **2013**, *14*, 507–515. [CrossRef]
41. Crossa, J.; Perez-Rodriguez, P.; Cuevas, J.; Montesinos-Lopez, O.; Jarquin, D.; de Los Campos, G.; Burgueno, J.; Gonzalez-Camacho, J.M.; Perez-Elizalde, S.; Beyene, Y.; et al. Genomic Selection in Plant Breeding: Methods, Models, and Perspectives. *Trends Plant Sci.* **2017**, *22*, 961–975. [CrossRef]
42. Barrero, L.S.; Willmann, M.R.; Craft, E.J.; Akther, K.M.; Harrington, S.E.; Garzon-Martinez, G.A.; Glahn, R.P.; Piñeros, M.A.; McCouch, S.R. Identifying genes associated with abiotic stress tolerance suitable for CRISPR/Cas9 editing in upland rice cultivars adapted to acid soils. *Plant Direct* **2022**, *6*, e469. [CrossRef] [PubMed]
43. Burbano-Erazo, E.; León-Pacheco, R.; Cordero-Cordero, C.; López-Hernández, F.; Cortés, A.; Tofiño-Rivera, A. Multi-Environment Yield Components in Advanced Common Bean (*Phaseolus vulgaris* L.) × Tepary Bean (*P. acutifolius* A. Gray) Interspecific Lines for Heat and Drought Tolerance. *Agronomy* **2021**, *11*, 1978. [CrossRef]
44. Buitrago-Bitar, M.A.; Cortés, A.J.; López-Hernández, F.; Londoño-Caicedo, J.M.; Muñoz-Florez, J.E.; Muñoz, L.C.; Blair, M.W. Allelic Diversity at Abiotic Stress Responsive Genes in Relationship to Ecological Drought Indices for Cultivated Tepary Bean, *Phaseolus acutifolius* A. Gray, and Its Wild Relatives. *Genes* **2021**, *12*, 556. [CrossRef] [PubMed]
45. Warschefsky, E.J.; Klein, L.L.; Frank, M.H.; Chitwood, D.H.; Londo, J.P.; von Wettberg, E.J.B.; Miller, A.J. Rootstocks: Diversity, Domestication, and Impacts on Shoot Phenotypes. *Trends Plant Sci.* **2016**, *21*, 418–437. [CrossRef]
46. Crossa, J.; Fritsche-Neto, R.; Montesinos-Lopez, O.A.; Costa-Neto, G.; Dreisigacker, S.; Montesinos-Lopez, A.; Bentley, A.R. The Modern Plant Breeding Triangle: Optimizing the Use of Genomics, Phenomics, and Enviromics Data. *Front. Plant Sci.* **2021**, *12*, 651480. [CrossRef]
47. Arenas, S.; Cortés, A.J.; Mastretta-Yanes, A.; Jaramillo-Correa, J.P. Evaluating the accuracy of genomic prediction for the management and conservation of relictual natural tree populations. *Tree Genet. Genomes* **2021**, *17*, 12. [CrossRef]
48. McCouch, S.; Navabi, K.; Abberton, M.; Anglin, N.L.; Barbieri, R.L.; Baum, M.; Bett, K.; Booker, H.; Brown, G.L.; Bryan, G.J.; et al. Mobilizing Crop Biodiversity. *Mol. Plant* **2020**, *13*, 1341–1344. [CrossRef]
49. Smale, M.; Jamora, N. Valuing genebanks. *Food Secur.* **2020**, *12*, 905–918. [CrossRef]

International Journal of
Molecular Sciences

MDPI

Review

Advances in Biological Control and Resistance Genes of Brassicaceae Clubroot Disease-The Study Case of China

Chaoying Zhang, Chunyu Du, Yuwei Li, Huiying Wang, Chunyu Zhang * and Peng Chen *

College of Plant Science, Huazhong Agricultural University, Wuhan 430070, China
* Correspondence: zhchy@mail.hzau.edu.cn (C.Z.); chenpeng@mail.hzau.edu.cn (P.C.)

Abstract: Clubroot disease is a soil-borne disease caused by *Plasmodiophora brassicae*. It occurs in cruciferous crops exclusively, and causes serious damage to the economic value of cruciferous crops worldwide. Although different measures have been taken to prevent the spread of clubroot disease, the most fundamental and effective way is to explore and use disease-resistance genes to breed resistant varieties. However, the resistance level of plant hosts is influenced both by environment and pathogen race. In this work, we described clubroot disease in terms of discovery and current distribution, life cycle, and race identification systems; in particular, we summarized recent progress on clubroot control methods and breeding practices for resistant cultivars. With the knowledge of these identified resistance loci and R genes, we discussed feasible strategies for disease-resistance breeding in the future.

Keywords: clubroot disease; *Plasmodiophora brassicae*; R gene

Citation: Zhang, C.; Du, C.; Li, Y.; Wang, H.; Zhang, C.; Chen, P. Advances in Biological Control and Resistance Genes of Brassicaceae Clubroot Disease-The Study Case of China. *Int. J. Mol. Sci.* **2023**, *24*, 785. https://doi.org/10.3390/ijms24010785

Academic Editor: Andrés J. Cortés

Received: 3 November 2022
Revised: 20 December 2022
Accepted: 21 December 2022
Published: 2 January 2023

1. Overview of Clubroot Disease

1.1. The Discovery and Distribution of Clubroot Disease

The pathogen causing clubroot disease in crucifer plants is called *Plasmodiophora brassicae*, and belongs to the genus *Plasmodiophora* in the phylum *Protozoa*. Clubroot disease has been found in continental European Brassica plants as early as the 13th century, and some reports suggest that its presence may be traceable to earlier than the Roman times [1]. In 1737, cruciferous clubroot diseases were officially reported in England from the west coast of the Mediterranean and southern Europe [2]. Early Scottish people generally believed that clubroot disease was caused by poor soil quality or unbalanced fertilizers [1]. In 1873, Russian biologist Michael Woronin first identified the clubroot pathogen and named it *Plasmodiophora brassicae* [3]. In the late 1960s, Woronin studied the relationship between the host and the pathogen and described the pathogen's life cycle and the interaction mode with the host [4]. From the late 19th and early 20th centuries, the clubroot disease was brought to Canada by European settlers [5]. In 2003, clubroot disease was officially reported on *B. rapa* and thereafter it became rapidly spread in Canada [6,7]. The incidence of clubroot disease in Asia started in Japan and has become a serious economic problem in Japan and Korea [8]. A variety of names have been given to clubroot disease from different countries and regions, indicating the diversity in the nature of the disease from both the pathogen side as well as the plant hosts [9]. In China, clubroot disease was first found in Taiwan and Fujian in the early 1920s [10], though in recent years it has expanded both to the north and to the west inland regions during agronomy development [11–13]. With the expansion of rapeseed cultivation and especially the modern farming activities, the incidence of clubroot disease has rapidly increased; provinces with the heaviest incidence include Sichuan, Hubei, Yunnan, and Anhui (Figure 1) [14].

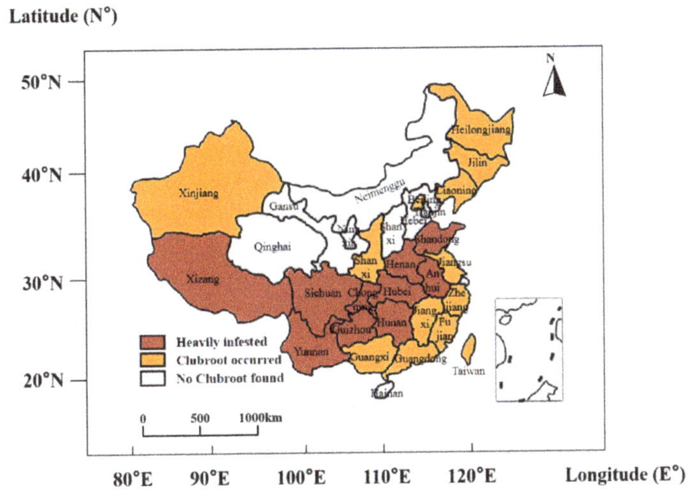

Figure 1. The distribution of clubroot disease in mainland China (Reprinted with permission from Ref. [14]).

1.2. Life History of *Plasmodiophora brassicae*

Clubroot disease occurs through a specialized parasitism of cruciferous plant roots. The causative pathogen is *Plasmodiophora brassicae*, and transmission of the disease occurs mainly through resting spores on crop residues from an infected field. The resting spores of *P. brassicae* can survive the winter and remain viable for a considerable time in a frozen state, thus posing great difficulties for disease control. The infestation of host roots by *P. brassicae* is generally divided into two stages: I. primary infection on root hairs and II. secondary infection in root cortical cells (Figure 2). The duration of these two phases varies slightly depending on the physiological subspecies of *P. brassicae* and the host [15]. A method of fluorescent probe-based confocal microscopy was used to investigate the root infection process of *P. brassicae* on *Arabidopsis* roots (Figure 2) [16]. During the primary infection (0–7 dpi, dpi = days post infection), resting spores germinated and produced primary mobile spores; the encapsulated primary mobile spores pierced the host cell wall to produce mononucleated protozoa in root epidermis (1 dpi). Mononuclear Plasmodium underwent mitosis and produced multinucleated zoospores (1–3 dpi), which was then accompanied by cytoplasmic cleavage to produce mononuclear secondary spores (3–4 dpi). Secondary free-living spores were released into root hairs or epidermal cells for primary infection (4–7 dpi). In root epidermal cells, the union of two haploid mononuclear secondary zoosporangium produces diploid mononuclear conidia (2n zygote), a process featuring an increase in chromatin volume and nucleus size. At 8 dpi, the presence of widespread mononuclear secondary plasmodium in root cortical cells marked the establishment of secondary infection. During 10–15 dpi, secondary plasmodium further developed binucleate, tetranucleate, and multinucleate forms, resulting in further occupation of roots by the pathogen and a dramatic increase in the root volume manifested as swelling symptoms. Upon 24 dpi, mononuclear resting spores could be found in cortical cells, marking the completion of one life cycle of *P. brassicae* (Figure 2C).

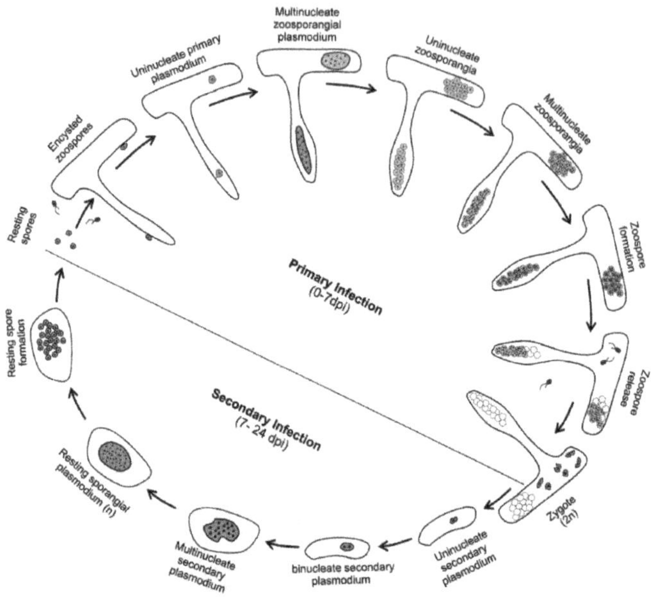

(A)

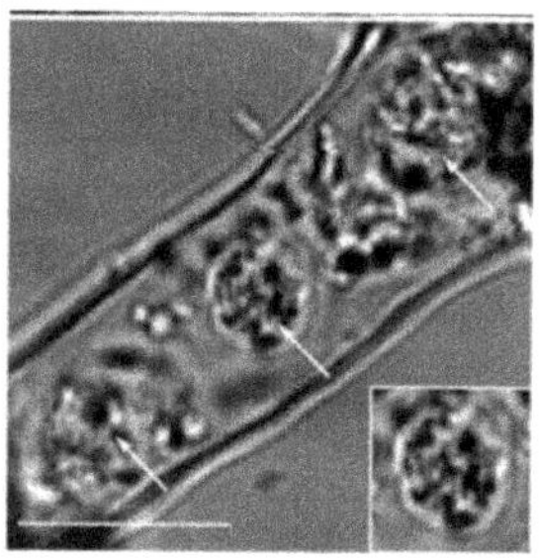

(B)

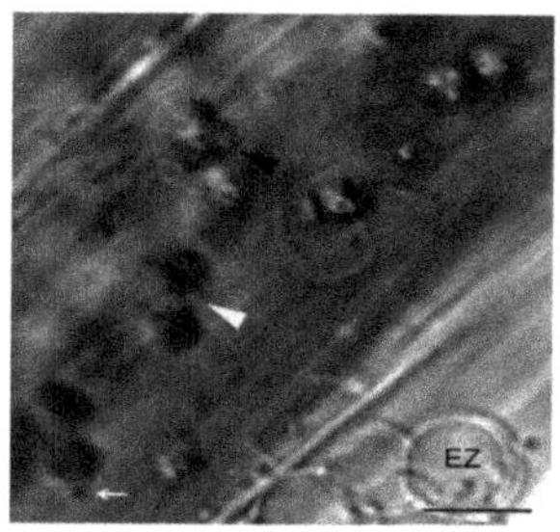

(C)

Figure 2. Diagram of the refined life cycle of *Plasmodiophora brassicae* (Reprinted with permission from Ref. [16]). (**A**) Complete life cycle of *Plasmodiophra brassicae*; (**B**) primary zoospore in root epidermis (white arrows); (**C**) secondary zoospore (white arrowhead) and a fusion of two zoospores to form a zygote (black arrowhead).

By comparison of the infestation process of *P. brassicae* on different host plants, it is now clear that *P. brassicae* host (e.g., rapeseed, cabbage) resistance is mainly determined by the secondary infection stage, i.e., during the cortical infestation, while non-host (e.g., rice, wheat, and barley) resistance acts primarily during the epidermal infestation stage [17]. Previous studies have provided important references for resolving the mechanisms of plant resistance and developing green and efficient prevention and control strategies against clubroot disease [18,19].

1.3. Identification of Physiological Races of P. brassicae

During an investigation of the pathogenesis of different hosts, researchers found that *P. brassicae* has a complex physiological race. That is, pathogens isolated from different regions might have different genetic backgrounds, which result in different disease phenotypes on a given host plant. Several taxonomy systems have been established for different *P. brassicae* species in the world, including (1) the Williams identification system, (2) the

European clubroot differential (ECD) system, and (3) the Sinitic Clubroot differentiation (SCD) system.

1.3.1. Williams Identification System

The Williams identification system was established in 1965, and it is still widely used (Table 1) [20]. This system uses four Brassica species, *Jersey Queen*, *Badger Shipper*, *Laurentian*, and *Wilhelmsburger*, and the classified *P. brassicae* are divided into 16 physiological races [20]. For example, when *Jersey Queen*, *Laurentian*, and *Wilhelmsburger* are susceptible (+) and *Badger Shipper* is resistant (−), the pathogen would be classified as race 1 in the Williams system.

Table 1. The Williams classification system [20].

Host Plant	Race															
	1	2	3	4	5	6	7	8	9	10	11	12	13	14	15	16
Jersey Queen	+	+	+	+	−	+	+	−	−	+	−	+	−	−	−	−
Badger Shipper	−	+	−	+	−	−	+	−	−	+	+	−	+	+	+	−
Laurentian	+	+	+	+	−	−	−	+	+	−	+	−	+	−	−	−
Wilhelmsburger	+	−	−	+	−	−	−	−	+	+	+	+	−	+	−	+

1.3.2. European Clubroot Differential System

The European Clubroot Differential (ECD) system was established in 1975 [21]. The system included 15 plant hosts which can be divided into three major groups: (1) ECD01–ECD05 was the *B. rapa* group (AA, 2n = 20). Within this group, ECD01–ECD04 were *rapifera* and ECD05 was *Pekinesis*, and they could be infected by all races and serve as susceptible controls. (2) ECD06–ECD10 were the *Brassica napus* L. (AACC, 2n = 28) group; and (3) ECD11–ECD15, the *Brassica oleracea* (CC, 2n = 18) group. These materials were collected from different regions of Europe by individual research groups, which served as representative host standards to classify different pathogen races. A binary recording calculation was used to dictate the race of pathogen based on the disease level from the three groups of hosts (Table 2). For example, if the first group's (*B. rapa* group) hosts are all susceptible, the second group's (*B. napus* group) are all resistant, and if only ECD15 of the third group is susceptible, then according to the binary calculation, 0 + 0 + 0 + 0 + 0 = 0/1 + 2 + 4 + 8 + 16 = 31/0 + 0 + 0 + 0 + 0 + 16 = 16, this pathogen race would be recorded as ECD 0/31/16.

1.3.3. The Sinitic Clubroot Differential (SCD) System

Although the ECD system includes three major types of *Brassica* species, there are still considerable regional differences in both the host and pathogen in Asian countries, such as China, Japan, and Korea. Therefore, a SCD (Sinitic Clubroot Differential) system was developed for pathogen race identification in mainland China (Table 3). Previous studies using the Williams system showed that the dominant *P. brassicae* race in of China was pathotype 4 [22]. However, due to the annual variation in pathogen populations and difficulty in standardization, a single-spore isolation method was developed recently by Zhang et al., who characterized *P. brassicae* strains isolated from nine different locations of Chinese cabbage cultivation field and obtained a total of 281 single-spore strains belonging to 15 disease types according to the Williams system, of which disease type 4 accounted for the largest proportion [23]. Bai et al. tested 42 species of *P. brassicae* on 12 different hosts, and developed the SCD system [24].

Table 2. The ECD classification system [22].

Differential Number	Differential Host	Binary Numbers	Decimal Numbers
	2n = 20 (*Brassica rapa* L. Sensu lato)		
ECD 01	ssp. *rapifera* line aaBBCC	2^0	1
ECD 02	ssp. *rapifera* line AAbbCC	2^1	2
ECD 03	ssp. *rapifera* line AABBcc	2^2	4
ECD 04	ssp. *rapifera* line AABBCC	2^3	8
ECD 05	ssp. *Pekinensis* line Granaat	2^4	16
	2n = 38 (*Brassica napus* L.)		
ECD 06	var. *napus* cv. Nevin line Dc101	2^0	1
ECD 07	var. *napus* cv. Giant line Dc119	2^1	2
ECD 08	var. *napus* line Dc128	2^2	4
ECD 09	var. *napus* cv. Clubrootresistance Dc129	2^3	8
ECD 10	var. *napus* cv. Wilhelmsburger Dc130	2^4	16
	2n = 18 (*Brassica oleracea* L.)		
ECD 11	var. *capitata* cv. Badger Shipper	2^0	1
ECD 12	var. *capitata* cv. Bindsachsener	2^1	2
ECD 13	var. *capitata* cv. Jersey Queen	2^2	4
ECD 14	var. *capitata* cv. Septa	2^3	8
ECD 15	var. *acephala* subvar. Laciniata cv. Verheul	2^4	16

Table 3. The Sinitic clubroot differential (SCD) system [24].

SCD	Pb1	Pb2	Pb3	Pb4	Pb5	Pb6	Pb7	Pb8	Pb9	Pb10	Pb11	Pb12	Pb13	Pb14
WIlliams classification system	2/4/7/11	4	4	4	2/4	4	4	4	4	4	4	4	4	4
H08	−	−	−	−	−	−	−	−	−	−	−	−	−	+
H03	−	−	−	−	−	−	−	−	+	+	+	−	+	+
H01	−	−	−	−	−	−	−	+	−	−	+	−	+	+
H04	−	−	−	−	−	−	+	−	−	+	+	+	+	+
H02	−	−	−	−	+	+	+	−	+	+	−	+	+	−
H05	−	−	−	+	−	+	−	−	+	+	−	+	+	−
H06	−	−	+	−	−	−	−	−	−	−	−	−	−	−
H07	−	+	−	−	−	−	−	−	−	−	−	−	−	−
H12	+	+	+	+	+	+	+	+	+	+	+	+	+	+

Note: Pb1–Pb14 are different physiological races of clubroot disease; H01–H08 are different resistant hosts of clubroot disease.

A total of 14 physiological races (Pb1–Pb14) were included in SCD system, with Pb1 as the dominant race [24]. Rui et al. improved the single-cell isolation protocol and combined the SCD system with the Williams system on strains collected from 11 provinces in China [25]. In terms of resistance breeding, disease-level phenotyping is critical. However, pathogens present in the field always constitute a population and this population is likely to change from year to year. Therefore, resistance developed towards a major pathotype may disappear upon long-term use. Since the pathogen could not be multiplied in vitro, it is critical to compare the disease phenotype based on single-spore method, i.e., a uniform genotype of the pathogen strain. Therefore, a scientific and efficient taxonomy system is very important, and this system must be in accordance with the regions to monitor changes possibly involved in field population. Different pathotypes of *P. brassicae* identified in the field can also facilitate resistance breeding in laboratory and greenhouse contexts.

1.4. Damage of Clubroot Disease

As mentioned above, *P. brassicae* infects the roots of cruciferous crops, and secondary infection results in swollen roots and the loss of root function, which in turn affects the development of the above-ground parts. *Brassica* L. is one of the most important genera in the cruciferous family, which include hundreds of agronomic crops such as rapeseed, Chinese cabbage, cabbage, shepherd's purse, radish, turnip, and many others. Many of them can be infected, but there are species that carry resistance genes and become immune to the disease. The area for cruciferous crop cultivation is expanding with modern agronomy techniques; additionally, the resting spores of *P. brassicae* can survive in soil for 8–12 years or even longer, letting mutations accumulate and contribute the appearance of new physiological races [26]. Warm temperature might be in favor of disease outburst, although so far no direct evidence has been reported. Modern farming measures also constitute a major reason for the rapid spreading of clubroot disease in recent years.

Since the clubroot disease infection can occur even in seedlings, the earlier the onset time, the more severe the disease outcome. At the beginning of the disease, there is no obvious phenotype on the aboveground parts, but the new leaf growth will be significantly inhibited. During the middle and late stages, the aboveground parts of the host plant will have stunted growth, yellowing and wilting starts to occur at the base of the leaves, and root galls of different sizes, shapes, and locations will be formed [27]. With the disease progression, the root xylem will be destroyed, the leaves and stems will eventually wilt, and the reduction of the absorption of water and nutrients by the roots will cause a great loss of yield and in extreme cases no harvest at all [28].

The size and location of the root galls/tumors is the basis for disease level grading. A four-level grading system of clubroot disease is proposed as follows: level 0-normal root, no tumor; level 1-no tumor on the main root, small tumor on lateral roots and fibrous roots; level 2-medium tumor on the main root, large tumor on some lateral roots; level 3-large tumor on the main root and lateral roots, enlargement of the basal part of stem, stunted plant growth [29,30]. In addition, Hu reported a disease-grading method especially developed for rapeseed using a scale of 1–4, which can better reflect and assess the severity of disease symptoms [31].

Due to the nature of soil-borne disease, temperature, soil pH, and humidity are significant factors affecting the germination of resting spores and therefore disease incidence. Studies have shown that soil temperatures of 18–25 °C, humidity of about 60%, and pH values of 5.4–6.5 are the optimum conditions for spore germination [32]. In accordance with the climate zone, Hubei, Hunan, Yunnan, Anhui, Sichuan, Jilin, Liaoning, northeast China, southwest China, and Shandong are the main sites of clubroot disease in China [33]. The disease affects an area of 3.2–4 million hm^2 per year in China, accounting for more than one-third of the total cultivation area of cruciferous crops. In an extreme year, the affected area can reach 9 million hm^2, with an average yield loss of 20%–30%, and in extreme cases a loss of more than 60% in the field. Therefore, clubroot disease has become a critical problem for the rapeseed industry and also a great threat for many vegetables; clubroot disease has demanded great attention during recent years as a bottleneck agricultural problem in China [34].

1.5. Control Measures of Clubroot Disease

The control of clubroot disease mainly follows the policy of "prevention-oriented, integrated control". It emphasizes the important role of agro-ecological control in disease management, while coordinating biological and chemical control techniques to ensure maximum socio-economic and ecological benefits. The current control measures for clubroot disease are listed below:

(1) Field management: *P. brassicae* is spatially aggregated in soil, with high incidence at entrances and field margins [35]. The viability and longevity of *P. brassicae* are closely related to soil properties, and it has been shown that alkaline addition in soil can reduce the germination rate of dormant spores, decrease root-hair infection, and

inhibit the maturation of sporangia and Zoosporangium [36]. Therefore, increasing soil pH with lime has been often used for the management of small acreage incidence [37–39]. In addition, it has been shown that high concentrations of calcium, boron, and magnesium have important effects on soil inoculum density [40]. High concentrations of calcium are involved in the induction of relevant defense compounds and the induction of host cell death by reducing dormant spore germination and sporangial development at the same time [41,42]. High concentrations of boron could slow down the development of *P. brassicae* by inhibiting its growth during primary infection stage [43]. Interestingly, clubroot incidence and severity were found to be affected by the level of total and individual glucosinolates between oilseed rape cultivars [44]. Since different crops have difference impact on agronomic residues on soil after growth seasons, crop rotation measures have also been used to avoid pathogen accumulation in open fields.

(2) Chemical control: chemical agents are used to inhibit the germination of resting spores. Pre-disease control is critical for disease prevention; current measures consist of pharmaceutical seed dressing, seedbed disinfection, soil fumigation, joint root irrigation, etc. A few conventional fungicides, such as SDD (sodium dimethyl dithiocarbamate), thiram, carbendazim, fluazinam, cyazofamid, and pentachloronitrobenzene, have shown to be effective. In addition, options are provided for combining different chemicals together to achieve better results, such as 58% metalaxyl mancozeb (1500 times dilution) and 75% chlorothalonil (1000 times dilution) for milder years [45].

(3) Biological control: the soil contains a large number of microorganisms; previous studies showed that *Trichoderma* and *Streptomyces* spp. can suppress *P. brassicae* in cauliflower both in greenhouse and in the field [46]. *Heteroconium chaetospira* is effective against the development of clubroot disease in cabbage at low to moderate soil moisture [47]. Application of formulated biocontrol agents including *Bacillus subtilis* and *Gliocladium catenulatum* could significantly reduce the incidence of clubroot in *Brassica napus* L. [48]. The endophytic fungus *Acremonium alternatum* was shown to suppress clubroot disease in cabbage and *Arabidopsis thaliana* [49].

(4) Breeding for disease-resistance cultivars: Field management can reduce disease incidence to some extent, but it requires a lot of labor and at the same time does not fundamentally solve the problem. Indeed, researchers have tried to quantify the abundance of clubroot pathogens using the qRT-PCR method, although in most cases the pathogen is present in a mixed population [50]. Comparative bioassays performed in growth chambers showed that resistance was under selection pressure, and the use of clubroot-resistant cultivars is recommended when *P. brassicae* DNA exceeds 1300 genes copies per gram soil [50]. Chemical control can be efficient but is costly and causes environmental pollution. Therefore, breeding for resistant cultivars could protect the plant and environment together from the disease; this is the most fundamental and effective way to prevent disease spreading. Resistant genes (R genes) could be identified from plant materials that are naturally immune. Breeding for disease resistance using R genes is not only very effective, but also in line with sustainable development strategies.

2. Plant Immune Pathways and R Genes

2.1. Plant Immune Response Pathways

Plants do not have specific immune cells or a somatic adaptive immune system, such as that of mammals. During evolution, plants have developed their own immune systems by recognizing invading pathogens (viruses, bacteria, and fungi) through various receptors on the cell surface as well as inside the cells [51–53]. Currently, the plant immune system is constituted by two pathways. (1) The primary immune pathway, also known as the PTI (pattern-triggered immunity) pathway, is activated by cell-surface pattern recognition receptors (PRR), and recognizes invading pathogenic microorganisms by microbe-associated molecular pattern (MAMP) or damage-associated molecular pattern

(DAMP). (2) The secondary immune response, also known as the ETI (effector-triggered immunity) pathway, is triggered by effectors released by pathogens; plants in turn can evolve resistance genes (R genes) that recognize the effectors and trigger host-immune responses [54] (Figure 3). The most commonly accepted model for PTI-ETI interaction is the "zigzag" model [55]. According to this model, PTI and ETI are temporally and spatially distinct and mediated by different factors, but they also interact on the molecular level and there are considerably overlap partners downstream of the ETI and PTI pathways. PTI is the front line of plant defense against pathogens and stimulates the basal defense, while ETI is an accelerated and amplified response of PTI and is generally more effective in preventing further transmission [56].

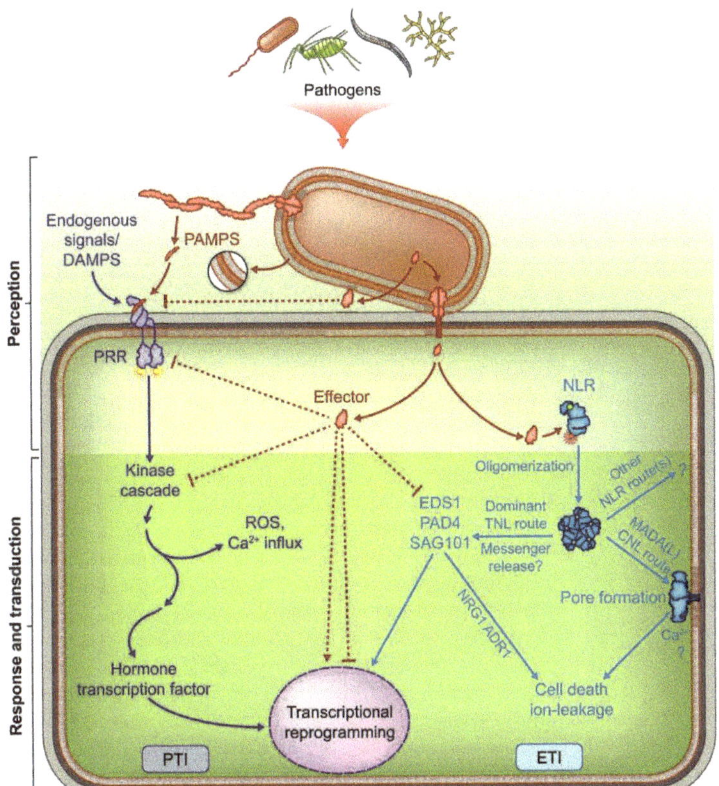

Figure 3. Schematic diagram of the plant immune system (Reprinted with permission from Ref. [3]).

2.2. *Disease-Resistance (R) Genes and the NBS–LRR Protein Family*

Most of the plant disease-resistance genes (R genes) identified so far encode proteins of the NBS–LRR (Nucleotide Binding Site–Leucine Rich Repeat) family, which are also known as NLR proteins as the major type for the plant R gene family. The NLR protein consists of three main components: the variable N-terminal structural domain, the NB (Nucleotide-Binding) structural domain, and the C-terminal conserved LRR (leucine-rich-repeat) structural domain [57,58]. Based on the characteristics of the N-terminal structural domain, NLRs are mainly divided into TNL with TIR (Toll-interleukin-1 receptor) at the N-terminal, and CNL with a CC (coiled-coil) structural domain at the N-terminal. The CNL class R genes were found in both dicotyledonous and monocotyledonous plants and significantly more than the TNL class. However, TNL class R genes were detected only in dicotyledons [59–61].

During plant immunity, NLR proteins act as intracellular immune recognition receptors, recognizing effectors released by pathogens and triggering immune responses [62–64]. The ways in which plant NLRs are involved in resistance are divided into direct and indirect effects. Typical of the "gene-for-gene" model is the interaction between the flax rust resistance fungal gene *AvrL567* and the L protein [65]. In rice, the Avr-Pita176 protein binds directly to the Pi-ta LRD region to initiate an immune response against rice blast fungus [66]. The TNL family member RPP1 (Recognition of Peronospora parasitica 1) in Arabidopsis directly and specifically recognizes the ATR1 (Arabidopsis thaliana Recognized 1) effector variant produced by the foliar oomycete pathogen *Hyaloperonospora arabidopsidis* (Hpa) to trigger an immune response [67].

However, most of the NLR proteins are bound to other host proteins before recognizing the effector (Table 4). The TNL protein RPS4 (resistance to *Pseudomonas syringae* 4) can interact specifically with the transcriptional activator bHLH84 and they mediate transcriptional regulation downstream of immunity [68]. RPS4 can also act in concert with RRS1 (resistance to *Ralstonia solanacearum* 1) to confer recognition of *Pseudomonas* AvrRps4 and *Ralstonia* PopP2 [69,70]. In AvrRps4-triggered resistance, RPS4 crosstalks with SNC1. While SPRF1 acts as a transcriptional repressor, its mutation activates SNC1 and lead to enhanced resistance [71]. The fact that RPS4 can interact with multiple proteins reflects the structural diversity of the protein, but the underlying mechanism regarding whether there is competition between multiple factors is not yet fully understood.

CNL protein RPM1 confers resistance to *Pseudomonas syringae* by recognizing the *Pseudomonas* effector Avrpm1 (ADP-ribosyltransferase) and AvrB through the phosphorylation of RIN4 during infection [72,73]. RPS2 (RESISTANT TO P. SYRINGAE2) is activated in Arabidopsis (At) RIN4 by the *Pseudomonas syringae* effector AvrRpt2, forming the AvrRpt2–RIN4–RPS2 defense-activation module [74]. CRT1 encodes a protein with ATPase activity and is an important mediator of defense signaling triggered by R proteins such as RPS2 [75]. Activation of the RPS5 protein requires PBS1 cleavage to trigger ADP–ATP exchange [76]. In other dicotyledons, the tobacco mosaic virus-resistant CNL protein NRG1 plays an important role in the recognition of the TNL proteins Roq1 and RPP1 [77]. The CNL protein Rx1 in potato interacts with NbGlK1 to regulate the binding affinity for DNA [78]. In monocotyledonous species, this indirect action-induced immune response is also prevalent. In barley, a series of MLA proteins (including MLA1, MLA6, and MLA10), which belong to CC-type NLRs, interact with RING-type E3 ligases and mediate the resistance to powdery mildew fungi (*Blumeria graminis*) [79,80]. In rice, Pigm genes encode a set of NLRs, including PigmR, which mediates broad-spectrum resistance. Additionally, PIBP1, a CNL protein containing an RNA-recognition structural domain (RRM), can interact with PigmR to accumulate in the nucleus in an NLR-dependent manner and directly bind target genes *OsWAK14* and *OsPAL1* A/T cis-acting elements of DNA to activate defense against rice plague [81].

In addition to this, the process of plant immunization is usually accompanied by a hypersensitivity response (HR) or local programmed cell death (PCD). Therefore, maintaining the homeostasis of plant NLR proteins is critical for balancing between immunity and growth [82–85]. The abovementioned studies fully demonstrate that NLR proteins have key roles in the disease-resistance pathways of different pathogens, both in monocotyledonous and dicotyledonous plants. By influencing the binding and possible recognition of effectors by NLR proteins and downstream helper NLRs, different circuits of immune response pathways involving phytohormones and transcriptional reprogramming are initiated and motivated, leading to resistance and morphological changes accompanied by with disease progression. In this sense, NLR-like R genes are the most important gene resources for disease-resistance breeding.

Table 4. NLR protein resistance in different crops and their intercrossing proteins.

	Host Plant	NLR Class	NLR Protein	NLR-Interacting Protein (Type)	Pathogen	Effector	Reference
Dicot	Arabidopsis	TNL	RPS4	bHLH84 (TF)	*Pseudomonas syringae*	AvrRps4	[68]
				RRS1 (Paired NLRs)	*Pseudomonas syringae*	AvrRps4, PopP2	[69]
					Ralstonia olanacearum		[70]
			SNC1	SRFR1 (TPR domain)	*Pseudomonas syringae*	AvrRps4	[71]
		CNL	RPS5	PBS1 (RLCK VII family kinase)	*Peronospora parasitica*	AvrPphB	[76]
			RPM1	HSP90.2 (Chaperone)	*Pseudomonas syringae*	AvrRpm1,AvrB	[72]
				RIN4 (Unknown)	*Pseudomonas syringae*	AvrRpm1,AvrB	[73]
			RPS2	CPR1 (E3 ligase (F-box))	*Pseudomonas syringae*	—	[84]
				RIN4 (Unknown)	*Pseudomonas syringae*	AvrRpt2	[74]
				CRT1 (ATPase activity)	*Pseudomonas syringae*	AvrRpt2	[75]
				MUSE13 (TRAF)	*Pseudomonas syringae*	AvrRpt2	[83]
	Tobacco	CNL	NRG1	EDS1 (Lipase-like)	*Tobacco mosaic virus*	P50	[77]
	Potato	CNL	Rx	GLK1 (TF)	*Potato virus X*	Coat protein	[78]
Monocot	Barley	CNL	MLA10	WRKY1,WRKY2 (TF)	*Blumeria graminis*	—	[80]
				MYB6 (TF)	*Blumeria graminis*	—	[85]
				MLR1 (E3 ligase (RING))	*Blumeria graminis*	—	[79]
			MLA1	MLR1 (E3 ligase (RING))	*Blumeria graminis*	AvrA1	
			MLA6	MLR1 (E3 ligase (RING))	*Blumeria graminis*	—	
	Rice	CNL	PIBP1	PigmR (TF)	*Magnaporthe oryzae*	—	[81]

3. Clubroot Resistance (CR) Genes and Resistance Breeding in *Brassica* Species

Brassica spp. have been domesticated and artificially selected over a long period of time from three diploid parents, *Brassica rapa* (AA, 2n = 20), *Brassica nigra* (BB, 2n = 16), and *Brassica oleracea* (CC, 2n = 18). The combination of the three allopolyploid species gives rise to diverse and rich members of the Brassica genus, including three tetraploid groups, *Brassica napus* L. (AACC), *Brassica juncea* (AABB), and *Brassica carinata* (BBCC) [86]. These hundreds of Brassica species share a high degree of genome structure similarity and duplication from ancestors, but also have accumulated mutations, gene loss, and functional divergence, leading to phenotypic differences in disease resistance.

In order to create a new germplasm with disease resistance, a parent with CR genes needs to be identified and crossed with an elite parent with other desirable agronomy traits such as a tight shape, high yield, and abiotic stress tolerance. Wide hybridization has been widely used to innovate CR cultivars using the crossing of species within the Brassica genus. Marker assisted selection (MAS) facilitates to narrow down the relevant genetic and chromosomal regions and pinpoint candidate genes. Traditional and molecular breeding are combined to accelerate the CR breeding process.

3.1. Progress on CR Loci Mapping and CR Gene Identification

B. oleracea, *Raphanus sativus* L. (RR, 2n = 18), and *B. rapa* ssp. *Rapifera* (AA, 2n = 20) are the main resource materials for clubroot resistance (CR) genes since they are immune to the clubroot pathogen. The most widely used materials for clubroot disease resistance are the ECD series of European turnip, especially ECD01, ECD02, ECD03, and ECD04 [87]. The CR loci of European turnips were mainly distributed on chromosomes from the A genome as quality traits. Some candidate CR genes have been identified by fine mapping and functional validation. Diederichsen and Sacristan artificially synthesized allotetraploids using ECD04 and *Brassica oleracea* var. *capitata Linnaeus*, and identified three CR dominant loci, *Pb-Bn1*, *Pb-Bn2*, and *Pb-Bn3* [88].

CR loci identified so far from the ECD series include *Crr1* (A08), *Crr2* (A01), *Crr3* (A03), *Crr4* (A06), *CRa* (A03), *CRb* (A03), *CRc* (A02), *CRk* (A02), *Rcr1* (A03), *PbBa3.1* (A03), *PbBa3.2* (A03), *PbBa3.3* (A03), and *PbBa8.1* (A08) (Figure 4). As shown in Figure 4, most of the CR loci were present on chromosome A03 of the A genome [89]. *Crr3*, *CRk*, *PbBa3.3*, *CRd*, and *BraCRc* loci are located on upper part of A03 and referred as the "A03-1 cluster"; *CRa*, *CRb*, *CRbki*, *BraA3P5X/G.CRa/bKato1.1*, *BraA3P5X/G.CRa/bKato1.2*, *Rcr1*, *Rcr2*, *Rcr4* and *BraCRa* are located on the lower arm of chromosome A03 and therefore named as "A03-2 cluster" [90]. Among all the CR loci and CR gene candidates, only *CRa* and *Crr1a* have been successfully cloned and functionally validated [91,92].

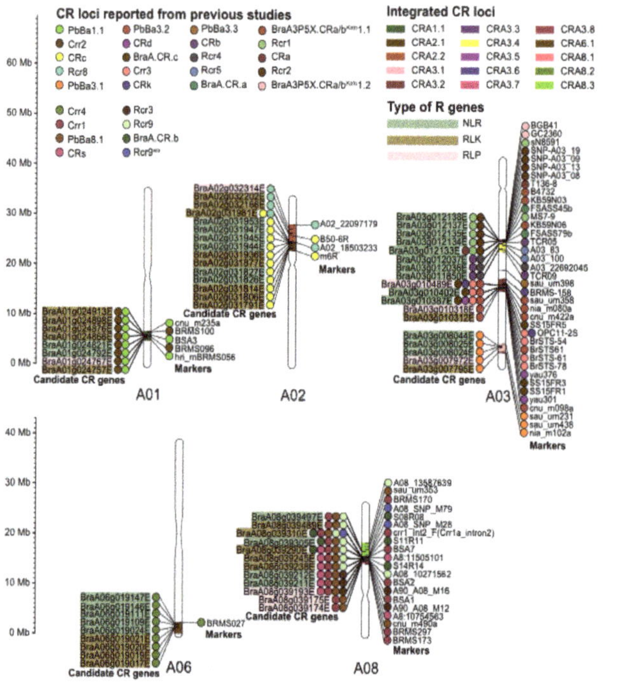

Figure 4. Physical mapping of CR loci (Reprinted with permission from Ref. [89]).

However, a single resistant variety cannot maintain stable resistance over a long time. Convergent breeding by aggregating disease-resistance genes from different sources in a single material is expected to improve the broad spectrum and persistence of disease resistance in varieties and become a more practical and effective breeding model [93]. In order to do this, more loci and CR genes need to be identified from different germplasms that are immune to the disease. The nature and relationship between these loci need to be evaluated before their utilization as gene resources for resistance breeding.

In 2010, Kamei et al. crossed Japanese radish (CR donor) with Chinese susceptible radish to construct a mapping population with 18 linkage groups; they used AFLP and SSR markers to identify a region of 554 Mb [94]. Matsumoto et al. obtained pure lines with high resistance by mounting three CR genes (*CRa*, *CRk*, and *CRc*), and demonstrated that disease resistance can be elevated by mounting CR genes [95]. By SNP mapping and RNA sequencing, Huang et al. identified two possible CR genes (*Bra019410* and *Bra019413*) from *Rcr2* loci in cabbage [96]. Recently, *Rsa10003637* and *RSA1005569/Rsa10025571* were identified as CR loci from radish; a significant correlation was found between the *Rsa10025569* locus and disease resistance in a BC1F1 population [97].

Arabidopsis, a model crop in the Cruciferae family, is a good model for analyzing the resistance mechanisms of *P. brassicae*. The earliest analyses of *Arabidopsis* resistance to clubroot disease were mainly performed with multiple metabolic pathways [98–100], and the observation of the natural response of *Arabidopsis* to clubroot disease in various locations [101–103]. Currently, in addition to the identification of the gene *RPB1*, located on chromosome 1, involved in clubroot disease resistance [104], Jubault et al. identified four additive QTL loci, *Pb-At5.23*, *Pb-At5.1*, *Pb-At1*, and *Pb-At4*, and all of these alleles for resistance were derived from the parent Bur-0 [105]. In addition to this, the homology of well-defined resistance genes on the *Arabidopsis* genome was used to further design designer markers for fine targeting [106].

In the typical Brassica radish, Kamei et al. identified *Crs1*, and found that the genomic region around *Crs1* and the genomes around *Crr3* on turnip (*B. rapa.*) share a common ancestor [94]. Gan et al. identified five QTL loci, *RsCr1*, *RsCr2*, *RsCr3*, *RsCr4*, and *RsCr5*, associated with clubroot disease resistance, with *RsCr1* being homologous to the well-defined locus *Crr1* [107]. Recently, Gan et al. identified a new locus *RsCr6* on chromosome 8 and screened for possible resistance candidate genes *R120263140* and *R120263070* [108].

The investigation of the resistance mechanism of different crops to clubroot disease can help us further understand the resistance loci as well as provide a solid theoretical basis for breeding against clubroot disease.

3.2. Genomic and Molecular Markers Associated with Clubroot Resistance

With the development of sequencing technologies, different omics have been used on combinations of different hosts and pathogens to understand plant–pathogen interactions [109,110]. Yu et al. performed QTL analysis on resistant cultivars to *P. brassicae* and mapped three QTLs on chromosomes A02, A03, and A08; one QTL, *Rcr4* on chromosome A03, was responsible for resistance to pathotypes 2, 3, 5, 6, and 8 in the Williams system [111]. The QTLs on chromosomes A02 and A08 were named *Rcr8* and *Rcr9* respectively, and two TNL genes were identified from genomic regions between *Bra020936* and *Bra020861* around *Rcr9* loci in *B. rapa* [111]. On the other hand, proteome and metabolome studies showed differential expression of proteins in lipid metabolism, plant defense, cell-wall repair, hormone production, and signal transduction in response to *P. brassicae* infestation [112].

Marker-assisted breeding has been extensively used for clubroot-resistance breeding. With more genomic information released, genomic sequence has been used more frequently for the development of new markers. Zheng et al. identified five molecular markers for pathotype identification, which can distinguish race P11 from P4, P7, and P9, and similarly P9 from P4, P7, and P11 [113]. Lei et al. validated the genetic stability of two co-dominant markers *CRaEX04-1* and *CRaEX04-3* associated with the *CRa* gene in cabbage using 57 resistant varieties and two genetic populations [114]. Jiang et al. studied the CR locus found in resistant "Kc84R" and identified *BnERF034* as one of the CR genes on chromosome A03 [115]. Indeed, many CR loci have been reported, including the ones recently identified by Wang et al., for two QTLs on A03 and A08, conferring resistance to pathotypes 3H, 3A, and 3D in turnip [116]. The development of markers and identification of genomic regions responsible for clubroot resistance laid an important foundation for marker-assisted breeding for the generation of resistant cultivars with durable resistance to clubroot disease.

3.3. Progress on CR Breeding for Clubroot Disease

In 2015, Gao et al. performed disease phenotyping on twenty germplasm of *Brassica napus* L; they found that Huayouza 9 and Huashuang 3 had strong resistance [117]. Using molecular marker-assisted selection, *PbBa8.1* locus of turnip ECD04 was transferred into the elite *B. napus* variety Huashuang 5 to create clubroot-resistant line H5R, which was immune to most of the race 4 pathogens in China [118]. A dominant CR gene CRd was successfully transferred from the self-incompatible line "85–74" to the conventional varieties "W3" and "Zhong Shuang 11", resulting in two new germplasms, "W3R" and "Zhong Shuang 11R", respectively [119]. In 2021, Li Qian et al. successfully developed the first hybrid oilseed rape variety "Huayouza 62R", by hybrid combination of a sterile line Huayouza 62R with a resistant cabbage donor parent Bing409R [80]. Huayouza 62R contains two disease-resistance loci, *PbBa8.1* and *CRb*, and showed excellent performance on field trial in disease areas [120]. Hou et al. used "Hua Resistant No. 5" as the source material and Ogu CMS (cytoplasmic male sterility) recovery line RF04 as the recipient, and created a kale-type spring oilseed rape immune to race 4 pathogens [121]. In an attempt to test the contribution of different CR genes and feasibility to promote the resistance by multiple CR genes in one germplasm, Nadil performed hybridization between kale type 409R containing *CRa* gene with kale type rape 305R containing *PaBa8.1*, and selected progeny with two CR loci. The plants with two CR loci displayed good additive effect for disease resistance, supporting a valid basis for the gene-mounting strategy for CR breeding [122].

In addition to breeding for clubroot disease resistance in oilseed rape, Sun Chaohui et al. developed an early maturing variety of cabbage "Anxiu" with resistance to clubroot disease in several trials in Shandong [123]. Yang et al. obtained a hybrid F1 of *B. oleracea* × *B. napus rape* carrying both clubroot disease-resistance genes and Ogura CMS-recovery genes through distant crosses and embryo rescue [124]. He et al. developed 14 new disease-resistant cabbage varieties using heterozygous crossbreeding and molecular marker-assisted selection techniques to meet the production needs of Yunnan cabbage [125]. A generation hybrid, "Jingchun CR3", with resistance to clubroot disease and tolerance to the shoots of cabbage was created by crossing two self-incompatible lines, CR1572 and CR1582, by Yu Yangjun et al. [126].

4. Conclusions

Clubroot disease is considered as a "cancer" for *Brassica* species; the fast spreading of the disease as well as the risk of losing resistance over time calls for a deeper understanding of the *Plasmodiophora* pathogen and the host pathways leading to disease resistance. In this review, we covered a basic background of the disease distribution and the pathogen's nature, with a greater focus on the plant ETI pathways and the roles of NLR proteins as R genes for clubroot disease. We strongly believe that with the identification and isolation of more CR (clubroot resistance) genes, more resistance materials could be developed to provide a better safeguard for the agronomy industry of *Brassica* species.

Author Contributions: C.Z. (Chaoying Zhang) and C.D. prepared figures and drafted the manuscript, and Y.L. and H.W. helped to revise the manuscript. C.Z. (Chunyu Zhang) and P.C. conceived the study, participated in its coordination. All authors have read and agreed to the published version of the manuscript.

Funding: This work was supported by the National Natural Science Foundation of China (U20A2034 and 31871659) and China Agriculture Research System (CARS-12) to CZ.

Data Availability Statement: Not applicable.

Conflicts of Interest: The authors declare no conflict of interest.

References

1. Watson, A.G.; Baker, K.F. Possible gene centers for resistance in the *genus Brassica* to *Plasmodiophora brassicae*. *Econ. Bot.* **1969**, *23*, 245–252. [CrossRef]
2. Suo, H.; Chen, Z.L.; Xu, H.; Song, B.; Fan, X.X.; Yuan, X.H.; He, C.Z. Research progress of cruciferae clubroot disease. *J. Anhui Agric. Sci.* **2015**, *43*, 115–117.
3. Anderson, A. Report on the disease of finger and toe in turnips Trans. *Highl. Agric. Soc. Scot. Ser.* **1855**, *6*, 118–140.
4. Woronin, M.S. *Plasmodiphora brassicae*, the cause of cabbage Hernia. *J. Sci. Bot.* **1978**, *11*, 548–574.
5. Desoignies, N.; Eickermann, M.; Delfosse, P.; Kremer, F.; Godart, N.; Hoffmann, L.; Legrève, A. First Report of *Plasmodiophora brasscae* on Rapeseed in the Grand Duchy of Luxembourg. *Plant Dis.* **2009**, *93*, 1220. [CrossRef] [PubMed]

6. Tewari, J.P.; Strelkov, S.E.; Orchard, D.; Hartman, M.; Lange, R.M.; Turkington, T.K. Identification of clubroot of crucifers on canola (*Brassica napus*) in Alberta. *Can. J. Plant Pathol.* **2005**, *27*, 143–144. [CrossRef]
7. Strelkov, S.E.; Hwang, S.F. Clubroot in the Canadian canola crop: 10 years into the outbreak. *Can. J. Plant Pathol* **2014**, *36*, 27–36. [CrossRef]
8. Hirai, M. Genetic analysis of clubroot resistance in *Brassica crops*. *Breed. Sci.* **2006**, *56*, 223–229. [CrossRef]
9. Dixon, G.R. The Occurrence and Economic Impact of *Plasmodiophora brassicae* and Clubroot Disease. *J. Plant Growth Regul.* **2009**, *28*, 194–202. [CrossRef]
10. Chai, A.L.; Xie, X.W.; Shi, Y.X.; Li, B.J. Special Issue: Research status of clubroot (*Plasmodiophora brassicae*) on cruciferous crops in China. *Can. J. Plant Pathol.* **2014**, *36*, 142–153. [CrossRef]
11. Huang, Q.W.; Ouyan, L.; Wang, Y.J. Occurrence and prevention of clubroot disease on cruciferous crops in Jiangxi. *Bull. Plant Prot.* **1955**, *8*, 1–4.
12. Li, H.M.; Pan, Y.H. Occurrence rules and control technology of clubroot disease on cruciferous crops. *J. Changjiang Veg.* **2018**, *9*, 15–16.
13. Zheng, J.; Wang, X.; Xiao, Y.; Wei, S.; Wang, D.; Huang, Y.; Wang, W.; Yang, H. Specific Genes Identified in Pathotype 4 of the Clubroot Pathogen *Plasmodiophora brassicae*. *Plant Dis.* **2019**, *103*, 495–503. [CrossRef]
14. Wang, Y.Y.; Yang, Z.Q.; Yang, Q.Y.; Zhang, C.Y. Genetic improvement and application of resistance to brassica napus clubroot. *J. Huazhong Agric. Univ.* **2021**, *40*, 1–5.
15. Schwelm, A.; Fogelqvist, J.; Knaust, A.; Julke, S.; Lilja, T.; Bonilla-Rosso, G.; Karlsson, M.; Shevchenko, A.; Dhandapani, V.; Choi, S.R.; et al. The *Plasmodiophora brassicae* genome reveals insights in its life cycle and ancestry of chitin synthases. *Sci. Rep.* **2015**, *5*, 11153. [CrossRef] [PubMed]
16. Liu, L.J.; Qin, L.; Zhou, Z.; Hendriks, W.; Liu, S.; Wei, Y. Refining the Life Cycle of *Plasmodiophora brassicae*. *Phytopathology* **2020**, *110*, 1704–1712. [CrossRef] [PubMed]
17. Liu, Y.; Li, D.L.; Yang, N.; Zhu, X.L.; Han, K.X.; Gu, R.; Bai, J.Y.; Wang, A.X.; Zhang, Y.W. Genome-Wide Identification and Analysis of CC-NBS-LRR Family in Response to Downy Mildew and Black Rot in Chinese Cabbage. *Int. J. Mol. Sci.* **2021**, *22*, 4266. [CrossRef] [PubMed]
18. McDonald, M.R.; Sharma, K.; Gossen, B.D.; Deora, A.; Feng, J.; Hwang, S.F. The Role of Primary and Secondary Infection in Host Response to *Plasmodiophora brassicae*. *Phytopathology* **2014**, *104*, 1078–1087. [CrossRef] [PubMed]
19. Pérez-López, E.; Hossain, M.M.; Tu, J.; Waldner, M.; Todd, C.D.; Kusalik, A.J.; Wei, Y.; Bonham-Smith, P.C. Transcriptome Analysis Identifies *Plasmodiophora brassicae* Secondary Infection Effector Candidates. *J. Eukaryot. Microbiol.* **2020**, *67*, 337–351. [CrossRef]
20. Williams, P.H. A system for the determination of races of *Plasmodiophora brassicae* that infect Cabbage and Rutabaga. *Phytopathology* **1966**, *56*, 624–626.
21. Shen, X.Q.; Nie, K.; Wu, Q.; Zhang, Y.G.; Meng, X.H. Preliminary report on identification of population differentiation of main physiological races of Chinese cabbage clubroot disease. *Chin. Veg.* **2009**, *8*, 59–62.
22. Buczacki, S.T.; Toxopeus, H.; Mattusch, P.; Johnston, T.D.; Dixon, G.R.; Hobolth, L.A. Study of physiologic specialization in Plasmodiophora brassicae: Proposals for attempted rationalization through an international approach. *Trans. Br. Mycol. Soc.* **1975**, *65*, 295–303. [CrossRef]
23. Zhang, J.; Wu, Y.; Feng, H.; Ge, W.J.; Liu, X.Y.; Lv, M.C.; Wang, Y.L.; Ji, R.Q. Acquisition and Identification of Monospore *Plasmodiophora brassicae* in Williams System. *Acta Hortic. Sin.* **2019**, *46*, 2415–2422.
24. Bai, P.Y. Establishment of Identification System for *Plasmodiophora brassicae* of *Brassica campestris*. Master's Thesis, Shenyang Agricultural University, Shenyang, China, 2019.
25. Rui, T.T.; Gao, Y.Q.; Li, X.J.; Shi, Y.X.; Xie, X.W.; Li, L.; Zhang, H.J.; Xu, W.J.; Chai, A.L.; Li, B.J. Isolation of Monospore Methylene Blue Agarose from *Brassica campestris* and Identification of *Plasmodiophora brassicae*. *Acta Hortic. Sin.* **2022**, *49*, 1290–1300.
26. Wang, J.; Huang, Y.; Hu, X.L.; Niu, Y.Z.; Li, X.L.; Liang, Y. Studies on symptoms, pathogen morphology and yield loss of *Brassica napus*. *Chin. J. Oil Crop Sci.* **2008**, *1*, 112–115.
27. Zahr, K.; Sarkes, A.; Yang, Y.; Ahmed, H.; Zhou, Q.; Feindel, D.; Harding, M.W.; Feng, J. *Plasmodiophora brassicae* in Its Environment: Effects of Temperature and Light on Resting Spore Survival in Soil. *Phytopathology* **2021**, *111*, 1743–1750. [CrossRef] [PubMed]
28. Zamani-Noor, N.; Brand, S.; Söchting, H.P. Effect of Pathogen Virulence on Pathogenicity, Host Range, and Reproduction of *Plasmodiophora brassicae*, the Causal Agent of Clubroot Disease. *Plant Dis.* **2022**, *106*, 57–64. [CrossRef]
29. Sun, B.Y.; Shen, X.Q.; Guo, H.F.; Zhou, Y.H. Research progress on clubroot disease and resistance breeding of cruciferous plants. *Chin. Veg.* **2005**, *4*, 34–37.
30. Si, J.; Li, C.Q.; Song, H.Y.; Ren, X.S.; Song, M.; Wang, X.J. Identification and Evaluation of Resistance of Cabbage to clubroot disease. *J. Southwest Univ.* **2009**, *31*, 26–30.
31. Hu, C.W.; Zhu, C.; Liu, X.Z.; Huang, T.C. Comparison of grading standards of *Brassica napus* clubroot disease and evaluation of resistance of varieties. *J. Anhui Agric. Sci.* **2020**, *48*, 27–29.
32. Chen, A.H. Occurrence and control of clubroot disease of cruciferous vegetables. *Fujian Agric. Sci. Technol.* **2018**, *11*, 26–29.
33. Li, J.P.; Chai, A.L.; Sun, R.F.; Li, B.J. New progress in research on clubroot disease of cruciferous vegetables. *Chin. Veg.* **2012**, *8*, 1–4.
34. Wang, J.; Huang, Y.; Li, X.L.; Li, H.Z. Research progress of cruciferae clubroot disease. *Chin. Bull. Bot.* **2011**, *37*, 153–158.
35. Botero-Ramirez, A.; Hwang, S.F.; Strelkov, S.E. *Plasmodiophora brassicae* Inoculum Density and Spatial Patterns at the Field Level and Relation to Soil Characteristics. *Pathogens* **2021**, *10*, 499. [CrossRef] [PubMed]

36. Struck, C.; Rüsch, S.; Strehlow, B. Control Strategies of Clubroot Disease Caused by *Plasmodiophora brassicae*. *Microorganisms* **2022**, *10*, 620. [CrossRef] [PubMed]
37. Feng, J.; Hwang, R.; Hwang, S.F.; Strelkov, S.E.; Gossen, B.D.; Zhou, Q.X.; Peng, G. Molecular characterization of a serine protease Pro1 from *Plasmodiophora brassicae* that stimulates resting spore germination. *Mol. Plant Pathol.* **2010**, *11*, 503–512. [CrossRef] [PubMed]
38. Niwa, R.; Nomura, Y.; Osaki, M.; Ezawa, T. Suppression of Clubroot Disease under Neutral pH Caused by Inhibition of Spore Germination of *Plasmodiophora brassicae* in the Rhizosphere. *Plant Pathol.* **2008**, *57*, 445–452. [CrossRef]
39. Tremblay, N.; Bélec, C.; Coulombe, J.; Godin, C. Evaluation of Calcium Cyanamide and Liming for Control of Clubroot Disease in Cauliflower. *Crop Prot.* **2005**, *24*, 798–803. [CrossRef]
40. Myers, D.F.; Campbell, R.N. Lime and the Control of Clubroot of Crucifers: Effects of pH, Calcium, Magnesium, and Their Interactions. *Phytopathology* **1985**, *75*, 670. [CrossRef]
41. Webster, M.A.; Dixon, G.R. Calcium pH and Inoculum Concentration Influencing Colonization by *Plasmodiophora brassicae*. *Mycol. Res.* **1991**, *95*, 64–73. [CrossRef]
42. Takahashi, H.; Takita, K.; Kishimoto, T.; Mitsui, T.; Hori, H. Ca2+ Is Required by Clubroot Resistant Turnip Cells for Transient Increases in PAL Activity That Follow Inoculation with *Plasmodiophora brassicae*. *J. Phytopathol.* **2002**, *150*, 529–535. [CrossRef]
43. Deora, A.; Gossen, B.D.; Walley, F.; McDonald, M.R. Boron Reduces Development of Clubroot in Canola. *Can. J. Plant Pathol.* **2011**, *33*, 475–484. [CrossRef]
44. Zamani-Noor, N.; Hornbacher, J.; Comel, C.J.; Papenbrock, J. Variation of Glucosinolate Contents in Clubroot-Resistant and -Susceptible *Brassica napus* Cultivars in Response to Virulence of *Plasmodiophora brassicae*. *Pathogens* **2021**, *10*, 563. [CrossRef] [PubMed]
45. Chai, H.P.; Kang, Y.J.; Wang, G.L.; Wang, P.S.; Zhang, B.Y. Study on Control Techniques of Cruciferae Vegetable Clubroot. *Mod. Agric. Sci. Technol.* **2017**, *15*, 98–100.
46. Cheah, L.H.; Veerakone, S.; Kent, G. Biological Control of Clubroot on Cauliflower with Trichoderma and *Streptomyces* spp. *New Zealand Plant Prot.* **2000**, *53*, 18–21. [CrossRef]
47. Narisawa, K.; Shimura, M.; Usuki, F.; Fukuhara, S.; Hashiba, T. Effects of Pathogen Density, Soil Moisture, and Soil pH on Biological Control of Clubroot in Chinese Cabbage by *Heteroconium chaetospira*. *Plant Dis.* **2005**, *89*, 285–290. [CrossRef]
48. Peng, G.; Mcgregor, L.; Lahlali, R. Potential biological control of clubfoot on canola and crucifer vegetable crops. *Plant Path.* **2011**, *60*, 566–574. [CrossRef]
49. Auer, S.; Jutta, L.M. Biological control of clubroot (*Plasmodiophora brassicae*) by the endophytic fungus Acremonium alternatum. *Endocyt. Cell. Res.* **2015**.
50. Wallenhammar, A.C.; Omer, Z.S.; Edin, E.; Jonsson, A. Influence of Soil-Borne Inoculum of *Plasmodiophora brassicae* Measured by qPCR on Disease Severity of Clubroot-Resistant Cultivars of Winter Oilseed Rape (*Brassica napus* L.). *Pathogens* **2021**, *10*, 433. [CrossRef]
51. Jones, J.D.G.; Vance, R.E.; Dangl, J.L. Intracellular innate immune surveillance devices in plants and animals. *Science* **2016**, *354*, 6316. [CrossRef]
52. Xia, Q.Z. Advances in plant immune system. *J. Huanggang Norm. Univ.* **2020**, *40*, 65–71.
53. Saur, I.M.L.; Panstruga, R.; Schulze-Lefert, P. NOD-like receptor-mediated plant immunity: From structure to cell death. *Nat. Rev. Immunol.* **2021**, *21*, 305–318. [CrossRef] [PubMed]
54. Song, W.; Forderer, A.; Yu, D.; Chai, J. Structural biology of plant defence. *New Phytol.* **2021**, 692–711. [CrossRef] [PubMed]
55. Jones, J.D.G.; Dangl, J.L. The plant immune system. *Nature* **2006**, *444*, 323–329. [CrossRef] [PubMed]
56. Ngou, B.P.M.; Jones, J.D.G.; Ding, P.T. Plant immune networks. *Trends Plant Sci.* **2022**, *27*, 255–273. [CrossRef]
57. Bernoux, M.; Zetzsche, H.; Stuttmann, J. Connecting the dots between cell surface- and intracellular-triggered immune pathways in plants. *Curr. Opin. Plant Biol.* **2022**, *69*, 2276. [CrossRef] [PubMed]
58. Takken, F.L.W.; Tameling, W.I.L. To Nibble at Plant Resistance Proteins. *Science* **2009**, *324*, 744–746. [CrossRef] [PubMed]
59. Tarr, D.E.; Alexander, H.M. TIR-NBS-LRR genes are rare in monocots: Evidence from diverse monocot orders. *BMC Res. Notes* **2009**, *2*, 197. [CrossRef] [PubMed]
60. Nepal, M.P.; Andersen, E.J.; Neupane, S.; Benson, B.V. Comparative Genomics of Non-TNL Disease Resistance Genes from Six Plant Species. *Genes* **2017**, *8*, 249. [CrossRef]
61. Wei, H.W.; Liu, J.; Guo, Q.W.; Pan, L.Z.; Chai, S.L.; Cheng, Y.; Ruan, M.Y.; Ye, Q.J.; Wang, R.Q.; Yao, Z.P.; et al. Genomic Organization and Comparative Phylogenic Analysis of NBS-LRR Resistance Gene Family in *Solanum pimpinellifolium* and *Arabidopsis thaliana*. *Evol. Bioinf.* **2020**, *16*, 1177. [CrossRef]
62. Belkhadir, Y.; Subramaniam, R.; Dangl, J.L. Plant disease resistance protein signaling: NBS-LRR proteins and their partners. *Curr. Opin. Plant Biol.* **2004**, *7*, 391–399. [CrossRef] [PubMed]
63. Takken, F.L.W.; Albrecht, M.; Tameling, W.I.L. Resistance proteins: Molecular switches of plant defence. *Curr. Opin. Plant Biol.* **2006**, *9*, 383–390. [CrossRef]
64. Bonardi, V.; Cherkis, K.; Nishimura, M.T.; Dangl, J.L. A new eye on NLR proteins: Focused on clarity or diffused by complexity? *Curr. Opin. Immunol.* **2012**, *24*, 41–50. [CrossRef] [PubMed]
65. Ellis, J.G.; Dodds, P.N.; Lawrence, G.J. Flax rust resistance gene specificity is based on direct resistance-avirulence protein interactions. *Annu. Rev. Phytopathol.* **2007**, *45*, 289–306. [CrossRef] [PubMed]

66. Jia, Y.; McAdams, S.A.; Bryan, G.T.; Hershey, H.P.; Valent, B. Direct interaction of resistance gene and avirulence gene products confers rice blast resistance. *EMBO J.* **2000**, *19*, 4004–4014. [CrossRef]
67. Ma, S.C.; Lapin, D.; Liu, L.; Sun, Y.; Song, W.; Zhang, X.X.; Logemann, E.; Yu, D.L.; Wang, J.; Jirschitzka, J.; et al. Direct pathogen-induced assembly of an NLR immune receptor complex to form a holoenzyme. *Science* **2020**, *370*, 1184. [CrossRef]
68. Xu, F.; Kapos, P.; Cheng, Y.T.; Li, M.; Zhang, Y.L.; Li, X. NLR-Associating Transcription Factor bHLH84 and Its Paralogs Function Redundantly in Plant Immunity. *PLoS Pathog.* **2014**, *10*, 1371. [CrossRef]
69. Sohn, K.H.; Segonzac, C.; Rallapalli, G.; Sarris, P.F.; Woo, J.Y.; Williams, S.J.; Newman, T.E.; Paek, K.H.; Kobe, B.; Jones, J.D.G. The Nuclear Immune Receptor RPS4 Is Required for RRS1(SLH1)-Dependent Constitutive Defense Activation in *Arabidopsis thaliana*. *PLoS Genet.* **2014**, *10*, 1371. [CrossRef]
70. Le Roux, C.; Huet, G.; Jauneau, A.; Camborde, L.; Tremousaygue, D.; Kraut, A.; Zhou, B.B.; Levaillant, M.; Adachi, H.; Yoshioka, H.; et al. A Receptor Pair with an Integrated Decoy Converts Pathogen Disabling of Transcription Factors to Immunity. *Cell* **2015**, *161*, 1074–1088. [CrossRef]
71. Kim, S.H.; Gao, F.; Bhattacharjee, S.; Adiasor, J.A.; Nam, J.C.; Gassmann, W. The Arabidopsis Resistance-Like Gene SNC1 Is Activated by Mutations in SRFR1 and Contributes to Resistance to the Bacterial Effector AvrRps4. *PLoS Pathog.* **2010**, *6*, 44. [CrossRef]
72. Choi, S.; Prokchorchik, M.; Lee, H.; Gupta, R.; Lee, Y.; Chung, E.H.; Cho, B.; Kim, M.S.; Kim, S.T.; Sohn, K.H. Direct acetylation of a conserved threonine of RIN4 by the bacterial effector HopZ5 or AvrBsT activates RPM1-dependent immunity in *Arabidopsis*. *Mol. Plant* **2021**, *14*, 1951–1960. [CrossRef] [PubMed]
73. Mackey, D.; Holt, B.F.; Wiig, A.; Dangl, J.L. RIN4 interacts with Pseudomonas syringae type III effector molecules and is required for RPM1-mediated resistance in *Arabidopsis*. *Cell* **2002**, *108*, 743–754. [CrossRef]
74. Alam, M.; Tahir, J.; Siddiqui, A.; Magzoub, M.; Shahzad-ul-Hussan, S.; Mackey, D.; Afzal, A.J. RIN4 homologs from important crop species differentially regulate the *Arabidopsis* NB-LRR immune receptor, RPS2. *Plant Cell Rep.* **2021**, *40*, 2341–2356. [CrossRef] [PubMed]
75. Kang, H.G.; Kuhl, J.C.; Kachroo, P.; Klessig, D.F. CRT1, an Arabidopsis ATPase that interacts with diverse resistance proteins and modulates disease resistance to turnip crinkle virus. *Cell Host Microbe* **2008**, *3*, 48–57. [CrossRef] [PubMed]
76. Ade, J.; DeYoung, B.J.; Golstein, C.; Innes, R.W. Indirect activation of a plant nucleotide binding site-leucine-rich repeat protein by a bacterial protease. *Proc. Natl. Acad. Sci. USA* **2007**, *104*, 2531–2536. [CrossRef]
77. Qi, T.; Seong, K.; Thomazella, D.P.T.; Kim, J.R.; Pham, J.; Seo, E.; Cho, M.J.; Schultink, A.; Staskawicz, B.J. NRG1 functions downstream of EDS1 to regulate TIR-NLR-mediated plant immunity in *Nicotiana benthamiana*. *Proc. Natl. Acad. Sci. USA* **2018**, *115*, 10979–10987. [CrossRef]
78. Townsend, P.D.; Dixon, C.H.; Slootweg, E.J.; Sukarta, O.C.A.; Yang, A.W.H.; Hughes, T.R.; Sharples, G.J.; Pålsson, L.O.; Takken, F.L.W.; Goverse, A.; et al. The intracellular immune receptor Rx1 regulates the DNA-binding activity of a Golden2-like transcription factor. *J. Biol. Chem.* **2018**, *293*, 3218–3233. [CrossRef]
79. Wang, T.; Chang, C.; Gu, C.; Tang, S.Y.; Xie, Q.; Shen, Q.H. An E3 Ligase Affects the NLR Receptor Stability and Immunity to *Powdery mildew*. *J. Plant Physiol.* **2016**, *172*, 2504–2515. [CrossRef]
80. Shen, Q.H.; Saijo, Y.; Mauch, S.; Biskup, C.; Bieri, S.; Keller, B.; Seki, H.; Ulker, B.; Somssich, I.E.; Schulze-Lefert, P. Nuclear activity of MLA immune receptors links isolate-specific and basal disease-resistance responses. *Science* **2007**, *315*, 1098–1103. [CrossRef]
81. Zhai, K.R.; Deng, Y.W.; Liang, D.; Tang, J.; Liu, J.; Yan, B.X.; Yin, X.; Lin, H.; Chen, F.D.; Yang, D.Y.; et al. RRM Transcription Factors Interact with NLRs and Regulate Broad-Spectrum Blast Resistance in Rice. *Mol. Cell* **2019**, *74*, 996. [CrossRef]
82. Li, X.; Kapos, P.; Zhang, Y.L. NLRs in plants. *Curr. Opin. Immunol.* **2015**, *32*, 114–121. [CrossRef] [PubMed]
83. Huang, S.; Chen, X.J.; Zhong, X.H.; Li, M.; Ao, K.V.; Huang, J.H.; Li, X. Plant TRAF Proteins Regulate NLR Immune Receptor Turnover. *Cell Host Microbe* **2016**, *20*, 271. [CrossRef] [PubMed]
84. Cheng, Y.T.; Li, Y.Z.; Huang, S.A.; Huang, Y.; Dong, X.N.; Zhang, Y.L.; Li, X. Stability of plant immune-receptor resistance proteins is controlled by SKP1-CULLIN1-F-box (SCF)-mediated protein degradation. *Proc. Natl. Acad. Sci. USA* **2011**, *108*, 14694–14699. [CrossRef] [PubMed]
85. Chang, C.; Yu, D.S.; Jiao, J.; Jing, S.J.; Schulze-Lefert, P.; Shen, Q.H. Barley MLA Immune Receptors Directly Interfere with Antagonistically Acting Transcription Factors to Initiate Disease Resistance Signaling. *Plant Cell* **2013**, *25*, 1158–1173. [CrossRef] [PubMed]
86. Li, L.; Luo, Y.; Chen, B.; Xu, K.; Zhang, F.; Li, H.; Huang, Q.; Xiao, X.; Zhang, T.; Hu, J.; et al. A Genome-Wide Association Study Reveals New Loci for Resistance to Clubroot Disease in *Brassica napus*. *Front. Plant Sci.* **2016**, *7*, 1483. [CrossRef] [PubMed]
87. Hasan, J.; Megha, S.; Rahman, H. Clubroot in Brassica: Recent advances in genomics, breeding, and disease management. *Genome* **2021**, *64*, 735–760. [CrossRef]
88. Manzanares-Dauleux, M.J.; Delourme, R.; Baron, F.; Thomas, G. Mapping of one major gene and of QTLs involved in resistance to clubroot in *Brassica napus*. *Theor. Appl. Genet.* **2000**, *101*, 885–891. [CrossRef]
89. Yang, Z.; Jiang, Y.; Gong, J.; Li, Q.; Dun, B.; Liu, D.; Yin, F.; Yuan, L.; Zhou, X.; Wang, H.; et al. R gene triplication confers European fodder turnip with improved clubroot resistance. *Plant Biotech. J.* **2022**, *20*, 1502–1517. [CrossRef]
90. Liu, S.T.; Wang, S.B.; Wang, R.H.; Zhang, Z.G.; Li, Q.Y.; Wang, L.H.; Zhao, Z.Z. Advances in Molecular Markers for Resistance to clubroot disease in Chinese Cabbage. *Shandong Agric. Sci.* **2022**, *54*, 156–164.

91. Ueno, H.; Matsumoto, E.; Aruga, D.; Kitagawa, S.; Matsumura, H.; Hayashida, N. Molecular characterization of the *CRa* gene conferring clubroot resistance in *Brassica rapa*. *Plant Mol. Biol.* **2012**, *80*, 621–629. [CrossRef]
92. Hatakeyama, K.; Suwabe, K.; Tomita, R.N.; Kato, T.; Nunome, T.; Fukuoka, H.; Matsumoto, S. Identification and Characterization of Crr1a, a Gene for Resistance to Clubroot Disease (*Plasmodiophora brassicae* Woronin) in *Brassica Rapa* L. *PLoS ONE* **2013**, *8*, e54745. [CrossRef] [PubMed]
93. Luo, M.; Jia, J.Z. Advances in Plant Genome Expressed Sequence Tags (EST) Program. *Prog. Biochem. Biophys.* **2001**, 494–497.
94. Kamei, A.; Tsuro, M.; Kubo, N.; Hayashi, T.; Wang, N.; Fujimura, T.; Hirai, M. QTL mapping of clubroot resistance in radish (*Raphanus sativus* L.). *Theor. Appl. Genet.* **2010**, *120*, 1021–1027. [CrossRef] [PubMed]
95. Matsumoto, E.; Ueno, H.; Aruga, D.; Sakamoto, K.; Hayashida, N. Accumulation of Three Clubroot Resistance Genes through Marker-assisted Selection in Chinese Cabbage (*Brassica rapa ssp pekinensis*). *J. Jpn. Soc. Hortic. Sci.* **2012**, *81*, 184–190. [CrossRef]
96. Huang, Z.; Peng, G.; Liu, X.J.; Deora, A.; Falk, K.C.; Gossen, B.D.; McDonald, M.R.; Yu, F.Q. Fine Mapping of a Clubroot Resistance Gene in Chinese Cabbage Using SNP Markers Identified from Bulked Segregant RNA Sequencing. *Front. Plant Sci.* **2017**, *8*, 1448. [CrossRef]
97. Wang, Q.B.; Wang, Y.P.; Qian, H.H.; Zhang, Z.Y.; Zhang, L. Evaluation of Germplasm and Development of Markers for Resistance to *Plasmodiophora brassicae* in Radish (*Raphanus sativus* L.). *Agronomy* **2022**, *12*, 554. [CrossRef]
98. Neuhaus, K.; Grsic-Rausch, S.; Sauerteig, S.; Ludwig-Muller, J. *Arabidopsis* plants transformed with nitrilase 1 or 2 in antisense direction are delayed in clubroot development. *J. Plant Physiol.* **2000**, *156*, 756–761. [CrossRef]
99. Brodmann, D.; Schuller, A.; Ludwig-Muller, J.; Aeschbacher, R.A.; Wiemken, A.; Boller, T.; Wingler, A. Induction of trehalase in *Arabidopsis* plants infected with the trehalose-producing pathogen *Plasmodiophora brassicae*. *Mol. Plant Microbe. Interact.* **2002**, *15*, 693–700. [CrossRef]
100. Siemens, J.; Keller, I.; Sarx, J.; Kunz, S.; Schuller, A.; Nagel, W.; Schmulling, T.; Parniske, M.; Ludwig-Muller, J. Transcriptome analysis of *Arabidopsis* clubroots indicate a key role for cytokinins in disease development. *Mol. Plant Microbe. Interact.* **2006**, *19*, 480–494. [CrossRef]
101. Denby, K.J.; Kumar, P.; Kliebenstein, D.J. Identification of Botrytis cinerea susceptibility loci in *Arabidopsis thaliana*. *Plant J.* **2004**, *38*, 473–486. [CrossRef]
102. Alix, K.; Lariagon, C.; Delourme, R.; Manzanares-Dauleux, M.J. Exploiting natural genetic diversity and mutant resources of *Arabidopsis thaliana* to study the *A. thaliana-Plasmodiophora brassicae* interaction. *Plant Breed.* **2007**, *126*, 218–221. [CrossRef]
103. Kover, P.X.; Cheverud, J. The genetic basis of quantitative variation in susceptibility of *Arabidopsis thaliana* to Pseudomonas syringae (Pst DC3000): Evidence for a new genetic factor of large effect. *New Phytol.* **2007**, *174*, 172–181. [CrossRef] [PubMed]
104. Fuchs, H.; Sacristán, M.D. Identification of a gene in *Arabidopsis thaliana* controlling resistance to clubroot (*Plasmodiophora brassicae*) and characterization of the resistance response. *Mol. Plant Microbe. Interact.* **1996**, *9*, 91–97. [CrossRef]
105. Jubault, M.; Lariagon, C.; Simon, M.; Delourme, R.; Manzanares-Dauleux, M.J. Identification of quantitative trait loci controlling partial clubroot resistance in new mapping populations of *Arabidopsis thaliana*. *Appl. Genet.* **2008**, *117*, 191–202. [CrossRef]
106. Saito, M.; Kubo, N.; Matsumoto, S.; Suwabe, K.; Tsukada, M.; Hirai, M. Fine mapping of the clubroot resistance gene, *Crr3*, in *Brassica rapa*. *Appl. Genet.* **2006**, *114*, 81–91. [CrossRef]
107. Gan, C.; Deng, X.; Cui, L. Construction of a high-density genetic linkage map and identification of quantitative trait loci associated with clubroot resistance in radish (*Raphanus sativus* L.). *Mol. Breed.* **2019**, *39*, 116. [CrossRef]
108. Gan, C.; Yan, C.; Pang, W.; Cui, L.; Fu, P.; Yu, X.; Qiu, Z.; Zhu, M.; Piao, Z.; Deng, X. Identification of Novel Locus *RsCr6* Related to Clubroot Resistance in Radish (*Raphanus sativus* L.). *Front. Plant Sci.* **2022**, *13*, 866211. [CrossRef]
109. Schwelm, A.; Ludwig-Müller, J. Molecular Pathotyping of *Plasmodiophora brassicae*-Genomes, Marker Genes, and Obstacles. *Pathogens* **2021**, *10*, 259. [CrossRef]
110. Ludwig-Müller, J. What Can We Learn from -Omics Approaches to Understand Clubroot Disease? *Int. J. Mol. Sci.* **2022**, *23*, 6293. [CrossRef]
111. Yu, F.; Zhang, X.; Peng, G.; Falk, K.C.; Strelkov, S.E.; Gossen, B.D. Genotyping-by-sequencing reveals three QTL for clubroot resistance to six pathotypes of *Plasmodiophora brassicae* in *Brassica rapa*. *Sci. Rep.* **2017**, *7*, 4516. [CrossRef]
112. Su, T.; Yu, S.; Wang, W.; Li, P.; Zhang, F.; Yu, Y.; Zhang, D.; Zhao, X. ITRAQ Analysis of Protein Profile during the Secondary Stage of Infection of *Plasmodiophora brassicae* in Chinese Cabbage (*Brassica rapa*). *J. Plant Pathol.* **2018**, *100*, 533–542. [CrossRef]
113. Zheng, J.; Wang, X.; Li, Q.; Yuan, S.; Wei, S.; Tian, X.; Huang, Y.; Wang, W.; Yang, H. Characterization of Five Molecular Markers for Pathotype Identification of the Clubroot Pathogen *Plasmodiophora brassicae*. *Phytopathology* **2018**, *108*, 1486–1492. [CrossRef] [PubMed]
114. Lei, T.; Li, N.; Ma, J.; Hui, M.; Zhao, L. Development of molecular markers based on CRa gene sequencing of different clubroot disease-resistant cultivars of Chinese cabbage. *Mol. Biol. Rep.* **2022**, *49*, 5953–5961. [CrossRef] [PubMed]
115. Jiang, X.; Su, Y.; Wang, M. Mapping of a novel clubroot disease resistance locus in *Brassica napus* and related functional identification. *Front. Plant Sci.* **2022**, *13*, 1014376. [CrossRef] [PubMed]
116. Wang, Z.; Megha, S.; Kebede, B.; Kav, N.N.V.; Rahman, H. Genetic and molecular analysis reveals that two major loci and their interaction confer clubroot resistance in canola introgressed from rutabaga. *Plant Genome* **2022**, *15*, e20241. [CrossRef] [PubMed]
117. Gao, Y.; Gao, F.; Li, S.J.; Peng, S.D.; Chen, S.W.; Lin, L.B. Screening of 20 *Brassica napus* resources resistant to clubroot disease. *J. Yunnan Agric. Univ.* **2015**, *30*, 346–350.

118. Zhan, Z.X.; Jiang, Y.F.; Zhu, Z.Y.; Zhang, C.S.; Yang, Q.Y.; Li, Q.; Hou, Z.K.; Gong, J.F.; Cheng, Y.G.; Wu, J.S.; et al. Development of Molecular Markers Closely Linked to Locus *PbBa8.1* and Breeding for Resistance to Clubroot in *Brassica napus*. *Acta Hortic. Sin.* **2015**, *37*, 766–771.
119. Jia, R. Improvement and Evaluation of Resistance to Clubroot Disease in *Brassica napus* Cultivars' W3 'and' Zhongshuang 11'. Master's Thesis, Shenyang Agricultural University, Shenyang, China, 2019.
120. Li, Q.; Nadil, S.; Zhou, Y.W.; Hhou, Z.K.; Gong, J.F.; Liu, Y.; Shang, Z.W.; Zhang, L.; Zhan, Z.X.; Chang, H.B.; et al. Breeding of a novel clubroot disease-resistant *Brassica napus* variety Huayouza 62R. *Acta Agric. Sin.* **2021**, *47*, 210–223. [CrossRef]
121. Hou, X.F.; Gu, Y.G.; Jia, D.H.; Li, Q.; Shi, B.X.; Miao, H.C.; Wang, Z.C.; Li, A. Breeding of a New *Brassica napus* Variety Huayuza 62R with Resistance to clubroot disease. *Mol. Breed.* **2021**, *19*, 4050–4056.
122. Nadil, S. Comparative Study the Response of *PbBa8.1* and *CRb* Pyramided *Brassica napus Lines* against Genetically Diverse Clubroot Pathogen *Plasmodiophora brassicae* Isolates. *J. Huazhong Agric. Univ.* **2020**, *6*, 208.
123. Sun, C.H.; Liu, Q.Q.; Duan, Y.X.; Liu, L.F.; Zhao, W.; Guo, R.B.; Sun, L.Q.; Cheng, F. Breeding of a New Chinese Cabbage Variety Anxiu with Resistance to clubroot disease. *China Cucurbits Veg.* **2022**, *35*, 102–104.
124. Yang, D.; Lv, F.X.; Li, C.J.; Xu, X.Z.; Hu, J.F.; Yang, H.L.; Lan, M.; Zhang, L.Q.; Dong, X.S.; He, J.M. Creation of an Interspecific Restorative Material for Ogura CMS of Cabbage with Resistance to clubroot disease. *Chin. Veg.* **2022**, *11*, 44–52.
125. He, J.M.; Lan, M.; Dao, J.R.; Hu, J.F.; Zhang, L.; Yang, H.L.; Hong, J.K. Innovation of Chinese Cabbage Breeding Technology and Breeding and Application of New Varieties. *Horticul.Res. Inst. Yunnan Acad. Agri. Sci.* **2022**.
126. Yu, Y.J.; Wang, H.H.; Su, T.B.; Zhang, F.L.; Zhang, D.S.; Zhao, Y.Y.; Yu, S.C.; Li, P.R.; Xin, X.Y.; Wang, J. A New Chinese Cabbage Variety 'Jingchun CR3' Resistant to clubroot disease and Bolting. *Acta. Hortic. Sin.* **2022**, *0518*, 1–2.

International Journal of
Molecular Sciences

MDPI

Review

MYB Transcription Factors Becoming Mainstream in Plant Roots

Zhuo Chen [1,2], Zexuan Wu [1,2], Wenyu Dong [1,2], Shiying Liu [1,2], Lulu Tian [1,2], Jiana Li [1,2] and Hai Du [1,2,*]

[1] College of Agronomy and Biotechnology, Chongqing Engineering Research Center for Rapeseed, Southwest University, Chongqing 400716, China
[2] Academy of Agricultural Sciences, Southwest University, Chongqing 400716, China
* Correspondence: haidu81@126.com; Tel.: +86-182-2348-0008

Abstract: The function of the root system is crucial for plant survival, such as anchoring plants, absorbing nutrients and water from the soil, and adapting to stress. MYB transcription factors constitute one of the largest transcription factor families in plant genomes with structural and functional diversifications. Members of this superfamily in plant development and cell differentiation, specialized metabolism, and biotic and abiotic stress processes are widely recognized, but their roles in plant roots are still not well characterized. Recent advances in functional studies remind us that *MYB* genes may have potentially key roles in roots. In this review, the current knowledge about the functions of *MYB* genes in roots was summarized, including promoting cell differentiation, regulating cell division through cell cycle, response to biotic and abiotic stresses (e.g., drought, salt stress, nutrient stress, light, gravity, and fungi), and mediate phytohormone signals. *MYB* genes from the same subfamily tend to regulate similar biological processes in roots in redundant but precise ways. Given their increasing known functions and wide expression profiles in roots, *MYB* genes are proposed as key components of the gene regulatory networks associated with distinct biological processes in roots. Further functional studies of *MYB* genes will provide an important basis for root regulatory mechanisms, enabling a more inclusive green revolution and sustainable agriculture to face the constant changes in climate and environmental conditions.

Keywords: MYB transcription factors; plant roots; development; biotic and abiotic stresses

Citation: Chen, Z.; Wu, Z.; Dong, W.; Liu, S.; Tian, L.; Li, J.; Du, H. MYB Transcription Factors Becoming Mainstream in Plant Roots. *Int. J. Mol. Sci.* **2022**, *23*, 9262. https://doi.org/10.3390/ijms23169262

Academic Editor: Karen Skriver

Received: 27 July 2022
Accepted: 16 August 2022
Published: 17 August 2022

1. Introduction

The plant root system provides the interface between plants and the complex soil environment, thereby being crucial for plant survival and crop productivity, such as absorbing nutrients and water from the soil, preventing lodging, and responding to biotic and abiotic stresses. In the past decades, plant root biology has attracted increasing attention as it has important implications for global food security under changing climate and environmental conditions [1–5]. The molecular basis of how the plant root system contributes to plant traits, such as lodging resistance, nutrient stress resistance, and biotic and abiotic stresses resistance are gradually clarified. It is commonly accepted that the root growth and development processes were regulated by complex mechanisms which include a series of transcription factors [6,7], such as *AtWOX5* controlling root apical meristem (RAM) division, *AtARF7/19* regulating the lateral roots (LRs) initiation, and *WEREWOLF* (*WER*/*AtMYB66*) determining the root hair (RH) cell formation [8–10].

The V-myb avian myeloblastosis viral oncogene homolog (MYB) transcription factors are commonly characterized by a highly conserved DNA-binding domain repeat (MYB domain) at the N-terminal, accompanied by a variable activation domain at the C-terminal [11,12]. MYB transcription factors (MYB TFs) are widely found throughout eukaryotic organisms and comprise a superfamily in land plant genomes, e.g., they account for ~13% of the 1500 transcription factors in the model plant *Arabidopsis thaliana* [13]. As compared with the large number in plant genomes, a few MYB TFs are generally present in unikonts, e.g., only three MYB TFs that are associated with cell cycle progression were reported in vertebrate animals including *Homo sapiens* and *Mus musculus* [14]; a single MYB TF that functions in cell cycle progression as well was present in invertebrates (such as *Drosophila melanogaster* and *Strongylocentrotus purpuratus*) [15]; similarly, only one MYB TF had been identified in cellular slime mold *Dictyostelium discoideum* [16]. In plants, based on the number of the highly conserved repeats (R) in the MYB domain, MYB proteins fall into four major families, namely MYB-related (R3/R1-MYB), 2R-MYB (R2R3-MYB), 3R-MYB (R1R2R3-MYB), and 4R-MYB (R1R2R2R1/R2-MYB) [17]. In general, the number of 3R-MYB and 4R-MYB families is very limited and conserved across plants, with most species possessing 3 and 2 members, respectively [17]. In contrast, numerous members are observed in MYB-related and 2R-MYB families in plants, especially in angiosperms [14,18]. Consistent with the small number, the functions of 3R-MYB and 4R-MYB families are conserved across plants and even eukaryotes. For example, the members of the 3R-MYB family in *Arabidopsis* play a similar role with their homologs in animals in cell cycle regulation [19–21]. On the contrary, the members of MYB-related and 2R-MYB families have diverse functions in many plant-specific processes [18,22]. To date, the major roles of *MYB*-related genes characterized are related to organ development, stress response, and circadian clock [18]. For example, CPC-like subfamily homologs are involved in cell fate determination [23], *AtMYBL* response to ABA and salt stresses [24], and *AtCCA1* with its homologs are involved in circadian rhythm [18]. Consistent with its vast number and phylogenetic classification, the 2R-MYB family has diverse functions in plants, which can be briefly summarized into three major biological processes: development and cell differentiation, biotic and abiotic stresses, and specialized metabolism [22]. For instance, the genes in the S15 subfamily are associated with cell fate determination [10], the genes in the S22 subfamily respond to drought and pathogen invasion [25], and those in the S6 subfamily participate in anthocyanin biosynthesis [26], etc. Due to inconvenient visualization, the root traits were generally neglected in plant genetics, biology, etc. Accordingly, to date, the functions of MYB proteins related to shoot attracted much more attention. However, an increasing number of *MYB* genes (*MYBs*) have been functionally characterized in roots, such as *AtMYB36* controlled LR primordium (LRP) development [27]; *AtMYB70* regulated root system development [28]; *AtMYB73/77* involved in LRs growth [29], etc. These results raised concerns about the roles of MYB TFs in roots.

In this review, the functions of *MYBs* in plant roots are described in detail, with a focus on their roles in regulating cell differentiation, coordinating cell cycle, responding to biotic and abiotic stresses, and transducing phytohormone signals. The function characteristics of MYB homologs in roots have been discussed from a subfamily perspective. Finally, the potential research directions of MYB TFs in roots for future work were prospected. To our knowledge, this is the first review of the functions of MYB TFs in roots.

2. MYB TFs Are Regulators of Root Cell Differentiation

The development of plant roots relies on highly precise regulation of cellular differentiation [7]. MYB TFs were widely recognized to regulate the differentiation of root cells, participating in the development processes of primary root (PR), LRs, and RHs.

The roles of MYB TFs in root cell differentiation were represented by *WER* (2R-MYB) and *CAPRICE* (*CPC*, R3-MYB) which act as the core factors in determining root epidermal cell fate. In *Arabidopsis*, the specification of hair and non-hair epidermal cells are position-dependent, with hair cells arising over clefts between the overlying cortical cells (Figure 1A). In cortical cells, JACKDAW (JKD), a plant-specific zinc finger protein, produces a signal that may bind to the leucine-rich repeat receptor-like kinase (LRR-RLK) SCRAMBLED (SCM) in epidermal cells [30]. Due to the larger contact surface, the epidermal cell located at the cleft between two overlying cortical cells received more signals, resulting in more SCM activation [30]. Subsequently, the activated SCM represses the expression of the *WER* gene, contributing to more CPC than WER, leading to hair cell (HC) formation

(Figure 1A) [31]. On the contrary, the epidermal cell on a single cortical cell has a little activated SCM because of the smaller contact surface, leading to the dominance of WER and thus developed as a non-hair cell (NHC) (Figure 1A) [31]. In NHC, a MYB protein WER, a bHLH protein GLABRA3 (GL3) or its homolog ENHANCER OF GLABRA3 (EGL3), and a WD40-repeat protein TRANSPARENT TESTA GLABRA1 (TTG1) form an activator protein complex MYB-bHLH-WD40 (MBW) to activate the expression of the downstream *GLABRA2* (*GL2*) gene that encodes a homeodomain protein [10,32–34]; subsequently, the GL2 protein represses a set of downstream genes such as *ROOT HAIR DEFECTIVE 6* (*RHD6*) and *RHD6-LIKE1* (*RSL1*) which are essential to the determination of RHs, leading to an NHC fate (Figure 1A) [35]. In HC, CPC protein competes with WER protein to form a repressor protein complex CPC-GL3/EGL3-TTG1 (MBW), which then inhibits the expression of the *GL2* gene [36,37]; thereby, the downstream genes of *GL2* (e.g., *RHD6* and *RSL1*) were normally expressed which then promote RH formation (Figure 1A).

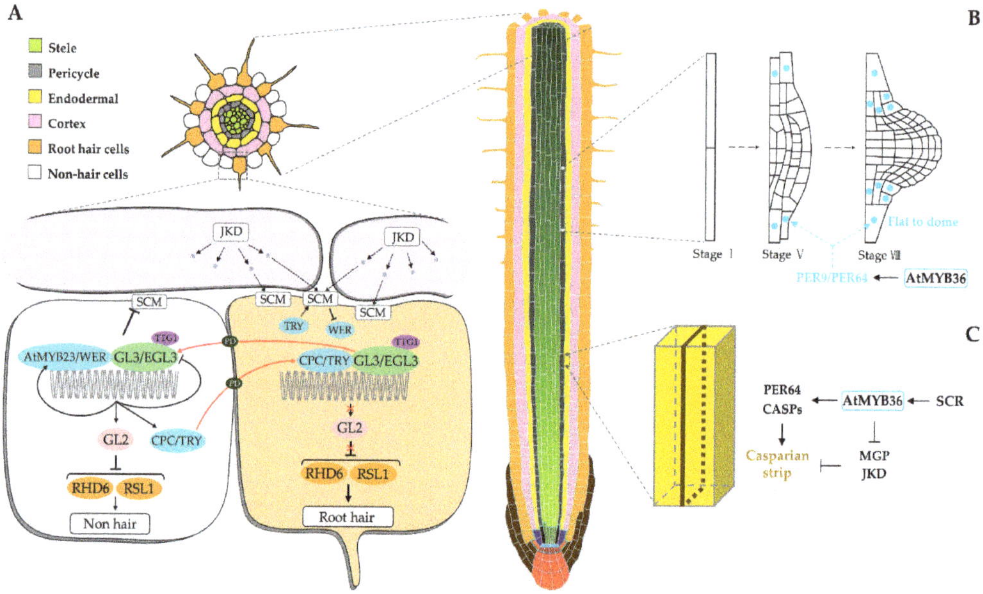

Figure 1. MYB transcription factors (MYB TFs) promote cell differentiation in *Arabidopsis* roots. (**A**) The model of *MYB* genes (*MYBs*) controls root epidermal cell specification. In non-hair cell (NHC), the WER-GL3/EGL3-TTG1 protein complex inhibits root hair (RH) formation by activating the *GL2* gene expression which then inhibits the expression of the downstream genes such as *RHD6* and *RSL1* to repress RH formation. In hair cell (HC), the CPC-GL3/EGL3-TTG1 complex cannot active *GL2* gene expression in HCs, thereby the downstream genes (e.g., *RHD6* and *RSL1*) can promote RH formation. The dominance of different MBW complexes in HCs and NHCs is determined by the position signals (grey dots) in NHCs and HCs and the lateral movements between NHCs and HCs (red arrows). Blue dots represent MYB proteins. (**B**) The model of the *AtMYB36* gene regulates lateral root development. Blue dots represent the expression position of the *AtMYB36* gene. (**C**) The model of *AtMYB36* regulates Casparian strip formation. *AtMYB36* promotes the formation of the Casparian strip by regulating the balance between proliferation and differentiation. Brown lines indicate the Casparian strip.

The competition between the two types of MYB proteins, WER and CPC, to form the MBW ternary protein complex is attributed to their close evolutionary relationship. The MYB domain of CPC protein and the R3 repeat in the MYB domain of WER protein shared a high degree of sequence similarity. Both of them contain the conserved motif DLx2Rx3Lx6Lx3R in the R3 repeat of the MYB domains that are involved in the interactions between MYB and bHLH proteins [18]. However, as compared to WER, the CPC protein lacks the transcription activation domain causing its opposite function in the RH formation process [38]. Accordingly, the R3 domain of the WER can functionally replace CPC, but the R3 domain of CPC cannot functionally replace that of WER [39]. In addition, the homologs of *WER* and *CPC* genes were widely demonstrated to act as similar roles in RH formation process. For example, the *AtMYB23* gene has a similar function to the *WER* gene to reinforce the NHC fate [40]; in *Arabidopsis*, there are six homologs of the *CPC* gene, including *TRIPTYCHON (TRY)*, *ENHANCER OF TRY AND CPC1 (ETC1)*, *ETC2*, *ETC3*, *TRICHOMELESS1 (TCL1)*, and *TCL2*, which act as positive regulators for RH formation [23,41,42]. In other plant species, the homologs of *WER/CPC* genes were reported to perform similar functions in RH cell differentiation. For example, overexpressing four homologs of the *WER* gene in *Brassica napus* (*BnMYB019*, *BnMYB189*, *BnMYB231*, and *BnMYB388*) in *Arabidopsis* could rescue the RH phenotype of the *wer* mutant to that of WT lines [43]; expressing the *Solanum lycopersicum SlTRY* gene and *Oryza sativa* (rice) *OsTCL1* gene in *Arabidopsis*, respectively, both enhanced RH differentiation and thus promoted RH formation [44,45]. These results show a new perspective on subfamily to understand the key roles of *WER* and *CPC* homologs in root epidermal cell differentiation.

Notably, a series of feedback exists in this process to enhance root epidermal cell fate specification. In NHC, in addition to the downstream *GL2* gene, the WER-GL3/EGL3-TTG1 complex also positively regulates the expression of the competitors of the WER protein, *CPC* and *TRY*, and even the homologs of *WER*, *AtMYB23* (Figure 1A) [40,46]. However, the WER-GL3/EGL3-TTG1 complex negatively regulates the expression of *SCM*, *GL3*, and *EGL3* genes in NHCs as well [47]. At the same time, the GL3 and EGL3 proteins can move from HC to NHC through plasmodesmata (PD) [47], resulting in more WER-GL3/EGL3-TTG1 complex in NHC that can upregulate more GL2 and CPC/TRY protein formation (Figure 1A). Subsequently, the GL2 protein controls the NHC formation, whereas the CPC and TRY proteins can move from NHC to HC through PD to promote HC formation [37,48]. Moreover, TRY in HC can upregulate the expression of the *SCM* gene, contributing to the HC fate [49]. Recently, phytosulfokine receptors (PSKRs) which belong to the LRR-RLK family and an O-fucosyltransferase protein, SPINDLY (SPY), were reported to participate in position-dependent root epidermal cell fate determination as well [50,51]. Together, both the position signals and various feedback pathways determine the dominance of different MBW complexes in HCs and NHCs which consequently specify the root epidermal cell fate.

Another *2R-MYB* gene, *AtMYB36*, is a key regulator of cell differentiation in *Arabidopsis* roots as well [27,52]. In endodermal cells, the *AtMYB36* gene is activated by SCARECROW (SCR) and then activates a series of genes such as *Casparian strip proteins* (*CASPs*), *Peroxidase 64* (*PER64*) and *Enhanced Suberin 1* (*ESB1*), as well as represses *JKD* and *MAGPIE* (*MGP*) involved in proliferation, resulting in Casparian strip formation that represents a successful transition from proliferation to differentiation in endodermal cells (Figure 1C) [53]. Notably, the JKD-involved formation of the Casparian strip in the endodermal cells is independent of its function in epidermal patterning [47]. Moreover, *AtMYB36* plays a similar role in LRP, as the size and shape of LRP are determined by the balance between cell proliferation and differentiation as well (Figure 1B). In this biological process, *AtMYB36* is expressed in the cells surrounding LRP where it activates a subset of peroxidase genes such as *PER9* and *PER64* to regulate the ROS balance, consequently promoting the transition from proliferation to differentiation at later stages of LRP development [27]. Together, *AtMYB36* regulates the formation of the Casparian strip in endodermal cells and LRP development by promoting the transition from proliferation to differentiation in a similar manner.

Taken together, the roles of *MYBs* in cell differentiation in RH development were well-known in many plant species, whereas their roles in other root cell differentiation processes remain to be explored in the future.

3. MYB TFs Regulate Root Growth and Development through Cell Cycle

Root growth and development is a dynamic balance between cell proliferation and cellular differentiation, which rely on the positive and/or negative regulation of cell cycle progression. Another well-known role of plant MYB TFs in roots is controlling the cell cycle.

Among the MYB superfamily, the *3R-MYBs* were well-known to regulate the cell cycle in both animals and plants. In animals, the MYB family is composed of three members, *MYB* (*C-MYB*), *MYBL1* (*A-MYB*), and *MYBL2* (*B-MYB*), which play important roles in controlling the cell cycle. The *B-MYB* gene is ubiquitously expressed in all cell types and activates the transcription of *CDC2* and *cyclin B1* in the G2/M phase; the *A-MYB* and *C-MYB* exhibit a tissue-specific expression profile and are also involved in the regulation of cell cycle [54]. In *Arabidopsis*, four of the five *3R-MYBs* are demonstrated to regulate cell cycle progression, except *AtMYB3R2* which is associated with circadian rhythms rather than cell cycle [19,20,55]. Among them, *AtMYB3R1* and *AtMYB3R4* (*AtMYB3R1/4*) act as transcriptional activators which are expressed in proliferating tissues, such as root tips and LRPs (Figure 2A). In *atmyb3r1/4* double mutants, the expression of G2/M-specific genes was significantly reduced, leading to short roots and defective cytokinesis in root epidermal cells [20,56]. In contrast, *AtMYB3R1*, *AtMYB3R3*, and *AtMYB3R5* (*AtMYB3R1/3/5*) act redundantly as transcriptional repressors in both proliferating and differentiated cells by inhibiting the expression of G2/M-specific genes directly (Figure 2A). In *atmyb3r1/3/5* triple mutants, the expression of G2/M-specific genes was significantly upregulated, leading to an increased size of root meristems and a longer length of PR [19]. Commonly, the above *AtMYB3Rs* all bind to the *cis*-acting mitosis-specific activator (MSA) element ("AACGG") in the promoter regions of several G2/M-specific genes (e.g., *AtKNOLLE*, *AtCYCB1*, and *AtCDC20*) to regulate the cell cycle (Figure 2A) [19,20]. However, the precise mechanism of how the *AtMYB3Rs* activate or repress the expression of downstream genes is still unclear. An explanation was that the AtMYB3Rs proteins can recruit different types of the DREAM (DP, RBR, E2F, and MuvB) complex which is well-known in the regulation of cell cycle [19]. In this process, the AtMYB3Rs indirectly recruit different E2F isoforms (E2FB or E2FC) through MuvB to form distinct DREAM complexes, such as AtMYB3R4-MuvB-E2FB and AtMYB3R3-MuvB-E2FC (Figure 2A,B) [57]. Consistently, E2FB functions as an activator while E2FC acts as a repressor in the regulation of cell cycle genes [58,59]. Notably, *AtMYB3R1* has dual roles in controlling the cell cycle, acting as an activator or a repressor in this process. In the *atmyb3r1* mutant, the expression of cell cycle genes underwent little change. However, enhanced downregulation and upregulation of G2/M-specific genes were observed when introducing *atmyb3r1* into *atmyb3r4* single mutant or *atmyb3r3/5* double mutants, respectively [19], demonstrating the dual roles of *AtMYB3R1* in the cell cycle. In RAM, the transcription of *AtMYB3R1* is downregulated by TSO1 (a core subunit of MuvB) rendering the cells unable to enter the differentiation process [60]. In other plants, *3R-MYB* homologs are demonstrated to perform a similar role in cell cycle progression. For instance, three *Nicotiana tabacum 3R-MYBs* (*NtmybA1*, *NtmybA2*, and *NtmybB*) and one rice *OsMYB3R-2* gene were reported to regulate the expression of cell cycle genes through binding to the MSA *cis*-element as well [61,62]. Recently, 225 *3R-MYBs* identified from 65 plant species appeared to be enriched for the MSA *cis*-element within their upstream promoter region, indicating a conserved functional involvement in cell cycle regulation [63]. Similarly, in animals, B-MYB is repressed by the DREAM complex in the G1/S phase while interacting with MuvB to activate the G2/M cell cycle genes [64,65]. These results suggest that *3R-MYBs* have undergone functional specialization during the evolution of plants.

The *2R-MYBs* also regulate the cell cycle in roots, and their regulatory mechanisms appear to be spatiotemporally specific that is distinct from the broad regulation manner of *3R-MYBs*. For example, the *AtMYB59* gene is specifically expressed in roots and functions in the S phase of the cell cycle progression (Figure 2A) [66]. In *AtMYB59*-overexpressing lines, about half of the mitotic cells in root tips are at metaphase, leading to shorter PR than wild-type (WT); in contrast, the *atmyb59* mutants show the opposite phenotype, indicating that *AtMYB59* inhibits root growth by disturbing the cell cycle [66]. Different from *AtMYB3Rs*,

AtMYB59 upregulates the expression of cell cycle genes (such as *AtCYCB1;1*) by binding to the MRE ("AACC"), MRE2 ("TATAACGGTTTTTT"), and ERE ("ATTTCAAA") *cis*-elements in the promoters instead of the MSA *cis*-element [66]. On the contrary, the *AtMYB56* (*BRAVO*) gene negatively regulates the expression of cell cycle genes (e.g., *AtCYCD2;2* and *AtCYCD3;3*) in quiescent center (QC) cells to ensure a low dividing activity (Figure 2B) [67]. Ectopic expression of *AtMYB56* resulted in the inhibition of root growth and the failure of root regeneration upon damage of stem cells [67]. Recently, the AtMYB56 protein has been reported to interact with the WOX5 protein to form a complex in QC cells (Figure 2B) [68]. In addition, an atypical 2R-MYB gene, *AtCDC5* (*cell division cycle 5*), is well-known for its role in regulating the cell cycle. This gene is predominantly expressed in proliferating cells and participates in the G2/M transition by upregulating the expression of the *AtCDKB1* gene (Figure 2A) [69]. Its binding site is the "CTCAGCG" motif in the promoters of target genes [69]. The root growth of the *AtCDC5*-RNAi plants was severely inhibited, and the *atcdc5* mutant was embryonic lethal, suggesting that *AtCDC5* is essential for cell cycle progression [69]. Moreover, these *2R-MYBs* commonly belong to the early derived subfamilies of the 2R-MYB family in plants, indicating that the ancient and important role of *2R-MYBs* in the cell cycle may be derived from *3R-MYBs*. To date, the knowledge regarding the roles of *2R-MYBs* in cell cycle progression is focused on the studies in *Arabidopsis*. However, as mentioned above, given the functional conservation characteristic of MYB homologs in diverse plant biological processes, it is foreseeable that *2R-MYBs* have an important role in root cell cycle progression in other plant species as well.

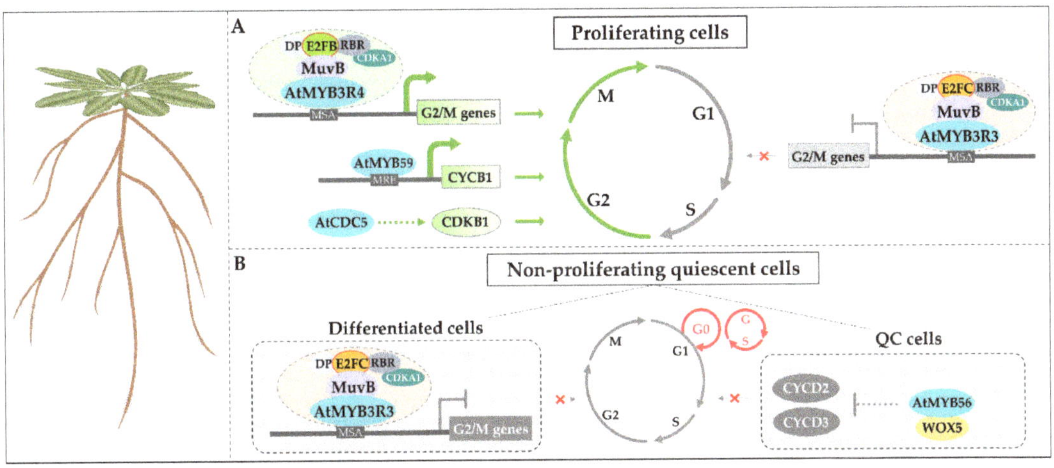

Figure 2. MYB TFs regulate the cell cycle in *Arabidopsis* roots. (**A**) In proliferating cells, AtMYB3R4 activates the expression of G2/M-specific genes by interacting with the DREAM complex containing E2FB at the G2/M phase, while AtMYB3R3 interacts with the DREAM complex containing E2FC to inhibit G2/M-specific gene expression at the G1/S phase. AtMYB59 activates *CYCB1* to promote G2/M transition in the root apical meristem (RAM). AtCDC5 participates in cell cycle progression by positively regulating *CDKB1* expression at the G2/M phase in RAM. (**B**) In non-proliferating quiescent cells, AtMYB3R3 interacts with the DREAM complex containing E2FC to inhibit the expression of G2/M-specific genes to maintain the quiescent state in differentiated cells; AtMYB56 represses the expression of *CYCD2* and *CYCD3* genes by interacting with WOX5 to maintain a low division rate of quiescent center cells. The blue dots represent MYB proteins.

Overall, the regulatory roles of *3R-MYBs* in root cell cycle progression are widely recognized in many plants showing a conserved mechanism across different plant species, while the roles of *2R-MYBs* in root cell cycle are still poorly understood, especially in non-model plants.

4. MYB TFs Function in Root System Architecture in Response to Biotic and Abiotic Stresses

Due to the immobility, plants face diverse volatile environments during the life cycle, whereas the root system plays a key role and shows strong plasticity in response to abiotic stress. To date, a mass of studies has demonstrated that plant MYB TFs are involved in diverse abiotic stress responses, such as drought, salt stress, nutrient stress, gravity, and light (Table 1). Most of these processes were achieved by regulating root growth and/or development, root morphology, and root system architecture (RSA).

Table 1. MYB gene functions in roots under biotic and abiotic stresses.

Species	Genes	Function Description	References
Arabidopsis thaliana	*AtMYB2*	Positively regulate LRs and RHs under P starvation	[70]
	AtMYB12	Maintain root growth under salt and drought stress	[71]
	AtMYB20	Promote PR growth under salt stress	[72]
	AtMYB30	Regulate root growth and development under salt stress	[73,74]
	AtMYB42	Protect root from ion toxicity under salt stress	[75]
	AtMYB59	Modulate RSA under nutrient stress and participate in the response to nematode infestation	[76–78]
	AtMYB60	Promote root growth under mild drought	[79]
	AtMYB62	Negative regulators of LRs growth under P starvation	[80]
	AtMYB73/77	Mediate the inhibition of LRs under UV-B light	[29]
	AtMYB96	Negatively regulate LRs under drought	[81]
	AtMYB88/124	Participate in PR and LRs gravitropism	[82]
	AtPHR1	A central regulator of P starvation	[83]
	AtPHL1/2	Dimerize with AtPHR1 to regulate P starvation responses	[84,85]
	AtHRS1	Regulate RHs and PR under P starvation	[86]
	AtHHO2	Promote LRs under P starvation	[87]
	CPC/TRY/ETC1	Positively regulate RHs development under P starvation	[88]
Glycine max	*GmMYB84*	Promote PR elongation under drought	[89]
Leymus chinensis	*LcMYB2*	Positively regulate root growth under drought	[90]
Malus×domestica	*MdFLP*	Promotes ARs in response to gravity	[91]
Oryza sativa	*OsMYB1*	Regulate LRs elongation under P starvation	[92]
	OsMYB2P-1	Positively regulate root growth under P starvation	[93]
	OsMYB4P	Promote PR growth under P starvation	[94]
	OsPHR1/2/3/4	A central regulator of P starvation	[95,96]
Panax notoginseng	*PnMYB2*	Regulate the resistance against the root rot	[97]
Populus ussuriensis	*PuMYB40*	Promote ARs formation under P starvation	[98]
Thinopyrum intermedium	*TiMYB2R-1*	Enhance the resistance to take-all disease	[99]
Triticum aestivum	*TaMpc1-D4*	Negative regulators with reduced root growth under drought	[100]
	TaPIMP2	Contribute to wheat resistance to root rot	[101]

PR, primary root; LRs, lateral roots; ARs, adventitious roots; RHs, root hairs; RSA, root system architecture; UV-B, ultraviolet B; P, phosphorus.

4.1. Drought

Drought is one of the most serious abiotic stresses affecting root growth. Under drought stress, the elongation of PR and the formation of LRs are adjusted to sustain the viability of plants. For example, as a typical adaptive response to environmental stress, PR growth can be promoted to absorb water deeper underground by sacrificing LRs development under drought conditions. MYB TFs have been reported frequently in drought resistance through modulating the RSA, including the length of PR and LRs as well as the number of LRs [25]. For example, in *Arabidopsis*, the *AtMYB96* gene was reported to negatively regulate LRs development and enhance drought resistance [81]; its overexpression lines showed significantly reduced LRs and enhanced drought resistance [81]. *AtMYB60*, a paralog of *AtMYB96*, promotes both PR and LRs growth to increase water uptake under mild drought stress [79]. In other plants, MYB TFs also regulate the RSA in response to drought. For instance, in *Glycine max*, *GmMYB84*-overexpressing lines exhibited enhanced drought resistance with a longer PR by controlling reactive oxygen species (ROS) balance [89]; in *Leymus chinensis*, *LcMYB2* improved plant drought resistance by promoting root growth as well [90]. However, MYB TFs play a negative role in drought response as well, e.g., overexpression of a wheat MYB gene *TaMpc1-D4* reduced the root length and repressed the expression of stress-related genes under drought stress [100].

4.2. Salt Stress

Salt stress is a major abiotic stress that adversely affects plant growth and development, significantly reducing crop productivity. Elevated soil salinity mainly causes ion toxicity and oxidative stress to roots, whereas MYB TFs are involved in these processes to respond to salt stress. *AtMYB30* enhances salt tolerance by improving alternative respiration which can maintain the root cellular redox homeostasis [73]. Moreover, *AtMYB30* links the reactive oxygen species (ROS) signaling and root cell elongation, and its mutant shows an increased cell length in root under H_2O_2 treatment [74]. Similarly, *AtMYB12* can upregulate the expression of ROS scavenging genes to maintain the root cellular ROS balance under both drought and salt stress conditions [71]. *AtMYB42* has been proven to participate in the regulation of ion toxicity. In this process, *AtMYB42* directly activates the expression of *salt overly sensitive 2* (*SOS2*), which plays a crucial role in regulating Na^+:K^+ homeostasis, resulting in the root system avoiding ion toxicity [75]. *AtMYB20* enhances salt tolerance by downregulating the expression of *type 2C serine/threonine protein phosphatases* (*PP2Cs*) that plays a negative role in ABA signaling [72]. The *AtMYB20*-overexpressing lines exhibit enhanced salt tolerance with a longer PR [72]. These results suggest that MYB TFs are involved in the regulation of diverse physiological processes to alleviate and even avoid the effects of salt stress, thereby ensuring root growth and development.

4.3. Nutrient Stress

Soil nutrient limitations, such as phosphorus (P), nitrogen (N), and potassium (K) starvation, are major abiotic stresses that affect plant growth and development and crop production. Consequently, roots have evolved a set of mechanisms to enhance nutrient acquisition, including changing RSA to enlarge root surface area to uptake nutrients from soil. Recently, an increasing number of *MYBs* has been reported to be involved in nutrient stress response processes by regulating the RSA, especially in response to P starvation. In *Arabidopsis*, a MYB-like gene *P Starvation Response 1* (*PHR1*), and its homologs *AtPHR1-like 1* (*AtPHL1*) and *AtPHL2* are recognized as central regulators of P starvation response (PSR) by directly regulating various *P starvation-induced* (*PSI*) genes, consequently affecting the P uptake and transport as well as modulating the RSA [83–85,102]. In *Oryza sativa*, four orthologs of *AtPHR1* (*OsPHR1/2/3/4*) function redundantly in a highly conserved manner to that of *AtPHR1* [95,96]. Two target genes of *AtPHR1*, *AtHRS1*, and *AtHHO2* (*HRS1 Homolog 2*) are MYB-like genes that participate in the regulation of RSA to enhance the adaptability to P starvation as well [103]. Under P starvation, *AtHRS1*-overexpression lines exhibited enhanced RHC development and shortened PR length, while *AtHHO2*-overexpression lines showed augmented LRs development [86,87]. Moreover, several *R3-MYBs*, *CPC*, *TRY*, and *ETC1* were well-known to play key roles in P starvation response by positively regulating RH development [88].

2R-MYBs participate in PSR as well. Under P starvation, *AtMYB2*-overexpression lines exhibited more and longer LRs and denser RHs [70]; the overexpression of *OsMYB2P-1* and *OsMYB4P* both enhanced the P starvation tolerance with enhanced root growth [93,94]. These three genes act as a positive regulator in PSR by activating the expression of downstream *PSI* genes, such as *miR399* and *P transporter* (*PHT*) genes. In contrast, *AtMYB62* acts as a negative regulator of the PSR by repressing the expression of *PSI* genes, such as *AtPHT1* and *AtACP5* [80]. Overexpression of the *AtMYB62* gene significantly decreased LRs length [80]. Rice 2R-MYB gene, *OsMYB1*, coordinately regulates the maintenance of P homeostasis and root development under P starvation conditions [92]. Moreover, MYB TFs were involved in adventitious root (AR) formation under P starvation, e.g., in *Populus ussuriensis*, *PuMYB40* controlled the P starvation resistance by promoting AR formation [98]. In addition, MYB TFs also participate in responding to other nutrient stress processes through modulating the RSA. Recently, *AtMYB59* has been reported to modulate the RSA in response to low K^+, low NO_3^-, or low calcium (Ca) stresses [76,77], suggesting its multiple roles in nutrient stress responses.

4.4. Light and Gravity

Light is an important signal that regulates root growth and development, especially ultraviolet B (UV-B) light which is a part of sunlight that markedly affects root morphology [104]. Recently, *AtMYB73* and *AtMYB77* genes were reported to directly interact with the UV-B photoreceptor UVR8 (UV Resistance Locus 8) to regulate root growth under UV-B light in *Arabidopsis* [29]. In this process, AtMYB73 and AtMYB77 proteins interact with auxin response factors to promote LRs growth and development; however, UVR8 can interact with AtMYB73/77 proteins in a UV-B-dependent manner to inhibit their DNA-binding activities, consequently inhibiting the LRs development [29]. The root system can change the growth direction in response to gravity stimulation, and *MYBs* are involved in this process. It was reported that *AtMYB88* and *AtMYB124 (AtFLP)* performed a redundant role in response to gravity stimulation by regulating the temporal–spatial expression patterns of *PIN-FORMED 3* (*PIN3*) and *PIN7* genes in gravity-sensing cells of primary and lateral roots [82]. After gravity stimulation (reorientation of 90°), the curvature of PR in the *atflp* mutant exhibited a defective gravity response. Moreover, *AtMYB88* and *AtFLP* functioned complementarily in establishing the gravitropic set-point angles, with the former functions in later stages while the latter in earlier stages [82]. Similarly, in *Malus×domestica*, *MdFLP*, an ortholog of *AtFLP*, regulated ARs gravitropism by directly binding to the promoters of *MdPIN3* and *MdPIN10* genes [91].

4.5. Biotic Stress

To date, many *MYBs* had been demonstrated to have an important role in biotic stress resistance in aerial parts, e.g., *AtMYB30* and *AtMYB72* participated in the response to pathogen attack in *Arabidopsis* leaf [105,106]; *AtMYB15*, *AtMYB34*, *AtMYB51*, and *AtMYB75* were involved in the resistance against insect herbivores [107]. In recent years, a few studies revealed the roles of *MYBs* in response to biotic stress in roots as well, such as pest and disease stress responses. In wheat, the *TaPIMP2* gene plays a positive role in defense responses to fungal pathogen *Bipolaris sorokiniana* infection by regulating the expression of defense-related genes, contributing to the host resistance to common root rot [101]. In *Panax notoginseng*, the *PnMYB2* gene has a positive role in root rot resistance caused by *Fusarium solani* pathogen through regulating JA signaling, disease-resistance-related genes, and photosynthesis [97]. The *Thinopyrum intermedium TiMYB2R-1* gene exhibits enhanced resistance to the take-all disease in wheat by upregulating the expression of defense-related genes [99]. In *Arabidopsis*, the *AtMYB59* gene was found to play an important role in the response to *Heterodera schachtii* infestation in roots [78]. Recently, based on transcriptome and co-expression network analyses in *Brassica rapa*, several *MYBs* were speculated to be involved in clubroot resistance caused by the soil-borne protist *Plasmodiophora brassicae* [108]. These studies demonstrated that *MYBs* are also involved in biotic stress resistance in roots, which is an important direction for future gene functional research on this gene family in plants.

5. The Regulatory Roles of MYB in Roots Are Mediated by Various Phytohormones

Phytohormones are involved in many biological processes in roots, such as root growth and development, root morphogenesis, stress response, and signal transduction. In fact, it was demonstrated that the roles of MYB TFs in roots are widely mediated by various phytohormones, including indole acetic acid (IAA), abscisic acid (ABA), gibberellin (GA), cytokinin (CTK), and brassinosteroid (BR). Meanwhile, MYB TFs are also involved in phytohormone response processes to regulate root growth and development and stress response by activating downstream phytohormone-related genes and/or acting as response factors to regulate various physiological biochemical processes in roots (Figure 3).

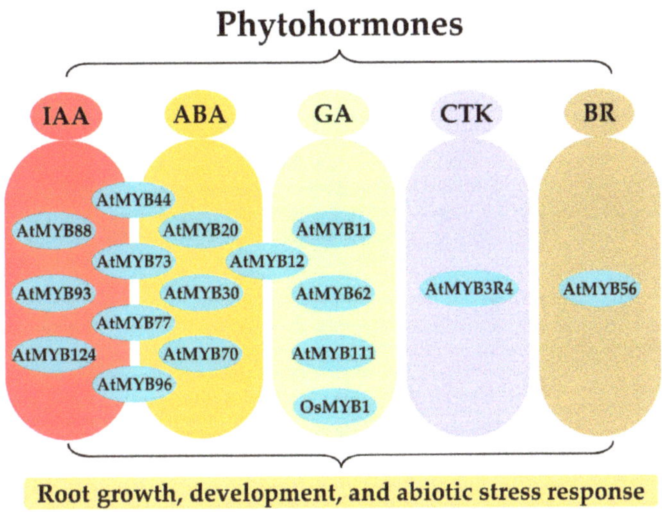

Figure 3. The regulatory roles of *MYBs* in roots are mediated by various phytohormones. IAA: indole acetic acid; ABA: abscisic acid; GA: gibberellin; CTK: cytokinin; BR: brassinosteroid. The blue dots represent MYB TFs. The *MYBs* crosslink different types of phytohormones indicating they can integrate different phytohormone signaling.

The IAA-mediated MYB signaling transduction pathways are widely involved in the regulation of diverse developmental and stress response processes in roots. In most cases, *MYBs* directly regulate the expression of IAA-related genes to modulate the RSA. For instance, in *Arabidopsis*, *AtMYB88* and *AtMYB124* participate in root gravitropism by regulating the transcription of *PIN3* and *PIN7* genes that encode auxin transporter proteins [82]; *AtMYB96* upregulates the expression of the *GH3* gene which encodes the auxin-conjugating enzyme to negatively regulate LRs growth [81]. In turn, IAA could induce the expression of *MYBs* to perform their biological function in roots. For example, the expression of *AtMYB96* could be induced by IAA leading to reduced LRs and enhanced drought resistance in *Arabidopsis* [81]; *AtMYB93* is a novel auxin-induced negative regulator in LRs development [109].

ABA is well-known for its key role in abiotic stress. Many MYB TFs are widely involved in the ABA signaling pathway to modulate RSA. For example, *AtMYB30* regulates root cell elongation depending on the ABA signaling [110]; *AtMYB20* modulates the RSA by downregulating the expression of *ABI1* and *PP2C* genes which encode the key regulators of the ABA signaling pathway [72]; *AtMYB12* modulates the RSA under salt and drought stresses by upregulating the ABA biosynthesis genes (e.g., *AtZEP*, *AtNCED*, *AtABA2*, and

AtAAO) [71]. In most cases, the roles of *MYBs* in roots were addressed by integrating signaling of different phytohormones, such as ABA-IAA. For instance, *AtMYB70* regulates the RSA through activating the expression of the *GH3* gene under ABA signaling [28]; *AtMYB44/73/77* promotes LRs growth by activating the auxin-responsive genes under ABA signaling [111]; *AtMYB96* is a molecular link that integrates ABA and IAA signaling in LRs growth under drought stress [81].

MYBs can regulate root development through other phytohormone signal pathways, such as GA, CTK, and BR. *AtMYB62* and *OsMYB1* function in the modulation of RSA under P starvation by changing the GA metabolism and signaling [80,92]. *AtMYB11*, *AtMYB12*, and *AtMYB111* positively regulate root growth under the presence of GA signaling by affecting the IAA content, suggesting that *AtMYB11/12/111*-mediated GA signals are integrated into the IAA signaling pathway [112]. In addition, CTK promotes the nuclear localization of AtMYB3R4 protein which is essential for cell division in both SAM and RAM [113]; *AtMYB56* acts as a switch to modulate the QC cells division by interacting with the BR-regulated transcription factor BRI1-EMS SUPPRESSOR 1 (BES1) [67]. Thus, the roles of MYB TFs in roots were generally mediated by phytohormone-mediated signaling pathways.

6. Conclusions and Prospects

The root system plays an essential role in plant anchorage, water, nutrient acquisition, and stress response. Therefore, genetic improvement of the root system has been recognized gradually as an effective strategy, which contributes to improving crop productivity under diverse growing conditions and thus enables a more inclusive green revolution. In the past decade, an increasing number of *MYBs* have been experimentally demonstrated to play critical roles in nearly all areas of root systems, such as PR, LRs, RHs, and ARs growth and development, and responding to diverse biotic and abiotic stresses. Among the MYB superfamily, *3R-MYBs* maintain their cell cycle regulatory roles in roots, whereas *1R-MYBs* and *2R-MYBs* function in root regulation in various ways, including cell cycle, cell differentiation, biotic and abiotic stresses, and phytohormone signaling pathways. The current understanding shows the fact that MYB TFs are becoming mainstream in plant roots, showing important application potential in molecular and genetic manipulation of root traits.

The MYB gene family is one of the largest transcription factor families in plant genome, with hundreds of members in most angiosperms. However, compared with its larger number, relatively limited members have been functionally characterized to date. The current knowledge regarding the functions of plant *MYBs* is mainly based on the studies in model plant *Arabidopsis*, these in non-model plants, especially in crops, remain rather unclear. Fortunately, an increasing number of sequenced genome and transcriptome data are available, which enabled genome-wide identification, systematic analysis, and overview of this gene family in diverse plant species. Consequently, the MYB superfamily has been widely identified and functionally verified in many plants, such as maize, soybean, and rapeseed. Based on the genome-wide expression analyses, a large proportion of *MYBs* have been proved to be preferentially and even specifically expressed in roots. Undoubtedly, much wider roles of *MYBs* in plant roots will be continuously recognized in the near future based on the efforts from both of the traditional and modern disciplines, such as Molecular Biology, Genetics, Cell Biology, Genomics, multi-omics, etc.

Author Contributions: Conceptualization, H.D.; methodology, formal analysis, Z.C.; software, validation, Z.W.; investigation, W.D.; data curation, S.L. and L.T.; writing—original draft preparation, H.D. and Z.C.; project administration, J.L.; writing—review and editing, supervision, funding acquisition, H.D. All authors have read and agreed to the published version of the manuscript.

Funding: This work was supported by the National Natural Science Foundation of China (32072094).

Institutional Review Board Statement: The study did not require ethical approval.

Informed Consent Statement: The study did not involve humans.

Data Availability Statement: The study did not report any data.

Acknowledgments: This work was mainly supported by the Du Hai lab in Southwest University.

Conflicts of Interest: The authors declare no conflict of interest.

References

1. Kohli, P.S.; Maurya, K.; Thakur, J.K.; Bhosale, R.; Giri, J. Significance of root hairs in developing stress-resilient plants for sustainable crop production. *Plant Cell Environ.* **2022**, *45*, 677–694. [CrossRef] [PubMed]
2. Lynch, J.P. Harnessing root architecture to address global challenges. *Plant J.* **2022**, *109*, 415–431. [CrossRef] [PubMed]
3. Calleja-Cabrera, J.; Boter, M.; Oñate-Sánchez, L.; Pernas, M. Root Growth Adaptation to Climate Change in Crops. *Front. Plant Sci.* **2020**, *11*, 544. [CrossRef] [PubMed]
4. Lombardi, M.; De Gara, L.; Loreto, F. Determinants of root system architecture for future-ready, stress-resilient crops. *Physiol. Plant.* **2021**, *172*, 2090–2097. [CrossRef]
5. Tracy, S.R.; Nagel, K.A.; Postma, J.A.; Fassbender, H.; Wasson, A.; Watt, M. Crop Improvement from Phenotyping Roots: Highlights Reveal Expanding Opportunities. *Trends Plant Sci.* **2020**, *25*, 105–118. [CrossRef]
6. Lee, J.-Y.; Colinas, J.; Wang, J.Y.; Mace, D.; Ohler, U.; Benfey, P.N. Transcriptional and posttranscriptional regulation of transcription factor expression in *Arabidopsis* roots. *Proc. Natl. Acad. Sci. USA* **2006**, *103*, 6055–6060. [CrossRef]
7. Motte, H.; Vanneste, S.; Beeckman, T. Molecular and Environmental Regulation of Root Development. *Annu. Rev. Plant Biol.* **2019**, *70*, 465–488. [CrossRef]
8. Forzani, C.; Aichinger, E.; Sornay, E.; Willemsen, V.; Laux, T.; Dewitte, W.; Murray, J.A. WOX5 Suppresses CYCLIN D Activity to Establish Quiescence at the Center of the Root Stem Cell Niche. *Curr. Biol.* **2014**, *24*, 1939–1944. [CrossRef]
9. Okushima, Y.; Fukaki, H.; Onoda, M.; Theologis, A.; Tasaka, M. ARF7 and ARF19 Regulate Lateral Root Formation via Direct Activation of *LBD/ASL* Genes in *Arabidopsis*. *Plant Cell* **2007**, *19*, 118–130. [CrossRef]
10. Lee, M.M.; Schiefelbein, J. WEREWOLF, a MYB-Related Protein in *Arabidopsis*, Is a Position-Dependent Regulator of Epidermal Cell Patterning. *Cell* **1999**, *99*, 473–483. [CrossRef]
11. Klempnauer, K.-H.; Gonda, T.J.; Bishop, J.M. Nucleotide sequence of the retroviral leukemia gene v-myb and its cellular progenitor c-myb: The architecture of a transduced oncogene. *Cell* **1982**, *31 Pt 1*, 453–463. [CrossRef]
12. Jiang, C.-K.; Rao, G.-Y. Insights into the Diversification and Evolution of R2R3-MYB Transcription Factors in Plants. *Plant Physiol.* **2020**, *183*, 637–655. [CrossRef] [PubMed]
13. Riechmann, J.L.; Heard, J.; Martin, G.; Reuber, L.; Jiang, C.; Keddie, J.; Adam, L.; Pineda, O.; Ratcliffe, O.J.; Samaha, R.R.; et al. *Arabidopsis* Transcription Factors: Genome-Wide Comparative Analysis Among Eukaryotes. *Science* **2000**, *290*, 2105–2110. [CrossRef] [PubMed]
14. Du, H.; Liang, Z.; Zhao, S.; Nan, M.-G.; Tran, L.-S.P.; Lu, K.; Huang, Y.-B.; Li, J.-N. The Evolutionary History of R2R3-MYB Proteins across 50 Eukaryotes: New Insights Into Subfamily Classification and Expansion. *Sci. Rep.* **2015**, *5*, 11037. [CrossRef] [PubMed]
15. Davidson, C.; Tirouvanziam, R.; Herzenberg, L.A.; Lipsick, J.S.; Jiggins, C.D.; Mavarez, J.; Beltrán, M.; McMillan, W.O.; Johnston, J.S.; Bermingham, E. Functional evolution of the vertebrate Myb gene family: B-Myb, but neither A-Myb nor c-Myb, complements Drosophila Myb in hemocytes. *Genetics* **2005**, *169*, 215–229. [CrossRef]
16. Stober-Grässer, U.; Brydolf, B.; Bin, X.; Grässer, F.; Firtel, R.A.; Lipsick, J.S. The Myb DNA-binding domain is highly conserved in *Dictyostelium discoideum*. *Oncogene* **1992**, *7*, 589–596.
17. Dubos, C.; Stracke, R.; Grotewold, E.; Weisshaar, B.; Martin, C.; Lepiniec, L. MYB transcription factors in *Arabidopsis*. *Trends Plant Sci.* **2010**, *15*, 573–581. [CrossRef] [PubMed]
18. Du, H.; Wang, Y.-B.; Xie, Y.; Liang, Z.; Jiang, S.-J.; Zhang, S.-S.; Huang, Y.-B.; Tang, Y.-X. Genome-Wide Identification and Evolutionary and Expression Analyses of MYB-Related Genes in Land Plants. *DNA Res.* **2013**, *20*, 437–448. [CrossRef]
19. Kobayashi, K.; Suzuki, T.; Iwata, E.; Nakamichi, N.; Suzuki, T.; Chen, P.; Ohtani, M.; Ishida, T.; Hosoya, H.; Mueller, S.; et al. Transcriptional repression by MYB 3R proteins regulates plant organ growth. *EMBO J.* **2015**, *34*, 1992–2007. [CrossRef] [PubMed]
20. Haga, N.; Kato, K.; Murase, M.; Araki, S.; Kubo, M.; Demura, T.; Suzuki, K.; Müller, I.; Voß, U.; Jürgens, G.; et al. R1R2R3-Myb proteins positively regulate cytokinesis through activation of *KNOLLE* transcription in *Arabidopsis thaliana*. *Development* **2007**, *134*, 1101–1110. [CrossRef]
21. Umeda, M.; Aki, S.S.; Takahashi, N. Gap 2 phase: Making the fundamental decision to divide or not. *Curr. Opin. Plant Biol.* **2019**, *51*, 1–6. [CrossRef] [PubMed]
22. Wu, Y.; Wen, J.; Xia, Y.; Zhang, L.; Du, H. Evolution and functional diversification of R2R3-MYB transcription factors in plants. *Hortic. Res.* **2022**, *9*, uhac058. [CrossRef] [PubMed]
23. Wang, S.; Chen, J.-G. Regulation of cell fate determination by single-repeat R3 MYB transcription factors in *Arabidopsis*. *Front. Plant Sci.* **2014**, *5*, 133. [CrossRef]

24. Zhang, X.; Ju, H.-W.; Chung, M.-S.; Huang, P.; Ahn, S.-J.; Kim, C.S. The R-R-type MYB-like transcription factor, AtMYBL, is involved in promoting leaf senescence and modulates an abiotic stress response in *Arabidopsis*. *Plant Cell Physiol.* **2010**, *52*, 138–148. [CrossRef] [PubMed]
25. Wang, X.; Niu, Y.; Zheng, Y. Multiple Functions of MYB Transcription Factors in Abiotic Stress Responses. *Int. J. Mol. Sci.* **2021**, *22*, 6125. [CrossRef]
26. Cao, Y.; Li, K.; Li, Y.; Zhao, X.; Wang, L. MYB Transcription Factors as Regulators of Secondary Metabolism in Plants. *Biology* **2020**, *9*, 61. [CrossRef]
27. Fernández-Marcos, M.; Desvoyes, B.; Manzano, C.; Liberman, L.M.; Benfey, P.N.; del Pozo, J.C.; Gutierrez, C. Control of *Arabidopsis* lateral root primordium boundaries by MYB 36. *New Phytol.* **2017**, *213*, 105–112. [CrossRef]
28. Wan, J.; Wang, R.; Zhang, P.; Sun, L.; Ju, Q.; Huang, H.; Lü, S.; Tran, L.-S.; Xu, J. MYB70 modulates seed germination and root system development in *Arabidopsis*. *iScience* **2021**, *24*, 103228. [CrossRef]
29. Yang, Y.; Zhang, L.; Chen, P.; Liang, T.; Li, X.; Liu, H. UV-B photoreceptor UVR8 interacts with MYB73/MYB77 to regulate auxin responses and lateral root development. *EMBO J.* **2020**, *39*, e101928. [CrossRef]
30. Hassan, H.; Scheres, B.; Blilou, I. JACKDAW controls epidermal patterning in the *Arabidopsis* root meristem through a non-cell-autonomous mechanism. *Development* **2010**, *137*, 1523–1529. [CrossRef]
31. Kwak, S.-H.; Schiefelbein, J. Regulated accumulation of the SCRAMBLED receptor and position-dependent cell type patterning in *Arabidopsis*. *Plant Signal. Behav.* **2009**, *4*, 332–335. [CrossRef] [PubMed]
32. Bernhardt, C.; Lee, M.M.; Gonzalez, A.; Zhang, F.; Lloyd, A.; Schiefelbein, J. The bHLH genes *GLABRA3 (GL3)* and *ENHANCER OF GLABRA3 (EGL3)* specify epidermal cell fate in the *Arabidopsis* root. *Development* **2003**, *130*, 6431–6439. [CrossRef] [PubMed]
33. Walker, A.R.; Davison, P.A.; Bolognesi-Winfield, A.C.; James, C.M.; Srinivasan, N.; Blundell, T.L.; Esch, J.J.; Marks, M.D.; Gray, J.C. The *TRANSPARENT TESTA GLABRA1* Locus, Which Regulates Trichome Differentiation and Anthocyanin Biosynthesis in *Arabidopsis*, Encodes a WD40 Repeat Protein. *Plant Cell* **1999**, *11*, 1337–1349. [CrossRef] [PubMed]
34. Masucci, J.; Rerie, W.; Foreman, D.; Zhang, M.; Galway, M.; Marks, M.D.; Schiefelbein, J. The homeobox gene GLABRA2 is required for position-dependent cell differentiation in the root epidermis of *Arabidopsis thaliana*. *Development* **1996**, *122*, 1253–1260. [CrossRef]
35. Shibata, M.; Sugimoto, K. A gene regulatory network for root hair development. *J. Plant Res.* **2019**, *132*, 301–309. [CrossRef]
36. Wada, T.; Kurata, T.; Tominaga, R.; Koshino-Kimura, Y.; Tachibana, T.; Goto, K.; Marks, M.D.; Shimura, Y.; Okada, K. Role of a positive regulator of root hair development, *CAPRICE*, in *Arabidopsis* root epidermal cell differentiation. *Development* **2002**, *129*, 5409–5419. [CrossRef]
37. Kurata, T.; Ishida, T.; Kawabata-Awai, C.; Noguchi, M.; Hattori, S.; Sano, R.; Nagasaka, R.; Tominaga, R.; Koshino-Kimura, Y.; Kato, T.; et al. Cell-to-cell movement of the CAPRICE protein in *Arabidopsis* root epidermal cell differentiation. *Development* **2005**, *132*, 5387–5398. [CrossRef]
38. Tominaga, R.; Iwata, M.; Okada, K.; Wada, T. Functional Analysis of the Epidermal-Specific MYB Genes *CAPRICE* and *WEREWOLF* in *Arabidopsis*. *Plant Cell* **2007**, *19*, 2264–2277. [CrossRef]
39. Wang, W.; Ryu, K.H.; Barron, C.; Schiefelbein, J. Root Epidermal Cell Patterning Is Modulated by a Critical Residue in the WEREWOLF Transcription Factor. *Plant Physiol.* **2019**, *181*, 1239–1256. [CrossRef]
40. Kang, Y.H.; Kirik, V.; Hulskamp, M.; Nam, K.H.; Hagely, K.; Lee, M.M.; Schiefelbein, J. The *MYB23* Gene Provides a Positive Feedback Loop for Cell Fate Specification in the *Arabidopsis* Root Epidermis. *Plant Cell* **2009**, *21*, 1080–1094. [CrossRef]
41. Wang, S.; Hubbard, L.; Chang, Y.; Guo, J.; Schiefelbein, J.; Chen, J.-G. Comprehensive analysis of single-repeat R3 MYB proteins in epidermal cell patterning and their transcriptional regulation in *Arabidopsis*. *BMC Plant Biol.* **2008**, *8*, 81. [CrossRef] [PubMed]
42. Tominaga, R.; Iwata, M.; Sano, R.; Inoue, K.; Okada, K.; Wada, T. *Arabidopsis CAPRICE-LIKE MYB 3* (*CPL3*) controls endoreduplication and flowering development in addition to trichome and root hair formation. *Development* **2008**, *135*, 1335–1345. [CrossRef] [PubMed]
43. Li, P.; Wen, J.; Chen, P.; Guo, P.; Ke, Y.; Wang, M.; Liu, M.; Tran, L.-S.P.; Li, J.; Du, H. MYB Superfamily in *Brassica napus*: Evidence for Hormone-Mediated Expression Profiles, Large Expansion, and Functions in Root Hair Development. *Biomolecules* **2020**, *10*, 875. [CrossRef] [PubMed]
44. Tominaga-Wada, R.; Nukumizu, Y.; Sato, S.; Wada, T. Control of Plant Trichome and Root-Hair Development by a Tomato (*Solanum lycopersicum*) R3 MYB Transcription Factor. *PLoS ONE* **2013**, *8*, e54019. [CrossRef] [PubMed]
45. Zheng, K.; Tian, H.; Hu, Q.; Guo, H.; Yang, L.; Cai, L.; Wang, X.; Liu, B.; Wang, S. Ectopic expression of R3 MYB transcription factor gene OsTCL1 in *Arabidopsis*, but not rice, affects trichome and root hair formation. *Sci. Rep.* **2016**, *6*, 19254. [CrossRef]
46. Schellmann, S.; Schnittger, A.; Kirik, V.; Wada, T.; Okada, K.; Beermann, A.; Thumfahrt, J.; Jürgens, G.; Hülskamp, M. TRIPTYCHON and CAPRICE mediate lateral inhibition during trichome and root hair patterning in *Arabidopsis*. *EMBO J.* **2002**, *21*, 5036–5046. [CrossRef]
47. Bernhardt, C.; Zhao, M.; Gonzalez, A.; Lloyd, A.; Schiefelbein, J. The bHLH genes *GL3* and *EGL3* participate in an intercellular regulatory circuit that controls cell patterning in the *Arabidopsis* root epidermis. *Development* **2005**, *132*, 291–298. [CrossRef]
48. Savage, N.S.; Walker, T.; Wieckowski, Y.; Schiefelbein, J.; Dolan, L.; Monk, N.A.M. A Mutual Support Mechanism through Intercellular Movement of CAPRICE and GLABRA3 Can Pattern the *Arabidopsis* Root Epidermis. *PLoS Biol.* **2008**, *6*, e235. [CrossRef]

49. Kwak, S.-H.; Schiefelbein, J. TRIPTYCHON, not CAPRICE, participates in feedback regulation of *SCM* expression in the *Arabidopsis* root epidermis. *Plant Signal. Behav.* **2014**, *9*, e973815. [CrossRef]
50. Kaufmann, C.; Stührwohldt, N.; Sauter, M. Tyrosylprotein sulfotransferase-dependent and -independent regulation of root development and signaling by PSK LRR receptor kinases in *Arabidopsis*. *J. Exp. Bot.* **2021**, *72*, 5508–5521. [CrossRef]
51. Mutanwad, K.V.; Zangl, I.; Lucyshyn, D. *Arabidopsis* O-fucosyltransferase SPINDLY regulates root hair patterning independently of gibberellin signaling. *Development* **2020**, *147*, dev192039. [CrossRef] [PubMed]
52. Kamiya, T.; Borghi, M.; Wang, P.; Danku, J.M.C.; Kalmbach, L.; Hosmani, P.S.; Naseer, S.; Fujiwara, T.; Geldner, N.; Salt, D.E. The MYB36 transcription factor orchestrates Casparian strip formation. *Proc. Natl. Acad. Sci. USA* **2015**, *112*, 10533–10538. [CrossRef] [PubMed]
53. Liberman, L.M.; Sparks, E.E.; Moreno-Risueno, M.A.; Petricka, J.J.; Benfey, P.N. MYB36 regulates the transition from proliferation to differentiation in the *Arabidopsis* root. *Proc. Natl. Acad. Sci. USA* **2015**, *112*, 12099–12104. [CrossRef] [PubMed]
54. Cicirò, Y.; Sala, A. MYB oncoproteins: Emerging players and potential therapeutic targets in human cancer. *Oncogenesis* **2021**, *10*, 19. [CrossRef]
55. Hanano, S.; Stracke, R.; Jakoby, M.; Merkle, T.; Domagalska, M.A.; Weisshaar, B.; Davis, S.J. A systematic survey in *Arabidopsis thaliana* of transcription factors that modulate circadian parameters. *BMC Genom.* **2008**, *9*, 182. [CrossRef]
56. Haga, N.; Kobayashi, K.; Suzuki, T.; Maeo, K.; Kubo, M.; Ohtani, M.; Mitsuda, N.; Demura, T.; Nakamura, K.; Jürgens, G.; et al. Mutations in *MYB3R1* and *MYB3R4* Cause Pleiotropic Developmental Defects and Preferential Down-Regulation of Multiple G2/M-Specific Genes in *Arabidopsis*. *Plant Physiol.* **2011**, *157*, 706–717. [CrossRef]
57. Magyar, Z.; Bögre, L.; Ito, M. DREAMs make plant cells to cycle or to become quiescent. *Curr. Opin. Plant Biol.* **2016**, *34*, 100–106. [CrossRef]
58. Sozzani, R.; Maggio, C.; Varotto, S.; Canova, S.; Bergounioux, C.; Albani, D.; Cella, R. Interplay between *Arabidopsis* Activating Factors E2Fb and E2Fa in Cell Cycle Progression and Development. *Plant Physiol.* **2006**, *140*, 1355–1366. [CrossRef]
59. del Pozo, J.C.; Diaz-Trivino, S.; Cisneros, N.; Gutierrez, C. The Balance between Cell Division and Endoreplication Depends on E2FC-DPB, Transcription Factors Regulated by the Ubiquitin-SCFSKP2A Pathway in *Arabidopsis*. *Plant Cell* **2006**, *18*, 2224–2235. [CrossRef]
60. Wang, W.; Sijacic, P.; Xu, P.; Lian, H.; Liu, Z. *Arabidopsis* TSO1 and MYB3R1 form a regulatory module to coordinate cell proliferation with differentiation in shoot and root. *Proc. Natl. Acad. Sci. USA* **2018**, *115*, E3045–E3054. [CrossRef]
61. Araki, S.; Ito, M.; Soyano, T.; Nishihama, R.; Machida, Y. Mitotic Cyclins Stimulate the Activity of c-Myb-like Factors for Transactivation of G2/M Phase-specific Genes in Tobacco. *J. Biol. Chem.* **2004**, *279*, 32979–32988. [CrossRef] [PubMed]
62. Ma, Q.; Dai, X.; Xu, Y.; Guo, J.; Liu, Y.; Chen, N.; Xiao, J.; Zhang, D.; Xu, Z.; Zhang, X.; et al. Enhanced Tolerance to Chilling Stress in *OsMYB3R-2* Transgenic Rice Is Mediated by Alteration in Cell Cycle and Ectopic Expression of Stress Genes. *Plant Physiol.* **2009**, *150*, 244–256. [CrossRef] [PubMed]
63. Feng, G.; Burleigh, J.G.; Braun, E.; Mei, W.; Barbazuk, W.B. Evolution of the 3R-MYB Gene Family in Plants. *Genome Biol. Evol.* **2017**, *9*, 1013–1029. [CrossRef] [PubMed]
64. Sadasivam, S.; Duan, S.; DeCaprio, J.A. The MuvB complex sequentially recruits B-Myb and FoxM1 to promote mitotic gene expression. *Genes Dev.* **2012**, *26*, 474–489. [CrossRef] [PubMed]
65. Sadasivam, S.; DeCaprio, J.A. The DREAM complex: Master coordinator of cell cycle-dependent gene expression. *Nat. Rev. Cancer* **2013**, *13*, 585–595. [CrossRef]
66. Mu, R.-L.; Cao, Y.-R.; Liu, Y.-F.; Lei, G.; Zou, H.-F.; Liao, Y.; Wang, H.-W.; Zhang, W.-K.; Ma, B.; Du, J.-Z.; et al. An R2R3-type transcription factor gene AtMYB59 regulates root growth and cell cycle progression in *Arabidopsis*. *Cell Res.* **2009**, *19*, 1291–1304. [CrossRef]
67. Vilarrasa-Blasi, J.; González-García, M.-P.; Frigola, D.; Fàbregas, N.; Alexiou, K.G.; Lopez-Bigas, N.; Rivas, S.; Jauneau, A.; Lohmann, J.; Benfey, P.N.; et al. Regulation of Plant Stem Cell Quiescence by a Brassinosteroid Signaling Module. *Dev. Cell* **2014**, *30*, 36–47. [CrossRef]
68. Betegón-Putze, I.; Mercadal, J.; Bosch, N.; Planas-Riverola, A.; Marquès-Bueno, M.; Vilarrasa-Blasi, J.; Frigola, D.; Burkart, R.C.; Martínez, C.; Conesa, A.; et al. Precise transcriptional control of cellular quiescence by BRAVO/WOX5 complex in *Arabidopsis* roots. *Mol. Syst. Biol.* **2021**, *17*, e9864. [CrossRef]
69. Lin, Z.; Yin, K.; Zhu, D.; Chen, Z.; Gu, H.; Qu, L.-J. AtCDC5 regulates the G2 to M transition of the cell cycle and is critical for the function of *Arabidopsis* shoot apical meristem. *Cell Res.* **2007**, *17*, 815–828. [CrossRef]
70. Baek, D.; Kim, M.C.; Chun, H.J.; Kang, S.; Park, H.C.; Shin, G.; Park, J.; Shen, M.; Hong, H.; Kim, W.-Y.; et al. Regulation of *miR399f* Transcription by AtMYB2 Affects Phosphate Starvation Responses in *Arabidopsis*. *Plant Physiol.* **2012**, *161*, 362–373. [CrossRef]
71. Wang, F.; Kong, W.; Wong, G.; Fu, L.; Peng, R.; Li, Z.; Yao, Q. AtMYB12 regulates flavonoids accumulation and abiotic stress tolerance in transgenic *Arabidopsis thaliana*. *Mol. Genet. Genom.* **2016**, *291*, 1545–1559. [CrossRef] [PubMed]
72. Cui, M.H.; Yoo, K.S.; Hyoung, S.; Nguyen, H.T.K.; Kim, Y.Y.; Kim, H.J.; Ok, S.H.; Yoo, S.D.; Shin, J.S. An *Arabidopsis* R2R3-MYB transcription factor, AtMYB20, negatively regulates type 2C serine/threonine protein phosphatases to enhance salt tolerance. *FEBS Lett.* **2013**, *587*, 1773–1778. [CrossRef] [PubMed]

73. Mabuchi, K.; Maki, H.; Itaya, T.; Suzuki, T.; Nomoto, M.; Sakaoka, S.; Morikami, A.; Higashiyama, T.; Tada, Y.; Busch, W.; et al. MYB30 links ROS signaling, root cell elongation, and plant immune responses. *Proc. Natl. Acad. Sci. USA* **2018**, *115*, E4710–E4719. [CrossRef] [PubMed]
74. Gong, Q.; Li, S.; Zheng, Y.; Duan, H.; Xiao, F.; Zhuang, Y.; He, J.; Wu, G.; Zhao, S.; Zhou, H.; et al. SUMOylation of MYB30 enhances salt tolerance by elevating alternative respiration via transcriptionally upregulating AOX1a in *Arabidopsis*. *Plant J.* **2020**, *102*, 1157–1171. [CrossRef]
75. Sun, Y.; Zhao, J.; Li, X.; Li, Y. E2 conjugases UBC1 and UBC2 regulate MYB42-mediated SOS pathway in response to salt stress in *Arabidopsis*. *New Phytol.* **2020**, *227*, 455–472. [CrossRef]
76. Du, X.-Q.; Wang, F.-L.; Li, H.; Jing, S.; Yu, M.; Li, J.; Wu, W.-H.; Kudla, J.; Wang, Y. The Transcription Factor MYB59 Regulates K^+/NO_3^- Translocation in the *Arabidopsis* Response to Low K^+ Stress. *Plant Cell* **2019**, *31*, 699–714. [CrossRef]
77. Fasani, E.; DalCorso, G.; Costa, A.; Zenoni, S.; Furini, A. The *Arabidopsis thaliana* transcription factor MYB59 regulates calcium signalling during plant growth and stress response. *Plant Mol. Biol.* **2019**, *99*, 517–534. [CrossRef]
78. Wiśniewska, A.; Wojszko, K.; Różańska, E.; Lenarczyk, K.; Kuczerski, K.; Sobczak, M. *Arabidopsis thaliana Myb59* Gene Is Involved in the Response to *Heterodera schachtii* Infestation, and Its Overexpression Disturbs Regular Development of Nematode-Induced Syncytia. *Int. J. Mol. Sci.* **2021**, *22*, 6450. [CrossRef]
79. Oh, J.E.; Kwon, Y.; Kim, J.H.; Noh, H.; Hong, S.-W.; Lee, H. A dual role for MYB60 in stomatal regulation and root growth of *Arabidopsis thaliana* under drought stress. *Plant Mol. Biol.* **2011**, *77*, 91–103. [CrossRef]
80. Devaiah, B.N.; Madhuvanthi, R.; Karthikeyan, A.S.; Raghothama, K.G. Phosphate Starvation Responses and Gibberellic Acid Biosynthesis are Regulated by the MYB62 Transcription Factor in *Arabidopsis*. *Mol. Plant* **2009**, *2*, 43–58. [CrossRef]
81. Seo, P.J.; Xiang, F.; Qiao, M.; Park, J.-Y.; Na Lee, Y.; Kim, S.-G.; Lee, Y.-H.; Park, W.J.; Park, C.-M. The MYB96 Transcription Factor Mediates Abscisic Acid Signaling during Drought Stress Response in *Arabidopsis*. *Plant Physiol.* **2009**, *151*, 275–289. [CrossRef] [PubMed]
82. Wang, H.-Z.; Yang, K.-Z.; Zou, J.-J.; Zhu, L.-L.; Xie, Z.D.; Morita, M.T.; Tasaka, M.; Friml, J.; Grotewold, E.; Beeckman, T.; et al. Transcriptional regulation of PIN genes by FOUR LIPS and MYB88 during *Arabidopsis* root gravitropism. *Nat. Commun.* **2015**, *6*, 8822. [CrossRef] [PubMed]
83. Rubio, V.; Linhares, F.; Solano, R.; Martín, A.C.; Iglesias, J.; Leyva, A.; Paz-Ares, J. A conserved MYB transcription factor involved in phosphate starvation signaling both in vascular plants and in unicellular algae. *Genes Dev.* **2001**, *15*, 2122–2133. [CrossRef] [PubMed]
84. Bustos, R.; Castrillo, G.; Linhares, F.; Puga, M.I.; Rubio, V.; Pérez-Pérez, J.; Solano, R.; Leyva, A.; Paz-Ares, J. A Central Regulatory System Largely Controls Transcriptional Activation and Repression Responses to Phosphate Starvation in *Arabidopsis*. *PLoS Genet.* **2010**, *6*, e1001102. [CrossRef] [PubMed]
85. Sun, L.; Song, L.; Zhang, Y.; Zheng, Z.; Liu, D. *Arabidopsis* PHL2 and PHR1 Act Redundantly as the Key Components of the Central Regulatory System Controlling Transcriptional Responses to Phosphate Starvation. *Plant Physiol.* **2016**, *170*, 499–514. [CrossRef]
86. Liu, H.; Yang, H.; Wu, C.; Feng, J.; Liu, X.; Qin, H.; Wang, D. Overexpressing *HRS1* Confers Hypersensitivity to Low Phosphate-Elicited Inhibition of Primary Root Growth in *Arabidopsis thaliana*. *J. Integr. Plant Biol.* **2009**, *51*, 382–392. [CrossRef]
87. Nagarajan, V.K.; Satheesh, V.; Poling, M.D.; Raghothama, K.G.; Jain, A. *Arabidopsis* MYB-Related HHO2 Exerts a Regulatory Influence on a Subset of Root Traits and Genes Governing Phosphate Homeostasis. *Plant Cell Physiol.* **2016**, *57*, 1142–1152. [CrossRef]
88. Chen, C.-Y.; Schmidt, W. The paralogous R3 MYB proteins CAPRICE, TRIPTYCHON and ENHANCER OF TRY AND CPC1 play pleiotropic and partly non-redundant roles in the phosphate starvation response of *Arabidopsis* roots. *J. Exp. Bot.* **2015**, *66*, 4821–4834. [CrossRef]
89. Wang, N.; Zhang, W.; Qin, M.; Li, S.; Qiao, M.; Liu, Z.; Xiang, F. Drought Tolerance Conferred in Soybean (*Glycine max* L.) by GmMYB84, a Novel R2R3-MYB Transcription Factor. *Plant Cell Physiol.* **2017**, *58*, 1764–1776. [CrossRef]
90. Zhao, P.; Hou, S.; Guo, X.; Jia, J.; Yang, W.; Liu, Z.; Chen, S.; Li, X.; Qi, D.; Liu, G.; et al. A MYB-related transcription factor from sheepgrass, LcMYB2, promotes seed germination and root growth under drought stress. *BMC Plant Biol.* **2019**, *19*, 564. [CrossRef]
91. Wang, Z.; Li, J.; Mao, Y.; Zhang, M.; Wang, R.; Hu, Y.; Mao, Z.; Shen, X. Transcriptional regulation of MdPIN3 and MdPIN10 by MdFLP during apple self-rooted stock adventitious root gravitropism. *BMC Plant Biol.* **2019**, *19*, 229. [CrossRef] [PubMed]
92. Gu, M.; Zhang, J.; Li, H.; Meng, D.; Li, R.; Dai, X.; Wang, S.; Liu, W.; Qu, H.; Xu, G. Maintenance of phosphate homeostasis and root development are coordinately regulated by MYB1, an R2R3-type MYB transcription factor in rice. *J. Exp. Bot.* **2017**, *68*, 3603–3615. [CrossRef] [PubMed]
93. Dai, X.; Wang, Y.; Yang, A.; Zhang, W.-H. *OsMYB2P-1*, an R2R3 MYB Transcription Factor, Is Involved in the Regulation of Phosphate-Starvation Responses and Root Architecture in Rice. *Plant Physiol.* **2012**, *159*, 169–183. [CrossRef] [PubMed]
94. Yang, W.T.; Baek, D.; Yun, D.-J.; Hwang, W.H.; Park, D.S.; Nam, M.H.; Chung, E.S.; Chung, Y.S.; Yi, Y.B.; Kim, D.H. Overexpression of OsMYB4P, an R2R3-type MYB transcriptional activator, increases phosphate acquisition in rice. *Plant Physiol. Biochem.* **2014**, *80*, 259–267. [CrossRef] [PubMed]
95. Guo, M.; Ruan, W.; Fangliang, H.; Huang, F.; Zeng, M.; Liu, Y.; Yu, Y.; Ding, X.; Wu, Y.; Wu, Z.; et al. Integrative Comparison of the Role of the PHOSPHATE RESPONSE1 Subfamily in Phosphate Signaling and Homeostasis in Rice. *Plant Physiol.* **2015**, *168*, 1762–1776. [CrossRef]

96. Ruan, W.; Guo, M.; Wu, P.; Yi, K. Phosphate starvation induced OsPHR4 mediates Pi-signaling and homeostasis in rice. *Plant Mol. Biol.* **2016**, *93*, 327–340. [CrossRef]
97. Qiu, B.; Chen, H.; Zheng, L.; Su, L.; Cui, X.; Ge, F.; Liu, D. An MYB Transcription Factor Modulates *Panax notoginseng* Resistance against the Root Rot Pathogen *Fusarium solani* by Regulating the Jasmonate Acid Signaling Pathway and Photosynthesis. *Phytopathology* **2022**, *112*, 1323–1334. [CrossRef]
98. Wang, H.; Pak, S.; Yang, J.; Wu, Y.; Li, W.; Feng, H.; Yang, J.; Wei, H.; Li, C. Two high hierarchical regulators, PuMYB40 and PuWRKY75, control the low phosphorus driven adventitious root formation in *Populus ussuriensis*. *Plant Biotechnol. J.* **2022**, *112*, 1323–1334. [CrossRef]
99. Liu, X.; Yang, L.; Zhou, X.; Zhou, M.; Lu, Y.; Ma, L.; Ma, H.; Zhang, Z. Transgenic wheat expressing Thinopyrum intermedium MYB transcription factor TiMYB2R-1 shows enhanced resistance to the take-all disease. *J. Exp. Bot.* **2013**, *64*, 2243–2253. [CrossRef]
100. Li, X.; Tang, Y.; Li, H.; Luo, W.; Zhou, C.; Zhang, L.; Lv, J. A wheat R2R3 MYB gene TaMpc1-D4 negatively regulates drought tolerance in transgenic *Arabidopsis* and wheat. *Plant Sci.* **2020**, *299*, 110613. [CrossRef]
101. Wei, X.; Shan, T.; Hong, Y.; Xu, H.; Liu, X.; Zhang, Z. TaPIMP2, a pathogen-induced MYB protein in wheat, contributes to host resistance to common root rot caused by Bipolaris sorokiniana. *Sci. Rep.* **2017**, *7*, 1754. [CrossRef] [PubMed]
102. Puga, M.I.; Rojas-Triana, M.; de Lorenzo, L.; Leyva, A.; Rubio, V.; Paz-Ares, J. Novel signals in the regulation of Pi starvation responses in plants: Facts and promises. *Curr. Opin. Plant Biol.* **2017**, *39*, 40–49. [CrossRef] [PubMed]
103. Li, Q.; Zhou, L.; Li, Y.; Zhang, D.; Gao, Y. Plant NIGT1/HRS1/HHO Transcription Factors: Key Regulators with Multiple Roles in Plant Growth, Development, and Stress Responses. *Int. J. Mol. Sci.* **2021**, *22*, 8685. [CrossRef] [PubMed]
104. Yang, Y.; Liu, H. Coordinated Shoot and Root Responses to Light Signaling in *Arabidopsis*. *Plant Commun.* **2020**, *1*, 100026. [CrossRef]
105. Raffaele, S.; Vailleau, F.; Léger, A.; Joubeés, J.; Miersch, O.; Huard, C.; Blée, E.; Mongrand, S.; Domergue, F.; Roby, D. A MYB Transcription Factor Regulates Very-Long-Chain Fatty Acid Biosynthesis for Activation of the Hypersensitive Cell Death Response in *Arabidopsis*. *Plant Cell* **2008**, *20*, 752–767. [CrossRef]
106. Van der Ent, S.; Verhagen, B.W.; Van Doorn, R.; Bakker, D.; Verlaan, M.G.; Pel, M.J.; Joosten, R.G.; Proveniers, M.C.; Van Loon, L.; Ton, J.; et al. MYB72 Is Required in Early Signaling Steps of Rhizobacteria-Induced Systemic Resistance in *Arabidopsis*. *Plant Physiol.* **2008**, *146*, 1293–1304. [CrossRef]
107. Cheong, Y.H.; Chang, H.-S.; Gupta, R.; Wang, X.; Zhu, T.; Luan, S. Transcriptional Profiling Reveals Novel Interactions between Wounding, Pathogen, Abiotic Stress, and Hormonal Responses in *Arabidopsis*. *Plant Physiol.* **2002**, *129*, 661–677. [CrossRef]
108. Yuan, Y.; Qin, L.; Su, H.; Yang, S.; Wei, X.; Wang, Z.; Zhao, Y.; Li, L.; Liu, H.; Tian, B.; et al. Transcriptome and Coexpression Network Analyses Reveal Hub Genes in Chinese Cabbage (*Brassica rapa* L. ssp. *pekinensis*) During Different Stages of *Plasmodiophora brassicae* Infection. *Front. Plant Sci.* **2021**, *12*, 650252. [CrossRef]
109. Gibbs, D.J.; Voß, U.; Harding, S.A.; Fannon, J.; Moody, L.A.; Yamada, E.; Swarup, K.; Nibau, C.; Bassel, G.W.; Choudhary, A.; et al. *At*MYB93 is a novel negative regulator of lateral root development in *Arabidopsis*. *New Phytol.* **2014**, *203*, 1194–1207. [CrossRef]
110. Sakaoka, S.; Mabuchi, K.; Morikami, A.; Tsukagoshi, H. MYB30 regulates root cell elongation under abscisic acid signaling. *Commun. Integr. Biol.* **2018**, *11*, e1526604. [CrossRef]
111. Zhao, Y.; Xing, L.; Wang, X.; Hou, Y.-J.; Gao, J.; Wang, P.; Duan, C.-G.; Zhu, X.; Zhu, J.-K. The ABA Receptor PYL8 Promotes Lateral Root Growth by Enhancing MYB77-Dependent Transcription of Auxin-Responsive Genes. *Sci. Signal.* **2014**, *7*, ra53. [CrossRef] [PubMed]
112. Tan, H.; Man, C.; Xie, Y.; Yan, J.; Chu, J.; Huang, J. A Crucial Role of GA-Regulated Flavonol Biosynthesis in Root Growth of *Arabidopsis*. *Mol. Plant* **2019**, *12*, 521–537. [CrossRef] [PubMed]
113. Yang, W.; Cortijo, S.; Korsbo, N.; Roszak, P.; Schiessl, K.; Gurzadyan, A.; Wightman, R.; Jönsson, H.; Meyerowitz, E. Molecular mechanism of cytokinin-activated cell division in *Arabidopsis*. *Science* **2021**, *371*, 1350–1355. [CrossRef] [PubMed]

International Journal of
Molecular Sciences

MDPI

Review

The miR393-Target Module Regulates Plant Development and Responses to Biotic and Abiotic Stresses

Jinjin Jiang [1], Haotian Zhu [1], Na Li [1], Jacqueline Batley [2,*] and Youping Wang [1,3,*]

1 Jiangsu Provincial Key Laboratory of Crop Genetics and Physiology, Yangzhou University, Yangzhou 225009, China
2 School of Biological Sciences, University of Western Australia, Perth, WA 6009, Australia
3 Joint International Research Laboratory of Agriculture and Agri-Product Safety, The Ministry of Education of China, Yangzhou University, Yangzhou 225009, China
* Correspondence: jacqueline.batley@uwa.edu.au (J.B.); wangyp@yzu.edu.cn (Y.W.)

Abstract: MicroRNAs (miRNAs), a class of endogenous small RNAs, are broadly involved in plant development, morphogenesis and responses to various environmental stresses, through manipulating the cleavage, translational expression, or DNA methylation of target mRNAs. miR393 is a conserved miRNA family present in many plants, which mainly targets genes encoding the transport inhibitor response1 (TIR1)/auxin signaling F-box (AFB) auxin receptors, and thus greatly affects the auxin signal perception, Aux/IAA degradation, and related gene expression. This review introduces the advances made on the miR393/target module regulating plant development and the plant's responses to biotic and abiotic stresses. This module is valuable for genetic manipulation of optimized conditions for crop growth and development and would also be helpful in improving crop yield through molecular breeding.

Keywords: miR393; target gene; plant development; stress response; auxin

Citation: Jiang, J.; Zhu, H.; Li, N.; Batley, J.; Wang, Y. The miR393-Target Module Regulates Plant Development and Responses to Biotic and Abiotic Stresses. *Int. J. Mol. Sci.* **2022**, *23*, 9477. https://doi.org/10.3390/ijms23169477

Academic Editors: Andrés J. Cortés and Hai Du

Received: 4 August 2022
Accepted: 19 August 2022
Published: 22 August 2022

1. Introduction

Small RNAs (sRNAs) are 20–30 nucleotide (nt) noncoding RNAs abundant in plants and animals. They were first reported in *Caenorhabditis elegans* [1], and classified into microRNAs (miRNAs), small interfering RNAs (siRNAs), and piwi-interacting RNAs (piRNAs) [2]. The 20–24 nt miRNAs influence plant growth and development, secondary metabolism and response to biotic and abiotic stresses through transcriptional and translational repression of specific target genes with complementary sites [3,4]. miRNAs can also silence target genes through DNA methylation and histone modification [5]. miR393 is a conserved miRNA family identified in many plant species which targets the genes encoding auxin receptors, Transport Inhibitor Response1 (TIR1) and Auxin Signaling F-box (AFB), as well as basic helix-loop-helix (bHLH), thus affecting the homeostasis of auxin signaling and regulating plant development and different stress responses [6–9]. Here, we review the molecular roles of miR393 and the *TIR1/AFB Auxin Receptor* (*TAAR*) gene family, as well as other putative target genes in plant growth, development, metabolism, and stress responses.

2. miR393 and the Target Genes Are Strongly Conserved in Plants

Based on the submitted miRNA sequences in miRBase (https://mirbase.org/, accessed on 10 May 2022), we find that miR393 is a highly conserved miRNA family by sequence alignment, with 21–22 nt in eudicots and monocotyledonous plants, including model plants (e.g., *Arabidopsis thaliana*, *Medicago truncatula*), main crops such as rice (*Oryza sativa*), maize (*Zea mays*), soybean (*Glycine max*), cotton (*Gossypium hirsutum*), rapeseed (*Brassica napus*), camelina (*Camelina sativa*), sorghum (*Sorghum bicolor*) and their relatives (e.g., *Arabidopsis lyrata*, *Aegilops tauschii*, *Brachypodium distachyon*) (Figure 1). Only one or two nucleotide variations exist in the first or last two sites that are not the 'seed' region (critical miRNA-target pairing region) of mature miR393 [10–12]. There are also some miRNAs classified as miR393 which are uncertain, for example miR393b differs greatly from the other members in soybean [12]. The perfect, matched target genes of miR393 predicted by psRNATarget (https://www.zhaolab.org/psRNATarget/, accessed on 10 May 2022) are *TIR1* and *AFBs* (*AFB1/2/3*). Other genes, such as *bHLHs*, *GRR1-like protein 1* (*GRH1*), and *Arabidopsis Response Regulators* (*ARRs*), are also reported as miR393 targets with a single or more mismatches by bioinformatic or degradome analysis, but most of these have not been correlated with miR393 through biological function analysis [13,14]. Similar to the conservation of miR393, the TIR1 and AFBs are also highly conserved in the model plants and main crops (Figure 2).

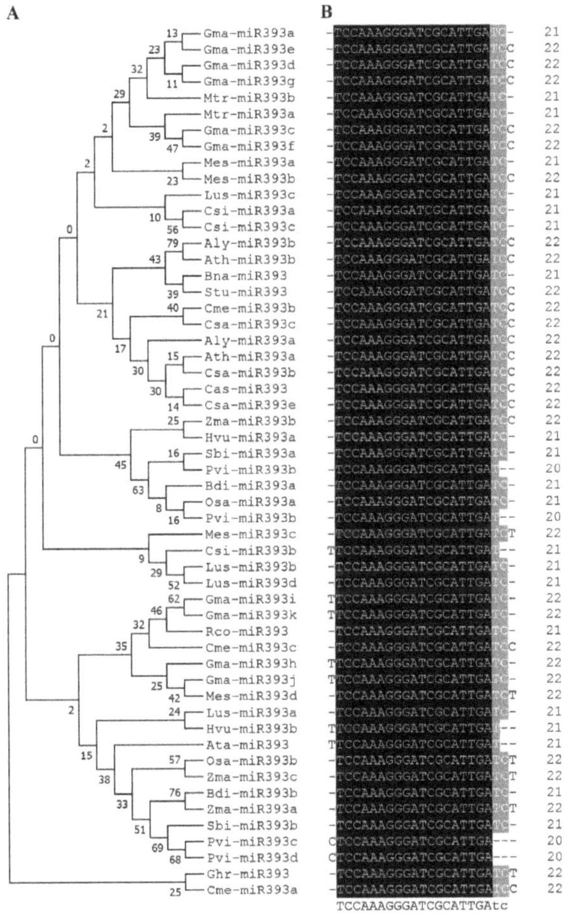

Figure 1. The conserved sequence similarity of miR393 members in different plants. (**A**) Phylogenetic relationships of pre-miR393s; (**B**) sequence similarity of mature miR393s. Aly, *Arabidopsis lyrata*; Ata, *Aegilops tauschii*; Ath, *Arabidopsis thaliana*; Bdi, *Brachypodium distachyon*; Bna, *Brassica napus*; Cas, *Camelina sativa*; Cme, *Cucumis melo*; Csa, *Cucumis sativus*; Csi, *Camellia sinensis*; Ghr, *Gossypium hirsutum*; Gma, *Glycine max*; Hvu, *Hordeum vulgare*; Lus, *Linum usitatissimum*; Mes, *Manihot esculenta*; Mtr, *Medicago truncatula*; Osa, *Oryza sativa*; Rco, *Ricinus communis*; Sbi, *Sorghum bicolor*; Stu, *Solanum tuberosum*; Zma, *Zea mays*; Pvi, *Panicum virgatum*.

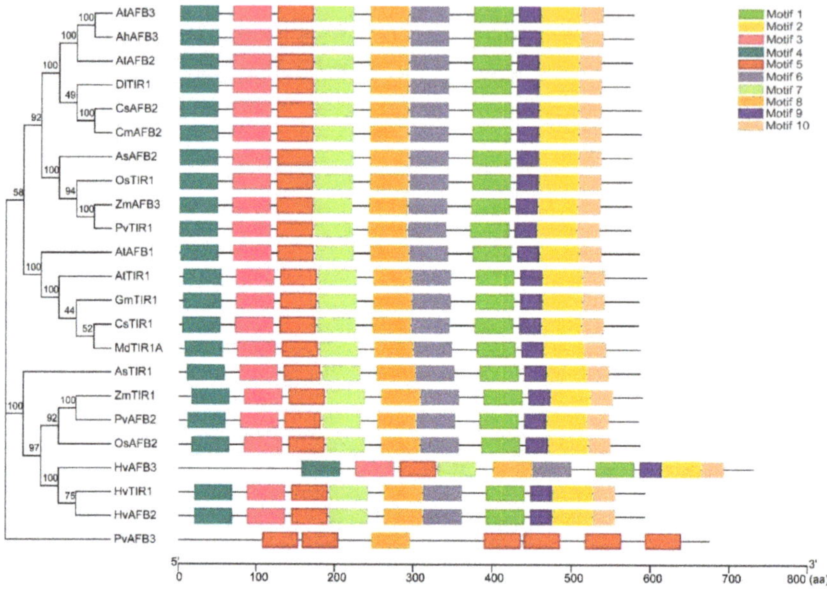

Figure 2. Phylogeny and structures of functionally characterized TIR1/AFBs in plants. Ah, *Arachis hypogaea*; At, *Arabidopsis thaliana*; As, *Agrostis stolonifera*; Cm, *Cucumis melo*; Cs: *Cucumis sativus*; Dl: *Dimocarpus longan*; Gm, *Glycine max*; Hv, *Hordeum vulgare*; Md, *Malus* × *domestica*; Os, *Oryza sativa*; Pv, *Panicum virgatum*; Zm, *Zea mays*.

miR393 members have distinctive expression patterns, and some of them are highly accumulated in the aerial organs of plants. For example, in *Arabidopsis*, miR393 was identified with greater abundance in leaves, stems, inflorescences and siliques, but was present at a very low level in roots [6]. Interestingly, miR393 was induced in *Arabidopsis* roots by nitrate, while the target genes *TIR1*, *AFB1*, *AFB2* and *bHLH77* were consistently repressed; however, *AFB3* was induced in response to nitrate [15]. Using promoter fusion lines, Parry et al. (2009) confirmed that miR393a and miR393b of *Arabidopsis* were expressed in primary root tips via β-glucuronidase (GUS) staining [7]. Similarly, the miR393b expression in rice was higher in aerial organs (e.g., leaves, shoots, booting panicles) but not expressed in roots, whereas miR393a was mainly expressed in the lateral root (LR) primordia of young seedlings and booting panicles [11]. The conserved targets of plant miR393s are the auxin receptors with F-box domains, TIR1 and AFBs, which function as components of the SKP/Cullin/F-box (SCF)-ubiquitin ligase complex and as Aux/IAA transcriptional repressors [16]. These auxin receptors associate with Cullin1 (CUL1) of SCF through the N-terminal H1 helix of the F-box domain [17]. Aux/IAAs recruit the TOPLESS (TPL) co-repressor to interact with both TPL and Auxin Response Factor (ARF) and, finally, repress the transcription of auxin-responsive genes [18]. Auxin is perceived by receptor complexes of TIR1/AFB and Aux/IAA, which could induce ubiquitination and 26S proteasome-mediated degradation of Aux/IAA, and release ARFs to activate auxin-responsive genes [19,20]. Thus, the miR393-TIR1/AFBs module is an important regulator that links auxin response with transcriptional regulation in plant development (Table 1).

Table 1. The functionally characterized miR393 and TIR1/AFBs in plants.

Plant Species	miRNA	Target Genes	Interactive Proteins	Regulated Genes	Functions	References
Arabidopsis thaliana	miR393a miR393b	*AtTIR1* *AtAFB1* *AtAFB2* *AtAFB3* *AtGRH1*	BDL/IAA12	*IAA1/3/7/12/19, GH3, DFL2, ARF1/9*	Regulates somatic embryogenesis, leaf shape and number, cotyledon epinasty, senescence, hypocotyl and root length, inflorescence height, and primary root length in response to nitrate. This module involves in hydrogen sulfide-mediated antibacterial resistance and immunity to *Pseudomonas syringae* pv. *tomato*. It also regulates glucosinolate and camalexin level to involve in plant resistance to biotroph and necrotroph pathogens. It regulates salt and osmotic stress responses through ABA signaling.	[6–10,15,21–31]
Oryza sativa	miR393a miR393b	*OsTIR1* *OsAFB2*	IAA1	*ARF, IAA6/9/20, miR390, AUX1, TB1*	Regulates tillering, flag leaf inclination, flowering and sensitivity to 2,4-D, primary and crown root growth, plant height, coleoptile elongation and stomatal development of submerged seeds, grain number and seed germination. The module is regulated by nitrogen and affects nitrogen-promoted tillering. It also plays roles in plant tolerance to salt, alkaline and drought stresses, as well as plant immunity to rice black streaked dwarf virus.	[11,32–39]
Zea mays	miR393	*ZmTIR1* *ZmAFB*			Might be involved in the internode elongation and development under maize ears; acts as a regulator in root development under short-term waterlogging.	[40,41]
Hordeum vulgare	miR393	*HvTIR1* *HvAFB2* *HvAFB3*		*ARF5, EPF1, SPCH, MUTE*	Regulates seedling growth, stomatal density and guard cell length. miR393 positively regulates length–width ratio of seeds and grain weight, and negatively regulates drought tolerance of barley. miR393 up-regulation alleviated aluminum-induced root inhibition and ROS-induced cell death, its down-regulation enhanced root sensitivity to aluminum stress.	[42–44]
Glycine max	miR393				Regulates root development and nodule formation of soybean and alfalfa. Inhibition of Gma-miR393 and overexpression of mutated GmTIR1C at the cleavage site increased the nodule number of soybean. Knockdown of Gma-miR393 enhanced plant susceptibility to *Phytophthora sojae*, repressed isoflavonoid biosynthetic gene expression in roots.	[12,45,46]
Arachis hypogaea	miR393	*AhAFB3*			Regulated by potassium and nitrogen deficiency, affects the primary and lateral root growth under nutrient deficiency.	[47]

Table 1. *Cont.*

Plant Species	miRNA	Target Genes	Interactive Proteins	Regulated Genes	Functions	References
Cucumis sativus	miR393	*CsTIR1* *CsAFB2*			CsTIR1 and CsAFB2 overexpression caused curling leaf and reduced stomata, poor seed germination, reduced plant height and seed size in tomato.	[48]
Camellia sinensis	miR393	*CsbHLH*			Negatively correlated with the biosynthesis of taste compounds, gallated catechin, caffeine, and theanine.	[49]
Dimocarpus longan	miR393	*DlTIR1*			miR393 was repressed and TIR1 was up-regulated under blue light.	[50]
Malus × *domestica*	miR393b	*MdTIR1A*			Negatively regulates adventitious root formation and auxin sensitivity of tobacco.	[51]
Cucumis melo	miR393	*CmAFB2*			miR393 overexpression delayed fruit ripening.	[52]
Agrostis stolonifera	miR393a	*AsAFB2* *AsTIR1*			Osa-miR393a inhibited AsAFB2/AsTIR1 expression; induced fewer and longer tillers, wider leaves and larger internodes; and improved plant tolerance to salt, drought and heat stresses.	[37]
Panicum virgatum	miR393	*PvAFB2* *PvAFB3* *PvTIR1*		*PvCOR47*, *PvICE1*, *PvRAV1*	Osa-miR393 improved cold tolerance and tillering of switchgrass. Pvi-miR393 was up-regulated by cold stress and down-regulated by auxin, the target genes were induced by cold stress with different patterns.	[53]

3. Functional Diversity of miR393 and TIR1/AFBs in Plant Development

3.1. Leaf Development

In *Arabidopsis*, an *miR393b* mutant exhibited increased leaf number, elongated leaves, higher cotyledon epinasty and earlier senescence, indicating that miR393b was necessary for normal leaf development and auxin-regulated leaf morphology [6,9]. In addition, it was shown that miR393 was necessary for the biogenesis of siTAARs from *TAARs*. *miR393b* mutation increased the expression level of primary auxin-responsive genes *IAA1/12/19* and *DFL2*, which were important for leaf development via regulating *TAAR* expression, auxin perception and auxin signaling homeostasis [6,9]. This finding was in agreement with a previous report that mimicry of miR393 (MIM393) resulted in down-curled and narrow leaves in *Arabidopsis* [21]. The four F-box proteins TIR1, AFB1, AFB2 and AFB3 redundantly regulate diverse aspects of *Arabidopsis* growth and development, such as cotyledon number, leaf size and shape, hypocotyl and root length, and inflorescence height, by interacting with BDL (BODENLOS)/IAA12 and regulating auxin responses throughout plant development [22]. miR393a and miR393b were partially redundant for proper leaf morphogenesis in *Arabidopsis*; both single and double mutants exhibited complex changes in *AUX/IAA* and *GH3* gene expression [23]. In rice, miR393a/b overexpression increased the flag leaf inclination and affected root morphology and plant height. The overexpression plants were resistant to exogenous 2, 4-D, with insensitive root and shoot growth under high auxin conditions. Similar leaf inclination was observed in *OsTIR1*- and *OsAFB2*-suppressed lines [11]. This study also confirmed that OsTIR1 and OsAFB2 interacted with OsIAA1 to regulate ARF activity and auxin signal transduction [11]. In cucumber, miR393 repressed the post-transcriptional expression of *CsTIR1* and *CsAFB2*; furthermore, the overexpression of these target genes in tomato resulted in leaf curling and reduced stomata formation [48]. In barley, miR393 regulated seedling growth, stomatal density and guard cell length by affecting the auxin signaling pathway and stomata development-related gene expression [42].

3.2. Root Development

The miR393-TIR1/AFBs can control root development in plants. Using single, double, triple and quadruple mutants of *TIR1*, *AFB1*, *AFB2* and *AFB3*, Parry et al. (2009) found that *AFB3* was more important than *AFB1* in the auxin response of *Arabidopsis* roots. *TIR1*, *AFB2*, and *AFB3*, but not *AFB1*, exhibited post-transcriptional regulation by miR393. TIR1 and AFB2 are dominant auxin receptors in seedling roots, which have been shown to be negatively regulated by miR393 and could interact with Aux/IAA proteins to regulate lateral root formation [7]. In a further study, miR393 was induced with exogenous IAA treatment. Overexpressing the miR393-resistant form of TIR1 (mTIR1) enhanced auxin sensitivity, inhibited primary root growth but promoted lateral root growth in *Arabidopsis*. Interestingly, miR393 expression was promoted in *35S:mTIR1* plants via a feedback regulation [10]. miR393 has been shown to regulate the lateral root system by modulating TIR1, which could bind the IAA proteins [6]. Furthermore, the ectopic overexpression of *Arabidopsis* miR393a reduced plant size and root length in tobacco [54]. In rice, miR393a is expressed mainly in the roots, especially in crown root tips and lateral root primordia, whereas miR393b is expressed in aerial tissues. The overexpression of rice miR393a/b increased the primary root length and reduced the crown root number in rice, but the root morphologies of *OsTIR1*- and *OsAFB2*-RNAi lines were similar to that of the wild type [11]. Under the inhibition of NAA, Xia et al. (2012) observed more roots and longer main roots in Osa-miR393 overexpression plants than wild-type rice, but no difference was observed when grown in water [32]. miR393a promoted primary root elongation of rice seeds germinated in air but inhibited coleoptile elongation and stomatal development of submerged rice seeds via modulating the auxin signaling during seed germination and seedling establishment [33]. In a further study, Osa-miR393 overexpression was shown to reduce the lateral root number and plant sensitivity to IAA, via repressing Osa-miR390 and *OsTIR*. The regulation of Osa-miR393 on root development could be induced by ABA and various stresses (e.g., drought, salt, heavy metal) [34]. The mutation of OsTIR1 and OsAFB2-5 improved primary root length, decreased adventitious root (AR) number and lateral root density, through interacting with OsIAA1 and regulating auxin responsive genes *OsIAA9/20* [35]. Recently, Li et al. (2021) found that down-regulated miR393 and up-regulated AFB3 might contribute to the primary and lateral root growth of peanut under potassium (K) deficiency, while induced miR393 and repressed AFB3 expression were observed in roots under nitrogen (N) deficiency [47].

3.3. Root Nodulation

miR393 and target-module-regulated auxin signaling also play important roles in the root nodulation of legume plants. Mao et al. (2013) analyzed the roles of miR393 and miR160 in regulating nodule formation of soybean, and found that miR393 overexpression reduced plant sensitivity to auxin, and the level of auxin sensitivity required for indeterminate nodule formation was higher than that of determinate nodules [45]. In common bean, 28 up-regulated miRNAs have been identified as aluminum toxicity nodule-responsive miRNA, including miR393, miR164 and miR170 [55]. Cai et al. (2017) proved that *GmTIR1/AFB3* were post-transcriptionally cleaved by miR393, and the inhibition of miR393 and overexpression of mutated *GmTIR1C* at the cleavage site increased the nodule number of soybean [12]. Li et al. (2021) confirmed that miR393b of apple (*Malus* × *domestica* Borkh.) rootstock repressed the expression of *MdTIR1A* and negatively regulated the adventitious root formation and auxin sensitivity of tobacco [51].

3.4. Branching and Internode Growth

Branching and internode growth are important for plant height, yield and light perception [56]. Osa-miR393 overexpression was shown to repress the target genes (*OsTIR1* and

OsAFB2), as well as an auxin transporter *OsAUX1* and a tillering inhibitor *OsTB1*, leading to increased tillers and early flowering in rice [32]. Osa-miR393 promoted rice tillering similar to the effect of nitrogen fertilizer, while Osa-miR393 mutation inhibited N-promoted tillering. The expression of OsIAA6 protein was positively regulated by Osa-miR393 and N nutrition, indicating that N-induced Osa-miR393 repressed *OsTIR1/AFB2*, alleviated axillary bud sensitivity to auxin and promoted tillering [36]. In rice, OsTIR1 and OsAFB2-5 could interact with OsIAA1 and redundantly regulate plant height and tillering number by affecting *OsIAA9/20* expression. The *Ostir1* and *Osafb2* mutants exhibited impaired plant height and increased tillering number compared to the wild type [35]. This finding was similar to a previous study that showed overexpression of miR393a/b reduced the plant height of rice [11]. However, the overexpression of miR393 target genes (*CsTIR1* and *CsAFB2*) from cucumber repressed the plant height of tomato [48]. Ectopic expression of Osa-miR393a in creeping bentgrass (*Agrostis stolonifera*) inhibited the expression of *AsAFB2* and *AsTIR1*, induced fewer but longer tillers, wider leaves and larger internodes [37]. Based on sRNA-seq and degradome analysis, Zhao et al. (2016) speculated that miR393 and the target genes might be involved in the internode elongation and development under maize ears [40]. The inhibition of miR393 in poplar was shown to promote phloem and xylem growth and increase lignin content through increasing the gene expression in the auxin signaling pathway, and the short tandem target mimic lines of miR393 (STTM393) showed more internodes and promoted growth compared with the control [57].

3.5. Flowering and Fruit Development

miR393 and its target genes can also control flowering, fruit and seed development. For example, in addition to the variations in root and leaf phenotype, the overexpression of mTIR1 also delayed the flowering of *Arabidopsis* [10]. miR393 was also shown to regulate silique length of *Arabidopsis* under different growth conditions, through affecting *ARF* and *IAA* expression and putative feedback regulation [6]. In rice, miR393 up-regulation led to early flowering, via repressing *OsTIR1* and *OsAFB2*, and auxin transporter *OsAUX1* [32]. OsTIR1 and OsAFB2 could interact with OsIAA1 to regulate auxin-responsive genes *OsIAA9/20*, and both *OsTIR1* and *OsAFB2* mutation impaired the grain number and germination of rice [35]. Bai et al. (2017) showed that miR393 up-regulated the length-width ratio of barley seeds, but both overexpression and mimicry of miR393 decreased the grain weight through targeting *TIR1/AFBs* in barley [43]. The overexpression of *CsTIR1* and *CsAFB2*, the target genes of miR393 in cucumber, impaired the seed size and seed germination in tomato [48]. The overexpression of Cme-miR393 repressed *CmAFB2* expression and delayed the fruit ripening of melon [52]. Shi et al. (2017) identified the miRNAs with reduced expression in peach fruit after NAA treatment, including miR393, miR156 and miR160. These miRNAs might control the fruit enlargement of peaches [58].

3.6. Secondary Metabolism and Other Roles

In additional to plant growth and development, the miR393/target gene module also plays a role in plant secondary metabolism, regenerability, and embryogenesis. Li et al. (2018) reported that blue light promoted the accumulation of flavonoids in embryogenic calli of longan, and miR393 expression was repressed, while *TIR1* was up-regulated under blue light [50]. Based on the profiling of the metabolome, transcriptome, degradome, and weighted correlation network analysis (WGCNA), Zhao et al. (2020) found that miR393-*bHLH* was negatively correlated with the biosynthesis of taste compounds (e.g., gallated catechin, caffeine, and theanine) in the tea plant (*Camellia sinensis*) [49]. miR393 inhibited the shoot regenerability and size of shoot apical meristem (SAM) in *Arabidopsis*, and similar regulation was shown in the *tir1-1* mutant [59]. Wojcik and Gaj (2016) proved that miR393 and TIR1/AFB controlled the somatic embryogenesis (SE) of *Arabidopsis* by affecting the explant sensitivity to auxin [24]. Omidvar et al. (2015) reported the miRNA expression pattern in a male-sterile tomato mutant and found that miR393 was down-regulated in the sterile line and might be involved in the anther development [25].

4. The Function of miR393 and TIR1/AFBs in Plant Abiotic Stress Response

4.1. Salt Stress

Salt stress induced miR393 expression in *Arabidopsis*, triggered stabilization of the Aux/IAA repressors and thus repressed the TIR1/AFB2-mediated auxin signaling. An *mir393ab* mutation enhanced lateral root number and root length during salinity, with an increased level of reactive oxygen species (ROS) and reduced ascorbate peroxidase (APX) activity compared to the wild genotype [26,27]. Chen et al. (2015) enhanced the salt tolerance of *A. thaliana* by overexpressing *mTIR1*, the transgenic lines displayed improved osmotic stress tolerance and accumulated more proline and anthocyanin. Compared with the wild type, the salt-stress-related genes were up-regulated, and the sodium content was reduced in mTIR1-overexpressing plants under salt stress [8]. Recently, Denver and Ullah (2019) found that the *mir393a*, *mir393b* and double mutant *mir393ab* of *Arabidopsis* were salt-sensitive, and miR393 regulated salt stress response through Receptor for Activated C Kinase 1A (RACK1A)-mediated ABA signaling [28]. Furthermore, the transgenic lines of tobacco overexpressing Ata-miR393a were less sensitive to IAA treatment and NaCl stress than the control [54]. Osa-miR393 expression changes under salt and alkali treatment, and the Osa-miR393 overexpression plants of rice and *Arabidopsis* were more sensitive to salinity and alkaline stresses [38]. In *Triticum aestivum*, miR393 was dramatically down-regulated under wounding treatment but was induced by salt stress [60].

4.2. Drought and Waterlogging

In plants, miR393 expression can be regulated by abiotic stresses, such as drought and waterlogging [61]. The lateral root growth of *Arabidopsis* overexpressing miR393-resistant AFB2 and TIR1 were found to be insensitive to ABA and osmotic stress, but the LRs in wild-type were significantly inhibited under ABA and PEG treatment. This indicated that miR393 inhibited LR development and regulated plant drought resistance by targeting *TIR1/AFB2* [29]. In addition to the roles in controlling tillering and flowering, Osa-miR393 could negatively regulate plant tolerance to salt and drought stresses, as well as plant sensitivity to auxin [32]. Lu et al. (2018) also reported that Osa-miR393 overexpression reduced the LR number and sensitivity to IAA by repressing Osa-miR390 and *OsTIR*. This regulation of Osa-miR393 on root development could be induced by ABA and various stresses (e.g., drought, salt, heavy metal) [34]. In maize, miR393 is a regulator in root development under short-term waterlogging, which could be induced under stress conditions in a waterlogging tolerant line but inhibited in a sensitive line [41]. In *Cynara cardunculus*, De Paola et al. (2012) found that Cca-miR393a was slightly up-regulated in salt-stressed leaves of globe artichoke, but not in the roots [62]. miR393 overexpression in barley enhanced drought sensitivity, while the knockdown of miR393 improved drought tolerance; this might be regulated through the ABA pathway [42]. miR393 was also down-regulated in different genotypes of *Sorghum bilolor* under drought stress [63]. The miR393 expression in wild *Ipomoea campanulata* and cultivated *Jacquemontia pentantha* was repressed under water deficit, but it was up-regulated in drought-stressed *Arabidopsis* [64]. In tomato, Sly-miR393 was up-regulated in above ground tissues of a drought-tolerant genotype and down-regulated in roots of a drought-sensitive genotype [14].

4.3. Temperature Stress

miR393 expression in *Arabidopsis* has also been shown to be induced by cold stress; *TIR1* was inhibited in response to low temperature [61]. Osa-miR393 could also regulate abiotic stress response of other monocotyledonous plants. For instance, Osa-miR393 ectopic expression improved the cold tolerance and tillering of switchgrass (*Panicum virgatum*). The cold responsive genes (*PvCOR47*, *PvICE1* and *PvRAV1*) were up-regulated, and the biomass and soluble sugar content were also increased in transgenic plants [53]. This study also showed that Pvi-miR393 of switchgrass was up-regulated by cold stress and down-regulated by auxin, while the predicted target genes (*PvAFB1*, *PvAFB2*, *PvAFB3* and *PvTIR1*) were induced by cold stress but in different patterns [53]. The overexpression of

Osa-miR393a in creeping bentgrass improved plant tolerance to salt, drought and heat stresses [37]. In wheat, miR393 could be up-regulated by salt and osmotic stresses but down-regulated under cold stress [65]. In banana, miR393-TIR1/AFB-triggered phasiRNAs were specifically enriched under cold stress [66]. After high-temperature (HT) treatment, Ghr-miR393 was up-regulated, but Ghr-novel-miR393b-3p and Ghr-novel-miR393c-3p were down-regulated in a HT-tolerant line of cotton compared with a HT-sensitive line [67].

4.4. Nutritional Stresses

In *Arabidopsis*, the miR393/AFB3 module controlled the root system architecture (RSA) in response to nitrate supply. miR393 and *AFB3* expression in roots were consistently regulated by nitrate, and *AFB3* could be induced by nitrate but then repressed by N metabolites from nitrate reduction and assimilation. Unlike the inhibited root development in wild type, both miR393-overexpression and *AFB3* mutation increased the primary root length in seedlings under three days of KNO_3 (5 mM) treatment [15]. Li et al. (2016) reported that nitrogen induced up-regulation of Osa-miR393 and down-regulation of target genes (*OsAFB2* and *OsTB1*) [36]. In peanut, miR393 and AFB3 expression were regulated by potassium and nitrogen deficiency, thus affecting the primary and lateral root growth [47]. Song et al. (2015) found that miR393 was up-regulated in leaves of *Chrysanthemum nankingense* under low-nitrogen conditions [68]. Lu et al. (2015) reported that miR393 was up-regulated in boron-deficient leaves of *Citrus sinensis* compared to the control, which might affect plant growth and development by repressing auxin signaling due to repressed TIR1 and AFB1/2 [69]. Under nitrogen stress, miR393 and the target genes were regulated coordinately in an N-stressed sensitive *Populus* clone (*Nanlin 895*) [70].

4.5. Metal Stresses

It has been speculated that miR393 has functions in plant response to metal stresses. He et al. (2014) reviewed the aluminum-responsive miRNAs in plants, including miR393 that was up-regulated in *M. truncatula* exposed to Al [71,72]. Ding and Zhu (2009) reviewed the miRNAs involved in plant adaptive response to copper and cadmium (Cd) stresses, of which, miR393 functioned in relieving Cd stress by repressing the target genes, *TIR1* and *bHLHs*, in *M. truncatula*, *Brassica napus* and rice [71,73–75]. In leaves, miR393 could be induced by metals such as Cd, hydrargyrum (Hg) and Al, and might play a role as a regulator in plant response to metal toxicity [71,76]. miR393 was up-regulated in 24 h aluminum toxicity roots from nitrate-fertilized common beans (*Phaseolus vulgaris*) [55]. Dmitriev et al. (2017) first reported that miR393, miR390 and miR319 were response to aluminum stress in flax, among which miR393 was up-regulated in resistant cultivars under Al stress [77]. The miR393 expression in the root apex of barley was inhibited by aluminum stress; the overexpression of miR393 alleviated the Al-induced root inhibition and reactive oxygen species (ROS)-induced cell death, while the inhibited miR393 expression enhanced the root sensitivity to Al stress [44].

5. The Role of miR393 and TIR1/AFBs in Biotic Stress Response

In *Arabidopsis*, the complementary strand of miR393 (miR393b*) has been identified as an Argonaute 2 (AGO2)-bound sRNA, which could target *MEMB12* encoding a SNARE protein localized in Golgi apparatus. *Pseudomonas syringae* pv. *tomato* (*Pst*) infection induced AGO2 and inhibited MEMB12 in a miR393b*-dependent manner. Both miR393b* overexpression and *memb12* mutation promoted the secretion of PR1, an antimicrobial pathogenesis-related protein in *Arabidopsis*, indicating that AGO2 and miR393/MEMB12 are important effectors or regulators in plant antibacterial immunity [30]. Ath-miR393 was induced during pattern-triggered immunity, and miR393 overexpression suppressed auxin signaling and inactivated ARF1/9 expression, increasing glucosinolate and decreasing camalexin levels, which were related to the plant resistance to biotrophic pathogens and susceptibility to necrotrophic pathogens, respectively. AtAFB1 overexpression could prevent inhibited auxin signaling by flg22, then reduce salicylic acid accumulation and cause plant susceptibility to biotrophs [78]. Zhao et al. (2012) isolated AGO-associated sRNA in *Arabidopsis* using immune precipitation and deep-sequencing, showing that miR393 was specifically enriched in bacterial-challenged plants [79]. The inoculation of *Burkholderia phytofirmans* PsJN on *A. thaliana* overexpressing miR393 did not increase the primary root length, fresh weight, or total chlorophyll content compared to the wild type. PsJN inoculation could not affect the target gene expression of miR393 in the wild *A. thaliana* at the four-leaf stage, but *AFB1/3* were up-regulated in the inoculated plants at the six-leaf stage [80]. Djami-Tchatchou and Dubery (2019) screened the miRNA expression pattern of bacterial lipopolysaccharide (LPS)-treated *Arabidopsis* leaf and callus, finding miR393 was repressed, but the target gene *LecRLK* was up-regulated, which might be related to the enhanced perception ability of LPS in *Arabidopsis* [81]. The overexpression and repression of miR393, respectively, suppressed and induced the expression of *Lectin Receptor-Like Kinases* (*LecRLK*) in *Arabidopsis* treated with LPS [82]. Shi et al. (2015) proved that miR393-mediated auxin signaling was involved in the hydrogen sulfide (H_2S)-mediated antibacterial resistance of *Arabidopsis* [31]. Sulfur dioxide (SO_2) pre-exposure of *Arabidopsis* resulted in the up-regulation of miR160, miR167 and miR393 and enhanced disease resistance against *Botrytis cinerea* [83].

miR393 can also control the biotic stress response in crops. In tobacco, miR393 was induced in leaves infiltrated with an oncogenic strain of *Agrobacterium tumefaciens* (C58), which may contribute to plant antibacterial resistance via repressing auxin signaling [84]. Recently, Nazari et al. (2017) found that miRNAs (miR393 and miR167) and flavonoids could be taken as biomarkers in the tobacco–*Agrobacterium* interaction, which were up-regulated or accumulated when treated with *Bacillus subtilis* [85]. In soybean, miR393 was up-regulated by soybean mosaic virus (SMV) infection, and the suppression of genes in auxin signaling pathways might be related to plant defense responses [86]. Supported with sRNA-seq data, Xu et al. (2015) found that miR393 was up-regulated in a soybean cyst nematode (SCN)-resistant line compared to a susceptible line of soybean, which exclusively target genes encoding TIR1, AFBs and ribosomal protein L20 [87]. Additionally, miR393 was also induced by *Phytophthora sojae* infection, knockdown of miR393 enhanced plant susceptibility to *P. sojae*, accompanied with repressed isoflavonoid biosynthetic gene expression in roots [46]. miR393 was a negative regulator of arbuscule formation in *O. sativa*, *S. lycopersicum*, and *M. truncatula*, by inhibiting the auxin receptor genes (*TIR1* and *AFBs*) and auxin perception in arbuscule-containing cells [88]. In rice, the overexpression of miR393 decreased the *TIR1* expression and increased plant susceptibility to Rice Black Streaked Dwarf Virus (RBSDV) [39]. sRNA-seq analysis revealed that the down-regulation of miR393 in maize might be related to the plant antiviral defense to synergistic infection [89]. The miR393 expression was significantly decreased in a sensitive cultivar (Hanatee) of cassava infected by *Colletotrichum gloeosporioides* but increased in resistant cultivar Huay Bong 60 [90].

Recently, miR393 was reported with expressional changes in fruits and vegetables under biotic stresses. In mulberry, miR393A was induced by phytoplasma infection, but miR393B was unaltered in infected leaves, and the two miR393 were regulated by different *cis*-acting elements [91]. Chand et al. (2017) identified 45 miRNAs responsive to immunity in garlic (*Allium sativum*), and transgenic plants overexpressing miR393, miR164a and miR168a showed enhanced resistance to *Fusarium oxysporum* f. sp. *cepae* [92]. Vinutha et al. (2020) found miR393 was up-regulated in tomato infected with Leaf Curl New Delhi Virus (ToLCNDV) [93].

6. Conclusions

As a miRNA family with conserved biological functions in plants, miR393 and its target genes TIR1/AFBs are broadly involved in the growth and development of leaf, root, branching, seed, secondary metabolism, as well as abiotic (e.g., salt, drought, cold, heavy metal) and biotic stress responses. So far, the regulatory network and putative applications of miR393/target module has been elaborated in *Arabidopsis*, rice, soybean, barley and cucumber. miR393s in these plants regulate auxin perception and signaling, Aux/IAA degradation and auxin-responsive gene expression, mainly by repressing auxin receptors TIR1 and AFBs. The miR393-TIR1/AFBs module is an important regulator links auxin response with transcriptional regulation in plant development; thus, we can harness this module as a valuable tool to manipulate crop traits for optimized yield and adaptability. This module may also be important in controlling development, agronomic trait, environmental adaptation of other plants such as tomato, tobacco, wheat and rapeseed, but the molecular mechanism of miR393/targets in these plants were barely reported except for the expressional variations under different growth conditions. On the other hand, it is still unclear whether there are other co-effectors or regulators of TIR1/AFBs, in addition to Aux/IAAs and CUL1. It is also intriguing that TIR1 and AFBs have different abilities to interact with IAAs and influence SCF assembly. For example, AFB1 in *Arabidopsis*, specialized later than other receptors, may play a unique role in Brassicaceae [94]. Additionally, the evolution and function of different miR393 members in crops remain unclear; many speculated members reported from degradome sequencing were not validated. How these miR393s and target genes regulate plant development and adaptation are unclear. A recent study showed that TIR1 and AFB2 were positive modulators of jasmonic acid (JA) homeostasis and AR formation in *Arabidopsis*, through controlling JA biosynthesis and conjunction [19]. This suggested that miR393-TIR1/AFBs might be involved in other hormone signaling pathways, and it is innovative to establish a connection between JA and miR393-auxin pathways. A detailed understanding of miR393 and different target genes will facilitate the design and utilization of this module in precise modification of agronomic traits and stress resistance in crops.

Author Contributions: J.J. drafted the manuscript; H.Z. and N.L. analyzed the sequence similarity; J.B. and Y.W. revised the manuscript. All authors have read and agreed to the published version of the manuscript.

Funding: This work was funded by the National Natural Science Foundations (31972963), the National Key Research and Development Program of China (2018YFE0108000), the Top Talent Support Program of Yangzhou University and the Jiangsu Qinglan Project, the Graduate Training Program for Innovation and Entrepreneurship (SJCX21_1602, KYCX22_3468), the Project of Special Funding for Crop Science Discipline Development (yzuxk202006), and the Priority Academic Program Development of Jiangsu Higher Education Institutions.

Institutional Review Board Statement: Not applicable.

Informed Consent Statement: Not applicable.

Data Availability Statement: Not applicable.

Conflicts of Interest: The authors declare no conflict of interest.

References

1. Lee, R.C.; Feinbaum, R.L.; Ambros, V. The *C. elegans* heterochronic gene *lin-4* encodes small RNAs with antisense complementarity to lin-14. *Cell* **1993**, *75*, 843–854. [CrossRef]
2. D'Ario, M.; Griffiths-Jones, S.; Kim, M. Small RNAs: Big impact on plant development. *Trends Plant Sci.* **2017**, *22*, 1056–1068. [CrossRef] [PubMed]
3. Xie, M.; Zhang, S.; Yu, B. microRNA biogenesis, degradation and activity in plants. *Cell. Mol. Life Sci.* **2015**, *72*, 87–99. [CrossRef] [PubMed]
4. Song, X.W.; Li, Y.; Cao, X.F.; Qi, Y.J. MicroRNAs and their regulatory roles in plant-environment interactions. *Annu. Rev. Plant Biol.* **2019**, *70*, 489–525. [CrossRef] [PubMed]
5. Khraiwesh, B.; Arif, M.A.; Seumel, G.I.; Ossowski, S.; Weigel, D.; Reski, R.; Frank, W. Transcriptional control of gene expression by microRNAs. *Cell* **2010**, *140*, 111–122. [CrossRef]
6. Si-Ammour, A.; Windels, D.; Arn-Bouldoires, E.; Kutter, C.; Ailhas, J.; Meins, F., Jr.; Vazquez, F. miR393 and secondary siRNAs regulate expression of the TIR1/AFB2 auxin receptor clade and auxin-related development of *Arabidopsis* leaves. *Plant Physiol.* **2011**, *157*, 683–691. [CrossRef]
7. Parry, G.; Calderon-Villalobos, L.I.; Prigge, M.; Peret, B.; Dharmasiri, S.; Itoh, H.; Lechner, E.; Gray, W.M.; Bennett, M.; Estelle, M. Complex regulation of the TIR1/AFB family of auxin receptors. *Proc. Natl. Acad. Sci. USA* **2009**, *106*, 22540–22545. [CrossRef]

8. Chen, Z.; Hu, L.; Han, N.; Hu, J.; Yang, Y.; Xiang, T.; Zhang, X.; Wang, L. Overexpression of a miR393-resistant form of *transport inhibitor response protein 1* (*mTIR1*) enhances salt tolerance by increased osmoregulation and Na^+ exclusion in *Arabidopsis thaliana*. *Plant Cell Physiol.* **2015**, *56*, 73–83. [CrossRef]
9. Windels, D.; Vazquez, F. miR393: Integrator of environmental cues in auxin signaling? *Plant Signal. Behav.* **2011**, *6*, 1672–1675. [CrossRef]
10. Chen, Z.H.; Bao, M.L.; Sun, Y.Z.; Yang, Y.J.; Xu, X.H.; Wang, J.H.; Han, N.; Bian, H.W.; Zhu, M.Y. Regulation of auxin response by miR393-targeted *transport inhibitor response protein 1* is involved in normal development in *Arabidopsis*. *Plant Mol. Biol.* **2011**, *77*, 619–629. [CrossRef]
11. Bian, H.; Xie, Y.; Guo, F.; Han, N.; Ma, S.; Zeng, Z.; Wang, J.; Yang, Y.; Zhu, M. Distinctive expression patterns and roles of the miRNA393/TIR1 homolog module in regulating flag leaf inclination and primary and crown root growth in rice (*Oryza sativa*). *New Phytol.* **2012**, *196*, 149–161. [CrossRef] [PubMed]
12. Cai, Z.; Wang, Y.; Zhu, L.; Tian, Y.; Chen, L.; Sun, Z.; Ullah, I.; Li, X. GmTIR1/GmAFB3-based auxin perception regulated by miR393 modulates soybean nodulation. *New Phytol.* **2017**, *215*, 672–686. [CrossRef] [PubMed]
13. Shen, E.; Zou, J.; Hubertus Behrens, F.; Chen, L.; Ye, C.; Dai, S.; Li, R.; Ni, M.; Jiang, X.; Qiu, J.; et al. Identification, evolution, and expression partitioning of miRNAs in allopolyploid *Brassica napus*. *J. Exp. Bot.* **2015**, *66*, 7241–7253. [CrossRef] [PubMed]
14. Candar-Cakir, B.; Arican, E.; Zhang, B. Small RNA and degradome deep sequencing reveals drought-and tissue-specific microRNAs and their important roles in drought-sensitive and drought-tolerant tomato genotypes. *Plant Biotechnol. J.* **2016**, *14*, 1727–1746. [CrossRef] [PubMed]
15. Vidal, E.A.; Araus, V.; Lu, C.; Parry, G.; Green, P.J.; Coruzzi, G.M.; Gutierrez, R.A. Nitrate-responsive miR393/AFB3 regulatory module controls root system architecture in *Arabidopsis thaliana*. *Proc. Natl. Acad. Sci. USA* **2010**, *107*, 4477–4482. [CrossRef] [PubMed]
16. Mockaitis, K.; Estelle, M. Auxin receptors and plant development: A new signaling paradigm. *Annu. Rev. Cell Dev. Biol.* **2008**, *24*, 55–80. [CrossRef]
17. Yu, H.; Zhang, Y.; Moss, B.L.; Bargmann, B.O.R.; Wang, R.H.; Prigge, M.; Nemhauser, J.L.; Estelle, M. Untethering the TIR1 auxin receptor from the SCF complex increases its stability and inhibits auxin response. *Nat. Plants* **2015**, *1*, 14030. [CrossRef]
18. Szemenyei, H.; Hannon, M.; Long, J.A. TOPLESS mediates auxin-dependent transcriptional repression during *Arabidopsis* embryogenesis. *Science* **2008**, *319*, 1384–1386. [CrossRef]
19. Lakehal, A.; Chaabouni, S.; Cavel, E.; Le Hir, R.; Ranjan, A.; Raneshan, Z.; Novak, O.; Pacurar, D.I.; Perrone, I.; Jobert, F.; et al. A molecular framework for the control of adventitious rooting by TIR1/AFB2-Aux/IAA-dependent auxin signaling in *Arabidopsis*. *Mol. Plant* **2019**, *12*, 1499–1514. [CrossRef]
20. Prigge, M.J.; Platre, M.; Kadakia, N.; Zhang, Y.; Greenham, K.; Szutu, W.; Pandey, B.K.; Bhosale, R.A.; Bennett, M.J.; Busch, W.; et al. Genetic analysis of the Arabidopsis TIR1/AFB auxin receptors reveals both overlapping and specialized functions. *Elife* **2020**, *9*, e54740. [CrossRef]
21. Todesco, M.; Rubiosomoza, I.; Pazares, J.; Weigel, D. A collection of target mimics for comprehensive analysis of microRNA function in *Arabidopsis thaliana*. *PLoS Genet.* **2010**, *6*, e1001031. [CrossRef] [PubMed]
22. Dharmasiri, N.; Dharmasiri, S.; Weijers, D.; Lechner, E.; Yamada, M.; Hobbie, L.; Ehrismann, J.S.; Jurgens, G.; Estelle, M. Plant development is regulated by a family of auxin receptor F box proteins. *Dev. Cell* **2005**, *9*, 109–119. [CrossRef] [PubMed]
23. Windels, D.; Bielewicz, D.; Ebneter, M.; Jarmolowski, A.; Szweykowska-Kulinska, Z.; Vazquez, F. miR393 is required for production of proper auxin signalling outputs. *PLoS ONE* **2014**, *9*, e95972. [CrossRef]
24. Wojcik, A.M.; Gaj, M.D. miR393 contributes to the embryogenic transition induced in vitro in *Arabidopsis* via the modification of the tissue sensitivity to auxin treatment. *Planta* **2016**, *244*, 231–243. [CrossRef] [PubMed]
25. Omidvar, V.; Mohorianu, I.; Dalmay, T.; Fellner, M. Identification of miRNAs with potential roles in regulation of anther development and male-sterility in 7B-1 male-sterile tomato mutant. *BMC Genom.* **2015**, *16*, 878. [CrossRef]
26. Sunkar, R.; Zhu, J.K. Novel and stress-regulated microRNAs and other small RNAs from *Arabidopsis*. *Plant Cell* **2004**, *16*, 2001–2019. [CrossRef]
27. Iglesias, M.J.; Terrile, M.C.; Windels, D.; Lombardo, M.C.; Bartoli, C.G.; Vazquez, F.; Estelle, M.; Casalongue, C.A. MiR393 regulation of auxin signaling and redox-related components during acclimation to salinity in *Arabidopsis*. *PLoS ONE* **2014**, *9*, e107678. [CrossRef]
28. Denver, J.B.; Ullah, H. miR393s regulate salt stress response pathway in *Arabidopsis thaliana* through scaffold protein RACK1A mediated ABA signaling pathways. *Plant Signal. Behav.* **2019**, *14*, 1600394. [CrossRef]
29. Chen, H.; Li, Z.; Xiong, L. A plant microRNA regulates the adaptation of roots to drought stress. *FEBS Lett.* **2012**, *586*, 1742–1747. [CrossRef]
30. Zhang, X.; Zhao, H.; Gao, S.; Wang, W.C.; Katiyar-Agarwal, S.; Huang, H.D.; Raikhel, N.; Jin, H. *Arabidopsis* Argonaute 2 regulates innate immunity via miRNA393*-mediated silencing of a Golgi-localized SNARE gene, *MEMB12*. *Mol. Cell* **2011**, *42*, 356–366. [CrossRef]
31. Shi, H.; Ye, T.; Han, N.; Bian, H.; Liu, X.; Chan, Z. Hydrogen sulfide regulates abiotic stress tolerance and biotic stress resistance in *Arabidopsis*. *J. Integr. Plant Biol.* **2015**, *57*, 628–640. [CrossRef] [PubMed]

32. Xia, K.; Wang, R.; Ou, X.; Fang, Z.; Tian, C.; Duan, J.; Wang, Y.; Zhang, M. *OsTIR1* and *OsAFB2* downregulation via *OsmiR393* overexpression leads to more tillers, early flowering and less tolerance to salt and drought in rice. *PLoS ONE* **2012**, *7*, e30039. [CrossRef] [PubMed]
33. Guo, F.; Han, N.; Xie, Y.; Fang, K.; Yang, Y.; Zhu, M.; Wang, J.; Bian, H. The miR393a/target module regulates seed germination and seedling establishment under submergence in rice (*Oryza sativa* L.). *Plant Cell Environ.* **2016**, *39*, 2288–2302. [CrossRef] [PubMed]
34. Lu, Y.; Feng, Z.; Liu, X.; Bian, L.; Xie, H.; Zhang, C.; Mysore, K.S.; Liang, J. MiR393 and miR390 synergistically regulate lateral root growth in rice under different conditions. *BMC Plant Biol.* **2018**, *18*, 261. [CrossRef]
35. Guo, F.; Huang, Y.; Qi, P.; Lian, G.; Hu, X.; Han, N.; Wang, J.; Zhu, M.; Qian, Q.; Bian, H. Functional analysis of auxin receptor OsTIR1/OsAFB family members in rice grain yield, tillering, plant height, root system, germination, and auxinic herbicide resistance. *New Phytol.* **2021**, *229*, 2676–2692. [CrossRef]
36. Li, X.; Xia, K.; Liang, Z.; Chen, K.; Gao, C.; Zhang, M. MicroRNA393 is involved in nitrogen-promoted rice tillering through regulation of auxin signal transduction in axillary buds. *Sci. Rep.* **2016**, *6*, 32158. [CrossRef]
37. Zhao, J.; Yuan, S.; Zhou, M.; Yuan, N.; Li, Z.; Hu, Q.; Bethea, F.G.; Liu, H.; Li, S.; Luo, H. Transgenic creeping bentgrass overexpressing Osa-miR393a exhibits altered plant development and improved multiple stress tolerance. *Plant Biotechnol. J.* **2019**, *17*, 233–251. [CrossRef]
38. Gao, P.; Bai, X.; Yang, L.; Lv, D.; Pan, X.; Li, Y.; Cai, H.; Ji, W.; Chen, Q.; Zhu, Y. *osa-MIR393*: A salinity- and alkaline stress-related microRNA gene. *Mol. Biol. Rep.* **2011**, *38*, 237–242. [CrossRef]
39. Zhang, H.; Tan, X.; Li, L.; He, Y.; Hong, G.; Li, J.; Lin, L.; Cheng, Y.; Yan, F.; Chen, J.; et al. Suppression of auxin signalling promotes rice susceptibility to rice black streaked dwarf virus infection. *Mol. Plant Pathol.* **2019**, *20*, 1093–1104. [CrossRef]
40. Zhao, Z.; Xue, Y.; Yang, H.; Li, H.; Sun, G.; Zhao, X.; Ding, D.; Tang, J. Genome-wide identification of miRNAs and their targets involved in the developing internodes under maize ears by responding to hormone signaling. *PLoS ONE* **2016**, *11*, e0164026. [CrossRef]
41. Liu, Z.; Kumari, S.; Zhang, L.; Zheng, Y.; Ware, D. Characterization of miRNAs in response to short-term waterlogging in three inbred lines of *Zea mays*. *PLoS ONE* **2012**, *7*, e39786. [CrossRef] [PubMed]
42. Yuan, W.; Suo, J.; Shi, B.; Zhou, C.; Bai, B.; Bian, H.; Zhu, M.; Han, N. The barley miR393 has multiple roles in regulation of seedling growth, stomatal density, and drought stress tolerance. *Plant Physiol. Biochem.* **2019**, *142*, 303–311. [CrossRef]
43. Bai, B.; Shi, B.; Hou, N.; Cao, Y.; Meng, Y.; Bian, H.; Zhu, M.; Han, N. microRNAs participate in gene expression regulation and phytohormone cross-talk in barley embryo during seed development and germination. *BMC Plant Biol.* **2017**, *17*, 150. [CrossRef] [PubMed]
44. Bai, B.; Bian, H.; Zeng, Z.; Hou, N.; Shi, B.; Wang, J.; Zhu, M.; Han, N. miR393-mediated auxin signaling regulation is involved in root elongation inhibition in response to toxic aluminum stress in barley. *Plant Cell Physiol.* **2017**, *58*, 426–439. [CrossRef]
45. Mao, G.; Turner, M.; Yu, O.; Subramanian, S. miR393 and miR164 influence indeterminate but not determinate nodule development. *Plant Signal. Behav.* **2013**, *8*, e26753. [CrossRef]
46. Wong, J.; Gao, L.; Yang, Y.; Zhai, J.; Arikit, S.; Yu, Y.; Duan, S.; Chan, V.; Xiong, Q.; Yan, J.; et al. Roles of small RNAs in soybean defense against *Phytophthora sojae* infection. *Plant J.* **2014**, *79*, 928–940. [CrossRef] [PubMed]
47. Li, L.; Li, Q.; Davis, K.E.; Patterson, C.; Oo, S.; Liu, W.; Liu, J.; Wang, G.; Fontana, J.E.; Thornburg, T.E.; et al. Response of root growth and development to nitrogen and potassium deficiency as well as microRNA-mediated mechanism in peanut (*Arachis hypogaea* L.). *Front. Plant Sci.* **2021**, *12*, 695234. [CrossRef]
48. Xu, J.; Li, J.; Cui, L.; Zhang, T.; Wu, Z.; Zhu, P.Y.; Meng, Y.J.; Zhang, K.J.; Yu, X.Q.; Lou, Q.F.; et al. New insights into the roles of cucumber TIR1 homologs and miR393 in regulating fruit/seed set development and leaf morphogenesis. *BMC Plant Biol.* **2017**, *17*, 130. [CrossRef]
49. Zhao, S.; Mi, X.; Guo, R.; Xia, X.; Liu, L.; An, Y.; Yan, X.; Wang, S.; Guo, L.; Wei, C. The biosynthesis of main taste compounds is coordinately regulated by miRNAs and phytohormones in tea plant (*Camellia sinensis*). *J. Agric. Food Chem.* **2020**, *68*, 6221–6236. [CrossRef]
50. Li, H.; Lin, Y.; Chen, X.; Bai, Y.; Wang, C.; Xu, X.; Wang, Y.; Lai, Z. Effects of blue light on flavonoid accumulation linked to the expression of miR393, miR394 and miR395 in longan embryogenic calli. *PLoS ONE* **2018**, *13*, e0191444.
51. Li, K.; Wei, Y.H.; Wang, R.H.; Mao, J.P.; Tian, H.Y.; Chen, S.Y.; Li, S.H.; Tahir, M.M.; Zhang, D. Mdm-MIR393b-mediated adventitious root formation by targeted regulation of *MdTIR1A* expression and weakened sensitivity to auxin in apple rootstock. *Plant Sci.* **2021**, *308*, 110909. [CrossRef] [PubMed]
52. Bai, S.; Tian, Y.; Tan, C.; Bai, S.; Hao, J.; Hasi, A. Genome-wide identification of microRNAs involved in the regulation of fruit ripening and climacteric stages in melon (*Cucumis melo*). *Hortic. Res.* **2020**, *7*, 106. [CrossRef] [PubMed]
53. Liu, Y.; Wang, K.; Li, D.; Yan, J.; Zhang, W. Enhanced cold tolerance and tillering in switchgrass (*Panicum virgatum* L.) by heterologous expression of *Osa-miR393a*. *Plant Cell Physiol.* **2017**, *58*, 2226–2240. [CrossRef]
54. Feng, X.M.; You, C.X.; Qiao, Y.; Mao, K.; Hao, Y.J. Ectopic overexpression of *Arabidopsis* AtmiR393a gene changes auxin sensitivity and enhances salt resistance in tobacco. *Acta Physiol. Plant.* **2010**, *32*, 997–1003. [CrossRef]
55. Mendoza-Soto, A.B.; Naya, L.; Leija, A.; Hernandez, G. Responses of symbiotic nitrogen-fixing common bean to aluminum toxicity and delineation of nodule responsive microRNAs. *Front. Plant Sci.* **2015**, *6*, 587. [CrossRef] [PubMed]

56. Wang, B.; Smith, S.M.; Li, J.Y. Genetic regulation of shoot architecture. *Annu. Rev. Plant Biol.* **2018**, *69*, 437–468. [CrossRef] [PubMed]
57. Chu, L.; He, X.; Shu, W.; Wang, L.; Tang, F. Knockdown of miR393 promotes the growth and biomass production in Poplar. *Front. Plant Sci.* **2021**, *12*, 714907. [CrossRef]
58. Shi, M.; Hu, X.; Wei, Y.; Hou, X.; Yuan, X.; Liu, J.; Liu, Y. Genome-wide profiling of small RNAs and degradome revealed conserved regulations of miRNAs on auxin-responsive genes during fruit enlargement in peaches. *Int. J. Mol. Sci.* **2017**, *18*, 2599. [CrossRef]
59. Wang, L.; Liu, Z.; Qiao, M.; Xiang, F. miR393 inhibits in vitro shoot regeneration in *Arabidopsis thaliana* via repressing *TIR1*. *Plant Sci.* **2018**, *266*, 1–8. [CrossRef]
60. Wang, B.; Sun, Y.F.; Song, N.; Wei, J.P.; Wang, X.J.; Feng, H.; Yin, Z.Y.; Kang, Z.S. MicroRNAs involving in cold, wounding and salt stresses in *Triticum aestivum* L. *Plant Physiol. Biochem.* **2014**, *80*, 90–96. [CrossRef]
61. Sunkar, R.; Chinnusamy, V.; Zhu, J.H.; Zhu, J.K. Small RNAs as big players in plant abiotic stress responses and nutrient deprivation. *Trends Plant Sci.* **2007**, *12*, 301–309. [CrossRef] [PubMed]
62. De Paola, D.; Cattonaro, F.; Pignone, D.; Sonnante, G. The miRNAome of globe artichoke: Conserved and novel micro RNAs and target analysis. *BMC Genom.* **2012**, *13*, 41. [CrossRef] [PubMed]
63. Hamza, N.B.; Sharma, N.; Tripathi, A.; Sanan-Mishra, N. MicroRNA expression profiles in response to drought stress in *Sorghum bicolor*. *Gene Expr. Patterns* **2016**, *20*, 88–98. [CrossRef] [PubMed]
64. Ghorecha, V.; Patel, K.; Ingle, S.; Sunkar, R.; Krishnayya, N.S. Analysis of biochemical variations and microRNA expression in wild (*Ipomoea campanulata*) and cultivated (*Jacquemontia pentantha*) species exposed to in vivo water stress. *Physiol. Mol. Biol. Plants* **2014**, *20*, 57–67. [CrossRef] [PubMed]
65. Gupta, O.P.; Meena, N.L.; Sharma, I.; Sharma, P. Differential regulation of microRNAs in response to osmotic, salt and cold stresses in wheat. *Mol. Biol. Rep.* **2014**, *41*, 4623–4629. [CrossRef] [PubMed]
66. Zhu, H.; Zhang, Y.; Tang, R.; Qu, H.; Duan, X.; Jiang, Y. Banana sRNAome and degradome identify microRNAs functioning in differential responses to temperature stress. *BMC Genom.* **2019**, *20*, 33. [CrossRef]
67. Chen, J.; Pan, A.; He, S.; Su, P.; Yuan, X.; Zhu, S.; Liu, Z. Different microRNA families involved in regulating high temperature stress response during cotton (*Gossypium hirsutum* L.) anther development. *Int. J. Mol. Sci.* **2020**, *21*, 1280. [CrossRef]
68. Song, A.; Wang, L.; Chen, S.; Jiang, J.; Guan, Z.; Li, P.; Chen, F. Identification of nitrogen starvation-responsive microRNAs in *Chrysanthemum nankingense*. *Plant Physiol. Biochem.* **2015**, *91*, 41–48. [CrossRef]
69. Lu, Y.B.; Qi, Y.P.; Yang, L.T.; Guo, P.; Li, Y.; Chen, L.S. Boron-deficiency-responsive microRNAs and their targets in *Citrus sinensis* leaves. *BMC Plant Biol.* **2015**, *15*, 271. [CrossRef]
70. Wang, X.; Li, X.; Zhang, S.; Korpelainen, H.; Li, C. Physiological and transcriptional responses of two contrasting *Populus* clones to nitrogen stress. *Tree Physiol.* **2016**, *36*, 628–642. [CrossRef]
71. Zhou, Z.S.; Huang, S.Q.; Yang, Z.M. Bioinformatic identification and expression analysis of new microRNAs from *Medicago truncatula*. *Biochem. Biophys. Res. Commun.* **2008**, *374*, 538–542. [CrossRef] [PubMed]
72. He, H.; He, L.; Gu, M. Role of microRNAs in aluminum stress in plants. *Plant Cell Rep.* **2014**, *33*, 831–836. [CrossRef] [PubMed]
73. Xie, F.L.; Huang, S.Q.; Guo, K.; Xiang, A.L.; Zhu, Y.Y.; Nie, L.; Yang, Z.M. Computational identification of novel microRNAs and targets in *Brassica napus*. *FEBS Lett.* **2007**, *581*, 1464–1474. [CrossRef]
74. Huang, S.Q.; Peng, J.; Qiu, C.X.; Yang, Z.M. Heavy metal-regulated new microRNAs from rice. *J. Inorg. Biochem.* **2009**, *103*, 282–287. [CrossRef] [PubMed]
75. Ding, Y.F.; Zhu, C. The role of microRNAs in copper and cadmium homeostasis. *Biochem. Biophys. Res. Commun.* **2009**, *386*, 6–10. [CrossRef]
76. Mendoza-Soto, A.B.; Sanchez, F.; Hernandez, G. MicroRNAs as regulators in plant metal toxicity response. *Front. Plant Sci.* **2012**, *3*, 105. [CrossRef]
77. Dmitriev, A.A.; Kudryavtseva, A.V.; Bolsheva, N.L.; Zyablitsin, A.V.; Rozhmina, T.A.; Kishlyan, N.V.; Krasnov, G.S.; Speranskaya, A.S.; Krinitsina, A.A.; Sadritdinova, A.F.; et al. miR319, miR390, and miR393 are involved in aluminum response in flax (*Linum usitatissimum* L.). *BioMed Res. Int.* **2017**, *2017*, 4975146.
78. Robert-Seilaniantz, A.; MacLean, D.; Jikumaru, Y.; Hill, L.; Yamaguchi, S.; Kamiya, Y.; Jones, J.D. The microRNA miR393 re-directs secondary metabolite biosynthesis away from camalexin and towards glucosinolates. *Plant J.* **2011**, *67*, 218–231. [CrossRef]
79. Zhao, H.; Lii, Y.; Zhu, P.; Jin, H. Isolation and profiling of protein-associated small RNAs. In *RNA Abundance Analysis*; Methods in Molecular Biology; Springer: Berlin/Heidelberg, Germany, 2012; Volume 883, pp. 165–176.
80. Zuniga, A.; Poupin, M.J.; Donoso, R.; Ledger, T.; Guiliani, N.; Gutierrez, R.A.; Gonzalez, B. Quorum sensing and indole-3-acetic acid degradation play a role in colonization and plant growth promotion of *Arabidopsis thaliana* by *Burkholderia phytofirmans* PsJN. *Mol. Plant Microbe Interact.* **2013**, *26*, 546–553. [CrossRef]
81. Djami-Tchatchou, A.T.; Dubery, I.A. Lipopolysaccharide perception leads to dynamic alterations in the microtranscriptome of *Arabidopsis thalian* a cells and leaf tissues. *BMC Plant Biol.* **2015**, *15*, 79. [CrossRef]
82. Djami-Tchatchou, A.T.; Dubery, I.A. miR393 regulation of lectin receptor-like kinases associated with LPS perception in *Arabidopsis thaliana*. *Biochem. Biophys. Res. Commun.* **2019**, *513*, 88–92. [CrossRef] [PubMed]
83. Xue, M.; Yi, H. Enhanced *Arabidopsis* disease resistance against *Botrytis cinerea* induced by sulfur dioxide. *Ecotoxicol. Environ. Saf.* **2018**, *147*, 523–529. [CrossRef] [PubMed]

84. Pruss, G.J.; Nester, E.W.; Vance, V. Infiltration with *Agrobacterium tumefaciens* induces host defense and development-dependent responses in the infiltrated zone. *Mol. Plant Microbe Interact.* **2008**, *21*, 1528–1538. [CrossRef] [PubMed]
85. Nazari, F.; Safaie, N.; Soltani, B.M.; Shams-Bakhsh, M.; Sharifi, M. *Bacillus subtilis* affects miRNAs and flavanoids production in Agrobacterium-Tobacco interaction. *Plant Physiol. Biochem.* **2017**, *118*, 98–106. [CrossRef]
86. Yin, X.; Wang, J.; Cheng, H.; Wang, X.; Yu, D. Detection and evolutionary analysis of soybean miRNAs responsive to soybean mosaic virus. *Planta* **2013**, *237*, 1213–1225. [CrossRef]
87. Xu, M.; Li, Y.; Zhang, Q.; Xu, T.; Qiu, L.; Fan, Y.; Wang, L. Novel miRNA and phasiRNA biogenesis networks in soybean roots from two sister lines that are resistant and susceptible to SCN race 4. *PLoS ONE* **2014**, *9*, e110051. [CrossRef]
88. Etemadi, M.; Gutjahr, C.; Couzigou, J.M.; Zouine, M.; Lauressergues, D.; Timmers, A.; Audran, C.; Bouzayen, M.; Becard, G.; Combier, J.P. Auxin perception is required for arbuscule development in arbuscular mycorrhizal symbiosis. *Plant Physiol.* **2014**, *166*, 281–292. [CrossRef]
89. Xia, Z.; Zhao, Z.; Gao, X.; Jiao, Z.; Wu, Y.; Zhou, T.; Fan, Z. Characterization of maize miRNAs in response to synergistic infection of maize chlorotic mottle virus and sugarcane mosaic virus. *Int. J. Mol. Sci.* **2019**, *20*, 3146. [CrossRef]
90. Pinweha, N.; Asvarak, T.; Viboonjun, U.; Narangajavana, J. Involvement of miR160/miR393 and their targets in cassava responses to anthracnose disease. *J. Plant Physiol.* **2015**, *174*, 26–35. [CrossRef]
91. Gai, Y.P.; Li, Y.Q.; Guo, F.Y.; Yuan, C.Z.; Mo, Y.Y.; Zhang, H.L.; Wang, H.; Ji, X.L. Analysis of phytoplasma-responsive sRNAs provide insight into the pathogenic mechanisms of mulberry yellow dwarf disease. *Sci. Rep.* **2014**, *4*, 5378. [CrossRef]
92. Chand, S.K.; Nanda, S.; Mishra, R.; Joshi, R.K. Multiple garlic (*Allium sativum* L.) microRNAs regulate the immunity against the basal rot fungus *Fusarium oxysporum* f. sp. *Cepae. Plant Sci.* **2017**, *257*, 9–21. [CrossRef] [PubMed]
93. Vinutha, T.; Vanchinathan, S.; Bansal, N.; Kumar, G.; Permar, V.; Watts, A.; Ramesh, S.V.; Praveen, S. Tomato auxin biosynthesis/signaling is reprogrammed by the geminivirus to enhance its pathogenicity. *Planta* **2020**, *252*, 51. [CrossRef] [PubMed]
94. Edger, P.P.; Hall, J.C.; Harkess, A.; Tang, M.; Coombs, J.; Mohammadin, S.; Schranz, M.E.; Xiong, Z.; Leebens-Mack, J.; Meyers, B.C.; et al. Brassicales phylogeny inferred from 72 plastid genes: A reanalysis of the phylogenetic localization of two paleopolyploid events and origin of novel chemical defenses. *Am. J. Bot.* **2018**, *105*, 463–469. [CrossRef] [PubMed]

International Journal of
Molecular Sciences

MDPI

Review

Applications of Artificial Intelligence in Climate-Resilient Smart-Crop Breeding

Muhammad Hafeez Ullah Khan [1,2], Shoudong Wang [1,2], Jun Wang [2], Sunny Ahmar [3], Sumbul Saeed [4], Shahid Ullah Khan [4], Xiaogang Xu [2], Hongyang Chen [2], Javaid Akhter Bhat [2] and Xianzhong Feng [1,2,*]

1 Key Laboratory of Soybean Molecular Design Breeding, Northeast Institute of Geography and Agroecology, Chinese Academy of Sciences, Changchun 130102, China
2 Zhejiang Lab, Hangzhou 310012, China
3 Institute of Biology, Biotechnology and Environmental Protection, Faculty of Natural Sciences, University of Silesia, Jagiellonska 28, 40-032 Katowice, Poland
4 National Key Laboratory of Crop Genetic Improvement, Huazhong Agricultural University, Wuhan 430070, China
* Correspondence: fengxianzhong@iga.ac.cn; Tel.: +86-431-8565-5051; Fax: +86-431-8554-2298

Abstract: Recently, Artificial intelligence (AI) has emerged as a revolutionary field, providing a great opportunity in shaping modern crop breeding, and is extensively used indoors for plant science. Advances in crop phenomics, enviromics, together with the other "omics" approaches are paving ways for elucidating the detailed complex biological mechanisms that motivate crop functions in response to environmental trepidations. These "omics" approaches have provided plant researchers with precise tools to evaluate the important agronomic traits for larger-sized germplasm at a reduced time interval in the early growth stages. However, the big data and the complex relationships within impede the understanding of the complex mechanisms behind genes driving the agronomic-trait formations. AI brings huge computational power and many new tools and strategies for future breeding. The present review will encompass how applications of AI technology, utilized for current breeding practice, assist to solve the problem in high-throughput phenotyping and gene functional analysis, and how advances in AI technologies bring new opportunities for future breeding, to make envirotyping data widely utilized in breeding. Furthermore, in the current breeding methods, linking genotype to phenotype remains a massive challenge and impedes the optimal application of high-throughput field phenotyping, genomics, and enviromics. In this review, we elaborate on how AI will be the preferred tool to increase the accuracy in high-throughput crop phenotyping, genotyping, and envirotyping data; moreover, we explore the developing approaches and challenges for multiomics big computing data integration. Therefore, the integration of AI with "omics" tools can allow rapid gene identification and eventually accelerate crop-improvement programs.

Keywords: artificial intelligence (AI); crop breeding; genomics; phenomics; envirotyping; big data

Citation: Khan, M.H.U.; Wang, S.; Wang, J.; Ahmar, S.; Saeed, S.; Khan, S.U.; Xu, X.; Chen, H.; Bhat, J.A.; Feng, X. Applications of Artificial Intelligence in Climate-Resilient Smart-Crop Breeding. *Int. J. Mol. Sci.* **2022**, 23, 11156. https://doi.org/10.3390/ijms231911156

Academic Editors: Andrés J. Cortés and Hai Du

Received: 14 August 2022
Accepted: 19 September 2022
Published: 22 September 2022

1. Introduction

Plant breeding is a time-honored tradition that continues the process of developing improved plant cultivars, which dates back to the dawn of agriculture. Humans began to discern degrees of excellence among the plants in their fields soon after the initial domestications of cereal grains, from which they stored the seeds of the best to grow new crops. Early plant-breeding technologies were the forerunners to such rudimentary selection strategies [1]. People from all across the world explored and cultivated nearly 7000 varieties of food plants during Breeding 1.0, which began 10–12 thousand years ago [2]. Breeding stage 2.0 started in the late nineteenth and early twentieth centuries when an inbreeding depression was discovered. During this time, many advances in plant breeding were made in the science of breeding itself, such as replicated field trials, controlled crossings, statistical analyses, formal experimental designs, hybrid breeding, pedigree-based estimates of breeding values, and precise yield measurement at scale (e.g., with multirow combines) [3]. Breeding 3.0 reached to about 30 years ago, when molecular markers and genomic data began to supplement phenotypic data [4]. Breeding 4.0 is rapidly approaching, with huge omics data and the rapid progress of informatic technologies [5]. Since the beginning of the plant breeding process, starting from crop domestication, new approaches mediated by different technological revolutions are being enriched in the science continuously to increase the pace, accuracy, and precision in plant breeding [6]. In the past decade, this science has already brought the green revolution by developing semi-dwarf, nutrient-responsive, and hybrid cultivars [7]. However, considering the population growth, decreasing arable land, and climate changes, these have demanded more precise, high-throughput approaches that can mediate to develop crop cultivars at a greater pace, with higher accuracy, and precision. In this regard, the field of artificial intelligence has recently emerged, which has been suggested to possess extraordinary potential to assist in breeding climate-resilient smart crops. Climate-resilient crops maintain or increase crop yields and quality under various climatic and environmental changes, and possess the ability to resist multiple biotic and abiotic stresses. To this end, climate-resilient smart crops will allow crops to address the interlinked challenges of food security and climate change [8].

The goal of artificial intelligence (AI) is to replicate some features of human intelligence using technology [9]. This discipline can be defined as a set of studies and techniques dealing with the computer science and mathematical aspects of statistical modeling, with significant economic and social implications, and the goal of developing technological systems capable of solving problems and performing tasks or duties normally performed by the human mind [10]. The increased interest in AI in the breeding world is due to the technological maturity attained, i.e., the ability to analyze large amounts of data in a short period of time to reveal unexpected linkages. The AI approach from a breeding standpoint, allows individuals to systematize information that is typically already available on the market in a disaggregated form, transforming data into breeding decisions, and thus only considering those tools that are useful to facilitate decision-making processes in crop breeding.

This paper presents how AI technologies help to solve problems of high-throughput phenotyping and gene functional analysis in current breeding practices, as well as tackle the current massive-data processing bottlenecks in both phenotyping and genotyping, and bring new avenues for future breeding to make envirotyping data extensively used in breeding.

2. AI Technologies Benefiting Crop Breeding

Artificial intelligence uses computers and technology to simulate the human mind's problem-solving and decision-making skills [11]. AI, often known as machine intelligence, is an area of computer science that focuses on developing and managing technology, which can learn to make decisions and carry out activities independently without the need for human effort [12]. AI is a broad term that encompasses a wide range of technologies; it is a catch-all word for any software or hardware component that helps with machine learning, computer vision, natural language comprehension, and natural language processing (NLP) [13]. Traditional complementary metal-oxide-semiconductor (CMOS) hardware and the same fundamental computational processes that drive traditional software are used in today's AI [14]. AI is the most rapidly emerging technology in computer science in today's digital world, and it creates intelligent computers that replicate the intellect of the human mind [15]. For instance, the deep neural network (DNN), artificial neural network (ANN), random forest (RF), and support vector machine (SVM) are a few examples of machine-learning algorithms, as well as advanced hi-tech equipment such as the internet of things (IoT) [16]. AI is a fascinating hi-tech system that provides an endless opportunity as far as its agricultural applications are considered; hence, this opens up new frontiers for digital breeding [17]. Future AI generations are projected to inspire new sorts of brain-inspired circuits and architectures capable of making data-driven judgments faster and more precisely than humans can [18]. Furthermore, artificial intelligence, big data, machine learning, and data analytics are all terms that appear often in current academic and corporate writings that deal with data [19].

Big data, machine learning, and AI are some of the terms used to characterize modern computer processes [20]. Big data is concerned with the use of huge data of diverse types and complex structures that cannot be handled well when analyzed through classical approaches [21]. In this context, the AI trains a computer to perform jobs that are beyond human efforts, especially by considering the time and labor involved, and which are typically involved in decision-making in a variety of situations [22]. Machine learning (ML) is a branch of AI in which computers discover relationships from massive training datasets. For environment and weather applications, a simple definition is: firstly, big data involves the collection of meteorological or Earth System-related measurements, as well as high spatial and temporal resolution Earth System model (ESM) outputs for analysis; secondly, ML is the refining or discovery of new linkages between locations, times, and quantities in the datasets (e.g., where sea surface temperature features aid the weather prediction for months over land regions); thirdly, AI is a means of providing automatic warnings and guidance to society in the event of oncoming weather extremes, based on the links discovered by machine learning [23]. The current ease for application of ML methods due to improved computing capabilities is aided in part by the unique usage of computer graphics processing units (GPUs), with GPU speed improving at a quicker rate than ordinary central processing units [24]. This is an innovative use of computer memory to make calculations both more efficient and considerably closer to the data storage location [23]. The main emphasis of employing AI in breeding is that it complements the work of the breeder by guaranteeing continuous farm monitoring. Indeed, with the automation of farms and the generalization of data, breeders may dedicate more time to higher-value jobs by spending less time in their buildings. AI saves time in data identification and processing, which is of considerable benefit. Breeders and technical advisors acquire confidence and reactivity, allowing them to act when it is most appropriate [25].

AI technology has been used to accelerate the process of breeding new plant varieties, such as high-throughput genomics and phenomics to advanced breeding [15,26–28]. Increasingly, ML methodology has been used in genomic prediction, genomic selection, and marker-assisted selection [27,28]. Many agricultural companies such as Monsanto and John Deere have already invested hundreds of millions of dollars to develop such technologies that can utilize extensive data on soil type, seed variety, and weather to help farmers reduce costs and enhance yields [29]. Many of the same data sources, such as weather forecasts and Google Maps, are used to fuel both of their businesses. In addition, they may access farm equipment data that are wirelessly sent to the cloud [30]. As part of a precision-farming experiment in Romania, companies like Nippon Electric Company, Limited (NEC; headquartered in Minato, Tokyo, Japan) and Dacom (headquarter in Santa Clara, USA) employed environmental sensors and huge data analytics tools to increase yields. The use of current technologies and information systems enhances the overall productivity of agriculture [31]. Due to the agricultural data sets' complexity, novel architecture and frameworks, algorithms, as well as the analytics face several obstacles in extracting the value and hidden information from this data [32]. The recent research on AI tools, including ML, deep learning, and predictive analysis intended toward increasing the planning, learning, reasoning, thinking, and action-taking abilities [33]. Plant Breeders are developing systems to aid in a better understanding of plant behavior under a variety of climatic situations [34]. Summit, the world's most powerful supercomputer, was recently unveiled with the potential to hold 27,000 GPUs, paving the way for a bright future. AI has the potential to be a game-changer in the near future for bringing an agricultural revolution and global food security [35].

3. AI Technologies Overcoming the Phenomics Bottlenecks

Plant phenomics has advanced rapidly in recent years providing the scope for precision breeding. This progress can be ascribed to an increase in the invention and availability of new technologies that allow for high-throughput phenotyping of complex plant features. In recent years, the use of AI in a variety of scientific fields has exploded. AI features, viz., computer vision, ML, and deep learning have been effectively integrated into non-invasive imaging procedures. Through the use of ML for robust picture analysis, this integration is steadily enhancing the efficiency of data gathering and analysis. Furthermore, AI has aided the development of software and tools for data gathering and management in field phenotyping. These include open-source devices and platforms that allow for community-driven research and data sharing, providing the enormous amounts of data needed for reliable phenotyping research [36] AI is used in three critical components of phenomic data management: algorithms and programs to convert sensory data into phenotypic information; model development to understand genotype-phenotype relationships with environmental interactions; database management to allow information and resources to be shared [36].

Experiments involving repeated trials in diverse conditions (considering the statistical need for an unbiased estimation) are required in order to screen plants for desirable features (such as grain size, abiotic stress tolerance, product quality, or yield potential). The measuring of individual plants in controlled conditions has been the subject of considerable phenotyping discussion; however, plant development under open-air circumstances is not accurately represented in controlled environments [37]. These things considered, a large gap has been seen practically regarding the performance of plants from lab-to-field [7]. The ongoing integration of AI into various technologies promises a development toward smarter, faster, and lower-cost solutions. In comparison to other imaging techniques, the integration of AI into the data management pipeline of tomography and thermography is on a smaller scale in the area of phenotyping image data analysis. Deep learning has been successfully used in the analysis of composite materials [38]; therefore, its application in the data analysis of these approaches is promising. Despite the fact that field phenotyping is a practical need in the crop breeding, still the high-throughput phenotyping under field conditions lags behind the indoor phenotypic facilities currently available. Thus, it needs more effort to develop such facilities to explore the practical aspects of phenomics. To increase their accuracy, AI technologies require a significant amount of data from numerous sources. This opens up the possibility of investing more in the customization of current technologies for field-data collecting, and the use of already available AI adaptable technology, such as smartphones, to boost the number and quality of data collected [39]. Smartphones have become common consumer items and the ease with which their sensors may be used suggests their great application in agriculture [36]. Advanced signal processing on smartphones must contend with constraints such as low battery life, limited computational power, and limited bandwidth [36]. The use of citizen science in data collecting alongside professional researchers has the potential to increase the amount of data collected [39]. The main purpose of using these approaches and technology is to offer the infrastructure for tracking how plants progress during the growing season and to make the data analysis, management, and use of results via AI methods easier [40].

Recent studies have showed that the phenotyping of crops through AI shows an improvement in crop phenotyping and predictions [41–47]. For example, Selvaraj et al. (2020) [41] reported that ML algorithms, viz., k-Nearest Neighbours (kNN), RF, and SVM revealed the best performance for root yield prediction in the cassava (*Manihot esculenta* Crantz), with the highest accuracy of R^2 = 0.67, 0.66, and 0.64, respectively. Moreover, AI-assisted high-throughput phenotyping systems have been successfully applied to: wheat and maize to identify the plant growth stage [42] and plant image segmentation [43]; oilseed crops for semantic segmentation of the crops and weeds [44]; the phenotyping of disease resistance of crops [45]; improvement of plant productivity [46,47].

4. Exploring the Potential of AI in Gene Function Analysis

The rapid development of high-throughput technologies in biological sciences has resulted in the generation of massive data in recent decades. Disciplines that attempt to collect and analyze enormous volumes of biological data are often referred to as "omics", which is used to indicate the total quantity of DNA contained in each cell of an organism, with an additional flavour of openness to big challenges [47]. "Omics" data has become too large and complicated to be analyzed visually or by using statistical correlations. This has incited the use of so-called Machine Intelligence or AI which manages large amounts of data that are insurmountable for human minds, while extracting information that goes beyond our current understanding of the system under investigation and, most importantly, improving automatically based on the training data [48].

AI is already being used extensively in plant genomics and also possesses more future applications for in-depth genome exploration. A number of ML tools and algorithms are available for different kinds of bioinformatics analysis, such as protein-coding gene identification, *cis*-regulatory element identification, gene expression, subcellular location, protein-protein interaction, gene ontology, metabolic pathways, phenotypes, and genomic prediction (as reviewed by Mahood et al. (2020) [49]). In the not-too-distant future, AI is likely to be used to address a variety of plant science genomics concerns.AI algorithms might potentially be used to address comparative genomic investigations or information transfer from a model plant to a crop of interest [50]. DeepBind [51] and DeepSEA [52] are two models that have been created in recent years to predict and analyze genetic features [26]. Various sorts of expressions or sequencing data analysis can be thought of, with the goal of predicting gene functions or the differential effects of gene expression on a trait [53].

Although a significant amount of genomic data was produced as a result of the fruitful breakthroughs of high-throughput sequencing technology, the enormous amount of data generated creates a huge problem for storage and examination of the data [26]. The AI technology of bioinformatics enables the measurement of simultaneous expressions of a large number of genes, or even each and every gene that is included in the genome under a wide range of situations [54,55]. All of this combines to give biologists a more "relevant" representation of their data and the ability to integrate it, which enables them to examine their genomic data, test and confirm their assumptions throughout the experimental cycle, and ultimately improve their research [56,57].

5. Linking of Crop Genome to Phenome with AI

Currently, modern breeding approaches are focused on linking the genotype with the crop phenotype accurately and precisely. In advanced breeding, linking the whole of the genome information to high-throughput phenotypes remains a massive challenge, and is impeding the optimal application of field phenotyping and omics [15]. Germplasm collection and mapping populations can efficiently differentiate the phenomics and genomics data through AI. Crop diversity, single nucleotide polymorphisms (SNPs) detection and selection, quantitative trait loci (QTL) analysis, genome-wide association study (GWAS) analysis, and genomic selection and sequences generate a large amount of data; AI can evaluate and link the phenomics and genomics data from these big data to improve the breeding approaches. AI related to a computation and training model can predict the gene functional analysis and high-throughput crop phenotyping and also predict the performance of yield and traits of the crop [46,47,50,58,59]. Therefore, the integration of AI with phenomics and genomics tools can allow for rapid gene identification associated with the crop phenotypes that eventually accelerate crop improvement programs. In Figure 1, we summarize how to apply AI technology to link high-throughput genomics and phenomics, which can result in the production of better breeding strategies.

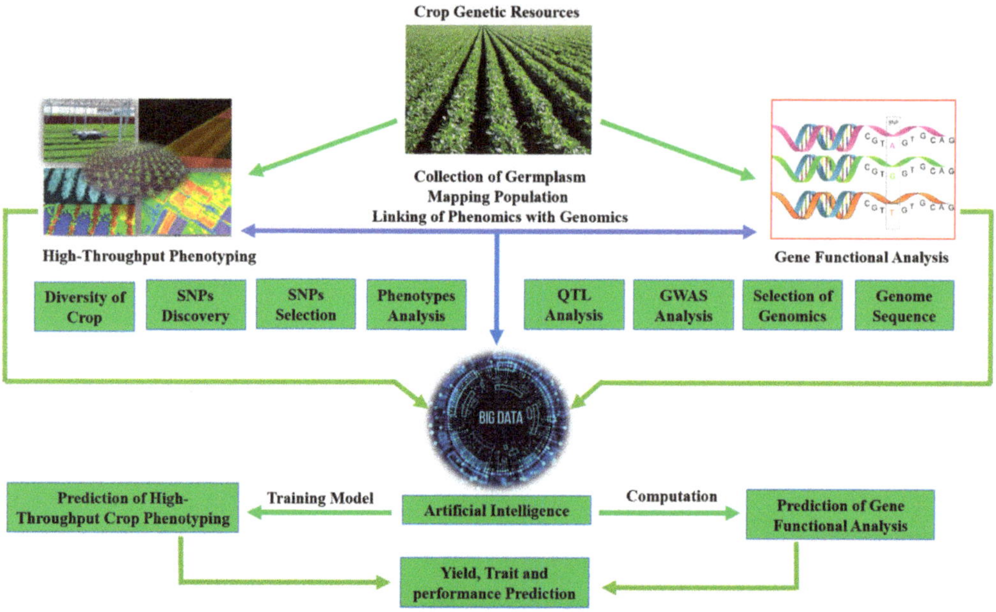

Figure 1. Artificial Intelligence used as a powerful tool for the prediction of high-throughput crop phenotyping and gene functional analysis in modern crop breeding. The high-throughput phenotypic and genotypic data were collected from large crop germplasm and breeding populations. The massive comprehensive database could integrate various resources with AI technology, such as phenotypic diversity of crops, SNPs polymorphisms, QTL analysis, GWAS analysis, genomics selection, and genome sequence. AI technologies are applied to predict the crop phenotype with whole genome prediction, the novel breeding strategies are produced through AI related to computation and training models.

Research on crop genomics is not only understanding the molecular mechanisms of phenotypes but also using technical data and bioinformatics techniques to analyse and understand the molecular mechanisms behind phenotypes [60]. To date, AI is a fascinating approach to bringing out these tasks inevitably [61]. AI approaches provide the platform to analyze huge, various, and useless datasets such as the generation of genome sequencing/photo imaging over conventional analytical strategies [15,62]. Recently, the AI approach has been explicitly employed in varied research fields of phenomics and genomics, such as: analysing genome assembly and genome-specific algorithms [26]; broad-range data analysis to mitigate multiplex biological complications in metabolomics, proteomics, genomics, transcriptomics, as well as systematic biology [62,63]; interpretation of gene expression cascades [64,65]; identification of significant SNPs in polyploid plants [66]; high-throughput crop stress phenotyping [41,67].

Scientists have employed AI and its developed models to modulate the flow of information from generic DNA to genetic-based phenotypes, to investigate the potential variants in natural populations [49]. More specifically, for breeders, AI will assist the further investigation of genetic loci to facilitate the agricultural output by triggering the genome algorithms and allowing high-throughput crop phenotyping in quantitative traits for open-field and controlled environments [49,68]. Additionally, AI can be cohesively combined with bioinformatics and genome sequencing analysis to interpret various molecular repertories such as transcription factor binding sites [69], long non-coding RNAs (lncRNAs) [70], microRNA (miRNAs), epistatic modifications, coding genes, targeted polyadenylation sites [71], as well as *cis*-regulatory elements (CREs) [49,72].

Various crop databases insert a huge amount of heterogeneous-related phenotypic and genotypic data (big data) recently providing insight into potential resources for breeders to untangle novel trait-identified candidate genes [73]. Luckily, AI provides a novel benchmark summary for analytical and computational methods for the integrated analysis of such enormous datasets based on the big-data spectrum [49,73]. In addition, employing AI to conclude the interrelations between candidate genes and CREs is a novel approach for categorizing and identifying previously unknown genes for significant crop improvements [74]. Furthermore, AI strategies have more potential for interpretation of the crop yield, variation in climatic assessment, high-throughput crop stress phenotyping, climate temperature, ultraviolent (UV) radiation, wind, and hail [26,73,75]. The role of AI is becoming more and more important in obtaining, analyzing, integrating, and managing genomic and phenomic data to increase agricultural climate resilience [76,77].

Next generation sequencing (NGS)-based genotyping methods have helped to improve gene-mapping resolution and gene identification and NGS-based genotyping for GWAS analysis has been used in crop improvement [68]. For example, in soybean, these kinds of studies have been widely used to identify genetic loci and candidate genes for seed weight [78], seed protein and oil contents [79], pod dehiscence [80], nitrogen fixation [81], soybean plant height and primary branches [81], agronomic traits [82], disease resistance [83], and tocopherol concentration [84]. Bulk segregant analysis (BSA) and its modified methodologies are currently used in many crops [85–90]. The NGS-based BSA is becoming a popular approach to identifying candidate genes for various traits, such as the soybean mosaic virus [91], charcoal rot resistance [92], flowering time [93], phytophthora resistance [94], and powdery-mildew resistance [95]. Recently, the deep-learning algorithm for BSA (DeepBSA) has been developed for QTL mapping and functional gene cloning in maize [96].

6. AI Making Envirotyping Data Accessible in Crop Breeding

Climate change has a great impact on the environment and crop production for the present and future. The concept of envirotyping is proposed as a third "typing" technology, accompanying phenotyping and genotyping to decode environmental influences on crop breeding [97]. Envirotyping plays a key role in crop modeling and the prediction of phenotypes through its efficient components, including the genotype-by-environment interaction (GEI), environmental signals, responsive genes, biotic and abiotic stresses, as well as integrative phenotyping [76]. Cortes et al. (2022) discuss the state of the emerging field of study known as "genome-environment associations", which combines ecological climatic data with evolutionary genomics (GEAs) [98]. The authors advocate for the community to begin collecting genomic estimated adaptive values (GEAVs) for genomic prediction (GP) and multi-dimensional ML models in order to take polygenic evolutionary adaptation into consideration. Xu et al. (2022) recently proposed an integrated genomic-enviromics prediction breeding scheme using integrated multiomics information, big data technology, and artificial intelligence [99].

Climate change, as well as the global population and pathogen pressure, have raised serious alarm over worldwide food security. In the coming years, strategies need to be developed to maximize the limited resources and their utilization for crop breeding and land management [100]. Climate-smart crops and climate-smart soils have been adapted to the environment for more effective breeding programs, and breeders may use this knowledge to generate new smart crops for the new climate. Genomic technologies together with high-throughput phenotyping are providing researchers and farmers with the information required to guide and notify the breeding methods for climate-smart breeding [76]. AI plays a vital role in integrating and manipulating this fast collecting abundance of data by conducting association studies to identify genomic targets which are related to adaptive climate-resilient traits [101]. Cortes et al. (2021) also put forward a roadmap to use ML, GP, and multi-trait gene editing approaches to capture novel abiotic stress tolerance variations from wild crop relatives to utilize these variants for bread drought-tolerant crops [102].

Breeders can use these data to adjust crops to their environment and they can be introduced via advanced selection or genome editing techniques [15]. Genomic and phenomic data will need to be integrated into comprehensive clade-specific databases and platforms, as well as accessible tools that breeders may utilize to inform the selection of climate-adapted characteristics, to effectively translate research into the field.

Our proposed breeding scheme (Figure 2) integrates genotypic, phenotypic, and envirotypic information to improve efficacy. The phenotypic data of crop plants both for indoor and outdoor environments are collected by high-throughput robotic systems [101] and the phenotypic information from various environments will transfer to a high-throughput phenotyping (HTP) server via Internet. The multiple datasets will take the genotypic, phenotypic, and envirotypic information together, and the G × E interaction (GEI) will quantify by multiple environments. AI technology, particularly ML and deep learning (DL), is used in cultivar selection for specific or multiple major environments. This approach will enable us to make decisions about the selected cultivars, and whether it is suitable for cultivation in limited environments or all major environments.

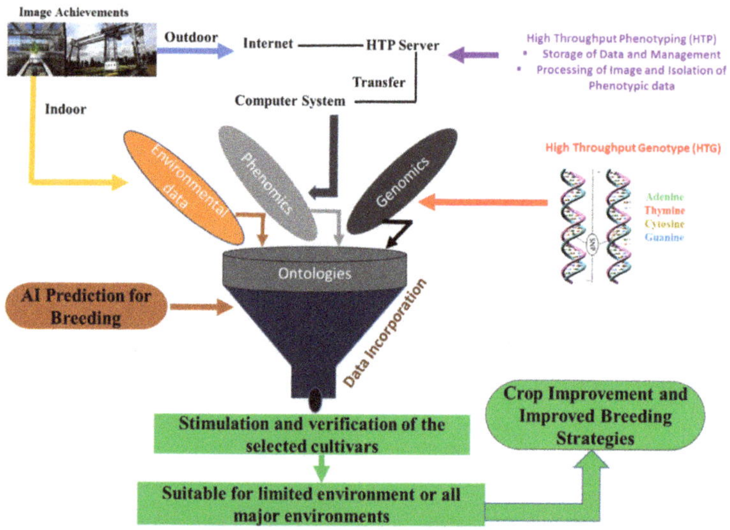

Figure 2. Integration and management of Genomics, Phenomics, and Enviromics data by artificial intelligence for crop-breeding improvement. The phenotypic data of crops are collected from both indoor and outdoor environments, the information of phenotypic, genotypic and environmental are combined together with AI technology. With mathematical modelling, logical deduction, and decision-making, the AI-assisted breeding system will simulate and verify the selected cultivars, whether it is suitable for cultivation in limited environments or all major environments.

7. Future Prospects of AI Breeding

In recent years, there has been significant growth in discussion regarding the importance of AI leading to debates about the applications of AI in the world. Plant breeding must be updated to take advantage of the digital revolution. Researchers and breeders must evaluate computer-generated suggestions against farmers' demands to be successful in their future work. Many sectors throughout the world, including agriculture breeding, are benefiting from greater profitability and have high economic growth rates as a result of the introduction and successful deployment of AI technologies. Furthermore, AI will focus on developing novel, human-centered techniques, assessing the application of robotic technology to a variety of industries and businesses around the world. AI will also transform the way different companies around the world expand and compete by representing new manufacturing concepts that will result in breeding profitability. To take full advantage of such prospects, most firms throughout the world will need to be more involved in the creation of various AI methods, such as putting human aspects at the centre of the nucleus. They will also concentrate on developing a variety of responsible AI machines with moral and ethical ideals, which will lead to positive outcomes as well as to enable individuals to perform tasks they are familiar with. The development of various AI systems will assist the global agricultural breeding sector in assuming the availability of symbolic structures, such as reasoning ability and knowledge existence. Furthermore, when AI achieves intelligence comparable to or higher than that of humans, there will be concerns about societal and political change.

These and other instances suggest that AI holds promise for some genomics applications. What does all of this have to do with plant genomics? One observation is that large-scale datasets, which are required to train the AI applications listed above, are not currently accessible for plants. In the not-too-distant future, AI is likely to be used to address a variety of plant science genomic concerns. It is possible that one area of plant genomics that may be addressed is how to deal with several species (such as wheat, soybean, rice, maize, tomato, and oilseed rape) at the same time, intercropping is becoming important for covering crops of mixed species. AI algorithms might potentially be used to address comparative genomics investigations or information transfer from a model plant to a crop of interest.

Farmers and breeders will be able to feed the data into cloud-based AI applications via portable devices, drones, and agriculture-equipment platforms making AI applications more widely accessible. The phenomics and genomics data obtained with ML and DL are accurate, but not good enough to totally rely on the technology to speed up breeding, which is still a tough, time-consuming, and costly process. When examined in genomes, epigenomics, transcriptomics, proteomics, metabolomics, and phenomics still provide minimal information.

Furthermore, plant scientists are finding amazing answers to most of these problems, changing our emphasis from algorithmic performance to new farming models that could enable a new agricultural revolution that is better for both humankind and the environment. AI will efficiently revolutionize the "omics" approaches and breeding management with high-technology methods. AI will develop a large number of genotypes and phenotypes that must be screened using model-based envirotyping for a wide range of adaptive genotypes and phenotypes, and segregating material must be developed and advanced using speed-breeding management or fast-generation advancement to shorten the breeding cycle and improve genetic gain.

Funding: This research was funded by the Zhejiang Lab (Grant No. 2021PE0AC04), Yazhou Bay Seed Lab (Grant No. B21HJ0101), and Jilin Province Science and Technology Development Plan Project (20210302005NC).

Institutional Review Board Statement: Not applicable.

Informed Consent Statement: Not applicable.

Data Availability Statement: No new data were created in this study. Data sharing is not applicable to this article.

Conflicts of Interest: The authors declare no conflict of interest.

References

1. Allard, R. *Plant Breeding*; Encyclopedia Britannica: Chicago, IL, USA, 2014.
2. Meyer, R.S.; DuVal, A.E.; Jensen, H.R. Patterns and processes in crop domestication: An historical review and quantitative analysis of 203 global food crops. *New Phytol.* **2012**, *196*, 29–48. [CrossRef] [PubMed]
3. Moran, P.; Smith, C. The correlation between relatives on the supposition of mendelian inheritance. *Trans. R. Soc. Edinb.* **1918**, *52*, 399–438.
4. Meuwissen, T.H.; Hayes, B.J.; Goddard, M.J.G. Prediction of total genetic value using genome-wide dense marker maps. *Genetics* **2001**, *157*, 1819–1829. [CrossRef]

5. Wallace, J.G.; Rodgers-Melnick, E.; Buckler, E.S. On the Road to Breeding 4.0: Unraveling the Good, the Bad, and the Boring of Crop Quantitative Genomics. *Annu. Rev. Genet.* **2018**, *52*, 421–444. [CrossRef] [PubMed]
6. Zargar, S.M.; Raatz, B.; Sonah, H.; Bhat, J.A.; Dar, Z.A.; Agrawal, G.K.; Rakwal, R. Recent advances in molecular marker techniques: Insight into QTL mapping, GWAS and genomic selection in plants. *J. Crop Sci. Biotechnol.* **2015**, *18*, 293–308. [CrossRef]
7. Bhat, J.A.; Yu, D.; Bohra, A.; Ganie, S.A.; Varshney, R.K. Features and applications of haplotypes in crop breeding. *Commun. Biol.* **2021**, *4*, 121266. [CrossRef]
8. Zohry, A.E.-H.; Ouda, S. Climate-Resilient Crops. In *Climate-Smart Agriculture: Reducing Food Insecurity*; Springer International Publishing: Cham, Germany, 2022; pp. 115–135. [CrossRef]
9. Yang, Y.; Siau, K.L. A qualitative research on marketing and sales in the artificial intelligence age. In Proceedings of the Thirteenth Midwest Association for Information Systems Conference, Saint Louis, MO, USA, 17–18 May 2018.
10. Amit, K. *Artificial Intelligence and Soft Computing: Behavioral and Cognitive Modeling of the Human Brain*; CRC Press: Boca Raton, FL, USA, 2018.
11. Jarrahi, M.H. Artificial intelligence and the future of work: Human-AI symbiosis in organizational decision making. *Bus. Horiz.* **2018**, *61*, 577–586. [CrossRef]
12. Dwivedi, Y.K.; Hughes, L.; Ismagilova, E.; Aarts, G.; Coombs, C.; Crick, T.; Duan, Y.; Dwivedi, R.; Edwards, J.; Eirug, A. Artificial Intelligence (AI): Multidisciplinary perspectives on emerging challenges, opportunities, and agenda for research, practice and policy. *Int. J. Inf. Manag.* **2021**, *57*, 101994. [CrossRef]
13. Li, Y.; Zhou, X.; Li, G. Bridging natural language and graphical user Interfaces. In *Artificial Intelligence for Human Computer Interaction: A Modern Approach*; Springer: Berlin/Heidelberg, Germany, 2021; pp. 463–493.
14. Kok, J.N.; Boers, E.J.; Kosters, W.A.; Van der Putten, P.; Poel, M. Artificial intelligence: Definition, trends, techniques, and cases. *Artif. Intell.* **2009**, *1*, 270–299.
15. Harfouche, A.L.; Jacobson, D.A.; Kainer, D.; Romero, J.C.; Harfouche, A.H.; Mugnozza, G.S.; Moshelion, M.; Tuskan, G.A.; Keurentjes, J.J.B.; Altman, A. Accelerating climate resilient plant breeding by applying next-generation artificial intelligence. *Trends Biotechnol.* **2019**, *37*, 1217–1235. [CrossRef]
16. Baduge, S.K.; Thilakarathna, S.; Perera, J.S.; Arashpour, M.; Sharafi, P.; Teodosio, B.; Shringi, A.; Mendis, P.J.A.i.C. Artificial intelligence and smart vision for building and construction 4.0: Machine and deep learning methods and applications. *Autom. Constr.* **2022**, *141*, 104440. [CrossRef]
17. Montesinos-López, O.A.; Montesinos-López, A.; Crossa, J.; Gianola, D.; Hernández-Suárez, C.M.; Martín-Vallejo, J. Multi-trait, multi-environment deep learning modeling for genomic-enabled prediction of plant traits. *G3 Genes Genomes Genet.* **2018**, *8*, 3829–3840. [CrossRef] [PubMed]
18. Xu, Y.; Liu, X.; Cao, X.; Huang, C.; Liu, E.; Qian, S.; Liu, X.; Wu, Y.; Dong, F.; Qiu, C.-W. Artificial intelligence: A powerful paradigm for scientific research. *Innovation* **2021**, *2*, 100179. [CrossRef]
19. Duan, Y.; Edwards, J.S.; Dwivedi, Y.K. Artificial intelligence for decision making in the era of Big Data–evolution, challenges and research agenda. *Int. J. Inf. Manag.* **2019**, *48*, 63–71. [CrossRef]
20. Cravero, A.; Sepúlveda, S. Use and adaptations of machine learning in big data—Applications in real cases in agriculture. *Electronics* **2021**, *10*, 552. [CrossRef]
21. Supriya, M.; Deepa, A. Machine learning approach on healthcare big data: A review. *Big Data Inf. Anal.* **2020**, *5*, 58–75. [CrossRef]
22. Berente, N.; Gu, B.; Recker, J.; Santhanam, R. Managing artificial intelligence. *MIS Q.* **2021**, *45*, 1433–1450.
23. Huntingford, C.; Jeffers, E.S.; Bonsall, M.B.; Christensen, H.M.; Lees, T.; Yang, H. Machine learning and artificial intelligence to aid climate change research and preparedness. *Environ. Res. Lett.* **2019**, *14*, 124007. [CrossRef]
24. Raschka, S.; Patterson, J.; Nolet, C. Machine learning in python: Main developments and technology trends in data science, machine learning, and artificial intelligence. *Information* **2020**, *11*, 193. [CrossRef]
25. Talaviya, T.; Shah, D.; Patel, N.; Yagnik, H.; Shah, M. Implementation of artificial intelligence in agriculture for optimisation of irrigation and application of pesticides and herbicides. *Artif. Intell. Agric.* **2020**, *4*, 58–73. [CrossRef]
26. Esposito, S.; Carputo, D.; Cardi, T.; Tripodi, P. Applications and trends of machine learning in genomics and phenomics for next-generation breeding. *Plants* **2020**, *9*, 34. [CrossRef]
27. Reinoso-Peláez, E.L.; Gianola, D.; González-Recio, O. Genome-enabled prediction methods based on machine learning. In *Genomic Prediction of Complex Traits*; Methods in Molecular Biology; Ahmadi, N., Bartholomé, J., Eds.; Humana: New York, NY, USA, 2022; Volume 2467. [CrossRef]
28. Crossa, J.; Pérez-Rodríguez, P.; Cuevas, J.; Montesinos-López, O.; Jarquín, D.; De Los Campos, G.; Burgueño, J.; González-Camacho, J.M.; Pérez-Elizalde, S.; Beyene, Y. Genomic selection in plant breeding: Methods, models, and perspectives. *Trends Plant Sci.* **2017**, *22*, 961–975. [CrossRef] [PubMed]
29. Faulkner, A.; Cebul, K.; McHenry, G. *Agriculture Gets Smart: The Rise of Data and Robotics*; Cleantech Agriculture Report; Cleantech Group: San Francisco, CA, USA, 2014.
30. Stergiou, C.; Psannis, K.E. Recent advances delivered by Mobile Cloud computing and internet of things for big data applications: A survey. *Int. J. Netw. Manag.* **2017**, *27*, e1930. [CrossRef]
31. El Bilali, H.; Allahyari, M.S. Transition towards sustainability in agriculture and food systems: Role of information and communication technologies. *Inf. Process. Agric.* **2018**, *5*, 456–464. [CrossRef]

32. Priya, R.; Ramesh, D.J.S.C.I. ML based sustainable precision agriculture: A future generation perspective. *Sustain. Comput. Inform.* **2020**, *28*, 100439. [CrossRef]
33. Shaw, J.; Rudzicz, F.; Jamieson, T.; Goldfarb, A. Artificial intelligence and the implementation challenge. *J. Med. Internet Res.* **2019**, *21*, e13659. [CrossRef]
34. Jeong, J.H.; Resop, J.P.; Mueller, N.D.; Fleisher, D.H.; Yun, K.; Butler, E.E.; Timlin, D.J.; Shim, K.M.; Gerber, J.S.; Reddy, V.R.; et al. Random Forests for Global and Regional Crop Yield Predictions. *PLoS ONE* **2016**, *11*, e0156571. [CrossRef]
35. Streich, J.; Romero, J.; Gazolla, J.G.F.M.; Kainer, D.; Cliff, A.; Prates, E.T.; Brown, J.B.; Khoury, S.; Tuskan, G.A.; Garvin, M. Can exascale computing and explainable artificial intelligence applied to plant biology deliver on the United Nations sustainable development goals? *Curr. Opin. Biotechnol.* **2020**, *61*, 217–225. [CrossRef]
36. Nabwire, S.; Suh, H.K.; Kim, M.S.; Baek, I.; Cho, B.K. Review: Application of Artificial Intelligence in Phenomics. *Sensors* **2021**, *21*, 4363. [CrossRef]
37. Yang, W.; Feng, H.; Zhang, X.; Zhang, J.; Doonan, J.H.; Batchelor, W.D.; Xiong, L.; Yan, J. Crop Phenomics and High-Throughput Phenotyping: Past Decades, Current Challenges, and Future Perspectives. *Mol. Plant* **2020**, *13*, 187–214. [CrossRef]
38. Bang, H.-T.; Park, S.; Jeon, H. Defect identification in composite materials via thermography and deep learning techniques. *Compos. Struct.* **2020**, *246*, 112405. [CrossRef]
39. Kumar, R.S.; Keerthana, G.; Murali, L.; Chidambaranathan, S.; Premkumar, C.; Mahaveerakannan, R. Enabling Artificial Intelligence and Cyber Security in Smart Manufacturing. In *Artificial Intelligent Techniques for Wireless Communication and Networking*; John Wiley & Sons, Ltd.: Hoboken, NJ, USA, 2022; pp. 269–286. [CrossRef]
40. Song, P.; Wang, J.; Guo, X.; Yang, W.; Zhao, C. High-throughput phenotyping: Breaking through the bottleneck in future crop breeding. *Crop J.* **2021**, *9*, 633–645. [CrossRef]
41. Selvaraj, M.G.; Valderrama, M.; Guzman, D.; Valencia, M.; Ruiz, H.; Acharjee, A. Machine learning for high-throughput field phenotyping and image processing provides insight into the association of above and below-ground traits in cassava (*Manihot esculenta Crantz*). *Plant Methods* **2020**, *16*, 87. [CrossRef] [PubMed]
42. Sadeghi-Tehran, P.; Sabermanesh, K.; Virlet, N.; Hawkesford, M.J. Automated method to determine two critical growth stages of wheat: Heading and flowering. *Front. Plant Sci.* **2017**, *8*, 252. [CrossRef] [PubMed]
43. Brichet, N.; Fournier, C.; Turc, O.; Strauss, O.; Artzet, S.; Pradal, C.; Welcker, C.; Tardieu, F.; Cabrera-Bosquet, L. A robot-assisted imaging pipeline for tracking the growths of maize ear and silks in a high-throughput phenotyping platform. *Plant Methods* **2017**, *13*, 96. [CrossRef] [PubMed]
44. Abdalla, A.; Cen, H.; Wan, L.; Rashid, R.; Weng, H.; Zhou, W.; He, Y. Fine-tuning convolutional neural network with transfer learning for semantic segmentation of ground-level oilseed rape images in a field with high weed pressure. *Comput. Electron. Agric.* **2019**, *167*, 105091. [CrossRef]
45. Mahlein, A.-K.; Kuska, M.T.; Thomas, S.; Wahabzada, M.; Behmann, J.; Rascher, U.; Kersting, K. Quantitative and qualitative phenotyping of disease resistance of crops by hyperspectral sensors: Seamless interlocking of phytopathology, sensors, and machine learning is needed! *Curr. Opin. Plant Biol.* **2019**, *50*, 156–162. [CrossRef]
46. Mochida, K.; Koda, S.; Inoue, K.; Hirayama, T.; Tanaka, S.; Nishii, R.; Melgani, F. Computer vision-based phenotyping for improvement of plant productivity: A machine learning perspective. *GigaScience* **2019**, *8*, giy153. [CrossRef]
47. Jung, J.; Maeda, M.; Chang, A.; Bhandari, M.; Ashapure, A.; Landivar-Bowles, J. The potential of remote sensing and artificial intelligence as tools to improve the resilience of agriculture production systems. *Curr. Opin. Biotechnol.* **2021**, *70*, 15–22. [CrossRef]
48. Grapov, D.; Fahrmann, J.; Wanichthanarak, K.; Khoomrung, S. Rise of deep learning for genomic, proteomic, and metabolomic data integration in precision medicine. *OMICS* **2018**, *22*, 630–636. [CrossRef]
49. Mahood, E.H.; Kruse, L.H.; Moghe, G.D. Machine learning: A powerful tool for gene function prediction in plants. *Appl. Plant Sci.* **2020**, *8*, e11376. [CrossRef] [PubMed]
50. Caudai, C.; Galizia, A.; Geraci, F.; Le Pera, L.; Morea, V.; Salerno, E.; Via, A.; Colombo, T. AI applications in functional genomics. *Comput. Struct. Biotechnol. J.* **2021**, *19*, 5762–5790. [CrossRef] [PubMed]
51. Alipanahi, B.; Delong, A.; Weirauch, M.T.; Frey, B.J. Predicting the sequence specificities of DNA-and RNA-binding proteins by deep learning. *Nat. Biotechnol.* **2015**, *33*, 831–838. [CrossRef]
52. Zhou, J.; Troyanskaya, O.G. Predicting effects of noncoding variants with deep learning–based sequence model. *Nat. Methods* **2015**, *12*, 931–934. [CrossRef] [PubMed]
53. Dasari, C.M.; Amilpur, S.; Bhukya, R. Exploring variable-length features (motifs) for predicting binding sites through interpretable deep neural networks. *Eng. Appl. Artif. Intell.* **2021**, *106*, 104485. [CrossRef]
54. Lemmon, E.M.; Lemmon, A.R. High-throughput genomic data in systematics and phylogenetics. *Annu. Rev. Ecol. Evol. Syst.* **2013**, *44*, 99–121. [CrossRef]
55. Huang, D.W.; Sherman, B.T.; Lempicki, R.A. Bioinformatics enrichment tools: Paths toward the comprehensive functional analysis of large gene lists. *Nucleic Acids Res.* **2009**, *37*, 1–13. [CrossRef] [PubMed]
56. Xia, E.H.; Li, F.D.; Tong, W.; Li, P.H.; Wu, Q.; Zhao, H.J.; Ge, R.H.; Li, R.P.; Li, Y.Y.; Zhang, Z.Z. Tea Plant Information Archive: A comprehensive genomics and bioinformatics platform for tea plant. *Plant Biotechnol. J.* **2019**, *17*, 1938–1953. [CrossRef] [PubMed]
57. Hu, H.; Scheben, A.; Edwards, D. Advances in integrating genomics and bioinformatics in the plant breeding pipeline. *Agriculture* **2018**, *8*, 75. [CrossRef]

58. Batley, J.; Edwards, D. The application of genomics and bioinformatics to accelerate crop improvement in a changing climate. *Curr. Opin. Plant Biol.* **2016**, *30*, 78–81. [CrossRef]
59. Wang, H.; Cimen, E.; Singh, N.; Buckler, E. Deep learning for plant genomics and crop improvement. *Curr. Opin. Plant Biol.* **2020**, *54*, 34–41. [CrossRef] [PubMed]
60. Altman, A.; Fan, L.; Foyer, C.; Cowling, W.; Mittler, R.; Qaim, M.; Weber, A.P.; Reynolds, M.; Varshney, R.K.; Fernie, A. Past and future milestones of plant breeding. *Trends Plant Sci.* **2021**, *26*, 530–538.
61. Williamson, H.F.; Brettschneider, J.; Caccamo, M.; Davey, R.P.; Goble, C.; Kersey, P.J.; May, S.; Morris, R.J.; Ostler, R.; Pridmore, T. Data management challenges for artificial intelligence in plant and agricultural research. *F1000Research* **2021**, *10*, 324. [CrossRef]
62. Libbrecht, M.W.; Noble, W.S. Machine learning applications in genetics and genomics. *Nat. Rev. Genet.* **2015**, *16*, 321–332. [CrossRef] [PubMed]
63. Xu, C.; Jackson, S.A. *Machine Learning and Complex Biological Data*; Springer: Berlin/Heidelberg, Germany, 2019.
64. Kwon, M.S.; Lee, B.T.; Lee, S.Y.; Kim, H.U. Modeling regulatory networks using machine learning for systems metabolic engineering. *Curr. Opin. Biotechnol.* **2020**, *65*, 163–170. [CrossRef]
65. Ni, Y.; Aghamirzaie, D.; Elmarakeby, H.; Collakova, E.; Li, S.; Grene, R.; Heath, L.S. A machine learning approach to predict gene regulatory networks in seed development in Arabidopsis. *Front. Plant Sci.* **2016**, *7*, 1936. [CrossRef]
66. Korani, W.; Clevenger, J.P.; Chu, Y.; Ozias-Akins, P. Machine learning as an effective method for identifying true single nucleotide polymorphisms in polyploid plants. *Plant Genome* **2019**, *12*, 180023. [CrossRef]
67. Zhao, J.; Bodner, G.; Rewald, B. Phenotyping: Using machine learning for improved pairwise genotype classification based on root traits. *Front. Plant Sci.* **2016**, *7*, 1864. [CrossRef]
68. Razzaq, A.; Kaur, P.; Akhter, N.; Wani, S.H.; Saleem, F. Next-Generation Breeding Strategies for Climate-Ready Crops. *Front. Plant Sci.* **2021**, *12*, 620420. [CrossRef]
69. Long, P.; Zhang, L.; Huang, B.; Chen, Q.; Liu, H. Integrating genome sequence and structural data for statistical learning to predict transcription factor binding sites. *Nucleic Acids Res.* **2020**, *48*, 12604–12617. [CrossRef]
70. Sun, L.; Liu, H.; Zhang, L.; Meng, J. lncRScan-SVM: A tool for predicting long non-coding RNAs using support vector machine. *PLoS ONE* **2015**, *10*, e0139654. [CrossRef] [PubMed]
71. Gao, X.; Zhang, J.; Wei, Z.; Hakonarson, H. DeepPolyA: A convolutional neural network approach for polyadenylation site prediction. *IEEE Access* **2018**, *6*, 24340–24349. [CrossRef]
72. Ernst, J.; Kellis, M. Chromatin-state discovery and genome annotation with ChromHMM. *Nat. Protoc.* **2017**, *12*, 2478–2492. [CrossRef]
73. Tong, H.; Nikoloski, Z. Machine learning approaches for crop improvement: Leveraging phenotypic and genotypic big data. *J. Plant Physiol.* **2021**, *257*, 153354. [CrossRef]
74. Li, Y.; Shi, W.; Wasserman, W.W. Genome-wide prediction of cis-regulatory regions using supervised deep learning methods. *BMC Bioinform.* **2018**, *19*, 202. [CrossRef] [PubMed]
75. Crane-Droesch, A. Machine learning methods for crop yield prediction and climate change impact assessment in agriculture. *Environ. Res. Lett.* **2018**, *13*, 114003. [CrossRef]
76. Marsh, J.I.; Hu, H.; Gill, M.; Batley, J.; Edwards, D. Crop breeding for a changing climate: Integrating phenomics and genomics with bioinformatics. *Theor. Appl. Genet.* **2021**, *134*, 1677–1690. [CrossRef]
77. Shen, Y.; Zhou, G.; Liang, C.; Tian, Z. Omics-based interdisciplinarity is accelerating plant breeding. *Curr. Opin. Plant Biol.* **2022**, *66*, 102167. [CrossRef]
78. Jing, Y.; Zhao, X.; Wang, J.; Teng, W.; Qiu, L.; Han, Y.; Li, W. Identification of the genomic region underlying seed weight per plant in soybean (*Glycine max* L. Merr.) via high-throughput single-nucleotide polymorphisms and a genome-wide association study. *Front. Plant Sci.* **2018**, *9*, 1392. [CrossRef]
79. Li, D.; Zhao, X.; Han, Y.; Li, W.; Xie, F. Genome-wide association mapping for seed protein and oil contents using a large panel of soybean accessions. *Genomics* **2019**, *111*, 90–95. [CrossRef]
80. Hu, D.; Kan, G.; Hu, W.; Li, Y.; Hao, D.; Li, X.; Yang, H.; Yang, Z.; He, X.; Huang, F.J.; et al. Identification of loci and candidate genes responsible for pod dehiscence in soybean via genome-wide association analysis across multiple environments. *Front. Plant Sci.* **2019**, *10*, 811. [CrossRef] [PubMed]
81. Torkamaneh, D.; Chalifour, F.-P.; Beauchamp, C.J.; Agrama, H.; Boahen, S.; Maaroufi, H.; Rajcan, I.; Belzile, F. Genome-wide association analyses reveal the genetic basis of biomass accumulation under symbiotic nitrogen fixation in African soybean. *Theor. Appl. Genet.* **2020**, *133*, 665–676. [CrossRef] [PubMed]
82. Sonah, H.; O'Donoughue, L.; Cober, E.; Rajcan, I.; Belzile, F. Identification of loci governing eight agronomic traits using a GBS-GWAS approach and validation by QTL mapping in soya bean. *Plant Biotechnol. J.* **2015**, *13*, 211–221. [CrossRef]
83. Zhao, X.; Han, Y.; Li, Y.; Liu, D.; Sun, M.; Zhao, Y.; Lv, C.; Li, D.; Yang, Z.; Huang, L. Loci and candidate gene identification for resistance to Sclerotinia sclerotiorum in soybean (*Glycine max* L. Merr.) via association and linkage maps. *Plant J.* **2015**, *82*, 245–255. [CrossRef]
84. Sui, M.; Jing, Y.; Li, H.; Zhan, Y.; Luo, J.; Teng, W.; Qiu, L.; Zheng, H.; Li, W.; Zhao, X. Identification of loci and candidate genes analyses for tocopherol concentration of soybean seed. *Front. Plant Sci.* **2020**, *11*, 539460. [CrossRef]
85. Giovannoni, J.J.; Wing, R.A.; Ganal, M.W.; Tanksley, S.D. Isolation of molecular markers from specific chromosomal intervals using DNA pools from existing mapping populations. *Nucleic Acids Res.* **1991**, *19*, 6553–6558. [CrossRef] [PubMed]

86. Michelmore, R.W.; Paran, I.; Kesseli, R.V. Identification of markers linked to disease-resistance genes by bulked segregant analysis: A rapid method to detect markers in specific genomic regions by using segregating populations. *Proc. Natl. Acad. Sci. USA* **1991**, *88*, 9828–9832. [CrossRef]
87. Yang, Z.; Huang, D.; Tang, W.; Zheng, Y.; Liang, K.; Cutler, A.J.; Wu, W. Mapping of quantitative trait loci underlying cold tolerance in Rice seedlings via high-throughput sequencing of pooled extremes. *PLoS ONE* **2013**, *8*, e68433. [CrossRef] [PubMed]
88. Takagi, H.; Abe, A.; Yoshida, K.; Kosugi, S.; Natsume, S.; Mitsuoka, C.; Uemura, A.; Utsushi, H.; Tamiru, M.; Takuno, S.; et al. QTL-seq: Rapid mapping of quantitative trait loci in rice by whole genome resequencing of DNA from two bulked populations. *Plant J.* **2013**, *74*, 174–183. [CrossRef]
89. Zhou, H.; Tang, K.; Li, G.; Liu, W.; Yu, H.; Yuan, X.; Yang, S.; Bhattacharyya, M.K.; Feng, X. A robust and rapid candidate gene mapping pipeline based on M2 populations. *Front. Plant Sci.* **2021**, *12*, 681816. [CrossRef]
90. Majeed, A.; Johar, P.; Raina, A.; Salgotra, R.K.; Feng, X.; Bhat, J.A. Harnessing the potential of bulk segregant analysis sequencing and its related approaches in crop breeding. *Front. Genet.* **2022**, *13*, 944501. [CrossRef] [PubMed]
91. Yang, Q.; Jin, H.; Yu, X.; Fu, X.; Zhi, H.; Yuan, F. Rapid identification of soybean resistance genes to soybean mosaic virus by SLAF-seq bulked segregant analysis. *Plant Mol. Biol. Rep.* **2020**, *38*, 666–675. [CrossRef]
92. da Silva, M.P.; Zaccaron, A.Z.; Bluhm, B.H.; Rupe, J.C.; Wood, L.; Mozzoni, L.A.; Mason, R.E.; Yingling, S.; Pereira, A. Bulked segregant analysis using next-generation sequencing for identification of genetic loci for charcoal rot resistance in soybean. *Physiol. Mol. Plant Pathol.* **2020**, *109*, 101440. [CrossRef]
93. Watanabe, S.; Tsukamoto, C.; Oshita, T.; Yamada, T.; Anai, T.; Kaga, A. Identification of quantitative trait loci for flowering time by a combination of restriction site–associated DNA sequencing and bulked segregant analysis in soybean. *Breed. Sci.* **2017**, 17013. [CrossRef]
94. Cheng, Y.; Ma, Q.; Ren, H.; Xia, Q.; Song, E.; Tan, Z.; Li, S.; Zhang, G.; Nian, H. Fine mapping of a Phytophthora-resistance gene RpsWY in soybean (*Glycine max* L.) by high-throughput genome-wide sequencing. *Theor. Appl. Genet.* **2017**, *130*, 1041–1051. [CrossRef]
95. Jiang, B.; Li, M.; Cheng, Y.; Cai, Z.; Ma, Q.; Jiang, Z.; Ma, R.; Xia, Q.; Zhang, G.; Nian, H. Genetic mapping of powdery mildew resistance genes in soybean by high-throughput genome-wide sequencing. *Theor. Appl. Genet.* **2019**, *132*, 1833–1845. [CrossRef]
96. Li, Z.; Chen, X.; Shi, S.; Zhang, H.; Wang, X.; Chen, H.; Li, W.; Li, L. DeepBSA: A deep-learning algorithm improves bulked segregant analysis for dissecting complex traits. *Molecular Plant* **2022**, *15*, 1418–1427. [CrossRef] [PubMed]
97. Xu, Y. Envirotyping for deciphering environmental impacts on crop plants. *TAG. Theor. Appl. Genet.* **2016**, *129*, 653–673. [CrossRef]
98. Cortés, A.J.; López-Hernández, F.; Blair, M.W. Genome–environment associations, an innovative tool for studying heritable evolutionary adaptation in orphan crops and wild relatives. *Front. Genet.* **2022**, *13*, 910386. [CrossRef]
99. Xu, Y.; Zhang, X.; Li, H.; Zheng, H.; Zhang, J.; Olsen, M.S.; Varshney, R.K.; Prasanna, B.M.; Qian, Q. Smart breeding driven by big data, artificial intelligence and integrated genomic-enviromic prediction. *Mol. Plant* 2022. [CrossRef]
100. Resende, R.T.; Piepho, H.P.; Rosa, G.J.M.; Silva-Junior, O.B.; E Silva, F.F.; de Resende, M.D.V.; Grattapaglia, D. Enviromics in breeding: Applications and perspectives on envirotypic-assisted selection. *Theor. Appl. Genet.* **2021**, *134*, 95–112. [CrossRef] [PubMed]
101. Shafiekhani, A.; Kadam, S.; Fritschi, F.B.; DeSouza, G.N. Vinobot and vinoculer: Two robotic platforms for high-throughput field phenotyping. *Sensors* **2017**, *17*, 214. [CrossRef] [PubMed]
102. Cortés, A.J.; López-Hernández, F. Harnessing crop wild diversity for climate change adaptation. *Genes* **2021**, *12*, 783. [CrossRef] [PubMed]

International Journal of
Molecular Sciences

Article

Whole Transcriptome Sequencing Unveils the Genomic Determinants of Putative Somaclonal Variation in Mint (*Mentha* L.)

Felipe López-Hernández * and Andrés J. Cortés *,†

Corporación Colombiana de Investigación Agropecuaria (AGROSAVIA)-CI La Selva, Rionegro 054048, Colombia
* Correspondence: llopez@agrosavia.co (F.L.-H.); acortes@agrosavia.co (A.J.C.)
† Secondary Address: Facultad de Ciencias Agrarias-Departamento de Ciencias Forestales, Universidad Nacional de Colombia-Sede Medellín, Medellín 050034, Colombia.

Abstract: Mint (*Mentha* L., Lamiaceae) is a strongly scented herb of the family Lamiaceae that is grown mostly by clonal propagation, making it a valuable species for the study of somaclonal variation and its phenotypic consequences. The recent introduction of a few species of mint in South America, followed by a presumably rampant propagation, make this region particularly ideal for studying the extent of somaclonal genetic diversity. Hence, the objective of this work was to offer a preliminary characterization of somaclonal genetically coding diversity of the mint in the northern Andes in order to address the question of whether somaclonal variants may have emerged despite relatively recent introductions in a region where mint is not native. A total of 29 clonally propagated specimens, collected in mint export farms in the province of Antioquia, a major region for mint production in the northwest Andes of Colombia, were genotyped using RNA sequencing (RNA-Seq). SNP calling was carried out from the leaves' transcriptome profiles of each plant by combining the GATK4 and TRINITY protocols, obtaining a total of 2033 loci across 912 transcripts with a minimum read depth of 20X and 4% of missing data. Unsupervised machine learning algorithms considered the *K*-means, AGNES and UPGMA approaches, all of which suggested three genetic clusters for *M. spicata* and a unique cluster for *M.* × *piperita*. The results indicate that at least two different origins of *M. spicata* reached the eastern region of the Antioquia province, clonally propagated in the locality ever since for local consumption and export. One of these ancestries had more population structure, possibly due to environmental or anthropological pressures that intervened in the fragmentation of this genetic group or to a higher somaclonal mutation rate. This work offers a first step into the study of the accumulation and transmission of presumably quasi-neutral somatic mutations at coding regions in an herbaceous clonally propagated scented species such as mint, likely favored by an expected population expansion after its Andean introduction. These ad hoc hypotheses warrant further study as part of future research.

Keywords: *M.* × *piperita* L.; *M. spicata*; RNA-Seq; comparative transcriptomics; crop biodiversity

Citation: López-Hernández, F.; Cortés, A.J. Whole Transcriptome Sequencing Unveils the Genomic Determinants of Putative Somaclonal Variation in Mint (*Mentha* L.). *Int. J. Mol. Sci.* **2022**, *23*, 5291. https://doi.org/10.3390/ijms23105291

Academic Editors: Jesús Osada and Satoshi Kishigami

Received: 3 April 2022
Accepted: 7 May 2022
Published: 10 May 2022

1. Introduction

The rate, extent and architecture of de novo mutations and somaclonal variations have been long-standing intriguing questions in molecular evolution [1]. Genomic features and chromosomal constraints are well-known determinants of hidden levels of genetic variation [2]. Unusual segregation rearrangements and high ploidy levels are also regarded as enhancers of the nucleotide diversity at evolutionary timescales [3]. Yet, somaclonal variations at shallower timeframes have remained elusive [4], partly due to their presumed rarity and difficulty targeting them in non-model species. Besides, they are often assumed neutral, a corollary of the neutral theory of molecular evolution [5], so their potential adaptive or co-opted, s.s. [6,7], values have seldom been tested.

Crops maintained by clonal propagation constitute unique experimental playgrounds to target somaclonal variants and explore their phenotypic consequences. In particular, crops grown for their non-reproductive organs have severe disruptions to their flowering and fruiting systems, making clonality a compulsory propagation strategy. Mint (*Mentha* L.), a strongly scented herb of the Lamiaceae family, is an example of this. Several phylogenetically intricate species [8] have been used for centuries for medicinal and savory purposes, including 30 species and hybrids that are distributed or introduced throughout the globe [9]. Beyond their use as herbs and spices and for pharmaceutical needs, distilling of essential oils (i.e., menthol) from commercial peppermint is now a major global economic commodity. Yet, its bio-economical potential remains enormous due to other secondary metabolites and novel uses from biomass waste.

Despite mint's cosmopolitan native range, the only subcontinent where they had to be introduced as part of the Colombian interchange was South America. Recent introductions of a few mint species [9], followed by their clonal propagation, make this region ideal for studying the extent of somaclonal genetic diversity. However, complex genomes, polyploidization and hybridization within the genus [8] make mandatory the use of transcriptomics to target somaclonal variation in allelic coding variants at higher resolution. Comparative works [10] suggest that the most efficient pipeline to reconstruct the matrix of allelic variants from RNA sequencing is the integration of the TRINITY algorithm as a de novo assembler [11] and the GATK4 protocol for SNP calling [12,13]. Hence, running the integrated TRINITY + GATK4 protocol has allowed recovering SNPs with 100% accuracy in cases such as with peaches and tangerines [10]. Yet, this protocol has not yet been tested for highly scented clonal herbs such as mint as a strategy to explore somaclonal variants in expressed regions. Given this research gap, this study's goal was to carry out a preliminary assessment of coding somaclonal diversity in mints from the northwest Andes using RNA sequencing (RNA-Seq) and the combined TRINITY + GATK4 protocols. Specifically, we wondered (1) whether somaclonal variation may have arisen despite relatively recent mint introductions to a region where the species is not native and (2) whether these variants may have any phenotypic effect. Exploring these questions offers a first step to unveiling the role of molecular evolution and molecular genetic resources in the plant improvement of clonally propagated crop species, just as envisioned by the International Journal of Molecular Science's special issue on Molecular Genetics and Plant Breeding. Diversity trends defined here will serve as null hypotheses to test underlying mechanistic molecular and biochemical processes.

2. Results

Variation across 29 specimens of mint sourced a matrix of allelic variants using the supertranscript and the hybrid protocol GATK4 + TRINITY for RNA-Seq. We used this SNP information to carry out a principal component analysis (PCA) as input for partitional and hierarchical clustering approaches. Additionally, we carried out a distance-based analysis using a UPGMA dendrogram. All these algorithms suggested a clear difference between *M. spicata* and *M. × piperita*, which validates the use of *M. × piperita* as a control group for the exploration of the somaclonal genetic diversity within *M. spicata*. The optimization in the number of clusters and the UPGMA dendrogram suggested the presence of three possible groups of *M. spicata* and one of *M. × piperita*.

2.1. The Commercial Protocol Recovered Greater Quantity and Quality of RNA

Among six RNA extraction protocols, the Qiagen® RNeasy Plant Mini Kit commercial kit allowed us to obtain the best concentration quality and A260/A280 and A260/230 absorbance ratios. In this sense, this commercial protocol was scaled to the 29 specimens (Table S2), conveying optimal RNA concentration and quality as follows: mean Nanodrop® and Qubit® concentrations of 922.021 ug/uL (IC: 236.52) and 94.79 ug/uL (CI: 8.78), and mean A260/280 and A260/230 ratios of 2.11 (CI: 0.016) and 2.02 (CI: 0.23).

2.2. Sequencing and Data Cleaning

The 29 genetic libraries built for RNA-Seq had a mean Qubit® concentration of 18.77 ug/uL (CI: 5.5431), a mean fragment size of 282.66 bp (CI: 3.099), and a quantification TapeStation® mean of 99.96 nM (CI: 29.165). For all specimens, electropherograms suggested fragment distributions with defined peaks and the absence of contaminants (Table S3). After the trimming for each specimen by the Trimmomatic algorithm, all samples had quality scores greater than 30 using 1.9 Illumina encoding (Figure S1A) without the presence of adapters (Figure S1D), high duplications percentage (Figure S1B) and GC deviation percentage to be expected for RNA-Seq data (Figure S1C).

2.3. Supertranscript of Mint as Reference for SNP Calling

A supertranscript was assembled as guiding reference for alignment of reads and SNP calling across all 29 specimens of mint (Table 1).

Table 1. Mapping statistics against the supertranscript. Mapping of each specimen's transcript profile used as reference the supertranscript and the GATK4 protocol. The total number of sequences per sample, mapped sequences, duplicates and purified sequences in the refinement of the protocol are shown in this table. For details on the sampling information of each specimen please refer to Materials and Methods.

Sample ID	Total Transcripts	Supplementary Transcripts	Duplicate Transcripts	Mapped Transcripts	Unmapped Duplicates	Mapped Ratio
MLV20-15244	13,595,799	438,550	10,261,916	13,431,369	3,169,453	0.988
MLV20-15245	13,893,704	422,388	9,887,386	13,560,195	3,672,809	0.976
MLV20-15246	19,286,261	628,197	14,805,807	18,976,542	4,170,735	0.984
MLV20-15247	13,264,697	388,203	9,755,435	12,984,582	3,229,147	0.979
MLV20-15248	14,873,365	403,064	10,434,595	14,174,092	3,739,497	0.953
MLV20-15249	11,814,041	375,642	8,206,501	11,568,380	3,361,879	0.979
MLV20-15250	12,501,455	260,480	9,220,160	11,900,514	2,680,354	0.952
MLV20-15251	12,138,901	348,593	8,715,938	11,859,700	3,143,762	0.977
MLV20-15252	15,738,616	474,121	11,982,280	15,212,909	3,230,629	0.967
MLV20-15253	13,479,991	429,102	9,136,387	13,252,005	4,115,618	0.983
MLV20-15254	14,800,126	468,816	10,804,162	14,579,876	3,775,714	0.985
MLV20-15255	15,818,146	444,333	11,731,011	15,593,508	3,862,497	0.986
MLV20-15256	15,727,298	509,431	11,699,111	15,532,158	3,833,047	0.988
MLV20-15257	21,828,779	704,838	17,154,954	21,564,209	4,409,255	0.988
MLV20-15258	14,676,563	452,082	10,462,672	14,444,502	3,981,830	0.984
MLV20-15259	14,468,839	474,835	10,691,456	14,284,865	3,593,409	0.987
MLV20-15260	15,790,191	453,092	9,426,258	12,650,256	3,223,998	0.801
MLV20-15261	9,381,824	276,062	5,290,265	8,045,530	2,755,265	0.858
MLV20-15262	15,512,116	493,130	10,567,253	15,326,192	4,758,939	0.988
MLV20-15263	20,188,093	592,807	14,099,354	19,789,887	5,690,533	0.98
MLV20-15264	12,867,595	441,468	8,952,424	12,227,682	3,275,258	0.95
MLV20-15265	7,777,046	231,316	4,725,157	7,386,866	2,661,709	0.95
MLV20-15266	20,521,417	686,319	15,064,734	20,332,633	5,267,899	0.991
MLV20-15267	20,002,666	669,043	14,947,013	19,630,558	4,683,545	0.981
MLV20-15268	24,788,824	718,176	19,701,023	24,412,087	4,711,064	0.985
MLV20-15269	19,153,299	556,314	14,032,872	18,839,624	4,806,752	0.984
MLV20-15270	11,730,613	372,805	8,680,285	11,611,304	2,931,019	0.99
MLV20-15271	16,841,331	556,417	11,787,644	16,511,475	4,723,831	0.98
MLV20-15272	19,610,849	640,811	14,233,830	19,274,477	5,040,647	0.983

De novo transcriptome assembly used TRINITY software from all trimmed *fastq*. It was composed of 509,754 transcripts (Figure S2A) with an average length of 557.9 bp, a minimum length of 178 bp, a maximum length of 12,186 bp and a GC percentage of 43.2%. The transcriptome presented splice isoforms that could increase the false-positive rate in calling allelic variants. Thus, using the transcriptome and trimmed *fastq*, we built a supertranscript collapsing regions of unique and common sequences between splice isoforms in a unique linear sequence (Davidson et al., 2017). This led to a supertranscript of 352,512 transcripts, with an average length of 472.2 pb, a minimum length of 201 bp, a maximum length of 15,765 bp and a GC percentage of 43.1% (Figure S2B).

2.4. A Total of 2033 SNP Markers Were Recovered in 912 Transcripts with Depth of 20X

We made a script to automate the process for SNP calling based on the GATK4 protocol (raw data, input files and bioinformatic scripts are available in https://github.com/FelipeLopez2019/RNAseq-SNP-Calling-GATK4-Mint). A mean of 15,588,705 reads (IC: 1,368,899) were obtained in all 29 specimens, of which 15,136,482 (IC: 1,403,825) were mapped to the supertranscript, with a mapping percentage of 96.81%. We obtained a high number of duplicated reads in the mapping, with a mean of 11,257,168 reads (IC: 1,166,846). In order to avoid confounding paralogous transcripts with allelic variants, duplicated reads were removed in the intermediate steps of the GATK4 protocol, obtaining a mean of 3,879,314 reads (IC: 293,552) without duplicates for all 29 specimens (Table 1). After that, we extracted a gVCF for each specimen using the function *HaplotypeCaller*, and we collapsed these gVCFs to a global VCF. This resulted in a matrix of allelic variants with 2033 loci in 912 transcripts for all specimens, with a minimum read depth of 20X and a maximum missing data percentage of 4%.

2.5. *M. spicata* Clones Were Distributed in Three Genetic Groups in the Northwest Andes

With the aim of reconstructing mint's genetic variability in the northwest Andes, we carried out clustering analyses following two families of algorithms, hierarchical and partitional methods, both of which used the principal components from the PCA analysis. Moreover, we performed visualization by means of unsupervised dendrogram clustering reconstructed using the UPGMA analysis. All methods suggested the presence of three genetic groups of *M. spicata* and a unique group for *M.* × *piperita*.

Specifically, from the 2033 loci distributed in 912 transcripts, we carried out a dimensional reduction to principal components. The first component explained 58.56% of the overall variance, the second component explained 22.27% of the variance and the third component accounted for 11.28%. We further performed clustering validation with the *NbClust* y *optCluster* algorithms. Both suggested a total of four clusters: three from *M. spicata* (spicata A, spicata B and spicata C) and one to *M.* × *piperita* (Figure 1A). All four clusters were recovered with the first two components, which together accounted for 80.83% of the overall variation, overpassing the 80% threshold. On the other hand, *optCluster* suggested that the best hierarchical algorithm to reconstruct the clustering was AGNES, and the best partitional algorithm was *K*-means. Both approaches tagged specimens identically. In the same way, the UPGMA approach suggested three groups for *M. spicata* (spicata A, spicata B and spicata C) and one for *M.* × *piperita* (Figure 1B).

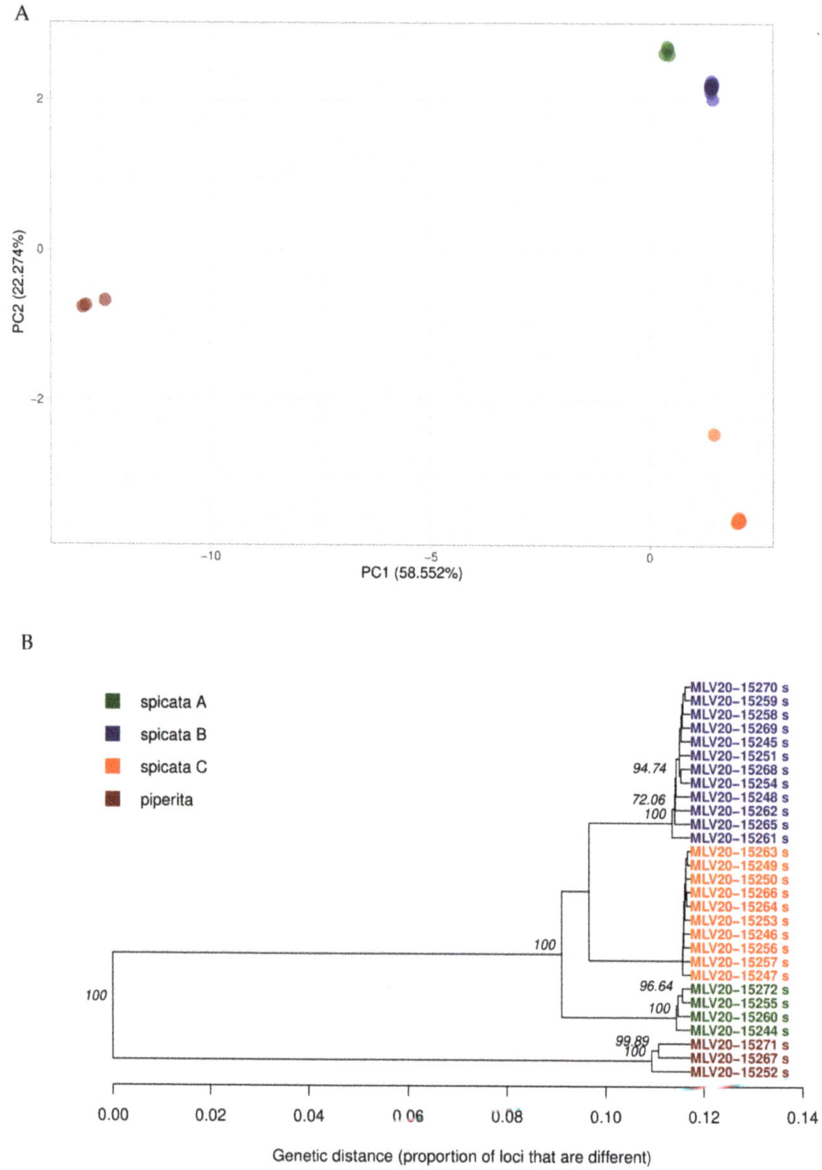

Figure 1. Overall diversity patterns in mint from the northwest Andes. (**A**) PCA analysis from 2033 loci distributed in 912 transcripts. The first component explained 58.56% of the variance, and the second component explained 22.27% of the variance (both totaling 80.83% of explained variance). Clustering validation was performed using the algorithms *NbClust* and *optCluster*, which suggested a total of four clusters: three of *M. spicata* (spicata A, spicata B and spicata C) and one of *M.* × *piperita*. All clusters were recovered by the first two components. (**B**) Dendrogram carried out by UPGMA analysis from 2033 variants distributed in 912 transcripts using Nei's distance and bootstrap as resampling method with 10,000 replicates. These results also suggested a total of four clusters: three from *M. spicata* (spicata A, spicata B and spicata C) and one for *M.* × *piperita*. Clusters under both approaches in (**A**,**B**) are fully concordant.

2.6. Gene Functionality of Polymorphic Transcripts

All 2033 variants distributed across the 912 transcripts spanned both *M. spicata* and *M.* × *piperita* (Figure S3). From all 912 transcripts, 96.10% retrieved hits in *BlastX* using the software *Blast2GO*. The 84.57% of these Blast *hits* were mapped, of which it was possible to annotate 94.27% with GO and enzyme codes (Table S4). Using the GO codes, we constructed three pathways to analyze the frequency of biological, cellular (Figure S4) and molecular (Figure S5) processes, as detailed below.

We first explored the top GO terms in each pathway using the high *Nodesores* and number of associated transcripts. In the biological pathway (Figure 2), we analyzed the GO terms in the *Nodescore* range from 35,822 to 1471.06 and the number of associated transcripts ranging from 365 to 428 sequences. The main GO terms were related to proteins of carbohydrate and energy metabolism (GO:0044237 and GO:0044238) and the synthesis of cysteine and methionine metabolism (GO:0008152 and GO:0009987) (Table S4). In the molecular pathway, we analyzed the GO terms in the *Nodescore* range of 626.86 to 704.47 and the number of associated transcripts ranging from 349 to 380 sequences. The main GO terms were related to proteins with catalytic activity, biosynthetic process and regulation of transcription (GO:0003824) (Table S4). On the other hand, for the cellular pathway, we analyzed the GO terms in the *Nodescore* range of 762 to 1224.97 and the number of associated transcripts ranging from 305 to 409 sequences. GO terms related to proteins linked in intracellular anatomical structure (GO:0005622) (Table S4).

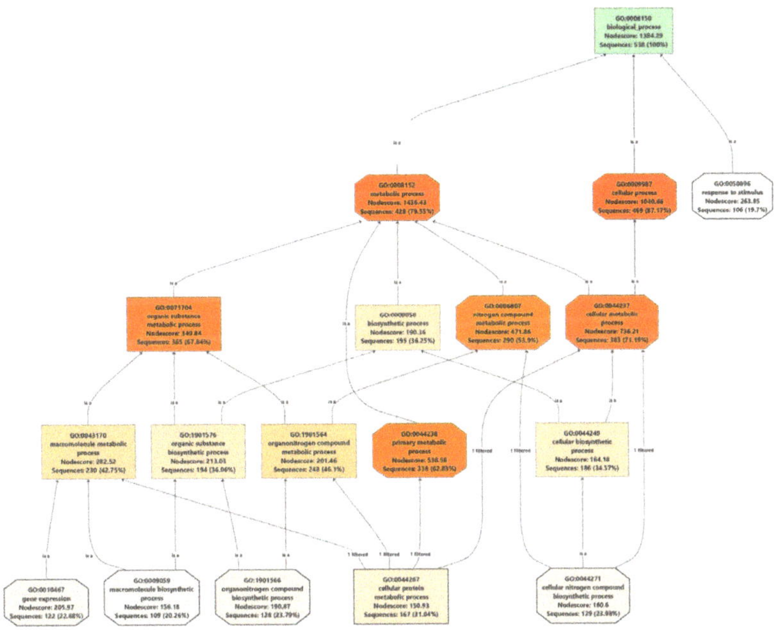

Figure 2. GO pathway to analyze polymorphic biological processes via GO codes from *Blast2GO*. GO terms ranged from 35,822 to 1471.06 and the number of associated transcripts ranging from 365 to 428 sequences. Main GO terms related to carbohydrate and energy metabolism (GO:0044237 and GO:0044238) and cysteine and methionine synthesis (GO:0008152 and GO:0009987). We also explored the enzyme codes of the 52 KEEG pathways associated with all polymorphic transcripts (Table 2, extended in Table S5). Of the KEEG pathways, 25% were related to carbohydrate metabolism, 19.23% of the KEEG pathways were related to amino acid metabolism, 11.54% of the KEEG pathways were related to energy metabolism and the other pathways were associated with less than 10% of the target queries. Main pathways of carbohydrate/amino acid synthesis were linked to glycolysis/gluconeogenesis (Figure S6), and cysteine/methionine metabolism (Figure S7) [14].

Table 2. Related KEEG pathways (52) across transcripts using enzyme codes in *Blas2GO* outputs.

KEGG Pathway	# of Transcripts	# of Enzymes	KEEG Label
Glycolysis/Gluconeogenesis	15	16	Carbohydrate metabolism
Cysteine and methionine metabolism	13	13	Amino acid metabolism
Carbon fixation in photosynthetic organisms	11	13	Energy metabolism
Methane metabolism	10	10	Energy metabolism
Pyruvate metabolism	9	12	Carbohydrate metabolism
Glycine, serine and threonine metabolism	9	9	Amino acid metabolism
Glyoxylate and dicarboxylate metabolism	9	9	Carbohydrate metabolism
Starch and sucrose metabolism	8	8	Carbohydrate metabolism
Galactose metabolism	7	8	Carbohydrate metabolism
Nitrogen metabolism	7	8	Energy metabolism
Citrate cycle (TCA cycle)	7	7	Carbohydrate metabolism
Oxidative phosphorylation	7	7	Energy metabolism
Tyrosine metabolism	7	7	Amino acid metabolism
Amino sugar and nucleotide sugar metabolism	6	7	Carbohydrate metabolism
Phenylalanine, tyrosine and tryptophan biosynthesis	6	7	Amino acid metabolism
Terpenoid backbone biosynthesis	6	7	Metabolism of terpenoids and polyketides
Carbon fixation pathways in prokaryotes	6	6	Energy metabolism
Glycerolipid metabolism	6	6	Lipid metabolism
Pentose phosphate pathway	6	6	Carbohydrate metabolism
Alanine, aspartate and glutamate metabolism	5	7	Amino acid metabolism
Tryptophan metabolism	5	5	Amino acid metabolism
Ubiquinone and other terpenoid-quinone synthesis	5	5	Metabolism of cofactors and vitamins
Ascorbate and aldarate metabolism	5	4	Carbohydrate metabolism
Glutathione metabolism	5	4	Metabolism of other amino acids
Phenylalanine metabolism	4	5	Amino acid metabolism
Phenylpropanoid biosynthesis	4	5	Biosynthesis of other secondary metabolites
alpha-Linolenic acid metabolism	4	4	Lipid metabolism
Fructose and mannose metabolism	4	4	Carbohydrate metabolism
O-Antigen nucleotide sugar biosynthesis	4	4	Glycan biosynthesis and metabolism
Porphyrin metabolism	4	4	Metabolism of cofactors and vitamins
Cyanoamino acid metabolism	3	4	Metabolism of other amino acids
Inositol phosphate metabolism	3	4	Carbohydrate metabolism
Pentose and glucuronate interconversions	3	4	Carbohydrate metabolism
Glycerophospholipid metabolism	3	3	Lipid metabolism
Selenocompound metabolism	3	3	Metabolism of other amino acids
Arginine biosynthesis	2	3	Amino acid metabolism
Carotenoid biosynthesis	2	2	Metabolism of terpenoids and polyketides
Drug metabolism—cytochrome P450	2	2	Xenobiotics biodegradation and metabolism
Drug metabolism—other enzymes	2	2	Xenobiotics biodegradation and metabolism
Fatty acid degradation	2	2	Lipid metabolism
Isoquinoline alkaloid biosynthesis	2	2	Biosynthesis of other secondary metabolites
Lysine degradation	2	2	Amino acid metabolism
Metabolism of xenobiotics by cytochrome P450	2	2	Xenobiotics biodegradation and metabolism
Nicotinate and nicotinamide metabolism	2	2	Metabolism of cofactors and vitamins
One carbon pool by folate	2	2	Metabolism of cofactors and vitamins
Propanoate metabolism	2	2	Carbohydrate metabolism
Steroid biosynthesis	2	2	Metabolism of terpenoids and polyketides

Table 2. *Cont.*

KEGG Pathway	# of Transcripts	# of Enzymes	KEEG Label
Styrene degradation	2	2	Xenobiotics biodegradation and metabolism
Sulfur metabolism	2	2	Energy metabolism
Thiamine metabolism	2	2	Metabolism of cofactors and vitamins
Tropane, piperidine and pyridine alkaloid synthesis	2	2	Biosynthesis of other secondary metabolites
Valine, leucine and isoleucine degradation	2	2	Amino acid metabolism

3. Discussion

Recent introductions of a few mint species in South America as part of the Colombian interchange, followed by presumably rampant clonal propagation, make this region ideal for studying the extent of putative somaclonal genetic diversity. In addition, the use of transcriptomics' profiles makes it possible to target variation in allelic coding variants at higher resolution while reducing the complexity of correctly aligning polymorphic non-coding regions and highly repetitive regions. Our results suggest that despite mint having relatively recent introductions to the northern Andes, where it is not native, coding variants were detected across all samples spanning a total of three groups for *M. spicata* and one group for *M.* × *piperita*, as well as putative somaclonal variation within clusters. The candidate somaclonal variants may be attributed to primary metabolic pathways, suggesting a likely predominant role of silent mutations, with sporadic co-opted variation as part of mint's colonization of the northern Andes (e.g., to shorter day length and unforeseen pressures by local populations of herbivores).

3.1. Origins and Extent of Genetic Clusters and Putative Somaclonal Variation in Mint

Gene ontology and KEGG analyses suggest a predominance of silent mutations, likely hidden from purifying selection [15]. However, novel mutational variants in defense transcripts may also speak for some co-opted (s.s. [6,7]) adaptive variants as part of human-mediated mint's colonization of the northern Andes, exposing introduced genotypes to shorter day lengths and unexpected antagonist biotic interactions. In order to discern the relative role of neutral and mutational load better, as part of a revisited neutral theory of molecular evolution [5] within a somaclonal framework, future studies should envision building a more compelling resource of somaclonal variance, as well as their timing of appearance. For instance, complete re-sequencing of the genome, with a careful phase reconstruction and purge of highly repetitive regions, could clarify both the fine-tuning of the three mint groups and the concrete metabolic pathway correlates beneath the candidate somaclonal variation.

3.2. Utility of RNA-Seq to Study Presumably Adaptive Variation

Capturing somaclonal variation from Whole Genome Re-sequencing (WGR) would be the most sensitive method. However, within the genus Mentha, there is only one genome annotated for horsemint (*M. longifolia*) with a genome size of 468,947 Mb, and *M. spicata* is an orphan species from a genomic perspective that does not yet have an annotated reference genome at the chromosome level. Producing this resource de novo would generate additional efforts, raising laboratory and computing costs. Meanwhile, genotyping based on reduced representation libraries, such as Genotyping-by-Sequencing (GBS), would not have the necessary resolution to capture somaclonal variability because it may be enriched in repetitive regions or conserved genes with little change to exhibit somaclonal segregation. This is why a more reliable alternative for a genus with complex genomes, repeated polyploidization and rampant hybridization [8] is transcriptomics, which nowadays is supported by modern RNA-Seq methodologies widely used to carry out expression analysis. In this sense, the present study was able to explore a de novo

RNA-Seq strategy using the TRINITY assembler to obtain a supertranscriptome [13], which served as a reference in the calling of variants.

Still, there are few studies testing different methods for the optimization of allelic variant calling from RNA-Seq. Therefore, we complemented the pipeline with recent recommendations for fruit tree species such as mandarin, in which authors reported a viable integration between the TRINITY and the GATK4 algorithms ultimately to increase the number of allelic variants captured while efficiently controlling for false positives. As a control species within the integrated pipeline, a peppermint accession was sampled under the same environmental conditions, given that it is also largely used to export from the northwestern Colombian province of Antioquia, east high plateau.

We were further able to confirm the requirement to bring the genotypes to a homogeneous greenhouse climate before sampling leaves for RNA-Seq sequencing. Otherwise, somaclonal variation may have been unbiased across transcripts and samples, precluding a systematic understanding of the different metabolic pathways prone to putative somaclonal variants. Preliminary greenhouse treatments enabled gene ontology and KEGG analyses to suggest that coding variation is more likely to appear in transcripts associated with primary metabolism (glycolysis/gluconeogenesis), synthesis of essential biomolecules and defense compounds (cysteine and methionine pathway) [14].

3.3. Caveats and Perspectives

Given the pilot nature of the present research to assess coding variation in a clonally propagated species such as mint, explicitly recognizing the study's limitations will guide future efforts to examine more rigorously some of the proposed ad hoc hypotheses. Among these, useful reference questions worth testing in more detail as null hypotheses in future research are: Did genotypes actually reach homogeneous acclimatization after the greenhouse treatment? What is the factual role of the presumed population expansion of mint linages after its Andean introduction? What is the cause of greater population structure in one of the *M. spicata* founding clades? Are putative somatic mutations genuinely quasi-neutral? What are their key phenotypic functional effects?

We must also acknowledge the fact that the assumption of the narrow founding mint's gene pool that reached the northern Andes, at most leading to three *M. spicata* clusters with two likely independent origins, deemed further consideration. Therefore, we encourage future evolutionary/phylogenetic studies to expand sampling beyond our study region in order to address this assumption more strongly.

From a more technical point of view, we had wished to be able to capture phased haplotypes. This is because haplotype divergence is fundamental to corroborate divergence among clonal organisms at the genomic level. After all, a null prediction is that haplotypes would accumulate mutations independently. Therefore, we invite future researchers to explicitly consider BAC library-cloning strategies as a pre-sequencing step that would enable accurate phase estimation and mutation tracing.

Last but not least, we would have preferred to carry out a full enrichment analysis as typically implemented in model species. Unfortunately, limitedly annotated resources for *M. spicata* precluded an accurate enrichment computation. Therefore, instead of reporting biased enrichment statistics, we opted for a detailed biological discussion of the major GO findings, which surprisingly did not include prominent flavor and flavonoid pathways. Still, we are looking forward to seeing better annotations for mint's genomic resources in the years to come. This may assist with novel uses of mint's clonal diversity.

4. Materials and Methods

4.1. Plant Material

A total of 38 specimens (Table S1) were sampled under protected conditions and free exposure from 14 mint export farms between 2019 and 2020 in the northwest Andes of Colombia, a major region for mint production. Only farms with clonally propagated mint were considered as a way to narrow down founder lineages. Mint production in these farms is mainly intended for exportation to the US fresh market.

According to previously standardized botanical traits [16], 36 of the 38 collected specimens were *M. spicata*, while the remaining two were *M.* × *piperita*. The identification of these specimens was carried out by farmers and specialized staff of the greenhouse at Universidad Católica de Oriente (UCO). Plant tissues (i.e., stems and roots) were sampled in situ and sent to UCO's greenhouse for clonal propagation and 9 months of acclimation before sampling leaves for RNA sequencing. Two additional specimens from UCO's collection were also included as species controls for *M. spicata* and *M.* × *piperita*.

4.2. RNA Extraction and Whole Transcriptome Sequencing

With the intention to recover the highest quantity and quality of mint's RNA, we carried out a comparative experiment of plant extraction protocols. A total of six different RNA extraction protocols and library preparation methods were tested at the Molecular Genetic Laboratory of AGROSAVIA in Tibaitatá's Research Station (Colombia) using two random samples of *M.* × *piperita* and *M. spicata*, in addition to UCO's reference controls for each species. The six protocols were: (1) In-house AGROSAVIA, (2) In-house AGROSAVIA modified in volumes and time, (3) In-house AGROSAVIA with variations for species with high content of polysaccharides and polyphenols, (4) In-house AGROSAVIA with Trizol, (5) Qiagen® RNeasy Plant Mini Kit (Hilden, Germany) and (6) Qiagen® RNeasy Plant Mini Commercial Kit with Trizol (Hilden, Germany). Of the six tested protocols, the Qiagen® RNeasy Plant Mini Kit commercial kit allowed to obtain a better quality and absorbance ratios of A260/A280 and A260/230.

Quantification of extracted cDNA was done by a spectrophotometry method using the Nanodrop® 2000 equipment (Thermo Fisher Scientific, United States) and by a fluorometric method using the Qubit® dsDNA HS fluorometer (Life Technologies, Sweden). Library construction was performed using the SureSelect Strand-Specific RNA® kit for multiplexed sequencing by Illumina®. Then, the libraries were quantified by the fluorometric method using the Qubit® dsDNA HS fluorometer. The concentration and fragment sizes of the cDNA libraries were evaluated using the TapeStation 4200® kit (Agilent Technologies, United States) and the High Sensitivity D1000 kit. DNA sequences were obtained using single-end Illumina 2500 Hiseq (Macrogen, South Korea).

4.3. Bioinformatics Processing

In order to clean the sequence data, we carried out a script to automate the software Trimmomatic (Bolger et al. 2014) with the main parameters: ILLUMINACLIP:TruSeq3-SE:2:30:10, SLIDINGWINDOW:4:20 and MINLEN:After that. We performed an analysis of quality from the trimmed *fastq* files in the software FastQC [17] using Illumina 1.9 encoding.

Comparative works [10] have suggested that the most efficient pipeline to reconstruct the matrix of allelic variants from RNA sequencing is the integration of the TRINITY algorithm [11] as a de novo assembler with a GATK [12] protocol for SNP calling. This way, we obtained a de novo transcriptome from the samples of mint by means of the software TRINITY [11] using the platform *Galaxy* v.2.9.1 [18]. A supertranscript was then built as a reference to map and identify allelic polymorphisms in the context of de novo assembly (i.e., without reference genome). The supertranscript was constructed by collapsing regions of unique and common sequences between splice isoforms into a single linear sequence [13]. We finally obtained a supertranscript using TRINITY to be used by GATK4 as a reference in the SNP calling protocol.

From all collected samples, total RNA was successfully extracted and bioinformatically processed for 29 specimens (Table 3) from leaves stored at −80°C using the standardized Qiagen® RNeasy Plant Mini Kit. Our ultimate aim was to identify allelic polymorphisms from these 29 samples of mint, targeting coding regions in an otherwise complex genome structure with presumably abundant polyploidization and hybridization signatures [8].

Table 3. *Mentha* spp. used for RNA-Seq and sampling localities in Colombia, Antioquia province.

Sequencing ID	ID UBV UCO	Species	Conditions	Municipality	Latitude	Longitude	Elevation
MLV20-15244_s	1	*M. spicata*	Open field	La Ceja	6.067500	−75.415056	2208
MLV20-15245_s	2	*M. spicata*	Open field	La Ceja	6.067500	−75.415056	2208
MLV20-15246_s	3	*M. spicata*	Protected cultivation	El Retiro	6.066111	−75.449528	2218
MLV20-15247_s	4	*M. spicata*	Open field	El Retiro	6.066111	−75.449528	2218
MLV20-15248_s	5	*M. spicata*	Protected cultivation	Medellin	6.198222	−75.511889	2554
MLV20-15249_s	6	*M. spicata*	Protected cultivation	Medellin	6.198222	−75.511889	2554
MLV20-15250_s	10	*M. spicata*	Open field	Marinilla	6.167222	−75.325861	2131
MLV20-15251_s	11	*M. spicata*	Open field	Marinilla	6.167222	−75.325861	2131
MLV20-15252_s	12	*M. piperita*	Protected cultivation	Rionegro	6.192972	−75.360972	2128
MLV20-15253_s	14	*M. spicata*	Open field	Rionegro	6.192972	−75.360972	2128
MLV20-15254_s	15	*M. spicata*	Open field	Rionegro	6.198111	−75.350639	2118
MLV20-15255_s	16	*M. spicata*	Open field	Rionegro	6.198111	−75.350639	2118
MLV20-15256_s	17	*M. spicata*	Protected cultivation	El Retiro	6.066111	−75.449528	2218
MLV20-15257_s	18	*M. spicata*	Protected cultivation	Medellin	6.198222	−75.511889	2554
MLV20-15258_s	19	*M. spicata*	Protected cultivation	Medellin	6.198222	−75.511889	2554
MLV20-15259_s	20	*M. spicata*	Protected cultivation	Medellin	6.198222	−75.511889	2554
MLV20-15260_s	21	*M. spicata*	Open field	Rionegro	6.067500	−75.415056	2208
MLV20-15261_s	22	*M. spicata*	Open field	Rionegro	6.067500	−75.415056	2208
MLV20-15262_s	24	*M. spicata*	Protected cultivation	La Unión	5.910333	−75.412278	2277
MLV20-15263_s	25	*M. spicata*	Protected cultivation	La Unión	5.910333	−75.412278	2277
MLV20-15264_s	26	*M. spicata*	Protected cultivation	San Vicente Ferrer	6.325194	−75.335694	2251
MLV20-15265_s	27	*M. spicata*	Protected cultivation	San Vicente Ferrer	6.325194	−75.335694	2251
MLV20-15266_s	28	*M. spicata*	Protected cultivation	San Vicente Ferrer	6.325194	−75.335694	2251
MLV20-15267_s	29	*M. piperita*	Protected cultivation	Carmen de Viboral	6.082287	−75.316147	2260
MLV20-15268_s	33	*M. spicata*	Protected cultivation	Carmen de Viboral	6.082287	−75.316147	2260
MLV20-15269_s	44	*M. spicata*	Protected cultivation	Guarne	6.213960	−75.417330	2200
MLV20-15270_s	45	*M. spicata*	Protected cultivation	Medellin	6.198222	−75.511889	2554
MLV20-15271_s	Reference piperita	*M. piperita*	Protected cultivation	-	-	-	2118
MLV20-15272_s	Reference spicata	*M. spicata*	Protected cultivation	-	-	-	2118

Therefore, we made a script to automate the process of SNP calling by the protocol GATK4 using the *HaplotypeCaller* function and the algorithm BWA-MEM [19,20]. The mapping statistics were computed using the *flagstat* function from the software *Samtools* v.1.9 [21] using the platform *Galaxy* v.2.9.1 [18]. Rather than relying on suboptimal phase imputation strategies poorly calibrated for mint RNA resources, we opted to carry out the analysis on more conservative allelic variants. The resultant SNP matrix was then filtered with a minimum depth of 20X and a stringent threshold of missing data of 4% using the software TASSEL v.5.2.78 [22]. All scripts, raw data and input files are available in https://github.com/FelipeLopez2019/RNAseq-SNP-Calling-GATK4-Mint.

4.4. Genetic Clustering and Putative Somaclonal Variation in Mint

From the identification of allelic polymorphisms consolidated in the SNP matrix, we carried out a prospective analysis of linear dimensional reduction by molecular principal components (PCA) using the *glPca* function of the R *adegenet* v.2.1.4 package [23]. After that, in order to assess the genetic substructure in the *M. spicata* and *M.* × *piperita* assemblages, we carried out a clustering analysis following a Data Science perspective [24]. Specifically, we ran a script of clustering validation using the function *OptCluster* from the R-package *OptCluster* [25], an optimized version of the function *clValid* [26]. The *OptCluster* function considered these methods: the partitional algorithms *K*-means [27,28] and *k*-medoids [29], and the hierarchical algorithms AGNES (AGglomerative NESting) and DIANA (Divisive analysis, [29]) with Genetic Algorithms and Cross-Entropy validation. We also ran another algorithm for clustering validation called *NbClust* [30]. Finally, we constructed a UPGMA dendrogram with 10,000 bootstrap replicates using the function *aboot* of the R-package *poppr* [31] and Nei's distance [32].

4.5. Gene Functionality of Polymorphic Transcripts

With the goal of obtaining the ontological analysis of polymorphic coding variation in mint, all polymorphic transcripts were analyzed by means of the software *Blast2GO* [33] using the algorithm *BlastX* with the *nr* database, threshold E-value of 1.0×10^{-6}, word size of 6, and 10 *hits* per transcript. Furthermore, we carried out the GO mapping from the hits of BLASTX to obtain functional labels using the GO Gene Annotation files and the *Uniprot* database. The annotation process was generated to select GO terms from the GO pool obtained in the previous step and assign them to the query sequence. We ran the annotation process using the taxonomy filter of 'flowering plants' [Magnoliopsida], a value of 55 as annotation cut-off, 5 as GO weight and a mean of convergence of 80% and 100% for computational analysis and experimental evidence, respectively. In order to sort the GO outputs in hierarchical pathways, we generated tree diagrams of biological, cellular and molecular processes using 150 as *Nodescore* filter, indicating the high associativity of GO ID with the mint transcripts. Finally, we carried out the KEEG pathway analysis using the *GO-EnzymeCode* function in *Blast2GO* (https://www.genome.jp/kegg/pathway.html, accessed on 30 March 2022) with the aim of exploring the metabolic pathways more prone to nucleotide diversity in mint.

5. Conclusions

Coding polymorphism in *M. spicata* from northwest South America likely traced back to two independent origins that were clonally propagated in the region ever since. Putative somaclonal variants were mostly found in primary metabolic pathways, indicating a plausible predominant role of silent mutations. However, co-opted adaptive variants may still be expected as part of mint's introduction and adaptation to the northern Andes, where it had to face shorter day length and unexpected herbivore pressures. These hypotheses on putative somatic mutations deem further study. Methodologically, the pipeline RNA-Seq + TRINITY + GATK4 + *Blast2GO* helps characterize somaclonal variation and associated metabolic pathways in complex genomes.

Supplementary Materials: The following supporting information can be downloaded at: https://www.mdpi.com/article/10.3390/ijms23105291/s1, **Table S1**. A total of 38 specimens were sampled under protected conditions and free exposure from 14 mint export farms during 2019 and 2020 in the northwest Andes (Antioquia province, Colombia). **Table S2**. RNA extraction was carried out across 29 specimens, retrieving an optimal state of RNA conservation and health, with a mean RNA concentration of 922.02 ug/uL (IC: 236.51) using Nanodrop®, a mean RNA concentration of 94.79 ug/uL (CI: 8.78) using Qubit®, mean A260/280 ratio of 2.110 (CI: 0.016), and mean A260/230 ratio of 2.015 (CI: 0.229). **Table S3**. Across all specimens, electropherogram suggested an ideal distribution of fragments with defined peaks, and the absence of contaminants. **Table S4**. *Blas2GO* outputs using all 912 transcripts derived from the supertranscript. Among these, 96.10% obtained hits in *BlastX* using the software *Blast2GO*. The 84.57% of these Blast hits were mapped, and the 94.27% of the mapped results were annotated with GO and Enzyme codes. **Table S5**. KEEG pathways associated to all transcripts from enzymes codes of *Blas2GO* outputs. **Figure S1**. *MultiFastQC* analysis for all specimens, with: (a) a quality score greater than 30 using 1.9 Illumina encoding, (b) a moderate duplications percentage, (c) a deviation of GC percentage to be expected in RNAseq data, and (d) absence of adapters. **Figure S2**. Supertranscript assembly. (a) De novo assembly of transcriptome using the software TRINITY from all trimmed *fastq*, composed of 509,754 transcripts with an average length of 557.9 bp, a minimum of 178 bp, and a maximum of 12,186 bp. (b) The supertranscript of mint presented 352,512 transcripts with an average length 472.2 pb, a minimum of 201 bp, and a maximum of 15,765 bp. **Figure S3**. Multidimensional Venn diagram depicting the overlap among transcript variants. **Figure S4**. GO pathway to analyze polymorphic cellular processes using GO codes from Blast2GO. **Figure S5**. GO pathway to analyze polymorphic molecular processes via GO codes from Blast2GO. **Figure S6**. Main KEGG pathway of carbohydrate metabolism for glycolysis/Gluconeogenesis synthesis using the enzymes codes from *Blast2GO* outputs. **Figure S7**. Main KEGG pathway of amino acid metabolism associated to cysteine and methionine synthesis.

Author Contributions: Conceptualization, F.L.-H. and A.J.C.; methodology, F.L.-H. and A.J.C.; data curation, F.L.-H.; formal analysis, F.L.-H.; visualization, F.L.-H.; writing—original draft preparation, F.L.-H. and A.J.C.; writing—review and editing, F.L.-H. and A.J.C.; supervision, A.J.C. All authors have read and agreed to the published version of the manuscript.

Funding: This research was funded by a grant from Sistema General de Regalías (SGR—Antioquia, Convocatoria regional cierre de brechas 805-2018), entitled "Desarrollo agroindustrial de las plantas aromáticas y condimentaría en el departamento de Antioquia mediante generación de valor agregado para segundos usos de interés industrial como contribución al cierre de brechas: aplicación en el sistema de producción de menta (*Mentha* spp.)" awarded to G. Franco as PI at AGROSAVIA under project number 62744.

Institutional Review Board Statement: Ethical review and approval were waived for this study because it did not involve humans or animals.

Informed Consent Statement: Not applicable because this study did not involve humans.

Data Availability Statement: Raw data, input files and bioinformatic scripts are available in https://github.com/FelipeLopez2019/RNAseq-SNP-Calling-GATK4-Mint. Data supporting reported results can be found is the Supplementary Materials. Any further details can be obtained per direct request to the corresponding authors.

Acknowledgments: We deeply thank to D. Castro-Restrepo, E. Castro and the staff at UCO's greenhouse for plant maintenance, acclimation and taxonomic guidance. Authors also acknowledge H.A. Cordoba-Novoa, and Y. Gómez-Vargas for assistance during RNA extraction and library preparation. Furthermore, we are in debt to G. Franco and L.M. Quintero for funding management and project coordination, as well as to C. Agudelo, E. Castro, B.M. Gaviria-Gutierrez, R. Navarro, E. Osorio, J.A. Rubiano-Rodríguez, and A. Tamayo-Velez for discussion while conceiving and drafting the original project proposal. Finally, we thank the valuable suggestions made by the reviewers, which enabled a better quality of this report, as well as International Journal of Molecular Science's Editorial Office for envisioning and inviting contribution to a timing Special Issue on Molecular Genetics and Plant Breeding, which advances the field of crop wild diversity [34,35] and its use for breeding [36,37] in tropical regions [38,39] and orphan species [40].

Conflicts of Interest: The authors declare no conflict of interest. The funders had no role in the design of the study, in the collection, analyses, or interpretation of data, in the writing of the manuscript, or in the decision to publish the results.

References

1. Ellegren, H.; Galtier, N. Determinants of Genetic Diversity. *Nat. Rev. Genet.* **2016**, *17*, 422–433. [CrossRef] [PubMed]
2. Wolf, J.B.W.; Ellegren, H. Making Sense of Genomic Islands of Differentiation in Light of Speciation. *Nat. Rev. Genet.* **2017**, *18*, 87–100. [CrossRef] [PubMed]
3. Abbott, R.; Albach, D.; Ansell, S.; Arntzen, J.W.; Baird, S.J.E.; Bierne, N.; Boughman, J.; Brelsford, A.; Buerkle, C.A.; Buggs, R.; et al. Hybridization and Speciation. *J. Evol. Biol.* **2013**, *26*, 229–246. [CrossRef] [PubMed]
4. Plomion, C.; Aury, J.M.; Amselem, J.; Leroy, T.; Murat, F.; Duplessis, S.; Faye, S.; Francillonne, N.; Labadie, K.; Le Provost, G.; et al. Oak Genome Reveals Facets of Long Lifespan. *Nat. Plants* **2018**, *4*, 440–452. [CrossRef] [PubMed]
5. Kimura, M. *Neutral Theory of Molecular Evolution*; Cambridge University Press: Cambridge, UK, 1983; ISBN 978-0-521-23109-1.
6. Gould, S.J.; Lewontin, R.C. The Spandrels of San Marco and the Panglossian Paradigm: A Critique of the Adaptationist Programme. *Proc. R. Soc. Lond. Ser. B Biol. Sci.* **1979**, *205*, 581–598. [CrossRef]
7. Barrett, R.D.H.; Hoekstra, H.E. Molecular Spandrels: Tests of Adaptation at the Genetic Level. *Nat. Rev. Genet.* **2011**, *12*, 767–780. [CrossRef]
8. Heylen, O.C.G.; Debortoli, N.; Marescaux, J.; Olofsson, J.K. A Revised Phylogeny of the *Mentha spicata* Clade Reveals Cryptic Species. *Plants* **2021**, *10*, 819. [CrossRef]
9. Vining, K.J.; Hummer, K.E.; Bassil, N.V.; Lange, B.M.; Khoury, C.K.; Carver, D. Crop Wild Relatives as Germplasm Resource for Cultivar Improvement in Mint (*Mentha* L.). *Front. Plant Sci.* **2020**, *11*, 1217. [CrossRef]
10. Zhao, Y.; Wang, K.; Wang, W.L.; Yin, T.T.; Dong, W.Q.; Xu, C.J. A High-Throughput SNP Discovery Strategy for RNA-Seq Data. *BMC Genom.* **2019**, *20*, 160. [CrossRef]
11. Grabherr, M.G.; Haas, B.J.; Yassour, M.; Levin, J.Z.; Thompson, D.A.; Amit, I.; Adiconis, X.; Fan, L.; Raychowdhury, R.; Zeng, Q.; et al. Full-Length Transcriptome Assembly from RNA-Seq Data without a Reference Genome. *Nat. Biotechnol.* **2011**, *29*, 644–652. [CrossRef]
12. McKenna, A.; Hanna, M.; Banks, E.; Sivachenko, A.; Cibulskis, K.; Kernytsky, A.; Garimella, K.; Altshuler, D.; Gabriel, S.; Daly, M.; et al. The Genome Analysis Toolkit: A MapReduce Framework for Analyzing next-Generation DNA Sequencing Data. *Genome Res.* **2010**, *20*, 1297–1303. [CrossRef] [PubMed]
13. Davidson, N.M.; Hawkins, A.D.K.; Oshlack, A. SuperTranscripts: A Data Driven Reference for Analysis and Visualisation of Transcriptomes. *Genome Biol.* **2017**, *18*, 148. [CrossRef] [PubMed]
14. Romero, L.C.; Aroca, M.Á.; Laureano-Marín, A.M.; Moreno, I.; García, I.; Gotor, C. Cysteine and Cysteine-Related Signaling Pathways in *Arabidopsis thaliana*. *Mol. Plant* **2014**, *7*, 264–276. [CrossRef] [PubMed]
15. Slotte, T.; Foxe, J.P.; Hazzouri, K.M.; Wright, S.I. Genome-Wide Evidence for Efficient Positive and Purifying Selection in *Capsella grandiflora*, a Plant Species with a Large Effective Population Size. *Mol. Biol. Evol.* **2010**, *27*, 1813–1821. [CrossRef]
16. Castro-Restrepo, D.; Díaz-García, J.; Serna-Betancur, R.; Martínez-Tobón, M.; Urrea, P.; Muñoz-Durango, K.; Sorio-Durango, E. *Cultivo y Producción de Plantas Aromáticas y Medicinales*; Castro-Restrepo, D., Ed.; Universidad Católica de Oriente: Rionegro, Columbia, 2013; ISBN 978-958-8385-74-7.
17. Andrews, S. FastQC: A Quality Control Tool for High Throughput Sequence Data. Available online: http://www.bioinformatics.babraham.ac.uk/projects/fastqc (accessed on 30 March 2022).
18. Afgan, E.; Baker, D.; Batut, B.; Van Den Beek, M.; Bouvier, D.; Ech, M.; Chilton, J.; Clements, D.; Coraor, N.; Grüning, B.A.; et al. The Galaxy Platform for Accessible, Reproducible and Collaborative Biomedical Analyses: 2018 Update. *Nucleic Acids Res.* **2018**, *46*, W537–W544. [CrossRef]
19. Li, H. Aligning Sequence Reads, Clone Sequences and Assembly Contigs with BWA-MEM. *arXiv* **2013**, arXiv:1303.3997.
20. Li, H.; Durbin, R. Fast and Accurate Long-Read Alignment with Burrows-Wheeler Transform. *Bioinformatics* **2010**, *26*, 589–595. [CrossRef]
21. Li, H.; Handsaker, B.; Wysoker, A.; Fennell, T.; Ruan, J.; Homer, N.; Marth, G.; Abecasis, G.; Durbin, R. The Sequence Alignment/Map Format and SAMtools. *Bioinformatics* **2009**, *25*, 2078–2079. [CrossRef]
22. Bradbury, P.J.; Zhang, Z.; Kroon, D.E.; Casstevens, T.M.; Ramdoss, Y.; Buckler, E.S. TASSEL: Software for Association Mapping of Complex Traits in Diverse Samples. *Bioinformatics* **2007**, *23*, 2633–2635. [CrossRef]
23. Jombart, T. Adegenet: A R Package for the Multivariate Analysis of Genetic Markers. *Bioinformatics* **2008**, *24*, 1403–1405. [CrossRef]
24. Alhusain, L.; Hafez, A.M. Nonparametric Approaches for Population Structure Analysis. *Hum. Genom.* **2018**, *12*, 25. [CrossRef] [PubMed]
25. Sekula, M.; Datta, S.; Datta, S. OptCluster: An R Package for Determining the Optimal Clustering Algorithm. *Bioinformation* **2017**, *13*, 101–103. [CrossRef] [PubMed]
26. Brock, G.; Pihur, V.; Datta, S.; Datta, S. ClValid: An R Package for Cluster Validation. *J. Stat. Softw.* **2008**, *25*, 1–22. [CrossRef]
27. Lloyd, S. Least Squares Quantization in PCM. *IEEE Trans. Inf. Theory* **1982**, *28*, 129–137. [CrossRef]
28. MacQueen, J.B. *Some Methods for Classification and Analysis of Multivariate Observations*; Lucien Marie Le Cam, J.N., Ed.; University of California: San Francisco, CA, USA, 1967; pp. 281–298.
29. Kaufman, L.; Rousseeuw, P.J. *Finding Groups in Data: An Introduction to Cluster Analysis*; John Wiley & Sons, Ltd.: Boca Raton, FL, USA, 2009; ISBN 0-471-73578-7.
30. Charrad, M.; Ghazzali, N.; Boiteau, V.; Niknafs, A. Nbclust: An R Package for Determining the Relevant Number of Clusters in a Data Set. *J. Stat. Softw.* **2014**, *61*, 1–36. [CrossRef]

31. Kamvar, Z.N.; Tabima, J.F.; Grünwald, N.J. Poppr: An R Package for Genetic Analysis of Populations with Clonal, Partially Clonal, and/or Sexual Reproduction. *PeerJ* **2014**, *2*, e281. [CrossRef]
32. Nei, M. *Molecular Evolutionary Genetics*; Columbia University Press: New York, NY, USA, 1987.
33. Götz, S.; García-Gómez, J.M.; Terol, J.; Williams, T.D.; Nagaraj, S.H.; Nueda, M.J.; Robles, M.; Talón, M.; Dopazo, J.; Conesa, A. High-Throughput Functional Annotation and Data Mining with the Blast2GO Suite. *Nucleic Acids Res.* **2008**, *36*, 3420–3435. [CrossRef]
34. Cortés, A.J.; López-Hernández, F. Harnessing Crop Wild Diversity for Climate Change Adaptation. *Genes* **2021**, *12*, 783. [CrossRef]
35. Cortés, A.J.; Cornille, A.; Yockteng, R. Evolutionary Genetics of Crop-Wild Complexes. *Genes* **2022**, *13*, 1. [CrossRef]
36. Guevara-Escudero, M.; Osorio, A.N.; Cortés, A.J. Integrative Pre-Breeding for Biotic Resistance in Forest Trees. *Plants* **2021**, *10*, 2022. [CrossRef]
37. Cortés, A.J.; López-Hernández, F.; Osorio-Rodriguez, D. Predicting thermal adaptation by looking into populations' genomic past. *Front. Genet.* **2020**, *11*, 564515. [CrossRef] [PubMed]
38. Blair, M.W.; Soler, A.; Cortés, A.J. Diversification and Population Structure in Common Beans (*Phaseolus vulgaris* L.). *PLoS ONE* **2012**, *7*, e49488. [CrossRef] [PubMed]
39. López-Hernández, F.; Cortés, A.J. Last-Generation Genome–Environment Associations Reveal the Genetic Basis of Heat Tolerance in Common Bean (*Phaseolus vulgaris* L.). *Front. Genet.* **2019**, *10*, 22. [CrossRef] [PubMed]
40. Cortés, A.J.; Restrepo-Montoya, M.; Bedoya-Canas, L.E. Modern Strategies to Assess and Breed Forest Tree Adaptation to Changing Climate. *Front. Plant Sci.* **2020**, *11*, 583323. [CrossRef] [PubMed]

International Journal of
Molecular Sciences

MDPI

Article

Evolution and Phylogeny of Soybean Mosaic Virus Based on 143 Complete Genomes

Hoseong Choi [1,†], Yeonhwa Jo [2,†], Soo Yeon Choi [3], Sang-Min Kim [3], Yu Mi Choi [4], Jin-Sung Hong [5], Bong Choon Lee [3,*] and Won Kyong Cho [2,*]

1 Plant Genomics and Breeding Institute, Seoul National University, Seoul 08826, Republic of Korea
2 College of Biotechnology and Bioengineering, Sungkyunkwan University, Suwon 16419, Republic of Korea
3 Crop Foundation Division, National Institute of Crop Science, Rural Development Administration, Wanju 55365, Republic of Korea
4 Agrobiodiversity Center, National Institute of Agricultural Sciences, Rural Development Administration, Jeonju 54874, Republic of Korea
5 Department of Applied Biology, Kangwon National University, Chuncheon 24341, Republic of Korea
* Correspondence: leebc21@korea.kr (B.C.L.); wonkyong@gmail.com (W.K.C.); Tel.: +82-63-238-5342 (B.C.L.); +82-31-290-7860 (W.K.C.)
† These authors contributed equally to this work.

Abstract: *Soybean mosaic virus* (SMV) of the genus *Potyvirus* is an important virus in cultivated soybeans. Here, we obtained 7 SMV genomes from soybean germplasms using RNA sequencing and conducted a comprehensive evolutionary and phylogenetic study of 143 SMV genomes derived from 10 plant species and 12 countries. The phylogenetic tree we constructed using coding DNA sequences revealed the existence of nine clades of SMV isolates/strains. Recombination analysis revealed 76 recombinant events and 141 recombinants in total. Clades 1 and 3 contain the most common SMV pathotypes, including G1 through G7, which are distributed worldwide. Clade 2 includes several Chinese SMV pathotypes. The SMV isolates were further divided into two groups. The SMV isolates in the first group, including clades 8 and 9, were identified from *Pinellia* and *Atractylodes* species, whereas those in the second group (clades 1 through 7) were mostly found in cultivated soybeans. The SMV polyprotein undergoes positive selection, whereas most mature proteins, except for the P1 protein, undergo negative selection. The P1 protein of SMV isolates in group 1 may be highly correlated with host adaptation. This study provides strong evidence that recombination and plant hosts are powerful forces driving the genetic diversity of the SMV genome.

Keywords: soybean mosaic virus; genome; evolution; phylogeny; diversity

Citation: Choi, H.; Jo, Y.; Choi, S.Y.; Kim, S.-M.; Choi, Y.M.; Hong, J.-S.; Lee, B.C.; Cho, W.K. Evolution and Phylogeny of Soybean Mosaic Virus Based on 143 Complete Genomes. *Int. J. Mol. Sci.* **2023**, *24*, 22. https://doi.org/10.3390/ijms24010022

Academic Editors: Andrés J. Cortés and Hai Du

Received: 17 October 2022
Revised: 6 December 2022
Accepted: 17 December 2022
Published: 20 December 2022

1. Introduction

Soybean mosaic virus (SMV) is a member of the genus *Potyvirus* in the family *Potyviridae* [1]. The main hosts of SMV are cultivated (*Glycine max* (L.) Merr.) and wild soybeans (*Glycine soya* Sieb. & Zucc.). SMV infects several plant species in the family Fabaceae [2]. In soybean plants, SMV infection results in foliar viral disease symptoms such as distorted and wrinkled leaves with mosaic or mottled patterns of color patches. In addition, SMV infection causes a reduction in seed size and the mottling of seed coats [3]. Moreover, dwarfism and a reduction in seed size and pod number were observed in soybean plants infected with SMV. Because it reduces the quality and quantity of soybean seed production, SMV is currently regarded as one of the most devastating viruses of cultivated soybeans [4].

SMV is usually transmitted by diverse aphids in a non-persistent manner [5]. It is also transmitted through seeds, with a seed transmission rate of up to 5% [6]. Therefore, spraying insecticides or planting virus-free seeds is useful for preventing the spread of SMV during soybean cultivation.

Several research groups have studied the interaction between SMV and soybeans by identifying and characterizing the resistance (*R*) genes against SMV and proteins that interact with known SMV strains [1]. As a result, many soybean cultivars that are resistant to diverse SMV strains with different pathogenicities have been identified. Of the known SMV resistance loci, the *Rsv1* locus that confers resistance to seven SMV strains (G1 through G7) was first identified in the PI96983 cultivar [7]. The dominant *Rsv1* gene, which is located on chromosome 13, is highly complex. For example, a previous study identified a cluster of *R* gene candidates composed of several NB-LRR (nucleotide-binding leucine-rich repeat) family genes [8].

The SMV genome consists of single-stranded positive-sense RNA. It is approximately 9000 nucleotides (nt) in length with a poly(A) tail at the 3′ end of its genome. The SMV genome has two open reading frames (ORFs), which encode a polyprotein and PIPO (pretty interesting *Potyviridae* ORF). The polyprotein precursor is processed into 10 functional proteins, namely, protein 1 (P1), helper component protease (HC-Pro), protein 3 (P3), 6 kDa protein 1 (6K1), cylindrical inclusion body (CI), 6K2, nuclear inclusion protein a-viral genome-linked protein (NIa-VPg), NIa-Pro, nuclear inclusion protein b (NIb), and coat protein (CP) by the cellular proteolytic activity of three viral-encoded proteinases.

Since the complete genome sequencing of the SMV strains G2 and G7 [9–11], many complete SMV genome sequences have been reported in diverse countries, including South Korea, China, the United States of America (USA), and Iran [12–15]. Furthermore, several high-throughput sequencing (HTS) technologies have facilitated the complete genome sequencing of known and novel viruses that infect a wide range of plant species [16].

Here, we obtained complete genome sequences for seven SMV isolates from soybean germplasms. After combining all available SMV genome sequences, we performed comprehensive analyses to elucidate the molecular evolution and phylogenetic relationships of the known SMV isolates/strains.

2. Results

2.1. Complete Genome Sequences of Seven SMV Isolates from Soybean Germplasms Obtained via RNA Sequencing

We examined viral infections in soybean germplasms using RT-PCR. Of these, we selected 10 soybean germplasms that were infected with SMV for RNA sequencing. Bioinformatic analyses identified 49 SMV-associated contigs ranging from 202 to 9592 bp in 4 of the 10 libraries. In addition, we obtained seven complete SMV genomes from four libraries, referred to as GR1, GR2, GR3, and GR4. In the GR1 library, two SMV isolates were identified. These isolates were referred to as GR1-1 and GR1-2, and both had genomes that were 9569 nucleotides (nt) long. We also obtained the complete genome sequences of three SMV isolates from the GR2 library. These isolates were referred to as GR2-1, GR2-2, and GR2-3, and all of them had genomes that were 9571 nt in length. We identified only single SMV isolates in the G3 and G4 libraries. These isolates were referred to as GR3 (9571 nt) and GR4 (9566 nt), respectively.

The SMV genome contains two ORFs and contains a poly (A) tail (Table 1). In the case of the SMV isolate GR1-1, ORF1 (positions 117 to 3066) encodes a polyprotein, whereas ORF2 (positions 2867 to 3094) encodes a PIPO protein. The BLASTN results showed that six SMV isolates (G1-1 and G1-2) had sequence similarity to the known SMV isolate GYGI-p (GenBank MT603834.1) identified from soybeans in Gyeonggi Province, Korea, with 99% coverage and 97.14% nucleotide identity. In addition, the SMV strain GR4 showed sequence similarity to SMV strain G5 (GenBank FJ640980.1) identified from soybeans in Korea, with 99% coverage and 99.27% nucleotide identity.

Table 1. Genetic diversity of 143 SMV genomes based on nucleotide sequences corresponding to individual proteins.

Protein	Sites	S	Eta	H	Hd	Pi	θw	Tajima's D
P1	1436	393	641	99	0.98 ± 0.003	0.11 ± 0.01	0.14 ± 0.03	−1.71
HC-Pro	1374	668	948	124	0.99 ± 0.001	0.07 ± 0.005	0.08 ± 0.02	−1.38
P3	1041	581	808	126	0.99 ± 0.001	0.06 ± 0.005	0.1 ± 0.02	−1.65
6K1	156	93	137	76	0.97 ± 0.005	0.06 ± 0.006	0.1 ± 0.02	−1.92
CI	1902	949	1335	131	0.99 ± 0.001	0.06 ± 0.004	0.09 ± 0.02	−1.55
6K2	159	79	114	66	0.95 ± 0.01	0.05 ± 0.004	0.08 ± 0.02	−1.78
NIa-VPg	576	264	360	103	0.99 ± 0.002	0.06 ± 0.004	0.08 ± 0.01	−1.38
NIa-Pro	729	335	461	112	0.99 ± 0.001	0.05 ± 0.004	0.08 ± 0.02	−1.70
NIb	1551	648	894	125	0.99 ± 0.001	0.06 ± 0.004	0.08 ± 0.02	−1.58
CP	873	365	483	118	0.99 ± 0.001	0.06 ± 0.004	0.08 ± 0.01	−1.41
Polyprotein	9803	4375	6181	142	0.99 ± 0.0008	0.06 ± 0.005	0.09 ± 0.02	−1.57
PIPO	225	92	106	72	0.96 ± 0.01	0.02 ± 0.003	0.07 ± 0.01	−2.07
Genome	10,044	8653	25,946	143	1.000 ± 0.0008	0.62 ± 0.01	0.18 ± 0.04	0.49

Sites: total number of sites; S: number of segregating sites; Eta: total number of mutations; H: number of haplotypes; Hd: haplotype diversity; Pi: nucleotide diversity; θw: Watterson's estimator of θ.

2.2. Examination of 143 Complete SMV Genomes Based on Country, Plant Host, and Year of Identification

We downloaded all SMV sequences from GenBank and only used SMV genomes with genome sequences that were over 9000 nt in length, meaning that the whole coding DNA sequence was covered. The selection process resulted in 136 SMV genome sequences from GenBank. We also included 7 SMV genome sequences from this study, which resulted in the analysis of a total of 143 SMV genome sequences. First, we examined the distribution of 143 SMV genomes by country (Figure 1A). The SMV genomes we studied were obtained from 12 countries: Brazil, Canada, China, Colombia, Germany, India, Iran, Japan, the Netherlands, Taiwan, and the USA. The majority of the SMV genomes were derived from two countries, China (56 genomes) and South Korea (55 genomes), followed by the USA (eight genomes) and Canada (five genomes). Next, we examined the distribution of the SMV genomes in plant hosts. The SMV genomes were identified in ten plant species: *Atractylodes macrocephala, Glycine max, Glycine soja*, other legumes, *Passiflora* species (including *Passiflora edulis*), *Pinellia pedatisecta, Pinellia ternata, Uraria crinita*, and *Vigna angularis* (Figure 1B). Of these, two plant species, *G. max* (100 genomes) and *G. soja* (31 genomes), were identified as major hosts for SMV.

Since the first SMV genome was reported in 1989, many SMV genomes have been sequenced (Figure 1C). Before 2000, only three complete genome sequences of three SMVs had been reported. However, the rapid development of sequencing technologies, such as HTS, has dramatically increased the number of complete SMV genome sequences. In 2006 and 2016, many SMV genomes were reported (24 and 18 genomes, respectively).

2.3. Phylogenetic Analysis of 143 SMV Genomes

We constructed a maximum likelihood (ML) phylogenetic tree using the 143 SMV coding DNA sequences (CDS) that we selected (Figure 2). We identified ten clades based on the ML phylogenetic tree (Figure 2). Three clades (clades 1, 2, and 3) were identified as major clades, and three (clades 8, 9, and 10) were identified as minor clades (Figure 1D). The SMV genomes in clade 1 were mostly derived from South Korea (28 genomes), whereas the SMV genomes in clade 2 were mostly derived from China (24 genomes) (Figure 3A). In clade 3, the majority of SMV genomes were derived from South Korea (27 genomes), followed by China (eight genomes), and Japan (five genomes). In clades 1 and 2, the SMV genomes were derived from four countries, whereas the SMV genomes in clade 3 were derived from ten countries. When the genomes were analyzed based on their identified plant hosts, the SMV genomes in clade 1 were identified as coming from *G. max* (22 genomes) and *G. soja* (14 genomes), whereas all the SMV genomes in clade 2 were derived from *G. max* (28 genomes) (Figure 3B). In clade 3, the SMV genomes came from six plant species. Among these, *G. max* (37 genomes) was the dominant plant host, followed by *G. soja* (14 genomes). The phylogenetic tree showed that most of the SMV genomes were closely related (Figure 2). However, 6 SMV genomes in clades 8, 9, and 10 were identified in China and were phylogenetically distant from the other 137 SMV genomes. The four SMV genomes in clade 8 were derived from *P. ternata* (three genomes) and *P. pedatisecta* (one genome), whereas the Am isolate in clade 9 and the Uraria isolate in clade 10 were found in *A. macrocephala* and *U. crinita*, respectively. During data analysis, we found that the SMV isolate Uraria (LC037232.1) from *Uraria crinita* in clade 10 showed strong sequence similarity to uraria mosaic virus (UMV) isolate OC (LC477217.1) [17]. Based on the molecular species demarcation for potyviruses, the SMV isolate Uraria should be revised to UMV isolate Uraria. Our results showed that 142 SMV isolates/strains in this study could be classified into 9 clades, and that the UMV isolate Uraria in this clade 10 could be used as an outgroup.

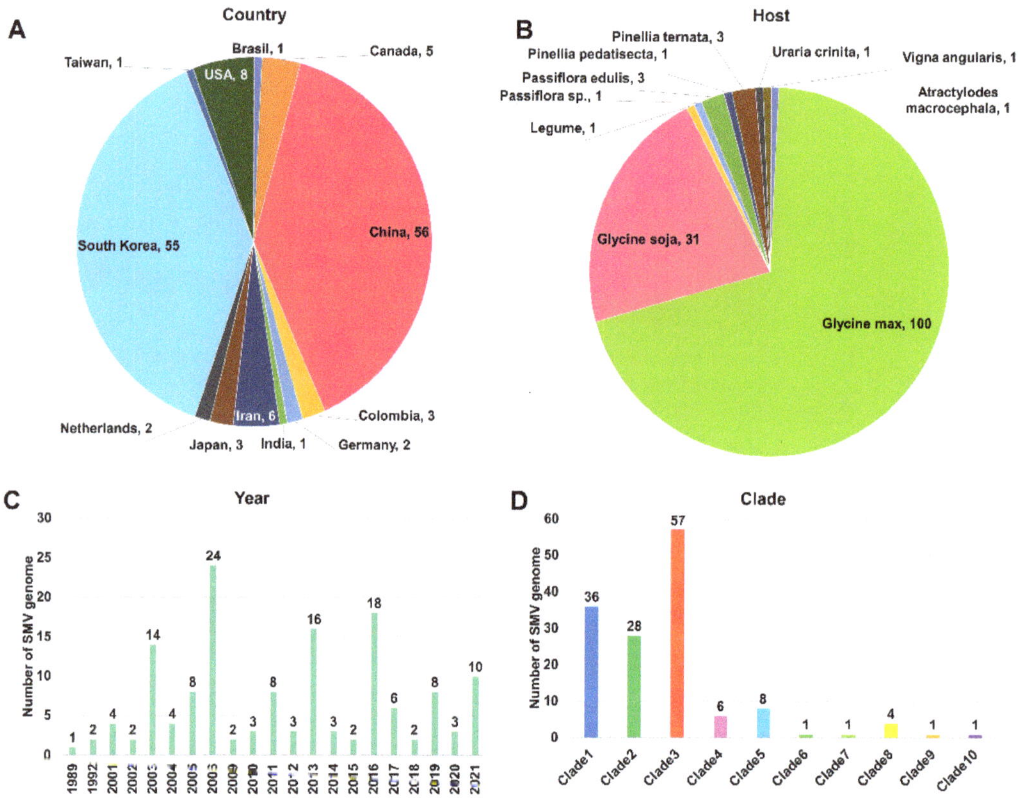

Figure 1. Information for 143 SMV genomes and the 10 identified clades. Pie charts display information for each country where SMV genomes were obtained (**A**) and the host plants from which SMV genomes were obtained (**B**). The bar graph shows the number of SMV genomes based on the year of identification (**C**). Information for the 10 clades of 143 SMV genomes that were identified using phylogenetic analysis (**D**).

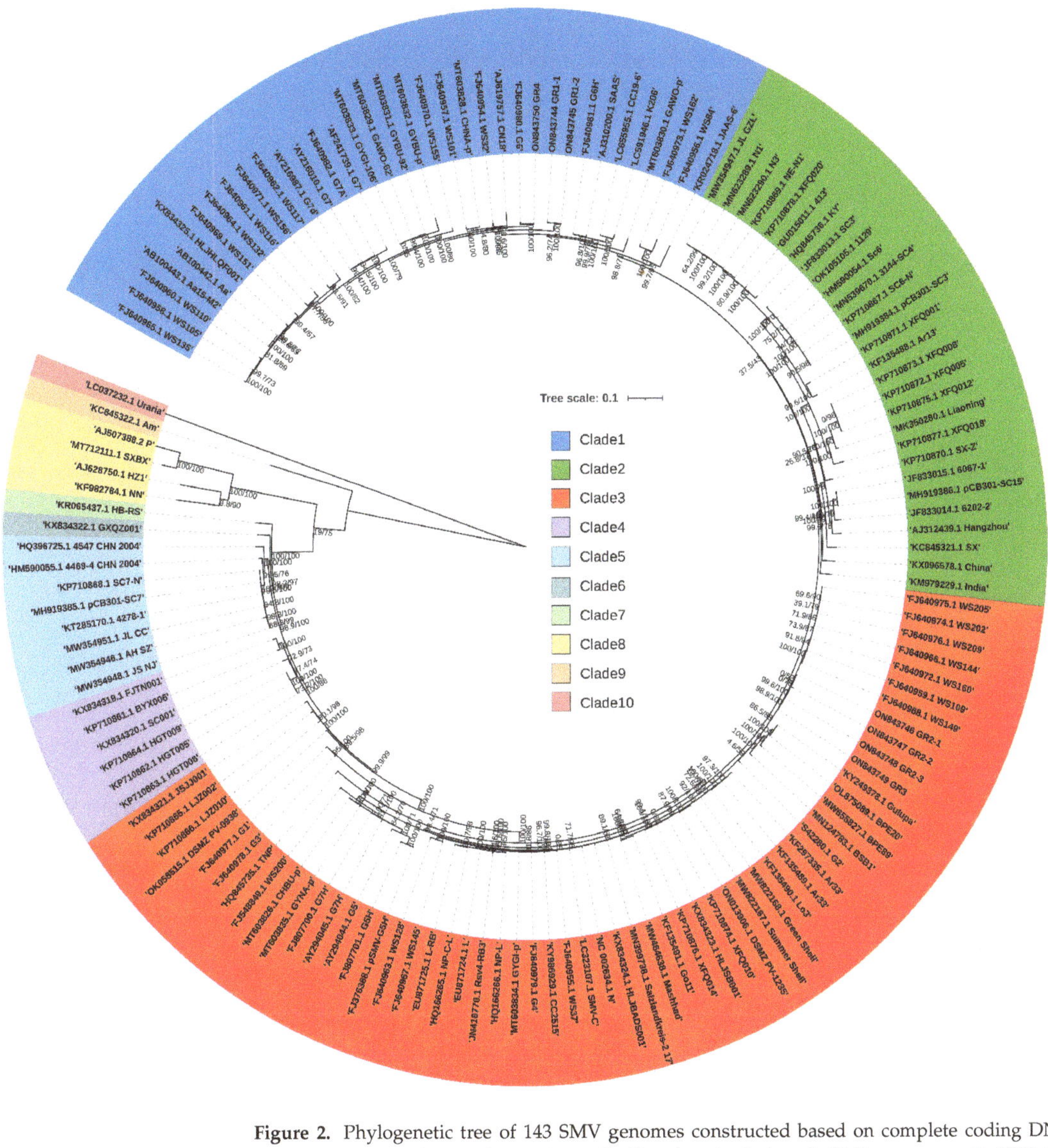

Figure 2. Phylogenetic tree of 143 SMV genomes constructed based on complete coding DNA sequences, generated using the maximum likelihood method. A total of 143 sequences with 9805 sites covering whole coding DNA sequences were used for phylogenetic tree construction using IQ-TREE with GTR + F+I + G4 for the best-fit model. The accession number and isolate (strain) name for each SMV genome are indicated. Both SH-aLRT and UFBoot support values are indicated on the branches. A total of 10 clades were found, and they are shown in 10 different colors. The isolate Uraria (GenBank LC037232.1) in clade 10 was used as an outgroup. Detailed information for individual SMV genomes can be found in Table S1.

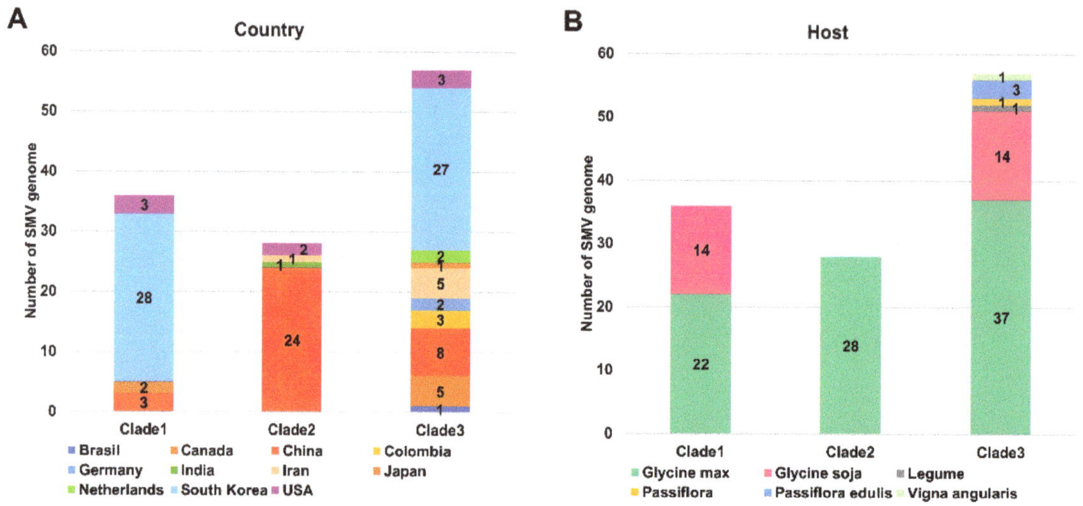

Figure 3. Composition of SMV isolates in the three major clades that became apparent when the isolates were analyzed based on their country of origin and host plant. The number of SMV isolates in each of the three major clades (clades 1 through 3) based on country of origin (**A**) and host plant (**B**) are indicated.

2.4. Recombination Analysis of 143 SMV Isolates/Strains

Recombination is an important process contributing to the genetic diversity of RNA viruses. To examine the recombination events in the 143 SMV isolates, all complete genome sequences were aligned and subjected to recombination analysis. First, we performed a network analysis using the SplitsTree4 program to examine the possible presence of recombinants in the 143 SMV genomes. The network tree showed many reticulated nodes, indicating that a high frequency of genetic recombination occurred among different SMV isolates/strains (Figure 4A). It was difficult to distinguish between several groups based on the network tree. However, the network tree revealed two major groups of SMV isolates/strains, designated as group 1 and group 2 (Figure 4B,C). Furthermore, we examined whether the seven known SMV strains G1–G7 clustered in different reticulated nodes (Figure 4C). We found that SMV strains G1 and G3 grouped together, and that SMV strains G2 and G4 also grouped together.

The aligned 143 SMV genome sequences were subjected to recombinant analysis using the RDP program (Table S2). A total of 76 recombinant events were identified. Of the 143 SMV isolates, 141 were predicted to be recombinants. The two non-recombinant SMV isolates were SMV isolate WS84 (FJ640956.1) from *G. soya* in Korea and SMV isolate Uraria (LC037232.1) from *Uraria crinita* in Taiwan, which were used as outgroups in the phylogenetic tree construction. Next, we determined whether the recombinant breakpoints were uniformly distributed across the SMV genome (Figure 5A). For this purpose, we calculated the breakpoints using a sliding window of 200 nt (Figure 5B). Most regions of the SMV genome showed a high number of breakpoints with strong statistical support (Figure 5C). For example, P1 had a relatively intact region, referred to as a cold spot, followed by P3. The regions with the highest number of breakpoints, referred to as hot spots, were found between the 6K1 and CI regions and between P1 and NIa-Pro (Figure 5B). In contrast, the recombination rate plot indicated by rho (4Ner) showed the highest value at the end of the 3′ region of the SMV genome, followed by the P1 region (Figure 5D). Except for these two regions, we did not observe any significant changes in the recombination rate across the SMV genome.

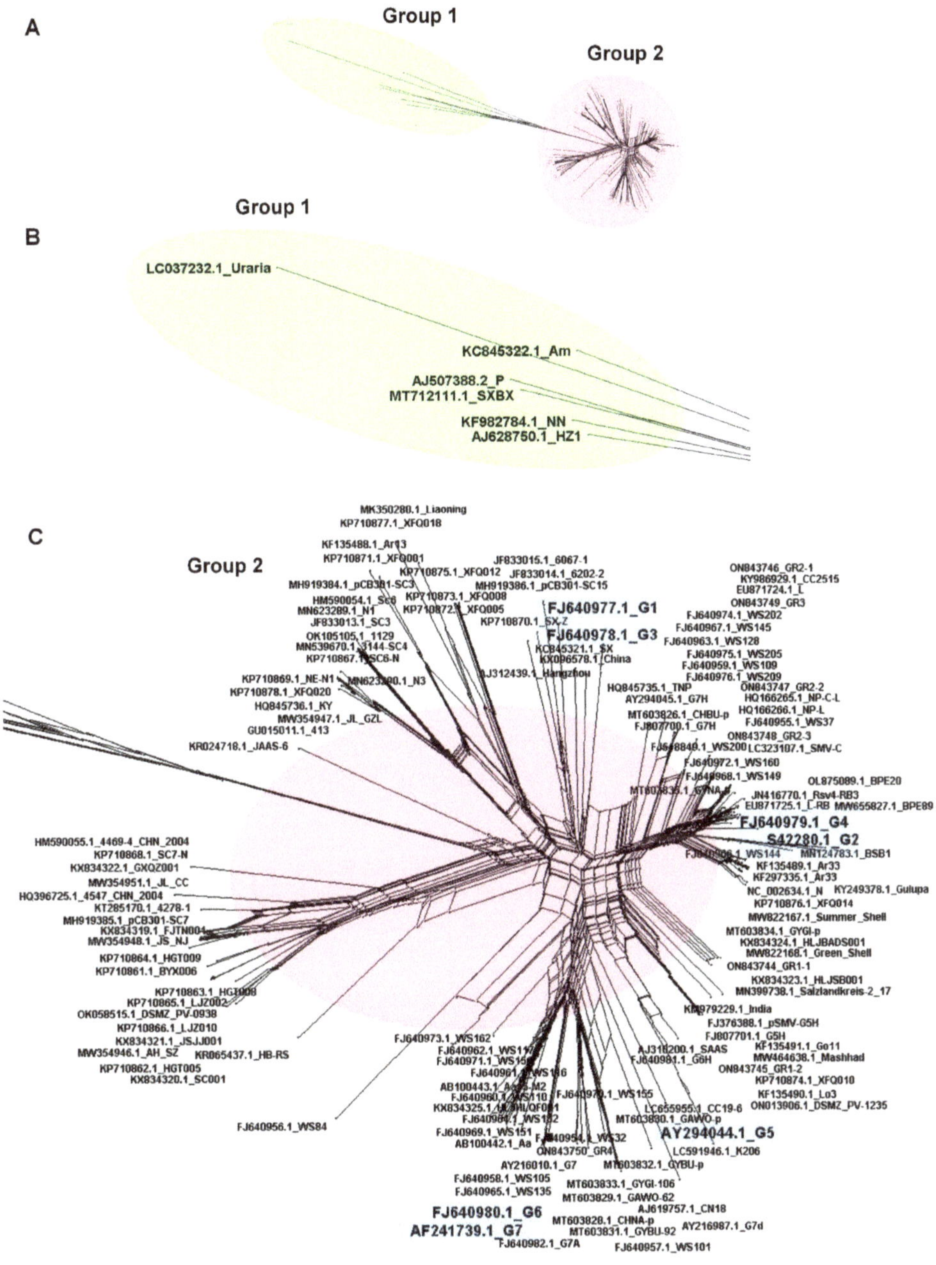

Figure 4. Phylogenetic network tree of 143 SMV genomes. The network tree was rooted using the complete genome sequences of 143 SMV genomes and contains 2 groups (Group 1 and Group 2) of SMV genomes (**A**). Group 1 is shown in a green oval in the network tree and was then further magnified (**B**). Group 2 is shown in a pink oval in the network and then was further magnified (**C**) The seven known SMV strains (G1–G7) are shown in the light blue colored box. The accession numbers and isolate names are also indicated. Using the nexus alignment file in the supplemental file S1, the network tree can be visualized in the SplitsTree program.

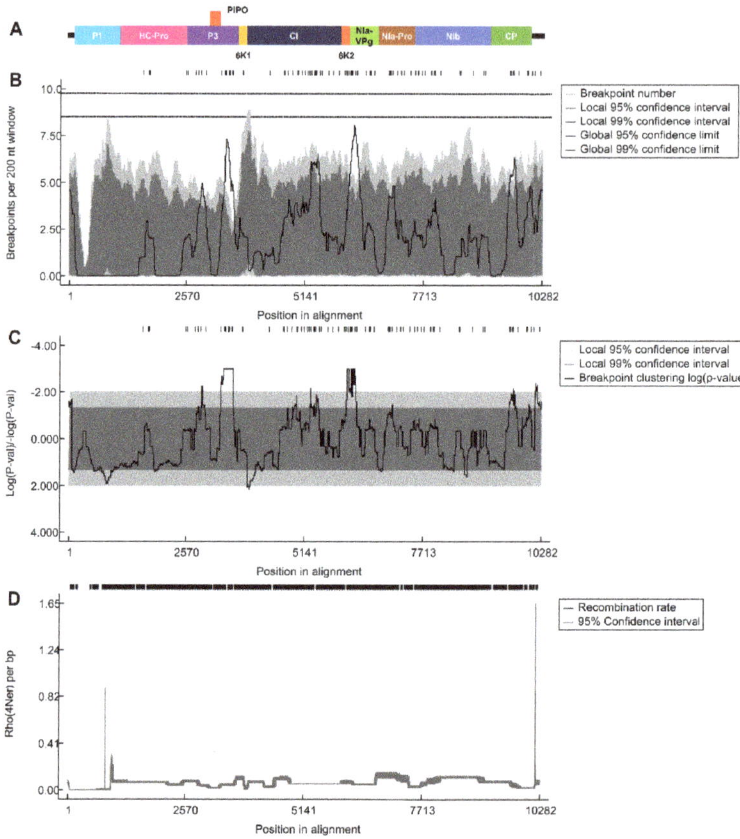

Figure 5. Recombination breakpoint distribution plots for the SMV genome. SMV genome organization showing individual protein regions (**A**). Recombination breakpoint distribution plot (**B**) and respective statistical log converted *p*-value (**C**). Locally and globally significant breakpoint clusters were identified using 95% and 99% permutation tests. Recombination rate plot (**D**).

Next, we tested whether recombination occurred frequently within the individual protein sequences. To this end, we conducted recombination analyses using nucleotide sequences corresponding to the individual proteins. We identified 41 recombination events and 125 recombinants in the polyprotein sequences, whereas no recombinant events were identified in PIPO (Figure 6). Among the mature proteins processed from the polyprotein, CI displayed the highest number of recombination events (9 events) and recombinants (17 recombinants). P1 and P3 had one and two recombinant events, respectively. However, the number of recombinants identified for P1 (20) and P3 (16) was very high. We identified two recombinant events and recombinants in both HC-Pro and NIb, whereas a single recombinant and recombination event was found in CP.

No recombinant events were found in five SMV proteins, namely, 6K1, 6K2, Nia-VPg, Nia-Pro, and PIPO. Furthermore, we examined the association between breakpoint distribution and specific genomic regions (protein-coding regions). Three protein regions, namely, CI, 6K2, and PIPO, showed high breakpoint levels compared to those of the other SMV genome regions.

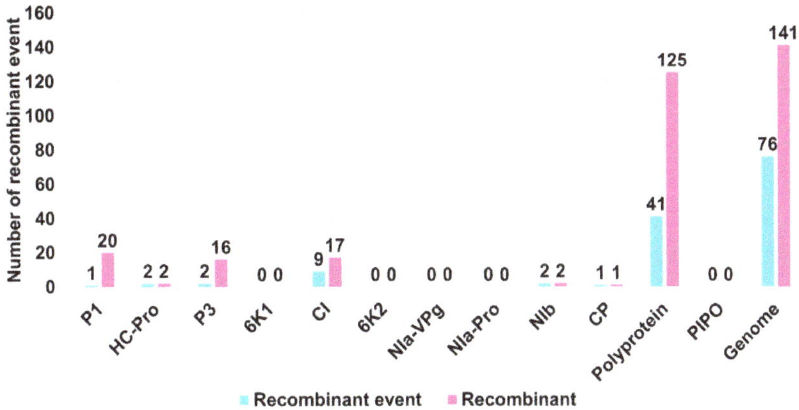

Figure 6. Number of recombinant events and recombinants in individual SMV proteins or genomes.

2.5. Genetic Diversity Analysis of 143 SMV Genomes

We also analyzed the genetic diversity of 143 SMV genomes, polyproteins, PIPOs, and 10 mature proteins processed from SMV polyproteins. The genetic diversity of the complete genome sequence was S = 8653 (number of segregating sites), Eta = 25946 (total number of mutations), Pi = 0.62182 ± 0.01438 (nucleotide diversity), θw = 0.18062 ± 0.04072 (Watterson's estimator of θ), and Tajima's D = 0.49152 (Table 1). The genetic diversity of the polyprotein was S = 4375, Eta = 6181, Pi = 0.06895 ± 0.00516, θw = 0.09303 ± 0.02100, and Tajima's D = −1.57646, and the genetic diversity of PIPO was S = 92, Eta = 106, Pi = 0.02977 ± 0.00394, θw = 0.07418 ± 0.01833, and Tajima's D = −2.07972. Of the ten mature proteins processed from the polyprotein, CI (Eta = 1335) had the highest number of mutations, whereas 6K2 had the lowest number of mutations. Haplotype analysis revealed that most SMV proteins exhibit a high level of haplotype diversity. CI (number of haplotypes, H = 131) showed the highest haplotype diversity (Hd = 0.9986 ± 0.0011), whereas 6K2 (H = 66) displayed the lowest haplotype diversity (Hd = 0.957 ± 0.011). According to the Pi value for nucleotide diversity, nucleotide diversity was the highest in P1 (Pi = 0.11369 ± 0.01319) and lowest in PIPO (Pi = 0.02977 ± 0.00394).

Next, we calculated the average number of segregated sites (θw) for each protein. Again, P1 (θw = 0.14546 ± 0.03353) showed the highest value of θw, followed by 6K1 (θw = 0.10768 ± 0.02658) and P3 (θw = 0.10100 ± 0.02311). PIPO (θw = 0.07418 ± 0.01833) had the lowest θw. To test the neutral mutation hypothesis, we calculated Tajima's D for individual SMV proteins [18]. All the 11 mature SMV proteins had negative Tajima's D values, ranging from −2.07972 (PIPO) to −1.38761 (HC-Pro). Only three proteins with small sizes (6K1, 6K2, and PIPO) showed statistically significant differences ($p < 0.05$) for Tajima's D values. Tajima's D value for the SMV genome was positive (0.49152), but this observation was not statistically supported ($p > 0.10$).

Next, we compared the nucleotide diversity (Pi) and the average number of segregating sites (θw) in the two SMV groups that were defined using network analysis (Tables S3 and S4). Group 1 contained 6 SMV genomes, whereas group 2 contained 137 SMV genomes. The nucleotide diversity values of all the SMV proteins in group 1 were always higher than those of group 2 and those of all 143 SMV genomes indicated by 'all' (Figure 7A). The nucleotide diversity for all 143 SMV genomes was 0.62182, which was much higher than that for group 1 (Pi = 0.21194) and group 2 (Pi = 0.056569). Among the SMV proteins, P1 (Pi = 0.40057) showed the highest nucleotide diversity in group 1, whereas PIPO (Pi = 0.02064) showed the lowest nucleotide diversity in group 2.

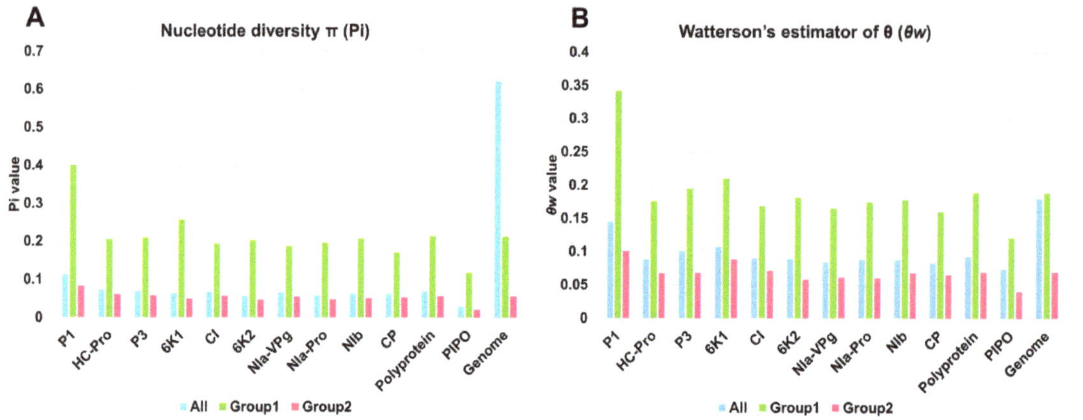

Figure 7. Genetic diversity of SMV proteins or genomes in different groups of SMV. Nucleotide diversity (Pi) (**A**) and average number of segregating sites (θw) (**B**) of individual SMV proteins or genomes in the three groups. 'All' indicates all 143 SMV genomes, and group 1 and group 2 include 6 and 137 SMV genomes, respectively.

The average number of segregating sites (θw) for all SMV proteins in group 1 was much higher than that in group 2 (Figure 7B). However, the difference in the average number of segregating sites between 'all' (Pi = 0.18062) and group 1 (Pi = 0.18856) was small. The average number of segregating sites for P1 (θw = 0.14546) in group 1 was the highest among the SMV proteins, whereas PIPO (θw = 0.0401) in group 2 had the lowest average number of segregating sites.

2.6. Estimation of the Nonsynonymous to Synonymous Rate Ratio of all SMV Protein Sequences

The ratio of nonsynonymous to synonymous substitutions (dN/dS), also known as Ka/Ks, is a useful measure of natural selection in protein-coding genes [19]. The average dN/dS ratio was greater than one for the SMV genome (1.004237), polyprotein (1.568258), and P1 (1.066121), whereas the average dN/dS ratio for the other ten mature SMV proteins was less than one (Table 2). The maximum dN/dS ratio was the highest for the polyprotein, followed by P1. In contrast, the maximum dN/dS ratio for NIa-VPg was the lowest among all examined SMV proteins. The number of nonsynonymous sites was much higher than that of synonymous sites for all examined proteins, except 6K2, in which the number of synonymous sites was approximately 1.2 times higher than that of nonsynonymous sites.

Next, we compared the average dN/dS ratio for all SMV proteins in three different groups, 'all', group 1, and group 2 (Table S5 and Figure 8). The average dN/dS ratios for the polyprotein and P1, followed by PIPO, were much higher than those for the other SMV proteins. The average dN/dS ratios for most SMV proteins were less than one except for P1 and polyprotein in 'all' and group 1. The average dN/dS ratios for P1 in 'all' and group 1 were 1.06612 and 0.64502, respectively. The average dN/dS ratios for polyprotein in 'all' and group 1 were 1.56825 and 1.02581, respectively. The average dN/dS ratios for PIPO in all three groups were comparable, ranging from 0.45433 (group 2) to 0.39788 (group 1). The lowest average dN/dS ratio was 0.01168 for NIa-Pro in group 2. The highest average dN/dS ratio was 1.56825 for polyprotein in group 1. Most SMV proteins, such as HC-Pro, P3, 6K1, CI, 6K2, NIa-VPg, NIa-Pro, Nib, and CP, had average dN/dS ratios of less than 0.2.

Table 2. Summary of synonymous site and nonsynonymous sites for individual SMV protein sequences.

Protein	Number of Sites	Average	Min	Max	Synonymous Site	Nonsynonymous Sites
P1	1436	1.066121	0	6.822368	99.16	356.84
HC-Pro	1374	0.050681	0	1.117647	294.73	1064.28
P3	1041	0.08778	0	1.326087	217.4	817.6
6K1	156	0.062452	0	1.373193	30.8	125.2
CI	1902	0.0342	0	1.173913	431.7	1467.3
6K2	159	0.023262	0	0.754717	32.24	26.76
Nia-VPg	576	0.025513	0	0.560976	121.22	439.78
Nia-Pro	729	0.012687	0	0.823529	147.79	536.21
Nib	1551	0.04367	0	2.2	288.16	1046.84
CP	873	0.0326	0	0.830508	169.69	619.31
Polyprotein	9803	1.568258	0	10	1702.53	6721.47
PIPO	225	0.429524	0	4.219048	47.26	174.74
Genome	10,044	1.004237	0	3.685714	1776.3	6854.7

Average, Min, and Max indicate the average, minimum, and maximum ratios of dN/dS, respectively, for the given protein sequences.

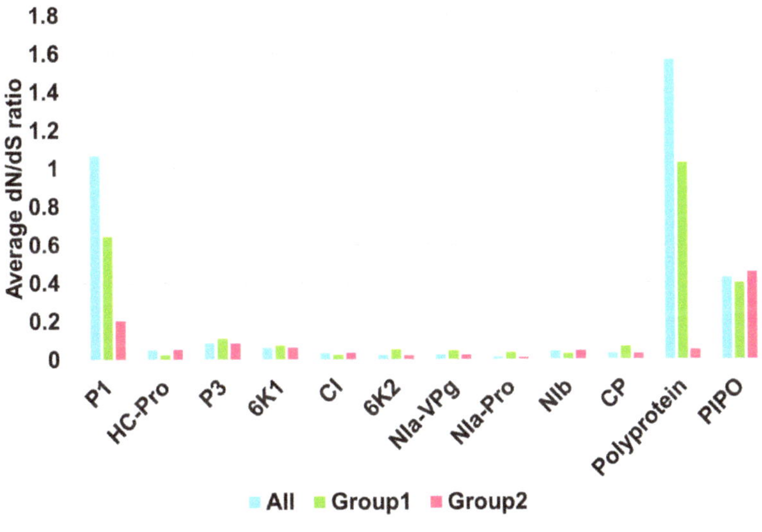

Figure 8. Average dN/dS ratio of SMV proteins in three groups. 'All' indicates the 143 SMV genomes, whereas group 1 and group 2 include 6 and 137 SMV genomes, respectively.

3. Discussion

In this study, we found seven complete SMV genome sequences from four soybean germplasms using RNA sequencing. Of the four soybean germplasms, two germplasms, GR1 and GR2, contained two and three SMV isolates, respectively. This result indicated that a single soybean seed can contain more than one SMV variant at a time. The nucleotide identity between the SMV isolates in GR1 was 99.68%, whereas that among the three SMV isolates in GR2 ranged from 99.87% to 99.26%. This result indicates that the SMV isolates within the same soybean seed have a high degree of sequence similarity. In RNA viruses, a viral quasispecies is defined as a viral population structure consisting of several viral variants (isolates) that play an important role in maintaining the genetic diversity of RNA viruses [20]. Seed transmission of SMV is well known; however, the seed transmission rate for SMV may vary [21]. Our previous study showed a high frequency of SMV mutations in soybean seed transcriptome data [22]. Thus, we showed that viral quasispecies contribute to SMV diversity within the soybean germplasm.

In RNA viruses, recombination, i.e., the exchange of genetic segments between two viral genomes, plays an important role in viral diversity and evolution [23]. As expected, the 143 SMV genomes showed a high number of recombinant events (76), and most SMV genomes (141) were predicted to be recombinants. Several previous studies have also revealed recombination events among SMV isolates/strains. Recombination events have also been identified between distinct SMV pathotypes [24]. A previous study revealed 19 SMV recombinant events in 44 SMV isolates, of which 30 were derived from Korea [12]. Another study revealed 32 recombination events in 83 SMV isolates, including 18 newly sequenced Chinese SMV isolates [13]. As the number of SMV genomes used for recombination analysis increases, the number of recombination events also seems to increase. Interestingly, a previous study suggested that two SMV isolates, 4547_CHN_2004 (HQ396725) and 4469-4_CHN_2004 (HM590055.1), could be derived from a recombination event between SMV and bean common mosaic virus (BCMV), and they were referred to as the SMV-R (recombination) strain [25]. Based on this result, we suggest that some SMV isolates in clade 5 containing two SMV-R strains might be recombinants between SMV and BCMV. Thus, the effect of recombination on the genetic diversity of SMV genomes may be much greater than expected. Moreover, we found no specific protein regions in which recombination occurred frequently. Recombination events occurred in most SMV genome regions covering the polyprotein, except for the P1 region, which was identified as a cold spot for recombination.

A previous study using 83 SMV genome sequences identified 4 major clades [13]. Phylogenetic analysis using SMV complete coding sequences revealed 9 clades that collectively contained 142 SMV isolates/strains. It is possible that a phylogenetic tree with a higher number of viral genome sequences has a higher number of clades.

Of the nine clades, clades 1, 2, and 3 contained a higher number of SMV isolates/strains than the other clades. We identified seven previously well-defined SMV strains (G1 through G7) based on pathotypes belonging to clade 1 (G6 and G7) and clade 3 (G1 through G5) [26]. SMV isolates in clades 1 and 3 were identified in diverse countries, and South Korea was a major source of these isolates. Moreover, SMV isolates in clade 1 and clade 3 were found in diverse plant species, including *G. max* and *G. soja*. These results suggest that SMV isolates in clades 1 and 3 are the most common SMV pathotypes, with a worldwide distribution. In particular, they have been found in South Korea and the USA. In contrast, SMV isolates in clade 2, derived mostly from *G. max* in China, were unique compared to those in the other clades. Two types of SMV strains have been identified in China. The first Chinese SMV type consists of 22 SMV SC strains isolated from soybean cultivars in southern China [27]. The second SMV Chinese type contains the N1–N3 strains, which were identified by evaluating soybean cultivars from northeast China [28]. Clade 2 contained the N1, N3, SC3, SC6, and SC6-N strains, which were previously identified as SMV strains from China. Therefore, the SMV isolates in clade 2 may be related to the well-known Chinese SMV pathotypes that originated in and are now distributed throughout China. In addition, 16 SMV isolates in clades 4 through 7 were derived from China and showed high sequence diversity when compared with other SMV isolates identified in *G. max* and *G. soja*. This result indicates that the wide geographical distribution of soybean cultivation in China may contribute to the genetic diversity of the SMV isolates. The most unique clades of the SMV isolates were clades 8 and 9. Clade 8 consists of four SMV isolates from two *Pinellia* species, whereas clade 9 contains a single SMV isolate from *Atractylodes macrocephala*. These results suggest that plant hosts are a powerful force driving mutations in the SMV genome.

Nucleotide diversity and Watterson estimator results for the nucleotide sequences corresponding to individual SMV proteins showed values that were all lower than one, indicating that all the SMV gene sequences analyzed were under negative selection. However, nucleotide diversity in the P1 region was much higher than that in other SMV protein regions, suggesting a high degree of nucleotide diversity in the P1 region. A comparison of the two groups of SMV isolates based on their plant hosts showed that the nucleotide diversity of SMV isolates in group 1, which were derived from *Pinellia* and *Atractylodes* species, was much higher than those in group 2, which were derived mostly from *G. max* and *G. soja*. The dN/dS ratio results for natural selection pressure revealed that most SMV protein sequences were negatively selected, except for P1, which showed positive selection in all analyzed SMV sequences. Our results are consistent with those of a recent study that used 104 complete SMV sequences [29]. However, the dN/dS ratios for P1 within group 1 and group 2 were less than one, indicating negative selection. Thus, we found that the P1 region of SMV isolates in group 1 contributes to the positive selection of the P1 protein. Moreover, a recent study based on the comparison of two different SMV isolates with high sequence variability in the P1 region suggested that the P1 protein of SMV might be related to viral adaptation to different hosts [30]. Both our results and that of previous studies suggest that positive selection and high sequence variability in the P1 protein of SMV isolates in group 1 might be highly correlated with host adaptation abilities.

The SMV polyprotein sequences in 'all' and group 1 showed positive selection. Again, the dN/dS ratio (1.568) for the polyprotein using all SMV sequences was significantly reduced in group 1 (1.025) after SMV isolates in group 1 were removed. Furthermore, the dN/dS ratio for the polyprotein was higher than that for the PIPO. Interestingly, the dN/dS ratios for PIPO in the three different groups were quite similar, suggesting the genetic conservation of PIPO sequences in strains derived from different plant species. Taken together, our results provide strong evidence that plant hosts such as *Pinellia* and *Atractylodes* species play important roles in maintaining the genetic diversity of the SMV genome.

4. Materials and Methods

4.1. Soybean Germplasms

Germplasms are defined as living genetic resources maintained as seeds and plant tissues. Before identifying soybean genetic resources resistant to SMV, we used RT-PCR to examine viral infections in 150 different soybean germplasms derived from the National Agrobiodiversity Center of the Rural Development Administration. In June 2020, soybean seeds, which were representative of the soybean germplasm, were planted in plastic nursery pots in a greenhouse at the National Institute of Crop Science in Wanju, Korea.

4.2. Total RNA Extraction and Library Zpreparation

We harvested mature trifoliate leaves from soybean seedlings and extracted total RNA using an RNeasy Plant Mini Kit (Qiagen, Hilden, Germany), according to the manufacturer's instructions. The extracted total RNA was used for reverse transcriptase-polymerase chain reaction (RT-PCR) analysis using SMV-specific primers. We selected 10 soybean germplasms that were infected with SMV. The quality of the extracted total RNAs was measured using a 2100 Bioanalyzer (Agilent Technologies, Santa Clara, CA, USA). For library preparation, we used total RNAs with an RNA integrity number (RIN) value greater than or equal to seven. After removing ribosomal RNA from the total RNA using TruSeq Stranded Total RNA with a Ribo-Zero Plant Kit (Illumina, San Diego, CA, USA), we generated 10 libraries for RNA sequencing using the TruSeq Stranded Total RNA LT Sample Prep Kit (Illumina).

4.3. RNA-Sequencing and Viral Genome Assembly

Ten libraries representing ten soybean germplasms were paired-end (2 × 101 bp) sequenced using a NovaSeq 6000 system (Macrogen, Seoul, Republic of Korea). We performed de novo transcriptome assembly using the raw sequence reads obtained from the Trinity assembler (version 2.13.2) with default parameters. The assembled contigs in each transcriptome were used for a BLASTX search against the viral reference protein database (21 August 2022) obtained from the National Center for Biotechnology Information (NCBI) using the DIAMOND program with an E-value of 1×10^{-5} as the cutoff [31]. Using the BLASTX results, we selected only virus-associated contigs. All virus-associated contigs were derived from SMV. Of the SMV-associated contigs, we selected viral contigs with sizes greater than 9000 bp to identify SMV sequences that covered all open reading frames (ORFs) using the NCBI ORFfinder [32].

4.4. Collection of SMV Genomes from GenBank

Using SMV as a query, we searched for and downloaded all SMV-associated sequences from GenBank in the NCBI database (11 August 2022). We obtained 667 SMV sequences from the GenBank database. In this study, we combined 667 sequences with 7 SMV genome sequences, resulting in a total of 674 sequences. Using reformat.sh from the BBMap suite, we deleted sequences of less than 9000 bp. After deleting redundant sequences, we obtained 143 SMV genome sequences.

4.5. Generation of Phylogenetic Trees

A total of 143 SMV genome sequences were aligned using MAFFT version 7 with the L-INS-i option [33]. Subsequently, we extracted only coding DNA sequences (CDSs) for 143 SMV genomes by deleting the 5′ and 3′ untranslated regions using the MEGA7 program [34]. After the alignment of the 143 coding DNA sequences, they were subjected to phylogenetic tree construction using IQ-Tree version 2.2.0 [35]. We used ModelFinder implemented in IQ-Tree for fast model selection [36]. Then, 2 fast approximate likelihood-based measures of branch support, SH-aLRT and ultrafast bootstrap, were conducted with 1000 bootstrap replicates. The generated maximum likelihood phylogenetic tree was imported into Figtree version 1.4.4 (http://tree.bio.ed.ac.uk/software/figtree/, (accessed on 1 August 2022)). The phylogenetic tree was modified using iTOL version 4 [37].

4.6. Recombination Analysis and Network Tree for 143 SMV Genomes

We conducted recombination analyses for 143 SMV genomes using RDP version 5.23 [38]. A full exploratory recombination scan using all seven available detection methods, namely RDP, GENECONV, Bootscan, Maxchi, Chimaera, SiSscan, and 3Seq, was used for the recombination analyses. Of the seven algorithms, recombinants predicted by at least five algorithms with a *p*-value less than 0.05 were regarded as true recombinants. We rechecked the identified recombinant events using all seven detection methods. An unrooted network tree based on the aligned 143 SMV genome sequences was generated using SplitsTree version 5.3.0 [38].

4.7. Analyses of Genetic Diversity and Selection Pressure for Individual SMV Proteins

After aligning the 143 SMV genome sequences, we manually extracted nucleotide sequences corresponding to individual SMV proteins. For example, the sequences of the two ORFs encoding polyproteins and PIPO were extracted first. Next, we obtained nucleotide sequences corresponding to the ten mature proteins cleaved from the SMV polyprotein. As a result, the nucleotide sequences of 12 SMV proteins were used for genetic diversity analysis using DnaSP version 6.12.03 [39]. The genetic diversity indices included the total number of sites, number of segregating sites (S), total number of mutations (Eta), number of haplotypes (H), haplotype diversity (Hd), nucleotide diversity (Pi), and Watterson's estimator of θ (θw).

We calculated the dN/dS ratio for the 12 individual SMV proteins to determine selection pressure using DnaSP. dN represents the average number of nonsynonymous substitutions per non-synonymous site, whereas dS represents the average number of synonymous substitutions per synonymous site. A dN/dS ratio of less than one indicates negative selection (purifying), whereas a dN/dS ratio greater than one indicates positive selection (diversifying). A dN/dS ratio of zero indicates neutral selection.

5. Conclusions

In this study, we obtained seven SMV genome sequences from soybean germplasms using RNA sequencing and combined all available SMV genome sequences, totaling 143 SMV genomes. Then we conducted a large-scale comprehensive evolutionary and phylogenetic study of these SMV genomes, which were identified from 10 plant species and 12 countries. A maximum likelihood phylogenetic tree constructed using the polyprotein sequences showed that the 142 SMV isolates/strains grouped into 9 clades. A recombination analysis revealed 76 recombinant events and 141 recombinants. Most regions of the SMV genome showed a high number of breakpoints, whereas P1 was identified as a cold spot. Clades 1 and 3 contain the most common SMV pathotypes, including G1 through G7, which are distributed worldwide, whereas clade 2 contains SMV pathotypes that are distributed in China and might have originated there. The SMV isolates were further divided into two groups. The SMV isolates in the first group were identified in *Pinellia* and *Atractylodes* species, whereas the second group was identified mostly in cultivated soybeans. The SMV polyprotein undergoes positive selection, whereas most mature proteins undergo negative selection. The only exception we found was the P1 protein. The P1 protein of SMV isolates in group 1 may be highly correlated with host adaptation. Taken together, our results provide strong evidence that recombination and plant hosts are powerful forces that drive mutations and genetic diversity in the SMV genome.

Supplementary Materials: The supporting information can be downloaded at: https://www.mdpi.com/article/10.3390/ijms24010022/s1.

Author Contributions: Data curation, H.C., Y.J., B.C.L. and W.K.C.; Formal analysis, H.C., Y.J., B.C.L. and W.K.C.; Funding acquisition, B.C.L.; Investigation, H.C., Y.J., S.Y.C., S.-M.K., J.-S.H., B.C.L. and W.K.C.; Project administration, B.C.L. and W.K.C.; Resources, Y.M.C., J.-S.H. and B.C.L.; Supervision, B.C.L. and W.K.C.; Writing—original draft, W.K.C.; Writing—review & editing, B.C.L. and W.K.C. All authors have read and agreed to the published version of the manuscript.

Funding: This research was funded by the "Cooperative Research Program for Agriculture, Science, & Technology Development" (Project No. PJ01416802) conducted by the Rural Development Administration, Republic of Korea, and was supported by the Korea Institute of Planning and Evaluation for Technology in Food, Agriculture and Forestry (IPET) through the Crop Viruses and Pests Response Industry Technology Development Program, funded by the Ministry of Agriculture, Food and Rural Affairs (MAFRA) (121055-2), and by a National Research Foundation of Korea (NRF) grant funded by the Korean government (MSIT) (No. NRF-2018R1D1A1B07043597). The funders had no role in study design, data collection and analysis, the decision to publish, or the preparation of the manuscript.

Institutional Review Board Statement: Not applicable.

Informed Consent Statement: Not applicable.

Data Availability Statement: The seven SMV genome sequences used in this study are publicly available in the NCBI GenBank database with the following accession numbers: ON843744–ON843750.

Acknowledgments: We would like to express our deepest gratitude to Jin Kyong Cho, Hyang Sook Kim, and Mi Kyong Kim for their support and commitment to soybean cultivation and harvest.

Conflicts of Interest: The authors declare no conflict of interest.

References

1. Liu, J.-Z.; Fang, Y.; Pang, H. The current status of the soybean-*Soybean mosaic virus* (SMV) pathosystem. *Front. Microbiol.* **2016**, *7*, 1906. [CrossRef] [PubMed]
2. Hajimorad, M.; Domier, L.; Tolin, S.; Whitham, S.; Saghai Maroof, M. *Soybean mosaic virus*: A successful potyvirus with a wide distribution but restricted natural host range. *Mol. Plant Pathol.* **2018**, *19*, 1563–1579. [CrossRef]
3. Domier, L.L.; Hobbs, H.A.; McCoppin, N.K.; Bowen, C.R.; Steinlage, T.A.; Chang, S.; Wang, Y.; Hartman, G.L. Multiple loci condition seed transmission of *Soybean mosaic virus* (SMV) and SMV-induced seed coat mottling in soybean. *Phytopathology* **2011**, *101*, 750–756. [CrossRef] [PubMed]
4. Wrather, J.; Anderson, T.; Arsyad, D.; Tan, Y.; Ploper, L.D.; Porta-Puglia, A.; Ram, H.; Yorinori, J. Soybean disease loss estimates for the top ten soybean-producing counries in 1998. *Can. J. Plant Pathol.* **2001**, *23*, 115–121. [CrossRef]
5. Domier, L.L.; Steinlage, T.A.; Hobbs, H.A.; Wang, Y.; Herrera-Rodriguez, G.; Haudenshield, J.S.; McCoppin, N.K.; Hartman, G.L. Similarities in seed and aphid transmission among *Soybean mosaic virus* isolates. *Plant Dis.* **2007**, *91*, 546–550. [CrossRef] [PubMed]
6. Jossey, S.; Hobbs, H.A.; Domier, L.L. Role of *Soybean mosaic virus*–encoded proteins in seed and aphid transmission in soybean. *Phytopathology* **2013**, *103*, 941–948. [CrossRef] [PubMed]

7. Kiihl, R.A.; Hartwig, E. Inheritance of reaction to *Soybean mosaic virus* in soybeans 1. *Crop. Sci.* **1979**, *19*, 372–375. [CrossRef]
8. Hayes, A.; Jeong, S.; Gore, M.; Yu, Y.; Buss, G.; Tolin, S.; Maroof, M.S. Recombination within a nucleotide-binding-site/leucine-rich-repeat gene cluster produces new variants conditioning resistance to *Soybean mosaic virus* in soybeans. *Genetics* **2004**, *166*, 493–503. [CrossRef]
9. Jain, R.; McKern, N.M.; Tolin, S.; Hill, J.; Barnett, O.; Tosic, M.; Ford, R.; Beachy, R.; Yu, M.; Ward, C. Confirmation that fourteen potyvirus isolates from soybean are strains of one virus by comparing coat protein peptide profiles. *Phytopathology* **1992**, *82*, 294–299. [CrossRef]
10. Jayaram, C.; Hill, J.H.; Miller, W.A. Nucleotide sequences of the coat protein genes of two aphid-transmissible strains of *Soybean mosaic virus*. *J. Gen. Virol.* **1991**, *72*, 1001–1003. [CrossRef]
11. Jayaram, C.; Hill, J.H.; Miller, W.A. Complete nucleotide sequences of two *Soybean mosaic virus* strains differentiated by response of soybean containing the Rsv resistance gene. *J. Gen. Virol.* **1992**, *73*, 2067–2077. [CrossRef] [PubMed]
12. Seo, J.-K.; Ohshima, K.; Lee, H.-G.; Son, M.; Choi, H.-S.; Lee, S.-H.; Sohn, S.-H.; Kim, K.-H. Molecular variability and genetic structure of the population of *Soybean mosaic virus* based on the analysis of complete genome sequences. *Virology* **2009**, *393*, 91–103. [CrossRef] [PubMed]
13. Zhou, G.-C.; Shao, Z.-Q.; Ma, F.-F.; Wu, P.; Wu, X.-Y.; Xie, Z.-Y.; Yu, D.-Y.; Cheng, H.; Liu, Z.-H.; Jiang, Z.-F. The evolution of *Soybean mosaic virus*: An updated analysis by obtaining 18 new genomic sequences of Chinese strains/isolates. *Virus Res.* **2015**, *208*, 189–198. [CrossRef] [PubMed]
14. Ahangaran, A.; Habibi, M.K.; Mohammadi, G.-H.M.; Winter, S.; García-Arenal, F. Analysis of *Soybean mosaic virus* genetic diversity in Iran allows the characterization of a new mutation resulting in overcoming Rsv4-resistance. *J. Gen. Virol.* **2013**, *94*, 2557–2568. [CrossRef]
15. Hajimorad, M.; Eggenberger, A.; Hill, J. Evolution of *Soybean mosaic virus*-G7 molecularly cloned genome in Rsv1-genotype soybean results in emergence of a mutant capable of evading Rsv1-mediated recognition. *Virology* **2003**, *314*, 497–509. [CrossRef]
16. Massart, S.; Olmos, A.; Jijakli, H.; Candresse, T. Current impact and future directions of high throughput sequencing in plant virus diagnostics. *Virus Res.* **2014**, *188*, 90–96. [CrossRef]
17. Nakasato, K.; Fujioka, S.; Sugawara, Y.; Ono, T.; Nishio, T.; Tsuda, S. First detection of two potyviruses, uraria mosaic virus and passiflora mosaic virus Y, from passionfruit in Japan. *J. Gen. Plant Pathol.* **2020**, *86*, 401–404. [CrossRef]
18. Tajima, F. Statistical method for testing the neutral mutation hypothesis by DNA polymorphism. *Genetics* **1989**, *123*, 585–595. [CrossRef]
19. Jeffares, D.C.; Tomiczek, B.; Sojo, V.; Reis, M.d. A beginners guide to estimating the non-synonymous to synonymous rate ratio of all protein-coding genes in a genome. In *Parasite Genomics Protocols*; Springer: Cham, Switzerland, 2015; pp. 65–90.
20. Domingo, E.; Perales, C. Viral quasispecies. *PLoS Genet.* **2019**, *15*, e1008271. [CrossRef]
21. Bowers, G., Jr.; Goodman, R. *Soybean mosaic virus*: Infection of soybean seed parts and seed transmission. *Phytopathology* **1979**, *69*, 569–572. [CrossRef]
22. Jo, Y.; Choi, H.; Bae, M.; Kim, S.-M.; Kim, S.-L.; Lee, B.C.; Cho, W.K.; Kim, K.-H. De novo genome assembly and single nucleotide variations for *Soybean mosaic virus* using soybean seed transcriptome data. *Plant Pathol. J.* **2017**, *33*, 478. [CrossRef] [PubMed]
23. Pérez-Losada, M.; Arenas, M.; Galán, J.C.; Palero, F.; González-Candelas, F. Recombination in viruses: Mechanisms, methods of study, and evolutionary consequences. *Infect. Genet. Evol.* **2015**, *30*, 296–307. [CrossRef] [PubMed]
24. Gagarinova, A.G.; Babu, M.; Strömvik, M.V.; Wang, A. Recombination analysis of *Soybean mosaic virus* sequences reveals evidence of RNA recombination between distinct pathotypes. *Virol. J.* **2008**, *5*, 143. [CrossRef] [PubMed]
25. Yang, Y.; Lin, J.; Zheng, G.; Zhang, M.; Zhi, H. Recombinant *Soybean mosaic virus* is prevalent in Chinese soybean fields. *Arch. Virol.* **2014**, *159*, 1793–1796. [CrossRef]
26. Cho, E.-K.; Goodman, R.M. Strains of *Soybean mosaic virus*: Classification based on virulence in resistant soybean cultivars. *Phytopathology* **1979**, *69*, 467–470. [CrossRef]
27. Li, K.; Yang, Q.; Zhi, H.; Gai, J. Identification and distribution of *Soybean mosaic virus* strains in southern China. *Plant Dis.* **2010**, *94*, 351–357. [CrossRef] [PubMed]
28. Wenqing, L.; Minghou, Z.; Peiwen, W.; Shuyi, X. Classification and distribution of strains of *Soybean mosaic virus* in northeast China. *Acta Phytophylacica Sin.* **1985**, *15*, 225–229.
29. Gao, L.; Wu, Y.; An, J.; Huang, W.; Liu, X.; Xue, Y.; Luan, X.; Lin, F.; Sun, L. Pathogenicity and genome-wide sequence analysis reveals relationships between *Soybean mosaic virus* strains. *Arch. Virol.* **2022**, *167*, 517–529. [CrossRef] [PubMed]
30. Mao, C.; Shan, S.; Huang, Y.; Jiang, C.; Zhang, H.; Li, Y.; Chen, J.; Wei, Z.; Sun, Z. The hypervariable N-terminal of *Soybean mosaic virus* P1 protein influences its pathogenicity and host defense responses. *Phytopathol. Res.* **2022**, *4*, 10. [CrossRef]
31. Buchfink, B.; Xie, C.; Huson, D.H. Fast and sensitive protein alignment using DIAMOND. *Nat. Methods* **2015**, *12*, 59–60. [CrossRef]
32. Wheeler, D.L.; Chappey, C.; Lash, A.E.; Leipe, D.D.; Madden, T.L.; Schuler, G.D.; Tatusova, T.A.; Rapp, B.A. Database resources of the national center for biotechnology information. *Nucleic Acids Res.* **2000**, *28*, 10–14. [CrossRef]
33. Katoh, K.; Standley, D.M. MAFFT multiple sequence alignment software version 7: Improvements in performance and usability. *Mol. Biol. Evol.* **2013**, *30*, 772–780. [CrossRef] [PubMed]
34. Kumar, S.; Stecher, G.; Tamura, K. MEGA7: Molecular evolutionary genetics analysis version 7.0 for bigger datasets. *Mol. Biol. Evol.* **2016**, *33*, 1870–1874. [CrossRef] [PubMed]

35. Minh, B.Q.; Schmidt, H.A.; Chernomor, O.; Schrempf, D.; Woodhams, M.D.; Von Haeseler, A.; Lanfear, R. IQ-TREE 2: New models and efficient methods for phylogenetic inference in the genomic era. *Mol. Biol. Evol.* **2020**, *37*, 1530–1534. [CrossRef] [PubMed]
36. Kalyaanamoorthy, S.; Minh, B.Q.; Wong, T.K.; Von Haeseler, A.; Jermiin, L.S. ModelFinder: Fast model selection for accurate phylogenetic estimates. *Nat. Methods* **2017**, *14*, 587–589. [CrossRef] [PubMed]
37. Letunic, I.; Bork, P. Interactive Tree of Life (iTOL) v4: Recent updates and new developments. *Nucleic Acids Res.* **2019**, *47*, W256–W259. [CrossRef] [PubMed]
38. Huson, D.H.; Bryant, D. Application of phylogenetic networks in evolutionary studies. *Mol. Biol. Evol.* **2006**, *23*, 254–267. [CrossRef]
39. Rozas, J.; Ferrer-Mata, A.; Sánchez-DelBarrio, J.C.; Guirao-Rico, S.; Librado, P.; Ramos-Onsins, S.E.; Sánchez-Gracia, A. DnaSP 6: DNA sequence polymorphism analysis of large data sets. *Mol. Biol. Evol.* **2017**, *34*, 3299–3302. [CrossRef]

International Journal of
Molecular Sciences

MDPI

Article

Viromes of 15 Pepper (*Capsicum annuum* L.) Cultivars

Yeonhwa Jo [1,†], Hoseong Choi [2,†], Jeong Hun Lee [3], Sang Hyun Moh [3,*] and Won Kyong Cho [1,*]

1 College of Biotechnology and Bioengineering, Sungkyunkwan University, Suwon 16419, Korea
2 Plant Genomics and Breeding Institute, Seoul National University, Seoul 08826, Korea
3 Plant Cell Research Institute of BIO-FD&C Co., Ltd., Incheon 21990, Korea
* Correspondence: shmoh@biofdnc.com (S.H.M.); wonkyong@gmail.com (W.K.C.); Tel.: +82-32-811-2027 (S.H.M.); +82-31-290-7860 (W.K.C.)
† These authors contributed equally to this work.

Abstract: Pepper (*Capsicum annuum* L.) plants produce berry fruits that are used as spices. Here, we examined the viromes of 15 pepper cultivars through RNA sequencing. We obtained 1,325 virus-associated contigs derived from 8 virus species. Bean broad wilt virus 2 (BBWV2) and cucumber mosaic virus (CMV) were identified as the major viruses infecting pepper plants, followed by potato virus Y, bell pepper endornavirus, and hot pepper endornavirus. The proportion of viral reads in each transcriptome ranged from 0.04% to 24.5%. BBWV2 was the dominant virus in seven cultivars, whereas CMV was dominant in five cultivars. All the bell pepper cultivars showed severe viral disease symptoms, whereas the commercially developed hot pepper cultivars were asymptomatic or had mild symptoms. In addition, 111 complete viral segments were obtained from 7 viruses. Based on the obtained viral genomes, the genetic relationship between the identified viruses and quasispecies of BBWV2 and CMV in each pepper plant was determined. Newly designed primers for nine viruses confirmed the results of RNA sequencing. Taken together, this study, for the first time, provides a comprehensive overview of viromes in 15 major pepper cultivars through RNA sequencing.

Keywords: *Capsicum annuum*; pepper; RNA sequencing; virome; virus

Citation: Jo, Y.; Choi, H.; Lee, J.H.; Moh, S.H.; Cho, W.K. Viromes of 15 Pepper (*Capsicum annuum* L.) Cultivars. *Int. J. Mol. Sci.* **2022**, 23, 10507. https://doi.org/10.3390/ijms231810507

Academic Editors: Andrés J. Cortés and Hai Du

Received: 9 August 2022
Accepted: 8 September 2022
Published: 10 September 2022

1. Introduction

The pepper plant *Capsicum annuum* L., a member of the nightshade family Solanaceae, is known to produce berry fruits [1]. Peppers are cultivated for their pungency and used as spices in many cuisines across the world [2]. The plant originated in Bolivia and was first cultivated in Mexico [3]. Many pepper cultivars spread worldwide during the Columbian Exchange; Portuguese traders introduced the plant to Asian countries. To date, five pepper species have been domesticated. The most commonly cultivated pepper groups are bell peppers, sweet peppers, and hot peppers.

In Korea, bell peppers and hot peppers are widely cultivated. Hot peppers are consumed as fresh fruits or are dried to produce pepper powder. Those hot peppers can be divided into red hot pepper and sweet pepper based on their spiciness. While most other countries use the term "paprika" for chili or hot peppers, in Korea, "paprika" is used interchangeably with bell pepper to refer to the large sweet peppers in a variety of colors such as green, orange, yellow, and red. The production of bell peppers, mostly consumed as fresh fruit, has been increasing in Korea. According to the Korea Seed and Variety Service registry, approximately 600 hot pepper and 60 bell pepper cultivars have been developed to date by more than 50 seedling companies. Most pepper cultivars grown in Korea are of the red hot pepper variety, used to produce red hot pepper powders, one of the most important spices in Korean cuisine, used in kimchi, pastes, and other side dishes [4].

Pepper plants are usually cultivated from April to October in Korea, and red hot pepper fruits are harvested from July to October. Diseases, such as phytophthora blight and anthracnose, and harmful insects, such as whitefly, oriental tobacco budworm, and thrip, cause serious damage to the quality and quantity of pepper production. Therefore, a variety of pesticides are regularly sprayed on pepper plants. However, excessive pesticide use poses a major risk to human and environmental health.

In addition to harmful bacteria, fungi, and insects, several viruses are also known to infect pepper plants. For example, DNA viruses belonging to the genus *Geminivirus* and RNA viruses with single- or double-stranded RNA genomes are major viruses that infect pepper plants [5]. Viroids such as columnea latent viroid and pepper chat fruit viroid also infect these plants [6].

In Korea, the most frequently identified viruses infecting pepper plants are the cucumber mosaic virus (CMV), broad bean wilt virus 2 (BBWV2), and tomato spotted wilt virus (TSWV) [7]. Hot pepper endornavirus (HPEV), pepper cryptic virus 1 (PCV1), and pepper cryptic virus 2 (PCV2) have also recently been identified [8]. However, unlike in other Asian countries, no DNA viruses or viroids infecting pepper plants have yet been identified in Korea.

While some farmers grow pepper seedlings, most purchase them from different nurseries. In many cases, nurseries cultivate multiple plant varieties within a limited space, such as in a greenhouse, potentially increasing the frequency of viral infections. In this study, we examined the viromes of 15 different pepper cultivars derived from 5 major nurseries in Korea, via RNA sequencing, followed by intensive bioinformatic analyses.

2. Results

2.1. Collection of Leaf Samples from 15 Different Pepper Cultivars

We collected 15 pepper cultivars from 5 nurseries in Korea (Table 1). The nurseries were located in Anyang (four cultivars), Hampyeong (five cultivars), Ilsan (four cultivars), Pusan (one cultivar), and Uljin (one cultivar). Of these, six were highly pungent red hot peppers, five were sweet peppers with a low level of pungency, and four were bell peppers.

Table 1. Summary of pepper cultivars used for virome study. A total of 15 different pepper cultivars derived from 5 nurseries indicated by region, were used for the virome study. Pepper cultivars were divided into three categories: red hot pepper, sweet pepper, and sweet paprika.

Index	Name	Category	Spicy	Region
P01	PR-9988	Red hot pepper	Hot	Hampyeong
P02	Mini paprika	Sweet paprika	Sweet	Ilsan
P03	TS-Monster	Red hot pepper	Hot	Hampyeong
P04	Boradori	Sweet pepper	Sweet	Ilsan
P05	Miin	Sweet pepper	Sweet	Ilsan
P06	Chegangtan	Red hot pepper	Hot	Pusan
P07	Dangjo	Sweet pepper	Sweet	Ilsan
P08	Ground cherry (Wrinkly)	Red hot pepper	Hot	Uljin
P09	Yellow paprika	Sweet paprika	Sweet	Anyang
P10	Vitamin	Sweet pepper	Sweet	Hampyeong
P11	Oi (Cucumber)	Sweet pepper	Sweet	Hampyeong
P12	Caltan	Red hot pepper	Hot	Anyang
P13	Cheongyang	Red hot pepper	Hot	Anyang
P14	Red paprika	Sweet paprika	Sweet	Anyang
P15	Green paprika	Sweet paprika	Sweet	Hampyeong

2.2. Viral Disease Symptoms in Different Pepper Cultivars

All the bell pepper plants showed severe viral disease symptoms such as stunting, mosaic pattern, and curling (Figure 1). These plants did not grow well and produced a small number of fruits. Among the hot pepper cultivars, P13 (Cheongyang) exhibited the strongest disease symptoms, including a mosaic pattern, mottling, chlorosis, and leaf distortion. However, the Cheongyang cultivar produced a high number of normal-quality pepper fruits. The remaining pepper cultivars showed mild viral disease symptoms or were asymptomatic (P01 and P03).

Figure 1. Images of 10 representative pepper cultivars. Few cultivars, such as P01 and P03, did not show any visible viral disease symptoms, while all paprika plants, such as P02 and P14, displayed very severe disease symptoms, such as stunting, and abnormal growth. Most pepper cultivars displayed mild symptoms, such as chlorosis, curling, mosaic, mottling, and vein banding.

2.3. Identification of Virus-Associated Contigs from 15 Different Pepper Cultivars

Regardless of the viral disease symptoms, we randomly selected individuals from different pepper cultivars. From each individual, we harvested five different leaf samples and extracted total RNA. We deleted ribosomes from these total RNAs and generated 15 different libraries representing individual pepper cultivars. By sequencing RNA followed by *de novo* transcriptome assembly using the Trinity program, we obtained 15 pepper transcriptomes. Each transcriptome was subjected to a BLASTX against the viral protein database. The identified virus-associated contigs were further subjected to a BLASTX against non-redundant protein databases to filter out non-viral sequences. Finally, 1325 virus-associated contigs were obtained. P14 contained the highest (213), and P01 showed the lowest (20) number of virus-associated contigs (Figure 2A).

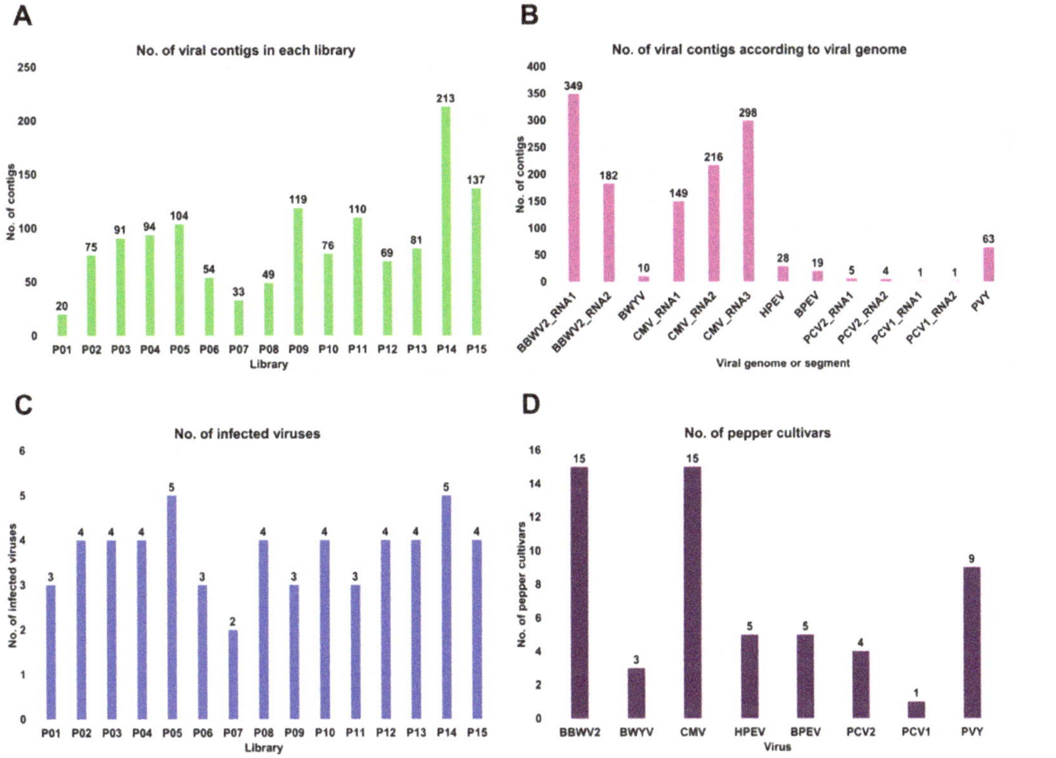

Figure 2. Number of identified virus-associated contigs and viruses from 15 different pepper cultivars: (**A**) number of viral contigs identified from each library; (**B**) number of viral contigs according to identified viral genome or segment; (**C**) number of viruses identified from each library; (**D**) number of pepper cultivars in which individual viruses were identified.

The identified virus-associated contigs were derived from eight RNA viruses, four of which (BBWV2, CMV, PCV2, and PCV1) were identified in this study and known to have multipartite genomes. For example, BBWV2, PCV2, and PCV1 had two RNA segments, whereas CMV had three. In total, 663 virus-associated contigs were derived from three CMV RNA segments: RNA1 (149 contigs), RNA2 (216 contigs), and RNA3 (298 contigs) (Figure 2B). A total of 531 virus-associated contigs were derived from two BBWV2 segments: RNA1 (349 contigs) and RNA2 (182 contigs). The other four RNA viruses, Beet western yellows virus (BWYV) (10 contigs), HPEV (28 contigs), bell pepper endornavirus (BPEV) (19 contigs), and potato virus Y (PVY) (63 contigs), were observed to have a monopartite genome.

All the examined pepper cultivars were coinfected with at least two viruses (Figure 2C). The number of infected viruses in each cultivar ranged from two (P07) to five (P05 and P14). The cultivars P01, P06, P09, and P11 were infected with three different viruses, whereas the remaining eight cultivars were infected with four different viruses.

We also examined the number of pepper cultivars infected with the individual virus (Figure 2D). BBWV2 and CMV were detected in 15 and 13 pepper cultivars, respectively. PVY infected nine cultivars, whereas BPEV and HPEV infected five cultivars. BWYV was detected in three cultivars, whereas only one cultivar (P14) was infected with PCV1.

The proportion of viral reads in each transcriptome ranged from 0.04% (P06) to 24.5% (P04) (Figure 3A). The proportion of viral reads for three cultivars (P01, P02, and P06) was less than 1%, whereas six cultivars (P04, P09, P10, P13, P14, and P15) had a viral read proportion of more than 20%. HPEV and PVY were predominant in P01 and P05, respectively, whereas PCV2 was the dominant virus in P02 (Figure 3B). BBWV2 was the dominant virus in seven cultivars (P03, P06, P07, P08, P10, P12, and P13), and CMV exhibited its dominance in five cultivars (P04, P09, P11, P14, and P15).

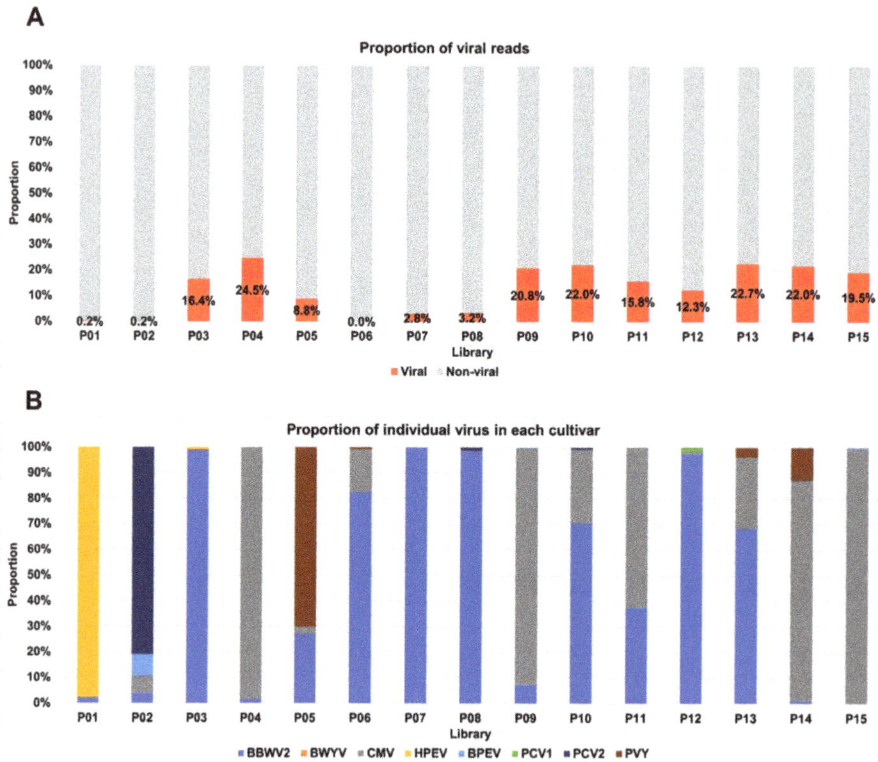

Figure 3. The proportion of viral reads in each library and the proportion of identified viruses in each library: (**A**) the proportion of viral reads (red color) as compared to non-viral reads (gray color); (**B**) the proportion of identified viruses in each library. Only viral reads assigned to the identified viruses were used for the calculation of virus proportion in each library.

2.4. Viral Genome Assembly

We obtained 111 complete viral segments for 7 viruses by *de novo* transcriptome assembly from 15 samples (Table S1). When the assembled viral genome covered all open reading frames (ORFs), we regarded it as the complete viral genome. Out of the eight identified viruses, four had multipartite genomes. Therefore, we examined the number of assembled segments for each virus (Figure 4). We obtained the highest number of BBWV2 RNA1 segments (24) followed by BBWV2 RNA2 (23) (Figure 4A). For CMV, the highest number of segments were obtained from CMV RNA3 (21), followed by CMV RNA1 (15) and CMV RNA2 (13) (Figure 4B). We obtained the same number of segments for each virus for PCV1 (1) and PCV2 (3; Figure 4C,D). We obtained the complete genomes for BPEV (2), BWYV (1), and PVY (4; Figure 5A). However, we did not obtain a complete genome sequence for HPEV.

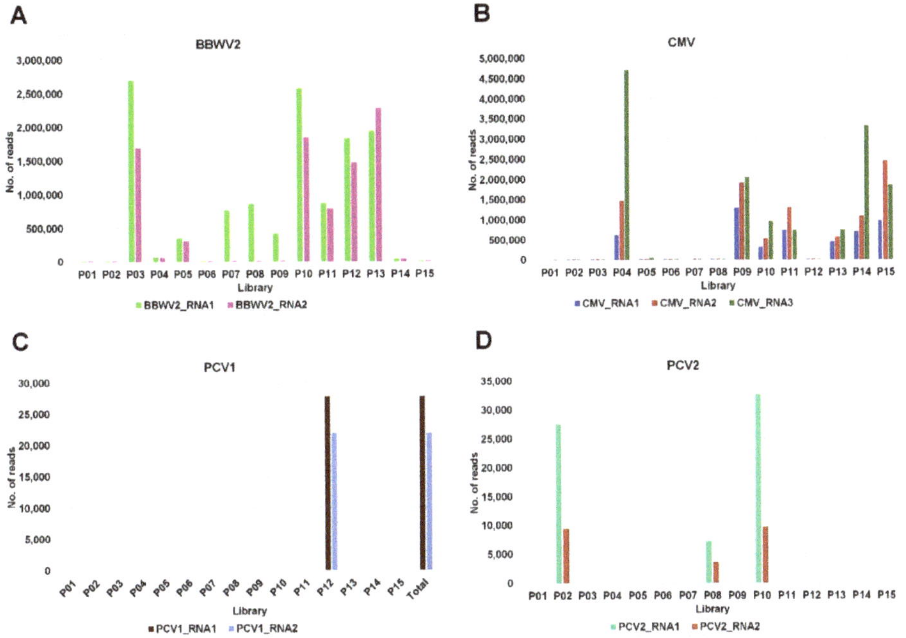

Figure 4. The number of viral reads in each library derived from four viruses composed of multiple viral segments. The number of viral reads in each library derived from BBWV2 composed of two RNA segments (**A**), CMV composed of three RNA segments (**B**), PCV1 composed of two RNA segments (**C**), and PCV2 composed of two RNA segments (**D**).

For BBWV2 and CMV, the complete genome sequences of several variants (isolates) were identified from the same pepper sample (Figure 5B,C). In each pepper sample, the number of BBWV2 RNA1 variants ranged from two to three, while the number of BBWV2 RNA2 variants ranged from six to three (Figure 5B). The numbers of variants of BBWV2 RNA1 and BBWV2 RNA2 were not consistent among the samples. For example, in P05, there were two variants of BBWV2 RNA1 and five variants of BBWV2 RNA2. We identified 20 variants of BBWV2 RNA1 and 16 variants of BBWV2 RNA2 from eight samples.

Similarly, we identified 36 CMV genome sequences, CMV RNA1 (9 sequences), CMV RNA2 (6 sequences), and CMV RNA3 (21 sequences) (Figure 5C). Interestingly, the number of variants for each CMV RNA segment varied among the pepper samples. The number of CMV RNA1 variants ranged from two to four, while the number of samples for CMV RNA2 variants ranged from two to three samples. For CMV RNA3, the number of variants in each sample ranged from two to five. However, the number of variants among the CMV RNA segments in the same sample was not consistent.

Based on the obtained viral genomes, we analyzed the phylogenetic relationships of the identified viral variants with known isolates for each virus. For BWYV, only a single variant, P03, was identified in this study. This variant showed sequence similarity to known BWYV isolates S14 and S19, which were identified in spinach in Japan. The phylogenetic tree using 11 BWYV isolates showed that BWYV variant P03 belonged to group A, possessing nine isolates of BWYV, whereas group B contained two BWYV isolates from France and the United States (Figure 6A).

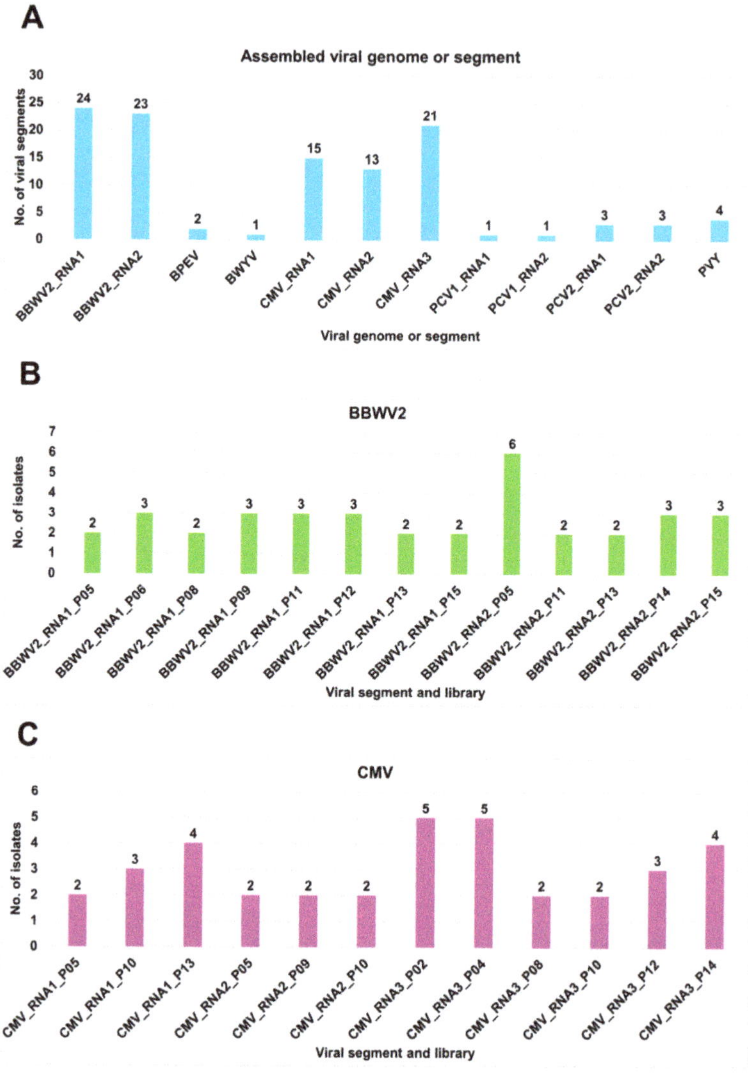

Figure 5. Number of assembled viral genomes: (**A**) total number of viral genomes or viral segments *de novo* assembled from RNA sequencing; (**B**) number of assembled viral variants for BBWV2 in each library; (**C**) number of assembled viral variants for CMV in each library. For viruses with multiple RNA segments, the number of assembled viral genomes is indicated by each viral segment.

Although we identified two alphaendornaviruses, HPEV and BPEV, only two complete genomes of the BPEV variants, P14 and P15, were obtained. The phylogenetic tree using 24 BPEV isolates demonstrated that P14 and P15 belonged to group A containing 18 BPEV isolates (Figure 6B). Most of the BPEV isolates in group A were closely related. In contrast, the six BPEV isolates in group B were distantly related.

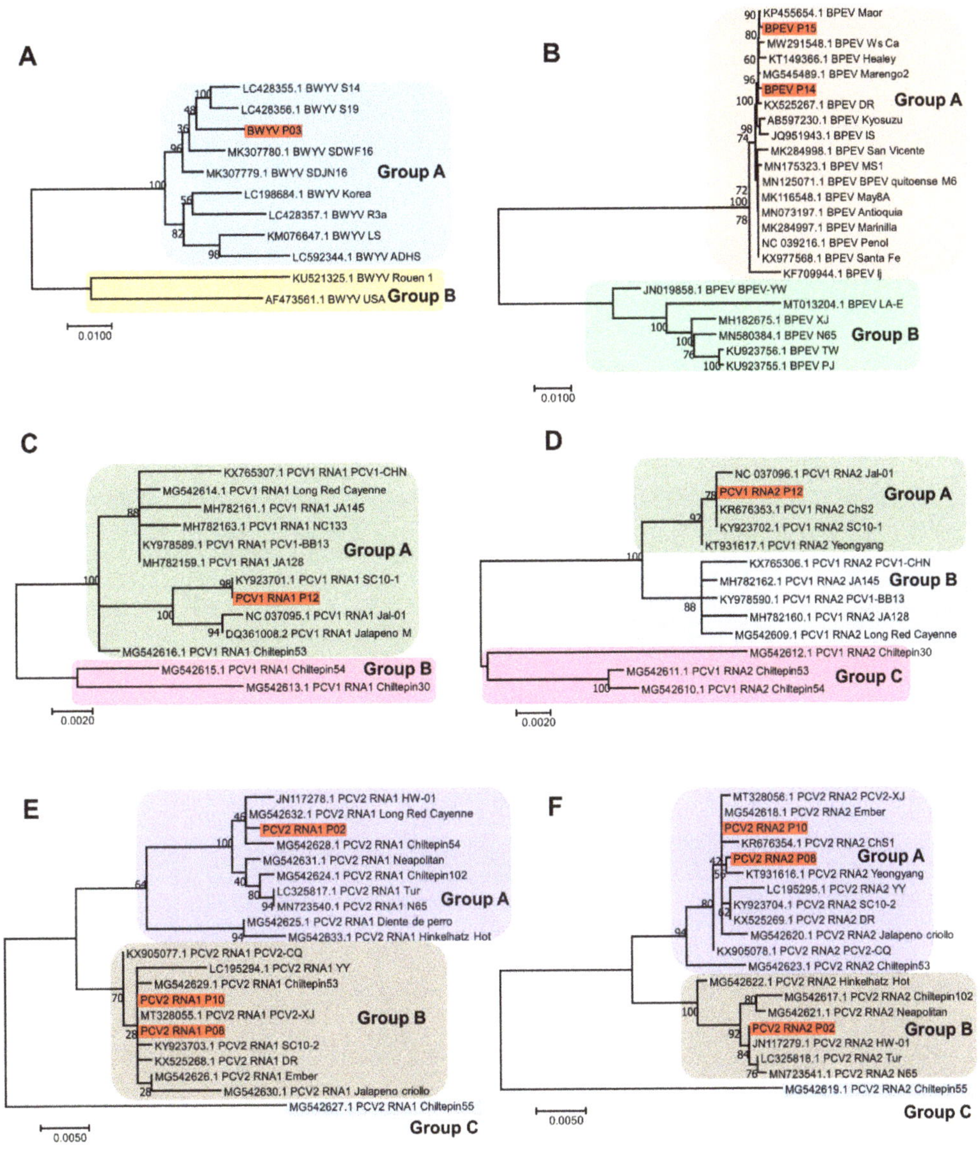

Figure 6. Phylogenetic tree of four viruses infecting pepper cultivars; BWYV, BPEV, PCV1, and PCV2: (**A**) phylogenetic tree of complete genome sequences of BWYV (**A**), BPEV (**B**), PCV1 RNA1 segment (**C**), PCV1 RNA2 segment (**D**), PCV2 RNA1 segment (**E**), and PCV2 RNA2 segment (**F**). Red-colored boxes indicate the assembled viral genome obtained from this study.

We identified a single PCV1 variant in the P12 library. The PCV1 genome consists of two RNA segments: RNA1 and RNA2. The phylogenetic tree based on the 13 PCV1 RNA1 sequences revealed two groups of PCV1 isolates (Figure 6C). RNA1 of PCV1 variant P12 showed strong sequence similarity to the known PCV1 isolate SC10-1 identified in Korea, and both the Korean isolates belonged to group A. A phylogenetic tree using 13 sequences of PCV1 RNA2 identified three groups of PCV1 isolates (Figure 6D). Group A included four isolates from Korea (P12, ChS2, SC10-1, and Yeongyang) and one isolate

(Jal-01) from the United States. The five isolates in group B were closely related. While the three PCV1 isolates in group C were identified from the same country (Mexico), they were distantly related.

For PCV2, composed of two RNA segments, we obtained three and two complete genome sequences for RNA1 and RNA2, respectively. We generated two phylogenetic trees based on the 21 PCV2 RNA1 and 20 PCV2 RNA2 sequences (Figure 6E,F). Both phylogenetic trees revealed the presence of three groups of PCV2 isolates. Interestingly, the PCV2 isolate Chiltepin55, identified in Mexico, was genetically distant from the other PCV2 isolates. Group A for PCV2 RNA1 contained 10 isolates, including the P02 isolate identified in this study, whereas group B contained 10 isolates, including variants P08 and P10 detected in this study (Figure 6E). For RNA2 of PCV2, group A contained 12 PCV2 isolates, including P08 and P10 found in this study, whereas 7 isolates belonged to group B, including P02 identified in this study (Figure 6F).

The genomes of four PVY variants were obtained in this study. According to a previous study, several strains of PVY have been identified [9]. We included nine known PVY strains for the construction of a phylogenetic tree of PVY based on a previous study [10]. The phylogenetic tree using 19 PVY isolates showed three groups, namely NTNa, NTNb, and N-Wi (Figure 7A). The four PVY variants identified in this study and two isolates from potatoes in Vietnam and Japan were grouped together with the three known PVY isolates belonging to the NTNa strain.

For BBWV2, which is composed of two RNA segments, two different phylogenetic trees were constructed based on the individual RNA segments obtained in this study (Figure 7B,C). The phylogenetic tree of BBWV2 RNA1 using the 24 sequences identified in this study showed two groups of BBWV2 isolates (Figure 7B). Group A contained 21 sequences, whereas group B contained 3 sequences from P01 and P15 samples. In general, variants from the same plant were observed to be closely related to each other. The phylogenetic tree based on the 23 sequences of BBWV2 RNA2 displayed three groups of BBWV2 variants (Figure 7C). P01, P15-1, P15-2, and P15-3 variants belonged to group A. Group A included 15 variants, whereas group B contained 6 variants. The P05 samples yielded six BBWV2 RNA2 sequences. Of these, four belonged to group A, and two belonged to group B.

We obtained 15, 13, and 21 sequences for CMV RNA1, RNA2, and RNA3, respectively. We constructed three phylogenetic trees based on individual RNA segments (Figure 7D–F), which were used to identify two groups of CMV variants. None of these groups contained the same list of CMV variants. In general, variants from the same sample were closely related. However, some variants from the same sample were observed to be distantly related to each other. For example, the four variants of CMV RNA1 from the P13 sample in group B were genetically different. In contrast, P04, P05-1, and P05-2 of CMV RNA2 as well as P04-1, P04-2, and P05 of CMV RNA3 were closely related, although they were derived from two different samples. CMV isolates can be classified into two major subgroups, I and II. Subgroup I can be further divided into IA and IB. We included RNA3 sequences of 13 known CMV isolates representing CMV subgroups for the phylogenetic tree construction based on a previous study [11]. Although 21 CMV RNA sequences in this study were divided into two groups (group A and group B), all the 21 CMV RNA3 variants were grouped together with the 5 CMV isolates belonging to subgroup IA (Figure 7F).

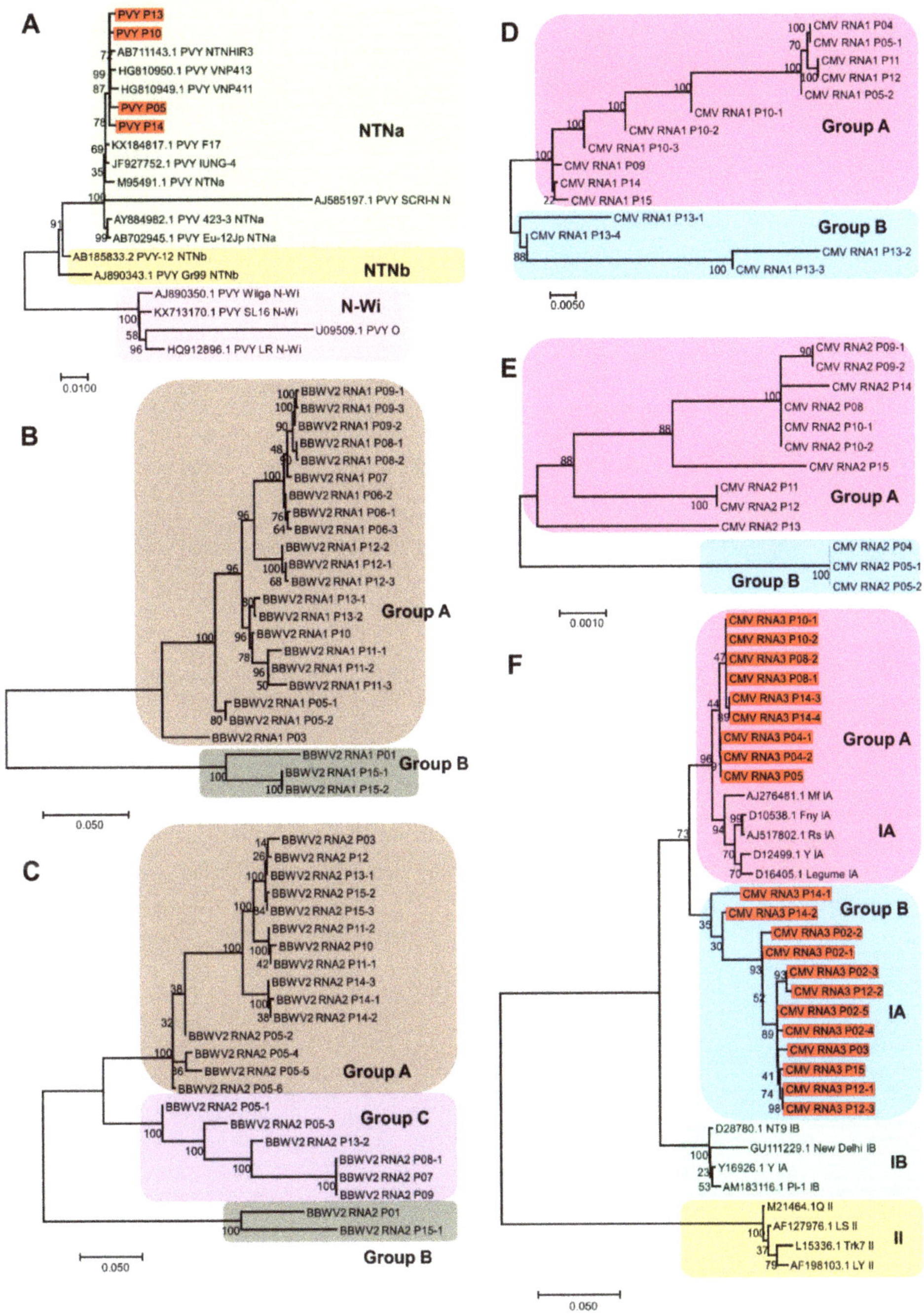

Figure 7. Phylogenetic tree of viruses, PVY, CMV and BBWV2, infecting pepper cultivars: phylogenetic tree of complete genome sequences of PVY (**A**), BBWV2 RNA1 segment (**B**), BBWV2 RNA2 segment (**C**). CMV RNA1 segment (**D**), CMV RNA2 segment (**E**), CMV RNA3 segment (**F**). The viral genome sequences in (**A**,**B**) derived from this study were indicated by red-colored boxes.

To confirm the results of RNA sequencing, we conducted RT-PCR using virus-specific primers. To detect viruses infecting pepper plants, we designed different primer pairs for nine viruses (Table 2). A primer pair amplifying a partial sequence of 18S rRNA for *C. annuum* (Ca18SrRNA) was used as a positive housekeeping gene for RT-PCR [12]. The designed primers can be used in three different methods: RT-PCR, real-time RT-PCR, and recombinase polymerase amplification (RPA). The same total RNAs used for RNA sequencing were used as templates for RT-PCR. Using a primer pair specific to 18S ribosomal RNA for pepper plants, we confirmed the quality of the total RNAs (Figure 8A). Using CMV-specific primers, we detected CMV in all the pepper samples, except P01 and P07 (Figure 8B). BBWV2 infected all the pepper plants, as revealed by RT-PCR (Figure 8C). BWYV was detected in three samples (P03, P04, and P05) (Figure 8D). In addition, we identified two alphaendornaviruses (BPEV and HPEV), which had similar genome organization and high sequence similarity. Therefore, we designed primer pairs that could selectively amplify the virus-specific sequences. Using the BPEV-specific primer pairs, we detected BPEV in the P02, P09, P13, P14, and P15 samples (Figure 8E), whereas HPEV was detected in the P01, P03, P04, and P05 samples (Figure 8F). Using the PCV1-specific primer pairs, we detected only PCV1 in the P12 sample (Figure 8G). In addition, we carried out RT-PCR to detect pepper mild mottle virus (PMMoV), pepper mottle virus (PepMoV), and TSWV, which were not identified with RNA sequencing (Figure 8H–J). The RT-PCR results revealed that none of the pepper samples was infected by PMMoV, PepMoV, or TSWV. These observations were consistent with the RNA sequencing results.

Table 2. Primer pairs used to detect viruses infecting pepper plants.

Viral Reference and Accession No.	Name	Position	Sequences	Size
18S rRNA of pepper EF564281	Ca18SrRNA-F1 Ca18SrRNA-R1		CCGGTCCGCCTATGGTGTGCACCGGTCGTC GCAGTTGTTCGTCTTTCATAAATCCAAGAA	285
Cucumber mosaic virus RNA3 D28780.1	CMVs3-1507F1 CMVs3-1661R1	1507–1536 1661–1632	TATTATGGTAAAAGGTTGTTGCTACCTGAT ACGGATAAGTCCGAGGAGGCAGAAACTTTA	155
Pepper mild mottle virus MN496154.1	PMMoV-5881F1 PMMoV-6059R1	5881–5910 6059–6030	CTGGTTTCAAAGTTTTCCGATATAATGCCG ACTGGCCCTAATGGCCACCGTCGCATCATC	179
Broad bean wilt virus 2 RNA2 JX183230.1	BBWVs2-2463F1 BBWVs2-2613R1	2463–2492 2613–2584	TGCAAATTCTTGACTCGCTGCCCCATATCG GAAACATCATCCCCTCTTGCCGAGTCCAAA	151
Tomato spotted wilt virus S segment KU179577.1	TSWV-S-2004F1 TSWV-S-2153R1	2004-2033 2153–5124	TAAGCAAGTTCTGCGAGTTTTGCCTGTTTT GAAGAAGGGGAAAGAGTATGCTGCTATACT	150
Beet western yellows virus MK307779.1	BWYV-3892F1 BWYV-4060R1	3892–3921 4060–4031	CAGAACTCCGGCTCCATCGCTTACGAGCTG GGATCCTGAATTGGTCCTCGGCGACGTCGT	169
Pepper mottle virus EU586122.1	PepMoV-8916F1 PepMoV-9089R1	8916–8945 9089–9060	ATGGTTTGGTGCATTGAAAATGGCACGTCC CATTTCTATATATGCCTCAGCCACATCAGA	174
Hot pepper endornavirus NC_027920.1	HPEV-13489F1 HPEV-13671R1	13489–13518 13671–13642	CTTTGATGCAAGTAAAGCAGACATATTGGC AGGTGAACATTTAATCTGTTAATAGCATGC	183
Bell pepper endornavirus NC_039216.1	BPEV-13478F1 BPEV-13665R1	13478–13507 13665–13636	GTGGCAACTTATTTTGATGCAGACAAAGCA ACATTTAGCCTGTTAATGGCATGTAGTTGC	188
Pepper cryptic virus 1 RNA2 NC_037096.1	PCV1s2-1081F1 PCV1s2-1243R1	1081–1110 1243–1214	CCGCAACAGGTACAACACAACGAAGGAAGA TTAGTCCTGATGACTGGAGGGAAGGTAACT	163

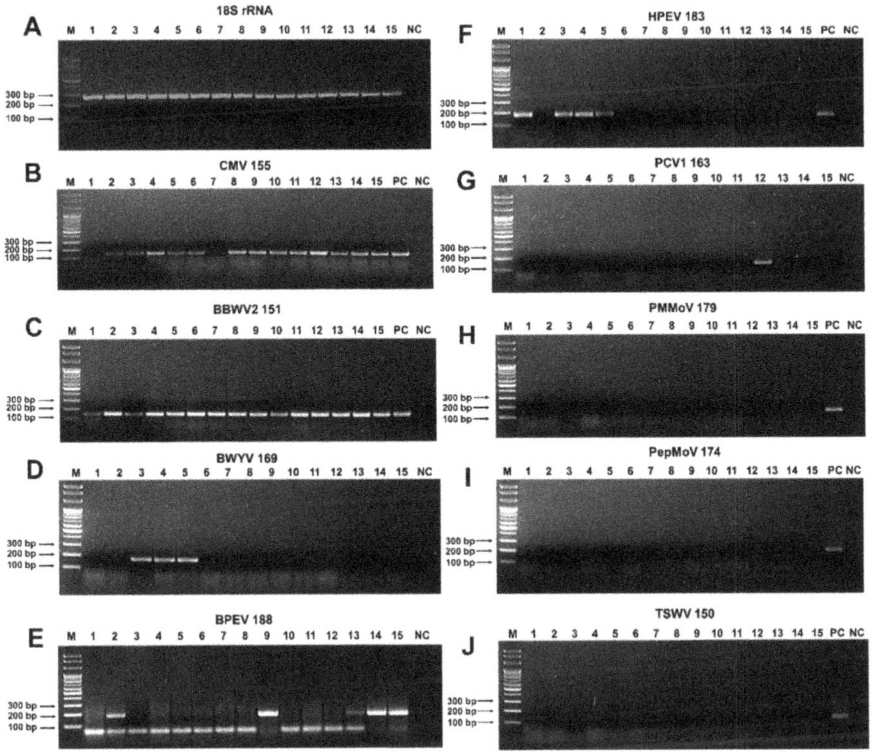

Figure 8. Detection of viruses infecting pepper plants by RT-PCR: (**A**) RT-PCR results using 18S rRNA specific primers to confirm the quality of extracted total RNAs. RT-PCR results using specific primers for CMV (**B**), BBWV2 (**C**), BWYV (**D**), BPEV (**E**), HPEV (**F**), PCV1 (**G**), PMMoV (**H**), PepMoV (**I**), and TSWV (**J**). C indicates positive control. The same total RNAs used for RNA sequencing were used for RT-PCR. The number with the virus name indicates the expected size of the RT-PCR amplicon. PC and NC indicate positive and negative controls, respectively.

3. Discussion

Viruses and viroids that infect pepper plants have been intensively examined in many countries [5,6,8]. However, viruses that infect different pepper cultivars have not yet been investigated. In this study, we, for the first time, revealed comprehensive viromes of 15 pepper cultivars cultivated in Korea, through sequencing.

We found that the most prominent viruses infecting pepper plants in Korea were BBWV2 and CMV. Both RNA sequencing and RT-PCR results revealed that all 15 cultivars were infected by BBWV2, while 13 cultivars were infected by CMV. Infection of BBWV2 has been widely reported in cultivated pepper plants in Korea [13]. However, there exists a limited number of studies reporting the infection of pepper plants by BBWV2 in other countries. For example, a previous study reported BBWV2 infection in pepper plants in the Czech Republic, using a double-antibody sandwich (DAS)–ELISA analysis [14]. We assume that either BBWV2 infection in pepper plants has not been intensively studied, or this virus is not a prominent source of infection in pepper plants in other countries.

In addition to hot pepper cultivars, we examined viruses infecting four bell pepper cultivars: P02, P09, P14, and P15. As expected, BPEV was identified in all the bell pepper cultivars. To our knowledge, this is the first study to report BPEV infection in bell pepper plants in Korea. In addition, we identified BPEV in a single hot pepper (Cheongyang cultivar, P13). Infection of hot peppers with BPEV has not been previously reported. This is the first study to report BPEV infection in hot pepper cultivars. HPEV was first identified in hot peppers in Korea [15] followed by hot peppers in China [16]. Interestingly, in this study, HPEV was identified in two hot pepper cultivars (P01 and P03) and two sweet pepper cultivars (P04 and P05) but not in bell pepper cultivars. Based on this result, we inferred that bell pepper cultivars could be the main hosts for BPEV in Korea, although there was an exception (P13). In addition, hot pepper and sweet pepper cultivars are the main hosts of HPEV. Both BPEV and HPEV are vertically transmitted via seeds [17]. Thus, it is likely that HPEV infecting sweet pepper cultivars could originate from breeding hot pepper cultivars. Both BPEV and HPEV belong to the genus *Alphaendornavirus* in the family *Endornaviridae* [18]. Their genome organization is reported to be quite similar to each other, and the known HPEV isolate has shown low sequence similarity (94% query cover and 72% identity) to the BPEV isolate Yolo Wonder. It was, therefore, difficult to distinguish between them without knowledge of their genome sequences.

According to a previous study in 2005, the most common viruses infecting pepper plants in Korea were CMV (19.8%) followed by PepMoV (13.4%), PMMoV (1.1%), BBWV2 (3.8%), and TMGMV (0.5%) [7]. Some viruses known to infect pepper plants, such as alfalfa mosaic virus (AMV), PVY, tobacco mosaic virus (TMV), and pepper vein chlorosis virus (PVCV), were not identified in our study. BBWV2 was the most prominent virus, followed by CMV; however, PepMoV, PMMoV, and TMGMV were not detected in this study. This is similar to our previous study, which also showed that CMV and BBWV2 were the two major viruses infecting overwintering pepper fruits [8]. Thus, CMV and BBWV2 might be the two major viruses infecting pepper plants in Korea.

PVY was initially identified in bell peppers (paprika) in Korea [19]. In our study, PVY was identified in sweet peppers (P05 and P10), hot peppers (P13), and bell peppers (P14 and P15). These results suggest that PVY could be an important virus infecting pepper plants owing to its presence in diverse pepper cultivars. Phylogenetic analyses revealed that the four PVY variants identified in this study were closely related to the PVY-infecting potato plants derived from Vietnam and Japan. Therefore, PVY variants infecting pepper plants might also be derived from potato plants. Generally, pepper and potato plants are cultivated during similar periods in Korea. Thus, aphid species identified as the main vectors of PVY can easily transmit PVY to the two plant species. In addition, we revealed that all the PVY variants in this study belong to the known PVY strain NTNa, which is one of the common PVY recombinants [9].

Due to differences in pepper samples and collection regions, other known viruses infecting pepper plants, such as PepMoV, PMMoV, and TMGMV, were not detected in this study. Moreover, in contrast to other Asian countries such as India [5] and Vietnam [6], we did not identify any DNA viruses and viroids, suggesting the importance of quarantine in preventing the introduction of these viruses and viroids to pepper plants in Korea.

In contrast to most previous studies that have used RT-PCR and ELISA assays to identify viruses infecting pepper plants in Korea [7], we used RNA sequencing to identify the known and unknown viruses infecting pepper plants. In a single study, we identified eight different viruses that infected pepper plants and provided a detailed viral composition for each pepper cultivar. Moreover, RNA sequencing enabled us to obtain 111 complete viral genome sequences.

It is noteworthy that the small number of reads derived from RNA sequencing might have been caused by cross-contamination from another library. For example, three reads for CMV RNA3 in the P01 cultivar and one read each for CMV RNA2 and CMV RNA3 could be cross-contaminants. The RT-PCR results demonstrated that CMV was not present in the P01 and P07 cultivars.

Due to the coinfection of different viruses in each cultivar, it was difficult to determine the major virus-causing disease symptoms. However, the proportion of viral reads in each library made it possible to detect the virus responsible for the disease symptoms in each cultivar. For example, while HPEV was the dominant virus in the P01 cultivar coinfected with HPEV and BBWV2, P01 (PR-9988 cultivar) did not display any disease symptoms. The viral abundance of BBWV2 was very low in P01. Recently, several pepper cultivars resistant to diseases such as phytophthora blight have been developed in Korea, and the disease-resistant cultivar is referred to as *Phytophthora* resistance (PR). In addition, we included several pepper cultivars that were resistant to anthracnose and viruses. For example, TS-Monster (P03) was developed for resistance against TSWV. Chegangtan (P06) is resistant to anthracnose. Caltan (P12) is resistant to CMV, TSWV, and anthracnose. Interestingly, most disease-resistant pepper cultivars showed no viral disease symptoms, despite being coinfected with several viruses.

Of the examined pepper cultivars, all four bell pepper cultivars displayed severe disease symptoms accompanied by improper growth and fruit production. To identify the major viruses causing disease symptoms in each bell pepper cultivar, we examined the dominant viruses based on the proportion of the viral reads. In P02, PCV2 was the most prominent virus, while CMV was the dominant virus in P09, P14, and P15. Based on this result, it was inferred that CMV may play a major role in the development of viral diseases in bell pepper plants. However, it was unclear whether PCV2 was the main virus causing disease symptoms in P02 because the PCV2-mediated disease symptoms have not been revealed to date. Moreover, BBWV2 and CMV may have coinfected the P02 cultivar. Although bell peppers are currently regarded as important horticultural plants in Korea, the viruses infecting bell peppers have not been intensively studied. In this study, we provided detailed information on viruses that infect major bell pepper cultivars.

BBWV2 is currently regarded as an important virus causing severe damage to pepper production in Korea. As expected, our results showed that BBWV2 was the dominant virus in 7 out of 15 cultivars: P03, P06, P07, P08, P10, P12, and P13. Therefore, it is important to prevent the spread of BBWV2 in pepper fields. Both CMV and BBWV2 can be transmitted by aphids and mechanically via sap. In addition, CMV is transmitted by seeds [20]; however, the seed transmission of BBWV2 has not been reported. Therefore, it may be of interest to examine the possible seed transmission of BBWV2 to prevent its spread in pepper plants.

Viral quasispecies is defined as population structure containing a large number of variant genomes caused by mutations [21]. In our pepper virome study, we revealed the viral quasispecies for BBWV2 and CMV based on the assembled viral genomes. To date, studies associated with viral quasispecies have been carried out by analyzing partial sequences of the target virus [22] or via RT-PCR followed by Sanger sequencing [23]. In contrast, we successfully identified viral quasispecies at the genome level through RNA sequencing. We showed that the viral quasispecies BBWV2 and CMV consisted of two and three RNA segments, respectively, according to their viral genome segment. We obtained the complete sequences for 24 BBWV2 RNA1 and 23 BBWV2 RNA2. The number of BBWV2 variants in each plant was similar, except for the P05 sample, which had two BBWV2 RNA1 variants and six BBWV2 RNA2 variants. For CMV, we obtained complete sequences for 15 RNA1, 13 RNA2, and 21 RNA3 variants. The number of CMV RNA variants in each pepper sample was different. For instance, none of the RNA1 and RNA2 variants were obtained from the P02 and P04 samples; however, five RNA3 variants were obtained from both samples. Phylogenetic analyses using the assembled BBWV2 and CMV variants revealed possible genetic relationships among the identified variants. In general, variants from the same sample were grouped, and some variants derived from different samples were closely related. For instance, the CMV RNA1 variants P04, P05-1, P11, P12, and P05-2 in group A were related. Similarly, at least 11 BBWV2 RNA2 variants in group A were closely related. These results suggest that the variants in the same group may have a common origin. In the case of CMV, all the variants in this study belonged to subgroup IA. The CMV isolates in subgroup I are tolerant to high temperatures and are frequently identified in regions with warmer climates [24]. Moreover, their viral disease symptoms are more severe than those in subgroup II [24]. The summer season in Korea is very hot and humid, and the disease symptoms caused by CMV in pepper plants were very severe, consistent with the results of a previous study [7].

In addition, other assembled virus genomes such as BWYV, BPEV, PCV1, and PCV2 obtained in this study were successfully employed for phylogenetic analyses to reveal genetic relationships with other known viral genomes.

The pepper plants used in this study originated from five different nurseries. Therefore, it is possible to track the origins of the identified viruses. However, we did not find any correlation between viral infections and the selected nurseries. Rather, most pepper cultivars displayed a cultivar-specific virome.

4. Materials and Methods

4.1. Plant Materials and Viral Disease Symptoms

Seedlings of 15 pepper cultivars were obtained from 5 nurseries (Table 1). The seedlings were grown in an open field. Leaves of a single pepper plant from each cultivar were harvested 1 month after transplantation and viral disease symptoms were observed. Collected pepper leaves were immediately frozen in liquid nitrogen.

4.2. Total RNA Extraction and Library Preparation

Frozen pepper leaf samples were ground using a mortar and pestle in liquid nitrogen. The ground powder (100 mg) was subjected to total RNA extraction using an RNeasy Plant Mini Kit (Qiagen, Hilden, Germany), according to the manufacturer's instructions. The quality of the extracted total RNAs was measured using a Technologies 2100 Bioanalyzer (Agilent, Santa Clara, CA, USA). The total RNAs with RNA Integrity Number (RIN) values greater than or equal to 7 were used for library preparation. The libraries for RNA sequencing were prepared by depleting ribosomal RNAs from the total RNAs using TruSeq Stranded Total RNA with a Ribo-Zero Plant kit (Illumina, San Diego, CA, USA), according to the manufacturer's instructions. A total of 15 libraries representing 15 pepper cultivars were prepared and, referred to as P01–P15.

4.3. RNA Sequencing and De Novo Transcriptome Assembly

The indexed libraries were paired-end-sequenced (150 bp $\times$ 2) using the Illumina HiSeq X system (Illumina). The raw RNA sequencing data were deposited in the sequence read archive (SRA) database in the National Center for Biotechnology Information (NCBI) with the respective accession numbers (Table S2). The raw data obtained as FASTQ files for each library were subjected to *de novo* transcriptome assembly using the Trinity assembler (version 2.13.2) with default parameters [25].

4.4. Identification of Viral Contigs and Calculation of Viral Reads

The assembled contigs from each library were subjected to a BLASTX search against the complete viral reference protein database derived from the NCBI using the DIAMOND program with an E-value of 1E-5 as cutoff [26]. BLASTX results in the DAA format were imported into the MEGAN6 program [27] to exclude the sequences derived from organisms other than viruses. Virus-associated contigs were extracted and subjected to a BLASTX search against NCBI's non-redundant protein database using the same conditions. The BLASTX results in the DAA format were imported into the MEGAN6 program to filter out the non-viral sequences. Based on the BLASTX results and taxonomy classification using MEGAN6, the viral contigs were classified according to the virus species. The sequences of the viruses composed of multiple segments were classified according to the viral segments.

4.5. Assembly of Viral Genome and Virus Genome Annotation

Viral sequences covering all ORFs for each virus species were selected. The sequences were subjected to the NCBI ORFfinder [28]. Individual ORF in each viral genome was annotated according to a known reference viral genome. The assembled viral genome was named after the library. Several viral genomes from the same library were identified for certain viruses. All the assembled viral genomes were deposited in the NCBI GenBank database with their respective accession numbers (Table S3).

4.6. Construction of Phylogenetic Trees

The assembled viral genomes for each virus species were subjected to BLASTN against the NCBI nucleotide database. The known viral genome sequences for each of the identified virus species were extracted. The viral genome sequences from this study and the known viral genome sequences from GenBank were aligned using MAFFT version 7 with the auto option [29]. The aligned sequences were subjected to TrimAl for automated sequence trimming using the option automated1 [30]. The trimmed sequences were imported into the MEGA7 program to construct a phylogenetic tree [31]. A phylogenetic tree for each virus species was generated using maximum likelihood with 1000 bootstrap replicates. In the case of BWYV, BPEV, PCV1, and PCV2, the complete genome sequences from this study, as well as the corresponding known viral genome sequences, were used for phylogenetic tree construction. In contrast, for the phylogenetic tree construction of CMV and BBWV2, each viral segment from this study was used.

4.7. Designing of Primers to Detect Viruses Infecting Pepper Plants via RT-PCR

Primers were designed for recombinase amplification assays using the PrimedRPA program, to detect a total of nine viruses, namely CMV, BBWV2, BWYV, BPEV, HPEV, PCV1, PMMoV, PepMoV, and TSWV [32]. RT-PCR was performed, and the primer pairs that specifically amplified the target virus species were selected. The designed primers had amplicon sizes less than 200 bp for all the PCR products. The selected primer pairs could be also used for three different methods: RT-PCR, real-time RT-PCR, and RPA. However, in this paper, we only present the RT-PCR results.

4.8. RT-PCR and Agarose Gel Electrophoresis

RT-PCR was performed using a SoGent DiaStar One-Step RT-PCR Kit (Solgent, Daejeon, Korea). The PCR reaction contained the following elements: 1 μL of template RNA (100 ng), 1 μL of the forward primer (10 pmol/μL), 1 μL of the reverse primer (10 pmol/μL), 6 μL of 5× Band Doctor, 6 μL of 5× buffer and 2 μL of the enzyme mix adjusted to a final volume of 30 μL using RNase-free water. RT-PCR was carried out using the following conditions: cDNA synthesis at 50 °C for 30 min, initial denaturation at 95 °C for 15 min, denaturation at 95 °C for 20 s, annealing at 60 °C for 40 s, and extension at 72 °C for 40 s. The denaturation-to-extension steps were repeated 34 times, with a total of 35 cycles continued to a final extension at 72 °C for 5 min. The samples were maintained at 4 °C. The PCR products were separated via gel electrophoresis on 1.5% agarose gel in the presence of a TBE buffer. The PCR products were stained using ethidium bromide and visualized under ultraviolet light.

5. Conclusions

In this study, we carried out a comprehensive virome study of 15 pepper cultivars derived from 5 nurseries based in Korea. All four bell pepper cultivars showed severe viral disease symptoms; however, most hot pepper cultivars were either asymptomatic or had mild viral disease symptoms. RNA sequencing followed by bioinformatic analyses identified 1325 virus-associated contigs that were derived from 8 different RNA viruses. The proportion of viral reads in most pepper cultivars was relatively high, ranging from 0.04% to 24.5%. We found that BBWV2 and CMV were the most frequently identified viruses, followed by PVY. BBWV2 was the dominant virus in seven cultivars, whereas CMV was the dominant virus in five cultivars. Moreover, we reported, for the first time, the infection of BPEV in four different bell pepper cultivars in Korea. Although the pepper plants in this study originated from different nurseries, most pepper cultivars displayed a cultivar-specific virome. A total of 111 complete viral sequences for the 7 viruses were obtained. Using the obtained viral genome sequences, we revealed the genetic relationship between the identified viral isolates and known isolates for each virus. Moreover, we revealed quasispecies of BBWV2 and CMV in several pepper cultivars. We developed RT-PCR primers specific for nine RNA viruses infecting pepper plants, confirming the RNA sequencing results. Taken together, this study revealed the viromes of 15 pepper cultivars through RNA sequencing.

Supplementary Materials: The supporting information can be downloaded at: https://www.mdpi.com/article/10.3390/ijms231810507/s1.

Author Contributions: Conceptualization, W.K.C.; data curation, Y.J. and H.C.; formal analysis, Y.J., H.C. and W.K.C.; funding acquisition, W.K.C.; investigation, Y.J., H.C. and W.K.C.; project administration, J.H.L., S.H.M. and W.K.C.; software, Y.J.; supervision, J.H.L. and W.K.C.; validation, H.C.; visualization, S.H.M.; writing—original draft preparation, W.K.C.; writing—review and editing, S.H.M. and W.K.C. All authors have read and agreed to the published version of the manuscript.

Funding: This research was funded by the Korea Institute of Planning and Evaluation for Technology in Food, Agriculture, and Forestry (IPET) through the Agri-Bio Industry Technology Development Program funded by the Ministry of Agriculture, Food, and Rural Affairs (MAFRA) (121054-2) and by the National Research Foundation of Korea (NRF) grant funded by the Korea government (MSIT) (No. NRF-2018R1D1A1B07043597).

Institutional Review Board Statement: Not applicable.

Informed Consent Statement: Not applicable.

Data Availability Statement: The raw data were deposited in the NCBI SRA database with the following accession numbers: SRR20337003–SRR20337017. The 111 viral genome sequences in this study were deposited in GenBank with the following accession numbers: ON759421–ON759531.

Acknowledgments: Won Kyong Cho would like to express our group's deepest gratitude to Hyang Sook Kim, Mi Kyong Kim, and Jin Kyong Cho for all their support with the current project.

Conflicts of Interest: The authors declare no conflict of interest.

References

1. Basu, S.K.; De, A.K. Capsicum: Historical and botanical perspectives. In *Capsicum*; CRC Press: Boca Raton, FL, USA, 2003; pp. 21–35.
2. Olatunji, T.L.; Afolayan, A.J. The suitability of chili pepper (*Capsicum annuum* L.) for alleviating human micronutrient dietary deficiencies: A review. *Food Sci. Nutr.* **2018**, *6*, 2239–2251. [CrossRef] [PubMed]
3. Kraft, K.H.; Brown, C.H.; Nabhan, G.P.; Luedeling, E.; Luna Ruiz, J.d.J.; Coppens d'Eeckenbrugge, G.; Hijmans, R.J.; Gepts, P. Multiple lines of evidence for the origin of domesticated chili pepper, Capsicum annuum, in Mexico. *Proc. Natl. Acad. Sci. USA* **2014**, *111*, 6165–6170. [CrossRef] [PubMed]
4. Kwon, D.Y.; Jang, D.-J.; Yang, H.J.; Chung, K.R. History of Korean gochu, gochujang, and kimchi. *J. Ethn. Foods* **2014**, *1*, 3–7. [CrossRef]
5. Jo, Y.; Choi, H.; Kim, S.-M.; Kim, S.-L.; Lee, B.C.; Cho, W.K. The pepper virome: Natural co-infection of diverse viruses and their quasispecies. *BMC Genom.* **2017**, *18*, 453. [CrossRef] [PubMed]
6. Choi, H.; Jo, Y.; Cho, W.K.; Yu, J.; Tran, P.-T.; Salaipeth, L.; Kwak, H.-R.; Choi, H.-S.; Kim, K.-H. Identification of viruses and viroids infecting tomato and pepper plants in vietnam by metatranscriptomics. *Int. J. Mol. Sci.* **2020**, *21*, 7565. [CrossRef] [PubMed]
7. Choi, G.-S.; Kim, J.-H.; Lee, D.-H.; Kim, J.-S.; Ryu, K.-H. Occurrence and distribution of viruses infecting pepper in Korea. *Plant Pathol. J.* **2005**, *21*, 258–261. [CrossRef]
8. Jo, Y.; Back, C.-G.; Kim, K.-H.; Chu, H.; Lee, J.H.; Moh, S.H.; Cho, W.K. Comparative Study of Metagenomics and Metatranscriptomics to Reveal Microbiomes in Overwintering Pepper Fruits. *Int. J. Mol. Sci.* **2021**, *22*, 6202. [CrossRef] [PubMed]
9. Green, K.J.; Brown, C.J.; Gray, S.M.; Karasev, A.V. Phylogenetic study of recombinant strains of Potato virus Y. *Virology* **2017**, *507*, 40–52. [CrossRef]
10. Glasa, M.; Hančinský, R.; Šoltys, K.; Predajňa, L.; Tomašechová, J.; Hauptvogel, P.; Mrkvová, M.; Mihálik, D.; Candresse, T. Molecular characterization of Potato Virus Y (PVY) using high-throughput sequencing: Constraints on full genome reconstructions imposed by mixed infection involving recombinant PVY strains. *Plants* **2021**, *10*, 753. [CrossRef]
11. Mrkvová, M.; Hančinský, R.; Predajňa, L.; Alaxin, P.; Achs, A.; Tomašechová, J.; Šoltys, K.; Mihálik, D.; Olmos, A.; Ruiz-García, A.B.; et al. High-Throughput Sequencing Discloses the Cucumber Mosaic Virus (CMV) Diversity in Slovakia and Reveals New Hosts of CMV from the Papaveraceae Family. *Plants* **2022**, *11*, 1665. [CrossRef]
12. Shen, L.; Liu, Z.; Yang, S.; Yang, T.; Liang, J.; Wen, J.; Liu, Y.; Li, J.; Shi, L.; Tang, Q.; et al. Pepper CabZIP63 acts as a positive regulator during Ralstonia solanacearum or high temperature high humidity challenge in a positive feedback loop with CaWRKY40. *J. Exp. Bot.* **2016**, *67*, 2439–2451. [CrossRef]
13. Kwak, H.-R.; Kim, M.-K.; Nam, M.; Kim, J.-S.; Kim, K.-H.; Cha, B.; Choi, H.-S. Genetic compositions of Broad bean wilt virus 2 infecting red pepper in Korea. *Plant Pathol. J.* **2013**, *29*, 274–284. [CrossRef]

14. Svoboda, J.; Leisova-Svobodova, L. First Report of Broad bean wilt virus-2 in Pepper in the Czech Republic. *Plant Dis.* **2013**, *97*, 1261. [CrossRef]
15. Lim, S.; Kim, K.H.; Zhao, F.; Yoo, R.H.; Igori, D.; Lee, S.-H.; Moon, J.S. Complete genome sequence of a novel endornavirus isolated from hot pepper. *Arch. Virol.* **2015**, *160*, 3153–3156. [CrossRef]
16. Zhang, Y.; Li, S.; Tan, R.; Zhang, D.; Zhang, S.; Liu, Y. First report of hot pepper endornavirus infecting pepper in China. *J. Plant Pathol.* **2020**, *102*, 229. [CrossRef]
17. Jo, Y.; Choi, H.; Yoon, J.-Y.; Choi, S.-K.; Cho, W.K. In silico identification of Bell pepper endornavirus from pepper transcriptomes and their phylogenetic and recombination analyses. *Gene* **2016**, *575*, 712–717. [CrossRef]
18. Valverde, R.A.; Khalifa, M.E.; Okada, R.; Fukuhara, T.; Sabanadzovic, S. ICTV virus taxonomy profile: Endornaviridae. *J. Gen. Virol.* **2019**, *100*, 1204–1205. [CrossRef] [PubMed]
19. Choi, H.-S.; Ko, S.-J.; Kim, M.-K.; Park, J.-W.; Lee, S.-H.; Kim, K.-H.; Were, H.-K.; Chois, J.-K.; Takanami, Y. Characteristics of Potato virus Y isolated from paprika in Korea. *Plant Pathol. J.* **2005**, *21*, 349–354. [CrossRef]
20. Ali, A.; Kobayashi, M. Seed transmission of Cucumber mosaic virus in pepper. *J. Virol. Methods* **2010**, *163*, 234–237. [CrossRef]
21. Andino, R.; Domingo, E. Viral quasispecies. *Virology* **2015**, *479–480*, 46–51. [CrossRef]
22. Park, S.-K.; Park, S.-H.; Yoon, J.-Y.; Park, J.-K.; Ryu, K.-H. A Subpopulation of RNA3 of Cucumber mosaic virus Quasispecies. *Plant Pathol. J.* **2003**, *19*, 210–216. [CrossRef]
23. Schneider, W.L.; Roossinck, M.J. Genetic diversity in RNA virus quasispecies is controlled by host-virus interactions. *J. Virol.* **2001**, *75*, 6566–6571. [CrossRef] [PubMed]
24. Li, N.; Yu, C.; Yin, Y.; Gao, S.; Wang, F.; Jiao, C.; Yao, M. Pepper crop improvement against cucumber mosaic virus (CMV): A review. *Front. Plant Sci.* **2020**, *11*, 598798. [CrossRef]
25. Grabherr, M.G.; Haas, B.J.; Yassour, M.; Levin, J.Z.; Thompson, D.A.; Amit, I.; Adiconis, X.; Fan, L.; Raychowdhury, R.; Zeng, Q.; et al. Full-length transcriptome assembly from RNA-Seq data without a reference genome. *Nat. Biotechnol.* **2011**, *29*, 644–652. [CrossRef]
26. Buchfink, B.; Xie, C.; Huson, D.H. Fast and sensitive protein alignment using DIAMOND. *Nat. Methods* **2015**, *12*, 59–60. [CrossRef]
27. Huson, D.H.; Beier, S.; Flade, I.; Górska, A.; El-Hadidi, M.; Mitra, S.; Ruscheweyh, H.-J.; Tappu, R. MEGAN community edition-interactive exploration and analysis of large-scale microbiome sequencing data. *PLoS Comput. Biol.* **2016**, *12*, e1004957. [CrossRef]
28. Wheeler, D.L.; Chappey, C.; Lash, A.E.; Leipe, D.D.; Madden, T.L.; Schuler, G.D.; Tatusova, T.A.; Rapp, B.A. Database resources of the national center for biotechnology information. *Nucleic Acids Res.* **2000**, *28*, 10–14. [CrossRef]
29. Katoh, K.; Standley, D.M. MAFFT multiple sequence alignment software version 7: Improvements in performance and usability. *Mol. Biol. Evol.* **2013**, *30*, 772–780. [CrossRef]
30. Capella-Gutiérrez, S.; Silla-Martínez, J.M.; Gabaldón, T. trimAl: A tool for automated alignment trimming in large-scale phylogenetic analyses. *Bioinformatics* **2009**, *25*, 1972–1973. [CrossRef]
31. Kumar, S.; Stecher, G.; Tamura, K. MEGA7: Molecular evolutionary genetics analysis version 7.0 for bigger datasets. *Mol. Biol. Evol.* **2016**, *33*, 1870–1874. [CrossRef]
32. Higgins, M.; Ravenhall, M.; Ward, D.; Phelan, J.; Ibrahim, A.; Forrest, M.S.; Clark, T.G.; Campino, S. PrimedRPA: Primer design for recombinase polymerase amplification assays. *Bioinformatics* **2019**, *35*, 682–684. [CrossRef] [PubMed]

International Journal of
Molecular Sciences

MDPI

Article

QTL Mapping and Candidate Gene Analysis for Seed Germination Response to Low Temperature in Rice

Nari Kim [1,†], Rahmatullah Jan [1,2,†], Jae-Ryoung Park [3], Saleem Asif [1], Dan-Dan Zhao [1], Eun-Gyeong Kim [1], Yoon-Hee Jang [1], Gyu-Hyeon Eom [1], Gang-Seob Lee [4,*] and Kyung-Min Kim [1,2,*]

1 Department of Applied Biosciences, Graduate School, Kyungpook National University, Daegu 41566, Korea; jennynari@hanmail.net (N.K.); rehmatbot@yahoo.com (R.J.); saleemasif10@gmail.com (S.A.); qx288mm@naver.com (D.-D.Z.); dkqkxk632@naver.com (E.-G.K.); uniunnie@naver.com (Y.-H.J.); znfnfn978@naver.com (G.-H.E.)
2 Coastal Agriculture Research Institute, Kyungpook National University, Daegu 41566, Korea
3 Crop Breeding Division, National Institute of Crop Science, Rural Development Administration, Wanju 55365, Korea; icd92@naver.com
4 Department of Agricultural Biotechnology, National Institute of Agricultural Sciences, Rural Development Administration, Jeonju 54874, Korea
* Correspondence: kangslee@korea.kr (G.-S.L.); kkm@knu.ac.kr (K.-M.K.); Tel.: +82-63-238-4714 (G.-S.L.); +82-53-950-5711 (K.-M.K.)
† These authors contributed equally to this work.

Abstract: Low temperature is a serious threat to the seed emergence of rice, which has become one of the main limiting factors affecting rice production in the world. It is of great significance to find the candidate genes controlling low-temperature tolerance during seed germination and study their functions for breeding new rice cultivars with immense low-temperature tolerance during seed germination. In the current experiment, 120 lines of the Cheongcheong Nagdong Double Haploid (CNDH) population were used for quantitative trait locus (QTL) analysis of low-temperature germinability. The results showed a significant difference in germination under low different temperature (LDT) (15 °C, 20 °C) conditions. In total, four QTLs were detected on chromosome 3, 6, and 8. A total of 41 genes were identified from all the four QTLs, among them, 25 genes were selected by gene function annotation and further screened through quantitative real-time polymerase chain reaction (qRT-PCR). Based on gene function annotation and level of expression under low-temperature, our study suggested the *OsGPq3* gene as a candidate gene controlling viviparous germination, ABA and GA signaling under low-temperature. This study will provide a theoretical basis for marker-assisted breeding and lay the basis for further mining molecular mechanisms of low-temperature germination tolerance in rice.

Keywords: low-temperature; seed germination; Cheongcheong Nagdong double haploid; quantitative trait locus; viviparous germination; marker-assisted breeding

Citation: Kim, N.; Jan, R.; Park, J.-R.; Asif, S.; Zhao, D.-D.; Kim, E.-G.; Jang, Y.-H.; Eom, G.-H.; Lee, G.-S.; Kim, K.-M. QTL Mapping and Candidate Gene Analysis for Seed Germination Response to Low Temperature in Rice. *Int. J. Mol. Sci.* **2022**, *23*, 7379. https://doi.org/10.3390/ijms23137379

Academic Editors: Andrés J. Cortés and Hai Du

Received: 9 June 2022
Accepted: 30 June 2022
Published: 2 July 2022

1. Introduction

Seeds are a means of plant survival and propagation; however, changes in the global environment influences their development and germination. Several main biotic and abiotic stress factors badly influence crop yield, especially rice. In addition to that, abiotic stresses such as high/low temperature, drought, salinity, and submergence directly or indirectly influence the physiological status and molecular mechanisms of rice which badly affect yield [1–4]. The impact of temperature on seed germination is associated with the biosynthesis and signaling of ABA and GA. ABA and GA are the major endogenous regulators that antagonistically control seed dormancy and germination in several plant species [5]. A low temperature during seed development enhances the accumulation of ABA and reduces GA [6,7]. Due to the low temperature, *NCED* (ABA metabolic) and

GA2ox2 (GA catabolic) genes express, which results in decreased ABA and GA level endogenously and directly promotes seed dormancy and reduces seed germination [8,9]. In contrast, it is reported that ABA contents reduces by cold imbibition and the level of ABA biosynthesis and signaling genes (*PYL, ABI4, ABI5, PP2C*) also change with cold imbibition [10]. Seed germination is directly associated with the induction or repression of key genes related to ABA (*NCED3, NCED6, NCED9, CYP707A2* and *ABI5*) and GA (*GA20-oxidase 4* and *GA3-oxidase 1*) biosynthesis and signaling pathways [11]. In addition, the phytohormones such as gibberellin (GA) and abscisic acid (ABA) play essential roles in the regulation of seed dormancy and germination [12,13]. Collectively, temperature regulates seed dormancy through mediating ABA and GA metabolism and signaling.

The ABA biosynthesis and signaling pathways during seed development and germination have been intensively studied until now. In a molecular base study, genes related to ABA biosynthesis, degradation and signaling were identified to play important roles in seed germination and development [14]. Central to ABA signaling in seeds are three core components: PYRABACTIN RESISTANCE/PYRABACTIN-LIKE/REGULATORY COMPONENTS OF ABA RECEPTORS (*PYR/PYL/RCAR*), PROTEIN PHOSPHATASE 2Cs (*PP2Cs*) and SNF1-RELATED PROTEIN KINASE 2s (*SnRK2s*; reviewed in [15]). The identification of *PYL/RCAR* family proteins verified that ABA receptors *PYL/RCAR* are essential ABA signaling components and predominantly function in seeds [16]. A study in rice has revealed that seeds expressing *OsPYL/RCAR* are hypersensitive to ABA during seed germination [17]. *PYLs* release protein phosphatase type 2C (*PP2C*) in the absence of ABA, *PP2C* (another essential component in ABA signaling) acts as a phosphatase [18]. *PP2C* suppresses downstream ABA signaling protein *SnRK2s* due to phosphorylation and inhibits the downstream ABA signaling network [19]. Therefore, *PP2C* acts as a negative regulator in the ABA signaling system [20]. However, in the presence of ABA, *PYR/PYL/RCAR* binds with ABA and *PP2C* to stop the phosphatase activity of *PP2C* which releases and enables the function of the *SnRK2* gene. Our candidate gene *OsGPq3* is also known as *OSK3* and *SnRK*. A recent report shows that *SnRK2* expresses in the nucleus during seed development and germination [21]. *SnRK2* is a positive regulator of downstream genes, which includes *bZIP, ABI5, ABI4,* and *ABI3* transcription factors, which are the main regulators of ABA-responsive genes [15]. *ABI3* gene expression induces seed dormancy while the *ABI3* mutant induces viviparous seed germination [22]. Viviparous 1 (*VP1*) is an *ABI3* ortholog in maize, shows a critical role in seed dormancy inhibition, promotes viviparous germination and reduces ABA sensitivity [23]. *ABI4* is a key component of the ABA signaling pathway and positively regulates primary seed dormancy by mediating the biosynthesis of ABA and GA. The GA levels of *ABI4* seeds are higher than that of wild rice, suggesting that *ABI4* represses GA biosynthesis [24]. Further study revealed that *ABI4* directly activates key GA catabolic gene *GA2ox7*, whereas GA can repress the expression of key ABA biosynthetic gene *NCED6* in an *ABI4*-dependent manner. In addition, GA can promote *ABI4* degradation while ABA stabilizes it [25]. *ABI5* is a major inducer of seed germination and post-germination growth [26]. Additionally, *SnRK2* plays a key role in the regulation of tiller enhancement. Recently, researchers suggested that the ABA receptor complex (*OsPYL/RCARs*) activates the *SnRK2* protein which phosphorylates tiller enhancer (*APC/CTE*) and interrupts the interaction between tiller enhancer and *OsPYL/RCAR* which subsequently stabilizes the ABA receptor complex (*OsPYL/RCARs*) [27]. In contrast, GA can reduce the level of *SnRK2s* and may promote *APC/CTE*-mediated degradation of *OsPYL/RCARs* [27]. This inference suggested that, GA can reduces ABA signaling by promoting the interaction between tiller enhancer and ABA receptors by reducing *SnRK2* activity which causes subsequent proteasomal degradation of *OsPYL/RCARs* [27].

In this study, a population of 120 lines of Cheongcheong/Nadong Double Haploid (CNDH), derived from two inbred rice lines, (Cheongcheong and Nagdong) was used as a QTL mapping population. The objective of this study was to: (i) analyze seed vigor of this population under LDT conditions; (ii) identify QTL responsible for seed germination under LDT conditions. This study will provide a basis for controlling viviparous germination, ABA and GA signaling during seed dormancy, seed germination and seed development and tiller number regulation.

2. Results

2.1. Phenotypic Evaluation of Germination Percentage under the LDT Conditions

In the present study, we evaluated the germination percentage (GP) of the CNDH population and their parental lines under LDT conditions as shown in (Figure 1a,b). Under the 15 °C temperature, Cheongcheong were not germinated until the 12th day while the Nagdong started germination after the 9th day and increased GP every day and the data were collected until 12 days. Under the 20 °C temperature, Cheongcheong initiated germination on the 7th day whereas Nagdong started germinating on the 3rd day. Our result shows that the Nagdong shows high germinability at both 15 °C and 20 °C as compared to Cheongcheong. In summary, both parental lines indicated a low GP under the 15 °C conditions compared with the 20 °C conditions and Nagdong showed a higher GP than Cheongcheong among parental lines (Table S1). On the basis of low and high germinability, we selected six lines that are; CNDH71, CNDH77, CNDH85, CNDH30, CNDH31, and CNDH51, from the CNDH population (Figure 1e,f). Among them, CNDH71, CNDH77, and CNDH85 showed 71%, 85%, and 80%, respectively, at 15 °C while 100% germinated at 20 °C. However, CNDH30, CNDH31, and CNDH51 were not germinated at 15 °C and at 20 °C their germination was 28%, 41% and 45%, respectively. The highest germinated lines, CNDH71, CNDH77, and CNDH85 had germinated on the 4th day, 3rd day, and 4th day, respectively, at 20 °C. CNDH30, CNDH31, and CNDH50 which showed low GP started germination on the 8th day, 7th day, and 7th day, respectively, at 20 °C. To statistically validate, our GP results of the CNDH population show normal frequency distribution under both conditions (Figure 1c,d). This experiment was repeated two times.

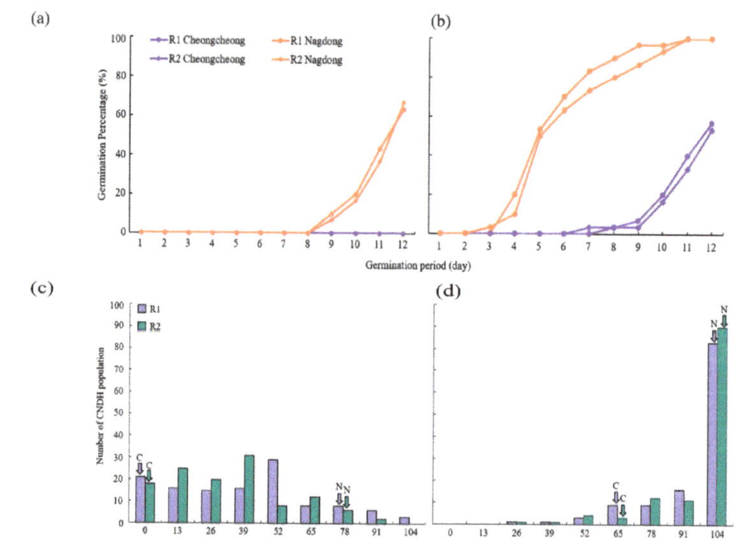

Figure 1. *Cont.*

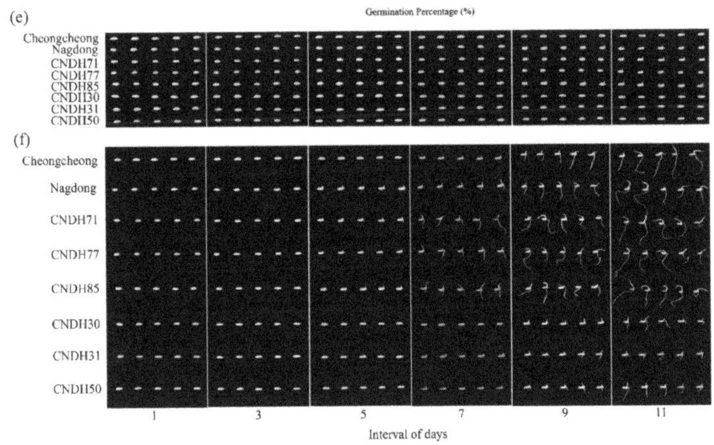

Figure 1. Representation of seed germination and frequency distribution. (**a**,**b**) Represent the GP of Cheongcheong and Nagdong on the daily basis under LDT conditions. (**c**,**d**) Shows frequency distribution for GP under the 15 °C and 20 °C conditions of CNDH population. (**e**,**f**) The seed germination pattern under 15 °C and 20 °C conditions. The highest GP in CNDH population is CNDH71, CNDH77, and CNDH85. The lowest GP in CNDH population is CNDH30, CNDH31, and CNDH50. All pictures were taken every two days after the seeds start germination. C; Cheongcheong, N; Nagdong, R1; replicate 1, and R2; replicate 2.

2.2. *Analysis of QTLs Associated with Seed Germination*

Phenotypic data for two repeat-experiment were collected to carry out the QTL mapping. Based on the two repeat-experiment data, four QTLs for seed germination under LDT conditions in the CNDH population were located on chromosomes 3, 6, and 8 (Table 1). The qGP6 and qGP6-1 were detected in RM528-RM20632 on chromosome 6 in both R1 and R2. RM7197-15063 on chromosome 3 was a region related to qGP3. The qGP8 was located at the RM23314-RM23178 on chromosome 8. The qGP3 was identified in RM7197-15063 on chromosome 3 and represented the highest LOD score of 3.36 and the phenotypic variation was 29%. The qGP8 was located on RM23314-RM23178 on chromosome 8 and showed 33% phenotypic variation with a 3.19 LOD score. The qGP6 and qGP6-1 were detected in RM528-RM20632 on chromosome 6 for two repeat-experiment and showed the LOD score of 2.64, and 2.94, respectively. The phenotypic variations were both 30%. Moreover, the alleles of a total of four QTLs associated with seed germination were derived from Cheongcheong. Two QTLs were overlapped on chromosome 6 at marker interval RM528-RM20632 in two repeat-experiments. Finally, the genes related to seed germination under LDT conditions were screened on all chromosomes (3, 6, and 8) (Figure 2).

Table 1. QTL associated with the germination percentage in the CNDH population.

Replicates	QTLs	Chr.	Marker Interval ^{z}	LOD	Additive Effect ^{y}	$R^{2\,x}$	Increasing Effect ^{w}
1	qGP6	6	RM528-RM20632	2.64	7.94	0.3	Cheongcheong
	qGP3	3	RM7197-RM15063	3.36	6.62	0.29	Cheongcheong
2	qGP6-1	6	RM528-RM20632	2.94	7.16	0.3	Cheongcheong
	qGP8	8	RM23314-RM23178	3.19	5.37	0.33	Cheongcheong

^{z} The markers which are in the significance threshold. ^{y} The positive values indicate the contribution from the mother plant. ^{x} The phenotypic variation. ^{w} The source of the allele generating an increase in assessed traits.

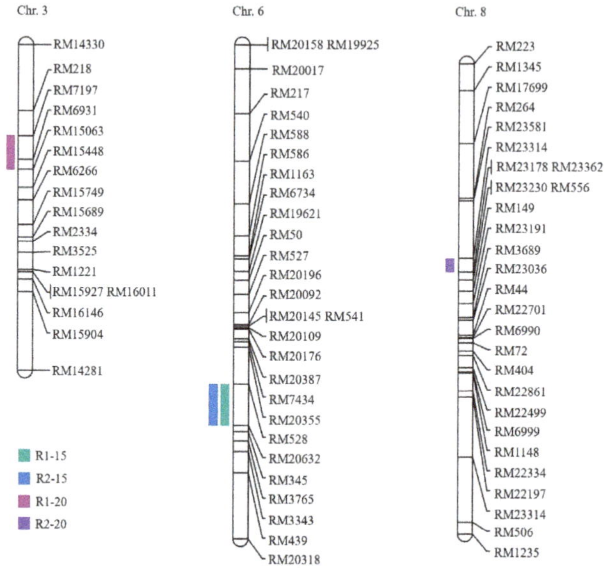

Figure 2. QTL mapping associated with seed germination in the CNDH population. The 4 QTLs were detected in RM7197-RM15063, RM528-RM20632, and RM23314-RM23178 on chromosome 3, 6, and 8, respectively. R1-15; replicate 1 at 15 °C, R2-15; replicate 2 at 15 °C, R1-20; replicate 2 at 20 °C, R2-20; replicate 2 at 20 °C.

2.3. Searching for Seed Germination Related Genes on the Basis of QTL Mapping

As a result of QTL mapping related to seed germination under LDT conditions, a total of four QTLs were detected in RM7197-RM15063, RM528-RM20632, and RM23314-RM23178 on chromosomes 3, 6, and 8, respectively. Depending on the detected four QTLs, genes related to seed germination under LDT conditions were identified using the NCBI database. There were 25, 7, and 9 related genes in RM7197-RM15063, RM528-RM20632, and RM23314-RM23178, respectively, were connected with cell function, embryo development, hormone, signaling, seed development, seed dormancy, and seed germination (Tables S2–S4). Among them, *Os03g0289100*, named *OsGPq3*, which is the *OSK3* (*SnRK*) was selected as the target gene (Figure 3). This gene, *OSK3* (*SnRK*) is involved in the ABA signaling pathway and is a regulator of *ABI3*, *ABI4*, and *ABI5* transcription factor, tiller enhancement, and viviparous germination.

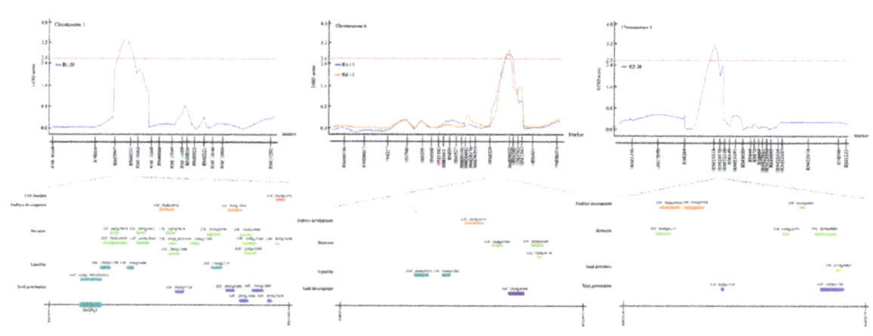

Figure 3. QTL analysis and physical mapping of seed germination-related genes. Genes associated with seed germination corresponding to cell function, embryo development, hormone, signaling, seed development, seed dormancy, and seed germination were identified in RM7197-RM15063, RM528-RM20632, and RM23314-RM23178 on chromosomes 3, 6, and 8, respectively. Among them, *OSK3* (*SnRK*) which plays a key role in the ABA signaling pathway was screened.

2.4. Relative Expression of Genes Related to Seed Germination under LDT

In this study, we predicted 25 related genes on different chromosomes and different loci. To narrow down these predicted genes, we further validated them through qRT-PCR (Figures 4 and 5). The results showed that the expression level of *OSK3* (*SnRK*), *PP2C*, and *PYL* was significantly regulated in parental lines under the LDT conditions (Figures 4 and 5). At 15 °C, *Os03g0289100* (*OSK3* or *SnRK*) gene was consistently downregulated until 96 h in Cheongcheong and highly upregulated after 96 h in Nagdong (Figures 4 and 5). On the other hand, at 20 °C it was significantly downregulated at 72 h and 96 h while upregulated after 120 h however, in Nagdong the overall expression was enhanced but, reduced non-significantly after 120 h (Figures 4 and 5). Another candidate *Os03g0292100* (*PP2C*) was significantly regulated by a low temperature (Figures 4 and 5). In Nagdong, the expression was consistently reduced and increased with time at 15 °C and 20 °C, respectively. While in Cheongcheong, after 48 h, the expression of *PP2C* was significantly reduced consistently. The expression level of the third candidate locus (*Os03g297600, PYL*) was significantly higher at 15 °C as compared to 20 °C in Cheongcheong except for 48 hrs. While in Nagdong, *PYL* was highly expressed after 12 h and 120 h at 15 °C and 20 °C, respectively.

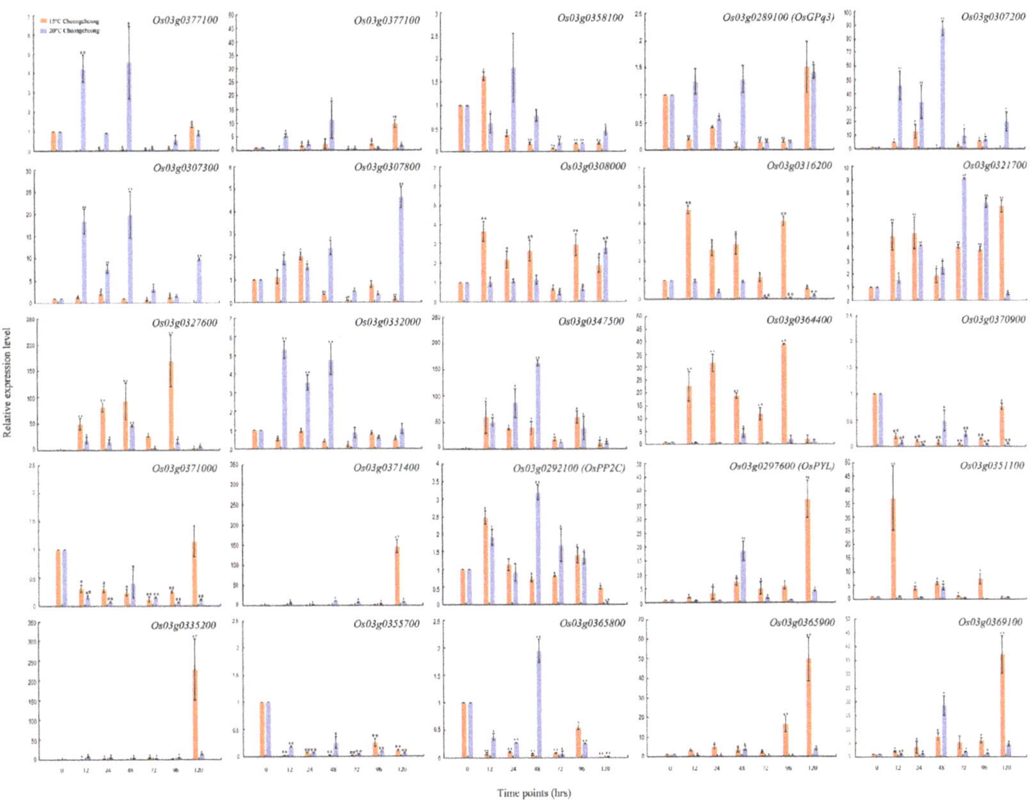

Figure 4. qRT-PCR data about the genes related to seed germination under LDT conditions in Cheongcheong parent. Relative expression levels were analyzed at 0, 12, 24, 48, 72, 96 and 120 h. Each time point is compared with 0 h under 15 °C and 20 °C conditions. Graph bars indicate mean ± standard deviation and asterisks show a significant difference (* $p < 0.05$, ** $p < 0.01$) analyzed by *t*-test.

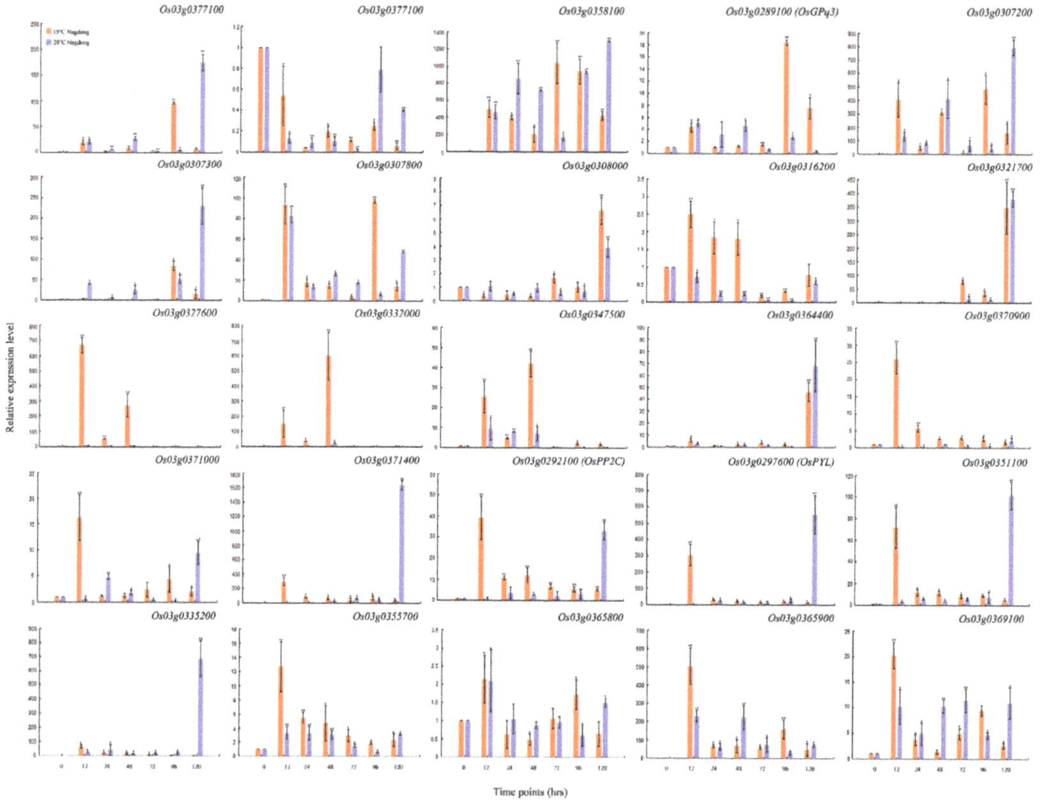

Figure 5. qRT-PCR data about the genes related to seed germination under LDT conditions in Nagdong parent. Relative expression levels were analyzed at 0, 12, 24, 48, 72, 96 and 120 h. Each time point is compared with 0 h under 15 °C and 20 °C conditions. Graph bars indicate mean ± standard deviation and asterisks show a significant difference (* $p < 0.05$, ** $p < 0.01$) analyzed by *t*-test.

2.5. Analysis of Phylogenetic Tree and Homology Sequence, and Protein Interaction

OsGPq3, which is an *OSK3* related to the ABA signaling pathway detected in RM7197-RM15063 on chromosome 3. The related genes for seed germination under LDT conditions represented cell function, embryo development, hormone, signaling, seed development, seed dormancy, and seed germination (Figure 6a). The result of phylogenetic tree analysis showed that *OsGPq3* had a genetic similarity among *Panicum hallii*, *Panicum virgatum*, *Setaria italica*, *Sorghum bicolor*, and *Zea mays* which were included in the Gramineae family (Figure 6b). The BLAST analysis using the NCBI database indicated that *OsGPq3* has highly similar sequences to the *OSK3* of *Setaria italica*, *Panicum hallii*, *Panicum virgatum*, *OSK4* of *Sorghum bicolor*, *Zea mays*, and SNF1 of *Zea mays* (Figure 6c). Furthermore, using the *OsGPq3* domain, we identified that *OsGPq3* interacted with 10 proteins (OS05T0405900-01, OsJ_18605, OS04T0311100-01, OsJ_06259, OS04T0382300-01, OS08T0379300-01, OSKbeta1, OJ1485_B09.7, OsJ_13432, and OS11T0586001-00) (Figure 6d).

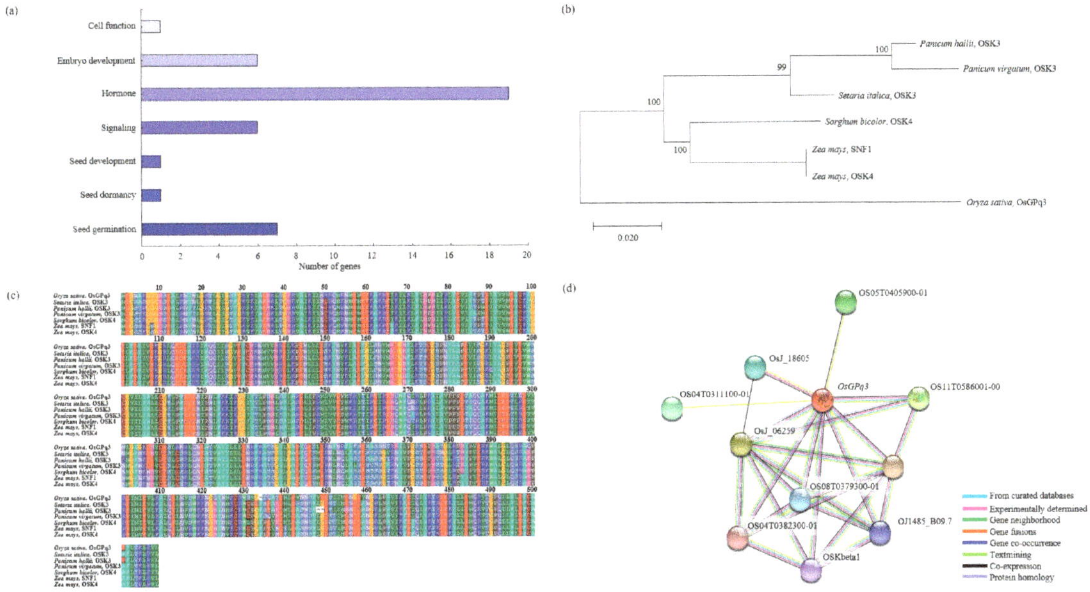

Figure 6. Sequence analysis of *OsGPq3*. (**a**) Related genes of seed germination were involved in cell function, embryo development, hormone, signaling, seed development, seed dormancy, and seed germination. (**b**) To analyze *OsGPq3* and the homology gene, the phylogenetic tree was used. (**c**) The multiple sequence alignment of *OsGPq3*; there is a high similarity among *Oryza sativa*, *Setaria italica*, *Panicum hallii*, *Panicum virgatum*, *Sorghum bicolor*, and *Zea mays*. The accession number of mentioned genes are present in Table S6. (**d**) *OsGPq3* interacts with OS05T0405900-01, OsJ_18605, OS04T0311100-01, OsJ_06259, OS04T0382300-01, OS08T0379300-01, OSKbeta1, OJ1485_B09.7, OsJ_13432, and OS11T0586001-00.

3. Discussion

Seed germination is the key factor in rice yield. Rice is a sensitive crop at low temperature compared with other main crops. In particular, the low temperature causes harmful influences on the overall process of rice growth [28]. Germination under low-temperature conditions is the main character related to the rice yield and quality and it is essential to verify and grow high-production rice under low-temperature conditions. Below 17 °C, rice is severely affected, mainly resulting in poor germination and seedling establishment, a severe reduction in growth, and lower yield [29]. In short, the germination of seeds is a crucial character to increase the rice yield, which is the main goal of the rice breeding system.

To find related genes, QTL analysis can be used to identify the practical alleles [30]. QTL analysis is an efficient way to identify genes related to germination which control different traits [31]. In the current study, QTL analysis was used for a 120 doubled haploid population from a cross between Cheongcheong (Indica variety) and Nagdong (Japonica variety). QTL analysis of seed germination under low temperature was used to detect target loci that play a vital role in these traits. The results analyzed the seed germination at 15 °C and 20 °C temperatures, which are the quantitative traits that exhibit continuous variation and follow the normal distribution under both the LDT conditions. Based on the two repeated experiment we detected four QTLs related to seed germination under the LDT conditions in the CNDH population, located in RM7197-RM15063, RM528-RM20632, RM23314-RM23178 on chromosome 3, 6, and 8, respectively. The qGP6 and qGP6-1 were detected in RM528-RM20632 on chromosome 6 in both the repeated experiment with LOD scores of 2.64 and 2.94, respectively. RM7197-15063 on chromosome 3 was a region related to qGP3 and showed the highest LOD score of 3.36. The qGP8 was located at the RM23314-RM23178 on chromosome 8 with an LOD score of 3.19. Among them, two QTLs were overlapped on chromosome 6 in the same marker interval RM528-RM20632 in both the repeated experiments. Li et al. and Hua et al. characterized qGP3-1 and qGP-6, respectively, associated with grain number per panicle [32,33]. It is also reported that the GA biosynthesis gene (*OsGA20ox1*) is located in qGP3 for the number of grains per panicle [34]. Consistent with our study, qGP6 was also reported as a major QTL related to seed germination percentage and increases germination by about 3%, and also detected in the RM528-RM340 marker interval [35]. Further reported studies evaluated that qGP6 on chromosome 6 was familiar with the region of qGW-6 for 1000 seed weight and sd6.1 for seed dormancy [36,37]. qGP8 was also previously reported associated with seed germination under osmotic stress detected in the RM6208-RM8264 marker interval [38]. However, the same QTL was detected in linked marker RM7027 associated with the grain number per panicle [39]. These inferences support our study that the four QTLs are linked to seed germination and yield. Among the detected QTLs, 41 germination-related genes were screened, and 25 candidates were identified on chromosome 3. These genes were screened based on gene function annotation.

To narrow down the most closely related genes, the expression level under the LDT conditions was evaluated through qRT-PCR (Figures 4 and 5). Among the 25 genes detected in the different locus, we selected three candidate genes (*SnRK/OSK, PP2C, PYL*) according to qRT-PCR validation. Due to previous reports, we assumed that these three genes (*SnRK/OSK, PP2C, PYL*) are significantly associated with seed germination under low temperature. All three genes are involved in ABA and GA signaling during seed germination under low temperature. ABA plays a key role in several developmental stages such as seed maturation and dormancy [40]. ABA and GA play the most crucial roles as phytohormones in mediating light and temperature-induced transition from seed dormancy establishment to seed germination. ABA promotes dormancy establishment during seed maturation and inhibits seed germination, while GA promotes seed germination [41]. Genetic analysis suggests that *PYR, PP2C,* and *SnRK* are important core components of the upstream signal transduction network that regulates the ABA-responsive process, including seed dormancy and seed germination reviewed in [42]. It is mentioned in the Introduction section that ABA physically combines with *PP2C* resulting in the dissociation of *SnRK* from the *SnRK* complex which directly phosphorylates *ABI3, ABI4,* and *ABI5* to mediate ABA responses [43]. Our result suggested that *PYR, PP2C,* and *SnRK* were significantly upregulated in Nagdong under both low-temperature conditions while Nagdong also showed high GP at both low-temperature conditions. Our results concluded that *PYR, PP2C,* and *SnRK* are closely related to seed germination under LDT conditions. These genes are also associated with viviparous germination discussed in the Introduction section. In our study we selected the *SnRK* gene for further future study; however, *PYR* and *PP2C* are also associated with the regulation of seed germination under low temperature.

One of the main functions of the *SnRK* gene is, to regulate ABA and GA mediated via the *ABI4* gene. A recent investigation determined that *ABI4* is significantly involved in GA and ABA antagonistic crosstalk. The *ABI4* enhances ABA biosynthesis and repress GA biosynthesis through the activation of ABA synthesizing and GA repressing genes [44]. This inference assumes that the expression of the *SnRK* gene can discourage viviparous germination via regulation of ABA biosynthesis and GA repression. In another study, it was found that the *SnRK* gene inhibits the tiller enhancer complex (APC/C^{TE}) which means that *SnRK* is associated with plant tiller number [27]. On the other hand, recently it was investigated via QTL analysis that the Auxin-related gene (*OsIAA17q5*) is closely related to the plant tiller number [45]. Our proposed model (Figure 7) shows that there is a significant association between the *SnRK*, GA, viviparous germination, and plant tiller pattern. However, molecularly it is poorly understood; therefore, further evaluation is needed to synchronize the crosstalk among the *SnRK*, GA regulation, viviparous germination and plant tiller number. In the next step of our experiment, we will further evaluate our identified gene through overexpression and genome editing with CRISPR-Cas9 technology. Further we will characterize the transgenic rice line by protein expression and ABA and GA signaling under low temperature.

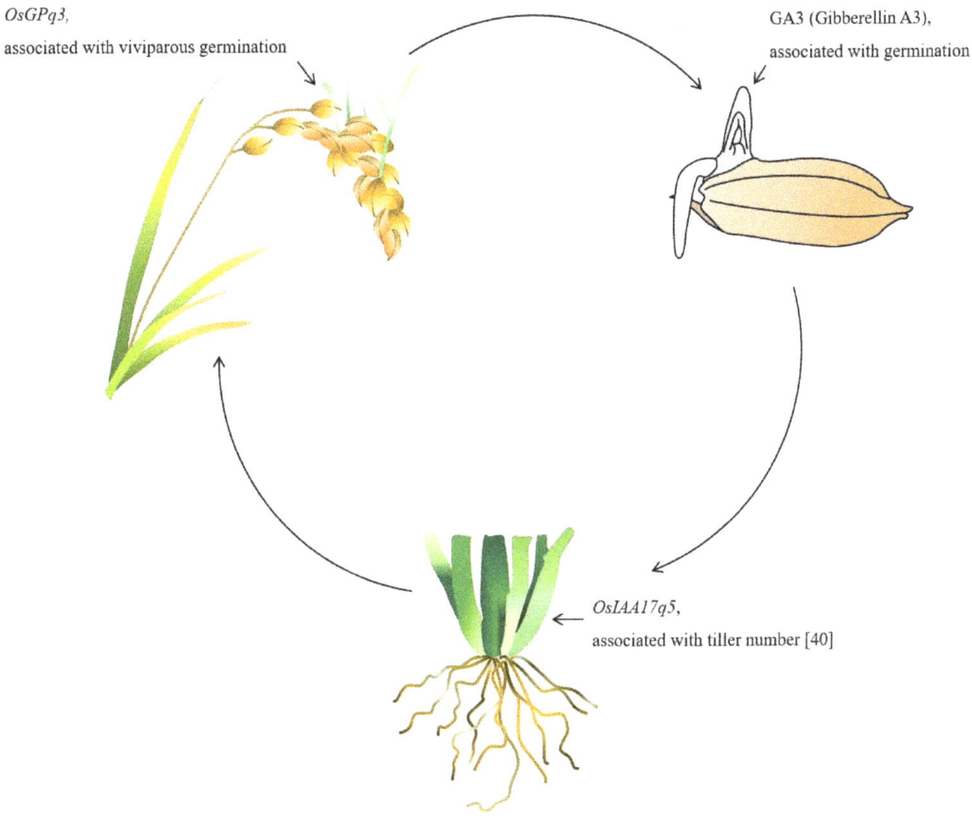

Figure 7. The purposed model of the *OsGPq3* function associated with the viviparous germination, GA regulation, and tiller number regulation.

4. Materials and Methods

4.1. Plant Materials and Preparation of Seeds for Germination Test

A total of 120 CNDH populations were used to analyze seed germination. The CNDH population which originated from another culture of F_1 hybrid was developed with doubled haploid from a cross between Cheongcheong and Nagdong. The CNDH population was created in 2005 was derived from crossbreeding between Cheongcheong (Indica variety) and Nagdong (Japonica variety). The seeds of the CNDH population were obtained in the experimental field at the Kyungpook National University in Gunwi. The seeds were sterilized with Spotak pesticide (Hankooksamgong, Seoul, South Korea) and put in an incubator at 33 °C for 3 days under dark conditions. The plants were transplanted 30 days after sowing in the experimental field. The planting density of the CNDH population and their parents were 30 × 15 cm. N, P_2O_5, and K_2O were applied at 9.0, 4.5, and 5.7 kg per 10 ha as a fertilizer to prevent diseases and pests (Park et al., 2021). After harvesting, only high-quality seeds were selected for germination.

4.2. Evaluation for Seed Germination

The 30 seeds of each CNDH population using three replicates per line were put into the 6 × 9 cm size plastic zipper bag. All plastic zipper bags were put into the beaker and then added distilled water until the whole seeds were immersed. The seeds were placed in an incubator at 20 °C and 15 °C, separately. The germinated seeds were counted every day until 12 days. The emergence of the radicle was considered the initiation of germination. When the radicle length reached approximately 2 mm, the seeds were considered germinated. The GP was calculated as GP (%) = (A total number of germinated seeds/A total number of seeds tested) × 100 [46].

4.3. QTL Analysis for GP

The CNDH population genetic map was created by using 778 SSR (simple sequence repeat) markers. Among 778 SSR markers, 423 SSR markers represented polymorphism. Through the PCR amplification, 222 SSR markers were selected and used for QTLs associated with seed germination. To analyze QTLs, Win QTL cartographer 2.5 software was used [47]. Composite interval mapping (CIM) was performed on a whole-genome scan and the LOD score was set at 2.5 [48]. To run this program, the genetic distance between markers, chromosome numbers, genotypic data, target trait values, and marker labels were required [49]. The detected QTLs were named the method proposed by McCouch [50].

4.4. Gene Information and Statistical Analysis Related to Seed Germination

As for QTL mapping, the identification of candidate genes is a primary factor in the analysis of QTL. To screen the candidate genes and create a physical map, RiceXpro (https://ricexpro.dna.affrc.go.jp/ (accessed on 3 June 2022)) and Rapdb (https://rapdb.dna.affrc.go.jp/ (accessed on 3 June 2022)) were utilized. Open Reading Frames (ORFs) were classified as functions associated with seed germination. Additionally, NCBI (https://www.ncbi.nlm.nih.gov/ (accessed on 3 June 2022)) and BioEdit 7.0 (https://bioedit.software.informer.com/7.0/ (accessed on 3 June 2022)) were used for homologous sequence analysis and STRING for the analysis of protein interactions related to genes. For statistical analysis, the mean and standard deviation were calculated and conducted in three replicates.

4.5. Prediction of the Related Genes

On the basis of QTL mapping, the genes related to seed germination under low temperature were screened using Rapdb (https://rapdb.dna.affrc.go.jp/ (accessed on 3 June 2022)) and RiceXpro (https://ricexpro.dna.affrc.go.jp/ (accessed on 3 June 2022)). Rapdb and RiceXpro selected all the ORFs in the target QTL region, and the agriGO tool (http://bioinfo.cau.edu.cn/agriGO/ (accessed on 3 June 2022)) was used to identify the function of all related genes by the gene ontology (GO) enrichment analysis. Based on the classification of gene functions, we searched for genes related to seed germination. For the multiple homologous sequence variation analyses of the related genes comparisons, NCBI (National Center for Biotechnology Information, Bethesda, MA, USA, https://www.ncbi.nlm.nih.gov/ (accessed on 3 June 2022)) and BioEdit 7.0 (https://bioedit.software.informer.com/7.0/ (accessed on 3 June 2022)) were used. Moreover, the MEGA 11 (https://www.megasoftware.net/ (accessed on 3 June 2022)) software was used for the phylogenetic tree. STRING (version 11.0) database (https://string-db.org/ (accessed on 3 June 2022)) was used for protein–protein interaction/association network.

4.6. Analysis of Related Gene Expression Level

To analyze the relative expression level of related genes, seed samples were collected at different times. The total RNA was extracted from seeds of Cheongcheong, Nagdong using the Trizol-based method. The quality and concentration of the total RNA was assessed by ultramicro spectrophotometer ND-2000 (Nanodrop, Wilmington, DE, USA). The RNA was diluted to make a 100 ng/µL concentration and then cDNA was synthesized using the UltraScript 2.0 cDNA Synthesis Kit (PCRBIOSYSTEMS, Wayne, PA, USA). For qRT-PCR, we used StepOnePlusTM RT-PCR System machine (Thermo Fisher Scientific, Seoul, Korea). For the reaction set up, 10 µL 2X Real-time PCR Master Mix (Including SYBR® Green I) (BIOFACT, Daejeon, Korea), 1 µL forward primer (20 pmol/µL), 1 µL reverse primer (20 pmol/µL), 100 ng of cDNA, and the remaining was used as nuclease-free water to make a final volume of 20 µL. The reaction conditions were as follows: polymerase activation at 95 °C for 10 min, denaturation and annealing at 95 °C for 15 s, and extension at 60 °C for 1 min. To calculate the mean and standard deviation, each reaction was performed in three replicates and *OsActin* was used as a housekeeping gene. The genes primer list is shown in (Table S5).

5. Conclusions

In summary, we carried out QTL analysis to identify the seed germination-related genes under LDT conditions. The results showed that four QTLs were detected in RM7197-RM15063, RM528-RM20632, and RM23314-RM23178 on chromosomes 3, 6, and 8, respectively. A total of 41 genes were screened on all chromosomes, among them, 25 related genes were selected by gene function annotation and further identified through qRT-PCR. Therefore, our study suggests that *OsGPq3* (*OSK/SnRK*) gene involved in viviparous germination, ABA signaling pathway, and tiller enhancement will provide a basis for further study associated with low-temperature germinability in rice.

Supplementary Materials: The following supporting information can be downloaded at: https://www.mdpi.com/article/10.3390/ijms23137379/s1.

Author Contributions: Conceptualization, N.K., R.J. and K.-M.K.; methodology, N.K.; software, D.-D.Z.; validation, J.-R.P., E.-G.K., Y.-H.J. and K.-M.K.; formal analysis, N.K.; investigation, N.K., S.A. and G.-H.E.; resources, K.-M.K.; data curation, K.-M.K.; writing—original draft preparation, N.K. and R.J.; writing—review and editing, N.K. and R.J.; visualization, E.-G.K.; supervision, G.-S.L. and K.-M.K. All authors have read and agreed to the published version of the manuscript.

Funding: This work was supported by a grant from the Rural Development Administration Agenda Program (Project No. PJ015608), RDA, Republic of Korea.

Institutional Review Board Statement: Not applicable.

Informed Consent Statement: Not applicable.

Data Availability Statement: The data presented in this study are available on request from the corresponding author.

Conflicts of Interest: The authors declare no conflict of interest.

References

1. Wani, S.H.; Kumar, V.; Shriram, V.; Sah, S.K. Phytohormones and their metabolic engineering for abiotic stress tolerance in crop plants. *Crop J.* **2016**, *4*, 162–176. [CrossRef]
2. Meena, K.K.; Sorty, A.M.; Bitla, U.M.; Choudhary, K.; Gupta, P.; Pareek, A.; Singh, D.P.; Prabha, R.; Sahu, P.K.; Gupta, V.K. Abiotic stress responses and microbe-mediated mitigation in plants: The omics strategies. *Front. Plant Sci.* **2017**, *8*, 172. [CrossRef] [PubMed]
3. Abhinandan, K.; Skori, L.; Stanic, M.; Hickerson, N.; Jamshed, M.; Samuel, M.A. Abiotic stress signaling in wheat–an inclusive overview of hormonal interactions during abiotic stress responses in wheat. *Front. Plant Sci.* **2018**, *9*, 734. [CrossRef] [PubMed]
4. Shakiba, E.; Edwards, J.D.; Jodari, F.; Duke, S.E.; Baldo, A.M.; Korniliev, P.; McCouch, S.R.; Eizenga, G.C. Genetic architecture of cold tolerance in rice (*Oryza sativa*) determined through high resolution genome-wide analysis. *PLoS ONE* **2017**, *12*, e0172133. [CrossRef]
5. Sohn, S.-I.; Pandian, S.; Kumar, T.S.; Zoclanclounon, Y.A.B.; Muthuramalingam, P.; Shilpha, J.; Satish, L.; Ramesh, M. Seed Dormancy and Pre-Harvest Sprouting in Rice—An Updated Overview. *Int. J. Mol. Sci.* **2021**, *22*, 11804. [CrossRef]
6. Kendall, S.L.; Hellwege, A.; Marriot, P.; Whalley, C.; Graham, I.A.; Penfield, S. Induction of dormancy in Arabidopsis summer annuals requires parallel regulation of DOG1 and hormone metabolism by low temperature and CBF transcription factors. *Plant Cell* **2011**, *23*, 2568–2580. [CrossRef]
7. He, H.; de Souza Vidigal, D.; Snoek, L.B.; Schnabel, S.; Nijveen, H.; Hilhorst, H.; Bentsink, L. Interaction between parental environment and genotype affects plant and seed performance in Arabidopsis. *J. Exp. Bot.* **2014**, *65*, 6603–6615. [CrossRef]

8. Footitt, S.; Douterelo-Soler, I.; Clay, H.; Finch-Savage, W.E. Dormancy cycling in Arabidopsis seeds is controlled by seasonally distinct hormone-signaling pathways. *Proc. Natl. Acad. Sci. USA* **2011**, *108*, 20236–20241. [CrossRef]
9. Finch-Savage, W.E.; Footitt, S. Seed dormancy cycling and the regulation of dormancy mechanisms to time germination in variable field environments. *J. Exp. Bot.* **2017**, *68*, 843–856. [CrossRef]
10. Weitbrecht, K.; Müller, K.; Leubner-Metzger, G. First off the mark: Early seed germination. *J. Exp. Bot.* **2011**, *62*, 3289–3309. [CrossRef]
11. Basbouss-Serhal, I.; Leymarie, J.; Bailly, C. Fluctuation of Arabidopsis seed dormancy with relative humidity and temperature during dry storage. *J. Exp. Bot.* **2016**, *67*, 119–130. [CrossRef] [PubMed]
12. Peng, J.; Harberd, N.P. The role of GA-mediated signalling in the control of seed germination. *Curr. Opin. Plant Biol.* **2002**, *5*, 376–381. [CrossRef]
13. Kucera, B.; Cohn, M.A.; Leubner-Metzger, G. Plant hormone interactions during seed dormancy release and germination. *Seed Sci. Res.* **2005**, *15*, 281–307. [CrossRef]
14. Ali, F.; Qanmber, G.; Li, F.; Wang, Z. Updated role of ABA in seed maturation, dormancy, and germination. *J. Adv. Res.* **2022**, *35*, 199–214. [CrossRef]
15. Nambara, E.; Okamoto, M.; Tatematsu, K.; Yano, R.; Seo, M.; Kamiya, Y. Abscisic acid and the control of seed dormancy and germination. *Seed Sci. Res.* **2010**, *20*, 55–67. [CrossRef]
16. Miao, C.; Xiao, L.; Hua, K.; Zou, C.; Zhao, Y.; Bressan, R.A.; Zhu, J.-K. Mutations in a subfamily of abscisic acid receptor genes promote rice growth and productivity. *Proc. Natl. Acad. Sci. USA* **2018**, *115*, 6058–6063. [CrossRef]
17. Kim, H.; Hwang, H.; Hong, J.-W.; Lee, Y.-N.; Ahn, I.P.; Yoon, I.S.; Yoo, S.-D.; Lee, S.; Lee, S.C.; Kim, B.-G. A rice orthologue of the ABA receptor, OsPYL/RCAR5, is a positive regulator of the ABA signal transduction pathway in seed germination and early seedling growth. *J. Exp. Bot.* **2012**, *63*, 1013–1024. [CrossRef]
18. Ma, Y.; Szostkiewicz, I.; Korte, A.; Moes, D.; Yang, Y.; Christmann, A.; Grill, E. Regulators of PP2C phosphatase activity function as abscisic acid sensors. *Science* **2009**, *324*, 1064–1068. [CrossRef]
19. Park, S.-Y.; Fung, P.; Nishimura, N.; Jensen, D.R.; Fujii, H.; Zhao, Y.; Lumba, S.; Santiago, J.; Rodrigues, A.; Chow, T.-f.F. Abscisic acid inhibits type 2C protein phosphatases via the PYR/PYL family of START proteins. *Science* **2009**, *324*, 1068–1071. [CrossRef]
20. Yoshida, T.; Nishimura, N.; Kitahata, N.; Kuromori, T.; Ito, T.; Asami, T.; Shinozaki, K.; Hirayama, T. ABA-hypersensitive germination3 encodes a protein phosphatase 2C (AtPP2CA) that strongly regulates abscisic acid signaling during germination among Arabidopsis protein phosphatase 2Cs. *Plant Physiol.* **2006**, *140*, 115–126. [CrossRef]
21. Nakashima, K.; Fujita, Y.; Kanamori, N.; Katagiri, T.; Umezawa, T.; Kidokoro, S.; Maruyama, K.; Yoshida, T.; Ishiyama, K.; Kobayashi, M. Three Arabidopsis SnRK2 protein kinases, SRK2D/SnRK2. 2, SRK2E/SnRK2. 6/OST1 and SRK2I/SnRK2. 3, involved in ABA signaling are essential for the control of seed development and dormancy. *Plant Cell Physiol.* **2009**, *50*, 1345–1363. [CrossRef] [PubMed]
22. Nambara, E.; Naito, S.; McCourt, P. A mutant of Arabidopsis which is defective in seed development and storage protein accumulation is a new abi3 allele. *Plant J.* **1992**, *2*, 435–441. [CrossRef]
23. Robichaud, C.; Sussex, I.M. The response of viviparous-1 and wild type embryos of Zea mays to culture in the presence of abscisic acid. *J. Plant Physiol.* **1986**, *126*, 235–242. [CrossRef]
24. Shu, K.; Zhang, H.; Wang, S.; Chen, M.; Wu, Y.; Tang, S.; Liu, C.; Feng, Y.; Cao, X.; Xie, Q. ABI4 regulates primary seed dormancy by regulating the biogenesis of abscisic acid and gibberellins in Arabidopsis. *PLoS Genet.* **2013**, *9*, e1003577. [CrossRef]
25. Shu, K.; Chen, Q.; Wu, Y.; Liu, R.; Zhang, H.; Wang, P.; Li, Y.; Wang, S.; Tang, S.; Liu, C. ABI 4 mediates antagonistic effects of abscisic acid and gibberellins at transcript and protein levels. *Plant J.* **2016**, *85*, 348–361. [CrossRef]
26. Lopez-Molina, L.; Mongrand, S.; Chua, N.-H. A postgermination developmental arrest checkpoint is mediated by abscisic acid and requires the ABI5 transcription factor in Arabidopsis. *Proc. Natl. Acad. Sci. USA* **2001**, *98*, 4782–4787. [CrossRef]
27. Lin, Q.; Wu, F.; Sheng, P.; Zhang, Z.; Zhang, X.; Guo, X.; Wang, J.; Cheng, Z.; Wang, J.; Wang, H. The SnRK2-APC/CTE regulatory module mediates the antagonistic action of gibberellic acid and abscisic acid pathways. *Nat. Commun.* **2015**, *6*, 7981. [CrossRef]
28. Najeeb, S.; Ali, J.; Mahender, A.; Pang, Y.; Zilhas, J.; Murugaiyan, V.; Vemireddy, L.R.; Li, Z. Identification of main-effect quantitative trait loci (QTLs) for low-temperature stress tolerance germination-and early seedling vigor-related traits in rice (*Oryza sativa* L.). *Mol. Breed.* **2020**, *40*, 10. [CrossRef]
29. Koseki, M.; Kitazawa, N.; Yonebayashi, S.; Maehara, Y.; Wang, Z.-X.; Minobe, Y. Identification and fine mapping of a major quantitative trait locus originating from wild rice, controlling cold tolerance at the seedling stage. *Mol. Genet. Genom.* **2010**, *284*, 45–54. [CrossRef]
30. Tanksley, S.; Nelson, J. Advanced backcross QTL analysis: A method for the simultaneous discovery and transfer of valuable QTLs from unadapted germplasm into elite breeding lines. *Theor. Appl. Genet.* **1996**, *92*, 191–203. [CrossRef]
31. Fujino, K.; Sekiguchi, H.; Matsuda, Y.; Sugimoto, K.; Ono, K.; Yano, M. Molecular identification of a major quantitative trait locus, qLTG3–1, controlling low-temperature germinability in rice. *Proc. Natl. Acad. Sci. USA* **2008**, *105*, 12623–12628. [CrossRef] [PubMed]
32. Li, Z.-K.; Luo, L.; Mei, H.; Wang, D.; Shu, Q.; Tabien, R.; Zhong, D.; Ying, C.; Stansel, J.; Khush, G. Overdominant epistatic loci are the primary genetic basis of inbreeding depression and heterosis in rice. I. Biomass and grain yield. *Genetics* **2001**, *158*, 1737–1753. [CrossRef] [PubMed]

33. Hua, J.; Xing, Y.; Xu, C.; Sun, X.; Yu, S.; Zhang, Q. Genetic dissection of an elite rice hybrid revealed that heterozygotes are not always advantageous for performance. *Genetics* **2002**, *162*, 1885–1895. [CrossRef] [PubMed]
34. Chen, L.; Bian, J.; Shi, S.; Yu, J.; Khanzada, H.; Wassan, G.M.; Zhu, C.; Luo, X.; Tong, S.; Yang, X. Genetic analysis for the grain number heterosis of a super-hybrid rice WFYT025 combination using RNA-Seq. *Rice* **2018**, *11*, 37. [CrossRef]
35. Wang, Z.-f.; Wang, J.-f.; Bao, Y.-m.; Wang, F.-h.; Zhang, H.-s. Quantitative trait loci analysis for rice seed vigor during the germination stage. *J. Zhejiang Univ. Sci. B* **2010**, *11*, 958–964. [CrossRef]
36. Wan, X.; Wan, J.; Weng, J.; Jiang, L.; Bi, J.; Wang, C.; Zhai, H. Stability of QTLs for rice grain dimension and endosperm chalkiness characteristics across eight environments. *Theor. Appl. Genet.* **2005**, *110*, 1334–1346. [CrossRef]
37. Li, C.; Zhou, A.; Sang, T. Genetic analysis of rice domestication syndrome with the wild annual species, Oryza nivara. *New Phytol.* **2006**, *170*, 185–194. [CrossRef]
38. Sabouri, A.; Sabouri, H.; De Ocampo, M. Genetic Analysis Seedling vigour under osmotic stress in Rice by QTL Mapping. *Russ. Agric. Sci.* **2012**, *38*, 423–429. [CrossRef]
39. WANG, S.-m.; CUI, G.-q.; Hui, W.; XIA, S.-s.; LI, Y.-f.; YANG, Z.-l.; LING, Y.-h.; ZHANG, C.-w.; HE, G.-h.; ZHAO, F.-m. Identification and QTL mapping of Z550, a rice backcrossed inbred line with increased grains per panicle. *J. Integr. Agric.* **2019**, *18*, 526–531. [CrossRef]
40. Umezawa, T.; Nakashima, K.; Miyakawa, T.; Kuromori, T.; Tanokura, M.; Shinozaki, K.; Yamaguchi-Shinozaki, K. Molecular basis of the core regulatory network in ABA responses: Sensing, signaling and transport. *Plant Cell Physiol.* **2010**, *51*, 1821–1839. [CrossRef]
41. Yan, A.; Chen, Z. The control of seed dormancy and germination by temperature, light and nitrate. *Bot. Rev.* **2020**, *86*, 39–75. [CrossRef]
42. Cutler, S.R.; Rodriguez, P.L.; Finkelstein, R.R.; Abrams, S.R. Abscisic acid: Emergence of a core signaling network. *Annu. Rev. Plant Biol.* **2010**, *61*, 651–679. [CrossRef] [PubMed]
43. Yu, F.; Li, M.; He, D.; Yang, P. Advances on post-translational modifications involved in seed germination. *Front. Plant Sci.* **2021**, *12*, 362. [CrossRef] [PubMed]
44. Shu, K.; Zhou, W.; Yang, W. APETALA 2-domain-containing transcription factors: Focusing on abscisic acid and gibberellins antagonism. *New Phytol.* **2018**, *217*, 977–983. [CrossRef] [PubMed]
45. Zhao, D.-D.; Park, J.-R.; Jang, Y.-H.; Kim, E.-G.; Du, X.-X.; Farooq, M.; Yun, B.-J.; Kim, K.-M. Identification of One Major QTL and a Novel Gene OsIAA17q5 Associated with Tiller Number in Rice Using QTL Analysis. *Plants* **2022**, *11*, 538. [CrossRef]
46. Azad, S.; Manik, M.R.; Hasan, S.; Matin, A. Effect of different pre-sowing treatments on seed germination percentage and growth performance of Acacia auriculiformis. *J. For. Res.* **2011**, *22*, 183–188. [CrossRef]
47. Yun, B.-W.; Kim, M.-G.; Handoyo, T.; Kim, K.-M. Analysis of rice grain quality-associated quantitative trait loci by using genetic mapping. *Am. J. Plant Sci.* **2014**, *2014*, 44442. [CrossRef]
48. Jan, R.; Aaqil Khan, M.; Asaf, S.; Park, J.-R.; Lee, I.-J.; Kim, K.-M. Flavonone 3-hydroxylase Relieves Bacterial Leaf Blight Stress in Rice via Overaccumulation of Antioxidant Flavonoids and Induction of Defense Genes and Hormones. *Int. J. Mol. Sci.* **2021**, *22*, 6152. [CrossRef]
49. Zhao, D.-D.; Son, J.-H.; Farooq, M.; Kim, K.-M. Identification of Candidate Gene for Internode Length in Rice to Enhance Resistance to Lodging Using QTL Analysis. *Plants* **2021**, *10*, 1369. [CrossRef]
50. McCough, S.R.; Doerge, R.W. QTL mapping in rice. *Trends Genet.* **1995**, *11*, 482–487. [CrossRef]

International Journal of *Molecular Sciences*

MDPI

Article

Resequencing of 410 Sesame Accessions Identifies SINST1 as the Major Underlying Gene for Lignans Variation

Senouwa Segla Koffi Dossou [1,2,†], Shengnan Song [1,†], Aili Liu [1], Donghua Li [1], Rong Zhou [1], Muez Berhe [1], Yanxin Zhang [1], Chen Sheng [1], Zhijian Wang [1], Jun You [1,*] and Linhai Wang [1,*]

1 Key Laboratory of Biology and Genetic Improvement of Oil Crops, Ministry of Agriculture and Rural Affairs, Oil Crops Research Institute of the Chinese Academy of Agricultural Sciences, Wuhan 430062, China
2 Laboratory of Plant Physiology and Biotechnologies, Faculty of Sciences, University of Lomé, Lomé 01BP 1515, Togo
* Correspondence: junyou@caas.cn (J.Y.); wanglinhai@caas.cn (L.W.); Tel.: +86-18607147952 (J.Y.); +86-15926338805 (L.W.)
† These authors contributed equally to this work.

Abstract: Sesame is a promising oilseed crop that produces specific lignans of clinical importance. Hence, a molecular description of the regulatory mechanisms of lignan biosynthesis is essential for crop improvement. Here, we resequence 410 sesame accessions and identify 5.38 and 1.16 million SNPs (single nucleotide polymorphisms) and InDels, respectively. Population genomic analyses reveal that sesame has evolved a geographic pattern categorized into northern (NC), middle (MC), and southern (SC) groups, with potential origin in the southern region and subsequent introduction to the other regions. Selective sweeps analysis uncovers 120 and 75 significant selected genomic regions in MC and NC groups, respectively. By screening these genomic regions, we unveiled 184 common genes positively selected in these subpopulations for exploitation in sesame improvement. Genome-wide association study identifies 17 and 72 SNP loci for sesamin and sesamolin variation, respectively, and 11 candidate causative genes. The major pleiotropic $SNP^{C/A}$ locus for lignans variation is located in the exon of the gene *SiNST1*. Further analyses revealed that this locus was positively selected in higher lignan content sesame accessions, and the "C" allele is favorable for a higher accumulation of lignans. Overexpression of $SiNST1^{C}$ in sesame hairy roots significantly up-regulated the expression of *SiMYB58, SiMYB209, SiMYB134, SiMYB276*, and most of the monolignol biosynthetic genes. Consequently, the lignans content was significantly increased, and the lignin content was slightly increased. Our findings provide insights into lignans and lignin regulation in sesame and will facilitate molecular breeding of elite varieties and marker-traits association studies.

Keywords: sesame; genome resequencing; lignan; genomic signatures; NST1; GWAS

Citation: Dossou, S.S.K.; Song, S.; Liu, A.; Li, D.; Zhou, R.; Berhe, M.; Zhang, Y.; Sheng, C.; Wang, Z.; You, J.; et al. Resequencing of 410 Sesame Accessions Identifies SINST1 as the Major Underlying Gene for Lignans Variation. *Int. J. Mol. Sci.* **2023**, *24*, 1055. https://doi.org/10.3390/ijms24021055

Academic Editors: Andrés J. Cortés and Hai Du

Received: 9 November 2022
Revised: 12 December 2022
Accepted: 13 December 2022
Published: 5 January 2023

1. Introduction

Sesame (*Sesamum indicum* L., 2n = 26) is a member of the Pedaliaceae family and one of the most important oilseed and industrial crops worldwide. It is primarily cultivated in subtropical and tropical regions in Asia, Africa, Latin America, and Central America [1]. Sesame seeds are promising sources of nutritional components and nutraceuticals for food and pharmacological industries [2,3]. It contains specific lignans, including sesamin, sesamolin, sesaminol, sesamol, etc., that have shown diverse physiological abilities such as anti-cancers, antioxidative, anti-diabetes, anti-inflammatory, anti-hypertensive, anti-proliferative, anti-melanogenesis, anti-depression and memory loss, anti-osteonecrosis, and anti-osteoporosis [4–9]. Although it is documented that sesame is the one of the most ancient oilseed crops known to humans, its origins are still a subject of debate, and its progenitor also is not determined [1,10–12]. Yu et al. [13] found that the domestication of sesame in China and India might have occurred independently. Other reports indicate that sesame was introduced in China in 2200 BC during the Han dynasty [14]. China is one of the leading countries in terms of sesame production and consumption (www.fao.org/statistics/en/, accessed on 22 November 2021) and has collected and preserved over 7000 accessions in its germplasms [15]. However, sesame's domestication, population structure, and genetic diversity remain largely unknown.

The release of the sesame reference genome via de novo assembly has allowed genomic studies on sesame biology [12]. The sesame reference genome quality has been improved via the resequencing of 430 recombinant inbred lines [16], which has provided a solid basis for investigating the molecular mechanisms controlling critical quality and agronomic traits of sesame such as high yield, high oil and lignans contents, seed coat color, plant architecture, and melatonin. Significant loci and candidate causative genes underlying these traits have been identified except for lignans variation [15,17–19]. The sesame-specific lignans seem to have evolved from defense and accommodation to biotic stresses [20–22], suggesting their underlying genes were positively selected during domestication and breeding [13]. Therefore, identifying adaptation-related genes may help understand sesame evolution [23]. The domestication and subsequent breeding of crops generate selective sweeps, eliminating or reducing variation amid the nucleotides adjoining a mutation in DNA [24,25]. Although large-scale genetic variations among diverse accessions of sesame have been reported [15], the detected genomic differentiation between the subpopulations was very low (F_{ST} = 0.02), and it was impossible to unravel regions under selective sweeps for exploitation in the crop improvement.

Lignans are a diversified group of phytochemicals that play essential roles in plants' interactions and adaptations to environmental conditions [26,27]. They are biosynthesized from phenylalanine via the phenylpropanoid pathway and mostly share the same precursors with lignin [26,28,29]. Most naturally occurring lignans are derived from the oxidative coupling of E-coniferyl alcohol moieties [26,28,30]. In sesame, the biosynthetic genes of the major lignans, sesamin and sesamolin have been identified, and their mechanisms of action elucidated [30,31]. Recently, two laccase (*SiLAC1* and *SiLAC39*) and two peroxidase (*SiPOD52* and *SiPOD63*) family genes were identified as the potential pinoresinol synthase genes [32]. However, the regulatory mechanisms of lignan biosynthesis in sesame are largely unknown. The regulation of monolignol biosynthesis is a complex molecular mechanism that involves diverse transcription factors and phytohormones, among which NST1/2 detain pivotal roles [33–35]. NST1 and NST2 are master switch regulators of secondary cell wall formation through interaction with diverse family genes, mostly Myb transcription factors [33–35]. Therefore, deciphering the complex regulatory network of lignan biosynthesis in sesame is of particular interest for biotechnological applications.

Whole-genome resequencing and population genomic analyses have considerably enhanced the understanding of crops' genetic variation and domestication history and facilitated marker-traits association studies [36–38]. In the present study, we resequenced a worldwide collection of sesame, including 316 accessions from various geographical regions in China and 94 accessions from twenty-eight countries. We investigated the population structure and genetic diversity and revealed key candidate genes under selective sweeps in subpopulations. In addition, the master regulatory gene of lignan and lignin biosynthesis was discovered based on a genome-wide association study (GWAS) and gene expression analyses and validated in sesame hairy roots.

2. Results

2.1. Genome-Wide Variation and Population Structure

We resequenced 410 sesame germplasm samples, including 316 collected from southern, middle, and northern China and 94 collected from 28 countries worldwide representing various geographical regions (Table S1). The population included 329 landraces and 81 modern cultivars. The resequencing generated a total of 3.533 Tb of raw data, with an average of 8.617 Mb per accession (Table S2). The average sequencing depth was 24.93×, and the genome coverage 96.08%. The GC content varied from 36.41 to 40.98% (Table S2). After mapping against the sesame reference genome, we obtained a final set of 5,385,583 single-nucleotide polymorphisms (SNPs) and 1,163,197 insertion-deletion (InDels). The SNPs and InDels are distributed unequally on the 13 chromosomes of the sesame genome (Figure 1a). Figures of 70.45, 10.70, and 4.45% of the SNPs were located in intergenic, intronic, and exonic regions, respectively (Figure 1c). Among the SNPs identified in coding sequences, 45.58% were synonymous SNPs and 52.79% non-synonymous SNPs (Figure 1b). The non-synonymous to synonymous SNPs ratio and the transition/transversion (Ts/Tv) ratio were 1.15 and 1.907, respectively (Table S3).

The SNP-based phylogenetic analysis classified the 410 sesame accessions into three major monophyletic clades which correlated with their geographical distribution (Figure 1d; Table S1). Group I (61 accessions), II (92 accessions), and III (257 accessions) included all the accessions from southern, northern, and central regions, respectively. Therefore, they were labeled SC (southern cluster, located south of 24° N latitude), NC (northern cluster, located north of 32–34° N latitude), and MC (central or middle cluster, located between SC and NC), respectively, to further identify selective sweeps. The landraces and modern cultivars exhibited admixture and were not separated on the phylogenetic tree (Figure 1d; Table S1). The principal component analysis (Figure 1e) and population structure analysis with different K-means levels (Figures 1f and S1) results were identical to the phylogenetic analysis results (Figures 1f and S1).

2.2. Genetic Diversity in Sesame Genomes

To explore genetic divergence among the sesame subpopulations, we performed a pairwise population differentiation (F_{ST}) across the genome between each group (Figure 2a). The F_{ST} value (F_{ST} = 0.068) between MC and NC groups is the smallest. The F_{ST} value between the SC and MC groups (F_{ST} = 0.156) was not too much different from that between the SC and NC groups (F_{ST} = 0.159). These results are consistent with the population structure and indicate that the MC and NC group populations have the closest distance and might evolve from the southern subpopulation. We then examined the genetic diversity by computing each group's nucleotide diversity (π). As we expected, the genetic diversity of the SC subpopulation ($\pi = 1.62 \times 10^{-3}$) is higher than that of the MC group ($\pi = 1.07 \times 10^{-3}$) and the NC group ($\pi = 1.09 \times 10^{-3}$) (Figure 2a). In addition, we explored Tajima's D values in each group and found a decrease in the average value from southern to northern regions (Figure 2c). We further investigated the linkage disequilibrium (LD) based on r^2 in the different groups (Figure 2b and Table S4). We found that the distance, when dropped to half of the maximum value, is 89.3 kb for the entire sesame population, which is slightly higher than that reported previously [15]. Finally, we investigated the directions of the gene flow between the subpopulations. The results indicated a "south to middle and north migration model" (Figure 2d,e).

2.3. Genomic Selected Signals in MC and NC Groups

To identify potential candidate genes for exploitation in sesame improvement, we searched for candidate selective genomic regions in MC and NC subpopulations using SC as a background population and adopting the criteria of $\log_2(\pi \text{ ratio}) > 2.5$ and $F_{ST} > 0.45$ (Figure 3a and Figure S3a). We identified 120 and 75 putative selective sweep regions comprising 968 and 571 genes for MC and NC groups, respectively (Tables S5 and S6).

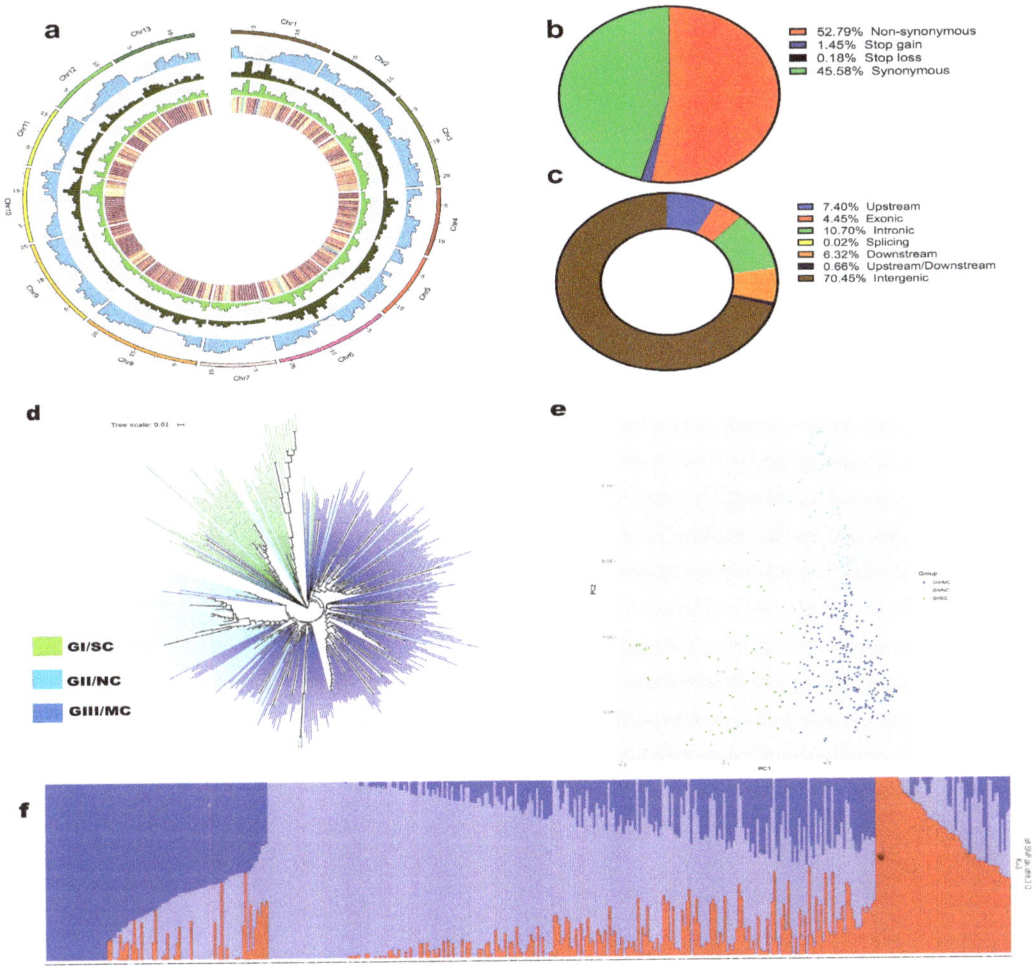

Figure 1. Population structure of sesame accessions in this study. (**a**) Chromosomal distribution of the genomic variants. From the outer to the inner are chromosomes, gene density, SNP density, InDels density, and nucleotide diversity. (**b**,**c**) Classification of the SNPs. (**d**) Phylogenetic tree of all accessions inferred from whole-genome SNPs filtered by linkage disequilibrium (LD) $r^2 < 0.05$; The three major groups (Southern cluster, SC; Northern cluster, NC; and middle or central cluster MC) are colored in purple, light blue, and blue, respectively. (**e**) Principal component analysis (PCA) plot for all the sesame accessions; dot colors correspond to the phylogenetic tree grouping. (**f**) Population stratification based on STRUCTURE for K = 3. Each color represents one ancestral population. Each accession is represented by a bar, and the length of each colored segment in the bar represents the proportion contributed by that ancestral population.

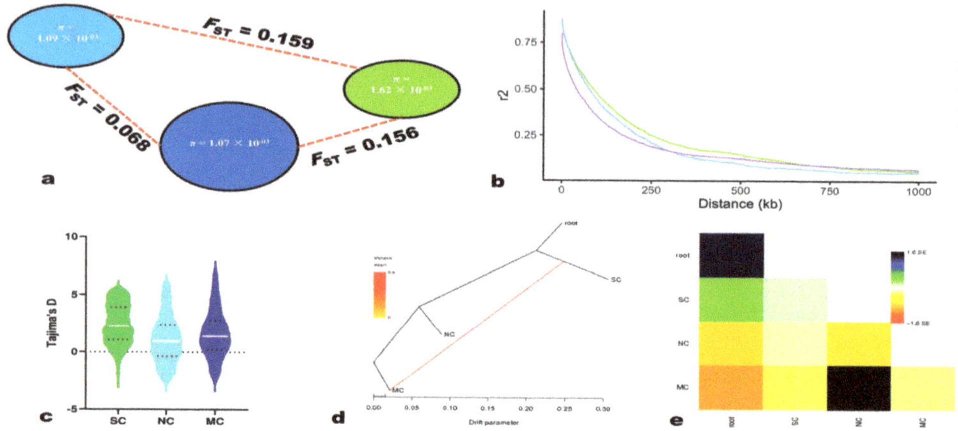

Figure 2. Genetic diversity and gene flow among the subpopulations. (**a**) The diversity (π) and genetic distance (F_{ST}) across the groups, where green, light blue, and blue colors indicate SC, NC, and MC, respectively; the radius of pie: genetic diversity; and dashed line length: F_{ST} value between two groups. SC, NC, and MC represent the subpopulations from the genetic diversity analysis. (**b**) Linkage disequilibrium (LD) differences between subpopulations. (**c**) Distribution of Tajima's D values. The white line indicates the mean. (**d**) Most frequently found maximum-likelihood trees for gene flow among the subpopulations by Treemix. (**e**) The residual fit values obtained from the maximum likelihood tree (**d**). The plot was constructed using the normalized residual covariances by dividing the residual covariance value between each pair of populations by the standard deviation between all sample pairs. On the right is the color ruler.

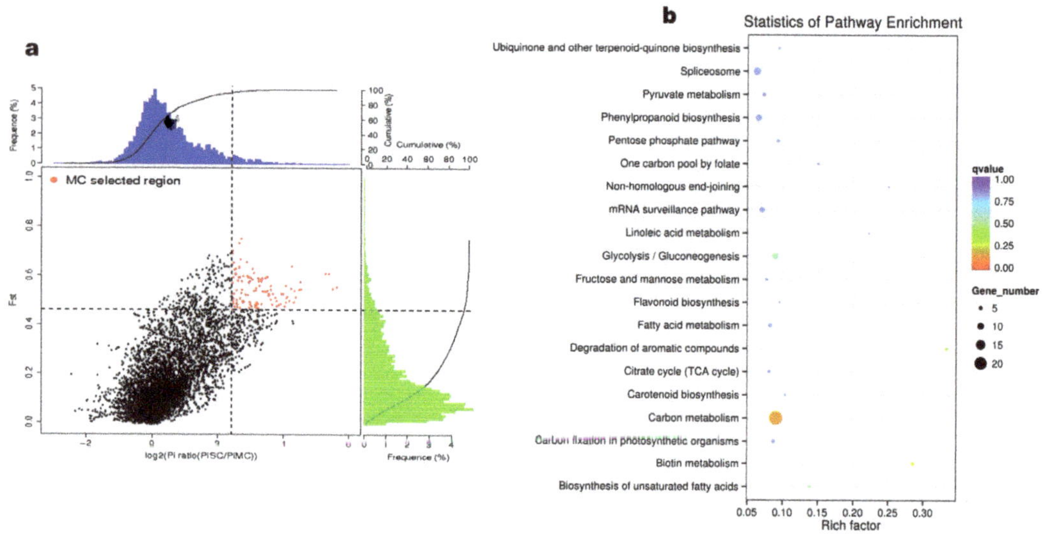

Figure 3. Selective sweeps and functional characterization of positively selected genes in MC group. (**a**) Selection signatures in MC subpopulation. Each red dot indicates a positive selectivegenomic region. (**b**) KEGG (Kyoto Encyclopedia of Genes and Genomes) analysis and enrichment results of genes detected under selective genomic regions in MC subpopulation. The color ruler and *p*-values are shown on the right.

Gene ontology (GO) enrichment analysis for genes selected in the MC subpopulation were mostly related to "structural molecule activity" (GO:0005198), "iron-sulfur cluster binding" (GO:0051536), "metal cluster binding" (GO:0051540), and "protein folding" (GO:0006457) (Figure S2). For the selected genes in NC, the GO terms "transferase activity, transferring phosphorus-containing groups" (GO:0016772), "hydrolase activity, acting on glycosyl bonds" (GO:0016798), "hydrolase activity, hydrolyzing O-glycosyl compounds" (GO:0004553), and "carbohydrate metabolic process" (GO:0005975) were predominant (Figure S3b). KEGG enrichment analysis assigned the MC-selected genes mainly to carbon metabolism, glycolysis/gluconeogenesis, and phenylpropanoid biosynthesis (Figure 3b). Meanwhile, the NC-selected genes were primarily assigned to ribosome, carbon metabolism, glycolysis/gluconeogenesis, phenylpropanoid biosynthesis, and purine and pyruvate metabolism (Figure S3c). We then constructed a Venn diagram among the two sets of selected genes and identified 184 genes (Figure S4). These genes encode diverse enzymes essential for plant development (Table S7).

2.4. Genome-Wide Associations for Lignans Variation in Sesame

Sesame contains two major and clinically important lignans, sesamin and sesamolin. To dissect the genetic basis of lignans variation in sesame, we cultivated all the 410 accessions in six different environments in Guangxi (GX, 2014), Wuhan (WH, 2014 and 2018), Hunan (HN, 2014), Shandong (SD, 2018), and Zhumadian (ZM, 2019). Seed samples from each location were analyzed by HPLC (high-performance liquid chromatography). The summary of the phenotypic data is presented in Table S8. In general, the sesamin and sesamolin contents varied from 0.06 to 10.65 mg/g and 0.05 to 4.77 mg/g, respectively. The coefficient of variation of sesamin ranged from 33.92 to 45.18%, while that of sesamolin ranged from 21.41 to 28.88%. The frequency distribution of the two traits and their correlations across the six environments are shown in Figure S5. As sesamin and sesamolin contents showed considerable variation, we fitted their values in the six environments to obtain a BLUP (best linear unbiased predictor) value, using a linear mixed model in the lme4 R package. To examine the influence of genetic variation on sesamin and sesamolin contents, we compared their respective values in the subpopulations (Figure 4a,b). We found no significant difference in the sesamolin content of the three groups (Figure 4b). However, the sesamin content in MC and SC was significantly ($p < 0.001$) higher than in NC (Figure 4a). We further investigated the population differentiation (F_{ST}) between high-sesamin (HS, value > 6 mg/g) and low-sesamin (LS, value < 1 mg/g) accessions. The F_{ST} analysis revealed significant divergent genomic regions comprising 2119 genes (Figure 4c; Tables S9 and S10). These genes were mainly related to metabolic processes, especially oxidoreductase activities and cellular processes (Table S10 and Figure S6). We searched for candidate selective regions for HS and LS content. We identified 41 and 65 putative selective genomic regions harboring 363 and 343 genes for HS and LS content, respectively (Tables S11–S14). KEGG annotation and enrichment analysis assigned the HS accessions selective genes mainly to plant-pathogen interaction, peroxisome, pentose and glucuronate interconversions, mRNA surveillance pathway, glycerolipid metabolism, and carbon metabolism (Figure S7a). While the LS accessions selective genes were principally assigned to metabolic pathways and biosynthesis of secondary metabolites (Figure S7b).

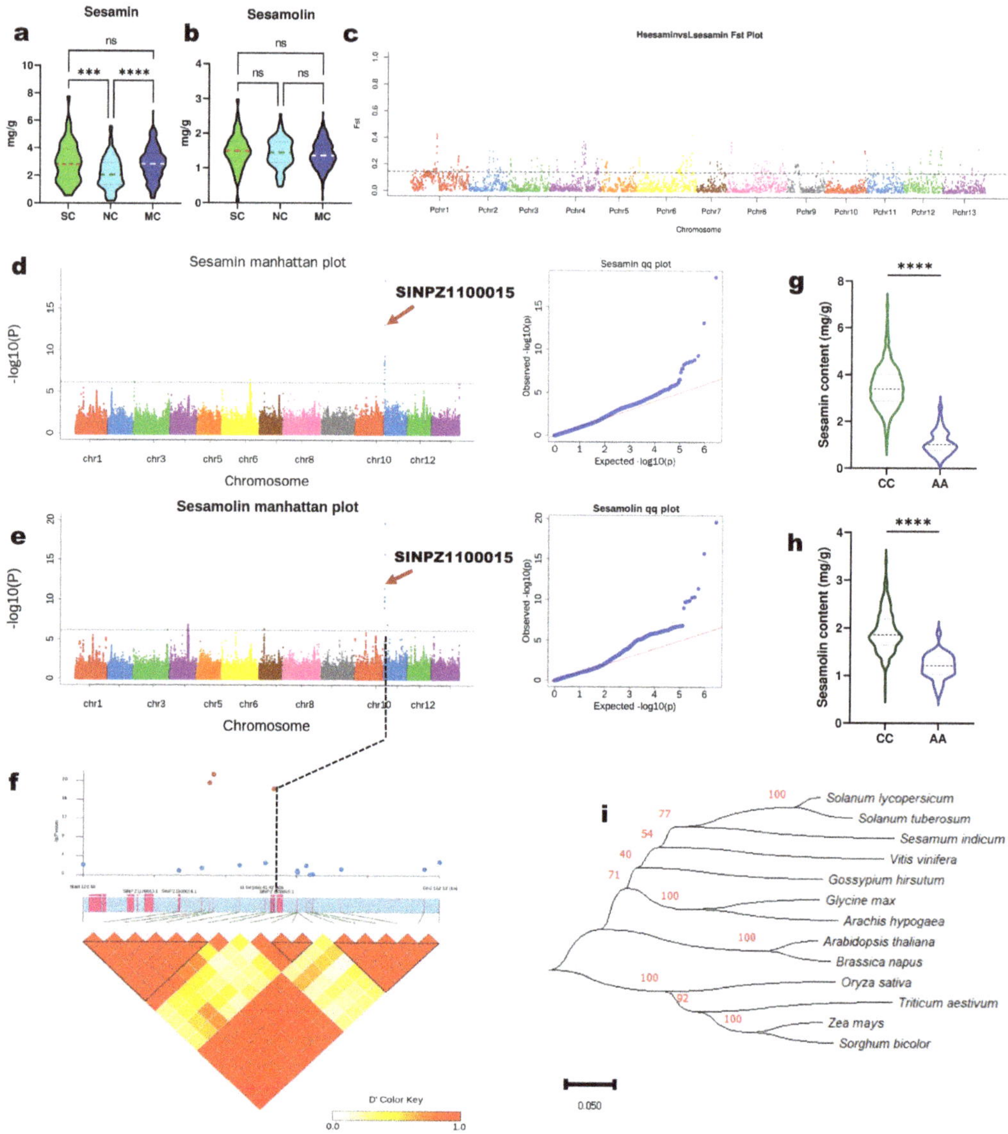

Figure 4. Genome-wide association for sesamin and sesamolin variation. (**a**,**b**) Boxplot of sesamin and sesamolin contents variation among the subpopulations, respectively. MC, NC, and MC indicate the different subpopulations. (**c**) Candidate selective genomic regions in high-sesamin content sesame accessions. F_{ST} stands for genetic differentiation. (**d**,**e**) Manhattan and QQ plots for GWAS analysis for sesamin and sesamolin, respectively. The dashed blue line indicates the threshold $-\log_{10}(p) = 6$. (**f**) LD (linkage disequilibrium) heat map around *SiNST1* (*SINPZ110015*). The dotted line from subfigures e and d to f show the position of *SiNST1* on the sesame genome. (**g**,**h**) Mining of favorable allele for sesamin and sesamolin contents, respectively; accessions with "C" allele, 362 and "A" allele, 34. (**i**) Phylogenetic analysis of *SiNST1* and NST1 from other species. *** $p < 0.001$; **** $p < 0.0001$, *t*-test.

We performed a GWAS on the two traits via the MLM (mixed linear model) implemented in the EMMAX (Efficient Mixed-Model Association eXpedited) software. The results showed that the model was adequate for analyzing these important traits (Figure 4d,e). In total, 17 and 72 significant SNP loci (threshold $-\log(p)$ value > 6) were identified for sesamin and sesamolin, respectively (Table S15). Interestingly, most of the loci locate at our previously detected QTL positions of the two lignans [39]. The major association signal (Chr11-142842) was pleiotropic for the two traits (Figure 4d,e) and is located in the exon of the gene *SINPZ1100015* (*SiNST1/SIN_1005755*), which encodes an NAC domain-containing protein 43. It is homologous to the *Arabidopsis* NST1, which is a master regulator of secondary cell wall deposition through the control of monolignol biosynthesis [33–35]. *SiNST1* has shown significant associations with variations in oil, seed coat thickness, and phytosterol contents in sesame [15,40]. In addition, we screened for candidate genes within the LD windows of the other associated loci. By combining gene function annotation and expression patterns in two contrasting sesamin content varieties, we selected ten additional candidate causative genes for lignans variation in sesame (Figure S8 and Table S16).

Interestingly, the locus of *SiNST1* exhibited significant divergence between HS and LS sesame accessions (Figure 4c and Table S10). The results of the regional association (LD) analysis of *SiNST1* $\pm$ 10 kb corroborate with the GWAS analysis (Figure 4f). *SiNST1* is primarily expressed in sesame seeds and capsules (Figure S9A). The non-synonymous SNP locus (Chr11-142842) consists of "C" allele change into "A" conducting amino acid change from Thr (threonine) to Lys (lysine). Superior allele analysis showed that the common allele "C" significantly increased ($p < 0.0001$) the sesamin and sesamolin contents than the variant allele "A" (Figure 4g,h). qRT-PCR analysis also showed that the "C" allele is favorable for lignans biosynthesis during seed development (Figure S9B). In sesame accessions with the "C" allele, *SiNST1* is highly expressed at the key stages (10 to 20 DPA) of sesamin and sesamolin biosynthesis [12,30,41]. While in sesame accessions with the "A" allele, *SiNST1* is mostly expressed at the early stages of seed development. Further, we searched for the coding sequence (CDS) of homologs of *SiNST1* in other crops via NCBI. The CDS of *SiNST1* consists of 376 amino acids (aa). Phylogenetic analysis clearly separated NST1 from monocots and dicots. On the phylogenetic tree, *SiNST1* clustered closely to NST1 in *Solanum tuberosum* and *S. lycopersicum* (Figure 4i).

2.5. *SiNST1 Is a Master Regulator of Lignans and Lignin Biosynthesis in Sesame*

Since NST1 is reported as a master regulator of lignans and lignin biosynthesis in plants by interacting with hormonal signals and MYB transcription factors, we explored the cis-acting elements in the promoter region of *SiNST1* and *PSS* (piperitol/sesamin synthase). The results showed that *SiNST1* contains mainly MYB binding, light responsiveness, and hormone-related elements (Table S17). *PSS* primarily contains MYB recognition elements and MeJA (methyl jasmonate) responsive elements (Table S18). We further performed a WGCNA (weighted gene correlation analysis) using the expression values (FPKM) of *SiNST1*, the 287 sesame MYB genes (*SiMYBs*) [42], and the 25 predicted CYP81Q1 genes [12], including *PSS*. As shown in Figure 5a, these genes were classified into four subgroups: blue; brown; grey; and turquoise modules. The blue module exhibited the highest positive correlation ($r = 0.31$) with sesamin and included *SiNST1* (Figure 5b). *SiNST1* appeared as the hub gene in this module and showed co-expression relationships with 85 *SiMYBs* and four CYP81Q1-related genes (Table S19). We found no direct interaction between *SiNST1* and *PSS*. *PSS* was classified into the turquoise module, which also exhibited a positive correlation ($r = 0.19$) with sesamin content (Figure 5b). The top ten genes in this module included seven MYBs and three CYP81Q1 genes (Table S20).

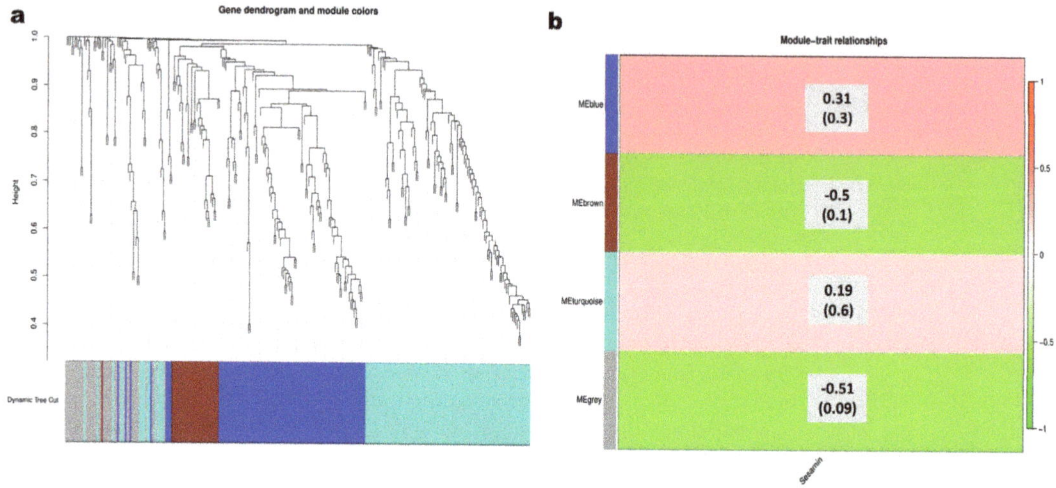

Figure 5. WGCNA results of *SiNST1*, *SiMYBs*, and predicted CYP81Q1-related genes in developing sesame seeds. (**a**) Classification of the analyzed genes in four modules represented by the blue, brown, grey, and turquoise colors. The genes with higher weight values have the most extended height on the dendrogram. (**b**) Heatmap of the correlation of WGCNA modules with sesamin content. The ordinate and abscissa indicate modules and the trait (sesamin), respectively. The green or red colors indicate the correlation coefficients (the top number in each cell). On the right is the color ruler. The bottom numbers (numbers in brackets) indicate the *p*-values.

To validate the potential function of *SiNST1*, we overexpressed *SiNST1*^{C} in sesame hairy roots (Figure S10). The *SiNST1*^{C}-overexpressing (OE) sesame hairy roots exhibited a significant increase ($p < 0.01$) in the content of sesamin and sesamolin (Figure 6e). As lignans and lignin share the same biosynthetic pathway, we also evaluated the lignin content in the sesame hairy roots. The results indicated a slight increase in the lignin content of OE compared to the control, but the difference was not significant (Figure 6d). To acquire insights into the regulatory mechanism of *SiNST1*, we investigated via qRT-PCR the expression levels of sesame monolignols biosynthetic genes, co-expressed *SiMYBs* predicted to involve lignin or phenylpropanoid biosynthesis, and *PSS*. The results showed that the *SiNST1* overexpression significantly influenced the expression of monolignol biosynthetic genes (Figure 6a). The expression levels of *SiPAL* (phenylalanine ammonia-lyase), *SiC4H* (cinnamate 4-hydroxylase), *SiC3H* (coumarate 3-hydroxylase), *SiCCR* (cinnamoyl-CoA reductase), *SiCOMT* (caffeic acid 3-O-methyltransferase), and *SiCCaAOMT* (caffeoyl-CoA O-methyltransferase), were significantly increased. Among the investigated MYB genes, *SiMYB58*, *SiMYB209*, *SiMYB134*, and *SiMYB276* exhibited a significant increase in their expression levels in OE (Figure 6b). The expression level of *PSS* in OE was also significantly increased (Figure 6c).

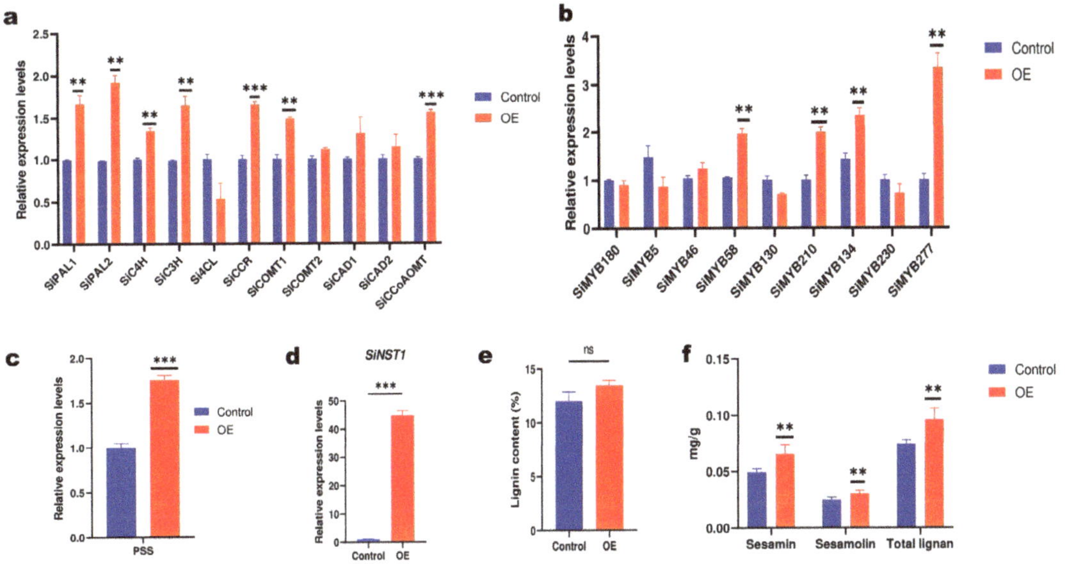

Figure 6. Functional validation of *SiNST1* in sesame hairy roots. (**a**) Expression analysis of monolignol biosynthetic genes. (**b**) Co-expressed *SiMYBs* with *SiNST1* predicted to involve in lignin biosynthesis. (**c**) piperitol/sesamin synthase, *PSS*. (**d**) *SiNST1*. (**e**) Lignin content. (**f**) Lignans content in transgenic sesame hairy roots. *** $p < 0.001$; ** $p < 0.01$; t-test.

3. Discussion

We provide comprehensive large-scale genome-wide polymorphism data in sesame, one of the most ancient oilseed crops with high economic, cultural, nutritional, and therapeutical values worldwide, especially in East Asia. The 410 sesame accessions resequencing data provide insights into the population structure and high throughput markers (5.38 and 1.16 million SNPs and InDels, respectively) for genotype selection and association studies to unveil key regulatory genes and dissect complex molecular mechanisms underlying agronomical and quality traits. The non-synonymous to synonymous SNPs ratio was 1.15, which is smaller than in soybean (*Glycine max* L., 1.61) [38], castor (*Ricinus communis* L., 1.39) [43], Tartary buckwheat (*Fagopyrum tataricum* L., 1.69) [44], and chickpea (*Cicer arietinum* L., 1.20) [45]. The transition/transversion (Ts/Tv) ratio of 1.907 for sesame is smaller than that for Tartary buckwheat (2.175) [44] but higher than that for maize (*Zea mays* L., 1.02) [46], tomato (*Solanum lycopersicum* L., 1.75) [47], and black gram (*Vigna mungo* L., 1.58) [48].

The domestication and evolution histories of sesame are unclear and are expected to be explored after the whole-genome sequencing of wild sesame plants [1]. Here, our investigations (phylogenetic and PCA analyses) show that sesame accessions were geographically different and exhibited a pattern of SC, MC, and NC. The landraces and modern cultivars exhibited admixture, which is consistent with the previous report by Wei et al. [15]. The population differentiation (F_{ST}), genetic diversity, and gene flow analyses indicated that the SC subpopulation is the oldest sesame population, and the MC and NC clades emanated from the SC group. The genetic distance (F_{ST}) between NC and MC groups was very low compared to their respective genetic distance with the SC group. The MC and NC subpopulations have relatively low genetic diversity (π) compared with the SC group. These results suggest that the SC subpopulation might be relatively close to wild sesame species. Moreover, the variation in genetic diversity suggests different intensities of genes selection during geographical differentiation. We identified 184 common genes under significant selective genomic regions in MC and NC groups. These genes encode diverse enzymes essential for plant development and may represent target genes for improving sesame quality traits and environmental adaptation. However, these potential environmental adaptation-related genome markers and candidate genes need to be confirmed via a Genome-Environment Association (GEA) and subsequently validated through functional genomics studies [49,50]. Otherwise, as the studied population contained majority accessions from China, our findings could imply that sesame introduction and domestication in China commenced in a locality in southern China and later extended to middle and northern China. Supportively, similar domestication events have been reported in soybean and Chinese castor [38,43]. This extensive genomic variation and substantiation of genetic differentiation amid regional subpopulations will favorize structured association mapping and the design of breeding programs that will seize the entire diversity in sesame [51–53].

Sesame lignans have become compounds of considerable scientific research interest due to their biological properties and huge importance [54–57]. Accordingly, the main current objective in sesame breeding is to create elite varieties containing high oil and lignans. The genetic variants governing natural variation in oil content and composition have been explored [15,58,59]. Selective sweeps analysis identified 363 positive selected genes in high-sesamin content sesame accessions. Functional analysis assigned these genes mostly in plant-pathogen interaction, peroxisome, pentose and glucuronate interconversions, mRNA surveillance pathway, glycerolipid metabolism, and carbon metabolism. These findings indicate that sesamin and sesamolin might be primarily involved in sesame biotic stress tolerance mechanisms. We found that the sesamin and sesamolin contents varied considerably among the six environments, and the content of sesamin significantly decreased from SC to NC. These results show genetic selection, domestication, and environmental conditions have significant impacts on sesame agronomic traits. We mainly focused on association analysis for sesamin and sesamolin variation and identified significantly associated loci and major candidate effect genes. The genetic polymorphism within the exonic region of the gene *SiNST1* may contribute substantially to lignan variation in sesame. Favorable allele mining provides important resources for marker-assisted selection. *SiNST1* encodes for an NAC domain-containing protein 43 and was positively selected in the high-sesamin content sesame accessions. Its homolog *AtNST1* in *Arabidopsis* is a master switches transcriptional regulator of SCW (secondary cell wall) formation by interacting with some MYB genes and hormonal signals [33–35]. Phylogenetic analysis clearly separated NST1 from monocots and dicots, indicating the gene might play different functions in the two categories of Angiosperm. On the phylogenetic tree, *SiNST1* clustered within dicots closely to NST1 in *Solanum tuberosum* and *S. lycopersicum*, indicating it might fulfill similar functions in sesame. *Cis*-acting elements analysis revealed that the promoter region of *SiNST1* mainly contains MYB binding, light responsiveness, and hormone-recognizing elements. These results suggest that *SiNST1* might regulate sesamin and sesamolin biosynthesis through interplays with MYB genes and phytohormones as per its homologous. Consistency, WGCNA revealed that *SiNST1* might interact with diverse *SiMYBs* in developing sesame seeds, confirming it might regulate diverse developmental processes in sesame. However, no direct interaction between *SiNST1* and *PSS* was noticed. *SiNST1* is also significantly associated with seed coat thickness, phytosterol content, and oil content in sesame [15,40]. The non-synonymous SNP locus in *SiNST1* consists of "C" allele change into "A" resulting in an amino acid change from Thr (threonine) to Lys (lysine). Sesame accession with the "C" allele accumulated significantly higher sesamin and sesamolin compared to accessions with the "A" allele. Supportively, gene expression analysis showed that $SiNST1^C$ is mainly expressed during lignan biosynthesis stages of 10 to 20 DPA, while $SiNST1^A$ is primarily expressed at seed coat development stages (early stages) [12,15,30]. These findings show that $SiNST1^C$ is a potential target for improving nutrient content in sesame, while $SiNST1^A$ may favorize seed coat thickness [15]. Together, our findings suggest that the thicker the sesame seed, the less oil and lignan may it contain. However, it should be noted that only 35 accessions with the "A" allele were contained in the studied population. Therefore, a future study is required to analyze the impact of nucleotide variability of *SiNST1* on sesame seed quality traits.

Since an efficient genetic transformation method for the sesame plant is not yet established, we overexpressed *SiNST1*^{C} in sesame hairy roots to investigate its regulatory mechanisms. The overexpression of *SiNST1*^{C} significantly induced the expression of *SiMYB58, SiMYB209, SiMYB134, SiMYB276,* and most monolignol biosynthetic genes. Consequently, the *SiNST1*^{C} -overexpressing sesame hairy roots exhibited an increase in lignan and lignin contents. However, the increase in the lignin content was not significant. These findings indicate that *SiNST1* plays a master switches function in regulating SCW formation and lignan biosynthesis in sesame by interacting with these four *SiMYBs*. Moreover, we identified ten other candidate genes for sesamin and sesamolin variation and seven *SiMYBs* that showed significant co-expression correlation with PSS in the turquoise module. These seven *SiMYBs* might be the key regulator of *PSS* expression. Therefore, further functional characterization of *SiNST1* and these genes in developing sesame seeds via subsequent knockout and overexpression is required when a genetic transformation protocol will be established to dissect this complex molecular regulatory network for biotechnological applications. Lignans and lignin are principally biosynthesized from the oxidative coupling of E-coniferyl alcohol, a critical reaction catalyzed by dirigent proteins (DIRs) [26,28,60,61]. Therefore, genome-wide characterization of the DIR gene family and identification of *SiDIRs* that may promote lignan accumulation in sesame are required for targeted balancing of the oxidation coupling in favor of lignan biosynthesis. In addition, identifying key quality-related candidate genes in wild sesame accessions will offer supplementary resources for sesame improvement. For instance, Wang et al. have found that the gene CYP92B14 promotes high sesamolin accumulation in wild allotetraploid sesame (*S. schinzianum*) [62]. This gene has to be functionally characterized for exploitation in genomic-assisted breeding of high sesamin and sesamolin content sesame varieties.

4. Materials and Methods

4.1. Plant Materials and Genome Sequencing

The 410 sesame accessions (329 landraces and 81 modern cultivars) were given by the National Mid-term Gene Bank, Oil crops Research Institute, Chinese Academy of Agricultural Sciences (OCRI, CAAS), Wuhan, China. The population is composed of 316 accessions from diverse geographical regions in China and 94 accessions from twenty-eight other countries (Table S1). Related to the type of variety, there were 329 and 81 landraces (local varieties) and modern cultivars (improved via breeding), respectively. Each accession sample was maintained via cultivation and self-pollination for over three generations prior to sequencing and phenotypes evaluation. A representative plant of each accession was chosen for the genome sequencing. Genomic DNA was extracted from leaves at the seedling stage via the CTAB method [63] and following the manufacturer's instructions. The sequencing library was constructed from high-quality DNA with an insert size of ~350 bp. The paired-end sequencing of each accession library was carried out on the Illumina HiSeqTM/MiSeqTM platform (Novogene Co., Ltd., Beijing, China).

4.2. Sequence Alignment, Variants Detection, and SNP Annotation

All the paired-end reads were mapped onto the sesame reference genome (version 2.0) [16] using the BWA (Burrows-Wheeler Aligner) software [64] with default parameters in order to locate the position of each read on the sesame genome. Next, we filtered out the low-quality (MQ < 20) reads with the SAMtools (v1.1) [65] program. We sorted and discarded PCR duplicates to construct a bam file index using Picard Tools (http://broadinstitute.github.io/picard/; v1.118, accessed on 2 March 2021). Finally, we called the genome variants (SNPs and InDels) with GATK (Genome Analysis Toolkit) [66], as described by Fan et al. [43]. To evaluate the quality of the identified SNPs, we randomly selected fifteen sesame accessions and repeated the whole-genome sequencing. The consistency of the results was approximately 99.8%.

The SNP annotation was conducted according to the sesame reference genome with ANNOVAR [67]. First, the SNPs were classified into upstream or downstream regions, intronic, exonic, intergenic, and splicing sites. Further, SNPs located in the exonic areas were grouped into synonymous and non-synonymous SNPs.

4.3. Population Genetic Analyses

The effective, high-quality SNPs were sorted using the criteria of Min depth $\geq$ 8x, MAF (Minor allele frequency) $\geq$0.05, and missing rate $\leq$0.2. Phylogenetic, PCA, structure, and LD analyses allow the examination of the evolutionary history or relationship between individuals of a population and characterize the population stratification and genetic variation. We constructed the approximate maximum-likelihood tree using the general time-reversible model in the TreeBeST (http://treesoft.sourceforge.net/treebest.shtml, accessed on 12 June 2021) software [68]. The population structure analysis was conducted in the Admixture program [69] with various K values. PCA (principal component analysis) was performed in the GCTA software [70]. The dataset used to conduct the PCA was screened for quality using the criteria of scaffold length >100 kb and (QUAL in the VCF file) <2000. Linkage disequilibrium (LD) decay analysis was performed in the PopLDdecay software [71].

To gain insight into the genetic diversity of the sesame population, we computed the genetic distance (F_{ST}) and nucleotide diversity (π) using VCFtools [72] with a 10 kb nonoverlapping window slid across each scaffold. The F_{ST} between clades was evaluated using the average value of all 10 kb sliding windows. The selective sweeps of geographical differentiation of SC subpopulation against other groups were revealed by considering both the F_{ST} value ($F_{ST} > 0.45$) and the π ratio ($\log_2(\pi \text{ ratio}) > 2.5$) in each sliding window, with a 5% cutoff [43]. The intergroup gene flow analysis was performed in Treemix [73], using a composite-likelihood approach. We defined the root via the root parameter and tested for many migration events to select the most frequent ones with significant ($p < 0.05$) gene flow.

4.4. Gene Ontology (GO) and KEGG Analyses

The gene annotation was conducted with PFAM terms using InterProScan 5 (http://www.ebi.ac.uk/Tools/pfa/iprscan5/, accessed on 19 September 2020) as per Hazzouri et al. [52]. Functional annotation and categorization of genes were achieved through gene ontology (GO) and KEGG analyses. The GO terms and the KEGG (Kyoto Encyclopedia of Genes and Genomes) analyses were carried out using AutoFact [74] and the Automated Annotation Server (KAAS; http://www.genome.jp/tools/kaas/, accessed on 9 August 2021), respectively. The ggplot2 package in R was used to sort the significant terms (p-value < 0.05) [75].

4.5. Phenotyping

The two primary lignans in sesame, sesamin and sesamolin, were investigated in six environmental conditions located in China, including Guangxi (GX, 2014), Wuhan (WH, 2014 and 2018), Hunan (HN, 2014), Shandong (SD, 2018), and Zhumadian (ZM, 2019). Wuhan belongs to the Yangtze river valley (southern China) with a warm and temperate climate. Zhumadian belongs to the Huang-Huai river valley (northern China) with a humid subtropical climate. Hunan and Guangxi are located at southern China with a subtropical humid monsoon and tropical monsoon climates, respectively. Shandong has a temperate monsoon climate. HPLC analysis of the sesamin and sesamolin contents in seed samples was carried out as previously reported [39,76]. The phenotypic data were analyzed with GraphPad.prism V. 9.0.0121 (GraphPad Software Inc., La Jolla, CA, USA).

4.6. GWAS Analysis and Candidate Genes Identification

The association analysis was carried out with the EMMAX (Efficient Mixed-Model Association eXpedited) program via the mixed linear model (MLM) [77], in order to uncover genetic markers that drive lignan variation in sesame. The significant association threshold was set at $-\log_{10}p > 6$. The Manhattan and QQ plots were constructed in R using the package "qq-man" [78]. The candidate genes in LD windows (±89 Kb) were selected through the integration of gene function annotation, non-synonymous SNPs, and gene expression analysis. Finally, we performed pairwise LD correlation analysis to confirm the major significant associated locus. The pairwise LD correlations analysis of the major candidate causative gene was carried out with LDBLockShow software [79].

4.7. Cis-Acting Elements and Phylogeny Analyses

Cis-acting elements prediction in the 2000 bp upstream promoter region of the targeted genes was achieved with PLANTCARE (http://bioinformatics.psb.ugent.be/webtools/plantcare/html/, accessed on 21 November 2021), in order to predict their respective potential functions. The protein sequence of *SiNST1* homolog in *Glycine max* (XM_003548125), *Arabidopsis thaliana* (NM_130243), *Brassica napus* (XM_013895547), *Triticum aestivum* (XM_044585881), *Oryza sativa* (XM_015793488), *Sorghum bicolor* (XM_002436372), *Zea mays* (XM_008650556), *Arachis hypogaea* (XM_025776021), *Gossypium hirsutum* (XM_016853342), *Vitis vinifera* (XM_002279509), *Solanum lycopersicum* (XM_004248375), and *S. tuberosum* (XM_015310959) were downloaded by blast from NCBI. These species were selected to represent monocots and dicots and based on previous studies [33–35]. The alignment and the phylogenetic analysis were performed with the MEGA X software [80].

4.8. WGCNA Analysis

Weighted gene co-expression network analysis (WGCNA) is a system biology method widely used to reveal correlation patterns among genes and identify candidate genes for specific agronomic traits. The WGCNA was conducted using the expression values (FPKM) of *SiMYBs* [42], predicted CYP81Q1 genes [12], and *SiNST1*. The expression values were extracted from available RNA-seq data of two sesame accessions (Zhongzhi16 and ZZM2748) with contrasting lignans content. The co-expression analysis was performed with the WGCNA package in R (version 4.0.2) [81].

4.9. Hairy Roots Transformation and Culture

The coding sequence of *SiNST1* was cloned (primers: F-AGAAGAGGGCTGGGTGGTT and R-GGATGACGAGTTGGGGCTAT) from Zhongfengzhi1 (harboring the favorable allele "C" at the locus Chr11-142842) and inserted into the pCAMBIA 1301S vector. The recombinant vector pCAMBIA 1301S-*SiNST1* was transformed to *Escherichia coli* DH5α (Weidi Biotechnology, Shanghai, China). After confirming that the transformation sequence was correct, a suitable amount of plasmid DNA was extracted from the previously transformed *E. coli* and transferred to *Agrobacterium rhizogenes* K599 for subsequent generating transgenic sesame hairy roots following previously described methods [82,83] with some modifications. Sesame sterile seedlings at two weeks old were wounded and used for the *Agrobacterium* infection for 10 min. Next, co-culture in the dark on MS solid medium for 48 h at 25 °C, followed by explants washing by MS liquid medium (containing 300 mg/mL cefotaxime) and sterile water. Then the explants were cultured on MS solid medium containing 200 mg/L timentin in a growth chamber to induce the hairy roots. The induced single sesame hairy root lines were discarded from explants after two weeks and grown on MS solid medium containing 200 mg/L timentin for detoxification. The positively transgenic lines were screened via PCR and grown on MS solid medium for two weeks under a 16/8 h light/dark photoperiod at 28 °C. Hairy roots were harvested, washed, and dried for sesamin, sesamolin, and lignin evaluation.

4.10. Lignans and Lignin Contents Analyses in the Sesame Hairy Roots

The extraction method of sesamin and sesamolin from the hairy roots was different. First, 0.3 g of crushed hairy roots was extracted with 5 mL absolute ethanol by shaking for 2 h. After centrifugation at 5000× *g* for 5 min, the supernatant was collected, and the ethanol was evaporated completely at 60 °C. Then, the residue was dissolved in 1 mL ethanol 80% by vortexing for 2 to 3 min. Finally, the extract was filtered and subjected to

HPLC analysis. The lignin (acid-soluble) content was evaluated as per Wu et al. via the Klason method [84].

4.11. *qRT-PCR Analysis*

Total RNA from the hairy root samples was extracted with the EASYspin Plus plant RNA Kit (Aidlab, Beijing, China) and, thereafter, reverse transcribed using the HIScript II first-strand cDNA synthesis Kit (Vazyme Biotechnology, Nanjing, China) following the manufacturer's instructions. The qRT-PCR was carried out with ChamQ™ SYBR1 qPCR Master Mix (Vazyme Biotech, Nanjing, China) on LightCycler480 (Roche, Switzerland) real-time PCR system. The histone H3.3 gene (*SINPZ1301428*) was used as an internal control to normalize the transcript levels [85]. Each gene was analyzed using three biological replicates under the same conditions. The expression levels of the genes were analyzed using the $2^{-\Delta\Delta CT}$ method [86]. The primers were designed with Primer3 (http://bioinfo.ut.ee/primer3-0.4.0/, accessed on 18 February 2022) and are listed in Additional file 1: Table S21.

4.12. *Statistical Analysis*

GraphPad Prism v9.0.0121 (GraphPad 159 Software Inc., La Jolla, CA, USA) was used for statistical analysis and graphs' construction. Statistical differences were performed by independent *t*-test at $p < 0.05$.

5. Conclusions

Overall, this study upgraded knowledge of genetic polymorphism in sesame through whole-genome resequencing of 410 worldwide accessions. Population structure and genetic diversity analyses classified sesame population into different geographic patterns, consistent with reported potential domestication of the crop from the southern to the middle and northern regions. The key selected genes in MC and NC subgroups were identified for exploitation in sesame improvement. Genotype and environmental conditions have significant impacts on lignan variation in sesame. We identified key loci and 11 candidate genes governing the major sesame lignans (sesamin and sesamolin) variation. Of them, *SiNST1* is the major underlying gene via a non-synonymous SNP (C/A). The "C" allele may favorize lignan and other quality traits accumulation, while the "A" may promote seed coat thickness. Functional characterization in sesame hairy roots revealed that *SiNST1* might function in synergetic action with *SiMYB58*, *SiMYB209*, *SiMYB134*, and *SiMYB276* to control monolignol biosynthetic gene expression at transcriptional levels. Our results may considerably contribute to sesame quality improvement via genomic-assisted breeding.

Supplementary Materials: The following supporting information can be downloaded at: https://www.mdpi.com/article/10.3390/ijms24021055/s1.

Author Contributions: Conceptualization, L.W. and J.Y.; methodology, formal analysis and data curation, S.S. and S.S.K.D.; validation and investigation, A.L., R.Z., C.S. and Z.W.; resources and visualization, Y.Z., D.L. and M.B.; writing—original draft preparation, S.S.K.D.; writing—review and editing, L.W., J.Y. and S.S.K.D.; supervision, project administration and funding acquisition, L.W. All authors have read and agreed to the published version of the manuscript.

Funding: This research was funded by the Agricultural Science and Technology Innovation Project of the Chinese Academy of Agricultural Sciences (CAAS-ASTIP-2016-OCRI), the Key Research Projects of Hubei province (2020BBA045), the Science and Technology Innovation Project of Hubei province (2021-620-000-001-035), the Fundamental Research Funds for Central Non-profit Scientific Institution (Y2022XK11, 1610172022010), the Open Project of Key Laboratory of Biology and Genetic Improvement of Oil Crops, Ministry of Agriculture and Rural Affairs, P.R. China (KF2022002), and China Agriculture Research System (CARS-14).

Institutional Review Board Statement: Not applicable.

Informed Consent Statement: Not applicable.

Data Availability Statement: The raw resequencing data reported in this paper have been deposited at the Genome Sequence Archive [87], National Genomics Data Center [88], China National Center for Bioinformation/Beijing Institute of Genomics, Chinese Academy of Sciences (GSA: CRA006452), publicly accessible at https://ngdc.cncb.ac.cn/gsa, accessed on 22 November 2021. The RNA-seq data is available at the National Center for Biotechnology Information (NCBI), Sequence Read Archive (SRA), accession PRJNA806786. Other datasets used and/or analyzed during the current study are available from the corresponding author on reasonable request.

Conflicts of Interest: The authors declare no conflict of interest.

References

1. Miao, H.; Zhang, H. *The Sesame Genome*; Miao, H., Zhang, H., Kole, C., Eds.; Compendium of Plant Genomes; Springer International Publishing: Cham, Switzerland, 2021; ISBN 978-3-319-98097-3.
2. Pathak, N.; Rai, A.K.; Kumari, R.; Bhat, K.V. Value addition in sesame: A perspective on bioactive components for enhancing utility and profitability. *Pharmacogn. Rev.* **2014**, *8*, 147–155. [CrossRef] [PubMed]
3. Dossou, S.S.K.; Xu, F.; You, J.; Zhou, R.; Li, D.; Wang, L. Widely targeted metabolome profiling of different colored sesame (*Sesamum indicum* L.) seeds provides new insight into their antioxidant activities. *Food Res. Int.* **2022**, *151*, 110850. [CrossRef] [PubMed]
4. Majdalawieh, A.F.; Massri, M.; Nasrallah, G.K. A comprehensive review on the anti-cancer properties and mechanisms of action of sesamin, a lignan in sesame seeds (*Sesamum indicum*). *Eur. J. Pharmacol.* **2017**, *815*, 512–521. [CrossRef] [PubMed]
5. Majdalawieh, A.F.; Dalibalta, S.; Yousef, S.M. Effects of sesamin on fatty acid and cholesterol metabolism, macrophage cholesterol homeostasis and serum lipid profile: A comprehensive review. *Eur. J. Pharmacol.* **2020**, *885*, 173417. [CrossRef]
6. Wu, M.S.; Aquino, L.B.B.; Barbaza, M.Y.U.; Hsieh, C.L.; De Castro-Cruz, K.A.; Yang, L.L.; Tsai, P.W. Anti-inflammatory and anticancer properties of bioactive compounds from *Sesamum indicum* L.—A review. *Molecules* **2019**, *24*, 4426. [CrossRef]
7. Deng, S.; Zhou, J.-L.; Fang, H.-S.; Nie, Z.-G.; Chen, S.; Peng, H. Sesamin Protects the Femoral Head from Osteonecrosis by Inhibiting ROS-Induced Osteoblast Apoptosis in Rat Model. *Front. Physiol.* **2018**, *9*, 1787. [CrossRef]
8. Ma, Z.P.; Zhang, Z.F.; Yang, Y.F.; Yang, Y. Sesamin promotes osteoblastic differentiation and protects rats from osteoporosis. *Med. Sci. Monit.* **2019**, *25*, 5312–5320. [CrossRef]
9. Zhao, Y.; Wang, Q.; Jia, M.; Fu, S.; Pan, J.; Chu, C.; Liu, X.; Liu, X.; Liu, Z. (+)-Sesamin attenuates chronic unpredictable mild stress-induced depressive-like behaviors and memory deficits via suppression of neuroinflammation. *J. Nutr. Biochem.* **2019**, *64*, 61–71. [CrossRef]
10. Bedigian, D. Evolution of sesame revisited: Domestication, diversity and prospects. *Genet. Resour. Crop Evol.* **2003**, *50*, 779–787.
11. Bedigian, D.; Seigler, D.S.; Harlan, J.R. Sesamin, sesamolin and the origin of sesame. *Biochem. Syst. Ecol.* **1985**, *13*, 133–139. [CrossRef]
12. Wang, L.; Yu, S.; Tong, C.; Zhao, Y.; Liu, Y.; Song, C.; Zhang, Y.; Zhang, X.; Wang, Y.; Hua, W.; et al. Genome sequencing of the high oil crop sesame provides insight into oil biosynthesis. *Genome Biol.* **2014**, *15*, R39. [CrossRef] [PubMed]
13. Yu, J.; Golicz, A.A.; Lu, K.; Dossa, K.; Zhang, Y.; Chen, J.; Wang, L.; You, J.; Fan, D.; Edwards, D.; et al. Insight into the evolution and functional characteristics of the pan-genome assembly from sesame landraces and modern cultivars. *Plant Biotechnol. J.* **2019**, *17*, 881–892. [CrossRef] [PubMed]
14. Qiu, Z.; Zhang, Y.; Bedigian, D.; Li, X.; Wang, C.; Jiang, H. Sesame Utilization in China: New Archaeobotanical Evidence from Xinjiang. *Econ. Bot.* **2012**, *66*, 255–263. [CrossRef]
15. Wei, X.; Liu, K.; Zhang, Y.; Feng, Q.; Wang, L.; Zhao, Y.; Li, D.; Zhao, Q.; Zhu, X.; Zhu, X.; et al. Genetic discovery for oil production and quality in sesame. *Nat. Commun.* **2015**, *6*, 8609. [CrossRef] [PubMed]
16. Wang, L.; Xia, Q.; Zhang, Y.; Zhu, X.; Zhu, X.; Li, D.; Ni, X.; Gao, Y.; Xiang, H.; Wei, X.; et al. Updated sesame genome assembly and fine mapping of plant height and seed coat color QTLs using a new high-density genetic map. *BMC Genom.* **2016**, *17*, 31. [CrossRef]
17. Cui, C.; Liu, Y.; Liu, Y.; Cui, X.; Sun, Z.; Du, Z.; Wu, K.; Jiang, X.; Mei, H.; Zheng, Y. Genome-wide association study of seed coat color in sesame (*Sesamum indicum* L.). *PLoS ONE* **2021**, *16*, e0251526. [CrossRef]
18. Zhou, R.; Dossa, K.; Li, D.; Yu, J.; You, J.; Wei, X.; Zhang, X. Genome-wide association studies of 39 seed yield-related traits in sesame (*Sesamum indicum* L.). *Int. J. Mol. Sci.* **2018**, *19*, 2794. [CrossRef]
19. Wang, X.; You, J.; Liu, A.; Qi, X.; Li, D.; Zhao, Y.; Zhang, Y.; Zhang, L.; Zhang, X.; Li, P. Variation in Melatonin Contents and Genetic Dissection of Melatonin Biosynthesis in Sesame. *Plants* **2022**, *11*, 2005. [CrossRef]
20. Horii, S.; Ishii, T. Effect of Arbuscular Mycorrhizal Fungi and Their Partner Bacteria on the Growth of Sesame Plants and the Concentration of Sesamin in the Seeds. *Am. J. Plant Sci.* **2014**, *05*, 3066–3072. [CrossRef]
21. Radhakrishnan, R.; Pae, S.B.; Shim, K.B.; Baek, I.Y. Penicillium sp. mitigates Fusarium-induced biotic stress in sesame plants. *Biotechnol. Lett.* **2013**, *35*, 1073–1078. [CrossRef]
22. Radhakrishnan, R.; Kang, S.M.; Baek, I.Y.; Lee, I.J. Characterization of plant growth-promoting traits of *Penicillium* species against the effects of high soil salinity and root disease. *J. Plant Interact.* **2014**, *9*, 754–762. [CrossRef]

23. Liu, Z.; Zhu, H.; Zhou, J.; Jiang, S.; Wang, Y.; Kuang, J.; Ji, Q.; Peng, J.; Wang, J.; Gao, L.; et al. Resequencing of 296 cultivated and wild lotus accessions unravels its evolution and breeding history. *Plant J.* **2020**, *104*, 1673–1684. [CrossRef] [PubMed]
24. Wu, D.; Liang, Z.; Yan, T.; Xu, Y.; Xuan, L.; Tang, J.; Zhou, G.; Lohwasser, U.; Hua, S.; Wang, H.; et al. Whole-Genome Resequencing of a Worldwide Collection of Rapeseed Accessions Reveals the Genetic Basis of Ecotype Divergence. *Mol. Plant* **2019**, *12*, 30–43. [CrossRef] [PubMed]
25. Maynard Smith, J.; Haigh, J. The hitch-hiking effect of a favourable gene. *Genet. Res. (Camb)* **2008**, *89*, 391–403. [CrossRef]
26. Hano, C.F.; Dinkova-Kostova, A.T.; Davin, L.B.; Cort, J.R.; Lewis, N.G. Lignans: Insights into Their Biosynthesis, Metabolic Engineering, Analytical Methods and Health Benefits. *Front. Plant Sci.* **2021**, *11*, 2020–2022. [CrossRef]
27. Markulin, L.; Drouet, S.; Corbin, C.; Decourtil, C.; Garros, L.; Renouard, S.; Lopez, T.; Mongelard, G.; Gutierrez, L.; Auguin, D.; et al. The control exerted by ABA on lignan biosynthesis in flax (*Linum usitatissimum* L.) is modulated by a Ca^{2+} signal transduction involving the calmodulin-like LuCML15b. *J. Plant Physiol.* **2019**, *236*, 74–87. [CrossRef]
28. Davin, L.B.; Lewis, N.G. An historical perspective on lignan biosynthesis: Monolignol, allylphenol and hydroxycinnamic acid coupling and downstream metabolism. *Phytochem. Rev.* **2003**, *2*, 257–288. [CrossRef]
29. Andargie, M.; Vinas, M.; Rathgeb, A.; Möller, E.; Karlovsky, P. Lignans of sesame (*Sesamum indicum* L.): A comprehensive review. *Molecules* **2021**, *26*, 883. [CrossRef]
30. Ono, E.; Nakai, M.; Fukui, Y.; Tomimori, N.; Fukuchi-Mizutani, M.; Saito, M.; Satake, H.; Tanaka, T.; Katsuta, M.; Umezawa, T.; et al. Formation of two methylenedioxy bridges by a *Sesamum* CYP81Q protein yielding a furofuran lignan, (+)-sesamin. *Proc. Natl. Acad. Sci. USA* **2006**, *103*, 10116–10121. [CrossRef]
31. Murata, J.; Ono, E.; Yoroizuka, S.; Toyonaga, H.; Shiraishi, A.; Mori, S.; Tera, M.; Azuma, T.; Nagano, A.J.; Nakayasu, M.; et al. Oxidative rearrangement of (+)-sesamin by CYP92B14 co-generates twin dietary lignans in sesame. *Nat. Commun.* **2017**, *8*, 2155. [CrossRef]
32. Zoclanclounon, Y.A.B.; Rostás, M.; Chung, N.J.; Mo, Y.; Karlovsky, P.; Dossa, K. Characterization of Peroxidase and Laccase Gene Families and In Silico Identification of Potential Genes Involved in Upstream Steps of Lignan Formation in Sesame. *Life* **2022**, *12*, 1200. [CrossRef] [PubMed]
33. Mitsuda, N.; Ohme-Takagi, M. NAC transcription factors NST1 and NST3 regulate pod shattering in a partially redundant manner by promoting secondary wall formation after the establishment of tissue identity. *Plant J.* **2008**, *56*, 768–778. [CrossRef] [PubMed]
34. Zhong, R.; Ye, Z.H. The *Arabidopsis* NAC transcription factor NST2 functions together with SND1 and NST1 to regulate secondary wall biosynthesis in fibers of inflorescence stems. *Plant Signal. Behav.* **2015**, *10*, 37–41. [CrossRef]
35. Zhong, R.; Richardson, E.A.; Ye, Z.H. Two NAC domain transcription factors, SND1 and NST1, function redundantly in regulation of secondary wall synthesis in fibers of *Arabidopsis*. *Planta* **2007**, *225*, 1603–1611. [CrossRef]
36. Gouin, A.; Legeai, F.; Nouhaud, P.; Whibley, A.; Simon, J.C.; Lemaitre, C. Whole-genome re-sequencing of non-model organisms: Lessons from unmapped reads. *Heredity* **2015**, *114*, 494–501. [CrossRef] [PubMed]
37. Martin, J.; Schackwitz, W.; Lipzen, A. Genomic sequence variation analysis by resequencing. *Methods Mol. Biol.* **2018**, *1775*, 229–239. [CrossRef]
38. Zhou, Z.; Jiang, Y.; Wang, Z.; Gou, Z.; Lyu, J.; Li, W.; Yu, Y.; Shu, L.; Zhao, Y.; Ma, Y.; et al. Resequencing 302 wild and cultivated accessions identifies genes related to domestication and improvement in soybean. *Nat. Biotechnol.* **2015**, *33*, 408–414. [CrossRef]
39. Xu, F.; Zhou, R.; Dossou, S.S.K.; Song, S.; Wang, L. Fine mapping of a major pleiotropic qtl associated with sesamin and sesamolin variation in sesame (*Sesamum indicum* L.). *Plants* **2021**, *10*, 1343. [CrossRef]
40. Wang, Z.; Zhou, Q.; Dossou, S.S.K.; Zhou, R.; Zhao, Y.; Zhou, W.; Zhang, Y.; Li, D.; You, J.; Wang, L. Genome-Wide Association Study Uncovers Loci and Candidate Genes Underlying Phytosterol Variation in Sesame (*Sesamum indicum* L.). *Agriculture* **2022**, *12*, 392. [CrossRef]
41. Dossou, S.S.K.; Xu, F.; Dossa, K.; Zhou, R.; Zhao, Y.; Wang, L. Antioxidant lignans sesamin and sesamolin in sesame (*Sesamum indicum* L.): A comprehensive review and future prospects. *J. Integr. Agric.* 2021. *in press*.
42. Mmadi, M.A.; Dossa, K.; Wang, L.; Zhou, R.; Wang, Y.; Cisse, N.; Sy, M.O.; Zhang, X. Functional characterization of the versatile MYB gene family uncovered their important roles in plant development and responses to drought and waterlogging in sesame. *Genes* **2017**, *8*, 362. [CrossRef]
43. Fan, W.; Lu, J.; Pan, C.; Tan, M.; Lin, Q.; Liu, W.; Li, D.; Wang, L.; Hu, L.; Wang, L.; et al. Sequencing of Chinese castor lines reveals genetic signatures of selection and yield-associated loci. *Nat. Commun.* **2019**, *10*, 3418. [CrossRef] [PubMed]
44. Zhang, K.; He, M.; Fan, Y.; Zhao, H.; Gao, B.; Yang, K.; Li, F.; Tang, Y.; Gao, Q.; Lin, T.; et al. Resequencing of global Tartary buckwheat accessions reveals multiple domestication events and key loci associated with agronomic traits. *Genome Biol.* **2021**, *22*, 23. [CrossRef] [PubMed]
45. Varshney, R.K.; Thudi, M.; Roorkiwal, M.; He, W.; Upadhyaya, H.D.; Yang, W.; Bajaj, P.; Cubry, P.; Rathore, A.; Jian, J.; et al. Resequencing of 429 chickpea accessions from 45 countries provides insights into genome diversity, domestication and agronomic traits. *Nat. Genet.* **2019**, *51*, 857–864. [CrossRef]
46. Batley, J.; Barker, G.; O'Sullivan, H.; Edwards, K.J.; Edwards, D. Mining for single nucleotide polymorphisms and insertions/deletions in maize expressed sequence tag data. *Plant Physiol.* **2003**, *132*, 84–91. [CrossRef]
47. Bhardwaj, A.; Dhar, Y.V.; Asif, M.H.; Bag, S.K. In Silico identification of SNP diversity in cultivated and wild tomato species: Insight from molecular simulations. *Sci. Rep.* **2016**, *6*, 38715. [CrossRef] [PubMed]

48. Raizada, A.; Souframanien, J. Transcriptome sequencing, de novo assembly, characterisation of wild accession of blackgram (*Vigna mungo* var. silvestris) as a rich resource for development of molecular markers and validation of SNPs by high resolution melting (HRM) analysis. *BMC Plant Biol.* **2019**, *19*, 358. [CrossRef]
49. Cortés, A.J.; López-Hernández, F.; Blair, M.W. Genome–Environment Associations, an Innovative Tool for Studying Heritable Evolutionary Adaptation in Orphan Crops and Wild Relatives. *Front. Genet.* **2022**, *13*, 910386. [CrossRef]
50. Kaushik, S.; Kaushik, S.; Sharma, D. Functional genomics. *Encycl. Bioinforma. Comput. Biol. ABC Bioinforma.* **2018**, *1–3*, 118–133. [CrossRef]
51. Hamblin, M.T.; Buckler, E.S.; Jannink, J.L. Population genetics of genomics-based crop improvement methods. *Trends Genet.* **2011**, *27*, 98–106. [CrossRef]
52. Hazzouri, K.M.; Flowers, J.M.; Visser, H.J.; Khierallah, H.S.M.; Rosas, U.; Pham, G.M.; Meyer, R.S.; Johansen, C.K.; Fresquez, Z.A.; Masmoudi, K.; et al. Whole genome re-sequencing of date palms yields insights into diversification of a fruit tree crop. *Nat. Commun.* **2015**, *6*, 8824. [CrossRef]
53. Teklu, D.H.; Shimelis, H.; Abady, S. Genetic Improvement in Sesame (*Sesamum indicum* L.): Progress and Outlook: A Review. *Agronomy* **2022**, *12*, 2144. [CrossRef]
54. Zhang, Y.; Liu, F.; Lin, Y.; Li, L.; Chen, M.; Ni, L. A Comprehensive Review on Distribution, Pharmacological Properties, and Mechanisms of Action of Sesamin. *J. Chem.* **2022**, *2022*, 4236525. [CrossRef]
55. Wei, P.; Zhao, F.; Wang, Z.; Wang, Q.; Chai, X.; Hou, G.; Meng, Q. Sesame (*Sesamum indicum* L.): A Comprehensive Review of Nutritional Value, Phytochemical Composition, Health Benefits, Development of Food, and Industrial Applications. *Nutrients* **2022**, *14*, 4079. [CrossRef] [PubMed]
56. Mohammed, S. An Overview on Nutritional Composition and Therapeutic Benefits of Sesame Seeds (*Sesamum indicum*). *Int. J. Res. Appl. Sci. Eng. Technol.* **2022**, *10*, 1119–1127. [CrossRef]
57. Kim, M.Y.; Kim, S.; Kim, J.I.; Oh, E.; Kim, S.W.; Lee, J.; Lee, E.; Lee, M.H. Antioxidant and Hepatoprotective Effects of Different Varieties of Sesame (*Sesamum indicum* L.) with Variation in Lignan Content. *J. Korean Soc. Food Sci. Nutr.* **2022**, *51*, 660–670. [CrossRef]
58. Wang, L.; Zhang, Y.; Li, D.; Dossa, K.; Wang, M.L.; Zhou, R.; Yu, J.; Zhang, X. Gene expression profiles that shape high and low oil content sesames. *BMC Genet.* **2019**, *20*, 45. [CrossRef] [PubMed]
59. Song, S.; You, J.; Shi, L.; Sheng, C.; Zhou, W.; Dossou, S.S.K.; Dossa, K.; Wang, L.; Zhang, X. Genome-wide analysis of nsltp gene family and identification of siltps contributing to high oil accumulation in sesame (*Sesamum indicum* L.). *Int. J. Mol. Sci.* **2021**, *22*, 5291. [CrossRef]
60. Vanholme, R.; De Meester, B.; Ralph, J.; Boerjan, W. Lignin biosynthesis and its integration into metabolism. *Curr. Opin. Biotechnol.* **2019**, *56*, 230–239. [CrossRef]
61. Paniagua, C.; Bilkova, A.; Jackson, P.; Dabravolski, S.; Riber, W.; Didi, V.; Houser, J.; Gigli-Bisceglia, N.; Wimmerova, M.; Budínská, E.; et al. Dirigent proteins in plants: Modulating cell wall metabolism during abiotic and biotic stress exposure. *J. Exp. Bot.* **2017**, *68*, 3287–3301. [CrossRef]
62. Wang, X.; Wang, S.; Lin, Q.; Lu, J.; Lv, S.; Zhang, Y.; Wang, X.; Fan, W.; Liu, W.; Zhang, L.; et al. The wild allotetraploid sesame genome provides novel insights into evolution and lignan biosynthesis. *J. Adv. Res.* 2022. *in press*. [CrossRef]
63. Murray, M.G.; Thompson, W.F. Rapid isolation of high molecular weight plant ONA. *Nucleic Acids Res.* **1980**, *8*, 2589–2598. [CrossRef]
64. Li, H.; Durbin, R. Fast and accurate short read alignment with Burrows-Wheeler transform. *Bioinformatics* **2009**, *25*, 1754–1760. [CrossRef] [PubMed]
65. Li, H.; Handsaker, B.; Wysoker, A.; Fennell, T.; Ruan, J.; Homer, N.; Marth, G.; Abecasis, G.; Durbin, R. The Sequence Alignment/Map format and SAMtools. *Bioinformatics* **2009**, *25*, 2078–2079. [CrossRef] [PubMed]
66. McKenna, A.; Hanna, M.; Banks, E.; Sivachenko, A.; Cibulskis, K.; Kernytsky, A.; Garimella, K.; Altshuler, D.; Gabriel, S.; Daly, M.; et al. The genome analysis toolkit: A MapReduce framework for analyzing next-generation DNA sequencing data. *Genome Res.* **2010**, *20*, 1297–1303. [CrossRef] [PubMed]
67. Wang, K.; Li, M.; Hakonarson, H. ANNOVAR: Functional annotation of genetic variants from high-throughput sequencing data. *Nucleic Acids Res.* **2010**, *38*, e164. [CrossRef] [PubMed]
68. Price, M.N.; Dehal, P.S.; Arkin, A.P. FastTree 2—Approximately maximum-likelihood trees for large alignments. *PLoS ONE* **2010**, *5*, e9490. [CrossRef]
69. Alexander, D.H.; Novembre, J.; Lange, K. Fast model-based estimation of ancestry in unrelated individuals. *Genome Res.* **2009**, *19*, 1655–1664. [CrossRef]
70. Reverter, A.; Fortes, M.R.S. *Genome-Wide Association Studies and Genomic Prediction*; Gondro, C., van der Werf, J., Hayes, B., Eds.; Methods in Molecular Biology; Humana Press: Totowa, NJ, USA, 2013; Volume 1019, ISBN 978-1-62703-446-3.
71. Zhang, C.; Dong, S.S.; Xu, J.Y.; He, W.M.; Yang, T.L. PopLDdecay: A fast and effective tool for linkage disequilibrium decay analysis based on variant call format files. *Bioinformatics* **2019**, *35*, 1786–1788. [CrossRef]
72. Danecek, P.; Auton, A.; Abecasis, G.; Albers, C.A.; Banks, E.; DePristo, M.A.; Handsaker, R.E.; Lunter, G.; Marth, G.T.; Sherry, S.T.; et al. The variant call format and VCFtools. *Bioinformatics* **2011**, *27*, 2156–2158. [CrossRef]
73. Pickrell, J.K.; Pritchard, J.K. Inference of population splits and mixtures from genome-wide allele frequency data. *Nat. Publ. Gr.* **2012**, *8*, e1002967.

74. Koski, L.B.; Gray, M.W.; Lang, B.F.; Burger, G. AutoFACT: An automatic functional annotation and classification tool. *BMC Bioinform.* **2005**, *6*, 151. [CrossRef]
75. Sun, X.; Wu, K.; Cook, D. PKgraph: An R package for graphically diagnosing population pharmacokinetic models. *Comput. Methods Programs Biomed.* **2011**, *104*, 461–471. [CrossRef]
76. Wang, L.; Zhang, Y.; Li, P.; Wang, X.; Zhang, W.; Wei, W.; Zhang, X. HPLC analysis of seed sesamin and sesamolin variation in a sesame germplasm collection in China. *JAOCS J. Am. Oil Chem. Soc.* **2012**, *89*, 1011–1020. [CrossRef]
77. Wang, S.-B.; Feng, J.-Y.; Ren, W.-L.; Huang, B.; Zhou, L.; Wen, Y.-J.; Zhang, J.; Dunwell, J.M.; Xu, S.; Zhang, Y.-M. Improving power and accuracy of genome-wide association studies via a multi-locus mixed linear model methodology. *Sci. Rep.* **2016**, *6*, 19444. [CrossRef] [PubMed]
78. Turner, S. qqman: An R package for visualizing GWAS results using Q-Q and manhattan plots. *J. Open Source Softw.* **2018**, *3*, 731. [CrossRef]
79. Dong, S.S.; He, W.M.; Ji, J.J.; Zhang, C.; Guo, Y.; Yang, T.L. LDBlockShow: A fast and convenient tool for visualizing linkage disequilibrium and haplotype blocks based on variant call format files. *Brief. Bioinform.* **2021**, *22*, bbaa227. [CrossRef]
80. Kumar, S.; Stecher, G.; Li, M.; Knyaz, C.; Tamura, K. MEGA X: Molecular Evolutionary Genetics Analysis across Computing Platforms. *Mol. Biol. Evol.* **2018**, *35*, 1547–1549. [CrossRef] [PubMed]
81. Langfelder, P.; Horvath, S. WGCNA: An R package for weighted correlation network analysis. *BMC Bioinform.* **2008**, *9*, 559. [CrossRef] [PubMed]
82. Jin, U.H.; Chun, J.A.; Han, M.O.; Lee, J.W.; Yi, Y.B.; Lee, S.W.; Chung, C.H. Sesame hairy root cultures for extra-cellular production of a recombinant fungal phytase. *Process Biochem.* **2005**, *40*, 3754–3762. [CrossRef]
83. Chun, J.A.; Lee, W.H.; Han, M.O.; Lee, J.W.; Yi, Y.B.; Goo, Y.M.; Lee, S.W.; Bae, S.C.; Cho, K.J.; Chung, C.H. Molecular and biochemical characterizations of dehydroascorbate reductase from sesame (*Sesamum indicum* L.) hairy root cultures. *J. Agric. Food Chem.* **2007**, *55*, 6067–6073. [CrossRef]
84. Wu, R.; Wang, L.; Wang, Z.; Shang, H.; Liu, X.; Zhu, Y.; Qi, D.; Deng, X. Cloning and expression analysis of a dirigent protein gene from the resurrection plant *Boea hygrometrica*. *Prog. Nat. Sci.* **2009**, *19*, 347–352. [CrossRef]
85. Wei, L.; Miao, H.; Zhao, R.; Han, X.; Zhang, T.; Zhang, H. Identification and testing of reference genes for Sesame gene expression analysis by quantitative real-time PCR. *Planta* **2013**, *237*, 873–889. [CrossRef] [PubMed]
86. Livak, K.J.; Schmittgen, T.D. Analysis of relative gene expression data using real-time quantitative PCR and the 2-ΔΔCT method. *Methods* **2001**, *25*, 402–408. [CrossRef] [PubMed]
87. Chen, T.; Chen, X.; Zhang, S.; Zhu, J.; Tang, B.; Wang, A.; Dong, L.; Zhang, Z.; Yu, C.; Sun, Y.; et al. The Genome Sequence Archive Family: Toward Explosive Data Growth and Diverse Data Types. *Genom. Proteom. Bioinforma.* **2021**, *19*, 578–583. [CrossRef] [PubMed]
88. Xue, Y.; Bao, Y.; Zhang, Z.; Zhao, W.; Xiao, J.; He, S.; Zhang, G.; Li, Y.; Zhao, G.; Chen, R. Database Resources of the National Genomics Data Center, China National Center for Bioinformation in 2022. *Nucleic Acids Res.* **2022**, *50*, D27–D38. [CrossRef]

International Journal of
Molecular Sciences

MDPI

Article

Identification of Quantitative Trait Locus and Candidate Genes for Drought Tolerance in a Soybean Recombinant Inbred Line Population

Wenqi Ouyang [1,†], Limiao Chen [1,†], Junkui Ma [2], Xiaorong Liu [2], Haifeng Chen [1], Hongli Yang [1], Wei Guo [1], Zhihui Shan [1], Zhonglu Yang [1], Shuilian Chen [1], Yong Zhan [3], Hengbin Zhang [3], Dong Cao [1,*] and Xinan Zhou [1,*]

1 Key Laboratory of Biology and Genetic Improvement of Oil Crops, Ministry of Agriculture and Rural Affairs, Oil Crops Research Institute of Chinese Academy of Agricultural Sciences, Wuhan 430062, China
2 The Industrial Crop Institute, Shanxi Academy of Agricultural Sciences, Taiyuan 030006, China
3 Crop Research Institute, Xinjiang Academy of Agricultural and Reclamation Science, Key Laboratory of Cereal Quality Research and Genetic Improvement, Xinjiang Production and Construction Crops, Shihezi 832000, China
* Correspondence: caodong@caas.cn (D.C.); zhouocri@sina.com (X.Z.)
† These authors contributed equally to this work.

Abstract: With global warming and regional decreases in precipitation, drought has become a problem worldwide. As the number of arid regions in the world is increasing, drought has become a major factor leading to significant crop yield reductions and food crises. Soybean is a crop that is relatively sensitive to drought. It is also a crop that requires more water during growth and development. The aim of this study was to identify the quantitative trait locus (QTL) that affects drought tolerance in soybean by using a recombinant inbred line (RIL) population from a cross between the drought-tolerant cultivar 'Jindou21' and the drought-sensitive cultivar 'Zhongdou33'. Nine agronomic and physiological traits were identified under drought and well-watered conditions. Genetic maps were constructed with 923,420 polymorphic single nucleotide polymorphism (SNP) markers distributed on 20 chromosomes at an average genetic distance of 0.57 centimorgan (cM) between markers. A total of five QTLs with a logarithm of odds (LOD) value of 4.035–8.681 were identified on five chromosomes. Under well-watered conditions and drought-stress conditions, one QTL related to the main stem node number was located on chromosome 16, accounting for 17.177% of the phenotypic variation. Nine candidate genes for drought resistance were screened from this QTL, namely *Glyma.16G036700*, *Glyma.16G036400*, *Glyma.16G036600*, *Glyma.16G036800*, *Glyma.13G312700*, *Glyma.13G312800*, *Glyma.16G042900*, *Glyma.16G043200*, and *Glyma.15G100700*. These genes were annotated as NAC transport factor, GATA transport factor, and BTB/POZ-MATH proteins. This result can be used for molecular marker-assisted selection and provide a reference for breeding for drought tolerance in soybean.

Keywords: soybean; drought; RIL; resequencing; QTLs

Citation: Ouyang, W.; Chen, L.; Ma, J.; Liu, X.; Chen, H.; Yang, H.; Guo, W.; Shan, Z.; Yang, Z.; Chen, S.; et al. Identification of Quantitative Trait Locus and Candidate Genes for Drought Tolerance in a Soybean Recombinant Inbred Line Population. *Int. J. Mol. Sci.* **2022**, *23*, 10828. https://doi.org/10.3390/ijms231810828

Academic Editors: Andrés J. Cortés and Hai Du

Received: 18 July 2022
Accepted: 10 September 2022
Published: 16 September 2022

Publisher's Note: MDPI stays neutral with regard to jurisdictional claims in published maps and institutional affiliations.

1. Introduction

Drought stress is one of the major environmental factors that cause changes in phenotypic, physiological, biochemical, and molecular levels in plants [1,2]. These changes adversely affect plant growth, plant development, and crop production. In total, 65% of global freshwater use is devoted to the growth stage of the plant. Severe drought would cause the termination of photosynthesis and disruption of the metabolism, and finally, lead to plant death [3,4]. The study of drought resistance in crops is very important. The identification of drought resistance genes and the exploration of drought-resistance mechanisms in plants is of vital importance as a means of breeding new varieties of drought-tolerant crops.

Soybean (*Glycine max*) is an important legume crop that can be processed into a variety of soybean oils or used to feed livestock and is a globally important cash and food crop [5]. As a crop with high water requirements [6], soybean is extremely sensitive to water deficits. Soybean requires irrigation of 1300–2200 g of water for every 1 g of seed formed at maturity [7], making it one of the most sensitive legume crops to water deficit [8]. Continued global drought has caused an approximately 40% loss in quality and yield of soybean [9]. Improving the drought resistance of soybean and selecting new drought-resistant soybean varieties is an important way to ensure high and stable yields in soybean production [10].

Conventional breeding enables the recombination of excellent drought-tolerant genes in high generations of crops [11,12], which is costly and time-consuming. The emergence of molecular marker-assisted breeding has become an important tool for studying genomic diversity and identifying domesticated selective regions, the key quantitative trait locus (QTL), and genes for important traits [13–15]. The QTL associated with grain yield under drought stress was identified using SSR markers in *Oryza sativa* L. [16]. With the rapid development of high-throughput sequencing technologies, QTL mapping and genome-wide association analysis (GWAS) have been widely used for the genetic analysis of drought tolerance traits in crops [17–19], including rice (*Oryza sativa* L.) [20,21], wheat (*Triticum aestivum* L.) [22,23], maize (*Zea mays* L.) [24], and soybean (*Glycine max*) [25]. The discovery of drought-resistance QTL and the screening of candidate drought-resistance genes based on agronomic traits associated with drought resistance in crops has greatly contributed to the development of drought-resistant crop breeding.

Drought tolerance is a complex quantitative trait. Many traits have changed under drought stress. Previous studies have indicated that some physiological traits, such as relative water content, relative electrical conductivity, and chlorophyll content, and some yield-related traits, such as plant height, number of nodes in main stems, leaf area, pod number, seed number, and seed weight, are affected by soil drought stress, in addition to the significant reduction in total dry matter and yield of soybean plants during the growth and flowering stages [26]. These traits can be considered indicators to judge the drought tolerance of the crop [27–29]. It is therefore necessary to have an evaluation of soybean phenotype, physiological, and yield-related traits for drought tolerance.

Some drought-tolerant phenotypic and physiologically related QTL have been identified in soybean, such as relative water content, the relative electrical conductivity of soybean leaves, and water use efficiency [23,27,30]. Drought is a complex multi-gene controlled quantitative trait [24], which influences plant height, plant weight, node number, and yield [31–35]. Hwang et al. [34] constructed five recombinant inbred line (RIL) populations to analyze loci controlling leaf wilting in soybean, and a total of seven stable QTLs were localized in the populations. A total of 136 soybean drought-tolerant lines were tested for SNPs between drought-tolerant and sensitive genotypes, and 13 genes associated with the number of nodes in main stems were identified [33]. By GWAS, Zhang et al. identified 53 QTLs in 19 soybean chromosomes, with two single nucleotide polymorphisms (SNPs) associated with plant height falling within the confidence interval of two QTLs in different water conditions [32]. In 373 soybean varieties, 31,260 SNPs were obtained, and 47 SNP loci associated with water use were located by GWAS [36]. The use of molecular markers to assist in breeding selection for identifying drought-tolerant QTL in soybean can greatly improve the efficiency of selection for drought-tolerant varieties of soybean [36].

In the present study, we constructed a high-density map by whole genome resequencing techniques (WGRS) of 162 soybean RIL lines generated by drought-tolerant cultivar Jindou21 and resequencing drought-sensitive cultivar Zhongdou33. Some agronomic traits (node number of the main stem (NNMS), chlorophyll content (CC), branches (BN), pull stem (PS), leaf area (LA), plant height (PH), biomass, seed weight per plant (SWPP), and maturity) and physiological traits (Relative water content of leave (RWCL) and Relative electric conductivity of leave (RECL)) were analyzed to identify QTLs and candidate genes for drought tolerance under well-watered conditions and drought stress conditions, by using a water-catch tank to simulate drought stress [37]. This research provides some valuable information for understanding the molecular basis and breeding for drought tolerance.

2. Results

2.1. Effect of Genotype and Water Status on Agronomic and Physiological Traits

The QQ chart was used to describe the distribution of 11 traits in soybean RILs. The results showed that the majority of traits show a normal distribution except PS and M under well-watered and drought-stress conditions (Figure 1). The drought condition was simulated by water-catch tank treatment. Field traits were measured at the beginning of the seventh week when there were the most significant differences in soil water contents between the two treatment conditions (Supplementary Figure S1). Therefore, mean square and significance tests were conducted using combined ANOVA of NNMS, CC, RWCL, BN, LA, PH, BIOMASS, SWPP, RWCL, and RECL to investigate the effects on agronomic and physiological traits under different genotypes and environments (Table 1). The results indicated that CC, RECL, PH, NNMS, BN, RWCL, biomass, and SWPP were highly significantly different ($p < 0.01$) among the RILs and water treatments. The interaction of different lines and water status has a highly significant effect on CC, LA, PH, NNMS, BN, RWCL, and BIOMASS ($p < 0.01$), a significant effect on SWPP ($p < 0.05$), and no significant effect on the trait RECL.

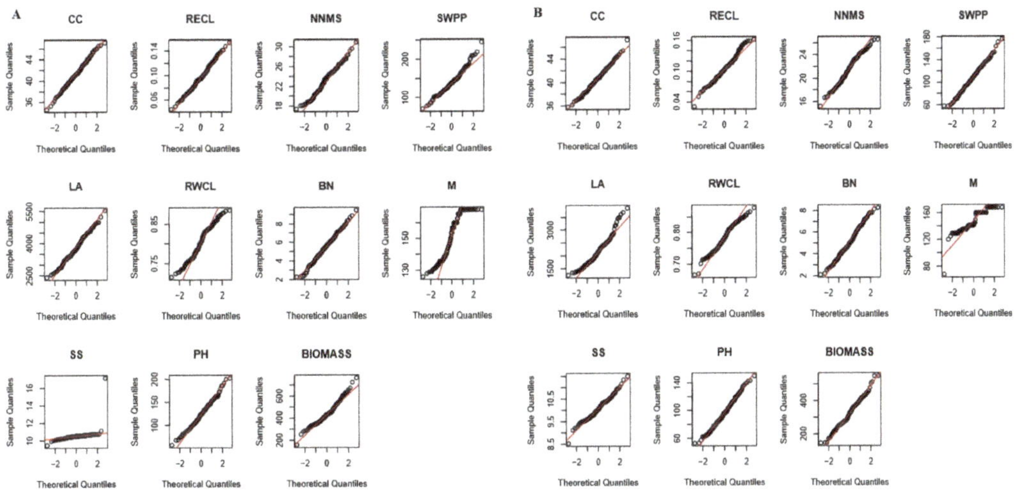

Figure 1. The QQ chart of normal distribution of 11 traits in 162 RILs under well-watered and drought stress conditions. (**A**) Distribution of traits under well-watered condition. (**B**) Distribution of traits under drought stress. LA, leaf area; PH, plant height; BN, branch number; CC, chlorophyll content; RWCL, relative water content of leaves; RECL, relative electric conductivity of leaves; NNMS, node number of main stem; BIOMASS, biomass; SWPP, seed weight per plant; PS, pull stem; M, maturity. The intercept of the red line is the mean and the slope is the standard deviation.

Table 1. Mean squares and significant tests after combined analysis of variance for nine agronomic traits under two water regimes.

Sources of Variation	DF	Seven Agronomic Traits							Two Physiological Traits	
		CC	LA	RECL	RWCL	PH	NNMS	BN	BIOMASS	SWPP
genotypes	161	22.707 **	1,158,244.424 **	0.002 **	0.006 **	3,295.922 **	33.930 **	8.982 **	37,381.694 **	2,779.981 **
Water status	1	61.573 **	622,894,650.534 **	0.013 **	0.096 **	171,580.293 **	669.760 **	55.676 **	3,034,425.977 **	228,060.573 **
genotypes × Water status	161	6.952 **	719,457.041 **	0.001	0.004 **	297.604 **	4.341 **	2.171 **	9,293.512 **	1,442.354 *

LA, leaf area; PH, plant height; BN, branch number; CC, chlorophyll content; RWCL, relative water content of leaves; RECL, relative electric conductivity of leaves; NNMS, node number of main stem; BIOMASS, biomass; SWPP, seed weight per plant. * represents significant difference at $p < 0.05$; ** represents significant difference at $p < 0.01$.

2.2. Traits Correlation and Principal Component Analysis

Figure 2 describes the correlation of seven agronomic traits and two physiological traits under well-watered and drought conditions. The correlation coefficients (r) indicate the degree of correlation between these traits. NNMS had a highly positive correlation with PH ($p < 0.001$, r > 0.6); similarly, NNMS and PH had a strong positive correlation with biomass ($p < 0.01$, r > 0.3) under well-watered drought stress conditions. In addition, RECL showed a significantly negative correlation with PH and biomass, respectively. Biomass had a positive correlation with SWPP under well-watered and drought stress conditions ($p < 0.001$, r > 0.6). Thus, NNMS and PH may be the key traits affecting biomass and SWPP under well-watered and drought-stress conditions.

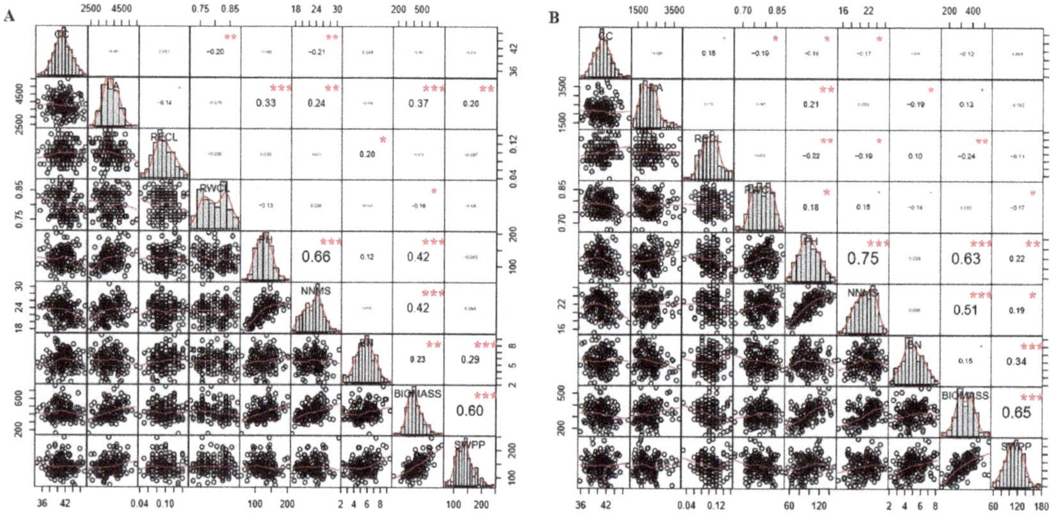

Figure 2. Pearson's correlation coefficients (r) describing associations of two physiological traits and seven agronomic traits of 162 soybean genotypes evaluated under well-watered condition and drought stress condition. (**A**) The correlation analysis among the traits under well-watered condition. (**B**) The correlation analysis among the traits under drought stress condition. LA, leaf area; PH, plant height; BN, branch number; CC, chlorophyll content; RWCL, relative water content of leaves; RECL, relative electric conductivity of leaves; NNMS, node number of main stem; BIOMASS, biomass; SWPP, seed weight per plant. The diagonal line shows the distribution of the nine traits. The bivariate scatter plot with fitted lines is displayed below the diagonal line. The correlation coefficient and significant difference are shown above the diagonal line, and the higher correlation coefficient is, the greater the numerical code. The red dot * represents significant difference at $p < 0.05$; the red dot ** represents significant difference at $p < 0.01$; the red dot *** represents significant difference at $p < 0.001$.

The relationships between genotypes and traits have been further investigated by principal component analysis of double-labeled plots under well-watered and drought stress conditions (Figure 3). The smaller the angle between the dimensional vectors of each trait in the plots, the higher the correlation between the traits. S151, S157, S161, and S152 had higher biomass that was mainly contributed by two traits NNMS and PH under drought stress conditions. S161 showed higher yields also mainly contributed by NNMS and PH.

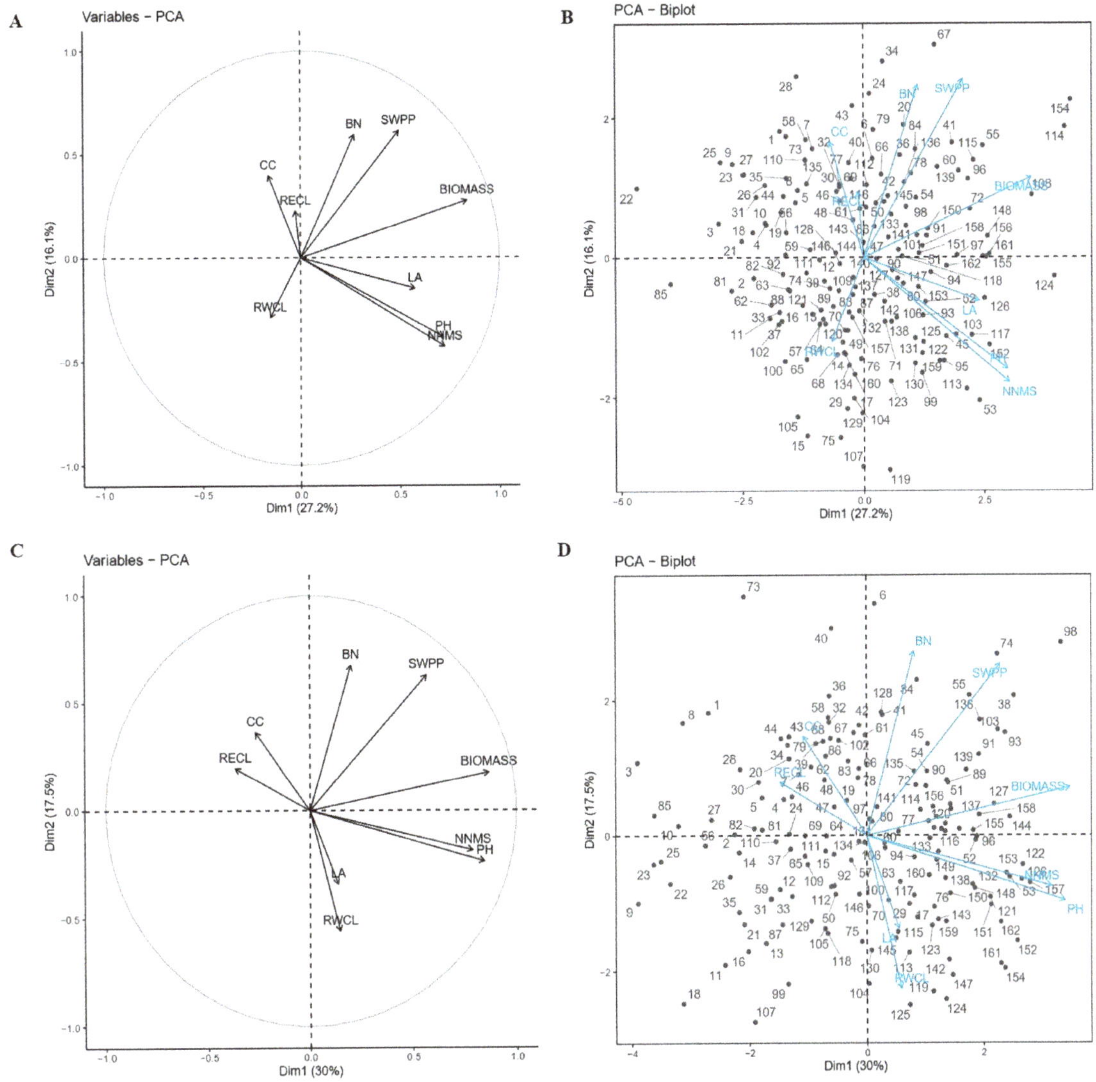

Figure 3. (**A**) The variables factor map of the nine traits under well-watered condition. (**B**) The principal component biplot displaying genotypic grouping under well-watered condition. (**C**) The variables factor map of the nine traits under drought stress condition. (**D**) The principal component biplot displaying genotypic grouping under drought-stress condition. LA, leaf area; PH, plant height; BN, branch number; CC, chlorophyll content; RWCL, relative water content of leaves; RECL, relative electric conductivity of leaves; NNMS, node number of main stem; BIOMASS, biomass; SWPP, seed weight per plant. The diagonal line shows the distribution of nine traits.

2.3. Linkage Mapping and QTL Analysis

The depth of the parental genomes was above 20X. The average coverage of the genomes was above 90%. The average depth of the offspring samples was 4.21X, with a coverage of above 92.86%. There was a total of 20 chromosomes in the high-density genetic map (Figure 4), containing 923,420 SNPs with a well-distributed linkage distance across the chromosomes. A total of 18 QTL associated with drought resistance were detected in 20 chromosomes under both normal and stress environments. Of these, under well-watered conditions, nine QTLs were localized on chromosomes 05, 08, 12, 13, 15, 16, 19, and 20, and nine QTLs were localized on chromosomes 02, 08, 11, 13, 15, 16, and 18 under drought stress (Table 2).

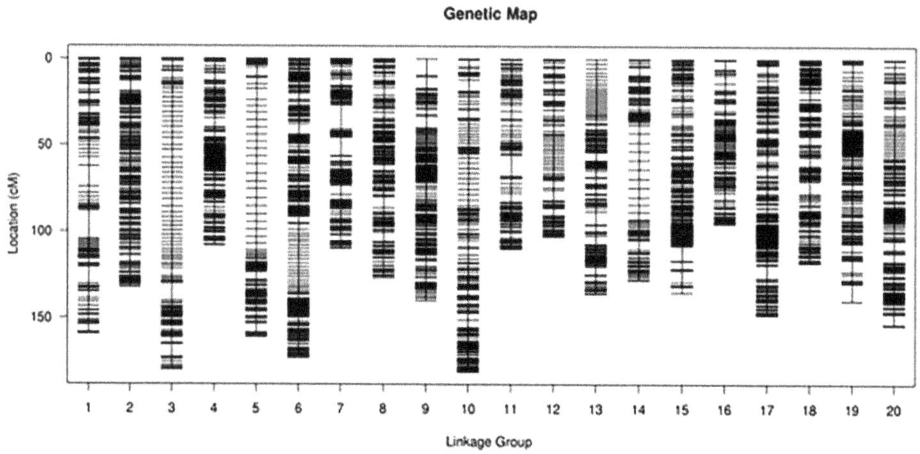

Figure 4. High-density linkage mapping of soybean RIL populations constructed by bin markers.

Table 2. Location and description of QTLs in RILs population derived from Jindou21 × Zhongdou33 grown under well-watered control (C) and drought stress (D) conditions.

Trait	Chr.	Start (cM)	End (cM)	LOD	Additive Effect	PVE (%)
LA-W	13	62.264	62.578	3.248	−143.48	5.436
LA-D	18	27.971	28.600	3.086	129.971	6.268
RECL-W	08	79.560	79.875	2.633	−0.005	5.874
RECL-D	15	39.392	39.706	4.744	0.007	9.220
RWCL-W	12	85.618	85.618	2.163	−0.008	3.757
RWCL-D	16	24.303	24.618	5.123	−0.014	10.063
PH-W	16	0.633	0.633	3.196	−8.135	8.187
PH-W	20	153.457	153.771	3.294	−8.326	8.577
PH-D	13	132.978	133.293	4.035	5.891	8.136
NNMS-W	16	13.769	13.769	8.681	−1.115	17.177
NNMS-D	16	13.769	13.769	7.510	−0.940	15.239
BN-W	19	113.603	113.603	4.209	0.386	7.454
BN-D	11	19.855	20.170	2.767	−0.307	5.311
BIOMASS-W	05	159.672	159.986	2.732	5.008	6.578
BIOMASS-W	15	81.000	81.250	2.955	−6.294	10.390
BIOMASS-D	11	89.689	90.318	3.930	4.915	7.6490
SWPP-D	02	104.790	105.355	2.230	4.457	3.846
SWPP-D	08	126.804	126.804	2.440	5.118	5.073

PVE, percentage of phenotypic variance explained by each QTL. LOD, logarithm of odds. LA, leaf area; PH, plant height; BN, branch number; CC, chlorophyll content; RWCL, relative water content of leaves; RECL, relative electric conductivity of leaves; NNMS, node number of main stem; BIOMASS, biomass; SWPP, seed weight per plant. The diagonal line shows the distribution of the nine traits; cM, centimorgan.

One QTL localized on chromosome 16 was identified for NNMS under well-watered and drought conditions, which may be a major QTL. It contributed 17.18% and 15.24% of the phenotypic variation, with maximum LOD values of 8.70 and 7.51 under well-watered and drought conditions, respectively (Figure 5).

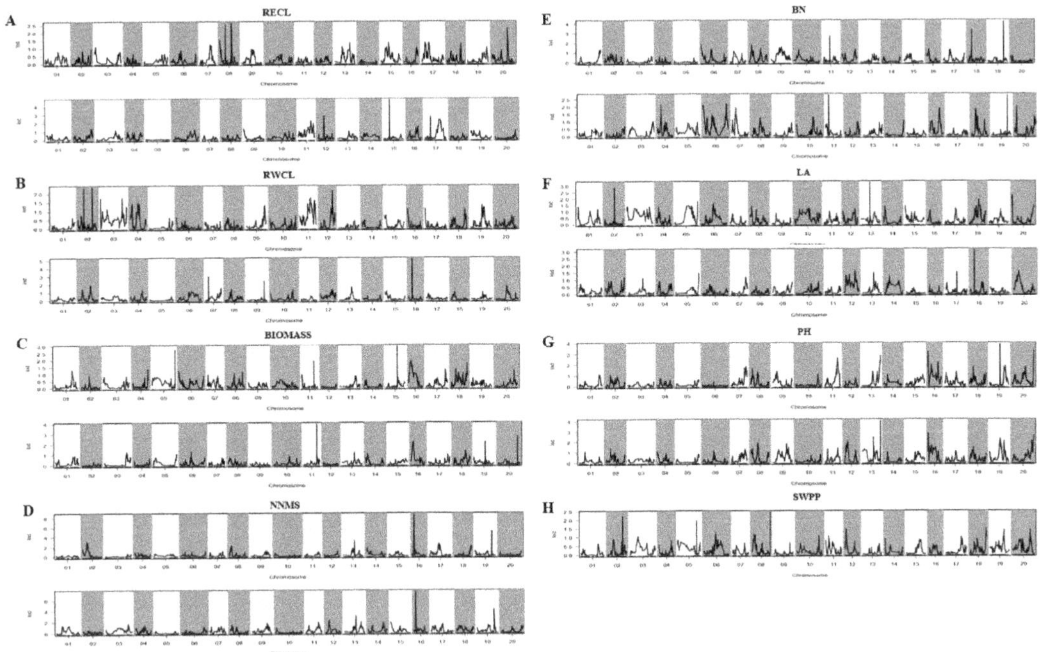

Figure 5. (**A**) RECL map positions (Mb) under well-watered condition and RECL map positions (Mb) under drought stress. (**B**) RWCL map positions (Mb) under well-watered condition and RWCL map positions (Mb) under drought stress. (**C**) BIOMASS map positions (Mb) under well-watered condition and BIOMASS map positions (Mb) under drought stress. (**D**) NNMS map positions (Mb) under well-watered condition and NNMS map positions (Mb) under drought stress. (**E**) BN map positions (Mb) under well-watered condition and BN map positions (Mb) under drought stress. (**F**) LA map positions (Mb) under well-watered condition and LA map positions (Mb) under drought stress. (**G**) PH map positions (Mb) under well-watered condition and PH map positions (Mb) under drought stress. (**H**) SWPP map positions (Mb) under drought stress.

Furthermore, two QTLs located on chromosomes 16 and 20 were identified for PH under well-watered conditions, and one was located on chromosome 13 under drought stress (PVE > 8%). One QTL identified for BN was located on chromosome 19 with a LOD value of 4.2 and 7.5% of PV under the well-watered condition. Under drought conditions, a QTL associated with RWCL located on chromosome 16 was identified, explaining up to 10% of the PV, while a QTL for RECL localized on chromosome 8 explained 9.2% of the phenotypic variation. For BIOMASS, two QTLs were identified on chromosomes 5 and 15, with LOD values above 2.5, and contributing 6.6 and 10.4%, respectively, under the well-watered condition, while one QTL localized on chromosome 11 had a LOD value of 3.9 and a PVE value of 7.6% under the drought condition. For SWPP, two QTLs located on two chromosomes were identified, explaining 3.85% and 5.07% of the PV with maximum LOD values of 2.23 and 2.44, respectively.

2.4. Prediction of Candidate Genes

Based on the results of the locus, 135 candidate genes of the major QTL for NNMS were identified under both drought and well-watered conditions (Table 3). The results showed that one QTL was located at 13.769 cM on chromosome 16, with 135 genes in the interval. The results of Gene Ontology (GO) enrichment and classification revealed that these genes are involved in the regulation of plant growth and development and response to a stimulus (Figure 6). These candidate genes are annotated as different kinases, transcription factors, and functional proteins (Table 3). The kinases mainly include mitogen-activated protein kinase (MAPK) and adenylyl-sulfate kinase (ASK). NAC, GATA transcription factor, ethylene-responsive transcription factor, and RAP-like transcription factors are also identified. In addition, some functional proteins, such as nucleoside-triphosphatase, BTB/POZ MATH (BPM) protein, and the plant flowering control gene FLOWERING LOCUS T (FT), were also annotated. It is worth noting that several candidate drought-tolerant genes were identified, such as NAC transport factor (*Glyma.16G042900* and *Glyma.16G043200*) and GATA transport factor (*Glyma.16G042300*), and BPM proteins, such as *Glyma.16G036700*, *Glyma.16G036400*, *Glyma.16G036600*, and *Glyma.16G036800*, under drought and well-watered conditions (Table 3). These gene families were involved in the regulation of drought stress [38–45].

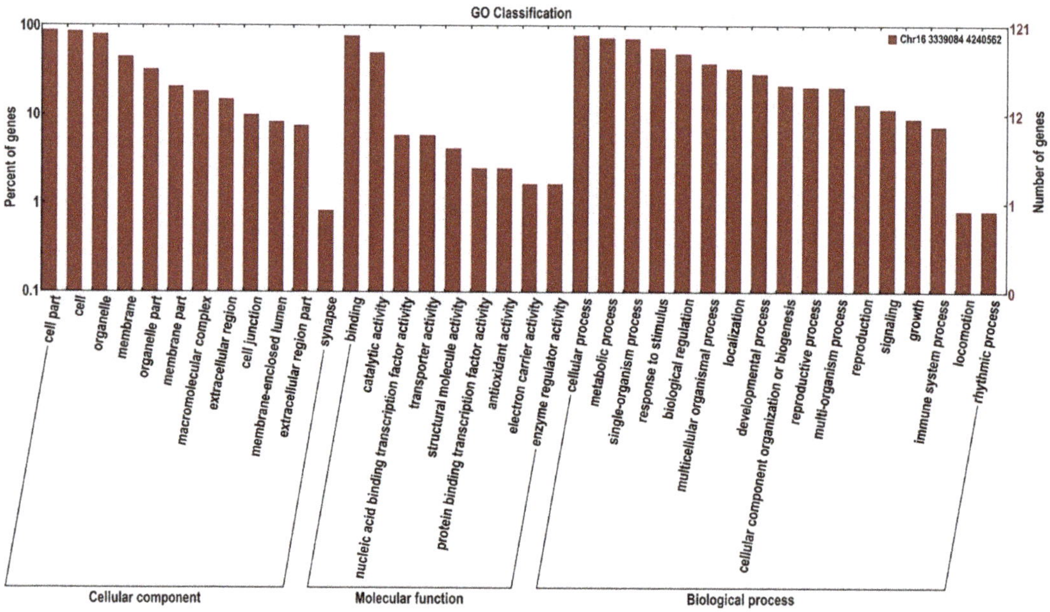

Figure 6. GO enrichment and classification of the candidate genes for NNMS under well-watered and drought conditions.

Table 3. Annotation and classification of candidate genes within NNMS-localized QTL intervals.

	Gene	Annotation	Annotation Homologous Genes in Arabidopsis	Function Annotation in Arabidopsis
Kinase	*Glyma.16G044000*	mitogen-activated protein kinase kinase kinase kinase 3-like	unknown	MAP3
	Glyma.16G040900	adenylyl-sulfate kinase 3	*AT3G03900*	ASK3
Transport Factor	*Glyma.16G042900*	NAC domain-containing protein 100	*AT5G18270*	NAC100
	Glyma.16G043200	NAC domain-containing protein 18	*AT3G04070*	NAC18
	Glyma.16G042300	GATA transcription factor 9-like	*AT4G32890*	GATA9
	Glyma.16G040000	ethylene-responsive transcription factor RAP2-11-like	*AT5G18560*	RAP2-11
Function Protein	*Glyma.16G039900*	importin-5	*AT5G19820*	IPO5
	Glyma.16G043700 *Glyma.16G043300* *Glyma.16G043400* *Glyma.16G043500*	nucleoside-triphosphatase-like	*AT5G18280*	APY2
	Glyma.16G039300 *Glyma.16G039400*	lysosomal beta glucosidase-like	*AT5G04885*	unknown
	Glyma.16G041200	glutamate dehydrogenase 1	*AT5G18170*	GDH1
	Glyma.16G043900	transmembrane and coiled-coil domain-containing protein	*AT5G06660*	unknown
	Glyma.16G036800	BTB/POZ and MATH domain-containing protein 5	*AT5G21010*	BPM5
	Glyma.16G036400 *Glyma.16G036700*	BTB/POZ and MATH domain-containing protein 4	*AT3G03740*	BPM4
	Glyma.16G038300	methionine synthase	*AT5G17920*	ATMS1
	Glyma.16G044400	transportin-3	*AT1G12930*	TNPO3
	Glyma.16G042000	short-chain type dehydrogenase/reductase-like	*AT3G03980*	unknown
	Glyma.16G037600	2-aminoethanethiol dioxygenase	*AT5G39890*	unknown
	Glyma.16G041900	inhibitor of growth protein 4-like	*AT1G54390*	ING2
	Glyma.16G036000	cation/calcium exchanger 4-like	*AT5G17860*	CAX7
	Glyma.16G036900	50S ribosomal protein L25-like isoform 1	*AT4G23620*	RIPL
	Glyma.16G040400	arabinogalactan peptide 14-like	*AT5G56540*	AGP14
	Glyma.16G040200	5&apos-adenylyl sulfate reductase-like 5-like	*AT3G03860*	APRL5
	Glyma.16G040900	adenylyl-sulfate kinase 1, chloroplastic-like	*AT3G03900*	APK3
	Glyma.16G040100	tubulin alpha-3 chain-like	*AT5G19780*	TUA5
	Glyma.16G041400	putative H/ACA ribonucleoprotein complex subunit 1-like	*AT3G03920*	unknown
	Glyma.16G039700	cation/H(+) antiporter 15-like	*AT1G05580*	CHX23
	Glyma.16G041300	hypothetical protein PRUPE_ppa015420mg	*AT3G42170*	BED zinc finger
	Glyma.16G036500	hypothetical protein MTR_7g013520	unknown	unknown
	Glyma.16G044300	omega-hydroxy palmitate O-feruloyl transferase-like	*AT1G65450*	unknown
	Glyma.16G039600	quinate hydroxycinnamoyl transferase-like	*AT2G19070*	SHT
	Glyma.16G038500	mavicyanin-like	*AT3G17675*	unknown
	Glyma.16G042500	glucan endo-1,3-beta-glucosidase 8-like	unknown	unknown
	Glyma.16G037200	Phosphoribosyl glycinamide formyl transferase	unknown	unknown
	Glyma.16G040800	conserved hypothetical protein	unknown	unknown
	Glyma.16G044200	protein FLOWERING LOCUS T-like	*AT1G65480*	FT
	Glyma.16G038400	nucleolar MIF4G domain-containing protein 1-like	*AT5G17930*	unknown
	Glyma.16G044600	acetyl-coenzyme A synthetase-like	*AT5G36880*	ACS
	Glyma.16G038200	ankyrin-3-like	*AT2G31820*	unknown

Under water-deficiency treatment, some candidate genes associated with drought resistance based on RECL were localized in chromosome 15 within 39.392–39.706 M, including a drought-resistance gene methionine sulfoxide reductase (MSR), *Glyma.15G100700*, which regulates chloroplasts. Four candidate genes were identified from 235 candidates with a drought stress response and PH on chromosome 13, annotated as PUB and NAC genes, respectively.

3. Discussion

3.1. The Character of the WGRS Approach to Identify QTL Makers and Candidate Region Analysis

The use of genetic maps is essential for finding important loci, precision mapping, and marker-assisted breeding [46]. Several genetic maps have been constructed for soybean based on molecular markers such as SSRs, ESTs, RFLP, and RAPD markers [47,48]. Using DNA markers, it was possible to identify regions on the genetic map that could be identified by the main genes, in accordance with standard mapping procedures. However, it is difficult to find the key locus associated with a specific trait by using non-specificity makes [49].

By advancing DNA sequencing technologies and applications, scientists have been able to improve plant breeding and aid in fine mapping processes in the past decade, as well as discover new types of molecular markers. Regarding detecting SNP markers and accurate genotyping, a high-density genetic map can be constructed with the next-generation sequencing technologies developed and the soybean reference genome sequence published [50,51]. The SNP molecular markers' identification by WGRS has been well used for studying drought resistance in many important crops [33,34].

In this study, the parental lines and 162 RILs were sequenced by WGRS to construct a high-density genetic map. Among them, the parental lines had a 20-fold sequencing depth and average genome coverage of 90% or more, and each RIL had a 4.21-fold sequencing depth and coverage of 92.86% or more. Ultimately, 32.84 Gbp of high-quality reads from Jindou 21, 33.3 Gbp of high-quality reads from Zhongdou33m and 748.99 Gbp of high-quality reads from their progeny were obtained. By utilizing bin markers and accurate genotypic data, it was possible to construct a high-density genetic map. An analysis of genotyping data showed that 4843 recombination bin markers represented 923,420 SNPs on 20 linkage groups. There was an average distance of 0.57 cM between adjacent bin markers on the linkage map, with a total length of 2,737.51 cM. The collinearity of the genetic maps and physical maps was good for each linkage group (Supplementary Figure S2). Here, we demonstrate that the WGRS strategy is an effective tool for detecting markers and building high-density linkage maps. The WGRS mapping enabled us to obtain a great number of genome-wide SNPs, which accurately reflect the genetic diversity and genomic diversity of soybean.

3.2. Yield-Related Traits Analysis

There have been many studies on the identification of drought tolerance in soybean varieties and the screening of drought-resistant germplasm resources, in which the identification of drought-tolerance indicators and traits in soybean varieties is a crucial step. In previous studies, drought-tolerance indicators mainly include yield traits, growth and development indicators, morphological indicators, and physiological and biochemical indicators [10,52]. Soybean drought-tolerance traits include leaf wilting [53,54], root morphology [53], yield under drought [54], etc. The drought treatment of soybean plants has also been shown to cause changes in the number of branches and main stem nodes and plant height [32,33].

In legumes, especially soybean, there are many studies on NNMS. NNMS is an important trait for soybean breeding. Soybean canopy and seed yield are determined by NNMS, which is one of the major plant agronomic traits [55]. Furthermore, there was a correlation between NNMS and other important agronomic traits, such as plant height, flowering, and maturity [56]. Li et al. reported that some QTLs associated with NNMS for plant density were identified using 144 four-way recombinant inbred lines (FW-RILs). The candidate genes were found on chr 06 and chr 19, named *Glyma.06G094400*, *Glyma.06G147600*, *Glyma.19G160800*, and *Glyma.19G161100* [57]. Fu et al. utilized 306 accessions from northeast China to identify 76 QTLs associated with NNMS for yield and identified 49 candidate genes [58]. Plant height is one of the main hot spots in plant abiotic stress. There are many studies on the association between plant height and drought tolerance. In soybean, six QTLs for drought tolerance associated with plant height have been identified (qPH2, qPH6, qPH7, qPH19-1, qPH19-2, and qPH19-3) [59,60].

In our study, there is a strong relationship between NNMS and PH ($p < 0.001$) under well-watered and drought-stress conditions. PH was also highly correlated with BIOMASS and Y traits under both conditions. The above result is consistent with earlier reports. Moreover, the correlation between NNMS and BIOMASS under the drought stress condition was enhanced compared to the well-watered condition. It is tempting to speculate that NNMS and PH are traits associated with drought tolerance.

3.3. Preliminary Analysis of the Potential Functions of Candidate Genes

In this study, the main QTL was localized with NNMS under both drought and well-watered conditions, and the candidate genes in this region were annotated with GATA, NAC transcription factors, and BPM proteins in Arabidopsis. In addition, under drought conditions, several candidate genes associated with RECL and PH were identified including MSR, PUB, and NAC genes.

Protein degradation is essential for plant growth and development. The BPM protein is part of the Cullin E3 ubiquitin ligase complex [39,61], and binds at least three transcription factor families—ERF/AP2 class I, the homologous cassette-leucine zip, and R2R3 MYB—to degrade target proteins by ubiquitination, which plays an important role in plant abiotic stress responses, especially drought resistance [39]. In this study, the main QTLs associated with NNMS have been identified. Four candidate genes, including *Glyma.16G036700*, *Glyma.16G036400*, *Glyma.16G036600*, and *Glyma.16G036800*, were annotated as BPM4-like proteins. Their homologous gene *AT3G03740* (BPM4) has been reported to interact with the transcription factor ERF/AP2 to regulate drought tolerance in Arabidopsis [62]. The plant U-box (PUB) gene family is also a major family of ubiquitin ligases in plants. They are involved in protein degradation pathways and physiological processes regulated by drought stress in plants [63,64]. For example, in Arabidopsis, PUB22 and PUB23 coordinate the regulation of drought signaling pathways through the ubiquitination of cytoplasmic RPN12a to enable plants to respond to water deficits [65]. In addition, the expression of PUB6 in soybean leaves and roots is induced by abscisic acid (ABA), high salinity, and polyethylene glycol (PEG), which play a negative regulatory role in drought tolerance [66]. Here, two candidate genes, *Glyma.13G312700* and *Glyma.13G312800,* associated with PH, were identified. Their homologs gene *PUB23*, encoding E3 ubiquitin protein ligase, negatively regulates drought tolerance by controlling the ABA receptor PYL9 in Arabidopsis [67].

The NAC (NAM, ATAF1/2, CUC2) protein family is a plant-specific transcription factor superfamily in most plants [40–42,68]. Wu et al. [42] revealed that PtrNAC72, a blocker of putrescine biosynthesis, may negatively regulate plant response to drought stress by acting as a deterrent to putrescine-related ROS homeostasis. Under dehydration stress conditions, ANAC096 cooperates with the bZIP-type transcription factors ABRE binding factor and ABRE binding protein (ABF/AREB) to support plants' survival [43]. In soybean, the overexpression of NAC085 decreases malondialdehyde content and increases superoxide dismutase, catalase, and ascorbate peroxidase activities under abiotic stresses [44]. In the study, *Glyma.16G042900* and *Glyma.16G043200* associated with NNMS were annotated as NAC100-like and NAC18-like, respectively. Their homologs genes were NAC87 (*AT5G18270*) and NAC18 (*AT3G04070*). In Arabidopsis, ding et al. identified dehydration stress memory response genes based on genome-wide RNA-Seq, in which NAC87 (*AT5G18270*) and NAC18 (*AT3G04070*) were associated [45].

MSR genes play an important role in plant stress resistance. MSR catalyzes the reduction of methionine sulfoxide to methionine residues. Reactive oxygen species (ROS) caused by biotic and abiotic stresses in plants lead to protein denaturation. MSR proteins can reduce plant damage during ROS disruption. For example, the pepper MSR2 is responsible for reducing oxidized porphobilinogen deaminase (PBGD), which can protect chlorophyll synthesis under drought conditions [38]. The MSR2 gene (*Glyma.15G100700*), which was screened by a QTL for RECL, could play an important role in drought tolerance in soybean. These candidate genes will require subsequent validation experiments to elucidate their drought-tolerance functions.

4. Materials and Methods

4.1. Plant Material and Growing Conditions

A RIL population developed from a cross between a drought-tolerant cultivar 'Jindou21' and the drought-sensitive cultivar 'Zhongdou33'. The parents and 160 RILs of F8 were grown in the field of the Industrial Crop Institute, Shanxi Academy of Agricultural Sciences, China in 2019.

4.2. Experimental Design and Drought Conditions

The accurate identification of drought tolerance in the field is difficult. The effect and feasibility of the collecting trough have been evaluated. Placing the collecting trough between soybean rows could reduce soil water content [37]. Here, the collecting troughs with a diameter of 25 cm were placed in the field in the vegetative period to collect rain. The method was used to simulate drought treatment to identify phenotypic and agronomic traits of the RIL lines. A soil moisture meter (FIELDSCOUT TDR 100, Campbell) was used to measure the soil water content under the control and drought stress conditions.

4.3. Measurement of Traits and Phenotyping

Drought-tolerance indicators were identified in the field and the greenhouse in 2019. The plant height, branch number, chlorophyll content, relative water content of leaves, the relative electric conductivity of leaves, the node number of the main stem, pull stem, and leaf area were measured at the R2 stage, whereas the biomass (including seeds), seed weight per plant, and maturity were measured when the plant was harvested at the R8 stage. Leaf area was determined using a portable leaf area meter (YMJ-D). Chlorophyll content was determined using a portable chlorophyll meter SPAD502. Stem strength was determined using a SUNDOO portable tester. Leaf relative water content and electrical conductivity were determined using the method in [69–71]. Each drought coefficient (DC) of 11 traits was calculated as the ratio of the individual trait under drought stress condition or well-watered conditions as shown in the equation below.

$$DC = \text{Trait in Drought Stress Condition} / \text{Trait in Well-watered Condition}$$

Five plants were measured per replicate of each trait (n = 3 biological replicates).

4.4. DNA Extraction, DNA Sequencing and SNP Identification

A total of 162 RILs, Jindou21, and Zhongdou33 genomic DNA were extracted from fresh young leaves using the Plant DNA Kit (D2485, Omega). In order to sequence each DNA sample, paired-end sequencing libraries were constructed with an insertion of 300–500 bp. An Illumina Hiseq 2000 system (Illumina Inc., San Diego, CA, USA) was used to sequence the libraries with a 150 bp (PE150) read length. Low-quality reads (quality score < 20e) were filtered out, and then raw reads were sorted to each sample according to barcode sequences. We used Samtools v 1.9 [72,73] to mark duplicates and GATK v 2.8.1 [74] to realign local elements and calibrate bases. With the default parameter settings, GATK and Samtools [73], SNP calling analysis was formed to produce a set of SNPs. There are eight segregation patterns based on polymorphic SNPs between parents (ab × cd, ef × eg, hk × hk, lm × ll, nn × np, aa × bb, ab × cc, and cc × ab). Only SNPs with the aa × bb pattern were chosen for further analysis.

4.5. Construction of Linkage Map

A total of 32.84 Gbp of clean data was obtained for parent Jindou21 and 33.3 Gbp for Zhongdou33. In total, 160 offspring had a total of 748.99 Gbp of data. Q30 in 162 RILs reached over 80%. A total of 1,913,252 SNPs were detected between the parents and were screened for redundancy.

To determine recombination breakpoints and construct a bin map of RILs, a slightly modified sliding window approach was employed [75]. The ratio of SNPs with 'Jindou21' and 'Zhongdou33' genotypes was calculated. According to Huang et al. [75], a physical bin map was constructed based on the recombination breakpoint position. Briefly, when RILs did not have recombination breakpoints within a 100 kb interval, these regions were combined into one bin. When recombination breakpoints did not occur within 100 kb intervals, the regions were combined into a single bin.

A genetic map with ultra-high density was constructed using bins as markers. Molecular markers were divided into linkage groups (LGs) based on their locations on the genome. To further confirm the robustness of markers in each LG, modified logarithms of odds (MLODs) were calculated between markers. Before ordering, markers with MLOD scores < 5 were filtered. For resolving genotyping errors within LGs, the HighMap [76] strategy was utilized. The SMOOTH error correction strategy is then applied based on parents' genotype contributions, and a k-nearest neighbor algorithm is used to impute missing genotypes. After applying the multipoint method of maximum likelihood, skewed markers were added to the map. Kosambi mapping function [77] was used to estimate map distances.

4.6. QTL Analysis

The effects of genotype and environment, as well as interaction effects between genotype and environment, were estimated using an analysis of variances (ANOVA). It was estimated that broad sense heritability is derived from the formula $H^2 = \sigma^2_g/(\sigma^2_g + \sigma^2_{ge}/n + \sigma^2_e/nr)$, where n delegates the number of environments, r delegates the number of replications, σ^2_g delegates the estimated genetic variance, σ^2_{ge} delegates variance for genotype-environment interaction, and σ^2_e delegates the experimental error. R package R/qtl was used to identify QTLs via composite interval mapping (CIM) [78]. LOD values were determined based on a 1000-permutation test. QTLs were called for LOD values of 2 and higher. A 1000-permutations test was used to determine LOD values. When LOD values exceeded 2, QTLs were identified.

4.7. Analysis of Phenotypic Data

The ANOVA analysis was performed using the SPSS 19 (George 2012) following a chi-squared test. To describe the magnitude of the relationships between physiological and agronomic traits, Pearson correlation coefficients (r) were calculated separately for the stress and non-stress treatments using R version 4.0.1 (2 July 2021). Correlation coefficients were calculated using the Pearson model. Normal distribution tests were plotted using the R package qqman, phenotypic correlation association plots were plotted using the R package Performance Analytics, and principal component analyses were plotted using the R packages FactoMineR, factoextra, and corrplot.

4.8. Potential Candidate Genes Prediction

The genes within QTLs were searched for potential candidates. The candidate genes were identified as those linked directly to drought stress or those that were associated with stress. A search for drought stress-related gene names and functions was carried out in Soybase (www.soybase.org, 13 June 2022), NCBI (https://www.ncbi.nlm.nih.gov/, 13 June 2022), and Phytozyme (https://phytozome.jgi.doe.gov, 13 June 2022).

5. Conclusions

A genetic map of 20 linkage groups with a total distance of 2737.51 cM was developed using 162 soybean recombinant inbred line populations and 923,420 SNP markers. The genetic map was used to identify drought-tolerant QTLs for traits in soybean. In this study, the main QTL on chromosome 16 was identified for NNMS under well-watered and drought conditions, which explained more than 10% phenotypic variation and had a LOD score larger than 6. Several of the candidate genes in this region were associated with NAC, BPM, and PUB proteins, which possibly enable plants to respond to drought stress, namely *Glyma.16G036700*, *Glyma.16G036400*, *Glyma.16G036600*, *Glyma.16G036800*, *Glyma.13G312700*, *Glyma.13G312800*, *Glyma.16G042900*, *Glyma.16G043200*, and *Glyma.15G100700*. The QTLs and candidate genes detected in this study could provide an important step toward clarifying the mechanisms of drought tolerance in soybean and further provide a theoretical basis for drought-tolerant breeding in soybean.

Supplementary Materials: The following supporting information can be downloaded at: https://www.mdpi.com/article/10.3390/ijms231810828/s1.

Author Contributions: X.Z., L.C., and D.C., experiment design; W.O. and L.C., writing the paper; J.M., X.L., H.Y. and Z.Y., experiment operation; W.G., Z.S., Y.Z. and H.Z., data analysis; S.C. and H.C., supplied reagents, materials, and analytical tools. All authors have read and agreed to the published version of the manuscript.

Funding: The work was funded by The National Natural Science Foundation of China (32171957); The Agricultural Science and Technology Innovation Program (ASTIP No. CAAS-ZDRW202201); The Opening Project of the Key Laboratory of Oil Crop Biology and Genetics Breeding, Ministry of Agriculture (KF2018007) National Genetically Modified Organisms Breeding Major Projects (2016ZX08004-002) and (2016ZX08004-005).

Institutional Review Board Statement: Not applicable.

Informed Consent Statement: Not applicable.

Data Availability Statement: The datasets used and/or analyzed during the current study are available from the corresponding author on reasonable request. However, some of the data are shown in Supplementary Materials.

Conflicts of Interest: The authors declare no conflict of interest.

Abbreviations

RIL	recombinant inbred line
QTL	quantitative trait locus
SNP	single nucleotide polymorphism
LOD	logarithm of odds
cM	centimorgan
WGRS	whole genome resequencing techniques
GWAS	genome-wide association analysis
DC	drought coefficient
LA	leaf area
PH	plant height
BN	branch number
CC	chlorophyll content
RWCL	relative water content of leaves
RECL	relative electric conductivity of leaves
NNMS	node number of main stem
BIOMASS	biomass
SWPP	seed weight per plant
PS	pull stem
M	maturity
GO	gene ontology
MAPK	mitogen-activated protein kinase
BPM	BTB/POZ math
FT	flowering locus t
ROS	reactive oxygen species
ASK	adenylyl-sulfate kinase
PEG	polyethylene glycol
MSR	methionine sulfoxide reductase
PBGD	porphobilinogen deaminase
LGs	linkage groups
MLODs	modified logarithms of odds
ANOVA	analysis of variances
CIM	composite interval mapping

References

1. Chaves, M.M.; Maroco, J.P.; Pereira, J.S. Understanding plant responses to drought—From genes to the whole plant. *Funct. Plant Biol.* **2003**, *30*, 239–264. [CrossRef] [PubMed]
2. Spinoni, J.; Naumann, G.; Carrao, H.; Barbosa, P.; Vogt, J. World drought frequency, duration, and severity for 1951–2010. *Int. J. Climatol.* **2014**, *34*, 2792–2804. [CrossRef]
3. Jaleel, C.A.; Gopi, R.; Sankar, B.; Gomathinayagam, M.; Panneerselvam, R. Differential responses in water use efficiency in two varieties of Catharanthus roseus under drought stress. *C. R. Biol.* **2008**, *331*, 42–47. [CrossRef] [PubMed]
4. Haile, F.J. Drought stress, insects, and yield loss. In *Biotic Stress and Yield Loss*; CRC Press: Boca Raton, FL, USA, 2000; pp. 131–148.
5. Fenta, B.A.; Beebe, S.E.; Kunert, K.J.; Burridge, J.D.; Barlow, K.M.; Lynch, J.P.; Foye, C.H. Field phenotyping of soybean roots for drought stress tolerance. *Agronomy* **2014**, *4*, 418–435. [CrossRef]
6. Dogan, E.; Kirnak, H.; Copur, O. Deficit irrigations during soybean reproductive stages and CROPGRO-soybean simulations under semi-arid climatic conditions. *Field Crops Res.* **2007**, *103*, 154–159. [CrossRef]
7. Chang, Y.; Song, S. Studies on the water requirment of soybeans and the effect of irrigation on their growth and yield. *Soybean Sci.* **1983**, *4*, 277–285.
8. Yang, P.H.; Li, G.Q.; Guo, L. Effect of drought stress on plasma membrane permeability of soybean varieties during flowering-poding stage. *Agric. Res. Arid Areas* **2003**, *21*, 127–130.
9. Specht, J.E.; Hume, D.J.; Kumudini, S.V. Soybean yield potential—A genetic and physiological perspective. *Crop Sci.* **1999**, *39*, 1560–1570. [CrossRef]
10. Manavalan, L.P.; Guttikonda, S.K.; Phan Tran, L.S.; Nguyen, H.T. Physiological and molecular approaches to improve drought resistance in soybean. *Plant Cell Physiol.* **2009**, *50*, 1260–1276. [CrossRef]
11. Gupta, P.K.; Kumar, J.; Mir, R.R. 4 Marker-assisted selection as a component of conventional plant breeding. *Plant Breed. Rev.* **2010**, *33*, 145.
12. Visarada, K.; Meena, K.; Aruna, C.; Srujana, S.; Saikishore, N.; Seetharama, N. Transgenic breeding: Perspectives and prospects. *Crop Sci.* **2009**, *49*, 1555–1563. [CrossRef]
13. Huang, X.; Kurata, N.; Wei, X.; Wang, Z.; Wang, A.; Zhao, Q.; Zhao, Y.; Liu, K. A map of rice genome variation reveals the origin of cultivated rice. *Nature* **2012**, *490*, 497–501. [CrossRef] [PubMed]
14. Jiao, Y.; Zhao, H.; Ren, L.; Song, W.; Zeng, B.; Guo, J.; Wang, B.; Liu, Z.; Chen, J.; Li, W. Genome-wide genetic changes during modern breeding of maize. *Nat. Genet.* **2012**, *44*, 812–815. [CrossRef] [PubMed]
15. Lin, T.; Zhu, G.; Zhang, J.; Xu, X.; Yu, Q.; Zheng, Z.; Lun, Y.; Li, S.; Wang, X. Genomic analyses provide insights into the history of tomato breeding. *Nat. Genet.* **2014**, *46*, 1220–1226. [CrossRef]
16. Bernier, J.; Kumar, A.; Ramaiah, V.; Spaner, D.; Atlin, G. A large-effect QTL for grain yield under reproductive-stage drought stress in upland rice. *Crop Sci.* **2007**, *47*, 507–516. [CrossRef]
17. Fan, Y.; Shabala, S.; Ma, Y.; Xu, R.; Zhou, M. Using QTL mapping to investigate the relationships between abiotic stress tolerance (drought and salinity) and agronomic and physiological traits. *BMC Genom.* **2015**, *16*, 43. [CrossRef]
18. Pham, A.T.; Maurer, A.; Pillen, K.; Brien, C.; Dowling, K.; Berger, B. Genome-wide association of barley plant growth under drought stress using a nested association mapping population. *BMC Plant Biol.* **2019**, *19*, 134. [CrossRef]
19. Varshney, R.; Hoisington, D.; Tyagi, A.; Tyagi, A.K. Advances in cereal genomics and applications in crop breeding. *Trends Biotechnol.* **2006**, *24*, 490–499. [CrossRef]
20. Sandhu, N.; Dixit, S.; Swamy, B.P.; Vikram, P.; Venkateshwarlu, C.; Catolos, M.; Kumar, A. Positive interactions of major-effect QTLs with genetic background that enhances rice yield under drought. *Sci. Rep.* **2018**, *8*, 1626. [CrossRef]
21. Panda, D.; Mishra, S.S.; Behera, P.K. Drought tolerance in rice: Focus on recent mechanisms and approaches. *Rice Sci.* **2021**, *28*, 119–132. [CrossRef]
22. Gupta, P.K.; Balyan, H.S.; Gahlaut, V. QTL analysis for drought tolerance in wheat: Present status and future possibilities. *Agronomy* **2017**, *7*, 5. [CrossRef]
23. Kirigwi, F.M.; Van Ginkel, M.; Brown-Guedira, G.; Gill, B.S.; Paulsen, G.M.; Fritz, A.K. Markers associated with a QTL for grain yield in wheat under drought. *Mol. Breed.* **2007**, *20*, 401–413. [CrossRef]
24. Hu, X.; Wang, G.; Du, X.; Zhang, H.; Xu, Z.; Wang, J. QTL analysis across multiple environments reveals promising chromosome regions associated with yield-related traits in maize under drought conditions. *Crop J.* **2021**, *9*, 759–766. [CrossRef]
25. Dossa, K.; Li, D.; Zhou, R.; Yu, J.; Wang, L.; Zhang, Y.; Xin, W.; Zhang, X. The genetic basis of drought tolerance in the high oil crop Sesamum indicum. *Plant Biotechnol. J.* **2019**, *17*, 1788–1803. [CrossRef] [PubMed]
26. Jha, P.K.; Kumar, S.N.; Inesa, A.V.M. Responses of soybean to water stress and supplemental irrigation in upper indo-Gangetic plain: Field experiment and modeling approach. *Field Crop Res.* **2018**, *219*, 76–86. [CrossRef]
27. Lopez, M.A.; Xavier, A.; Rainey, K.M. Phenotypic variation and genetic architecture for photosynthesis and water use efficiency in soybean (*Glycine max* L. Merr). *Front. Plant Sci.* **2019**, *10*, 680. [CrossRef]
28. Hufstetler, E.V.; Boerma, H.R.; Carter, T.E.; Earl, H.J. Genotypic variation for three physiological traits affecting drought tolerance in soybean. *Crop Sci.* **2007**, *47*, 25–35. [CrossRef]
29. Long, S.P.; Zhu, X.G.; Naidu, S.L.; Ort, D.R. Can improvement in photosynthesis increase crop yields? *Plant Cell Environ.* **2006**, *29*, 315–330. [CrossRef]

30. Wang, W.; Jiang, W.; Liu, J.; Li, Y.; Gai, J.; Li, Y. Genome-wide characterization of the aldehyde dehydrogenase gene superfamily in soybean and its potential role in drought stress response. *BMC Genom.* **2017**, *18*, 518. [CrossRef]
31. Zhang, Y.; Liu, Z.; Wang, X.; Li, Y.; Li, Y.; Gou, Z.; Zhao, X.; Hong, H.; Ren, H.; Qi, X.; et al. Identification of genes for drought resistance and prediction of gene candidates in soybean seedlings based on linkage and association mapping. *Crop J.* **2022**, *10*, 830–839. [CrossRef]
32. Chen, L.; Fang, Y.; Li, X.; Zeng, K.; Chen, H.; Zhang, H.; Yang, H.; Cao, D.; Hao, Q.; Yuan, S.; et al. Identification of soybean drought-tolerant genotypes and loci correlated with agronomic traits contributes new candidate genes for breeding. *Plant Mol. Biol.* **2020**, *102*, 109–122. [CrossRef] [PubMed]
33. Yan, C.; Song, S.; Wang, W.; Wang, C.; Li, H.; Wang, F.; Li, S.; Sun, X. Screening diverse soybean genotypes for drought tolerance by membership function value based on multiple traits and drought-tolerant coefficient of yield. *BMC Plant Biol.* **2020**, *20*, 321. [CrossRef] [PubMed]
34. Hwang, S.; King, C.A.; Ray, J.D.; Cregan, P.B.; Chen, P.; Carter, T.E.; Li, Z.; Abdel-Haleem, H.; Matson, K.W.; Schapaugh, W., Jr.; et al. Confirmation of delayed canopy wilting QTLs from multiple soybean mapping populations. *Theor. Appl. Genet.* **2015**, *128*, 2047–2065. [CrossRef] [PubMed]
35. Kaler, A.S.; Ray, J.D.; Schapaugh, W.T.; King, C.A.; Purcell, L.C. Genome-wide association mapping of canopy wilting in diverse soybean genotypes. *Theor. Appl. Genet.* **2017**, *130*, 2203–2217. [CrossRef] [PubMed]
36. Zhang, Y.; Zhang, H.; Fu, J.; Du, Y.Y.; Qu, J.; Song, Y.; Wang, P.W. The GmXTH1 gene improves drought stress resistance of soybean seedlings. *Mol. Breed.* **2022**, *42*, 1–18. [CrossRef]
37. Ma, J.K.; Liu, X.R.; Zhou, W.; Liu, X. Application of collecting trough method on identification of soybean drought resistance. *Soybean Sci.* **2016**, *35*, 766–771.
38. Kim, J.S.; Park, H.M.; Chae, S.A.; Lee, T.H.; Hwang, D.J.; Oh, S.D.; Park, J.; Song, D.; Pan, C.; Choi, D.; et al. The pepper *MSRB2* gene confers drought tolerance in rice through the protection of chloroplast-targeted genes. *PLoS ONE* **2014**, *9*, e90588. [CrossRef]
39. Chen, L.; Lee, J.H.; Weber, H.; Tohge, T.; Witt, S.; Roje, S.; Hellmann, H. Arabidopsis BPM proteins function as substrate adaptors to a cullin3-based E3 ligase to affect fatty acid metabolism in plants. *Plant Cell* **2013**, *25*, 2253–2264. [CrossRef] [PubMed]
40. Puranik, S.; Sahu, P.P.; Srivastava, P.S.; Prasad, M. NAC proteins: Regulation and role in stress tolerance. *Trends Plant Sci.* **2012**, *17*, 369–381. [CrossRef]
41. Olsen, A.N.; Ernst, H.A.; Leggio, L.L.; Skriver, K. DNA-binding specificity and molecular functions of NAC transcription factors. *Plant Sci.* **2005**, *169*, 785–797. [CrossRef]
42. Wu, H.; Fu, B.; Sun, P.P.; Xiao, C.; Liu, J.H. A NAC transcription factor represses putrescine biosynthesis and affects drought tolerance. *Plant Physiol.* **2016**, *172*, 1532–1547. [CrossRef] [PubMed]
43. Xu, Z.Y.; Kim, S.Y.; Hyeon, D.Y.; Kim, D.H.; Dong, T.; Park, Y.; Hwang, I. The Arabidopsis NAC transcription factor ANAC096 cooperates with bZIP-type transcription factors in dehydration and osmotic stress responses. *Plant Cell* **2013**, *25*, 4708–4724. [CrossRef] [PubMed]
44. Nguyen, K.H.; Mostofa, M.G.; Li, W.; Van Ha, C.; Watanabe, Y.; Le, D.T.; Tran, L.S.P. The soybean transcription factor GmNAC085 enhances drought tolerance in Arabidopsis. *Environ. Exp. Bot.* **2018**, *151*, 12–20. [CrossRef]
45. Ding, Y.; Liu, N.; Virlouvet, L.; Riethoven, J.J.; Fromm, M.; Avramova, Z. Four distinct types of dehydration stress memory genes in *Arabidopsis thaliana*. *BMC Plant Biol.* **2013**, *13*, 229. [CrossRef]
46. Shang, J.; Li, N.; Xu, Y.; Ma, S.; Wang, J. Construction of a high-density genetic map for watermelon (*Citrullus lanatus* L.) based on large-scale SNP discovery by specific length amplified fragment sequencing (SLAF-seq). *Sci. Hortic.* **2016**, *203*, 38–46. [CrossRef]
47. Xia, Z.; Tsubokura, Y.; Hoshi, M.; Hanawa, M.; Yano, C.; Okamura, K.; Ahmed, T.A.; Anai, T.; Watanabe, S.; Hayashi, M.; et al. An integrated high-density linkage map of soybean with RFLP, SSR, STS, and AFLP markers using a single F_2 population. *DNA Res.* **2007**, *14*, 257–269. [CrossRef]
48. Xu, Y.; Li, P.; Yang, Z.; Xu, C. Genetic mapping of quantitative trait loci in crops. *Crop J.* **2017**, *5*, 175–184. [CrossRef]
49. Colasuonno, P.; Marcotuli, I.; Gadaleta, A.; Soriano, J.M. From genetic maps to QTL cloning: An overview for durum wheat. *Plants* **2021**, *10*, 315. [CrossRef]
50. Schmutz, J.; Cannon, S.B.; Schlueter, J.; Ma, J.; Mitros, T.; Nelson, W.; Hyten, D.L.; Song, Q.; Thelen, J.; Cheng, J.; et al. Genome sequence of the palaeopolyploid soybean. *Nature* **2010**, *463*, 178–183. [CrossRef]
51. Liu, Y.; Du, H.; Li, P.; Shen, Y.; Peng, H.; Liu, S.; Zhou, G.; Zhang, H.; Liu, Z.; Shi, M.; et al. Pan-genome of wild and cultivated soybeans. *Cell* **2020**, *182*, 162–176. [CrossRef]
52. Abdel-Haleem, H.; Carter, T.E.; Purcell, L.C.; King, C.A.; Ries, L.L.; Chen, P.; Boerma, H.R. Mapping of quantitative trait loci for canopy-wilting trait in soybean (*Glycine max* L. Merr). *Theor. Appl. Genet.* **2012**, *125*, 837–846. [CrossRef] [PubMed]
53. Chen, Q.S.; Zhang, Z.C.; Liu, C.Y.; Xin, D.W.; Qiu, H.M.; Shan, D.P.; Hu, G.H. QTL analysis of major agronomic traits in soybean. *Agric. Sci. China* **2007**, *6*, 399–405. [CrossRef]
54. Oya, T.; Nepomuceno, A.L.; Neumaier, N.; Farias, J.R.B.; Tobita, S.; Ito, O. Drought tolerance characteristics of Brazilian soybean cultivars—evaluation and characterization of drought tolerance of various Brazilian soybean cultivars in the field. *Plant Prod. Sci.* **2004**, *7*, 129–137. [CrossRef]
55. Fahim, A.M.; Liu, F.; He, J.; Wang, W.; Xing, G.; Gai, J. Evolutionary QTL-allele changes in main stem node number among geographic and seasonal subpopulations of Chinese cultivated soybeans. *Mol. Genet. Genom.* **2021**, *296*, 313–330. [CrossRef]

56. Fahim, A.M.; Pan, L.; Li, C.; He, J.; Xing, G.; Wang, W.; Zhang, F.; Li, N.; Gai, J. QTL-allele system of main stem node number in recombinant inbred lines of soybean (*Glycine max*) using association versus linkage mapping. *Plant Breed.* **2021**, *140*, 870–883. [CrossRef]
57. Li, W.X.; Wang, P.; Zhao, H.; Sun, X.; Yang, T.; Li, H.; Hou, Y.; Liu, C.; Siyal, M.; Raja Veesar, R.; et al. QTL for main stem node number and its response to plant densities in 144 soybean FW-RILs. *Front. Plant Sci.* **2021**, *12*, 666796. [CrossRef]
58. Fu, M.; Wang, Y.; Ren, H.; Du, W.; Yang, X.; Wang, D.; Cheng, Y.; Zhao, J.; Gai, J. Exploring the QTL–allele constitution of main stem node number and its differentiation among maturity groups in a Northeast China soybean population. *Crop Sci.* **2020**, *60*, 1223–1238. [CrossRef]
59. Yan, H.; Wang, H.; Cheng, H.; Hu, Z.; Chu, S.; Zhang, G.; Yu, D. Detection and fine-mapping of SC7 resistance genes via linkage and association analysis in soybean. *J. Integr. Plant Biol.* **2015**, *57*, 722–729. [CrossRef]
60. Ren, H.; Han, J.; Wang, X.; Zhang, B.; Yu, L.; Gao, H.; Hong, H.; Sun, R.; Tian, Y.; Qi, X.; et al. QTL mapping of drought tolerance traits in soybean with SLAF sequencing. *Crop J.* **2020**, *8*, 977–989. [CrossRef]
61. Škiljaica, A.; Lechner, E.; Jagić, M.; Majsec, K.; Malenica, N.; Genschik, P.; Bauer, N. The protein turnover of Arabidopsis BPM1 is involved in regulation of flowering time and abiotic stress response. *Plant Mol. Biol.* **2020**, *102*, 359–372. [CrossRef]
62. Weber, H.; Hellmann, H. *Arabidopsis thaliana* BTB/POZ-MATH proteins interact with members of the ERF/AP2 transcription factor family. *FEBS J.* **2009**, *276*, 6624–6635. [CrossRef] [PubMed]
63. Trujillo, M. News from the PUB: Plant U-box type E3 ubiquitin ligases. *J. Exp. Bot.* **2018**, *69*, 371–384. [CrossRef] [PubMed]
64. Ryu, M.Y.; Cho, S.K.; Hong, Y.; Kim, J.; Kim, J.H.; Kim, G.M.; Yang, S.W. Classification of barley U-box E3 ligases and their expression patterns in response to drought and pathogen stresses. *BMC Genom.* **2019**, *20*, 326. [CrossRef] [PubMed]
65. Cho, S.K.; Ryu, M.Y.; Song, C.; Kwak, J.M.; Kim, W.T. Arabidopsis PUB22 and PUB23 are homologous U-Box E3 ubiquitin ligases that play combinatory roles in response to drought stress. *Plant Cell* **2008**, *20*, 1899–1914. [CrossRef]
66. Wang, N.; Liu, Y.; Cai, Y.; Tang, J.; Li, Y.; Gai, J. The soybean U-box gene GmPUB6 regulates drought tolerance in *Arabidopsis thaliana*. *Plant Physiol. Biochem.* **2020**, *155*, 284–296. [CrossRef]
67. Zhao, J.; Zhao, L.; Zhang, M.; Zafar, S.A.; Fang, J.; Li, M.; Li, X. Arabidopsis E3 ubiquitin ligases PUB22 and PUB23 negatively regulate drought tolerance by targeting ABA receptor PYL9 for degradation. *Int. J. Mol. Sci.* **2017**, *18*, 1841. [CrossRef]
68. Olsen, A.N.; Ernst, H.A.; Leggio, L.L.; Skriver, K. NAC transcription factors: Structurally distinct, functionally diverse. *Trends Plant Sci.* **2005**, *10*, 79–87. [CrossRef]
69. Larbi, A.; Mekliche, A. Relative water content (RWC) and leaf senescence as screening tools for drought tolerance in wheat. *Options Méditerr. Sér. A Sémin. Méditerr.* **2004**, *60*, 193–196.
70. Liu, F.H.; Guo, Y.; Gu, D.M.; Gang, X.; Zheng, C.; Chen, S.Y. Salt tolerance of transgenic plants with BADH cDNA. *Yi Chuan Xue Bao* **1997**, *24*, 54–58.
71. Zhou, R.; Chen, H.F.; Wang, X.Z.; Zhang, X.; Shan, Z.; Wu, X.; Qiu, D.Z.; Wu, J. QTL analysis of yield, yield components, and lodging in soybean. *Acta Agron. Sin.* **2009**, *35*, 821–830. [CrossRef]
72. Danecek, P.; Bonfield, J.K.; Liddle, J.; Marshall, J.; Ohan, V.; Pollard, M.O.; Whitwham, A.; Keane, T.; McCarthy, S.A.; Li, H. Twelve years of SAMtools and BCFtools. *Gigascience* **2021**, *10*, 008. [CrossRef] [PubMed]
73. Li, H.; Handsaker, B.; Wysoker, A.; Fennell, T.; Ruan, J.; Homer, N.; Marth, G.; Abecasis, G.; Durbin, R. The sequence alignment/map (SAM) format and SAMtools. *Bioinformatics* **2009**, *25*, 2078–2079. [CrossRef] [PubMed]
74. McKenna, A.; Hanna, M.; Banks, E.; Sivachenko, A.; Cibulskis, K.; Kernytsky, A.; Garimella, K.; Altshuler, D.; Gabriel, S.; Daly, M.; et al. The Genome Analysis Toolkit: A MapReduce framework for analyzing next generation DNA sequencing data. *Genome Res.* **2010**, *20*, 1297–1303. [CrossRef] [PubMed]
75. Huang, X.; Feng, Q.; Qian, Q.; Wang, L.; Wang, A.; Guan, J.; Fan, D.; Weng, Q.; Huang, T.; Dong, G.; et al. High-throughput genotyping by whole-genome resequencing. *Genome Res.* **2009**, *19*, 1068–1076. [CrossRef]
76. Liu, D.; Ma, C.; Hong, W.; Huang, L.; Liu, M.; Liu, H.; Zeng, H.; Deng, D.; Xin, H.; Song, J.; et al. Construction and analysis of high-density linkage map using high-throughput sequencing data. *PLoS ONE* **2014**, *9*, e98855. [CrossRef]
77. Kosambi, D. The estimation of map distances from recombination values. *Ann. Eugen.* **1943**, *12*, 172–175. [CrossRef]
78. Broman, K.W.; Wu, H.; Sen, S.; Churchill, G.A. R/qtl: QTL mapping in experimental crosses. *Bioinformatics* **2013**, *19*, 889–890. [CrossRef]

International Journal of
Molecular Sciences

MDPI

Article

Development and Validation of SNP and InDel Markers for Pod-Shattering Tolerance in Soybean

Jeong-Hyun Seo [1,†], Sanjeev Kumar Dhungana [1,†], Beom-Kyu Kang [1], In-Youl Baek [1], Jung-Sook Sung [1], Jee-Yeon Ko [1], Chan-Sik Jung [1], Ki-Seung Kim [2] and Tae-Hwan Jun [3,4,*]

1 Department of Southern Area Crop Science, National Institute of Crop Science, Rural Development Administration, Miryang 50424, Korea; next0501@korea.kr (J.-H.S.); sanjeev@korea.kr (S.K.D.); hellobk01@korea.kr (B.-K.K.); baekiy@korea.kr (I.-Y.B.); sjs31@korea.kr (J.-S.S.); kjeeyeon@korea.kr (J.-Y.K.); jung100@korea.kr (C.-S.J.)
2 Innovative Technology Department, FarmHannong, Ltd., Nonsan 33010, Korea; leehan26@snu.ac.kr
3 Department of Plant Bioscience, Pusan National University, Miryang 50463, Korea
4 Life and Industry Convergence Research Institute, Pusan National University, Miryang 50463, Korea
* Correspondence: thjun76@pusan.ac.kr
† These authors contributed equally to this work.

Abstract: Pod-shattering causes a significant yield loss in many soybean cultivars. Shattering-tolerant cultivars provide the most effective approach to minimizing this loss. We developed molecular markers for pod-shattering and validated them in soybeans with diverse genetic backgrounds. The genes *Glyma.16g141200*, *Glyma.16g141500*, and *Glyma.16g076600*, identified in our previous study by quantitative trait locus (QTL) mapping and whole-genome resequencing, were selected for marker development. The whole-genome resequencing of three parental lines (one shattering-tolerant and two shattering-susceptible) identified single nucleotide polymorphism (SNP) and/or insertion/deletion (InDel) regions within or near the selected genes. Two SNPs and one InDel were converted to Kompetitive Allele-Specific PCR (KASP) and InDel markers, respectively. The accuracy of the markers was examined in the two recombinant inbred line populations used for the QTL mapping, as well as the 120 varieties and elite lines, through allelic discrimination and phenotyping by the oven-drying method. Both types of markers successfully discriminated the pod shattering-tolerant and shattering-susceptible genotypes. The prediction accuracy, which was as high as 90.9% for the RILs and was 100% for the varieties and elite lines, also supported the accuracy and usefulness of these markers. Thus, the markers can be used effectively for genetic and genomic studies and the marker-assisted selection for pod-shattering tolerance in soybean.

Keywords: Kompetitive Allele-Specific PCR; soybean; pod shattering tolerance; single nucleotide polymorphism; insertion/deletion; candidate gene; molecular marker

Citation: Seo, J.-H.; Dhungana, S.K.; Kang, B.-K.; Baek, I.-Y.; Sung, J.-S.; Ko, J.-Y.; Jung, C.-S.; Kim, K.-S.; Jun, T.-H. Development and Validation of SNP and InDel Markers for Pod-Shattering Tolerance in Soybean. *Int. J. Mol. Sci.* **2022**, *23*, 2382. https://doi.org/10.3390/ijms23042382

Academic Editors: Andrés J. Cortés and Hai Du

Received: 10 January 2022
Accepted: 18 February 2022
Published: 21 February 2022

1. Introduction

Soybean is an economically important crop worldwide. The productivity of soybean is affected by various biotic and abiotic factors, including pod-shattering. The characterization of soybean productive units and the use of growth simulation models can help to identify some of the related factors of soybean productivity [1,2]. Pod-shattering is the opening of the mature pod along the dorsal or ventral suture with the dispersal of the seeds [3]. While it is essential for seed dispersal in wild species, in domesticated crops it can cause significant yield loss [4]. These losses vary from negligible to significant, depending on the harvest delay, the environmental conditions, and/or the tolerance of the variety [5–7].

Linkage mapping, commonly known as the quantitative trait locus (QTL) analysis [8], as well as genome-wide association studies (GWAS) [9] are mapping techniques that locate genomic regions associated with the quantitative variations for target traits. In soybean, the major genomic regions associated with the pod-shattering tolerance are on chromosome

16. These regions were identified through linkage mapping using diverse molecular markers, including restriction fragment length polymorphism markers [10], simple-sequence repeat (SSR) markers [11–13], and high-density single nucleotide polymorphism (SNP) markers [14]. A few minor QTLs have been mapped on chromosomes 2, 5, 10, 14, and 19 by linkage mapping [10,11]; on chromosomes 1, 4, 6, 8, 9, 11, 17, 18, and 20 through GWAS [15]; and on chromosomes 1, 5, 8, and 14 through linkage mapping using specific locus-amplified fragment sequencing [16]. We previously mapped major and minor QTLs on chromosome 16 and reported candidate genes for the pod-shattering tolerance in soybean [17].

Although pod-shattering is more than 90% heritable [18], pod-shattering in the field is greatly affected by diverse environmental factors, including humidity and temperature. To address the problems inherent in field evaluations, most pod-shattering assessments in soybeans are made using the oven-dry method [15,19,20]. However, the conventional method for evaluating pod-shattering (i.e., keeping the mature pods at room temperature for a few days to equalize the moisture content among samples) is time-consuming and labor-intensive. Moreover, this technique is applicable only for mature pods (R8 stage) at later growth stages. Therefore, an efficient and accurate selection technique, such as molecular markers that can be applied at all growth stages, is needed to determine the pod-shattering tolerance versus susceptibility in soybean breeding programs.

Among various molecular markers, Kompetitive Allele-Specific PCR (KASP) and insertion/deletion (InDel) markers have been used widely for many crops because they are simple, reproducible, stable, accurate, fast, and are low-cost [21–24]. KASP assays for the marker-assisted selection (MAS) of the egusi trait in the watermelon [24], the flowering gene in lentils [22], and the white tip nematode (*Aphelenchoides besseyi*) in rice [25] have been developed and applied to diverse genetic backgrounds. In soybean, KASP assays have been employed for a few traits, including soybean cyst nematode resistance [26,27], frogeye leaf spot resistance [28], and *Fusarium graminearum* resistance [29]. However, there are few reports of selection markers for the pod-shattering tolerance in soybean [30,31]. We designed SNP and/or InDel markers and validated them in populations with diverse genetic backgrounds. These markers could be used by MAS for initial germplasm selection and for developing soybean cultivars with a pod-shattering tolerance.

2. Results

2.1. Development of KASP Markers

For KASP markers, we selected the SNPs or InDel regions in the candidate genes of the major QTL [17] by comparing the DNA sequences of the tolerant parent Daewonkong (DW) and the susceptible parents Tawonkong (TW) and Saeolkong (SO). Of the five candidate genes from within the major QTL, three genes had five SNPs. For the two genes *Glyma.16g141200* and *Glyma.16g141500* (Supplementary Figures S1 and S2) on chromosome 16, these SNPs had no other polymorphic variations within 100 bp upstream or downstream.

The first marker, *KASP-PS-1* for *Glyma.16g141200*, had a single bp deletion (TC > T) at 29,916,524 bp on chromosome 16. The tolerant parent DW and the reference genome (reference, version) had a "C" allele in the target region, whereas the susceptible parents TW and SO had the deletion (Table 1, Supplementary Figure S1). The second marker, *KASP-PS-2* for *Glyma.16g141500*, had a SNP variant (T > G) at 29,966,815 bp on chromosome 16. The tolerant parent DW and the reference genome had the "T" allele in the target region, whereas the susceptible parents had the "G" allele. We developed the KASP marker with allele-specific primers and common primers in each SNP (Table 1, Supplementary Figure S2).

Table 1. Summary of SNP information on the candidate genes of the major quantitative trait locus [17].

Gene Model	SNP Position (bp)	Polymorphic Site	SNP Information	Amino Acid Change
Glyma.16g141200	29,916,524	3′ UTR	Deletion (TC > T)	-
Glyma.16g141500	29,964,216	5′ UTR	A > C	-
Glyma.16g141500	29,966,815	5′ UTR	T > G	-
Glyma.16g141600	29,970,894	Coding region	T > C Nonsynonymous	(Asp > Gly)
Glyma.16g141600	29,970,957	Coding region	T > C Nonsynonymous	(Asn > Ser)

2.2. Development of InDel Markers

Previously, we reported a novel candidate gene *Glyma.16g076600* related to abscisic acid (ABA) catabolism [17]. There were four SNPs, one insertion, and one deletion in the target region, causing amino acid changes (Table 2). This gene had a translational stop codon (nonsense) mutation, which would stop protein expression from the gene. We designated the InDel marker for the 18-bp insertion region of the candidate gene *Glyma.16g076600* (Supplementary Figure S3). DW had the additional sequence differences (the DW-specific allele) compared to the reference genome TW and SO (Table 2).

Table 2. Summary of SNP/InDel information on the candidate gene of the minor quantitative trait locus [17].

Gene Model	Position (bp)	Polymorphism Site	Codon Change	Amino Acid Change
Glyma.16g076600	7,775,892	Coding region	C > T Nonsynonymous	(Glu > Lys)
	7,775,945	Coding region	T > A Nonsynonymous	(Lys > Met)
	7,775,948	Coding region	A > G Nonsynonymous	(Ile > Thr)
	7,775,970	Coding region	Insertion (18 bp)	Stop gained and disruptive inframe insertion
	7,776,045	Coding region	C > T Nonsynonymous	(Met > Ile)
	7,777,575	Coding region	Deletion (3 bp) Nonsynonymous	Inframe deletion (Asn > -)

2.3. Validation of KASP and InDel Markers in RIL Populations

To validate the KASP markers in RILs, we genotyped the lines with two KASP markers and determined the pod-shattering ratio in 154 RILs from the DW × TW (DT population) and 153 RILs from the DW × SO (DS population). In the scatter plot, the RILs, shown by the blue spot, had a tolerance allele from DW, while the RILs indicated by the red spot had a susceptibility allele (Figures 1 and 2). The results showed that the pod-shattering-tolerant parent DW and the susceptible parents TW and SO were successfully distinguished by both markers. Similarly, the tolerant lines and susceptible lines were clearly distinguished according to their genotypes. Among the RILs, 56 lines had the tolerance allele and 88 lines had the susceptibility allele, and the others were undetermined or had a heterozygous allele in the DT population. Similarly, 72 lines had the tolerance allele, 78 lines had the susceptibility allele, and the others had a heterozygous allele in the DS population (Figures 1 and 2).

In the DT population, the lines with the tolerance genotype had 1.6%, 5.1%, and 9.0% pod-shattering ratios for 24, 48, and 72 h of oven-drying, respectively. Those with the susceptible genotype had high ratios of 25.9%, 55.0%, and 75.1% for the same periods of drying, respectively. Similarly, in the DS population, 4.9%, 12.3%, and 18.3% of the pods of tolerant lines were shattered after 24, 48, and 72 h, respectively. Shattered pods of susceptible lines were 44.0%, 80.4%, and 86.3% for the same periods of drying, respectively (Table 3).

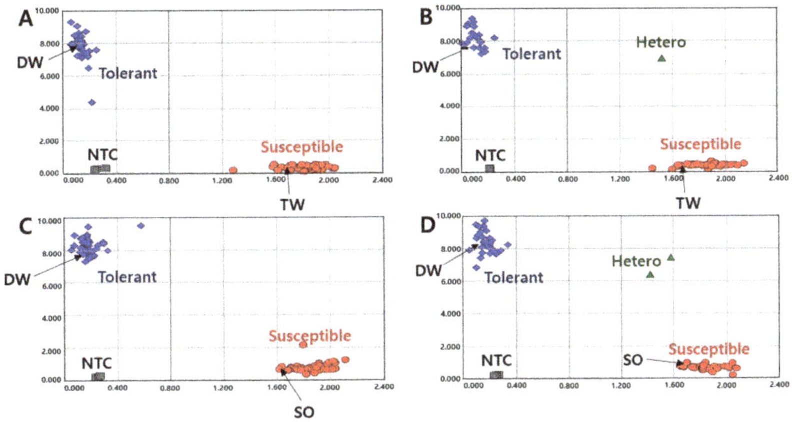

Figure 1. Allelic discrimination of RIL populations using the *KASP-PS-1* marker. The upper two ((**A**): 91 RILs and 2 parents, (**B**): 63 RILs and 2 parents because up to 96 genotypes were accommodated in a single experimental set-up) figures for 154 RILs of DT combination (DW × TW population) and the lower two ((**C**): 91 RILs and 2 parents, (**D**): 62 RILs and 2 parents) figures for 153 RILs of DS combination (DW × SO population). NTC: No template control, DW: Daewonkong, TW: Tawonkong, and SO: Saeolkong.

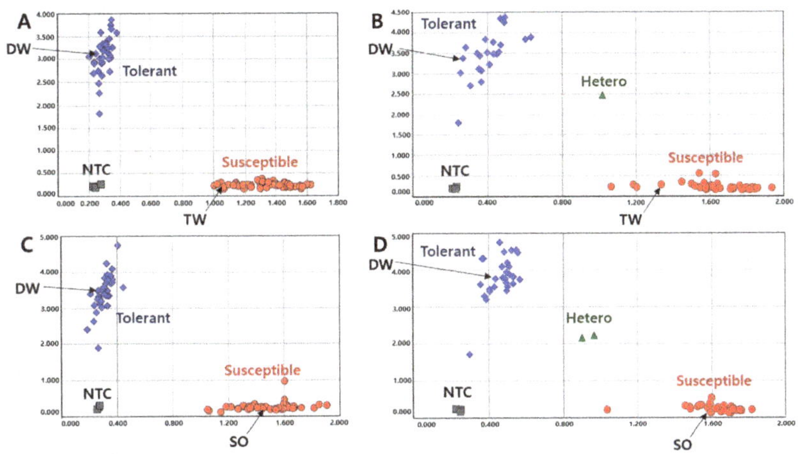

Figure 2. Allelic discrimination of RIL populations using the *KASP-PS-2* marker. The upper two figures ((**A**): 91 RILs and 2 parents, (**B**): 63 RILs and 2 parents because up to 96 genotypes were accommodated in a single experimental set-up) for 154 RILs of DT combination (DW × TW population) and the lower two figures ((**C**): 91 RILs and 2 parents, (**D**): 62 RILs and 2 parents) for 153 RILs of DS combination (DW × SO population). NTC: No template control, DW: Daewonkong, TW: Tawonkong, and SO: Saeolkong.

As expected for the InDel marker in the RILs, the tolerant parent and susceptible parents gave different product sizes (DW, 144 bp; TW and SO, 126 bp) for the InDel marker (Supplementary Figure S4). Among the RILs, 69 lines had the tolerance allele, 74 lines had the susceptibility allele, and the others had undetermined or heterozygous alleles in the DT population. Similarly, 74 lines had the tolerance allele, 76 lines had the susceptibility allele, and the others had undetermined or heterozygous alleles in the DS population (Supplementary Figure S4).

Table 3. Phenotypic (pod-shattering ratio) variations in the recombinant inbred line populations according to the genotypes.

Marker Type	Population	Genotype	Pod Shattering Ratio (%)		
			24 h	48 h	72 h
KASP	DT	P1	1.6a	5.1a	9.0a
		P2	25.9b	55.0b	75.1b
	DS	P1	4.9a	12.3a	18.3a
		P2	44.0b	80.4b	86.3b
InDel	DT	P1	7.4a	20.3a	27.9a
		P2	24.7b	49.5b	69.0b
	DS	P1	7.6a	19.7a	26.4a
		P2	43.0b	75.9b	81.2b

DT: Daewonkong × Tawonkong, DS: Daewonkong × Saeolkong. 'P1' indicates that the target allele is the same as the tolerant parent 'Daewonkong', 'P2' indicates that the target allele is the same as susceptible parents 'Tawonkong' and 'Saeolkong'. Pod-shattering ratio is the mean value at the different drying periods (24, 48, and 72 h). Different letters within the same column indicate significant differences ($p < 0.05$).

In the DT population, 7.4%, 20.3%, and 27.9% of the pod-shattering ratios were found for the genotypes with the DW allele after 24, 48, and 72 h of drying, respectively. However, the lines with the TW or SO allele produced 24.7%, 49.5%, and 69.0% of the pod-shattering ratios for the same periods of drying, respectively (Table 3). Likewise, in the DS population, 7.6%, 19.7%, and 26.4% of the pod-shattering ratios were observed in the tolerant lines, whereas the susceptible lines gave 43.0%, 75.9%, and 81.2% of the pod-shattering ratios after 24, 48, and 72 h of oven-drying, respectively (Table 3).

2.4. Validation of KASP and InDel Markers in Diverse Varieties and Elite Lines

A total of 120 varieties and elite lines with diverse genetic backgrounds were tested to validate the KASP (Figure 3) and InDel (Supplementary Figure S5) markers. Although the efficiency of the InDel marker was lower than that of KASP, both markers successfully distinguished the pod-shattering-tolerant and -susceptible genotypes. In the KASP marker analysis, 1.4%, 3.6%, and 5.7% of the pod-shattering ratios were seen in the tolerant genotypes after 24, 48, and 72 h of oven-drying, respectively, while 44.6%, 76.2%, and 85.1% of the pod-shattering ratios were observed in the susceptible genotypes, respectively (Table 4). In the case of the InDel marker, the varieties and elite lines with the tolerant genotype showed 4.7%, 11.6%, and 14.4% of the pod-shattering ratios after 24, 48, and 72 h of oven-drying, respectively. However, the lines representing the susceptible genotype had 15.0%, 26.1%, and 30.7% pod-shattering ratios after 24, 48, and 72 h of oven-drying, respectively (Table 4).

Table 4. Phenotypic (pod-shattering ratio) variations in the 120 varieties and elite lines according to the genotypes.

Marker Type	Genotype	Pod-Shattering Ratio (%)		
		24 h	48 h	72 h
KASP	P1	1.4a	3.6a	5.7a
	P2	44.6b	76.2b	85.1b
InDel	P1	4.7a	11.6a	14.4a
	P2	15.0b	26.1b	30.7b

'P1' indicates that the target allele is the same as that of the tolerant parent 'Daewonkong', 'P2' indicates that the target allele is the same as that of the susceptible parents 'Tawonkong' and 'Saeolkong'. Pod-shattering ratio is the mean value at the different drying periods (24, 48, and 72 h). Different letters followed by the values within the same column indicate significant differences ($p < 0.05$).

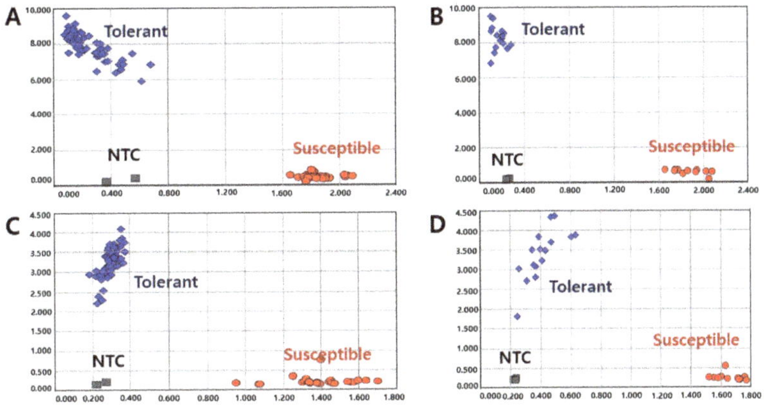

Figure 3. Allelic discrimination of 120 varieties and elite lines using *KASP-PS-1* ((**A**): 93 and (**B**): 27) and *KASP-PS-2* ((**C**): 93 and (**D**): 27) markers. NTC: No template control.

2.5. Association between Allelic Combination and Phenotypes

To determine the associations between the allelic combinations obtained with KASP and InDel markers (A, tolerant; B, susceptible) and phenotypic variations, we divided 307 RILs from the two populations into four groups; i.e., the tolerant genotypes in both markers (AA); the tolerant genotypes with the KASP marker, but also the susceptible genotypes with the InDel marker (AB); the susceptible genotypes with the KASP marker, but also the tolerant genotypes with the InDel marker (BA); and the susceptible genotypes with both markers (BB). As a result, we discovered significant differences in the pod-shattering ratio depending on the allelic combinations of RILs. The RILs with the allele type AA had 2.4%, 7.2%, and 12.2% of the pod-shattering ratios after 24, 48, and 72 h of oven-drying, respectively, while allele type AB had 10.0%, 20.6%, and 26.1% of the pod-shattering ratios after 24, 48, and 72 h of oven-drying, respectively. Allele type BA had 24.0%, 61.3%, and 75.1% of the pod-shattering ratios after 24, 48, and 72 h of oven-drying, respectively, and allele type BB had 37.2%, 68.5%, and 81.7% of the pod-shattering ratios after 24, 48, and 72 h of oven-drying, respectively (Figure 4A). The varieties and elite lines also showed similar patterns of pod-shattering after 24, 48, and 72 h of oven-drying to that of RILs (AA, 0.2%, 0.5%, and 1.3%; AB, 1.8%, 4.6%, and 7.2%; BA, 24.3%, 60.9%, and 71.8%; and BB, 48.5%, 79.1%, and 87.7%) (Figure 4B).

The prediction accuracy of the KASP marker was very high, with 81.7% to 88.2% in the RIL populations and 91.3% to 95.5% in the varieties and elite lines. Considering both types of markers in the strains with tolerant genotypes, for the RIL populations, the prediction accuracy was 83.4% to 90.9%, and for the varieties and elite lines, the prediction accuracy was 100%, which was higher than the KASP marker alone (Table 5).

2.6. Haplotype Analysis of Candidate Genes

The results of the haplotype analysis for *Glyma.16g076600*, *Glyma.16g141200*, and *Glyma.16g141500* are shown in Figure 5. For *Glyma.16g076600*, there were four SNPs, one insertion (18 bp) at 7,775,970 bp that was used to develop the InDel marker, and one deletion in the target region (Table 2, Figure 5, and Supplementary Figure S3). *Glyma.16g141200* had a single bp deletion (TC > T) at 29,916,524 bp. *Glyma.16g141500* had an SNP variant (T > G) at 29,966,815 bp that was considered while developing KASP markers (Table 1, Figure 5, and Supplementary Figures S2 and S3). The haplotype analysis of the 120 varieties and elite lines showed that the InDel and SNPs were divided into four haplotypes (Haps), with the maximum pod-shattering variations of 48.3, 78.6, and 86.4 at 24, 48, and 72 h of oven-drying, respectively, between Haps 1 and 4 (Figure 5B).

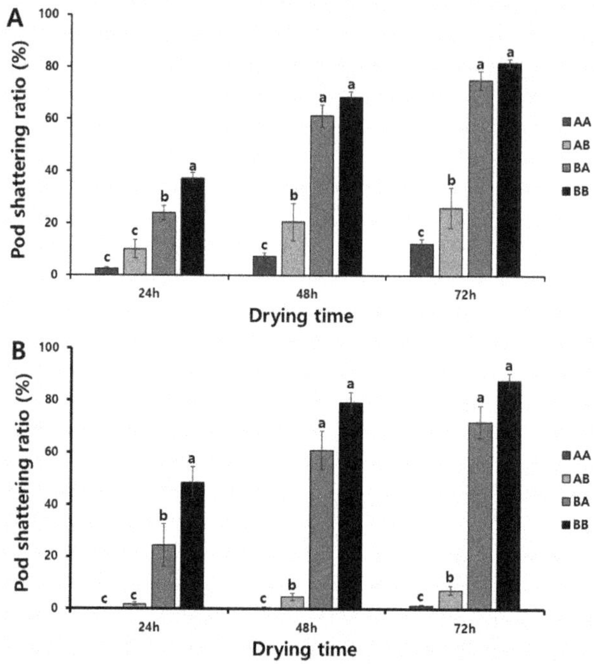

Figure 4. Phenotypic variation for pod-shattering ratio according to their genotypes: (**A**) two RIL populations and (**B**) varieties and elite lines. AA: tolerant genotypes in both markers; AB: tolerant genotypes in KASP marker but susceptible genotype in InDel marker; BA: susceptible genotypes in KASP marker but tolerant genotypes in InDel marker; and BB: susceptible in both markers. Standard errors are used to depict error bars. The letters (a–c) above bar diagrams represent significant differences at $p < 0.05$ by Duncan's multiple range test.

Table 5. The prediction accuracy between the genotypes and phenotypes of recombinant inbred lines (RILs), varieties, and elite lines for three drying periods.

Genotypes	RIL Populations						Varieties and Elite Lines					
	24 h		48 h		72 h		24 h		48 h		72 h	
	Tol	Sus	Tol	Sus	Tol	Sus	Tol	Sus	Tol	Sus	Tol	Sus
X	116	12	101	27	82	46	84	4	80	8	73	12
Y	31	135	4	162	1	165	4	27	0	31	0	31
Accuracy (%)	86.0		88.2		81.7		91.3		95.5		92.9	
AA	101	7	90	18	73	35	22	0	22	0	22	0
BB	22	109	2	129	1	130	0	26	0	26	0	26
Accuracy (%)	88.4		90.9		83.4		100		100		100	

Genotypes, X: tolerant genotype for KASP marker; Y: susceptible genotype for KASP marker; AA: tolerant genotype in both types of markers; and BB: susceptible genotype in both types of markers. Tol: number of tolerant lines (shattered pods $\leq$ 10%) and Sus: number of susceptible lines (shattered pods > 10%). Accuracy: prediction accuracy is the possibility that tolerant genotypes showed a tolerant phenotype and susceptible genotypes showed a susceptible phenotype.

A

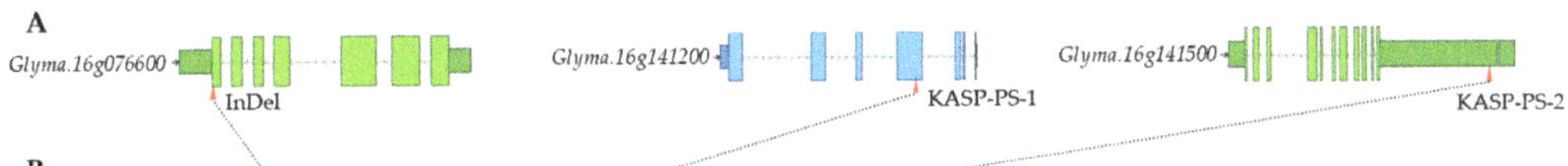

B

Gene	*Glyma.16g076600*	*Glyma.16g141200*	*Glyma.16g141500*	Number of genotypes	Pod shattering ratio		
Position	7775970	29916524	29966815				
Allele	-/CCGAGTGATATTTAATTC	C/-	T/G		24 h	48 h	72 h
Reference	-	C	T				
Hap1	CCGAGTGATATTTAATTC	C	T	23	0.2^{c}	0.5^{d}	1.3^{d}
Hap2	-	C	T	65	1.8^{c}	4.6^{c}	7.2^{c}
Hap3	CCGAGTGATATTTAATTC	-	G	5	24.3^{b}	60.9^{b}	71.8^{b}
Hap4	-	-	G	27	48.5^{a}	79.1^{a}	87.7^{a}

C

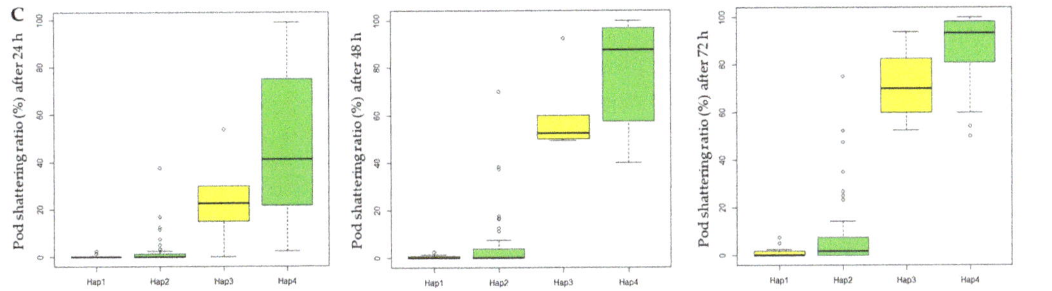

Figure 5. Haplotype analysis of *Glyma.16g076600*, *Glyma.16g141200*, and *Glyma.16g141500*. (**A**) Schematic representation of gene structure and InDel and SNPs positions. (**B**) Results of haplotype analysis. Hap: Haplotype. Superscript letters (a-d) represent significant differences at $p < 0.05$. (**C**) Haplotype variation analysis. Colored boxes indicate the pod-shattering ratio of soybean genotypes in each Hap.

3. Discussion

Pod-shattering causes a significant yield loss in soybeans. The selection of elite breeding lines with good agronomic traits, including pod-shattering tolerance, is essential for soybean breeding programs around the world. Molecular markers offer an effective and efficient approach for studying pod-shattering, as well as for breeding programs.

Several studies have identified molecular markers targeting genes on chromosome 16, the major genomic region associated with the pod-shattering phenotype [10,17,32]. Three InDel markers (SRM0, SRM1, and SRM2) were developed for the *qPDH1* locus [13]. The one SNP variation in the coding region of *Glyma16g25580* (Wm82.a1.v2) was developed as a Pdh1 Simple Probe real-time PCR assay for predicting pod-shattering [4]. The SNPs within *Glyma.16g141600* on chromosome 16 also have significant associations with pod-shattering [32]. However, for a highly accurate selection of soybean genotypes, we need additional molecular markers for pod-shattering tolerance [31]. Therefore, we wanted to confirm the candidate genes and develop additional allele-specific markers related to the pod-shattering tolerance from an elite cultivar DW with strong tolerance to pod-shattering.

It has been reported that the inheritances of pod shattering in soybean are controlled by additive gene action [33,34], which is the effect of multiple genes with a linear or additive fashion [35]. Previous studies have suggested that the genetic variation in pod-shattering in soybean is controlled by partial dominance [36], dominant epistasis [37], dominance, or partial dominance [38]. Bhor et al. [39] found that two major genes with inhibitory epistasis were responsible for the mode of inheritance of pod-shattering. A single major gene may not be enough to differentiate soybean genotypes for pod-shattering. Han et al. [40] reported no difference between soybean cultivars Heihe 43 (tolerant pod-shattering) and Heihe 18 (susceptible to pod-shattering) using the major pod-shattering tolerance gene *SHATTERING1-5* [41]. However, the cultivars were successfully differentiated by *Glyma.05g225900* and *Glyma.05g227400*. Therefore, the identifications of additional genes and the use of molecular markers linked with the genes would be highly useful for genetic studies and breeding programs in soybean. In the present study, we developed KASP markers for *Glyma.16g141200* and *Glyma.16g141500*, as well as an InDel marker for *Glyma.16g076600*, which are all candidate genes for pod-shattering that we identified within the QTLs detected previously [17]. *Glyma.16g076600* is a novel candidate gene found in our previous study [17]. In addition, we evaluated the efficiency of the markers using soybean genotypes with diverse genetic backgrounds that are tolerant or susceptible to pod-shattering.

The functions of the three candidate genes involved in the pod-shattering mechanisms have been studied previously. It has been suggested that pod-shattering involves a bZIP transcription factor related to *Glyma.16g141500* (bZIP transcription factor bZIP117), located in a major pod-shattering QTL (*qPDH1*) in soybean [14]. The bZIP proteins play a role in the expression of the proline dehydrogenase (ProDH) gene in *Arabidopsis* [42]. Kavi Kishor et al. [43] reported that proline is an important source of the cell wall matrix. Proline-based signaling proteins are also important for the cell wall of *Arabidopsis thaliana* [44]. The increased expression of ProDH2 in senescent leaves and in the abscission zone of floral organs implicates a bZIP transcription factor in pod-shattering [45].

The pod-shattering tolerance in soybean may result from excessive secondary cell wall deposition and lignification in the outer part of the suture [41]. This is supported by a study in common vetch, in which high-throughput RNA sequencing was used to evaluate global transcriptome variations associated with the pod ventral sutures of accessions for the shattering-susceptible and shattering-resistant strains [46]. A similar study demonstrated that the shattering-resistant varieties of common vetch did not have an abscission zone in the bundle sheath; thus, this favors the development of strong sutures [47]. The expression of a gene associated with stem bolting in *Arabidopsis* indicates that members of the bZIP family may be involved in secondary cell wall thickening [48]. Moreover, several TGA-type bZIP genes may control the expression of the genes involved in abscission [49]. The upregulation of 11 bZIP genes in the abscission zones of *Elymus nutans* supports their role as positive regulators in abscission [50]. Similar to the involvement of *Glyma.16g141500* in abscission, protein phosphatase 2C, which is related to *Glyma.16g141200*, may also function in abscisic acid signal transduction in *Arabidopsis thaliana* [51]. *Glyma.16g076600*, a homolog of *AT4G19230* in *Arabidopsis*, is a member of the CYP707A gene family that encodes a protein related to ABA catabolism [52]. Plant hormones, such as ABA, may regulate silique dehiscence in *Arabidopsis* and *Brassica* [53].

We evaluated the KASP and InDel markers for the pod-shattering phenotypes using the oven-drying method in 307 RILs and 120 varieties and elite lines. The prediction accuracy of the KASP marker was very high, with 81.7% to 88.2% in the RIL populations, and 91.3% to 95.5% in the varieties and elite lines. The lower prediction accuracy in the RIL populations may be due to a recombination in the RILs (although they were $F_{7:8}$), and these results were similar to the results of Kim et al. [31]. The prediction accuracy of this study is comparable to a previous study [31]. The slightly higher values obtained in the study [31] might be due to their use of fewer RILs of the shattering-tolerant breeding lines. Using both types of markers could increase the prediction accuracy in soybean lines, compared to using KASP markers alone. Adding markers from the additional candidate genes could increase the prediction accuracy. The haplotype analysis of candidate genes revealed distinct groups of soybean genotypes for the pod-shattering tolerance. The three genes used here as markers for pod-shattering in soybean, as well as the KASP and InDel markers developed in this study, could be used in MAS to improve the pod-shattering tolerance in soybean breeding programs.

4. Materials and Methods

A summary of the study design has been provided as a schematic diagram (Supplementary Figure S6).

4.1. Plant Materials

The QTLs and candidate genes for pod-shattering tolerance were identified using two RIL populations ($F_{7:8}$) derived from the cross between a pod-shattering-tolerant cultivar Daewonkong (DW) and the pod-shattering-susceptible cultivars, Tawonkong (TW) and Saeolkong (SO), as described previously [17,54] (Supplementary Figure S7). Populations consisting of 154 RILs (DW × TW; DT), 153 RILs (DW × SO; DS), and 120 varieties and elite lines were used to validate the KASP and InDel markers for the candidate genes *Glyma.16g141200*, *Glyma.16g141500*, and *Glyma.16g076600*. Two RIL populations, as well as 120 soybean varieties and elite lines used in this study, were developed in our institute according to the cultivation methods of the Agricultural Science Technology Standards for Investigation (Rural Development Administration, Jeonju, Korea) and they comply with the guidelines and legislation of Korea. The varieties and elite lines were selected based on their economic values and cultivation practices in Korea. The panel of the 120 varieties and elite lines includes 86 pod-shattering-tolerant and 34 pod-shattering-susceptible soybeans. All the plant materials are deposited at the National Institute of Crop Science, Rural Development Administration, Miryang, Korea, and can be obtained on reasonable request after meeting the institutional obligations.

4.2. Evaluation of Pod-Shattering Tolerance

The pod-shattering tolerance for 400 genotypes (307 RILs, 120 varieties, and elite lines) was calculated, as described previously [17,54]. In brief, mature pods were harvested at the R8 stage, stored at room temperature for a week, oven-dried (40 °C) for 24, 48, or 72 h, and the number of shattered pods were counted. The pod-shattering ratio = (number of shattered pods/total number of pods) × 100 (%).

4.3. DNA Extraction from the Parental Lines and Whole-Genome Resequencing

Genomic DNA was extracted from the young trifoliate leaves of three parental cultivars (Daewonkong, Tawonkong, and Saeolkong) of the RIL populations using a DNA extraction kit (Exgene Plant SV Miniprep Kit; GeneAll, Seoul, Korea) following the manufacturer's instructions. The concentration and quality of the extracted DNA were determined with a NanoDrop 2000 spectrophotometer (Thermo Fisher Scientific, Waltham, MA, USA). The Illumina Hi-seq 4000 whole-genome resequencing of the parental cultivars (Macrogen Inc., Seoul, Korea) revealed the SNP and/or InDel regions, as compared to the reference genome of *Glycine max* (Wm82.a2.v1) [55]. The sequences were visualized using the Golden Helix Genome Browser v 3.0.0 software (Golden Helix, Bozeman, MT, USA).

4.4. Molecular Marker Development and Genotyping

The genomic DNA from 307 RILs of two populations, as well as 120 varieties and elite lines, was extracted, as described above. The KASP markers for the target SNPs were developed with allele-specific primers and common primers. The KASP markers were genotyped by using the ABI7300 system (ABI, Foster City, CA, USA) and were analyzed with 7300 system SDS RQ study software (Thermo Fisher Scientific, Waltham, MA, USA). The PCR (polymerase chain reaction) conditions for genotyping with the KASP marker were as follows: 94 °C for 15 min; 10 cycles of 94 °C for 20 s, 61 °C for 1 min; and 26 cycles of 94 °C for 20 s; followed by 55 °C for 1 min. The InDel marker for the target sequence was generated using the Primer3 program (http://bioinfo.ut.ee/primer3-0.4.0, Accessed 19 October 2021). The InDel marker was genotyped by capillary electrophoresis (QIAxcel, Qiagen, Hilden, Germany). The PCR conditions for genotyping with the InDel marker were as follows: 95 °C for 5 min; 35 cycles of 95 °C for 20 s, 55 °C for 1 min, 72 °C for 1 min; followed by 72 °C for 5 min.

4.5. Determination of Prediction Accuracy

The prediction accuracy was evaluated following the method by Kim et al. [31]. The susceptibility accuracy was calculated as the proportion of the lines with susceptible phenotypes, compared to the lines with susceptibility alleles. Similarly, the tolerance accuracy was calculated as the proportion of the lines with resistant phenotypes compared to the lines with resistance alleles. The prediction accuracy was calculated as the mean of susceptible accuracy and tolerant accuracy.

4.6. Haplotype Analysis

The haplotype analysis for *Glyma.16g076600*, *Glyma.16g141200*, and *Glyma.16g141500* was carried out to assess genetic variations among soybean genotypes using the methods by Zhang et al. [56] with some modifications. In 120 varieties and elite lines panels, one InDel and two SNPs markers were used for the haplotype analysis. The average score and genotype count were determined from the pod-shattering ratio, and the haplotypes that were significantly associated with pod shattering were identified.

4.7. Statistical Analysis

R software V 4.1.1 (R Foundation for Statistical Computing, Vienna, Austria) was used to determine the analysis of variance and compare the phenotypic variations among the genotypes of RIL populations, varieties, and elite lines.

5. Conclusions

Two KASP markers and one InDel marker were developed based on our previous studies on QTL mapping and the analysis of candidate genes for pod-shattering tolerance using two RIL populations. The whole-genome resequencing of three parental lines revealed SNP and/or InDel regions based on comparisons to the Williams_82 reference genome (ver. 2.1). The markers were validated in 307 RILs, 120 varieties, and elite lines through allelic discrimination. The validity of the markers in these lines was further supported by the pod-shattering ratios and the high prediction accuracy of these results. The consistency of the results in diverse genetic backgrounds indicates that the markers can be used to select pod-shattering tolerant genotypes. These markers may also provide insights into the molecular mechanisms underlying the pod-shattering tolerance in soybean.

Supplementary Materials: The following supporting information can be downloaded at: https://www.mdpi.com/article/10.3390/ijms23042382/s1.

Author Contributions: Conceptualization: J.-H.S., C.-S.J. and T.-H.J.; Methodology: J.-H.S.; Investigation: B.-K.K. and S.K.D.; Data Curation: B.-K.K. and J.-Y.K.; Resources: B.-K.K., I.-Y.B. and J.-S.S.; Writing-Original Draft Preparation: J.-H.S., S.K.D. and T.-H.J.; Writing-Review and Editing: J.-H.S., S.K.D., K.-S.K. and T.-H.J.; Funding Acquisition: J.-H.S. All authors have read and agreed to the published version of the manuscript.

Funding: This research was funded by the National Institute of Crop Science, Rural Development Administration (No. PJ01186803) of the Republic of Korea.

Institutional Review Board Statement: Not applicable.

Informed Consent Statement: Not applicable.

Data Availability Statement: The data sets generated in this study are included in this published article and its Supplementary Materials.

Conflicts of Interest: The authors declare no conflict of interest.

References

1. Rey, L.; Lopez, Y.; Jaramillo, A.; White, J.W. Evaluation of a Growth Simulation Model Applied to Soybean Genotypes (*Glycine Max* L. Merr) under Tropical Conditions. *Corpoica Cienc. Tecnol. Agropecu.* **1996**, *1*, 16. [CrossRef]
2. Painii Montero, V.F.; Santillán Muñoz, O.B.; Montes Escobar, K.; Garces Fiallos, F.R. Characterization of Soybean Productive Units in the Ecuadorian Coast. *Cienc. Tecnol. Agropecu.* **2020**, *21*, 1–20. [CrossRef]
3. Nilmani, B.; Dhirendra, K.; Shrivastava, A.N. Studies on the Factors Affecting Pod Shattering in Soybean. *Indian J. Genet. Plant Breed.* **2013**, *73*, 270–277. [CrossRef]
4. Funatsuki, H.; Suzuki, M.; Hirose, A.; Inaba, H.; Yamada, T.; Hajika, M.; Komatsu, K.; Katayama, T.; Sayama, T.; Ishimoto, M.; et al. Molecular Basis of a Shattering Resistance Boosting Global Dissemination of Soybean. *Proc. Natl. Acad. Sci. USA* **2014**, *111*, 17797–17802. [CrossRef] [PubMed]
5. Tukamuhabwa, P.; Dashiell, K.E.; Rubaihayo, P.; Nabasirye, M. Determination of Field Yield Loss and Effect of Environment on Pod Shattering in Soybean. *Afr. Crop Sci. J.* **2002**, *10*, 203–209.
6. Zhang, L.; Bellaloui, N. Effects of Planting and Maturity Dates on Shattering Patterns under Early Soybean Production System. *Am. J. Plant Sci.* **2012**, *3*, 119–124. [CrossRef]
7. Zhang, Q.; Tu, B.; Liu, C.; Liu, X. Pod Anatomy, Morphology and Dehiscing Forces in Pod Dehiscence of Soybean (*Glycine max* (L.) Merrill). *Flora* **2018**, *248*, 48–53. [CrossRef]
8. Miles, C.; Wayne, M. Quantitative Trait Locus (QTL) Analysis. *Nat. Educ.* **2008**, *1*, 208.
9. Visscher, P.M.; Andrew, T.; Nyholt, D.R. Genome-Wide Association Studies of Quantitative Traits with Related Individuals: Little (Power) Lost but Much to Be Gained. *Eur. J. Hum. Genet.* **2008**, *16*, 387–390. [CrossRef]
10. Bailey, M.A.; Mian, M.A.R.; Carter, T.E.; Ashley, D.A.; Boerma, H.R. Pod Dehiscence of Soybean: Identification of Quantitative Trait Loci. *J. Hered.* **1997**, *88*, 152–154. [CrossRef]
11. Kang, S.-T.; Kwak, M.; Kim, H.-K.; Choung, M.-G.; Han, W.-Y.; Baek, I.-Y.; Kim, M.Y.; Van, K.; Lee, S.-H. Population-Specific QTLs and Their Different Epistatic Interactions for Pod Dehiscence in Soybean [*Glycine max* (L.) Merr.]. *Euphytica* **2009**, *166*, 15–24. [CrossRef]
12. Funatsuki, H.; Ishimoto, M.; Tsuji, H.; Kawaguchi, K.; Hajika, M.; Fujino, K. Simple Sequence Repeat Markers Linked to a Major QTL Controlling Pod Shattering in Soybean. *Plant Breed.* **2006**, *125*, 195–197. [CrossRef]
13. Suzuki, M.; Fujino, K.; Nakamoto, Y.; Ishimoto, M.; Funatsuki, H. Fine Mapping and Development of DNA Markers for the QPDH1 Locus Associated with Pod Dehiscence in Soybean. *Mol. Breed.* **2010**, *25*, 407–418. [CrossRef]
14. Gao, M.; Zhu, H. Fine Mapping of a Major Quantitative Trait Locus That Regulates Pod Shattering in Soybean. *Mol. Breed.* **2013**, *32*, 485–491. [CrossRef]
15. Hu, D.; Kan, G.; Hu, W.; Li, Y.; Hao, D.; Li, X.; Yang, H.; Yang, Z.; He, X.; Huang, F.; et al. Identification of Loci and Candidate Genes Responsible for Pod Dehiscence in Soybean via Genome-Wide Association Analysis across Multiple Environments. *Front. Plant Sci.* **2019**, *10*, 811. [CrossRef] [PubMed]
16. Han, J.; Han, D.; Guo, Y.; Yan, H.; Wei, Z.; Tian, Y.; Qiu, L. QTL Mapping Pod Dehiscence Resistance in Soybean (*Glycine max* L. Merr.) Using Specific-Locus Amplified Fragment Sequencing. *Theor. Appl. Genet.* **2019**, *132*, 2253–2272. [CrossRef] [PubMed]
17. Seo, J.-H.; Kang, B.-K.; Dhungana, S.K.; Oh, J.-H.; Choi, M.-S.; Park, J.-H.; Shin, S.-O.; Kim, H.-S.; Baek, I.-Y.; Sung, J.-S.; et al. QTL Mapping and Candidate Gene Analysis for Pod Shattering Tolerance in Soybean (*Glycine max*). *Plants* **2020**, *9*, 1163. [CrossRef]
18. Parker, T.A.; Berny Mier y Teran, J.C.; Palkovic, A.; Jernstedt, J.; Gepts, P. Pod Indehiscence Is a Domestication and Aridity Resilience Trait in Common Bean. *New Phytol.* **2020**, *225*, 558–570. [CrossRef]
19. Kang, S.T.; Kim, H.K.; Baek, I.Y.; Chung, M.G.; Han, W.Y.; Shin, D.C.; Lee, S.-H. Genetic Analysis of Pod Dehiscence in Soybean. *Korean J. Crop Sci.* **2005**, *50*, 281–285.
20. Tukamuhabwa, P.; Rubaihayo, P.; Dashiell, K.E. Genetic Components of Pod Shattering in Soybean. *Euphytica* **2002**, *125*, 29–34. [CrossRef]
21. Zhao, S.; Li, A.; Li, C.; Xia, H.; Zhao, C.; Zhang, Y.; Hou, L.; Wang, X. Development and Application of KASP Marker for High Throughput Detection of AhFAD2 Mutation in Peanut. *Electron. J. Biotechnol.* **2017**, *25*, 9–12. [CrossRef]
22. Wang, D.; Yang, T.; Liu, R.; Li, N.; Wang, X.; Sarker, A.; Zhang, X.; Li, R.; Pu, Y.; Li, G.; et al. RNA-Seq Analysis and Development of SSR and KASP Markers in Lentil (Lens culinaris Medikus Subsp. culinaris). *Crop J.* **2020**, *8*, 953–965. [CrossRef]
23. Hechanova, S.L.; Bhattarai, K.; Simon, E.V.; Clave, G.; Karunarathne, P.; Ahn, E.-K.; Li, C.-P.; Lee, J.-S.; Kohli, A.; Hamilton, N.R.S.; et al. Development of a Genome-Wide InDel Marker Set for Allele Discrimination between Rice (*Oryza sativa*) and the Other Seven AA-Genome Oryza Species. *Sci. Rep.* **2021**, *11*, 8962. [CrossRef] [PubMed]
24. Paudel, L.; Clevenger, J.; McGregor, C. Refining of the Egusi Locus in Watermelon Using KASP Assays. *Sci. Hortic.* **2019**, *257*, 108665. [CrossRef]
25. Devran, Z.; Göknur, A. Development and Validation of a SNP-Based KASP Assay for Rapid Identification of Aphelenchoides besseyi Christie, 1942. *Crop Prot.* **2020**, *136*, 105235. [CrossRef]
26. Shi, Z.; Liu, S.; Noe, J.; Arelli, P.; Meksem, K.; Li, Z. SNP Identification and Marker Assay Development for High-Throughput Selection of Soybean Cyst Nematode Resistance. *BMC Genom.* **2015**, *16*, 314. [CrossRef]
27. Tran, D.T.; Steketee, C.J.; Boehm, J.D.; Noe, J.; Li, Z. Genome-Wide Association Analysis Pinpoints Additional Major Genomic Regions Conferring Resistance to Soybean Cyst Nematode (Heterodera glycines Ichinohe). *Front. Plant Sci.* **2019**, *10*, 401. [CrossRef]

28. Pham, A.-T.; Harris, D.K.; Buck, J.; Hoskins, A.; Serrano, J.; Abdel-Haleem, H.; Cregan, P.; Song, Q.; Boerma, H.R.; Li, Z. Fine Mapping and Characterization of Candidate Genes That Control Resistance to Cercospora sojina K. Hara in Two Soybean Germplasm Accessions. *PLoS ONE* **2015**, *10*, e0126753. [CrossRef]
29. Cheng, P.; Gedling, C.R.; Patil, G.; Vuong, T.D.; Shannon, J.G.; Dorrance, A.E.; Nguyen, H.T. Genetic Mapping and Haplotype Analysis of a Locus for Quantitative Resistance to Fusarium graminearum in Soybean Accession PI 567516C. *Theor. Appl. Genet.* **2017**, *130*, 999–1010. [CrossRef]
30. Miranda, C.; Culp, C.; Škrabišová, M.; Joshi, T.; Belzile, F.; Grant, D.M.; Bilyeu, K. Molecular Tools for Detecting Pdh1 Can Improve Soybean Breeding Efficiency by Reducing Yield Losses Due to Pod Shatter. *Mol. Breed.* **2019**, *39*, 27. [CrossRef]
31. Kim, J.-M.; Kim, K.-H.; Jung, J.; Kang, B.K.; Lee, J.; Ha, B.-K.; Kang, S. Validation of Marker-Assisted Selection in Soybean Breeding Program for Pod Shattering Resistance. *Euphytica* **2020**, *216*, 166. [CrossRef]
32. Lee, J.S.; Kim, K.R.; Ha, B.-K.; Kang, S. Identification of SNPs Tightly Linked to the QTL for Pod Shattering in Soybean. *Mol. Breed.* **2017**, *37*, 54. [CrossRef]
33. Tiwari, S.; Bhatnagar, P. Pod Shattering as Related to Other Agronomic Attributes in Soybean. *Trop. Agric.* **1991**, *68*, 102–103.
34. Umar, F.A.; Mohammed, M.S.; Oyekunle, M.; Usman, I.S.; Ishaq, M.N.; Dachi, S.N. Estimates of Combining Ability for Resistance to Pod Shattering in Soybean (*Glycine max* (L.) Merrill) Genotypes. *J. Plant Breed. Crop Sci.* **2017**, *9*, 217–223. [CrossRef]
35. Plomin, R.; DeFries, J.C.; McClearn, G.E.; McGuffin, P. *Behavioral Genetics*; Worth Publishers: New York, NY, USA, 2001.
36. Tsuchiya, T.; Sunada, K. Breeding Studies on Pod Shattering in Soybeans. II. Methods of Testing for Shattering and Varietal Differences. *Soybean Abstr.* **1980**, *3*, 1706.
37. Tukamuhabwa, P.; Rubaihayo, P.R.; Dashiell, K.; Adipala, E. Inheritance to Pod Shattering in Soybean. *Afr. J. Crop Sci.* **2000**, *8*, 203–211.
38. Tiwari, S.; Bhatnagar, P. Consistent Resistance for Pod Shattering in Soybean Varieties. *Indian J. Agric. Sci.* **1993**, *63*, 173–174.
39. Bhor, T.J.; Chimote, V.P.; Deshmukh, M.P. Inheritance of Pod Shattering in Soybean (*Glycine max* (L.) Merrill). *Electron. J. Plant Breed.* **2014**, *5*, 671–676.
40. Han, D.; Han, J.; Jiang, S.; Su, B.; Zhang, B.; Liu, Z.; Yan, H.; Qiu, L.-J. Shattering-Resistance of an Elite Soybean Variety 'Heihe 43' and Identification of Shattering-Resistant Genes. *Euphytica* **2021**, *217*, 120. [CrossRef]
41. Dong, Y.; Yang, X.; Liu, J.; Wang, B.-H.; Liu, B.-L.; Wang, Y.-Z. Pod Shattering Resistance Associated with Domestication Is Mediated by a NAC Gene in Soybean. *Nat. Commun.* **2014**, *5*, 3352. [CrossRef]
42. Satoh, R.; Fujita, Y.; Nakashima, K.; Shinozaki, K.; Yamaguchi-Shinozaki, K. A Novel Subgroup of BZIP Proteins Functions as Transcriptional Activators in Hypoosmolarity-Responsive Expression of the ProDH Gene in Arabidopsis. *Plant Cell Physiol.* **2004**, *45*, 309–317. [CrossRef] [PubMed]
43. Kavi Kishor, P.B. Role of Proline in Cell Wall Synthesis and Plant Development and Its Implications in Plant Ontogeny. *Front. Plant Sci.* **2015**, *6*, 544. [CrossRef] [PubMed]
44. Ihsan, M.Z.; Ahmad, S.J.N.; Shah, Z.H.; Rehman, H.M.; Aslam, Z.; Ahuja, I.; Bones, A.M.; Ahmad, J.N. Gene Mining for Proline Based Signaling Proteins in Cell Wall of Arabidopsis thaliana. *Front. Plant Sci.* **2017**, *8*, 233. [CrossRef] [PubMed]
45. Funck, D.; Eckard, S.; Müller, G. Non-Redundant Functions of Two Proline Dehydrogenase Isoforms in Arabidopsis. *BMC Plant Biol.* **2010**, *10*, 70. [CrossRef]
46. Dong, R.; Dong, D.; Luo, D.; Zhou, Q.; Chai, X.; Zhang, J.; Xie, W.; Liu, W.; Dong, Y.; Wang, Y.; et al. Transcriptome Analyses Reveal Candidate Pod Shattering-Associated Genes Involved in the Pod Ventral Sutures of Common Vetch (*Vicia sativa* L.). *Front. Plant Sci.* **2017**, *8*, 649. [CrossRef] [PubMed]
47. Dong, D.; Yan, L.; Dong, R.; Liu, W.; Wang, Y.; Liu, Z. Evaluation and Analysis of Pod Dehiscence Factors in Shatter-Susceptible and Shatter-Resistant Common Vetch. *Crop Sci.* **2017**, *57*, 2770–2776. [CrossRef]
48. Ehlting, J.; Mattheus, N.; Aeschliman, D.S.; Li, E.; Hamberger, B.; Cullis, I.F.; Zhuang, J.; Kaneda, M.; Mansfield, S.D.; Samuels, L.; et al. Global Transcript Profiling of Primary Stems from Arabidopsis thaliana Identifies Candidate Genes for Missing Links in Lignin Biosynthesis and Transcriptional Regulators of Fiber Differentiation. *Plant J.* **2005**, *42*, 618–640. [CrossRef]
49. Tucker, M.L.; Whitelaw, C.A.; Lyssenko, N.N.; Nath, P. Functional Analysis of Regulatory Elements in the Gene Promoter for an Abscission-Specific Cellulase from Bean and Isolation, Expression, and Binding Affinity of Three TGA-Type Basic Leucine Zipper Transcription Factors. *Plant Physiol.* **2002**, *130*, 1487–1496. [CrossRef]
50. Zhao, Y.; Zhang, J.; Zhang, Z.; Xie, W. Elymus Nutans Genes for Seed Shattering and Candidate Gene-Derived EST-SSR Markers for Germplasm Evaluation. *BMC Plant Biol.* **2019**, *19*, 102. [CrossRef]
51. Meyer, K.; Leube, M.P.; Grill, E. A Protein Phosphatase 2C Involved in ABA Signal Transduction in Arabidopsis thaliana. *Science* **1994**, *264*, 1452–1455. [CrossRef]
52. Saito, S.; Hirai, N.; Matsumoto, C.; Ohigashi, H.; Ohta, D.; Sakata, K.; Mizutani, M. Arabidopsis CYP707A s Encode (+)-Abscisic Acid 8′-Hydroxylase, a Key Enzyme in the Oxidative Catabolism of Abscisic Acid. *Plant Physiol.* **2004**, *134*, 1439–1449. [CrossRef]
53. Jaradat, M.R.; Ruegger, M.; Bowling, A.; Butler, H.; Cutler, A.J. A Comprehensive Transcriptome Analysis of Silique Development and Dehiscence in Arabidopsis and Brassica Integrating Genotypic, Interspecies and Developmental Comparisons. *GM Crops Food* **2014**, *5*, 302–320. [CrossRef] [PubMed]
54. Seo, J.H.; Kang, B.K.; Kim, H.T.; Kim, H.S.; Choi, M.S.; Oh, J.H.; Shin, S.O.; Baek, I.Y.; Kwak, D.Y. Variation in Pod Shattering in a RIL Population and Selection for Pod Shattering Tolerance in Soybean [*Glycine max* (L.) Merr]. *Korean J. Crop Sci.* **2019**, *64*, 414–421. [CrossRef]

55. Schmutz, J.; Cannon, S.B.; Schlueter, J.; Ma, J.; Mitros, T.; Nelson, W.; Hyten, D.L.; Song, Q.; Thelen, J.J.; Cheng, J.; et al. Genome Sequence of the Palaeopolyploid Soybean. *Nature* **2010**, *463*, 178–183. [CrossRef] [PubMed]
56. Zhang, H.; San, M.L.; Jang, S.-G.; Lee, J.-H.; Kim, N.-E.; Lee, A.-R.; Park, S.-Y.; Cao, F.-Y.; Chin, J.-H.; Kwon, S.-W. Genome-Wide Association Study of Root System Development at Seedling Stage in Rice. *Genes* **2020**, *11*, 1395. [CrossRef]

International Journal of
Molecular Sciences

MDPI

Article

Genome-Wide Association Analysis for Hybrid Breeding in Wheat

Monika Mokrzycka [1], Stefan Stojałowski [2], Mirosław Tyrka [3], Przemysław Matysik [4], Barbara Żmijewska [4], Rafał Marcinkowski [4], Urszula Woźna-Pawlak [5], Róża Martofel [5], Michał Rokicki [5,6], Monika Rakoczy-Trojanowska [7] and Paweł Krajewski [1,*]

1 Institute of Plant Genetics, Polish Academy of Science, Strzeszyńska 34, 60-479 Poznań, Poland
2 Department of Plant Genetics, Breeding and Biotechnology, West Pomeranian University of Technology Szczecin, Słowackiego 17, 71-434 Szczecin, Poland
3 Department of Biotechnology and Bioinformatics, Faculty of Chemistry, Rzeszow University of Technology, Powstańców Warszawy 6, 35-959 Rzeszów, Poland
4 Plant Breeding Strzelce Group IHAR Ltd., Główna 20, 99-307 Strzelce, Poland
5 Poznań Plant Breeding Ltd., Kasztanowa 5, 63-004 Tulce, Poland
6 Plant Breeding and Acclimatization Institute, Radzików, 05-870 Błonie, Poland
7 Department of Plant Genetics, Breeding and Biotechnology, Institute of Biology, Warsaw University of Life Sciences, Nowoursynowska 159, 02-776 Warszawa, Poland
* Correspondence: pkra@igr.poznan.pl

Abstract: Disclosure of markers that are significantly associated with plant traits can help develop new varieties with desirable properties. This study determined the genome-wide associations based on DArTseq markers for six agronomic traits assessed in eight environments for wheat. Moreover, the association study for heterosis and analysis of the effects of markers grouped by linkage disequilibrium were performed based on mean values over all experiments. All results were validated using data from post-registration trials. GWAS revealed 1273 single nucleotide polymorphisms with biologically significant effects. Most polymorphisms were predicted to be modifiers of protein translation, with only two having a more pronounced effect. Markers significantly associated with the considered set of features were clustered within chromosomes based on linkage disequilibrium in 327 LD blocks. A GWAS for heterosis revealed 1261 markers with significant effects.

Keywords: genome-wide association study; wheat hybrid breeding

Citation: Mokrzycka, M.; Stojałowski, S.; Tyrka, M.; Matysik, P.; Żmijewska, B.; Marcinkowski, R.; Woźna-Pawlak, U.; Martofel, R.; Rokicki, M.; Rakoczy-Trojanowska, M.; et al. Genome-Wide Association Analysis for Hybrid Breeding in Wheat. *Int. J. Mol. Sci.* **2022**, *23*, 15321. https://doi.org/10.3390/ijms232315321

Academic Editors: Andrés J. Cortés and Hai Du

Received: 19 October 2022
Accepted: 27 November 2022
Published: 5 December 2022

1. Introduction

Wheat (*Triticum aestivum* L.) is an important source of calories and proteins for humans worldwide [1]. This crop is cultivated on approximately 216 million hectares and yields over 765 million tons of grain [2]. For the last 60 years, the average yield of wheat has been increased 3-fold to the current level of approx. 3.5 t/ha [2,3], but the progress in this regard has been relatively slow in recent years. Norman Borlaug's "Green Revolution" initiated this continuous progress, which aimed to release new highly productive cultivars for developing countries. The use of heterosis and hybrid wheat breeding is a possible option for maintaining or boosting yield productivity in modern wheat cultivars. The possibility of the commercial use of wheat F1 hybrids was suggested as early as 1963 by Briggle [4]. Wheat hybrids may significantly outperform traditional cultivars regarding yield level (usually by under 5%, but in particular situations by more than 40%; [5]) because of the effect of heterosis and can reveal higher stability in various environments due to their heterozygotic nature [6–8]. Moreover, the certified hybrid seeds desired by producers protect the interests of breeders [9].

The availability of high-throughput genotyping techniques, such as genotyping by sequencing (GBS) or single nucleotide polymorphism (SNP) arrays, has accelerated global genetic analyses of large genome species, such as wheat [10–14]. The results of genome-wide association studies (GWAS) can be applied to the selection of the most valuable genotypes (genomic selection). GWAS was recently used to analyze anther extrusion in wheat [15,16]. Association studies based on various types of populations for the wheat plant traits analyzed in this report were also performed; we refer to a number of them in Discussion, Sections 3.1–3.3. More complex analyses focusing on the multi-trait requirements of hybrid breeding have not yet been reported. In the breeding of traditional cultivars, the target of GWAS is relatively clear, and the objective of genomic selection is to identify genotypes with alleles associated with increased values of yield-related traits that are frequently affected by environmental changes [17–20]. The application of GWAS for genomic selection in hybrid breeding procedures is complex. The successful release of a cultivar results from selecting a proper maternal line (seed parent), paternal line (pollen parent), and prediction of the heterosis effect revealed by the final hybrid. The challenge is that a desirable trait in one of these three components can be undesirable in another. Furthermore, the best performance of the final hybrid should be achieved when numerous loci of high-yielding parents remain heterozygotic. Genomic selection in hybrid breeding should be adjusted to meet these requirements.

In addition to the agronomic performance of related traits, the condition of heterosis breeding success is the appropriate flowering biology of the parental components. Different elements of flowering biology determine the suitability of male and female wheat parents for hybrid breeding. Efficient anther extrusion and a high quantity of released pollen over an extended period are considered the most important traits desired in male candidates for the successful production of hybrid wheat seeds. For female parents as pollen recipients, more significant is the ability to open flowers in the time necessary to get pollinated for extended periods, coupled with hairy stigmas. Considering the plant height, it is recommended that the pollen parent be taller than the seed parent because the pollen of wheat is much less mobile than the pollen of typical wind-pollinated plants such as maize and rye [3,7,16,21–23].

Despite several studies, the processes of flower opening and anther extrusion, a very complex and dynamic phenomenon combining anatomical, physiological, and biochemical parameters, are still not fully understood or defined [24–26]. Nevertheless, the variation and heritability of flowering-related traits are considered moderate to high, which provides the opportunity to select desirable parental components [7,27].

In the presented work, we use the data presented in [28], which was applied to the analysis of the genetic structure of a large pool of 509 wheat varieties and breeding lines. On adding to them the phenotypic observations obtained in two independent series of field trials, we presume that by using GWAS we will identify chromosomal regions associated with six agronomically important traits that are significant for wheat hybrid breeding.

To verify this presumption, we set the following objectives: (1) to apply the genotyping data [28] consisting of SNP marker observations determined by the DArTseq platform for GWAS, and (2) to identify chromosomal regions associated with six traits within a set of wheat genotypes across different environments.

2. Results

2.1. Phenotypic Characteristics of the Accessions

Six phenotypic traits were assessed in eight environments. The distributions of genotypic means (BLUPs obtained in individual experiments; Table S1) were approximately normal; a slight skewness to the right was observed for time of flowering—FT (Figure S1). Two phenological traits, time of heading—HT and FT, were positively correlated with each other and with plant height—PH (p-value < 0.05; Figure S1). In contrast, for PH, a negative correlation was observed with yield-related traits (number of kernels per spike—KN, weight of kernels per spike—KW, and thousand kernel weight—TKW). KW was positively correlated with TKW.

In ANOVA, all fixed effects, i.e., effects of the year (Y), localization (L), and Y × L interaction, were significant (at least at a p-value < 0.05, but most of them at p-value < 0.001). The variance components for random effects of genotypes (G) were much larger than those for G × Y and G × L interactions for traits HT, FT, and PH, which was reflected by the large heritability for these traits (88–90%; Table 1). The variance components were in the same order for yield-related traits, and therefore, heritability was smaller (38–54%).

Table 1. Mean values, coefficients of variation (CV), variance components (with standard errors) for genotypes (G) and their interactions with year (Y) and localization (L), and broad-sense heritability coefficients for the six traits.

Trait	Mean	Minimum	Maximum	CV	Variance Component ± s.e.				Heritability (%)
					G	G × Y	G × L	Error	
HT	25.45	10.5	34.44	13.8	6.06 ± 0.42	0.91 ± 0.07	0.36 ± 0.03	1.56 ± 0.03	90.41
FT	31.67	17.87	39.95	11.74	4.04 ± 0.32	1.39 ± 0.11	0.46 ± 0.04	1.38 ± 0.03	80.77
PH	81.5	42.66	127.73	17.42	32.52 ± 2.37	5.92 ± 0.55	2.41 ± 0.31	22.15 ± 0.42	86.79
KN	62.24	27.69	103.9	15.88	14.05 ± 1.73	9.68 ± 1.12	11.44 ± 1.03	63.5 ± 1.2	54.60
KW	2.53	1.03	4.33	19.49	0.018 ± 0.003	0.028 ± 0.003	0.023 ± 0.002	0.14 ± 0.003	38.79
TKW	41.87	23.86	55.96	12.03	4.36 ± 0.56	5.41 ± 0.42	2.25 ± 0.18	9.72 ± 0.18	52.97

A comparison of the distributions of genotypic means in particular experiments showed that accessions headed and flowered earlier and were shorter in 2018 than in 2019 (Figure 1A), and that the yield parameters were higher in 2018. The weather conditions, as characterized by monthly precipitation and temperatures, differed slightly between the two growing seasons. In 2017/2018, winter temperatures were lower, and summer precipitation was slightly higher than in 2018/2019 (Figure 1B).

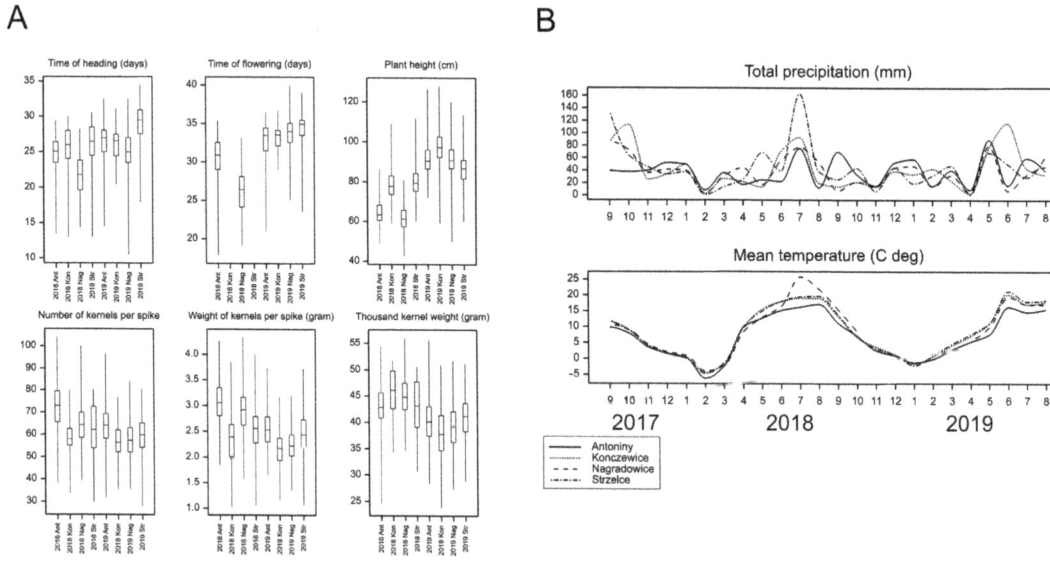

Figure 1. (**A**) Distributions of genotypic means in experiments; Ant—Antoniny, Kon—Konczewice, Nag—Nagradowice, Str—Strzelce; (**B**) Observations of weather parameters in 2017–2019 at four locations of experiments (ten-day summaries).

Stability analysis in the AMMI model revealed that, in general, the reactions of genotypes to conditions in the four locations were more similar in 2019 than in 2018 (Figure S2A). In 2018, for traits HT, PH, KN, and TKW, a considerable difference existed between reactions to conditions in (Antoniny, Nagradowice) vs. (Konczewice, Strzelce). The patterns of the interactions for localization were similar for HT and TKW. Genotypes showed continuous distributions of interactions. For FT accessions, PHR_6 (Akteur), STH_148 (UKR 12), and for KW accessions, STH_65 (STH 9025), STH_167 (Cornea ost.) showed outlying instability. The largest significant correlations existed between instability variances for genotypes estimated for pairs of traits: HT and FT and KW and TKW (Figure S2B,C); correlations for trait pairs (PH, FT) and (KW, KN) were statistically significant but very small. The ranks of genotypes concerning instability variances for all traits are presented in Table S2; they show that the most stable genotypes for one trait were usually much less stable for the other.

2.2. Association Analysis

GWAS revealed 7603 SNPs with polymorphisms significantly associated with at least one phenotypic trait (only SNPs with more than five accessions in each of the two homozygous classes; BH corrected p-value < 0.05). The characteristics of the allelic substitution effects for these loci are shown in Table 2.

Table 2. Characteristics of allelic substitution effects for SNPs associated significantly with phenotypic traits.

Trait	Number of Associations	Negative Significant Effects		Positive Significant Effects		2th Percentile	
		Min	Max	Min	Max	Lower	Upper
HT	4206	−1.57	−0.0024	0.0003	1.49	−0.68	0.72
FT	4108	−1.30	−0.0027	0.0014	1.13	−0.61	0.64
PH	1999	−4.24	−0.0006	0.0013	3.18	−1.72	1.63
KN	1858	−4.29	−0.0015	0.0113	3.27	−1.36	1.54
KW	738	−0.09	−0.0001	0.0002	0.12	−0.05	0.05
TKW	1908	−1.41	−0.0002	0.0001	1.19	−0.66	0.61

By considering also the criterion related to the size of the substitution effect being in the second lower or upper percentile, the number of associated markers was reduced to 1273, and the number of significant associations was reduced to 1344 (Figure 2, Table 3 and Table S3).

The mean linkage disequilibrium (LD) with other SNPs in the region ±5 Mb was determined for significant SNPs. A detailed analysis of LD has been provided in [28] (a heatmap of LD on chromosome 1A in Figure S2A, exemplary LD clusters in Figure S2B, and the plots of LD vs. physical distance between markers with 0–20 Mb distance intervals in Figure 5A of [28]). The percentage of associations with a small mean LD (<0.01) in this interval varied from 5% for PH to 20% for KW (Table 3). The significant SNPs were clustered into small groups of no more than five markers within chromosomes based on LD (Table 3). The highest number of LD clusters was on chromosome 3A for the two phenological traits, on chromosome 2B for PH, on 2B and 6B for KN, on 7B for KW, and on 2A for TKW (Figure 3).

The fraction of associations showing interaction with the environment was the lowest for PH and KN and the highest for KW. Associations were unevenly distributed in the wheat subgenomes. Most of the associated SNPs mapped to the A and B subgenomes (38.4% and 45.01%, respectively), whereas only 14.14% were derived from D, and 2.44% remained unmapped (Table 3, Figure 2); however, the proportion of the relative number of SNPs was similar: 8.99% from the A subgenome, 10.14% from B, 8.45% from D and 11.07 from Un. For five traits, most associations were detected in subgenome A, and only for PH were most associations detected in subgenome B. More than 50% of the associated SNPs for all the traits were located in genes. Most SNPs associated with traits were predicted to modify protein translation, and only two SNPs significantly influenced the protein structure. The first SNP (1130302 | F | 0-44 | TG), which was associated with FT, altered the splicing donor site in TraesCS6A02G085100 (Arabidopsis RPT3 ortholog), which is responsible for ATP-dependent degradation of ubiquitinated proteins. The second SNP (1090593 | F | 0-44 | CT) was associated with TKW and introduced a STOP codon in the gene TraesCS4B02G086500, annotated with "carbohydrate metabolic process" and "polygalacturonase activity" GO terms.

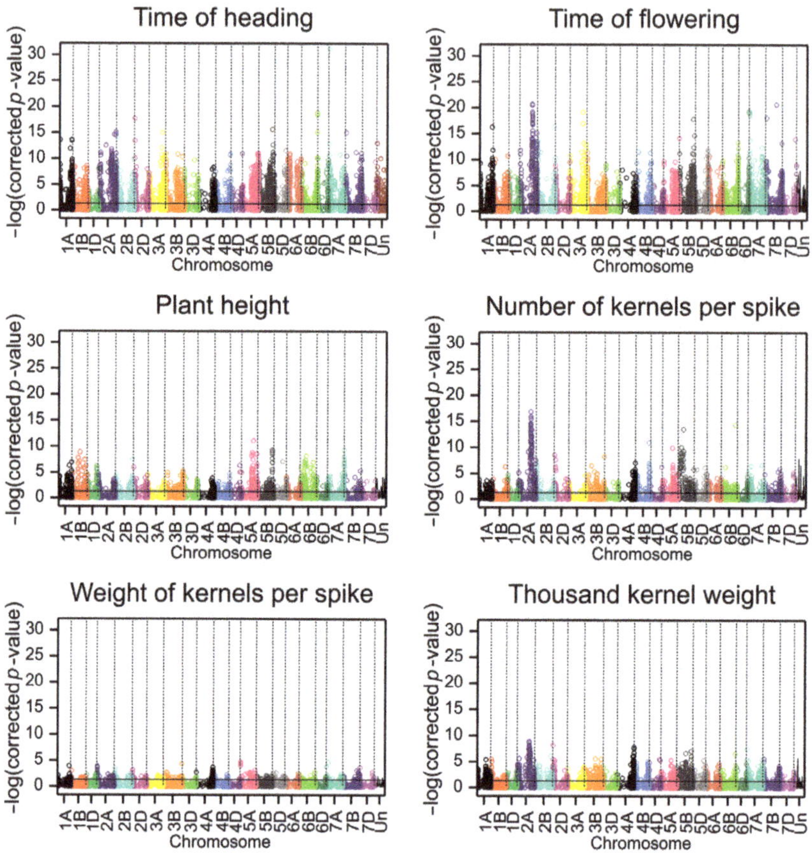

Figure 2. Manhattan plots for associations of SNPs with studied phenotypic traits. Points represent *p*-values for associations corrected by the Benjamini–Hochberg method. The black line visualizes the critical significance level of 0.05.

Allelic substitution effects for different traits were correlated (Figure S3). However, not all correlations were of the same size and sign as those of the traits themselves. For example, allelic effects for PH were negatively correlated with effects for phenological traits (positive correlation for traits), and effects for KW were positively correlated with phenological traits (negative correlation for traits).

SNPs associated with traits were divided into three sets corresponding to plant phenology (HT or FT, set 1), PH (set 2), and kernel properties (KN or KW or TKW, set 3), with the numbers of associations in these sets being 592, 391, and 429, respectively. There were 26, 99, and 26 SNPs common to sets 1 and 2, 1 and 3, and 2 and 3, respectively, and the 12 SNPs common for all three sets (Figure 4, Table S3). For pairs of traits, the largest number of common significant SNPs was recorded for HT and FT (312), which is consistent with a large correlation between these two traits reported above.

Table 3. GE interaction and distribution of trait-associated SNPs in linkage blocks, subgenomes, and genes.

Trait	Number of Associations	Number of SNPs with Mean LD <0.01	Number of SNP Clusters	With GE Interaction (% out of Significant)	% in Subgenome			% in Genes	Number of SNPs Affecting Protein Translation			
					A	B	D		High	Low	Moderate	Modifier *
HT	444	71	220	66.0	47.1	36.9	13.3	55.4	0	57	45	342 (144)
FT	460	78	238	66.1	43.7	34.8	18.7	56.7	1	64	38	357 (158)
PH	391	21	123	38.1	20.2	63.2	13.3	59.6	0	52	66	273 (115)
KN	170	25	94	31.2	44.1	32.4	21.8	55.9	0	18	20	132 (57)
KW	51	10	36	68.6	41.2	39.2	19.6	54.9	0	7	6	38 (15)
TKW	272	22	115	61.4	55.1	36.4	5.9	58.5	1	23	29	219 (106)

* Total number of markers indicated by VEP [29] (in brackets: number of markers assigned to specific genes).

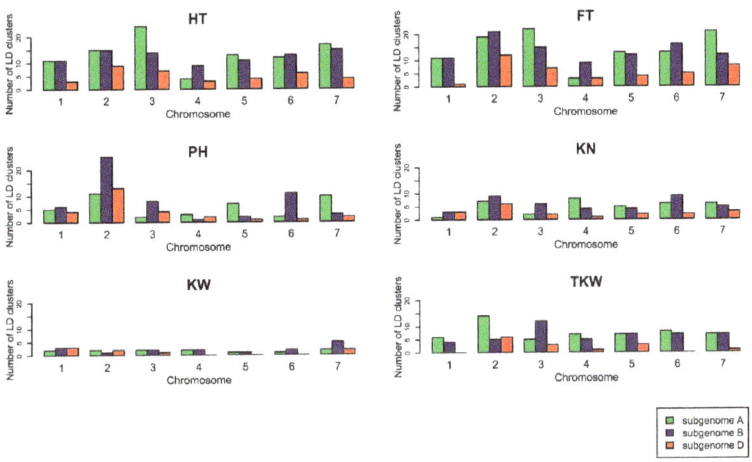

Figure 3. Number of LD clusters containing associated SNPs by chromosomes for each trait.

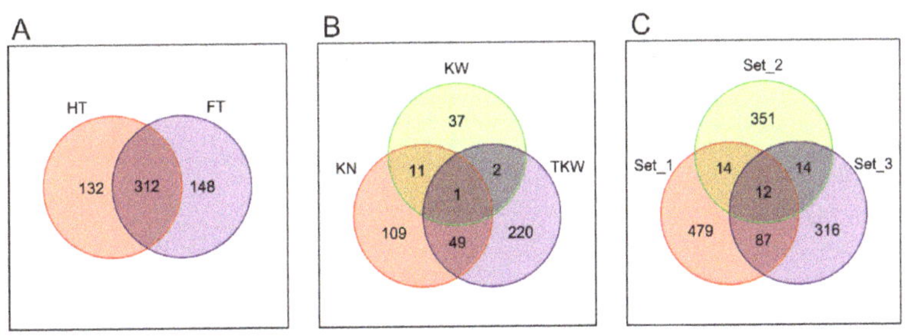

Figure 4. The number of common and specific SNPs in (**A**,**B**) association sets for traits and (**C**) association set 1 (HT or FT), set 2 (PH) and set 3 (KN, KW or TKW).

Of the 26 markers that belonged to sets 1 and 2, 17 were assigned to genes, with 5 being modifiers and 4 substitutions having moderate effects (Table 4). There were three markers with concordant effects on earliness and plant height, with the modifying effect of substitutions. Two of these (2253029 | F | 0-10 | CT and 1237800 | F | 0-13 | CG) were associated with genes (Table S3). Out of 12 SNPs common for sets 1, 2 and 3, eleven were associated with KN and one with TKW. The same direction of effect could be seen for FT and KW, while the opposite direction was seen for KN and PH.

Table 4. Markers with concordant effects for sets 1 and 2 (in bold) and assigned to genes. Types of translation effects caused by substitution are provided (LOW, MDR—moderate, or MFI—modifier).

Marker	Gene	Effect	Interpro Description
2253029 | F | 0-10 | CT (negative)	TraesCS2A02G482200	MFI	NAD-dependent epimerase/dehydratase; NAD(P)-binding domain superfamily
3938110 | F | 0-10 | CG	TraesCS2B02G045100	LOW	NB-ARC;P-loop containing nucleoside triphosphate hydrolase; Leucine-rich repeat domain superfamily
1237800 | F | 0-13 | CG (positive)	TraesCS2B02G164700	MFI	F-box-like domain superfamily
4910338 | F | 0-15 | GC	TraesCS2B02G475700	MDR	Zf-FLZ domain; Zf-FLZ domain
2322929 | F | 0-10 | AT	TraesCS2B02G490600	MFI	Ribonuclease H-like superfamily; Exonuclease, RNase T/DNA polymerase III; Ribonuclease H superfamily
2322355 | F | 0-40 | CG	TraesCS2B02G521100	LOW	Glycosyl transferase;1,3-beta-glucan synthase subunit FKS1-like, domain-1
1238701 | F | 0-16 | GA	TraesCS2D02G127300	LOW	F-box domain;F-box-like domain superfamily
1675478 | F | 0-15 | TG	TraesCS3A02G506600	LOW	NAD(P)-binding domain superfamily; NAD(P)-binding domain superfamily
7350269 | F | 0-8 | TC	TraesCS3A02G517700	LOW	
1049114 | F | 0-20 | AG	TraesCS3D02G511900	MFI	Ubiquitin-like domain; Ubiquitin domain; Ubiquitin-like domain superfamily
1675534 | F | 0-35 | AC	TraesCS3D02G513900	MDR	UDP-glucuronosyl/UDP-glucosyltransferase
2252787 | F | 0-19 | CT	TraesCS3D02G529700	MFI	Coenzyme Q-binding protein COQ10, START domain; START-like domain superfamily
7352843 | F | 0-43 | AT	TraesCS3D02G531000	MDR	Transcription initiation factor IIA, gamma subunit; Transcription factor IIA, helical; Transcription factor IIA, beta-barrel; Transcription initiation factor IIA, gamma subunit, C-terminal
7352096 | F | 0-35 | GC	TraesCS3D02G541900	LOW	Uncharacterised conserved protein UCP015417, vWA
7353078 | F | 0-15 | CA	TraesCS3D02G542800	MDR	Oxoglutarate/iron-dependent dioxygenase; Non-haem dioxygenase N-terminal domain; Isopenicillin N synthase-like
3024403 | F | 0-23 | CT	TraesCS3D02G543100	LOW	DnaJ domain; Tetratricopeptide-like helical domain superfamily; Tetratricopeptide repeat-containing domain
1009915 | F | 0-65 | CG	TraesCSU02G059500	LOW	Leucine-rich repeat, cysteine-containing subtype; SKP1/BTB/POZ domain superfamily; BTB/Kelch-associated; Leucine-rich repeat domain superfamily
1056528 | F | 0-9 | TG (negative) *	-	MFI	-

* marker not in the gene but showing congruent association with HT, FT, and PH.

2.3. Marker LD Clusters

Clusters of markers significantly associated with HT, FT (set 1), and with PH (set 2) (Table 3) were considered for LD-based GWAS. Combinations of their genotypes, represented by at least 25 accessions from the investigated collection, were considered for further analysis. The characteristics of the 327 marker clusters are presented in Table S4. Among them, we attempted to identify the effects of genotypic combinations that were higher than the separate allele effects of particular single markers. Examples of genotypic combinations with large negative effects on FT, HT, and PH are shown in Figure 5. For marker LD block no. 21, which clusters markers "1207903 | F | 0-56 | AT_4992730 | F | 0-46 | CT", the variant "T/T T/T" was present in 26 accessions and was related to early flowering (effect of −2.9 days) and low plants (effect of −3.8 cm) in comparison to the average of all accessions. One of these markers was assigned as an upstream modifier of the gene TraesCS1B02G028100, annotated as involved in "serine-type endopeptidase inhibitor activity." Other interesting marker clusters with higher effects contributed by haplotypes were identified, for example, no. "291", 2260931 | F | 0-40 | CT_1004422 | F | 0-41 | AG. We noted that 76% of the genotypic variants in marker LD blocks were fully homozygous. Therefore, their information was equivalent to the knowledge of haplotypes.

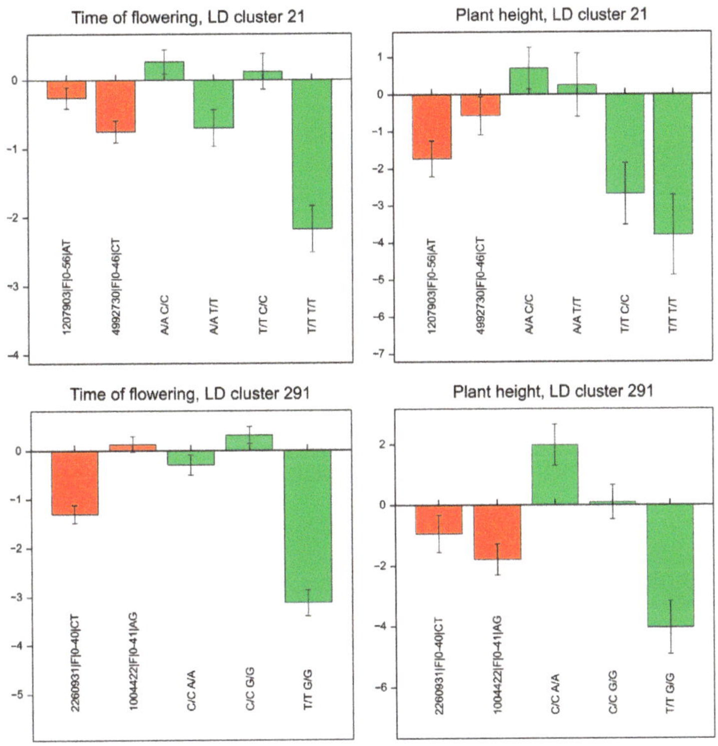

Figure 5. Effects found in the association analysis: red—allelic effects of individual SNPs, averaged over eight experiments; green—haplotype effects (deviations from the general mean). Bars represent standard errors of effects. Illustrated are the results for two marker clusters with effects concordant for earliness (FT) and plant height (PH).

2.4. Heterosis

Searching for heterosis of SNP polymorphisms (on the set of SNPs with more than five heterozygous lines; BH corrected p-value < 0.05) revealed 1261 SNPs with significant effects (Table 5 and Table S5), with the largest number of effects for phenological traits and no effects for KW. Significant SNPs were evenly distributed over the subgenomes, with 53–80% of SNPs in the genes (Figure 6). None of the SNPs revealed a high translation effect, and the majority of the predicted effects were classified as low or modifying protein translation.

Table 5. Characteristics of heterosis effects for the five traits (KW was omitted due to the lack of significant effects).

Trait	Number of Significant Effects	% in Genome			% in Genes	Number SNPs with Protein Translation Effect (in Genes)		
		A	B	D		Low	Moderate	Modifier *
HT	437	33.9	28.6	35.5	57.7	106	52	275 (94)
FT	393	35.4	28.0	35.4	57.0	96	45	249 (83)
PH	324	35.5	31.8	28.7	52.8	52	41	229 (78)
KN	62	37.1	32.3	29.0	62.9	18	3	41 (18)
TKW	45	42.2	20.0	37.8	68.9	13	2	30 (16)

* Total number of markers indicated by VEP [29] (in brackets: number of markers assigned to specific genes).

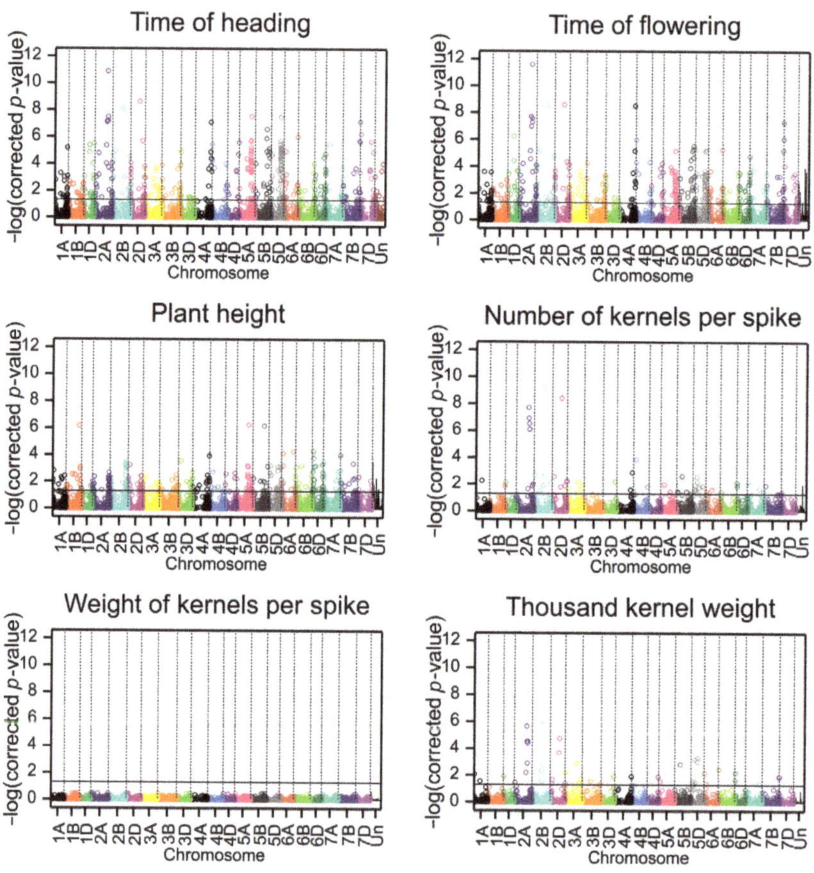

Figure 6. Manhattan plots for heterosis effects of SNPs. Points represent p-values for associations corrected by the Benjamini–Hochberg method. The black line visualizes the critical significance level of 0.05.

Association sets for traits (HT, FT), (PH), (KN, KW, and TKW) based on heterosis effects contained 480, 324, and 83 markers, respectively. The latter (yield-related) set contained 52 markers assigned to genes; the heterosis effects in this set were: from −4.50 to 3.09 grains for KN, from −0.12 to 0.09 g for KW, and from −2.09 to 4.13 g for TKW. The highest negative effect for KN (−4.50 grains) was for marker 3024735 | F | 0-10 | GA (TraesCS7B02G481400).

The correlations of heterosis effects were similar to those of additive effects (Figure S4).

2.5. Allelic Substitution Effects vs. Heterosis Effects

Allelic effects were correlated with heterosis effects for markers with high heterozygosity (Figure 7). The range of additive effects was smaller than that of heterosis effects for all traits.

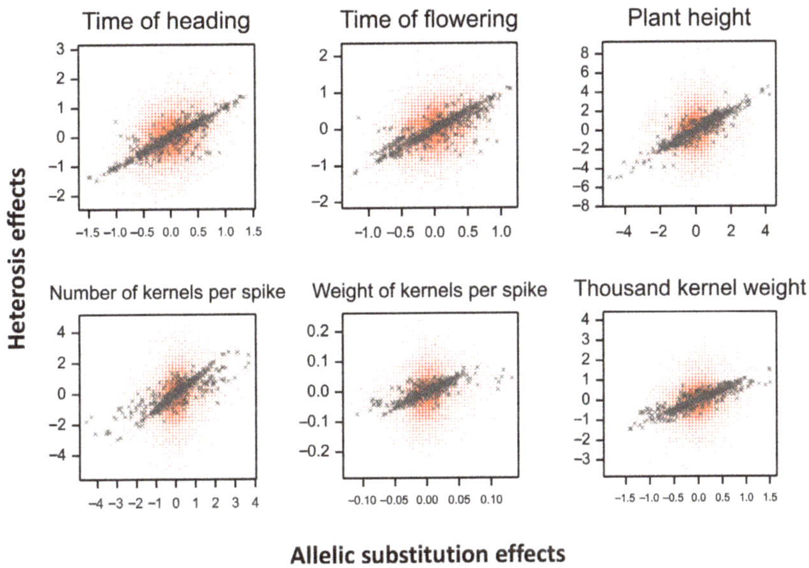

Figure 7. Heterosis effects vs. additive effects for six phenotypic traits; gray points for markers with heterozygosity > 0.5.

One marker, 1238701 | F | 0-16 | GA (TraesCS2D02G127300), belonged to sets 1 and 2 with respect to additive effects (effects for HT −0.75, for PH 2.47), and to set 3 with respect to heterosis effects (effect 2.25 for TKW).

2.6. GO Annotation of Associated Markers

GO terms (biological processes) represented in sets of genes assigned to SNPs in association sets 1, 2, and 3 for allelic effects and heterosis effects are shown in Table S6A and Table S6B, respectively. Most frequently (16–26%), markers associated with both additive and heterosis effects were involved in redox processes and protein phosphorylation. The next most common GO terms were regulation of transcription (DNA-templated), carbohydrate metabolic process, and transmembrane transport. Responses to auxins were identified in groups of genes responsible for plant height additive effects more frequently than proteolysis.

2.7. Validation of Results Using Post-Registration Trial Data

Data from post-registration trials (PRT) were used to validate the GWAS results obtained in the HYBRE experiments. The general varietal means of HT, PH, and TKW from the two series of experiments were correlated (0.84, 0.86, and 0.76, respectively; *p*-value < 0.001; Figure 8A). Allelic effects from GWAS were also correlated (0.29, 0.42, and 0.40, respectively; *p*-value < 0.001; Figure 8B). However, only three SNPs were significantly associated with traits in both series of trials: one for HT (3024420 | F | 0-9 | CT, modifier, intergenic), and two for PH (1087592 | F | 0-37 | GA, modifier, downstream of TraesCS5A02G277900 and 1126438 | F | 0-22 | CT, modifier, intergenic). The heterosis effects were also correlated (Figure 8C).

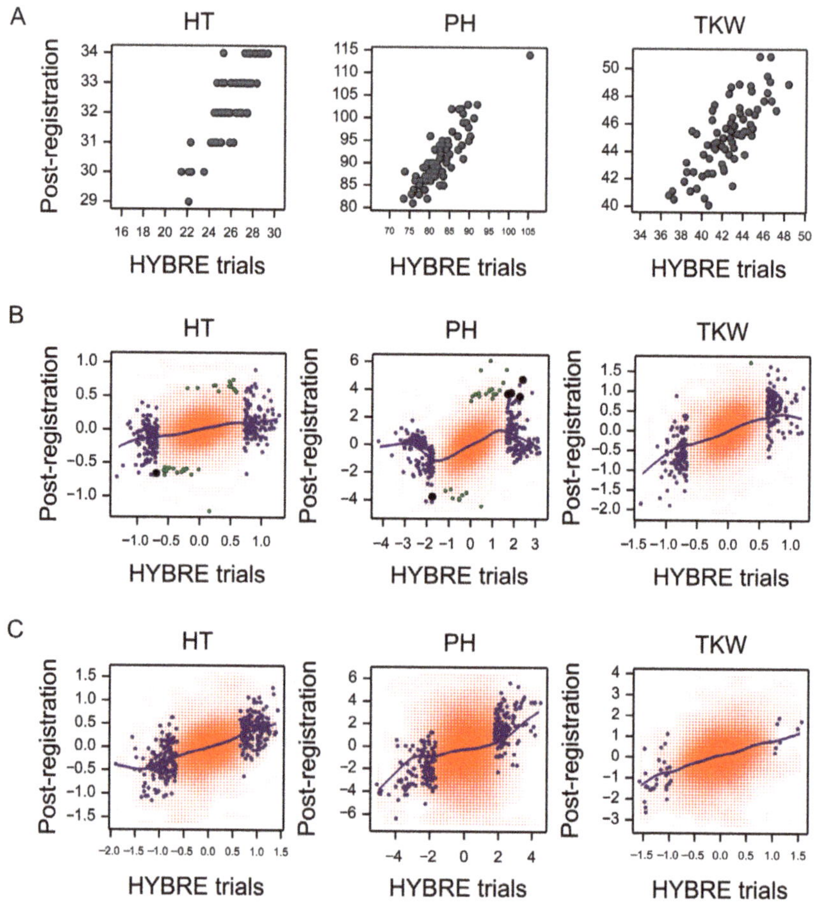

Figure 8. (**A**) Scatterplots of varietal means observed in post-registration experiments (PRT) vs. means observed in HYBRE trials, for 75 varieties occurring in both series of experiments. (**B**) Density plots of allelic effects for markers estimated in post-registration experiments (PRT) vs. effects estimated in HYBRE trials in the set of 75 varieties. (**C**) Density plots of heterosis effects estimated in PRT vs. HYBRE. (**B**,**C**) Blue—significant in HYBRE, green—significant in PRT, and black—significant in both analyses. The blue lines are trend lines determined by smooth regression using the thin plate method.

3. Discussion

The main goal of hybrid wheat breeding is to exploit the heterosis effect. Final success depends on several factors, but yield is a crucial goal. Without a significant increase in productivity offered by hybrid cultivars, they will not be the choice of farmers; as an alternative, they may apply less expensive seeds of well-performing classic cultivars. The cost of hybrid seed production is also important. It depends on numerous factors related to plant morphology and the biology of flowering [3,7,30]. The cross-pollination of wheat plants is based on the wind, but it is significantly less effective than that of typical open-pollinated plants, such as rye and maize. Waines and Hedge [31] noticed that singular wheat pollen could be found as far as 1000 m from the plantation, but hybrid seeds are usually undetectable beyond 20–30 m. The distance between parental lines of wheat allowing for economically sufficient effectiveness of seed production is limited to 2–3 m, and larger values can be applied when the pollen parent is taller than the seed parent. The reduction of plant height in wheat (highly recommended for maternal components) can be achieved by utilizing properly selected reduced height (Rht) genes.

In this study, we present a multifaceted statistical analysis aimed at providing information on a broad population of cultivars and breeding strains, which can primarily be used in wheat hybrid breeding. We investigated other morphological and phenological traits significant for efficient seed production and chosen traits related to wheat productivity, aiming to verify whether our non-standard genome-wide association study methods have potential application in hybrid breeding of wheat.

The analysis of variance showed that genotypic variability, measured in relation to variabilities caused by genotype by year and genotype by localization interactions, and, consequently, broad-sense heritability, was higher for phenological traits and for plant height than for yield-related traits. This indicates the possibility of selecting valuable parental forms independently, to some extent, from the target environment. Additionally, stability analysis (AMMI) showed that, although a given genotype may be subjected to various degrees of environmental variance for various traits, some correlation exists among instability variances for traits related to hybrid component selection.

However, our association analysis was performed assuming that SNP variant effects can also interact with environmental conditions. This aspect of GWAS analysis is often addressed by performing separate analyses for each environment (cf. [32–34]), mainly because of the limited options presented by the data analysis software. Based on the mixed model developed by Malosetti et al. [35], our approach allowed for the explicit testing of markers by environment interaction effects, which provided a conclusion on the lower environmental variability of SNP effects for plant height than for phenology. We also performed GWAS analysis using marker LD clusters. The most promising marker clusters were those for which the effects of genotypic combinations were higher than the allelic effects of individual markers. Furthermore, concordant effects of earliness and plant height are required for the practical importance of the haplotype. The applied procedure allowed the identification of groups of markers with a high impact on phenotypic traits, which could be expected for individual polymorphisms. Thus, we demonstrated the possibility of using the phenomenon of non-allelic interactions in LD-informed hybrid breeding. Moreover, as hybrid breeding is based on the effects of heterozygous materials, another association analysis was performed with respect to heterosis. It aimed to identify genomic loci for which heterozygotes could be more profitable than homozygotes, especially for yield-related traits. Although the results obtained in this way are far less valuable than those that could be gained from observations of hybrids, this is undoubtedly progress compared to the methodological approaches used to date in this type of research.

Validation of GWAS results can be done by an independent experimental study concerning another pool of genotypes. In our case, the pool covers most of the accessions that are practically interesting for project stakeholders. Another form of validation can be based on a real breeding process; this has been started by passing our results to breeders, who have already performed selected crosses and are currently assessing the value of hybrids by their standard procedures. In this report, to verify the results of GWAS, publicly available data from independent, post-registration trials performed in a wider set of environments were used. The mean BLUPs for genotypes from the HYBRE experiments were correlated with the general means from the PRT. Moreover, substantial correlations between allelic substitution effects obtained from GWAS on both datasets were obtained; the same was true for heterosis effects. The effects concerning plant height were found to be the most correlated. This means that the data obtained in the HYBRE experiment are representative, and it can be assumed that new breeding creations can be evaluated in limited field experiments rather than in large experimental systems.

3.1. *Flowering-Related QTLs and Genes*

The expected female component of the hybrid should contain long stigmatic hairs that are fully extruded and receptive for long periods [7]. Wheat stigma remains receptive for up to 13 days after anthesis; however, it is usually highly receptive for no more than 3 days after anthesis [21]. Effective anther extrusion of the pollen parent is crucial for the successful setting of hybrid seeds. Recently, Denisow et al. [36] showed that the anther extrusion plays a much more important role in the contribution to the final amount of pollen available for cross-pollination than previously thought. Flowering generally begins at the center of a wheat spike and proceeds in both directions (up and down). Within each spikelet, the primary floret opens first, followed by the secondary, tertiary, etc. The first two florets of the spikelet produce the most abundant anthers, filled with the most viable pollen [37]. Thus, it is suggested that when these highly effective florets of the male component begin flowering, the female parent should be ready to receive pollen. The optimal difference between the flowering times of the seed parent and pollen parent is 2–3 days [30]. The phenotypic efficiency of some of the SNP markers indicated in our analyses met these criteria. The studied set of wheat accessions had been focused on genotypes cultivated in Europe [28]. Therefore, phenotypic variation was, in some way, limited. Despite this, some of the selected SNP markers revealed additive effects ranging from 1–1.6, corresponding to a difference from 2 to over 3 days of flowering time of homozygotic wheat genotypes, which are potential parental components of hybrids.

Markers associated with flowering time are necessary to synchronize the flowering of hybrid wheat components. Eight QTLs for heading time have been reported [38]. One of them, QHD_7A_psr_ParW670_CFLN17, corresponding to a relatively short region 710–719 Mbp on 7A, colocalized with the 3025631 | F | 0-10 | GA marker found in our study. The regulation of flowering has been extensively studied at the gene level. Flowering is mainly regulated by vernalization (VRN1, VRN2, VRN3, and VRN-D4), photoperiod (Ppd-A1, Ppd-B1 Ppd, and Ppd-D1), and earliness per se (eps) genes [39]. Some of these genes showed pleiotropic effects, i.e., Ppd-A1 increased TKW and yield. Similarly, Ppd-B1 was associated with a high kernel number [40]. Eps genes interact with Ppd1 and are associated with spikelet number [41]. In wheat, a homolog of the Arabidopsis early-flowering 3 (ELF3) gene was identified as a candidate gene for Eps-Am1 [42]. Additional genes affecting flowering time in wheat were reviewed by Zhang et al. [39].

Searching Ensembl Plants for wheat orthologs of 204 Arabidopsis thaliana genes related to flowering [43] resulted in a list of 617 wheat genes. Of these, 31 were assigned to some of the SNPs analyzed in this study. Five SNPs were associated with FT or HT, with a negative effect of the ALT allele; these were orthologs of the A. thaliana genes UGT87A2, MFT, FRI, AGL57, TT16, and MAF4.

3.2. *Plant Height-Related QTLs and Genes*

Some Rht loci have been suggested to negatively affect anther extrusion [44,45]. The insertion of Rht genes was crucial for the success of the "green revolution" in wheat. Replacement of some widely used Rht loci with alternative variants more neutral for the flowering process may be recommended in hybrid breeding [46]. For example, the Rht1 and Rht24 genes reduce plant height, but Rht1 simultaneously reduces anther extrusion, whereas Rht24 does not have such adverse effects [30]. All wheats are assumed to be monomorphic for Rht-A1a [47]. Rht24 occurs at relatively high frequencies in European and Chinese wheat cultivars and was mapped to the same region as Rht14, Rht16, and Rht18 [48,49].

QTLs and candidate genes for PH were collected from the catalog of gene symbols for wheat ([47] with updates), and 44 microarray probes reported for PH [50] were mapped to separate physically linked regions in the Chinese Spring genome V 1.0 (Table S7A). We found eight DArTseq markers overlapping the four regions identified with these microarray probes. Additionally, DarTseq markers were found in the physical regions of QHt.nau-2D, Rht8, Rht13, and Rht22 [51–53].

Information on known genes responsible for wheat PH was used to search for homologous and paralogous genes (with identity to target > 50%). Sequences (TraesCS4A02G271000, TraesCS4B02G043100, and TraesCS4D02G040400) corresponding to gibberellic acid insensitive (GAI) Rht-A1, Rht-B1, and Rht-D1 were mapped to the 4AL, 4BS, and 4DS chromosome arms [54,55]. These GAI loci were highly conserved, and a search with blastn revealed no additional candidate homologous genes.

The remaining Rht genes were sensitive to GA. Two known genes from this group are GA2-β-dioxygenase (GA2oxA9) and AP2-D (Q gene), which correspond to Rht18 and Rht23 loci, respectively [50,56]. Two GA2oxA9 homologs and three paralogs were found, and five DArTseq markers coincided with the GA2oxA9 homolog located on the 6B chromosome. Two sequences were used to identify AP-2 homologs. Five AP-2 homologs were found, three of which were located in proximity (<5 Mbp) to 14 DArTseq markers associated with PH (Table S7A).

We identified 224 genes (containing 313 SNPs) assigned to DNA sequences related to PH using the text search in the description of genes by GO terms and Interpro features for the texts "gibberellin," "auxin," "cytochrome," "kaurene," "kaurenoic," and "DELLA." Most detected hits were for "cytochrome," as in [50]. Of these, only 12 genes (with 14 SNPs) associated with PH showed noticeable negative or positive phenotypic effects. Three of them were orthologs of the A. thaliana genes CYP85A1, CYP85A2, CYP84A4, CYP84A1, and CYP734A1.

3.3. Spike Traits Related to QTLs and Genes

The positions of significant SNPs for grain weight, number per spike, and TKW were compared with those reported in previous studies (Table S7B). Maintaining a higher KN is an important breeding target for stress-tolerant lines [57]. A total of 142 regions were associated with KN, and 11 overlapped (±1 Mb) with the DArTseq markers reported in our study. Only precise QTLs spanning regions not exceeding 5 Mb were included in the search for common positions with SNP markers identified in our study. Ninety-two markers associated with kernel weight have been described in the literature, and only one marker had a position congruent with the SNP identified in our study. Out of the 123 SNPs or QTL described as responsible for TKW, five SNP markers were identified in previously described regions. TaGW8-B1 is associated with agronomic traits in bread wheat cultivars. The TaGW8-B1a allele increases TKW and spikelet number per spike and provides higher yields than cultivars with the TaGW8-B1b allele [58]. KASP markers for TKW have also been developed [59–62].

3.4. SNP Translation Effects

Among the identified polymorphisms, two SNPs had high translational effects. The first identified polymorphism, 1090593 | F | 0-44 | CT, was associated with TKW and located in the gene TraesCS4B02G086500, annotated with the GO term "polygalacturonase activity". Polygalacturonases are hydrolyzing enzymes implicated in a wide range of plant developmental processes, such as cell elongation, organ abscission, fruit ripening, microspore release, and pollen tube growth [63,64]. Members of two out of the five clades, C and F, are also expressed in grasses during root and seed development [65]. They were detected in the outer pericarp and intermediate layers of grains of the related species Aegilops tauschii [66]. Ye et al. [67] found three putative candidate genes for QTL mapped on chromosome 4A in wheat; two (TraesCS4A02G229600 and TraesCS4A02G229700) were orthologous to the Arabidopsis gene At2g43860, coding for polygalacturonase. SNP 1090593 | F | 0-44 | CT was localized in the same homologous group but on chromosome 4B and had a high translational level, resulting in a STOP codon instead of a lysine codon. It is predicted that this substitution resulted in protein shortening (by 40 residues) and an altered C-terminus (the structures of wild-type and altered proteins, Figure S5).

The second SNP (1130302 | F | 0-44 | TG) was located in gene TraesCS6A02G085100 (Arabidopsis RPT3 ortholog) with a high translational effect and annotated with the GO term "protein ubiquitination" and was associated with FT. Protein ubiquitination is a sophisticated system of post-translational modification in all eukaryotes and has been demonstrated to play a key role in various plant developmental stages and processes, such as seed dormancy and germination, root growth, flowering time control, self-incompatibility, chloroplast development, and several abiotic stress responses [68]. The SNP 1130302 | F | 0-44 | TG was positioned in the intron and localized in the 5′UTR without any direct impact on the protein sequence. In general, cis-acting elements present in UTRs are essential for post-transcriptional control, including alternative polyadenylation, riboswitching, short-peptide translation, nonsense-mediated decay, and alternative splicing [69]. However, the T/G mutation found in our study may disrupt splicing, possibly resulting in the retention of this intron. Intron retention in the 5′UTR affects gene expression (by alternative splicing, alternative polyadenylation, and protein translation). Such an effect has been shown, inter alia, in Arabidopsis ZIF2 [70], EF1α-A3 [71] and rice OsmiR156h [72]. It has been postulated that UTR-related regulation of gene expression helps plants adapt to environmental fluctuations [69]. Flowering time is a major factor in climatic adaptation [73–75].

4. Materials and Methods

4.1. Germplasm Resources

The study was conducted using 509 wheat varieties and breeding lines. A set of 277 European varieties, registered mainly in Germany, Poland, and the United Kingdom, was used. Advanced breeding lines were represented by 232 accessions from the ongoing programs of the Plant Breeding Strzelce (STH) and Poznań Plant Breeding (PHR) companies. The genetic characteristics of the studied resources were provided by Tyrka et al. [28].

4.2. Field Phenotyping

Field experiments were conducted during two vegetation seasons, 2017/18 and 2018/19. Trials were established at four experimental stations of two breeding companies (STH and PHR) located in Strzelce (GPS: 52.3149° N, 19.4025° E), Kończewice (GPS: 53.1848° N, 18.5637° E), Nagradowice (GPS: 52.3178° N, 17.1529° E), and Leszno/Antoniny (GPS: 51.8586° N, 16.5902° E). Two replications of the studied objects were performed at each location. They were distributed within three randomized blocks. Each plot had an area of 1 m^2. Cultivation conditions in brief: Sowing rate: 250 kernels per m^2. Fertilization: 45 kg ha^{-1} of P_2O_5 and 80 kg ha^{-1} of K_2O prior sawing followed by 150 kg of N per hectare applied in 2 doses during the vegetation season. Plant protection: only herbicides and insecticides were applied (no treatments against fungi and lodging). The following traits were analyzed in the field experiments:

Time of heading (GS55)—HT (number of days since May 1)

Time of flowering (GS65)—FT (number of days since May 1)

Plant height—PH (average height of plants on a plot in cm)

Number of kernels per spike—KN (assessed on 10 randomly collected spikes in each plot)

Weight of kernels per spike—KW (assessed on 10 randomly collected spikes in each plot; results in grams)

Thousand kernel weight—TKW (average from three random samples per plot, each containing 100 kernels)

To validate the results obtained from the HYBRE experiments, publicly available data from Polish state post-registration trials (www.coboru.gov.pl, accessed on 23 June 2019 [76]) from 2015–2018 were used. Only data on HT, PH, and TKW were available in the official trial reports among the six traits studied in our research.

4.3. Genotyping

A detailed description of genotyping and genotypic data processing was provided by Tyrka et al. [28]. Briefly, for each genotype, DNA was isolated from 15–20 bulked 2-week-old seedlings. DNA concentration and purity were determined using a NanoDrop spectrophotometer (Thermo Fisher Scientific, Waltham, MA, USA), and DNA quality was assessed using 1.5% agarose gel electrophoresis. The DNA was stored at −20 °C and diluted to a working concentration of 50 ng/μL for subsequent wheat DArTseq 1.0 genotyping completed by Diversity Arrays Technology (Bruce, Australia). Only selected high-quality data for 13,499 SNPs were taken into account in this research (dominant markers of the Silico-DArT type were omitted).

4.4. Data Analysis

Phenotypic data were analyzed using a linear mixed model (LMM) with fixed effects of the year (Y), location (L), Y × L interaction, and random effects of genotypes (G), G × Y interaction, and G × L interaction. Broad-sense heritability was estimated by the method described by Cullis et al. [77]. The additive main effects and multiplicative interaction (AMMI) analysis was performed as described by Gauch [78]. Genome-wide association study (GWAS), using SNP polymorphisms and genotypic means (BLUPs) obtained in individual experiments, was performed by a method that allows for the interaction of genetic effects with the environment, developed by van Eeuwijk et al. [79] and Malosetti et al. [35]. This is based on a linear mixed model with the population structure estimated by eigenanalysis of the kinship matrix (see [28]) and the compound symmetry variance-covariance model used for environmental variation. Marker data used for these analyses were coded as follows: 0, reference (REF) homozygote; 1, heterozygote; and 2, alternative (ALT) homozygote. Standard GWAS considers the additive marker effects, but also non-additive effects might explain an important proportion of the variation in complex traits ([80]). Therefore, GWAS for heterosis effects, estimated as the difference between the mean for heterozygotes and the mean for two homozygotes, was performed based on mean values over all experiments. To do this, marker data coded with 0, 1, and 2 values in the model matrix (used in GWAS for allelic substitution effects) were recoded to values of 0 for homozygotes (of both types) and 1 for heterozygotes [80]. The population structure was represented in the model by the eigenanalysis scores. The effects of marker linkage disequilibrium (LD) clusters were analyzed using mean trait values over all experiments and analysis of covariance, with the genotypic combinations in these clusters as classifying factors and eigenscores as covariates. p-values for allelic substitution effects and heterosis effects were corrected for multiple testing using the Benjamini–Hochberg (BH) method. Only SNP data with more than five accessions present in both homozygous classes were used during the genome-wide association analyses. An SNP effect was considered statistically significant if the BH-corrected p-value was lower than 0.05.

Statistical analyses were performed using Genstat for Windows 19th edition [81]. Visualizations of results were performed using Genstat 19 or R software. The annotation of SNP markers with respect to genomic positions, neighboring genes, their Gene Ontology classification, and SNP translation effects [29] used in this study were the same as those used by Tyrka et al. [28].

5. Conclusions

In conclusion, by employing an appropriate experimental and statistical approach, we generated a set of SNP markers that can be used in breeding practice to predict the earliness and height of plants to assess whether a given genotype will be particularly good as a maternal (early and short plant) or paternal form (late and tall plant). The estimated heritability of phenological traits and plant height was relatively high; therefore, selection based on the markers indicated during this study may be successful. In contrast, the heritability of yield-related traits in our field trials was relatively low; nevertheless, it remained within the range reported by other authors [82–85].

The main achievements of the presented work are:

1. The successful parallel selection of homozygous parental genotypes (based on traits regulated by additive genes controlling phenology and plant height) and components revealing a high heterosis effect (the choice based on the highest values of spike-yield-related traits revealed by heterozygous genotypes) using GWAS.
2. Demonstrating that the application of clustered markers for the choice of genotypes in multi-feature processes may be more efficient than classic selection based on single marker polymorphism (SNP)
3. Validation of the GWAS results using post-registration trial data.

Supplementary Materials: The following supporting information can be downloaded at: https://www.mdpi.com/article/10.3390/ijms232315321/s1.

Author Contributions: Conceptualization, M.M., S.S., M.T., M.R.-T. and P.K.; Data curation, M.M.; Formal analysis, M.M. and P.K.; Funding acquisition, M.R.-T.; Methodology, M.M., S.S., M.T., M.R.-T. and P.K.; Resources, P.M., B.Ż., R.M. (Rafał Marcinkowski), U.W.-P., R.M. (Róża Martofel) and M.R.; Visualization, M.M. and P.K.; Writing—original draft, M.M., S.S., M.T., M.R.-T. and P.K.; Writing—review & editing, M.M., S.S., M.T., M.R.-T. and P.K. All authors have read and agreed to the published version of the manuscript.

Funding: This work was supported by the Polish National Centre for Research and Development (project HYBRE, no. BIOSTRATEG3/343665/6/NCBR/2017).

Institutional Review Board Statement: Not applicable.

Informed Consent Statement: Not applicable.

Data Availability Statement: The data presented in this study are available in Supplementary Materials.

Acknowledgments: Some computations were performed with the use of resources of Poznań Supercomputing and Networking Center (www.psnc.pl, accessed on 28 November 2022).

Conflicts of Interest: The authors declare no conflict of interest.

Abbreviations

AMMI	Additive main-effects and multiplicative interaction
BLUP	Best linear unbiased prediction
FT	Time of flowering
GBS	Genotyping by Sequencing
GS	Growth stage (BBCH scale)
GWAS	Genome-wide association studies
HYBRE	The project acronym and the name of the consortium
HT	Time of heading
KN	Number of kernels per spike
KW	Weight of kernels per spike
PH	Plant height
PRT	Post-registration trials
SNP	Single Nucleotide Polymorphism
TKW	Thousand kernel weight
VEP	Variant Effect Predictor

References

1. Poutanen, K.S.; Kårlund, A.O.; Gómez-Gallego, C.; Johansson, D.P.; Scheers, N.M.; Marklinder, I.M.; Eriksen, A.K.; Silventoinen, P.C.; Nordlund, E.; Sozer, N.; et al. Grains—A major source of sustainable protein for health. *Nutr. Rev.* **2022**, *80*, 1648–1663. [CrossRef] [PubMed]
2. FAOSTAT. Production domain. In *Crops*; FAO: Rome, Italy, 2019.
3. El Hanafi, S.; Bendaou, N.; Kehel, Z.; Sanchez-Garcia, M.; Tadesse, W. Phenotypic evaluation of elite spring bread wheat genotypes for hybrid potential traits. *Euphytica* **2020**, *216*, 168. [CrossRef]

4. Briggle, L.W.; Reitz, L.P. *Classification of Triticum Species and of Wheat Varieties Grown in the United States*; U.S. Department of Agriculture: Washington, DC, USA, 1963.
5. Fabrizius, M.A.; Busch, R.H.; Khan, K.; Huckle, L. Genetic diversity and heterosis of spring wheat crosses. *Crop Sci.* **1998**, *38*, 1108–1112. [CrossRef]
6. Krystkowiak, K.; Adamski, T.; Surma, M.; Kaczmarek, Z. Relationship between phenotypic and genetic diversity of parental genotypes and the specific combining ability and heterosis effects in wheat (*Triticum aestivum* L.). *Euphytica* **2009**, *165*, 419–434. [CrossRef]
7. Whitford, R.; Fleury, D.; Reif, J.C.; Garcia, M.; Okada, T.; Korzun, V.; Langridge, P. Hybrid breeding in wheat: Technologies to improve hybrid wheat seed production. *J. Exp. Bot.* **2013**, *64*, 5411–5428. [CrossRef] [PubMed]
8. Mühleisen, J.; Piepho, H.-P.; Maurer, H.P.; Longin, C.F.; Reif, J.C. Yield stability of hybrids versus lines in wheat, barley, and triticale. *Theor. Appl. Genet.* **2014**, *127*, 309–316. [CrossRef] [PubMed]
9. Creech, C. Making the Case for Certified Wheat Seed. CropWatch. 2017. Available online: https://cropwatch.unl.edu/2017/making-case-certified-wheat-seed (accessed on 28 November 2022).
10. Gupta, P.K.; Mir, R.R.; Mohan, A.; Kumar, J. Wheat Genomics: Present Status and Future Prospects. *Int. J. Plant Genom.* **2008**, 896451. [CrossRef]
11. Wang, S.C.; Wong, D.; Forrest, K.; Allen, A.; Chao, S.; Huang, B.E.; Maccaferri, M.; Salvi, S.; Milner, S.G.; Cattivelli, L.; et al. Characterization of polyploid wheat genomic diversity using a high-density 90000 single nucleotide polymorphism array. *Plant Biotechnol.* **2014**, *12*, 787–796. [CrossRef]
12. Liu, J.D.; He, Z.H.; Rasheed, A.; Wen, W.E.; Yan, J.; Zhang, P.Z.; Wan, Y.X.; Zhang, Y.; Xie, C.J.; Xia, X.C. Genome-wide association mapping of black point reaction in common wheat (*Triticum aestivum* L.). *BMC Plant Biol.* **2017**, *17*, 220. [CrossRef]
13. Liu, J.; Xu, Z.; Fan, X.; Zhou, Q.; Cao, J.; Wang, F.; Ji, G.; Yang, L.; Feng, B.; Wang, T.A. Genome-Wide Association Study of Wheat Spike Related Traits in China. *Front. Plant Sci.* **2018**, *9*, 1584. [CrossRef]
14. Li, F.J.; Wen, W.E.; He, Z.H.; Liu, J.D.; Jin, H.; Cao, S.H.; Geng, H.W.; Yan, J.; Zhang, P.Z.; Wan, Y.X.; et al. Genome-wide linkage mapping of yield related traits in three Chinese bread wheat populations using high-density SNP markers. *Theor. Appl. Genet.* **2018**, *131*, 1903–1924. [CrossRef] [PubMed]
15. Muqaddasi, Q.H.; Lohwasser, U.; Nagel, M.; Börner, A.; Pillen, K.; Röder, M.S. Genome-Wide Association Mapping of Anther Extrusion in Hexaploid Spring Wheat. *PLoS ONE* **2016**, *11*, e0155494. [CrossRef]
16. Adhikari, A.; Basnet, B.R.; Crossa, J.; Dreisigacker, S.; Camarillo, F.; Bhati, P.K.; Jarquin, D.; Manes, Y.; Ibrahim, A.M.H. Genome-Wide Association Mapping and Genomic Prediction of Anther Extrusion in CIMMYT Hybrid Wheat Breeding Program via Modeling Pedigree, Genomic Relationship, and Interaction With the Environment. *Front. Genet.* **2020**, *11*, 586687. [CrossRef] [PubMed]
17. Gahlaut, V.; Jaiswal, V.; Singh, S.; Balyan, H.S.; Gupta, P.K. Multi-Locus Genome Wide Association Mapping for Yield and Its Contributing Traits in Hexaploid Wheat under Different Water Regimes. *Sci. Rep.* **2019**, *9*, 19486. [CrossRef] [PubMed]
18. Jamil, M.; Ali, A.; Gul, A.; Ghafoor, A.; Napar, A.A.; Ibrahim, A.M.H.; Naveed, N.H.; Yasin, N.A.; Mujeeb-Kazi, A. Genome-wide association studies of seven agronomic traits under two sowing conditions in bread wheat. *BMC Plant Biol.* **2019**, *19*, 149. [CrossRef]
19. Li, F.; Wen, W.; Liu, J.; Zhang, Y.; Cao, S.; He, Z.; Rasheed, A.; Jin, H.; Zhang, C.; Yan, J.; et al. Genetic architecture of grain yield in bread wheat based on genome-wide association studies. *BMC Plant Biol.* **2019**, *19*, 168. [CrossRef]
20. Tsai, H.-Y.; Janss, L.L.; Andersen, J.R.; Orabi, J.; Jensen, J.D.; Jahoor, A.; Jensen, J. Genomic prediction and GWAS of yield, quality and disease-related traits in spring barley and winter wheat. *Sci. Rep.* **2020**, *10*, 3347. [CrossRef]
21. De Vries, A.P. Flowering biology of wheat, particularly in view of hybrid seed production—A review. *Euphytica* **1971**, *20*, 152–170. [CrossRef]
22. Longin, F.; Mühleisen, J.; Maurer, H.P.; Zhang, H.; Gowda, M.; Reif, J. Hybrid breeding in autogamous cereals. *Theor. Appl. Genet.* **2012**, *125*, 1087–1096. [CrossRef]
23. Langer, S.M.; Longin, C.F.H.; Würschum, T. Phenotypic evaluation of floral and flowering traits with relevance for hybrid breeding in wheat (*Triticum aestivum* L.). *Plant Breed.* **2014**, *133*, 433–441. [CrossRef]
24. Okada, T.; Ridma, J.E.A.; Jayasinghe, M.; Nansamba, M.; Baes, M.; Warner, P.; Kouidri, A.; Correia, D.; Nguyen, V.; Whitford, R.; et al. Unfertilized ovary pushes wheat flower for cross-pollination. *J. Exp. Bot.* **2018**, *69*, 399–412. [CrossRef] [PubMed]
25. Selva, C.; Riboni, M.; Baumann, U.; Würschum, T.; Whitford, R.; Tucker, M.R. Hybrid breeding in wheat: How shaping floral biology can offer new perspectives. *Funct. Plant Biol.* **2020**, *47*, 675–694. [CrossRef] [PubMed]
26. Zajączkowska, U.; Denisow, B.; Łotocka, B.; Dołkin-Lewko, A.; Rakoczy-Trojanowska, M. Spikelet movements, anther extrusion and pollen production in wheat cultivars with contrasting tendencies to cleistogamy. *BMC Plant Biol.* **2021**, *21*, 136. [CrossRef]
27. Hayes, J.D. Hybrid Wheat—Results and Problems, by A. A. Pickett. 259 pp. Berlin, Hamburg: Paul Parey. DM 96.00 (paperback). ISBN 3 489 53510 3. *J. Agric. Sci.* **1993**, *121*, 294. [CrossRef]
28. Tyrka, M.; Mokrzycka, M.; Bakera, B.; Tyrka, D.; Szeliga, M.; Stojałowski, S.; Matysik, P.; Rokicki, M.; Rakoczy-Trojanowska, M.; Krajewski, P. Evaluation of genetic structure in European wheat cultivars and advanced breeding lines using high-density genotyping-by sequencing approach. *BMC Genom.* **2021**, *22*, 81. [CrossRef] [PubMed]
29. McLaren, W.; Gil, L.; Hunt, S.E.; Riat, H.S.; Ritchie, G.R.; Thormann, A.; Flicek, P.; Cunningham, F. The Ensembl Variant Effect Predictor. *Genome Biol.* **2016**, *17*, 122. [CrossRef]

30. Gupta, D.R.; Avila, C.S.R.; Win, J.; Soanes, D.M.; Ryder, L.S.; Croll, D.; Bhattacharjee, P.; Hossain, M.S.; Mahmud, N.U.; Mehebub, M.S.; et al. Cautionary Notes on Use of the MoT3 Diagnostic Assay for Magnaporthe oryzae Wheat and Rice Blast Isolates. *Phytopathology* **2019**, *109*, 504–508. [CrossRef] [PubMed]
31. Waines, J.G.; Hegde, S.G. Intraspecific Gene Flow in Bread Wheat as Affected by Reproductive Biology and Pollination Ecology of Wheat Flowers. *Crop Sci.* **2003**, *43*, 451–463. [CrossRef]
32. Racedo, J.; Gutiérrez, L.; Perera, M.F.; Ostengo, S.; Pardo, E.M.; Cuenya, M.I.; Welin, B.; Castagnaro, A.P. Genome-wide association mapping of quantitative traits in a breeding population of sugarcane. *BMC Plant Biol.* **2016**, *16*, 142. [CrossRef]
33. Mwadzingeni, L.; Shimelis, H.; Rees, D.J.G.; Tsilo, T.J. Genome-wide association analysis of agronomic traits in wheat under drought-stressed and non-stressed conditions. *PLoS ONE* **2017**, *12*, e0171692. [CrossRef]
34. Mourad, A.M.I.; Sallam, A.; Belamkar, V.; Wegulo, S.; Bowden, R.; Jin, Y.; Mahdy, E.; Bakheit, B.; El-Wafaa, A.A.; Poland, J.; et al. Genome-Wide Association Study for Identification and Validation of Novel SNP Markers for Sr6 Stem Rust Resistance Gene in Bread Wheat. *Front. Plant Sci.* **2018**, *9*, 380. [CrossRef] [PubMed]
35. Malosetti, M.; Ribaut, J.M.; van Eeuwijk, F.A. The statistical analysis of multi-environment data: Modeling genotype-by-environment interaction and its genetic basis. *Front. Physiol.* **2013**, *4*, 44. [CrossRef] [PubMed]
36. Denisow, B.; Masierowska, M.; Winiarczyk, K.; Rakoczy-Trojanowska, M. The pollen dispersal ability for cross-pollination in winter wheat (*Triticum aestivum* L.) is related to anther extrusion capability rather than to pollen output. *S. Afr. J. Bot.* **2022**, *148*, 283–292. [CrossRef]
37. Allan, R.E. Wheat. In *Hybridization of Crop Plants*; Agronomy Books. 3; Fehr, W.R., Hadley, H.H., Eds.; Iova State University: Ames, IA, USA, 1980; Available online: https://lib.dr.iastate.edu/agron_books/3 (accessed on 28 November 2022).
38. Wilkinson, P.A.; Winfield, M.O.; Barker, G.L.A.; Allen, A.M.; Burridge, A.; Coghill, J.A.; Burridge, A.; Edwards, K.J. CerealsDB 2.0: An integrated resource for plant breeders and scientists. *BMC Bioinform.* **2013**, *13*, 219. [CrossRef]
39. Zhang, J.; She, M.; Yang, R.; Jiang, Y.; Qin, Y.; Zhai, S.; Balotf, S.; Zhao, Y.; Anwar, M.; Alhabbar, Z.; et al. Yield-Related QTL Clusters and the Potential Candidate Genes in Two Wheat DH Populations. *Int. J. Mol. Sci.* **2021**, *22*, 11934. [CrossRef]
40. Arjona, J.; Royo, C.; Dreisigacker, S.; Ammar, K.; Villegas, D. Effect of Ppd-A1 and Ppd-B1 Allelic Variants on Grain Number and Thousand Kernel Weight of Durum Wheat and Their Impact on Final Grain Yield. *Front. Plant Sci.* **2018**, *9*, 888. [CrossRef]
41. Zikhali, M.; Griffiths, S. The effect of Earliness per se (Eps) genes on flowering time in bread wheat. In *Advances in Wheat Genetics: From Genome to Field, Proceedings of the 12th International Wheat Genetics Symposium, Yokohama, Japan, 8–14 September 2013*; Yasunari, O., Shigeo, T., Hirokazu, H., Eds.; Springer: Tokyo, Japan, 2015; pp. 339–345. [CrossRef]
42. Wang, J.; Wen, W.; Hanif, M.; Xia, X.; Wang, H.; Liu, S.; Liu, J.; Yang, L.; Cao, S.; He, Z. TaELF3-1DL, a homolog of ELF3, is associated with heading date in bread wheat. *Mol. Breed.* **2016**, *36*, 161. [CrossRef]
43. Peng, F.Y.; Hu, Z.; Yang, R.-C. Genome-Wide Comparative Analysis of Flowering-Related Genes in Arabidopsis, Wheat, and Barley. *Int. J. Plant Genom.* **2015**, 874361. [CrossRef]
44. Lu, Q.; Lillemo, M.; Skinnes, H.; He, X.; Shi, J.; Ji, F.; Dong, Y.; Bjørnstad, A. Anther extrusion and plant height are associated with Type I resistance to Fusarium head blight in bread wheat line 'Shanghai-3/Catbird'. *Theor. Appl. Genet.* **2013**, *126*, 317–334. [CrossRef]
45. Boeven, P.H.G.; Longin, C.F.H.; Leiser, W.L.; Kollers, S.; Ebmeyer, E.; Würschum, T. Genetic architecture of male floral traits required for hybrid wheat breeding. *Theor. Appl. Genet.* **2016**, *129*, 2343–2357. [CrossRef]
46. Würschum, T.; Liu, G.; Boeven, P.H.G.; Longin, C.F.H.; Mirdita, V.; Kazman, E.; Zhao, Y.; Reif, J.C. Exploiting the Rht portfolio for hybrid wheat breeding. *Theor. Appl. Genet.* **2018**, *131*, 1433–1442. [CrossRef] [PubMed]
47. McIntosch, R.A.; Yamazaki, Y.; Devos, K.M.; Dubcovsky, J.; Rogers, W.J.; Appels, R. Catalogue of gene symbols for wheat. In Proceedings of the 10th International Wheat Genetics Symposium, Paestrum, Italy, 1–6 September 2003; Pogna, N.E., Romano, N., Pogna, E.A., Galterio, G., Eds.; Instituto Sperimentale per la Cerealicoltura: Rome, Italy, 2003; pp. 1–34.
48. Tian, X.; Wen, W.; Xie, L.; Fu, L.; Xu, D.; Fu, C.; Wang, D.; Chen, X.; Xia, X.; Chen, Q.; et al. Molecular Mapping of Reduced Plant Height Gene Rht24 in Bread Wheat. *Front. Plant Sci.* **2017**, *8*, 998. [CrossRef]
49. Würschum, T.; Langer, S.M.; Longin, C.F.H.; Tucker, M.R.; Leiser, W.L. A modern Green Revolution gene for reduced height in wheat. *Plant J.* **2017**, *92*, 892–903. [CrossRef] [PubMed]
50. Zanke, C.D.; Ling, J.; Plieske, J.; Kollers, S.; Ebmeyer, E.; Korzun, V.; Argillier, O.; Stiewe, G.; Hinze, M.; Neumann, K.; et al. Whole Genome Association Mapping of Plant Height in Winter Wheat (*Triticum aestivum* L.). *PLoS ONE* **2014**, *9*, e113287. [CrossRef]
51. Korzun, V.; Röder, M.; Ganal, M W.; Worland, A.J.; Law, C.N. Genetic analysis of the dwarfing gene (Rht8) in wheat. Part I. Molecular mapping of Rht8 on the short arm of chromosome 2D of bread wheat (*Triticum aestivum* L.). *Theor. Appl. Genet.* **1998**, *96*, 1104–1109. [CrossRef]
52. Ellis, M.H.; Rebetzke, G.J.; Azanza, F.; Richards, R.A.; Spielmeyer, W. Molecular mapping of gibberellin-responsive dwarfing genes in bread wheat. *Theor. Appl. Genet.* **2005**, *111*, 423–430. [CrossRef]
53. Peng, Z.S.; Li, X.; Yang, Z.J.; Liao, M.L. A new reduced height gene found in the tetraploid semi-dwarf wheat landrace Aiganfanmai. *Genet. Mol. Res.* **2011**, *10*, 2349–2357. [CrossRef]
54. Ellis, H.; Spielmeyer, W.; Gale, R.; Rebetzke, J.; Richards, A. "Perfect" markers for the Rht-B1b and Rht-D1b dwarfing genes in wheat. *Theor. Appl. Genet.* **2002**, *105*, 1038–1042. [CrossRef]
55. Pearce, S.; Saville, R.; Vaughan, S.P.; Chandler, P.M.; Wilhelm, E.P.; Sparks, C.A.; Al-Kaff, N.; Korolev, A.; Boulton, M.I.; Phillips, A.L.; et al. Molecular characterization of Rht-1 dwarfing genes in hexaploid wheat. *Plant Physiol.* **2011**, *157*, 1820–1831. [CrossRef]

56. Xu, T.; Bian, N.; Wen, M.; Xiao, J.; Yuan, C.; Cao, A.; Zhang, S.; Wang, X.; Wang, H. Characterization of a common wheat (*Triticum aestivum* L.) high-tillering dwarf mutant. *Theor. Appl. Genet.* **2017**, *130*, 483–494. [CrossRef]
57. Alemu, A.; Suliman, S.; Hagras, A.; Thabet, S.; Al-Abdallad, A.; Abdelmula, A.A.; Tadesse, W. Multi-model genome-wide association and genomic prediction analysis of 16 agronomic, physiological and quality related traits in ICARDA spring wheat. *Euphytica* **2021**, *217*, 205. [CrossRef]
58. Yan, X.; Zhao, L.; Ren, Y.; Dong, Z.; Cui, D.; Chen, F. Genome-wide association study revealed that the TaGW8 gene was associated with kernel size in Chinese bread wheat. *Sci. Rep.* **2019**, *9*, 2702. [CrossRef] [PubMed]
59. Yang, J.; Zhou, Y.; Wu, Q.; Chen, Y.; Zhang, P.; Zhang, Y.; Cao, T. Molecular characterization of a novel TaGL3–5A allele and its association with grain length in wheat (*Triticum aestivum* L.). *Theor. Appl. Genet.* **2019**, *132*, 1799–1814. [CrossRef] [PubMed]
60. Yang, J.; Zhou, Y.; Zhang, Y.; Hu, W.; Wu, Q.; Chen, Y.; Zhao, H. Cloning, characterization of TaGS3 and identifcation of allelic variation associated with kernel traits in wheat (*Triticum aestivum* L.). *BMC Genet.* **2019**, *20*, 98. [CrossRef] [PubMed]
61. Yang, L.; Zhao, D.; Meng, Z.; Xu, K.; Yan, J.; Xia, X.; Zhang, Y. QTL mapping for grain yield-related traits in bread wheat via SNP-based selective genotyping. *Theor. Appl. Genet.* **2019**, *133*, 857–872. [CrossRef] [PubMed]
62. Daba, S.D.; Tyagi, P.; Brown-Guedira, G.; Mohammadi, M. Genome-wide association studies to identify loci and candidate genes controlling kernel weight and length in a historical United States wheat population. *Front. Plant Sci.* **2018**, *9*, 1045. [CrossRef]
63. Hadfield, K.A.; Bennett, A.B. Polygalacturonases: Many genes in search of a function. *Plant Physiol.* **1998**, *117*, 337–343. [CrossRef]
64. Babu, Y.; Bayer, M. Plant Polygalacturonases Involved in Cell Elongation and Separation-The Same but Different? *Plants* **2014**, *3*, 613–623. [CrossRef]
65. Yang, Y.; Yu, Y.; Liang, Y.; Anderson, C.T.; Cao, J. A Profusion of Molecular Scissors for Pectins: Classification, Expression, and Functions of Plant Polygalacturonases. *Front. Plant Sci.* **2018**, *9*, 1208. [CrossRef]
66. Chateigner-Boutin, A.L.; Suliman, M.; Bouchet, B.; Alvarado, C.; Lollier, V.; Rogniaux, H.; Guillon, F.; Larré, C. Endomembrane proteomics reveals putative enzymes involved in cell wall metabolism in wheat grain outer layers. *J. Exp. Bot.* **2015**, *66*, 2649–2658. [CrossRef]
67. Ye, X.; Li, J.; Cheng, Y.; Yao, F.; Long, L.; Wang, Y.; Wu, Y.; Li, J.; Wang, J.; Jiang, Q.; et al. Genome-wide association study reveals new loci for yield-related traits in Sichuan wheat germplasm under stripe rust stress. *BMC Genom.* **2019**, *20*, 640. [CrossRef] [PubMed]
68. Shu, K.; Yang, W. E3 Ubiquitin Ligases: Ubiquitous Actors in Plant Development and Abiotic Stress Responses. *Plant Cell Physiol.* **2017**, *58*, 1461–1476. [CrossRef] [PubMed]
69. Srivastava, A.K.; Lu, Y.; Zinta, G.; Lang, Z.; Zhu, J.K. UTR-Dependent Control of Gene Expression in Plants. *Trends Plant Sci.* **2018**, *23*, 248–259. [CrossRef] [PubMed]
70. Remy, E.; Cabrito, T.R.; Batista, R.A.; Hussein, M.A.M.; Teixeira, M.C.; Athanasiadis, A.; Sá-Correia, I.; Duque, P. Intron Retention in the 5′UTR of the Novel ZIF2 Transporter Enhances Translation to Promote Zinc Tolerance in Arabidopsis. *PLoS Genet.* **2014**, *10*, e1004375. [CrossRef]
71. Chung, B.Y.; Simons, C.; Firth, A.E.; Brown, C.M.; Hellens, R.P. Effect of 5′UTR introns on gene expression in Arabidopsis thaliana. *BMC Genom.* **2006**, *7*, 120. [CrossRef]
72. Zhao, M.; Liu, B.; Wu, K.; Ye, Y.; Huang, S.; Wang, S.; Wang, Y.; Han, R.; Liu, Q.; Fu, X.; et al. Regulation of OsmiR156h through Alternative Polyadenylation Improves Grain Yield in Rice. *PLoS ONE* **2015**, *10*, e0126154. [CrossRef] [PubMed]
73. Jung, C.; Muller, A.E. Flowering time control and applications in plant breeding. *Trends Plant Sci.* **2009**, *14*, 563–573. [CrossRef]
74. Gaudinier, A.; Blackman, B.K. Evolutionary processes from the perspective of flowering time diversity. *New Phytol.* **2020**, *225*, 1883–1898. [CrossRef]
75. Wu, C.-C.; Wei, F.-J.; Chiou, W.-Y.; Tsai, Y.-C.; Wu, H.-P.; Gotarkar, D.; Wei, Z.H.; Lai, M.H.; Hsing, Y.I.C. Studies of rice Hd1 haplotypes worldwide reveal adaptation of flowering time to different environments. *PLoS ONE* **2020**, *15*, e0239028. [CrossRef]
76. Research Centre for Cultivar Testing. Available online: http://www.coboru.pl (accessed on 23 June 2019).
77. Cullis, B.R.; Smith, A.B.; Coombes, N.E. On the design of early generation variety trials with correlated data. *J. Agric. Biol. Environ. Stat.* **2006**, *11*, 381–393. [CrossRef]
78. Gauch, H.G. *Statistical Analysis of Regional Yield Trials: AMMI Analysis of Factorial Designs*; Elsevier: Amsterdam, The Netherlands, 1992.
79. van Eeuwijk, F.A.; Bink, M.C.A.M.; Chenu, K.; Chapman, S.C. Detection and use of QTL for complex traits in multiple environments. *Curr. Opin. Plant Biol.* **2010**, *13*, 193–205. [CrossRef] [PubMed]
80. Vidotti, M.S.; Lyra, D.H.; Morosini, J.S.; Granato, Í.S.C.; Quecine, M.C.; Azevedo, J.L.D.; Fritsche-Neto, R. Additive and heterozygous (dis)advantage GWAS models reveal candidate genes involved in the genotypic variation of maize hybrids to Azospirillum brasilense. *PLoS ONE* **2019**, *14*, e0222788. [CrossRef] [PubMed]
81. VSN International. Genstat for Windows. 19th ed. Hemel Hempstead. Web Page. 2017. Available online: Genstat.co.uk (accessed on 28 November 2022).
82. Kashif, M.; Khaliq, I. Heritability, correlation and path coefficient analysis for some metric traits in wheat. *Int. J. Agric. Biol.* **2004**, *6*, 138–142.
83. Nukasani, V.; Potdukhe, N.R.; Bharad, S.; Deshmukh, S.; Shinde, S.M. Genetic variability, correlation and path analysis in wheat. *J. Wheat Res.* **2013**, *5*, 48–51.

84. Guo, Z.; Zhao, Y.; Röder, M.S.; Reif, J.C.; Ganal, M.W.; Chen, D.; Schnurbusch, T. Manipulation and prediction of spike morphology traits for the improvement of grain yield in wheat. *Sci. Rep.* **2018**, *8*, 14435. [CrossRef]
85. Mangini, G.; Gadaleta, A.; Colasuonno, P.; Marcotuli, I.; Signorile, A.M.; Simeone, R.; De Vita, P.; Mastrangelo, A.M.; Laidò, G.; Pecchioni, N.; et al. Genetic dissection of the relationships between grain yield components by genome-wide association mapping in a collection of tetraploid wheats. *PLoS ONE* **2018**, *13*, e0190162. [CrossRef]

International Journal of
Molecular Sciences

MDPI

Article

Identification and Validation of Quantitative Trait Loci Mapping for Spike-Layer Uniformity in Wheat

Kunyu Zhou [1,2,†], Yu Lin [1,2,†], Xiaojun Jiang [1,2], Wanlin Zhou [1,2], Fangkun Wu [1,2], Caixia Li [2], Yuming Wei [1,2] and Yaxi Liu [1,2,*]

1 State Key Laboratory of Crop Gene Exploration and Utilization in Southwest China, Chengdu 611130, China; zhoukunyu@stu.sicau.edu.cn (K.Z.); linyu@stu.sicau.edu.cn (Y.L.); jiangxj999@gmail.com (X.J.); zhouwanlin@stu.sicau.edu.cn (W.Z.); 2018112004@stu.sicau.edu.cn (F.W.); ymwei@sicau.edu.cn (Y.W.)
2 Triticeae Research Institute, Sichuan Agricultural University, Chengdu 611130, China; licaixia@sicau.edu.cn
* Correspondence: liuyaxi@sicau.edu.cn; Tel.: +86-28-86290952; Fax: +86-28-82650350
† These authors contributed equally to this work.

Abstract: Spike-layer uniformity (SLU), the consistency of the spike distribution in the vertical space, is an important trait. It directly affects the yield potential and appearance. Revealing the genetic basis of SLU will provide new insights into wheat improvement. To map the SLU-related quantitative trait loci (QTL), 300 recombinant inbred lines (RILs) that were derived from a cross between H461 and Chinese Spring were used in this study. The RILs and parents were tested in fields from two continuous years from two different pilots. Phenotypic analysis showed that H461 was more consistent in the vertical spatial distribution of the spike layer than in Chinese Spring. Based on inclusive composite interval mapping, four QTL were identified for SLU. There were two major QTL on chromosomes 2BL and 2DL and two minor QTL on chromosomes 1BS and 2BL that were identified. The additive effects of *QSlu.sicau-1B*, *Qslu.sicau-2B-2*, and *QSlu.sicau-2D* were all from the parent, H461. The major QTL, *QSlu.sicau-2B-2* and *QSlu.sicau-2D*, were detected in each of the conducted trials. Based on the best linear unbiased prediction values, the two loci explained 23.97% and 15.98% of the phenotypic variation, respectively. Compared with previous studies, the two major loci were potentially novel and the two minor loci were overlapped. Based on the kompetitive allele-specific PCR (KASP) marker, the genetic effects for *QSlu.sicau-2B-2* were validated in an additional RIL population. The genetic effects ranged from 26.65% to 32.56%, with an average value of 30.40%. In addition, *QSlu.sicau-2B-2* showed a significant ($p < 0.01$) and positive influence on the spike length, spikelet number, and thousand kernel weight. The identified QTL and the developed KASP marker will be helpful for fine-mapping these loci, finally contributing to wheat breeding programs in a marker-assisted selection way.

Keywords: QTL mapping; spike-layer uniformity; kompetitive allele-specific PCR marker; yield traits; recombinant inbred lines; wheat

Citation: Zhou, K.; Lin, Y.; Jiang, X.; Zhou, W.; Wu, F.; Li, C.; Wei, Y.; Liu, Y. Identification and Validation of Quantitative Trait Loci Mapping for Spike-Layer Uniformity in Wheat. *Int. J. Mol. Sci.* **2022**, *23*, 1052. https://doi.org/10.3390/ijms23031052

Academic Editors: Andrés J. Cortés and Hai Du

Received: 19 December 2021
Accepted: 15 January 2022
Published: 19 January 2022

1. Introduction

Wheat (*Triticum aestivum* L.) is one of the most important crops in the world, and it provides 20% of the global human population with calories [1,2]. To feed the world's growing population, an annual rate increase of 2.4% in wheat productivity is expected. As the current annual rate of 0.9% is well below expectations, it is urgent to increase wheat productivity. Breeding high-yield wheat cultivars is one of the most effective strategies to increase the total food production.

Compared with conventional breeding methods, marker-assisted selection is an economic and efficient strategy to breed cultivars. Marker-assisted selection is a tool for breeders, allowing the possibility to select desirable traits more directly using DNA markers that are linked to genes/loci of interest [3]. Quantitative trait loci (QTL) mapping can provide usable molecular markers that are linked to a gene of interest. Up to now, QTL mapping has been widely applied to discover the locations of loci underlying traits in plants [4–13]. With the development of high-throughput sequencing technology, single nucleotide polymorphism (SNP) arrays, including the 9K, 55K, 90K, 660K, and 820K arrays, have been successfully used for constructing high-density genetic maps of wheat [14–18]. Based on this, QTL mapping has been used for detection within relatively narrow genetic intervals [14,15,19]. The availability of high-throughput SNPs and high-quality reference genomes combine to make the identification of QTL more precise. The identified QTL-linked markers, via dense genetic maps, could be directly used in wheat improvement through molecular assisted selection (MAS) and further accelerating fine mapping and map-based gene cloning.

Spike-layer uniformity (SLU), the consistency of the spike distribution in the vertical space, is determined by the different plant heights (PH) between the main stem and tillers. A uniform spike-layer ensures, not only uniform grain sizes, but also a synchronized maturation time and facilitates harvesting [20]. In rice, ideotype plants with a high yield potential are suggested to have uniformity in the panicle layer [20]. Several genes have been demonstrated to control SLU and affect yield [21,22]. In wheat, the SLU is a quantitative trait and directly affects the yield potential and appearance [23,24]. It has relatively high heritability and is affected by both genotypes and environmental factors [23,24]. A previous study found that the genotype "Jing411" has a higher SLU value than "Kenong9204" [23]. This indicated that "Jing411" has a more consistent vertical spatial distribution. To identify the wheat genotypes with uniform spike-layers and loci that are associated with this trait, a natural population, including 225 diverse bread wheats, was also used in a previous study. In previous studies, QTL for SLU have been identified on chromosomes 1A, 1B, 1D, 2A, 2B, 2D, 3A, 3B, 4A, 4B, 5A, 5B, 6A, 6B, 6D, 7A, and 7D [23,24]. Among these, only one locus, *qSlu-4A*, was detected in three environments; the others were identified in one or two environments. To breed wheat with consistent spike heights, it is necessary to detect stable QTL of SLU among multiple environments. These expected QTL were reliable in the marker-assisted selection of wheat improvement.

Spike-layer uniformity is an important trait that directly affects wheat yield potential and appearance. Uncovering the genetic basis of SLU in wheat will provide a new insight into wheat improvement. In this study, the SLU of H461 was more consistent than that of Chinese Spring. The derived recombinant inbred line (RIL) population was used to evaluate the SLU from two consecutive years from two different environments. Based on the high-density map that was constructed by the 55K SNP array, QTL mapping was performed for this trait in multiple environments and using the best linear unbiased prediction (BLUP) values. The genetic effects of the objective locus were validated in a different background using an additional RIL population. Further analysis was performed to reveal the genetic relationships between the SLU and several yield traits.

2. Results

2.1. Phenotypic Characterization of Spike-Layer Uniformity

A significant ($p < 0.01$) difference was identified for the SLU among genotypes, environments, and genotype–environment interactions (Table 1). In all four trials, the SLU of H461 ranged from 0.66 to 0.83, while the SLU of Chinese Spring ranged from 0.18 to 0.24 (Figure 1, Table 2). The spike-layer uniformity of H461 was significantly higher than that of Chinese Spring in all four trials (Table 2). Based on the BLUP values, the SLU of H461 was 0.80, which was 3.6 times higher than that of Chinese Spring. The results revealed the great potential of wheat spike-layer improvement using the material H461. In all the trials, transgressive segregation was observed in the RIL population (Figure 1, Table 2). The broad-sense heritability of SLU was 0.88, indicating that this trait was majorly controlled by genetic factors. Significant ($p < 0.01$) correlations were observed among all four environments with correlation coefficients ranging from 0.62 to 0.79 (Supplementary Table S1), which indicated that the SLUs that were obtained from the four trials were strongly correlated.

Table 1. Analysis of variance for spike-layer uniformity across the four environments.

Variable	DF [a]	Type III Sum of Square	Mean Square	F Value	Significance [b]
Environment	3	3.76	1.25	141.77	**
Replication	4	0.02	0.005	0.52	ns
Genotype	298	32.51	0.11	12.35	**
Environment × Genotype	885	10.29	0.01	1.32	**

[a] DF, degrees of freedom. [b] **, significant at $p < 0.01$; ns, not significant.

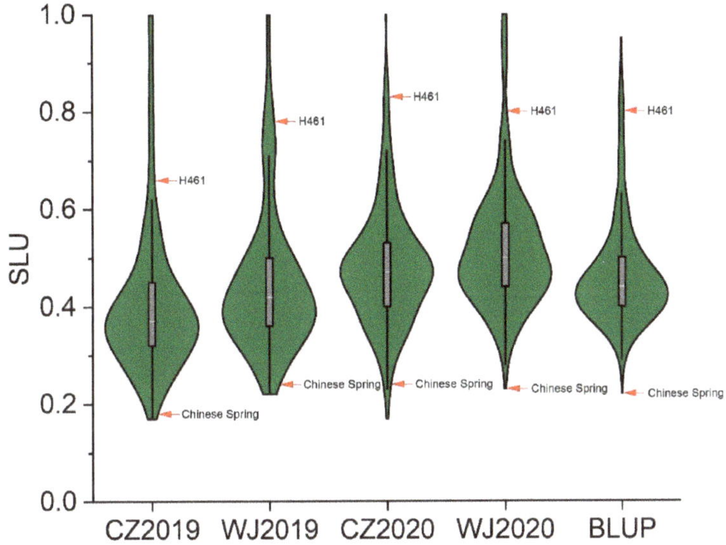

Figure 1. Violin plots for spike-layer uniformity distributions of the H461 × Chinese Spring recombinant inbred line population in different environments. CZ2019, CZ2020, WJ2019, and WJ2020 denote field experiments in Chongzhou (2019, 2020) and Wenjiang (2019, 2020). BLUP, best linear unbiased prediction.

Table 2. Phenotypic variation and the broad-sense heritability of spike-layer uniformity in different environments.

Environment [a]	Parents			Recombinant Inbred Line (RIL) Population					
	H461	Chinese Spring	Difference [b]	Mean	Min	Max	SD [c]	CV (%) [d]	Heritability
CZ2019	0.66	0.18	0.49 **	0.41	0.17	1.00	0.15	37.61	
WJ2019	0.78	0.24	0.53 **	0.45	0.22	1.00	0.15	32.84	
CZ2020	0.83	0.24	0.59 **	0.48	0.17	1.00	0.12	25.11	
WJ2020	0.80	0.23	0.57 **	0.52	0.28	1.00	0.13	24.34	
BLUP	0.80	0.22	0.58 **	0.58	0.29	0.95	0.11	18.69	0.88

[a] CZ2019, CZ2020, WJ2019, and WJ2020 denote field experiments in Chongzhou (2019, 2020) and Wenjiang (2019, 2020). BLUP, best linear unbiased prediction. [b] **, significant at $p < 0.01$. [c] SD, standard deviation. [d] CV, coefficient of variation.

2.2. *Identification of QTL Conferring Spike-Layer Uniformity*

A total of four QTL were identified for the SLU (Figure 2, Table 3). Among them, two major QTL were located on chromosomes 2B (*QSlu.sicau-2B-2*) and 2D (*QSlu.sicau-2D*), and two minor QTL were located on chromosomes 1B (*QSlu.sicau-1B*) and 2B (*QSlu.sicau-2B-1*). *QSlu.sicau-2B-2* (flanked by AX-108770043 and AX-108927717) was detected in all four trials, as well as in the BLUP values. This locus was located on the long arm of chromosome 2B at 165.70–166.40 cM and explained 13.89–23.97% of the phenotypic variation. The other major locus, *QSlu.sicau-2D,* was detected in all the trials, as well in the BLUP values. It was located on the long arm of chromosome 2D at 230.67–235.60 cM and explained

12.93–15.98% of the phenotypic variation. The minor locus *QSlu.sicau-1B* (flanked by AX-109490843 and AX-111619113) was detected in CZ2020, as well as in the BLUP values and was located on the short arm of chromosome 1B at 6.87–7.22 cM, explaining 3.43–3.55% of the phenotypic variation. The other minor locus, *QSlu.sicau-2B-1* (flanked by AX-111487903 and AX-111457622), was detected in WJ2019 as well as in the BLUP values, explaining 2.05–3.91% of the phenotypic variation. This locus was on the long arm of chromosome 2B at 137.91–139.55 cM. With the exception of *QSlu.sicau-2B-1*, alleles for the higher SLUs at these loci were all from the parent, H461.

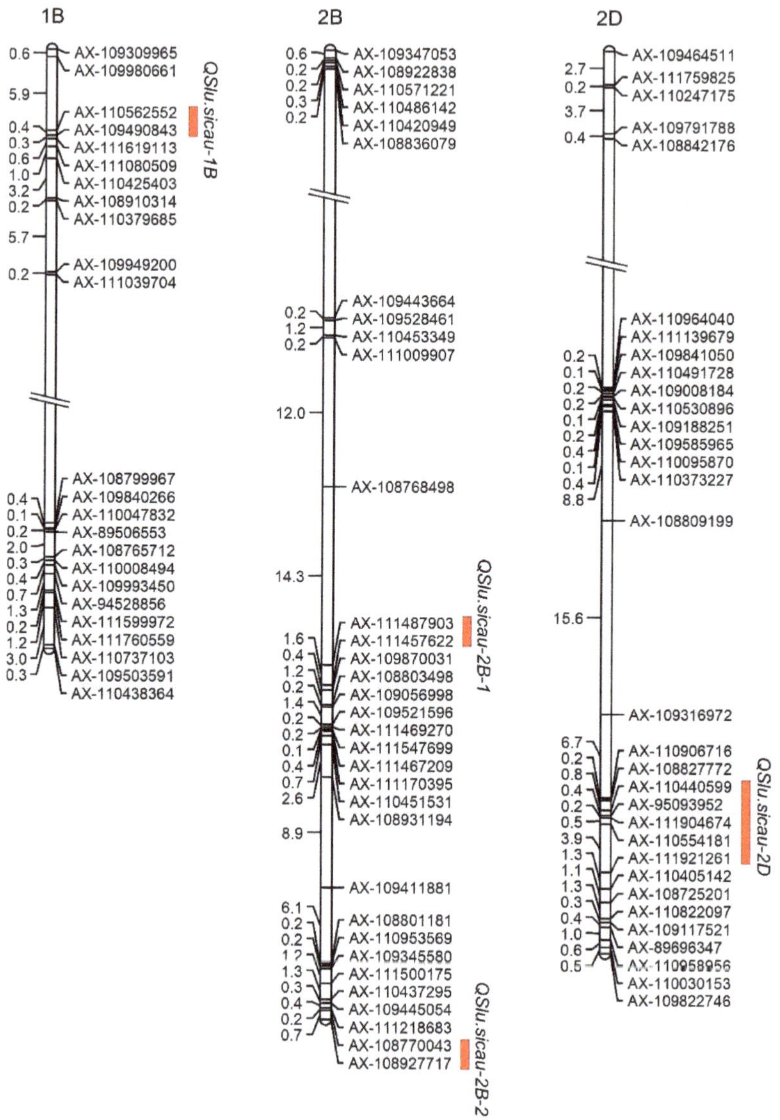

Figure 2. Chromosomal locations of QTL for spike-layer uniformity in the H461 × Chinese Spring recombinant inbred line population. The red bar points to the genetic interval of QTL. The numbers on the left side of the chromosomes point to the genetic distance (cM) between the markers. The marker's names are shown on the right side of chromosomes

Table 3. Quantitative trait loci for spike-layer uniformity that were identified in different environments and using BLUP values.

QTL	Environment [a]	Chromosome	Genetic Distance (cM)	Physical Positions (Mb)	Flanking Marker	LOD [b]	PVE (%) [c]	AE [d]
QSlu.sicau-1B	CZ2020	1B	6.87–7.22	572.94–577.58	AX-109490843 and AX-111619113	3.73	3.43	0.024
	BLUP	1B	6.87–7.22	572.94–577.58	AX-109490843 and AX-111619113	5.96	3.55	0.024
QSlu.sicau-2B-1	WJ2019	2B	137.91–139.55	755.21–777.78	AX-111487903 and AX-111457622	6.17	3.91	−0.034
	BLUP	2B	137.91–139.55	755.21–777.78	AX-111487903 and AX-111457622	3.47	2.05	−0.018
QSlu.sicau-2B-2	CZ2019	2B	165.70–166.40	798.21–802.29	AX-108770043 and AX-108927717	17.66	17.98	0.070
	WJ2019	2B	165.70–166.40	788.13–792.19	AX-108770043 and AX-108927717	28.76	21.21	0.078
	CZ2020	2B	165.70–166.40	788.13–792.20	AX-108770043 and AX-108927717	14.06	13.89	0.048
	WJ2020	2B	165.70–166.40	788.13–792.21	AX-108770043 and AX-108927717	16.12	16.73	0.055
	BLUP	2B	165.70–166.40	788.13–792.22	AX-108770043 and AX-108927717	32.70	23.97	0.062
QSlu.sicau-2D	CZ2019	2D	231.72–235.60	565.41–640.84	AX-110554181 and AX-111921261	13.61	13.67	0.062
	WJ2019	2D	230.67–231.02	633.66–634.41	AX-110440599 and AX-95093952	18.98	12.93	0.061
	CZ2020	2D	231.72–235.60	565.41–640.84	AX-110554181 and AX-111921261	14.46	14.88	0.050
	WJ2020	2D	231.72–235.60	565.41–640.84	AX-110554181 and AX-111921261	14.84	15.39	0.053
	BLUP	2D	230.67–231.02	633.66–634.41	AX-110440599 and AX-95093952	23.75	15.98	0.051

[a] CZ2019, CZ2020, WJ2019, and WJ2020 denote field experiments in Chongzhou (2019, 2020) and Wenjiang (2019, 2020). BLUP, best linear unbiased prediction. [b] LOD, the logarithm of odds score. [c] PVE, the percentage of phenotypic variance explained by individual QTL. [d] AE, additive effect. Positive values mean alleles from H461 increase the trait scores, and negative values mean alleles from Chinese Spring increase the trait scores.

2.3. Validation of the Major and Stable QTL, with a Different Genetic Background and Its Effect on Yield-Related Traits

The tightly linked SNP, AX-108927717, was converted into a kompetitive allele-specific PCR (KASP) marker (KASP-7717) to validate the major and stable locus, *QSlu.sicau-2B-2*, with a different genetic background (Table 4). The KASP marker was polymorphic, between H461 and the other two parents (Chinese Spring and CM42). In the RILs that were derived from H461 and CM42, the SLU of the group carrying the "aa" allele was significantly ($p < 0.01$) higher than that of the group carrying the "AA" allele (Figure 3). The differences between the two groups were 32.56%, 31.98%, and 26.65% for CZ2019, WJ2019, and BLUP, respectively, with an average value of 30.40% (Figure 3). In addition, the spike length (SL) of the group carrying the "aa" allele was significantly ($p < 0.05$) higher than that of the group carrying the "AA" allele. The group carrying the "aa" allele had a significantly ($p < 0.01$) higher spikelet number (SN) and thousand kernel weight (TKW) than the group carrying the "AA" allele. Thus, *QSlu.sicau-2B-2* showed a significant and positive influence on SL, SN, and TKW, but it did not affect PH (Table 5). These results indicated that this locus has great potential in wheat yield improvement.

Table 4. The primers of the kompetitive allele-specific marker for alleles of *Qslu.sicau-2B-2* that were used in this study.

SNP	Primer [a]	Primer sequence (5′ to 3′)	Allele [b]
AX-108927717	FAM	TTGGAATGTCTCCATCCCAC	aa
	HEX	TTGGAATGTCTCCATCCCAG	AA
	Common reverse	CCTCTCCTATATCTGGCTTCTGTTG	

[a] FAM probe sequence of the forward primer is GAAGGTGACCAAGTTCATGCT, and the HEX probe sequence of the reverse primer is GAAGGTCGGAGTCAACGGATT. [b] aa, lines carrying the homozygous alleles from H461; AA, lines carrying the homozygous alleles from the non-H461 parents.

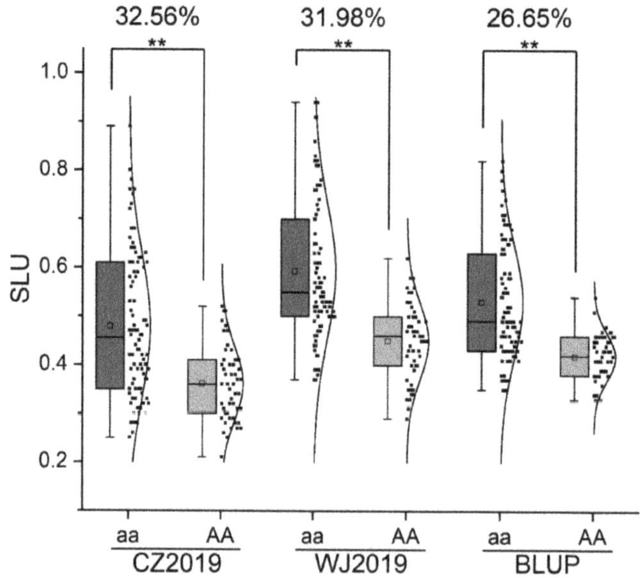

Figure 3. Effects of *Qslu.sicau-2B-2* in the H461 × CM42 recombinant inbred line population. Aa, lines carrying the homozygous alleles from H461. AA, lines carrying the homozygous alleles from the non-H461 parents. CZ2019 and WJ2019 denote the field experiments in 2019 in Chongzhou and Wenjiang, respectively. BLUP, best linear unbiased prediction. **, significant at $p < 0.01$.

Table 5. Effects of *QSlu.sicau-2B-2* on the yield-related traits in the H461 × HC42 recombinant inbred line population.

Environment [a]	Allele	Plant Height (cm)	Spike Length (cm) [b]	Spikelet Number	Thousand Kernel Weight (g)
CZ2019	aa	84.22	13.60 *	22.11 **	50.15 **
	AA	84.22	13.23	20.81	45.84
WJ2019	aa	82.01	13.45 **	20.92 **	50.15 **
	AA	82.01	12.80	19.74	45.84
BLUP	aa	83.10	13.50 *	21.48 **	47.58 **
	AA	81.25	13.05	20.35	44.60

[a] CZ2019 and WJ2019 denote field experiments in 2019 in Chongzhou and Wenjiang, respectively. BLUP, best linear unbiased prediction. [b] *, significant at $p < 0.05$. **, significant at $p < 0.01$.

2.4. Candidate Genes for QSlu.sicau-2B-2

Based on the Chinese Spring Reference Genome Sequence RefSeq v2.1 [25,26], *QSlu.sicau-2B-2* was located on the long arm of chromosome 2B at 798.21–802.29 Mb. A total of 49 predicted genes were identified in the interval. After homologous analysis, a total of 47 genes were homologous with rice. There were two homologous genes of rice that have been reported regarding their functions, *GSTU5* [27] and *GSTU45* [27]. Expression analysis of these genes is shown in Figure S1. *TraesCS2B03G1529700*, which encodes thioredoxin, showed the highest expression, from the seedling stage to the dough stage (Table S2). The expressions of eight genes, including *TraesCS2B03G1522100*, *TraesCS2B03G1522700*, *TraesCS2B03G1524300*, *TraesCS2B03G1526400*, *TraesCS2B03G1531500*, *TraesCS2B03G1531600*, and *TraesCS2B03G1534400*, were shown to be highly expressed in all developmental stages.

3. Discussion

The uniformity of the spike layer significantly affects the plant architecture and grain yield in wheat. The high heritability of SLU in the present study indicated that genetic factors strongly determined this trait (Table 1). The high heritability for SLU was also observed in the population that included 20 wheat cultivars/lines [28]. Previous studies also confirmed that genetic factors mainly controlled SLU [22,28,29]. Thus, it is necessary to understand the genetic basis of SLU for MAS to be applied in wheat breeding.

The genetic basis of SLU is poorly understood in wheat. In rice, a few genes have been demonstrated to control plant uniformity, including *DWT1*, *DWL2*, and *OsPIP5K1* [21,22]. *DWT1* encodes a *WOX* transcription factor and interacts with *OsPIP5K1* to coordinate rice uniformity [21,22]. *DWL2* has the same redundant functions as *DWT1* and acts with *OsPIP5K1* to coordinate uniformity in rice [21]. To the best of our knowledge, only two studies on QTL identification for SLU have been reported [23,24]. A total of 19 QTL have been detected and are located on chromosomes 1B, 1D, 2A, 2B, 2D, 4A, 4B, 5A, 5B, 6B, 7A, and 7D. Based on the RIL population consisting of 188 lines, 16 QTL were detected on chromosomes 1B, 1D, 2A (2 QTL), 2B (2 QTL), 4A, 4B, 5A, 5B, 6B (2 QTL), 7A (2 QTL), and 7D (2 QTL), explaining 5.04–18.43% phenotypic variation. Of those, one (*qSlu-4A*) could be detected across three environments and two (*qSlu-4B* and *qSlu-6B.1*) across two environments [23]. In the present study, only four QTL were detected for SLU (Figure 2, Table 3). This is due to the loci that were detected in only one environment being removed in this study. It is well known that QTL that are detected in one environment are not useful in breeding. Due to the significant differences between parents, the large population, and abundant polymorphism markers, two stable and major QTL were detected in this study. Recently, using a set of 225 diverse spring wheat accessions, 14 marker-trait associations for SLU were detected on chromosomes 1A, 1B, 2D, 3A, 3B, 4A, 4B, 5A, 5B, 6A, 6B, 6D, and 7A [24]. Only the SNP, SNP_10924, was significantly associated with the SLU in two environments, explaining up to 6% of the phenotypic variation [24]. These results indicated that detecting a more stable and major QTL for SLU is necessary.

Based on the Chinese Spring Reference Sequence RefSeq v2.1 [25,26], QTL were compared with previous studies [23,24]. In this study, one minor-effect QTL, *QSlu.sicau-1B*, was identified on chromosome 1B at 572.94–577.58 Mb. In a previous study, the marker SNP_11184 was located on the short arm of chromosome 1B and was significantly associated with SLU [24]; it was far from *QSlu.sicau-1B*. In another study, the locus *qSlu-1BL* was identified on chromosome 1B in the interval between wPt-2315 and AX-108850061 [23]. Due to the unavailable physical position of the DArT marker wPt-2315, the physical position of this marker was referred to by the adjacent marker, AX-111595814. Thus, *qSlu-1BL* is probably located on chromosome 1B at 575.35–587.26 Mb, overlapping with *QSlu.sicau-1B*. *QSlu.sicau-2B-1* was another minor-effect QTL. This QTL was located on chromosome 2B at 755.21–777.78 Mb, overlapping with *qSlu-2B.2* (flanked by AX-111041164 and AX-94403958) [23]. The locus *qSlu-2B.2* was located on chromosome 2B at 735.06–779.53 Mb. *QSlu.sicau-2B-2* was a major and stable QTL with the highest effect. This QTL was located on the long arm of chromosome 2B at 798.21–802.29 Mb. It is far from *qSlu-2B.1* (chromosome 2B at 725.48-725.76 Mb) and *qSlu-2B.2* (chromosome 2B at 735.06-779.53 Mb), indicating that *QSlu.sicau-2B-2* is a potential novel QTL for SLU. *QSlu.sicau-2D* is another QTL with a major effect. This QTL is located on the long arm of chromosome 2D at 565.41–640.84 Mb. One of the SLU-related QTL on the short arm of chromosome 2D was reported previously [24], indicating *QSlu.sicau-2D* as another potential novel QTL. A tightly linked marker of *QSlu.sicau-2B-2* was designed to be a KASP marker that was applied in wheat molecular breeding programs (Table 4). The KASP marker, KASP-7717, was genotyped in the HC42 population. The locus *QSlu.sicau-2B-2* showed a great effect in a different background with an average difference of 30.40% (Figure 3). This result indicated that *QSlu.sicau-2B-2* in MAS for plant architecture improvement in wheat is practicable.

Spike-layer uniformity is a vital plant architecture trait and affects the yield potential in crops with tillers, such as rice, barley, and wheat. Previous studies have found that the SLU and yield potential were positively correlated [22,29–31]. It has been demonstrated that DWT1 regulates intra-panicle uniformity and coordinates panicle and internodes in rice [22]. In wheat, knowledge of the genetic relationships between the SLU and yield-related traits is limited. A previous study found that the SLU was significantly negatively correlated with PH, TKW, and the yield per plant [23]. In the present study, *QSlu.sicau-2B-1*, located on chromosome 2B at 755.21–777.78 Mb, was overlapped with the interval of *QSl.sicau-2B-1* (Tables 3 and S3). The additive effects of these two QTL were both from Chinese Spring. This indicated that this locus could regulate both the SLU and SL and not the lowest tiller height (LTH) or PH. *QSlu.sicau-2B-2*, located on chromosome 2B at 798.21–802.29 Mb, was overlapped with the interval of *QGns.sicau-2B* [14]. This locus was also overlapped with the interval of *QSl.sicau-2B-2* (Tables 3 and S3). The additive effects of these three QTL were both from H461. This indicated that this locus could regulate SLU and coordinate the grain number per spikelet and SL. In the RIL population that was derived from the crossing of H461 and CM42, *QSlu.sicau-2B-2* also had a positive influence on SL, SN, and TKW, but did not affect PH (Table 5). This finding was inconsistent with previous results [23] indicating that the mechanism of *QSlu.sicau-2B-2* may be different. This locus could improve the uniformity of the spike layer and several yield-related traits, demonstrating that applications of *QSlu.sicau-2B-2* in wheat yield improvement by MAS are feasible and necessary. A total of 49 genes were located in the interval of *QSlu.sicau-2B-2*. Of those, eight showed high expression in the wheat development stages (from the seedling stage to the dough stage, Figure S1 and Table S2). *TraesCS2B03G1529700* encodes chaperone protein DnaJ (Table S2). A previous study also found that the gene that encodes chaperone protein DnaJ was associated with wheat spike development, seed development, and the grain yield [24]. Combined with expression analyses, *TraesCS2B03G1529700* may be the most critical candidate gene for this locus. To validate it, qRT-PCR and transgene testing should be conducted in further studies.

4. Materials and Methods

4.1. Plant Materials, Trial Design, and Phenotype Evaluation

A total of 300 RILs (F_8) that were derived from the cross between H461, and Chinese Spring (denoted as the HCS population) were used to map the SLU-related QTL in this study [14]. In addition, 200 RILs (F_8) that were derived from the crossing of H461 and CM42 (denoted as the HC42 population) were used to validate the genetic effects of the major QTL. The SLU of H461 was significantly higher than that of Chinese Spring and CM42.

The HCS population was evaluated in Chongzhou (30°32′ N, 103°38′ E) in 2019 (CZ2019) and 2020 (CZ2020), and in Wenjiang (30°43′ N, 103°51′ E) in 2019 (WJ2019) and 2020 (WJ2020). The HC42 population was evaluated in CZ2019 and WJ2019. Chongzhou and Wenjiang have yellow–brown soil and paddy soil, respectively. The plant dates and meteorological data, including the temperature and rainfall, of these four trials are presented in Table S4. The experiment was set as an incomplete block design with two repetitions for each environment. Each line contained 15 plants that were grown in a single row for each repetition. The row length was 150 cm and the row spacing was 30 cm. Standard irrigations, fertilization management, and pest and disease preventions were performed throughout the developmental period. The lowest tiller height (LTH), PH, and SL of the HCS population were evaluated in all four environments. The LTH, PH, SL, SN, and TKW of the HC42 population were also assessed. Each plant's lowest and highest productive tiller was used to evaluate the LTH and PH, respectively. The LTH and PH were measured from the ground level to the tip of the spike, excluding awns. The SL was measured from the base to the tip of the spike, excluding awns. The SN was counted as the number of fertile spikelets per spike. TKW was measured as twice the weight of 500 kernels, harvested in the middle row at the maturation stage. The SLU was calculated as SL/(PH-LTH+SL), as shown in Figure S2.

4.2. Phenotypic Data Analysis

The average values of each conducted trial were calculated using SPSS 20.0 (IBM Corp., Armonk, NY, USA). The BLUP values were calculated across all the conducted trials using SAS 9.4 (SAS Institute Inc., Cary, NC, USA), as reported previously [32,33]. The broad-sense heritability of SLU was calculated as $H^2 = Vg/(Vg + Vge/n + Ve/nr)$; where, Vg, Vge, and Vg are the estimates of variances of genotype, genotype × environment interaction, and random error variances, respectively, while n and r are the number of environments and replicates [34]. The Student's *t*-test was used to evaluate the SLU differences between H461 and Chinese Spring.

4.3. Genotyping, Map Construction, QTL Mapping

The HCS population was genotyped using the wheat 55K SNP array containing 53,063 markers from our previous study [14]. The high-density linkage map of the HCS population contained 21,197 SNPs (represented by 3087 bin markers), spanning 3792.71 cM in total, as described in our previous study [14]. The linkage map contained 21 groups corresponding to 21 wheat chromosomes, respectively. QTL mapping for SLU was detected using the inclusive composite interval mapping [35] in IciMapping 4.2 [36]. To increase the authenticity and reliability of the QTL detected, the threshold was set as LOD scores ≥3, and the loci that were detected in only one environment or only using BLUP values were removed [37–39]. QTL were named according to the rules that were established by the International Rules of Genetic Nomenclature (http://wheat.pw.usda.gov/ggpages/wgc/98/Intro.htm (accessed on 12 September 2021)). For example, "Slu" and "sicau" represent "spike-layer uniformity" and "Sichuan Agricultural University," respectively. The flanking markers of the QTL were mapped on the Chinese Spring Reference Sequence RefSeq v2.1 to obtain their chromosomal and physical locations [25,26].

4.4. Marker Development, QTL Validation, and Candidate Genes Prediction

SNP probe sequences flanking the target QTL were used to convert them into KASP markers. KASP primers were designed and analyzed following the procedure that was described by Suzuki et al. (1991), as in our previous report [40,41]. The amplification reactions were performed in a total volume of 10 μL, containing 5 μL of 2 × KASP Mastermix (JasonGen, Beijing, China), 1.4 μL of KASP Assay Mix (mixture of 0.168 μmol of FAM primer, 0.168 μmol of HEX primer, and 0.42 μmol common reverse primer), 2.6 μL of DNase/RNase-free water, and 1 μL of genomic DNA (concentration of ~100 ng/μL). The amplifications were carried out using the CFX96Touch^{TM} real-time PCR detection system (BioRad, Hercules, California, USA) with the following steps: 95 °C for 15 min, 10 touchdown cycles for 65–55 °C (decreasing by 1 °C per cycle) for 60 s, and 30 cycles for 94 °C for 20 s and 57 °C for 60 s.

For the HCM population, 141 randomly selected lines were genotyped using the designed KASP marker. The RILs were classed into two groups based on the alleles that were present in the parents and their lines. The lines carrying homozygous alleles from H461 and non-H461 parents were denoted as "aa" and "AA", respectively. The Student's *t*-test was used to calculate the significances of the traits between the two groups. Based on Chinese Spring Reference Sequence RefSeq v2.1, the predicted genes that were located in the physical interval of the QTL were determined using homologous analysis by KOBAS 3.0 [42] and funRiceGenes [43] using rice (*Oryza sativa*) as the background species. In silico gene expression analysis was also performed to identify candidate genes from the Wheat Expression Browser (http://www.wheat-expression.com/ (accessed on 7 January 2022)).

5. Conclusions

The wheat line H461 was more consistent in the vertical spatial distribution of the spike layer than Chinese Spring. The spike-layer uniformity of 300 RILs that were derived from these two parents was evaluated in multiple environments. Combined with a previously constructed high-density genetic map, two major-effect QTL were successfully identified on the long arms of chromosomes 2B and 2D, respectively. There were two minor-effect QTL that were identified on the short arm of chromosome 1B and the long arm of chromosome 2B. A comparison analysis that was based on a physical map of Chinese Spring indicated that the two major loci, *QSlu.sicau-2B-2* and *QSlu.sicau-2D*, were probably novel. The effect of *QSlu.sicau-2B-2* was validated in an additional RIL population using a developed KASP marker. *QSlu.sicau-2B-2* also positively affected SL, SN, and TKW. The results indicated that this QTL and KASP marker could be used in MAS for new wheat cultivar breeding with a scientifically reasonable spike-layer distribution.

Supplementary Materials: The following are available online at https://www.mdpi.com/article/10.3390/ijms23031052/s1.

Author Contributions: Conceptualization, Y.L. (Yaxi Liu).; methodology, K.Z. and Y.L. (Yu Lin); software, Y.L. (Yu Lin); validation, K.Z. and X.J.; formal analysis, K.Z., Y.L. (Yu Lin) and X.J.; investigation, W.Z., F.W. and C.L.; resources, Y.L. (Yaxi Liu); data curation, C.L. and Y.L. (Yaxi Liu); writing—original draft preparation, K.Z., Y.L. (Yu Lin) and X.J.; writing—review and editing, Y.L. (Yu Lin), Y.W. and Y.L. (Yaxi Liu); visualization, Y.L. (Yu Lin); supervision, Y.W. and Y.L. (Yaxi Liu); project administration, C.L.; funding acquisition, Y.L. (Yaxi Liu). All authors have read and agreed to the published version of the manuscript.

Funding: This study was funded by the Key Research and Development Program of Sichuan Province (2021YFN0034 and 2021YFG0028) and the National Natural Science Foundation of China (31771794).

Institutional Review Board Statement: Not applicable.

Informed Consent Statement: Not applicable.

Data Availability Statement: The data presented in this study are available on request from the corresponding author.

Conflicts of Interest: The authors declare no conflict of interest.

References

1. Chaves, M.S.; Martinelli, J.A.; Wesp-Guterres, C.; Graichen, F.A.S.; Brammer, S.P.; Scagliusi, S.M.; Da Silva, P.R.; Wiethoelter, P.; Montan Torres, G.A.; Lau, E.Y.; et al. The importance for food security of maintaining rust resistance in wheat. *Food Secur.* **2013**, *5*, 157–176. [CrossRef]
2. Sinchire, D.B.M.; Álvarez, L.S.J.; Valdivieso, J.I.B.; Mora, E.D.C. Oat and wheat forage production under hydroponic and conventional systems. *Cienc. Tecnol. Agropecu.* **2020**, *21*, 1–16. [CrossRef]
3. Henkrar, F.; Udupa, S. Marker assisted selection in plant breeding. *Moroccan J. Agric. Sci.* **2020**, *1*, 237–247.
4. Hu, X.; Wang, G.; Du, X.; Zhang, H.; Xu, Z.; Wang, J.; Chen, G.; Wang, B.; Li, X.; Chen, X. QTL analysis across multiple environments reveals promising chromosome regions associated with yield-related traits in maize under drought conditions. *Crop J.* **2020**, *9*, 759–766. [CrossRef]
5. Wang, Z.; Hu, H.; Jiang, X.; Tao, Y.; Lin, Y.; Wu, F.; Hou, S.; Liu, S.; Li, C.; Chen, G. Identification and validation of a novel major quantitative trait locus for plant height in common wheat (*Triticum aestivum* L.). *Front Genet.* **2020**, *11*, 1314. [CrossRef]
6. Zhao, M.; Ma, Z.; Wang, L.; Tang, Z.; Mao, T.; Liang, C.; Gao, H.; Zhang, L.; He, N.; Fu, L. SNP-based QTL mapping for panicle traits in the japonica super rice cultivar Liaoxing 1. *Crop J.* **2020**, *8*, 769–780. [CrossRef]
7. Zhou, H.; Luo, W.; Gao, S.; Ma, J.; Zhou, M.; Wei, Y.; Zheng, Y.; Liu, Y.; Liu, C. Identification of loci and candidate genes controlling kernel weight in barley based on a population for which whole genome assemblies are available for both parents. *Crop J.* **2020**, *9*, 854–861. [CrossRef]
8. Zhu, X.; Leiser, W.L.; Hahn, V.; Würschum, T. Identification of seed protein and oil related QTL in 944 RILs from a diallel of early-maturing European soybean. *Crop J.* **2020**, *9*, 238–247. [CrossRef]
9. Song, J.; Li, B.; Cui, Y.; Zhuo, C.; Gu, Y.; Hu, K.; Wen, J.; Yi, B.; Shen, J.; Ma, C.; et al. QTL mapping and diurnal transcriptome analysis identify candidate genes regulating brassica napus flowering time. *Int. J. Mol. Sci.* **2021**, *22*, 7559. [CrossRef] [PubMed]
10. He, J.; Zhang, D.; Chen, X.; Li, Y.; Hu, M.; Sun, S.; Su, Q.; Su, Y.; Li, S. Identification of QTLs and a candidate gene for reducing pre-harvest sprouting in *Aegilops tauschii-Triticum aestivum* chromosome segment substitution lines. *Int. J. Mol. Sci.* **2021**, *22*, 3729. [CrossRef] [PubMed]
11. Arif, M.A.R.; Attaria, F.; Shokat, S.; Akram, S.; Waheed, M.Q.; Arif, A.; Boenier, A. Mapping of QTLs associated with yield and yield related traits in durum wheat (*Triticum durum* Desf.) under irrigated and drought conditions. *Int. J. Mol. Sci.* **2020**, *21*, 2372. [CrossRef] [PubMed]
12. Talukder, Z.I.; Underwood, W.; Ma, G.; Seiler, G.J.; Misar, C.G.; Cai, X.; Qi, L. Genetic dissection of phomopsis stem canker resistance in cultivated sunflower using high density SNP linkage map. *Int. J. Mol. Sci.* **2020**, *21*, 1497. [CrossRef] [PubMed]
13. Hina, A.; Cao, Y.; Song, S.; Li, S.; Sharmin, R.A.; Elattar, M.A.; Bhat, J.A.; Zhao, T. High-resolution mapping in two ril populations refines major "QTL hotspot" regions for seed size and shape in soybean (*Glycine max* L.). *Int. J. Mol. Sci.* **2020**, *21*, 1040. [CrossRef] [PubMed]
14. Lin, Y.; Jiang, X.; Hu, H.; Zhou, K.; Wang, Q.; Yu, S.; Yang, X.; Wang, Z.; Wu, F.; Liu, S. QTL mapping for grain number per spikelet in wheat using a high-density genetic map. *Crop J.* **2021**, *9*, 1108–1114. [CrossRef]
15. Cui, F.; Zhang, N.; Fan, X.; Zhang, W.; Zhao, C.; Yang, L.; Pan, R.; Chen, M.; Han, J.; Zhao, X.; et al. Utilization of a Wheat660K SNP array-derived high-density genetic map for high-resolution mapping of a major QTL for kernel number. *Sci. Rep.* **2017**, *7*, 3788. [CrossRef] [PubMed]
16. Winfield, M.O.; Allen, A.M.; Burridge, A.J.; Barker, G.L.; Benbow, H.R.; Wilkinson, P.A.; Coghill, J.; Waterfall, C.; Davassi, A.; Scopes, G.; et al. High-density SNP genotyping array for hexaploid wheat and its secondary and tertiary gene pool. *Plant Biotechnol. J.* **2016**, *14*, 1195–1206. [CrossRef]
17. Wang, S.; Wong, D.; Forrest, K.; Allen, A.; Chao, S.; Huang, B.E.; Maccaferri, M.; Salvi, S.; Milner, S.G.; Cattivelli, L.; et al. Characterization of polyploid wheat genomic diversity using a high-density 90,000 single nucleotide polymorphism array. *Plant Biotechnol. J.* **2014**, *12*, 787–796. [CrossRef]
18. Cavanagh, C.R.; Chao, S.; Wang, S.; Huang, B.E.; Stephen, S.; Kiani, S.; Forrest, K.; Saintenac, C.; Brown-Guedira, G.L.; Akhunova, A.; et al. Genome-wide comparative diversity uncovers multiple targets of selection for improvement in hexaploid wheat landraces and cultivars. *Proc. Natl. Acad. Sci. USA* **2013**, *110*, 8057–8062. [CrossRef] [PubMed]
19. Wang, Z.; Liu, Y.; Shi, H.; Mo, H.; Wu, F.; Lin, Y.; Gao, S.; Wang, J.; Wei, Y.; Liu, C.; et al. Identification and validation of novel low-tiller number QTL in common wheat. *Theor. Appl. Genet.* **2016**, *129*, 603–612. [CrossRef]
20. Ma, L.; Bao, J.; Guo, L.; Zeng, D.; Li, X.; Ji, Z.; Xia, Y.; Yang, C.; Qian, Q. Quantitative trait loci for panicle layer uniformity identified in doubled haploid lines of rice in two environments. *J. Integr. Plant Biol.* **2009**, *51*, 818–824. [CrossRef]
21. Fang, F.; Ye, S.; Tang, J.; Bennett, M.J.; Liang, W. DWT1/DWL2 act together with OsPIP5K1 to regulate plant uniform growth in rice. *New Phytol.* **2020**, *225*, 1234–1246. [CrossRef]
22. Wang, W.; Li, G.; Zhao, J.; Chu, H.; Lin, W.; Zhang, D.; Wang, Z.; Liang, W. DWARF TILLER1, a WUSCHEL-related homeobox transcription factor, is required for tiller growth in rice. *PLoS Genet.* **2014**, *10*, e1004154. [CrossRef]
23. Zhao, C.; Zhang, N.; Wu, Y.; Sun, H.; Liu, C.; Fan, X.; Yan, X.; Xu, H.; Ji, J.; Cui, F. QTL for spike-layer uniformity and their influence on yield-related traits in wheat. *BMC Genet.* **2019**, *20*, 23. [CrossRef] [PubMed]
24. Malik, P.; Kumar, J.; Sharma, S.; Sharma, R.; Sharma, S. Multi-locus genome-wide association mapping for spike-related traits in bread wheat (*Triticum aestivum* L.). *BMC Genom.* **2021**, *22*, 597. [CrossRef] [PubMed]

25. International Wheat Genome Sequencing Consortium (IWGSC). Shifting the limits in wheat research and breeding using a fully annotated reference genome. *Science* **2018**, *361*, 7191. [CrossRef]
26. Zhu, T.; Wang, L.; Rimbert, H.; Rodriguez, J.C.; Deal, K.R.; De Oliveira, R.; Choulet, F.; Keeble-Gagnère, G.; Tibbits, J.; Rogers, J.; et al. Optical maps refine the bread wheat Triticum aestivum cv. Chinese Spring genome assembly. *Plant J.* **2021**, *107*, 303–314. [CrossRef]
27. Jain, M.; Ghanashyam, C.; Bhattacharjee, A. Comprehensive expression analysis suggests overlapping and specific roles of rice glutathione S-transferase genes during development and stress responses. *BMC Genom.* **2010**, *11*, 73. [CrossRef]
28. Yao, W. Studies on inheritance of spike layer uniformity of wheat. *Seed* **2000**, *3*, 19–21.
29. Hu, Y. The difference of the spike-layer architecture and its relation to yield in winter wheat cultivars. *Seed* **2001**, *119*, 19–21.
30. Li, J.; Yan, Q. Studies on the relationship between the evenness degree and yield characters of hybrid early rice. *Hunan Agric. Sci.* **2005**, *1*, 14–17.
31. Zhao, D.H.; Yan, Q.Q.; Wang, J.R.; Wan, H.Q. Study on relationship between population uniformity and yield characters of late hybrid rice. *Guangxi Agric. Sci.* **2005**, *36*, 205–207.
32. Liu, Y.; Lin, Y.; Gao, S.; Li, Z.; Ma, J.; Deng, M.; Chen, G.; Wei, Y.; Zheng, Y. A genome-wide association study of 23 agronomic traits in Chinese wheat landraces. *Plant J.* **2017**, *91*, 861–873. [CrossRef] [PubMed]
33. Piepho, H.P.; Möhring, J.; Melchinger, A.E.; Büchse, A. BLUP for phenotypic selection in plant breeding and variety testing. *Euphytica* **2008**, *161*, 209–228. [CrossRef]
34. Smith, S.E.; Kuehl, R.O.; Ray, I.M.; Hui, R.; Soleri, D. Evaluation of simple methods for estimating broad-sense heritability in stands of randomly planted genotypes. *Crop Sci.* **1998**, *38*, 1125–1129. [CrossRef]
35. Li, H.; Hearne, S.; Banziger, M.; Li, Z.; Wang, J. Statistical properties of QTL linkage mapping in biparental genetic populations. *Heredity* **2010**, *105*, 257–267. [CrossRef]
36. Meng, L.; Li, H.; Zhang, L.; Wang, J. QTL IciMapping: Integrated software for genetic linkage map construction and quantitative trait locus mapping in biparental populations. *Crop J.* **2015**, *3*, 269–283. [CrossRef]
37. Deng, M.; Wu, F.; Zhou, W.; Li, J.; Shi, H.; Wang, Z.; Lin, Y.; Yang, X.; Wei, Y.; Zheng, Y.; et al. Mapping of QTL for total spikelet number per spike on chromosome 2D in wheat using a high-density genetic map. *Genet. Mol. Biol.* **2019**, *42*, 603–610. [CrossRef]
38. Rice, J.P.; Saccone, N.L.; Corbett, J. Model-based methods for linkage analysis. *Adv. Genet.* **2008**, *60*, 155–173. [CrossRef]
39. Lebrec, J.; Putter, H.; Houwelingen, J.C. Score test for detecting linkage to complex traits in selected samples. *Genet Epidemiol* **2004**, *27*, 97–108. [CrossRef] [PubMed]
40. Lin, Y.; Jiang, X.; Tao, Y.; Yang, X.; Wang, Z.; Wu, F.; Liu, S.; Li, C.; Deng, M.; Ma, J.; et al. Identification and validation of stable quantitative trait loci for grain filling rate in common wheat (*Triticum aestivum* L.). *Theor. Appl. Genet.* **2020**, *133*, 2377–2385. [CrossRef] [PubMed]
41. Suzuki, Y.; Sekiya, T.; Hayashi, K. Allele-specific polymerase chain reaction: A method for amplification and sequence determination of a single component among a mixture of sequence variants. *Anal. Biochem.* **1991**, *192*, 82–84. [CrossRef]
42. Bu, D.; Luo, H.; Huo, P.; Wang, Z.; Zhang, S.; He, Z.; Wu, Y.; Zhao, L.; Liu, J.; Guo, J.; et al. KOBAS-i: Intelligent prioritization and exploratory visualization of biological functions for gene enrichment analysis. *Nucleic. Acids Res.* **2021**, *49*, W317–W325. [CrossRef] [PubMed]
43. Yao, W.; Li, G.; Yu, Y.; Ouyang, Y. FunRiceGenes dataset for comprehensive understanding and application of rice functional genes. *Gigascience* **2017**, *7*, 119. [CrossRef] [PubMed]

International Journal of
Molecular Sciences

MDPI

Article

PIF4 Promotes Expression of *HSFA2* to Enhance Basal Thermotolerance in *Arabidopsis*

Jiaheng Yang [1,†], Xiao Qu [1,†], Li Ji [1,2], Guanhui Li [1], Chen Wang [1], Changyu Wang [1], Yan Zhang [1], Lanjie Zheng [1], Wanchen Li [2,*] and Xu Zheng [1,*]

1 State Key Laboratory of Wheat and Maize Crop Science, Center for Crop Genome Engineering, Longzi Lake Campus, College of Agronomy, Henan Agricultural University, Zhengzhou 450046, China; hnndyjh@163.com (J.Y.); 18839770781@163.com (X.Q.); 18482109281@163.com (L.J.); liguanhui2175@163.com (G.L.); wcgfly@163.com (C.W.); 15688179707@163.com (C.W.); yan000zhang@163.com (Y.Z.); zhenglanjie@henau.edu.cn (L.Z.)

2 Key Laboratory of Biology and Genetic Improvement of Maize in Southwest Region, Ministry of Agriculture, Maize Research Institute, Sichuan Agricultural University, Chengdu 611130, China

* Correspondence: aumdyms@sicau.edu.cn (W.L.); zhengxu@henau.edu.cn (X.Z.)

† These authors contributed equally to this work.

Abstract: Heat stress (HS) seriously restricts the growth and development of plants. When plants are exposed to extreme high temperature, the heat stress response (HSR) is activated to enable plants to survive. Sessile plants have evolved multiple strategies to sense and cope with HS. Previous studies have established that PHYTOCHROME INTERACTING FACTOR 4 (PIF4) acts as a key component in thermomorphogenesis; however, whether PIF4 regulates plant thermotolerance and the molecular mechanism linking this light transcriptional factor and HSR remain unclear. Here, we show that the overexpression of *PIF4* indeed provides plants with a stronger basal thermotolerance and greatly improves the survival ability of *Arabidopsis* under severe HS. Via phylogenetic analysis, we identified two sets (six) of PIF4 homologs in wheat, and the expression patterns of the PIF4 homologs were conservatively induced by heat treatment in both wheat and *Arabidopsis*. Furthermore, the PIF4 protein was accumulated under heat stress and had an identical expression level. Additionally, we found that the core regulator of HSR, HEAT SHOCK TRANSCRIPTION FACTOR A2 (HSFA2), was highly responsive to light and heat. Followed by promoter analysis and ChIP-qPCR, we further found that PIF4 can bind directly to the G-box motifs of the *HSFA2* promoter. Via effector–reporter assays, we found that PIF4 binding could activate *HSFA2* gene expression, thereby resulting in the activation of other HS-inducible genes, such as heat shock proteins. Finally, the overexpression of *PIF4* led to a stronger basal thermotolerance under non-heat-treatment conditions, thereby resulting in an enhanced tolerance to severe heat stress. Taken together, our findings propose that PIF4 is linked to heat stress signaling by directly binding to the *HSFA2* promoter and triggering the HSR at normal temperature conditions to promote the basal thermotolerance. These functions of PIF4 provide a candidate direction for breeding heat-resistant crop cultivars.

Keywords: heat stress; PIF4; thermotolerance; HSFA2; wheat

Citation: Yang, J.; Qu, X.; Ji, L.; Li, G.; Wang, C.; Wang, C.; Zhang, Y.; Zheng, L.; Li, W.; Zheng, X. PIF4 Promotes Expression of *HSFA2* to Enhance Basal Thermotolerance in *Arabidopsis*. *Int. J. Mol. Sci.* **2022**, *23*, 6017. https://doi.org/10.3390/ijms23116017

Academic Editors: Hai Du, Andrés J. Cortés and Dilantha Fernando

Received: 14 April 2022
Accepted: 26 May 2022
Published: 27 May 2022

1. Introduction

As sessile organisms, plants have evolved multiple strategies to sense and cope with many forms of abiotic stress [1,2]. Due to the rising atmospheric CO_2 concentration [3], high temperatures have been one of the major forces of the suppression of crop yields [4].

In general, the response of plants to temperature can be divided into two forms: acclimation and the tolerance response [5,6]. The definitions of warm temperatures and heat stress (HS) depend on the natural living conditions of specific species [5]. The optimum temperature for *Arabidopsis* is 20–22 °C, with 27–29 °C being considered to be a high ambient temperature [6]. Sessile plants can change their architecture in response to high ambient temperatures (27–29 °C) through elongated hypocotyls and petioles, thinned leaves, and hyponastic growth and early flowering, and these developmental processes have been collectively named thermomorphogenesis [7–9]. Temperatures above 30 °C are considered to be HS conditions, and temperatures above 36 °C are considered to be severe HS conditions [10]. HS negatively impacts diverse aspects of plant growth and development [1,2,11,12]. For example, HS-induced protein misfolding results in protein denaturation, plasma membrane destabilization, and reactive oxygen species (ROS) accumulation, causing plant cell death [13–15]. When plants suffer from heat stress, the heat stress response (HSR) is induced via H_2O_2-induced mitogen-activated protein kinases (MAPKs) and Ca^{2+}-dependent calmodulin (CaM3) [16–21]. Heat shock transcription factors (HSFs) are the master regulators of the heat stress response and are conserved among eukaryotes [22]. In *Arabidopsis*, there are 21 HSFs that can be grouped into three classes: HSFA, HSFB, and HSFC [23]. When plant cells sense heat stress, the HSFA1s act as master activation regulators of various *HSR* genes that include not only *heat shock proteins* (*HSPs*) but also TFs, such as *HSFA2*, *HSFA3*, *HSFA7a*, *HSFA7b*, *HSFB1*, *HSFB2a*, *HSFB2b*, and dehydration-responsive element binding 2A (*DREB2A*) [11,13,22,24–26].

Previous studies have established that *HSFAs* and *DREB2A* are positive regulators of HSR and that they can enhance and maintain the expression of *HSR* genes and amplify the heat stress response to enhance thermotolerance in plants [27–31]. HSFA1 directly targets *DREB2A*, further activating *HSFA3* expression to maintain the heat stress response for longer periods of time [29,30]. Increasing data indicate that *HSFA2* is the most highly heat-induced *HSF* in *Arabidopsis* and that the acquired thermotolerance is completely abolished in *hsfa2* knockout mutants [11,28,32]. *HSFA2* is also related to HS memory in plants [28]. When plants are exposed to a mild primary HS, H3K4me2 and H3K4me3 are remarkably enriched at the loci of HS memory genes (*APX2*, *HSP18.2*, and *HSP22*), leading to a higher and longer expression of the HS memory genes, which are key regulators of acquired thermotolerance [32–34]. After mild primary HS, HSFA3 binds to HSFA2 and then forms heteromeric complexes that are essential for the accumulation of H3K4me2 and H3K4me3, thereby resulting in enhanced HS memory [21]. HSFA2 also establishes a heritable feedback loop with the H3K27me3 demethylase RELATIVE OF EARLY FLOWERING 6 (REF6), ensuring transgenerational thermomemory in *Arabidopsis* [35].

HSPs function as molecular chaperones that are responsible for assisting stress-induced misfolded protein folding to maintain cell homeostasis [11,36]. Under normal temperature conditions, HSP70 and HSP90 negatively regulate the nuclear localizations and activities of HSFA1s, whereas heat stress induces the accumulation of unfolded proteins that interact competitively with HSP70/HSP90, thereby causing HSFA1s to be released from the HSP70/HSP90 complex and become active [22,37,38]. Plants have evolved multiple molecular mechanisms to activate the heat stress response; however, there is little information on the linking of the HSR and light signaling transcription factors, and the underlying regulatory mechanism remains unclear.

Previous studies have established that PHYTOCHROME INTERACTING FACTOR 4 (PIF4), a bHLH transcription factor, acts as a key component in thermomorphogenesis [39]. The transcription of PIF4 is induced by high ambient temperatures [40], thereby enhancing the accumulation of PIF4 protein. The increased PIF4 protein induces hypocotyl growth by activating the expression of auxin biosynthetic genes such as *YUCCA8 (YUC8)*, *CYP79B2*, and *TAA1*; auxin signaling genes such as *IAA19* and *IAA29*; growth-promoting genes such as *ATHB2* and *LNGs*; and brassinosteroid biosynthetic genes such as *BES1* and *BZR1* [41]. The evening complex (EC), which consists of *ELF3*, *ELF4*, and *LUX*, mediates the warm-temperature activation of *PIF4* mRNA expression [42]. EC represses *PIF4* and *PIF5* expression by directly binding to their promoters [43,44]. Very recently, PIF4 was shown to play a critical role in mediating *Arabidopsis* leaf senescence induced by heat stress (42 °C) [45].

Nevertheless, previous studies have established that PIF4 is a core hub of thermomorphogenesis when plants are exposed to high ambient temperatures [40,41,46–50]. However, whether PIF4 also regulates HSR when plants are subjected to severe HS and the mechanistic details underlying this proceeding remains unclear. Here, we demonstrate that the high-level overexpression of *PIF4* significantly enhances the survival rate of *Arabidopsis* seedlings after severe HS. PIF4 binds directly to the *HSFA2* promoter and positively regulates its expression, leading to a stronger capacity to withstand HS and the enhanced basal thermotolerance of plants. Additionally, the heat responses of PIFs are conserved in wheat; this may provide a candidate direction for breeding heat-resistant crop cultivars to ensure food security.

2. Results

2.1. Phylogenetic Analysis of PIFs in Dicot and Monocot Plants

To investigate whether the PIF proteins are conserved across plants and their application values, we performed phylogenetic analysis on the PIFs in dicot (*Arabidopsis thaliana, Solanum lycopersicum,* and *Glycine max*) and monocot plants (*Triticum aestivum, Zea mays,* and the *Oryza sativa Japonica Group*). The proteins containing at least one motif (APA or APB) and a bHLH domain were selected for phylogenetic tree construction.

The phylogenetic tree is presented in Figure 1. The tree shows that all of the PIF genes in the plants could be classified into four groups: group PIF2/3/6, group PIF7/8, group PIF1/4/5, and OsPIL15, which is consistent with previous studies [51,52]. The protein domains were highly conserved in dicot and monocot plants, indicating the evolutionary and functional conservation among PIF gene family members.

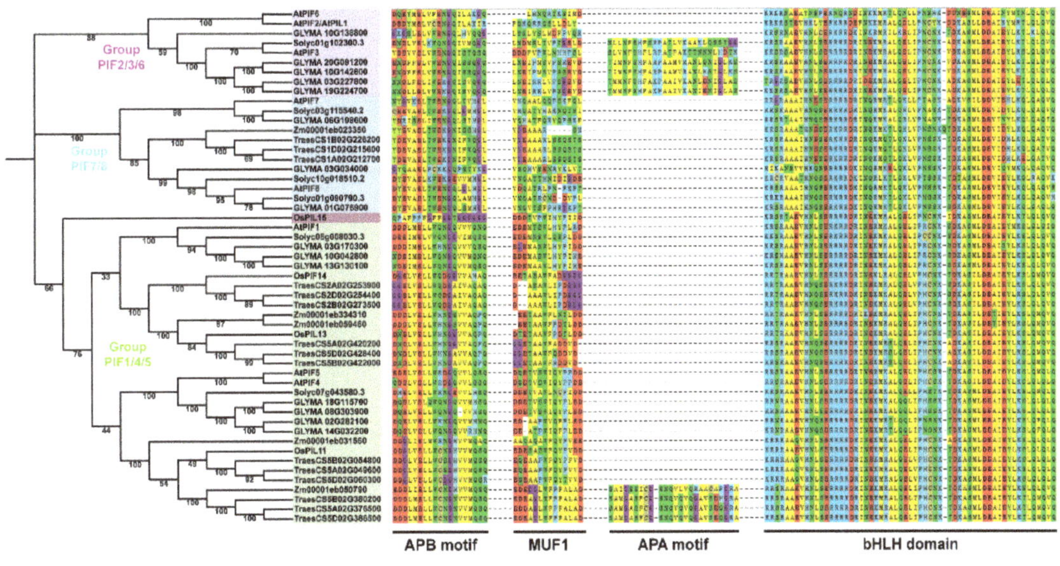

Figure 1. A neighbor-joining (NJ) phylogenetic tree of PIFs in *Arabidopsis thaliana, Solanum lycopersicum, Glycine max, Triticum aestivum, Zea mays,* and the *Oryza sativa Japonica Group.* The full-length protein sequences were used for the construction of a phylogenetic tree using MEGAX. The tree showed four major groups, which are indicated with different backgrounds. Multiple sequence alignment and motif analysis were performed using MEGAX and MEME.

2.2. PIFs Positively Regulated Thermotolerance under Severe HS

The PIFs were found to be responsible for thermomorphogenesis, regardless of whether or not it contributed to thermotolerance. To investigate whether phytochrome interacting factors (PIFs) regulate thermotolerance in plants, we first performed thermotolerance phenotype assays to test the survival rate of WT and *pifq* (*pif1/3/4/5* quadruple mutant) after severe HS.

As the PIFs were accumulated under dark conditions, we performed the thermotolerance phenotype assays using seedlings that had been grown in darkness for 4 days. Then, we treated the seedlings at 45 °C for 1 h and allowed them to recover at 22 °C for 3 days under long day conditions (16 h light/8 h darkness). Compared to WT, the *pifq* seedlings exhibited remarkably weaker thermotolerance (Figure 2a). Consistently, the survival rate data revealed that severe HS-induced plant death was enhanced in the *pifq* mutant (5.05%) (Figure 2b). These results suggest that PIFs positively regulate thermotolerance under severe HS in plants.

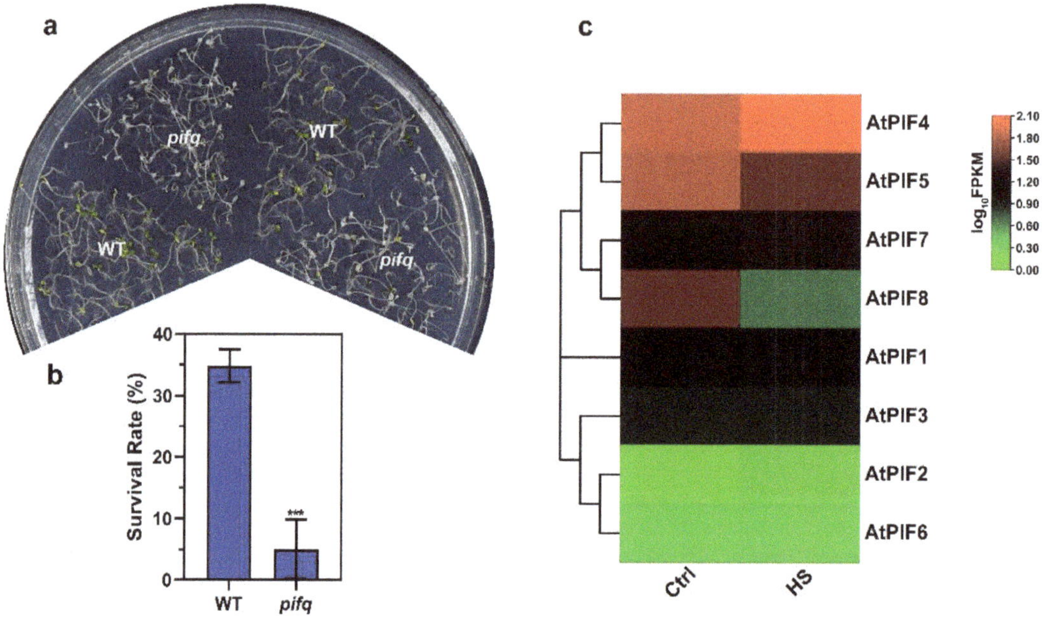

Figure 2. The pifq (pif1/3/4/5) mutant exhibits reduced thermotolerance. (**a**) Thermotolerance phenotypes of 4-day-old pifq seedlings grown in darkness were treated at 45 °C for 1 h and were allowed to recover at 22 °C for 3 d under long day conditions (16 h light/8 h darkness). (**b**) Quantification of the survival rates as indicated in (**a**). Values are the means ± SD of three independent biological replicates. Statistical differences calculated using Student's *t*-test are shown. Asterisks indicate the significant differences (*** $p < 0.001$). (**c**) Heat maps of the expression of the PIF genes under heat stress. The FPKM values of each gene are normalized by log10.

Previous studies showed that there were eight members in the *Arabidopsis PIF* family: *PIF1*, *PIF2*, *PIF3*, *PIF4*, *PIF5*, *PIF6*, *PIF7*, and *PIF8* [53,54]. To determine which PIF gene was induced by heat stress, we next checked the PIF expression pattern on 7-day-old WT seedlings under normal temperature and HS conditions (37 °C for 1 h). The heat map of the *PIFs* under normal temperature and HS conditions revealed that *PIF4* was up-regulated under HS, but that the other *PIFs* were not (Figure 2c). Taken together, these observations led us to investigate the roles of *PIF4* under severe HS further.

2.3. Spatio-Temporal Expression Analysis of PIFs in Arabidopsis and Triticum

The transcriptome data were downloaded from public databases, and the expression patterns of the 8 *AtPIFs* and 15 *TaPIFs* in different tissues (root, hypocotyl/shoot, cotyledon/leaves, spike, and seed) were characterized (Figure 3). The result revealed that the *AtPIFs* and *TaPIFs* exhibited similar expression patterns in different tissues. The PIFs were described as the master negative regulator of photomorphogenesis in previous studies [53,54]. We found that the expression levels of *AtPIFs* and *TaPIFs* in the hypocotyl and cotyledon or in leaves and shoots were remarkably higher than they were in roots and seeds or in grains (Figure 3). This indicated that the function of PIFs on photomorphogenesis is conserved among *Arabidopsis* and *Triticum*.

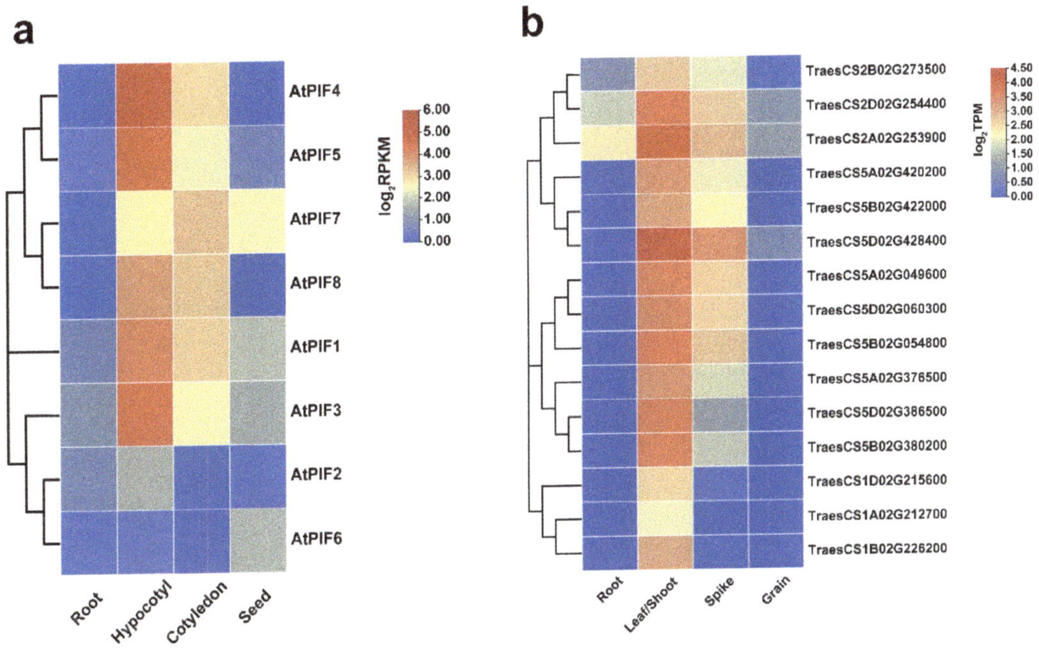

Figure 3. Expression analysis of *PIFs* in different *Arabidopsis* and *Triticum* tissues. (**a**) Hierarchical clustering of expression profiles of *AtPIF* genes in different tissues. The RPKM values of each gene are normalized by log2. (**b**) Hierarchical clustering of the expression profiles of *TaPIF* genes in different tissues. The TPM values of each gene are normalized by log2.

2.4. TaPIF4 Exhibited Similar Transcription Patterns in Response to Heat Stress

The phylogenetic tree showed that the *AtPIF4* homolog genes in wheat were probably *TraesCS5A02G049600*, *TraesCS5D02G060300*, *TraesCS5B02G054800* and *TraesCS5A02G376500*, *TraesCS5D02G386500*, and *TraesCS5B02G380200*, which were in the clade and are closely associated with *AtPIF4* (Figure 1).

We then performed RNA-seq on 7-day-old Col-0 seedlings and 2-week-old AK58 wheat cultivar seedlings treated with different abiotic stresses. The data showed that the expression of *TraesCS5A02G049600*, *TraesCS5D02G060300*, and *TraesCS5B02G054800* were significantly up-regulated under HS conditions (Figure 4b), consistent with *AtPIF4* (Figure 4a). Interestingly, *AtPIF4* was only up-regulated in response to HS but down-regulated in response to other types of abiotic stress, such as Mannitol and NaCl (Figure 4a). *TraesCS5A02G049600* and *TraesCS5D02G060300* were up-regulated under NaCl stress. Moreover, *TraesCS5A02G376500*, *TraesCS5D02G386500*, and *TraesCS5B02G380200* were induced significantly by the cold treatment.

These observations suggest that the *AtPIF4* homolog genes in wheat exhibited similar transcription patterns in response to heat stress but diverse responses to other abiotic stresses, such as NaCl and cold.

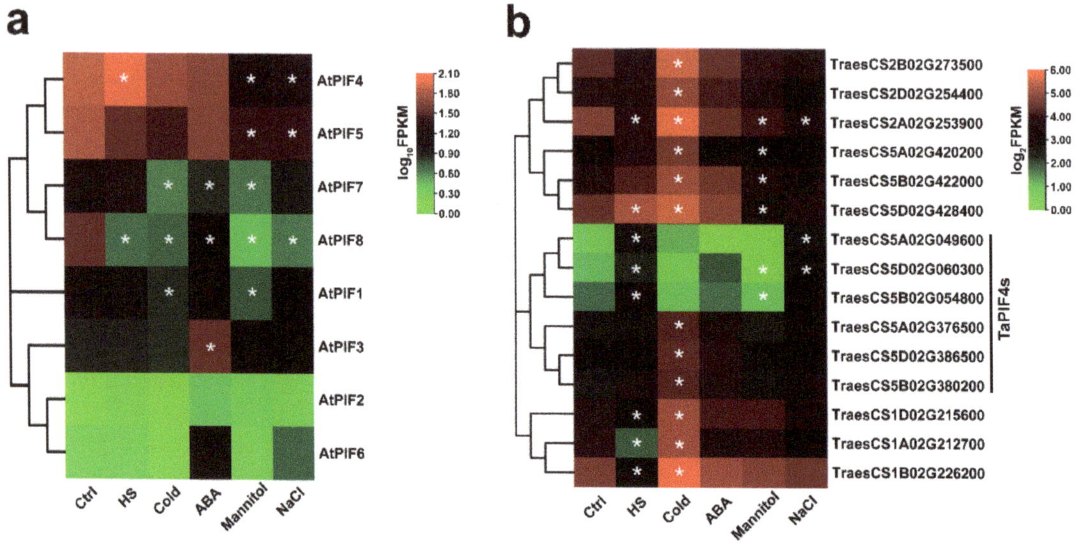

Figure 4. Expression analysis of the PIFs of Arabidopsis and Triticum under different abiotic stress treatments. (**a**) Hierarchical clustering of the expression profiles of AtPIF genes under different abiotic stress treatments. The FPKM values of each gene are normalized by log10. (**b**) Hierarchical clustering of the expression profiles of TaPIF genes under different abiotic stress treatments. The FPKM values of each gene are normalized by log2. Statistical differences calculated by DESeq2 are shown. Asterisks indicate the significant difference (* $p < 0.05$).

2.5. PIF4 Was a Key Factor for Plant Thermotolerance under Severe HS

Previous studies have well established that PIF4 is a core hub of thermomorphogenesis and photomorphogenesis in plants [6,55]. In darkness, PIF4 promotes skotomorphogenic development, thereby leading to rapid hypocotyl elongation [56]. The photomorphogenesis phenotype assay showed that pifq and the PIF4 loss-of-function mutant (pif4-1) exhibited a shorter hypocotyl compared to the WT under red light; in contrast, the overexpressed PIF4 (PIF4-OE) exhibited a longer hypocotyl (Figure S1), consistent with previous studies [53,54,57].

We wondered whether PIF4 was involved in thermotolerance regulation. Next, we performed the thermotolerance phenotype assay using WT, pif4-1, and PIF4-OE. Under severe HS conditions (Figure 5a), the pif4-1 plants exhibited weaker acquired thermotolerance compared to WT, and on the contrary, the PIF4-OE showed stronger thermotolerance (Figure 5b), which was evident in the lower survival rate of pif4-1 (6.12%) and the higher survival rate of PIF4-OE (74.49%) compared to WT (34.69%) (Figure 5c). In summary, these results demonstrate that PIF4 is not only a core hub of thermomorphogenesis but also a key factor for plant thermotolerance under severe HS conditions.

2.6. Heat Stress Induced the Accumulation of PIF4

We found that PIF4 expression was induced by heat stress (Figure 2c), and we wondered whether the post-transcription of PIF4 was regulated by the severe HS. To rule out the expression effect, we chose constitutive promoter-driven PIF4. Additionally, the qPCR data showed that the expression levels of PIF4 were not induced by heat shock treatment in the *35S:PIF4-TAP/WT* seedlings and that even the expression value of PIF4 after HS for 2 h was down-regulated (Figure S2). Next, the immunoblot assay was performed to detect PIF4 protein accumulation using *35S:PIF4-TAP/WT* seedlings under 45 °C for 0.5, 1, and 2 h. The TAP-tag consisted of His, Myc, and Flag, so the *35S:PIF4-TAP/WT* seedlings were first detected by using the anti-Myc primary antibody (Figure S3). After the severe

HS treatment, the total proteins were extracted with lysis buffer and then immunoblotted with an anti-Myc primary antibody and a horseradish peroxidase-conjugated anti-mouse IgG secondary antibody. Remarkably, the abundance of PIF4 proteins increased with the HS treating time and remained unchanged for up to 2 h, indicating that HS induces PIF4 accumulation (Figure 6). Taken together, severe heat stress induces the post-transcriptional accumulation of PIF4, and this might be critical for the plant response to severe heat stress.

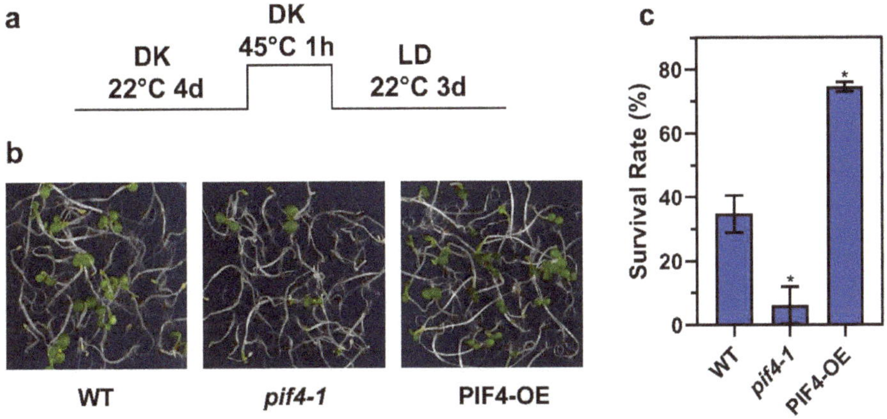

Figure 5. *PIF4* overexpression enhances the thermotolerance of *Arabidopsis*. (**a**) Col-0, *pif4-1*, and PIF4-OE seedlings were grown in darkness for 4 d and then treated at 45 °C for 1 h and recovered at 22 °C for 3 d under long day conditions (16 h light/8 h darkness). (**b**) The thermotolerance phenotypes of Col-0, *pif4-1*, and PIF4-OE were shown after HS treatment, as indicated in (**a**). (**c**) Quantification of the survival rates indicated in (**a**). Values are the means ± SD of three independent biological replicates. Statistical differences calculated by Student's *t*-test are shown. Asterisks indicate the significant differences (* $p < 0.05$).

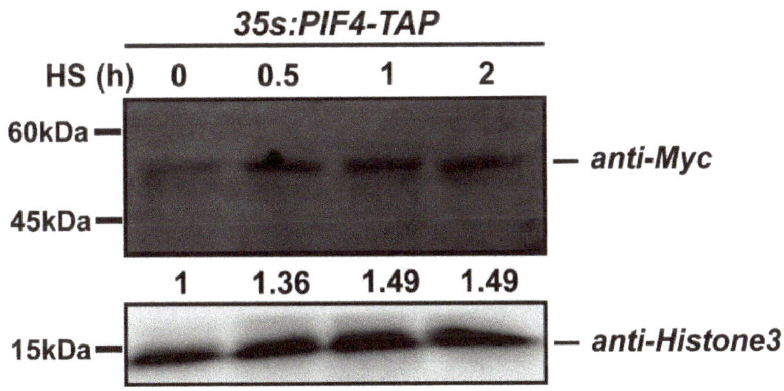

Figure 6. The protein accumulation of PIF4 under heat stress conditions. Detection of the PIF4 protein levels in 4-day-old *35S:PIF4-TAP/WT* seedlings grown in darkness and in the seedlings transferred to 45 °C for 0.5, 1, and 2 h. The proteins were detected by immunoblot analysis using an anti-Myc antibody.

2.7. Overexpression of PIF4 Enhanced the Basal Expression of HS-Inducible Genes

Heat shock factors (HSFs) are known as key regulators in the response to heat stress. Here, we found that light signaling factor PIF4 was responsible for the plant heat stress response. Additionally, there were 21 HSFs in Arabidopsis, and we wondered if there could be an HSF that could be regulated by both light and heat signaling. According to this hypothesis, an HSF could work as a linker to connect light to thermotolerance. To investigate which HSF responds to both light and heat, we checked the expression profiles of all 21 HSFs in darkness and white light (WL, 16 h light/8 h darkness) and under high temperatures (16 h light/8 h darkness, 37 °C for 1 h). With the results, we had a heat map to indicate the expression patterns, and there were four HSFs that could be induced by light and heat, HSFA2, HSFB2a/b, and HSFB4. Unsurprisingly, HSFA2 was induced the most in heat stimuli, as it was previously determined to be the main regulator of acquired thermotolerance (Figure 7a). For the HSFBs, although they were induced by heat, they were not involved in plant thermotolerance regulation [58,59]. As such, HSFA2 might work as a linker to connect the PIF4 to thermotolerance regulation.

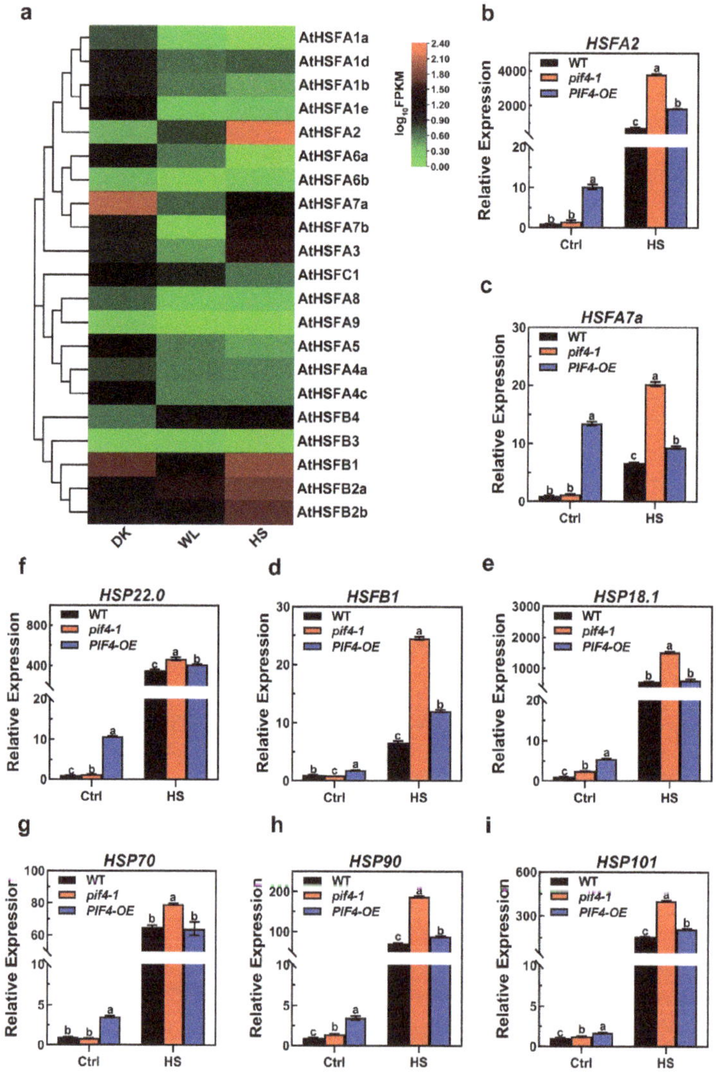

Figure 7. The expression patterns of heat stress (HS)-inducible genes under severe heat stress conditions. (**a**) Heat maps of the expression of the AtHSF genes in 7-day-old WT seedlings under DK,

WL (16 h light/8 h darkness), and HS (16 h light/8 h darkness, 37 °C for 1 h). The FPKM values of each gene are normalized by log10. (**b–i**) Relative expression levels of HS-inducible genes in overexpressed WT, pif4-1 mutant, and PIF4-OE lines under control (Ctrl) and HS conditions (45 °C for 1 h). Gene expression levels are expressed relative to the value in non-heat-treated Col-0 seedlings. Error bars indicate the mean ± SD. Actin2 was used as a reference gene. Statistical differences were calculated by one-way ANOVA. Different letters above each bar indicate statistically significant differences as determined by Tukey's multiple testing methods ($p < 0.05$).

The opposite thermotolerance phenotype of the pif4-1 mutants and PIF4-OE prompted us to explore the differences in the transcriptional levels of the heat stress (HS)-inducible gene associated with basal and acquired thermotolerance between WT, pif4-1, and PIF4-OE.

Notably, we also observed that compared to WT and pif4-1, the transcriptional levels of all of the HS-inducible genes were significantly higher in PIF4-OE under non-heat conditions (Figure 7b–i). Together, these data indicate that the basal thermotolerance of PIF4-OE is stronger than that WT and pif4-1 and demonstrate that these are probably the critical reasons why PIF4-OE has a higher survival rate than WT under severe HS.

Surprisingly, the qRT-PCR results showed that the expression levels of HSFA2, HSFA7a, HSFB1, HSP22.0, HSP90, and HSP101 were significantly up-regulated in both pif4-1 and PIF4-OE compared to those in WT plants under severe HS. Moreover, the transcriptional levels of the HS-inducible genes in pif4-1 were even higher than PIF4-OE under severe HS (Figure 7). These results showed that PIF4 regulated the expression of HSR genes under severe HS. Nevertheless, the weaker thermotolerance of pif4-1 indicates that PIF4 not only regulates the transcription of HSR genes but that it also might impact the post-transcriptional levels or modification of HSR proteins. The presence or absence of PIF4 might affect the function of these HSR genes under severe HS.

2.8. PIF4 Directly Bound to the Promoter of HSFA2 and Promoted Its Gene Expression

It was previously demonstrated that PIF4 directly activates or represses the expression of downstream genes via binding to G-box (CACGTG) and/or PBE-box (CACATG) motifs in their promoters (Figure 8a) [60–62]. Next, we analyzed the promoter structures of the above HS-inducible genes. The results showed that the promoter regions of HSFA2, HSP90, and HSP101 harbored 1 to 3 G-box motifs (Figure 8b). This observation indicated that PIF4 might bind directly to the loci of HSFA2, HSP90, and HSP101 to regulate their transcription. HSFA2 was the most highly heat-induced HSF, representing the master regulator, and played a critical role in the acquired thermotolerance of Arabidopsis [28,31]. Moreover, the expression of HSFA2 was highly responsive to light and heat (Figure 7a).

We next performed ChIP-qPCR assays using 7-day-old WT and *35s:PIF4-TAP* seedlings grown at 22 °C in darkness to investigate whether PIF4 could directly regulate the expression of HSFA2 via binding to the G-box motifs under non-heat conditions. The specific primers were designed to detect the enrichment of the HSFA2 promoter of the "P1" and "P2" regions harboring G-box motifs (Figure 8c), and the enrichment in the TA3 promoter was used as a negative control [63]. The results showed that the "P1" and "P2" regions of the HSFA2 promoter were remarkably enriched in the ChIP DNA samples from *35s:PIF4-TAP* seedlings but not in the WT seedlings compared to the TA3 promoter region (Figure 8c).

To further investigate whether the binding of PIF4 in the loci of the HSFA2 promoter is responsible for regulating its expression, we performed an effector–reporter assay in Nicotiana benthamiana leaves to confirm this hypothesis. As expected, 35s:PIF4-GFP but not 35s:GFP elevated the LUC activity of P_{HSFA2}:LUC remarkably (Figure 8d,e). Interestingly, the promotion of PIF4 to the HSFA2 promoter was abolished in $P_{HSFA2mGbox}$:LUC, which is the G-box mutant of P_{HSFA2}:LUC reporters, confirming that the positive role of PIF4 on HSFA2-driven transcriptional activation is reliant on G-box motifs (Figure 8d,e). Collectively, these results suggest that PIF4 directly binds to the G-box motifs in the HSFA2 promoter to promote its expression under non-heat conditions, leading to enhanced basal thermotolerance in PIF4-OE.

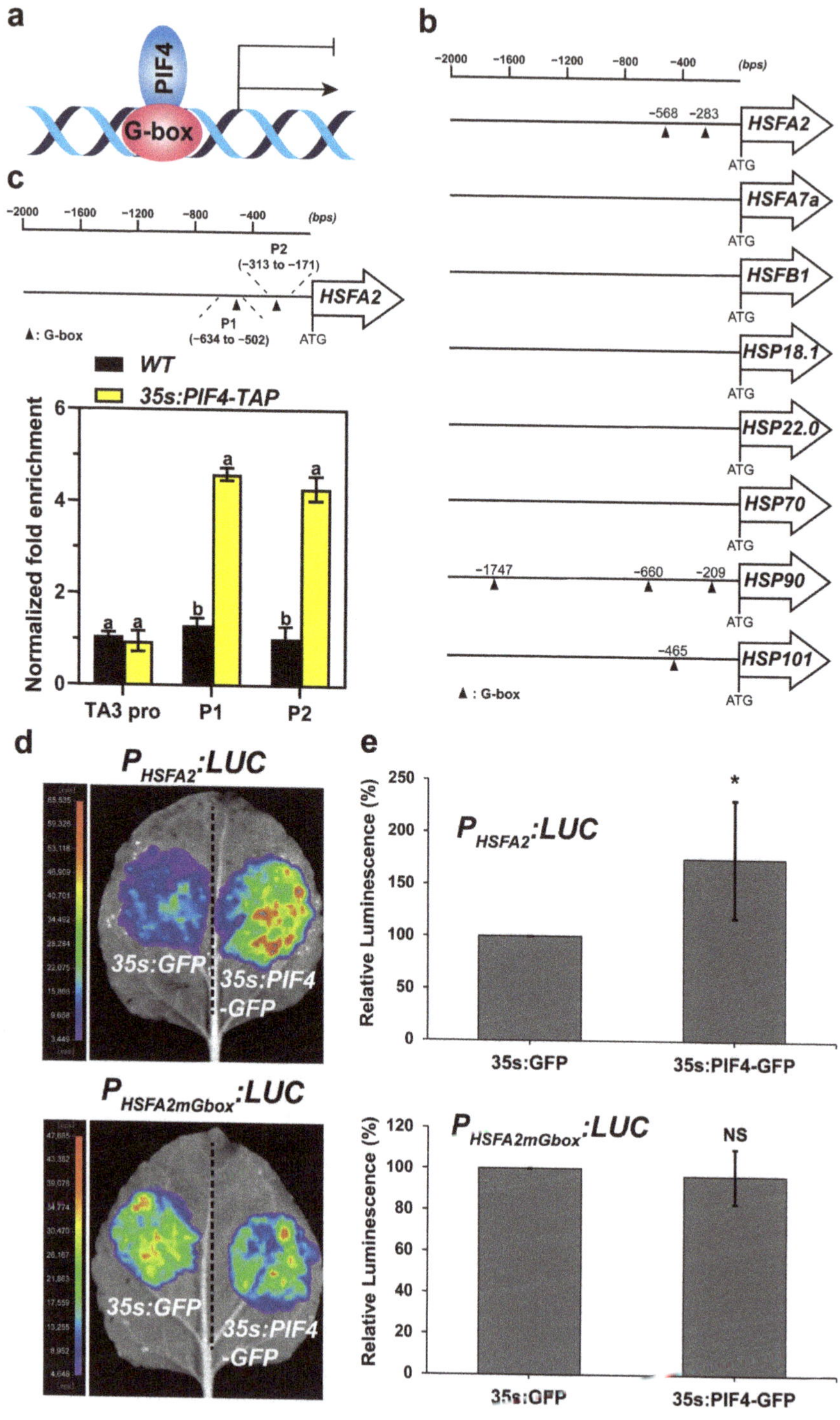

Figure 8. PIF4 directly binds to the G-box motifs in the HSFA2 promoter. (**a**) PIF4 directly binds to the G-box motif of gene promoters to activate or repress their expression. (**b**) Schematic representation of

the HSR gene promoters with the location of the G-box motif. The triangle indicates the G-box motif. (**c**) Schematic representation of the HSFA2 promoter with the location of two G-box motifs. P1 and P2 represent the respective primer positions used for ChIP-qPCR. ChIP-qPCR analysis of PIF4 binding to the HSFA2 promoter. ChIP assays were performed on 7-day-old WT and *35s:PIF4-TAP* seedlings grown at 22 °C in darkness. The protein–DNA complexes were immunoprecipitated using anti-Myc beads. ChIP DNA was quantified by qRT-PCR with primers specific to the G-box motifs (P1 and P2) and TA3 promoter (negative control). The ChIP-qPCR values of these regions were normalized to TUB2. The enrichment level in WT was set to 1. Values are the means ± SD of four technical replicates. Statistical differences were calculated by one-way ANOVA. Different letters above each bar indicate statistically significant differences as determined by Tukey's multiple testing methods ($p < 0.05$). (**d**) Activation of the HSFA2 promoter by PIF4 in Nicotiana benthamiana leaves. A reporter vector P_{HSFA2}:LUC containing the HSFA2 promoter (1500 bp upstream of the start codon) driving LUC. $P_{HSFA2mGbox}$:LUC represents the G-box mutation of P_{HSFA2}:LUC, and 35s:GFP (negative control) and 35s:PIF4 were used as effector constructs. (**e**) The quantification of the luciferase activity for the samples shown in (**b**,**d**). All the 35s:GFP controls were set at 100%. The ratio represents the relative luminescence intensity of the 35s:PIF4 effector compared to the 35s:GFP effector in the same reporter. Data are means of relative luminescence ratios and error bars represent SD of relative luminescence ratios for at least four independent biological replicates. Asterisks indicate statistically significant differences (* $p < 0.05$, NS, no significance).

3. Discussion

In recent years, global warming has caused frequent occurrences of extreme weather, which have had a detrimental impact on plant growth and development throughout the whole plant life cycle [64]. As sessile organisms, plants cannot migrate to avoid extreme temperatures; therefore, multiple strategies have evolved to sense and cope with heat stress in plants [6,8,21]. In general, extreme temperatures display two forms: high ambient temperatures (27~29 °C) and heat stress (≥30 °C) [5,6]. It is worth noting that temperatures above 36 °C are considered to be severe HS for Arabidopsis [6]. Phytochrome interacting factors (PIFs) have been reported to play a critical role in seedling development, hypocotyl elongation, and thermomorphogenesis [55,60]. In addition to this, our data showed that PIF4 can be involved in thermotolerance regulation and how it functions.

3.1. The Expression Patterns of PIF4 in Response to Heat Stress Were Highly Conserved in Arabidopsis and Triticum

Previous studies have demonstrated that PIFs contain two highly conserved motifs: the active PHYB binding (APB) and active PHYA binding (APA) motifs, which are necessary for interaction with PHYB and PHYA [53,65,66]. MUF1 is reported to play a critical role in the transcriptional activation of PIFs [67]. Furthermore, the basic/helix–loop–helix domain is highly conserved in bHLH transcription factors [68]. We next performed a BLASTP search using Ensemble Plants (http://plants.ensembl.org/index.html, accessed on 11 April 2022) and chose the sequences with an e-value less than 1×10^{-10}. The MUF1 and bHLH domain sequences were defined by MEME and were consistent with previous studies [51,68]. The phylogenetic tree showed that all of the PIF genes of plants could be classified into three groups: PIF2/3/6, PIF7/8, and PIF1/4/5, consistent with previous studies [51,52]. The highly conserved protein domains in the dicot and monocot plants indicated the evolutionary and functional conservation among PIF gene family members (Figure 1). Meanwhile, the homology analysis of the PIFs in Arabidopsis and Triticum showed that there are two clusters of AtPIF4 homologs in wheat (Figure 1). We found that the expression levels of the AtPIFs and TaPIFs in the hypocotyl and cotyledon or in leaves and shoots were remarkably higher than in roots and seeds or grains (Figure 3). This indicated that the function of PIFs in photomorphogenesis is conserved among Arabidopsis and Triticum.

In this study, the seedlings lacking PIF genes (pifq mutants) caused a drastic reduction in the survival rate under severe HS, and PIF4 expression was induced by heat stress (Figure 2). Furthermore, AtPIF4 and the predicted TaPIF4s (TraesCS5A02G049600, TraesCS5D02G060300, TraesCS5B02G054800, TraesCS5A02G376500, and TraesCS5B02G380200) exhibited similar transcription patterns in response to heat stress but diverse responses to other types of abiotic stress, such as NaCl, cold, and ABA (Figure 4), indicating that the function of PIF4s was conserved in response to heat stress.

3.2. PIF4 Was Essential for Plants in Response to the Severe Heat Stress (45 °C)

Previous studies have established that PIF4 acts as a key component in thermomorphogenesis, which changes the morphology architecture of plants in response to moderately high ambient temperature [40,50,69]. In more detail, PIF4 can activate the expression of auxin biosynthetic genes such as YUCCA8 (YUC8), CYP79B2, and TAA1; auxin signaling genes such as IAA19 and IAA29; growth-promoting genes such as ATHB2 and LNGs; and brassinosteroid biosynthetic genes such as BES1 and BZR1 in response to high ambient temperatures [39,42,46,48,49,70,71]. Recent studies have also shown that PIF4 and PIF5 mediate heat-stress-induced leaf senescence [45]. Nevertheless, the molecular mechanisms of PIF4 underlying severe heat stress (45 °C) remain unclear.

To clarify the role of PIF4 in the heat stress response of plants, we investigated the thermotolerance phenotype of pif4-1 and PIF4-OE under severe HS. It is known that PIF proteins accumulate in darkness when plants are exposed to light and when the phytochromes interact with PIFs, leading to their rapid turnover via ubiquitination and phosphorylation and the mediation of the PIF degradation via the 26S proteasome pathway [65,72–76]. As such, we performed thermotolerance phenotype assays using seedlings grown in darkness for 4 days. The seedlings were then treated at 45 °C for 1 h and recovered at 22 °C for 3 days under long day conditions (16 h light/8 h darkness).

Then, we found that the death of the seedlings with induced heat stress was remarkably reduced in PIF4-OE and significantly enhanced in pif4-1 compared to WT (Figure 5a), which was evident in the lower survival rate of pif4-1 (6.12%) and higher survival rate of PIF4-OE (74.49%) compared to WT (34.69%) (Figure 5b). Furthermore, to investigate PIF4 regulation at the post-transcriptional level by severe HS, we detected the protein accumulation of PIF4 under 45 °C for 0.5, 1, and 2 h. Additionally, the PIF4 protein levels increased as the HS treatment time was prolonged (Figure 6). Heat promoted the transformation of PHYB from Pfr to Pr, which might inhibit PIF4 protein degradation [77,78]. The increased PIF4 protein accumulation might play a critical role in the response of plants to severe heat stress.

3.3. PIF4 Directly Bound to the Promoter of HSFA2 to Active Its Expression

Recent studies have indicated that HSFA2 and HSFA7a are positive regulators of HSR and that they maintain HSR gene expression, leading to plants having enhanced thermotolerance [25,28]. It was found that HSFA2 and HSFA7a could directly bind to the heat stress elements of HSPs to activate their expression and gain a stronger and longer HSR for the heat acclimation response [25,28]. It is worth noting that HSPs function as molecular chaperones and are involved in the HS-induced unfolded protein response to maintain plant cell homeostasis [11,36]. Moreover, HSFB1 repressed the expression of the heat-stress-induced HSFs but is necessary for the acquired thermotolerance [58].

In the current study, we found that the transcriptional levels of HS-inducible marker genes were significantly higher in PIF4-OE compared to in WT and pif4-1 under non-heat conditions (Figure 7). These results showed that the transcriptional levels of HS-inducible genes were regulated by PIF4. Furthermore, these data also indicated that the higher transcriptional levels of HSR genes might result in the stronger basal thermotolerance of PIF4-OE than WT and pif4-1. Additionally, it was probably the critical reason for the higher survival rate of PIF4-OE than WT under severe HS.

Surprisingly, the qRT-PCR results showed that compared to WT, the transcriptional levels of HSFA2, HSFA7a, HSFB1, HSP22.0, HSP90, and HSP101 were significantly up-regulated in both pif4-1 and PIF4-OE after severe HS treatment, and all of the transcriptional levels of the HS-inducible genes in pif4-1 were higher than PIF4-OE under severe HS (Figure 7). Nevertheless, the higher transcriptional levels of HS-inducible genes in pif4-1 did not rescue its heat-resistant phenotype (Figure 5). This result suggests that PIF4 not only regulates the transcription of HSR genes but that it also might mediate the post-transcriptional regulation of HSR proteins. The presence or absence of PIF4 may determine the positive function of these HSR genes under severe HS, and this field requires further investigation in the future.

Due to PIF4 directly binding to the G-box (CACGTG) and/or PBE-box (CACATG) motifs of downstream genes to activate or repress their expression [60–62], we next investigated the possibility of PIF4 binding to the promoters of these HS-inducible genes. The promoter sequence analysis showed that HSFA2, HSP90, and HSP101 harbor one to three G-box motifs in their promoter regions (Figure 8b). HSFA2 is considered to be the most highly heat-induced HSF and plays a critical role in the acquired thermotolerance of Arabidopsis. Next, we performed the ChIP-qPCR assay to confirm this hypothesis. The results showed that the "P1" and "P2" regions of the HSFA2 promoter were remarkably enriched in the ChIP DNA samples from the *35s:PIF4-TAP* seedlings but not in the WT seedlings compared to the TA3 promoter region (Figure 8c), indicating that PIF4 can bind directly to the loci of HSFA2 in vivo. Furthermore, effector–reporter assays showed that the binding of PIF4 to the G-box motifs elevated the transcriptional levels of HSFA2 remarkably (Figure 8d,e).

Taking these results together, we proposed a hypothetical model that elucidates the molecular function of PIF4 in response to severe heat stress. PIF4 can directly bind to the G-box motifs of the HSFA2 promoter to activate its gene expression, thereby resulting in the activation of other HS-inducible genes, such as heat shock proteins. These lead to a stronger basal thermotolerance under non-heat-treatment conditions, resulting in higher tolerance to severe heat stress (Figure 9).

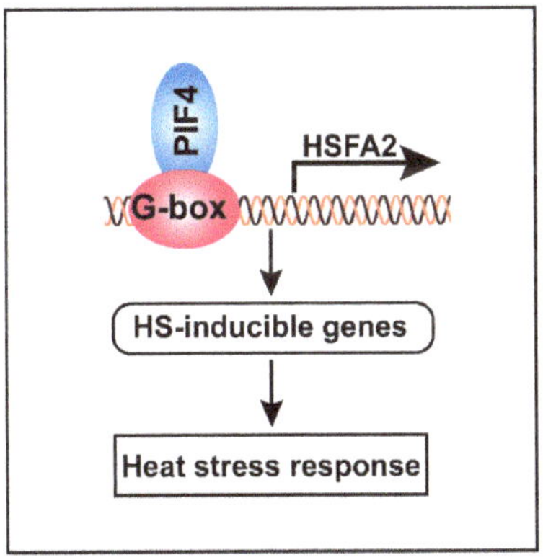

Figure 9. Schematic model showing how PIF4 modulates the heat stress response in plants. PIF4 directly binds to the G-box motifs to promote *HSFA2* gene expression, thereby resulting in the activation of the HS-inducible genes. These leads to a stronger basal thermotolerance under non-heat-treatment conditions, thereby resulting in enhanced tolerance to severe heat stress.

The elevated thermotolerance of crops is critical for ensuring food security. Here, we found that the overexpression of PIF4 could provide a stronger basal thermotolerance for plants, greatly improving the survival ability of Arabidopsis under severe HS, and the AtPIF4 homolog genes in wheat exhibited similar transcription patterns in response to heat stress. Our findings establish a molecular mechanism of PIF4 in mediating the heat stress response and provide a candidate direction for breeding heat-resistant wheat cultivars. Although our research yielded a model for PIF4 modulating the heat stress response to cope with severe HS rather than thermomorphogenesis, a number of questions remain unclearly answered. These include why pif4-1 had stronger transcriptional levels of HS-inducible genes but the thermotolerance was still weaker. Other questions include whether the post-transcriptional HSR genes in pif4-1 or PIF4-OE were different and whether PIF4 could directly bind to other HS-inducible transcription factors (TFs) or HSP promoters. Does the overexpression of TaPIF4s improve the thermotolerance of wheat under severe HS? All of the above questions await further investigation in the future.

4. Materials and Methods

4.1. Plant Materials, Growth Conditions, and Heat Treatments

All of the plants used in this study were of Columbia-0 background. The *pif4-1* (SALK_140393C), *pifq*, and *PIF4-OE* used in this research were described previously [57,79,80]. The *35s:PIF4-TAP* (*His/Myc/Flag*) transgenic line was provided by Dr. Hongtao Liu [81]. All of the *Arabidopsis* seeds were sterilized and then incubated at 4 °C for 3 days in the dark. Next, seeds were sowed on $\frac{1}{2}$ Murashige and Skoog (MS) medium containing 0.8% agar and 1% sucrose (pH 5.7).

For the thermotolerance assays, the seedlings were grown in darkness for 4 days and then treated at 45 °C for 1 h and recovered at 22 °C for 3 days under long day conditions (16 h light/8 h darkness). For the photomorphogenesis assays, the seedlings were grown under continuous red light (15 μmol m^{-2} s^{-1}) for 4 days.

4.2. RNA-Seq Analysis

WT (Col-0) *Arabidopsis* were grown in normal conditions (dark or long day (16 h light/8 h darkness)) for 6 days, and AK-58 wheat cultivar seedlings were grown under normal conditions for 2 weeks and then treated with HS (37 °C), cold (5 °C), ABA (20 μM), Mannitol (200 mM), and NaCl (150 mM) before sampling. Each treatment was performed with two biological replicates. Samples were sequenced on the Illumina platform by Biomarker Technologies (Beijing, China). The expression levels were calculated using the FPKM method [82]. The FPKM values were normalized by log2 or log10 for heat maps. The heat maps were generated using TBtools [83].

To analyze the expression patterns of the *PIFs* in different tissues of *Arabidopsis* and *Triticum*, the transcriptomic data of *Arabidopsis* and Chinese Spring were downloaded from the *Arabidopsis* eFP Browser (http://bar.utoronto.ca/efp/cgi-bin/efpWeb.cgi, accessed on 11 April 2022) [84] and ExpVIP (http://www.wheat-expression.com/, accessed on 11 April 2022), respectively. The expression levels of *Arabidopsis* and Chinese Spring were calculated with the RPKM and TPM methods, respectively. The RPKM and TPM values were normalized by log2 for the heat maps. The heat maps were generated using TBtools [83].

4.3. Protein Extraction and Immunoblot Assays

The total proteins were extracted with lysis buffer (50 mM Tris-HCl pH 7.5, 150 mM NaCl, 1 mM EDTA, 10% (*vol/vol*) glycerol, 0.1%NP-40, 1 mM PMSF, and 1× complete Protease Inhibitor Mixture (Roche)). Samples were centrifuged at 4 °C (12,000× *g* for 10 min). Then, the supernatants were collected into new centrifuge tubes. Each protein sample was denatured at 95 °C for 10 min and then separated on 10% (*w*/*v*) SDS PAGE gels. The proteins were transferred to a polyvinylidene fluoride (PVDF) membrane (Millipore) that was the same size. Then, the membrane was blocked with 5% skimmed milk powder. The proteins were immunoblotted with anti-Myc primary antibody (M20002, Abmart, 1: 5000 (*v*/*v*)) and horseradish peroxidase-conjugated anti-mouse IgG secondary antibody (AB0102, Abways, 1: 10,000 [*v*/*v*]). The chemiluminescence signals were captured with a Li-Cor/Odyssey system. The protein band intensities were calculated with ImageJ (https://imagej.nih.gov/ij/, accessed on 11 April 2022). The relative intensities were calculated with Histone3 (CY6587, Abways, 1: 5000 (*v*/*v*)) control. Immunoblot experiments were repeated for three biological replicates, essentially with the same conclusions, and the representative result is shown in Figure 6.

4.4. RNA Extraction and Quantitative Real-Time PCR Analysis

The total RNA was extracted with TRNzol Universal Reagent (DP424, TIANGEN). The quality of the RNA samples was detected by agarose gel electrophoresis using Nanodrop one for concentration determination. The RNA was reverse transcribed into cDNA with the PrimeScript™II 1st Strand cDNA Synthesis Kit (Takara, Kusatsu City, Japan, 6210A). qRT-PCR was performed using 2×TSINGKE Master qPCR Mix (SYBR Green I) (TSINGKE, TSE201) with the LightCycler® 480 System (Roche, Basel, Switzerland). A two-step qPCR amplification program was used as follows: 95 °C for 1 min followed by 40 cycles of 95 °C for 10 s and 60 °C for 30 s. The cycle threshold (CT) values were calculated using the $2^{-\Delta\Delta CT}$ method. Each assay was performed in three technical replicates, and the relative expression levels were normalized to Actin2. The primer sequences used for qRT-PCR are listed in Supplemental Table S1.

4.5. Effector–Reporter Assays

To detect whether the PIF4 protein impacts *HSFA2 transcription* depends on whether or not G-box motifs are present. We constructed two different reporters for the effector–reporter assays. The P_{HSFA2}*:LUC* reporter contained the *HSFA2* promoter driving LUC. The $P_{HSFA2mGbox}$*:LUC* reporter indicated both the G-box mutations of P_{HSFA2}*:LUC*. *35s:GFP* (negative control) and *35s:PIF4* were used as effectors. These constructs were transformed into *Agrobacterium tumefaciens* (strain GV3101). The GV3101 cultures harboring different reporters were cultured in liquid LB medium (50 mg/mL kanamycin and rifampicin) to an $OD600_{nm}$ between 0.5 and 0.6. Then, *Agrobacterium* were collected after centrifugation (4000× *g* at 25 °C for 10 min) and resuspended using the infiltration buffer (10 mM MES pH 5.6, 150 μM Acetosyringone, and 10 mM $MgCl_2$) to a final $OD600_{nm}$ between 0.8 and 1.0. Then, *Agrobacterium* harboring *35s:GFP* or *35s:PIF4* were mixed in a 1:1 ratio and then infiltrated into *Nicotiana* leaves. The infiltrated *Nicotiana* was grown under normal conditions for 2 days. The *Nicotiana* leaves were infiltrated with 1mM D-fluorescein potassium salt before the luciferase activity was detected. LUC signaling was captured with the NightShade LB985 Plant Imaging System. The luciferase activity for each group was calculated using IndiGO software. All of the *35s:GFP* controls were set at 100%. The ratio represents the relative luminescence intensity of the *35s:PIF4* effector to the *35s:GFP* effector in the same reporter. Each assay was performed in at least four independent biological replicates, and representative results are shown.

4.6. ChIP-qPCR Assays

ChIP assays have been described in previous studies [85]. In brief, 7-day-old WT and *35s:PIF4-TAP* seedlings were crosslinked with 1% formaldehyde under vacuum. After 10 min, crosslinking was stopped by adding glycine to a final concentration of 125 mM. The seedlings were rinsed with water five times, and excessive moisture was removed with absorbent paper. The tissues were ground with 10 mL extraction buffer I (10 mM Tris-HCl pH 8, 0.4 M sucrose, 10 mM $MgCl_2$, 5 mM BME, 1×protease inhibitor (Beyotime, Jiangsu, China, P1015), and 0.1 mM PMSF). The homogenate was filtered through a strainer (300 mesh) into a 15 mL falcon tube. Then, the filtered solution was centrifuged at 4000 rpm at 4 °C for 20 min. The supernatant was removed, and the pellet was resuspended in 1 mL extraction buffer II (10 mM Tris-HCl pH 8, 0.25 M sucrose, 1% Triton X-100, 10 mM $MgCl_2$, 5 mM BME, 1mM EDTA, 1×protease inhibitor, and 0.1 mM PMSF). After centrifugation (12,000× *g* at 4 °C for 10 min), the supernatant was removed, and the pellet was resuspended in 300 μL of extraction buffer III (10 mM Tris-HCl pH 8, 1.7 M sucrose, 0.15% Triton X-100, 2 mM $MgCl_2$, 5 mM BME, 1 mM EDTA, 1×protease inhibitor, and 0.1 mM PMSF). The solution was then laid on the top of a clean 300 μL of extraction buffer 3 and then centrifuged at 16,000× *g* at 4 °C for 1 h. The supernatant was removed, the chromatin pellet was resuspended in 200 μL nuclei lysis buffer (50 mM Tris-HCl pH 8, 1% SDS, 10 mM EDTA, 1×protease inhibitor, and 0.1 mM PMSF), and the tissues were then sonicated at 4 °C to 250~500 bp genomic DNA fragments. After centrifugation (14,000 rpm at 4 °C for 5 min), the supernatant was removed to a new tube, and 1.8 mL of ChIP dilution buffer (16.7 mM Tris-HCl pH 8, 1.1% Triton X-100, 1.2 mM EDTA, 167 mM NaCl, 1×protease inhibitor, and 0.1 mM PMSF) was added to dilute the 1% SDS to 0.1% SDS. The protein–DNA complex was immunoprecipitated by anti-Myc magnetic beads (Beyotime, P2118) and incubated while rotating at 4 °C overnight. The magnetic beads were attached to a magnet, and washed with 1 mL of low-salt wash buffer (20 mM Tris-HCl pH 8, 150 mM NaCl, 0.1% SDS, 1% TritonX-100, and 2 mM EDTA) two times, washed with high-salt wash buffer (20 mM Tris-HCl pH 8, 500 mM NaCl, 0.1% SDS, 1% TritonX-100, and 2 mM EDTA) two times, washed with LiCl wash buffer (10 mM Tris-HCl pH 8, 0.25 M LiCl, 1% NP40, 1% sodium deoxycholate, and 1 mM EDTA) two times, and, finally, washed with TE buffer (10 mM Tris-HCl pH 8, 1 mM EDTA) one time; each washing was performed at 4 °C for 5 min. After washing, the immunocomplexes were eluted from the magnetic beads twice with 250 μL of elution buffer (0.1 M $NaHCO_3$ and 1% SDS); then, 20 μL 5 M NaCl was added to the eluate and reverse crosslinked at 65 °C for at least 6 h or overnight. Then, 10 μL of 0.5 M EDTA, 20 μL of 1 M Tris-HCl pH 6.5, and 2 μL of 10 mg/mL proteinase K were added to the eluate, and it was incubated for 1 h at 45 °C. The DNA was extracted using an equal volume of phenol/chloroform. The purified DNA was analyzed by qRT-PCR, which is described in Materials and Methods Section 4.4. The ChIP-qPCR values were normalized to TUB2. Additionally, the TA3 promoter was used as a negative control. Four technical replicates were performed for each ChIP-qPCR experiment. The primer sequences used for ChIP-qPCR are listed in Supplemental Table S2.

4.7. Homology Analysis of PIFs in Arabidopsis and Triticum

To further clarify the homology relationship of the PIFs in Arabidopsis and Triticum, the PIF protein sequences of Arabidopsis and Triticum were downloaded and compared. To download the sequences, we performed a BLASTP search using Ensemble Plants (http://plants.ensembl.org/index.html, accessed on 11 April 2022) and chose the sequences with e-values less than 1×10^{-10}. Full-length protein alignments were performed by using MUSCLE [86], and phylogenetic trees were constructed using MEGA-X (https://www.megasoftware.net/, accessed on 11 April 2022). A total of 1000 bootstrap replications were performed using the neighbor-joining method. The tree was next modified by iTOL (https://itol.embl.de/, accessed on 11 April 2022). The MUF1 and bHLH domain sequences were defined using MEME (https://meme-suite.org/meme/tools/meme-chip, accessed on 11 April 2022).

4.8. Statistical Analysis

Statistical differences were calculated by Student's *t*-test or one-way ANOVA. Asterisks indicate the significant differences (* $p < 0.05$). Different letters above each bar indicate statistically significant differences determined by Tukey's multiple testing methods ($p < 0.05$).

5. Conclusions

In summary, we propose a hypothetical model that elucidates the molecular function of PIF4 in response to severe heat stress. PIF4 can directly bind to the G-box motifs of the HSFA2 promoter to activate its gene expression, thereby resulting in the activation of other HS-inducible genes, such as heat shock proteins. These lead to a stronger basal thermotolerance under non-heat-treatment conditions, thereby resulting in enhanced tolerance to severe heat stress. Meanwhile, the ATPIF4 homolog genes in wheat exhibited similar transcription patterns in response to heat stress. Our findings establish a molecular mechanism of PIF4 in mediating the heat stress response and provide a candidate direction for breeding heat-resistant wheat cultivars.

Supplementary Materials: The following are available online at https://www.mdpi.com/article/10.3390/ijms23116017/s1.

Author Contributions: J.Y. and X.Z. conceived and designed this research; J.Y. and X.Q. performed the experiment; J.Y., X.Q. and L.J. analyzed the data; J.Y. wrote the manuscript; G.L., C.W. (Chen Wang), C.W. (Changyu Wang), Y.Z., L.Z., W.L. and X.Z. performed writing—review and editing. All authors have read and agreed to the published version of the manuscript.

Funding: This work was supported by the Talents Project of Henan Agricultural University (No. 30601733) and the International Training Program for high-level Talents of Henan Province (No. 30602056).

Acknowledgments: We thank Hongtao Liu (Institute of Plant Physiology and Ecology, Shanghai Institutes for Biological Sciences, Chinese Academy of Sciences) for kindly providing the *35s:PIF4-TAP* seeds. We thank Jianping Yang (College of Agronomy, Henan Agricultural University) for kindly providing the *pif4-1* and PIF4-OE seeds.

Conflicts of Interest: The authors declare no conflict of interest.

References

1. Zhang, H.; Zhu, J.; Gong, Z.; Zhu, J.K. Abiotic stress responses in plants. *Nat. Rev. Genet.* **2022**, *23*, 104–119. [CrossRef] [PubMed]
2. Zhu, J.K. Abiotic stress signaling and responses in plants. *Cell* **2016**, *167*, 313–324. [CrossRef] [PubMed]
3. Wheeler, T.; von Braun, J. Climate change impacts on global food security. *Science* **2013**, *341*, 508–613. [CrossRef] [PubMed]
4. Zhao, C.; Liu, B.; Piao, S.; Wang, X.; Lobell, D.; Huang, Y.; Huang, M.; Yao, Y.; Bassu, S.; Ciais, P.; et al. Temperature increase reduces global yields of major crops in four independent estimates. *Proc. Natl. Acad. Sci. USA* **2017**, *114*, 9326–9331. [CrossRef]
5. Yeh, C.H.; Kaplinsky, N.J.; Hu, C.; Charng, Y.Y. Some like it hot, some like it warm: Phenotyping to explore thermotolerance diversity. *Plant Sci.* **2012**, *195*, 10–23. [CrossRef]
6. Perrella, G.; Bäurle, I.; van Zanten, M. Epigenetic regulation of thermomorphogenesis and heat stress tolerance. *New. Phytol.* **2022**, *234*, 1144–1160. [CrossRef]
7. Quint, M.; Delker, C.; Franklin, K.A.; Wigge, P.A.; Halliday, K.J.; van Zanten, M. Molecular and genetic control of plant thermomorphogenesis. *Nat. Plants* **2016**, *2*, 15190. [CrossRef]
8. Casal, J.; Balasubramanian, S. Thermomorphogenesis. *Annu. Rev. Plant Biol.* **2019**, *70*, 321–346. [CrossRef]
9. Gil, K.; Park, C. Thermal adaptation and plasticity of the plant circadian clock. *New. Phytol.* **2019**, *221*, 1215–1229. [CrossRef]
10. Wahid, A.; Gelani, S.; Ashraf, M.; Foolad, M. Heat tolerance in plants: An overview. *Environ. Exp. Bot.* **2007**, *61*, 199–223. [CrossRef]
11. Jacob, P.; Hirt, H.; Bendahmane, A. The heat-shock protein/chaperone network and multiple stress resistance. *Plant Biotechnol. J.* **2017**, *15*, 405–414. [CrossRef] [PubMed]
12. Ding, Y.; Shi, Y.; Yang, S. Molecular regulation of plant responses to environmental temperatures. *Mol. Plant* **2020**, *13*, 544–564. [CrossRef] [PubMed]
13. Ohama, N.; Sato, H.; Shinozaki, K.; Yamaguchi-Shinozaki, K. Transcriptional regulatory network of plant heat stress response. *Trends Plant Sci.* **2017**, *22*, 53–65. [CrossRef] [PubMed]
14. Mittler, R.; Finka, A.; Goloubinoff, P. How do plants feel the heat? *Trends Biochem. Sci.* **2012**, *37*, 118–125. [CrossRef]
15. Hasanuzzaman, M.; Nahar, K.; Alam, M.M.; Roychowdhury, R.; Fujita, M. Physiological, biochemical, and molecular mechanisms of heat stress tolerance in plants. *Int. J. Mol. Sci.* **2013**, *14*, 9643–9684. [CrossRef]
16. Neill, S.J.; Desikan, R.; Clarke, A.; Hurst, R.D.; Hancock, J.T. Hydrogen peroxide and nitric oxide as signalling molecules in plants. *J. Exp. Bot.* **2002**, *53*, 1237–1247. [CrossRef]
17. Fichman, Y.; Mittler, R. Rapid systemic signaling during abiotic and biotic stresses: Is the ROS wave master of all trades? *Plant J.* **2020**, *102*, 887–896. [CrossRef]
18. Kovtun, Y.; Chiu, W.; Tena, G.; Sheen, J. Functional analysis of oxidative stress-activated mitogen-activated protein kinase cascade in plants. *Proc. Natl. Acad. Sci. USA* **2000**, *97*, 2940–2945. [CrossRef]
19. Link, V.; Sinha, A.; Vashista, P.; Hofmann, M.; Proels, R.; Ehness, R.; Roitsch, T. A heat-activated MAP kinase in tomato: A possible regulator of the heat stress response. *FEBS Lett.* **2002**, *531*, 179–183. [CrossRef]
20. Sangwan, V.; Örvar, B.; Beyerly, J.; Hirt, H.; Dhindsa, R. Opposite changes in membrane fluidity mimic cold and heat stress activation of distinct plant MAP kinase pathways. *Plant J.* **2002**, *31*, 629–638. [CrossRef]
21. Shekhawat, K.; Almeida-Trapp, M.; García-Ramírez, G.; Hirt, H. Beat the heat: Plant- and microbe-mediated strategies for crop thermotolerance. *Trends Plant Sci.* **2022**, *21*, 1360–1385. [CrossRef] [PubMed]
22. Ohama, N.; Kusakabe, K.; Mizoi, J.; Zhao, H.; Kidokoro, S.; Koizumi, S.; Takahashi, F.; Ishida, T.; Yanagisawa, S.; Shinozaki, K.; et al. The transcriptional cascade in the heat stress response of arabidopsis is strictly regulated at the level of transcription factor expression. *Plant Cell* **2016**, *28*, 181–201. [CrossRef] [PubMed]
23. Nover, L.; Bharti, K.; Döring, P.; Mishra, S.; Ganguli, A.; Scharf, K. Arabidopsis and the heat stress transcription factor world: How many heat stress transcription factors do we need? *Cell Stress Chaperones* **2001**, *6*, 177–189. [CrossRef]

24. Liu, H.; Liao, H.; Charng, Y. The role of class A1 heat shock factors (HSFA1s) in response to heat and other stresses in Arabidopsis. *Plant Cell Environ.* **2011**, *34*, 738–751. [CrossRef]
25. Yoshida, T.; Ohama, N.; Nakajima, J.; Kidokoro, S.; Mizoi, J.; Nakashima, K.; Maruyama, K.; Kim, J.; Seki, M.; Todaka, D.; et al. Arabidopsis HsfA1 transcription factors function as the main positive regulators in heat shock-responsive gene expression. *Mol. Genet. Genom.* **2011**, *286*, 321–332. [CrossRef]
26. Liu, H.; Charng, Y. Common and distinct functions of Arabidopsis class A1 and A2 heat shock factors in diverse abiotic stress responses and development. *Plant Physiol.* **2013**, *163*, 276–290. [CrossRef]
27. Sakuma, Y.; Maruyama, K.; Qin, F.; Osakabe, Y.; Shinozaki, K.; Yamaguchi-Shinozaki, K. Dual function of an Arabidopsis transcription factor DREB2A in water-stress-responsive and heat-stress-responsive gene expression. *Proc. Natl. Acad. Sci. USA* **2006**, *103*, 18822–18827. [CrossRef] [PubMed]
28. Charng, Y.; Liu, H.; Liu, N.; Chi, W.; Wang, C.; Chang, S.; Wang, T. A heat-inducible transcription factor, HsfA2, is required for extension of acquired thermotolerance in Arabidopsis. *Plant Physiol.* **2007**, *143*, 251–262. [CrossRef]
29. Schramm, F.; Larkindale, J.; Kiehlmann, E.; Ganguli, A.; Englich, G.; Vierling, E.; von Koskull-Döring, P. A cascade of transcription factor DREB2A and heat stress transcription factor HsfA3 regulates the heat stress response of Arabidopsis. *Plant J.* **2008**, *53*, 264–274. [CrossRef]
30. Yoshida, T.; Sakuma, Y.; Todaka, D.; Maruyama, K.; Qin, F.; Mizoi, J.; Kidokoro, S.; Fujita, Y.; Shinozaki, K.; Yamaguchi-Shinozaki, K. Functional analysis of an Arabidopsis heat-shock transcription factor HsfA3 in the transcriptional cascade downstream of the DREB2A stress-regulatory system. *Biochem. Biophys. Res. Commun.* **2008**, *368*, 515–521. [CrossRef]
31. Nishizawa-Yokoi, A.; Nosaka, R.; Hayashi, H.; Tainaka, H.; Maruta, T.; Tamoi, M.; Ikeda, M.; Ohme-Takagi, M.; Yoshimura, K.; Yabuta, Y.; et al. HsfA1d and HsfA1e involved in the transcriptional regulation of HsfA2 function as key regulators for the Hsf signaling network in response to environmental stress. *Plant Cell Physiol.* **2011**, *52*, 933–945. [CrossRef] [PubMed]
32. Friedrich, T.; Oberkofler, V.; Trindade, I.; Altmann, S.; Brzezinka, K.; Lämke, J.; Gorka, M.; Kappel, C.; Sokolowska, E.; Skirycz, A.; et al. Heteromeric HSFA2/HSFA3 complexes drive transcriptional memory after heat stress in Arabidopsis. *Nat. Commun.* **2021**, *12*, 3426. [CrossRef] [PubMed]
33. Shekhawat, K.; Saad, M.; Sheikh, A.; Mariappan, K.; Al-Mahmoudi, H.; Abdulhakim, F.; Eida, A.; Jalal, R.; Masmoudi, K.; Hirt, H. Root endophyte induced plant thermotolerance by constitutive chromatin modification at heat stress memory gene loci. *EMBO Rep.* **2021**, *22*, e51049. [CrossRef]
34. Liu, H.; Lämke, J.; Lin, S.; Hung, M.; Liu, K.; Charng, Y.; Bäurle, I. Distinct heat shock factors and chromatin modifications mediate the organ-autonomous transcriptional memory of heat stress. *Plant J.* **2018**, *95*, 401–413. [CrossRef] [PubMed]
35. Liu, J.; Feng, L.; Gu, X.; Deng, X.; Qiu, Q.; Li, Q.; Zhang, Y.; Wang, M.; Deng, Y.; Wang, E.; et al. An H3K27me3 demethylase-HSFA2 regulatory loop orchestrates transgenerational thermomemory in Arabidopsis. *Cell Res.* **2019**, *29*, 379–390. [CrossRef] [PubMed]
36. Swindell, W.; Huebner, M.; Weber, A. Transcriptional profiling of Arabidopsis heat shock proteins and transcription factors reveals extensive overlap between heat and non-heat stress response pathways. *BMC Genom.* **2007**, *8*, 125. [CrossRef] [PubMed]
37. Yamada, K.; Fukao, Y.; Hayashi, M.; Fukazawa, M.; Suzuki, I.; Nishimura, M. Cytosolic HSP90 regulates the heat shock response that is responsible for heat acclimation in Arabidopsis thaliana. *J. Biol. Chem.* **2007**, *282*, 37794–37804. [CrossRef]
38. Hahn, A.; Bublak, D.; Schleiff, E.; Scharf, K. Crosstalk between Hsp90 and Hsp70 chaperones and heat stress transcription factors in tomato. *Plant Cell* **2011**, *23*, 741–755. [CrossRef]
39. Martínez, C.; Espinosa-Ruíz, A.; de Lucas, M.; Bernardo-García, S.; Franco-Zorrilla, J.; Prat, S. PIF4-induced BR synthesis is critical to diurnal and thermomorphogenic growth. *EMBO Rep.* **2018**, *37*, e99552. [CrossRef]
40. Koini, M.; Alvey, L.; Allen, T.; Tilley, C.; Harberd, N.; Whitelam, G.; Franklin, K. High temperature-mediated adaptations in plant architecture require the bHLH transcription factor PIF4. *Curr. Biol.* **2009**, *19*, 408–413. [CrossRef]
41. Xu, Y.; Zhu, Z. PIF4 and PIF4-Interacting Proteins: At the Nexus of Plant Light, Temperature and Hormone Signal Integrations. *Int. J. Mol. Sci.* **2021**, *22*, 10304. [CrossRef] [PubMed]
42. Qiu, Y. Regulation of PIF4-mediated thermosensory growth. *Plant Sci.* **2020**, *297*, 110541. [CrossRef] [PubMed]
43. Nusinow, D.; Helfer, A.; Hamilton, E.; King, J.; Imaizumi, T.; Schultz, T.; Farré, E.; Kay, S. The ELF4-ELF3-LUX complex links the circadian clock to diurnal control of hypocotyl growth. *Nature* **2011**, *475*, 398–402. [CrossRef]
44. Nozue, K.; Covington, M.; Duek, P.; Lorrain, S.; Fankhauser, C.; Harmer, S.; Maloof, J. Rhythmic growth explained by coincidence between internal and external cues. *Nature* **2007**, *448*, 358–361. [CrossRef]
45. Li, N.; Bo, C.; Zhang, Y.; Wang, L. Phytochrome Interacting Factors PIF4 and PIF5 promote heat stress induced leaf senescence in Arabidopsis. *J. Exp. Bot.* **2021**, *72*, 4577–4589. [CrossRef]
46. Franklin, K.; Lee, S.; Patel, D.; Kumar, S.; Spartz, A.; Gu, C.; Ye, S.; Yu, P.; Breen, G.; Cohen, J.; et al. Phytochrome-interacting factor 4 (PIF4) regulates auxin biosynthesis at high temperature. *Proc. Natl. Acad. Sci. USA* **2011**, *108*, 20231–20235. [CrossRef]
47. Kumar, S.; Lucyshyn, D.; Jaeger, K.; Alós, E.; Alvey, E.; Harberd, N.; Wigge, P. Transcription factor PIF4 controls the thermosensory activation of flowering. *Nature* **2012**, *484*, 242–245. [CrossRef]
48. Oh, E.; Zhu, J.; Wang, Z. Interaction between BZR1 and PIF4 integrates brassinosteroid and environmental responses. *Nat. Cell Biol.* **2012**, *14*, 802–809. [CrossRef]
49. Sun, J.; Qi, L.; Li, Y.; Chu, J.; Li, C. PIF4-mediated activation of YUCCA8 expression integrates temperature into the auxin pathway in regulating arabidopsis hypocotyl growth. *PLoS Genet.* **2012**, *8*, e1002594. [CrossRef]

50. Lee, S.; Wang, W.; Huq, E. Spatial regulation of thermomorphogenesis by HY5 and PIF4 in Arabidopsis. *Nat. Commun.* **2021**, *12*, 3656. [CrossRef]
51. Possart, A.; Xu, T.; Paik, I.; Hanke, S.; Keim, S.; Hermann, H.; Wolf, L.; Hiß, M.; Becker, C.; Huq, E.; et al. Characterization of Phytochrome Interacting Factors from the Moss Physcomitrella patens Illustrates Conservation of Phytochrome Signaling Modules in Land Plants. *Plant Cell* **2017**, *29*, 310–330. [CrossRef] [PubMed]
52. Li, W.; Liu, Y.; Wang, W.; Liu, J.; Yao, M.; Guan, M.; Guan, C.; He, X. Phytochrome-interacting factor (PIF) in rapeseed (*Brassica napus* L.): Genome-wide identification, evolution and expression analyses during abiotic stress, light quality and vernalization. *Int. J. Biol. Macromol.* **2021**, *180*, 14–27. [CrossRef] [PubMed]
53. Leivar, P.; Monte, E. PIFs: Systems integrators in plant development. *Plant Cell* **2014**, *26*, 56–78. [CrossRef] [PubMed]
54. Leivar, P.; Quail, P. PIFs: Pivotal components in a cellular signaling hub. *Trends Plant Sci.* **2011**, *16*, 19–28. [CrossRef] [PubMed]
55. Xu, D. Multifaceted Roles of PIF4 in Plants. *Trends Plant Sci.* **2018**, *23*, 749–751. [CrossRef] [PubMed]
56. Xu, X.; Paik, I.; Zhu, L.; Huq, E. Illuminating Progress in Phytochrome-Mediated Light Signaling Pathways. *Trends Plant Sci.* **2015**, *20*, 641–650. [CrossRef] [PubMed]
57. Leivar, P.; Monte, E.; Oka, Y.; Liu, T.; Carle, C.; Castillon, A.; Huq, E.; Quail, P. Multiple phytochrome-interacting bHLH transcription factors repress premature seedling photomorphogenesis in darkness. *Curr. Biol.* **2008**, *18*, 1815–1823. [CrossRef]
58. Ikeda, M.; Mitsuda, N.; Ohme-Takagi, M. Arabidopsis HsfB1 and HsfB2b act as repressors of the expression of heat-inducible Hsfs but positively regulate the acquired thermotolerance. *Plant Physiol.* **2011**, *157*, 1243–1254. [CrossRef]
59. Olmo, R.; Cabrera, J.; Díaz-Manzano, F.; Ruiz-Ferrer, V.; Barcala, M.; Ishida, T.; García, A.; Andrés, M.; Ruiz-Lara, S.; Verdugo, I.; et al. Root-knot nematodes induce gall formation by recruiting developmental pathways of post-embryonic organogenesis and regeneration to promote transient pluripotency. *New Phytol.* **2020**, *227*, 200–215. [CrossRef]
60. Pham, V.; Kathare, P.; Huq, E. Phytochromes and phytochrome interacting factors. *Plant Physiol.* **2018**, *176*, 1025–1038. [CrossRef]
61. Zhang, Y.; Mayba, O.; Pfeiffer, A.; Shi, H.; Tepperman, J.; Speed, T.; Quail, P. A quartet of PIF bHLH factors provides a transcriptionally centered signaling hub that regulates seedling morphogenesis through differential expression-patterning of shared target genes in Arabidopsis. *PLoS Genet.* **2013**, *9*, e1003244. [CrossRef] [PubMed]
62. Hornitschek, P.; Lorrain, S.; Zoete, V.; Michielin, O.; Fankhauser, C. Inhibition of the shade avoidance response by formation of non-DNA binding bHLH heterodimers. *EMBO J.* **2009**, *28*, 3893–3902. [CrossRef] [PubMed]
63. Nawkar, G.; Kang, C.; Maibam, P.; Park, J.; Jung, Y.; Chae, H.; Chi, Y.; Jung, I.; Kim, W.; Yun, D.; et al. HY5, a positive regulator of light signaling, negatively controls the unfolded protein response in Arabidopsis. *Proc. Natl. Acad. Sci. USA* **2017**, *114*, 2084–2089. [CrossRef] [PubMed]
64. Lippmann, R.; Babben, S.; Menger, A.; Delker, C.; Quint, M. Development of Wild and Cultivated Plants under Global Warming Conditions. *Curr. Biol.* **2019**, *29*, R1326–R1338. [CrossRef]
65. Al-Sady, B.; Ni, W.; Kircher, S.; Schäfer, E.; Quail, P. Photoactivated phytochrome induces rapid PIF3 phosphorylation prior to proteasome-mediated degradation. *Mol. Cell* **2006**, *23*, 439–446. [CrossRef]
66. Khanna, R.; Huq, E.; Kikis, E.; Al-Sady, B.; Lanzatella, C.; Quail, P. A novel molecular recognition motif necessary for targeting photoactivated phytochrome signaling to specific basic helix-loop-helix transcription factors. *Plant Cell* **2004**, *16*, 3033–3044. [CrossRef]
67. Dalton, J.; Bätz, U.; Liu, J.; Curie, G.; Quail, P. A Modified Reverse One-Hybrid Screen Identifies Transcriptional Activation Domains in PHYTOCHROME-INTERACTING FACTOR 3. *Front. Plant Sci.* **2016**, *7*, 881. [CrossRef]
68. Toledo-Ortiz, G.; Huq, E.; Quail, P. The Arabidopsis basic/helix-loop-helix transcription factor family. *Plant Cell* **2003**, *15*, 1749–1770. [CrossRef]
69. Zhao, H.; Bao, Y. PIF4: Integrator of light and temperature cues in plant growth. *Plant Sci.* **2021**, *313*, 111086. [CrossRef]
70. Hwang, G.; Zhu, J.; Lee, Y.; Kim, S.; Nguyen, T.; Kim, J.; Oh, E. PIF4 promotes expression of LNG1 and LNG2 to induce thermomorphogenic growth in Arabidopsis. *Front. Plant Sci.* **2017**, *8*, 1320. [CrossRef]
71. Zhu, J.; Oh, E.; Wang, T.; Wang, Z. TOC1-PIF4 interaction mediates the circadian gating of thermoresponsive growth in Arabidopsis. *Nat. Commun.* **2016**, *7*, 13692. [CrossRef] [PubMed]
72. Shen, H.; Moon, J.; Huq, E. PIF1 is regulated by light-mediated degradation through the ubiquitin-26S proteasome pathway to optimize photomorphogenesis of seedlings in Arabidopsis. *Plant J.* **2005**, *44*, 1023–1035. [CrossRef] [PubMed]
73. Shen, Y.; Khanna, R.; Carle, C.; Quail, P. Phytochrome induces rapid PIF5 phosphorylation and degradation in response to red-light activation. *Plant Physiol.* **2007**, *145*, 1043–1051. [CrossRef]
74. Oh, E.; Yamaguchi, S.; Kamiya, Y.; Bae, G.; Chung, W.; Choi, G. Light activates the degradation of PIL5 protein to promote seed germination through gibberellin in Arabidopsis. *Plant J.* **2006**, *47*, 124–139. [CrossRef] [PubMed]
75. Lorrain, S.; Allen, T.; Duek, P.; Whitelam, G.; Fankhauser, C. Phytochrome-mediated inhibition of shade avoidance involves degradation of growth-promoting bHLH transcription factors. *Plant J.* **2008**, *53*, 312–323. [CrossRef] [PubMed]
76. Qi, L.; Liu, S.; Li, C.; Fu, J.; Jing, Y.; Cheng, J.; Li, H.; Zhang, D.; Wang, X.; Dong, X.; et al. Phytochrome-interacting factors interact with the aba receptors PYL8 and PYL9 to orchestrate aba signaling in darkness. *Mol. Plant* **2020**, *13*, 414–430. [CrossRef]
77. Jung, J.; Domijan, M.; Klose, C.; Biswas, S.; Ezer, D.; Gao, M.; Khattak, A.; Box, M.; Charoensawan, V.; Cortijo, S.; et al. Phytochromes function as thermosensors in Arabidopsis. *Science* **2016**, *354*, 886–889. [CrossRef]
78. Legris, M.; Klose, C.; Burgie, E.; Rojas, C.; Neme, M.; Hiltbrunner, A.; Wigge, P.; Schäfer, E.; Vierstra, R.; Casal, J. Phytochrome B integrates light and temperature signals in Arabidopsis. *Science* **2016**, *354*, 897–900. [CrossRef]

79. Huai, J.; Zhang, X.; Li, J.; Ma, T.; Zha, P.; Jing, Y.; Lin, R. SEUSS and PIF4 Coordinately Regulate Light and Temperature Signaling Pathways to Control Plant Growth. *Mol. Plant* **2018**, *11*, 928–942. [CrossRef]
80. de Lucas, M.; Davière, J.; Rodríguez-Falcón, M.; Pontin, M.; Iglesias-Pedraz, J.; Lorrain, S.; Fankhauser, C.; Blázquez, M.; Titarenko, E.; Prat, S. A molecular framework for light and gibberellin control of cell elongation. *Nature* **2008**, *451*, 480–484. [CrossRef]
81. Ma, D.; Li, X.; Guo, Y.; Chu, J.; Fang, S.; Yan, C.; Noel, J.; Liu, H. Cryptochrome 1 interacts with PIF4 to regulate high temperature-mediated hypocotyl elongation in response to blue light. *Proc. Natl. Acad. Sci. USA* **2016**, *113*, 224–229. [CrossRef] [PubMed]
82. Florea, L.; Song, L.; Salzberg, S. Thousands of exon skipping events differentiate among splicing patterns in sixteen human tissues. *F1000Research* **2013**, *2*, 188. [CrossRef] [PubMed]
83. Chen, C.; Chen, H.; Zhang, Y.; Thomas, H.; Frank, M.; He, Y.; Xia, R. TBtools: An Integrative Toolkit Developed for Interactive Analyses of Big Biological Data. *Mol. Plant* **2020**, *13*, 1194–1202. [CrossRef] [PubMed]
84. Winter, D.; Vinegar, B.; Nahal, H.; Ammar, R.; Wilson, G.; Provart, N. An "Electronic Fluorescent Pictograph" browser for exploring and analyzing large-scale biological data sets. *PLoS ONE* **2007**, *2*, e718. [CrossRef]
85. Zhang, X.; Clarenz, O.; Cokus, S.; Bernatavichute, Y.; Pellegrini, M.; Goodrich, J.; Jacobsen, S. Whole-genome analysis of histone H3 lysine 27 trimethylation in Arabidopsis. *PLoS Biol.* **2007**, *5*, e129. [CrossRef]
86. Edgar, R. MUSCLE: A multiple sequence alignment method with reduced time and space complexity. *BMC Bioinform.* **2004**, *5*, 113. [CrossRef]

International Journal of
Molecular Sciences

MDPI

Article

AtEAU1 and AtEAU2, Two EAR Motif-Containing ABA Up-Regulated Novel Transcription Repressors Regulate ABA Response in *Arabidopsis*

Na Zhang [1,2,†], Siyu Chen [1,†], Adnan [2], Xutong Wang [1], Saddam Hussain [2], Yuxin Cheng [2], Yingying Li [2], Yuan Yuan [2], Chen Wang [2], Rao Lin [2], Huiyuan Zhang [1], Jiachen Wang [1], Tianya Wang [2] and Shucai Wang [1,*]

1 Laboratory of Plant Molecular Genetics and Crop Gene Editing, School of Life Sciences, Linyi University, Linyi 276000, China
2 Key Laboratory of Molecular Epigenetics of MOE, Northeast Normal University, Changchun 130024, China
* Correspondence: wangshucai@lyu.edu.cn
† These authors contributed equally to this work.

Abstract: EAR (Ethylene-responsive element binding factor-associated Amphiphilic Repression) motif-containing transcription repressors have been shown to regulate plant growth and development, and plant responses to plant hormones and environmental stresses including biotic and abiotic stresses. However, the functions of most EAR-motif-containing proteins remain largely uncharacterized. The plant hormone abscisic acid (ABA) also plays important roles in regulating plant responses to abiotic stresses via activation/repression of ABA-responsive genes. We report here the identification and functional characterization of two ABA-responsive EAR motif-containing protein genes, *AtEAU1* (*Arabidopsis thaliana EAR motif-containing ABA Up-regulated 1*) and *AtEAU2*. Quantitative RT-PCR results show that the expressions of *AtEAU1* and *AtEAU2* were increased by ABA treatment, and were decreased in the ABA biosynthesis mutant *aba1-5*. Assays in transfected *Arabidopsis* protoplasts show that both AtEAU1 and AtEAU2 were specifically localized in the nucleus, and when recruited to the promoter region of the reporter gene by a fused DNA binding domain, repressed reporter gene expression. By using T-DNA insertion mutants and a gene-edited transgene-free mutant generated by CRISPR/Cas9 gene editing, we performed ABA sensitivity assays, and found that ABA sensitivity in the both *ateau1* and *ateau2* single mutants was increased in seedling greening assays. ABA sensitivity in the *ateau1 ateau2* double mutants was also increased, but was largely similar to the *ateau1* single mutants. On the other hand, all the mutants showed a wild type response to ABA in root elongation assays. Quantitative RT-PCR results show that the expression level of *PYL4*, an ABA receptor gene was increased, whereas that of *ABI2*, a PP2C gene was decreased in the *ateau1* and *ateau1* single, and the *ateau1 ateau2* double mutants. In summary, our results suggest that *AtEAU1* and *AtEAU2* are ABA-response genes, and AtEAU1 and AtEAU2 are novel EAR motif-containing transcription repressors that negatively regulate ABA responses in *Arabidopsis*, likely by regulating the expression of some ABA signaling key regulator genes.

Keywords: ABA; transcription factor; EAR motif-containing protein; AtEAUs; *Arabidopsis*

Citation: Zhang, N.; Chen, S.; Adnan; Wang, X.; Hussain, S.; Cheng, Y.; Li, Y.; Yuan, Y.; Wang, C.; Lin, R.; et al. AtEAU1 and AtEAU2, Two EAR Motif-Containing ABA Up-Regulated Novel Transcription Repressors Regulate ABA Response in *Arabidopsis*. *Int. J. Mol. Sci.* **2022**, *23*, 9053. https://doi.org/10.3390/ijms23169053

Academic Editor: Bartolome Sabater

Received: 28 June 2022
Accepted: 8 August 2022
Published: 13 August 2022

1. Introduction

The plant hormone ABA (abscisic acid) regulates plant tolerance to abiotic stresses such as drought, salt, cold, and low oxygen, in most of the cases, via regulating the expression of downstream ABA response genes [1–8]. Under stress conditions, levels of ABA increase in plant cells, therefore ABA molecules are able to bind to the PYR1 (Pyrabactin Resistance 1)/PYL(PYR1-like)/RCAR (Regulatory Component of ABA Receptor) receptors [5,9–13]. The binding of ABA allows the PYR1/PYL/RCAR receptors to interact with the A-group PP2Cs (PROTEIN PHOSPHATASE 2C) phosphatases, which under normal conditions, bind to the SnRKs (NONFERMENTING 1 (SNF1)-RELATED

PROTEIN KINASES) kinases and inhibit their activities [2,3,6,14]. Once released from A-group PP2Cs phosphatases, SnRKs kinases are self-activated, and then phosphorylate the ABF (ABA-responsive element-binding protein)/AREB(ABRE-binding factor)/ABI5 (ABA INSENSITIVE 5)-type bZIP (basic region leucine zipper) transcription factors. Phosphorylated ABF/AREB/ABI5 transcription factors are then able to regulate the expression of ABA-responsive genes, resulting in altered responses of plants to ABA and abiotic stresses [2,6,7,14–17].

In addition to the ABF/AREB/ABI5 transcription factors, some of the transcription factors from known transcription factor families encoded by ABA response genes are also able to regulate plant responses to ABA and abiotic stresses. For example, the expression of both B3 transcription factor gene *ABI3* and AP2 (APETALA2) transcription factor gene *ABI4* are induced by ABA, and ABI3 and ABI4 positively regulate ABA response in *Arabidopsis* [18,19]. On the other hand, expression of the bHLH transcription factor gene *bHLH129* is down-regulated by ABA, and bHLH129 negatively regulates ABA response in *Arabidopsis* [6]. In addition to transcription factors from known transcription factor families, some novel transcription factors have also been identified from unknown function ABA response genes. As an example, transcription repressor family genes *AITRs* (*ABA-induced transcription repressors*) were identified from unknown function ABA response genes, and AITRs negatively regulate ABA and abiotic tresses responses in *Arabidopsis* [6,7]. In particular, we found that some AITRs are EAR (ERF-associated amphiphilic repression) motif-containing proteins that contain a fully conserved LxLxL EAR motif [6].

The EAR motifs were first identified in the class II ERFs (ethylene-responsive factors) and some of the C2H2 family transcription factors that confer transcriptional repression activities [20], and then in other transcription repressors such as Aux/IAA proteins and ovate family proteins (OFPs) [21,22]. The EAR motifs were initially identified as a (L/F)DLN(L/F)xP amino acid signature [20]; further analysis of the EAR motifs in the class II ERFs, the C2H2 family proteins, and some other EAR motif-containing proteins have refined the amino acid signatures as DLNxxP and LxLxL, and transcriptional repression mediated by EAR motif-containing proteins has been considered as the main transcriptional repression form in plants [23].

Consistent with the importance of EAR motif-containing in regulating gene expression, accumulated evidence suggests that EAR motif-containing proteins are involved in the regulation of plant growth and development, and plant response to plant hormones and environmental stresses, including biotic and abiotic stresses, in most of the cases, via functioning as transcription repressors or recruiting co-repressors to repress the expression of the downstream target genes [22–31]. As examples, the transcription repressor OFP1 regulates cell elongation in *Arabidopsis* via directly repressing the expression of the plant hormone GA (Gibberellin) biosynthesis gene *GA20ox1* (*GA20-oxidase 1*) [22], KIX8 (KINASE-INDUCIBLE DOMAIN INTERACTING 8) and KIX9 represses *Arabidopsis* leaf growth via recruiting the co-repressor TOPLESS to form a repressor protein complex [32], whereas ERF7 regulates ABA and drought responses in *Arabidopsis* via interacting with the histone deacetylase HDA19, the protein kinase PKS3, and AtSin3, a homolog of a human global transcription corepressor [25].

Considering the importance of EAR motif-containing proteins, a genome-wide search of the LxLxL or DLNxxP EAR motif-containing proteins was performed in 71 different plant species, and more than 20,000 EAR motif-containing proteins were identified, with 556 encoded by 411 genes in *Arabidopsis*, but most of them are functionally uncharacterized [31]. On the hand, as the EAR motif-containing proteins show a high amino acid sequence diversity [31], it is very likely that some of the EAR motif-containing proteins may still remain unidentified. As an example, we found that some of the AITRs have a fully conserved LxLxL EAR motif [7], but none of them are identified in the genome-wide search of EAR motif-containing proteins [31]. Therefore we first tried to identify novel EAR motif-containing proteins that may involve in the regulation of ABA response by searching EAR motifs in proteins encoded by unknown function ABA-responsive genes from available transcriptome datasets, and we successfully identified SlEAD1 (*Solanum lycopersicum* EAR motif-containing ABA down-regulated 1) as a negative regulator of ABA response in tomato [33].

Here, by using an opposite strategy, i.e., identifying ABA response genes from unknown function EAR motif-containing protein genes [31], and searching for protein homologs, we identified AtEAU1 (*Arabidopsis thaliana* EAR motif-containing ABA Up-regulated 1) and AtEAU2 as novel regulators of ABA response in *Arabidopsis*.

2. Results

2.1. AtEAU1 and AtEAU2 Are Closely Related EAR Motif-Containing Proteins

A protein encoded by gene At1g78170 is among the identified EAR motif-containing proteins [31]. As a protein encoded by gene At1g22250 is closely related to the protein encoded by gene At1g78170, and the expression of these two genes is induced by ABA treatment (see next section for details), we named them *AtEAU1* (*Arabidopsis thaliana EAR motif-containing ABA Up-regulated 1*) and *AtEAU2*, respectively.

Protein homolog identification on Phytozome (https://phytozome-next.jgi.doe.gov, accessed on 1 September 2016) show that the next closely related protein to AtEAU1 and AtEAU2 is the protein encoded by gene At4g08910, which was also identified as an EAR motif-containing protein [31]. However, phylogenetic analysis shows that the most closely related protein identified from other plant species including soybean, tomato, rice and poplar are more closely related to At4g08910, and they formed a clade, whereas AtEAU1 and AtEAU2 formed another clade (Figure 1A).

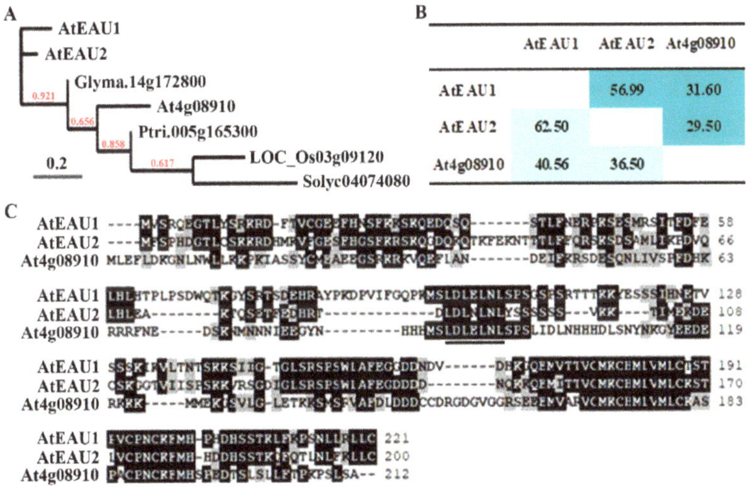

Figure 1. AtEAUs in *Arabidopsis*. (**A**) Phylogenetic analysis of AtEAU1, AtEAU2, and their closely related proteins. The full-length amino acid sequences of the AtEAU1 and AtEAU2 and their closely related proteins from *Arabidopsis* and other several different plants were obtained from phytozome (https://phytozome-next.jgi.doe.gov, accessed on 1 September 2016), and subjected to phylogenetic analysis by using the "One Click" mode with default settings on phylogeny (http://www.phylogeny.fr/simple_phylogeny.cgi, accessed on 1 September 2016). Numbers above the branches indicate the branch support values. (**B**) Amino acid identity and similarity of AtEAU1, AtEAU2 and At4g08910. The full-length amino acid sequences of AtEAU1, AtEAU2, and At4g08910 were used for SIAS (http://imed.med.ucm.es/Tools/sias.html, accessed on 1 September 2016) assay. The amino acid identity percentage was shaded in light blue, and the similarity percentage in blue. (**C**) Sequence alignment of AtEAUs and At4g08910. Black shades indicate the identical amino acids and gray shades indicate the similar amino acids. Underline indicates the LxLxL motif, one of the two known EAR motifs.

Consistent with the phylogenetic tree, amino acid identity and similarity assays show that AtEAU1 and AtEAU2 shared ~57% amino acid identity and ~62% similarity, respectively, but sharing between At4g08910 and AtEAU1 or AtEAU2 was only about 30% and 40%, respectively (Figure 1B). On the other hand, even though At4g08910 shared relatively low amino acid identity and similarity with AtEAU1 or AtEAU2, it has a fully conserved LxLxL EAR motif presented in AtEAU1 or AtEAU2 (Figure 1C).

2.2. The Expression of AtEAU1 and AtEAU2 Are Up-Regulated by ABA and They Share Similar Expression Patern

Since AtEAU1 and AtEAU2 are closely related, but showed low amino acid identity and similarity with At4g08910, and are in a different clade than At4g08910 in the phylogenetic analysis (Figure 1), we characterized only AtEAU1 and AtEAU2 in this study.

Expression of *AtEAU1* and *AtEAU2* in response to exogenous ABA treatment was examined using the Col wild-type *Arabidopsis* seedlings. Seedlings of the Col wild type were treated with ABA, and RNA isolated was used for qRT-PCR analysis. As shown in Figure 2A, the expression levels of *AtEAU1* and *AtEAU2* increased about 2.5 and 1.5 fold, respectively, in seedlings treated with ABA compared with mock-treated seedlings. We also examined the expression of *AtEAU1* and *AtEAU2* in seedlings of the *aba1-5*, an ABA biosynthesis mutant in the *Ler* wild type background [34]. We found that the expression level of *AtEAU1* and *AtEAU2* decreased to about 1/3 and 1/2 of that in the *Ler* wild type seedlings, respectively (Figure 2B). These results suggest that *AtEAU1* and *AtEAU2* are ABA response genes.

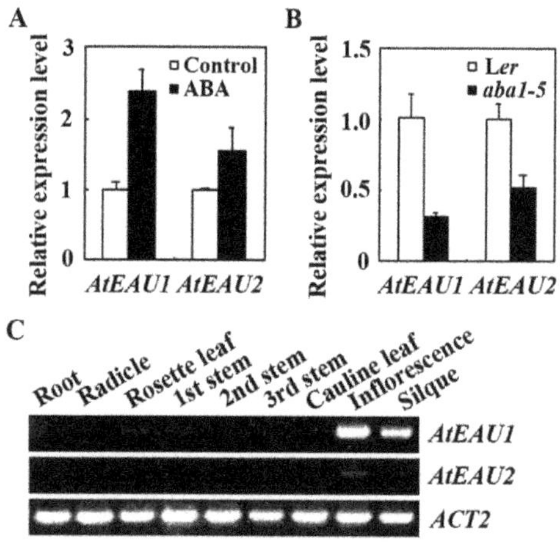

Figure 2. Expression of *AtEAUs* in seedlings treated with ABA and in different tissues and organs. (**A**) Expression of *AtEAU1* and *AtEAU2* in ABA-treated seedlings. Twelve-day-old seedlings of the Col wild type *Arabidopsis* were treated with 50 μM ABA for 4 h, RNA was then isolated and subjected to quantitative RT-PCR (qRT-PCR) to examine the expression of *AtEAU1* and *AtEAU2*. *ACT2* was used as an inner reference gene for qRT PCR, and the expression levels of *AtEAU1* and *AtEAU2* in mock-treated control seedlings were set as 1. Data represent the mean ± SD of three replicates. (**B**) Expression of *AtEAU1* and *AtEAU2* in the *aba1-5* ABA biosynthesis mutant seedlings. RNA was isolated from 12-day-old seedlings of the *Ler* wild type and the *aba1-5* mutant, and qRT-PCR was used to examine the expression of *AtEAU1* and *AtEAU2*. *ACT2* was used as an inner reference gene for qRT-PCR, and expression of *AtEAU1* and *AtEAU2* in the *Ler* wild-type seedlings was set as 1. Data represent the mean ± SD of three replicates. (**C**) Expression pattern of *AtEAU1* and *AtEAU2*. Roots, radicle, rosette leaves, stems including 1st, 2nd, and 3rd stem from bottom to top of the inflorescence, cauline leaves, inflorescences, and siliques were collected from 5-week-old plants, RNA was then isolated and subjected to RT-PCR analysis to examine the expression of *AtEAU1* and *AtEAU2*. *ACT2* was used as a control for RT-PCR analysis.

We further examined the expression pattern of *AtEAU1* and *AtEAU2*. Different tissues and organs were collected from adult *Arabidopsis* plants, and RNA was isolated and used for RT-PCR analysis. As shown in Figure 2C, *AtEAU1* and *AtEAU2* shared a largely similar expression pattern, i.e., both have relative higher express levels in inflorescence and siliques. However, the expression of *AtEAU1* was also detectable in rosette leaves and stems, whereas *AtEAU2* was not, possibly due to its relatively low expression level (Figure 2C).

2.3. AtEAU1 and AtEAU2 Function as Transcription Repressors

As AtEAU1 has been identified as EAR motif-containing protein [31], and AtEAU2 also has a fully conserved LxLxL EAR motif, we thus examined if they may function as transcription repressors by using transfection assays in *Arabidopsis* protoplasts.

We first examined the protein subcellular localization of AtEAU1 and AtEAU2. Plasmids of the *AtEAU1-GFP* and *AtEAU2-GFP* constructs were transfected into *Arabidopsis* protoplasts isolated from leaves of 3–4 weeks old Col wild type plants, and GFP fluorescence in the protoplasts was observed under a confocal microscope. As shown in Figure 3A, GFP fluorescence for both AEAU1-GFP and AtEAU2-GFP was observed only in the nucleus, suggesting that AEAU1 and AtEAU2 are nucleus proteins.

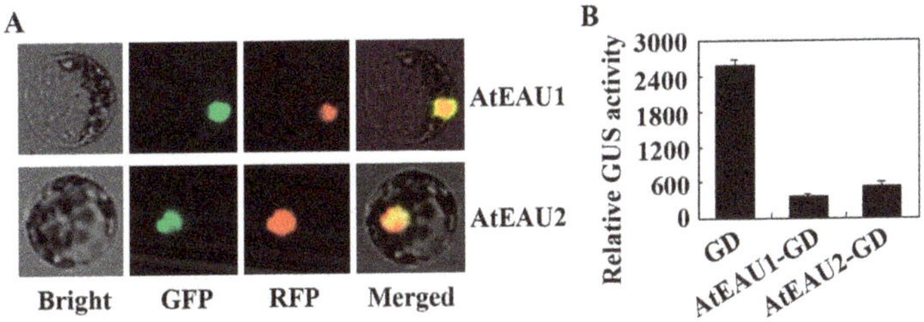

Figure 3. Protein subcellular localization and transcription activity of AtEAUs. (**A**) Protein subcellular localization of AtEAU1 and AtEAU2. Plasmid DNA of *AtEAU1-GFP* and *AtEAU2-GFP* was co-transfected with nucleus indicator *NLS-RFP* into protoplasts isolated from the Col wild-type *Arabidopsis*; transfected protoplasts were incubated for 20–22 h in darkness and then examined under a fluorescence microscope. Left panel: bright-field channel; middle panel: GFP channel; right panel: merged image. (**B**) Transcriptional activities of AtEAU1 and AtEAU2. Plasmid DNA of the *AtEAU1-GD* and *AtEAU2-GD* was cotransfected with a plasmid of the *LD-VP* activator gene, and the *LexA-Gal4:GUS* reporter gene into protoplasts isolated from the Col wild-type *Arabidopsis*, and the transfected protoplasts were incubated for 20–22 h in darkness before GUS activity was assayed. Cotransfection of the *GD* plasmid DNA was used as a control. Data represent the mean ± SD of three replicates.

We then examined if AtEAU1 and AtEAU2 may able to repress reporter gene expression in transfected protoplasts when recruited to the promoter of the reporter gene by a fused DNA binding domain. Plasmids of the reporter construct *LexA-Gal4:GUS* and the transcriptional activator construct *LD-VP* were co-transfected with the effect construct *AtEAU1-GD* or *AtEAU2-GD* into *Arabidopsis* protoplasts. Co-transfection of effector construct *GD* was used as a control. As the transcriptional activator LD-VP is able to bind to the *LexA* promoter via the fused LD domain and activate the expression of the *LexA-Gal4:GUS* reporter gene, whereas the AtEAU1 or AtEAU2 proteins are able to bind to the *Gal4* promoter via the fused GD domain, expression levels of the reporter gene will be decreased if AtEAU1 and AtEAU2 function as transcription repressors. As shown in Figure 3B, compared with cotransfection of *GD*, GUS activity was dramatically reduced when *AtEAU1-GD* or *AtEAU2-GD* was cotransfected, suggesting that AtEAU1 and AtEAU2 are transcription repressors.

2.4. Generation of Single and Double Mutants for Gene AtEAU1 and AtEAU2

In order to examine if AtEAU1 and AtEAU2 may involve in the regulation of ABA response in *Arabidopsis*, we isolated/generated single and double mutants for *AtEAU1* and *AtEAU2*. Since there is 1 T-DNA insertion line available for gene *AtEAU1*, and 2 are available for *AtEAU2* from ABRC, we obtained the seeds and isolated single homozygous mutants *ateau1-1* (SAIL_197_H05), *ateau2-1* (SAIL_642_C12), and *ateau2-2* (SAIL_1242_E02) (Figure 4A). By crossing *ateau1-1* and *ateau2-1* mutants and examining F2 plants, we obtained *ateau1 ateau2* double mutants.

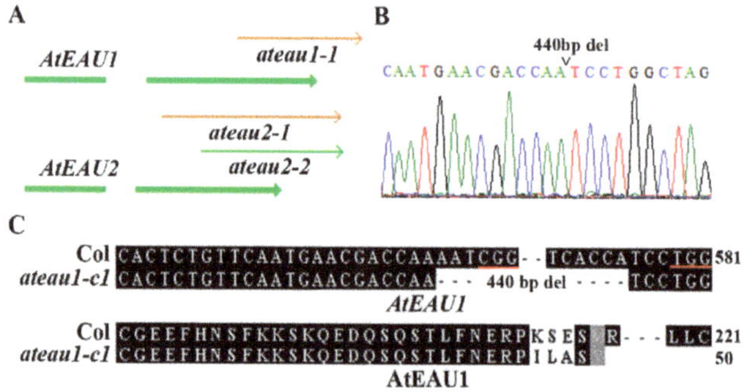

Figure 4. Isolation of the T-DNA insertion mutants for *AtEAUs* and generation of the gene-edited mutant for *AtEAU1*. (**A**) The T-DNA insertion position of the *ateau1-1*, *ateau2-1*, and *ateau2-2* single mutants. Seeds of the T-DNA lines SAIL_197_H05, SAIL_642_C12, and SAIL_1242_E02 were obtained from ABRC, and homozygous mutants were identified by genotyping. (**B**) Gene editing status of *AtEAU1* in the *ateau1-c1* mutant. DNA was isolated from leaves of the T2 transgene-free plants, and subjected to PCR amplification of the genome sequence of *AtEAU1*. The PCR products were recovered and sequenced, and the sequencing results obtained were compared with *AtEAU1* genome sequence to check the editing status. Open arrowhead indicates the position where the 440 bp fragment was deleted. (**C**) Alignment of the nucleotide sequences (up panel) of *AtEAU1* and amino acid sequences (low panel) of AtEAU1 in the Col wild type and the *ateau1-c1* mutant. In the nucleotide sequence alignment, numbers beside sequences indicate the relative position of the nucleotides to the start codon, and underlines indicate the PAM sites. In the amino acid alignment, *AtEAU1* sequences in the mutant were subjected to ORFfinder (https://www.ncbi.nlm.nih.gov/orffinder/, accessed on 1 September 2016) for ORF analysis, and the predicted amino acid sequences were aligned with AtEAU1 amino acid sequence. Numbers beside the sequences indicate the total amino acids in the corresponding proteins.

Considering that only one T-DNA insertion mutant was obtained for *AtEAU1*, we decided to generate an additional mutant by using CRISPR/Cas9 gene-editing of *AtEAU1*. Two different target sequences were selected and used to generate CRISPR/Cas9 construct. By transforming the Col wild-type *Arabidopsis*, examining gene editing status in T1 transgenic plants, and isolating Cas9-free homozygous mutants from T2 plants, we successfully obtained a transgene-free homozygous fragment deletion mutant, namely *ateau1-c1*. Sequencing results show that there is a 440bp fragment deletion in the *ateau1-c1* mutant (Figure 4B). The deletion led to substitutions of a few amino acids and a premature stop in *AtEUA1*, resulting in a short amino acid sequence of AtEAU1 with only 50 amino acids in the *ateau1-c1* mutant (Figure 4C).

2.5. ABA Sensitivity Is Increased in the Single and Double Mutants of AtEAU1 and AtEAU2

By using the mutants obtained, we examined the role of AtEAU1 and AtEAU2 in regulating of ABA response in *Arabidopois* using seedling greening assay. Seeds of the Col wild type, the *ateau1* and *ateau2* single and the *ateau1 ateau2* double mutants were plated on 1/2 MS plates at the presence of 1 µM ABA or solvent alone as a control, and seedlings with green cotyledons were counted 14 days after the transfer of the plates into the growth room. The percentage of green cotyledons was then calculated.

We found that, on the control plates, seedlings for all the plants turned to green, however, on the ABA-containing plates, the green seeding rate for the Col wild type was nearly 95%, whereas that for both the *ateau1-1* and *ateau1-c1* single mutants was about 60%, and that for the *ateau2* single mutants was less than 20% (Figure 5A). To our surprise, we found that the green seeding rate for the *ateau1 ateau2* double mutants was about 60%, similar to the *ateau1-1* and *ateau1-c1* single mutants (Figure 5A).

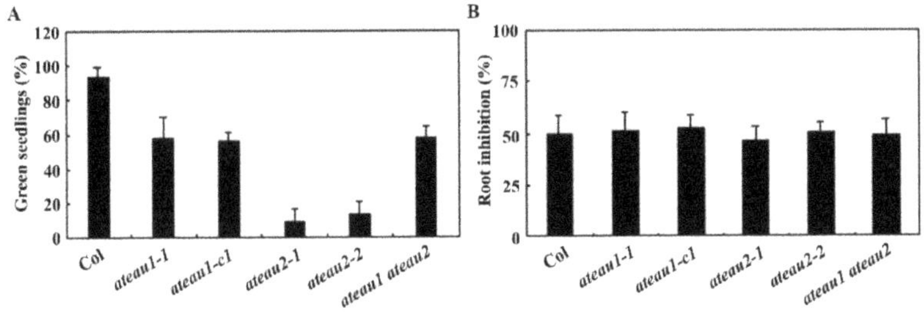

Figure 5. Loss-of-function of *AtEAUs* enhanced ABA response in *Arabidopsis*. (**A**) Cotyledon greening of the Col wild type *Arabidopsis*, the *ateau1* and *ateau2* single, and the *ateau1 ateau2* double mutants in response to ABA treatment. Sterilized seeds were sown on 1/2 MS plates containing 1 μM ABA or solvent control, respectively. The plates were kept for 2 days at 4 °C in darkness, and then transferred to a growth room. Fourteen days after the transfer, seedlings with green cotyledons were counted and the percentage of green cotyledons was then calculated. Data represent the mean ± SD of three replicates. (**B**) Root elongation of the Col wild type, the *ateau1* and *ateau2* single, and the *ateau1 ateau2* double mutants in response to ABA treatment. Sterilized seeds were sown on 1/2 MS plates, kept for 2 days at 4 °C in darkness, then transferred to a growth room and grown vertically for 2 days. Seedlings with similar root length were selected and transferred to plates containing 5 μM ABA or solvent control, respectively. New elongated root length was measured, and percentage of inhibition was then calculated. Data represent the mean ± SD of 10 seedlings.

We also examined response of the mutants to ABA by using root elongation inhibition assays. Sterilized seeds of the Col wild type, the *ateau1* and *ateau2* single and the *ateau1 ateau2* double mutants were plated on 1/2 MS plates and grown vertically for 2 days, then seedlings with root at the same length were transferred to 1/2 MS plates at the presence or absence of 5 μM ABA and grown for 9 additional days, then new elongated root length was measured and the percentage of inhibition was calculated. We found that root elongation of all the seedlings, including the Col wild type, the *ateau1* and *ateau2* single, and the *ateau1 ateau2* double mutants was inhibited by about 50% (Figure 5B).

2.6. *Expression Levels of Some ABA Signaling Genes Are Altered in the Single and Double Mutants of AtEAU1 and AtEAU2*

Having shown that *AtEAU1* and *AtEAU2* are ABA response genes, and ABA sensitivity was increased in the *ateau1* and *ateau2* single and the *ateau1 ateau2* double mutants, we wanted to further examine how AtEAU1 and AtEAU2 may regulate plant response to ABA. We examined the expression of ABA single key regulator genes in seedlings of the Col wild type, the *ateau1-1* and *ateau2-1* single, and the *ateau1 ateau2* double mutants, and found that the expression level of the ABA receptor gene *PYL4* was increased, but the PP2C gene *ABI2* was decreased in the seedlings of the *ateau1-1* and *ateau2-1* single and the *ateau1 ateau2* double mutants when compared with that in the seedlings of the Col wild type plants (Figure 6).

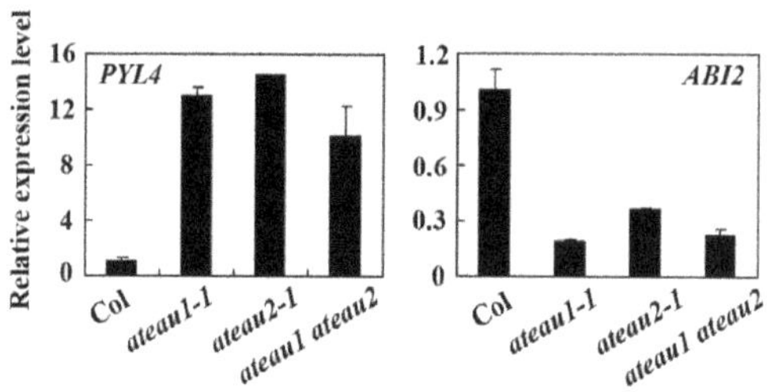

Figure 6. Expression of *PYL4* and *ABI2* in mutants of *AtEAUs*. RNA was isolated from 12-day-old seedlings of the Col wild type, the *ateau1-1*, *ateau2-1* single, and the *ateau1 ateau2* double mutants, and used for qRT-PCR analysis. Expression of *ACT2* was used as an inner reference gene, and expression of *PYL4* and *ABI2* in the seedlings of the Col wild-type plants was set as 1. Data represent the mean ± SD of three replicates.

3. Discussion

EAR motif-containing-protein mediated transcriptional repression has been considered to be the main transcriptional repression form in plants [23], consistent with this, EAR motif-containing proteins are involved in the regulation of plant growth and development, as well as plant response to hormones and environmental stresses [22–26,28–31]. So far it has also been shown that EAR motif-containing proteins are involved in the regulation of hormone responses including ethylene, jasmonate, auxin, and strigolactone [20,23,26–28]. Some of the EAR motif-containing proteins, including ERF7, AITRs, and SlEAD1 have been shown to regulate ABA response in plants [6,25,33]. We provide evidence here that AtEAU1 and AtEAU2 are ABA responsive novel EAR motif-containing transcription repressors that regulating ABA response in *Arabidopsis*.

First, AtEAU1 has previously reported to be an EAR motif-containing protein [31], and our bioinformatics analysis identified AtEAU2 as its closely related EAR motif-containing protein (Figure 1). Second, we found that the expression levels of *AtEAU1* and *AtEAU2* were increased in response to exogenously ABA treatment, but decreased in the ABA biosynthesis mutant *aba1-5* (Figure 2). Third, consistent with the fact that AtEAU1 and AtEAU2 are closely related and both have a fully conserved LxLxL EAR motif (Figure 1), both AtEAU1 and AtEAU2 were specifically localized in the nucleus in *Arabidopsis* protoplast assays, and functioned as transcription repressors (Figure 3). Last but not least, ABA sensitivity was increased in the *ateau1* and *ateau2* single mutants in green seedling assays (Figure 5). These results show that *AtEAU1* and *AtEAU2* are ABA responsive EAR motif-containing protein genes, and AtEAU1 and AtEAU2 function as transcription repressors to regulate ABA response in *Arabidopsis*. However, in root elongation assays, no difference was observed between the seedlings of the Col wild type and the mutants, suggesting the AtEAU1 and AtEAU2 may not regulate ABA sensitivity at root elongation stage.

Considering that both *AtEAU1* and *AtEAU2* are ABA response genes (Figure 2), and AtEAU1 and AtEAU2 shared higher amino acid identity and similarity (Figure 1), and repressed reporter gene expression in transfected protoplasts (Figure 3), we expect that they may have redundant functions, therefore we generated *ateau1 ateau2* double mutant, and tested whether that was the case. However, we found that the *ateau2* single mutants display greater sensitivity to ABA than the *ateau1* single mutants, but sensitivity in the *ateau1 ateau2* double mutant is largely similar to the *ateau1* single mutants (Figure 5), and expression level changes of *PYL4* and *ABI2* in the *ateau1 ateau2* double mutant are also largely similar to those in the *ateau1* single mutants (Figure 6). These results suggest that AtEAU1 and AtEAU2 may function sequentially to regulate ABA response in *Arabidopsis*, and AtEAU1 may function down stream of AtEAU2. However, more experiments are required to examine if that is the case.

Consistent with the key roles played by ABA in regulating plant abiotic stress responses via regulating the expression of downstream ABA response genes, changes in the expression of ABA-signaling key regulator genes, including the PYR1/PYL/RCAR receptor genes, the PP2Cs phosphatase genes, the SnRKs kinase genes, and ABF/AREB/ABI5-type bZIP transcription factor genes all affected plant responses to abiotic stresses [1–7]. On the other hand, changes in the expression of some downstream ABA response genes also affected plant abiotic stress response, for example, the R2R3 MYB transcription gene *MYB44*, the heat shock factor gene *HSFA6b*, and the bHLH transcription factor gene *bHLH112* are all ABA responsive genes involved in the regulation of plant responses to abiotic stresses [35–37]. Consistent with the importance of EAR motif-containing proteins in regulating transcriptional repression in plants [23], some EAR motif-containing proteins such as ERF7, AITRs, and SlEAD1 have been shown to be involved in regulating ABA responses [6,25,33]. So far we have used two different strategies to identify EAR motif-containing proteins; i.e., searching EAR motifs in proteins encoded by ABA-responsive genes with unknown functions, and identifying ABA response genes from unknown function EAR motif-containing protein genes [31]. By using the first strategy, we have successfully identified SlEAD1 as a novel negative regulator of ABA response in tomato [33], and here by using the second strategy, we identified AtEAU1 and AtEAU2 as novel regulators of ABA response in *Arabidopsis*, indicating that both strategies are practical. Since most of the EAR motif-containing proteins that have been identified are functionally uncharacterized, and some of the EAR motif-containing proteins may have not been identified due to their sequence diversity [31,33], more EAR motif-containing ABA response regulator should be identified by using the two strategies mentioned above, and some of the genes may be used for plant breeding to improve plant abiotic stress tolerance.

Nevertheless, we identified the EAR motif-containing protein genes *AtEAU1* and *AtEAU2* as ABA responsive genes, we showed that AtEAU1 and AtEAU2 function as transcription repressors, and we found that AtEAU1 and AtEAU2 negatively regulate ABA responses in *Arabidopsis*, possibly by regulating the expression of the ABA signaling key regulator genes *PYL4* and *ABI2*.

4. Materials and Methods

4.1. Bioinformatics Analysis

Homologs of AtEAU1 (At1g78170) in *Arabidopsis*, soybean, rice, tomato, and poplar were identified on Phytozome (https://phytozome-next.jgi.doe.gov, accessed on 1 September 2016) under the term Protein Homologs, and homologs were then used to search on Phytozome again to make sure their homolog protein is AtEUA1. The full-length amino acid sequences of AtEAU1 and its protein homologs were used for phylogenetic analysis on Phylogeny (http://www.phylogeny.fr/simple_phylogeny.cgi, accessed on 1 September 2016) by using the "One Click" mode with default settings, and for sequence alignment using Bioedit 7.2. Amino acid sequence identity and similarity assays of AtEAU1, AtEAU2 and At4g08910 were analyzed on SIAS (http://imed.med.ucm.es/Tools/sias.html, accessed on 1 September 2016).

4.2. Plant Materials and Growth Conditions

The Columbia-0 (Col) ecotype *Arabidopsis* (*Arabidopsis thaliana*) was used as a wild-type for plant transformation, protoplast isolation, and as a control for ABA response analysis. Seeds of SAIL_197_H05, a T-DNA insertion line for *AtEAU1*, and SAIL_642_C12 and SAIL_1242_E02, T-DNA insertion lines for *AtEAU2* (At1g22250) obtained from the ABRC (Arabidopsis Biological Resource Center) are all in Col background, and were used to isolate the *ateau1-1* and *ateau2-1* and *ateau2-2* mutants, respectively. The *ateau1 ateau2-1* double mutant was generated by crossing *ateau1-1* and *ateau2-1* single mutants, and identifying homozygous double mutants in the F2 generation. The *Ler* wild type and the *aba1-5* mutant in *Ler* background [34] were used to examine the expression of *AtEAUs*.

To generate plants for protoplast isolation and plant transformation, seeds of the Col wild type were sown directly into soil pots, germinated and grown in a growth room. To generate seedlings for RNA isolation and ABA response assays, seeds of the Col wild-type, the *ateau1* and *ateau2* single and *ateau1 ateau2* double mutants, the *Ler* wild type, and the *aba1-5* mutant were sterilized for 10 min with 25% (*v*/*v*) bleach, washed with sterilized water four times, and plated on 0.6% (*w*/*v*) phytoagar (PlantMedia) solidified 1/2 MS (Murashige and Skoog) plates with vitamins (Plant Media) and 1% (*w*/*v*) sucrose. The plates were kept in darkness at 4 °C for 2 days, and then transferred to a growth room. The growth conditions in the growth room were set at 22 °C, and a 16 h light/8 h dark long-day condition with photon density at ~120 μmol m^{-2} s^{-1}.

4.3. RNA Isolation and RT-PCR

To examine the expression changes of *AtEAU1* and *AtEAU2* in response to ABA treatment, 12-day-old seedlings of the Col wild-type were treated for 4 h with 50 μM ABA or solvent methanol, and then RNA was isolated. To examine the expression levels of *AtEAU1* and *AtEAU2* in the *aba1-5* mutant seedlings, RNA was isolated from 12-day-old seedlings of the *Ler* wild type and the *aba1-5* mutant. To examine the expression of ABA signaling genes in the *ateau1-1* and *ateau2-1* single and *ateau1 ateau2* double mutants, RNA was isolated from seedlings of 12-day-old Col wild-type, *ateau1-1* and *ateau2-1* single and *ateau1 ateau2* double mutants. To examine the expression pattern of *AtEAUs*, roots, radicle, rosette leaves, first stems, second stems, third stems, cauline leaves, inflorescences, and siliques were collected from 5-week-old plants, and used for RNA isolation.

Total RNA isolated was used for cDNA synthesis as described previously [38], and then used to examine the expression of *AtEAUs* and ABA signaling genes by using RT-PCR/qRT-PCR. The primers used for *AtEAU1* were 5′-TTCATGCACCCACACGATCA-3′ and 5′-AGGTTCTTCTCTACAAAAAGCCTAA-3′; for *AtEAU2* they were 5′-TCAAAAG GAGCAAACAAGGAGA-3′ and 5′-GGCTTTATCAACATGGCGCT-3′. *ACT2* (*ACTIN2*) was used as a control gene for RT-PCR and an inner reference gene for qRT-PCR. The primers for *PYL4*, *ABI2*, and *ACT2* were as reported previously [39–41].

4.4. Construsts

The reporter gene *LexA-Gal4:GUS*, the control effector gene *GD*, the transcription activator gene *LD-VP*, and the nucleus indicator gene *NLS-RFP* used for *Arabidopsis* protoplast transfection were as described previously [21,22,42]. To generate *GD* and *GFP*-fused *AtEAU1* and *AtEAU2* constructs used for *Arabidopsis* protoplast transfection, the full-length ORF (open reading frame) sequences of *AtEAU1* and *AtEAU2* were RT-PCR amplified by using RNA isolated from 12-day-old seedlings of the Col wild type, digested with proper enzymes, and cloned under the control of the CaMV *35S* promoter, and placed in frame with a C-terminal GD or GFP tag into *pUC19* vector [21,22]. To generate CRISPR/Cas9 construct for gene editing of *AtEAU1*, exon sequences of *AtEAU1* were used for CRISPRscan (http://www.crisprscan.org/?page=sequence, accessed on 1 September 2016) to identify potential target sequences. Selected target sequences were evaluated for potential offtarget on Cas-OFFinder (http://www.rgenome.net/cas-offinder/, accessed on 1 September 2016). Two specific target sequences, i.e., 5′-TTCAATGAACGACCAAAAT (CGG)-3′ and 5′-GACTAAGTCGATCACCATCC (TGG)-3′, were selected. The target sequences were cloned into the *pHEE401E* vector as described previously in [43]. The primers used to generate the CRISPR/-Cas9 construct for editing *AtEAU1* were DT1-BsF (*AtEAU1*), 5′-ATATATGGTCTCGATTGTTCAATGAACGACCAAAAT GTT-3′, DT1-F0 (*AtEAU1*), 5′-TGTTCAATGAACGACCAAAATGTTTTAGAGCTAGAAATAGC-3′, DT2-R0 (*AtEAU1*), 5′-AACGGATGGTGATCGACTTAGTCAATCTC TTAGTCGACTCTAC-3′, and DT2-BsR (*AtEAU1*), 5′-ATTATTGGTCTCGAAACGGATGGTGATCGACTTAGTC-3′. The U626-IDF and U629-IDR primers used for colony PCR and sequencing of the CRISPR/Cas9 constructs generated were described previously [44].

4.5. Plant Transformation and Transgenic Plants Selection

About 5-week-old Col wild-type plants with several mature flowers were transformed with the CRISPR/Cas9 constructs to generate the *ateau1-c1* single mutant by using the floral dip method [45]. Transgenic plants generated were selected by plating the T1 seeds collected on 1/2 MS plates with 30 μg/mL Hygromycin and 100 μg/mL Carbenicillin. Gene editing status in the T1 plants was examined by PCR amplification and sequencing the genomic sequence of *AtEAU1*. T2 seeds were collected from gene-edited T1 plants, germinated directly in soil pots, and used to identify Cas9-free homozygous mutant as described in the next section.

4.6. DNA Isolation and PCR

To examine gene editing status of *AtEAU1*, DNA was isolated from leaves of the T1 transgenic plants or the Cas9-free T2 plants, and used for PCR amplification of the genomic sequences of *AtEAU1*. To isolate transgene-free mutants, DNA was isolated from leaves of the T2 offspring of the gene-edited T1 plants, and subjected to PCR amplification of the *Cas9* gene fragment by using the primers described previously [44].

4.7. Plasmid DNA Isolation, Protoplast Isolation and Transfection

Plasmid DNA of the reporter and effector genes was isolated by using the Gold Hi Endo Free Plasmid Maxi Kit (CWBIO) according to the manufacturers' instructions. Protoplasts were isolated from rosette leaves collected from 3–4-week-old Col wild-type plants, and the isolated protoplasts were transfected as described previously [6,38,46,47]. For subcellular location assays, plasmid DNA of *AtEAU1-GFP* and *AtEAU2-GFP* were transfected into protoplasts. For transcriptional activity assays, plasmid DNA of the reporter gene *LexA-Gal4:GUS*, the activator gene *LD-VP*, and the effector genes *GD*, *AtEAU1-GD*, or *AtEAU2-GD* were cotransformed into protoplasts. The transfected protoplasts were incubated in darkness at room temperature for 20–22 h, then GFP fluorescence was observed under an Olympus FV1000 confocal microscope, and GUS activities were measured using a Synergy™ HT microplate reader.

4.8. ABA Sensitivity Assays

Assays of ABA-inhibited cotyledon greening and root elongation were performed as described previously [6,48–50]. For cotyledon greening assays, green seedlings on plates containing 1 μm ABA or solvent alone were counted 14 days after the plates were transferred to the growth room. For root elongation assays, 2-day-old seedlings grown on vertically placed plates were transferred to plates containing 5 μm ABA or solvent as a control; the length of new elongated roots was measured 9 days after the transfer. All the experiments were repeated at least three times.

Author Contributions: Conceptualization, S.W.; investigation, N.Z., S.C., A., X.W., S.H., Y.C., Y.L., Y.Y., C.W., R.L., H.Z. and J.W.; data curation, N.Z., S.C., T.W. and S.W.; writing—original draft preparation, S.W. and N.Z.; writing—review and editing, S.W., N.Z. and S.C.; project administration, S.W.; funding acquisition, S.W. All authors have read and agreed to the published version of the manuscript.

Funding: This research was funded by National Natural Science Foundation of China, grant number 32071938, and a startup funding from Linyi University, grant number LYDX2019BS039.

Institutional Review Board Statement: Not applicable.

Informed Consent Statement: Not applicable.

Data Availability Statement: All data were obtained were presented in this article.

Acknowledgments: We thank all our lab members in both Linyi University and Northeast Normal University for their helpful discussion and suggestion.

Conflicts of Interest: The authors declare no conflict of interest.

References

1. Fujita, Y.; Nakashima, K.; Yoshida, T.; Katagiri, T.; Kidokoro, S.; Kanamori, N.; Umezawa, T.; Fujita, M.; Maruyama, K.; Ishiyama, K. Three SnRK2 protein kinases are the main positive regulators of abscisic acid signaling in response to water stress in *Arabidopsis*. *Plant Cell Physiol.* **2009**, *50*, 2123–2132. [CrossRef]
2. Guo, J.; Yang, X.; Weston, D.J.; Chen, J.G. Abscisic acid receptors: Past, present and future. *J. Integr. Plant Biol.* **2011**, *53*, 469–479. [CrossRef] [PubMed]
3. Yoshida, T.; Mogami, J.; Yamaguchi-Shinozaki, K. ABA-dependent and ABA-independent signaling in response to osmotic stress in plants. *Curr. Opin. Plant Biol.* **2014**, *21*, 133–139. [CrossRef] [PubMed]
4. Park, S.Y.; Peterson, F.C.; Mosquna, A.; Yao, J.; Volkman, B.F.; Cutler, S.R. Agrochemical control of plant water use using engineered abscisic acid receptors. *Nature* **2015**, *520*, 545–548. [CrossRef] [PubMed]
5. Zhao, Y.; Chan, Z.; Gao, J.; Xing, L.; Cao, M.; Yu, C.; Hu, Y.; You, J.; Shi, H.; Zhu, Y.; et al. ABA receptor PYL9 promotes drought resistance and leaf senescence. *Proc. Natl. Acad. Sci. USA* **2016**, *113*, 1949–1954. [CrossRef]
6. Tian, H.; Chen, S.; Yang, W.; Wang, T.; Zheng, K.; Wang, Y.; Cheng, Y.; Zhang, N.; Liu, S.; Li, D.; et al. A novel family of transcription factors conserved in angiosperms is required for ABA signalling. *Plant Cell Environ.* **2017**, *40*, 2958–2971. [CrossRef] [PubMed]
7. Chen, S.; Zhang, N.; Zhou, G.; Hussain, S.; Ahmed, S.; Tian, H.; Wang, S. Knockout of the entire family of AITR genes in *Arabidopsis* leads to enhanced drought and salinity tolerance without fitness costs. *BMC Plant Biol.* **2021**, *21*, 137. [CrossRef] [PubMed]
8. González-Guzmán, M.; Gómez-Cadenas, A.; Arbona, V. Abscisic acid as an emerging modulator of the responses of plants to low oxygen conditions. *Front. Plant Sci.* **2021**, *12*, 661789. [CrossRef] [PubMed]
9. Ma, Y.; Szostkiewicz, I.; Korte, A.; Moes, D.; Yang, Y.; Christmann, A.; Grill, E. Regulators of PP2C phosphatase activity function as abscisic acid sensors. *Science* **2009**, *324*, 1064–1068. [CrossRef] [PubMed]
10. Park, S.Y.; Fung, P.; Nishimura, N.; Jensen, D.R.; Fujii, H.; Zhao, Y.; Lumba, S.; Santiago, J.; Rodrigues, A.; Chow, T. Abscisic acid inhibits type 2C protein phosphatases via the PYR/PYL family of START proteins. *Science* **2009**, *324*, 1068–1071. [CrossRef]
11. Santiago, J.; Rodrigues, A.; Saez, A.; Rubio, S.; Antoni, R.; Dupeux, F.; Park, S.Y.; Marquez, J.A.; Cutler, S.R.; Rodriguez, P.L. Modulation of drought resistance by the abscisic acid receptor PYL5 through inhibition of clade A PP2Cs. *Plant J.* **2009**, *60*, 575–588. [CrossRef]
12. Gonzalez-Guzman, M.; Pizzio, G.A.; Antoni, R.; Vera-Sirera, F.; Merilo, E.; Bassel, G.W.; Fernández, M.A.; Holdsworth, M.J.; Perez-Amador, M.A.; Kollist, H.; et al. *Arabidopsis* PYR/PYL/RCAR receptors play a major role in quantitative regulation of stomatal aperture and transcriptional response to abscisic acid. *Plant Cell* **2012**, *24*, 2483–2496. [CrossRef] [PubMed]
13. Antoni, R.; Gonzalez-Guzman, M.; Rodriguez, L.; Peirats-Llobet, M.; Pizzio, G.A.; Fernandez, M.A.; De Winne, N.; De Jaeger, G.; Dietrich, D.; Bennett, M.J.; et al. PYRABACTIN RESISTANCE1-LIKE8 plays an important role for the regulation of abscisic acid signaling in root. *Plant Physiol.* **2013**, *161*, 931–941. [CrossRef]
14. Dong, T.; Park, Y.; Hwang, I. Abscisic acid: Biosynthesis, inactivation, homoeostasis and signalling. *Essays Biochem.* **2015**, *58*, 29–48. [PubMed]
15. Gosti, F.; Beaudoin, N.; Serizet, C.; Webb, A.A.; Vartanian, N.; Giraudat, J. ABI1 protein phosphatase 2C is a negative regulator of abscisic acid signaling. *Plant Cell* **1999**, *11*, 1897–1909. [CrossRef]
16. Fujii, H.; Verslues, P.E.; Zhu, J.K. Identification of two protein kinases required for abscisic acid regulation of seed germination, root growth, and gene expression in *Arabidopsis*. *Plant Cell* **2007**, *19*, 485–494. [CrossRef]
17. Umezawa, T.; Nakashima, K.; Miyakawa, T.; Kuromori, T.; Tanokura, M.; Shinozaki, K.; Yamaguchi-Shinozaki, K. Molecular basis of the core regulatory network in ABA responses: Sensing, signaling and transport. *Plant Cell Physiol.* **2010**, *51*, 1821–1839. [CrossRef] [PubMed]
18. Finkelstein, R.R.; Gampala, S.S.; Rock, C.D. Abscisic acid signaling in seeds and seedlings. *Plant Cell* **2002**, *14*, S15–S45. [CrossRef] [PubMed]
19. Chandrasekaran, U.; Luo, X.; Zhou, W.; Shu, K. Multifaceted signaling networks mediated by Abscisic Acid Insensitive 4. *Plant Commun.* **2020**, *1*, 100040. [CrossRef]
20. Ohta, M.; Matsui, K.; Hiratsu, K.; Shinshi, H.; Ohme-Takagi, M. Repression domains of class II ERF transcriptional repressors share an essential motif for active repression. *Plant Cell* **2001**, *13*, 1959–1968. [CrossRef] [PubMed]
21. Tiwari, S.B.; Hagen, G.; Guilfoyle, T.J. Aux/IAA proteins contain a potent transcriptional repression domain. *Plant Cell* **2004**, *16*, 533–543. [CrossRef]
22. Wang, S.; Chang, Y.; Guo, J.; Chen, J.G. *Arabidopsis* Ovate family protein 1 is a transcriptional repressor that suppresses cell elongation. *Plant J.* **2007**, *50*, 858–872. [CrossRef]
23. Kagale, S.; Rozwadowski, K. EAR motif-mediated transcriptional repression in plants: An underlying mechanism for epigenetic regulation of gene expression. *Epigenetics* **2011**, *6*, 141–146. [CrossRef] [PubMed]
24. McGrath, K.C.; Dombrecht, B.; Manners, J.M.; Schenk, P.M.; Edgar, C.I.; Maclean, D.J.; Scheible, W.R.; Udvardi, M.K.; Kazan, K. Repressor- and activator-type ethylene response factors functioning in jasmonate signaling and disease resistance identified via a genome-wide screen of *Arabidopsis* transcription factor gene expression. *Plant Physiol.* **2005**, *139*, 949–959. [CrossRef] [PubMed]
25. Song, C.P.; Agarwal, M.; Ohta, M.; Guo, Y.; Halfter, U.; Wang, P.; Zhu, J.K. Role of an *Arabidopsis* AP2/EREBP-type transcriptional repressor in abscisic acid and drought stress responses. *Plant Cell* **2005**, *17*, 2384–2396. [CrossRef] [PubMed]

26. Chini, A.; Fonseca, S.; Fernandez, G.; Adie, B.; Chico, J.M.; Lorenzo, O.; García-Casado, G.; López-Vidriero, I.; Lozano, F.M.; Ponce, M.R.; et al. The JAZ family of repressors is the missing link in jasmonate signalling. *Nature* **2007**, *448*, 666–671. [CrossRef] [PubMed]
27. Guilfoyle, T.J.; Hagen, G. Auxin response factors. *Curr. Opin. Plant Biol.* **2007**, *10*, 453–460. [CrossRef]
28. Jiang, L.; Liu, X.; Xiong, G.; Liu, H.; Chen, F.; Wang, L.; Meng, X.; Liu, G.; Yu, H.; Yuan, Y.; et al. DWARF 53 acts as a repressor of strigolactone signalling in rice. *Nature* **2013**, *504*, 401–405. [CrossRef]
29. Zhai, Q.Z.; Zhang, X.; Wu, F.M.; Feng, H.L.; Deng, L.; Xu, L.; Zhang, M.; Wang, Q.; Li, C. Transcriptional mechanism of jasmonate receptor COI1-mediated delay of flowering time in *Arabidopsis*. *Plant Cell* **2015**, *27*, 2814–2828. [CrossRef]
30. Liang, Y.; Ward, S.; Li, P.; Bennett, T.; Leyser, O. SMAX1-LIKE7 signals from the nucleus to regulate shoot development in *Arabidopsis* via partially EAR motif-independent mechanisms. *Plant Cell* **2016**, *28*, 1581–1601. [CrossRef]
31. Yang, J.; Liu, Y.; Yan, H.; Tian, T.; You, Q.; Zhang, L.; Xu, W.; Su, Z. PlantEAR: Functional analysis platform for plant EAR motif-containing proteins. *Front. Genet.* **2018**, *9*, 590. [CrossRef] [PubMed]
32. Gonzalez, N.; Pauwels, L.; Baekelandt, A.; De Milde, L.; Van Leene, J.; Besbrugge, N.; Heyndrickx, K.S.; Cuéllar Pérez, A.; Durand, A.N.; De Clercq, R.; et al. A repressor protein complex regulates leaf growth in *Arabidopsis*. *Plant Cell* **2015**, *27*, 2273–2287. [CrossRef]
33. Wang, W.; Wang, X.; Wang, Y.; Zhou, G.; Wang, C.; Hussain, S.; Adnan; Lin, R.; Wang, T.; Wang, S. SlEAD1, an EAR motifcontaining ABA down-regulated novel transcription repressor regulates ABA response in tomato. *GM Crops Food* **2020**, *11*, 275–289. [CrossRef] [PubMed]
34. Ton, J.; Mauch-Mani, B. Beta-amino-butyric acid-induced resistance against necrotrophic pathogens is based on ABA-dependent priming for callose. *Plant J.* **2004**, *38*, 119–130. [CrossRef] [PubMed]
35. Jung, C.; Seo, J.S.; Han, S.W.; Koo, Y.J.; Kim, C.H.; Song, S.I.; Nahm, B.H.; Choi, Y.D.; Cheong, J.J. Overexpression of AtMYB44 enhances stomatal closure to confer abiotic stress tolerance in transgenic *Arabidopsis*. *Plant Physiol.* **2008**, *146*, 623–635. [CrossRef]
36. Liu, Y.; Ji, X.; Nie, X.; Qu, M.; Zheng, L.; Tan, Z.; Zhao, H.; Huo, L.; Liu, S.; Zhang, B.; et al. *Arabidopsis* AtbHLH112 regulates the expression of genes involved in abiotic stress tolerance by binding to their E-box and GCG-box motifs. *New Phytol.* **2015**, *207*, 692–709. [CrossRef]
37. Huang, Y.C.; Niu, C.Y.; Yang, C.R.; Jinn, T.L. The heat stress factor HSFA6b connects ABA signaling and ABA-mediated heat responses. *Plant Physiol.* **2016**, *172*, 1182–1199. [CrossRef]
38. Wang, X.; Wang, X.; Hu, Q.; Dai, X.; Tian, H.; Zheng, K.; Wang, X.; Mao, T.; Chen, J.G.; Wang, S. Characterization of an activation-tagged mutant uncovers a role of GLABRA2 in anthocyanin biosynthesis in *Arabidopsis*. *Plant J.* **2015**, *83*, 300–311. [CrossRef]
39. Ding, Z.J.; Yan, J.Y.; Li, G.X.; Wu, Z.C.; Zhang, S.Q.; Zheng, S.J. WRKY41 controls *Arabidopsis* seed dormancy via direct regulation of ABI3 transcript levels not downstream of ABA. *Plant J.* **2014**, *79*, 810–823. [CrossRef]
40. Liu, S.; Hu, Q.; Luo, S.; Yang, X.; Wang, X.; Wang, S. Expression of wild type PtrIAA14.1, a polpar Aux/IAA gene causes morphological changes in *Arabidopsis*. *Front. Plant Sci.* **2015**, *6*, 388. [CrossRef]
41. Ma, Q.; Xia, Z.; Cai, Z.; Li, L.; Cheng, Y.; Liu, J.; Nian, H. GmWRKY16 enhances drought and salt tolerance through an ABA-mediated pathway in *Arabidopsis thaliana*. *Front. Plant Sci.* **2019**, *9*, 1979. [CrossRef]
42. Lee, Y.J.; Kim, D.H.; Kim, Y.-W.; Hwang, I. Identification of a signal that distinguishes between the chloroplast outer envelope membrane and the endomembrane system in vivo. *Plant Cell* **2001**, *13*, 2175–2190. [CrossRef]
43. Wang, Z.P.; Xing, H.L.; Dong, L.; Zhang, H.Y.; Han, C.Y.; Wang, X.C.; Chen, Q.J. Egg cell-specific promoter-controlled CRP-SPR/Cas9 efficiently generates homozygous mutants for multiple target genes in *Arabidopsis* in a single generation. *Genome Biol.* **2015**, *16*, 144. [CrossRef]
44. Chen, S.; Zhang, N.; Zhang, Q.; Zhou, G.; Tian, H.; Hussain, S.; Ahmed, S.; Wang, T.; Wang, S. Genome editing to integrate seed size and abiotic stress tolerance traits in *Arabidopsis* reveals a role for DPA4 and SOD7 in the regulation of inflorescence architecture. *Int. J. Mol. Sci.* **2019**, *20*, 2695. [CrossRef]
45. Clough, S.J.; Bent, A.F. Floral dip: A simplified method for Agrobacterium-mediated transformation of *Arabidopsis* thaliana. *Plant J.* **1998**, *16*, 735–743. [CrossRef]
46. Wang, S.; Tiwari, S.B.; Hagen, G.; Guilfoyle, T.J. Auxin Response Factor7 restores the expression of auxin-responsive genes in mutant *Arabidopsis* leaf mesophyll protoplasts. *Plant Cell* **2005**, *17*, 1979–1993. [CrossRef]
47. Dai, X.; Zhou, L.; Zhang, W.; Cai, L.; Guo, H.; Tian, H.; Schiefelbein, J.; Wang, S. A single amino acid substitution in the R3 domain of GLABRA1 leads to inhibition of trichome formation in *Arabidopsis* without affecting its interaction with GLABRA3. *Plant Cell Environ.* **2016**, *39*, 897–907. [CrossRef]
48. Guo, J.; Wang, J.; Xi, L.; Huang, W.D.; Liang, J.; Chen, J.G. RACK1 is a negative regulator of ABA responses in *Arabidopsis*. *J. Exp. Bot.* **2009**, *60*, 3819–3833. [CrossRef]
49. Tian, H.; Guo, H.; Dai, X.; Cheng, Y.; Zheng, K.; Wang, X.; Wang, S. An ABA down-regulated bHLH transcription repressor gene, bHLH129 regulates root elongation and ABA response when overexpressed in *Arabidopsis*. *Sci. Rep.* **2015**, *5*, 17587. [CrossRef]
50. Zheng, K.; Wang, Y.; Wang, S. The non-DNA binding bHLH transcription factor PACLOBUTRAZOL RESISTANCES are involved in the regulation of ABA and salt responses in *Arabidopsis*. *Plant Physiol. Biochem.* **2019**, *139*, 239–245. [CrossRef]

International Journal of
Molecular Sciences

MDPI

Article

Global DNA Methylation and mRNA-miRNA Variations Activated by Heat Shock Boost Early Microspore Embryogenesis in Cabbage (*Brassica oleracea*)

Congcong Kong [1,†], Henan Su [1,2,†], Siping Deng [1], Jialei Ji [1], Yong Wang [1], Yangyong Zhang [1], Limei Yang [1], Zhiyuan Fang [1] and Honghao Lv [1,*]

[1] Key Laboratory of Biology and Genetic Improvement of Horticultural Crops, Ministry of Agriculture, Institute of Vegetables and Flowers, Chinese Academy of Agricultural Sciences, Beijing 100081, China; 13121238399@163.com (C.K.); 18810835083@163.com (H.S.); dengsiping@163.com (S.D.); jijialei@caas.cn (J.J.); wangyong03@caas.cn (Y.W.); zhangyangyong@caas.cn (Y.Z.); yanglimei@caas.cn (L.Y.); fangzhiyuan@caas.cn (Z.F.)

[2] Institute of Horticulture, Henan Academy of Agricultural Sciences, Zhengzhou 450002, China

* Correspondence: lvhonghao@caas.cn

† These authors contributed equally to this study.

Citation: Kong, C.; Su, H.; Deng, S.; Ji, J.; Wang, Y.; Zhang, Y.; Yang, L.; Fang, Z.; Lv, H. Global DNA Methylation and mRNA-miRNA Variations Activated by Heat Shock Boost Early Microspore Embryogenesis in Cabbage (*Brassica oleracea*). *Int. J. Mol. Sci.* **2022**, *23*, 5147. https://doi.org/10.3390/ijms23095147

Academic Editors: Andrés J. Cortés and Hai Du

Received: 25 February 2022
Accepted: 4 May 2022
Published: 5 May 2022

Publisher's Note: MDPI stays neutral with regard to jurisdictional claims in published maps and institutional affiliations.

Abstract: Microspore culture, a type of haploid breeding, is extensively used in the cultivation of cruciferous crops such as cabbage. Heat shock (HS) treatment is essential to improve the embryo rate during the culture process; however, its molecular role in boosting early microspore embryogenesis (ME) remains unknown. Here we combined DNA methylation levels, miRNAs, and transcriptome profiles in isolated microspores of cabbage '01-88' under HS (32 °C for 24 h) and normal temperature (25 °C for 24 h) to investigate the regulatory roles of DNA methylation and miRNA in early ME. Global methylation levels were significantly different in the two pre-treatments, and 508 differentially methylated regions (DMRs) were identified; 59.92% of DMRs were correlated with transcripts, and 39.43% of miRNA locus were associated with methylation levels. Significantly, the association analysis revealed that 31 differentially expressed genes (DEGs) were targeted by methylation and miRNA and were mainly involved in the reactive oxygen species (ROS) response and abscisic acid (ABA) signaling, indicating that HS induced DNA methylation, and miRNA might affect ME by influencing ROS and ABA. This study revealed that DNA methylation and miRNA interfered with ME by modulating key genes and pathways, which could broaden our understanding of the molecular regulation of ME induced by HS pre-treatment.

Keywords: microspore culture; DNA methylation; small RNA; differentially expressed genes

1. Introduction

Cabbage (*Brassica oleracea* var. *capitata*), a member of the cruciferous family, is an important vegetable crop. Most commercial cabbage cultivars are F_1 hybrids to ensure high uniformity and yield [1]. To produce hybrids, high-generation inbred lines are needed, and the parental inbred lines are usually produced through a labor-intensive traditional method-system selection [2,3]. Microspore culture is an effective technique to produce doubled haploid (DH) lines for F_1 hybrid production, which accelerates the breeding process significantly [4,5].

Microspore embryogenesis (ME) is a complex and interesting process in which isolated microspores are induced to form haploid embryos in vitro, and pre-treatment is essential for embryo production [6]. As the first 8 h of culture at 32 °C was found to be critical in determining the developmental fate of the isolated microspores in *B. napus*, most molecular-level research to date has focused on this period [7,8]. Breakthrough progress has also been made in isolated microspore culture of *B. oleracea* vegetables in recent years.

In cabbage, the ME was significantly enhanced at 32.5 °C for 24 h [9]. In broccoli, the combination of cold pre-treatment (4 °C) for 1 or 2 days and heat shock (32.5 °C) for 1 day significantly enhanced microspore embryogenesis efficiency compared with cold pre-treatment alone [10]. These studies suggested that heat shock (HS) plays an important role in *B. oleracea*. For HS stress, the activation of heat shock proteins (HSPs) were directly linked to heat-responsive pathways [11]. HSPs were initially thought to have a prominent role in microspore induction, but it was found that uninduced microspores exposed to HS also accumulated HSPs as well [12]. Currently, the view is that HSPs may have a role more directly related to stress tolerance [13]. However, microspore culture of cabbage mainly focused on the optimization of culture conditions, while few studies have been carried out on the mechanism of heat shock-induced ME.

In plants, epigenetic modifications play a crucial role in responding quickly to environmental stress. DNA methylation, representing an epigenetic mechanism, regulates gene expression, which is significant for seed/embryo development, gametophyte development, and stress responses [14–17]. DNA methylation can be categorized into three contexts: symmetric CG and CHG (where H = A, T, or C) and asymmetric CHH [18,19]. Methylation modification is stable and heritable, but it may change in response to development and the environment [20]. Recently, it has been found that cellular reprogramming in ME involved in DNA methylation levels decrease with the activation of cell proliferation, and subsequently increases with embryonic differentiation in some plant species, like *Brassica napus* [21], *Hordeum vulgare* [22], and *Quercus suber* [23]. A recent study also demonstrated that disrupted genome methylation in response to high temperature (39−41 °C in the daytime and 29−31 °C at night) has distinct effects on microspore abortion and anther indehiscence in cotton, and transcriptome analysis indicated that high temperature disturbs sugar and ROS metabolism via disrupting DNA methylation [24]. For *B. napus*, studies have shown treatment on microspores at 32 °C for 6 h triggered DNA hypomethylation, and the total number of differentially methylated region-related genes at 32 °C for 6 h was about twice as high as that at 18 °C for 6 h [25]. Heat pre-treatment is one of the most important stresses for ME induction [13], apart from this, culture of isolated microspores of *B. oleracea* at 25 °C has been proposed as an optimal system to study the gametophytic development in vitro [9].

Additionally, microRNA (miRNA), a class of small non-coding RNAs approximately 18–25 nt in length, is another important epigenetic regulator that post-transcriptionally represses gene expression in a wide variety of eukaryotic organisms [26,27]. As a major type of small RNAs, miRNA participates in the regulation of diverse plant developmental processes such as seed germination [28], embryogenesis [29], leaf growth [30], and fruit development and ripening [31]. In other reports, the negative regulation of miRNAs through target mRNA degradation or translational repression is crucial for responding to biotic and abiotic stresses [32,33]. To investigate the role of miRNAs in barley ME development, Bélanger et al. [34] detected that the abundance of 51 of these miRNAs differed significantly during microspore development under thermal, osmotic, and starvation stresses, and degradome analysis revealed that the transition of microspore to embryogenesis pathway involved miRNA-directed regulation of members of the ARF, SPL, GRF, and HD-ZIPIII transcription factor families. Through small RNA sequencing of buds in cotton, Chen et al. [35] found that expression abundance of miR156 under high temperature stress was 10% higher than the total expression abundance under normal temperature at microspore release stage, which was the main regulators responding to HT stress with positive or negative regulation patterns. Although implicated in an array of embryonic developmental processes, miRNA functions during ME in cabbage remain unclear. Thus, identifying miRNAs and their targets and elucidating their regulatory mechanisms are critical to understand ME in cabbage.

In summary, DNA methylation and miRNA regulation have been shown play an important role in ME development in some plant species, but the epigenetic modification in *B. oleracea* microspore toward embryoids is not clear. In this study, we have performed the first integrated genome-wide analysis of DNA methylation, miRNAs, and mRNA transcriptional activity, using HS stress (32 °C for 24 h) and normal temperature stress (25 °C for 24 h) microspores from cabbage accession '01-88'. We constructed two DNA libraries, six small RNA libraries, and six mRNA libraries, and the objective were to analyze the landscape of methylome distribution, differentially expressed miRNAs (DERs) and differentially expressed genes (DEGs) involved in cabbage ME development. The work performed in this study will broadens our understanding of the molecular regulation of ME induced by HS stress.

2. Results

2.1. Microspore Collection and Culture

The isolated microspores from the highly embryogenic inbred line '01-88' were subjected to HS (32 °C) and non-HS (25 °C) pre-treatment for 24 h, respectively. Then after 3 weeks of culture at 25 °C, embryoids were produced only from isolated microspores under 32 °C HS for 24 h (Figure 1A). The average number of embryoids per bud was 47 after 32 °C HS treatment, compared with zero embryoids per bud under 25 °C normal temperature treatment (Figure 1C), and the difference in the embryoids rates between the 32 °C and 25 °C treatments was significant ($p < 0.01$).

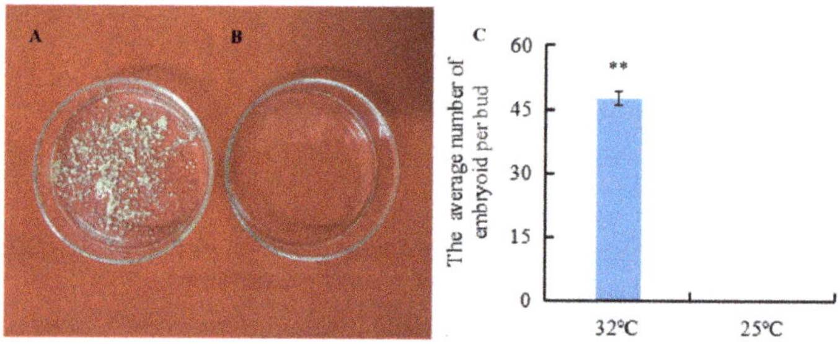

Figure 1. Embryoids produced by isolated microspores under the 32 °C and 25 °C pre-treatments. (**A**) Embryoids produced by isolated microspores cultured for 3 weeks after 24 h treatment at 32 °C; (**B**) Embryoids produced by isolated microspores cultured for 3 weeks after 24 h treatment at 25 °C; (**C**) The embryoid rates after the 32 °C and 25 °C treatments. Error bars represent standard error of the mean. Significant differences has been marked with asterisks (** $p < 0.01$).

2.2. The DNA Methylation Landscapes of Microspores Treated at Two Temperatures

Genome wide profiling of DNA methylation using bisulphite sequencing (BS-Seq) was performed in microspores treated with heat temperature at 32 °C for 24 h (HT32) and normal temperature at 25 °C for 24 h (NT25), respectively. The two libraries generated a total of 104,763,044 (HT32) and 133,801,556 (NT25) raw reads. Following filtering, 54,750,572 clean reads (52.26%) in HT32 and 69,437,722 clean reads (51.90%) in NT25 were successfully and uniquely aligned to the '02-12' cabbage reference genome (BRAD, http://brassicadb.cn, 12 June 2021). These data were used to retrieve the methylation level for each CG, CHG, and CHH site. A total of 32,351,452 cytosine methylation levels (mCs) (60.8% at CG sites, 31.0% at CHG sites, and 8.2% at CHH sites) and 42,291,691 mCs (60.4% at CG sites, 30.9% at CHG sites, and 8.7% at CHH sites) were identified in HT32 and NT25, respectively (Figure 2). Fisher's test were used to calculate the significance of the CG, CHG, and CHH methylation numbers and the total methylation numbers between the two pre-treatments, respectively, and the p values were all less than 0.01. The total mCs were slightly different between the HT32 (21.15%) and NT25 (21.60%) in the whole cytosine sites; The overall genomic methylation degree of the mCGs was higher in the HT32 samples (69.31%) than in the NT25 samples (68.95%); For the mCHGs, the methylation level showed identical between the HT32 (35.32%) and NT25 (35.32%), whereas for the mCHH, the methylation level was lower for HT32 (9.41%) than for NT25 (9.97%) (Figure S1).

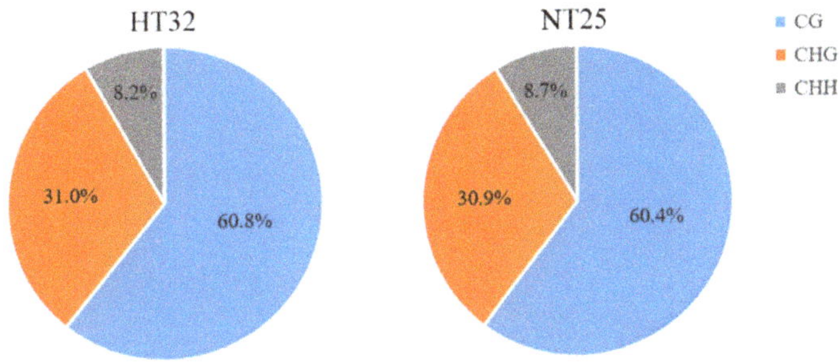

Figure 2. The cytosine methylation level (mC) percentage for each sequence context in HT32 and NT25.

To show the genome-wide DNA methylation landscape of the two samples, DNA methylation maps in the three contexts were represented using a Circos histogram (Figure 3). The outermost circle is a scale presentation based on the corresponding chromosome length, the outer and inner three circles are the background of methylation of CG, CHG, and CHH in the chromosome intervals of HT32 and NT25, and the darker the color, the higher the background level of methylation. We observed some different methylation peaks, and the methylation levels of mCHH in HT32 and NT25 could be clearly distinguished in some local positions of chromosomes, e.g., 3–5 Mb, 14–16 Mb and 25–30 Mb on chromosome 1. Further, we calculated the percentage of DNA methylation levels in each context throughout the nine chromosomes (Figure S2); these nine chromosomes showed similar methylation patterns to the genome.

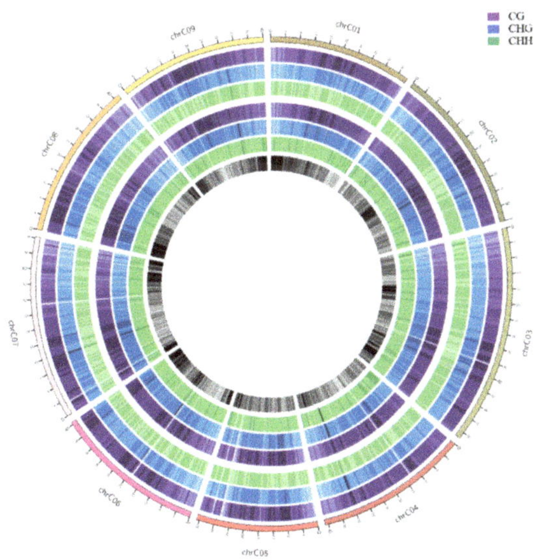

Figure 3. Genomic methylation levels of HT32 and NT25. The first circle represents chromosome name and scale. The second to fourth circles (purple, blue and green) represent the methylation background display of CG, CHG and CHH in the corresponding chromosome intervals of HT32 groups, respectively; The fifth to seventh circles (purple, blue and green) represents the methylation background display of CG, CHG and CHH in the corresponding chromosome intervals of NT25 group, respectively. The eighth circle represents the number of genes in the corresponding interval, and the darker the color, the more genes there are in that region.

2.3. DNA Methylation Levels in Different Genomic Regions

We analyzed the differential methylation levels of gene bodies (first, internal, and last exons and introns) and their 2 kb up- and downstream regions (distal, intermediate, and proximal promoter; downstream) between HT32 and NT25 (Figure 4). In general, the methylation levels of CG, CHG, and CHH were extremely low at the transcriptional start site (TSS) and transcriptional end site (TES), and the CG, CHG, and CHH methylation levels were higher in the flanking regions than in the gene bodies (Figure 4). Moreover, the CG, CHG, and CHH methylation levels were slightly higher in HT32 than in NT25, except CG in the flanking area, the number of mCs in HT32 and NT25 was similar. Notably, the frequency of mCs in the CG context was higher than that in the CHH and CHG contexts in the region of analysis. These distinct methylation levels revealed that microspores induction under 32 °C treatment could have higher CG, CHG, and CHH methylation in the distal promoter region.

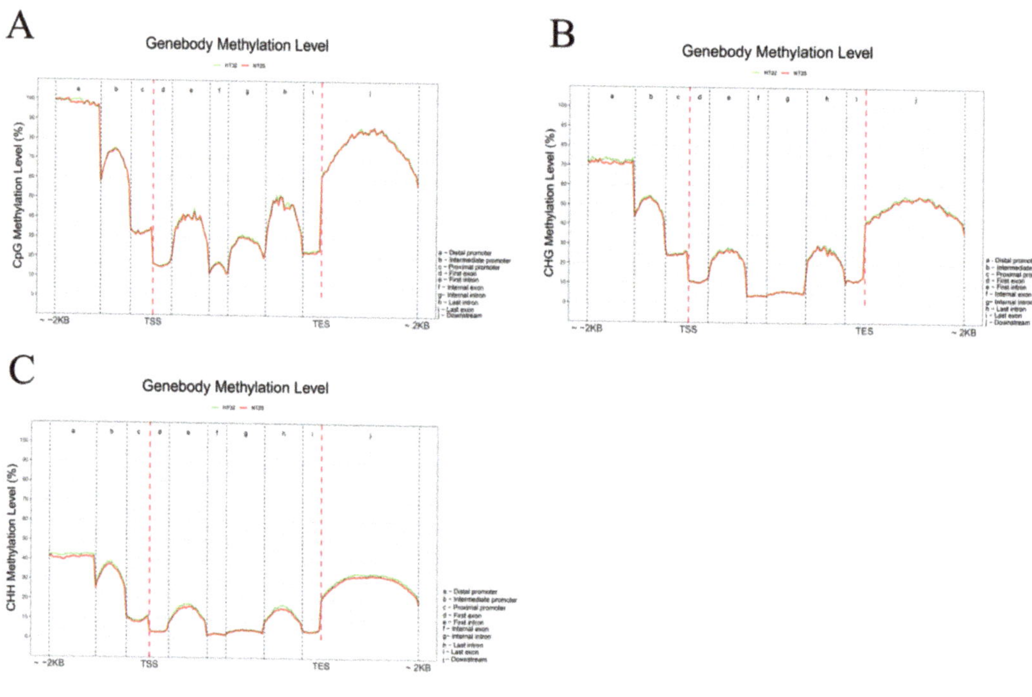

Figure 4. Comparative analysis of DNA methylation levels in different genomic regions between HT32 and NT25. (**A**) CG methylation level. (**B**) CHG methylation level. (**C**) CHH methylation level. Peak distribution in different genomic regions. Methylation density is the ratio of methylation peaks to the corresponding sequence lengths.

2.4. Identification of Differentially Methylated Genomic Regions (DMRs) and DMR-Associated Genes

To investigate specific DNA methylation in cabbage ME, the genomic regions associated with hypermethylation or hypomethylation were profiled. In total, 1038 differentially methylated genomic regions (DMRs) ($p < 0.05$) were detected in the two treatments. The number of hypermethylated and hypomethylated regions was 414 and 624, respectively, all of which were mapped to the chromosomes (Table S1). As DNA methylation was associated with gene expression, DMRs in the promoter, introns, exons, and downstream regions were defined as DMR-associated genes. In total, 508 DMR-associated genes were identified, which included 212 hypermethylated genes and 296 hypomethylated genes in HT32. GO enrichment analysis was conducted on these DMR-associated genes. We found 15, 10, and 25 terms in cellular components (CC), molecular functions (MF), and biological processes (BP), respectively (Figure 5A). The GO terms of the integral component of membranes in CC, carbohydrate metabolic in BP, and protein binding in MF were most enriched terms. Moreover, the DMR−associated genes were enriched in the KEGG pathways database to identify pathways that were responsive to early microspore embryogenesis. In total, 25 pathways were detected in our study, which included five branches: organismal systems, metabolism, genetic information processing, environmental information processing, and cellular processes. The most abundant pathway in each branch was the endocrine system, amino acid metabolism, translation, signal transduction, and transport and catabolism, respectively (Figure 5B).

2.5. Association Analysis with the DNA Methylome and mRNA Profiles

To explore the influence of DNA methylation on the expression of genes, we analyzed the relationship between DMRs and differentially expressed genes (DEGs, $|\log2FC| > 1$ and $p < 0.05$) on a genome-wide scale. In transcriptome data, a total of 7226 DEG (4312 upregulated genes and 2914 downregulated genes) transcripts were identified in HT32 and NT25 (Table S2, Figure S3). GO enrichment analysis was conducted on the DEGs. We found that 52, 99, and 136 functional groups in CC, MF and BP, respectively (Figure S4A). Furthermore, the GO terms of the integral component of membrane in CC, the translation in BP, and the structural constituent of ribosome in MF were most enriched terms. These DEGs were involved in certain embryogenesis-related processes, including the response to abscisic acid (ABA), embryo development ending in seed dormancy, ABA-activated signaling pathway, and response to cytokinin (CTK). Furthermore, 2761 DEGs were associated with the corresponding KEGG pathway, and they were primarily enriched in ribosome, plant-pathogen interaction, flavonoid biosynthesis, stilbenoid, diarylheptanoid and gingerol biosynthesis, ribosome biogenesis in eukaryotes, and galactose metabolism. Among them, 329 and 281 DEGs were enriched in ribosome, plant-pathogen interaction (Figure S4B).

The general observation was that the hypermethylated DMR-associated genes exhibited lower levels of transcript abundance, and hypomethylated DMR-associated genes exhibited higher levels of transcript abundance relative to all genes. In HT32 and NT25, 59.92% of all DMRs were associated with transcripts, which included 622 transcripts related to the DMRs. Box-plots showing differential expression levels of genes associated with DMRs are displayed (Figure 6A). For hyper-DMRs between HT32 and NT25, 54 transcripts were downregulated, and 64 transcripts were upregulated (Figure 6B). Some of the DMR-associated genes were negatively correlated with the transcript abundance. For the hypo-DMRs, 53 transcripts were downregulated, and 71 transcripts were upregulated (Figure 6B). The results suggested that these DMRs played a role in the regulation of gene expression and might contribute to the differential temperature stress responses.

2.6. Validation of Gene Expression

We validated the expression levels of nine genes from the genes that were negatively regulated by microRNA and methylation in HT32 and NT25 using quantitative real-time PCR (qRT-PCR). We found that Bol019334, Bol018627, Bol044765, and Bol037384 were upregulated, while Bol007881, Bol010698, Bol035641, Bol016979, and Bol042216 were downregulated in HT32 (Figure 7). All genes were significant except gene Bol044765 ($p < 0.01$). Overall, the qRT-PCR results were in concordance with results obtained from RNA-seq data analyses, indicating the high quality of our data. The primers used for qRT-PCR are listed in Table S3.

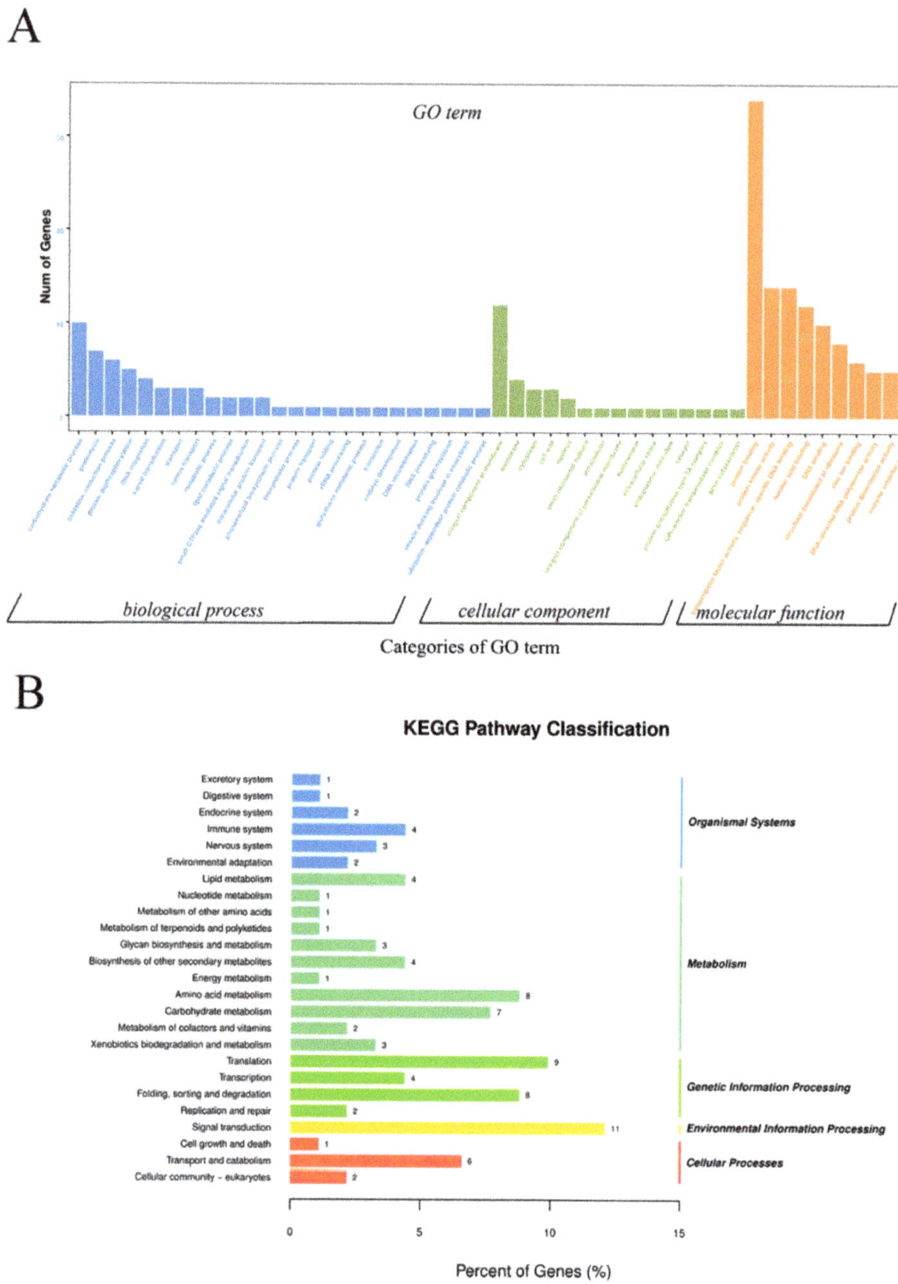

Figure 5. GO and KEGG enrichment of DMR-associated genes. (**A**) GO enrichment of DMR-associated genes. (**B**) KEGG enrichment of DMR-associated genes.

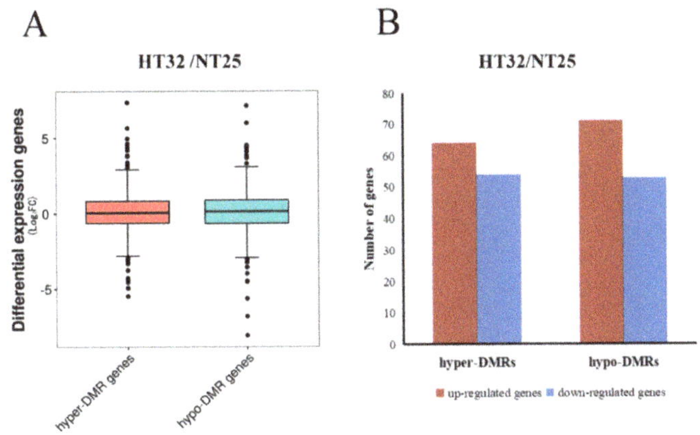

Figure 6. Correlation between differential methylation and level of transcript abundance. (**A**) Differential expression levels of genes associated with hyper/hypo DMRs. (**B**) The number of up/down-regulated genes associated with hyper/hypo DMRs.

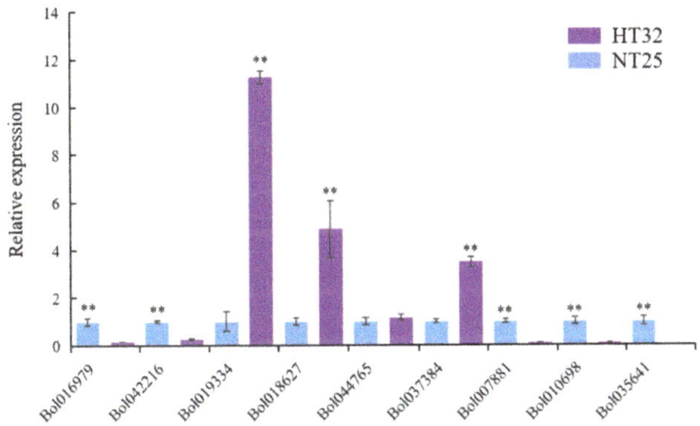

Figure 7. The qRT-PCR validation and expression analysis of 9 genes. The gene expression was normalized using *actin* and expressed relative to gene expression in HT32 sample (negative control). Error bars represent standard error of the mean. Significant differences has been marked with asterisks (** $p < 0.01$).

2.7. Association Analysis with the DNA Methylome and miRNA Profiles

To examine the relationship between the extent of DNA methylation and miRNA abundance, we performed small RNA−seq analysis. Six independent small RNA libraries (HT32-1, -2, -3 and NT25-1, -2, -3) were constructed to sequence the small RNAs and generated a total of about 12.17–18.43 million raw reads (Table S4). In total, 870 miRNAs (including the putative miRNAs) were found in cabbage, and the predicted targets of 63 differentially expressed miRNAs (DERs) were analyzed (Table S5). Among the 63 differentially expressed miRNAs, 31 miRNAs were up-regulated, and the remaining 32 miRNAs were downregulated in HT32. For 31 up-regulated miRNAs, the number of normalized reads of aly-miR156h-3p_L+1R+2_1ss11CA was very high (5.73-fold), followed by PC-5p-495929_3, bra-miR156b-p3. Most miRNA members of the same family had similar expression profiles. For example, eight miR2592 and four miR156 family members were significantly up-regulated. Three miR394 family members were found to be significantly down-regulated. In addition, 475 predicted targets were found (Table S6). GO enrichment analysis was conducted on the predicted targets, and they were primarily enriched in meristem initiation, cell differentiation, seed development, and primary shoot apical meristem specification (Figure S5A). According to the KEGG pathway analysis (Figure S5B), miRNA targets were primarily enriched in plant hormone signal transduction. Then, we identified the interaction network of miRNA and target genes involved in the plant hormone signal transduction pathway (Figure 8). We found that one miRNA could target several genes, and most miRNA members of the same family targeted the same genes, e.g., bolmiR172a targeted 10 genes, and ath-miR394a, bna-miR394a, and aly-miR394a-59 targeted Bol024053.

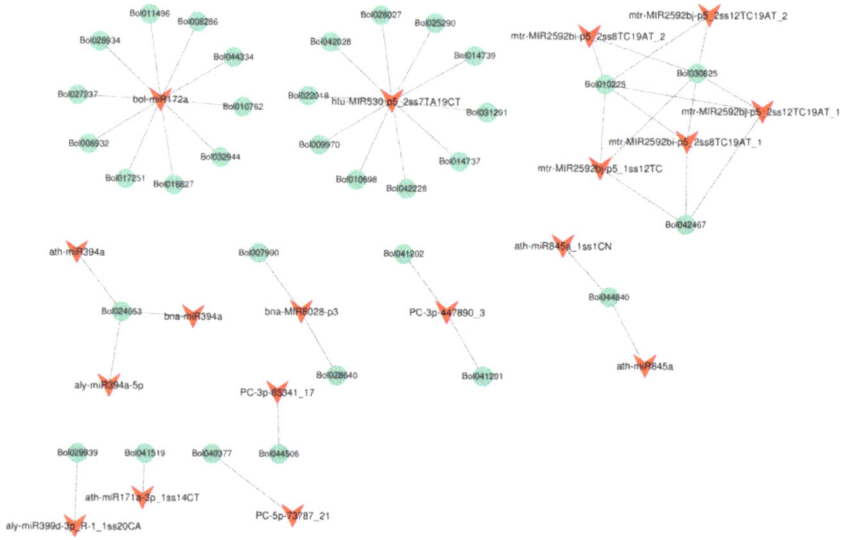

Figure 8. The interaction network of miRNA and target genes in plant hormone signal transduction pathway (ko04075).

Next, we investigated the presence of mCs in the miRNA loci. In total, 39.43% of miRNA loci were associated with mCs. Interestingly, there was no CHG locus in miRNAs associated with mCs. The proportion of CG was slightly higher as compared to the CHH-context mCs in the two treatments (Figure 9A). For the mCGs, the methylation level was higher for HT32 than for NT25. In addition, we estimated the abundance of miRNA in differential DMR-associated genes and found that the abundance of miRNAs was significantly enriched in hypo-DMRs (Figure 9B). These results suggest that miRNAs participate in alteration of DNA methylation levels in different treatments.

2.8. *Association Analysis of the DNA Methylation, miRNA and mRNA in Cabbage Microspore Embryogenesis*

To investigate the effects of DNA methylation and miRNA on gene expression, we analyzed DEGs targeted by both DMR and DER in HT32 and NT25. In total, 31 DEGs were targeted by both methylation and miRNA (Table S7). These target genes encoded proteins involved in diverse cellular processes, including phenylpropanoid biosynthesis, carotenoid biosynthesis, plant hormone signal transduction, and more (Table S4). Many studies have found that DNA methylation and miRNA negatively regulated gene expression [36]. Although 19 of the 31 genes were targeted by both DMR and DER, but their gene expression has not completely negative correlation with DNA methylation and miRNA regulation, which were relatively complex, and these DEGs might not be completely regulated by DNA methylation and miRNA or only one of them. Therefore, we selected the remaining

12 key genes whose expression levels were negatively correlated with DNA methylation and miRNA regulation (Figure 10). In hypomethylated regions, miRNAs including bna-miR6028, bol-miR172a, and ptc-miR6478 in cabbage were downregulated under 32 °C stress, and the target genes Bol024133 (TIR-NBS-LRR family), Bol044765 (exostosin family), Bol018627 (ethylene-responsive transcription factor 1B-like, AP2), and Bol019334 (peroxidase CB) were upregulated. In hypermethylated regions, miRNAs including ath-miR164, bra-miR164, htu-miR530, and mtr-miR2592 in cabbage were upregulated under 32 °C stress, and the target genes Bol035641 (kinesin motor protein-related), Bol010698 (SNF1 kinase homolog 10), Bol016979 (unknown), Bol042216 (DNA-binding), Bol017014 (atypical CYS HIS rich thioredoxin 5), Bol021292 (glutamate-ammonia ligase), and Bol007881 (plant vesicle-associated membrane) were downregulated. These genes were associated with the ROS response, hormone signal transduction, and defense response, indicating that the ME mechanism during heat shock induction may be similar to the plant–pathogen interaction.

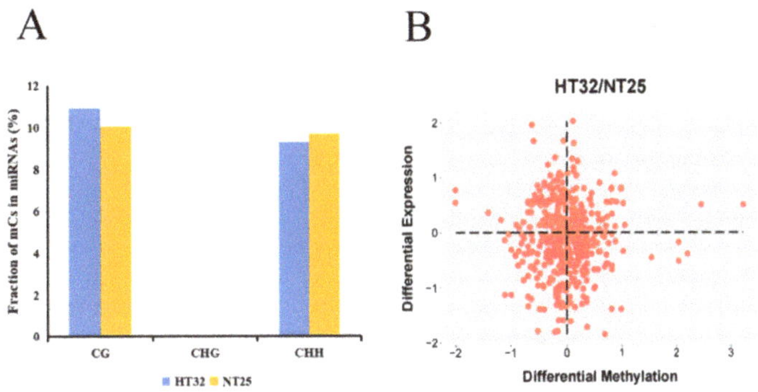

Figure 9. Correlation between differential methylation and level of miRNA abundance. (**A**) Fraction of different mCs present in miRNAs. (**B**) Differential expression levels of miRNAs associated with hyper/hypo DMRs.

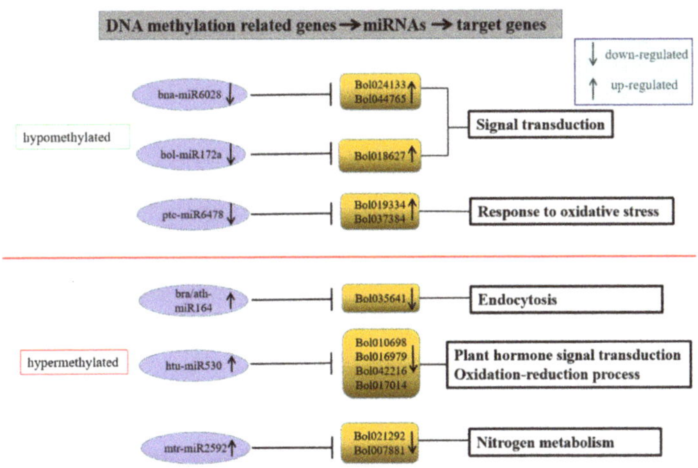

Figure 10. Genes with negative correlation with miRNA and methylome.

3. Discussion

3.1. Global Analysis of HS Boosting ME in B. oleracea

In recent years, a large number of studies have suggested that DNA methylation plays a crucial role in the regulation of seed development, fertility, embryogenesis, and stress adaption [37–40]. The epigenetic marks, such as DNA methylation, may be attributed to the phenotypic consequences such as embryogenesis from isolated microspores. Therefore, it is important to investigate the DNA methylomes of isolated microspore under 32 °C treatment for 24 h to understand the epigenetic regulation of microspore embryogenesis. In this study, the global DNA methylation pattern was profiled in HT32 and NT25. A comparison of HT32 and NT25 indicated that the genomic cytosine methylation levels showed few differences, including a slightly lower methylation level at mCHH and a higher level at mCG in HT32. Hypomethylation is considered a common feature associated with adaptive responses to various stresses [41–43]. In this study, there were 124 hypomethylated DMR-associated DEGs in HT32 and NT25, and these genes could be critical in ME of *B. oleracea*.

Recently, high throughput sequencing has been widely applied to identify genome−wide miRNAs involved in plant development and environmental stress responses [44,45]. However, the comprehensive identification of miRNAs and their targets involved in cabbage microspore embryogenesis has not been conducted. In this study, 63 differentially expressed miRNAs between isolated microspores subjected to 32 °C (24 h) and 25 °C (24 h) were identified, and 18 differentially expressed miRNAs with more than a three-fold relative change between isolated microspores subjected to 32 °C (24 h) and 25 °C (24 h) were identified. Previous studies demonstrated that diverse miRNAs exist in plants and they might have potential regulatory roles in somatic embryogenesis and seed development. In maize, miR156 regulates embryogenic callus differentiation [46], and it is also responsible for cotyledon embryo development in larch [47]. Overexpression of csi-miR156a significantly enhanced the initial phases of SE induction in preserved citrus embryogenic callus [48]. Both miR156 and miR172 are master regulators of phase transition and seed germination in plants [49]. Auxin homeostasis is important for embryo development and is mediated by the action of miR165/166, miR164, and miR160 [50]. In our study, the expression level of bra-miR156b and aly-miR156h was 4.15 and 5.73, and bra-miR164e was only identified at 32 °C after 24 h. We speculated that these miRNAs might be involved in the microspore embryogenesis process. miRNAs and target genes jointly regulate plant growth and development. During somatic embryogenesis, miRNA genes function in various cellular, physiological, and developmental processes to mediate stress adaptation, developmental regulation, and hormone responses [51]. In this study, miRNA target genes were primarily enriched in plant hormone signal transduction, which is consistent with previous research results [50,51].

A correlation between miRNA abundance and DNA methylation has been established in some studies [52,53]. However, whether miRNA and DNA methylation regulate gene expression in microspore embryogenesis under 32 °C heat-shock stress in cabbage is unknown. In this study, we identified 17 miRNAs and 31 target genes that were differentially methylated under 32 °C HS stress. Several pairs of genes that were negatively regulated by miRNA and methylation may play an important role in ME. Six target genes were involved in the oxidation-reduction process; the reactive oxygen intermediates produced as signaling molecules by these gene products, along with each gene's role in oxidation-reduction reactions, could control cabbage response to various processes including programmed cell death, abiotic stress response, and systemic signaling [54].

In the induction process of ME in some plants, HSPs could be related to the acquisition of embryogenic potential by microspores, but some scholars hold different views in *B. napus*. Seguí-Simarro et al. [12] found that uninduced microspores exposed to HS also accumulated HSPs. Zhao et al. [55] found that there was no specific expression of HSP during microspore culture in colchicin-treated *B. napus*, suggesting that HSP may not be necessary for ME. In our mRNA data, there was no significant difference in HSP expression between the two samples, suggesting that HSPs could not be necessary in *B. oleracea*, as it was in *B. napus*.

In miRNA data, bol-miR172a was downregulated to target AP2. In *Arabidopsis*, miR172 associates with AGO1 or AGO10 and acts to silence specific target gene expression; miR172 targets the transcription factor gene AP2, which regulates juvenile to adult-phase transition, floral organ identity, and flowering time [56–58]. Interestingly, the expression level of AP2 was 14.95, and the DNA methylation level was hypomethylated in HT32 and NT25. We identified those genes with altered expression levels through transcriptional silencing mechanisms including DNA methylation and miRNA cleavage, and these changes may be correlated with the embryogenesis induced by heat shock treatment.

3.2. Heat Shock-Induced DNA Methylation and miRNA Associated–Genes Invovled in the ROS Response and ABA Signal Transduction

According to previous studies [59,60], ME was related to ROS response. Among the 31 DEGs targeted by methylation and miRNA, seven genes could be found to be closely related to early ME (Table 1). According to the GO and KEGG annotation information (Figure S4), two genes (Bol037384 and Bol01933, peroxidases) were involved in the response to oxidative stress and could remove ROS. The two genes were negatively regulated by the same miRNA targeting and methylation in HT32, but were upregulated in the transcriptome. We hypothesized that HS mediated ROS metabolism disrupted DNA methylation and miRNA, leading to protective function in ME.

Table 1. Seven DEGs targeted by methylation and miRNA.

Target Genes	mRNA Regulation	Methylation Regulation	Location of DMR	miRNA Name	miRNA Regulation	*p* Value	Description
Bol007667	up	hypermethylation	promoter	PC-5p-73787_21	down	<0.05	plant U-box 9
Bol012681	up		downstream	bna-MIR6028-p3	down	<0.05	calmodulin-binding protein
Bol017014	down		downstream	htu-MIR530-p5_2ss7TA19CT	up	<0.05	atypical CYS HIS rich thioredoxin 5
Bol016003	up	hypomethylation	downstream	PC-5p-70200_22	up	<0.05	glycine-rich protein 2B
Bol045794	up		downstream	htu-MIR530-p5_2ss7TA19CT	up	<0.05	cyclic nucleotide-gated channel 6
Bol037384	up		downstream	ptc-miR6478_R-2	down	<0.05	peroxidase CB
Bol019334	up		downstream	ptc-miR6478_R-2	down	<0.05	peroxidase superfamily protein

There are other reports that ME is strongly affected by ABA regulation [61,62]. In our study, Bol007667 (plant U-box 9, PUB9) was involved in the ABA-activated signaling pathway, and Bol016003 (glycine-rich protein 2B) was involved in the ABA response. Reynolds et al. [61] showed that the mechanism of ABA in ME was related to the *EcMt* gene (early cysteine-labeled metallothionein gene). During this process, calcium is involved in ABA signal transduction to initiate the EcMt gene; thus, ME could have a possible relationship with calmodulin [62]. In HT32 and NT25, one atypical CYS HIS sulfur-rich thioredoxin gene (Bol017014) and two calcium-related genes (Bol012681 and Bol045794) were identified, which were regulated by methylation and miRNA.

It is generally believed that methylation and miRNA located at promoters and genes had a greater influence on gene expression. However, the regulation of methylation and miRNA on genes is very complex, which occurs in different regions of genes and may have different functions [63,64]. Therefore, the expression of these genes could be affected by methylation or miRNA and played a regulatory role in ME, which was worthy of further study.

Overall, it would be interesting to study the changes of methylation levels in cabbage microspore under HS. These results could broaden our understanding of the molecular regulation of ME induced by HS pre-treatment, and have a significance for producing DH lines from F_1 hybrids, which accelerates the breeding process significantly.

4. Conclusions

To study the regulatory roles of DNA methylation and miRNA in early ME, we combined DNA methylation levels, miRNAs, and transcriptome profiles in isolated microspores of cabbage '01-88' under HS and normal temperature. For global DNA methylation, 508 DMRs were identified. The association analysis found 59.92% of DMRs were correlated with transcripts and 39.43% of miRNA locus were associated with mCs. Furthermore, a total of 17 miRNAs and 31 target DEGs were found to be differentially methylated. These DEGs were mainly involved in ROS response and ABA signaling, and provided evidence that epigenetic regulation play a crucial role in ME activated by HS.

5. Materials and Methods

5.1. *Plant Materials, Treatments, and Sample Collection*

The microspore culture of cabbage accession '01-88' (excellent in the embryo yield under microspore culture) was carried out at the Institute of Vegetables and Flowers, Chinese Academy of Agricultural Sciences, Beijing, China. The microspore isolation and culture procedures were performed following Lv [65] and were slightly modified. Briefly, buds with a length of 3–3.5 mm and an anther-to-petal length ratio of about 3:2 were selected. These buds surfaces were sterilized by 75% ethanol and 8% sodium hypochlorite, then transferred to test tubes with round bottom containing a small amount of B5 medium (3.21 g B5 and 130 g sucrose in 1 L dH_2O, pH 5.9) and grounded with a glass rod. The isolated microspores were filtered, centrifuged, and resuspended in the NLN-13 medium (1.77 g NLN and 130 g sucrose in 1 L dH_2O, pH 5.9). Subsequently, a 2 mL microspore suspension adjusted to 1×10^5 cfu/mL was added to each petri dish (60 × 15 mm), and were incubated at heat temperature at 32 °C for 24 h (HT32) and normal temperature at 25 °C for 24 h (NT25), respectively.

A biological replicate required 80 buds, and 26 petri dishes of microspore suspension could be obtained, 20 of which were centrifuged to collect microspores for sequencing, and 6 petri dishes were incubated at 25 °C in the dark for 3 weeks until embryoids were produced. T test was used to evaluate the number of embryoids produced under the two treatments, with three replicates per treatment, and the significance difference was $p < 0.05$. Small RNA sequencing and RNA sequencing were performed with three biological replicates per treatment, and both were harvested in the same sample. Bisulfite sequencing processed genomic DNA isolated from HT32 and NT25 (pooled in equal quantity from the two independent biological replicates).

5.2. *Genomic Methyl Cytosine Library Construction and Bisulphite Sequencing (BS-Seq)*

Total DNA was extracted using a Hi-Fast Plant Genomic DNA Kit (GeneBetter Biotech, Beijing, China) following the manufacturer's procedure. The quality of DNA was evaluated by agarose gel electrophoresis and a spectrophotometer (Thermo Fisher Scientific Inc., Wilmington, DE, USA). The genomic DNA samples were fragmented by sonication to a size of 100–300 bp. The DNA fragments were end repaired and the 3′-end added a single 'A' nucleotide. Subsequently, the Accel-NGS Methyl-Seq DNA Library Kit (Vazyme Biotech Co, Nanjing, China) was utilized for attaching adapters to single−stranded DNA fragments. The DNA fragments were subjected to bisulfite conversion using a ZYMO EZ DNA Methylation-Gold kit (NEB, Ipswich, MA, USA). After desalting, size selection and PCR amplification, the two libraries were sequenced by the pair-end 2 × 150 bp sequencing on an Illumina Hiseq 4000 platform (LC Bio, Hangzhou, China).

5.3. *Bioinformatic Analysis*

In-house cutadapt and Perl scripts were used to remove reads with adapter contamination, low quality bases, and undetermined bases [66]. Then, sequence quality was verified using FastQC (http://www.bioinformatics.babraham.ac.uk/projects/fastqc/, accessed on 12 June 2021). Reads that passed quality control were mapped to the reference genome using WALT [67]. After alignment, the reads were further deduplicated using samtool [68].

For each cytosine site (or guanine corresponding to a cytosine on the opposite strand) in the *B. oleracea* genome sequence (http://brassicadb.cn, 12 June 2021). The DNA methylation level was determined by the ratio of the number of reads supporting C (methylated) to that of the total reads (methylated and unmethylated) using in-house Perl scripts and MethPipe [69]. Genomic methylation levels of HT32 and NT25 was visualized by circos software [70]. DMRs were calculated using the R package-MethylKit with default parameters (1000 bp slide windows, 500 bp overlap, $p < 0.05$). Multiple testing correction (FDR < 0.05) was followed to test each window. According to the read coverage of methylated C in each sample $\geq$30, the level ratio of average methylation was >15, the fold change was >2 or <0.5, and the DMR with gene annotation was retained to filter and screen the DMRs [71].

5.4. Small RNA Library Construction, Sequencing Analysis, and Prediction of miRNA Target Genes

Total RNAs were isolated separately from the microspores in HT32 and NT25 using Trizol reagent (Invitrogen, GeneBetter Biotech, Beijing, China) following the manufacturer's protocol. Each treatment had 3 replicates, which were HT32-1, -2, -3 and NT25-1, -2, -3. To construct the six small RNA libraries (HT32-1, -2, -3 and NT25-1, -2, -3), small RNA (18 to 30 nt in length) was fractionated by polyacrylamide gel electrophoresis and ligated to the 5′ and 3′ RNA adapters. Reverse transcription and PCRs were performed to obtain sufficient single-stranded cDNA. Finally, these six small RNA libraries were sequenced by Illumina Hiseq2500 platform (LC Bio, Hangzhou, China).

Raw reads were analyzed according to procedures described in a previous study [72] by ACGT101-miR v4.2 software package (LC Sciences, Houston, TX, USA) to remove adapter dimers and low-quality reads (the reads with more than one base with a Q-value lower than 20). The clean small RNA reads were matched to the GenBank database (https://www.ncbi.nlm.nih.gov/genbank/, accessed on 3 July 2021), Rfam database (https://rfam.xfam.org/, 3 July 2021) and Repbase database (https://www.girinst.org/repbase/) to identify and remove rRNA, scRNA, snRNA and tRNA by blastn (E-value < 10^{-5}).

Subsequently, unique sequences with a length in 18–25 nt were mapped to specific species precursors in miRBase 21.0 (ftp://mirbase.org/pub/mirbase/CURRENT/, accessed on 10 July 2021) by a BLAST search to identify known miRNAs and novel 3p- and 5p-derived miRNAs. Length variation at the 3′ and 5′ ends and one mismatch inside of the sequence were allowed in the alignment. The annotated miRNAs were determined as known miRNAs. The unmapped sequences were BLASTed against the specific genomes, and the hairpin RNA structures containing sequences were predicted from the flank 120-nt sequences using RNAfold software (http://rna.tbi.univie.ac. at/cgi-bin/RNAfold.cgi, accessed on 13 July 2021) [73].

Differentially expressed miRNAs (DERs) was analysed based on normalized deep-sequencing counts by *T* test and the significance difference was the FDR adjusted $p < 0.05$. The fold change and p value for each miRNA was calculated based on the normalized expression. To predict the genes targeted by DERs, the putative target sites of miRNA were identified using the psRNA Target program (http://plantgrn.noble.org/psRNATarget/, accessed on 12 June 2021) with default parameters, the predicted target genes were evaluated based on complementarity scoring and maximum expectation [74]. The *B. oleracea* genome database (http://brassicadb.cn, accessed on 12 June 2021) was used as the sequence library for target searches. The gene functions targeted by miRNAs were clarified using GO (http://www.geneontology.org/, accessed on 15 July 2021) and KEGG (http://www.kegg.jp/kegg, accessed on 12 June 2021) annotation. Significantly enriched GO terms and KEGG pathways were identified using hypergeometric tests and a bonferroni correction with $p < 0.05$ as a threshold [73]. Additionally, the miRNA-pathway interaction network was constructed using Cytoscape v2.8.3 software (http://www.cytoscape.org/, accessed on 15 July 2021).

5.5. mRNA Sequencing, Assembly, and Annotation

The quantity of total RNA was determined using a NanoDrop 2000 (Thermo Fisher Scientific Inc., Wilmington, DE, USA), and the integrity was assessed by an Agilent 2100 (Agilent Technologies Co., Ltd., Beijing, China) with RIN number >7.0. Six cDNA libraries (HT32-1, -2, -3 and NT25-1, -2, -3) were prepared and sequencing was performed using the Illumina HiSeq 4000 platform (LC Bio, Hangzhou, China). The low quality reads containing sequencing adaptors, sequencing primers, and/or nucleotides with a quality score (Q < 20) were removed. Then, clean reads were mapped to the reference genome (http://brassicadb.cn, 12 June 2021) by HISAT [75]. Mapped reads were assembled using String Tie [76], and their expression levels were calculated by the FPKM method. Genes differential expression analysis was performed by edgeR between two samples. The genes with the parameter of the FDR adjusted $p < 0.05$ and absolute fold change ≥ 2 were considered differentially expressed genes (DEGs). All DEGs were mapped to GO terms and KEGG pathways in the respective databases. Significantly enriched GO terms and KEGG pathways were identified using hypergeometric tests and a Bonferroni correction with $p < 0.05$ as a threshold [77].

5.6. qRT−PCR Analysis

The transcript levels of genes were identified using a real−time quantitative polymerase chain reaction (qRT-PCR). Isolated micropores were extracted using TRIzol reagent (Invitrogen, GeneBetter Biotech, Beijing, China) following the manufacturer's protocol. A Revert Aid First Strand cDNA Synthesis Kit (Vazyme Biotech Co, Nanjing, China) was used to reverse transcribe RNA to cDNA. The qRT-PCR reactions were conducted as reported previously, and *actin* (GenBank accession number XM_013731369.1; qRT-PCR forward primer: 5′-CCAGAGGTCTTGTTCCAGCCATC-3′, reverse primer: 5′-GTTCCACCAC TGAGCACAATGTTAC-3′) was used as the internal reference gene in cabbage; the test was repeated three times [78]. We calculated the relative expression levels of the genes using the $2^{-\Delta\Delta Ct}$ method [79]. Significance was determined using a two-tailed Student's *T* test ($p < 0.05$). The primer sequences of 9 genes were shown in Table S2.

5.7. Association of miRNA, Transcriptome, and Methylation Sequencing Data

These association analyses in miRNA, transcriptome, and methylation were based on small RNA-Seq, RNA-Seq and MeDIP-Seq sequencing data. We filtered the raw datas to reduce the influence of sequencing error as described previously of the standard analysis. Firstly, the methylation regions of miRNA and mRNA were sorted out, and the methylation levels of miRNA and mRNA were annotated respectively. Then, we correlated miRNA target genes and differentially methylated regions (1000 bp slide windows, 500 bp overlap, FDR adjusted $p < 0.05$) with mRNA data ($|\log2FC| > 1$, FDR adjusted $p < 0.05$). We screened and sorted the data and regulatory relationships with significant differences among the three groups. Lastly, we performed GO and KEGG enrichment analyses of the associated core gene sets (hypergeometric tests and a bonferroni correction with $p < 0.05$ as a threshold).

Supplementary Materials: The following supporting information can be downloaded at: https://www.mdpi.com/article/10.3390/ijms23095147/s1.

Author Contributions: Formal analysis, C.K., S.D., J.J. and Y.W.; Funding acquisition, Y.W., Y.Z. and L.Y.; Project administration H.L. and Z.F.; Writing—original draft, C.K. and H.S. All authors have read and agreed to the published version of the manuscript.

Funding: This work was financially supported by grants from the Science and Technology Innovation Program of the Chinese Academy of Agricultural Sciences (CAAS−ASTIP−IVFCAAS), China Agriculture Research System of MOF and MARA (CARS−23) and Major Science and Technology Projects of Inner Mongolia Autonomous Region (2021ZD0001).

Institutional Review Board Statement: Not applicable.

Informed Consent Statement: Informed consent was obtained from all subjects involved in the study.

Data Availability Statement: The raw reads of our BS-Seq, RNA-seq and miRNA-seq data in this work were deposited in the Sequence Read Archive under accession numbers PRJNA810282.

Conflicts of Interest: The authors declare that they have no conflict of interest.

References

1. Ji, J.; Yang, L.; Fang, Z.; Zhuang, M.; Zhang, Y.; Lv, H.; Liu, Y.; Li, Z. Complementary transcriptome and proteome profiling in cabbage buds of a recessive male sterile mutant provides new insights into male reproductive development. *J. Proteom.* **2018**, *179*, 80–91. [CrossRef] [PubMed]
2. Bosemark, N.O. Genetics and breeding. In *The Sugar Beet Crop*; Springer: Dordrecht, The Netherlands, 1993; pp. 67–119.
3. Vanous, K.; Vanous, A.; Frei, U.K.; Lübberstedt, T. Generation of maize (*Zea mays*) doubled haploids via traditional methods. *PGPB* **2017**, *2*, 147–157. [CrossRef]
4. Su, H.; Han, F.; Yang, L.; Zhuang, M.; Zhang, Y.; Wang, Y.; Li, Z.; Fang, Z.; Lv, H. Studies on optimization of Cabbage Isolated Microspore culture conditions and Generation of Embryoids from high-generation inbred lines. *China Veg.* **2018**, *4*, 30–36. (In Chinese)
5. Ferrie, A.; Caswell, K. Isolated microspore culture techniques and recent progress for haploid and doubled haploid plant production. *Plant Cell Tiss. Org.* **2011**, *104*, 301–309. [CrossRef]
6. Pechan, P.M.; Bartels, D.; Schell, D.C.W.; Schell, J. Messenger-RNA and protein changes associated with induction of Brassica microspore embryogenesis. *Planta* **1991**, *184*, 161–165. [CrossRef]
7. Cordewener, J.H.G.; Busink, R.; Traas, J.A.; Custers, J.B.M.; Hans, J.M. Induction of microspore embryogenesis in *Brassica napus* is accompanied by specific changes in protein synthesis. *Planta* **1994**, *195*, 50–56. [CrossRef]
8. Cordewener, J.H.G.; Hause, G.; Görgen, E.; Busink, R.; Hause, B.; Dons, H.J.M.; Van Lammeren, A.A.M.; Van Lookeren Campagne, M.M.; Pechan, P. Changes in synthesis and localization of members of the 70−kDa class of heat−shock proteins accompany the induction of embryogenesis in *Brassica napus* L. microspores. *Planta* **1995**, *196*, 747–755. [CrossRef]
9. Yuan, S.; Su, Y.; Liu, Y.; Fang, Z.; Yang, L.; Zhuang, M.; Zhang, Y.; Sun, P. Effects of pH, MES, arabinogalactan−proteins on microspore cultures in white cabbage. *Plant Cell Tiss. Org.* **2012**, *110*, 69–76. [CrossRef]
10. Yuan, S.; Liu, Y.; Fang, Z.; Yang, L.; Zhuang, M.; Zhang, Y.; Sun, P. Effect of combined cold pre-treatment and heat shock on microspore cultures in broccoli. *Plant Breed.* **2009**, *130*, 80–85. [CrossRef]
11. López-Hernández, F.; Cortés, A.J. Last-generation genome-environment associations reveal the genetic basis of heat tolerance in common bean (*Phaseolus vulgaris* L.). *Front. Genet.* **2019**, *10*. [CrossRef]
12. Seguí-Simarro, J.M.; Testillano, P.S.; Risueo, M.C. Hsp70 and hsp90 change their expression and subcellular localization after microspore embryogenesis induction in *Brassica napus* L. *J. Struct Biol.* **2003**, *142*, 379–391. [CrossRef]
13. Seguí-Simarro, J.M.; Nuez, F. How microspores transform into haploid embryos: Changes associated with embryogenesis induction and microspore−derived embryogenesis. *Physiol. Plantarum.* **2008**, *134*, 1–12. [CrossRef] [PubMed]
14. Naydenov, M.; Baev, V.; Apostolova, E.; Gospodinova, N.; Sablok, G.; Gozmanova, M.; Yahubyan, G. High-temperature effect on genes engaged in DNA methylation and affected by DNA methylation in *Arabidopsis*. *Plant Physiol. Bioch.* **2015**, *87*, 102–108. [CrossRef] [PubMed]
15. Jullien, P.E.; Susaki, D.; Yelagandula, R.; Higashiyama, T.; Berger, F. DNA methylation dynamics during sexual reproduction in *Arabidopsis thaliana*. *Curr. Biol.* **2012**, *22*, 1825–1830. [CrossRef]
16. He, X.J.; Chen, T.; Zhu, J.K. Regulation and function of DNA methylation in plants and animals. *Cell Res.* **2011**, *21*, 442–465. [CrossRef]
17. Saze, H.; MittelstenScheid, O.; Paszkowski, J. Maintenance of CpG methylation is essential for epigenetic inheritance during plant gametogenesis. *Nat. Genet.* **2003**, *34*, 65–69. [CrossRef]
18. Henderson, I.R.; Jacobsen, S.E. Epigenetic inheritance in plants. *Nature* **2007**, *447*, 418–424. [CrossRef]
19. Lister, R.; Pelizzola, M.; Dowen, R.H.; Hawkins, R.D.; Hon, G.; Tonti-Filippini, J.; Nery, J.R.; Lee, L.; Ye, Z.; Ngo, Q.; et al. Human DNA methylomes at base resolution show widespread epigenomic differences. *Nature* **2009**, *462*, 315–322. [CrossRef]
20. Grafi, G. Epigenetics in plant development and response to stress. *Biochim. Biophys. Acta.* **2011**, *1809*, 351–352. [CrossRef]
21. Solís, M.T.; Rodríguez-Serrano, M.; Meijón, M.; Cañal, M.J.; Cifuentes, A.; Risueño, M.C.; Testillano, P.S. DNA methylation dynamics and MET1a-like gene expression changes during stress-induced pollen reprogramming to embryogenesis. *J. Exp. Bot.* **2012**, *63*, 6431–6444. [CrossRef]
22. El-Tantawy, A.A.; Solís, M.T.; Risueño, M.C.; Testillano, P.S. Changes in DNA methylation levels and nuclear distribution patterns after microspore reprogramming to embryogenesis in barley. *Cytogenet. Genome Res.* **2014**, *143*, 200–208. [CrossRef] [PubMed]
23. Rodríguez-Sanz, H.; Manzanera, J.A.; Solís, M.T.; Gómez-Garay, A.; Pintos, B.; Risueño, M.C.; Testillano, P.S. Early markers are present in both embryogenesis pathways from microspores and immature zygotic embryos in cork oak, *Quercus suber* L. *BMC Plant Biol.* **2014**, *14*, 224. [CrossRef] [PubMed]
24. Ma, Y.; Min, L.; Wang, M.; Wang, C.; Zhao, Y.; Li, Y.; Fang, Q.; Wu, Y.; Xie, S.; Ding, Y.; et al. Disrupted Genome Methylation in Response to High Temperature Has Distinct Affects on Microspore Abortion and Anther Indehiscence. *Plant Cell* **2018**, *30*, 1387–1403. [CrossRef] [PubMed]

25. Li, J.; Huang, Q.; Sun, M.; Zhang, T.; Li, H.; Chen, B.; Xu, K.; Gao, G.; Li, F.; Yan, G.; et al. Global DNA methylation variations after short−term heat shock treatment in cultured microspores of *Brassica napus* cv. *Topas*. *Sci. Rep.* **2016**, *6*, 38401. [CrossRef]
26. Tang, G.; Reinhart, B.J.; Bartel, D.P.; Zamore, P.D. A biochemical framework for RNA silencing in plants. *Genes Dev.* **2003**, *17*, 49–63. [CrossRef]
27. Reinhart, B.; Weinstein, E.; Rhoades, M.; Bartel, B.; Bartel, D. MicroRNAs in plants. *Genes Dev.* **2002**, *16*, 1616–1626. [CrossRef]
28. Hu, J.; Jin, J.; Qian, Q.; Huang, K.; Ding, Y. Small RNA and degradome profiling reveals miRNA regulation in the seed germination of ancient eudicot Nelumbo nucifera. *BMC Genom.* **2016**, *17*, 684. [CrossRef]
29. Gao, C.; Wang, P.; Zhao, Z.; Zhao, C.; Xia, H.; Hou, L.; Ju, Z.; Zhang, Y.; Li, C.; Wang, X. Small RNA profiling and degradome analysis reveal regulation of microRNA in peanut embryogenesis and early pod development. *BMC Genom.* **2017**, *18*, 220. [CrossRef]
30. Lukasik, A.; Pietrykowska, H.; Paczek, L.; Szweykowska-Kulinska, Z.; Zielenkiewicz, P. High-throughput sequencing identification of novel and conserved miRNAs in the *Brassica oleracea* leaves. *BMC Genom.* **2013**, *14*, 801. [CrossRef]
31. Gao, C.; Ju, Z.; Cao, D.; Zhai, B.; Qin, G.; Zhu, H.; Fu, D.; Luo, Y.; Zhu, B. MicroRNA profiling analysis throughout tomato fruit development and ripening reveals potential regulatory role of RIN on microRNAs accumulation. *Plant Biotechnol. J.* **2015**, *13*, 370–382. [CrossRef]
32. Bartel, D. MicroRNAs: Genomics, biogenesis, mechanism, and function. *Cell* **2004**, *116*, 281–297. [CrossRef]
33. Jones−Rhoades, M.; Bartel, D.; Bartel, B. MicroRNAs and their regulatory roles in plants. *Annu. Rev. Plant Biol.* **2006**, *57*, 19–53. [CrossRef] [PubMed]
34. Bélanger, S.; Baldrich, P.; Lemay, M.A.; Marchand, S.; Esteves, P.; Meyers, B.C.; Belzile, F. The commitment of barley microspores into embryogenesis correlates with miRNA-directed regulation of members of the SPL, GRF and HD-ZIPIII transcription factor families. *Plant Direct* **2020**, *4*, e00289. [CrossRef]
35. Chen, J.; Pan, A.; He, S.; Su, P.; Yuan, X.; Zhu, S.; Liu, Z. Different microRNA families involved in regulating high temperature stress response during cotton (*Gossypium hirsutum* L.) anther development. *IJMS* **2020**, *21*, 1280. [CrossRef] [PubMed]
36. Han, F.; Zhang, X.; Xing, L.; Su, H.; Kong, C.; Fang, Z.; Yang, L.; Mu, Z.; Zhang, Y.; Liu, Y. Comparative Analysis of Genome Wide DNA Methylation Profiles for the Genic Male Sterile Cabbage Line 01-20S and Its Maintainer Line. *Genes* **2017**, *8*, 159. [CrossRef] [PubMed]
37. An, Y.C.; Goettel, W.; Han, Q.; Bartels, A.; Liu, Z.; Xiao, W. Dynamic Changes of Genome—Wide DNA Methylation during Soybean Seed Development. *Sci. Rep.* **2017**, *7*, 12263. [CrossRef] [PubMed]
38. Hu, J.; Chen, X.; Zhang, H.; Ding, Y. Genome-wide analysis of DNA methylation in photoperiod and thermo−sensitive male sterile rice Peiai 64S. *BMC Genom.* **2015**, *16*, 102. [CrossRef]
39. Xiao, W.; Custard, K.D.; Brown, R.C.; Lemmon, B.E.; Harada, J.J.; Goldberg, R.B.; Fischer, R.L. DNA methylation is critical for *Arabidopsis* embryogenesis and seed viability. *Plant Cell* **2006**, *18*, 805–814. [CrossRef]
40. Jain, M.; Moharana, K.C.; Shankar, R.; Kumari, R.; Garg, R. Genome wide discovery of DNA polymorphisms in rice cultivars with contrasting drought and salinity stress response and their functional relevance. *Plant Biotechnol. J.* **2014**, *12*, 253–264. [CrossRef]
41. Wang, M.; Qin, L.; Xie, C.; Li, W.; Yuan, J.; Kong, L.; Yu, W.; Xia, G.; Liu, S. Induced and constitutive DNA methylation in a salinity-tolerant wheat introgression line. *Plant Cell Physiol.* **2014**, *55*, 1354–1365. [CrossRef]
42. Choi, C.S.; Sano, H. Abiotic-stress induces demethylation and transcriptional activation of a gene encoding a glycerophosphodieste rase-like protein in tobacco plants. *Mol. Genet. Genom.* **2007**, *277*, 589–600. [CrossRef] [PubMed]
43. Angers, B.; Castonguay, E.; Massicotte, R. Environmentally induced phenotypes and DNA methylation: How to deal with unpredictable conditions until the next generation and after. *Mol. Ecol.* **2010**, *19*, 1283–1295. [CrossRef]
44. Chávez–Hernández, E.C.; Alejandri-Ramírez, N.D.; Juárez-González, V.T.; Dinkova, T.D. Maize miRNA and target regulation in response to hormone depletion and light exposure during somatic embryogenesis. *Front. Plant Sci.* **2015**, *6*, 555. [CrossRef] [PubMed]
45. Khraiwesh, B.; Zhu, J.K.; Zhu, J. Role of miRNAs and siRNAs in biotic and abiotic stress responses of plants. *Biochim. Biophys. Acta.* **2012**, *1819*, 137–148. [CrossRef] [PubMed]
46. Yang, X.; Wang, L.; Yuan, D.; Lindsey, K.; Zhang, X. Small RNA and degradome sequencing reveal complex miRNA regulation during cotton somatic embryogenesis. *J. Exp. Bot.* **2013**, *64*, 1521–1536. [CrossRef] [PubMed]
47. Zhang, J.; Zhang, S.; Han, S.; Wu, T.; Li, V.; Li, W.; Qi, L. Genome-wide identification of microRNAs in larch and stage-specific modulation of 11 conserved microRNAs and their targets during somatic embryogenesis. *Planta* **2012**, *236*, 647–657. [CrossRef]
48. Long, J.M.; Liu, C.Y.; Feng, M.Q.; Liu, M.Q.; Wu, X.M.; Guo, W.W. miR156-SPL modules regulate induction of somatic embryogenesis in citrus callus. *J. Exp. Bot.* **2018**, *69*, 2979–2993. [CrossRef]
49. Li, C.; Zhang, B. MicroRNAs in control of plant development. *J. Cell Physiol.* **2016**, *231*, 303–313. [CrossRef]
50. Martin, R.C.; Liu, P.P.; Goloviznina, N.A.; Nonogaki, H. microRNA, seeds, and Darwin? Diverse function of miRNA in seed biology and plant responses to stress. *J. Exp. Bot.* **2010**, *61*, 2229–2234. [CrossRef]
51. Siddiqui, Z.H.; Abbas, Z.K.; Ansari, M.W.; Khan, M.N. The role of miRNA in somatic embryogenesis. *Genomics* **2019**, *111*, 1026–1033. [CrossRef]
52. Garg, R.; Chevala, V.N.; Shankar, R.; Jain, M. Divergent DNA methylation patterns associated with gene expression in rice cultivars with contrasting drought and salinity stress response. *Sci. Rep.* **2015**, *5*, 14922. [CrossRef]

53. Zhang, H.; Zhao, H.; Wu, S.; Huang, F.; Wu, K.; Zeng, X.; Chen, X.; Xu, P.; Wu, X. Global Methylation Patterns and Their Relationship with Gene Expression and Small RNA in Rice Lines with Different Ploidy. *Front. Plant Sci.* **2016**, *7*, 1002. [CrossRef] [PubMed]
54. Mittler, R. Oxidative stress, antioxidants and stress tolerance. *Trends Plant Sci.* **2002**, *7*, 405–410. [CrossRef]
55. Zhao, J.; Newcomb, W.; Simmonds, D. Heat-shock proteins 70 kDa and 19 kDa are not required for induction of embryogenesis of *Brassica napus* L. cv. *Topas microspores*. *Plant Cell Physiol.* **2003**, *44*, 1417–1421. [CrossRef] [PubMed]
56. Aukerman, M.J.; Sakai, H. Regulation of flowering time and floral organ identity by a MicroRNA and its APETALA2-like target genes. *Plant Cell* **2003**, *15*, 2730–2741. [CrossRef]
57. Chen, X. A microRNA as a translational repressor of APETALA2 in *Arabidopsis* flower development. *Science* **2004**, *303*, 2022–2025. [CrossRef] [PubMed]
58. Lauter, N.; Kampani, A.; Carlson, S.; Goebel, M.; Moose, S.P. microRNA172 down-regulates glossy15 to promote vegetative phase change in maize. *Proc. Natl. Acad. Sci. USA* **2005**, *102*, 9412–9417. [CrossRef]
59. Ikeda-Iwai, M.; Umehara, M.; Satoh, S.; Kamada, H. Stress-induced somatic embryogenesis in vegetative tissues of *Arabidopsis thaliana*. *Plant J.* **2010**, *34*, 107–114. [CrossRef]
60. Żur, I.; Dubas, E.; Golemiec, E.; Szechyńska-Hebda, M.; Gołębiowska, G.; Wędzony, M. Stress-related variation in antioxidative enzymes activity and cell metabolism efficiency associated with embryogenesis induction in isolated microspore culture of triticale (x Triticosecale Wittm.). *Plant Cell Rep.* **2009**, *28*, 1279–1287. [CrossRef]
61. Reynolds, T.L.; Crawford, R.L. Changes in abundance of an abscisic acid-responsive, early cysteine-labeled metallothionein transcript during pollen embryogenesis in bread wheat (*Triticum aestivum*). *Plant Mol. Biol.* **1996**, *32*, 823. [CrossRef]
62. Reynolds, T.L. Effects of calcium on embryogenic induction and the accumulation of abscisic acid, and an early cysteine-labeled metallothionein gene in androgenic microspores of *Triticum aestivum*. *Plant Sci.* **2000**, *150*, 201–207. [CrossRef]
63. Jones, P.A. Functions of DNA methylation: Islands, start sites, gene bodies and beyond. *Nat. Rev. Genet.* **2012**, *13*, 484–492. [CrossRef] [PubMed]
64. Xiao, M.; Li, J.; Li, W.; Wang, Y.; Wu, F.; Xi, Y.; Zhang, L.; Ding, C.; Luo, H.; Li, Y.; et al. MicroRNAs Activate Gene Transcription Epigenetically as an Enhancer Trigger. *RNA Biol.* **2017**, *14*, 1326–1334. [CrossRef]
65. Lv, H.; Wang, Q.; Yang, L.; Fang, Z.; Liu, Y.; Zhuang, M.; Zhang, Y.; Yang, Y.; Xie, B.; Wang, X. Breeding of cabbage (*Brassica oleracea* var. capitata) with fusarium wilt resistance based on microspore culture and marker-assisted selection. *Euphytica* **2014**, *200*, 465–473. [CrossRef]
66. Martin, M. Cutadapt removes adapter sequences from high-throughput sequencing reads. *EMBnet J.* **2011**, *17*, 10–12. [CrossRef]
67. Chen, H.; Smith, A.D.; Chen, T. WALT: Fast and accurate read mapping for bisulfite sequencing. *Bioinformatics* **2016**, *32*, btw490. [CrossRef]
68. Li, H.; Handsaker, B.; Wysoker, A.; Fennell, T.; Ruan, J.; Homer, N.; Marth, G.; Abecasis, G.; Durbin, R. The Sequence Alignment/Map format and SAMtools. *Bioinformatics* **2009**, *25*, 2078–2079. [CrossRef]
69. Song, Q.; Decato, B.; Hong, E.E.; Zhou, M.; Smith, A.D. A Reference Methylome Database and Analysis Pipeline to Facilitate Integrative and Comparative Epigenomics. *PLoS ONE* **2013**, *8*, e81148. [CrossRef]
70. Krzywinski, M.; Schein, J.; Birol, I.; Connors, J.; Gascoyne, R.; Horsman, D.; Marra, M.A. Circos: An information aesthetic for comparative genomics. *Genome Res.* **2009**, *19*, 1639–1645. [CrossRef]
71. Akalin, A.; Kormaksson, M.; Sheng, L.; Garrett-Bakelman, F.E.; Figueroa, M.E.; Melnick, A.; Mason, C.E. Methylkit: A comprehensive R package for the analysis of genome-wide DNA methylation profiles. *Genome Biol.* **2012**, *13*, R87. [CrossRef]
72. Jeyaraj, A.; Zhang, X.; Hou, Y.; Shangguan, M.; Gajjeraman, P.; Li, Y. Genome-wide identification of conserved and novel micrornas in one bud and two tender leaves of tea plant (*Camellia sinensis*) by small rna sequencing, microarray-based hybridization and genome survey scaffold sequences. *BMC Plant Biol.* **2017**, *17*, 212. [CrossRef] [PubMed]
73. Zhao, S.; Wang, X.; Yan, X.; Guo, L.; Mi, X.; Xu, Q.; Zhu, J.; Wu, A.; Lin, L.; Wei, C. Revealing of MicroRNA Involved Regulatory Gene Networks on Terpenoid Biosynthesis in Camellia sinensis in Different Growing Time Points. *J. Agric. Food Chem.* **2018**, *66*, 12604–12616. [CrossRef] [PubMed]
74. Dai, X.; Zhao, P.X. psRNATarget: A plant small RNA target analysis server. *Nucleic Acids Res.* **2011**, *39*, W155–W159. [CrossRef] [PubMed]
75. Kim, D.; Langmead, B.; Salzberg, S.L. HISAT: A fast spliced aligner with low memory requirements. *Nat. Methods* **2015**, *12*, 357–360. [CrossRef]
76. Pertea, M.; Pertea, G.M.; Antonescu, C.M.; Chang, T.C.; Mendell, J.T.; Salzberg, S.L. StringTie enables improved reconstruction of a transcriptome from RNA-seq reads. *Nat. Biotechnol.* **2015**, *33*, 290–295. [CrossRef]
77. Su, H.; Xing, M.; Liu, X.; Fang, Z.; Yang, L.; Zhuang, M.; Zhang, Y.; Wang, Y.; Lv, H. Genome-wide analysis of HSP70 family genes in cabbage (*Brassica oleracea* var. *capitata*) reveals their involvement in floral development. *BMC Genom.* **2019**, *20*, 369. [CrossRef]
78. Ren, J.; Liu, Z.; Chen, W.; Xu, H.; Feng, H. Anthocyanin Degrading and Chlorophyll Accumulation Lead to the Formation of Bicolor Leaf in Ornamental Kale. *Int. J. Mol. Sci.* **2019**, *20*, 603. [CrossRef]
79. Schmittgen, T.D.; Livak, K.J. Analyzing real-time PCR data by the comparative CT method. *Nat. Protoc.* **2008**, *3*, 1101–1108. [CrossRef]

International Journal of
Molecular Sciences

MDPI

Article

Mapping and Validation of *BrGOLDEN*: A Dominant Gene Regulating Carotenoid Accumulation in *Brassica rapa*

Lei Zhang [1,2,3,†], Shifan Zhang [1,†], Yun Dai [1], Shaoxing Wang [1,2], Chenggang Wang [2], Fei Li [1], Hui Zhang [1], Guohu Chen [2], Lingyun Yuan [2], Jinfeng Hou [2], Xiaoyan Tang [2], Shidong Zhu [2], Rifei Sun [1], Guoliang Li [1,*] and Shujiang Zhang [1,*]

[1] Institute of Vegetables and Flowers, Chinese Academy of Agricultural Sciences, Beijing 100081, China
[2] Vegetable Genetics and Breeding Laboratory, College of Horticulture, Anhui Agricultural University, Hefei 230036, China
[3] Beijing Academy of Agricultural and Forestry Sciences, Key Laboratory of Biology and Genetic Improvement of Horticultural Crops (North China), Beijing Key Laboratory of Vegetable Germplasm Improvement, Beijing 100097, China
* Correspondence: liguoliang@caas.cn (G.L.); zhangshujiang@caas.cn (S.Z.); Tel.: +86-130-5152-4685 (G.L.); +86-137-1868-1292 (S.Z.)
† These authors contributed equally to this work.

Abstract: In plants, the accumulation of carotenoids can maintain the balance of the photosystem and improve crop nutritional quality. Therefore, the molecular mechanisms underlying carotenoid synthesis and accumulation should be further explored. In this study, carotenoid accumulation differed significantly among parental *Brassica rapa*. Genetic analysis was carried out using the golden inner leaf '1900264' line and the light−yellow inner leaf '1900262' line, showing that the golden inner leaf phenotype was controlled by a single dominant gene. Using bulked−segregant analysis sequencing, *BraA09g007080.3C* encoding the ORANGE protein was selected as a candidate gene. Sequence alignment revealed that a 4.67 kb long terminal repeat insertion in the third exon of the *BrGOLDEN* resulted in three alternatively spliced transcripts. The spatiotemporal expression results indicated that *BrGOLDEN* might regulate the expression levels of carotenoid−synthesis−related genes. After transforming *BrGOLDEN* into *Arabidopsis thaliana*, the seed−derived callus showed that $BrGOLDEN_{Ins}$ and $BrGOLDEN_{Del}$ lines presented a yellow color and the $BrGOLDEN_{Ldel}$ line presented a transparent phenotype. In addition, using the yeast two−hybrid assay, $\mathrm{BrGOLDEN_{Ins}}$, $\mathrm{BrGOLDEN_{Ldel}}$, and $\mathrm{Brgolden_{wt}}$ exhibited strong interactions with BrPSY1, but $\mathrm{BrGOLDEN_{Del}}$ did not interact with BrPSY1 in the split−ubiquitin membrane system. In the secondary and 3D structure analysis, $\mathrm{BrGOLDEN_{Del}}$ was shown to have lost the PNFPSFIPFLPPL sequences at the 125 amino acid position, which resulted in the α−helices of $\mathrm{BrGOLDEN_{Del}}$ being disrupted, restricting the formation of the 3D structure and affecting the functions of the protein. These findings may provide new insights into the regulation of carotenoid synthesis in *B. rapa*.

Keywords: carotenoid; *BrGOLDEN*; *Brassica rapa*; spatiotemporal expression; Y2H

Citation: Zhang, L.; Zhang, S.; Dai, Y.; Wang, S.; Wang, C.; Li, F.; Zhang, H.; Chen, G.; Yuan, L.; Hou, J.; et al. Mapping and Validation of *BrGOLDEN*: A Dominant Gene Regulating Carotenoid Accumulation in *Brassica rapa*. *Int. J. Mol. Sci.* **2022**, 23, 12442. https://doi.org/10.3390/ijms232012442

Academic Editor: Pedro Martínez-Gómez

Received: 10 September 2022
Accepted: 13 October 2022
Published: 18 October 2022

1. Introduction

Carotenoids play a fundamental role in human nutrition by maintaining human health and mitigating a range of diseases [1]. They are derived from isoprenoids, the largest class of natural pigments in plants. To date, more than 750 natural carotenoid molecules have been identified from animals, plants, and microorganisms [2]. In plants, carotenoids protect the photosystem from damage by dissipating excess light energy released by the photosynthetic mechanism [3]. Carotenoids can also serve as precursors for the phytohormone strigolactone and abscisic acid biosynthesis [4–7]. In edible fleshy fruit, such as cantaloupe, melon, and citrus, carotenoid volatile derivatives affect the aroma and flavor [8]. Moreover, the activity of reactive oxygen species and free radicals can be inhibited by carotenoids to maintain cell structure and normal life activities [9]. Therefore, improving the carotenoid content in plants has attracted increasing attention, especially the content of carotenoids in fruit, vegetables, and grains, which are the main components in the human diet. In recent years, considerable details of the plant carotenoid biosynthetic pathway have been elucidated [10–12].

Carotenoids can be produced in all differentiated plastids and mainly accumulate in the chloroplasts of green tissues and chromoplasts of petals and fruit [13,14]. In plants, many core catalytic reaction genes and enzymes for carotenoid synthesis have been identified (Figure 7e) [2]. Carotenoids are formed from the condensation of the 5−carbon precursors isopentenyl diphosphate and dimethylallyl diphosphate, which are produced via the plastidial 2−C−methyl−D−erythritol 4−phosphate (MEP) pathway in plastids [15]. The first committed step in the carotenoid pathway occurs when phytoene synthase (*PSY*) condenses two molecules of geranylgeranyl diphosphate (*GGPP*) to form 15−*cis*−phytoene [16]. *PSY* is considered the rate−limiting enzyme that determines carotenoid synthesis. 15−*Cis*−phytoene is transformed into trans−lycopene under the catalysis of phytoene desaturase (*PDS*) and ζ−carotene desaturase (*ZDS*). The next step can follow two paths: α−carotene is formed via lycopene ε−cyclase (*ε−LCY*) and β−carotene is formed via *β−LCY* [17–20]. Subsequently, α− and β−carotene are catalyzed by heme−containing cytochrome P450−type hydroxylases (*CYP*) and non−heme β−ring hydroxylases (*BCH*), respectively, to hydroxylate the distal ring structure to generate yellow lutein and zeaxanthin. Zeaxanthin can be epoxidized by zeaxanthin epoxidase (*ZEP*) to produce violaxanthin, which can be reversed by violaxanthin de−epoxidase (*VDE*). This interconversion forms the xanthophyll cycle. Finally, neoxanthin is produced by neoxanthin synthase (*NXS*), completing the main biosynthetic pathway of carotenoids [21–23]. These carotenoids are indispensable for their role in catalyzing the production of strigolactone, caprolactone, β−citraurin, volatiles, and scents. Thus far, in *Arabidopsis thaliana* [24] and many horticultural plants, such as *Daucus carota* [25], *Solanum lycopersicum* [26,27], *Solanum tuberosum* [28], *Capsicum annuum* [29], and *Brassica rapa* [30], there has been great progress in the study of carotenoid metabolism.

Chinese cabbage (*B. rapa*) is a leafy vegetable belonging to the *Brassica* genus in the cruciferous family [31]. Its cultivated area and consumption rank first among all vegetables in China. The inner leaves of Chinese cabbage are rich in carotenoids, which are responsible for the diverse colors of Chinese cabbage leaves. The abnormal accumulation of carotenoids in Chinese cabbage with orange heads is mostly controlled by a series of recessive genes [32]. Matsumoto et al. (1998) developed rapid fragment length polymorphism (RFLP) markers for *orange−yellow* gene (*Oy*) pigmentation in Chinese cabbage and mapped *Oy* on linkage group 1. Genetic analysis has indicated that *Oy* is a recessive gene [33]. Using bulked segregant analysis (BSA)−seq and sequence characterized amplified region (SCAR) markers, localization analysis of the orange inner leaf trait of Chinese cabbage was carried out and 110 Chinese cabbage samples were verified using the three developed SCAR markers. The candidate gene was located on the R9 linkage group [34]. Feng et al. (2012) used 600 individual plants from the F2 population to map and develop new markers closely related to the genes for the orange color on the inner leaves of Chinese cabbage. Using linkage analysis, a gene was located at the end of A09 and the distance between the markers *syau19* and *syau15* was found to be 4.6 cM [32]. In addition, the *orange* locus was finely mapped in the F_2 S_4 populations and *ORF1*, a candidate gene, encodes a carotenoid isomerase, involved in the isomerization of carotenoids. Sequencing analysis has revealed insertions/deletions in the *ORF1* promoter region between the two parents. Therefore, *CRTISO* (*carotene isomerase*) is the most likely candidate gene for *Bror* [35]. Lee et al. (2014) found a mutation in a gene encoding *CRTISO* in an orange−colored Chinese cabbage and verified several molecular markers that could distinguish *CRTISO* genotypes. This was the first time that *CRTISO* has been identified as a candidate gene in the OC inner leaves of Chinese cabbage [36]. In subsequent studies, genetic analysis and fine mapping were conducted in the orange inner leaves of Chinese cabbage from different parents, indicating that *BrCRTISO* is a single recessive candidate gene. In the parental material, *BrCRTISO* has many Indels and single nucleotide polymorphisms (SNPs) in the promoter and coding regions, resulting in orange inner leaves in Chinese cabbage [37–39].

To date, many recessive genes regulating carotenoid accumulation in *B. rapa* have been discovered, but no dominant gene controlling this trait has been found. In this study, *BrGOLDEN*, a dominant gene, was mapped and cloned to regulate carotenoid accumulation in *B. rapa*. Tri−crossed hybrid lines were constructed through hybridization between the *B. rapa* line with golden inner leaves and the non−golden highly inbred line. Carotenoid pigments were identified by high−performance liquid chromatography (HPLC) in the leaves of different *B. rapa* lines. In addition, the promoter and coding sequences of the candidate genes were cloned and transformed into *A. thaliana* to verify their functions. The interaction proteins were screened by Yeast Two−Hybrid (Y2H), which revealed that *PSY* played an important role in regulating carotenoids and different transcripts of *BrGOLDEN* were selected. Meanwhile, quantitative real−time (qRT)−PCR was used to analyze the expression of the candidate genes and genes in the carotenoid regulation pathways, indicating how *BrGOLDEN* influences other genes. This study lays a foundation for the elucidation of the molecular mechanisms of carotenoid biosynthesis in the golden inner leaves of the *B. rapa* line, which may help determine the carotenoid regulation pathways.

2. Results

2.1. Phenotypic Evaluation and Genetic Analysis

Line '1900264′, a commercial *B. rapa* variety, is a male−sterile hybrid line (P1, called F_1, parental information is unknown) with golden inner leaves (Figure 1a–c). Line '1900262′ (P2) is a highly inbred *B. rapa* line that has non−golden inner leaves and no significant carotenoid accumulation in the short stem tissue (Figure 1d–f). The morphology and growth of the two Chinese cabbage lines are similar in the seedling and rosette stages (Figure 1g–j). However, compared with line '1900264′, line '1900262′ is an early heading material, with a higher degree of heading and relatively curly leaves (Figure 1k,l). The tri−crossed hybrid lines were obtained through '1900264′ × '1900262′. Then, the inner leaf phenotypes of the tri−crossed hybrid line were investigated and counted and the golden and non−golden phenotypes conformed to a 1:1 segregation ratio at the Chi−squared test (χ^2 test) level (Table 1). Therefore, the golden inner leaf phenotype was controlled by a pair of dominant nuclear genes, which was named *BrGOLDEN*. The phenotypic investigation also showed that the 'golden circle' appeared in Chinese cabbage with golden inner leaves; we speculate that the 'golden circle' in short stem tissue may be linked to the golden inner leaf phenotype (Figure 1a–f).

Table 1. Genetic analyses of the tri−crossed hybrid populations.

Populations	Generations	Total Plants	Non−Golden Inner Leaves	Golden Inner Leaves	Expected Ratio	χ^2 Test (Chi−Squared Test)
Tri−crossed hybrid lines	P_1 (1900262, aa genotype)	20	20			
	P_2 (1900264, Aa genotype)	20		20		
	$P_2 \times P_1$	151	80	71	1:1	0.54

2.2. Comparison of Microstructure and the Carotenoid Component

The microstructures of the short stems and the inner leaves at the maturity stage were analyzed by observing microsection images of the '1900264′ and '1900262′ lines. In line '1900264′, more chromoplasts were aggregated in the xylem cells of the short stem, but fewer chromoplasts were produced in the parenchyma cells (Figure 2a) and, in the inner leaves, the golden chromoplasts were distributed around all cells (Figure 2c). However, this distribution of chromoplasts was not observed in the cells of the short stem and the inner leaves in the '1900262′ line, whose inner leaves were the non−golden phenotype (Figure 2b,d).

Figure 1. Morphological observation of two *Brassica rapa* lines. Sectional view of the '1900264′ line (**a**) and '1900262′ line (**d**) at the mature stage (14 weeks). Phenotypes of short stems in the '1900264′ line (**b**,**c**) and the '1900262′ line (**e**,**f**) at the mature stage (14 weeks). Phenotype of the '1900264′ line grown for 4 (**g**), 6 (**i**), and 10 (**k**) weeks. Phenotype of the '1900262′ line grown for 4 (**h**), 6 (**j**), and 10 (**l**) weeks.

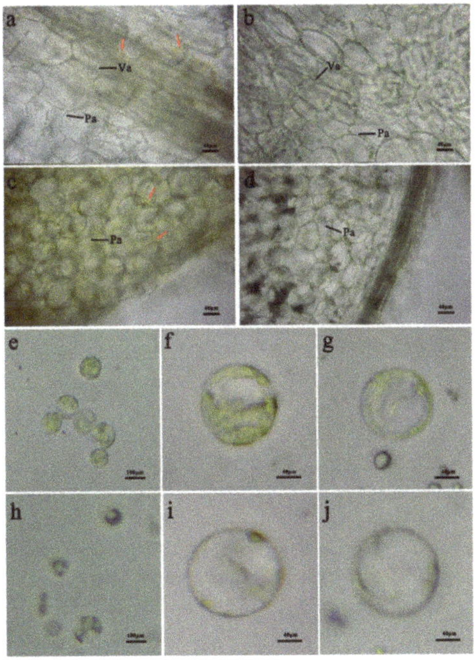

Figure 2. Microstructure of the inner leaves and short stems in the '1900262′ and '1900264′ lines. Microstructure of the short stem cells in the '1900264′ (**a**) and '1900262′ lines (**b**). Microstructure of the inner leaf cells in the '1900264′ (**c**) and '1900262′ lines (**d**). Inner leaf protoplasts of the '1900264′ (**e**) and '1900262′ lines (**h**) under 10× light microscope. Inner leaf protoplasts of the '1900264′ (**f**) and '1900262′ lines (**i**) under 40× light microscope. Short stem protoplasts of the '1900264′ (**g**) and '1900262′ lines (**j**) under 40× light microscope. The point of the red arrow indicates the chromoplasts. Va: vascular bundle; Pa: parenchyma cell.

The protoplasts were separated from the short stem and inner leaf tissues of the parent lines and observed under a light microscope. Many chromoplasts were distributed in the short stem and inner leaf cells of the '1900264′ line (Figure 2e–g). Chromoplasts were not observed in the short stems of the non−golden inner leaves of the '1900262′ line, but a few chromoplasts were found in the protoplasts of the inner leaves (Figure 2h–j). This may be because the '1900262′ line has light yellow inner leaves and a low carotenoid content; thus, the presence of a few chromoplasts was observed. Thus, the chromoplasts contained in these cells may be related to the color of the short stems and the inner leaves in the golden *B. rapa* line.

To analyze the carotenoid compositions of the inner leaves, the 5 cm long inner leaves of the two parent lines ('1900264′ and '1900262′ lines) at the mature stage were used to detect carotenoid components and content. Using the liquid chromatography−tandem mass spectrometry (LC−MS/MS) platform, 20 and 46 carotenoid components were subjected to absolute and relative quantification, respectively (Table S1). Among them, eight components were primarily detected and the contents of the other components were relatively low. Eight carotenoid components were detected. The total carotenoid content of the '1900264′ line was 3.3−times that of the '1900262′ line (Figure 3). Except for neoxanthin, violaxanthin, and antheraxanthin, the contents of the other five carotene components differed between the '1900264′ and '1900262′ lines. The '1900264′ line was rich in β−carotene and lutein, which accounted for 35.2% and 56.2% of the total carotenoid content, respectively. Compared with the '1900262′ line, the content of β−carotene increased by 13.6−fold. Interestingly, colorless (E/Z)−phytoene, as the upstream product of carotenoid biosynthesis [40], was detected only in the '1900264′ line.

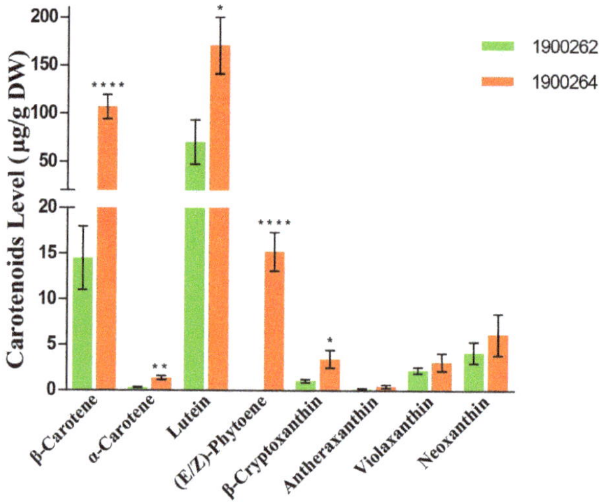

Figure 3. Comparison of carotenoid content in the inner leaves of the '1900262′ and '1900264′ lines. Data are the average ± SE of three biological replicates. Asterisk (*) indicates that there is a statistical difference, * indicates $p \leq 0.05$; ** indicates $p \leq 0.01$; **** indicates $p \leq 0.0001$.

2.3. Primary Mapping for the Golden Inner Leaf Phenotype by BSA−seq

The BSA method can rapidly identify markers linked to any specific gene or genomic region [41]. Forty golden and non−golden individuals were selected from the tri−crossed hybrid lines to construct the golden pool (GP) and non−golden pool (NGP), respectively. After resequencing, quality control, assembly, and alignment with the *B. rapa* reference genome V3.0 (http://brassicadb.org, accessed on 10 February 2021), the GP contained 179.87 Mb reads (98.39% coverage) and the NGP contained 180.56 Mb reads (98.41% coverage) (Table S2). Using association analysis of the Euclidean distance (ED) algorithm [42] and the △index algorithm [43], the intersection association area corresponding to all SNPs and Indels was obtained. Finally, *BrGOLDEN* was primarily located in the 2.2 Mb region of A09 (Figure 4).

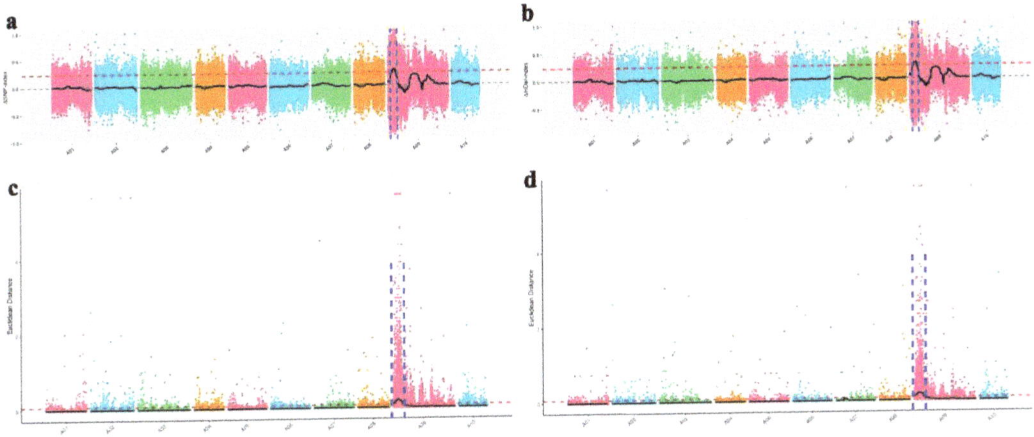

Figure 4. Genetic mapping of the golden inner leaf phenotype in *B. rapa*. Identification of the *BrGOLDEN* candidate region through the △SNP index (**a**) and △InDel index; (**b**) association analysis method. *BrGOLDEN* candidate regions identified by Euclidean distance (ED) and SNP (**c**) or ED and InDel (**d**) association analysis. The higher the index value or ED value, the better the association effect.

2.4. Verifying the BrGOLDEN Candidate Gene

Based on the BSA−seq mapping results, the candidate region of 2.2 Mb was identified in combination with the *B. rapa* and *A. thaliana* databases. A total of 532 genes was annotated within the candidate region (Table S4). The *BraA09g007080.3C* at the anterior end of A09 was homologous to *AT5G61670* (*AtOR*) in *A. thaliana*. In cauliflower, *BoOR* is a dominant gene regulating carotene accumulation [44]. Similar to *AtOR*, *BraA09g007080.3 C* contained a DnaJ−like zinc finger domain. In a previous study, *AtOR* interacted directly with *PSY* (phytoene synthase), which acts as a positive posttranscriptional regulator to control carotenoid biosynthesis [45,46]. Thus, *BraA09g007080.3C* is likely the candidate gene *BrGOLDEN* associated with inner leaf color.

To analyze the *BraA09g007080.3C* sequences of possible candidate genes in the '1900264′ and '1900262′ lines, the gDNA and CDS were amplified and sequenced using specific primers (Table S4). In line '1900262′, the gDNA of the *Brgolden* was 1648 bp, contained eight exons and seven introns, and two transcripts were obtained from it. $Brgolden_{wt}$ was 918 bp and encoded a protein with 305 amino acids. $Brgolden_{Lins}$ (Lins, large insert transcript), with the second intron (61 bp) fully inserted, results in a frameshift mutation that prematurely terminates translation and, presumably, this transcript may be non−functional (Figure 5a). The reverse transcription (RT)−PCR results showed that $Brgolden_{wt}$ was more abundant than $Brgolden_{Lins}$ in the '1900262′ line (Figure 5b). In line '1900264′, sequence analysis of *BrGOLDEN* showed that a large fragment of 4.67 kb was inserted at 558 bp and the insertion event led to the production of three alternative splicing transcripts. During the transcription process, only 21 bp remained at the 5′ end of the large insertion fragment. $BrGOLDEN_{Ins}$ (Ins, insert transcript) contained an additional 18 bp insertion in the third exon. $BrGOLDEN_{Del}$ (Del, deletion transcript) contained a 39 bp deletion in the third exon and $BrGOLDEN_{Ldel}$ (Ldel, large deletion transcript) contained a 126 bp large deletion in the third and fourth exons (Figure 5a,c). The relative expression level of $BrGOLDEN_{Del}$ was the most abundant in the '1900264′ line (Figure 5b). Amino acid sequence alignment showed that these transcripts were highly homologous and all contained four conserved C××C×××G motifs (Figure S1). Thus, we propose that *BraA09g007080.3C* is the candidate gene for golden and non−golden inner leaf color in Chinese cabbage.

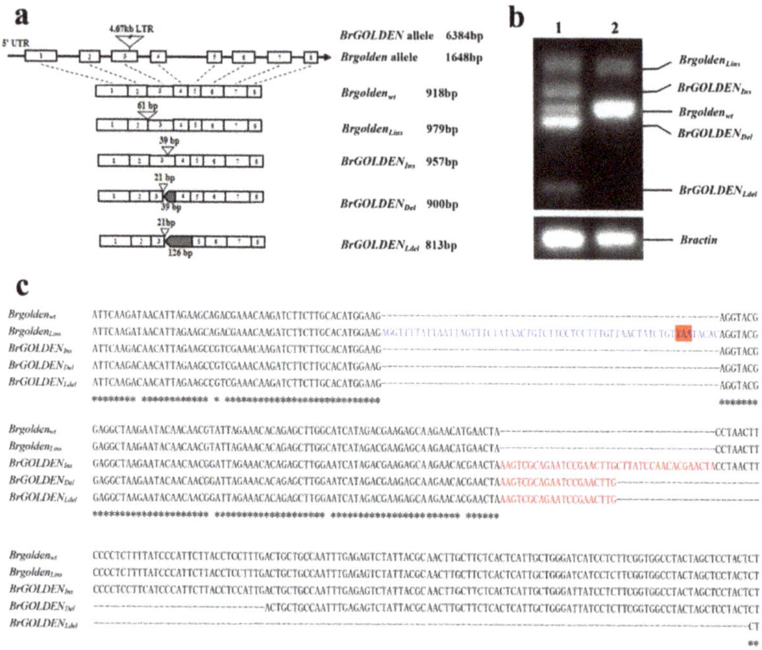

Figure 5. *BrGOLDEN* structure analysis. (**a**) The *BrGOLDEN* with the 4.67 kb LTR insertion site and different transcripts. Open boxes with numbers represent exons and solid lines represent introns. Triangles indicate insertion sites and grey boxes indicate deletion sites. *Brgolden$_{Lins}$*, large insertion; *BrGOLDEN$_{Ins}$*, insertion; *BrGOLDEN$_{Del}$*, deletion; *BrGOLDEN$_{Ldel}$*, large deletion. (**b**) RT−PCR amplification of *BrGOLDEN* in the '1900264′ (lane 1) and '1900262′ lines (lane 2) cDNAs. *Bractin* served as an internal control. (**c**) Nucleotide sequence alignment of five transcript fragments. The blue words represent the transcripts resulting from the complete insertion of the second intron in the '1900262′ line and the red box indicates the translation end site. The red words and dotted lines indicate the splicing patterns of the three transcripts unique to the '1900264′ line.

The *OR* homologous gene *BrGOLDEN* in *B. rapa* was selected as a candidate gene for regulating carotenoid accumulation. OR is unique to plants and there were obvious differences among various plant families [44]. To analyze the evolutionary relationships of *BrGOLDEN* in horticultural crops, the sequences were aligned in different species. A highly conserved Cys−rich DnaJ domain was present in Cruciferae, Cucurbitaceae, and Solanaceae, among others. Phylogenetic analysis showed that *B. rapa* (*BraA09g007080.3C*) was most closely related to *Brassica juncea*, *Brassica napus*, *Brassica oleracea*, and *Brassica carinata* (Figure 6).

2.5. Relative Expression Level Analysis of BrGOLDEN and Carotenoid Synthesis Pathway Genes

To investigate whether *BrGOLDEN* promoted carotenoid accumulation by affecting the expression of carotenoid−synthesis genes, the relative expression levels were analyzed, including the *BrGOLDEN/Brgolden* and 12 carotenoid synthesis pathway genes (*ZDS*, *CR-TISO*, *VDE*, *LCYB*, *ZEP*, *DXS*, *PSY*, *GGPPS11.1*, *GGPPS11.2*, *BCH2*, *CYP97C1*, and *CYP97A3*) (Figure 7e). The proteins encoded by these genes were extensively involved in carotenoid synthesis and catalytically produced carotenoid components were also significantly detected. RNA was extracted and reverse transcribed into cDNA from four tissues (the outer leaves (OL), inner leaves (IL), short stems (S), and roots (R)) and in three periods (seedling, rosette, and heading stages) in the '1900262′ and '1900264′ lines. Compared with *Brgolden*, the relative expression levels of *BrGOLDEN* were not significantly higher, in the four tissues from the three different periods (Figure S2a). Compared with IL, S, and R, the relative expression levels of *BrGOLDEN* and *Brgolden* in OL were highest (Figure S2a,b). Twelve key genes of the carotenoid synthesis pathway were analyzed at maturity. Compared with line '1900262′, the relative expression level of *DXS* was significantly higher in the roots of line '1900264′ (Figure 7a). However, these genes were not significantly different in the short stems of either line (Figure 7b). In particular, the relative expression levels of four genes (*ZEP*, *DXS*, *CYP97C1*, and *PSY*) were higher in the inner leaves of the '1900264′ line than in the '1900262′ line (Figure 7c). In addition, the relative expression levels of *ZEP* were similar between lines '1900262′ and '1900264′ but were significantly higher than the other 11 genes in the outer leaves (Figure 7d). Previous studies found similar transcript levels of carotenoid biosynthesis genes in cauliflower and *A. thaliana* [47,48]. Thus, *BrGOLDEN* may promote carotenoid accumulation by regulating the expression of genes in the carotenoid synthesis pathway in *B. rapa*.

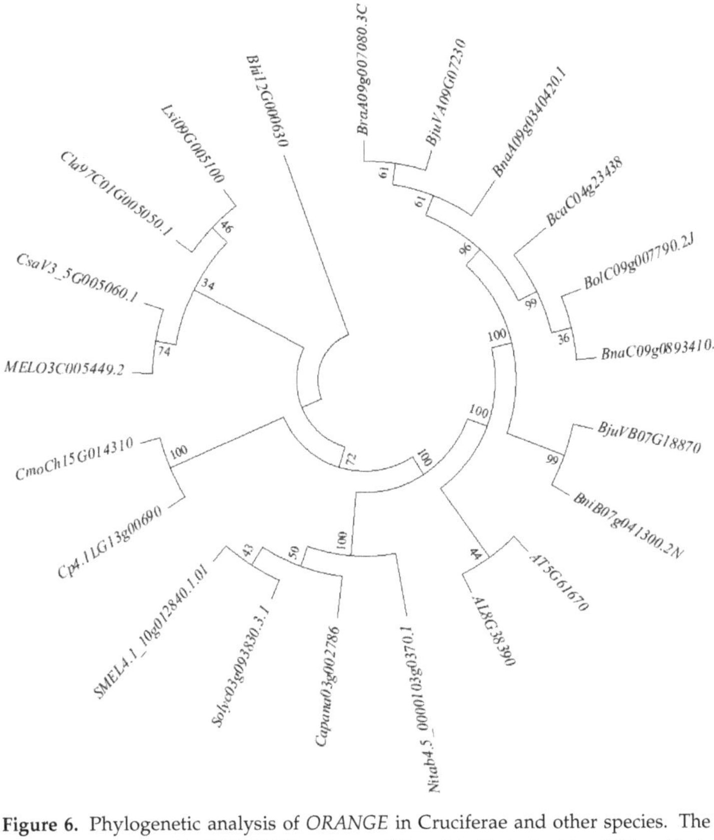

Figure 6. Phylogenetic analysis of *ORANGE* in Cruciferae and other species. The phylogenetic tree was constructed in MEGA 7.0 software (Mega Limited, Auckland, New Zealand) using the neighbor–joining method. The bootstrap replication values were determined from 1000 replicates.

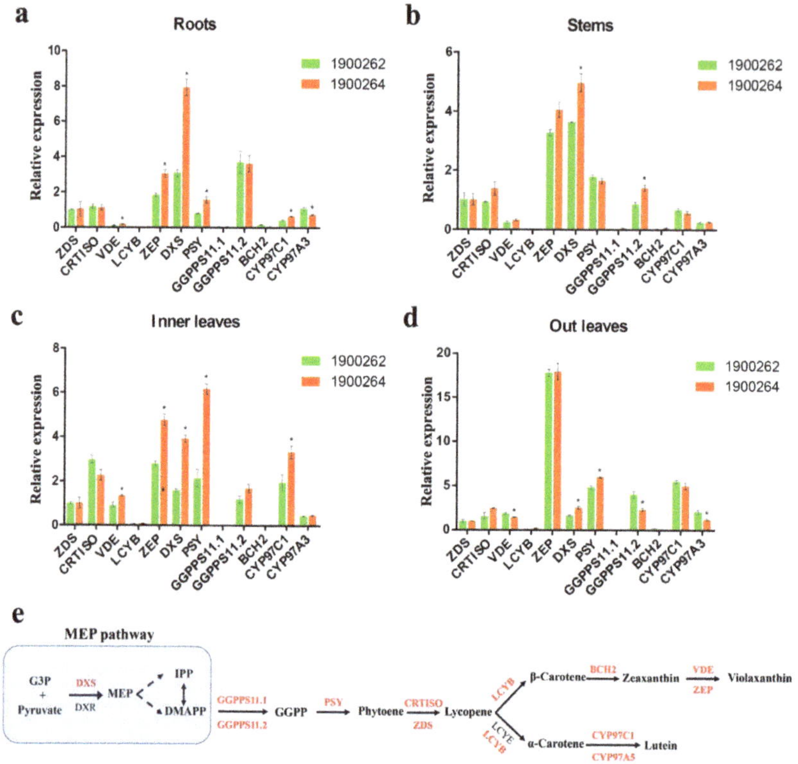

Figure 7. Expression of carotenoid biosynthetic genes in lines '1900262' and '1900264'. Relative expression levels of carotenoid−synthesis−related genes in the roots (**a**), short stems (**b**), inner leaves (**c**), and outer leaves (**d**) of lines '1900262' and '1900264'. *ZDS*, *BraA05g040310.3C*; *CRTISO*, *BraA09g063710.3C*; *VDE*, *BraA06g005700.3C*; *LCYB*, *BraA03g017230.3C*; *ZEP*, *BraA07g016880.3C* and *BraA07g016890.3C*; *DXS*, *BraA01g021140.3C*; PSY, *BraA02g006890.3C*; *GGPPS11.1*, *BraA08g021280.3C*; *GGPPS11.2*, *BraA01g002000.3C*; *BCH2*, *BraA10g010930.3C*; *CYP97C1*, *BraA07g021850.3.5C*; *CYP97A3*, *BraA08g009810.3C*. An asterisk (*) indicates that there is a statistical difference, $p \leq 0.05$, $n = 3$. *Bractin* served as a control. (**e**) Outline of the carotenoid biosynthetic pathway. Red words indicate genes validated in the experiment.

2.6. Brgolden Exhibits Distinct Expression Patterns

The *cis*−acting element of the *Brgolden* promoter was predicted and transgenic plants with a β−galactosidase (GUS) reporter driven by the *Brgolden* promoter ($Pro_{Brgolden}$:GUS) were generated in *A. thaliana*. Basic elements, such as central promoter elements (TATA−box) and enhancer elements (CAAT−box) necessary for eukaryotes, were included in the *Brgolden* promoter. In addition, there were several specific regulatory elements, including five light−responsive−related elements (GATA−motif, Box 4, TCT−motif, I−box, and G−box), two types of hormone−responsive *cis*−elements (TCA−element and ABRE), elements required for anaerobic induction (ARE), and a specific element of the roots (motif I) and seeds (RY−element) (Table 2). The GUS staining results of the three−week−old transgenic *A. thaliana* plants showed that GUS was expressed in the leaves, leaf veins, and petioles, but less so in the roots (Figure 8a,b). At the mature stage, there was higher GUS activity in the calyx, stigma, seed pods, and the junctions between the seed pods and seed stalks, whereas there was less GUS expression in the petals (Figure 8c–f). In summary, the *Brgolden* promoter may be a constitutive promoter driving *GUS* expression in different tissues.

Table 2. Analysis of *cis*−acting elements in the *Brgolden* promoter.

Component Name	Core Sequence	Numbers	Predictive Function
MBS	CAACTG	2	MYB binding site involved in drought-inducibility
RY-element	CATGCATG	1	Cis-acting regulatory element involved in seed-specific regulation
GATA-motif	GATAGGA	1	Part of a light-responsive element
TCA-element	CCATCTTTTT	1	Cis-acting element involved in salicylic acid responsiveness
Motif I	gGTACGTGGCG	1	Cis-acting regulatory element root specific
Box 4	ATTAAT	1	Part of a conserved DNA module involved in light responsiveness
ARE	AAACCA	4	Cis-acting regulatory element essential for the anaerobic induction
ABRE	ACGTG	3	Cis-acting element involved in the abscisic acid responsiveness
TATA-box	TATATA	19	Core promoter element around −30 of transcription start
TATA-box	TATA	15	
TATA-box	TATAAAA	5	
TATA-box	TATTTAAA	1	
TATA-box	TATAA	3	
TATA-box	ccTATAAAaa	1	
TCT-motif	TCTTAC	2	Part of a light-responsive element
CAAT-box	CCAAT	9	Common cis-acting element in promoter and enhancer regions
CAAT-box	CAAT	10	
I-box	GTATAAGGCC	1	Part of a light-responsive element
G-box	CACGTG	2	Cis-acting regulatory element involved in light responsiveness
G-box	TACGTG	1	
	motif_sequence	10	Short-function
W box	TTGACC	1	Cis-acting regulatory element essential for the pathogen induction

Figure 8. β−Galactosidase (GUS) staining of different tissues of $Pro_{Brgolden}$:GUS transgenic *Arabidopsis thaliana*. Three−week−old seedlings of $Pro_{Brgolden}$: GUS (**a**) and wild−type lines (**b**). Mature silique of $Pro_{Brgolden}$: GUS (**c**) and wild−type lines (**d**). Flowers of $Pro_{Brgolden}$: GUS (**e**) and wild−type lines (**f**).

2.7. Ectopic Overexpression of *BrGOLDEN* Increases Carotenoid Accumulation in *A. thaliana* Calluses

The function of *BrGOLDEN* was analyzed by constructing 35S:*BrGOLDEN* overexpressing (OE) vectors and transforming them into the *A. thaliana* wild type (*Col*−0). From the obtained transgenic lines, one OE representative line of each alternative splicing transcript (35S:$BrGOLDEN_{Ins}$, 35S:$BrGOLDEN_{Del}$, and 35S:$BrGOLDEN_{Ldel}$) was selected for the observation of phenotypic changes. All transgenic *A. thaliana* mutants were similar to the wild type in plant morphology and leaf phenotype (Figure 9a), which was consistent with the results of the wild type and $BoOR_{MUT}$ *B. oleracea* [47]. This also indicates that carotenoid composition and content are in a dynamic balance in the leaf tissue for optimal photosynthesis [49]. Compared with the leaves, non−green tissues, such as calluses, tend to respond phenotypically to an increased carotenoid content [50]. In recent years, calluses have been used to rapidly characterize the function of genes involved in carotenoid biosynthesis and accumulation [51]. Therefore, calluses were induced from the seeds of these overexpressing *A. thaliana* lines and the wild type (Figure 9b). Under the same conditions, the calluses induced by 35S:$BrGOLDEN_{Ldel}$ and the wild type were nearly transparent or white, while the calluses of $BrGOLDEN_{Ins}$ and $BrGOLDEN_{Del}$ were yellow, inferring that there was an accumulation of carotenoids. Thus, the function of *BrGOLDEN* was determined by different spliced transcripts.

Figure 9. Phenotypic observations of plants and seed−derived calluses of *A. thaliana* lines overexpressing *BrGOLDEN* lines. (**a**) Rosette stage phenotypes of 35 S:*BrGOLDEN* transgenic and wild−type *A. thaliana* seedlings at four weeks. (**b**) Representative images from seed−derived calluses of *BrGOLDEN* overexpressing lines and the wild type.

2.8. Identification of the Target Proteins Involved in BrGOLDEN Interactions

PSY is the rate−limiting enzyme in the carotenoid synthesis pathway and its activity greatly affects carotenoid accumulation. Previous studies have shown that the N−terminus of OR physically interacts with PSY and positively regulates the PSY level and enzymatic activity, while the C−terminus forms a dimeric structure [46,52]. Treatment with the protein synthesis inhibitor cycloheximide (CHX), OR, was found to stabilize PSY, greatly reduce the protein turnover rate, and significantly increase carotenoid levels [53]. In this study, $Brgolden_{wt}$ was a double−transmembrane protein (Figure 10a). Due to the different splicing modes of the N−terminus of the three alternative splicing transcripts, the integrity of the N−terminal and the transmembrane domain was destroyed (Figure S3). Therefore, the split−ubiquitin membrane Y2H system was used to verify whether $Brgolden_{wt}$,

BrGOLDEN$_{Ins}$, and BrGOLDEN$_{Del}$ interacted with BrPSY1. The interaction between BrGOLDEN$_{Ldel}$ and BrPSY1 was verified using Y2H based on the nuclear interaction system (Figure 10b). Interestingly, although BrGOLDEN$_{Ins}$, BrGOLDEN$_{Ldel}$, and Brgolden$_{wt}$ exhibited strong interactions with BrPSY1, BrGOLDEN$_{Del}$ did not interact with BrPSY1 in the split−ubiquitin membrane system. In addition, after BrGOLDEN$_{Del}$−PBT3−N and BrPSY1−pPR3−N co−transformed yeast strains through the membrane interaction system, the negative and positive controls had the expected results. The strain on the DDO plate grew well, indicating that it was successfully transferred into the host cells. However, they did not grow on either the QDO or TDO plates supplemented with 5 mM 3AT, which was verified with replication. There was no interaction between BrGOLDEN$_{Del}$−PBT3−N and BrPSY1−pPR3−N in the split−ubiquitin membrane system. The amino acid deletion of BrGOLDEN$_{Del}$ played an important role in its interaction with BrPSY1.

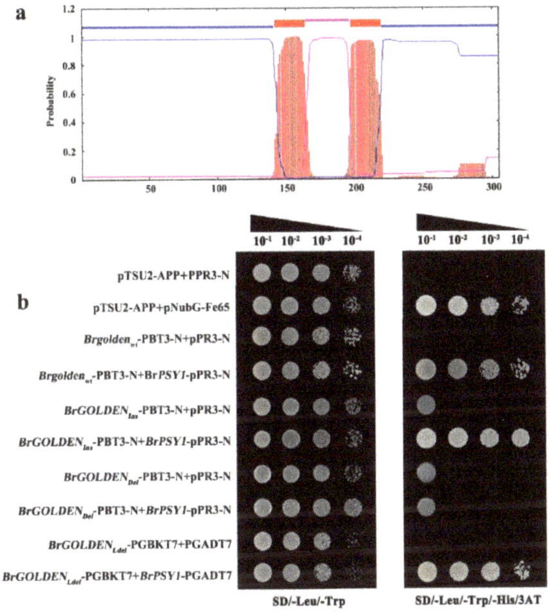

Figure 10. Interactions of BrPSY1 with BrGOLDEN. (**a**) Prediction of transmembrane helices in Brgolden$_{wt}$. The red line represents the transmembrane. The blue line represents the inside. The pink line represents the outside. Positions 142–164 and 197–219 amino acids represent transmembrane helices. The abscissa indicates the amino acid site. (**b**) Co−transformation of prey and bait vectors in various combinations validates protein interactions. Co−transformation of pPR3−N and pTSU2−App vectors was used as a negative control and co−transformation of pTSU2−APP and pNubG−Fe65 vectors was used as a positive control. Dilutions (1–10^{-4}) of different concentrations of the saturated cultures were spotted onto SD/−Leu/−Trp/−His/−3AT plates.

Compared with Brgolden$_{wt}$, the amino acid sequence of BrGOLDEN$_{Del}$ lost the PNFPSFIPFLPPL sequence at position 125 amino acids and was replaced by the sequence KSQNPNL (Figure S1). The protein secondary structures of BrGOLDEN$_{Del}$ and Brgolden$_{wt}$ were predicted by Phyre2 (Figure S4) and included 11 β−strands and seven and nine α−helices, respectively. Due to sequence deletions, the important α−helix of the secondary structure of BrGOLDEN$_{Del}$ was disrupted. The missing protein structure is located on the surface of the 3D structure, which plays an important role in the interactions between BrGOLDEN and BrPSY1. Using 3D structure prediction analysis, the Brgolden$_{wt}$ structure was predicted successfully and BrGOLDEN$_{Del}$ was not, which indicated that the loss of the sequences at position 125 amino acids limited its function. In addition, by analyzing the

3D structure of $Brgolden_{wt}$, this protein may be a tryptophan RNA−binding attenuator protein inhibitory protein with three molecular functions (binding, protein binding, and identical protein binding) and three cellular components (intracellular anatomical structure, cytoplasm, and cellular anatomical entity) [54]. This study further verified that the 3D structure of the protein was the most direct factor affecting the functions of the protein. Due to the mutation of several amino acids in $BrGOLDEN_{Del}$, the 3D structure of this protein was greatly changed, affecting the interaction between $BrGOLDEN_{Del}$ and BrPSY1.

3. Discussion

The synthesis of carotenoids is very important in plants and the various 40−carbon compounds produced by metabolic pathways provide plants with red, orange, and yellow colors and give them ornamental value. The varied inner leaf colors of Chinese cabbage are attributed to different carotenoid compositions and concentrations. Due to the nutritional and health−related properties of carotenoids, research on carotenoid metabolism in Chinese cabbage inner leaves is an active field. At the same time, carotenoids are widely involved in growth and development processes, such as photosynthesis and plant stress. Studies have shown that carotenoids are synthesized in chromoplasts and are mainly found in non−green tissues [55]. In this study, a dominant gene *BrGOLDEN*, controlling carotenoid accumulation, was identified for the first time in the Chinese cabbage line with golden inner leaves. The results showed that *BraA09g007080.3C*, which was homologous to *AtOR*, was considered a candidate gene for *BrGOLDEN*. ORANGE is an important protein that stabilizes the rate−limiting enzyme PSY for carotenoid synthesis, which converts GGPPs to phytone [56,57]. As PSY protein level is crucial for overall pathway activity and carotenoid amounts, a tight control of PSY proteostasis is expected. Studies have showed that OR can also regulate the plastid−localized protein degradation machinery, namely the Clp protease complex, to maintain PSY proteostasis and fine−tune carotenogenesis [46,53].

In a previous study, the homolog *BoOR* of *BrGOLDEN* was reported in *B. oleracea* [44]. Comparing the sequences of *BrGOLDEN* with *BoOR*, an insertion event and three alternative splicing transcripts of *BrGOLDEN* were very similar to the *BoOR* mutant (Figures S5 and S6). In *B. oleracea*, a large fragment from C06 was inserted into *BoOR* in C09 and an extra 16 bp was generated, resulting in a 4686 bp fragment insertion. Unlike *BoOR*, *BrGOLDEN* retained only the entire insertion fragment sequence (4670 bp). RT−PCR was used to detect regulatory elements, including the *CaMV* 35S promoter, *FMV* 35S promoter, *NOS* promoter, *NOS* terminator, and *CaMV* 35S terminator, according to the testing method for genetically modified new varieties (Ministry of Agriculture and Rural Affairs of the People's Republic of China) and none of the target amplification products were detected in the parents (Figure S7). Thus, it was inferred that *BrGOLDEN* was not transformed into *B. rapa* by transgenic technology. Rather, it was transferred into *B. rapa* through distant hybridization between *B. rapa* and *B. oleracea*.

The large insertion found was a long terminal repeat (LTR) belonging to transposons, which are common in plants. Some studies have been reported in *A. thaliana* [58], *B. rapa* [59], and *Oryza sativa* [60], and the proportion of transposon−related sequences in each genome accounts for 10%, 39.5%, and 46%, respectively. LTRs are the most abundant transposon type in the *B. rapa* genome, accounting for 27.1% [59]. In animals and plants, transposons are extremely rich in functions, mainly mediating the formation of new genes, gene disruption, and regulation of gene expression or activity. The effects of transposons on genes are not only manifested at the transcriptional level but also at the post−transcriptional level [61]. These functions provide a theoretical basis into the regulation of carotenoid metabolism by *BrGOLDEN*.

The expression level of *Brgolden* in golden inner leaf Chinese cabbage was not more significant than that of *BrGOLDEN*. It is speculated that the accumulation of carotenoids is due to the new function acquired by the *BrGOLDEN* mutation, rather than the expression level. In plants, the *ORANGE* mutant exerts its functions through a variety of pathways. The differentiation of membranous chromoplasts can be induced by $AtOR^{His}$ or $BoOR_{mut}$, but not by *AtOR* or *BoOR* [48,62]. Alternatively, *ORANGE* enhanced the accumulation and preservation of β−carotene and lutein in the chromoplasts, turning the tissue golden [48]. Similar results were observed in Chinese cabbage. Interestingly, in line '1900264', the outer leaf tissue with high expression of the *BrGOLDEN* appeared green, similar to the control. The short stems with lower expression showed an obvious golden color, forming a 'golden circle' structure. Overexpression of the *PSY* encodes a key rate−limiting enzyme for carotenoid production in *A. thaliana*, but the effect of *PSY* overexpression on carotenoid accumulation in leaves is extremely small [50]. This may be because the composition and content of carotenoids are highly conserved in the green tissues, thus, limiting chromoplast formation [49]. We speculate that these two genes may limit their functions through the same pathway in green tissue.

ORANGE can promote membranous chromoplast formation and carotenoid accumulation in many plants [44,48,63]. For example, the overexpression of $BoOR_{MUT}$ in potato tubers produces orange tissue with increased carotenoid content. The mutation of a single amino acid in the *OR* ($CmOR^{His}$) of melon can lead to a change in callus color and an increase in carotenoid content in *A. thaliana* and can specifically induce the formation of membranous chromoplasts [64]. In our study, *BrGOLDEN* may be derived from the cauliflower orange curd mutant but with some differences from $BoOR_{MUT}$, mainly reflected in the size of the LTR fragment and SNP variation in the transcripts. $BrGOLDEN_{Ins}$ and $BrGOLDEN_{Del}$ could act alone to turn the callus golden in the induced *A. thaliana* callus and the deletion of the large fragment of $BrGOLDEN_{Ldel}$ limited this function. It may exert other effects by interacting with *BrPSY1* in the nucleus or acting through a *BrPSY1*−independent pathway. PSY is a vital enzyme in carotenoid synthesis and the overexpression of *AtOR* and $AtOR^{His}$ did not cause changes in the transcription levels of *PSY* and other enzyme genes in *A. thaliana*. Previous studies have demonstrated that *OR* may post−transcriptionally regulate the protein level of PSY [48]. This provides a reasonable explanation for the promotion of carotenoid accumulation in *B. rapa* by *BrGOLDEN*. Interestingly, in our study, $BrGOLDEN_{Del}$ did not interact with BrPSY1 in the membrane system, suggesting that it may promote carotenoid accumulation through a PSY1−independent pathway. These results contribute to diversity of molecular mechanisms of carotenoid synthesis and accumulation and also provide new insight for the improvement and marker−assisted selection (MAS) of carotenoid−rich Chinese cabbage varieties.

4. Materials and Methods

4.1. Plant Materials and Callus Induction

The '1900264' line, a commercial *B. rapa* variety, is an F_1 cytoplasmic sterile hybrid line with golden inner leaves and a 'golden circle' in the short stem tissue. Line '1900262' is a highly inbred *B. rapa* line with non−golden inner leaves and no significant carotenoid accumulation in the short stem tissue. The tri−crossed hybrid lines were obtained from the cross '1900264' × '1900262'. All lines were cultivated in the Shunyi Base of the Chinese Academy of Agricultural Sciences. Plant samples at different stages were collected for DNA, RNA, and carotenoid analysis. The samples were frozen in liquid nitrogen for later use.

The *A. thaliana* (Col−0) was used for function verification. The *A. thaliana* seeds were plated on Murashige and Skoog (MS) medium containing 0.8% agar and 3% sucrose. The seeds were vernalized at 4 °C for 3D and then transferred to a constant temperature incubator at 22 °C with a photoperiod of 16 h light/8 h dark. After 10 d of growth, the seedlings were transplanted to the substrate until flowering.

To induce calluses, *A. thaliana* seeds were sterilized with 5% (*v*/*v*) NaClO for 15 min, washed 5–6 times with sterilized ddH_2O and sown on seed−derived callus (SDC) medium (4.33 g/L MS basal salts/pH to 5.8 by 1 M KOH, 0.1% (*v*/*v*) Gamborg's vitamin, 3% (*w*/*v*) sucrose, 0.5 mg/L 2,4−D, 2 mg/L indole−3−acetic acid, 0.5 mg/L 2−isopentenyladenine, 0.4% (*w*/*v*) agar) as previously described [50]. The seeds were germinated under 16 h of light/8 h of dark at 26 °C for 5 d and grown in the dark for 2 weeks in a growth incubator. Callus samples were collected, frozen in liquid nitrogen, and stored at −80 °C.

4.2. Genetic Analysis of the Golden Inner Leaf Phenotype

For genetic analysis, the non−golden inner leaf line '1900262′ (P_1) was crossed with the golden inner leaf line '1900264′ (P_2) to generate the tri−crossed hybrid lines. Phenotype characterization was performed for each generation (P_1, P_2, the tri−crossed hybrid lines) and the segregation ratios of the tri−crossed hybrid lines were analyzed using the chi−square (χ^2) test. All phenotypes and samples were analyzed at the Shunyi Base, Chinese Academy of Agricultural Sciences.

4.3. Protoplast Separation, Microstructure Observation, and Carotenoid Analysis

In this study, the microstructure was observed by isolating protoplasts and producing free−hand slices. Protoplasts were isolated by vacuum infiltration for 30 min and gently shaking for 3 h, with 1.5 g tissue in 10 mL of a solution of 0.4 M mannitol, 1.5% (*w*/*v*) cellulose, 0.4% (*w*/*v*) macerozyme (enzymes were manufactured by Yakult Pharmaceutical Industry), 20 mM KCl, MES (pH 5.7), and 10 mM $CaCl_2$. Samples were centrifuged at 100× *g* for 2 min and resuspended with 8% mannitol for observation [65]. At the mature stage, tissue sections of the inner leaves and short stems of the two parent materials were made. Sections were mounted in water under a coverslip, observed, and photographed under a light microscope (Carl Zeiss AG, Oberkochen, Germany).

The carotenoid content in the parental lines ('1900264′ and '1900262′) was detected using the AB Sciex QTRAP 6500 LC−MS/MS platform. In the heading stage (14 weeks), Chinese cabbage inner leaves (~5 cm) were freeze−dried to a constant weight. The dried samples were then homogenized and powdered in a mill. The content of carotenoid components in Chinese cabbage was determined using previously described methods [66]. Three biological replicates were tested for each sample.

4.4. BSA−seq Mapping

The tri−crossed hybrid lines, comprising 151 individuals, were derived from a cross between lines '1900262′ and '1900264′. From the populations, equal amounts of DNA were pooled from 40 plants with a non−golden inner leaf phenotype, which constituted the non−golden pool (NGP) and from 40 plants with a golden inner leaf phenotype, which constituted the golden pool (GP). Two mixed pools and two parent pools were used for association analyses. Approximately 40× genome sequences for each pool were generated using the Illumina HiSeq 2500 platform (Illumina, San Diego, American) [67]. The *B. rapa* V3.0 and *A. thaliana* V10.0. were used as the reference genomes. The △SNP index and △Indel index of each position were calculated for sliding window analysis [68,69]. The SNPs and InDels were detected with GATK [70]. Gene functions were annotated with BLAST software (EMBL-EBI, Hinxton, UK) [71].

4.5. Cloning and Sequencing of Candidate Genes

The full−length *BrGOLDEN* sequence and CDS sequence were amplified from the gDNA and cDNA of lines '1900262′ and '1900264′ using the primers GOLDEN−F/R (Table S4). The PCR product was constructed for the T1 vector via the pEASY−T1 Cloning Kit (TransGen, Beijing, China). A single colony was selected and sent to Sangon Biotech (Sangon, Shanghai, China) for sequencing. Sequencing results were compared at http://abc.gao-lab.org/, accessed on 5 May 2021. Subsequently, transmembrane domain prediction of gene−coding proteins was performed (https://services.healthtech.dtu.dk/service.php?TMHMM-2.0, accessed on 12 May 2021). Protein structures were predicted on Phyre^2 (http://www.sbg.bio.ic.ac.uk/phyre2/html/page.cgi?id=index, accessed on 12 May 2021). Sequence alignment of ORANGE protein sequences was performed in Cruciferae and other species. The phylogenetic tree was constructed in MEGA 7.0 software (Mega Limited, Auckland, New Zealand) and the neighbor−joining method, Poisson correction, and pairwise deletion method were used [72].

4.6. *Quantitative Reverse–Transcription PCR*

The relative expression levels of genes were analyzed via qRT–PCR. The total RNA from the seedling, rosette, and heading stages of both Chinese cabbage lines was extracted using a Plant Total RNA Mini Kit (Gene Better, Beijing, China) following the manufacturer's instructions. The RNA quality was assessed using an RNA 6000 Nano Chip on an Agilent 2100 Bioanalyzer (http://www.chem.agilent.com, accessed on 22 April 2021). One microgram of total RNA was converted into cDNA using a reverse transcription kit (TransGen, Beijing, China). The qRT–PCR was performed using SYBR Green PCR Master Mix (Vazyme, Nanjing, China). The relative expression levels of all genes were calculated using the $2^{-\Delta\Delta CT}$ method using *Bractin* as a control [73]. Three technical replicates were performed for each experiment. The primers were designed using Premier 5 (Table S4).

4.7. *Construction of $Pro_{Brgolden}$:GUS Fusion Vector and Histochemical Staining of Transgenic Plants*

Based on the upstream sequence of *Brgolden* in the *B. rapa* genome database (http://brassicadb.cn, accessed on 9 September 2022), primers (Table S4) were designed to amplify the promoter sequence (~1.6 kb upstream) and ligated with the Easy–T1 vector. Prediction of promoter *cis*–acting elements was performed at Plant CARE [74]. After sequencing, the correct sequence was found by blasting in NCBI. The $Pro_{Brgolden}$:GUS expression vector was constructed by restriction enzymes *Sca* I and *BamH* I and then ligated with T4 ligase (TSINGKE, Beijing, China). The recombinant vectors were then transformed into *A. thaliana Col–0* using the floral–dip method [75]. *Arabidopsis thaliana* from different tissues was selected for histochemical staining, as previously described [76].

4.8. *Generation of Transgenic Plants Overexpressing BrGOLDEN*

The complete coding region of *BrGOLDEN* was amplified from the cDNA by PCR using a pair of primers with a homologous arm (Table S4). The PCR product was inserted into the pCAMBIA1305 vector digested with *BamH* I and *Xba* I using a Trelief SoSoo Cloning Kit (TSINGKE, Beijing, China) to generate pCAMBIA1305–$BrGOLDEN_{Ins}$, pCAMBIA1305–$BrGOLDEN_{Ldel}$, and pCAMBIA1305–$BrGOLDEN_{Del}$. The recombinant plasmid was then transferred into *E. coli* DH5α and sent to Sangon Biotechnology for sequencing to verify the correct insertion. We subsequently introduced the overexpression construct into *Agrobacterium tumefaciens* strain GV3101. pCAMBIA1305–$BrGOLDEN_{Ins}$, pCAMBIA1305–$BrGOLDEN_{Ldel}$, and pCAMBIA1305–$BrGOLDEN_{Del}$ were then mixed to transform *A. thaliana* [75].

4.9. *Y2H Assay*

The full–length coding sequences of $BrGOLDEN_{Ldel}$ and *BrPSY1* were cloned into pGBKT7 and pGADT7, respectively, following the Yeastmaker Yeast Transformation System 2 user manual (Clontech, Mountain View, CA, USA). Both plasmids were co–transformed into the golden yeast strain. The resultant strains were grown on plates for 3D at 30 °C. The interaction was tested via growth assays on media lacking leucine, tryptophan, and histidine but containing 3AT.

Interactions between BrPSY1 and $BrGOLDEN_{Ins}$, $BrGOLDEN_{Ldel}$, and $Brgolden_{wt}$ were investigated in yeast with the DUAL membrane starter system (Dual–Systems Biotech). Full–length coding sequences of $BrGOLDEN_{Ins}$, $BrGOLDEN_{Ldel}$, and $Brgolden_{wt}$ were cloned into the PBT3–N vector and *BrPSY1* was cloned into the pPR3–N vector. The primers are described in Table S4. The constructed bait vector and *BrPSY1* were co–transformed into yeast strain NMY51. All strains were grown in different media: (1) SD medium lacking Trp and Leu; (2) QDO (SD medium lacking Trp, Leu, His, and Ade); and (3) QDO + 3AT (QDO with 5 mM 3–amino–1,2,4–triazole). The empty pPR3–N and PBT3 vectors were co–transformed into the NMY51 strain and tested for autoactivation with QDO medium.

4.10. Statistical Analysis

GraphPad Prism 8 (San Diego, CA, USA) and Microsoft Office Excel 2010 software (Redmond, WA, USA) were used to analyze the data from carotenoid content and qRT−PCR. IBM SPSS 25.0 (Armonk, NY, USA) was used to evaluate the significant differences ($p < 0.05$). Multiple *t*-tests were used for mean comparisons. The average standard deviations were adopted to indicate the measured values.

Supplementary Materials: The following supporting information can be downloaded at: https://www.mdpi.com/article/10.3390/ijms232012442/s1.

Author Contributions: Conceptualization, L.Z., G.L. and S.Z. (Shujiang Zhang); Methodology, L.Z. and S.Z. (Shifan Zhang); Project administration, G.C., L.Y., J.H., X.T., S.Z. (Shidong Zhu) and R.S.; Supervision, Y.D., S.W., C.W., F.L. and H.Z.; Writing—original draft, L.Z.; Writing—review and editing, G.L. and S.Z. (Shujiang Zhang). All authors have read and agreed to the published version of the manuscript.

Funding: This work was performed at the Key Laboratory of Biology and Genetic Improvement of Horticultural Crops, Ministry of Agriculture, Beijing, China. This work was funded by the National Natural Science Foundation of China (32102373), the Beijing Natural Science Foundation (6212030), the China Agriculture Research System (CARS−23-A−14), the Beijing Joint Research Program for Germplasm Innovation and New Variety Breeding (G20220628003), and the Technology Innovation Program of the Chinese Academy of Agricultural Sciences (IVF-BRF2022002).

Data Availability Statement: The data that support the results are included in this article and its Supplementary Materials. Other relevant materials are available from the corresponding author upon reasonable request.

Conflicts of Interest: The authors declare no conflict of interest.

References

1. Bradshaw, J.E.; Pande, B.; Bryan, G.J.; Hackett, C.A.; McLean, K.; Stewart, H.E.; Waugh, R. Interval mapping of quantitative trait loci for resistance to late blight [*Phytophthora infestans* (Mont.) de Bary], height and maturity in a tetraploid population of potato (*Solanum tuberosum* subsp. *tuberosum*). *Genetics* **2004**, *168*, 983–995. [CrossRef]
2. Walter, M.H.; Strack, D. Carotenoids and their cleavage products: Biosynthesis and functions. *Nat. Prod. Rep.* **2011**, *28*, 663–692. [CrossRef]
3. Demmig-Adams, B.; Adams, W.W. Antioxidants in photosynthesis and human nutrition. *Science* **2002**, *298*, 2149–2153. [CrossRef]
4. Milborrow, B.V. The pathway of biosynthesis of abscisic acid in vascular plants: A review of the present state of knowledge of ABA biosynthesis. *J. Exp. Bot.* **2001**, *52*, 1145–1164. [CrossRef]
5. Bouvier, F.; Suire, C.; Mutterer, J.; Camara, B. Oxidative remodeling of chromoplast carotenoids: Identification of the carotenoid dioxygenase CsCCD and CsZCD genes involved in crocus secondary metabolite biogenesis. *Plant Cell* **2003**, *15*, 47–62. [CrossRef]
6. Fester, T.; Hause, B.; Schmidt, D.; Halfmann, K.; Schmidt, J.; Wray, V.; Hanse, G.; Strack, D. Occurrence and localization of apocarotenoids in arbuscular mycorrhizal plant roots. *Plant Cell Physiol.* **2002**, *43*, 256–265. [CrossRef]
7. Giuliano, G.; Al-Babili, S.; von Lintig, J. Carotenoid oxygenases: Cleave it or leave it. *Trends Plant Sci.* **2003**, *8*, 145–149. [CrossRef]
8. Lewinsohn, E.; Sitrit, Y.; Bar, E.; Azulay, Y.; Ibdah, M.; Meir, A.; Yosef, E.; Zamir, D.; Tadmor, Y. Not just colors—carotenoid degradation as a link between pigmentation and aroma in tomato and watermelon fruit. *Trends Food Sci. Technol.* **2005**, *16*, 407–415. [CrossRef]
9. Lu, S.; Li, L. Carotenoid metabolism: Biosynthesis, regulation, and beyond. *J. Integr. Plant Biol.* **2008**, *50*, 778–785. [CrossRef]
10. Hirschberg, J. Carotenoid biosynthesis in flowering plants. *Curr. Opin. Plant Biol.* **2001**, *4*, 210–218. [CrossRef]
11. Taylor, M.; Ramsay, G. Carotenoid biosynthesis in plant storage organs: Recent advances and prospects for improving plant food quality. *Physiol. Plant.* **2005**, *124*, 143–151. [CrossRef]
12. DellaPenna, D.; Pogson, B.J. Vitamin synthesis in plants: Tocopherols and carotenoids. *Annu. Rev. Plant Biol.* **2006**, *57*, 711–738. [CrossRef]
13. Li, L.; Yuan, H. Chromoplast biogenesis and carotenoid accumulation. *Arch. Biochem. Biophys.* **2013**, *539*, 102–109. [CrossRef]
14. Cazzonelli, C.I.; Pogson, B.J. Source to sink: Regulation of carotenoid biosynthesis in plants. *Trends Plant Sci.* **2010**, *15*, 266–274. [CrossRef]
15. Rodriguez-Concepcion, M. Supply of precursors for carotenoid biosynthesis in plants. *Arch. Biochem. Biophys.* **2010**, *504*, 118–122. [CrossRef]
16. Brown, A.F.; Yousef, G.G.; Chebrolu, K.K.; Byrd, R.W.; Everhart, K.W.; Thomas, A.; Reid, R.W.; Parkin, I.A.P.; Sharpe, A.G.; Oliver, R.; et al. High-density single nucleotide polymorphism (SNP) array mapping in *Brassica oleracea*: Identification of QTL associated with carotenoid variation in broccoli florets. *Theor. Appl. Genet.* **2014**, *127*, 2051–2064. [CrossRef]

17. Cunningham, F.X., Jr.; Chamovitz, D.; Misawa, N.; Gantt, E.; Hirschberg, J. Cloning and functional expression in Escherichia coli of a cyanobacterial gene for lycopene cyclase, the enzyme that catalyzes the biosynthesis of beta-carotene. *FEBS Lett.* **1993**, *328*, 130–138. [CrossRef]
18. Cunningham, F.X., Jr.; Pogson, B.; Sun, Z.; McDonald, K.A.; DellaPenna, D.; Gantt, E. Functional analysis of the beta and epsilon lycopene cyclase enzymes of Arabidopsis reveals a mechanism for control of cyclic carotenoid formation. *Plant Cell* **1996**, *8*, 1613–1626. [CrossRef]
19. Pecker, I.; Gabbay, R.; Cunningham, F.X., Jr.; Hirschberg, J. Cloning and characterization of the cDNA for lycopene beta-cyclase from tomato reveals decrease in its expression during fruit ripening. *Plant Mol. Biol.* **1996**, *30*, 807–819. [CrossRef]
20. Ronen, G.; Cohen, M.; Zamir, D.; Hirschberg, J. Regulation of carotenoid biosynthesis during tomato fruit development: Expression of the gene for lycopene epsilon-cyclase is down-regulated during ripening and is elevated in the mutant Delta. *Plant J. Cell Mol. Biol.* **1999**, *17*, 341–351. [CrossRef]
21. Nisar, N.; Li, L.; Lu, S.; Khin, N.C.; Pogson, B.J. Carotenoid Metabolism in Plants. *Mol. Plant* **2015**, *8*, 68–82. [CrossRef]
22. Rodriguez-Concepcion, M.; Boronat, A. Breaking new ground in the regulation of the early steps of plant isoprenoid biosynthesis. *Curr. Opin. Plant Biol.* **2015**, *25*, 17–22. [CrossRef]
23. Hermanns, A.S.; Zhou, X.; Xu, Q.; Tadmor, Y.; Li, L. Carotenoid Pigment Accumulation in Horticultural Plants. *Hortic. Plant J.* **2020**, *6*, 343–360. [CrossRef]
24. Ruiz-Sola, M.A.; Rodriguez-Concepcion, M. Carotenoid biosynthesis in Arabidopsis: A colorful pathway. *Arab. Book* **2012**, *10*, e0158. [CrossRef]
25. Clotault, J.; Peltier, D.; Berruyer, R.; Thomas, M.; Briard, M.; Geoffriau, E. Expression of carotenoid biosynthesis genes during carrot root development. *J. Exp. Bot.* **2008**, *59*, 3563–3573. [CrossRef]
26. Fujisawa, M.; Nakano, T.; Shima, Y.; Ito, Y. A Large-Scale Identification of Direct Targets of the Tomato MADS Box Transcription Factor RIPENING INHIBITOR Reveals the Regulation of Fruit Ripening. *Plant Cell* **2013**, *25*, 371–386. [CrossRef]
27. Wu, M.B.; Xu, X.; Hu, X.W.; Liu, Y.D.; Cao, H.H.; Chan, H.E.; Gong, Z.H.; Yuan, Y.J.; Luo, Y.Q.; Feng, B.H.; et al. SlMYB72 Regulates the Metabolism of Chlorophylls, Carotenoids, and Flavonoids in Tomato Fruit(1). *Plant Physiol.* **2020**, *183*, 854–868. [CrossRef]
28. Fraser, P.D.; Truesdale, M.R.; Bird, C.R.; Schuch, W.; Bramley, P.M. Carotenoid Biosynthesis during Tomato Fruit Development (Evidence for Tissue-Specific Gene Expression). *Plant Physiol.* **1994**, *105*, 405–413. [CrossRef]
29. Wang, Q.; Huang, X.-Q.; Cao, T.-J.; Zhuang, Z.; Wang, R.; Lu, S. Heteromeric Geranylgeranyl Diphosphate Synthase Contributes to Carotenoid Biosynthesis in Ripening Fruits of Red Pepper (*Capsicum annuum* var. conoides). *J. Agric. Food Chem.* **2018**, *66*, 11691–11700. [CrossRef]
30. Paul, P.; Dhandapani, V.; Li, X.; Choi, S.R.; Hur, Y.; Lim, Y.P. Identification of candidate genes involved in the biosynthesis of carotenoids in *Brassica rapa*. *Hortic. Environ. Biotechnol.* **2014**, *55*, 342–351. [CrossRef]
31. Zhang, L.; Dai, Y.; Yue, L.X.; Chen, G.H.; Yuan, L.Y.; Zhang, S.F.; Li, F.; Zhang, H.; Li, G.L.; Zhu, S.D.; et al. Heat stress response in Chinese cabbage (*Brassica rapa* L.) revealed by transcriptome and physiological analysis. *PeerJ* **2022**, *10*, e13427. [CrossRef]
32. Feng, H.; Li, Y.; Liu, Z.; Liu, J. Mapping of or, a gene conferring orange color on the inner leaf of the Chinese cabbage (*Brassica rapa* L. ssp pekinensis). *Mol. Breed.* **2012**, *29*, 235–244. [CrossRef]
33. Matsumoto, E.; Yasui, C.; Ohi, M.; Tsukada, M. Linkage analysis of RFLP markers for clubroot resistance and pigmentation in Chinese cabbage (*Brassica rapa* ssp. pekinensis). *Euphytica* **1998**, *104*, 79–86. [CrossRef]
34. Zhang, F.; Wang, G.; Wang, M.; Liu, X.; Zhao, X.; Yu, Y.; Zhang, D.; Yu, S. Identification of SCAR markers linked to or, a gene inducing beta-carotene accumulation in Chinese cabbage. *Euphytica* **2008**, *164*, 463–471. [CrossRef]
35. Zhang, J.; Li, H.; Zhang, M.; Hui, M.; Wang, Q.; Li, L.; Zhang, L. Fine mapping and identification of candidate Br-or gene controlling orange head of Chinese cabbage (*Brassica rapa* L. ssp. pekinensis). *Mol. Breed.* **2013**, *32*, 799–805. [CrossRef]
36. Lee, S.; Lee, S.C.; Byun, D.H.; Lee, D.Y.; Park, J.Y.; Lee, J.H.; Lee, H.O.; Sung, S.H.; Yang, T.J. Association of molecular markers derived from the *BrCRTISO1* gene with prolycopene-enriched orange-colored leaves in *Brassica rapa* [corrected]. *Theor. Appl. Genet.* **2014**, *127*, 179–191. [CrossRef]
37. Li, P.; Zhang, S.; Zhang, S.; Li, F.; Zhang, H.; Liu, X.; Wu, J.; Wang, X.; Sun, R. Carotenoid identification and molecular analysis of carotenoid isomerase-encoding *BrCRTISO*, the candidate gene for inner leaf orange coloration in Chinese cabbage. *Mol. Breed.* **2015**, *35*, 72. [CrossRef]
38. Su, T.; Yu, S.; Zhang, J.W.F.; Yu, Y.; Zhang, D.; Zhao, X.; Wang, W. Loss of Function of the Carotenoid Isomerase Gene *BrCRTISO* Confers Orange Color to the Inner Leaves of Chinese Cabbage (*Brassica rapa* L. ssp pekinensis). *Plant Mol. Biol. Report.* **2015**, *33*, 648–659. [CrossRef]
39. Zou, C.L.; Zheng, Y.; Wang, P.; Zhang, X.; Wang, Y.H.; Liu, Z.Y.; Feng, H. Fine mapping and characterization of the or gene in Chinese cabbage (*Brassica rapa* L. ssp pekinensis). *Genet. Mol. Res.* **2016**, *15*. [CrossRef]
40. Yuan, H.; Zhang, J.; Nageswaran, D.; Li, L. Carotenoid metabolism and regulation in horticultural crops. *Hortic. Res.* **2015**, *2*, 15036. [CrossRef]
41. Michelmore, R.W.; Paran, I.; Kesseli, R.V. Identification of markers linked to disease-resistance genes by bulked segregant analysis: A rapid method to detect markers in specific genomic regions by using segregating populations. *Proc. Natl. Acad. Sci. USA* **1991**, *88*, 9828–9832. [CrossRef] [PubMed]
42. Hill, J.T.; Demarest, B.L.; Bisgrove, B.W.; Gorsi, B.; Su, Y.C.; Yost, H.J. MMAPPR: Mutation Mapping Analysis Pipeline for Pooled RNA-seq. *Genome Res.* **2013**, *23*, 687–697. [CrossRef] [PubMed]

43. Fekih, R.; Takagi, H.; Tamiru, M.; Abe, A.; Natsume, S.; Yaegashi, H.; Sharma, S.; Sharma, S.; Kanzaki, H.; Matsumura, H. MutMap plus: Genetic Mapping and Mutant Identification without Crossing in Rice. *PLoS ONE* **2013**, *8*, e68529. [CrossRef]
44. Lu, S.; Van Eck, J.; Zhou, X.; Lopez, A.B.; O'Halloran, D.M.; Cosman, K.M.; Conlin, B.J.; Paolillo, D.J.; Garvin, D.F.; Vrebalov, J.; et al. The cauliflower or gene encodes a DnaJ cysteine-rich domain-containing protein that mediates high levels of beta-carotene accumulation. *Plant Cell* **2006**, *18*, 3594–3605. [CrossRef] [PubMed]
45. Sun, T.; Zhou, F.; Huang, X.-Q.; Chen, W.-C.; Kong, M.-J.; Zhou, C.-F.; Zhuang, Z.; Li, L.; Lu, S. ORANGE Represses Chloroplast Biogenesis in Etiolated Arabidopsis Cotyledons via Interaction with TCP14. *Plant Cell* **2019**, *31*, 2996–3014. [CrossRef]
46. Zhou, X.; Welsch, R.; Yang, Y.; Alvarez, D.; Riediger, M.; Yuan, H.; Fish, T.; Liu, J.; Thannhauser, T.W.; Li, L. Arabidopsis OR proteins are the major posttranscriptional regulators of phytoene synthase in controlling carotenoid biosynthesis. *Proc. Natl. Acad. Sci. USA* **2015**, *112*, 3558–3563. [CrossRef]
47. Li, L.; Paolillo, D.J.; Parthasarathy, M.V.; DiMuzio, E.M.; Garvin, D.F. A novel gene mutation that confers abnormal patterns of beta-carotene accumulation in cauliflower (*Brassica oleracea* var. botrytis). *Plant J.* **2001**, *26*, 59–67. [CrossRef]
48. Yuan, H.; Owsiany, K.; Sheeja, T.E.; Zhou, X.; Rodriguez, C.; Li, Y.; Welsch, R.; Chayut, N.; Yang, Y.; Thannhauser, T.W.; et al. A Single Amino Acid Substitution in an ORANGE Protein Promotes Carotenoid Overaccumulation in Arabidopsis. *Plant Physiol.* **2015**, *169*, 421–431. [CrossRef]
49. Domonkos, I.; Kis, M.; Gombos, Z.; Ughy, B. Carotenoids, versatile components of oxygenic photosynthesis. *Prog. Lipid Res.* **2013**, *52*, 539–561. [CrossRef]
50. Maass, D.; Arango, J.; Wuest, F.; Beyer, P.; Welsch, R. Carotenoid Crystal Formation in Arabidopsis and Carrot Roots Caused by Increased Phytoene Synthase Protein Levels. *PLoS ONE* **2009**, *4*, e6373. [CrossRef]
51. Bai, C.; Rivera, S.M.; Medina, V.; Alves, R.; Vilaprinyo, E.; Sorribas, A.; Canela, R.; Capell, T.; Sandmann, G.; Christou, P.; et al. An in vitro system for the rapid functional characterization of genes involved in carotenoid biosynthesis and accumulation. *Plant J.* **2014**, *77*, 464–475. [CrossRef] [PubMed]
52. Chayut, N.; Yuan, H.; Ohali, S.; Meir, A.; Sa'ar, U.; Tzuri, G.; Zheng, Y.; Mazourek, M.; Gepstein, S.; Zhou, X.; et al. Distinct Mechanisms of the ORANGE Protein in Controlling Carotenoid Flux. *Plant Physiol.* **2017**, *173*, 376–389. [CrossRef] [PubMed]
53. Welsch, R.; Zhou, X.; Yuan, H.; Alvarez, D.; Sun, T.; Schlossarek, D.; Yang, Y.; Shen, G.; Zhang, H.; Rodriguez-Concepcion, M.; et al. Clp Protease and OR Directly Control the Proteostasis of Phytoene Synthase, the Crucial Enzyme for Carotenoid Biosynthesis in Arabidopsis. *Mol. Plant* **2018**, *11*, 149–162. [CrossRef] [PubMed]
54. Shevtsov, M.B.; Chen, Y.L.; Gollnick, P.; Antson, A.A. Crystal structure of Bacillus subtilis anti-TRAP protein, an antagonist of TRAP/RNA interaction. *Proc. Natl. Acad. Sci. USA* **2005**, *102*, 17600–17605. [CrossRef] [PubMed]
55. Zhou, M.; Deng, L.; Guo, S.; Yuan, G.; Li, C.; Li, C. Alternative transcription and feedback regulation suggest that SlIDI1 is involved in tomato carotenoid synthesis in a complex way. *Hortic. Res.* **2022**, *9*, uhab045. [CrossRef] [PubMed]
56. Ruiz-Sola, M.A.; Coman, D.; Beck, G.; Barja, M.V.; Colinas, M.; Graf, A.; Welsch, R.; Ruetimann, P.; Buehlmann, P.; Bigler, L.; et al. Arabidopsis GERANYLGERANYL DIPHOSPHATE SYNTHASE 11 is a hub isozyme required for the production of most photosynthesis-related isoprenoids. *New Phytol.* **2016**, *209*, 252–264. [CrossRef]
57. Camagna, M.; Grundmann, A.; Bar, C.; Koschmieder, J.; Beyer, P.; Welsch, R. Enzyme Fusion Removes Competition for Geranylgeranyl Diphosphate in Carotenogenesis. *Plant Physiol.* **2019**, *179*, 1013–1027. [CrossRef]
58. Kaul, S.; Koo, H.L.; Jenkins, J.; Rizzo, M.; Rooney, T.; Tallon, L.J.; Feldblyum, T.; Nierman, W.; Benito, M.I.; Lin, X.Y. Analysis of the genome sequence of the flowering plant *Arabidopsis thaliana*. *Nature* **2000**, *408*, 796–815.
59. Wang, X.; Wang, H.; Wang, J.; Sun, R.; Wu, J.; Liu, S.; Bai, Y.; Mun, J.-H.; Bancroft, I.; Cheng, F. The genome of the mesopolyploid crop species *Brassica rapa*. *Nat. Genet.* **2011**, *43*, 1035–1039. [CrossRef]
60. Matsumoto, T.; Wu, J.Z.; Kanamori, H.; Katayose, Y.; Fujisawa, M.; Namiki, N.; Mizuno, H.; Yamamoto, K.; Antonio, B.A.; Baba, T.; et al. The map-based sequence of the rice genome. *Nature* **2005**, *436*, 793–800.
61. Macia, A.; Blanco-Jimenez, E.; Garcia-Perez, J.L. Retrotransposons in pluripotent cells: Impact and new roles in cellular plasticity. *Biochim. Biophys. Acta-Gene Regul. Mech.* **2015**, *1849*, 417–426. [CrossRef] [PubMed]
62. Paolillo, D.J.; Garvin, D.F.; Parthasarathy, M.V. The chromoplasts of Or mutants of cauliflower (*Brassica oleracea* L. var. botrytis). *Protoplasma* **2004**, *224*, 245–253. [CrossRef] [PubMed]
63. Yazdani, M.; Sun, Z.X.; Yuan, H.; Zeng, S.H.; Thannhauser, T.W.; Vrebalov, J.; Ma, Q.Y.; Xu, Y.M.; Fei, Z.J.; Van Eck, J.; et al. Ectopic expression of ORANGE promotes carotenoid accumulation and fruit development in tomato. *Plant Biotechnol. J.* **2019**, *17*, 33–49. [CrossRef] [PubMed]
64. Tzuri, G.; Zhou, X.J.; Chayut, N.; Yuan, H.; Portnoy, V.; Meir, A.; Sa'ar, U.; Baumkoler, F.; Mazourek, M.; Lewinsohn, E.; et al. A 'golden' SNP in CmOr governs the fruit flesh color of melon (*Cucumis melo*). *Plant J.* **2015**, *82*, 267–279. [CrossRef]
65. Mathur, J.; Koncz, C. Establishment and maintenance of cell suspension cultures. *Methods Mol. Biol.* **1998**, *82*, 27–30.
66. Fantini, E.; Falcone, G.; Frusciante, S.; Giliberto, L.; Giuliano, G. Dissection of Tomato Lycopene Biosynthesis through Virus-Induced Gene Silencing. *Plant Physiol.* **2013**, *163*, 986–998. [CrossRef]
67. Meynert, A.M.; Ansari, M.; FitzPatrick, D.R.; Taylor, M.S. Variant detection sensitivity and biases in whole genome and exome sequencing. *BMC Bioinform.* **2014**, *15*, 247. [CrossRef]
68. Islam, M.S.; Zeng, L.; Thyssen, G.N.; Delhom, C.D.; Kim, H.J.; Li, P.; Fang, D.D. Mapping by sequencing in cotton (*Gossypium hirsutum*) line MD52ne identified candidate genes for fiber strength and its related quality attributes. *Theor. Appl. Genet.* **2016**, *129*, 1071–1086. [CrossRef]

69. Wang, N.; Liu, Z.; Zhang, Y.; Li, C.; Feng, H. Identification and fine mapping of a stay-green gene (Brnye1) in pakchoi (*Brassica campestris* L. ssp chinensis). *Theor. Appl. Genet.* **2018**, *131*, 673–684. [CrossRef]
70. McKenna, A.; Hanna, M.; Banks, E.; Sivachenko, A.; Cibulskis, K.; Kernytsky, A.; Garimella, K.; Altshuler, D.; Gabriel, S.; Daly, M.; et al. The Genome Analysis Toolkit: A MapReduce framework for analyzing next-generation DNA sequencing data. *Genome Res.* **2010**, *20*, 1297–1303. [CrossRef]
71. Altschul, S.F.; Madden, T.L.; Schaffer, A.A.; Zhang, J.; Zhang, Z.; Miller, W.; Lipman, D.J. Gapped BLAST and PSI-BLAST: A new generation of protein database search programs. *Nucleic Acids Res.* **1997**, *25*, 3389–3402. [CrossRef] [PubMed]
72. Kumar, S.; Stecher, G.; Tamura, K. MEGA7: Molecular Evolutionary Genetics Analysis Version 7.0 for Bigger Datasets. *Mol. Biol. Evol.* **2016**, *33*, 1870–1874. [CrossRef] [PubMed]
73. Livak, K.J.; Schmittgen, T.D. Analysis of relative gene expression data using real-time quantitative PCR and the 2(-Delta Delta C(T)) Method. *Methods* **2001**, *25*, 402–408. [CrossRef]
74. Lescot, M.; Dehais, P.; Thijs, G.; Marchal, K.; Moreau, Y.; Van de Peer, Y.; Rouze, P.; Rombauts, S. PlantCARE, a database of plant cis-acting regulatory elements and a portal to tools for in silico analysis of promoter sequences. *Nucleic Acids Res.* **2002**, *30*, 325–327. [CrossRef]
75. Hong, B.; Tong, Z.; Ma, N.; Kasuga, M.; Yamaguchi-Shinozaki, K.; Gao, J.-P. Expression of the Arabidopsis DREB1A gene in transgenic chrysanthemum enhances tolerance to low temperature. *J. Hortic. Sci. Biotechnol.* **2006**, *81*, 1002–1008. [CrossRef]
76. Zhang, J.; Liu, Z.; Liang, J.; Wu, J.; Cheng, F.; Wang, X. Three genes encoding AOP2, a protein involved in aliphatic glucosinolate biosynthesis, are differentially expressed in *Brassica rapa*. *J. Exp. Bot.* **2015**, *66*, 6205–6218. [CrossRef] [PubMed]

International Journal of
Molecular Sciences

MDPI

Article

Comparative Study of Starch Phosphorylase Genes and Encoded Proteins in Various Monocots and Dicots with Emphasis on Maize

Guowu Yu [1,2,†], Noman Shoaib [1,†], Ying Xie [1], Lun Liu [1], Nishbah Mughal [1], Yangping Li [2], Huanhuan Huang [1], Na Zhang [3], Junjie Zhang [4], Yinghong Liu [5], Yufeng Hu [2], Hanmei Liu [4,*] and Yubi Huang [1,2,*]

1 National Demonstration Center for Experimental Crop Science Education, College of Agronomy, Sichuan Agricultural University, Chengdu 611130, China; 13863@sicau.edu.cn (G.Y.); nomanshoaib@stu.sicau.edu.cn (N.S.); xieying89@sicau.edu.cn (Y.X.); liulun@stu.sicau.edu.cn (L.L.); nishbahmughal97@gmail.com (N.M.); huanghuanhuan@sicau.edu.cn (H.H.)
2 State Key Laboratory of Crop Gene Exploration and Utilization in Southwest China, Sichuan Agricultural University, Chengdu 611130, China; ypli@sicau.edu.cn (Y.L.); 13958@sicau.edu.cn (Y.H.)
3 College of Science, Sichuan Agricultural University, Chengdu 611130, China; 72008@sicau.edu.cn
4 College of Life Science, Sichuan Agricultural University, Ya'an 625014, China; junjiezhang@sicau.edu.cn
5 Maize Research Institute, Sichuan Agricultural University, Chengdu 611130, China; 13964@sicau.edu.cn
* Correspondence: 13175@sicau.edu.cn (H.L.); 10024@sicau.edu.cn (Y.H.)
† These authors contributed equally to this work.

Abstract: Starch phosphorylase (PHO) is a multimeric enzyme with two distinct isoforms: plastidial starch phosphorylase (PHO1) and cytosolic starch phosphorylase (PHO2). PHO1 specifically resides in the plastid, while PHO2 is found in the cytosol. Both play a critical role in the synthesis and degradation of starch. This study aimed to report the detailed structure, function, and evolution of genes encoding PHO1 and PHO2 and their protein ligand-binding sites in eight monocots and four dicots. "True" orthologs of PHO1 and PHO2 of *Oryza sativa* were identified, and the structure of the enzyme at the protein level was studied. The genes controlling PHO2 were found to be more conserved than those controlling PHO1; the variations were mainly due to the variable sequence and length of introns. Cis-regulatory elements in the promoter region of both genes were identified, and the expression pattern was analyzed. The real-time quantitative polymerase chain reaction indicated that PHO2 was expressed in all tissues with a uniform pattern of transcripts, and the expression pattern of PHO1 indicates that it probably contributes to the starch biosynthesis during seed development in *Zea mays*. Under abscisic acid (ABA) treatment, PHO1 was found to be downregulated in *Arabidopsis* and *Hordeum vulgare*. However, we found that ABA could up-regulate the expression of both PHO1 and PHO2 within 12 h in *Zea mays*. In all monocots and dicots, the 3D structures were highly similar, and the ligand-binding sites were common yet fluctuating in the position of aa residues.

Keywords: comparative study; cytosolic starch phosphorylase; plastidial starch phosphorylase; starch phosphorylase

Citation: Yu, G.; Shoaib, N.; Xie, Y.; Liu, L.; Mughal, N.; Li, Y.; Huang, H.; Zhang, N.; Zhang, J.; Liu, Y.; et al. Comparative Study of Starch Phosphorylase Genes and Encoded Proteins in Various Monocots and Dicots with Emphasis on Maize. *Int. J. Mol. Sci.* **2022**, *23*, 4518. https://doi.org/10.3390/ijms23094518

Academic Editors: Andrés J. Cortés and Hai Du

Received: 27 February 2022
Accepted: 12 April 2022
Published: 20 April 2022

1. Introduction

Maize (*Zea mays*) is one of the three major cereal crops globally and the largest in food crops, with a yield second to rice (*Oryza sativa*). The major part of the yield is known as starch, which is the most important form of reserve polysaccharide synthesized in cellular organelles in plants called plastids [1]. Starch biosynthesis involves a set of enzymes including ADP-glucose pyrophosphorylase (AGPase; EC 2.7.7.27), starch synthases (SSs; EC 2.4.1.21), branching enzymes (SBEs; EC 2.4.1.18), starch debranching enzymes [DBEs, 3.2.1.41; mainly isoamylases (ISAs)] and starch phosphorylase (PHO; EC 2.4.1.1) in the cereal endosperm [2,3]. In recent times, PHO has been investigated comprehensively due to its involvement in starch biosynthesis [4–9].

The PHO is found to be involved in the transfer of glucosyl unit from glucose-1-phosphate (Glc-1-P) to growing α-1-4 linked glucan chain in a reversible reaction that depends on the available substrate level [10]. In plants, two distinct forms of PHO have been reported [11]: plastidial starch phosphorylase (PHO1) and cytosolic starch phosphorylase (PHO2). These two forms have significant sequence similarities to each other but largely differ in their molecular sizes, substrate specificities physiological roles, and intracellular localization [12]. PHO1 resides in the plastid [13] and PHO2 in the cytosol. PHO1 plays an important role only in starch degradation due to the high Pi/Glc-1-P ratio found in vivo [14]. However, no direct evidence is available to support this. Moreover, several starch-degrading enzymes have been identified [15,16], including glucan water dikinase (GWD; EC 2.7.9.4) and phosphoglucan water dikinase (PWD; EC 2.7.9.5), which produces the branched and linear glucans from the starch [17]. However, α-amylase (EC 3.2.1.1) is thought to be the first enzyme to initiate starch degradation. These findings make the PHO1 an indirect-acting or a regulatory candidate that can influence the activity of other participating enzymes. The exact role of PHO1 is still in debatable. Moreover, PHO1 is reported to be involved in starch metabolism in various monocots and dicots, including *Zea mays* [18], *Oryza sativa* [19] *Triticum aestivum* [4], *Hordeum vulgare* [20], and *Solanum tuberosum* [21]. PHO2 can degrade the branched glucans and can even attack the starch granules [22]. In addition, it can use the degraded products from the starch and maintain the Glc-1-P level in the cytosol [23].

These two enzymes are highly similar to each other, however they differ in the presence of transit peptides in the N-terminal and an extra stretch of 78–82 amino acid (aa) residues (L80 domain) in the middle region of PHO1. The presence of this extra peptide can be responsible for varied biochemical activities [9]. Previously, L80 insertion is found to be an active substrate for proteasomes in *Ipomoea batatas* [8]. The PHO1 in *Solanum tuberosum* maintains its intact structure in young tubers, however steadily degrades into smaller peptides as the tubers become mature [13]. Both the intact (105 kDa) and degraded (55 kDa) forms of PHO1 have been reported in *Vigna radiata*. Interestingly, both forms were identified to make active catalytic complexes [24]. In contrast, the PHO1 of cereals, including *Oryza sativa* [19,25], *Zea mays* [18], *Hordeum vulgare* [26], and *Triticum aestivum* [10] is not degraded by the proteasome, even though they all have a PEST motif in L80 insertion. A detailed comparative study would be required to uncover these differences to reach a conclusion. No major differences exist between PHO1 and PHO2 in terms of binding Glc-1-P and Pi (inorganic phosphate). However, these can be distinguished by their affinity to bind the glucan substrate (PHO1 has a low and PHO2 has a high affinity for highly branched polysaccharides) [19,27]. The L80 domain in PHO1 was found to sterically prevent the binding of the enzyme to the branched polysaccharide substrate [28]. Quantitative data about PHO abundance and activity is lacking and as of yet only *Hordeum vulgare* PHO1 crystal structure has been characterized [26].

Several reports on microarray based analysis combined with promoter sequences indicated that *PHO* may play a role under different stresses. However, the regulation pattern found variables among the main cereal crops, especially in *Arabidopsis thaliana*, where *PHO1* seemed to play a role in the tolerance of abiotic stress rather than the degradation of starch [29]. A study on *PHO1* of *Hordeum vulgare* suggested no effect of ABA on the expression of *PHO1*, although there was up-regulation of *PHO2* [20]. In contrast, in *Arabidopsis thaliana* significant down-regulation for both isoforms were detected. Considerable information about PHO and its gene is available. However, detailed information about gene diversity, structure variation, function, evolution, and PHO responding to phytohormones is unclear in some important crops, including *Zea mays*. In recent years, significant progress has been made by scientists in sequencing plant genomes. The availability of genome data is essential to answer the biological questions through comparative analysis [30]. Therefore, the present study was carried out using the PHO1 and PHO2 genes of *Oryza sativa* as a reference (as these are well characterized and closely related to *Zea mays*) to identify and characterize its orthologs from eight important monocots and four important dicots. The obtained results are presented in this communication, including "true" orthologs of *PHO1* and *PHO2* genes, evolution of gene structure for both isoforms across mentioned species, synteny and collinearity analysis, identification of conserved and variable domains with subtle differences between both isoforms and among the monocots and dicots, promotor analysis, and expression analysis with the effect of ABA treatment on the expression of both isoforms of PHO in *Zea mays*.

2. Results

2.1. Gene Sequence and Structure

2.1.1. Identification of Orthologs of PHO1 and PHO2

The "true" orthologs of the PHO1 and the PHO2-encoding genes of *O. sativa* were identified from the 12 different species using the criteria mentioned in the Materials and Methods section. Thus, the genome, coding DNA sequence (CDS), and cDNA of PHO1 and PHO2 from the mentioned orthologs were obtained and are presented in Table 1 (PHO1) and Table 2 (PHO2). In all examined species, a single ortholog for each PHO1 and PHO2 was found. The similarity of cDNA of PHO1 in all mentioned orthologs with that of *O. sativa* (used as reference) ranged from 78.1% to 81.7% in monocots and 42.2% to 65.0% in dicots. Likewise, the similarity of cDNA of PHO2 ranged from 80.4% to 81.7% in monocots and 62.9% to 65.9% in dicots. The CDS similarity of PHO1 ranged from 90.4% to 91.9% in monocots and 80.5% to 81.6% in dicots. Likewise, the CDS similarity of PHO2 ranged from 86.0% to 93.5% in monocots and 82.8% to 85.1% in dicots. The similarity index of gene sequences was low compared with that of cDNA and CDS; it ranged from 49.2% to 63.0% in monocots and 45.9% to 46.3% in dicots; for PHO2 it ranged from 36.7% to 62.4% in monocots and 11.8% to 47.6% in dicots.

Table 1. Details of cDNAs, CDS, genes and protein sequences of PHO1 in different monocots and dicots with respect to *O. sativa* PHO1.

Species PHO1	cDNA		CDS		Gene		* Protein	
	Length (bp)	% Identical	Length (bp)	% Identical	Length (bp)	% Identical	Length (aa)	% Identical
S. bicolor	3445	79.0	2958	91.9	6536	62.6	985	86.2
Z. mays	3523	78.1	2955	91.9	9022	49.2	984	85.8
S. italica	3385	80.9	2943	92.4	6909	62.8	980	87.2
P. halli	3351	80.7	29.64	91.9	6671	63.0	987	85.8
B. distachyon	3410	81.7	2937	92.5	7498	62.9	978	87.5
T. aestivum	3312	80.0	2917	90.4	7461	58.4	971	84.3
O. sativa	3408	100	2937	100	7239	100	978	100
H. vulgare	3305	77.3	2907	90.6	7441	61.1	968	84.9
A. thaliana	3180	64.5	2889	81.5	5008	45.9	962	69.3
S. lycopersicum	3332	65.0	2901	80.9	8718	46.3	966	67.0
S. tuberosem	4942	42.2	2901	81.6	4942	42.2	966	67.4
C. annuum	3456	63.2	2937	80.5	8233	45.9	978	68.5

* Accession ids for *S. bicolor*, XP_021306483.1; *Z. mays*, NP_001296783.1; *S. italica*, XP_004981704.1; *P. halli*, XP_023795811.1; *B. distachyon*, XP_003559211.1; *T. aestivum*, ACC59201.1; *O. sativa*, XP_015631420.1; *H. vulgare*, KAE8783983.1; *A. thaliana*, Q9LIB2.1; *S. lycopersicum*, NP_001362574.1; *S. tuberosem*, NP_001275215.1; *C. annuum*, XP_016569840.1.-indicates absence of data.

2.1.2. Gene Structure Comparison

The gene structures for PHO1 (*PHO1*) and PHO2 (*PHO2*) from different species were compared using *O. sativa* as a reference. The length of the genes was a major difference, which ranged from 2.92 Kb to 9.02 Kb in *PHO1* and 5.24 Kb to 34.54 Kb in *PHO2*. *PHO1* was the longest for *Z. mays* and smallest for *S. tuberosum* (Table 1). Likewise, *PHO2* was the longest for *S. lycopersicum* and smallest for *Arabidopsis. thaliana* (Table 2). The variation in gene size was primarily attributed to the variations in size and number of introns. While comparing cDNA, introns were found to be lacking in 1.828–5.499 Kb in *PHO1* and

2.072–12.201 Kb in *PHO2*. cDNAs for both *PHO1* and *PHO2* were longer in dicots than in monocots. In CDS comparison, only marginal differences were observed for both *PHO1* and *PHO2* in monocots and dicots except for the *PHO1* of *A. thaliana*, where the CDS length was the shortest (compared with CDS of *PHO1* in all species) and *PHO2* of *P. hallii*, where the CDS length was the longest (compared with CDS *PHO1* in all species). Therefore, the detected cDNA variations were due to the length of untranslated regions (UTRs) in the cDNAs.

Table 2. Details of cDNAs, CDS, gene, and protein sequences of PHO2 in different monocots and dicots with respect to *O. sativa* PHO2.

Species PHO2	cDNA		CDS		Gene		* Protein	
	Length (bp)	% Identical	Length (bp)	% Identical	Length (bp)	% Identical	Length (aa)	% Identical
S. bicolor	3167	80.8	2517	92.5	6227	60.6	838	89.5
Z. mays	2868	80.4	2517	92.6	12,006	36.7	838	89.9
S. italica	2937	81.7	2514	92.4	6312	61.6	837	89.2
P. halli	2979	81.2	2664	86.0	5902	59.4	897	83.2
B. distachyon	2909	82.5	2502	93.4	6891	62.4	833	90.3
T. aestivum	2856	80.4	2499	93.2	6114	62.2	832	90.3
O. sativa	3036	100	2526	100	6275	100	841	100
H. vulgare	2857	79.4	2499	93.5	5569	59.6	832	90.4
A. thaliana	3168	62.9	2526	85.1	5240	47.6	841	77.6
S. lycopersicum	2908	65.4	2514	84.9	15,109	23.7	837	77.1
S. tuberosem	2833	65.9	2517	85.1	13,343	29.7	838	77.8
C. annuum	2965	65.0	2544	82.8	34,548	11.8	847	75.0

* Accession ids for *S. bicolor*, XP_002458773.1; *Z. mays*, NP_001151625.1; *S. italica*, XP_004970602.1; *P. halli*, XP_025814328.1; *B. distachyon*, XP_003564622.1; *T. aestivum*, Q9LKJ3.1; *O. sativa*, XP_015621275.1; *H. vulgare*, KAE8796869.1; *A. thaliana*, CAB61943.1; *S. lycopersicum*, XP_004246972.1; *S. tuberosem*, NP_001275118.1; *C. annuum*, XP_016542003.1.-indicates absence of data.

The number of introns and exons also displayed some differences among *PHO1* and *PHO2* of the examined species (Figure 1). Each *PHO1* had 15 exons and 14 introns in both monocots and dicots (except for the *PHO1* of *S. tuberosum*, which had seven exons and six introns). However, each *PHO2* had 15 exons with 14 introns (except for *PHO2* of *S. bicolor*, which had 16 introns) in monocots and 16 exons with 15 introns in dicots.

The difference in the size of introns and exons was also observed. The introns in *PHO1* were generally shorter than those of introns in *PHO2* where these were found to be larger in size ranging from 77 to 4638 bp (except for the first intron of *PHO1* in monocots). The intron 11 of the *PHO1* of *Z. mays* was found to be the longest among all *PHO1* of monocots and dicots (2092 bp). Likewise, the intron 13 of the *PHO2* of *S. tuberosum* was found to be the longest among all *PHO2* of monocots and dicots (4638 bp). The variations in the exon length were found to be relatively low, ranging from 77 to 476 bp for *PHO1* and 77 to 492 bp for *PHO2*. The Exon 12 of the *PHO1* in *S. lycopersicum* was the shortest (77 bp), whereas the exon 15 was the longest (476 bp). Likewise, the exon 4 of *PHO2* was the shortest (77 bp) and the exon 15 of *PHO2* was the longest (492 bp) in *O. sativa*.

Each type of intron phase was found in the *PHO1* of all monocots and dicots. However, only intron phase 0 and intron phase 2 were found in the *PHO2* of all monocots and dicots (Figure 1). An identical intron phase pattern was observed in both *PHO1* and the *PHO2* (except the PHO1 of *S. bicolor*). For both *PHO1* and *PHO2* phases, phase 0 was the most dominant in all the examined species, ranging from 71.42–78.57% for *PHO1* and 78.57–80% for *PHO2*. The average sequence similarity for exons in *PHO1* was found to be higher in monocots (82.84–85.31%) than in dicots (33.18–69.50%). However, the average sequence similarity for introns was low in both monocots (36.15–39.40%) and dicots (23.23–33.47%) (Figure 2). For *PHO2* exons, the similarity ranged from 84.50% to 86.17% in monocots and 19.21% to 31.75% in dicots; whereas, for introns, it ranged from 41.1% to 50.1% in monocots (except for *S. italica* where this was 72.8%) and 20% to 32.1% in dicots (Tables S1 and S2).

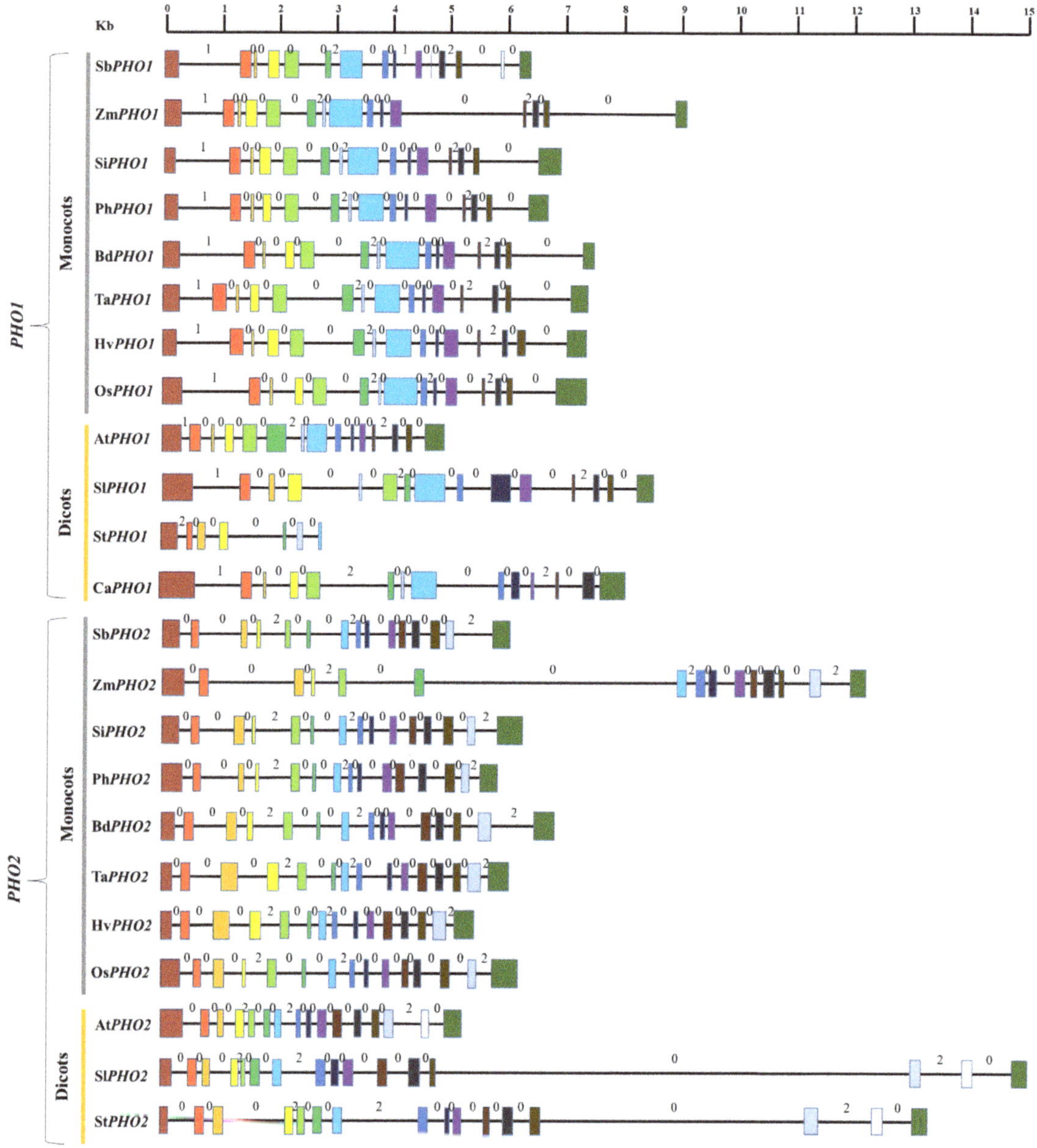

Figure 1. Gene structure of PHO1 and PHO2 from the translation start to stop sites in various monocots and dicots. Symbols include: Sb, *Sorghum bicolor;* Zm, *Zea mays;* Si, *Setaria italica;* Ph, *Panicum hallii;* Bd, *Brachypodium distachyon;* Ta, *Triticum aestivum;* Hv, *Hordeum vulgare;* Os, *Oryza sativa;* At, *Arabidopsis thaliana;* Sl, *Solanum lycopersicum;* St, *Solanum tuberosum;* and Ca, *Capsicum annuum.* Solid boxes indicate the exons, and lines indicate the introns. Colors represent the similarities among respective exons as a reference to *O. sativa.* The values of zero, one, and two marked above each intron are intron phases.

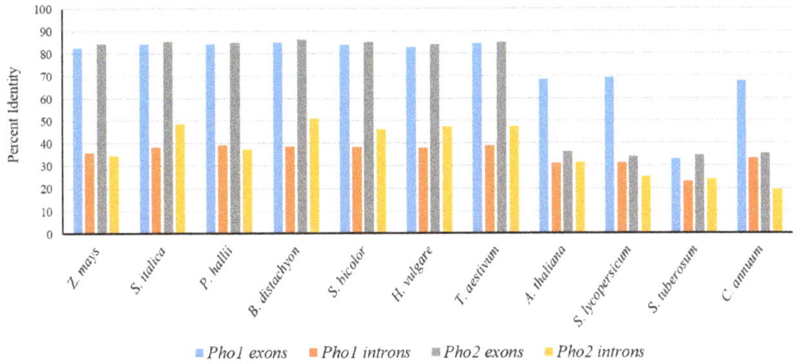

Figure 2. Percent identity of exons and introns in *PHO1* and *PHO2* of monocots and dicots with respect to exons and introns of *O. sativa*.

2.1.3. Synteny and Collinearity Analysis

The synteny conservation analysis was carried out using the chunk of 31 genes with 15 genes flanking on either side of the *PHO1* and *PHO2* of *O. sativa* on chromosome number 3 and 1, respectively. For *PHO2*, some degree of synteny conservation was observed for *Z. mays*, *S. italica*, *T. aestivum*, *H. vulgare*, *B. distachyon*, and *S. bicolor*. However, for *PHO1*, some degree of synteny conservation for only *Z. mays*. and *P. hallii* was observed. The synteny of 12 of the 30 genes flanking on the *PHO1* of *O. sativa* was shared on the parallel chromosome 1 in *Z. mays* and chromosome 9 in *P. hallii*. Similarly, the synteny of 16 of the 30 genes flanking on *O. sativa PHO2* was shared on the parallel chromosome 3 in *Z. mays*, 16 of the 30 genes on chromosome V in *S. italica*, 19 of the 30 genes on chromosome 3A in *T. aestivum*, 17 of the 30 genes on chromosome 3H in *H. vulgare*, 19 of the 30 genes on chromosome 2 in *B. distachyon*, and 21 of the 30 genes on chromosome 3 in *S. bicolor*. The microsynteny analysis in terms of the nucleotide for *PHO1* and *PHO2* has already been described earlier in the gene structure comparison section (Figures S1 and S2).

2.1.4. SSRs and Retro-Elements in PHO1 and PHO2

Both the *PHO1* and the *PHO2* were found to have the SSRs; but with variations in numbers and sequences of bp (Tables S3 and S4). No retrotransposons (LINEs and LTR elements), SINEs and transposons were identified in any of *PHO1* and *PHO2* of monocots and dicots.

2.1.5. Promoter Analysis of PHO1 and PHO2

Cis-regulatory elements were identified through promotor analysis of about 1 kb upstream region of both *PHO1* and *PHO2* in each of the monocots and dicots. These regulatory elements (e.g., ABRE and MBS) apparently responded to hormones such as abscisic acid (ABA) and abiotic stress. Regulatory elements that induced the tissue-specific expression such as "endosperm expression" were also identified (Figure 3). TATA box and CAAT elements were common in both *PHO1* and *PHO2*. The effect of ABA regulatory elements was further analyzed in both *PHO1* and *PHO2* of *Z. mays* by employing the ABA treatments to cDNA (elaborated in the expression analysis section).

2.2. Protein Sequence and Structure

2.2.1. Comparison of Protein Sequence

An extra stretch of about 78–82 aa residues was found in the middle region of PHO1 and not in PHO2; while for both PHO1 and PHO2, the length of the protein sequences was almost similar in all monocots and dicots, ranging 962–985 aa in PHO1 and 832–897 aa in PHO2 (Tables 1 and 2). The extra peptide in the middle region of PHO1 was designated as the "L80 domain". The comparison of aa sequences of both the PHO1 and the PHO2 in all monocots and dicots concerning the PHO1 and PHO2 of *O. sativa* showed higher similarity for PHO2 (89.2–90.4% in monocots and 75.0–77.8% in dicots) than for PHO1 (84.3–87.5% in monocots and 67.0–69.3% in dicots). PHO2 is found to be more conserved than PHO1.

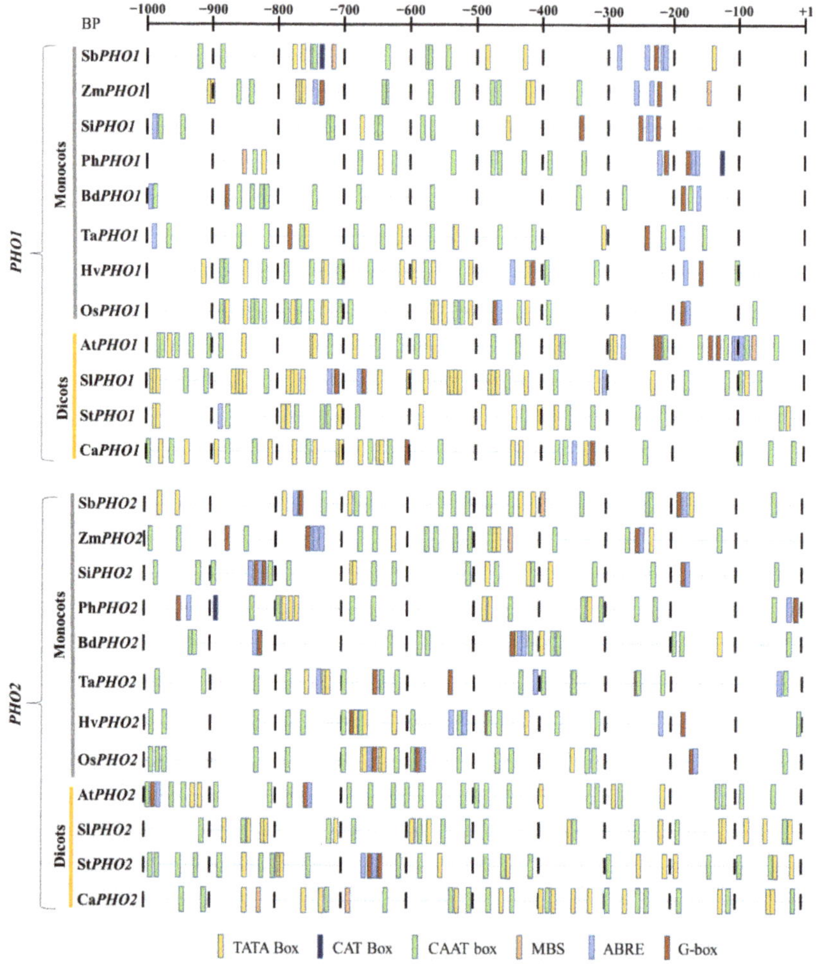

Figure 3. Representative figure showing regulatory elements identified in the 1 kb upstream region of *PHO1* and *PHO2*. Different color bars represent the major regulatory elements. Symbols include: Sb, *Sorghum bicolor;* Zm, *Zea mays;* Si, *Setaria italica;* Ph, *Panicum hallii;* Bd, *Brachypodium distachyon;* Ta, *Triticum aestivum;* Hv, *Hordeum vulgare;* Os, *Oryza sativa;* At, *Arabidopsis thaliana;* Sl, *Solanum lycopersicum;* St, *Solanum tuberosum;* and Ca, *Capsicum annuum.*

2.2.2. Sequence Variation in PHO1 and PHO2

Among all monocots and dicots, PHO1 was longer in *S. bicolor* than in other species. PHO1 had two conserved domains (CDs): GT-35-Glycogen-Phosphorylase (97–950 aa) and a unique L80 domain (500–600 aa). The L80 domain was hyper-variable and predicted to consist of a PEST motif (a motif of negatively charged aa residues, which could act as a substrate for proteasomes) [31,32]. Similarly, the N-terminal (1–97 aa) was also hyper-variable, although the linker segments were conserved (Figure 4). Insertions, deletions, and mismatches were the main reasons for variations. Each of the PHO1 was found with the insertion of a variable transit peptide of up to 50 amino acids, which lacked in PHO2 except for the PHO2 of *P. halli* (Figure S3). The transit peptide contained several structural features common to transit peptides of chloroplast proteins, however, contained a surprisingly large number of histidine residues.

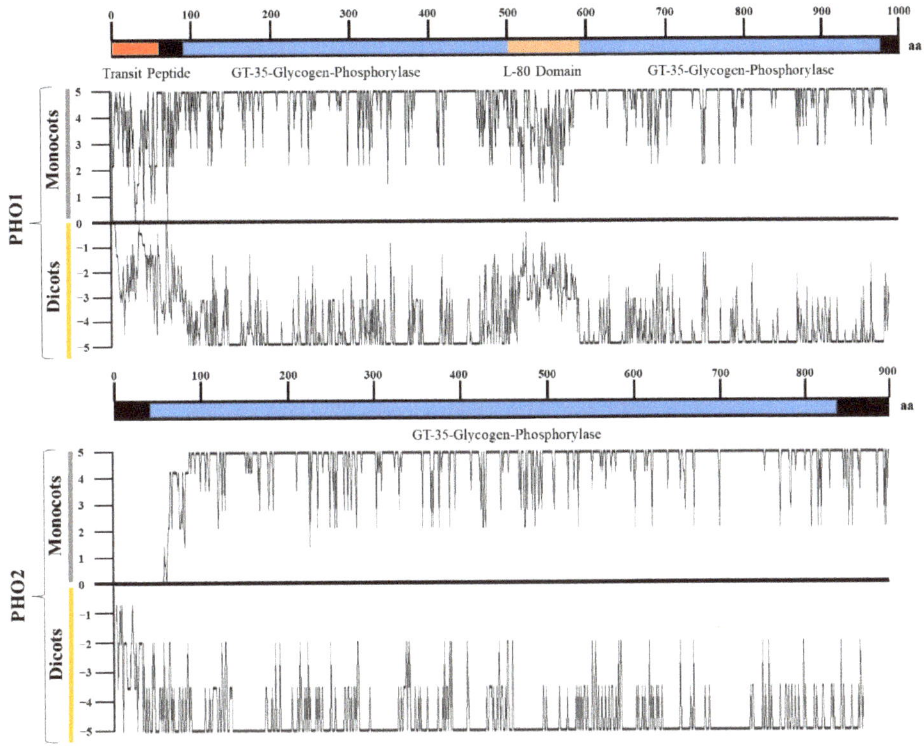

Figure 4. Amino acids sequence similarity of PHO1 and PHO2 among various monocots and dicots with respect to the consensus sequence. A value position of zero indicates the consensus sequence. Similar amino acids are plotted on scales 1–5 in monocots and −1 to −5 in dicots.

PHO2 was relatively longer in *P. hallii* than in any other species. Unlike the lower conservation for PHO1, aa residues in PHO2 were more conserved for each of the 12 examined species. However, just like PHO1, PHO2 also possessed hyper-variable N-terminal residues in all species (Figure 4). PHO2 was also associated with the GT-35-Glycogen-Phosphorylase domain (45–830 aa), however unlike PHO1, it lacked the L80 domain in the middle region. Sequence variations were also observed for PHO2.

2.2.3. Three Dimensional (3D) Structure Comparison of PHO1 and PHO2

The three-dimensional (3D) structures of PHO1 and PHO2 for all 12 species were generated using the PHO1 of *H. vulgare* (PDB id: 5lr8.1.A) and the PHO2 of *A. thaliana* (PDB id: 4beq.1.A) as templates, respectively. All generated structures had a high level of confidence, as evident from the following observations (Table S5): (a) relative to allowed and disallowed regions, a high proportion of aa residues were present in the favored regions of the Ramachandran plot; (b) the value of G-factor was within the standard range for all comparisons; (c) the quality factor values calculated using ERRAT and VERIFY3D were also high; and (d) the Qmean and Dfire energy values (Swiss-model assessment–based values) were negative, suggesting the favorable energy environment for given aa residues. Both PHO1 and PHO2 were in homo-dimeric forms and almost had similar structures, except the L80 domain arranged outside the main core (Figure 5).

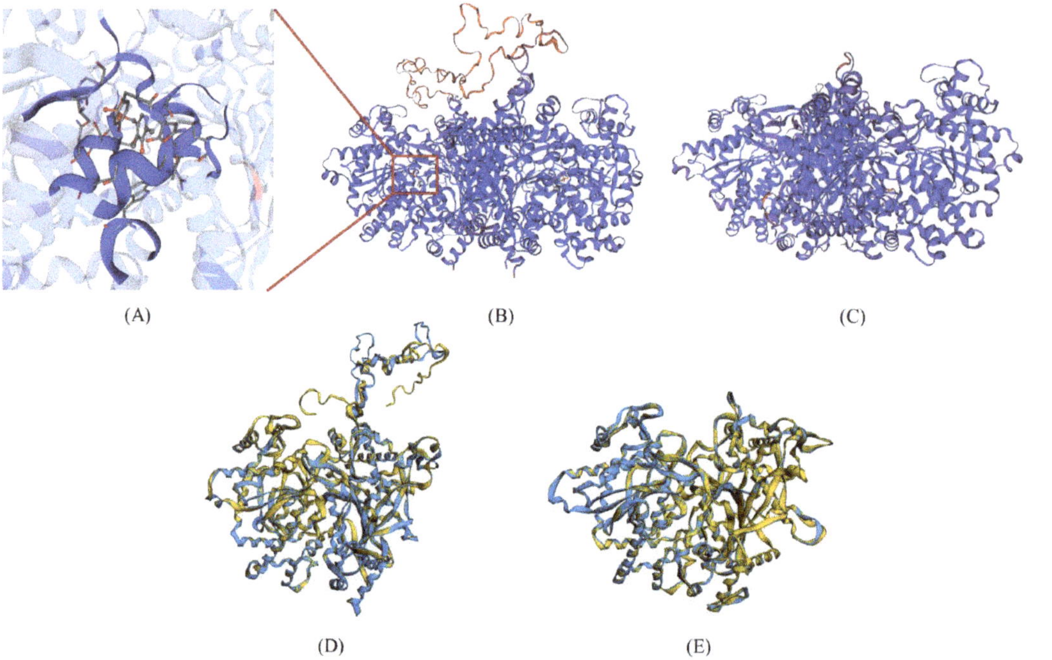

Figure 5. The 3D structures of maize PHO1 and PHO2: (**A**) Amino acid residues and their location involved in the binding of a ligand (PLP); (**B**) Homo-dimeric 3D structure of the PHO1 of *Z. mays*. The L80 domain (orange) arranged outside the main protein core; (**C**) Homo-dimeric 3D structure of the PHO2 of *Z. mays*; (**D**) Superimposed structures of the predicted PHO1 of *Z. mays* (yellow-colored) over the PHO1 of *H. vulgare* (blue colored); (**E**) Superimposed structures of the predicted PHO2 of *Z. mays* (yellow-colored) over the PHO2 of *A. thaliana* (blue colored).

Different values for various parameters obtained after superimposition are mentioned in Table S6. A high similarity level was observed upon pairwise alignment of 3D structures of PHO1 and PHO2 from each of the 12 species with the reference structure of the *O. sativa* ranging from 85.23% to 94.34% for PHO1 and 89.12% to 94.34% for PHO2. In addition, a high similarity level of intrinsic physical and chemical properties was also observed, ranging from 56.56% to 97.94% for PHO1 and 70.45% to 99.27% for PHO2 in each of the mentioned species. For PHO1, the root mean square deviation (RMSD) value was 2.04–5.68 Å in monocots and 2.10–3.87 Å in dicots. For PHO2 this was 0.04–1.63 Å in monocots and 0.25–0.37 Å in dicots. The main protein core regions of PHO1 and PHO2 were favorably superimposed to the reference structure except for the L80 domains in PHO1 (Figure 5, Figures S4 and S5 and Table S7).

2.2.4. Ligand Binding Sites Analysis

The aa residues constituting the ligand-binding sites were identified only in PHO1 except for the PHO1 of *C. annuum*, *S. tuberosum*, *B. distachyon*, and *S. lycopersicum*. However, no ligand binding sites were identified for PHO2 except in C. annum (Figure 5 and Table S8). The ligand-binding sites were confined to the residues in the GT-35-glycogen-phosphorylase domain and were nearer to the C-terminal. Generally, one to two clusters of ligands were predicted for the binding of pyridoxal-5-phosphate (PLP) molecules. For PHO1, aa residues ranging from 15 to 16 and from 14 to 15 were involved in the binding of the ligands in monocots and in dicot, respectively (similar for PHO2). In all monocots and dicots, the ligand-binding sites were common (~99%), however, they fluctuated in the position of aa residues.

2.3. Phylogenetic Analysis

Initially, three separate trees were constructed for phylogenetic analysis based on aa sequences using PHO1, PHO2, and PHO1 + PHO2 for each. All trees were mostly identical. Only the tree based on PHO1 + PHO2 was performed (Figure 6). The tree had two major clades (PHO1 and PHO2), each with two main clusters including eight monocots and four dicots for both PHO1 and PHO2. The PHO1 and the PHO2 of *Z. mays* shared closer relationships with *S. italica*, *S. bicolor*, and *P. hallii* among monocots and were grouped in a subcluster.

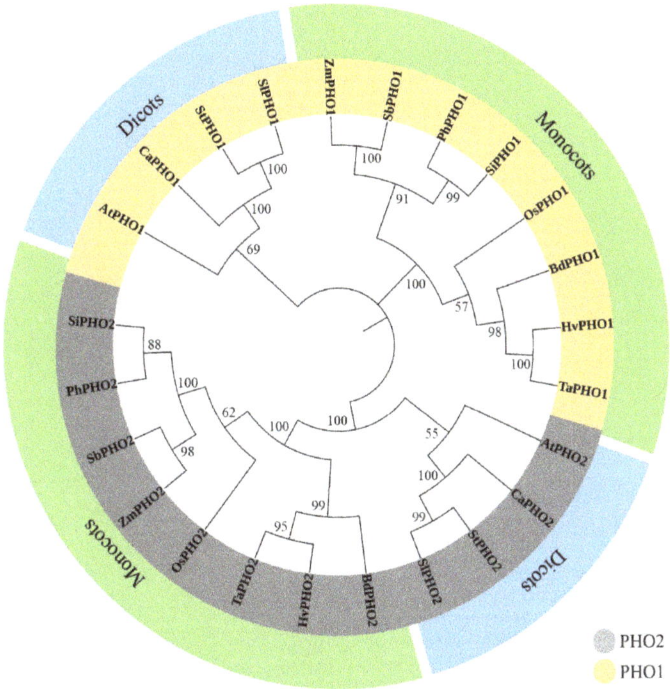

Figure 6. A phylogenetic tree constructed by the neighbor-joining method using the amino acid sequences of PHO1 and PHO2 to depict the relationship among monocots and dicots. The bootstrap value is calculated based on 1000 replications and displayed on each node. Symbols include: Sb, *Sorghum bicolor*; Zm, *Zea mays*; Si, *Setaria italica*; Ph, *Panicum hallii*; Bd, *Brachypodium distachyon*; Ta, *Triticum aestivum*; Hv, *Hordeum vulgare*; Os, *Oryza sativa*; At, *Arabidopsis thaliana*; Sl, *Solanum lycopersicum*; St, *Solanum tuberosum*; and Ca, *Capsicum annuum*. The branch length indicates the magnitude of genetic changes.

2.4. Expression Analysis of PHO1 and PHO2

The results based on the microarray analysis showed that the expression pattern of PHO1 and PHO2 remained similar among all monocots and dicots. However, the expression pattern in the PHO1 of *Z. mays* was much more complex. In *Z. mays*, real-time quantitative polymerase chain reaction (RT-qPCR) analysis indicated that the expression level of PHO1 was the highest in the endosperm and the lowest in the anthers. Unlike PHO1, the expression level of PHO2 was the highest in the embryo and the lowest in the pollens. The results indicated that the expression level of PHO1 was low in the early stages of seed development, highest at 12 days after pollination, and then decreased until the end of the seed development. However, the expression level of PHO2 remained the same throughout the seed development and peaked at 24 DAF (Figure 7).

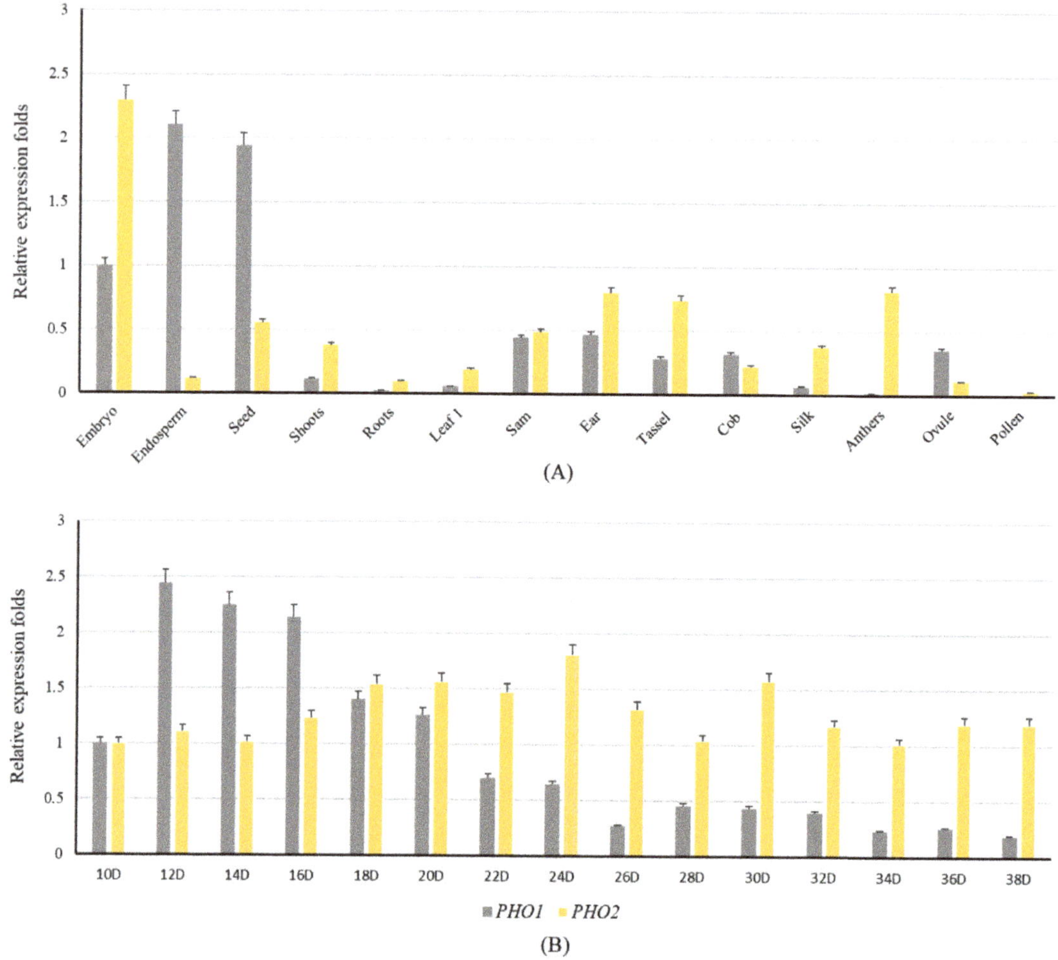

Figure 7. Distribution and expression pattern of the *PHO1* and *PHO2* transcripts of *Z. mays:* (**A**) Tissue-specific expression levels of the *PHO1* and *PHO2* of *Z. mays*. The relative expression pattern is shown and the transcript level in the embryo is used as control. The leaf1 is the first leaf harvested from the mature plant when it is flowering. The 15 DAP seeds and dissected endosperm were used to analyze the transcript level; (**B**) Relative expression levels of the *PHO1* and *PHO2* of *Z. mays* in different developmental stages of the endosperm. The transcript level in 10 DAP endosperm was used as a control.

Effect of ABA Treatment on the Level of Expression of PHO1 and PHO2

Responses of PHO1 and PHO2 under ABA treatment varied considerably among the main cereal crops. Based on microarray data under different stress conditions, PHO1 was found to be downregulated by ABA; however, ABA only significantly altered the expression of PHO2 in the main cereal crops. Unlike the PHO1 of *H. vulgare*, the PHO1 of *Z. mays* was significantly up-regulated under ABA treatment just as PHO2 (Figure 8).

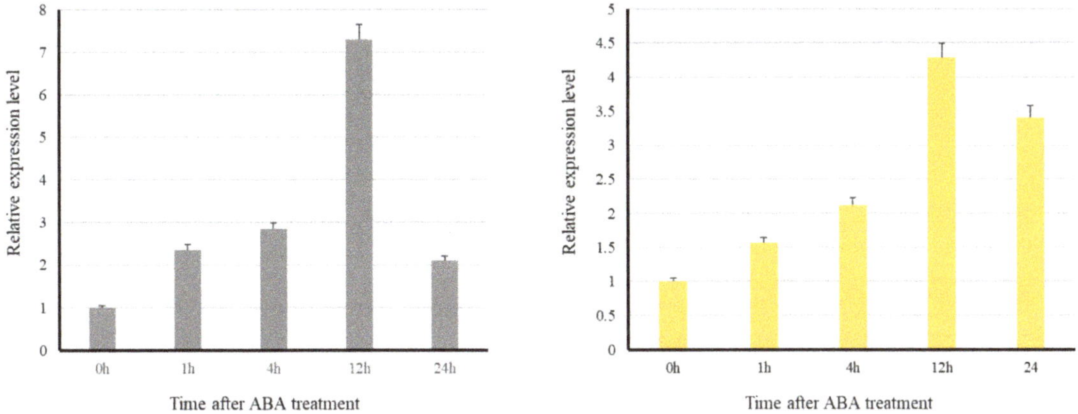

(A) (B)

Figure 8. Expression level of the *PHO1* and *PHO2* of maize under ABA treatment. The 15 DAP middle seed of corn cob was used to treat with ABA in MS medium. Furthermore, 0 h indicates no ABA treatment and is used as a control. (**A**) Effect on *PHO1* transcripts expression. (**B**) Effect on *PHO2* transcripts expression.

3. Discussion

Among both isoforms, PHO1 has been intensively investigated due to its main role in the growth and development of plants. In the present study we isolated the *PHO1* and *PHO2* of *Z. mays* and characterized their gene and protein structures together with evolutionary analysis in several other monocots and dicots. We analyzed their expression pattern based on RT-qPCR and microarray data and demonstrated the effect of ABA on the expression of *PHO1* and *PHO2* transcripts. Our results demonstrated the conservation of both isoforms of PHO, not only at the protein level, but also at the gene level in all the monocots and dicots that were characterized in this study. At the protein level, the conservation was evident by estimating similarity, identity, sequence coverage, presence of GT-35-glycogen-phosphorylase domain, and the 3D structure analysis across all the examined monocots and dicots. At the gene level, the exons were found to be more conserved than the introns in both isoforms of PHO across all demonstrated species. The expression of *PHO1* was found throughout the endosperm development suggesting that it could contribute with starch biosynthetic genes (SSs, SBEs and DBEs) during the seed development in *Z. mays*, as the availability of *PHO1* correlates with these enzymes [33–35], while, *PHO2* was found to be primarily expressed during seed germination. Various motifs were identified in the promoter regions of starch phosphorylase genes. The expression level of both isoforms of PHO under ABA treatment could be upregulated in *Z. mays* as suggested by the experimental results. In contrast, the expression level under different stress conditions varied much among several species.

PHO2 is found to be more conserved than PHO1 at both protein and gene levels. The higher similarity in PHO2 might be due to its functional conservation, which was nearly reported as similar in various species. Moreover, the variations in PHO1 could be attributed to its varying functions [8,28,36]. The regulatory and catalytic domains were found to be conserved in both PHO1 and PHO2 of all monocots and dicots. The ligand-binding sites were found to be mostly common for PHO, suggesting that their catalytic properties were closely related. Previously, it was described that a large portion of both of the PHO proteins shared by various species was similar and could be related to the function of phosphorylase [37]. Our results indicated that the PHO1 of the monocots was different from the PHO1 of dicots and PHO2 of all monocots and dicots in terms of an extra peptide in the middle (L80 region) and a transit peptide in the N-terminus. The addition of the extra peptide in the middle of PHO1 made it a differentially acting enzyme and could be critical in functional depiction [28]. Moreover, these regions were highly variable across the PHO1 of all monocots. The exact fate of this region has not been characterized yet. A specialized PEST motif [32] and various phosphorylation sites [31,38] have been described in this region of PHO1, which makes it critical in terms of structure, function, and regulation. Previously, phosphorylation of the starch biosynthetic enzymes was reported as the key modification for the regulation and activity of various enzymes [39]. The 3D modeling suggested the outside arrangements of the L80 domain from the main protein core and was found to restrict the activity of PHO1 for smaller polysaccharide substrates and sterically hinder the large polysaccharides, such as starch or glycogen [26]. Other than direct actions on the surface of the substrate, the L80 domain could also be involved in the phosphorylation-dependent complex formation [1]. Site-directed point mutations on phosphorylation sites, cleavage, or even complete removal of this L80 domain could increase the affinity of PHO1 for the larger molecules. Other than the L80 region and transit peptide in PHO1, both isoforms were mostly identical indicating that they might evolve from a single gene and the L80 domain was an additional product of an evolutionary process. The superimposition of predicted dimeric structures of PHO1 over the PHO1 of *H. vulgare* (only known structure of dimeric PHO1) and PHO2 over the PHO2 of *A. thaliana* (only known structure of dimeric PHO1) showed the accuracy of the structures and confirmed the identity of the true orthologs identified in the study.

The sequences of genes for PHO1 and PHO2 of monocots and dicots varied and were found to be more conserved in monocots. The number of introns and exons of *PHO1* remained constant during evolution, except in *S. tuberosum* (seven exons and six introns). An additional exon and intron were found in the *PHO2* of dicots. After the divergence of monocots and dicots from a common ancestor [40], exon loss occurred in monocots [41], which was supported by the fact that intron was longer in monocots than in dicots. The divergence of monocots and dicots was also apparent in our evolutionary tree. This is reasonable as storage organs for starch domestication patterns could vary in monocots and dicots. Higher variations in dicots suggested higher conservation in monocots than in dicots.

The lack of shared gene content (synteny) and gene order (collinearity) in most of the monocots and dicots, except in *Z. mays*, *S. bicolor*, *P. hallii* and *S. italica*, suggested that the chromosomes carrying the genes for PHO1 had undergone a massive reshuffling during evolution. The retention of some degree of synteny and collinearity in *S. bicolor*, *S. italica*, and *P. hallii* was understandable as these were more related to *Z. mays* compared with other evaluated species [42]. However, the lack of synteny and collinearity among the PHO1 of *H. vulgare*, *T. aestivum*, and *B. distachyon* was surprising as extensive conservation was reported in grasses [43], as seen in PHO2 of various species. The available literature on synteny and microsynteny [44,45] also suggested that synteny at the whole-genome level was conserved among grasses, however, it was missing in the genomic regions [44–47].

Several *cis*-regulatory elements have been prospected previously in the promotor region [32,37]. We also identified the elements responsible for the response to abiotic stress, tissue-specific expression, and light. The light-responsive element might have an important role as the starch synthesis took place during the daytime. Thus, a tight regulation pathway was required to control the response to light. The response elements in the upstream region of *PHO1* and *PHO2* might be responsible for the high expression level during the endosperm development. The prospected response elements in the present study need to be validated using wet-lab to determine the functional significance.

Microarray-based data have indicated the uniform pattern of *PHO1* transcripts in various tissues and seed developmental stages among *H. vulgare*, *O. sativa*, and *A. thaliana*. The higher expression of *PHO1* in the developing endosperm during the anthesis and grain filling period, and the lower expression in vegetative tissues, suggested its role in starch metabolism in *Z. mays*. Similar results were reported in *T. aestivum* [4], *O. sativa* [48], and *H. vulgare* [37]. The expression analysis of the *PHO2* of *Z. mays* suggested that it could mainly involve the degradation and maturation of starch, as the expression pattern was found to be relatively higher in the lateral stages of endosperm development. These results are agreeable with the PHO2 of *H. vulgare* [37] and *T. aestivum* [4] and it was shown that the expression of PHO2 correlates with its activity and it may degrade the reserve starch in plant organs [4]. In addition, it has been suggested previously that it may also be involved in metabolizing the products of starch degradation from chloroplast in cells with intact plastid, and regulates the cytosolic Glc-1-P level. The expression mode of *PHO1* was similar in the reported cereal crops, indicating that *PHO1* might play the identical role/activity in the metabolic pathways in each of the examined species, as the availability of *PHO1* correlates with the expression levels of other starch biosynthetic enzymes in higher plants [35]. The expression pattern of both isoforms are found to be relatively low in leaves for each of the examined species. Early speculation was that phosphorolysis was the major route of degradation in chloroplast [49], however, it is now clear that many chloroplast and starch containing plastids possess amylases in addition to PHO and can degrade the starch hydrolytically as well as phosphorolytically [50]. The importance of *PHO1* and *PHO2* in leaves have not been properly characterized. However, a study on the knockout mutant of *A. thaliana* (lack of *PHO1*) have found no alteration in starch and sugar metabolism and suggested that hydrolysis, rather than phosphorylsis, may be the major route of degradation in the chloroplast [51]. In addition, they show that phosphorolysis is not necessary for a normal metabolism in most leaf cells.

All studied *PHO1* and *PHO2* are proposed to have the regions (in promotors) that have functions in ABA-induced expression, but a study on PHO1 of *H. vulgare* suggested no effect of ABA on the expression of PHO1 but up-regulation of PHO2 [20]. However, in *A. thaliana* significant down-regulation for both isoforms were detected. In contrast, it is found that ABA can up-regulate the expression of both the *PHO1* and *PHO2* in *Zea mays*. The effect of abiotic stresses on the expression of PHO has not been determined for various crops and could be worthy of investigation. For *PHO1* and *PHO2* in maize, ABA-induced expression changes might be the same according to phylogenetic analysis and aa residue alignments.

4. Materials and Methods

4.1. Identification of "True" Orthologs of PHO1 and PHO2 Encoding Genes

Full-length cDNA and protein sequences of PHO in *O. sativa* were used as a reference in tBLASTx to identify "true" orthologs of PHO1 and PHO2 encoding genes in various monocots: *Zea mays* (*Z. mays*), *Setaria italica* (*S. italica*), *Panicum hallii* (*P. hallii*), *Brachypodium distachyon* (*B. distachyon*), *Oryza sativa* (*O. sativa*), *Hordeum vulgare* (*H. vulgare*) and dicots; *Arabidopsis thaliana* (*A. thaliana*), *Solanum lycopersicum* (*S. lycopersicum*), *Solanum tuberosum* (*S. tuberosum*), and *Capsicum annuum* (*C. annuum*) according to the criteria described in [52]. Briefly, the identified orthologs were based on a high level of sequence identities (more than 80% for monocots and more than 70% for dicots) and query the coverage along the PHO1 and PHO2 protein length, the presence of same domains and motifs as found in the original query sequence, and the conservation of sequence intervals between domains and motifs. Retrieved gene sequences were used to identify the full-length gene sequences from various databases, including NCBI (https://www.ncbi.nlm.nih.gov/, accessed on 5 January 2022), EMBL (https://www.embl.org//, accessed on 5 January 2022), and Ensembl (http://plants.ensembl.org/index.html/, accessed on 5 January 2022).

4.2. Gene Structure Analysis

Genomic DNA and CDS were used to determine the intron–exon junctions in full-length gene sequences. The intron phases (0, 1 and 2) were marked according to their position relative to the reading frame: intron insertion between two codons (phase 0), intron insertion after the first base pair of the codon (phase 2), or intron insertion after the second base pair of the codon (phase 3). Both exon and intron regions were employed to measure the GC content by GC calculator (https://www.sciencebuddies.org/, accessed on

1 January 2022). To evaluate the simple sequence repeats (SSRs) and retro-elements, a repeat masker 4.0.5 (http://www.repeatmasker.org/,accessed on 1 January 2022) was used with the default setting. The cis-regulatory response elements were identified in 1 kb upstream genomic regions of the translation start site using the PlantCare bioinformatics tool [53]. Following the criteria described by [54], the response elements on the sense and nonsense strand showing the matrix value of >5 was accepted.

4.3. Protein Sequence Analysis

All orthologs for PHO1 and PHO2 were used in multiple sequence alignment to generate the consensus amino acid (aa) sequence through DNAMAN software with default settings (http://www.dnaman.com, accessed on 1 January 2022). An aa present in all orthologs at a specific position was used in consensus sequence generation. Maize aa was used in consensus sequence if a different aa was found at the position. Insertions were also added to the consensus sequence that was relative to reference species [55]. All the aa sequences of orthologs were compared with the consensus sequence to find out the sequence similarity. A scale of 0–5 was used for the similarity index for the monocots and dicots. As such, zero value indicated the lack of similarity, whereas the value of five in monocots and the value of five in dicots suggested the conservation of aa compared with aa in the consensus sequence. CDD analysis was carried out for depicting various domains and motifs in the consensus sequence (http://www.ncbi.nlm.nih.gov/Structure/cdd/wrpsb.cg, accessed on 1 January 2022).

4.4. 3D Structure Analysis of the PHO1 and the PHO2

The Swiss model of the automated mode was used to generate the 3D structures of PHO1 and PHO2 using their aa sequences. The generated 3D structures were subjected to geometric and energetic verifications, and for this purpose, the following servers were used; structure analysis and verification server (SEVES) (http://nihserver.mbi.ucla.edu/SAVES, accessed on 1 January 2022), PROCHECK to check the aa in favored regions relative to other regions [56], ERRAT to evaluate the digits of non-bonded interactions among different atoms [57], VERIFY3D to evaluate the 3D structure compatibility with its aa sequence [58] and a structure assessment tool in the Swiss-model server. To confirm the PHO1 and PHO2 3D structures, the FATCAT server was used by superimposing the 3D structures of PHO1 and PHO2 from each of the mentioned species on the 3D structures of the PHO1 of *H. vulgare* and the PHO2 of *A. thaliana*, respectively.

4.5. Ligand-Binding Site Analysis

Scrutiny for the ligand-binding sites was conducted for both PHO1 and PHO2 using the 3DligandSite (http://www.sbg.bio.ic.ac.uk/3dligandsite/advanced.cgi, accessed on 5 January 2022) online software [59]. Since no output was available when PHO2 from the mentioned species was used, no analysis was carried out for PHO2.

4.6. Phylogenetic, Synteny and Collinearity Analysis

MEGA7 [60] was employed to analyze the phylogenetic relationship using the aa sequences of PHO1 and PHO2. The unrooted phylogenetic tree was constructed by the neighbor-joining method of distance matrix [61] with default settings. Synteny and collinearity of PHO1 and PHO2 were studied with Genomicus [62] (https://www.genomicus.bio.ens.psl.eu/, accessed on 1 January 2022) using blocks of genes associated with the PHO1 and PHO2 of *O. sativa*.

4.7. Preparation of Plant Material

Mo17 inbred line of *Z. mays* was selected to identify the expression level and the effect of ABA treatment on the expression level of PHO1 and PHO2. The inbred line was grown under the normal field conditions at Chongzhou Research Base of Sichuan Agricultural

University, China. The materials were collected at 2-day intervals and preserved at −80 °C for experiments.

4.8. Expression Analysis

In silico expression analysis of the genes for PHO1 and PHO2 of the main cereal crops (*O. sativa*, *H. vulgare*, and *A. thaliana*) was carried out using the "Genevestigator" microarray database. In the case of *Z. mays*, the expression analysis in different tissues was carried out using the expression level data determined by RT-qPCR. The extraction of RNA, cDNA synthesis, and expression analysis were carried out as described by Jian M. [37]. The *GAPDH* gene encoding glycerinaldehyde-phosphate dehydrogenase was used as internal reference. The following primers were used to measure the transcript level of *PHO1*: Fwd: CTAACAGGACAATATGCA; Rev: GCTTCATTGGCCTTGGCA. The following primers were used to measure the transcript level of *PHO2*: Fwd: ATGAGTGCGGCGGCGAA; Rev: CTTGGCAATGCGCTGCTTCA.

4.9. ABA Treatment Analysis

The middle seed of corn cob 15 days after self-pollination of Mo17 was placed in a liquid MS medium containing 50 µmol/L ABA to slowly shake (50 rpm) for 24 h. At the same time, the seeds were cultured in the MS medium without ABA as the control. Some seeds were taken out at indicated time points (0, 1, 4, 12, and 24 h), quickly frozen with liquid nitrogen, and stored for an RT-qPCR experiment.

Supplementary Materials: The following supporting information can be downloaded at: https://www.mdpi.com/article/10.3390/ijms23094518/s1.

Author Contributions: G.Y., Y.H. (Yubi Huang) and N.S. designed the experiments and wrote the article. L.L., N.M. and H.H. analyzed the structure of the PHO1 and PHO2 genes. G.Y. and N.S. analyzed the regulatory elements and amino acids sequence similarity of PHO1 and PHO2. Y.X., Y.L. (Yinghong Liu) and J.Z. performed the phylogenetic tree analysis. Y.L. (Yangping Li), N.Z. and Y.H. (Yufeng Hu) helped to perform and analyze the RT-qPCR. H.L. and Y.H. (Yubi Huang) read and approved the contents. All authors have read and agreed to the published version of the manuscript.

Funding: This work was supported by the National Key Research and Development Program of China (2021YFF1000304), the Natural Science Foundation of China (No: 31501322 and 31971960), Postdoctoral Special Foundation of Sichuan Province (No: 03130104), Overseas Scholar Science and Technology Activities Project Merit Funding (No: 00124300).

Institutional Review Board Statement: Not applicable.

Informed Consent Statement: Not applicable.

Data Availability Statement: Not applicable.

Acknowledgments: We would like to thank Hui Li for her technical support. This work was partly supported by the Maize Research Institute, Sichuan Agricultural University.

Conflicts of Interest: The authors declare that the research was conducted in the absence of any commercial or financial relationships that could be construed as a potential conflict of interest.

References

1. Tetlow, I.J.; Wait, R.; Lu, Z.; Akkasaeng, R.; Bowsher, C.G.; Esposito, S.; Kosar-Hashemi, B.; Morell, M.K.; Emes, M.J. Protein phosphorylation in amyloplasts regulates starch branching enzyme activity and protein-protein interactions. *Plant Cell* **2004**, *16*, 694–708. [CrossRef] [PubMed]
2. Crofts, N.; Nakamura, Y.; Fujita, N. Critical and speculative review of the roles of multi-protein complexes in starch biosynthesis in cereals. *Plant Sci.* **2017**, *262*, 1–8. [CrossRef] [PubMed]
3. James, M.G.; Denyer, K.; Myers, A.M. Starch synthesis in the cereal endosperm. *Curr. Opin. Plant Biol.* **2003**, *6*, 215–222. [CrossRef]
4. Schupp, N.; Ziegler, P. The relation of starch phosphorylases to starch metabolism in wheat. *Plant Cell Physiol.* **2004**, *45*, 1471–1484. [CrossRef]

5. Dauvillee, D.; Chochois, V.; Steup, M.; Haebel, S.; Eckermann, N.; Ritte, G.; Ral, J.P.; Colleoni, C.; Hicks, G.; Wattebled, F.; et al. Plastidial phosphorylase is required for normal starch synthesis in Chlamydomonas reinhardtii. *Plant J.* **2006**, *48*, 274–285. [CrossRef]
6. Young, G.H.; Chen, H.M.; Lin, C.T.; Tseng, K.C.; Wu, J.S.; Juang, R.H. Site-specific phosphorylation of L-form starch phosphorylase by the protein kinase activity from sweet potato roots. *Planta* **2006**, *223*, 468–478. [CrossRef]
7. Nakamura, Y.; Ono, M.; Utsumi, C.; Steup, M. Functional interaction between plastidial starch phosphorylase and starch branching enzymes from rice during the synthesis of branched maltodextrins. *Plant Cell Physiol.* **2012**, *53*, 869–878. [CrossRef]
8. Lin, Y.C.; Chen, H.M.; Chou, I.M.; Chen, A.N.; Chen, C.P.; Young, G.H.; Lin, C.T.; Cheng, C.H.; Chang, S.C.; Juang, R.H. Plastidial starch phosphorylase in sweet potato roots is proteolytically modified by protein-protein interaction with the 20S proteasome. *PLoS ONE* **2012**, *7*, e35336. [CrossRef]
9. Hwang, S.K.; Singh, S.; Cakir, B.; Satoh, H.; Okita, T.W. The plastidial starch phosphorylase from rice endosperm: Catalytic properties at low temperature. *Planta* **2016**, *243*, 999–1009. [CrossRef]
10. Tickle, P.; Burrell, M.M.; Coates, S.A.; Emes, M.J.; Tetlow, I.J.; Bowsher, C.G. Characterization of plastidial starch phosphorylase in *Triticum aestivum* L. endosperm. *J. Plant Physiol.* **2009**, *166*, 1465–1478. [CrossRef]
11. Tsai, C.Y.; Nelson, O.E. Two additional phosphorylases in developing maize seeds. *Plant Physiol.* **1969**, *44*, 159–167. [CrossRef] [PubMed]
12. Subasinghe, R.M.; Liu, F.; Polack, U.C.; Lee, E.A.; Emes, M.J.; Tetlow, I.J. Multimeric states of starch phosphorylase determine protein-protein interactions with starch biosynthetic enzymes in amyloplasts. *Plant Physiol. Biochem.* **2014**, *83*, 168–179. [CrossRef] [PubMed]
13. Brisson, N.; Giroux, H.; Zollinger, M.; Camirand, A.; Simard, C. Maturation and subcellular compartmentation of potato starch phosphorylase. *Plant Cell* **1989**, *1*, 559–566. [CrossRef] [PubMed]
14. Newgard, C.B.; Hwang, P.K.; Fletterick, R.J. The family of glycogen phosphorylases: Structure and function. *Crit. Rev. Biochem. Mol. Biol.* **1989**, *24*, 69–99. [CrossRef] [PubMed]
15. Zeeman, S.C.; Delatte, T.; Messerli, G.; Umhang, M.; Stettler, M.; Mettler, T.; Streb, S.; Reinhold, H.; Kotting, O. Starch breakdown: Recent discoveries suggest distinct pathways and novel mechanisms. *Funct. Plant Biol.* **2007**, *34*, 465–473. [CrossRef]
16. Orzechowski, S. Starch metabolism in leaves. *Acta Biochim. Pol.* **2008**, *55*, 435–445. [CrossRef]
17. Fettke, J.; Eckermann, N.; Kotting, O.; Ritte, G.; Steup, M. Novel starch-related enzymes and carbohydrates. *Cell. Mol. Biol.* **2007**, *52*, OL883-904.
18. Yu, Y.; Mu, H.H.; Wasserman, B.P.; Carman, G.M. Identification of the maize amyloplast stromal 112-kD protein as a plastidic starch phosphorylase. *Plant Physiol.* **2001**, *125*, 351–359. [CrossRef]
19. Satoh, H.; Shibahara, K.; Tokunaga, T.; Nishi, A.; Tasaki, M.; Hwang, S.K.; Okita, T.W.; Kaneko, N.; Fujita, N.; Yoshida, M.; et al. Mutation of the plastidial alpha-glucan phosphorylase gene in rice affects the synthesis and structure of starch in the endosperm. *Plant Cell* **2008**, *20*, 1833–1849. [CrossRef]
20. Ma, J.; Jiang, Q.T.; Zhang, X.W.; Lan, X.J.; Pu, Z.E.; Wei, Y.M.; Liu, C.; Lu, Z.X.; Zheng, Y.L. Structure and expression of barley starch phosphorylase genes. *Planta* **2013**, *238*, 1081–1093. [CrossRef]
21. Albrecht, T.; Greve, B.; Pusch, K.; Kossmann, J.; Buchner, P.; Wobus, U.; Steup, M. Homodimers and heterodimers of Pho1-type phosphorylase isoforms in *Solanum tuberosum* L. as revealed by sequence-specific antibodies. *Eur. J. Biochem.* **1998**, *251*, 343–352. [CrossRef] [PubMed]
22. Steup, M.; Robenek, H.; Melkonian, M. In-vitro degradation of starch granules isolated from spinach chloroplasts. *Planta* **1983**, *158*, 428–436. [CrossRef] [PubMed]
23. Rathore, R.S.; Garg, N.; Garg, S.; Kumar, A. Starch phosphorylase: Role in starch metabolism and biotechnological applications. *Crit. Rev. Biotechnol.* **2009**, *29*, 214–224. [CrossRef] [PubMed]
24. Ko, Y.T.; Chang, J.Y.; Lee, Y.T.; Wu, Y.H. The identification of starch phosphorylase in the developing mungbean (*Vigna radiata* L.). *J. Agric. Food Chem.* **2005**, *53*, 5708–5715. [CrossRef]
25. Hwang, S.K.; Koper, K.; Satoh, H.; Okita, T.W. Rice Endosperm Starch Phosphorylase (Pho1) Assembles with Disproportionating Enzyme (Dpe1) to Form a Protein Complex That Enhances Synthesis of Malto-oligosaccharides. *J. Biol. Chem.* **2016**, *291*, 19994–20007. [CrossRef]
26. Cuesta-Seijo, J.A.; Ruzanski, C.; Krucewicz, K.; Meier, S.; Hagglund, P.; Svensson, B.; Palcic, M.M. Functional and structural characterization of plastidic starch phosphorylase during barley endosperm development. *PLoS ONE* **2017**, *12*, e0175488. [CrossRef]
27. Tsai, C.Y.; Nelson, O.E. Phosphorylases I and II of Maize Endosperm. *Plant Physiol.* **1968**, *43*, 103–112. [CrossRef]
28. Mori, H.; Tanizawa, K.; Fukui, T. A chimeric alpha-glucan phosphorylase of plant type L and H isozymes. Functional role of 78-residue insertion in type L isozyme. *J. Biol. Chem.* **1993**, *268*, 5574–5581. [CrossRef]
29. Zeeman, S.C.; Thorneycroft, D.; Schupp, N.; Chapple, A.; Weck, M.; Dunstan, H.; Haldimann, P.; Bechtold, N.; Smith, A.M.; Smith, S.M. Plastidial alpha-glucan phosphorylase is not required for starch degradation in Arabidopsis leaves but has a role in the tolerance of abiotic stress. *Plant Physiol.* **2004**, *135*, 849–858. [CrossRef]
30. Young, N.D.; Debelle, F.; Oldroyd, G.E.; Geurts, R.; Cannon, S.B.; Udvardi, M.K.; Benedito, V.A.; Mayer, K.F.; Gouzy, J.; Schoof, H.; et al. The Medicago genome provides insight into the evolution of rhizobial symbioses. *Nature* **2011**, *480*, 520–524. [CrossRef]

31. Hwang, S.K.; Koper, K.; Okita, T.W. The plastid phosphorylase as a multiple-role player in plant metabolism. *Plant Sci.* **2020**, *290*, 110303. [CrossRef] [PubMed]
32. Shoaib, N.; Liu, L.; Ali, A.; Mughal, N.; Yu, G.; Huang, Y. Molecular Functions and Pathways of Plastidial Starch Phosphorylase (PHO1) in Starch Metabolism: Current and Future Perspectives. *Int. J. Mol. Sci.* **2021**, *22*, 10450. [CrossRef]
33. Liu, F.; Makhmoudova, A.; Lee, E.A.; Wait, R.; Emes, M.J.; Tetlow, I.J. The amylose extender mutant of maize conditions novel protein-protein interactions between starch biosynthetic enzymes in amyloplasts. *J. Exp. Bot.* **2009**, *60*, 4423–4440. [CrossRef] [PubMed]
34. Tetlow, I.J.; Davies, E.J.; Vardy, K.A.; Bowsher, C.G.; Burrell, M.M.; Emes, M.J. Subcellular localization of ADPglucose pyrophosphorylase in developing wheat endosperm and analysis of the properties of a plastidial isoform. *J. Exp. Bot.* **2003**, *54*, 715–725. [CrossRef] [PubMed]
35. Morell, M.K.; Blennow, A.; Kosar-Hashemi, B.; Samuel, M.S. Differential expression and properties of starch branching enzyme isoforms in developing wheat endosperm. *Plant Physiol.* **1997**, *113*, 201–208. [CrossRef]
36. Nakamura, Y.; Ono, M.; Sawada, T.; Crofts, N.; Fujita, N.; Steup, M. Characterization of the functional interactions of plastidial starch phosphorylase and starch branching enzymes from rice endosperm during reserve starch biosynthesis. *Plant Sci.* **2017**, *264*, 83–95. [CrossRef]
37. Ma, J.; Jiang, Q.T.; Zhao, Q.Z.; Zhao, S.; Lan, X.J.; Dai, S.F.; Lu, Z.X.; Liu, C.; Wei, Y.M.; Zheng, Y.L. Characterization and expression analysis of waxy alleles in barley accessions. *Genetica* **2013**, *141*, 227–238. [CrossRef]
38. Walley, J.W.; Shen, Z.; Sartor, R.; Wu, K.J.; Osborn, J.; Smith, L.G.; Briggs, S.P. Reconstruction of protein networks from an atlas of maize seed proteotypes. *Proc. Natl. Acad. Sci. USA* **2013**, *110*, E4808–E4817. [CrossRef]
39. Yu, G.; Lv, Y.; Shen, L.; Wang, Y.; Qing, Y.; Wu, N.; Li, Y.; Huang, H.; Zhang, N.; Liu, Y.; et al. The Proteomic Analysis of Maize Endosperm Protein Enriched by Phos-tag(tm) Reveals the Phosphorylation of Brittle-2 Subunit of ADP-Glc Pyrophosphorylase in Starch Biosynthesis Process. *Int. J. Mol. Sci.* **2019**, *20*, 986. [CrossRef]
40. Chaw, S.M.; Chang, C.C.; Chen, H.L.; Li, W.H. Dating the monocot-dicot divergence and the origin of core eudicots using whole chloroplast genomes. *J. Mol. Evol.* **2004**, *58*, 424–441. [CrossRef]
41. Roy, S.W.; Gilbert, W. The evolution of spliceosomal introns: Patterns, puzzles and progress. *Nat. Rev. Genet.* **2006**, *7*, 211–221. [CrossRef] [PubMed]
42. Salse, J.; Piegu, B.; Cooke, R.; Delseny, M. New in silico insight into the synteny between rice (*Oryza sativa* L.) and maize (*Zea mays* L.) highlights reshuffling and identifies new duplications in the rice genome. *Plant J.* **2004**, *38*, 396–409. [CrossRef]
43. Pfeifer, M.; Martis, M.; Asp, T.; Mayer, K.F.; Lubberstedt, T.; Byrne, S.; Frei, U.; Studer, B. The perennial ryegrass GenomeZipper: Targeted use of genome resources for comparative grass genomics. *Plant Physiol.* **2013**, *161*, 571–582. [CrossRef] [PubMed]
44. Sorrells, M.E.; La Rota, M.; Bermudez-Kandianis, C.E.; Greene, R.A.; Kantety, R.; Munkvold, J.D.; Miftahudin; Mahmoud, A.; Ma, X.; Gustafson, P.J.; et al. Comparative DNA sequence analysis of wheat and rice genomes. *Genome Res.* **2003**, *13*, 1818–1827. [CrossRef] [PubMed]
45. Ahn, S.; Anderson, J.A.; Sorrells, M.E.; Tanksley, S.D. Homoeologous relationships of rice, wheat and maize chromosomes. *Mol. Gen. Genet.* **1993**, *241*, 483–490. [CrossRef]
46. Tarchini, R.; Biddle, P.; Wineland, R.; Tingey, S.; Rafalski, A. The complete sequence of 340 kb of DNA around the rice Adh1-adh2 region reveals interrupted colinearity with maize chromosome 4. *Plant Cell* **2000**, *12*, 381–391. [CrossRef] [PubMed]
47. Vision, T.J. Gene order in plants: A slow but sure shuffle. *New Phytol.* **2005**, *168*, 51–60. [CrossRef]
48. Hwang, S.K.; Nishi, A.; Satoh, H.; Okita, T.W. Rice endosperm-specific plastidial alpha-glucan phosphorylase is important for synthesis of short-chain malto-oligosaccharides. *Arch. Biochem. Biophys.* **2010**, *495*, 82–92. [CrossRef]
49. Stitt, M.; Wirtz, W.; Gerhardt, R.; Heldt, H.; Spencer, C.; Walker, D.; Foyer, C. A comparative study of metabolite levels in plant leaf material in the dark. *Planta* **1985**, *166*, 354–364. [CrossRef]
50. da Silva, P.M.; Eastmond, P.J.; Hill, L.M.; Smith, A.M.; Rawsthorne, S. Starch metabolism in developing embryos of oilseed rape. *Planta* **1997**, *203*, 480–487. [CrossRef]
51. Lao, N.T.; Schoneveld, O.; Mould, R.M.; Hibberd, J.M.; Gray, J.C.; Kavanagh, T.A. An Arabidopsis gene encoding a chloroplast-targeted β-amylase. *Plant J.* **1999**, *20*, 519–527. [CrossRef] [PubMed]
52. Dhaliwal, A.K.; Mohan, A.; Gill, K.S. Comparative analysis of ABCB1 reveals novel structural and functional conservation between monocots and dicots. *Front. Plant Sci.* **2014**, *5*, 657. [CrossRef] [PubMed]
53. Lescot, M.; Dehais, P.; Thijs, G.; Marchal, K.; Moreau, Y.; Van de Peer, Y.; Rouze, P.; Rombauts, S. PlantCARE, a database of plant cis-acting regulatory elements and a portal to tools for in silico analysis of promoter sequences. *Nucleic Acids Res.* **2002**, *30*, 325–327. [CrossRef] [PubMed]
54. Chen, X.; Wang, Z.; Wang, J.; Wang, M.; Zhao, L.; Wang, G. Isolation and characterization of *Brittle2* promoter from *Zea Mays* and its comparison with *Ze19* promoter in transgenic tobacco plants. *Plant Cell* **2007**, *88*, 11–20. [CrossRef]
55. Navarro, C.; Moore, J.; Ott, A.; Baumert, E.; Mohan, A.; Gill, K.S.; Sandhu, D. Evolutionary, Comparative and Functional Analyses of the Brassinosteroid Receptor Gene, BRI1, in Wheat and Its Relation to Other Plant Genomes. *PLoS ONE* **2015**, *10*, e0127544. [CrossRef]
56. Laskowski, R.A.; Rullmannn, J.A.; MacArthur, M.W.; Kaptein, R.; Thornton, J.M. AQUA and PROCHECK-NMR: Programs for checking the quality of protein structures solved by NMR. *J. Biomol. NMR* **1996**, *8*, 477–486. [CrossRef]

57. Colovos, C.; Yeates, T.O. Verification of protein structures: Patterns of nonbonded atomic interactions. *Protein Sci.* **1993**, *2*, 1511–1519. [CrossRef]
58. Eisenberg, D.; Luthy, R.; Bowie, J.U. VERIFY3D: Assessment of protein models with three-dimensional profiles. *Methods Enzymol.* **1997**, *277*, 396–404. [CrossRef]
59. Wass, M.N.; Kelley, L.A.; Sternberg, M.J. 3DLigandSite: Predicting ligand-binding sites using similar structures. *Nucleic Acids Res.* **2010**, *38*, W469–W473. [CrossRef]
60. Kumar, S.; Stecher, G.; Tamura, K. MEGA7: Molecular Evolutionary Genetics Analysis Version 7.0 for Bigger Datasets. *Mol. Biol. Evol.* **2016**, *33*, 1870–1874. [CrossRef]
61. Saitou, N.; Nei, M. The neighbor-joining method: A new method for reconstructing phylogenetic trees. *Mol. Biol. Evol.* **1987**, *4*, 406–425. [CrossRef] [PubMed]
62. Muffato, M.; Louis, A.; Poisnel, C.E.; Roest Crollius, H. Genomicus: A database and a browser to study gene synteny in modern and ancestral genomes. *Bioinformatics* **2010**, *26*, 1119–1121. [CrossRef] [PubMed]

International Journal of
Molecular Sciences

MDPI

Article

Genome-Wide Characterization of High-Affinity Nitrate Transporter 2 (NRT2) Gene Family in *Brassica napus*

Run-Jie Du [1,2], Ze-Xuan Wu [1,2], Zhao-Xi Yu [1,2], Peng-Feng Li [1,2], Jian-Yu Mu [1,2], Jie Zhou [1,2], Jia-Na Li [1,2] and Hai Du [1,2,*]

1 College of Agronomy and Biotechnology, Southwest University, Chongqing 400716, China; dddrj12345@email.swu.edu.cn (R.-J.D.); w1045123799@126.com (Z.-X.W.); yzx20010120@gmai.com (Z.-X.Y.); pengfengli17@126.com (P.-F.L.); mujianyu728@126.com (J.-Y.M.); zj893422105@163.com (J.Z.); ljn1950@swu.edu.cn (J.-N.L.)
2 Academy of Agricultural Sciences, Southwest University, Chongqing 400716, China
* Correspondence: haidu81@126.com or dh20130904@swu.edu.cn; Tel.: +86-1822-348-0008

Abstract: Nitrate transporter 2 (NRT2) plays an essential role in Nitrogen (N) uptake, transport, utilization, and stress resistance. In this study, the NRT2 gene family in two sequenced *Brassica napus* ecotypes were identified, including 31 genes in 'Zhongshuang11' (*BnaZSNRT2s*) and 19 in 'Darmor-*bzh*' (*BnaDarNRT2s*). The candidate genes were divided into three groups (Group I–III) based on phylogenetic analyses, supported by a conserved intron-exon structure in each group. Collinearity analysis revealed that the large expansion of *BnaZSNRT2s* attributed to allopolyploidization of ancestors *Brassica rapa* and *Brassica oleracea*, and small-scale duplication events in *B. napus*. Transcription factor (TF) binding site prediction, *cis*-element analysis, and microRNA prediction suggested that the expressions of *BnaZSNRT2s* are regulated by multiple factors, and the regulatory pattern is relatively conserved in each group and is tightly connected between groups. Expression assay showed the diverse and differentiated spatial-temporal expression profiles of *BnaZSNRT2s* in Group I, but conserved patterns were observed in Group II/III; and the low nitrogen (LN) stress up-regulated expression profiles were presented in Group I–III, based on RNA-seq data. RT-qPCR analyses confirmed that *BnaZSNRT2.5A-1* and *BnaZSNRT2.5C-1* in Group II were highly up-regulated under LN stress in *B. napus* roots. Our results offer valid information and candidates for further functional *BnaZSNRT2s* studies.

Keywords: *Brassica napus*; NRT2; gene family; evolution; nitrate; expression profile

Citation: Du, R.-J.; Wu, Z.-X.; Yu, Z.-X.; Li, P.-F.; Mu, J.-Y.; Zhou, J.; Li, J.-N.; Du, H. Genome-Wide Characterization of High-Affinity Nitrate Transporter 2 (NRT2) Gene Family in *Brassica napus*. *Int. J. Mol. Sci.* **2022**, *23*, 4965. https://doi.org/10.3390/ijms23094965

Academic Editor: Frank M. You

Received: 7 April 2022
Accepted: 28 April 2022
Published: 29 April 2022

1. Introduction

Nitrogen (N), mainly in the form of nitrate, is a crucial element for plant development and a vital factor affecting diverse plant bioprocesses, such as photosynthesis and protein synthesis [1,2]. Thus, it is substantially required by crops to form agricultural production [3–6]. Normally, sufficient N conditions can protect crop yield and quality in variable environments [7]. However, the distributions of biologically available N, including nitrate, ammonium, and small peptides, are uneven in natural and agricultural land worldwide [8]. In this case, large amounts of fertilizer were generally applied in production, and approximately 60% of the annual fertilizer consumption is N, which ensures more than 40% of the population's basic food needs [6,9]. On the other hand, excessive N fertilizer application brings a series of environmental damages, such as acid rain [10], water eutrophication [11,12], the greenhouse effect [13,14], and poor soil fertility [15]. Hence, analyses concerning increasing the efficiency in N absorption and utilization perform crucial roles in crop yield, quality, and even environmental protection.

NITRATE TRANSPORTER 2 (NRT2) homologous proteins are typical high-affinity nitrate transporters in the plant kingdom and are responsible for the nitrate uptake process in plants. In general, NRT2 proteins (NRT2s) have a typical membrane topology connected by a cytosolic loop, including 1 MFS domain that exhibits dual affinities for nitrate [16] and 12 transmembrane domains [17], which are usually located on the cell plasma membrane [18]. In *Arabidopsis*, there are 7 NRT2 genes (*AtNRT2s*) that were divided into three groups [6,19]; with Group I containing 5 members while Groups II/III only include 1 member each. In Group I, *AtNRT2.1* and *AtNRT2.2* play vital roles in nitrate uptake in roots [20]; *AtNRT2.4* acts a key role in nitrate uptake under low nitrogen (LN) stress in both shoots and roots [21]; *AtNRT2.6* is involved in biotic and abiotic stresses [22]; whereas *AtNRT2.3* remains to be functional elucidated yet. In Group II, *AtNRT2.5* plays an essential role in severe nitrogen starvation response. The expression of *AtNRT2.5* was highly induced after a long N starvation, and then it acted as the major transporter for high-affinity nitrate uptake [23]. In Group III, *AtNRT2.7* is expressed in tonoplast and contributes to the N accumulation in seeds [24]. Notably, except for *AtNRT2.7*, all of the other AtNRT2 transporters could interact with *AtNAR2.1*, enhancing the nitrate uptake capacity of *AtNRT2s* [25]. Similarly, the homologs of *AtNRT2s* in other plants were widely demonstrated to perform numerous roles in N uptake, transport, and utilization processes across developmental stages. For instance, in rice, *OsNRT2.3a* plays a key role in increasing N use efficiency and field [26]. In wheat, *TaNRT2.5* is expressed in the embryo and shell and acts in nitrate accumulation in seeds [27]. In maize, *ZmNRT2.1* regulates nitrate uptake along the root axis [28]. Together, *NRT2* homologs play key roles in nitrate uptake and even utilization in plants. Thus, systematically identifying the NRT2 gene families in plant genomes and exploring their roles involved in nitrate utilization processes may contribute to promoting the N utilization efficiency (NUE) and crop yields without resorting to excessive N fertilizer.

Brassica napus (*B. napus*; $A_nA_nC_nC_n$, 2n = 38) is a significant oil crop worldwide which was an allopolyploid produced by *Brassica rapa* (*B. rapa*; A_nA_n, n = 10) and *Brassica oleracea* (*B. oleracea*; C_nC_n, n = 9) ~7500 years ago [29,30]. So far, the genomes of 2 *B. napus* ecotypes have been sequenced, namely 'Darmor-*bzh*' and 'Zhongshuang 11' (ZS11). *B. napus* rely on the amount of N in the production, which needs more N fertilizer to produce a unit of yield than wheat and rice [31,32]. However, the NUE of *B. napus* is much lower than wheat and rice, causing a mass of N loss [33,34]. Given the important roles in N utilization-related processes, identifying the NRT2 encoding genes at a genome-wide level and exploring their roles in N uptake and utilization processes in *B. napus* has potential research significance and application value, aiming to improve the NUE and even the yield and quality of *B. napus*.

In our study, we conducted global analyses of the NRT2 gene family in 2 ecotypes of *B. napus* genomes ('Darmor-*bzh*' and 'ZS11') at the genome-wide level, accompanied by a series of bioinformatics assays of the candidates, including sequence structure, phylogenetic relationship, chromosomal location, collinearity relationship, gene duplication, regulatory mechanism prediction, etc. Then, we analyzed the spatial-temporal expression profiles of candidates in 52 *B. napus* (ZS11) tissues/organs across distinct developmental stages. Moreover, the LN stress expression patterns of the candidates in *B. napus* (ZS11) seedling roots were analyzed by RNA-Seq and RT-qPCR methods, respectively. Our findings constitute the first step toward further research on the molecular functions of *NRT2s* in *B. napus*.

2. Results

2.1. Identification of NRT2 Genes in B. napus

To identify the NRT2s in *B. napus* genome, we performed preparatory BLASTP and Tblastn searches using the protein sequences of *Arabidopsis* NRT2 protein (AtNRT2s) [35] as queries. Two sequenced genome databases of *B. napus* varieties in GENOSCOPE (Darmor–*bzh*, http://www.genoscope.cns.fr/brassicanapus/, accessed on 19 August 2014) [30] and BnPIR (Zhongshuang 11, ZS11, http://cbi.hzau.edu.cn/bnapus/, accessed on 10 September 2020) [29] were used. After excluding the redundant sequences, the remainders were further verified by SMART (http://smart.embl-heidelberg.de/, accessed on 26 October 2020) and

ExPASy (https://web.expasy.org/compute_pi/, accessed on 23 February 2022) online software to ensure the candidates contain the typical sequence features of this gene family. Finally, we obtained 19 candidate genes from the 'Darmor-*bzh*' genome (*BnaDarNRT2s*) and 31 candidates from the 'ZS11' genome (*BnaZSNRT2s*) with relative complete functional domains. Then, we named them based on the *Arabidopsis* homologous *NRT2s* and their chromosome locations in *B. napus*, such as the four homologs of *AtNRT2.1* gene in A_n subgenome were named as *BnaZSNRT2.1A-1* to *BnaZSNRT2.1A-4* whereas these in C_n were named as *BnaZSNRT2.1C-1* to *BnaZSNRT2.1C-4* (Table 1 and Table S1).

Table 1. Features of the 31 NRT2 genes (BnaZSNRT2s) identified in Brassica napus Zhongshuang 11 (ZS11) ecotype.

Gene Name	Genome ID	Chromosome	Protein Length (aa)	CDS Length (bp)	DNA Length (bp)	pI	Molecular Weight (kDa)	Subcellular Localization	
								Predicted by Cell-PLoc2.0	Predicted by WoLF PSORT
BnaZSNRT2.6A-1	BnaA01G0234100ZS	chrA01	546	1641	2624	7.23	59.39	Cell membrane	plasmalemma
BnaZSNRT2.6A-2	BnaA01G0234200ZS	chrA01	541	1626	1721	8.80	58.48	Cell membrane	plasmalemma
BnaZSNRT2.7A-1	BnaA02G0054200ZS	chrA02	484	1455	1525	8.14	52.06	Cell membrane	plasmalemma
BnaZSNRT2.3A-1	BnaA02G0096600ZS	chrA02	502	1509	6544	9.06	53.76	Cell membrane	plasmalemma
BnaZSNRT2.1A-1	BnaA06G0047500ZS	chrA06	530	1593	2163	9.07	57.70	Cell membrane	plasmalemma
BnaZSNRT2.1A-2	BnaA06G0047600ZS	chrA06	530	1593	1804	9.03	57.73	Cell membrane	plasmalemma
BnaZSNRT2.6A-3	BnaA06G0186600ZS	chrA06	543	1632	2940	6.89	59.12	Cell membrane	plasmalemma
BnaZSNRT2.6A-4	BnaA06G0186700ZS	chrA06	538	1617	1720	9.04	58.37	Cell membrane	plasmalemma
BnaZSNRT2.5A-1	BnaA08G0276500ZS	chrA08	499	1500	1773	9.01	54.17	Cell membrane	plasmalemma
BnaZSNRT2.1A-3	BnaA08G0300800ZS	chrA08	530	1593	2129	8.79	57.61	Cell membrane	plasmalemma
BnaZSNRT2.2A-1	BnaA09G0667700ZS	chrA09	506	1521	1983	9.02	54.83	Cell membrane	plasmalemma
BnaZSNRT2.1A-4	BnaA09G0667800ZS	chrA09	529	1590	1860	8.90	57.34	Cell membrane	plasmalemma
BnaZSNRT2.4A-1	BnaA10G0160100ZS	chrA10	527	1584	2368	8.90	57.61	Cell membrane	plasmalemma
BnaZSNRT2.3A-2	BnaA10G0160300ZS	chrA10	536	1611	1964	9.14	58.21	Cell membrane	plasmalemma
BnaZSNRT2.6C-1	BnaC01G0301400ZS	chrC01	546	1641	2618	7.66	59.40	Cell membrane	plasmalemma
BnaZSNRT2.6C-2	BnaC01G0301600ZS	chrC01	541	1626	1721	8.85	58.56	Cell membrane	plasmalemma
BnaZSNRT2.7C-1	BnaC02G0063100ZS	chrC02	502	1509	1579	7.61	53.45	Cell membrane	plasmalemma
BnaZSNRT2.3C-1	BnaC02G0116100ZS	chrC02	536	1611	2105	9.04	58.06	Cell membrane	plasmalemma
BnaZSNRT2.6C-3	BnaC03G0602800ZS	chrC03	538	1617	1720	9.13	58.13	Cell membrane	plasmalemma
BnaZSNRT2.6C-4	BnaC03G0603000ZS	chrC03	154	465	1109	9.86	17.09	Cell membrane	cytoplasm
BnaZSNRT2.6C-5	BnaC03G0603300ZS	chrC03	162	489	17582	6.91	17.55	Chlo Cyto	E.R.
BnaZSNRT2.6C-6	BnaC03G0603600ZS	chrC03	184	555	652	9.93	19.86	Cell membrane	chloroplast
BnaZSNRT2.1C-1	BnaC05G0059500ZS	chrC05	475	1428	1806	9.03	51.83	Cell membrane	plasmalemma
BnaZSNRT2.1C-2	BnaC05G0059600ZS	chrC05	530	1593	2059	9.03	57.76	Cell membrane	plasmalemma
BnaZSNRT2.8C-1	BnaC08G0033200ZS	chrC08	113	342	342	6.02	12.19	Cell membrane	cytoplasm
BnaZSNRT2.1C-3	BnaC08G0033300ZS	chrC08	395	1188	1579	9.00	43.15	Cell membrane	chloroplast
BnaZSNRT2.5C-1	BnaC08G0220400ZS	chrC08	498	1497	1799	9.11	54.13	Cell membrane	plasmalemma
BnaZSNRT2.2C-1	BnaC08G0532700ZS	chrC08	502	1509	1967	8.91	54.40	Cell membrane	plasmalemma
BnaZSNRT2.1C-4	BnaC08G0532800ZS	chrC08	529	1590	1915	8.95	57.38	Cell membrane	plasmalemma
BnaZSNRT2.4C-1	BnaC09G0443000ZS	chrC09	527	1584	4336	8.90	57.52	Cell membrane	plasmalemma
BnaZSNRT2.3C-2	BnaC09G0443100ZS	chrC09	567	1704	1987	9.08	61.87	Cell membrane	plasmalemma

Abbreviations: aa, amino acids; CDS, coding sequence; pI, isoelectric point; E.R., endoplasmic reticulum; Chlo, chloroplast; Cyto, Cytoplasm.

As shown in Table 1 and Table S1, the length of BnaDarNRT2 proteins (BnaDarNRT2s) and BnaZSNRT2 proteins (BnaZSNRT2s) ranged from 154 to 506 amino acids and 113 to

567 amino acids. The average value was 427.16 and 472.26, and the standard deviation (SD) was 99.54 and 126.73, respectively. Isoelectric point (pI) of BnaDarNRT2s and BnaZSNRT2s varied from 7.54 to 9.93 and 6.02 to 9.93, with an average value of 8.94 and 8.65, and the SD was 0.57 and 0.87, respectively. Their molecular weight (MW) varied from 17.09 to 54.83 kDa and 12.19 to 61.87 kDa, with the average value being 46.44 and 51.25, and the SD 10.7 and 13.75, respectively. Subcellular localization analysis showed that nearly all 19 BnaDarNRT2s and 30 BnaZSNRT2s were located on the cell membrane. Only BnaZSNRT2.6C-5 was located on chloroplast/cytoplasm, which suggested their potential function features in the nitrate utilization process.

2.2. Phylogenetic and Sequence Structure Analysis of B. napus NRT2 Gene Family

To investigate the phylogenetic relationship of the candidate *NRT2s*, a neighbor-joining (NJ) phylogenetic tree of the 19 BnaDarNRT2s, 31 BnaZSNRT2s, and 7 AtNRT2s was generated based on the multi-alignment of the whole-length protein sequences (Figure 1A). Due to technical reasons (having no common information site(s) between sequences), *BnaZSNRT2.8C-1* was excluded from the phylogenetic tree because of severe sequence deletion. According to the topology and bootstrap values of the NJ tree, the candidates were separated into three groups: Group I–III. Group I was the largest, which contained 15 *BnaDarNRT2s*, 26 *BnaZSNRT2s*, and 5 *AtNRT2s*; Group II/III both contained 2 *BnaDarNRT2s*, 2 *BnaZSNRT2s*, and 1 *AtNRT2s*. The number of *BnaDarNRT2s* and *BnaZSNRT2s* in Group II and III was equal, while the number of homologs between these two ecotypes was quite different in Group I. Excepting for 11 homologous gene pairs, there are 4 *BnaDarNRT2s* and 15 *BnaZSNRT2s* having non-homologs in Group I. In general, the *AtNRT2s* have homologs in both ecotypes, and the *AtNRT2s* in Group I have more homologs in 'ZS11' than in 'Darmor-*bzh*', excepting *AtNRT2*.6, which only has *BnaZSNRT2s* homologs. Subsequently, we analyzed the sequence identity and similarity of the full-length DNA, CDS, and protein sequences of each homologous gene pair (Table S2). The results showed that the sequence identity and similarity in Group II/III were very high, with the protein sequence identity ranging from 84.6–100% in each group, while that was decreased in Group I (ranging from 62.10–100%). These results indicated that the gene number, sequence features, and even functions of the homologous gene pairs in Group II and III were highly conserved during evolution, while those in Group I may undergo functional diversification.

Sequence structure feature observed 8 relatively conserved intron insertion sites in terms of conserved insertion site and phase, namely intron "1" to "8" (Figure 1C). The intron number of *NRT2s* varies in Group I–III or even within a group, while members of Group I had 1 to 5 introns, Group II had 2 introns, and Group III had 1 intron. Although the number of introns was diverse in the three groups, the intron insertion site and phase were generally conserved within each group. Members of Group II contained intron "4" and "7", which were conserved in this group. Similarly, the intron patterns were also completely conserved in group III, members of which contained intron "7". By contrast, the intron number, insertion site, and phase in Group I were diverse and relatively less conserved, which contained 6 relative conserved intron sites, "1–3", "5", "6", and "8". Among them, intron "5" (39/46, ~85%) and "6" (37/46, ~80%) were highly conserved in this group, with several aberrations due merely to the severe sequence missing and flanking sequence diversity. In contrast, the rest introns ("1", "2", "3", and "8") were only conserved within several genes. Together, the exon-intron structures of Group II and III were highly conserved in comparison with that of Group I. The intron "7" was shared and completely conserved in Group II and III, implying the close relationship between these two groups. Moreover, almost all introns were located outside of the transmembrane domains (TMs), except introns "5" and "8" in Group I.

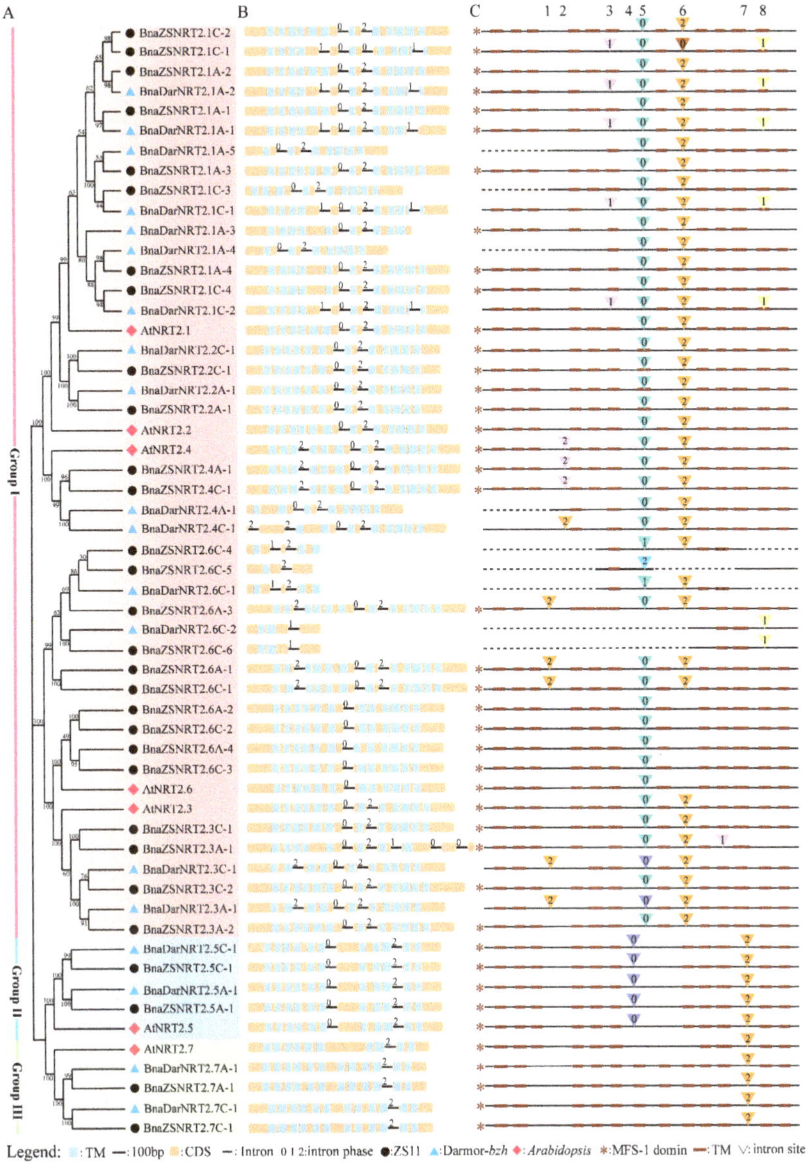

Figure 1. Phylogenetic tree and sequence structures of the candidate *NRT2s* in *B. napus* and *Arabidopsis*. (**A**) Neighbor-joining (NJ) phylogenetic tree of *NRT2s* in *B. napus* and *Arabidopsis*. Different background colors represent different groups. (**B**) Gene structures of *NRT2s* in *B. napus* and *Arabidopsis*. Exons are shown by yellow boxes, transmembrane (TM) domains are shown by blue boxes, and the lines between the colored boxes correspond to the introns. Numbers 0, 1, and 2 represent introns phase 0, 1, and 2, respectively. (**C**) Intron insertion patterns of *NRT2s* in *B. napus* and *Arabidopsis*. The top numbers (1–8) indicate the orders of the 8 conserved intron sites. The triangles represent different intron insertion sites. Each column (triangle with the same color and number) represents the same intron insertion site and phase. The dashed lines represent the severely missing sequences of candidates.

The protein domain prediction using HMMER online software (https://www.ebi.ac.uk/Tools/hmmer/, accessed on 19 November 2021) showed that 42 of the 50 members (84%) of BnaDarNRT2s and BnaZSNRT2s have 8 to 12 TMs (Figure 1B). Most members of Group I (34/41, ~82.93%) had 8 to 12 TMs, all members of Group II possessed 11 TMs, and nearly all of Group III had 12 TMs except AtNRT2.7 (10 TMs). The MFS-1 domain existed in most candidates (Figure 1C), which was located in the middle region, covering nearly all TMs. Moreover, the distribution and characteristics of the MFS-1 and TM domains were highly conserved within each group, especially in Groups II and III (Figure S1). Our results demonstrated that the protein domains were relatively conserved in the NRT2 gene family, and they were highly conserved in the same group.

2.3. Chromosomal Location and Collinearity Relationship in BnaZSNRT2s

As shown in Figure 2A, the 31 *BnaZSNRT2s* were scattered on 12 of the 19 *B. napus* chromosomes. There are 5, 4, and 4 *BnaZSNRT2s* located on chromosomes C_n08, A_n06, and C_n03, respectively; each of the last 9 chromosomes contained 2 *BnaZSNRT2s*. The numbers of *BnaZSNRT2s* on A_n-subgenome (14 genes) and C_n-subgenome (17 genes) showed a biased trend, with more genes on C_n-subgenome.

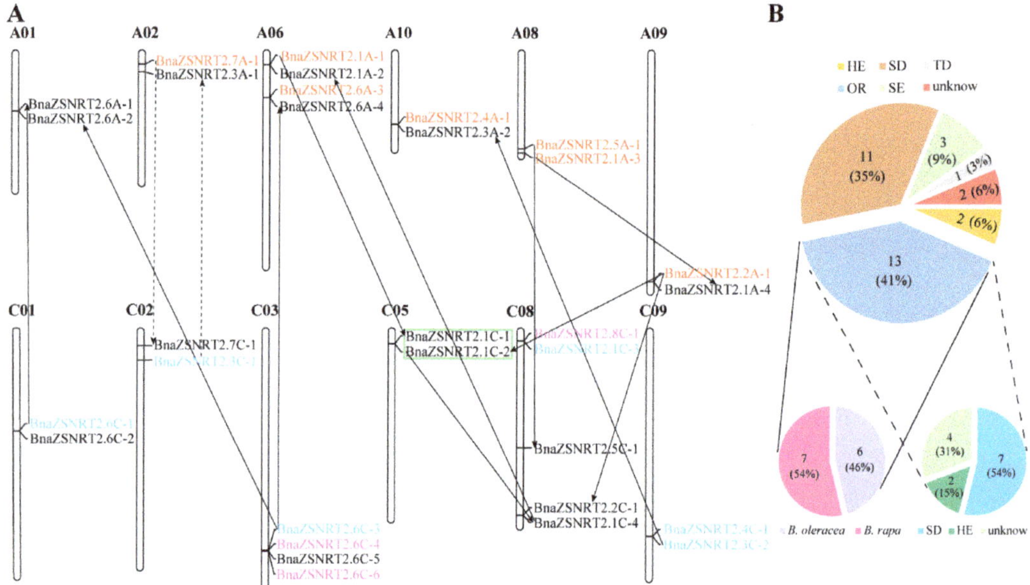

Figure 2. Chromosome distribution and duplication of *BnaZSNRT2s*. (**A**) 31 *BnaZSNRT2s* were mapped on 12 chromosomes in *B. napus*. The genes in orange and blue originated from *B. rapa* and *B. oleracea*, respectively; genes in purple were involved in segmental exchange (SE) events; the genes with green frames were tandem duplication (TD) pairs. The black and dashed lines with an arrow represent the duplication direction of genes involved in segmental duplication (SD) and homologous exchange (HE) events, respectively. (**B**) The big pie chart represents the percentage of *BnaZSNRT2s* derived from SD, SE, HE, TD, and orthologous region (OR) events, respectively; the left small pie chart represents the percentage of *BnaZSNRT2s* from OR and then experienced small duplication events in *B. napus*, and the right small pie chart represents the percentage of *BnaZSNRT2s* involved in OR events from *B. rapa* and *B. oleracea*.

The number of *NRT2s* in *B. napus* genome is larger than in the other species reported, e.g., *Arabidopsis* (7 genes) [35], poplar (6 genes) [36], barley (10 genes) [37]. This may because *B. napus* ($A_nA_nC_nC_n$, 2n = 38) was newly originated from the hybridization event between *Brassica rapa* (A_nA_n, n = 10) and *Brassica oleracea* (C_nC_n, n = 9) ~7500 years ago [30], and Brassicaceae species underwent a whole-genome triplication (WGT) event [38]. Therefore, in theory, *B. rapa*, *B. oleracea*, and *B. napus* genomes may have 21, 21, and 42 *NRT2s* expanded from the 7 *AtNRT2s*. In fact, only 14, 14, 19, and 31 genes were identified in *B. rapa*, *B. oleracea*, 'Darmor-*bzh*', and 'ZS11', respectively (Table S3), indicating that many *NRT2s* may have been lost during evolution. Theoretically, 5 *AtNRT2s* in Group I should expand to 30 *BnaZSNRT2s*/*BnaDarNRT2s*, and 1 *AtNRT2* in Group II/III should expand to 6 homologs in *B. napus*, respectively. In fact, 27, 2, and 2 genes were identified in Group I−III in the 'ZS11'ecotype, and 15, 2, and 2 genes were identified in Group I−III in the 'Darmor-*bzh*' ecotype, indicating that 'ZS11' retained 12 genes more than 'Darmor-*bzh*' in Group I, and the two ecotypes both lost 4 genes in Group II/III.

Collinearity relationship analysis found that 29 of the 31 (~93.55%) *BnaZSNRT2s* had the collinearity relationship in *B. napus*, *B. rapa*, and/or *B. oleracea*, except for *BnaZSNRT2.6C-2* and *BnaZSNRT2.6C-5* (Figure 2B, Table S4). Among the 29 *BnaZSNRT2s*, 13 genes (~44.83%) were inherited from allopolyploidy between *B. rapa* and *B. oleracea*, including 7 genes (~24.14%) were inherited from *B. rapa* and 6 genes (~20.69%) were from *B. oleracea*; the last 16 (~55.17%) *BnaZSNRT2s* were originated from other duplication events within *B. napus* genome, including 11 (68.75%) genes from segmental duplication (SD), 3 (18.75%) genes from the segmental exchange (SE), and 2 (12.5%) genes from homologous exchange (HE) events. Only 1 tandem duplication (TD) event was identified (*BnaZSNRT2.1C-1*/*BnaZSNRT2.1C-2*). Moreover, all the 3 genes from the SE event were from A_n-subgenome, which replaced the genes on C_n-subgenome in 'ZS11'. This demonstrated bias retention for genes derived from *B. rapa* in *B. napus* after allopolyploidy. Furthermore, 9 (69.23%) of the 13 *BnaZSNRT2s* derived from allopolyploidy have undergone small-scale duplications in the *B. napus* genome as well, including 2 genes that experienced two SD events, 5 genes underwent one SD event, and the other 2 genes underwent one HE events, implying that the larger number of *BnaZSNRT2s* expansion might mainly attribute to allopolyploidy and subsequent SD events in *B. napus*. Additionally, the sequence similarity and identity of the full-length DNA, CDS, and protein sequences of the 8 duplicated gene pairs in *B. napus* were very high, and the sequence identities were on average ~85.74%, ~91.49%, and ~94.25%, respectively (Table S5). This indicated that the duplicated genes are functionally redundant.

Taken together, our results proved that 'ZS11' showed a higher retention rate than 'Darmor-*bzh*'; allopolyploidy and SD events (24/29, ~83%) mainly contributed to the massive expansion of *BnaZSNRT2s* with these derived from *B. rapa* were inclined to be reserved in *B. napus* genome. In the following sections, we will focus on the features of the candidate genes in the native variety ZS11 ecotype.

2.4. Potential Regulatory Mechanism in the Promoter Regions of BnaZSNRT2s

The transcription factor (TF) binding sites in the promoter sequences (−1500bp) of the 31 *BnaZSNRT2s* were predicted by the online PlantTFDB software, and then a regulatory network was generated (Figure 3A). A total of 292 TF binding sites were predicted among 29 *BnaZSNRT2s*, excepting *BnaZSNRT2.7A-1* and *BnaZSNRT2.1C-2* (Table S6), which belonged to 21 TF families, with Dof (50 sites), bZIP (36 sites), MYB (29 sites) and B3 (22 sites) families contained most sites (Figure 3B). In Group I, the candidates belong to 21 TF families that might regulate 26 *BnaZSNRT2s* with 5 TF families may only regulate 1 gene respectively (e.g., G2like, GRAS, SRS), whereas the last 16 TF families (e.g., Dof, MYB, B3) regulate multi-genes respectively. In Group II, B3, Dof, and C2H2 families might regulate *BnaZSNRT2.5C-1*; SBP and Dof might regulate *BnaZSNRT2.5A-1*. In Group III, both Dof and AP2 might regulate *BnaZSNRT2.7C-1*. The complicated regulatory network by TF in Group I might indicate the diverse expression profiles of *BnaZSNRT2s*.

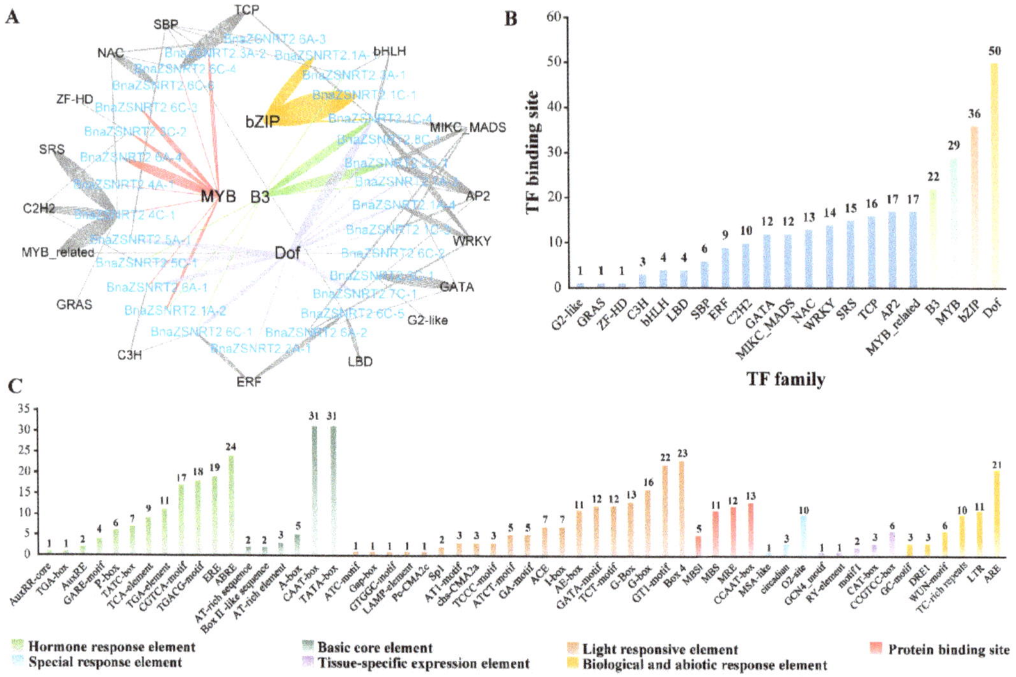

Figure 3. Transcription factor (TF) binding network and *cis*-element analysis in the promoter regions of *BnaZSNRT2s*. (**A**) The potential TF binding network of the 31 *BnaZSNRT2s* predicted by the PlantTFDB tool. (**B**) The TF gene families with potential binding sites in the promoter regions of the *BnaZSNRT2s*. (**C**) The *cis*-elements in the promoter regions of the *BnaZSNRT2s*. The ordinate represents the number of *BnaZSNRT2s*.

To understand the potential regulatory mechanism, we subsequently predicted *cis*-acting elements on the −1500bp upstream promoter regions of the 31 *BnaZSNRT2s* by PlantCARE online software. A total of 2941 *cis*-acting elements were found, classified into 56 types (Table S7). Except for the common basic core elements (e.g., A-box, TATA-box) and light-responsive elements (e.g., ACE), the rest were divided into five major groups, including hormone response element, protein binding site, special response element, tissue-specific expression element, and biological and abiotic response elements (Figure 3C). In the hormone response element group, 12 types of *cis*-elements were predicted, and 35, 24, and 19 *BnaZSNRT2s* might be involved in MeJA (TGACG-motif, CGTCA-motif), abscisic acid (ABRE), and ethylene (ERE) responsive processes. In the protein binding site group, all types of *cis*-acting elements were related to MYB binding site. In the special response element group, 10 *BnaZSNRT2s* might be involved in zein metabolism regulation (O2-site). In the tissue-specific expression element group, 9 *BnaZSNRT2s* might be related to meristem activation and expression (CCGTCC-box, CAT-box). In the biotic and abiotic-response *cis*-element group, 21, 11, and 10 *BnaZSNRT2s* might be involved in anoxia (ARE), low temperature (LTR), and defense and stress (TC-rich repeats) responsive processes. Moreover, we found that the predicted *cis*-acting elements in each group of the NRT2 gene family were similar *in B. napus* (Table S7). In Groups II and III, 53% (9/17) and 70% (12/17) of the *cis*-acting elements existed in all members, respectively (Table S7). In Group I, 31% (17/55) of the *cis*-acting elements were presented in the majority of candidates (Table S7). These results indicate a complex regulatory network of *BnaZSNRT2s* responding to multi-factors and a relatively conserved regulation mechanism in each group, especially in Groups II and III.

In conclusion, our results suggested that *BnaZSNRT2s* may closely respond to plant hormones and abiotic stresses and may be regulated by many TF family members.

2.5. Potential miRNAs Targets of BnaZSNRT2s

We predicted the potential miRNA targets in the CDS sequences of the candidate 31 BnaZSNRT2s by psRNATarget online software (Table S8, Figure 4). Accordingly, 21 miRNAs were found to have potential targets in 28 *BnaZSNRT2s*, except for *BnaZSNRT2.6C-5*, *BnaZSNRT2.6C-6*, and *BnaZSNRT2.8C-1*. In Group II, miR6030 might target both *BnaZSNRT2.5A-1* and *BnaZSNRT2.5C-1*; meanwhile, the latter might also be the target of miR395. In Group III, miR164 might target both *BnaZSNRT2.7A-1* and *BnaZSNRT2.7C-1*, and the latter might be the targets of miR166 and miR6031 as well. In Group I, a relatively complex regulation by multi-miRNAs was observed, where 16 BnaZSNRT2s might be targeted by 3 to 5 miRNAs, and miR397, miR164, miR172, miR395, and miR167 might target 10, 8, 8, 7, and 7 BnaZSNRT2s in this group, respectively. Moreover, single miRNA tends to target multiple BnaZSNRT2s in Group I–III, and the type of miRNA regulating each group was relatively conserved, indicating the relatively conserved expression and function in the same group. Additionally, miR164 might target 2 genes in Group III and 8 genes in Group I; miR395 might target 1 gene in Group II and 7 genes in Group I. The regulation of NRT2 homologs by miRNAs in Group I is more complicated than in Group II/III, implicating the diverse expression profile and function of Group I members.

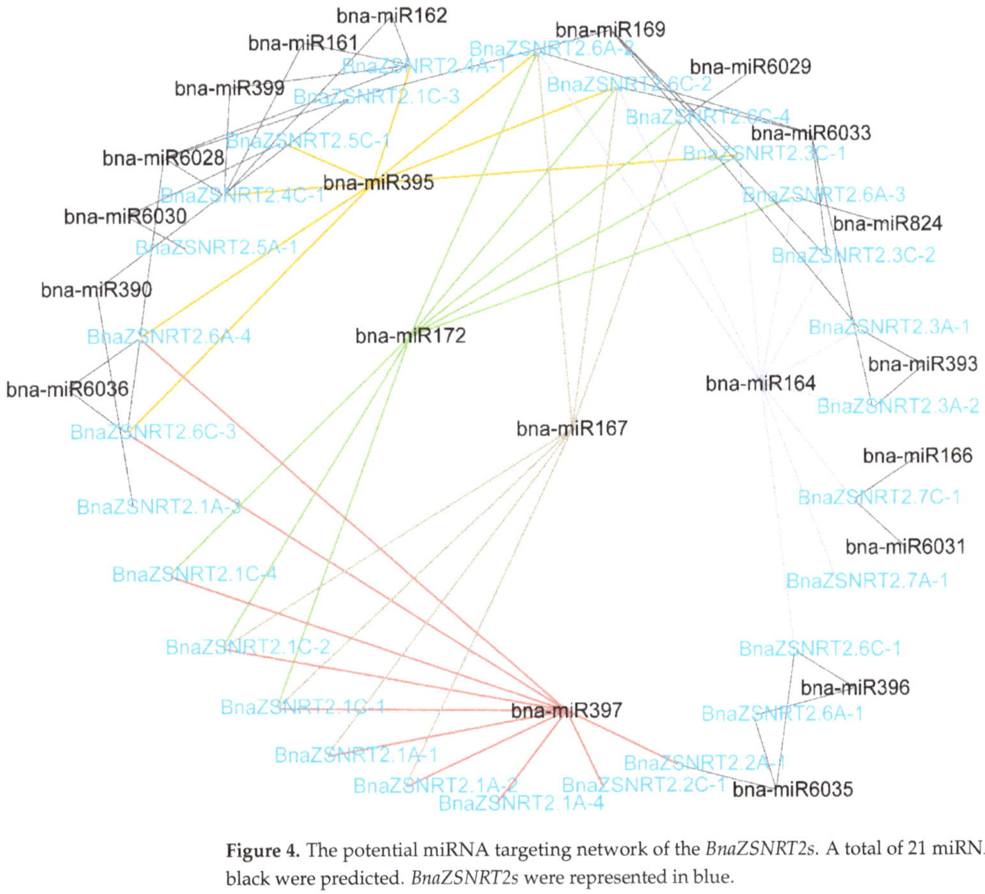

Figure 4. The potential miRNA targeting network of the *BnaZSNRT2s*. A total of 21 miRNAs with black were predicted. *BnaZSNRT2s* were represented in blue.

2.6. Spatial-Temporal Expressions of BnaZSNRT2s in Different Developmental Stages of B. napus

To study the expression patterns of *BnaZSNRT2s* in extensive tissues/organs across distinct developmental stages in *B. napus*, a public RNA-seq dataset BnTIR (http://yanglab.hzau.edu.cn, accessed on 10 September 2020) including 52 samples was applied. Except for 14 genes with no detectable expression levels (TPM < 1) in all tissues investigated, the remaining 17 genes belonging to Group I–III have obvious preferential expression profiles in 'ZS11'. The genes in the three groups generally have distinct expression profiles (Figure 5A). In general, the homologs in the same group had conserved expression patterns, implicating their potentially functional conservation and redundancy. In Group I, nearly all of the 13 genes were commonly highly expressed in roots (except *BnaZSNRT2.6A-4*) and showed uneven expression levels in shoot tissues with 4 genes (*BnaZSNRT2.6A-2*, *BnaZSNRT2.6C-2*, *BnaZSNRT2.6A-4*, and *BnaZSNRT2.6C-3*) having higher expression levels in the late developmental stages of silique and seed tissues as well. This implicated that the homologs in Group I might perform diverse functions in *B. napus*. In contrast, the expression profiles were highly conserved in the last two groups. In Group II, the 2 members (*BnaZSNRT2.5A-1* and *BnaZSNRT2.5C-1*) were highly expressed in leaf and silique tissues with a gradually increased in leaf (Figure 5B). Similarly, the 2 members of Group III (*BnaZSNRT2.7A-1* and *BnaZSNRT2.7C-1*) were highly expressed in leaf, silique, and seed tissues, and the expression levels were stable in leaf instead of silique and seed tissues (Figure 5B). The similar expression patterns between Group II and III indicated their close relationship and even similar functions. Given the general biological functions of the organs in the plant, we speculated that members of Group I may be related to the N uptake and transport in roots, while members of Group II and III may involve in N transport, storage, and/or accumulation in *B. napus*. Additionally, the Pearson correlation coefficient of 6 (75%) sister pairs was ≥0.8 (Table S5), indicating their expression conservation and even functional redundancy.

2.7. Expression Profile of BnaZSNRT2s under LN Stress

To discover the potential functions of *BnaZSNRT2s* in the N utilization process in *B. napus*, an LN stress RNA-seq dataset (PRJNA612634) was applied in this study. Excepting 13 *BnaZSNRT2s* that were not expressed (FPKM < 1) in all samples, the last 18 expressed *BnaZSNRT2s* belong to three groups (Figure 6A). Consistent with the large number, up to 14 members in Group I had detectable transcript accumulation (FPKM ≥ 1) in the dataset, which was preferentially expressed in roots. Their expressions were significantly up-regulated under LN treatment after 3-, 5-, and 12-days (Figure 6B). Moreover, their expression patterns were somewhat different: *BnaZSNRT2.6A-2*, *BnaZSNRT2.4A-1*, *BnaZSNRT2.6C-2*, and *BnaZSNRT2.4C-1* were obviously preferentially up-regulated after 12 days LN treatment in roots; *BnaZSNRT2.1A-2*, *BnaZSNRT2.1A-3*, *BnaZSNRT2.1C-3*, and *BnaZSNRT2.8C-1* were significantly up-regulated after 5- and 12-days LN treatments; and the rest 6 members were up-regulated after 3-, 5- and 12-days LN treatment. This implicated the expression profile diversification of the homologs in Group I. Consistent with the spatial-temporal expression profile (Figure 5), the genes in Group II (*BnaZSNRT2.5A-1* and *BnaZSNRT2.5C-1*) and III (*BnaZSNRT2.7A-1* and *BnaZSNRT2.7C-1*) had highly conserved expression patterns under LN treatments respectively. The expressions of *BnaZSNRT2.7A-1* and *BnaZSNRT2.7C-1* in Group III were slightly up-regulated after 3-, 5- and 12-days LN treatment (Figure 6B). Whereas the expression levels of *BnaZSNRT2.5A-1* and *BnaZSNRT2.5C-1* in Group II were dramatically increased in roots after 12-days of LN treatment. Their expression levels were both peaked in leaf and roots after 12-day LN treatment, showing long-term N deficiency expression profiles. Notably, *BnaZSNRT2.5A-1* and *BnaZSNRT2.5C-1* showed very low expression levels in normal conditions in roots (Figure 5A), but their expressions were significantly up-regulated in roots under LN stress (Figure 6A), indicating their potential roles in roots in response to LN stress. To confirm the LN-induced expression profiles of *BnaZSNRT2.5A-1* and *BnaZSNRT2.5C-1* obtained from the RNA-seq dataset, an RT-qPCR assay was further applied. As shown in Figure 6C, the expression profiles of *BnaZSNRT2.5A-1* and *BnaZSNRT2.5C-1* were similar and were significantly up-regulated after long-term LN treatments, supporting the credibility of RNA-seq analysis. Notably, a peak expression level was observed at 7 days under LN treatment in the RT-qPCR assay, suggesting their expression trend under the LN stress condition.

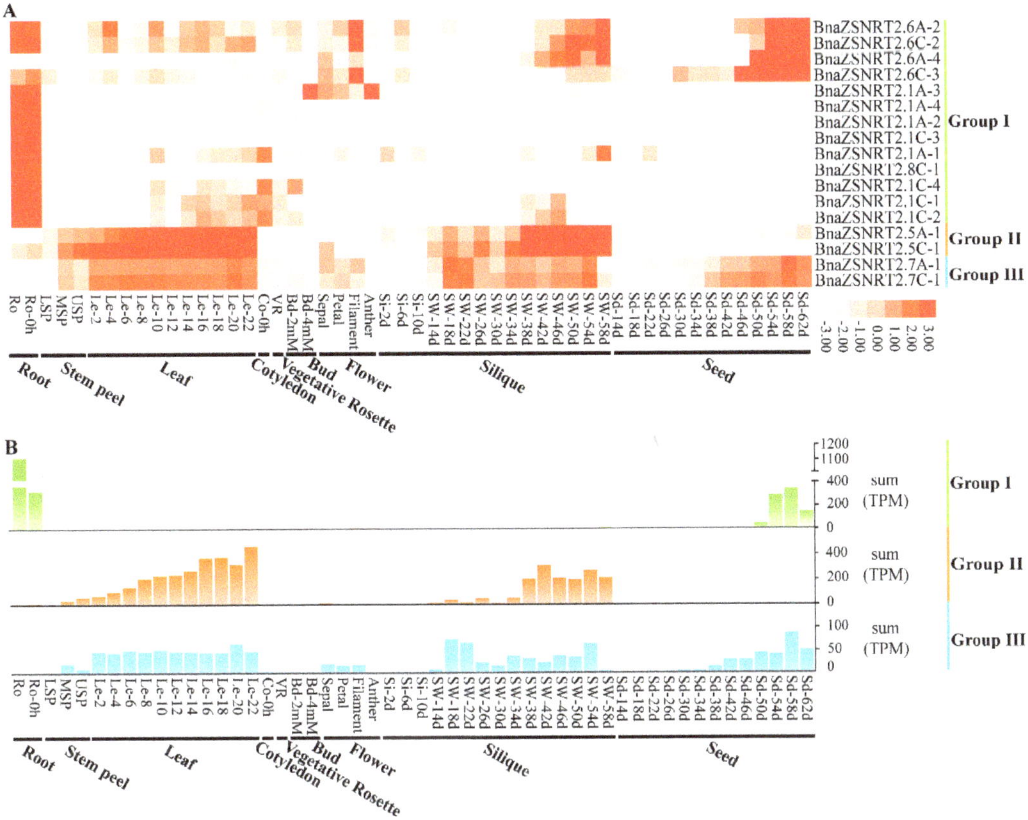

Figure 5. Expression pattern of *BnaZSNRT2s* in *B. napus* at different developmental stages. (**A**) The expression profile of *BnaZSNRT2s* in 52 tissues. (**B**) The sum of expression levels in Group I–III. The ordinate represents the sum of expression levels (TPM). "Ro" = root, "LSP" = lower stem peel, "MSP" = middle stem peel, "USP" = upper stem peel, "Le" = leaf, "Co" = cotyledon, "VR" = vegetative rosette, "Bd" = bud, "Si" = silique, "SW" = silique wall, "Sd" = seed; "h," and "d," indicate hour and day, respectively.

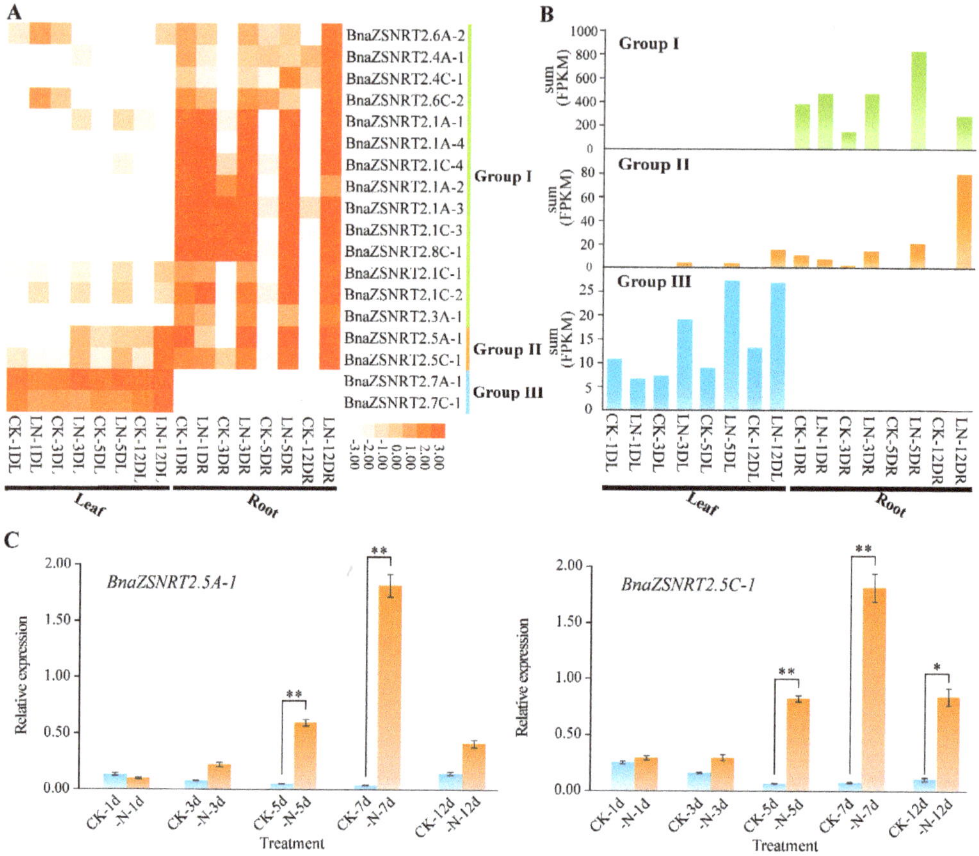

Figure 6. Expression pattern of *BnaZSNRT2s* under low nitrogen (LN) stress treatments. (**A**) The expression profiles of *BnaZSNRT2s* in 'ZS11' seedling leaf and root tissues under LN stress treatment based on RNA-Seq dataset. "CK" = normal N condition, "LN" = Low N; "D", "L", and "R" indicate day, leaf, and root, respectively. (**B**) The sum of expression levels in each group. The ordinate represents the sum of expression levels (FPKM) corresponding to each tissue investigated. (**C**) Expression levels of *BnaZSNRT2.5A-1* and *BnaZSNRT2.5C-1* under LN treatments by RT-qPCR method. The reference genes were *BnaActin7* (GenBank accession no. AF024716) and *BnaUBI* (GenBank accession no. NC027770). Error bars represent the standard deviation of three independent experiments. *: Significant difference (0.05 > *p* > 0.01); **: Extremely significant difference.

3. Discussion

3.1. A Conserved Patterns of NRT2s Intron Insertion Patterns and Phylogeny

Numerous studies have indicated that the intron insertion patterns were commonly conserved in each gene family or subfamily in plants [39–41]. In this study, we identified 19 *BnaDarNRT2s* and 31 *BnaZSNRT2s* in two *B. napus* ecotypes. A total of 8 intron insertion sites were observed in *BnaDarNRT2s*, *BnaZSNRT2s*, and *AtNRT2s*, with "5" and "6" were nearly absolutely conserved in Group I, "4" and "7" were completely conserved in Group II, and "7" were conserved in Group III (Figure 1C). To our knowledge, this is the first time to focus on the intron insertion patterns of this family in plants. To further confirm our results in *Arabidopsis* and *B. napus*, we expand the analyses to 21 plant species, consisting of chlorophyte, bryophyte, basic angiosperm, monocots, and eudicots (Figure 7, Table S9). We found that the intron patterns of this gene family generally tracked with accepted taxonomy across viridiplantae in an evolutionary context. For the 8 intron insertion sites, especially the highly conserved "4", "5", "6", and "7" sites, none were distributed in chlorophyte. The intron "5" was first found in Bryophyte and was conserved in *P. patens* and *S. fallax* and investigated. Introns "5", "6", and "7" were observed as early as in basic angiosperm *A. trichopoda* and *C. kanehirae*; all of the 4 sites were observed in angiosperms while all of the investigated dicots genes had introns, and introns "4", "5", "6" and "7" were widely observed. However, only 24.1% (7/29) of monocot genes had introns with 5 genes sharing intron "7". Notably, intron "4" was only distributed in Brassicaceae, indicating a new origin in this lineage. Overall, the intron insertion sites might be firstly derived after the divergence of chlorophyte and embryophyte, and intron "5" might be an ancestor and widely distributed in embryophyte except for monocots. An obvious bias trend between monocots and dicots was observed according to the divergence of intron sites.

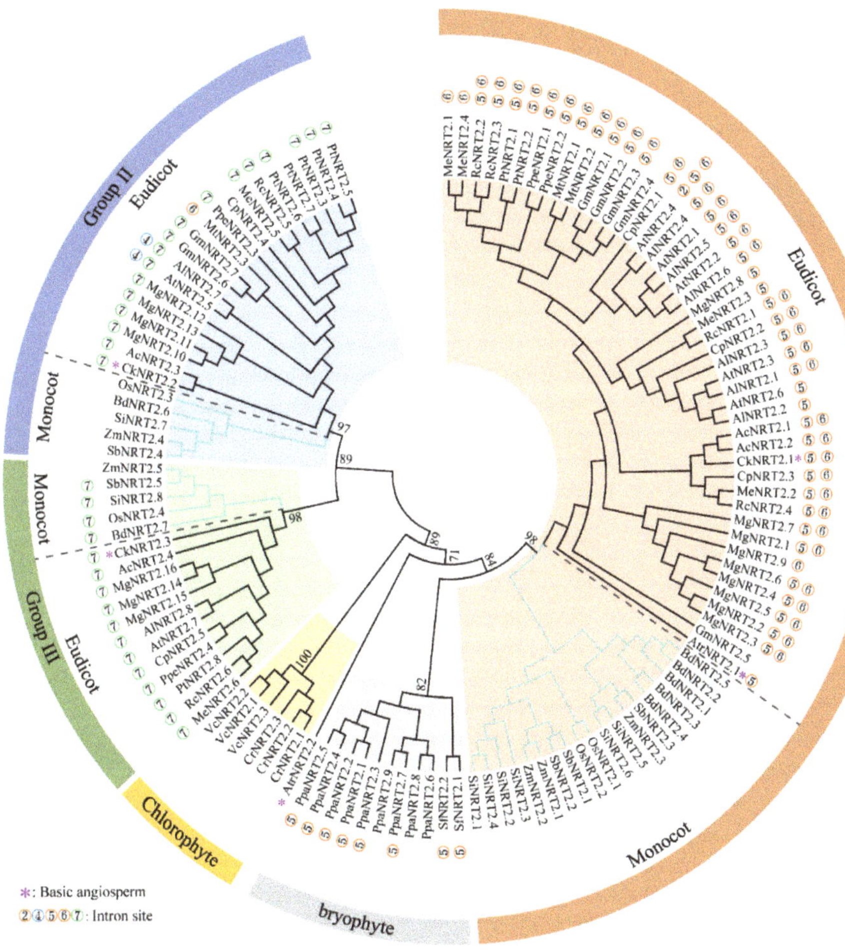

Figure 7. Phylogenetic analysis of NRT2 gene families in 22 plants. Different groups were colored by a special background. Monocot and eudicot within Groups I–III were divided by dashed lines. The number beside the branch indicates the bootstrap value of each group from 1000 replicates.

To date, the NRT2 gene family has been identified in numerous plant genomes, such as *Brachypodium* [42], rice [43], barley [37], *M. esculenta* [44], and *A. thaliana* [35]. These studies generally classified this gene family into three groups, namely I−III. To gain insights into the evolutionary mechanisms and sequence features of the NRT2 gene family in the plant kingdom, we further identified the candidates in 22 plant genomes in Phytozome v13 ranging from chlorophyte to angiosperms, including *C. reinhardtii, V. carterigenomes, P. patens, S. fallax, A. trichopoda, C. kanehirae, A. coerulea, A. lyrata, A. thaliana, C. papaya, G. max, M. esculenta, M. guttatus, M. truncatula, P. persica, P. trichocarpa, R. communis, B. distachyon, O. sativa, S. bicolor, S. italica* and *Z. mays*. The *NRT2s* were observed in all of these species, with the number ranging from 2 (*A. trichopod* and *S. fallax*) to 16 (*M. guttatus*) (Table S9). Consistent with previous studies [45,46], based on the systematic identification and phylogenetic analyses, we divided this gene family into five groups, including Group I (homologs of *AtNRT2.1−AtNRT2.4* and *AtNRT2.6*), Group II (homologs of *AtNRT2.5*), Group III (homologs of *AtNRT2.7*), chlorophyte-specific group (Group IV), and bryophyte-specific group (Group V) (Figure 7). Interestingly, we found that the intron distribution trend was highly consistent with the phylogenetic relationship of this gene family in viridiplantae: the genes with both introns "5" and "6" in angiosperms except monocots were observed in Group I. The genes with both introns "4" and "7" were clustered in Group II, and those with only intron "7" were in Group III. Notably, though the genes in dicots and monocots were commonly clustered into each of the angiosperm-specific groups (I−III), members of these two lineages shared the same intron pattern only in Group III, and the monocot genes in the last two groups were generally intron-less instead. Overall, our analyses indicated that the intron patterns might reflect the phylogeny of *NRT2s* even in an evolutionary context.

3.2. Expression and Function Characteristics of NRT2s in Plants

In this study, we revealed that the genes in Group I have a wide and less conserved expression pattern in *B. napus,* most of which were expressed in roots, whereas those in Groups II and III have a higher and conserved expression profile in the acrial part, especially in leaves and siliques (Figure 5). A similar situation was observed in many other eudicots [19,35–37,42,44,47,48]. For example, in Group I, many members were expressed abundantly in roots and a few other organs, such as *MtNRT2.1* in *M. truncatula* [47], *MeNRT2.1−MeNRT2.4* in *M. esculenta* [44], and *AtNRT2.1, AtNRT2.3, AtNRT2.4* and *AtNRT2.6* in *A. thaliana* [35]; In Group II, the genes generally showed strong transcripts in shoot tissues (e.g., stems and leaves) and weak expression levels in roots, such as poplar *PtNRT2.5A* and *PtNRT2.5B* [36], *M. truncatula MtNRT2.3* [47], and *M. esculenta MeNRT2.5* [44]; while members of Group III were specially expressed in shoots, such as *PtNRT2.7* [36] and *AtNRT2.7* [35]. Notably, the expression profiles of this gene family in monocots are different from that of dicots, especially those of Group II and III that showed higher expression levels in both roots and leaves (e.g., *HvNRT2.10* [37], *BdNRT2.5* in *B. distachyon* [42], *OsNRT2.3* in rice [48]; *BdNRT2.7* [42] and *OsNRT2.4* [48]), whereas the expression pattern of Group I was similar to eudicots, such as *HvNRT2.2−HvNRT2.9* in barley [37] and *OsNRT2.1/OsNRT2.2* in rice [48]. Together, the spatial-temporal expression profile of *NRT2s* in Group I was generally similar between monocots and eudicots, whereas those of Group II and III were distinct.

Consistent with their critical roles in nitrate transporters, the *NRT2s* generally responded positively to external nitrate or N deficiency in numerous plants. For example, *AtNRT2.1/AtNRT2.4/AtNRT2.6, MeNRT2.1−2.4, HvNRT2.2−2.9* and *OsNRT2.1−2.2* in Group I, and *AtNRT2.5, MeNRT2.5, HvNRT2.1,* and *OsNRT2.3* in Group II were up-regulated under low nitrate or LN treatment in roots [35,37,43,44]. At the same time, *AtNRT2.7* in Group III was intensively induced by limited nitrate supply in shoots [35]. Similarly, we found that the *BnaZSNRT2s* in Group I and II were significantly up-regulated in roots under LN treatment, whereas these in Group III were positively induced in leaves under LN treatment (Figure 6). Notably, in contrast to the situation in Group I and III, the genes in

Group II showed an obvious inverse trend between the spatial-temporal (Figure 5) and LN stress (Figure 6) expression profiles which were up-regulated in roots under LN treatments instead of leaves. Moreover, the LN stress expression profile of this gene family is generally consistent with their known functions in plants. For instance, members of Group I were widely proved to function in roots under nitrate or N starvation, e.g., *MeNRT2.2* [44], *CmNRT2.1* [49], and *AtNRT2.1* [50]. *OsNRT2.3a* in Group II acted as a long-distance nitrate transport from roots to shoots under low nitrate treatment [51], while members of Group III mainly functioned in seeds, such as wheat *TaNRT2.5* [52] and *AtNRT2.7* [24]. These results suggested a spatial-temporal collaborative role of this gene family in the root and shoot tissues in N utilization in plants.

4. Materials and Methods

4.1. Identification of NRT2 Genes in Plants

The *AtNRT2s* were acquired from the TAIR database (http://www.arabidopsis.org, accessed on 11 July 2019). To identify the *NRT2s* in the *B. napus* genome, BLASTP and Tblastn searches were performed in GENOSCOPE (http://www.genoscope.cns.fr/brassicanapus/, accessed on 19 August 2014) and BnPIR (http://cbi.hzau.edu.cn/bnapus/, accessed on 10 September 2020) databases respectively, using the *AtNRT2s* as the queries with a low-stringency criterion (cutoff $p < 0.1$). After deleting the redundant sequences, the remainders were confirmed by HMMER (https://www.ebi.ac.uk/Tools/hmmer/, accessed on 19 November 2021) and SMART (http://smart.embl-heidelberg.de/, accessed on 26 October 2020) tools to ensure they had the characteristic domains of the NRT2 family. DNA, cDNA, and encoding protein sequences of the candidate genes were acquired from the GENOSCOPE and BnPIR database. We also identified the *NRT2s* in *C. reinhardtii*, *V. carterigenomes*, *P. patens*, *S. fallax*, *A. trichopoda*, *C. kanehirae*, *B. rapa*, *B. oleracea*, *A. coerulea*, *A. lyrata*, *C. papaya*, *G. max*, *M. esculenta*, *M. guttatus*, *M. truncatula*, *P. persica*, *P. trichocarpa*, *R. communis*, *B. distachyon*, *O. sativa*, *S. bicolor*, *S. italica* and *Z. mays* from Phytozome v13 database (https://phytozome-next.jgi.doe.gov/, accessed on 22 November 2011) [53] by the same method. The physicochemical properties and subcellular localization analyses of candidates were predicted by ExPASy online tool (http://www.expasy.org/tools/, accessed on 23 February 2022) [54] and by Cell-PLoc-2 (http://www.csbio.sjtu.edu.cn/bioinf/Cell-PLoc-2/, accessed on 28 June 2010) and WoLF PSORT (https://www.genscript.com/wolf-psort.html, accessed on 21 May 2007), respectively.

4.2. Phylogenetic and Sequence Structure Analysis of B. napus NRT2 Gene Family

To discover the evolutionary relationship of the NRT2 family in *B. napus* and *Arabidopsis*, the protein sequences of candidates in *B. napus* (*BnaDarNRT2s* and *BnaZSNRT2s*) and *Arabidopsis* (*AtNRT2s*) were applied to perform a multiple sequence alignment analyses by MAFFT version 7 tool with default parameters (https://mafft.cbrc.jp/alignment/server/, accessed on 06 September 2017). Based on multiple sequence alignment analyses, a phylogenetic tree was generated by MEGA version 7 [55] using the NJ method with the following major parameters: Poisson correction, bootstrap with 1000 replicates, and pairwise deletion. The sequence structures of *BnaDarNRT2s*, *BnaZSNRT2s*, and *AtNRT2s* were analyzed by a Gene Structure Display Server (GSDS) 2.0 (http://gsds.cbi.pku.edu.cn/, accessed on 10 December 2014) [56], using the cDNA and protein sequences of candidates. The domains and TM regions of BnaDarNRT2s, BnaZSNRT2s, and AtNRT2s were predicted by the HMMER tool (https://www.ebi.ac.uk/Tools/hmmer/, accessed on 19 November 2021), and then were visualized by Weblogo online software (http://weblogo.threeplusone.com/, accessed on 06 June 2004). The intron insertion sites and phases of the candidates were obtained by comparing the DNA and cDNA sequences of each gene manually using MEGA version 7.

4.3. Chromosomal Location and Collinearity Relationship of BnaZSNRT2s

The chromosome location information of candidate *BnaZSNRT2s* was acquired from the BnPIR database. The chromosome map of candidates was drawn by MapChart v2.32 software. The collinearity relationship of *BnaZSNRT2s*, *BrNRT2s*, *BoNRT2s*, and *AtNRT2s* was calculated by the CoGe online tool (https://genomevolution.org/coge/, accessed on 04 February 2008). The duplication events of *BnaZSNRT2s* were identified in the previous study [57].

4.4. The Potential Regulatory Mechanism in the Promoter Regions of BnaZSNRT2s

The −1500 bp upstream promoter sequence of *BnaZSNRT2s* was used to predict the potential TFs regulation information by the PlantTFDB database (http://planttfdb.gao-lab.org/, accessed on 24 October 2016) with a threshold *p*-value $< 10^{-6}$. The potential *cis*-elements in the promoter regions of *BnaZSNRT2s* (−1500 bp upstream sequence) were predicted by PlantCARE online software (http://bioinformatics.psb.ugent.be/webtools/plantcare/html/, accessed on 11 September 2000). The regulation network of *BnaZSNRT2s* was generated by Cytoscape 3.8.2 software [58]. The potential miRNAs regulating sites of *BnaZSNRT2s* were analyzed by the psRNATarget website (expectation ≤ 6) (http://plantgrn.noble.org/psRNATarget/, accessed on 30 April 2018).

4.5. Spatial-Temporal and LN Stress Expression Profile Analysis of BnaZSNRT2s by RNA-Seq Data

To explore the temporal-spatial expression patterns of *BnaZSNRT2s* at different developmental stages in *B. napus*, the RNA-Seq dataset in BnTIR (http://yanglab.hzau.edu.cn/, accessed on 10 September 2020) was obtained. The heatmap was generated by Cluster 3.0 [59] and Java Treeview software [60] according to log2-transformed data. The genes with FPKM < 1 were speculated to be pseudogenes or specifically expressed genes under other conditions, and thus were not included in the heatmap in this study. Similarly, to explore the LN-inductive expression patterns of *BnaZSNRT2s*, an LN stress RNA-Seq dataset of 'ZS11' ecotype seedling roots and leave was acquired from NCBI (BioProject: PRJNA612634). The heatmap was generated by Cluster 3.0 [59] and Java Treeview software [60] based on the log2-transformed data of candidates.

4.6. RT-qPCR Analysis of BnaZSNRT2s under LN Conditions

Seeds of ZS11 were acquired from the College of Agriculture and Biotechnology, Southwest University. The treatments of the seeds and seedlings referred to a previous study [61]. The normal nutrient solution comprised of 0.5 mM K_2SO_4, 0.25 mM KH_2PO_4, 325 μM $MgSO_4$, 50 μM NaCl, 8 μM H_3BO_3, 0.4 μM $MnSO_4$, 0.4 μM $ZnSO_4$, 0.4 μM $CuSO_4$, 0.1 μM Na_2MoO_4, 40 μM Fe-EDDHA, 10 μM $C_2H_4N_4$, 1.8 mM $Ca(NO_3)_2$ and 0.2 μM $(NH_4)_2SO_4$ (control, CK). For the LN treatment, 1.8 mM $Ca(NO_3)_2$ and 0.2 mM $(NH_4)_2SO_4$ were replaced by 0.09 mM $Ca(NO_3)_2$ and 0.001 mM $(NH_4)_2SO_4$ [62]. The pH of the solution was 5.8. The root tissues were reaped at 1, 3, 5, 7, and 12 days after the treatments, which were promptly frozen in liquid nitrogen, and then stored at −80 °C for the purpose of RNA isolation. The RT-qPCR analysis method referred to our previous study [61], using the GoScriptTM Reverse Transcription Mix, Oligo(dT) kit (Promega, Beijing, China), and Taq Pro Universal SYBR qPCR Master Mix (Vazyme, Nanjing, China). The primers were designed by Primer Premier 5 software, and the primer sequences were presented in Table S10. Ultimately, we acquired the data (mean ± standard deviation) of all three independent repeated trials and calculated the relative expression of *BnaZSNRT2s* by the $2(-\Delta\Delta Ct)$ method. Standard errors were presented by Error bars from three independent repeated trials. Expression level Differences of *BnaZSNRT2s* were assessed by a One-way ANOVA test (* $p < 0.05$; ** $p < 0.01$) using Excel 2016.

Supplementary Materials: The following supporting information can be downloaded at: https://www.mdpi.com/article/10.3390/ijms23094965/s1.

Author Contributions: Conceptualization, H.D.; methodology, R.-J.D.; software, Z.-X.W.; validation, Z.-X.Y., and P.-F.L.; formal analysis, J.-Y.M.; investigation, J.Z.; writing—original draft preparation, H.D. and R.-J.D.; writing—review and editing, H.D.; supervision, H.D.; project administration, J.-N.L.; funding acquisition, H.D. All authors have read and agreed to the published version of the manuscript.

Funding: This work was supported by the National Key Research and Development Program of China (2018YFD1000900).

Institutional Review Board Statement: The study did not require ethical approval.

Informed Consent Statement: The study did not involve humans.

Data Availability Statement: The study did not report any data.

Conflicts of Interest: The authors declare no conflict of interest.

References

1. Evans, J.R.; Clarke, V.C. The nitrogen cost of photosynthesis. *J. Exp. Bot.* **2019**, *70*, 7–15. [CrossRef] [PubMed]
2. Frink, C.R.; Waggoner, P.E.; Ausubel, J.H. Nitrogen fertilizer: Retrospect and prospect. *Proc. Natl. Acad. Sci. USA* **1999**, *96*, 1175–1180. [CrossRef] [PubMed]
3. Zhao, X.Q.; Shen, R.F. Aluminum-Nitrogen Interactions in the Soil-Plant System. *Front. Plant. Sci.* **2018**, *9*, 807. [CrossRef] [PubMed]
4. Vitousek, P.M.; Howarth, R.W. Nitrogen limitation on land and in the sea: How can it occur? *Biogeochemistry* **1991**, *13*, 87–115. [CrossRef]
5. Alvarez, J.M.; Vidal, E.A.; Gutierrez, R.A. Integration of local and systemic signaling pathways for plant N responses. *Curr. Opin. Plant Biol.* **2012**, *15*, 185–191. [CrossRef]
6. Wang, Y.Y.; Cheng, Y.H.; Chen, K.E.; Tsay, Y.F. Nitrate Transport, Signaling, and Use Efficiency. *Annu. Rev. Plant. Biol.* **2018**, *69*, 85–122. [CrossRef]
7. Robertson, G.P.; Vitousek, P.M. Nitrogen in Agriculture: Balancing the Cost of an Essential Resource. *Annu. Rev. Environ. Resour.* **2009**, *34*, 97–125. [CrossRef]
8. Gutiérrez, R.A. Systems biology for enhanced plant nitrogen nutrition. *Science* **2012**, *336*, 1673–1675. [CrossRef]
9. Fixen, P.E.; West, F.B. Nitrogen fertilizers: Meeting contemporary challenges. *Ambio* **2002**, *31*, 169–176. [CrossRef]
10. Ju, X.T.; Xing, G.X.; Chen, X.P.; Zhang, S.L.; Zhang, L.J.; Liu, X.J.; Cui, Z.L.; Yin, B.; Christie, P.; Zhu, Z.L.; et al. Reducing environmental risk by improving N management in intensive Chinese agricultural systems. *Proc. Natl. Acad. Sci. USA* **2009**, *106*, 3041–3046. [CrossRef]
11. Khan, F.A.; Ansari, A.A. Eutrophication: An Ecological Vision. *Bot. Rev.* **2005**, *71*, 449–482. [CrossRef]
12. Yang, X.E.; Wu, X.; Hao, H.L.; He, Z.L. Mechanisms and assessment of water eutrophication. *J. Zhejiang Univ. Sci. B* **2008**, *9*, 197–209. [CrossRef] [PubMed]
13. Sieling, K.; Kage, H. Apparent fertilizer N recovery and the relationship between grain yield and grain protein concentration of different winter wheat varieties in a long-term field trial. *Eur. J. Agron.* **2021**, *124*, 126246. [CrossRef]
14. Dimkpa, C.O.; Fugice, J.; Singh, U.; Lewis, T.D. Development of fertilizers for enhanced nitrogen use efficiency—Trends and perspectives. *Sci. Total Environ.* **2020**, *731*, 139113. [CrossRef] [PubMed]
15. Liu, J.; Diamond, J. China's environment in a globalizing world. *Nature* **2005**, *435*, 1179–1186. [CrossRef] [PubMed]
16. Yan, N. Structural Biology of the Major Facilitator Superfamily Transporters. *Annu. Rev. Biophys.* **2015**, *44*, 257–283. [CrossRef] [PubMed]
17. Galvan, A.; Fernández, E. Eukaryotic nitrate and nitrite transporters. *Cell. Mol. Life Sci.* **2001**, *58*, 225–233. [CrossRef]
18. Tong, J.; Walk, T.C.; Han, P.; Chen, L.; Shen, X.; Li, Y.; Gu, C.; Xie, L.; Hu, X.; Liao, X.; et al. Genome-wide identification and analysis of high-affinity nitrate transporter 2 (NRT2) family genes in rapeseed (Brassica napus L.) and their responses to various stresses. *BMC Plant Biol.* **2020**, *20*, 464. [CrossRef]
19. Castro-Rodríguez, V.; Cañas, R.A.; de la Torre, F.N.; Pascual, M.B.; Avila, C.; Cánovas, F.M. Molecular fundamentals of nitrogen uptake and transport in trees. *J. Exp. Bot.* **2017**, *68*, 2489–2500. [CrossRef]
20. Guan, M.; Chen, M.; Cao, Z. NRT2.1, a major contributor to cadmium uptake controlled by high-affinity nitrate transporters. *Ecotoxicol. Environ. Saf.* **2021**, *218*, 112269. [CrossRef]
21. Kiba, T.; Feria-Bourrellier, A.B.; Lafouge, F.; Lezhneva, L.; Boutet-Mercey, S.; Orsel, M.; Brehaut, V.; Miller, A.; Daniel-Vedele, F.; Sakakibara, H.; et al. The Arabidopsis nitrate transporter NRT2.4 plays a double role in roots and shoots of nitrogen-starved plants. *Plant Cell* **2012**, *24*, 245–258. [CrossRef] [PubMed]

22. Dechorgnat, J.; Patrit, O.; Krapp, A.; Fagard, M.; Daniel-Vedele, F. Characterization of the Nrt2.6 gene in Arabidopsis thaliana: A link with plant response to biotic and abiotic stress. *PLoS ONE* **2012**, *7*, e42491. [CrossRef] [PubMed]
23. Lezhneva, L.; Kiba, T.; Feria-Bourrellier, A.B.; Lafouge, F.; Boutet-Mercey, S.; Zoufan, P.; Sakakibara, H.; Daniel-Vedele, F.; Krapp, A. The Arabidopsis nitrate transporter NRT2.5 plays a role in nitrate acquisition and remobilization in nitrogen-starved plants. *Plant J.* **2014**, *80*, 230–241. [CrossRef] [PubMed]
24. Chopin, F.; Orsel, M.; Dorbe, M.F.; Chardon, F.; Truong, H.N.; Miller, A.J.; Krapp, A.; Daniel-Vedele, F. The Arabidopsis ATNRT2.7 nitrate transporter controls nitrate content in seeds. *Plant Cell* **2007**, *19*, 1590–1602. [CrossRef] [PubMed]
25. Kotur, Z.; Mackenzie, N.; Ramesh, S.; Tyerman, S.D.; Kaiser, B.N.; Glass, A.D.M. Nitrate transport capacity of the Arabidopsis thaliana NRT2 family members and their interactions with AtNAR2.1. *New Phytol.* **2012**, *194*, 724–731. [CrossRef]
26. Chen, J.; Liu, X.; Liu, S.; Fan, X.; Zhao, L.; Song, M.; Fan, X.; Xu, G. Co-Overexpression of OsNAR2.1 and OsNRT2.3a Increased Agronomic Nitrogen Use Efficiency in Transgenic Rice Plants. *Front. Plant Sci.* **2020**, *11*, 1245. [CrossRef]
27. Liu, R.; Jia, T.; Cui, B.; Song, J. The expression patterns and putative function of nitrate transporter 2.5 in plants. *Plant Signal. Behav.* **2020**, *15*, 1815980. [CrossRef]
28. Lupini, A.; Mercati, F.; Araniti, F.; Miller, A.J.; Sunseri, F.; Abenavoli, M.R. NAR2.1/NRT2.1 functional interaction with NO3(-) and H(+) fluxes in high-affinity nitrate transport in maize root regions. *Plant Physiol. Biochem.* **2016**, *102*, 107–114. [CrossRef]
29. Song, J.M.; Guan, Z.; Hu, J.; Guo, C.; Yang, Z.; Wang, S.; Liu, D.; Wang, B.; Lu, S.; Zhou, R.; et al. Eight high-quality genomes reveal pan-genome architecture and ecotype differentiation of Brassica napus. *Nat. Plants* **2020**, *6*, 34–45. [CrossRef]
30. Chalhoub, B.; Denoeud, F.; Liu, S.; Parkin, I.A.; Tang, H.; Wang, X.; Chiquet, J.; Belcram, H.; Tong, C.; Samans, B.; et al. Plant genetics. Early allopolyploid evolution in the post-Neolithic Brassica napus oilseed genome. *Science* **2014**, *345*, 950–953. [CrossRef]
31. Stahl, A.; Vollrath, P.; Samans, B.; Frisch, M.; Wittkop, B.; Snowdon, R.J. Effect of breeding on nitrogen use efficiency-associated traits in oilseed rape. *J. Exp. Bot.* **2019**, *70*, 1969–1986. [CrossRef] [PubMed]
32. Bouchet, A.S.; Laperche, A.; Bissuel-Belaygue, C.; Baron, C.; Morice, J.; Rousseau-Gueutin, M.; Dheu, J.E.; George, P.; Pinochet, X.; Foubert, T.; et al. Genetic basis of nitrogen use efficiency and yield stability across environments in winter rapeseed. *BMC Genet.* **2016**, *17*, 131. [CrossRef] [PubMed]
33. Sorin, C.; Leport, L.; Cambert, M.; Bouchereau, A.; Mariette, F.; Musse, M. Nitrogen deficiency impacts on leaf cell and tissue structure with consequences for senescence associated processes in Brassica napus. *Bot. Stud.* **2016**, *57*, 11. [CrossRef] [PubMed]
34. Avice, J.C.; Etienne, P. Leaf senescence and nitrogen remobilization efficiency in oilseed rape (Brassica napus L.). *J. Exp. Bot.* **2014**, *65*, 3813–3824. [CrossRef]
35. Orsel, M.; Krapp, A.; Daniel-Vedele, F. Analysis of the NRT2 nitrate transporter family in Arabidopsis. Structure and gene expression. *Plant Physiol.* **2002**, *129*, 886–896. [CrossRef]
36. Bai, H.; Euring, D.; Volmer, K.; Janz, D.; Polle, A. The nitrate transporter (NRT) gene family in poplar. *PLoS ONE* **2013**, *8*, e72126. [CrossRef]
37. Guo, B.; Li, Y.; Wang, S.; Li, D.; Lv, C.; Xu, R. Characterization of the Nitrate Transporter gene family and functional identification of HvNRT2.1 in barley (Hordeum vulgare L.). *PLoS ONE* **2020**, *15*, e0232056. [CrossRef]
38. Cheng, F.; Wu, J.; Wang, X. Genome triplication drove the diversification of Brassica plants. *Hortic. Res.* **2014**, *1*, 14024. [CrossRef]
39. Wen, J.; Guo, P.; Ke, Y.; Liu, M.; Li, P.; Wu, Y.; Ran, F.; Wang, M.; Li, J.; Du, H. The auxin response factor gene family in allopolyploid Brassica napus. *PLoS ONE* **2019**, *14*, e0214885. [CrossRef]
40. Du, H.; Liang, Z.; Zhao, S.; Nan, M.G.; Tran, L.S.; Lu, K.; Huang, Y.B.; Li, J.N. The Evolutionary History of R2R3-MYB Proteins Across 50 Eukaryotes: New Insights Into Subfamily Classification and Expansion. *Sci. Rep.* **2015**, *5*, 11037. [CrossRef]
41. Guo, P.; Wen, J.; Yang, J.; Ke, Y.; Wang, M.; Liu, M.; Ran, F.; Wu, Y.; Li, P.; Li, J.; et al. Genome-wide survey and expression analyses of the GRAS gene family in Brassica napus reveals their roles in root development and stress response. *Planta* **2019**, *250*, 1051–1072. [CrossRef] [PubMed]
42. Wang, J.; Hüner, N.; Tian, L. Identification and molecular characterization of the Brachypodium distachyon NRT2 family, with a major role of BdNRT2.1. *Physiol. Plant.* **2019**, *165*, 498–510. [CrossRef] [PubMed]
43. Cai, C.; Wang, J.Y.; Zhu, Y.G.; Shen, Q.R.; Li, B.; Tong, Y.P.; Li, Z.S. Gene structure and expression of the high-affinity nitrate transport system in rice roots. *J. Integr. Plant Biol.* **2008**, *50*, 443–451. [CrossRef] [PubMed]
44. You, L.; Wang, Y.; Zhang, T.; Zhu, Y.; Ren, N.; Jiang, X.; Zhou, Y. Genome-wide identification of nitrate transporter 2 (NRT2) gene family and functional analysis of MeNRT2.2 in cassava (Manihot esculenta Crantz). *Gene* **2022**, *809*, 146038. [CrossRef]
45. Slot, J.C.; Hallstrom, K.N.; Matheny, P.B.; Hibbett, D.S. Diversification of NRT2 and the origin of its fungal homolog. *Mol. Biol. Evol.* **2007**, *24*, 1731–1743. [CrossRef]
46. von Wittgenstein, N.J.; Le, C.H.; Hawkins, B.J.; Ehlting, J. Evolutionary classification of ammonium, nitrate, and peptide transporters in land plants. *BMC Evol. Biol.* **2014**, *14*, 11. [CrossRef]
47. Pellizzaro, A.; Clochard, T.; Planchet, E.; Limami, A.M.; Morère-Le Paven, M.C. Identification and molecular characterization of Medicago truncatula NRT2 and NAR2 families. *Physiol. Plant.* **2015**, *154*, 256–269. [CrossRef]
48. Feng, H.; Yan, M.; Fan, X.; Li, B.; Shen, Q.; Miller, A.J.; Xu, G. Spatial expression and regulation of rice high-affinity nitrate transporters by nitrogen and carbon status. *J. Exp. Bot.* **2011**, *62*, 2319–2332. [CrossRef]
49. Gu, C.; Song, A.; Zhang, X.; Wang, H.; Li, T.; Chen, Y.; Jiang, J.; Chen, F.; Chen, S. Cloning of chrysanthemum high-affinity nitrate transporter family (CmNRT2) and characterization of CmNRT2.1. *Sci. Rep.* **2016**, *6*, 23462. [CrossRef]

50. Remans, T.; Nacry, P.; Pervent, M.; Girin, T.; Tillard, P.; Lepetit, M.; Gojon, A. A central role for the nitrate transporter NRT2.1 in the integrated morphological and physiological responses of the root system to nitrogen limitation in Arabidopsis. *Plant Physiol.* **2006**, *140*, 909–921. [CrossRef]
51. Tang, Z.; Fan, X.; Li, Q.; Feng, H.; Miller, A.J.; Shen, Q.; Xu, G. Knockdown of a rice stelar nitrate transporter alters long-distance translocation but not root influx. *Plant Physiol.* **2012**, *160*, 2052–2063. [CrossRef]
52. Li, W.; He, X.; Chen, Y.; Jing, Y.; Shen, C.; Yang, J.; Teng, W.; Zhao, X.; Hu, W.; Hu, M.; et al. A wheat transcription factor positively sets seed vigour by regulating the grain nitrate signal. *New Phytol.* **2020**, *225*, 1667–1680. [CrossRef] [PubMed]
53. Goodstein, D.M.; Shu, S.; Howson, R.; Neupane, R.; Hayes, R.D.; Fazo, J.; Mitros, T.; Dirks, W.; Hellsten, U.; Putnam, N.; et al. Phytozome: A comparative platform for green plant genomics. *Nucleic Acids Res.* **2012**, *40*, D1178–D1186. [CrossRef] [PubMed]
54. Artimo, P.; Jonnalagedda, M.; Arnold, K.; Baratin, D.; Csardi, G.; de Castro, E.; Duvaud, S.; Flegel, V.; Fortier, A.; Gasteiger, E.; et al. ExPASy: SIB bioinformatics resource portal. *Nucleic Acids Res.* **2012**, *40*, W597–W603. [CrossRef] [PubMed]
55. Kumar, S.; Stecher, G.; Tamura, K. MEGA7: Molecular Evolutionary Genetics Analysis Version 7.0 for Bigger Datasets. *Mol. Biol. Evol.* **2016**, *33*, 1870–1874. [CrossRef]
56. Hu, B.; Jin, J.; Guo, A.Y.; Zhang, H.; Luo, J.; Gao, G. GSDS 2.0: An upgraded gene feature visualization server. *Bioinformatics* **2015**, *31*, 1296–1297. [CrossRef]
57. Li, P.; Wen, J.; Chen, P.; Guo, P.; Ke, Y.; Wang, M.; Liu, M.; Tran, L.P.; Li, J.; Du, H. MYB Superfamily in Brassica napus: Evidence for Hormone-Mediated Expression Profiles, Large Expansion, and Functions in Root Hair Development. *Biomolecules* **2020**, *10*, 875. [CrossRef]
58. Shannon, P.; Markiel, A.; Ozier, O.; Baliga, N.S.; Wang, J.T.; Ramage, D.; Amin, N.; Schwikowski, B.; Ideker, T. Cytoscape: A software environment for integrated models of biomolecular interaction networks. *Genome Res.* **2003**, *13*, 2498–2504. [CrossRef]
59. de Hoon, M.J.; Imoto, S.; Nolan, J.; Miyano, S. Open source clustering software. *Bioinformatics* **2004**, *20*, 1453–1454. [CrossRef]
60. Saldanha, A.J. Java Treeview–extensible visualization of microarray data. *Bioinformatics* **2004**, *20*, 3246–3248. [CrossRef]
61. Zhou, J.; Zhou, H.J.; Chen, P.; Zhang, L.L.; Zhu, J.T.; Li, P.F.; Yang, J.; Ke, Y.Z.; Zhou, Y.H.; Li, J.N.; et al. Genome-Wide Survey and Expression Analysis of the KT/HAK/KUP Family in Brassica napus and Its Potential Roles in the Response to K(+) Deficiency. *Int. J. Mol. Sci.* **2020**, *21*, 9487. [CrossRef] [PubMed]
62. Liu, K.H.; Huang, C.Y.; Tsay, Y.F. CHL1 is a dual-affinity nitrate transporter of Arabidopsis involved in multiple phases of nitrate uptake. *Plant Cell* **1999**, *11*, 865–874. [CrossRef] [PubMed]

International Journal of
Molecular Sciences

MDPI

Article

Genome-Wide Identification and Analysis of Ariadne Gene Family Reveal Its Genetic Effects on Agronomic Traits of *Brassica napus*

Sumbal Wahid †, Meili Xie *,†, Sehrish Sarfraz, Jie Liu, Chuanji Zhao, Zetao Bai, Chaobo Tong, Xiaohui Cheng, Feng Gao * and Shengyi Liu

Key Laboratory of Biology and Genetics Improvement of Oil Crops, Oil Crops Research Institute of Chinese Academy of Agricultural Sciences, Ministry of Agriculture and Rural Affairs, Wuhan 430062, China; sumbalwahid@gmail.com (S.W.); sehrishsarfraz04@gmail.com (S.S.); whjiejiel@163.com (J.L.); zhaochuanji@caas.cn (C.Z.); baizetao_2005@163.com (Z.B.); tongchaobo@126.com (C.T.); cxh5495@163.com (X.C.); liusy@oilcrops.cn (S.L.)
* Correspondence: xiemeili@caas.cn (M.X.); gaofeng01@caas.cn (F.G.)
† These authors contributed equally to this work.

Abstract: E3 ligases promote protein ubiquitination and degradation, which regulate every aspect of eukaryotic life. The Ariadne (ARI) proteins of RBR (ring between ring fingers) protein subfamily has been discovered as a group of potential E3 ubiquitin ligases. Only a few available research studies show their role in plant adaptations processes against the external environment. Presently, the functions of ARI proteins are largely unknown in plants. Therefore, in this study, we performed genome-wide analysis to identify the *ARI* gene family and explore their potential importance in *B. napus*. A total of 39 *ARI* genes were identified in the *B. napus* genome and were classified into three subfamilies (A, B and C) based on phylogenetic analysis. The protein–protein interaction networks and enrichment analysis indicated that *BnARI* genes could be involved in endoreduplication, DNA repair, proteasome assembly, ubiquitination, protein kinase activity and stress adaptation. The transcriptome data analysis in various tissues provided us an indication of some *BnARI* genes' functional importance in tissue development. We also identified potential *BnARI* genes that were significantly responsive towards the abiotic stresses. Furthermore, eight *BnARI* genes were identified as candidate genes for multiple agronomic traits through association mapping analysis in *B. napus*; among them, *BnaA02g12100D*, which is the ortholog of *AtARI8*, was significantly associated with ten agronomic traits. This study provided useful information on *BnARI* genes, which could aid targeted functional research and genetic improvement for breeding in *B. napus*.

Keywords: E3 ligases; Ariadne (ARI) proteins; ubiquitination; *AtARI8*; phylogenetic analysis; association mapping analysis; agronomic traits; *B. napus*

Citation: Wahid, S.; Xie, M.; Sarfraz, S.; Liu, J.; Zhao, C.; Bai, Z.; Tong, C.; Cheng, X.; Gao, F.; Liu, S. Genome-Wide Identification and Analysis of Ariadne Gene Family Reveal Its Genetic Effects on Agronomic Traits of *Brassica napus*. *Int. J. Mol. Sci.* **2022**, *23*, 6265. https://doi.org/10.3390/ijms23116265

Academic Editors: Andrés J. Cortés and Hai Du

Received: 5 May 2022
Accepted: 29 May 2022
Published: 3 June 2022

1. Introduction

Ubiquitination is one of the important mechanisms that prepare the cell response toward internal and external stimuli during plant life [1]. It is the process of protein degradation in which the ubiquitin-26S proteasome system (UPS) is the main player [2]. The targeted proteins are first bound to the multiple ubiquitin (Ub) proteins. Ub is a 76-aminoacid protein that is a highly conserved housekeeping protein in all the eukaryotes. A series of enzymes, including ubiquitin-protein ligases (E3), ubiquitin-activating enzymes (E1) and ubiquitin-conjugating enzymes (E2 or Ubc), are used to transfer Ub to a target protein. Firstly, the Ub protein is activated by interacting with E1 and forms a E1-Ub molecule at the expense of ATP molecules. This activated Ub molecule is transferred to E2 through the interaction of E1-Ub with E2 and forms an intermediate complex, E2-Ub. Then, the E3 covalently attaches the Ub with the target protein by interacting with the

E2-Ub intermediate complex and target protein. In this way, many targeted proteins are bound to Ub proteins binds by the repetition of the above-mentioned process, and hence are recognized and degraded by the 26S proteasome [1–3]. About 6% of *A. thaliana* proteome constitutes this proteasome system [4]. Among E1, E2 and E3, E3 are most abundant in eukaryotes and 1400 E3s are predicted in *A. thaliana* [5]. By far, E3s are of more importance because they are diverse, directly interacting with the conjugated Ub and binding to the target protein, therefore maintaining specificity [6]. These E3 ligases maintain the cellular response by targeting cell division, signaling, immune responses and DNA repair [7].

In general, E3 ligases are divided into two groups, each of which has either a HECT domain or a RING (really interesting new gene) domain [7,8]. The RING domain containing E3s is usually identified with a cysteine-histidine rich ring motif with Zinc atoms [7]. It is one of the most detected domains in *A. thaliana* [9] and it has been involved in protein–protein interaction processes [10]. In Arabidopsis and rice genomes, 426 and 425 RING type E3 ligases have been identified, respectively [6,11]. The RING E3 ligases were responsive towards dehydration, cold and heat treatment in rice leaf tissue [12]. AIRP2 (ABA insensitive RING protein 2) is also a E3 ring type ligase which was involved in Abscisic-Acid-mediated drought tolerance in Arabidopsis [13]. RING E3 ligases also play an important role in the plasticity of the flowering time of plants; for example, the photoperiodic pathways are regulated through ubiquitination and, hence, timely floral induction is ensured [14].

One of the largest group of proteins that contain the RING domain is RBR (ring between ring fingers) proteins, identified by the presence of their RING1-IBR-RING2 domains (RBR supra domain) [15]. It constitutes RING1 at the N-terminal, an IBR (in between ring), and RING2 at the C-terminal. E3 ligase activity is seen in several RBR proteins [15]. The plant RBR proteins consist of Ariadne (ARI proteins), ARA54, Plant II and Helicase subfamilies [16]. The best known RBR protein is Parkin, which is involved in autosomal recessive familial Parkinson's disease [17,18], and the Ariadne family shares its RBR domain similarity [17,19]. The RING1 is a typical RING finger with a C3HC4 signature of conserved cysteine and histidine residues, the IBR domain has a typical C6HC signature, while the RING2 is shorter than the canonical RING structure. In addition to the RBR domain, ARI proteins also contain the Ariadne domain at the C-terminus [17,20].

The *ARI* gene family has been identified in human [21–23], Drosophila [24], mouse [25], and Arabidopsis [9]. However, the mechanism and functions of ARI proteins are largely unknown, for only limited literature is available regarding the functional importance of these genes. Some ARI proteins are involved in E2 enzyme interactions in Drosophila [24] and human [23]. Ubiquitination activity is only proven in *ARI8* in *A. thaliana* [6]. *AtARI12* was involved in UV- B pathway through interaction with Constitutively Photomorphogenic 1 (COP1), which is a key component of the light signaling pathway [26,27]. The overexpression of a soybean Ariadne-like gene caused aluminum tolerance in Arabidopsis [28]. The orthologous gene of *AtARI7* in *Hypericum perforatum* was associated with apospory [29].

Brassica napus, an allotetraploid plant generated through hybridization between *Brassica rapa* and *Brassica oleracea* [30], is a major worldwide crop due its industrial value and oil production. Like other crops, *B. napus* productivity has been affected by several abiotic stresses including low temperature, drought, salinity, and so on [31–33]. Therefore, it is important to improve stress adaptation by identifying stress-related genes in *B. napus*. To date, genome-resource and various transcriptome datasets have been published [34,35]. However, *ARI* gene family members and their potential importance have not been investigated in *B. napus*. Therefore, a genome-wide study was designed to find out the number of genes in this family. The evolutionary relationships, gene architectures, conserved motifs, gene duplications, and protein–protein interactions in *B. napus* were all investigated in this work. Diverse tissues and environmental stressors transcriptome data was used to understand the expression patterns of *ARI* genes in *B. napus*. Furthermore, we also investigated the genetic variations (SNPs) in the *BnARI* gene family and associated them with several agronomic traits in the natural population of *B. napus*. This research enriched our knowledge of *BnARI* genes and laid the groundwork for future functional research and genetic breeding of *B. napus*.

2. Results

2.1. Identification of ARI Genes in B. napus

We identified 39 *ARI* genes in the *B. napus* genome by querying 16 *A. thaliana* ARI protein sequences (Table 1 and Table S1). All the BnARI proteins contained the RBR and Ariadne domains. The detailed information of *BnARI* genes is summarized in Table 1. The protein length varied from 369 to 672 amino acids and the molecular weight ranged from 40.5 to 75.1 kDa. The isoelectric points increased from 4.7 to 6.48 (Table 1). Based on CELLO server prediction, 37 ARI proteins were localized in the nucleus (Table 1).

Table 1. Characteristics of the *ARI* genes in *B. napus* (pI, isoelectric point; MW, molecular weight).

ID	Classification	pI	MW (kDa)	Amino acids	Chromosome	Start	End	Duplication Type	Exon Number	Subcellular Localization
BnaC01g04040D	subfamily A	5.44	68.30577	598	C01	2107232	2110597	WGD or Segmental	6	Nuclear
BnaA01g02770D		5.49	67.70018	592	A01	1363570	1366772	WGD or Segmental	6	Nuclear
BnaA07g03810D		5.5	66.84028	583	A07	3464224	3466892	WGD or Segmental	7	Nuclear
BnaC07g05410D		5.39	67.01159	582	C07	8591383	8594140	WGD or Segmental	7	Nuclear
BnaA09g02140D		5.38	61.03602	525	A09	1051875	1053449	WGD or Segmental	1	Nuclear
BnaA02g29030D		5.28	55.36126	481	A02	21265416	21266987	WGD or Segmental	3	Nuclear
BnaC09g01530D		5.45	56.70796	485	C09	836916	838508	WGD or Segmental	3	Nuclear
BnaC02g37060D		5.07	55.77174	485	C02	40014335	40015918	WGD or Segmental	3	Nuclear
BnaC02g04010D	subfamily B	5.53	58.47465	510	C02	2040175	2041821	WGD or Segmental	2	Nuclear
BnaC05g04050D		5.12	63.11043	556	C05	1976038	1981190	WGD or Segmental	15	Nuclear
BnaA02g34990D		5.83	62.01988	541	A02_random	24816	26477	WGD or Segmental	2	Nuclear
BnaA10g03930D		5.11	63.19455	559	A10	2088735	2093862	WGD or Segmental	15	Nuclear
BnaA03g03500D		5	56.69741	496	A03	1688116	1689720	WGD or Segmental	2	Nuclear
BnaC03g04990D		5.02	59.7861	526	C03	2416072	2417649	WGD or Segmental	1	Nuclear
BnaA04g18230D		4.99	63.39872	556	A04	14658509	14662929	WGD or Segmental	15	Nuclear
BnaC03g17530D		4.9	63.69074	558	C03	8960359	8965066	WGD or Segmental	15	Nuclear
BnaC04g42310D		4.92	63.61489	558	C04	42886627	42891037	WGD or Segmental	15	Nuclear
BnaA08g31720D		5.08	62.69513	554	A08_random	2100605	2104866	WGD or Segmental	15	Nuclear
BnaA03g14490D		4.85	63.39747	555	A03	6669544	6674154	WGD or Segmental	15	Nuclear
BnaC08g01560D		5.03	62.76486	555	C08	1211774	1215865	WGD or Segmental	15	Nuclear
BnaA02g12100D		5.06	62.48365	548	A02	6326435	6331272	WGD or Segmental	16	Nuclear
BnaC02g45230D		5.05	64.28373	565	C02_random	956624	961488	WGD or Segmental	15	Nuclear
BnaAnng26970D		4.96	57.65184	499	Ann_random	30922371	30924172	Tandem	2	Nuclear
BnaAnng26980D		6.48	51.01917	440	Ann_random	30926298	30927929	Tandem	2	Nuclear
BnaC04g42500D		5.09	63.14806	548	C04	43066664	43068633	Tandem	2	Nuclear
BnaA03g14710D		5.83	60.14362	522	A03	6761186	6762872	WGD or Segmental	2	Nuclear
BnaC04g42490D		5.39	66.50624	574	C04	43060145	43062382	WGD or Segmental	3	Nuclear
BnaC05g04020D		5.18	61.60797	529	C05	1967024	1969723	WGD or Segmental	14	Nuclear
BnaA10g03910D		5.28	60.45582	522	A10	2073034	2075988	WGD or Segmental	15	Nuclear
BnaC03g50610D	subfamily C	4.87	75.1735	657	C03	35066895	35072312	WGD or Segmental	3	Nuclear
BnaA02g33880D		5.15	51.08088	446	A02	24242394	24244388	WGD or Segmental	2	Extracellular
BnaC02g42600D		5.31	51.23706	447	C02	45189629	45191512	WGD or Segmental	2	Extracellular
BnaA06g22860D		4.7	61.10942	532	A06	16009537	16013955	WGD or Segmental	3	Nuclear
BnaC03g50620D		4.82	42.64821	369	C03	35073684	35074888	Tandem	2	Nuclear
BnaC02g01010D		5.61	54.80394	483	C02	436156	437604	WGD or Segmental	1	Nuclear
BnaA06g18370D		4.99	50.6674	447	A06	10552571	10553998	WGD or Segmental	2	Nuclear
BnaA10g22750D		5.06	74.29611	672	A10	15235470	15237795	WGD or Segmental	3	Nuclear
BnaAnng05970D		5.38	54.8181	478	Ann_random	6115791	6117224	WGD or Segmental	1	Nuclear
BnaC09g47300D		6.05	73.81982	665	C09	46717181	46719562	WGD or Segmental	4	Nuclear

2.2. Phylogenetic Analysis of BnARI Proteins

To determine the evolutionary relationship between *BnARI* and *AtARI* genes, a phylogenetic tree based on the NJ (neighbor-joining) method with 1000 bootstrap replications was constructed using their protein sequences. A total of 16 *AtARIs* and 39 *BnARIs* were clustered into three subfamilies (Figure 1, Table 1). In addition, in each subfamily, the *BnARIs* were clustered with their closest homologous gene in *A. thaliana*. Based on the previous nomenclature system used in *A. thaliana*, these subfamilies were named subfamily A, B and C, respectively (Figure 1, Table 1). Subfamily B was the largest and included 21 *ARI* genes, while Subfamily C and A contained 10 and 8 *ARI* genes, respectively.

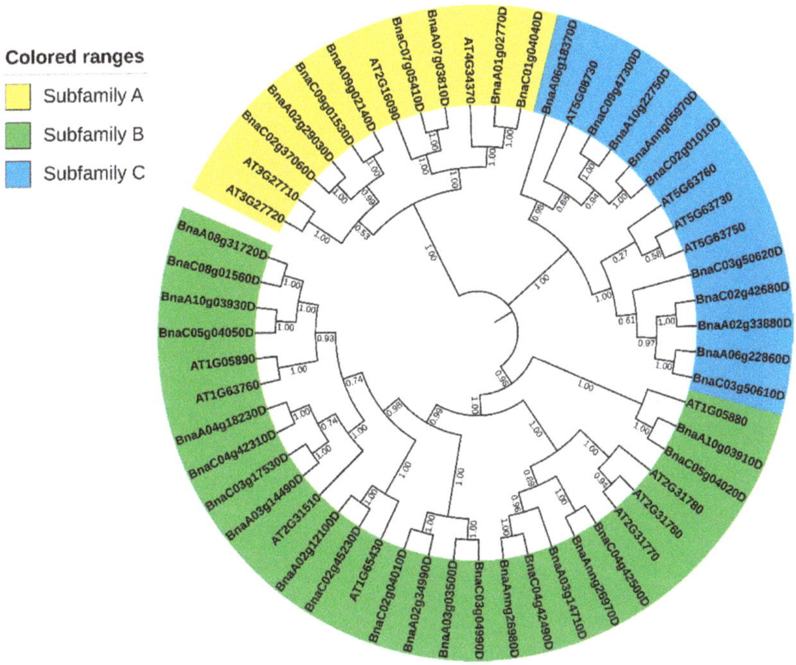

Figure 1. Phylogenetic analysis of ARI proteins in *B. napus* and *A. thaliana*. All ARI proteins are grouped into three subfamilies, and each subfamily is represented by a different color. The number on the branches shows the bootstrap values.

2.3. Chromosomal Location and ARI Genes Duplication in B. napus

In *B. napus*, 33 *ARI* genes were unevenly distributed on the 19 chromosomes; the remaining six genes were located on random chromosomes (Table 1, Figure 2). A total of 20 and 19 *ARIs* were located on the A and C subgenomes, respectively. Chromosome C03 had the most *ARIs* (4 genes) while chromosomes A01, A04, A07, A09, C01, C07 and C08 had a single *ARI* gene. Chromosomes A05, A08, and C06 did not contain any *ARI* genes.

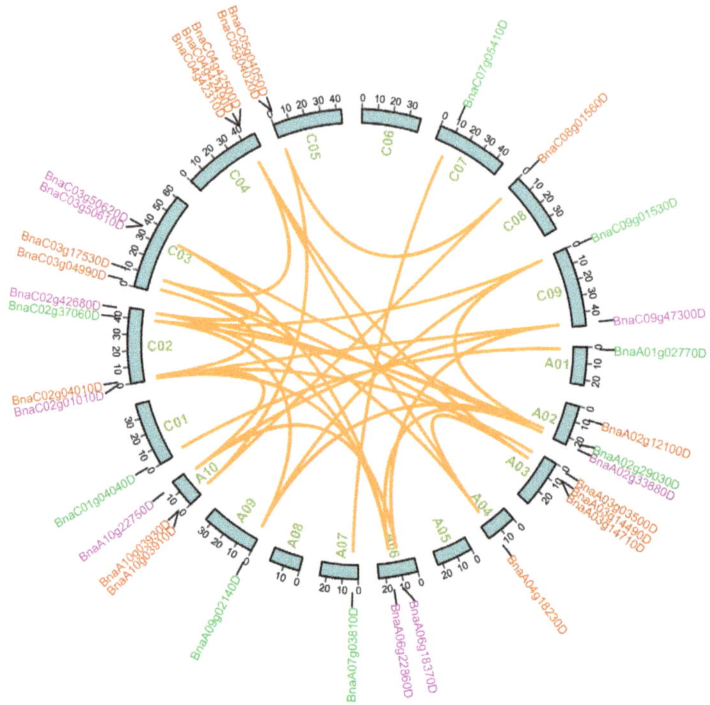

Figure 2. The chromosomal locations and duplicated genes analysis of *ARI* genes in *B. napus*. The locations of all the chromosomal *BnARI* genes are represented on different chromosomes, excluding the random fragment chromosomes. The different colors mean different *BnARI* subfamilies genes. Subfamily A, subfamily B and subfamily C are indicated by green, brown and purple color, respectively. The orange lines are used to highlight the duplicated *BnARI* gene pairs.

The expansion of the gene family is contributed by duplication events in the plant genome [36]. Therefore, the duplication events were analyzed for *ARI* genes. Based on BLAST and MCScanX analysis, the results showed that 35 out of 39 *ARIs* resulted from whole-genome duplication (WGD) or segmental duplication, while four genes were derived from tandem duplication (Table 1). Furthermore, 17 duplicated *ARI* pairs between subgenomes were found in *B. napus* (Figure 2; Table S2). To determine the selection pressure on these duplicated pairs during the evolution, the ratios of non-synonymous to synonymous substitutions (Ka/Ks) and divergence time were determined. The Ka/Ks value ranged from 0.038 to 0.035 with an average of 0.268 between gene pairs. The Ka/Ks ratio was significantly less than 1. The estimated divergence time of the duplicated pairs ranged from 1.509 Mya to 10.101 MYA with an average of 3. 877 MYA time period. The results indicated that *BnARIs* were under strong purifying selection (Table S2).

2.4. Gene Structure and Conserved Motif Analysis of BnARIs

The exons and introns were examined to obtain insights into structural evolution in the *BnARI* gene family. The gene structure information was retrieved from the "*Darmor*-bzh" genome file in the BnaOmics database (Available online: https://bnaomics.ocri-genomics.net, accessed on 6 February 2022) using TBtool software. On average, each gene contained six exons but the number of exons of *BnARIs* greatly varied and ranged from 1 to 15 (Table 1, Figure 3). All the three subfamilies possessed a different number of exons; for example, subfamily C had the lowest number of exons (one–four), subfamily A contained up to seven exons while the subfamily B exhibited diverse exon number variations ranging from 1 to 15 (Table 1, Figure 3).

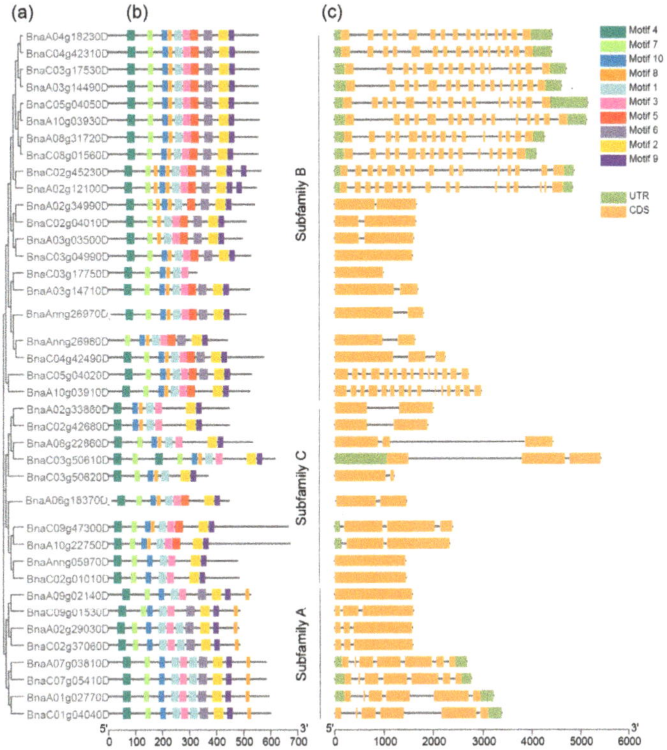

Figure 3. The phylogenetic relationship, conserved motifs and gene structure of *ARI* genes in *B. napus*. (**a**) The phylogenetic relationship of BnARI proteins. (**b**) The motif composition of BnARI proteins. Motifs (1–10) are shown in different colored boxes. (**c**) Gene structures of the BnARI genes. Orange boxes represent the CDS whereas green boxes represent UTR.

Furthermore, we extracted the protein data of BnARIs from the Protein FASTA file of *B. napus* in the BnaOmics database (Available online: https://bnaomics.ocri-genomics.net, accessed on 6 February 2022) to investigate the motif composition of all the *BnARIs*. The online MEME sever [37] was used to perform the motif analysis. Motifs 1, 5 and 8 were annotated as IBR; 2 and 6 were annotated as Ariadne domain; and motif 7 was annotated as Zinc finger (Table S3). Motifs 1, 2 and 7 were present in all the *BnARIs* except *BnaA02g33880D* and *BnaC02g42680D* which do not contain motif 7 (Figure 3). Both Ariadne domains (motifs 2 and 6) were present in subfamily A and B, whereas motif 6 was absent in subfamily C. Subfamily B and C both contained motif 1, 7, 8, 5 and 2 in principle but the only difference was subfamily C do not contain motif 6. The remaining motifs, 3, 4, 9, 10, were unknown domains but these were distributed evenly within the same subfamily (Figure 3). The motif arrangement in each subfamily also verified the phylogeny classification (Figures 1 and 3).

2.5. Cis-Elements and Protein Interaction Analysis of BnARI Genes

To determine the potential function of these *BnARIs*, we further analyzed the cis-elements in their promoter regions and proteins that could interact with them by using online public databases [38,39]. The 2kb upstream region of *BnARIs* were retrieved from the "*Darmor*-bzh" genome file and were analyzed for cis-elements through the PlantCARE database. These promoters were mainly enriched in growth-related and stress-related elements (Figure 4, Tables S4 and S5). ABRE (an element responsive to abscisic acid), CGTCA-motif (an element responsive to methyl jasmonic acid), ERE (an element response to ethylene), TCA-element (an element responsive to salicylic acid), P-box, GARE, AuxRR (an element responsive to auxin), TATC-box (an element responsive to gibberellin) and TGA-element (an element responsive to auxin) were hormone-related cis-regulatory elements and were listed according to their abundance (Figure 4, Table S4). Other growth-related elements included Circadian, O2-site (zein-metabolism-responsive element), CAT-box (a meristem-related element), MBSI (flavonoid-synthesizing element), HD-Zip (cell-differentiation element) and GCN4_motifs (an endosperm-development element); these elements are also presented according to their abundance (Figure 4, Table S4).

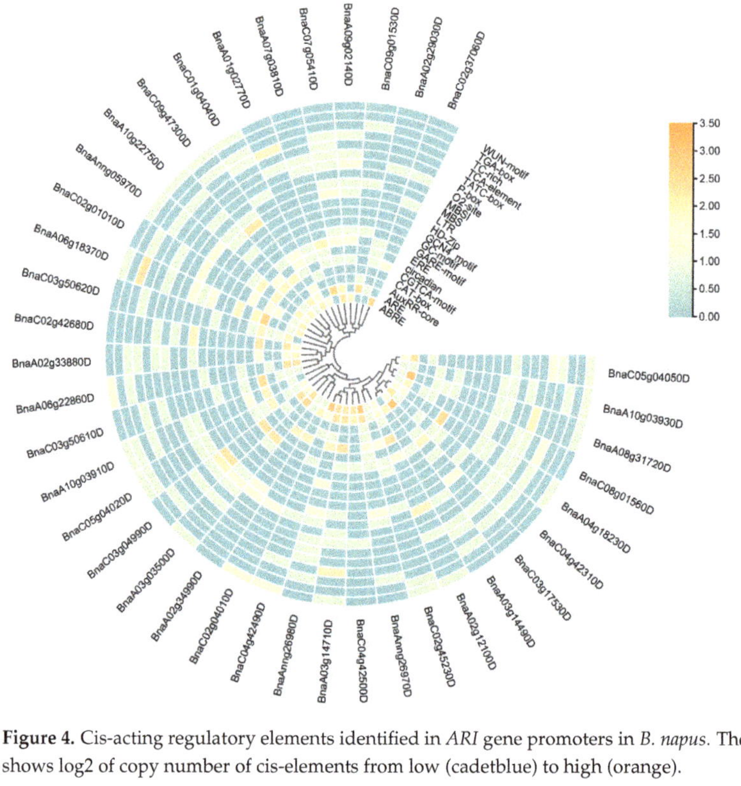

Figure 4. Cis-acting regulatory elements identified in *ARI* gene promoters in *B. napus*. The color bar shows log2 of copy number of cis-elements from low (cadetblue) to high (orange).

Stress-related elements were predominant in *ARIs* promoters and these included ARE (an element that responds to anaerobic induction), LTR (an element responsive to low temperature), MBS (the MYB binding site for drought stress), WUN-motif (an element responsive to wounds), TC-rich (a defense-responsive element), and GC-motif (an anoxic-specific element) (Figure 4, Table S4). Among all the promoter cis-elements, ARE was most abundant, whereas 33/39 of *BnARIs* contained ABREs (Figure 4, Table S4).

In subfamily A, a large number of stress- and hormone-related cis-acting elements were identified in the promoters of *BnARI* genes: ARE, ABRE and ERE elements had maximum copies and were detected in 5, 5 and 6 of the 8 *BnARI* gene promoters, respectively (Figure 4, Tables S4 and S5). The subfamily B members were enriched with stress-related element such as ARE, LTR, MBS and the WUN motif and at least 12 of 21 genes contain these elements. All 21 genes of this subfamily had the ARE motif, with an average of three copies. In 10 genes of subfamily C, CGTCA (on average, three copies), ARE (on average, two copies) and ABRE (on average, two copies) motif were found in 9, 10 and 9 genes, respectively. Stress-related cis-acting elements were relatively more abundant in number in all *BnARI* promoters (Figure 4, Tables S4 and S5). These results implied the importance of *BnARI* genes in stress adaptations.

To elucidate the functional role of *BnARIs*, we predicted the protein networks based on known protein interactions in *A. thaliana* (Figure 5, Table S6). By using 16 AtARI proteins as a query, a total of 1323 proteins were identified in the protein interactive database of *A. thaliana* that were homologous to 4714 proteins in *B. napus*. Most of the BnARI proteins interacted with each other (Figure 5a, Table S6). To further access the functional categories of these interacting proteins, Gene Ontology (GO) and Kyoto Encyclopedia of Genes and Genomes (KEGG) enrichment analyses were performed (Figure 5b, Table S6). The KEGG pathway showed that the *BnARIs* participated in the endoreduplication, post-replication repair, growth, proteasome assembly, protein transport, protein kinase activity and stress-related processes. The detailed GO analysis revealed that the aforementioned pathways were consistent biological processes terms; in the cellular component terms, mainly, the ubiquitin ligases complexes, spliseosomal complexes, autophagosome and nucleosomes were enriched (Table S6). Meanwhile, under molecular function category, protein kinase activity, cyclin-dependent protein serine/threonine kinase regulator activity, RNA polymerase activity, U5/U6 snRNA binding, Atg8 activating enzyme activity and ubiquitin ligase activity terms were enriched (Table S6).

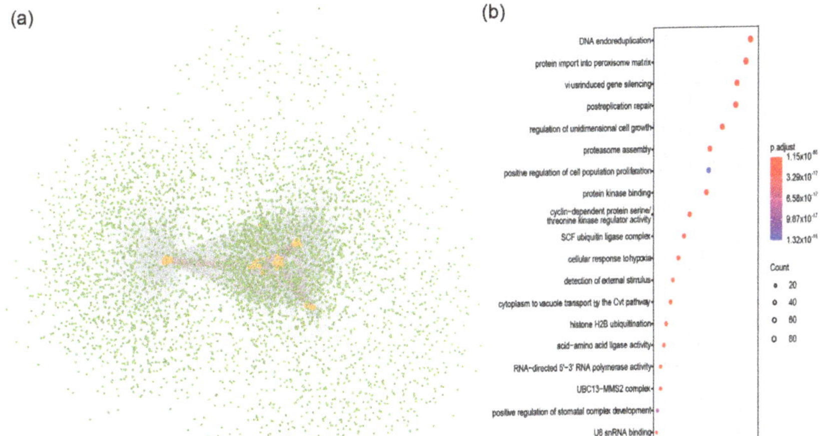

Figure 5. The ARI-protein interaction-network analysis in *B. napus*. (**a**) Protein–protein interaction network of BnARI proteins. The BnARI proteins are shown by orange circles, the green circles represent proteins that interacted with BnARI proteins. The red lines indicate the interaction between BnARI proteins, and the gray lines indicate the interaction between BnARI and other proteins. (**b**) KEGG pathway analysis of proteins interacted with BnARI proteins. The color bar shows the value of *p*.adjust from low (red) to high (purple).

2.6. Expression Patterns of BnARI Genes in Different Tissues and under Abiotic Stresses

We utilized the 12 tissues (root, leaf, bud, silique, stamen, new petal, blossomy petal, wilting petal, stem, sepal, ovule and pericarp) in a *B. napus* cultivar "ZS11" from our lab resources [34], to analyze the expression patterns of *BnARI* genes to predict their potential functions. We observed variable gene expressions across these tissues (Figure 6, Table S7). The results showed that the majority of genes were expressed in root and bud tissues. A total of 5 genes showed maximum expression of 70–161 FPKM in the root tissue. These genes included *BnaC01g04040D* (an ortholog of *AtARI1*) of subfamily A, *BnaC02g42680D* (an ortholog of *AtARI13*), *BnaAnng05970D* (an ortholog of *AtARI16*), *BnaC03g50620D* (an ortholog of *AtARI14*) of subfamily C and *BnaC05g04050D* (an ortholog of *AtARI5*) of subfamily B with FPKM values 72.6495, 78.5711, 152.675 and 161.35, respectively (Table S7). Two genes, *BnaA10g22750D* and *BnaAnng05970D* (an ortholog of *AtARI16*), showed predominant expression in the ovule (>60 FPKM) and one gene *BnaA01g02770D* (an ortholog of *AtARI1*) was highly expressed in the new petal tissue (60 FPKM). A total of 13 *BnARI* genes were moderately expressed among different tissues and their FPKM values ranged between 5 and 20; whereas 15 *BnARIs* were expressed below the threshold (>5) in all the tissues (Table S7). The low-expressed genes in all the tissues could be pseudogenes (Figure 6, Table S7).

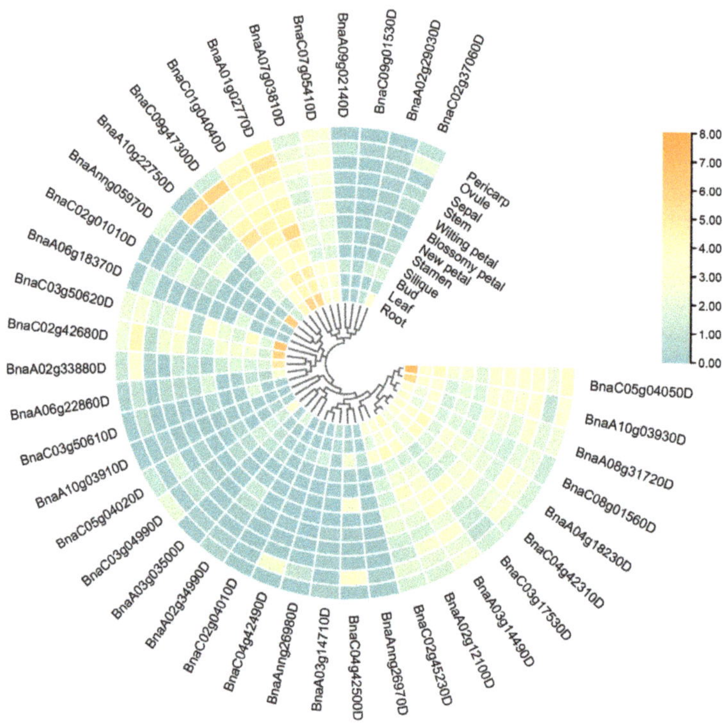

Figure 6. Expression profile of *ARI* genes in different tissues of *B. napus*. Heatmap was generated by taking log2 fold of FPKM values. The color bar shows relative expression from low (cadetblue) to high (orange).

There was expression variation among different tissues between the subfamilies (Figure 6, Table S7). In subfamily A, the genes were highly expressed in root > new petal > ovule. Subfamily B is the largest family of *BnARIs*, but the majority of the genes are moderately expressed in all the tissues (up to 17 FPKM). The predominant expression was observed in root > bud > pericarp (Figure 6, Table S7). All the *BnARIs* in silique tissue and orthologs of *AtARI12* in all the tissues had no FPKM greater than the threshold value; whereas, in subfamily C, all the *BnARIs* showed an FPKM below threshold in leaf, blossomy petal, stem, sepal and pericarp (Figure 6, Table S7). The predominant expression (>37) of remaining genes was observed in root > ovule > petals. We also observed the expression divergence between the duplicated genes of the same *AtARI* gene at expression level and tissue specificity; for example, in subfamily C, *AtARI16* has two duplicated pairs in *B. napus*: *BnaAnng05970D*/*BnaC02g01010D* and *BnaC09g47300D*/*BnaA10g22750D*. There was tissue-specific expression not only between each pair but also between both pairs (Figure 6, Table S7). In subfamily A, *BnaC02g37060D* (ortholog of *AtARI3*) was expressed in root tissue only, while the remaining three orthologs did not show expression in any tissue (Figure 6, Table S7). In subfamily B, the duplicated pair *BnaC04g42500D*/*BnaAnng26970D* (ortholog of *AtARI9*): the former partner was expressed in the ovule, new petal and bud (up to 13 FPKM) whereas the latter partner did not show expression in any tissue (Table S7). All these results represented clues for functional differentiation in the *ARI* gene family in *B. napus*.

Moreover, according to predominant gene expression and the diverse expression patterns of duplicated gene pairs among various tissues, five *BnARI* duplicated genes (*BnaC01g04040D*, *BnaA01g02770D*; *BnaC05g04050D*, *BnaA10g03930D*; *BnaA02g12100D*, *BnaC02g45230D*; *BnaA02g33880D*, *BnaC02g42680D*; *BnaA10g22750D*, *BnaC09g47300D*) from the three subfamilies were selected for qRT-PCR analysis (Figure 7, Table S8). The expression pattern results verified the correctness of RNA-seq data. Expression divergence was observed in all the cases among the duplicated pairs. Most of the genes showed high expression in root tissue (Figure 7). These candidate genes must be investigated for their importance in root growth in future studies.

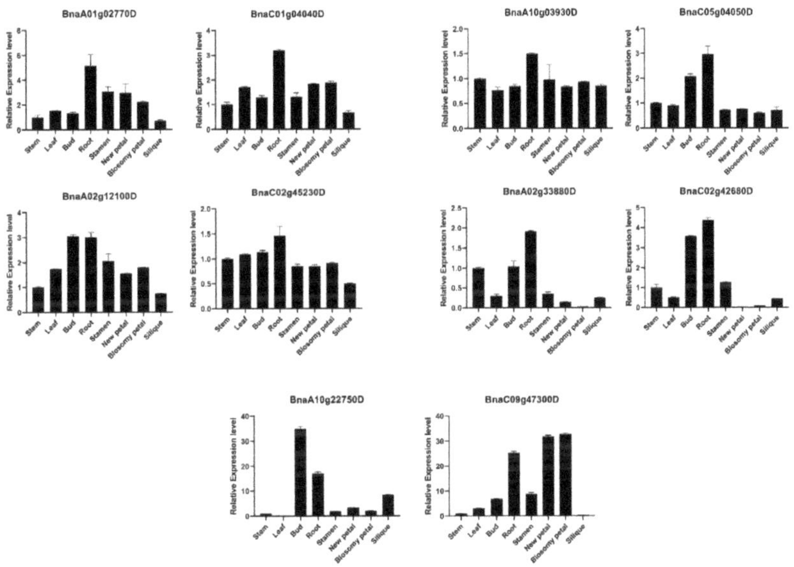

Figure 7. A qRT-PCR expression analysis of 10 *ARI* genes in eight different tissues of *B. napus*. The $2^{-\Delta\Delta CT}$ method was used to analyze the results. The error bars indicate the standard error of the mean of three biological replicates for every tissue.

We also investigated the expression pattern of *BnARIs* under abiotic stresses that included dehydration, cold, ABA and salinity treatments. The transcriptome data for this analysis was retrieved from a previous study by referring to Zhang [40]. The majority of gene expressions were upregulated when the stresses were applied (Figure 8, Table S9). Thirteen genes showed a significant change in expression compared to the control ($\geq$two folds) (Figure 8, Table S9). The expression of only one ortholog of *AtARI5*, *BnaC03g04990D*, was upregulated two folds after a 1 h and 8 h dehydration, 4 h and 24 h cold stress and 4 h saline treatment. One ortholog of *AtARI3*, *BnaC02g37060D*, displayed increased expression (3.5 fold) in 8 h dehydration stress while its expression was downregulated (−2.5 fold) under 4 h ABA treatment (Figure 8, Table S9). The orthologs of *AtARI13* displayed a significant response towards all the stress conditions. The expression of *AtARI13* orthologs (*BnaC03g50610D*, *BnaA02g33880D*, *BnaC02g42680D*, *BnaA06g22860D*) was upregulated

(2–5 fold) after 24 h cold treatment (Figure S7, Table S9). The expression analysis shows the importance of these *ARI* genes during adaptation in *B. napus*.

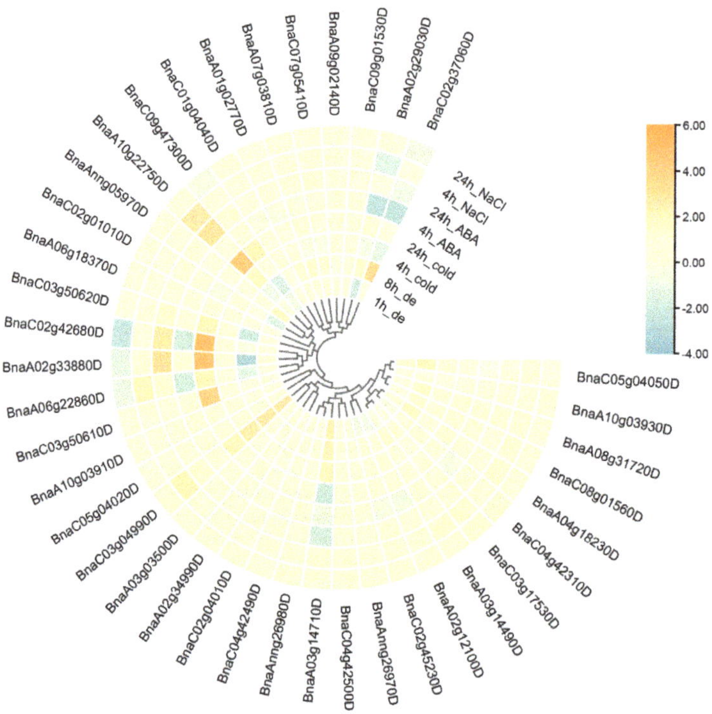

Figure 8. Expression profile of *ARI* genes under abiotic stress conditions in *B. napus*. The expression data is processed by comparing the control with each treated sample and calculating the log2 fold change to generate the heatmap. The color bar represents relative expression levels from low (cadetblue) to high (orange).

2.7. Functional Importance of BnARIs Using Association Mapping Analysis in Natural Population

We utilized the 324 natural population accession data of our lab to find the potential effects of *BnARIs* in the agronomic traits of *B. napus*. We first analyzed their genetic variations (SNPs) and then performed association mapping analysis in natural population (Figure 9, Table S10) [41]. On average, ~40 SNPs were detected per *ARI* gene and SNP density was higher in the A subgenome (50 SNPs), whereas the SNP density was lower in the C subgenome (24 SNPs). Moreover, the average density of SNPs among the three subfamilies was in the following order: subfamily B (46) > subfamily B (42) > subfamily A (19) (Table S10). One gene, *BnaA01g02770D* from subfamily A; six genes, *BnaA10g03930D*, *BnaA04g18230D*, *BnaA03g14490D*, *BnaA02g12100D*, *BnaC02g45230D* from subfamily B; and one gene, *BnaAnng05970D* from subfamily C showed greater than 80 SNPs. There was variation in SNP number between the duplicated pairs; for example, *BnaC01g04040D*/*BnaA01g02770D* had 13/88 SNPs, *BnaA02g12100D*/*BnaA03g14490D* had 161/95 SNPs, *BnaA02g33880D*/*BnaC02g42680D* had 18/48, while *BnaA10g03930D*/*BnaC04g42500D* had 110/0 SNPs, respectively (Table S10). The SNP annotation results showed that a total of 750 SNPs were detected in the exon regions and (274) 44% of SNPs resulted in missense mutations in *BnARI* genes. These SNPs could lay the foundation of the functional importance of the *ARI* gene in *B. napus*.

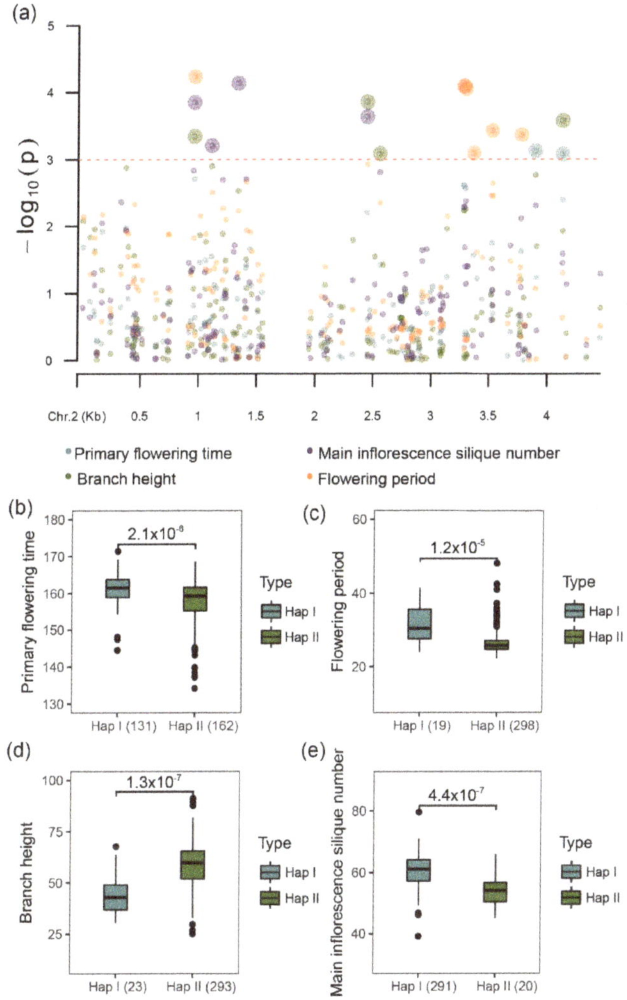

Figure 9. Association analysis of genetic variations in *BnaA02g12100D* in 324 worldwide collections of *B. napus* population with several agronomic traits. (**a**) Manhattan plot of *BnaA02g12100D* with primary flowering time, flowering period, branch height and main inflorescence silique number ($p < 0.001$). (**b**–**e**) The haplotype analysis of *BnaA02g12100D* genetic variation for primary flowering time, flowering period, branch height and main inflorescence silique number.

In order to examine the effect of *ARI* genes on the final phenotype of the *B. napus* plant, association mapping analysis for primary flowering time (PFT), full flowering time (FFT1), final flowering time (FFT2), early flowering stage (EFS), late-flowering stage (LFS), flowering period (FP), branch number (BN), branch height (BH), plant height (PH), main inflorescence length (MIL), main inflorescence silique number (MISN), and main inflorescence silique density (MISD) were conducted. Finally, eight *BnARI* genes were identified as candidate genes for multiple agronomic traits (Table S10). Interestingly, *BnaA02g12100D* (an ortholog of *AtARI8, AT1G65430*) was significantly associated with nearly all traits used ($p < 0.001$) (Table S10). For flowering time, flowering period, branch height and main inflorescence silique number, the population were clearly grouped into two haplotypes, and the t-test results showed that significant differences were observed between two haplotype groups ($p < 1 \times 10^{-5}$) (Figure 9). Moreover, we analyzed the proteins that interacted with *BnaA02g12100D* (Table S10). The GO enrichment analysis showed that they were not only involved in ubiquitination (GO:0019005, GO:0000151, GO:0034450, GO:0034450, and so on), but also participated in multicellular organism development (GO:0007275), the negative regulation of flower development (GO:0009910), vegetative phase change (GO:0010050), the regulation of circadian rhythm (GO:0042752), meristem maintenance (GO:0010073), cellular response to auxin stimulus (GO:0071365), the maintenance of meristem identity (GO:0010074), developmental growth (GO:0048589), and the regulation of meristem development (GO:0048509) (Table S11). All these processes were related with the regulation of endogenous hormone and the development of meristem, which could finally affect flowering time and plant architecture. Overall, the results suggested that *ARI* genes could affect the agronomic traits of *B. napus*.

3. Discussion

The ring E3 ubiquitin ligases are widely investigated for their role in plant adaptation and development [13,14,28,42]. Since the discovery of Parkin protein that causes juvenile parkinsonism, the RBR subclass of RING-containing E3 ligases has been recognized as an essential group of proteins [18]. The plant RBR family, on the other hand, has received little attention. The RBR family has four subfamilies named as Ariadne (ARI proteins), ARA54, Plant II and Helicase in plants [16]. The Ariadne subfamily shares domain (IBR) similarity with the Parkin protein of humans. However, the functions of this subfamily are largely unidentified in plants. The Ariadne gene family is not separately explored in detail in plants except for *A. thaliana*. We performed a genome-wide investigation of Ariadne (ARI proteins) and attempted to anticipate the potential functions of this gene family in *B. napus*.

After its separation from the Arabidopsis lineage, the genus Brassica experienced genome triplication, followed by interspecific hybridization between *B. rapa* and *B. oleracea*, resulting in the allotetraploid *B. napus* [30]. As a result of these duplication events, the genome size was expanded in *B. napus* during the evolution [30] with the expectation of six genes for one *A. thaliana* gene in the *B. napus*. We identified a total of 39 *ARI* genes in *B. napus* as compared with 16 *AtARI* genes (Table 1). The expansion of this gene family was only about 2.5 fold more than the ancestor *A. thaliana*, representing gene loss [43]. We did not obtain any ortholog against two *AtARI* genes (*AtARI6* (*At1g63760*), *AtARI4* (*At3g27720*) which were considered as pseudogenes [9]). Many studies showed that WGD or segmental duplication play a critical role during the expansion of gene families [44–46]. We obtained similar results in *BnARI* genes: all the genes were produced from the WGD or segmental duplication except four genes *BnaAnng26970D*, *BnaAnng26980D*, *BnaC04g42500D*, *BnaA03g14710D* and *BnaC03g50620D*, which were produced as a result of tandem duplication events (Table 1). The ka/Ks ratio was less than one for all duplicated pairs, suggesting purifying selection during the evolution. There was variation in Ka/Ks for the duplicated pairs, suggesting that they had evolved at different rates of evolution. The average divergence time for all the duplicated pairs was about 4 MYA, suggesting recent duplications in the *BnARI* gene family, which was also consistent with the evolutionary process of *B. napus*.

The evolution and differentiation of genes was deduced by comparing protein homologies, gene structure and motif combinations. We used this way to classify the *BnARI* gene family. Using Arabidopsis ARI proteins as a reference, these ARI proteins were clustered into three subfamilies (A, B and C). This classification was further confirmed by gene structure and motif analysis within the *BnARI* gene-family members. The number of exon and introns within subfamily C was conserved but exon and intron number was not conserved within subfamily A and B. Then, the investigation of motifs revealed that the coding sequence was highly conserved within the subfamilies, suggesting that the coding sequences were conserved in subfamily A, B and C.

Based on Cello server prediction, almost all the proteins were localized to a nuclear region, and these results were supported by a previous study [9], except for *BnaA02g33880D* and *BnaC02g42680D*, which were localized to an extracellular region. The possible explanation for this observation was that only these two proteins did not contain motif 7, which encoded for the Ring finger (Figure 3, Table S3). However, further functional studies are required to decode the mechanism.

To determine the possible function of *BnARI* genes, we investigated the cis-elements in their promoters that could influence their expression pattern [47] (Figure 4, Tables S4 and S5). In the RNA-seq data of twelve tissues, which included the root, leaf, bud, silique, stamen, new petal, blossomy petal, wilting petal, stem, sepal, ovule and pericarp of blooming of *B. napus* [34], interestingly, the highest gene expression was observed in the root tissue (Figure 6, Table S7). Mostly, genes were expressed in the root and bud tissue and the fewest genes were expressed in the silique tissue. We observed differential expression patterns of all the genes among different tissues, as discussed in the Results section (Figure 6, Table S7). Expressional divergence was observed between the duplicated gene pairs as well, suggesting a clue for sub/neo functionalization and pseudogenization, like in other polyploid crops [48]. Plants face several kinds of external stimuli including light, drought, high temperature, influencing their growth and development. Therefore, they evolve many strategies to cope with these challenges [49]. In the promoters of *BnARIs*, we detected various type of stress-related cis-elements (Figure 8, Table S4) that could predict their importance in abiotic stress. In the RNA-seq data for four stress treatments (dehydration, cold, ABA and salinity), subfamily C, especially all the orthologs (*BnaC03g50610D*, *BnaA02g33880D*, *BnaC02g42680D*, *BnaA06g22860D*) of *AT5G63750*, (*AtARI13*) showed significant responses towards the stresses applied (Figure 8, Table S9). The previous studies also showed the importance of *ARI* genes against abiotic stress in plants [26–28]. The gene ontology (GO) and KEGG pathway analyses predicted the involvement of *BnARIs* in endoreduplication (polyploidy), DNA damage, kinase activity, proteasome assembly, ubiquitination and stress mechanisms (Figure 5, Table S6). E3 ligases are also known to be involved in the aforementioned processes [7].

We also investigated genetic variations (SNPs) for *BnARIs* in the natural *B. napus* population [41] (Table S10). The greater number of SNPs in the genomic regions suggested that a lot of variations occurred in *BnARIs* during the evolution. The SNP density was much higher in the A subgenome than the C subgenome (Table S10) and these results were consistent with the *SR* and *GATA* gene families in *B. napus* [45,50]. In this study, a total of 271 missense mutations could help in the functional differentiation of these *BnARI* genes. We further used these genetic variations and performed association analysis with different agronomic traits to predict the importance of *ARI* genes in *B. napus* (Figure 9). Among all the genes, *BnaA02g12100D* (ortholog of *ARI8*, *AT1G65430*) was significantly associated with nearly all the traits used (threshold 3) ($p < 0.001$) (Figure 9, Table S10). This gene has a high and more diverse expression pattern than the other duplicated partner (*BnaC02g45230D*) in our transcriptome studied data (Figure 6, Table S7). In addition, its expression variation was observed in all the stresses as compared with the control. Based on the GO enrichment analysis for its interacted protein, it could be involved in diverse biological processes including ubiquitination, phosphorylation of proteins, meristem development and response to hormones (Table S11), which eventually influence the phenotype and adaptation of plants. According to a previous study, when expression analysis was carried out among *RBR* genes in 79 developmental stages of *A. thaliana*, *ARI8* (ortholog of *BnaA02g12100D*) was the predominantly expressed gene and its expression was specifically detected to be high in the mature pollen stage [16]. Among the 103 tissues of several developmental stages in *B. napus*, the expression of *BnaA02g12100D* in mature anther was specifically high [35]. Expression specificity in the male gametophyte could influence plant fitness and phenotype by causing the proteins to become ubiquitinated [51,52]. Among all the AtARI proteins, ubiquitin activity was only proven in *AtARI8* [6]. Ubiquitination is one of the critical mechanism that controlled the photoperiodic pathway in flowering regulation [14]. Likewise, the E3 ligases were involved in the regulation of some important flowering time genes, for example, constans, constitutive photomorphogenic 1, and target of early activation tagged 2 [53]. The knockout mutants of their orthologous genes in animals were lethal [24]. Nevertheless, the knockout studies for *ARI* genes are not available in any plants and the role of *ARI* genes in development and stress tolerance should be further explored. Therefore, these findings can provide a better clue to understand the significance of *BnARIs* in phenotypic variation and adaptation.

4. Materials and Methods

4.1. Identification of ARI Gene Family in B. napus

To identify the *BnARI genes*, we performed the BLASTP search in *B. napus* at BnaOmics database (Available online: https://bnaomics.ocri-genomics.net, accessed on 6 February 2022), 16 AtARI protein sequences were used as queries with e-value 1×10^{-5}. The NCBI Conserved Domain Database (Available online: https://www.ncbi.nlm.nih.gov/cdd, accessed on 6 February 2022) [54] and Pfam database (Available online: http://pfam.xfam.org/, accessed on 6 February 2022) [55] were used for verification of candidate genes that contained IBR and Ariadne domain. The redundant genes were removed manually.

The genome data files of "*Darmor*-bzh" were used to collect sequence information for all *BnARIs*, including ID, CDS, proteins, chromosomal position at BnaOmics database (Available online: https://bnaomics.ocri-genomics.net, accessed on 6 February 2022) [30]. The ExPasy tool (Available online: http://www.expasy.org/, accessed on 6 February 2022) was used to compute the peptide length (aa), molecular weight (MW), and isoelectric point (PI) of each BnARI protein. CELLO v2.5 (Available online: http://cello.life.nctu.edu.tw/, accessed on 6 February 2022) [56] was used to predict the subcellular localization of BnARI proteins.

4.2. Phylogenetic Analysis of BnARI Family

Multiple sequence alignments of ARI proteins from *A. thaliana* and *B. napus* were performed by using the ClustalW v2 [57] programme to acquire insights into the phylogenetic relationships between ARI family members. The neighbor-joining (NJ) technique was used with 1000 bootstrap replications in the MEGA v11 [58] to create the phylogenetic tree. Furthermore, iTOL v6.5.2 (Available online: https://itol.embl.de/, accessed on 12 February 2022) was used to visualize the tree. According to homology, *ARI* genes were further subdivided into subfamilies.

4.3. Chromosomal Distribution, Duplication Status, Ka/Ks Ratio

The physical positions of the *BnARI* genes on chromosomes were identified using the *B. napus* genome annotation file in TBtools programme, v1.098 [59]. BLASTP, with e-value of 1×10^{-10} and MCScanX [60], was used to evaluate the duplication patterns, such as segmental and tandem duplications. The TBtools programme v1.098 [59] was used to visualize the chromosomal location and duplicated *BnARI* genes.

The ratios of synonymous substitution rate (ks) and non-synonymous substitution rate (ka) of homologous *BnARI* gene pairs were calculated using TBtools programme (v1.098) [59]. $T = Ks/2R$, where R is 1.5×10^{-8} synonymous substitutions per site per year, was used to calculate divergence [61]. Ka/Ks ratio less than one indicated purifying selection, whereas Ka/Ks ratio greater than one indicated positive selection.

4.4. Identification of Conserved Motifs and Gene Structure in BnARI Gene Family

The conserved motifs in the BnARI proteins were analyzed using Multiple Expectation Maximization for Motif Elicitation (MEME 5.4.1) [37]. The parameters were organized into 10 motifs with widths between 6–50 amino acids and the remaining options were left at default settings. The Pfam database (Available online: http://pfam.xfam.org/search, accessed on 15 February 2022) was used to annotate the discovered motifs. The gene-structure information was obtained from *B. napus* genome files. The gene and motif structures were visualized using the TBtools programme, v1.098 [59].

4.5. Identification of Cis-Acting Regulatory Elements and Protein–Protein Interaction in BnARI Gene Family

To find the cis-acting regulatory elements in the *BnARI* genes promoters, 2 kb upstream each gene was extracted and examined through PlantCARE database (Available online: http://bioinformatics.psb.ugent.be/webtools/plantcare/html/, accessed on 15 February 2022) [38]. The number of cis-acting elements were converted to log2 transformation and were visualized using TBtools programme, v1.098 [59].

To determine the protein–protein interactions (PPIs) of BnARIs, first PPIs of the AtARI proteins were downloaded from STRING database (Available online: https://www.string-db.org/, accessed on 15 February 2022) [39], the BnARI-interacting proteins were predicted based on the homologs in *A. thaliana*, and the network was visualized using Cytoscape [62]. Using the clusterProfiler in R [63], the genes that interacted with the BnARI proteins were selected for gene ontology and KEGG enrichment analysis.

4.6. Expression Analysis of BnARI Genes in Different Tissues and under Abiotic Stress

We used transcriptome data from twelve tissues of "ZS 11" (root, leaf, bud, silique, stamen, new petal, blossomy petal, wilting petal, stem, sepal, ovule and pericarp) [34] and four treatments (dehydration, cold, ABA and salinity) to detect the expression of *BnARI* genes [40]. The FPKM values were transformed to log2 folds, and TBtools was used to produce heatmaps of all data.

4.7. RNA Extraction and Reverse Transcription-Quantitative PCR (qRT PCR)

The *B. napus* v. "ZS 11" was grown in Oil Crop Research Institute (OCRI) Wuhan fields. Three samples for each tissue (leaf, stem, root, petal, sepal, carpel, and stamen) were collected at the blooming stage. The 21 samples were immediately put in liquid nitrogen before storing at −80 °C. The Invitrogen TRIZOL Reagent (Thermofisher, Waltham, MA, USA) was used to isolate total RNA. A TaKaRa reverse transcription kit (Prime ScriptTMRT reagent Kit, TaKaRa, Beijing, China) was used to make first-strand complementary DNA (cDNA). Furthermore, qRT-PCR primers for potential *BnARI* genes were constructed to test the expression pattern (Table S8). The qRT-PCR was performed using Bio Supermix (Bio-rad, Hercules, CA, USA) according to the manufacturer's instructions, with reaction steps as follows: 95 °C for 3 min; 40 cycles of 95 °C for 15 s; 56 °C for 15 s, followed by 65 °C for 5 s and 95 °C for 5 s in three biological replicates. Ten genes of 5 duplicated pairs were analyzed for expression by qRT-PCR. The *B. napus*—β actin gene (*AF111812*) was used as internal reference. The $2^{-\Delta\Delta Ct}$ method [64] was used to compute the relative expression.

4.8. Functional Significance of BnARIs by Using Association Mapping in Natural Population

To determine the functional significance of *BnARIs* at the population level, we investigated the natural genetic variations (SNPs) in them using 324 accessions, which were collected from around the world [41]. The SNPs were obtained and annotated through SnpEff programme v4.11 [65]. Primary flowering time (PFT), full flowering time (FFT1), final flowering time (FFT2), early flowering stage (EFS), late-flowering stage (LFS), flowering period (FP), branch number (BN), branch height (BH), plant height (PH), main inflorescence length (MIL), main inflorescence silique number (MISN), and main inflorescence silique density (MISD) were selected as studied traits [41]. EMMAX [66] was used to perform the association mapping analysis between genetic variations and agronomic traits. CMplot and ggplot2 in R (Available online: https://mirrors.tuna.tsinghua.edu.cn/CRAN/, accessed on 15 February 2022) were used to draw the Manhattan plot and the boxplot.

4.9. Statistical Analysis

The statistical analysis was performed by R (v3.6) and means of two continuous normally distributed variables were compared by independent samples Student's T-test. A value of $p < 0.05$ was considered significant.

5. Conclusions

In this study, the detailed investigation and characterization of Ariadne genes was carried out. All 39 *ARI* genes were classified into three subfamilies. Gene architectures and motifs were comparable across genes from the same subfamily. The presence of cis-acting regulatory elements in the *ARI* genes promoters, as well as their expression patterns in diverse tissues and under various environmental conditions, and the protein interaction analysis demonstrated that they could play an important role in development and stress tolerance. Furthermore, genetic variations in *BnARI* genes provided diverse potential influences on agronomic traits. In summary, this study supplied useful information on *BnARI* genes, and it will aid further functional research and genetic improvement for breeding in *B. napus*.

Supplementary Materials: The following are available online at https://www.mdpi.com/article/10.3390/ijms23116265/s1.

Author Contributions: M.X., S.W. and S.L. designed the research. Formal analysis, S.W., M.X. and S.S.; Funding acquisition, S.L., C.T.; Investigation, S.W., M.X., S.S., J.L.; Data curation, X.C., Z.B.; Project administration, M.X. and F.G.; Resources, C.Z.; S.W. and S.S. wrote the manuscript; Review and editing, S.L., M.X., S.W., S.S., Z.B. and F.G. All authors have read and agreed to the published version of the manuscript.

Funding: This research was funded by the National Key Research and Development Program of China (2018YFE0108000), Central Public-interest Scientific Institution Basal Research Fund, grant number 2021-2060302-061-027 No: 2021-2060302-061-029, the Agricultural Science and Technology Innovation Program of Chinese Academy of Agricultural Sciences (CAAS-ASTIP-2013-OCRI), China Agriculture Research System of MOF and MARA (CARS-12), and the Young Top-notch Talent Cultivation Program of Hubei Province.

Institutional Review Board Statement: Not applicable.

Informed Consent Statement: Not applicable.

Data Availability Statement: The corresponding data are shown in Supplementary Materials.

Acknowledgments: We thank Zhixian Qiao of the Analysis and Testing Center at IHB for technical support in RNA-seq analysis.

Conflicts of Interest: The authors declare no conflict of interest.

References

1. Chen, L.; Hellmann, H. Plant E3 ligases: Flexible enzymes in a sessile world. *Mol. Plant* **2013**, *6*, 1388–1404. [CrossRef] [PubMed]
2. Smalle, J.; Vierstra, R.D. The ubiquitin 26S proteasome proteolytic pathway. *Annu. Rev. Plant Biol.* **2004**, *55*, 555–590. [CrossRef] [PubMed]
3. Hellmann, H.; Estelle, M. Plant development: Regulation by protein degradation. *Science* **2002**, *297*, 793–797. [CrossRef] [PubMed]
4. Vierstra, R.D. The ubiquitin–26S proteasome system at the nexus of plant biology. *Nat. Rev. Mol. Cell Biol.* **2009**, *10*, 385–397. [CrossRef] [PubMed]
5. Mazzucotelli, E.; Belloni, S.; Marone, D.; De Leonardis, A.M.; Guerra, D.; Di Fonzo, N.; Cattivelli, L.; Mastrangelo, A.M. The E3 ubiquitin ligase gene family in plants: Regulation by degradation. *Curr. Genom.* **2006**, *7*, 509–522. [CrossRef] [PubMed]
6. Stone, S.L.; Hauksdóttir, H.; Troy, A.; Herschleb, J.; Kraft, E.; Callis, J. Functional analysis of the RING-type ubiquitin ligase family of Arabidopsis. *Plant Physiol.* **2005**, *137*, 13–30. [CrossRef] [PubMed]
7. Sharma, B.; Joshi, D.; Yadav, P.K.; Gupta, A.K.; Bhatt, T.K. Role of ubiquitin-mediated degradation system in plant biology. *Front. Plant Sci.* **2016**, *7*, 806. [CrossRef]
8. Callis, J. The ubiquitination machinery of the ubiquitin system. *Arab. Book Am. Soc. Plant Biol.* **2014**, *12*, e0174. [CrossRef]
9. Mladek, C.; Guger, K.; Hauser, M.T. Identification and characterization of the ARIADNE gene family in Arabidopsis. A group of putative E3 ligases. *Plant Physiol.* **2003**, *131*, 27–40. [CrossRef]
10. Borden, K.L.B. RING domains: Master builders of molecular scaffolds? *J. Mol. Biol.* **2000**, *295*, 1103–1112. [CrossRef]
11. Lim, S.D.; Yim, W.C.; Moon, J.-C.; Kim, D.S.; Lee, B.-M.; Jang, C.S. A gene family encoding RING finger proteins in rice: Their expansion, expression diversity, and co-expressed genes. *Plant Mol. Biol.* **2010**, *72*, 369–380. [CrossRef] [PubMed]
12. Lim, S.D.; Hwang, J.G.; Jung, C.G.; Hwang, S.G.; Moon, J.C.; Jang, C.S. Comprehensive analysis of the rice RING E3 ligase family reveals their functional diversity in response to abiotic stress. *DNA Res.* **2013**, *20*, 299–314. [CrossRef] [PubMed]

13. Cho, S.K.; Ryu, M.Y.; Seo, D.H.; Kang, B.G.; Kim, W.T. The Arabidopsis RING E3 ubiquitin ligase AtAIRP2 plays combinatory roles with AtAIRP1 in abscisic acid-mediated drought stress responses. *Plant Physiol.* **2011**, *157*, 2240–2257. [CrossRef]
14. Piñeiro, M.; Jarillo, J.A. Ubiquitination in the control of photoperiodic flowering. *Plant Sci.* **2013**, *198*, 98–109. [CrossRef] [PubMed]
15. Eisenhaber, B.; Chumak, N.; Eisenhaber, F.; Hauser, M.T. The ring between ring fingers (RBR) protein family. *Genome Biol.* **2007**, *8*, 209. [CrossRef]
16. Marín, I. Diversification and specialization of plant RBR ubiquitin ligases. *PLoS ONE* **2010**, *5*, e11579. [CrossRef]
17. Marin, I.; Ferrús, A. Comparative genomics of the RBR family, including the Parkinson's disease–related gene Parkin and the genes of the Ariadne subfamily. *Mol. Biol. Evol.* **2002**, *19*, 2039–2050. [CrossRef]
18. Beasley, S.A.; Hristova, V.A.; Shaw, G.S. Structure of the Parkin in-between-ring domain provides insights for E3-ligase dysfunction in autosomal recessive Parkinson's disease. *Proc. Natl. Acad. Sci. USA* **2007**, *104*, 3095–3100. [CrossRef]
19. Morett, E.; Bork, P. A novel transactivation domain in parkin. *Trends Biochem. Sci.* **1999**, *24*, 229–231. [CrossRef]
20. Marín, I.; Lucas, J.I.; Gradilla, A.-C.; Ferrús, A. Parkin and relatives: The RBR family of ubiquitin ligases. *Physiol. Genom.* **2004**, *17*, 253–263. [CrossRef]
21. Marteijn, J.A.F.; van Emst, L.; Erpelinck-Verschueren, C.A.J.; Nikoloski, G.; Menke, A.; de Witte, T.; Löwenberg, B.; Jansen, J.H.; van der Reijden, B.A. The E3 ubiquitin-protein ligase Triad1 inhibits clonogenic growth of primary myeloid progenitor cells. *Blood* **2005**, *106*, 4114–4123. [CrossRef] [PubMed]
22. Ardley, H.C.; Tan, N.G.S.; Rose, S.A.; Markham, A.F.; Robinson, P.A. Features of the parkin/ariadne-like ubiquitin ligase, HHARI, that regulate its interaction with the ubiquitin-conjugating enzyme, Ubch7. *J. Biol. Chem.* **2001**, *276*, 19640–19647. [CrossRef] [PubMed]
23. Duda, D.M.; Olszewski, J.L.; Schuermann, J.P.; Kurinov, I.; Miller, D.J.; Nourse, A.; Alpi, A.F.; Schulman, B.A. Structure of HHARI, a RING-IBR-RING ubiquitin ligase: Autoinhibition of an Ariadne-family E3 and insights into ligation mechanism. *Structure* **2013**, *21*, 1030–1041. [CrossRef] [PubMed]
24. Aguilera, M.; Oliveros, M.; Martínez-Padrón, M.; Barbas, J.A.; Ferrús, A. Ariadne-1: A vital Drosophila gene is required in development and defines a new conserved family of ring-finger proteins. *Genetics* **2000**, *155*, 1231–1244. [CrossRef]
25. Itier, J.-M.; Ibáñez, P.; Mena, M.A.; Abbas, N.; Cohen-Salmon, C.; Bohme, G.A.; Laville, M.; Pratt, J.; Corti, O.; Pradier, L. Parkin gene inactivation alters behaviour and dopamine neurotransmission in the mouse. *Hum. Mol. Genet.* **2003**, *12*, 2277–2291. [CrossRef]
26. Lang-Mladek, C.; Xie, L.; Nigam, N.; Chumak, N.; Binkert, M.; Neubert, S.; Hauser, M.T. UV-B signaling pathways and fluence rate dependent transcriptional regulation of ARIADNE12. *Physiol. Plant.* **2012**, *145*, 527–539. [CrossRef]
27. Xie, L.; Lang-Mladek, C.; Richter, J.; Nigam, N.; Hauser, M.T. UV-B induction of the E3 ligase ARIADNE12 depends on CONSTITUTIVELY PHOTOMORPHOGENIC 1. *Plant Physiol. Biochem.* **2015**, *93*, 18–28. [CrossRef]
28. Zhang, X.; Wang, N.; Chen, P.; Gao, M.; Liu, J.; Wang, Y.; Zhao, T.; Li, Y.; Gai, J. Overexpression of a soybean ariadne-like ubiquitin ligase gene GmARI1 enhances aluminum tolerance in Arabidopsis. *PLoS ONE* **2014**, *9*, e111120. [CrossRef]
29. Schallau, A.; Arzenton, F.; Johnston, A.J.; Hähnel, U.; Koszegi, D.; Blattner, F.R.; Altschmied, L.; Haberer, G.; Barcaccia, G.; Bäumlein, H. Identification and genetic analysis of the APOSPORY locus in *Hypericum perforatum* L. *Plant J.* **2010**, *62*, 773–784. [CrossRef]
30. Chalhoub, B.; Denoeud, F.; Liu, S.; Parkin, I.A.P.; Tang, H.; Wang, X.; Chiquet, J.; Belcram, H.; Tong, C.; Samans, B.; et al. Early allopolyploid evolution in the post-neolithic *Brassica napus* oilseed genome. *Science* **2014**, *345*, 950–953. [CrossRef]
31. Ayyaz, A.; Miao, Y.; Hannan, F.; Islam, F.; Zhang, K.; Xu, J.; Farooq, M.A.; Zhou, W. Drought tolerance in *Brassica napus* is accompanied with enhanced antioxidative protection, photosynthetic and hormonal regulation at seedling stage. *Physiol. Plant.* **2021**, *172*, 1133–1148. [CrossRef] [PubMed]
32. Lei, Y.A.N.; Shah, T.; Cheng, Y.; Yan, L.Ü.; Zhang, X.; Zou, X. Physiological and molecular responses to cold stress in rapeseed (*Brassica napus* L.). *J. Integr. Agric.* **2019**, *18*, 2742–2752.
33. Lohani, N.; Jain, D.; Singh, M.B.; Bhalla, P.L. Engineering multiple abiotic stress tolerance in canola, *Brassica napus*. *Front. Plant Sci.* **2020**, *3*. [CrossRef] [PubMed]
34. Li, Y.; Dong, C.; Hu, M.; Bai, Z.; Tong, C.; Zuo, R.; Liu, Y.; Cheng, X.; Cheng, M.; Huang, J.; et al. Identification of flower-specific promoters through comparative transcriptome analysis in *Brassica napus*. *Int. J. Mol. Sci.* **2019**, *20*, 5949. [CrossRef] [PubMed]
35. Chao, H.; Li, T.; Luo, C.; Huang, H.; Ruan, Y.; Li, X.; Niu, Y.; Fan, Y.; Sun, W.; Zhang, K.; et al. Brassicaedb: A gene expression database for brassica crops. *Int. J. Mol. Sci.* **2020**, *21*, 5831. [CrossRef]
36. Cannon, S.B.; Mitra, A.; Baumgarten, A.; Young, N.D.; May, G. The roles of segmental and tandem gene duplication in the evolution of large gene families in *Arabidopsis thaliana*. *BMC Plant Biol.* **2004**, *4*, 10. [CrossRef]
37. Bailey, T.L.; Boden, M.; Buske, F.A.; Frith, M.; Grant, C.E.; Clementi, L.; Ren, J.; Li, W.W.; Noble, W.S. MEME Suite: Tools for motif discovery and searching. *Nucleic Acids Res.* **2009**, *37*, 202–208. [CrossRef]
38. Lescot, M.; Déhais, P.; Thijs, G.; Marchal, K.; Moreau, Y.; Van De Peer, Y.; Rouzé, P.; Rombauts, S. PlantCARE, a database of plant cis-acting regulatory elements and a portal to tools for in silico analysis of promoter sequences. *Nucleic Acids Res.* **2002**, *30*, 325–327. [CrossRef]
39. Szklarczyk, D.; Gable, A.L.; Lyon, D.; Junge, A.; Wyder, S.; Huerta-Cepas, J.; Simonovic, M.; Doncheva, N.T.; Morris, J.H.; Bork, P. STRING v11: Protein–protein association networks with increased coverage, supporting functional discovery in genome-wide experimental datasets. *Nucleic Acids Res.* **2019**, *47*, D607–D613. [CrossRef]

40. Zhang, Y.; Ali, U.; Zhang, G.; Yu, L.; Fang, S.; Iqbal, S.; Li, H.; Lu, S.; Guo, L. Transcriptome analysis reveals genes commonly responding to multiple abiotic stresses in rapeseed. *Mol. Breed.* **2019**, *39*, 158. [CrossRef]
41. Tang, M. *Population Genome Variations and Subgenome Asymmetry in Brassica napus L.*; Huazhong Agricultural University: Wuhan, China, 2019.
42. Cho, S.K.; Ryu, M.Y.; Kim, J.H.; Hong, J.S.; Oh, T.R.; Kim, W.T.; Yang, S.W. RING E3 ligases: Key regulatory elements are involved in abiotic stress responses in plants. *BMB Rep.* **2017**, *50*, 393. [CrossRef] [PubMed]
43. Albalat, R.; Cañestro, C. Evolution by gene loss. *Nat. Rev. Genet.* **2016**, *17*, 379–391. [CrossRef]
44. Sehrish, S.; Sumbal, W.; Xie, M.; Zhao, C.; Zuo, R.; Gao, F.; Liu, S. Genome-Wide Identification and Characterization of SET Domain Family Genes in *Brassica napus* L. *Int. J. Mol. Sci.* **2022**, *23*, 1936. [CrossRef] [PubMed]
45. Xie, M.; Zuo, R.; Bai, Z.; Yang, L.; Zhao, C.; Gao, F.; Cheng, X.; Huang, J.; Liu, Y.; Li, Y. Genome-wide characterization of Serine/Arginine-Rich gene family and its genetic effects on agronomic traits of *Brassica napus*. *Front. Plant Sci.* **2022**, *13*, 829668. [CrossRef] [PubMed]
46. Wu, Y.; Ke, Y.; Wen, J.; Guo, P.; Ran, F.; Wang, M.; Liu, M.; Li, P.; Li, J.; Du, H. Evolution and expression analyses of the MADS-box gene family in *Brassica napus*. *PLoS ONE* **2018**, *13*, e0200762. [CrossRef] [PubMed]
47. Oudelaar, A.M.; Higgs, D.R. The relationship between genome structure and function. *Nat. Rev. Genet.* **2021**, *22*, 154–168. [CrossRef]
48. Chaudhary, B.; Flagel, L.; Stupar, R.M.; Udall, J.A.; Verma, N.; Springer, N.M.; Wendel, J.F. Reciprocal silencing, transcriptional bias and functional divergence of homeologs in polyploid cotton (Gossypium). *Genetics* **2009**, *182*, 503–517. [CrossRef]
49. Meena, K.K.; Sorty, A.M.; Bitla, U.M.; Choudhary, K.; Gupta, P.; Pareek, A.; Singh, D.P.; Prabha, R.; Sahu, P.K.; Gupta, V.K. Abiotic stress responses and microbe-mediated mitigation in plants: The omics strategies. *Front. Plant Sci.* **2017**, *8*, 172. [CrossRef]
50. Zhu, W.; Guo, Y.; Chen, Y.; Wu, D.; Jiang, L. Genome-wide identification, phylogenetic and expression pattern analysis of GATA family genes in *Brassica napus*. *BMC Plant Biol.* **2020**, *20*, 543. [CrossRef]
51. Borges, F.; Gomes, G.; Gardner, R.; Moreno, N.; McCormick, S.; Feijó, J.A.; Becker, J.D. Comparative transcriptomics of Arabidopsis sperm cells. *Plant Physiol.* **2008**, *148*, 1168–1181. [CrossRef]
52. Haerizadeh, F.; Wong, C.E.; Bhalla, P.L.; Gresshoff, P.M.; Singh, M.B. Genomic expression profiling of mature soybean (Glycine max) pollen. *BMC Plant Biol.* **2009**, *9*, 25. [CrossRef] [PubMed]
53. Pavicic, M.; Mouhu, K.; Wang, F.; Bilicka, M.; Chovanček, E.; Himanen, K. Genomic and phenomic screens for flower related RING type ubiquitin E3 ligases in Arabidopsis. *Front. Plant Sci.* **2017**, *8*, 416. [CrossRef] [PubMed]
54. Lu, S.; Wang, J.; Chitsaz, F.; Derbyshire, M.K.; Geer, R.C.; Gonzales, N.R.; Gwadz, M.; Hurwitz, D.I.; Marchler, G.H.; Song, J.S. CDD/SPARCLE: The conserved domain database in 2020. *Nucleic Acids Res.* **2020**, *48*, D265–D268. [CrossRef] [PubMed]
55. Mistry, J.; Chuguransky, S.; Williams, L.; Qureshi, M.; Salazar, G.A.; Sonnhammer, E.L.L.; Tosatto, S.C.E.; Paladin, L.; Raj, S.; Richardson, L.J. Pfam: The protein families database in 2021. *Nucleic Acids Res.* **2021**, *49*, D412–D419. [CrossRef]
56. Yu, C.; Chen, Y.; Lu, C.; Hwang, J. Prediction of protein subcellular localization. *Proteins Struct. Funct. Bioinform.* **2006**, *64*, 643–651. [CrossRef]
57. Larkin, M.A.; Blackshields, G.; Brown, N.P.; Chenna, R.; McGettigan, P.A.; McWilliam, H.; Valentin, F.; Wallace, I.M.; Wilm, A.; Lopez, R. Clustal W and Clustal X version 2.0. *Bioinformatics* **2007**, *23*, 2947–2948. [CrossRef]
58. Tamura, K.; Stecher, G.; Kumar, S. MEGA11: Molecular evolutionary genetics analysis version 11. *Mol. Biol. Evol.* **2021**, *38*, 3022–3027. [CrossRef]
59. Chen, C.; Xia, R.; Chen, H.; He, Y. TBtools, a toolkit for biologists integrating various biological data handling tools with a user-friendly interface. *bioRxiv* **2018**, 289660. [CrossRef]
60. Wang, Y.; Tang, H.; DeBarry, J.D.; Tan, X.; Li, J.; Wang, X.; Lee, T.; Jin, H.; Marler, B.; Guo, H. MCScanX: A toolkit for detection and evolutionary analysis of gene synteny and collinearity. *Nucleic Acids Res.* **2012**, *40*, e49. [CrossRef]
61. Wei, L.; Zhu, Y.; Liu, R.; Zhang, A.; Zhu, M.; Xu, W.; Lin, A.; Lu, K.; Li, J. Genome wide identification and comparative analysis of glutathione transferases (GST) family genes in *Brassica napus*. *Sci. Rep.* **2019**, *9*, 9196. [CrossRef]
62. Shannon, P.; Markiel, A.; Ozier, O.; Baliga, N.S.; Wang, J.T.; Ramage, D.; Amin, N.; Schwikowski, B.; Ideker, T. Cytoscape: A software environment for integrated models of biomolecular interaction networks. *Genome Res.* **2003**, *13*, 2498–2504. [CrossRef] [PubMed]
63. Yu, G.; Wang, L.-G.; Han, Y.; He, Q.-Y. clusterProfiler: An R package for comparing biological themes among gene clusters. *OMICS J. Integr. Biol.* **2012**, *16*, 284–287. [CrossRef]
64. Livak, K.J.; Schmittgen, T.D. Analysis of relative gene expression data using real-time quantitative PCR and the $2^{-\Delta\Delta CT}$ method. *Methods* **2001**, *25*, 402–408. [CrossRef] [PubMed]
65. Cingolani, P.; Platts, A.; Wang, L.L.; Coon, M.; Nguyen, T.; Wang, L.; Land, S.J.; Lu, X.; Ruden, D.M. A program for annotating and predicting the effects of single nucleotide polymorphisms, SnpEff: SNPs in the genome of *Drosophila melanogaster* strain w1118; iso-2; iso-3. *Fly* **2012**, *6*, 80–92. [CrossRef] [PubMed]
66. Kang, H.M.; Sul, J.H.; Service, S.K.; Zaitlen, N.A.; Kong, S.; Freimer, N.B.; Sabatti, C.; Eskin, E. Variance component model to account for sample structure in genome-wide association studies. *Nat. Genet.* **2010**, *42*, 348–354. [CrossRef]

International Journal of
Molecular Sciences

MDPI

Article

Conservation and Divergence of the Trihelix Genes in Brassica and Expression Profiles of *BnaTH* Genes in *Brassica napus* under Abiotic Stresses

Cuiping Zhang, Lijing Lu, Ruolin Gong, Xing Su, Fengbo Liu, Ru Zhang * and Jihong Hu *

State Key Laboratory of Crop Stress Biology for Arid Areas, College of Agronomy, Northwest A&F University, Xianyang 712100, China

* Correspondence: zhangru0819@yeah.net (R.Z.); hujh05@nwafu.edu.cn (J.H.)

Abstract: Trihelix (TH) proteins are a family of plant-specific transcription factors that play a role in light response and are extensively involved in plant growth and development, as well as in various stress responses. However, the function of *TH* genes in *Brassica napus* (*B. napus*) remains unclear, as does the evolution and differentiation pattern of *TH* genes in Brassica plants. Here, we identified a total of 455 *TH* genes in seven species, including six Brassica species and *Arabidopsis*, which were grouped into five clades, GT-1, GT-2, GTγ, SH4, and SIP1, each with 69, 142, 44, 55, and 145 members, respectively. The types and distributions of motifs of the TH proteins and the structures of the *TH* genes are conserved in the same subgroup, and some variations in certain amino acid residues occur in *B. napus* when inheriting motifs from *Brassica rapa* (*B. rapa*) and *Brassica oleracea* (*B. oleracea*). Collinearity analysis revealed that the massive expansion of *TH* genes in tetraploid species was attributed to the hetero-tetraploidization of diploid ancestors and gene duplication events within the tetraploid species. Comparative analysis of the membership numbers of five subgroups in different species revealed that the GT-2 and SIP1 genes underwent significant expansion during evolution, possibly to support the better adaptation of plants to their environments. The differential expression of the *BnaTH* genes under five stresses indicates that the *BnaTH* genes are involved in plant responses to stresses such as drought, cold, and heat. The presence of different stress-responsive *cis*-elements in the upstream promoter region of the genes indicated that *BnaTH* genes have the potential to cope with variable environments. Meanwhile, qRT-PCR analyses also confirmed that five *TH* genes respond to different abiotic stresses. Our results provide information and candidates for further studies on the role of *TH* genes in stress resistance of *B. napus*.

Keywords: trihelix; transcription factor; brassica; evolution; abiotic stress; drought

Citation: Zhang, C.; Lu, L.; Gong, R.; Su, X.; Liu, F.; Zhang, R.; Hu, J. Conservation and Divergence of the Trihelix Genes in Brassica and Expression Profiles of *BnaTH* Genes in *Brassica napus* under Abiotic Stresses. *Int. J. Mol. Sci.* **2022**, *23*, 15766. https://doi.org/10.3390/ijms232415766

Academic Editors: Andrés J. Cortés and Hai Du

Received: 31 October 2022
Accepted: 9 December 2022
Published: 12 December 2022

1. Introduction

Trihelix (TH) transcription factors, as a key regulatory factor, were originally called GT elements due to their specific binding to GT elements, which is required for light-responsive transcription [1,2]. TH factors contain a conserved trihelix DNA-binding domain characterized by a typical trihelix (helix–loop–helix–loop–helix) structure, which binds specifically to GT elements in the promoters of target genes [3,4] and plays a pivotal role in a variety of developmental processes and environmental stress responses. Based on its conserved domains and evolutionary relationships, the TH gene family is divided into five clades, namely GT-1, GT-2, SH4, SIP1, and GTγ. Studies have revealed that GT factors, which have a trihelix structure that resembles the Myb/SANT-LIKE DNA-binding domain, likely originated from an MYB-like gene carrying only one repeat [3]. However, the recognition sequences of the GT transcription factor and the MYB protein are different, which may be caused by the insertion of gaps between the two helices [3].

In recent years, with the completion of the genome assembly of various species, the genome-wide identification of the TH gene family has been reported in many species. In total, 30 and 31 *TH* genes were identified in *Arabidopsis* and rice, respectively [2,5]. In addition, 31, 35, 94, 52, 56, 52, 20, and 36 *TH* genes were identified in *Fagopyrum esculentum*, *Moso bamboo*, *Triticum aestivum*, *Gossypium arboreum*, *Populus trichocarpa*, *B. rapa*, *Chrysanthemum*, and *Solanum lycopersicum*, respectively [6–13]. Recent studies have revealed that TH factors not only respond to light stress and regulate the expression of light-responsive genes but are also involved in abiotic stress responses [2,14]. For instance, the TH family gene *ShCIGT* mediates cold tolerance and drought tolerance by interacting with SnRK1 in tomato [15]. Overexpression or knockout of the *AST1* in *Arabidopsis* could positively influence tolerance to salt, osmotic, and drought stress [16]. *BnSIP1-1* of *B. napus* has been proved to be capable of coping with various abiotic stresses, and the overexpression of *BnSIP1-1* can improve the germination rate of *B. napus* under osmotic and salt stresses [17]. *Arabidopsis*, AtGT2L, as a Ca^{2+}-dependent CaM-binding protein, is strongly induced to express under cold and salt stresses [18]. However, the function of TH family genes in Brassica species and their roles in abiotic stresses have not been systematically reported. Thus, it is essential to systematically characterize the TH family genes in Brassica species, to explore their evolutionary relationships among species, and to further investigate their roles under abiotic stresses.

B. napus (AnAnCnCn, 2n = 38), a member of the Brassica species, is an allopolyploid that was formed by hybridization between *B. rapa* (AnAn, n = 10) and *B. oleracea* (CnCn, n = 9) about 7500 years ago [19,20]. In addition, the Brassica genus also includes *Brassica nigra* (*B. nigra*; BnBn, n = 8), along with two allotetraploids, *Brassica juncea* (*B. juncea*; AnAnBnBn, 2n = 36) and *Brassica carinata* (*B. carinata*; BnBnCnCn, 2n = 34) [21,22]. The evolutionary relationships of these six Brassica species are known as the classic U's triangle model [23]. Studies have confirmed that the Brassica species *B. nigra*, *B. rapa*, *B. oleracea*, *B. juncea*, *B. carinata*, and *B. napus* have undergone a whole-genome triplication (WGT) since their divergence from the *Arabidopsis* species [24,25]. The genome underwent chromosomal rearrangement, gene loss, and the divergence of retained paralogues after WGT, with substantial gene loss occurring in an asymmetric and reciprocal manner [26]. A large number of over-retained genes after WGT belongs to the gene families that underlie adaptation to environmental factors such as salt, cold, osmotic stress, light, wounding, pathogen defense, and so on [24]. To date, whole genome sequencing of six species in the U's model has been completed. Plant growth and development are accompanied by adaptation to the environment, which is always complex and changeable. Therefore, exploring genes such as *TH* genes, and identifying their roles in abiotic stresses, will provide insights into improving crop yields and environmental adaptation.

B. napus is an important oilseed crop, and the production of rapeseed is second only to that of soybeans in terms of the worldwide production of oilseed crops. However, with continuous changes in the global climate, environmental factors such as drought and low temperatures have become the main limiting factors for rapeseed production, resulting in a decline in rapeseed yields and quality. Although many breeding works have used heterosis to screen and breed varieties with high yields and oil content, the research into abiotic stress in rapeseed is still relatively underdeveloped. Therefore, it is urgent to cultivate stress-resistant *B. napus* varieties capable of adapting to the changing environment in order to ensure yield and quality.

In our study, we identified and characterized the TH transcription factors in six Brassica species and expression patterns in *B. napus* under different abiotic stresses. The evolution of the TH gene family was investigated in Brassica genus. To establish the role of the *TH* genes in rapeseed, we evaluated their expression patterns in different tissues and under different abiotic stresses, including drought, cold, and heat. Our research provides new insights into enhancing stress resistance in *B. napus* with the goal of improving yield and quality.

2. Results

2.1. Identification of TH Genes in Seven Species

A total of 455 *TH* genes was identified in seven species, with 29, 51, 49, 54, 92, 86, and 94 in *Arabidopsis*, *B. nigra*, *B. rapa*, *B. oleracea*, *B. juncea*, *B. carinate*, and *B. napus*, respectively. All *TH* genes were subsequently renamed according to their species and chromosomal locations (Table S1), yielding *AtTH-1* to *-29*, *BniTH-1* to *-51*, *BraTH-1* to *-49*, *BolTH-1* to *-54*, *BjuTH-1* to *-92*, *BolTH-1* to *-86*, and *BraTH-1* to *-94*. Next, the physicochemical properties of the TH proteins, including protein length, molecular weight (MW), theoretical isoelectric point (pI), instability index (II), aliphatic index (AI), and the grand average of hydropathicity (GRAVY), were calculated to provide further information (Table S2). The length of TH proteins varied from 109 (BolTH-35) to 1149 (BnaTH-19) amino acids (aa), with an MW range of 12,940.84 (BolTH-35) –27,364.53 (BolTH-18) kDa. The pI values were between 4.65 (AtTH-18) and 10.83 (BniTH-44), and the II ranged from 28.75 (BraTH-4) to 86.3 (BcaTH-32). The AI can vary from 38.53 (BolTH-35) to 87 (BcaTH-51). All the TH proteins presented GRAVY values of <0, ranging from −0.357 (BcaTH-42) to −1.358 (BcaTH-51), implying their hydrophilic characterization. Subcellular localization revealed that nearly all the TH proteins (389, 85.5%) were located in the nucleus. This result is consistent with their roles as transcription factors. In addition, small proportions of the proteins are located in other regions, such as the chloroplast (22, 4.8%), cytoplasm (34, 7.5%), peroxisome (8, 1.8%), endoplasmic reticulum (1, 0.2%), and vacuole (1, 0.2%).

2.2. Phylogenetic Analysis and Classification of TH Genes

To uncover the phylogenetic relationships of the *TH* genes, we constructed an unrooted phylogenetic tree based on the multiple sequence alignment of 455 TH proteins among the seven species, namely *Arabidopsis*, *B. nigra*, *B. juncea*, *B. carinate*, *B. rapa*, *B. napus*, and *B. oleracea* (Figure 1). These *TH* genes were classified into five clades, namely GT- 1, GT-2, GTγ, SH4, and SIP1 [11], containing 69, 142, 44, 55, and 145 *TH* genes, respectively. The SIP1 clades contained the most *TH* members (145, 31.9%), while the GTγ clades had the fewest *TH* genes (44, 9.7%). These results indicated that the number of *TH* genes is uneven across the five subgroups.

2.3. Gene Structure and Protein Conservative Domain Analysis

Through MEME domain analysis, we identified 10 distinct motifs for the structural diversification of 455 *TH* genes, which we refer to as motifs 1 to 10 (Figure S1 and Table S3). Overall, similar motif compositions and distributions were found in the same subgroups, whereas they differed significantly among the subgroups. All subgroup members contain about three motifs, except the GT-2 subgroup, which has about 10 motifs. All TH proteins contain motif 1, whereas motif 3 is almost unique to GT-2. Furthermore, the GT-2 clade with the most abundant motif types, contains all of the motifs except for motif 4 and motif 9, and most members contain two of motif 1, two of motif 2, and two of motif 3. In addition, motif 4 which has the longest sequence, is mostly present in GT-1 and only appears a few times in SIP1. These results suggested that subgroup-specific motifs might play pivotal roles in specific functions. Gene structure analysis showed that the exon lengths of genes in the same subgroup were diverse, but the number was almost always conserved (Figure S1). For example, most GTγ and SIP1 members contain one exon, and GT-1 and SH4 members contain 2–5 exons. In contrast, the GT-2 subfamily has a more complex exon number and length than the others.

In order to explore the gene structural changes in *B. napus* during evolution, we performed structural analysis for *B. napus* with its diploid progenitors *B. rapa* and *B. oleracea*. We separately predicted ten motifs in these three species and compared their amino acid residues (Figure 2, Table S4). We found that most motifs in *B. napus* were similar to those in *B. rapa* and *B. oleracea*. In particular, the amino acid residues of motif 1 and motif 6 were almost identical to their corresponding sequences in the progenitors, indicating that these two motifs were highly conserved during rapeseed evolution. In addition, we found that some motifs in *B. napus* were more likely to be derived from *B. rapa*. For example, motif 5 in *B. napus* is more similar to motif 4 in *B. rapa* than motif 5 in *B. oleracea*, and motif 4 in *B. napus* is more likely to have originated from motif 6 in *B. rapa*. Furthermore, the orthologous among the three species have similar gene structures and motif features (Figure 2 and Table S5). For instance, *BnaTH-57* has similar exons with its homologs *BraTH-28* and *BolTH-18*; additionally, the composition and distribution of motifs in protein sequences and the amino acid residues in the motifs are identical. Moreover, there is a unique motif 9 in *B. napus*, which may have been newly formed during *B. napus* evolution.

Figure 1. Phylogenetic tree using 455 TH proteins from 7 species, including *Arabidopsis*, *B. carinate*, *B. juncea*, *B. napus*, *B. oleracea*, *B. nigra*, *B. oleracea*, and *B. rapa*, marked as yellow pentagram, purple triangle, pink triangle, red pentagram, purple pentagram, green triangle, and purple pentagram, respectively. The clades of group GT-1, GT-2, GTγ, SH4, and SIP1 are distinguished by different colors.

Moreover, we found that some BraTH motifs originated from only one species, such as BnaTH-11, which has the orthologs BraTH-14 and BolTH-13, whose motifs appear to be derived from BolTH-13 only, as the corresponding motif cannot be found in BraTH-14.

However, most BnaTH motifs have more complex origins, and they may be derived from both *B. rapa* and *B. oleracea*, such as, BnaTH-51 and BnaTH-8. Our analysis of members of the SIP1 subgroup in *B. napus* and its diploid ancestors revealed that some BnaTHs inherited motifs from ancestors with motif insertions and deletions (Figure S2). For example, there are six motifs in BnaTH-28, while its homologs BolTH-20 and BraTH-25 have seven and five, respectively, and there are six motifs in BnaTH-46 compared to four in the homolog BolTH-2. These results implied that the evolution of *TH* genes in *B. napus* is intricate and diverse, and that they have inherited genetic material from ancestral species while undergoing a number of mutations in order to meet autogenous growth and development.

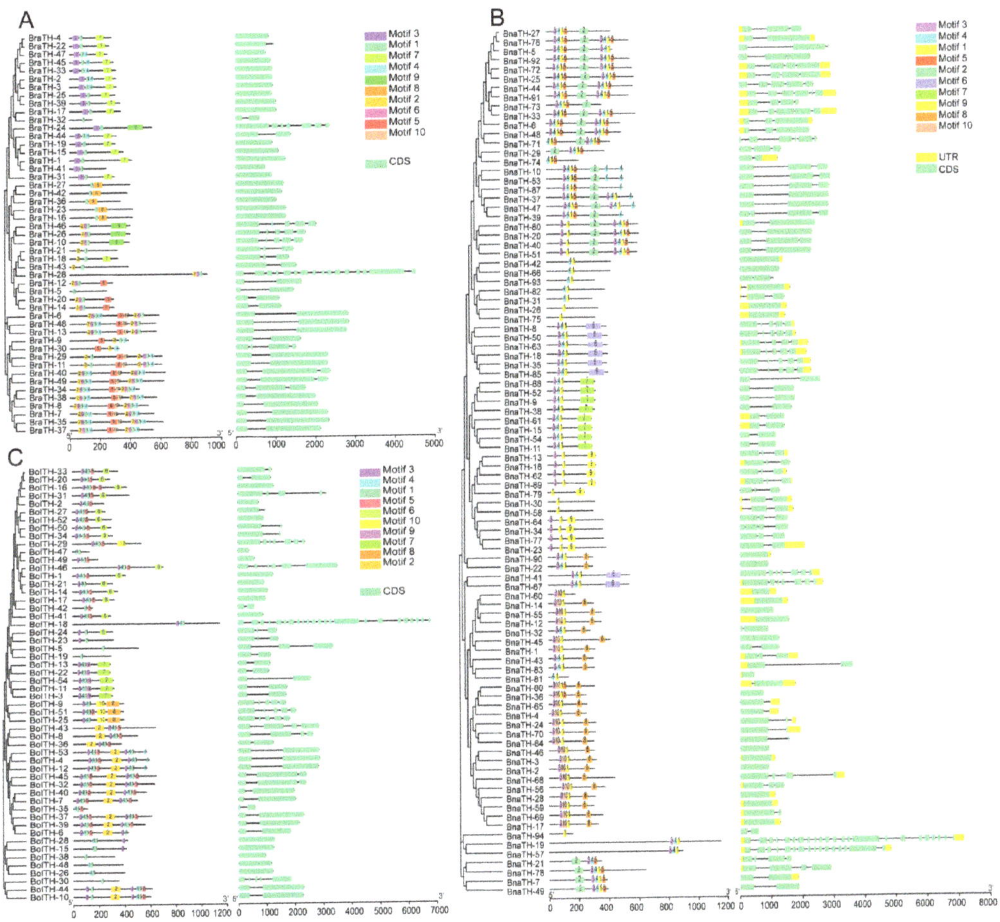

Figure 2. Comparative analysis of conserved domains and gene structures of *BraTH* genes, *BolTH* genes, and *BnaTH* genes. The colored boxes with numbers indicate the different conserved motifs identified by MEME. The scale bar at the bottom estimates the lengths of the motifs, exons, and introns. (**A**) Phylogenetic relationships, conserved protein motifs, and gene structures of 49 *BraTH* genes. (**B**) Phylogenetic relationships, conserved protein motifs, and gene structures of 94 *BnaTH* genes. (**C**) Phylogenetic relationships, conserved protein motifs, and gene structures of 54 *BolTH* genes.

2.4. Cis-Acting Elements in the Promoter Region of TH Genes

In order to better understand the potential regulatory roles of *TH* genes, the *cis*-acting elements in the 1500 bp upstream region of *TH* gene promoters were predicted using PlantCARE software. Twenty-seven distinct *cis*-acting elements were identified and grouped into three main categories (Tables S6 and S7). The first one was engaged in light responsiveness, such as Box 4 elements, I-boxes, TCT motif, ATC motif, GATA motif, MRE, G-box, SP1, AE-box, ATCT-motif, MRE, TCCC-motif, chs-CMA1a, and GT1-motif; the second was associated with stress responses, including drought stress (MBS), anaerobic induction (ARE), and low-temperature stress (LTR); the third was involved in plant growth and development, including circadian (circadian control), meristem expression (CAT-box), and phytohormone responses, such as abscisic acid (ABRE), methyl jasmonate (CGTCA-motif and TGACG-motif), auxin (TGA element), salicylic acid (TCA element), and so on (Tables S6 and S7). A total of 99.3% (442/445) of genes contained light-responsive elements, with *BcaTH-15* containing the most light-responsive elements (24) and *BnaTH-66* having the least (1). Among the 168 genes featuring drought-responsive elements, *BcaTH-74* contains the most numbers of MBS (4). In addition, the most abundant of these 168 genes are members of the SIP1 subgroup (53, 31.5%), followed by the GT-1 subgroup (39, 23.2%); meanwhile, the SH4 subgroup (17, 10.1%) had the fewest members (Table S7). There are 220 genes involved in low temperature response elements; of these, members of the SIP1 subgroup account for the largest proportion (75, 34.1%), followed by GT-2 (55, 25.0%). *BolTH-43* contains 9 low-temperature-response elements, which is the most among the 220 genes. In addition, there are a large number of other elements in the *TH* genes, such as methyl-jasmonate (MeJA)-responsive elements, abscisic-acid (ABA)-responsive elements, salicylic-acid (SA)-responsive elements, auxin-responsive elements, gibberellin-responsive elements, and ethylene (Eth)-responsive elements, indicating that they may be involved in phytohormone signaling pathways (Table S7). These results suggested that *TH* genes play an important role in plant growth and response to stress resistance. Moreover, compared with other subgroups, GT-2 and SIP1 have more of these elements, implying that the gene expansions of the subgroups GT-2 and SIP1 may be associated with environmental adaptation.

2.5. Chromosomal Distribution and Gene Duplication Events of the TH Family

Location information of *TH* genes from seven species was obtained according to the genome annotations (Table S1). Some genes were not accurately mapped to chromosomes because of the incomplete assembly of certain genomes. Overall, *TH* genes are unevenly distributed on the chromosomes of the respective species (Figure 3). We found that the numbers and distributions of *TH* genes located in the C subgenome of *B. napus* and *B. oleracea* were remarkably consistent. Additionally, the *TH* genes located on the A subgenome of the *B. napus* and *B. rapa* genome also have high consistency, but the similarity is relatively low. In addition, the number of *TH* genes in *B. napus* is slightly lower than that on the corresponding chromosomes of *B. rape* and *B. oleracea*. This may be due to the loss of some functionally redundant *TH* genes during the evolution of *B. napus*, or the incomplete assembly of chromosomes.

The intra-species synteny relationships of *TH* genes were analyzed using the genomic information of the seven species (Figure 4, Table S8). A total of 393 paralogous pairs was identified, including 6, 90, 131, 85, 25, 29, and 27 pairs in *Arabidopsis*, *B. carinate*, *B. juncea*, *B. napus*, *B. nigra*, *B. oleracea*, and *B. rapa*, respectively. These paralogous gene pairs contained 358 *TH* genes, and most of them belong to subgroup GT-2 (110, 30.7%), followed by subgroup SIP1 (106, 29.6%). Overall, compared with the diploid species, the tetraploid species experienced more segmental duplication events. In addition, we detected nine pairs of tandem duplicated genes with 17 *TH* genes, which mainly belonged to subgroup GT-2 (Table S9).

The nonsynonymous rate (*Ka*), synonymous rate (*Ks*), and evolutionary constraint (*Ka/Ks*) ratio of duplicated gene pairs of *TH* genes were calculated (Table S8). The *Ka/Ks* ratios of all paralogous gene pairs were less than 1; 383 (98.5%) pairs were less than 0.5, and 162 (41.2%) pairs were less than 0.2, which might have encountered extremely strong purification selection after triplication, resulting in the severe inhibition of the functional differentiation of *TH* genes.

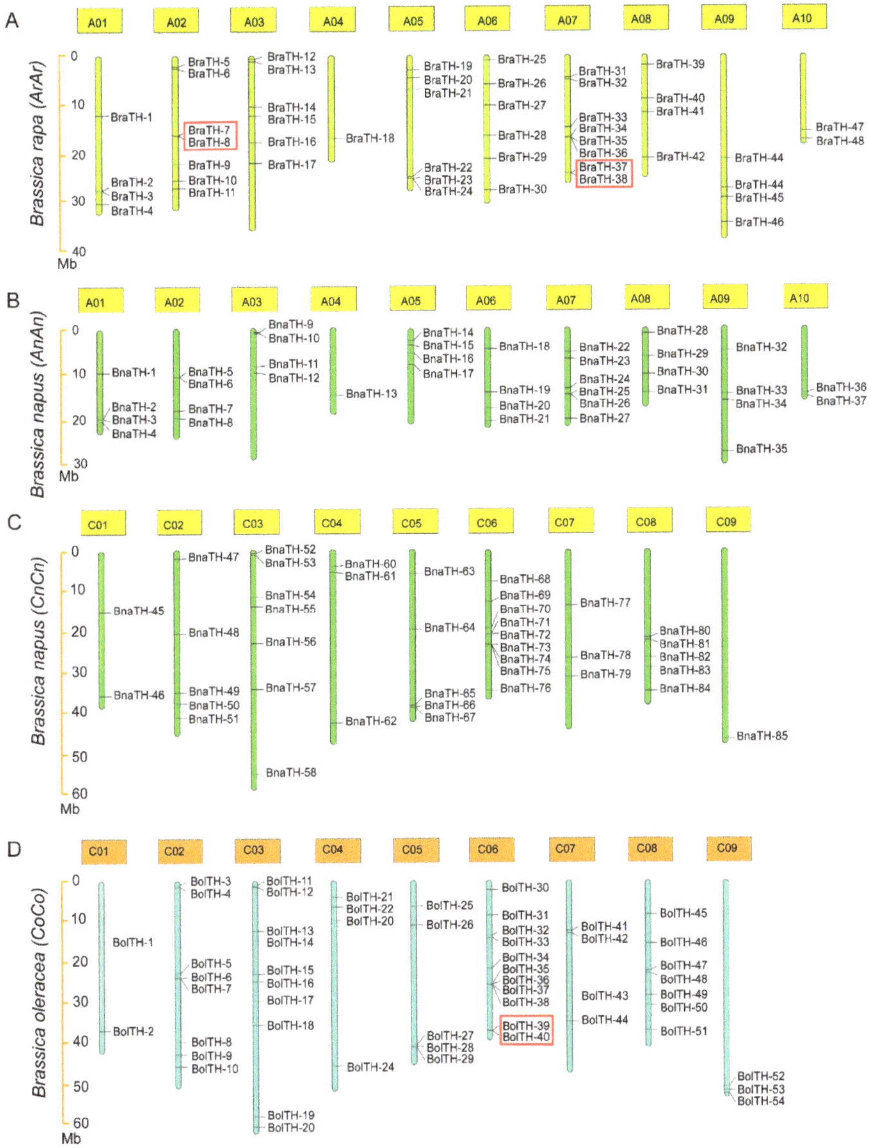

Figure 3. Chromosomal distribution of *BraTH* genes, *BolTH* genes, and *BnaTH* genes. The colored bars represent chromosomes, and the chromosome numbers are shown at the top of the bars. *TH* genes are labeled at the right of the chromosomes. The scale on the left side indicates the chromosome length (Mb). (**A**) Chromosomal localization of *BraTH* genes in *B. rapa*. (**B**) Chromosomal localization of *BnaTH* genes in A subgenome of *B. napus*. (**C**) Chromosomal localization of *BnaTH* genes in C subgenome of *B. napus*. (**D**) Chromosomal localization of *BolTH* genes in *B. oleracea*. Genes in the red box are tandem duplication genes.

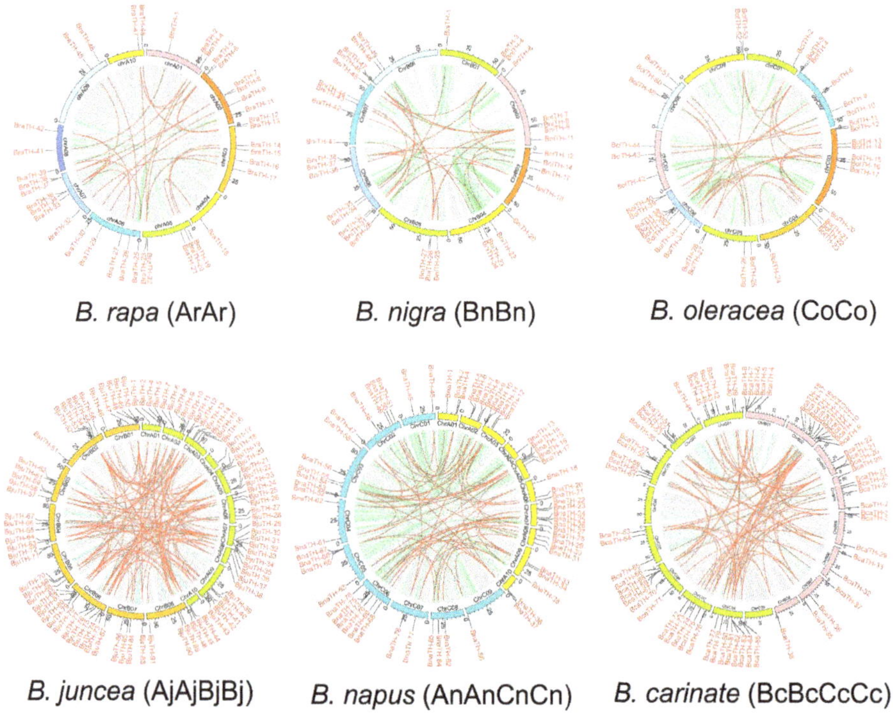

Figure 4. The collinearity of *TH* genes in *B. rapa*, *B. nigra*, *B. oleracea*, *B. juncea*, *B. napus*, and *B. carinate*. The green line indicates all synteny blocks between each chromosome and the red line indicates duplicated *TH* gene pairs. The chromosome number is shown at the top of each chromosome. The scale marked on the chromosome indicates the chromosome length (Mb).

2.6. Evolutionary and Collinearity Analysis of TH Genes in Brassica

In order to trace the evolution of the TH gene family of Brassica, we analyzed the orthologous relationship between *Arabidopsis* and six Brassica species, and between the tetraploid Brassica species and their diploid ancestors (Figure 5, Table S5). A total of 655 pairs of *TH* genes showed collinearity, among which 245 pairs of orthologous genes were composed of 27 *Arabidopsis TH* genes and corresponding Brassica *TH* genes (Figure 6). The orthologous gene pairs between *Arabidopsis* and *B. rapa*, *B. oleracea*, and *B. napus* are 30, 31, and 45, respectively, indicating that some redundant genes were lost during the allo-tetraploidization of *B. napus*. In contrast, *TH* genes lost the most during the formation of the tetraploid *B. carinate*, while *B. juncea* lost the least (Figure 6). In addition, compared with the numbers of *TH* genes in tetraploid species and the numbers of orthologous genes between tetraploid species and their diploid ancestors, we observed that the *TH* genes in *B. napus* had the most variation in tetraploid species, which might differentiate into more functional genes. The number of orthologous gene pairs between *B. napus* and *B. rapa* and between *B. napus* and *B. oleracea* was 53 and 64, respectively, which may suggest that *TH* genes from *B. oleracea* are more inclined to be retained in *B. napus*.

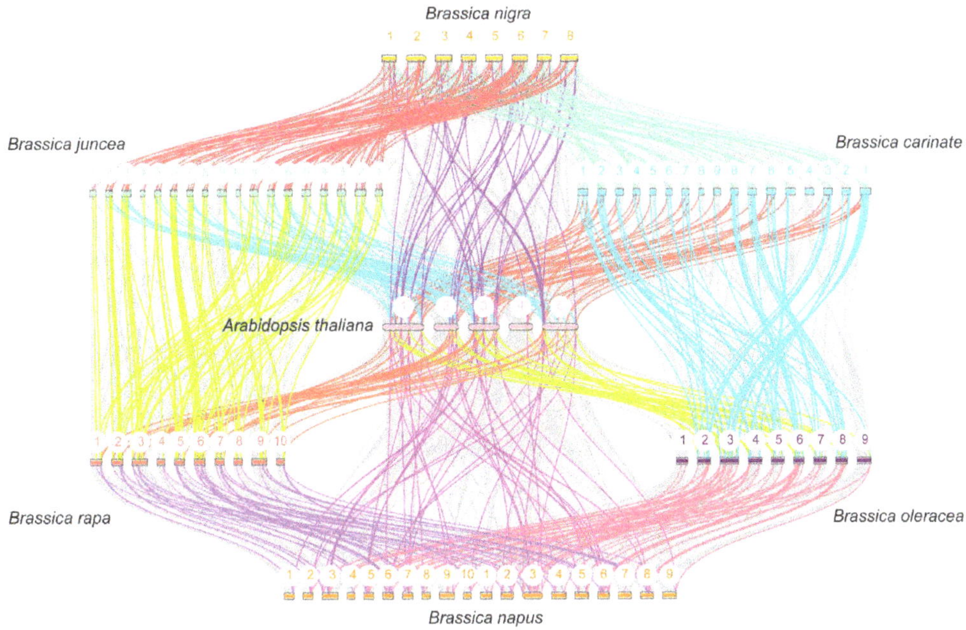

Figure 5. The collinearity analysis of *TH* genes between *Arabidopsis* and six Brassica species and between tetraploid and their diploid ancestors. The grey lines in the background indicate the collinear blocks in the genomes of the two species connected by the grey lines, while the colored lines highlight the syntenic *TH* gene pairs. The colored bars represent the chromosomes of the different species. The chromosome number is labeled at the top of each chromosome.

To evaluate the expansion of different clades, we investigated the relationships between *TH* genes and clades in each species (Figure 7). Overall, the GT-2 and SIP1 clades hold the largest number of *TH* genes, especially in tetraploid. However, the proportion of genes in different subgroups does not seem to change much in each species. As shown in Figure 6, the orthologous genes between *B. rapa* and *B. napus* are 37 and 52, respectively, and the orthologous genes of *B. oleracea* and *B. napus* are 42 and 59, respectively, but there are 94 *TH* genes in *B. napus*. In addition, by calculating the difference between the total *TH* genes in diploid species and the inherited *TH* genes, we found that many of the missing genes in *B. napus* belonged to the GT-2 subgroup, and most of the missing genes in *B. carinate* belonged to SIP1, but almost no genes were discarded in *B. juncea*. Compared with other subgroups, the genes in the GT-2 and SIP1 subgroups might produce more genes with new functions through duplication events after allotetraploidy. These genes with new functions may promote the growth and development of plants or improve the adaptability to their environment.

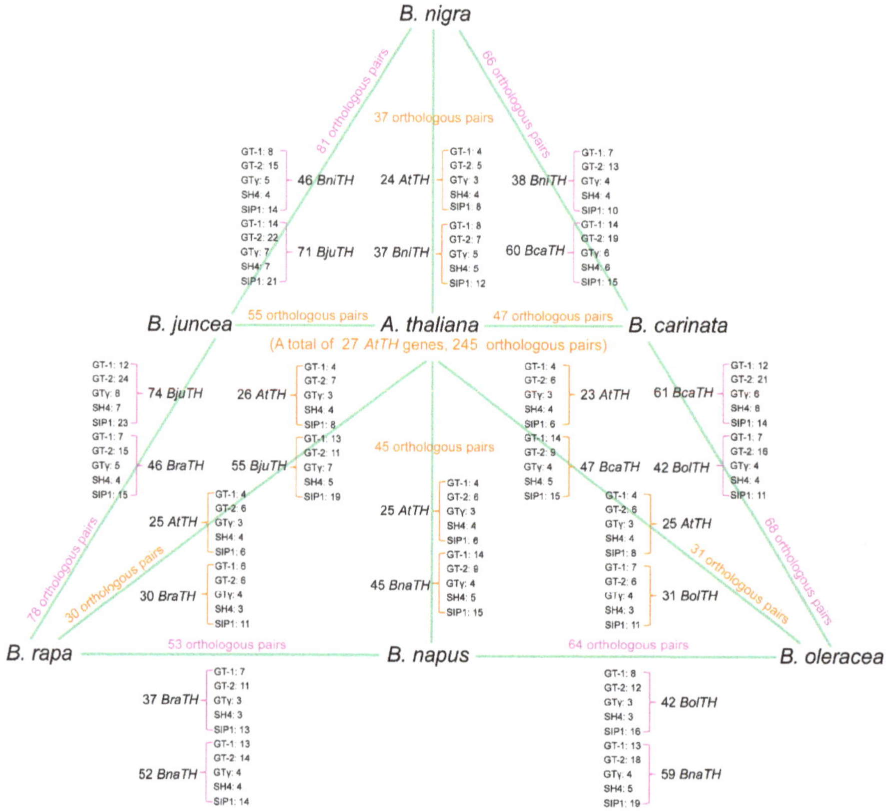

Figure 6. Numbers of orthologous gene pairs among different species and subfamily of orthologous genes. Orange represents the orthologous gene pair between *Arabidopsis* and Brassica species. Magenta represents the orthologous gene pair between Brassica species.

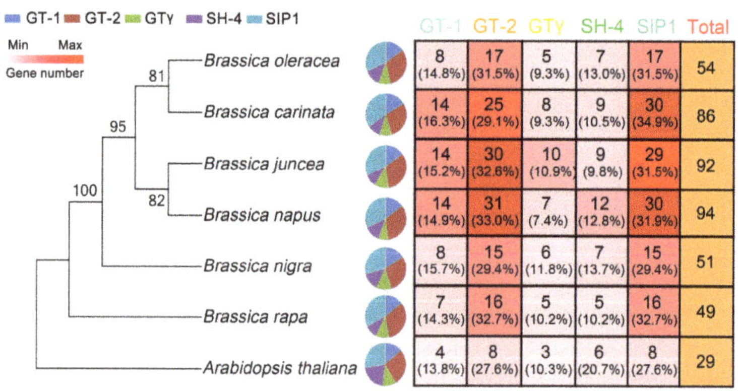

Figure 7. Clade distribution of *TH* genes in *Arabidopsis*, *B. rapa*, *B. nigra*, *B. oleracea*, *B. juncea*, *B. napus*, and *B. carinate*. The species tree constructed on the basis of the evolutionary relationships between the seven species is shown on the left. The subgroup distribution of *TH* genes in each species is shown in the middle pie chart. The counts and percentages of *TH* gene subgroup members in each species are given in the table on the right.

2.7. Interaction Analysis of BnaTH Genes

There are 21 *BnaTH* genes with interaction relationships, most of which belong to the GT-2 subgroup (14, 66.7%) (Figure S3). In the protein interaction network, we identified two hub genes, namely *BnaTH-94* and *BnaTH-41*, which can interact with eleven and eight *BnaTH* genes, respectively. In total, 28 miRNAs were found to target 35 *BnaTH* genes, and 73 pairs of interaction relationships were generated, among which the GT-2 (19, 54.3%) and SIP1 (12, 34.3%) subgroups accounted for the vast majority. Most miRNAs can target 1–3 *BnaTH* genes, but the miR396a can regulate 16 *BnaTH* genes (Figure S3). In addition, *BnaTH-19* can be regulated by nine miRNAs, which indicates that the function of this gene may be diverse.

2.8. Expression Patterns of BnaTH Genes in Different Tissues and under Abiotic Stresses

In order to investigate the putative functions of *TH* genes in plant growth and development, we analyzed the expression patterns of *BnaTH* genes in nine different tissues (cotyledon, leaf stem, root, petal, sepal, pollen, ovule, silique, and seed) of *B. napus* and at different developmental stages of buds, seeds, and siliques (Figure 8). The results revealed that the expression profiles of the *TH* genes were diverse, but that, to a certain extent, the genes of the same subgroup displayed similar expression patterns. Overall, almost every tissue has gene expressions from five different subfamilies, but, in the same subfamily, the expression profiles are not exactly the same, indicating the different functions of the genes in the same subfamily. Seven genes (*BnaTH-76*, *BnaTH-27*, *BnaTH-33*, *BnaTH-25*, *BnaTH-73*, *BnaTH-29*, and *BnaTH-72*) belonging to the GT-2 subgroup are highly expressed in the cotyledons and leaves, which may mean that they play a role in the light responses of plants and in photosynthesis. With the growth of siliques, the expression of three genes (*BnaTH-89*, *BnaTH-19*, and *BnaTH-57*) in siliques also increased gradually, which may indicate involvement in the silique. We found that *BnaTH-49* was highly expressed in ovules and seeds, but weakly expressed in other tissues. Moreover, it was observed that the expression of *BnaTH-49* decreased with the maturation of seeds, which suggested that this gene might have negative regulation with seed maturation. In addition, a large number of *BnaTH* genes is abundantly expressed in stems, roots, buds, and most other tissues, which indicates that *TH* genes are generally involved in plant growth and development.

To determine the response of *BnaTH* genes to abiotic stresses, the expression levels of *BnaTH* genes in *B. napus* under drought stress, cold stress, heat stress, salt stress, and ABA treatment were analyzed (Figure 9). Expression profile analysis revealed that many *BnaTH* genes could be induced under different stresses. Most *TH* genes were significantly up- or down-regulated under heat treatment, implying that *TH* genes are generally responsive to heat stress and that they have different regulatory networks. Some genes showed similar expression patterns under different stress treatments. For example, the expression levels of *BnaTH-85*, *BnaTH-26*, *BnaTH-78*, *BnaTH-92*, and *BnaTH-4* were significantly increased after drought treatment and salt stress, indicating that these genes may be associated with multiple stress response processes. In addition, we found that some homologous genes showed similar expression patterns, such as *BnaTH-2* and *BnaTH-56*, while others had opposite expression profiles, such as *BnaTH-5* and *BnaTH-27*. Overall, most of the genes that can respond to abiotic stresses belong to GT-2 and SIP1 clades, which suggests that the GT-2 and SIP1 clade genes may be significantly expanded to adapt to the environment.

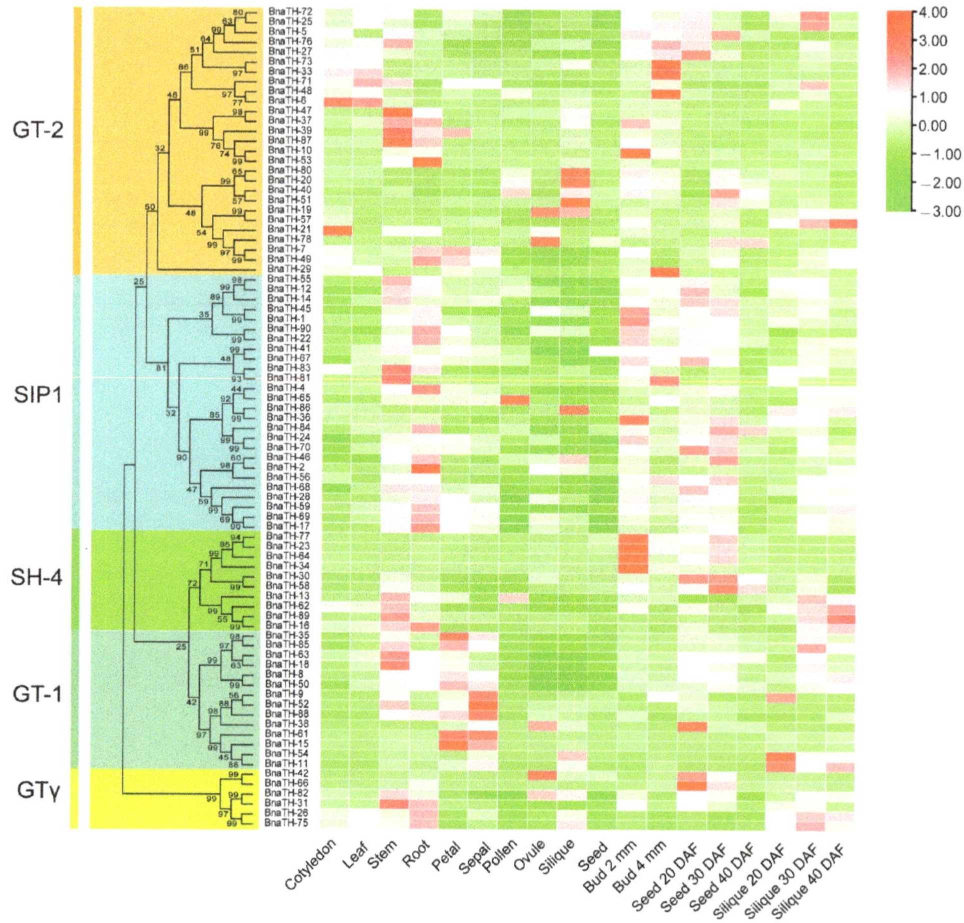

Figure 8. Expression pattern of *BnaTH* in different tissues and different developmental stages. The subgroup distribution and phylogenetic tree of BnaTH are shown on the left. FPKM values of *BnaTH* genes transformed by log2 were used in TBtools to construct heat maps. Expression levels are depicted by the different colors on the scale. Red and green represent high and low expression levels respectively.

To validate the expression patterns of *TH* genes under drought, cold, and heat stresses, qRT-PCRs were used to detect the expression levels of five randomly selected *TH* genes (Figure 10). The differential expression profiles of these selected genes under different stresses imply that they have different functions. For example, under cold treatment, the expression of *BnaTH-26* decreased significantly at 1 h, and remained at low levels at 6 and 12 h, while the *BnaTH-83* continued to increase. The expression of *BnaTH-74* decreased gradually with the prolonged drought treatment, while *BnaTH-83* showed the opposite expression pattern. All five of the selected genes responded to heat stress. Moreover, interestingly, except for *BnaTH-85*, the expression levels of the genes decreased first and then increased after heat treatment. Thus, all five genes were induced by various abiotic stresses with differential expression levels.

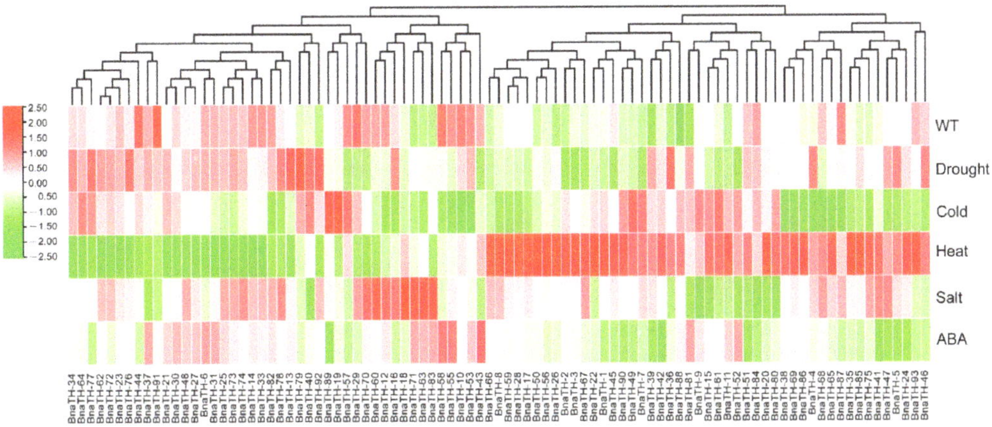

Figure 9. Expression patterns of *BnaTH* under different stresses. FPKM values of *BnaTH* genes transformed by log2 were used in TBtools to construct heat maps. Expression levels are depicted by the different colors on the scale. Red and green represent high and low expression levels respectively.

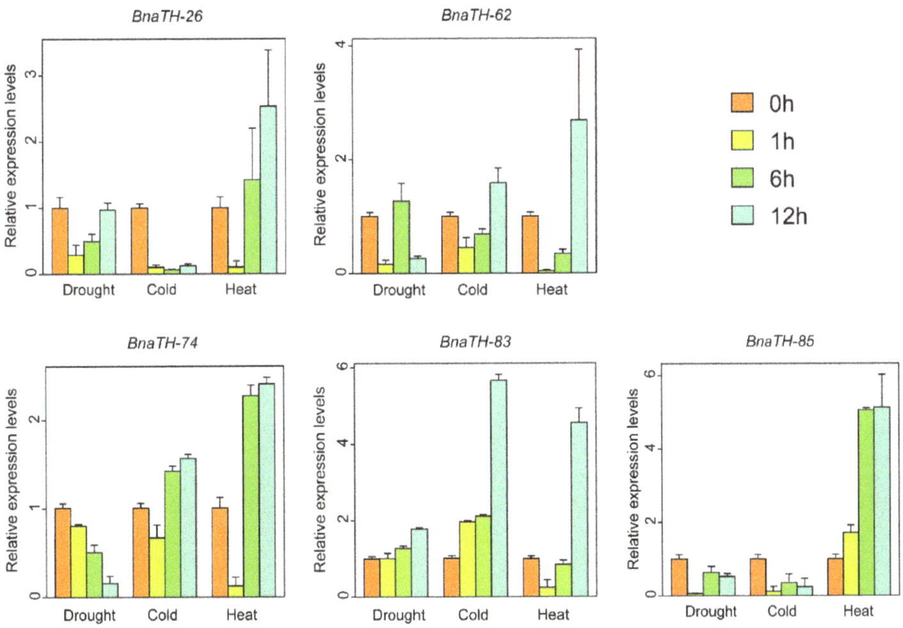

Figure 10. Expression profiles of five random selected *TH* genes in response to drought, cold, and heat treatments. Leaves were collected at 0 h, 1 h, 6 h, and 12 h after different treatments. Data were normalized to the *BnACTIN* gene. Vertical bars indicate standard deviations.

3. Discussion

As sessile organisms, plants must endure various abiotic stresses, including drought, cold, high temperatures, salt, and other environmental factors. Transcription factor TH has been found to be involved in regulating plant growth and development, as well as responses to biotic and abiotic stresses. With the release of the genome sequences of many plants, genome-wide identification of *TH* genes has been carried out in many species. However, other than for around 10 reported *TH* genes (Table S10), the identification and the function of Brasscia *TH* genes, especially in *B. napus*, have not been well reported [5,16–18,27–33]. In this study, a total of 455 *TH* genes was identified in six Brassica species and *Arabidopsis*, and the phylogenetic relationship, structure, *cis*-acting elements, chromosome distribution, gene duplication events, and collinearity were investigated. Furthermore, the expression patterns of these *TH* genes in *B. napus* were analyzed using RNA-seq with the highly sensitive method of qRT-PCRs [34,35]. Through comparative genomics, we explored the evolution of the *TH* gene family in Brassica, and present comprehensive information about this gene family, which provides clues for the functional study of *TH* genes in *B. napus*.

3.1. The Retention and Loss of TH Genes in Brassica during Evolution

Previous studies have proposed that the genome of Brassica originated from a genome with a similar structure to *Arabidopsis* through three rearrangement variants and inherited from a common hexaploid ancestor [25,36]. After whole-genome duplication or triplication (WGD/WGT), plants generally tend to restore the number of genes to the diploid level through gene loss [37]. However, in Brassica, this is more suitable for collinear genes in conserved syntenic regions. It is estimated that 60% of the genome is lost after triplication, and the number of genes before triplication is almost recovered [26]. Massive gene loss and the frequent recombination of triplicated genomic blocks occurred in the Brassica genome after WGT, resulting in complex mosaics of triplicated ancestral genomic blocks in the A and C genomes [26]. *B. napus*, as a relatively young amphidiploid, has not undergone significant chromosome rearrangement since the genome fusion of the progenitors A and C. However, homologous recombination events between the two corresponding genomes are universal in newly shaped *B. napus* lines, and low levels of homologous recombination have been recognized in established canola cultivars [38–40]. Through comparative mapping of *B. napus* and *Arabidopsis*, three collinear fragments were usually identified in each diploid genome for every region of *Arabidopsis* studied [41,42].

In this study, 49 and 54 *TH* genes were identified in *B. rapa* and *B. oleracea*, respectively, while 94 *TH* genes were identified in *B. napus*. This situation also been observed in two other allotetraploids, *B. carinate* and *B. juncea*, indicating that some redundant *TH* genes were lost in the process of evolution. In order to explore the reasons for the loss of *TH* genes in tetraploid species, we took the number of *TH* genes in *Arabidopsis* as a reference, and first compared and analyzed the *TH* genes in diploid plants. Furthermore, we found that 20 *TH* genes in *Arabidopsis* have only one homologous gene in *B. rapa*, and five *TH* genes have two homologous genes in *B. rapa*. In *B. oleracea*, 19 *TH* genes correspond to a homologous gene with *Arabidopsis*, and 12 *TH* genes correspond evenly to six *Arabidopsis TH* genes. However, we found a large number of *AtTH* genes with more than one homologous gene in *B. nigra*; among these, 12 *AtTH* genes only found one homologous gene in *B. nigra*, but 11 *AtTH* genes found two homologous genes in *B. nigra*, and even *AtTH-24* corresponded to three homologous genes (*BniTH-9*, *-27*, *-46*) in *B. nigra*. This indicated that the genome of Brassica ancestors lost a large number of *TH* genes after WGT, and the number may tend to recover to its pre-tripling level. In addition, *B. nigra* retained more *TH* genes during the evolution of Brassica diploid species from hexaploid ancestors. In the collinearity analysis of *Arabidopsis* and Brassica tetraploid species, we found that most *TH* genes in *Arabidopsis* only found one or two homologous genes in each species. However, in every species of Brassica, a high proportion of *TH* segmental duplication genes was recognized, especially in tetraploid species, and most of these genes were from subgroups GT-2 and SIP1. To sum up, segmental duplication events may drive the expansion of *TH* genes in the evolution of Brassica species, and these duplicated genes may support plants to better adapt to various environments.

3.2. A Conserved Pattern of Motifs and TH Gene Structures

We confirmed that the conserved motifs and gene structures of 455 TH family members were consistent with the classifications of subfamilies. These subfamily members with similar gene structures and motif distributions may have arisen from gene duplication events and, therefore, they may have similar functions. Alternatively, these genes may have undergone pseudogenization, neofunctionalization, or subfunctionalization during evolution. We found that some motifs are present in all subgroups, while some motifs are almost unique to subfamilies, and these subgroup-specific motifs may perform a distinct function. In addition, our comparative analysis of the 10 conserved motifs of the TH proteins in *B. rapa*, *B. oleracea*, and *B. napus* revealed that almost all motifs in *B. napus* could find their counterparts in *B. rapa* or *B. oleracea*, but a small number of amino acid residues in these sequences were mutated, suggesting that *B. napus* inherited motifs from ancestral diploids in a relatively conserved state during hetero-tetraploidization.

3.3. Expression and Function Characteristics of TH Genes in B. napus

In the present study, we revealed the broad expression patterns of *TH* genes in *B. napus*; their transcripts expressed in almost all tissues and universally participated in the growth and development of *B. napus*, based on the RNA-seq data. Changes in *TH* gene expression patterns under abiotic stresses such as drought and osmotic stresses have been reported [16]. In our study, we determined the expression patterns of *TH* genes in *B. napus* seedlings exposed to five stresses to elucidate the roles of *TH* genes in response to adverse environmental conditions. The results showed that *TH* genes could be induced by various stresses, and the expression levels of some genes were significantly up-regulated after stresses; in contrast, *TH* genes could only be weakly induced by ABA. The genes *BnaTH-8*, *BnaTH-59*, and *BnaTH-28* were all abundantly expressed under heat stress, whereas the genes *BnaTH-4*, *BnaTH-80*, and *BnaTH-67* were differentially expressed under drought, cold, and salt stresses, respectively, indicating that these genes are involved in many biological processes in *B. napus* and respond to different stresses. The interaction relationships of BnaTH proteins and miRNA targeted *BnaTH* genes also revealed that *BnaTHs* regulated plant development and stress responses. In addition, the involvement of *cis*-acting elements in the expression of genes at different stages of plant growth and development and under different stresses has been widely reported. In our study, we identified a large number of stress-related *cis*-acting elements, such as MBS (drought), LTR (low temperature stress), and phytohormone response elements such as abscisic acid (ABRE), methyl jasmonate (CGTCA-motif and TGACG-motif), auxin (TGA element), salicylic acid (TCA element). Some phytohormones are important for plant stress resistance, such as ABA, which mediates the physiological response to stress induced by dehydration and salt stresses, suggesting that some hormone response elements not only affect plant growth and development but also have the potential to respond to stresses. Further qRT-PCR analysis verified that some genes were up-regulated under drought, cold, and heat stress, but most of them showed weak changes at the initial stage and significant differences at 12 h after the stress treatment (Figure 10).

4. Materials and Methods

4.1. Plant Materials

Seeds of *B. napus* Zhongshuang 9 (ZS9) were cultured in an illuminated incubator with a 16 h/8 h light dark cycle and a temperature of 22 °C. The seedlings were subjected to various stresses (drought, cold, and heat treatments) at the three-leaf stage. For the drought stress treatments, the seedlings were watered with a Hoagland nutrient solution containing 18% polyethylene glycol (PEG)-6000 (*w/v*). In the cold stress treatment, plants were kept in an illuminated incubator at 4 °C with the other conditions unchanged. Seedlings were also grown in an oven at 40 °C for heat stress treatment. We collected leaves from the plants treated with drought, cold, and heat stresses at 0 h, 1 h, 6 h, and 12 h after the beginning of treatment. After collection, all samples were frozen immediately in liquid nitrogen and stored at −80 °C for RNA extraction.

4.2. Identification and Characterization of the Trihelix (TH) Genes

We downloaded the entire genome sequence information of six Brassica species from the Brassica database (BRAD version 3.0; http://brassicadb.cn/, accessed on 18 July 2022) [43]. Additionally, the *Arabidopsis* genome information was acquired from Ensembl (http://plants.ensembl.org/, accessed on 18 July 2022) [44]. The Hidden Markov Model (HMM) profile (PF13837) downloaded from the Pfam database (http://pfam.xfam.org/, accessed on 18 July 2022) was utilized to search the trihelix domain (E-value < 1×10^{-3}) by means of HMMER v3.0 software (http://hmmer.janelia.org/, accessed on 18 July 2022) [45,46]. We constructed seven species-specific HMM profiles, using the extracted protein domain sequences with E-value < 1×10^{-28} in the first search result. Then, we performed a second scan of the protein sequence (E-value < 1×10^{-3}) to identify the *TH* genes based on the newly created species-specific HMM profiles. The output putative TH protein sequences were further authenticated via the online websites Pfam (http://pfam.xfam.org/, accessed on 19 July 2022) and SMART (https://smart.embl.de/, accessed on 19 July 2022) [47,48]. The basic physical and chemical characteristics of the candidate TH proteins, including molecular weight (MW), theoretical isoelectric point (pI), the aliphatic index (AI), the instability index (II), and the grand average of hydropathicity (GRAVY), were calculated using the ProtParam program (https://web.expasy.org/protparam/, accessed on 20 July 2022) [49]. In addition, the subcellular localization of TH proteins was predicted using WOLF PSORT (https://wolfpsort.hgc.jp/, accessed on 20 July 2022) [50].

4.3. Phylogenetic Analysis and Classification

Multiple sequence alignment of the AtTH, BcaTH, BjuTH, BnaTH, BniTH, BolTH, and BraTH protein sequences was performed using the ClustalX v2.1 program. Then, an unrooted phylogenetic tree was constructed via the neighbor-joining (NJ) method using MEGA 7.0 software with 1000 bootstrap repetitions [51]. The phylogenetic tree was visualized using Evolview software (https://evolgenius.info//evolview-v2/#login, accessed on 20 July 2022) [52].

4.4. Gene Structure and Conserved Motif Analysis

The exon-intron structure information was attained from the annotation file (gff3) of the sequenced genome. The online software MEME (https://meme-suite.org/, accessed on 21 July 2022) [53] was used to investigate the conserved motifs of TH proteins; the parameters were set with the maximum number of motifs as 10 and the motif width as 5–200 amino acids. TBtools was utilized to visualize the map of the exon–intron structures and motifs [54].

4.5. Analysis of Cis-Acting Elements in the Promoter Region

The PlantCARE (https://bioinformatics.psb.ugent.be/webtools/plantcare/html/, accessed on 22 July 2022) database was utilized to calculate the *cis*-acting elements within the 1500 bp sequence upstream of the transcription start site of the *TH* genes [55]. GSDS online software (http://gsds.gao-lab.org/, accessed on 22 July 2022) was employed to visualize the distribution map of *cis*-acting elements [56].

4.6. Chromosomal Distribution, Gene Duplication (Collinear), and Evolution of TH Genes

The chromosome location data for the *TH* genes were obtained from the genomic annotation file GFF3 of the seven species, then MapChart software [57] was used to locate the distribution of the *TH* genes on the chromosome, respectively. The analysis of the intra-species collinearity of *TH* genes in seven species was performed using collinear scanning toolkits (McScanX) [58], and the collinearity of gene pairs was presented using Circos v0.55 [59]. Moreover, McScanX software was used to analyze and plot collinearity between species.

4.7. Ka, Ks, and Ka/Ks Analysis

The synonymous (*Ks*) and nonsynonymous (*Ka*) substitution rates, and the selection pressure *Ka/Ks* ratio of homologous gene pairs and tandem array genes (adjacent *TH* genes on individual chromosomes), were calculated using TBtools [54].

4.8. Interaction Analysis of BnaTHs

The interaction of BraTH proteins was predicted using the STRING website (https://cn.string-db.org/, accessed on 15 September 2022). The potential miRNA-regulating sites of *BnaTH* genes were analyzed using psRNATarget [60]. The interaction relationships of BnaTH was visually edited using Cytoscape 3.8.2 [61].

4.9. RNA Isolation and qRT-PCR Analysis

Total RNA was extracted from leaves treated with different abiotic stresses using an RNA Prep Pure Plant Kit (Tiangen Biochemical Technology (Beijing, China) Co., Ltd.: DP201101X). The concentration and purity of the RNA were determined using a Nanodrop micro spectrophotometer (Thermo Scientific, Waltham, MA, USA) and agarose gel electrophoresis. The first strand of cDNA derived from mRNA was synthesized using HiScript®Q RT SuperMix (Vazyme, Nanjing, China). Quantitative real-time PCR (qRT-PCR) was carried out by SYBR-green fluorescence with a QuantStudio™ Real-Time PCR System (Thermo Fisher Scientific). Five *TH* genes were randomly selected to validate the expression patterns under drought, cold, and heat stresses (Table S11). The data were normalized by the internal control gene *BnACTIN* (*BnaA03g55890D*) and the $2^{-\Delta\Delta CT}$ analysis method was utilized to calculate relative expression levels [62].

4.10. Expression Profile Analysis

To explore the spatiotemporal expression patterns of rapeseed under different tissues and stresses, the relative FPKM value (fragments per kilobase of transcript per million fragments mapped) of RNA-seq data, PRJNA389508 (seedlings treated with five stress conditions) and PRJNA311067 (seeds harvested at different times after pollination), were retrieved in the BrassicaEDB website (https://brassica.biodb.org/downloads, accessed on 20 August 2022), and the expression levels of different tissues (cotyledon, leaf, stem, root, petal, sepal, etc.) were acquired from the *Brassica napus* information resource (http://yanglab.hzau.edu.cn/BnIR/expression_zs11, accessed on 20 August 2022) [63–66]. The FPKM values were log2 transformed and the heat map of hierarchical clustering was visualized using TBtools software [54].

5. Conclusions

In summary, a total of 455 *TH* genes was identified in six Brassica species and *Arabidopsis*. The evolutionary history of the *TH* genes was explored by integrating phylogenetic analysis, conserved motif identification, gene duplication event analysis, collinearity analysis between six Brassica species and *Arabidopsis*, and comparative genomic analysis. Many of the missing genes in *B. napus* belonged to the GT-2 subgroup, and most of the missing genes in *B. carinate* belonged to SIP1, but almost no genes were discarded in *B. juncea*. In addition, some mutations occurred in the motifs of TH proteins following hetero-tetraploidization in *B. napus*, with changes in the sequences of their amino acid residues. Expression pattern analysis revealed that *BnaTH* genes have different expression patterns, which can be induced by various stresses. Further transgenic validation or CRISPR/Cas9 knock out of these *BnaTH* genes could reveal their roles in stresses. In this study, a systematic analysis of *TH* genes in Brassica species, including their origin and evolution, and the function of *BnaTH* genes, was carried out to provide insights for further studies on the function of *TH* genes in *B. napus*.

Supplementary Materials: The following supporting information can be downloaded at: https://www.mdpi.com/article/10.3390/ijms232415766/s1.

Author Contributions: J.H., C.Z. and R.Z. conceived and designed this study. C.Z. and R.Z. performed the data analysis. C.Z., L.L., R.G., X.S. and F.L. prepared the samples and performed experiments. C.Z. drafted the manuscript. J.H. and R.Z. revised and edited the manuscript. All authors have read and agreed to the published version of the manuscript.

Funding: This study was supported by the Start-Up Funds of Northwest A&F University (Z1090222020).

Institutional Review Board Statement: Not applicable.

Informed Consent Statement: Not applicable.

Data Availability Statement: The transcriptome data are available from the online website BrassicaEDB (https://brassica.biodb.org/downloads accessed on 20 August 2022, PRJNA389508 and PRJNA311067) and Brassica napus information resource (http://yanglab.hzau.edu.cn/BnIR/expression_zs11, accessed on 20 August 2022).

Conflicts of Interest: The funders had no role in the design of the study; in the collection, analyses, or interpretation of data; in the writing of the manuscript; or in the decision to publish the results.

References

1. Green, P.J.; Kay, S.A.; Chua, N.H. Sequence-specific interactions of a pea nuclear factor with light-responsive elements upstream of the rbcS-3A gene. *EMBO J.* **1987**, *6*, 2543–2549. [CrossRef] [PubMed]
2. Kaplan-Levy, R.N.; Brewer, P.B.; Quon, T.; Smyth, D.R. The trihelix family of transcription factors—Light, stress and development. *Trends Plant Sci.* **2012**, *17*, 163–171. [CrossRef] [PubMed]
3. Nagano, Y. Several features of the GT-factor trihelix domain resemble those of the Myb DNA-binding domain. *Plant Physiol.* **2000**, *124*, 491–494. [CrossRef] [PubMed]
4. Nagano, Y.; Inaba, T.; Furuhashi, H.; Sasaki, Y. Trihelix DNA-binding protein with specificities for two distinct cis-elements: Both important for light down-regulated and dark-inducible gene expression in higher plants. *J. Biol. Chem.* **2001**, *276*, 22238–22243. [CrossRef] [PubMed]
5. Gao, M.J.; Lydiate, D.J.; Li, X.; Lui, H.; Gjetvaj, B.; Hegedus, D.D.; Rozwadowski, K. Repression of seed maturation genes by a trihelix transcriptional repressor in *Arabidopsis* seedlings. *Plant Cell* **2009**, *21*, 54–71. [CrossRef]
6. Ma, Z.; Liu, M.; Sun, W.; Huang, L.; Wu, Q.; Bu, T.; Li, C.; Chen, H. Genome-wide identification and expression analysis of the trihelix transcription factor family in tartary buckwheat (*Fagopyrum tataricum*). *BMC Plant Biol.* **2019**, *19*, 344. [CrossRef]
7. Cheng, X.; Xiong, R.; Yan, H.; Gao, Y.; Liu, H.; Wu, M.; Xiang, Y. The Trihelix family of transcription factors: Functional and evolutionary analysis in Moso bamboo (*Phyllostachys edulis*). *BMC Plant Biol.* **2019**, *19*, 154. [CrossRef]
8. Xiao, J.; Hu, R.; Gu, T.; Han, J.; Qiu, D.; Su, P.; Feng, J.; Chang, J.; Yang, G.; He, G. Genome-wide identification and expression profiling of trihelix gene family under abiotic stresses in wheat. *BMC Genom.* **2019**, *20*, 287. [CrossRef]
9. Mo, H.; Wang, L.; Ma, S.; Yu, D.; Lu, L.; Yang, Z.; Yang, Z.; Li, F. Transcriptome profiling of *Gossypium arboreum* during fiber initiation and the genome-wide identification of trihelix transcription factors. *Gene* **2019**, *709*, 36–47. [CrossRef]
10. Wang, Z.; Liu, Q.; Wang, H.; Zhang, H.; Xu, X.; Li, C.; Yang, C. Comprehensive analysis of trihelix genes and their expression under biotic and abiotic stresses in *Populus trichocarpa*. *Sci Rep.* **2016**, *6*, 36274. [CrossRef]
11. Wang, W.; Wu, P.; Liu, T.; Ren, H.; Li, Y.; Hou, X. Genome-wide analysis and expression divergence of the Trihelix family in *Brassica rapa*: Insight into the evolutionary patterns in plants. *Sci. Rep.* **2017**, *7*, 6463. [CrossRef]
12. Song, A.; Wu, D.; Fan, Q.; Tian, C.; Chen, S.; Guan, Z.; Xin, J.; Zhao, K.; Chen, F. Transcriptome-wide identification and expression profiling analysis of chrysanthemum Trihelix transcription factors. *Int. J. Mol. Sci.* **2016**, *17*, 198. [CrossRef]
13. Yu, C.; Cai, X.; Ye, Z.; Li, H. Genome-wide identification and expression profiling analysis of trihelix gene family in tomato. *Biochem. Biophys. Res. Commun.* **2015**, *468*, 653–659. [CrossRef]
14. Qin, Y.; Ma, X.; Yu, G.; Wang, Q.; Wang, L.; Kong, L.; Kim, W.; Wang, H.W. Evolutionary history of trihelix family and their functional diversification. *DNA Res.* **2014**, *21*, 499–510. [CrossRef]
15. Yu, C.; Song, L.; Song, J.; Ouyang, B.; Guo, L.; Shang, L.; Wang, T.; Li, H.; Zhang, J.; Ye, Z. *ShCIGT*, a Trihelix family gene, mediates cold and drought tolerance by interacting with SnRK1 in tomato. *Plant Sci.* **2018**, *270*, 140–149. [CrossRef]
16. Xu, H.; Shi, X.; He, L.; Guo, Y.; Zang, D.; Li, H.; Zhang, W.; Wang, Y. *Arabidopsis thaliana* Trihelix transcription factor AST1 mediates salt and osmotic stress tolerance by binding to a novel AGAG-Box and some GT motifs. *Plant Cell Physiol.* **2018**, *59*, 946–965. [CrossRef]
17. Luo, J.; Tang, S.; Mei, F.; Peng, X.; Li, J.; Li, X.; Yan, X.; Zeng, X.; Liu, F.; Wu, Y.; et al. *BnSIP1-1*, a Trihelix family gene, mediates abiotic stress tolerance and ABA signaling in *Brassica napus*. *Front. Plant Sci.* **2017**, *8*, 44. [CrossRef]
18. Xi, J.; Qiu, Y.; Du, L.; Poovaiah, B.W. Plant-specific trihelix transcription factor AtGT2L interacts with calcium/calmodulin and responds to cold and salt stresses. *Plant Sci.* **2012**, *185–186*, 274–280. [CrossRef]

19. Chalhoub, B.; Denoeud, F.; Liu, S.; Parkin, I.A.P.; Tang, H.; Wang, X.; Chiquet, J.; Belcram, H.; Tong, C.; Samans, B.; et al. Plant genetics. Early allopolyploid evolution in the post-Neolithic *Brassica napus* oilseed genome. *Science* **2014**, *345*, 950–953. [CrossRef]
20. Hu, J.; Chen, B.; Zhao, J.; Zhang, F.; Xie, T.; Xu, K.; Gao, G.; Yan, G.; Li, H.; Li, L.; et al. Genomic selection and genetic architecture of agronomic traits during modern rapeseed breeding. *Nat. Genet.* **2022**, *54*, 694–704. [CrossRef]
21. Kang, L.; Qian, L.; Zheng, M.; Chen, L.; Chen, H.; Yang, L.; You, L.; Yang, B.; Yan, M.; Gu, Y.; et al. Genomic insights into the origin, domestication and diversification of *Brassica juncea*. *Nat. Genet.* **2021**, *53*, 1392–1402. [CrossRef] [PubMed]
22. Song, X.; Wei, Y.; Xiao, D.; Gong, K.; Sun, P.; Ren, Y.; Yuan, J.; Wu, T.; Yang, Q.; Li, X.; et al. *Brassica carinata* genome characterization clarifies U's triangle model of evolution and polyploidy in Brassica. *Plant Physiol.* **2021**, *186*, 388–406. [CrossRef] [PubMed]
23. Nagaharu, U. Genome analysis in Brassica with special reference to the experimental formation of *B. napus* and peculiar mode of fertilization. *Jpn. J. Bot.* **1935**, *7*, 389–452.
24. Wang, X.; Wang, H.; Wang, J.; Sun, R.; Wu, J.; Liu, S.; Bai, Y.; Mun, J.-H.; Bancroft, I.; Cheng, F.; et al. The genome of the mesopolyploid crop species *Brassica rapa*. *Nat. Genet.* **2011**, *43*, 1035–1039. [CrossRef] [PubMed]
25. Lysak, M.A.; Koch, M.A.; Pecinka, A.; Schubert, I. Chromosome triplication found across the tribe Brassiceae. *Genome Res.* **2005**, *15*, 516–525. [CrossRef]
26. Liu, S.; Liu, Y.; Yang, X.; Tong, C.; Edwards, D.; Parkin, I.A.P.; Zhao, M.; Ma, J.; Yu, J.; Huang, S.; et al. The Brassica oleracea genome reveals the asymmetrical evolution of polyploid genomes. *Nat. Commun.* **2014**, *5*, 3930. [CrossRef]
27. Luo, J.; Jiang, W.; Tang, S.; Mei, F.; Yan, X.; Zeng, X.; Wu, G. BnSIP1-1 involves in light response and regulation of endogenous hormones and flowering time of *Brassica Napus*. *J. Plant Growth Regul.* **2021**, *40*, 2049–2057. [CrossRef]
28. Breuer, C.; Kawamura, A.; Ichikawa, T.; Tominaga-Wada, R.; Wada, T.; Kondou, Y.; Muto, S.; Matsui, M.; Sugimoto, K. The trihelix transcription factor GTL1 regulates ploidy-dependent cell growth in the *Arabidopsis* trichome. *Plant Cell.* **2009**, *21*, 2307–2322. [CrossRef]
29. Yoo, C.Y.; Pence, H.E.; Jin, J.B.; Miura, K.; Gosney, M.J.; Hasegawa, P.M.; Mickelbart, M.V. The *Arabidopsis* GTL1 transcription factor regulates water use efficiency and drought tolerance by modulating stomatal density via transrepression of SDD1. *Plant Cell.* **2010**, *22*, 4128–4141. [CrossRef]
30. Völz, R.; Kim, S.K.; Mi, J.; Mariappan, K.G.; Guo, X.; Bigeard, J.; Alejandro, S.; Pflieger, D.; Rayapuram, N.; Al-Babili, S.; et al. The Trihelix transcription factor GT2-like 1 (GTL1) promotes salicylic acid metabolism, and regulates bacterial-triggered immunity. *PLoS Genet.* **2018**, *14*, e1007708. [CrossRef]
31. Wang, X.H.; Li, Q.T.; Chen, H.W.; Zhang, W.K.; Ma, B.; Chen, S.Y.; Zhang, J.S. Trihelix transcription factor GT-4 mediates salt tolerance via interaction with TEM2 in *Arabidopsis*. *BMC Plant Biol.* **2014**, *14*, 339. [CrossRef]
32. Li, X.; Qin, G.; Chen, Z.; Gu, H.; Qu, L.J. A gain-of-function mutation of transcriptional factor PTL results in curly leaves, dwarfism and male sterility by affecting auxin homeostasis. *Plant Mol. Biol.* **2008**, *66*, 315–327. [CrossRef]
33. Kaplan-Levy, R.N.; Quon, T.; O'Brien, M.; Sappl, P.G.; Smyth, D.R. Functional domains of the PETAL LOSS protein, a trihelix transcription factor that represses regional growth in *Arabidopsis thaliana*. *Plant J.* **2014**, *79*, 477–491. [CrossRef]
34. Marguerat, S.; Bähler, J. RNA-seq: From technology to biology. *Cell Mol. Life Sci.* **2010**, *67*, 569–579. [CrossRef]
35. Auer, P.L.; Doerge, R.W. Statistical design and analysis of RNA sequencing data. *Genetics* **2010**, *185*, 405–416. [CrossRef]
36. Town, C.D.; Cheung, F.; Maiti, R.; Crabtree, J.; Haas, B.J.; Wortman, J.R.; Hine, E.E.; Althoff, R.; Arbogast, T.S.; Tallon, L.J.; et al. comparative genomics of *Brassica oleracea* and *Arabidopsis thaliana* reveal gene loss, fragmentation, and dispersal after polyploidy. *Plant Cell* **2006**, *18*, 1348–1359. [CrossRef]
37. Sankoff, D.; Zheng, C.; Zhu, Q. The collapse of gene complement following whole genome duplication. *BMC Genom.* **2010**, *11*, 313. [CrossRef]
38. Udall, J.A.; Quijada, P.A.; Osborn, T.C. Detection of chromosomal rearrangements derived from homologous recombination in four mapping populations of *Brassica napus* L. *Genetics* **2005**, *169*, 967–979. [CrossRef]
39. Sharpe, A.G.; Parkin, I.A.; Keith, D.J.; Lydiate, D.J. Frequent nonreciprocal translocations in the amphidiploid genome of oilseed rape (*Brassica napus*). *Genome* **1995**, *38*, 1112–1121. [CrossRef]
40. Parkin, I.A.; Sharpe, A.G.; Keith, D.J.; Lydiate, D.J. Identification of the A and C genomes of amphidiploid *Brassica napus* (oilseed rape). *Genome* **1995**, *38*, 1122–1131. [CrossRef]
41. Parkin, I.A.; Lydiate, D.J.; Trick, M. Assessing the level of collinearity between *Arabidopsis thaliana* and *Brassica napus* for *A. thaliana* chromosome 5. *Genome* **2002**, *45*, 356–366. [CrossRef] [PubMed]
42. Cavell, A.C.; Lydiate, D.J.; Parkin, I.A.; Dean, C.; Trick, M. Collinearity between a 30-centimorgan segment of *Arabidopsis thaliana* chromosome 4 and duplicated regions within the *Brassica napus* genome. *Genome* **1998**, *41*, 62–69. [CrossRef] [PubMed]
43. Chen, H.; Wang, T.; He, X.; Cai, X.; Lin, R.; Liang, J.; Wu, J.; King, G.; Wang, X. BRAD V3.0: An upgraded Brassicaceae database. *Nucleic Acids Res.* **2021**, *50*, D1432–D1441. [CrossRef] [PubMed]
44. Cunningham, F.; Allen, J.E.; Allen, J.; Alvarez-Jarreta, J.; Amode, M.R.; Armean, I.M.; Austine-Orimoloye, O.; Azov, A.G.; Barnes, I.; Bennett, R.; et al. Ensembl 2022. *Nucleic Acids Res.* **2022**, *50*, D988–D995. [CrossRef] [PubMed]
45. El-Gebali, S.; Mistry, J.; Bateman, A.; Eddy, S.R.; Luciani, A.; Potter, S.C.; Qureshi, M.; Richardson, L.J.; Salazar, G.A.; Smart, A.; et al. The Pfam protein families database in 2019. *Nucleic Acids Res.* **2019**, *47*, D427–D432. [CrossRef]
46. Eddy, S.R. Profile hidden Markov models. *Bioinformatics* **1998**, *14*, 755–763. [CrossRef]
47. Mistry, J.; Chuguransky, S.; Williams, L.; Qureshi, M.; Salazar, G.A.; Sonnhammer, E.L.L.; Tosatto, S.C.E.; Paladin, L.; Raj, S.; Richardson, L.J.; et al. Pfam: The protein families database in 2021. *Nucleic Acids Res.* **2021**, *49*, D412–D419. [CrossRef]

48. Letunic, I.; Khedkar, S.; Bork, P. SMART: Recent updates, new developments and status in 2020. *Nucleic Acids Res.* **2021**, *49*, D458–D460. [CrossRef]
49. Artimo, P.; Jonnalagedda, M.; Arnold, K.; Baratin, D.; Csardi, G.; de Castro, E.; Duvaud, S.; Flegel, V.; Fortier, A.; Gasteiger, E.; et al. ExPASy: SIB bioinformatics resource portal. *Nucleic Acids Res.* **2012**, *40*, W597–W603. [CrossRef]
50. Horton, P.; Park, K.J.; Obayashi, T.; Fujita, N.; Harada, H.; Adams-Collier, C.J.; Nakai, K. WoLF PSORT: Protein localization predictor. *Nucleic Acids Res.* **2007**, *35*, W585–W587. [CrossRef]
51. Kumar, S.; Stecher, G.; Tamura, K. MEGA7: Molecular evolutionary genetics analysis version 7.0 for bigger datasets. *Mol. Biol. Evol.* **2016**, *33*, 1870–1874. [CrossRef]
52. He, Z.; Zhang, H.; Gao, S.; Lercher, M.J.; Chen, W.H.; Hu, S. Evolview v2: An online visualization and management tool for customized and annotated phylogenetic trees. *Nucleic Acids Res.* **2016**, *44*, W236–W241. [CrossRef]
53. Bailey, T.L.; Boden, M.; Buske, F.A.; Frith, M.; Grant, C.E.; Clementi, L.; Ren, J.; Li, W.W.; Noble, W.S. MEME SUITE: Tools for motif discovery and searching. *Nucleic Acids Res.* **2009**, *37*, W202–W208. [CrossRef]
54. Chen, C.; Chen, H.; Zhang, Y.; Thomas, H.R.; Frank, M.H.; He, Y.; Xia, R. TBtools: An integrative toolkit developed for interactive analyses of big Biological Data. *Mol. Plant* **2020**, *13*, 1194–1202. [CrossRef]
55. Lescot, M.; Déhais, P.; Thijs, G.; Marchal, K.; Moreau, Y.; Van de Peer, Y.; Rouzé, P.; Rombauts, S. PlantCARE, a database of plant cis-acting regulatory elements and a portal to tools for in silico analysis of promoter sequences. *Nucleic Acids Res.* **2002**, *30*, 325–327. [CrossRef]
56. Hu, B.; Jin, J.; Guo, A.Y.; Zhang, H.; Luo, J.; Gao, G. GSDS 2.0: An upgraded gene feature visualization server. *Bioinformatics* **2015**, *31*, 1296–1297. [CrossRef]
57. Voorrips, R.E. MapChart: Software for the graphical presentation of linkage maps and QTLs. *J. Hered.* **2002**, *93*, 77–78. [CrossRef]
58. Wang, Y.; Tang, H.; Debarry, J.D.; Tan, X.; Li, J.; Wang, X.; Lee, T.H.; Jin, H.; Marler, B.; Guo, H.; et al. MCScanX: A toolkit for detection and evolutionary analysis of gene synteny and collinearity. *Nucleic Acids Res.* **2012**, *40*, e49. [CrossRef]
59. Krzywinski, M.; Schein, J.; Birol, I.; Connors, J.; Gascoyne, R.; Horsman, D.; Jones, S.J.; Marra, M.A. Circos: An information aesthetic for comparative genomics. *Genome Res.* **2009**, *19*, 1639–1645. [CrossRef]
60. Dai, X.; Zhuang, Z.; Zhao, P.X. psRNATarget: A plant small RNA target analysis server (2017 release). *Nucleic Acids Res.* **2018**, *46*, W49–W54. [CrossRef]
61. Shannon, P.; Markiel, A.; Ozier, O.; Baliga, N.S.; Wang, J.T.; Ramage, D.; Amin, N.; Schwikowski, B.; Ideker, T. Cytoscape: A software environment for integrated models of biomolecular interaction networks. *Genome Res.* **2003**, *13*, 2498–2504. [CrossRef] [PubMed]
62. Livak, K.J.; Schmittgen, T.D. Analysis of relative gene expression data using real-time quantitative PCR and the $2^{-\Delta\Delta CT}$ method. *Methods* **2001**, *25*, 402–408. [CrossRef] [PubMed]
63. Wang, P.; Yang, C.; Chen, H.; Luo, L.; Leng, Q.; Li, S.; Han, Z.; Li, X.; Song, C.; Zhang, X.; et al. Exploring transcription factors reveals crucial members and regulatory networks involved in different abiotic stresses in *Brassica napus* L. *BMC Plant Biol.* **2018**, *18*, 202. [CrossRef] [PubMed]
64. Wan, H.; Cui, Y.; Ding, Y.; Mei, J.; Dong, H.; Zhang, W.; Wu, S.; Liang, Y.; Zhang, C.; Li, J.; et al. Time-Series Analyses of Transcriptomes and proteomes reveal molecular networks underlying oil accumulation in canola. *Front. Plant Sci.* **2016**, *7*, 2007. [CrossRef]
65. Liu, D.; Yu, L.; Wei, L.; Yu, P.; Wang, J.; Zhao, H.; Zhang, Y.; Zhang, S.; Yang, Z.; Chen, G.; et al. BnTIR: An online transcriptome platform for exploring RNA-seq libraries for oil crop *Brassica napus*. *Plant Biotechnol. J.* **2021**, *19*, 1895–1897. [CrossRef]
66. Chao, H.; Li, T.; Luo, C.; Huang, H.; Ruan, Y.; Li, X.; Niu, Y.; Fan, Y.; Sun, W.; Zhang, K.; et al. BrassicaEDB: A Gene Expression Database for Brassica Crops. *Int. J. Mol. Sci.* **2020**, *21*, 5831. [CrossRef]

International Journal of *Molecular Sciences*

MDPI

Article

De Novo Transcriptome Analysis of *R. nigrum* cv. Aldoniai in Response to Blackcurrant Reversion Virus Infection

Ingrida Mažeikienė, Ana Dovilė Juškytė, Vidmantas Bendokas * and Vidmantas Stanys

Lithuanian Research Centre for Agriculture and Forestry, Institute of Horticulture, Kaunas str. 30, 54333 Babtai, Lithuania
* Correspondence: vidmantas.bendokas@lammc.lt; Tel.: +370-37-555253

Abstract: The most damaging pathogen in blackcurrant plantations is mite-transmitted blackcurrant reversion virus (BRV). Some *Ribes* species have an encoded genetic resistance to BRV. We performed RNA sequencing analysis of BRV-resistant blackcurrant cv. Aldoniai to evaluate the molecular mechanisms related to the BRV infection response. The RNA of virus-inoculated and mock-inoculated microshoots was sequenced, and the transcriptional changes at 2- and 4-days post inoculation (dpi) were analyzed. The accumulation and expression of BRV RNA1 were detected in infected plants. In total, 159,701 transcripts were obtained and 30.7% were unigenes, annotated in 7 databases. More than 25,000 differentially expressed genes (DEGs) according to FPKM were upregulated or downregulated. We observed 221 and 850 DEGs at 2 and 4 dpi, respectively, in BRV-infected microshoots related to the stress response. The proportion of upregulated DEGs at 4 dpi was about 3.5 times higher than at 2 dpi. Pathways of the virus defense response were activated, and key candidate genes were identified. The phenylpropanoid and the cutin, suberine, and wax biosynthesis pathways were activated in infected plants. Our comparative de novo analysis of the *R. nigrum* transcriptome provides clues not only for understanding the molecular BRV resistance mechanisms but also for breeding BRV-tolerant genotypes.

Keywords: BRV infection; biotic stress; blackcurrant; defense response; RNA-Seq; de novo transcriptome

Citation: Mažeikienė, I.; Juškytė, A.D.; Bendokas, V.; Stanys, V. De Novo Transcriptome Analysis of *R. nigrum* cv. Aldoniai in Response to Blackcurrant Reversion Virus Infection. *Int. J. Mol. Sci.* **2022**, *23*, 9560. https://doi.org/10.3390/ijms23179560

Academic Editors: Andrés J. Cortés and Hai Du

Received: 28 July 2022
Accepted: 21 August 2022
Published: 24 August 2022

1. Introduction

Several viruses can cause diseases in *Ribes* spp. Plant infections by viruses result in a range of symptoms, damage host plants, and adversely affect the economic value of blackcurrant plantations [1]. Blackcurrant reversion virus (BRV) is specific for currants and is the most damaging virus in *Ribes* spp. plantations worldwide. It is transmitted by the biological vector *Cecidophyopsis* spp. (Acari: *Eriophyidae*) in a persistent manner and mechanically through agrotechnical means and grafting. Atypical to other nepoviruses, BRV is not transmitted through seeds or pollen [2]. This viruscauses blackcurrant reversion disease (BRD) in European (E) or Russian (R) forms with several morphological distortions; the worst case is complete yield loss due to plants' infertility [1,3,4].

BRV transmission that uses mechanical inoculation with sap or purified virions is complicated. The new method of transmission of BRV from BRV-positive *Ribes* to BRV-negative *Ribes* in vitro through root soaking in the inoculum was created earlier by [5].

BRV is a small (approximately 29 nm in diameter) icosahedral *Nepovirus* [6]. BRD etiology has been studied very sparsely; however, the positive-sense single-stranded RNAs of BRV have been characterized [7–9]. BRV is bipartite and has three isometric virion components: M with RNA1 (7711 bp), B with RNA2 (4600 bp), and T with satRNA (1432 bp), with all of them containing a single open reading frame (ORF). RNA2 encodes 55 or 54 kDa coat protein (CP). BRV particles with 54 kDa CP are suitable for mechanical inoculation through leaf abrasion [6].

Ribes nigrum cv. Aldoniai is a new cultivar that was created at the LAMMC in 2017. Aldoniai is distinguished as a high-yielding, self-pollinating, gall-mite-resistant and BRV-resistant cultivar [10]. The cultivar combines various blackcurrant genotypes and is a descendant of *R. dikuscha* Fisch., *R. nigrum* spp. *sibiricum* Pavl., and *R. nigrum* spp. *europaeum* Pavl. Resistance to BRV in cv. Aldoniai was inherited from *R. dikuscha*. Additionally, this cultivar has pyramidic resistance to gall mite and BRV according to previous studies involving molecular markers [11].

To survive, plants have evolved multiple sophisticated and complex regulatory mechanisms to defend themselves against virus infection, including gene silencing pathways, hormone-mediated signaling pathways, and metabolism regulation [12]. Studies of the molecular basis of BRV resistance have focused mainly on qualitative resistance and some molecular markers related to resistance to BRV were suggested [13,14]. Although resistance to BRV in *Ribes* spp. appears to be a quantitative trait controlled by a dominant gene [15,16], the genetic control of resistance is complex and remains unknown.

Hence, it is necessary to attain a comprehensive understanding of the molecular mechanisms of resistance to BRV in *Ribes* for novel blackcurrant breeding programs and acceleration of the selection process. Recently, RNA-sequencing (RNA-Seq) technology and digital gene expression analysis have provided new and rapid approaches for detecting differences in the transcriptomes of non-model plants [17–19]. Global investigation of gene expression during BRV infection will help to elucidate the mechanisms of BRV resistance in *Ribes* plants. RNA-seq, de novo transcriptome assembly, and flavonoid gene analysis of blackcurrant berries were studied [18]. Transcriptome analysis of blackcurrant's response to biotic stress has not been performed, and data analysis of genetic studies of the single genus *Ribes* from the order Saxifragales is complicated due to a lack of a possibility of comparison to genetic data of other reference genera in Eudicots (*Rosids* and *Asterids*) [18,20].

In this study, we used next-generation sequencing approaches on the BRV-resistant blackcurrant cultivar to analyze the responses to BRV infection at the transcriptomic level. We investigated the differences in gene expression between virus-infected and mock-inoculated samples at different time points following BRV infection (at 2 and 4 days after infection). Our results showed that some plant defense and virus resistance genes were differentially expressed between the infected plants and control. Our study provides insight into the molecular mechanism of blackcurrant's response/resistance to BRV infection and advances our understanding of plant–virus interactions.

2. Results

2.1. Quality Control of BioProject Data

Twelve cDNA libraries of four experiment treatments C_2, V_2, C_4, and V_4 were generated for RNA-Seq of *R. nigrum* cv. Aldoniai to study genes and pathways related to resistance to BRV. More than 305 million high-quality clean reads were produced from microshoot RNA and were used for performing de novo transcriptome assembly of *R. nigrum*. The raw data obtained by NGS in this study is available from the NCBI database (accession PRJNA797914 of BioProject). In total, 92.9 Gbases of raw reads of SRA data per BioProjec were generated (Table 1). Clean reads accounted for 9.85%, comprising 6.3 to 8.9 Gbases among samples with a Q20 percentage of 97.02–97.55% and a GC percentage of 42.61–43.60%.

More than 159,701 transcripts were assembled and a total of 48,966 non-redundant unigenes were generated. A total of 15,220 unigenes (31.08%) had a sequence length of up to 500 bp and 13,431 unigenes (27.43%) had a length of 501–1000 bp. In addition, 9266 unigenes (18.92%) had a length of 1001–2000 bp, and 11,049 unigenes (22.56%) had a sequence longer than 2000 bp. As a result, the number of transcripts assembled in this project increased as the sequence length increased; however, the percent of the assembled number of unigenes decreased despite the increasing number of transcripts for a given length in bp (Figure 1).

Table 1. Summary data of quality control (QC) of reads of the genome *R. nigrum* cv. Aldoniai.

Sample Library ID	Raw Reads	Raw Bases, G	Clean Reads	Clean Bases, G	Error Rate, %	Q20	GC_pct, %
C_2_1	28,086,156	8.4	27,637,344	8.3	0.03	97.45	43.53
C_2_2	24,977,381	7.5	24,594,501	7.4	0.03	97.51	43.32
C_2_3	24,061,978	7.2	23,627,353	7.1	0.03	97.55	43.36
V_2_1	27,826,308	8.3	27,343,527	8.2	0.03	97.52	43.57
V_2_2	24,844,092	7.5	24,476,571	7.3	0.03	97.28	43.49
V_2_3	26,612,831	8.0	26,276,865	7.9	0.03	97.41	43.47
C_4_1	23,144,738	6.9	22,864,209	6.9	0.03	97.50	43.60
C_4_2	21,330,139	6.4	21,028,990	6.3	0.03	97.18	43.27
C_4_3	30,235,731	9.1	29,812,238	8.9	0.03	97.25	43.36
V_4_1	23,810,909	7.1	23,507,940	7.1	0.03	97.21	42.93
V_4_2	28,857,614	8.7	28,387,016	8.5	0.03	97.58	42.61
V_4_3	25,987,937	7.8	25,614,722	7.7	0.03	97.02	42.80

Error rate (%)—base error rate of whole sequencing; Q20—the percentage of bases whose Q Phred values are greater than 20; GC pct (%)—the percentage of G and C base numbers of the total bases.

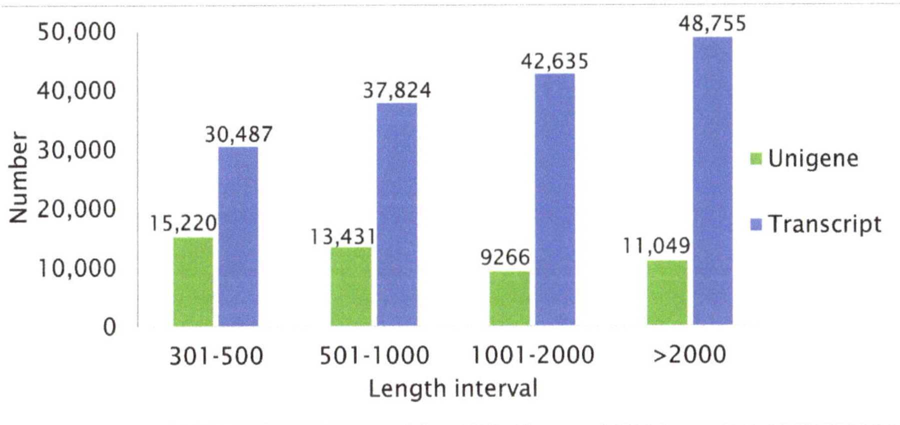

Figure 1. Length distribution of the assembled transcripts and unigenes from the transcriptome of *R. nigrum* cv. Aldoniai.

2.2. Annotation of Unigenes in R. nigrum Transcriptome

All assembled unigenes (48,966) obtained by sequencing of the *R. nigrum* transcriptome were annotated into seven databases: GO, KO, KOG, NR, NT, PFAM, and SwissProt (Figure 2A). The largest number of unigenes 24,497 (50.02%) in the de novo assembled genome were identified and functionally annotated by the NR database. The annotation data according to the databases GO, NT, PFAM, and SwissProt were similar, and 37.85%, 39.84%, 37.85%, and 37.48% of the unigenes were identified, respectively. Finally, the least number of functionally annotated genes were identified according to the data of KO and KOG, with 18.12% and 10.12%, respectively.

The highest matching percent of 23.0% in the *R. nigrum* unigene sequences was with the genes of *Vitis vinifera* according to the NR database (Figure 2B). The assembled unigenes showed a 6.4% phylogenetic identity with the genes of *Quercus suber*, and 6.3% with the genes of *Camellia sinensis*. We established a lower genetic identity between blackcurrant and *Actinidia chinensis* and *Juglans regia*, which was 3.6% in both cases. About half (47.0%) of the assembled unigenes did not match the protein sequences of other plant species. These unigenes, with lengths of 150–450 bp, were specific and unique to *R. nigrum* and have not been identified yet. Further, their protein sequences do not match with proteins from other plant species.

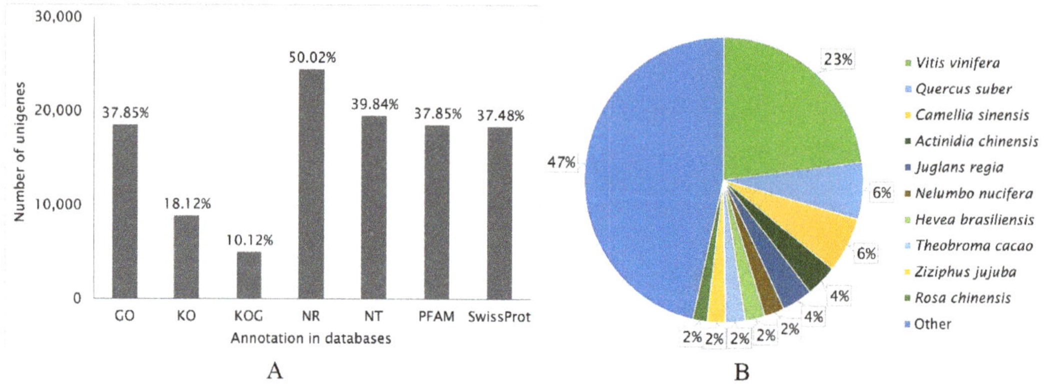

Figure 2. Number of unigenes of blackcurrant cv. Aldoniai annotated into different databases (**A**) and the identity (round value) of plant species to *R. nigrum* gene annotation according to the NR database (**B**).

The functions of *R. nigrum* unigenes were annotated using GO, KEGG, and KOG analysis data (Figure 3). Using GO analysis, 80,470 unigenes were grouped into 3 functional categories (biological process, cellular component, and molecular function) and assigned to 43 functional groups (Figure 3A). In total, 40,335 unigenes identified in the *R. nigrum* microshoot transcriptome were related to genes involved in 26 various biological processes. In total, 11,133 and 10,081 unigenes were involved in cellular and metabolic processes, respectively. In total, 4135 and 3852 unigenes were involved in biological regulation and regulation of biological process. In total, 3199 unigenes were involved in localization and 2395 unigenes were involved in response to stimulus. Fewer of the identified unigenes were involved in signaling (1279) and interspecies interaction between organisms (1162), 40 of which were associated with the response to virus. We also identified unigenes related to the multicellular organism (541), development (444), multi-organism (387), negative regulation (327), reproduction (295), and other processes, as shown in Figure 3A. In total, 17,358 unigenes were divided into 5 subcategories in the cellular component category. In total, 22,777 unigenes were involved in molecular function and they were distributed into 12 subcategories, in which unigenes from the binding (10,243) and catalytic activity (8384) subcategories dominated.

The functions of *R. nigrum* unigenes were predicted and classified by searching the KOG database (Figure 3B). In total, 4957 unigenes were classified into 25 groups composed of ancestral proteins. *R. nigrum* unigenes were assigned in all groups via KOG. Three dominant groups of unigenes in *R. nigrum* emerged according to the KOG classification data. Post-translational modification, protein turnover, and chaperones group (O) consisted of 671 unigenes (13.54%); general function prediction only group (R) included 634 unigenes (12.79%); and the translation, ribosomal structure, and biogenesis group (J) included 545 unigenes (10.99%). However, the smallest groups were extracellular structures (W) and cell motility (N), with seven and two unigenes, respectively.

All annotated unigenes (9382) were classified into five major metabolic pathways: 1033 in cell process (A), 984 in environmental information processing (B), 1824 in genetic information processing (C), 3911 in metabolism (D), and 1630 in organismal systems (E) via the KEGG database (Figure 3C). The five metabolic pathways were further divided into 33 subcategories. In total, 41.69% of the unigenes were annotated to the metabolism pathways. A maximum number of unigenes of 927 were related to signal transduction metabolism, 775 unigenes were related to translation, and 689 unigenes were related to carbohydrate metabolism.

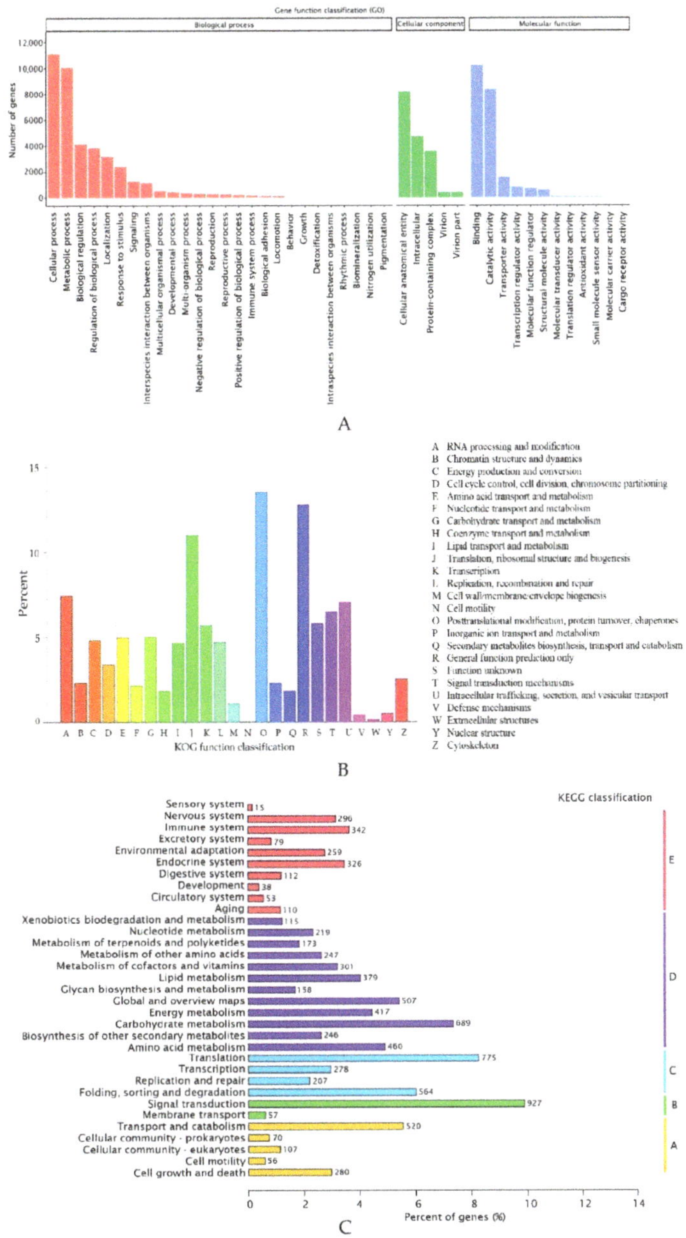

Figure 3. Characteristics of blackcurrant unigenes based on Gene Ontology (GO) categorization (**A**), Clusters of Orthologous Group (KOG) classification (**B**), and Kyoto Encyclopedia of Genes and Genomes (KEGG) pathways classification (**C**). The *x*-axis is the names of the 25 KOG group, and the *y*-axis is the percentage of genes annotated under this group of the total annotated genes. The *y*-axis is the names of the KEGG pathways, and the *x*-axis is the number of genes annotated in the pathway and the ratio between the number in this pathway and the total number of annotated genes. The KEGG metabolic pathway genes are divided into 5 branches: A: Cellular Processes, B: Environmental Information Processing, C: Genetic Information Processing, D: Metabolism, E: Organismal Systems.

2.3. *Differentially Expressed Genes (DEGs) to Virus Defense Response in R. nigrum cv. Aldoniai Microshoots*

Using the de novo assembled transcriptome as a reference, the genes expressed in the virus-inoculated and mock-inoculated (control) groups after 2 and 4 dpi were identified. In total, 48,092 genes after 2 dpi and 47,334 genes after 4 dpi were evaluated (Figure 4). The padj ≤ 0.05 and log2 ratio ≥ 1 were used to identify DEGs in the virus-inoculated samples in comparison to control plants. DEGs increased in virus-infected plants by approximately 4 times during the post-inoculation period from 2 to 4 days. In total, at 2 dpi, 221 DEGs were identified, of which 104 were downregulated and 117 were upregulated in response to BRV infection (Figure 4A and Table S1). The fold change in gene expression varied between one and six. A significant increase in DEGs (850 genes) in virus-inoculated samples was determined at 4 dpi, with 445 genes being downregulated and 405 being upregulated (Figure 4B and Table S1). The expression level remained unchanged (false) for 99.54% (47,871) of the genes at 2 dpi and 98.20% (46,484) of the genes at 4 dpi.

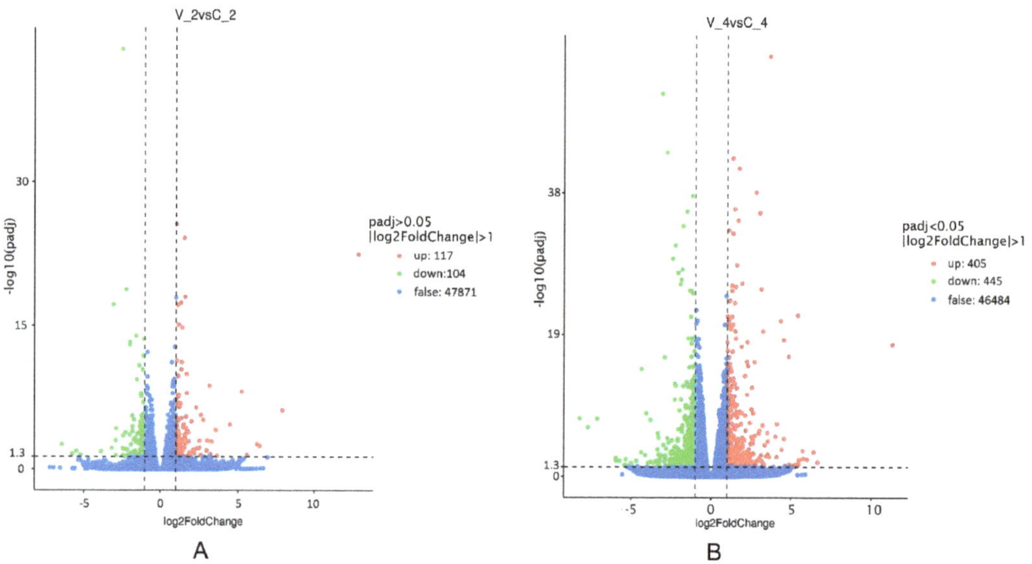

Figure 4. The number of differentially expressed genes (DEGs) in response to BRV infection at 2 (**A**) and 4 dpi (**B**) in *R. nigrum* cv. Aldoniai. The threshold is normally set as: [log2 (Fold Change)] > 1 and q-value < 0.005. The *x*-axis shows the fold change in the gene expression between different samples, and the *y*-axis shows the statistical significance of the differences. Statistically significant differences are represented by red or green dots.

Plants in this in vitro inoculation assay were infected with inoculum with three different BRV isolates [5]. The expression of viral polyprotein encoded by RNA1 (cluster-12591.29271) was detected in virus-inoculated plants. The log2FoldChange (log2FC) in the microshoots was 12.85 (q value 3.4704×10^{-23}) at 2 dpi and 11.29 (q value 2.0349×10^{-18}) at 4 dpi (Figure 4 and Table S1). Three polyproteins of BRV with genetic diversity in strand RNA2 were identified in the transcriptomes of infected blackcurrant in Trinity data (Figure S1), but clean reads were not annotated according to databases, and the expression of virus RNA2 in microshoots of blackcurrant was not observed.

The enriched gene pathways identified by GO (functional categories grouping) are presented in Figure 5. Twenty functional pathways with reliable expression of specific genes were identified in each post-inoculation period. However, the structure of enriched gene pathways was partially different between the studied periods. Specifically, for the activated pathways, the DEG characteristics are presented in the Supplementary Materials

(Table S2). Statistically significantly upregulated DEGs (gene ratio 19:65) at 2 dpi were identified in the molecular function oxidoreductase activity (Figure 5A). Meanwhile, at 4 dpi, the oxidoreductase activity function genes were downregulated at a ratio of 54:256, and the biological process response to stress was upregulated at a gene ratio of 26:204 (Figure 5B).

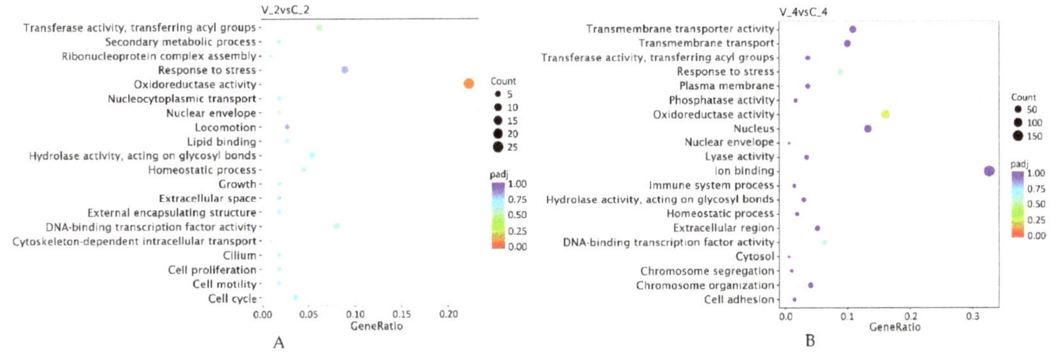

Figure 5. Enriched gene pathways identified according to GO involving differentially expressed genes (DEGs) following biotic stress (virus) inoculation at 2 (**A**) and 4 dpi (**B**) in *R. nigrum* cv. Aldoniai.

The KEGG enrichment scattered plot shows the DEGs enrichment analysis results in twenty KEGG pathways. The significant interactions of 13 genes at 2 dpi and 17 genes at 4 dpi show genes were involved in certain biological functions via KEGG analysis (Figure 6 and Table S3). Two days after BRV inoculation, unigenes in the phenylpropanoid biosynthesis were upregulated (ratio 7:23). Meanwhile, unigenes in the protein processing in the endoplasmic reticulum pathway (gene ratio 4:13) and sesquiterpenoid and triterpenoid biosynthesis (gene ratio 2:13) showed significant reductions in the expression levels. Although the plant hormone signal transduction pathway (gene ratio 3:13) had separate reliably expressed genes, the pathway itself was not statistically reliable. Four days after BRV infection, five genes with upregulation in phenylpropanoid biosynthesis were appointed; however, there was also negative expression of seven genes in this pathway. Negative signaling in the cutin, suberine, and wax biosynthesis (gene ratio 5:103) was observed in this period also.

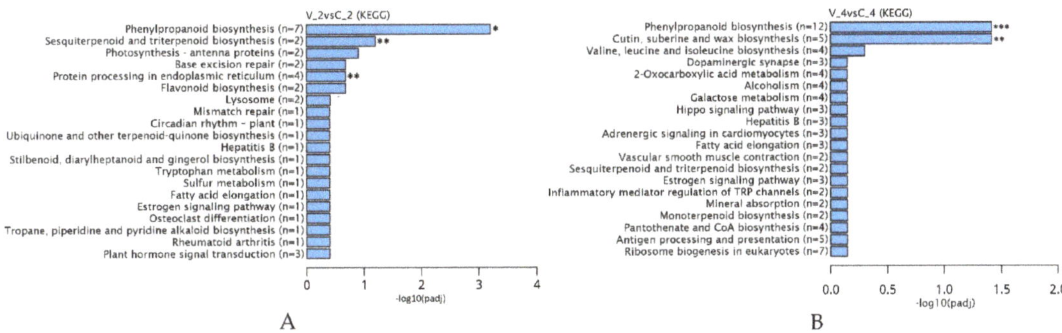

Figure 6. (**A**,**B**) Kyoto Encyclopedia of Genes and Genomes (KEGG) pathway enrichment analysis of up- and downregulated genes (n = x). All pathways were enriched using the criterion $p < 0.05$. *—significant upregulation, **—significant downregulation, ***—significant up- and downregulation of the pathways.

3. Discussion

3.1. Transcriptome of R. nigrum Microshoots

We conducted de novo assembly of transcriptomes of BRV-inoculated and mock-inoculated microshoots in BRV-resistant blackcurrant cv. Aldoniai. *R. nigrum* belongs to the Saxifragales order and *Grossulariaceae* family; molecular-genetic studies of plants from this family are very sparce. The nutritional and medicinal values of currant fruits are indisputable, and have stimulated research on the synthesis of secondary metabolites in fruits. Several genetic studies related to fruit ripening and secondary metabolite synthesis have been performed based on RNA transcripts [18,21,22]. Our main attention in this study was concentrated on the ecological aspect of plant cultivation since BRV pathogen and BRD cannot be treated and cause great economic losses. Consequently, the development of virus-resistant plants is a relevant task for breeders. Although studies of the genetically determined disease resistance of the *Ribes* genus have been carried out for a long time [3,15], a system of molecular marker usage was offered to breeders [11,14,23]. The resistance or tolerance of individual *Ribes* spp. genotypes of BRV is determined genetically, and markers are created, but the host plant–pathogen interaction at the genetic level has not been studied. De novo transcriptome assembly using NGS was performed for the first time in BRV-infected plants according to the methodology under in vitro conditions created earlier [5]; this method opened possibilities to study specific and precise host–virus interactions in *Ribes* spp.

Aldoniai is the new BRV-resistant Lithuanian cultivar [10] with the genealogy of interspecific hybrids. This cultivar was used as a model because the virus may be observed in plant material up to 6–8 days after inoculation and significant expression changes in *pathogenesis-related 1* (*PR1*) were detected at 2 and 4 dpi as stated in previous studies. This gene was used as a molecular marker to indicate blackcurrant's defense response to BRV infection [5,24]. Therefore, the transcriptomes of cv. Aldoniai microshoots (mock- and BRV-inoculated) at 2 and 4 dpi were analyzed (Table 1). Raw reads from 21,330,139 to 30,235,731 were received in the mRNA of microshoots while 168,239,621 raw reads in ripe fruits of blackcurrants cv. Ben Hope were counted [18]. The *Ribes* spp. genome lacked an available reference genome sequence; therefore, de novo assembly of the Illumina reads was carried out for cv. Aldoniai using Trinity software and 159,701 transcripts were assembled (Figure 1). This number of transcripts is higher than in cv. Ben Hope fruits (145,906 transcripts) but lower in comparison to *R. nigrum* var. *sibiricum* (186,129 transcripts) [18]. The Siberian subspecies are one of the ancestors of the cv. Aldoniai. In total, 48,966 non-redundant unigenes were generated in our study and only half of them were annotated in the NR database. Among the homologous sequences of other plant species, the matching degree with *V. vinifera* was the highest (Figure 2). Similar results were obtained while studying fruit transcriptomes and performing phylogenetic analysis of plants [18]. According to the time of plant species divergence and comparative analysis of plant transcriptomes, *Ribes* spp. (Saxifragales) diverged about 117 million years ago from the Vitales and most genes related to biological processes and molecular functions etc. are characterized by high genetic diversity [18,25]. Therefore, they are not identified in the NCBI Blast database at the nucleotide and amino acid levels. Conservative genes are better recognized in the biological cellular and metabolic processes or molecular functions binding and catalytic activity (Figure 3).

3.2. Virus Defense Response to BRV in R. nigrum

In our study, expression differences were observed in 47,871 genes. DEGs related to virus infection with a significant value of log2FC increased at 4 dpi in comparison to 2 dpi in BRV-infected blackcurrants (Figure 4).

BRV infection was approved by the presence of virus RNA1 and RNA2 in the transcriptome of inoculated samples (Figure S1). According to the data of a previous study [5], inoculum with three different BRV isolates was used in our experiment. RNA2 strand is responsible for translation of three proteins: X, movement protein (MP), and coat protein

(CP). However, the expression of polyprotein from RNA2 did not occur in virus-inoculated samples at 2 or 4 dpi. An analysis of DEGs showed that polyprotein from RNA1 of BRV were strongly upregulated in virus-infected *R. nigrum* microshoots.

RNA1 encodes five proteins: N-terminus protease co-factor, followed by nucleotide-binding protein or helicase, putative VPg, cysteine protease (CysPro), and RNA-dependent RNA polymerase [9]. Different genes of the virus are recognized as plant virus avirulence (*Avr*) factors in the viral Avr-R system [26]. Only polyprotein of BRV RNA1 showed significant expression (Figure 4 and Table S1) (log2FC 12.85 and 11.29 at 2 and 4 dpi, respectively) and could be recognized as an *Avr* gene in blackcurrant's plant–pathogen interaction pathway. We assume that the genes of BRV RNA1 can activate the chemical defense in *R. nigrum*. A different genetic response of the blackcurrants was observed due to BRV-induced stress at 2 and 4 dpi (Figure 5). A change in the reduction–oxidation status in plant cells is one of the earliest responses to biotic stress. The production of reactive oxygen species (ROS) and nitric oxide (NO) and their compounds synergistically activates the hypersensitive response (HR) in model plants [27]. In our case, genes from the oxidoreductase activity pathway (GO:0016491) were the first upregulated genes in the virus-infected samples of *R. nigrum* at 2 dpi (Figure 5, Table S2). At 4 dpi, oxidoreductase activity genes were suppressed; however, at this moment, the biological process response to stress (GO:0006950) was significantly upregulated in the microshoots. We propose that at 4 dpi, the microshoots of cv. Aldoniai reacted to artificially induced stress.

Some plant compound elicitors (benzoic acid, chitosan, salicylic acid, etc.) that activate the defense in plants were studied [26]. Synthesis of the signal-mediating phytohormones: salicylic acid (SA) (at 2 dpi), jasmonic acid (JA), and ethylene (at 4 dpi) was observed in the plant hormone signal transduction pathway (ko:04075) in BRV-infected plants (Figures S2 and S3). These phytohormones stimulate plant responses to different biotic and abiotic stresses, leading to activation of systemic acquired resistance (SAR) and induced systemic resistance (ISR) [28,29]. It was established that both SAR and ISR were activated in BRV-infected microshoots.

Various biosynthetic pathways can be activated in plants depending on the elicitor compound stimulation [30]. Several of the upregulated genes from the phenolic compound biosynthesis pathway identified in BRV-infected microshoots were related to the activation of different defense-related enzymes (peroxidases) (Table S2). The phenylpropanoid biosynthesis pathway (ko:00940) was significantly upregulated at both 2 and 4 dpi in our experiment (Figure 6). The synthesis of some compounds, including coumarine, cinnamaldehyde, caffeoyl quinic acid, and lignin, was increased at 2 dpi (Figure S4). The synthesis of ferulic acid, sinapis acid, caffeoyl-aldehyde, caffeoyl-alcohol, 5-Hydroxy-coniferaldehyde, and 5-Hydroxy-coniferyl alcohol was also increased at 4 dpi while the synthesis of cinnamic acid was suppressed. Some lignin synthesis genes in *R. nigrum*-infected samples were up- and downregulated at 4 dpi (Figure S5). Consequently, the responses to virus infection in blackcurrants appeared through the synthesis of phenylpropanoids, which also play important roles in the response to biotic stress in other plants [31].

Cuticular wax biosynthesis and its roles in plant disease resistance are well established [32]. Negative signaling of the genes in the cutin, suberine, and wax biosynthesis pathway (ko:00073) in virus-inoculated plants was observed at 4 dpi (Figure S6). Suppression of the synthesis of fatty acids in the plant defense system must be related to the production of signaling molecules JA (oxylipin) for defense regulation, with remodeling of the membrane lipid composition and defense signaling [33]. Long-chain fatty acid synthesis suppression has a negative effect on the biosynthesis of the plant cuticle and the generation of bioactive molecules, including sphingolipids. Therefore, BRV infection leads to the suppression of the ko:00073 pathway, thereby reducing the plant's resistance, and thus facilitating the entry of other pathogens such as bacteria and fungi into the plant cells.

In this study, we reported a transcriptome analysis that includes data on blackcurrant's response to an economically important and very harmful pathogen: BRV. Such data is necessary for gene identification and functional analysis to help improve gene resources.

These resources will enable investigations of the stress response, defense response, and metabolic processes in plants beyond the *R. nigrum*–BRV interaction pathways analyzed in this study. Our results clarified that blackcurrant cv. Aldoniai (resistant to BRV interspecific hybrid) has a complex of genetically determined defense mechanisms aimed at BRV infection, which were significantly activated 2 and 4 days after viral infection.

4. Materials and Methods

4.1. RNA Isolation, cDNA Library Preparation, and Sequencing

Plants of cv. Aldoniai were inoculated with BRV under in vitro conditions at the Department of Orchard Plant Genetics and Biotechnology of the Institute of Horticulture, LAMMC according to the methodology described by [5]. Control samples of mock- (C) and virus-inoculated (V) microshoots were collected at 2- and 4-days post inoculation (dpi) (sample library ID C_2 and V_2, C_4 and V_4) (Table 1). Three individual microshoots from each treatment were collected to provide biological replicates. Samples were immediately frozen in liquid nitrogen and stored until RNA extraction with a GeneJET Plant RNA Purification Mini Kit (Thermo Scientific, Vilnius, Lithuania) according to the manufacturer's instructions. The concentration and quality of the RNA were measured with an Implen GmbH spectrophotometer (Implen, Munich, Germany).

4.2. De Novo Transcriptome Analysis

Qualified RNA, with 4 μg per sample, was sent to Novogene (Cambridge, UK) for mRNA library preparation (poly-A enrichment). The next-generation sequencing (NGS) was conducted on an Illumina 6000 NovaSeq PE150 (6 Gb raw data per sample) platform. In total, the data of 36 replicates (3 biological replicates per treatment and 3 techniques) were used in the de novo transcriptome analysis of *R. nigrum* cv. Aldoniai.

4.3. Functional Annotation of Unigenes

Low-quality raw reads were filtered as follows: reads containing adaptors were removed, N > 10% and with Q score of over 50% bases below 5 (Table 1). The de novo transcriptome (absence of a reference genome) was reconstructed using Trinity [34]. Assembled transcripts were clustered with the command-line software program Corset [35], and redundancy was removed. The longest transcripts of each cluster were selected as unigenes.

4.4. Gene Functional Annotation

Functional annotation of all assembled unigenes was performed using seven public databases: Non-Redundant Protein Sequence Database (NR), Nucleotide Sequence Database (NT), Gene Ontology (GO), Swiss-Prot, Kyoto Encyclopedia of Genes and Genomes (KEGG), Pfam, and Clusters of Orthologous Groups for Eukaryotic Complete Genomes (KOG). The genes successfully annotated in GO were grouped into three main domains: Biological Process (BP), Cellular Component (CC), and Molecular Function (MF). Genes annotated in KOG were divided into functional groups. The genes successfully annotated in KEGG were classified according to the KEGG pathway they joined in.

4.5. Gene Expression Analysis

To compare the gene expression levels under different conditions, the FPKM (Fragments Per Kilobase of transcript per Million mapped reads) values were used. For biological replicates, the final FPKM was the mean value. The FPKM values were transformed to log2FoldChange (log2FC); log2FC ≥ 1 was considered to have significant expression.

Supplementary Materials: The supporting information can be downloaded at: https://www.mdpi.com/article/10.3390/ijms23179560/s1.

Author Contributions: Conceptualization, I.M. and A.D.J.; methodology, I.M., A.D.J. and V.S.; software, I.M. and V.B.; investigation, A.D.J. and I.M.; resources, V.S.; data curation, I.M.; writing—original draft preparation, I.M. and V.B.; writing—review and editing, V.S. and A.D.J.; visualization, I.M. All authors have read and agreed to the published version of the manuscript.

Funding: This research received no external funding.

Institutional Review Board Statement: Not applicable.

Informed Consent Statement: Not applicable.

Data Availability Statement: This study did not report any data.

Conflicts of Interest: The authors declare no conflict of interest.

References

1. Šutic, D.D.; Ford, R.E.; Tošic, M.T. Virus diseases of small fruits. In *Handbook of Plant Virus Diseases*; Šutic, D.D., Ford, R.E., Tošic, M.T., Eds.; CRC Press: Boca Raton, FL, USA, 1999; pp. 433–475.
2. Dolan, A.; MacFarlane, S.A.; McGavin, W.J.; Brennan, R.M.; McNicol, J.W. Blackcurrant reversion virus: Validation of an improved diagnostic test, accelerating testing in breeding and certification of blackcurrants. *J. Berry Res.* **2011**, *1*, 201–208. [CrossRef]
3. Adams, A.N.; Thresh, J.M. Reversion of black currant. In *Virus Diseases of Small Fruits*; Converse, R.H., Ed.; United States Department of Agriculture: Washington, DC, USA, 1987; pp. 133–136.
4. Susi, P. Black currant reversion virus, a mite-transmitted nepovirus. *Mol. Plant Pathol.* **2004**, *5*, 167–173. [CrossRef] [PubMed]
5. Juškytė, A.D.; Mažeikienė, I.; Stanys, V. An effective method of *Ribes* spp. inoculation with blackcurrant reversion virus under in vitro conditions. *Plants* **2022**, *11*, 1635. [CrossRef] [PubMed]
6. Seitsonen, J.J.T.; Susi, P.; Lemmetty, A.; Butcher, S.J. Structure of the mite-transmitted *Blackcurrant reversion nepovirus* using electron cryo-microscopy. *Virology* **2008**, *378*, 162–168. [CrossRef]
7. Latvala-Kilby, S.; Lehto, K. The complete nucleotide sequence of RNA2 of blackcurrant reversion nepovirus. *Virus Res.* **1999**, *65*, 87–92. [CrossRef]
8. Pacot-Hiriart, C.; Latvala-Kilby, S.; Lehto, K. Nucleotide sequence of black currant reversion associated nepovirus RNA1. *Virus Res.* **2001**, *79*, 145–152. [CrossRef]
9. Moročko-Bičevska, I.; Stalažs, A.; Lācis, G.; Laugale, V.; Baļķe, I.; Zuļģe, N.; Strautiņa, S. *Cecidophyopsis* mites and blackcurrant reversion virus on *Ribes* hosts: Current scientific progress and knowledge gaps. *Ann. Appl. Biol.* **2021**, *180*, 26–43. [CrossRef]
10. Mažeikienė, I.; Stanys, V.; Juškytė, A.D.; Sasnauskas, A.; Šikšnianas, T. Juodojo serbento veislės 'Aldoniai' ir 'Didikai'. *Sodininkystė Daržininkystė* **2017**, *36*, 3–14.
11. Mazeikiene, I.; Juskyte, A.D.; Stanys, V. Application of marker-assisted selection for resistance to gall mite and Blackcurrant reversion virus in *Ribes* genus. *Zemdirbyste* **2019**, *106*, 359–366. [CrossRef]
12. Ding, S.W. RNA-based antiviral immunity. *Nat. Rev. Immunol.* **2010**, *10*, 632–644. [CrossRef]
13. Brennan, R.; Jorgensen, L.; Hackett, C.; Woodhead, M.; Gordon, S.; Russell, J. The development of a genetic linkage map of blackcurrant (*Ribes nigrum* L.) and the identification of regions associated with key fruit quality and agronomic traits. *Euphytica* **2008**, *161*, 19–34. [CrossRef]
14. Brennan, R.M.; Jorgensen, L.; Gordon, S.; Loades, K.; Hackett, C.; Russell, J. The development of a PCR-based marker linked to resistance to the blackcurrant gall mite (*Cecidophyopsis ribis* Acari: *Eriophyidae*). *Theor. Appl. Genet.* **2009**, *118*, 205–211. [CrossRef] [PubMed]
15. Anderson, M.M. Resistance to gall mite (*Phytoptus ribis* Nal.) in the Eucoreosma section of *Ribes*. *Euphytica* **1971**, *20*, 422–426. [CrossRef]
16. Keep, E.; Knight, V.H.; Parker, J.H. Progress in the integration of characters in gall mite resistant black currants. *J. Hortic. Sci.* **1982**, *57*, 189–196. [CrossRef]
17. Deng, S.; Ma, J.; Zhang, L.; Chen, F.; Sang, Z.; Jia, Z.; Ma, L. De novo transcriptome sequencing and gene expression profiling of *Magnolia wufengensis* in response to cold stress. *BMC Plant Biol.* **2019**, *19*, 321. [CrossRef] [PubMed]
18. Thole, V.; Bassard, J.E.; Ramírez-González, R.; Trick, M.; Ghasemi Afshar, B.; Breitel, D.; Hill, L.; Foito, A.; Shepherd, L.; Freitag, S.; et al. RNA-seq, de novo transcriptome assembly and flavonoid gene analysis in 13 wild and cultivated berry fruit species with high content of phenolics. *BMC Genom.* **2019**, *20*, 995. [CrossRef] [PubMed]
19. He, D.; Zhang, J.; Zhang, X.; He, S.; Xie, D.; Liu, Y.; Li, C.; Wang, Z.; Liu, Y. Development of SSR markers in *Paeonia* based on de novo transcriptomic assemblies. *PLoS ONE* **2020**, *15*, e0227794. [CrossRef]
20. Simpson, M.G. Diversity and classification of flowering plants: Eudicots. In *Plant Systematics*, 3rd ed.; Simpson, M.G., Ed.; MA Academic Press: Burlington, NJ, USA, 2019; pp. 275–448. ISBN 978-0-12-812628-8.
21. Jarret, D.A.; Morris, J.; Cullen, D.W.; Gordon, S.L.; Verrall, S.R.; Milne, L.; Hedley, P.E.; Allwood, J.W.; Brennan, R.M.; Hancock, R.D. A transcript and metabolite atlas of blackcurrant fruit development highlights hormonal regulation and reveals the role of key transcription factors. *Front. Plant Sci.* **2018**, *9*, 1235. [CrossRef]

22. Starkevič, P.; Ražanskienė, A.; Starkevič, U.; Kazanavičiūtė, V.; Denkovskienė, E.; Bendokas, V.; Šikšnianas, T.; Rugienius, R.; Stanys, V.; Ražanskas, R. Isolation and analysis of anthocyanin pathway genes from *Ribes* genus reveals MYB gene with potent anthocyanin-inducing capabilities. *Plants* **2020**, *9*, 1078. [CrossRef]
23. Mazeikiene, I.; Bendokas, V.; Stanys, V.; Siksnianas, T. Molecular markers linked to resistance to the gall mite in blackcurrant. *Plant Breed.* **2012**, *131*, 762–766. [CrossRef]
24. Juškytė, A.D.; Mažeikienė, I.; Stanys, V. Putative genes of pathogenesis-related proteins and coronatine-insensitive protein 1 in *Ribes* spp. *Plants* **2022**, *11*, 355. [CrossRef] [PubMed]
25. One Thousand Plant Transcriptomes Initiative. One thousand plant transcriptomes and the phylogenomics of green plants. *Nature* **2019**, *574*, 679–685. [CrossRef] [PubMed]
26. Huang, C. From player to pawn: Viral avirulence factors involved in plant immunity. *Viruses* **2021**, *13*, 688. [CrossRef]
27. Matika, D.E.F.; Loake, G.J. Redox regulation in plant immune function. *Antioxid. Redox Signal.* **2014**, *21*, 1373–1388. [CrossRef] [PubMed]
28. Alazem, M.; Lin, N.S. Roles of plant hormones in the regulation of host-virus interactions. *Mol. Plant Pathol.* **2015**, *16*, 529–540. [CrossRef] [PubMed]
29. Kamle, M.; Borah, R.; Bora, H.; Jaiswal, A.K.; Singh, R.K.; Kumar, P. Systemic acquired resistance (SAR) and induced systemic resistance (ISR): Role and mechanism of action against phytopathogens. In *Fungal Biotechnology and Bioengineering*; Hesham, A.L., Upadhyay, R., Sharma, G., Manoharachary, C., Gupta, V., Eds.; Springer: Cham, Switzerland, 2020; pp. 457–470. ISBN 978-3-030-41870-0. [CrossRef]
30. Thakur, M.; Sohal, B.S. Role of elicitors in inducing resistance in plants against pathogen infection: A review. *Int. Sch. Res. Not.* **2013**, *2013*, 762412. [CrossRef] [PubMed]
31. Vogt, T. Phenylpropanoid biosynthesis. *Mol. Plant* **2010**, *3*, 2–20. [CrossRef]
32. Wang, X.; Kong, L.; Zhi, P.; Chang, C. Update on cuticular wax biosynthesis and its roles in plant disease resistance. *Int. J. Mol. Sci.* **2020**, *21*, 5514. [CrossRef]
33. Raffaele, S.; Leger, A.; Roby, D. Very long chain fatty acid and lipid signaling in the response of plants to pathogens. *Plant Signal. Behav.* **2009**, *4*, 94–99. [CrossRef]
34. Grabherr, M.G.; Haas, B.J.; Yassour, M.; Levin, J.Z.; Thompson, D.A.; Amit, I.; Adiconis, X.; Fan, L.; Raychowdhury, R.; Zeng, Q.; et al. Full-length transcriptome assembly from RNA-Seq data without a reference genome. *Nat. Biotechnol.* **2011**, *29*, 644–652. [CrossRef]
35. Davidson, N.M.; Oshlack, A. Corset: Enabling differential gene expression analysis for *de novo* assembled transcriptomes. *Genome Biol.* **2014**, *15*, 410. Available online: http://genomebiology.com/2014/15/8/410 (accessed on 5 July 2022). [PubMed]

International Journal of
Molecular Sciences

Article

Integrated Transcriptome and Proteome Analysis Provides Insight into the Ribosome Inactivating Proteins in *Plukenetia volubilis* Seeds

Guo Liu *, Zhihua Wu, Yan Peng, Xiuhua Shang and Liqiong Gao

Research Institute of Fast-Growing Trees, Chinese Academy of Forestry, 30 Mid Renmin Avenue, Zhanjiang 524022, China
* Correspondence: cercliug@caf.ac.cn

Abstract: *Plukenetia volubilis* is a highly promising plant with high nutritional and economic values. In our previous studies, the expression levels of ricin encoded transcripts were the highest in the maturation stage of *P. volubilis* seeds. The present study investigated the transcriptome and proteome profiles of seeds at two developmental stages (Pv-1 and Pv-2) using RNA-Seq and iTRAQ technologies. A total of 53,224 unigenes and 6026 proteins were identified, with functional enrichment analyses, including GO, KEGG, and KOG annotations. At two development stages of *P. volubilis* seeds, 8815 unique differentially expressed genes (DEGs) and 4983 unique differentially abundant proteins (DAPs) were identified. Omics-based association analysis showed that ribosome-inactivating protein (RIP) transcripts had the highest expression and abundance levels in Pv-2, and those DEGs/DAPs of RIPs in the GO category were involved in hydrolase activity. Furthermore, 21 RIP genes and their corresponding amino acid sequences were obtained from libraries produced with transcriptome analysis. The analysis of physicochemical properties showed that 21 RIPs of *P. volubilis* contained ricin, the ricin_B_lectin domain, or RIP domains and could be divided into three subfamilies, with the largest number for type II RIPs. The expression patterns of 10 RIP genes indicated that they were mostly highly expressed in Pv-2 and 4 transcripts encoding ricin_B_like lectins had very low expression levels during the seed development of *P. volubilis*. This finding would represent valuable evidence for the safety of oil production from *P. volubilis* for human consumption. It is also notable that the expression level of the Unigene0030485 encoding type I RIP was the highest in roots, which would be related to the antiviral activity of RIPs. This study provides a comprehensive analysis of the physicochemical properties and expression patterns of RIPs in different organs of *P. volubilis* and lays a theoretical foundation for further research and utilization of RIPs in *P. volubilis*.

Keywords: *Plukenetia volubilis*; transcriptome; proteomics; ribosome-inactivating proteins; physicochemical property; expression pattern

Citation: Liu, G.; Wu, Z.; Peng, Y.; Shang, X.; Gao, L. Integrated Transcriptome and Proteome Analysis Provides Insight into the Ribosome Inactivating Proteins in *Plukenetia volubilis* Seeds. *Int. J. Mol. Sci.* **2022**, *23*, 9562. https://doi.org/10.3390/ijms23179562

Academic Editors: Andrés J. Cortés and Hai Du

Received: 25 July 2022
Accepted: 21 August 2022
Published: 24 August 2022

1. Introduction

Ribosome-inactivating proteins (RIPs) are toxic N-glycosidase enzymes found in most plant species and distributed across multiple organs, with the function of defense against fungal or viral infections [1,2]. Ricin toxin (RT), belonging to the RIP family of toxins, is a protein with highly toxic properties, present in abundance in the seeds of castor bean (*Ricinus communis*) [3–5]. The members of the RIP family are classified as type I RIPs, which include a single polypeptide chain (~30 kDa) with RNA-*N*-glycosidase enzyme activity, and type II RIPs that include a lectin domain (ricin toxin B_chain, RTB) and a glycosidase domain (ricin toxin A_chain, RTA) [6]. These two chains are connected by a single interchain disulfide bond [7]. Pokeweed antiviral protein (PAP) and trichosanthin (TCS) belong to the subfamily of type I RIP, but abrin and RT are typical type II RIPs [8]. Ricin has attracted interest mainly due to its cytotoxicity to mammalian cells and potential applications as biological weapons, which makes it to be considered a moderate threat by the US Center for Disease Control and Prevention (CDC) [6]; moreover, ricin also can be used as a research tool to study its intracellular transport and therapeutic effect on tumors [9,10] and AIDS (acquired immune deficiency syndrome) [10,11]. These applications are attributed to the toxicity of RT, which is a thousand times higher than that of arsenic and 2–3 times higher than that of cobra venom. Ricin is a protein with lectin domains, exhibiting hemagglutination activity [12].

Plukenetia volubilis Linneo (Euphorbiaceae) is an underutilized oilseed crop with high nutritional values, native to the Amazon basin of South America, and has been traditionally utilized by indigenous Incas since pre-Hispanic times [13,14]. The mature seeds contain approximately 35.0~60.0% lipids, of which more than 90.0% are polyunsaturated fatty acids (PUFAs). The PUFAs in the seeds of *P. volubilis* comprised approximately 35.2~50.8% α-linolenic acid (C18:3n-3, ω-3, ALA) and approximately 33.4~41.0% linoleic acid (C18:2n-6, ω-6, LA), which are the two fatty acids essential for humans and must be obtained from the diet [15]. The seeds also contain approximately 26.6~31.6% protein and have antioxidant properties due to the presence of phenols, carotenoids, and tocopherols [16]. Due to the excellent nutritional composition in its seeds, especially lipid, *P. volubils* has gained increasing popularity and awareness in the global markets in recent years [17]. Although cold-pressed edible oil from *P. volubilis* seeds has been shown to not contain toxins or harmful substances for health and thus is safe for human consumption, the roasted seeds from *P. volubilis* are not approved by the European Union (EU) due to a lack of relevant knowledge about the composition and content of alkaloids in these seeds [18]. Sruchamnong et al. [19] found that the fresh leaves and seeds of *P. volubilis* contained alkaloids, saponins, and possibly lectins, the main groups of naturally occurring plant toxins, which belong to the main classes of secondary metabolites and can be found in many parts of plants, including seeds, bark, leaves, and roots. These phytotoxins might become unstable under heat treatment, suggesting that the long-term consumption of large quantities of fresh seeds and leaves of *P. volubilis* should be avoided [18,19]. In addition, Liu et al. [20] and Wang et al. [21] performed the transcriptome analysis of five and two developmental stages of *P. volubilis* seeds, respectively. Surprisingly, the amounts of RIPs transcripts were the highest at the maturity stage or the fast oil accumulation stage, and the expression levels of those transcripts were greater than 10^6. Specifically, in the seeds of *R. communis* seeds, 10 RIP genes, including 4 type I and 6 type II, also showed the highest expression levels in stages approaching maturity [22]. Therefore, knowledge of the physicochemical properties and expression patterns of RIP genes in *P. volubilis* is important in attempts to ensure a sustained supply of polyunsaturated fatty acids (PUFAs) without posing any threat to human health.

Considering the current global challenges, it is absolutely imperative to ensure food security, mitigate climate change, and alleviate malnutrition. Therefore, underutilized crops may help improve agricultural resilience, eliminate the need for external inputs, build climate resilience, facilitate dietary diversification, and improve income opportunities for farmers [14]. *P. volubilis* is a promising new crop in this regard with great potential for further domestication, and their seed oil has an excellent composition and good sensory acceptability [16]. The planting processes and requirements of *P. volubilis* are relatively simple, so it is a well-established plant that has numerous potential applications in gastronomy, cosmetics, and medicine.

According to the dynamic change of the fatty acids in the development of *P. volubilis* seeds [23] and the expression trend of RIP genes in the transcriptome of five and two developmental stages [20,21], a combination of transcriptomic and proteomics were used to identify differentially expressed genes (DEGs) and differentially abundant proteins (DAPs) in two developmental stages of *P. volubilis* seeds in the present study. Subsequently, RIP genes, physicochemical properties of ribosome-inactivating proteins, and expression patterns of RIP genes in *P. volubilies* were explored. The construction of transcriptional regulatory networks would help understand the gene functions. These results will lay a theoretical foundation for further utilization of RIPs in *P. volubilis*.

2. Results

2.1. Sequencing, Assembly, and Annotation

To determine changes in gene expression across the transcriptome at two developmental stages of *P. volubilis* seeds, RNA sequencing analysis was performed using RNA samples including Pv-1 (10 days after pollination (DAP)) and Pv-2 (100 DAP). A total of 227.63 million reads were generated from the two libraries (an average of 37.94 million), encompassing 34.14 Gb of sequence data (an average of 5.69 Gb). After stringent quality assessment and data filtering, a total of 227.48 million high-quality reads (an average of 37.91 million) were selected for further analysis. The Q20 percentage reached more than 98.29%, and the GC content ranged from 44.36% to 45.71%. The de novo transcriptome assembly was carried out using the Trinity software (2.8.6). As a result, 54.25 M bases were assembled, and 53,224 unigenes were obtained, with an average length of 1019 bp and an N50 of 1834 bp. Results of the BUSCO analysis showed that by the assembly, 91.60% of the conserved single-copy orthologous genes, including 77.01% complete (C) and 14.58% fragment (F) genes, were retrieved. The high percentage of completeness (C) and low percentages of both fragments (F) and missing sequences (M) (8.40%) indicated that the transcriptome assembly had a good representation of the transcriptome of *P. volubilis* seeds.

Using ESTScan 2.0b software with a cut-off E-value of 1×10^{-5}, a total of 40,172 unigenes were annotated (Figure S1a), of which 38,331 (72.02%), 20,497 (51.02%), 30,859 (57.98%), 24,366 (45.78%), and 36,410 (68.41%) unigenes showed significant similarities to known genes in the NR, GO, SwissProt, KOG, and KEGG databases, respectively. Approximately 34.71% (13,942) of unigenes could be assigned to a homolog in all five databases (Figure S1a). As shown in Figure S1b, a large number of unigenes in *P. volubilis* showed close identities to the genes in other plant species. The highest number of homologous genes in *P. volubilis* (13,982 unigenes, 36.48% of the total) was identified in *R. communis*, followed by *Brassica napus* (3898, 10.17%), and *Jatropha curcas* (3130, 8.17%).

According to the gene ontology (GO) annotation (Figure S2), the sub-categories metabolic process (GO:0008152, 13,208 unigenes, 64.44%) and cellular process (GO:0009987, 11,466 unigenes, 55.94%) were most prominently represented in the category of biological process (BP). For the category of molecular function (MF), catalytic activity (GO:0003824, 10,762 unigenes, 52.51%) and binding (GO:0005488, 9075 unigenes, 44.27%) were the most highly represented GO terms. For the cellular component (CC) category, however, the most frequently represented terms were cell (GO:0005623, 8939 unigenes, 43.61%) and cell part (GO:0044464, 8935 unigenes, 43.60%). The KOG annotation results (Figure S3) showed that the largest number of unigenes was clustered into the functional category of "general function prediction only" (5010 unigenes, 20.56%), followed by "post-translational modification", "protein turnover", "chaperones" (3273 unigenes, 13.43%), and "signal transduction mechanisms" (2971 unigenes, 12.19%). Around 1033 (4.24%) unigenes were assigned to the cluster of "lipid transport and metabolism". Based on the KEGG pathway assignment, 30,859 unigenes were assigned to a total of 133 KEGG pathways (Table S1), with the majority of these unigenes involved in the pathways of metabolic pathways (ko01100, 4404 unigenes, 12.10%) and biosynthesis of secondary metabolites (ko01110, 2461 unigenes, 6.76%). Among the 14 KEGG pathways associated with lipid metabolism, 161 unigenes were mapped to the pathways of the glycerolipid metabolism (ko00561), followed by glycerophospholipid metabolism (ko00564; 159 unigenes) and fatty acid degradation (ko00071; 132 unigenes).

2.2. Enrichment Analysis of DEGs Based on GO and KEGG Pathways

We identified mRNAs with a fold change $\geq$2 and a false discovery rate (FDR) < 0.05 as significant DEGs. In total, a set of 8815 DEGs (2188 upregulated and 6627 downregulated) showed $\geq$2-fold changes or different gene expression levels between the two libraries. Of 18 DEGs with RPKM values > 1000 in the Pv-1 (Table S2), 3 DEGs encode a proline-rich protein that is a structural constituent of cell wall proteins (Unigene0006989, Unigene0007377, and Unigene0009962), and 2 DEGs encode chlorophyll a/b-binding protein that may be involved in pigment biosynthesis or the assembly of thylakoid membranes (Unigene0000529 and Unigene0002244). It is noteworthy that among 27 DEGs with RPKM values > 1000 in the Pv-2 (Table S3), the 4 most abundant transcripts (Unigene0000208, Unigene0000207, Unigene0000209, and Unigene0000172) and Unigene0012386 code for ricin-like protein and the ribosome-inactivating protein gelonin, respectively, indicating the toxicity of raw seeds of *P. volubilis*. Transcripts encoding 2S albumin-like (Unigene0003223, Unigene0049069, Unigene0046600, Unigene0011183, and Unigene0048736), legumin-like (Unigene0048787, Unigene0010858, and Unigene0010856) and vicilin-like proteins (Unigene0049803 and Unigene0044374) were also abundant; these proteins are specifically synthesized during seed maturation. Other abundant transcripts of the *Oleosin* genes (Unigene0044871, Unigene0028565, and Unigene0050906) are highly abundant in land plants and could positively regulate total seed oil accumulation [24,25].

GO enrichment analysis was carried out, and based on primary biological functions, DEGs were classified into three main categories, including BP (1476 terms; 3987 unigenes), MF (224 terms; 3878 unigenes), and CC (574 terms; 2336 unigenes) (Table S4). Among the top 20 GO enrichment terms (Figure 1a), catalytic activity (GO:0003824) contained the largest number of unigenes (2861; 606/2255, up-/downregulated, respectively), followed by membrane part (GO:0044425; 766; 183/583, up-/downregulated, respectively), and intrinsic component of membrane (GO:0031224; 722; 170/552, up-/downregulated, respectively). In particular, three unigenes (Unigene0000207, Unigene0000208, and Unigene0000209) encode a ricin-like protein and one unigene encodes the ribosome-inactivating protein gelonin (Unigene0012388) with extremely high expression levels at the Pv-2 stage were enriched in hydrolase activity (GO:0016787).

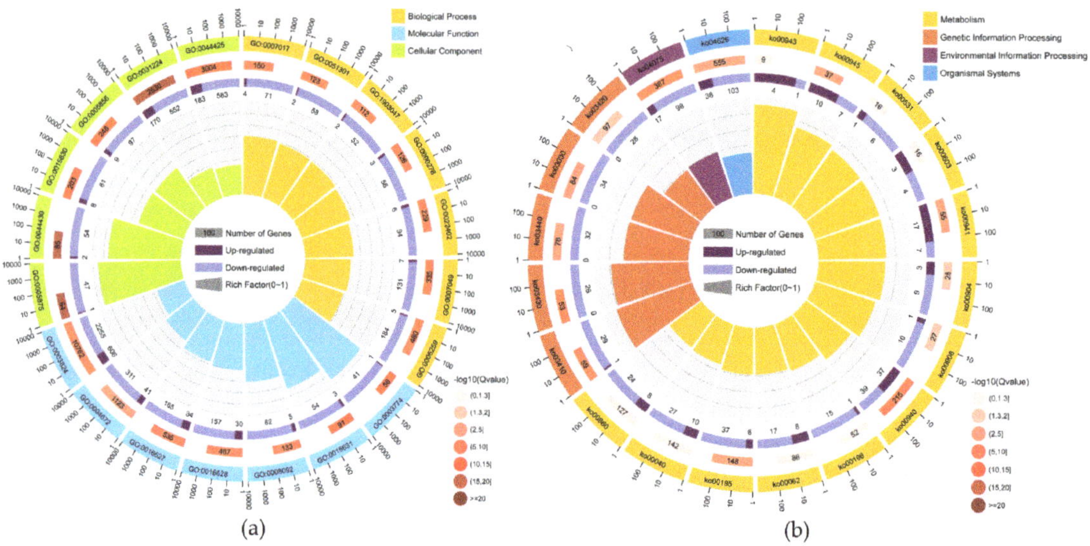

Figure 1. Top 20 of GO function (**a**) and KEGG pathway (**b**) enrichment of DEGs in the transcriptome of two developmental stages of *P. volubilis* seeds. The first circle: top 20 GO terms/KEGG pathways. The number outside of the circle is the coordinate ruler of the gene number. Different colors represent different GO terms/KEGG A classes; the second circle: the number and Q value of the GO term/KEGG pathway in the background gene. The more genes, the longer the bars, and the smaller the *Q*-value, the redder the color; the *Q*-value is the *p* value corrected by the FDR method. The third circle: a bar chart of the proportion of up and down-regulated genes, dark purple represents the proportion of up-regulated genes, and light purple represents the proportion of down-regulated genes; the specific values are shown below. Fourth circle: RichFactor value of each KEGG pathway.

To further determine whether the metabolic pathways were involved during the two developmental stages of seeds, 129 KEGG pathways were represented by DEGs (1928 unigenes) (Table S5). The pathway represented by the largest number of unigenes (749) was the metabolic pathway (Ko01100). The top 20 enriched KEGG pathways are shown in Figure 1b, among which pathways of Photosynthesis (ko00195), Photosynthesis-antenna proteins (ko00196), and Porphyrin and chlorophyll metabolism (ko00860) are closely related to plant photosynthesis.

2.3. Protein Identification and Functional Annotation

The comparative proteome analysis of Pv-1 and Pv-2 was carried out by iTRAQ technology for the comprehensive transcriptome study. A total of 113,636 total spectra, 56,818 spectra, 28,409 unique spectra, 34,825 peptides, 26,510 unique peptides, and 6026 proteins were identified (Figure 2a). The analysis of the characteristics of peptides showed that 4700 (78.00%) peptides were related to at least 2 unique peptides (Figure 2b). The mass of peptides ranged from 1000 to 4000 Da, and the peptides within the range of 1000–3000 Da accounted for 96.89% (27,525) of total peptides (Figure 2c). The average length of the identified peptides was 11.99 amino acids, which was within the reasonable range (Figure 2d).

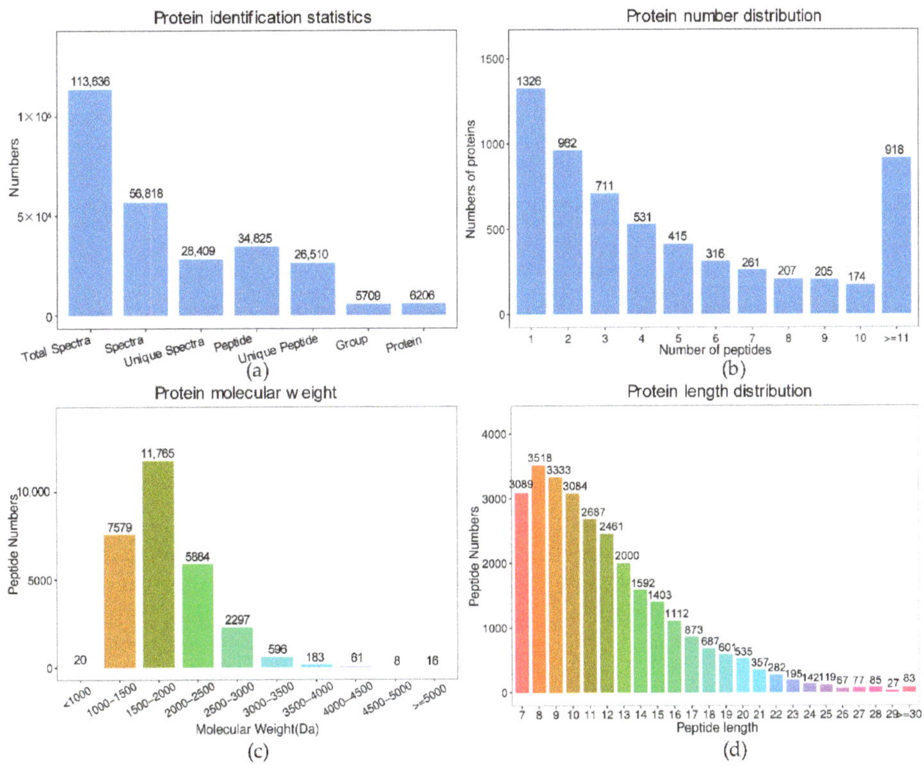

Figure 2. (**a**) The identification results of proteins in *P. volubilis* seeds; (**b**) Number distribution of all identified unique proteins. The abscissa is the number of peptides, and the ordinate is the number of proteins in this peptide number; (**c**) Molecular weight (Da) of all identified proteins; (**d**) Length distribution of all identified protein.

The changes in the structure of protein species and average masses in seeds of *P. volubilis* at two growth stages can be used to study protein function. Out of a total of 6026 identified proteins, 4336 (71.95%), 2613 (43.36%), and 4534 (75.24%) proteins could be annotated in GO, KEGG, and KOG databases, respectively (Figure S4). Altogether, 5375 (89.20%) proteins were successfully annotated in these three public databases. Based on GO categorization, 4336 proteins were classified into a total of 2348 GO terms, including BP (1536 terms), CC (245 terms), and MF (567 terms) (Table S6). In the BP category, metabolic process (GO:0008152) and cellular process (GO:0009987) were the predominant terms, followed by organic substance metabolic process (GO:0071704). In the CC category, cell (GO:0005623) and cell part (GO:0044464) were the predominant terms, followed by intracellular (GO:0005622) and intracellular part (GO:0044424). In the MF category, however, catalytic activity (GO:0003824) and binding (GO:0005488) were the main distributed terms, followed by organic cyclic compound binding (GO:0097159) and heterocyclic compound binding (GO:1901363). A total of 2613 proteins were assigned to 128 pathways in the KEGG database (Table S7). The most represented pathways included the metabolic pathway (ko01100, 1173 proteins), secondary metabolite biosynthesis (ko01110, 645 proteins), and ribosome (ko03010, 257 proteins). Notably, 14 pathways, such as fatty acid degradation (ko00071), glycerolipid metabolism (ko00561), fatty acid biosynthesis (ko00061, 35 proteins), etc., were closely linked to lipid biosynthesis and metabolism and took place during the seed germination of *P. volubilis*. In total, 4534 proteins were assigned to COG classes (Figure S5). Among the 25 COG categories, "General function prediction only" represented the largest group (763 proteins), followed by "Post-translational modification", "protein turnover", and "chaperones" (712 proteins). The "Lipid transport and metabolism" category contained 221 proteins.

2.4. Enrichment Analysis of DAPs Based on GO and KEGG Pathways

A total of 4983 proteins were identified as DAPs at two developmental stages of *P. volubilis* seeds (fold change > 1.2, p-value < 0.05), of which 169 DAPs were upregulated and 4814 DAPs were downregulated (Table S8). Consistent with the RNA-seq data, the number of upregulated DAPs was lower than that of downregulated DAPs. It is particularly noteworthy that based on the relative quantification, the number of ricin-like proteins (Unigene0000207 and Unigene0000209) was the highest in Pv-2, followed by legumin proteins (Unigene0048787, Unigene0010856, and Unigene0010858). However, the quantity of β-glucosidase (Unigene0018636) in the Pv-1 was the highest, followed by the MLP-like protein (Unigene0051229), and ricin (Unigene0000207). The proteins closely involved in fatty acid biosynthesis, such as oil body-associated proteins (Unigene0048411 and Unigene0032502), Oleosins (Unigene0028565 and Unigene0050906), FAD3 (Unigene0043398), FAD2 (Unigene0034350), SAD (Unigene0011486), and long-chain Acyl-CoA synthetase (Unigene0026569), were all significantly up-regulated in the fast oil accumulation stage.

GO functional annotation and KEGG pathway enrichment analysis were used to compare the changes in annotated DAPs. A total of 3619 DAPs were enriched in 2256 GO terms, including 1483 terms in BP, 234 terms in CC, and 539 terms in MF (Table S9). For the BP category, DAPs were mainly related to the metabolic process (GO:0008152, 2476 DAPs), the cellular process (GO:0009987, 2021 DAPs), and the organic substance metabolic process (GO:0071704, 1881 DAPs). In the CC category, DAPs were mainly associated with the cell (GO:0005623), cell part (GO:0044464), intracellular (GO:0005622), and intracellular part (GO:0044424). In the MF category, however, DAPs were mainly associated with catalytic activity (GO:0003824), binding (GO:0005488), organic cyclic compound binding (GO:0097159), and heterocyclic compound binding (GO:1901363). The KEGG enrichment analysis results showed that DAPs were assigned to 127 pathways, of which 96 were metabolic pathways (Table S10). In enrichment of the top 20 pathways based on Q-value (Figure S6), the most enriched pathways included Spliceosome (ko03040), Plant-pathogen interaction (ko04626), and Ascorbate and aldarate metabolism (ko00053). A total of 23 and 35 DAPs were enriched in the phosphatidylinositol signaling system (ko04070) and the plant MAPK signaling pathway (ko04016), respectively, among the first 20 pathways, which were related to signal transduction [26,27].

2.5. Association Analysis of Transcriptome and Proteome Data

A global correlation analysis was performed between the proteins and their corresponding transcripts to explore the consistency between the transcriptome and proteome data [28]. Association analysis revealed that 5877 proteins matched the transcripts, of which 4974 were DAPs, 1950 were DEGs, and 1687 were both DAPs and DEGs (Figure 3a). Similar to a previous study on *Drosophila melanogaster*, only a moderate correlation (R = 0.49) was observed between transcriptome and proteome data.

Quantitative and enrichment analyses of genes and proteins were performed in each region of the nine-quadrant map, which was drawn based on changes in the expression of genes and proteins at the transcriptome and proteome levels. As shown in Figure 3b, the transcripts/proteins (146) concentrated at the center of the plot (quadrant 5) were NDEGs/NDAPs, respectively. The proteins in quadrant 4 (1443) and quadrant 6 (17) were DAPs whose corresponding genes were not differentially expressed. The proteins in quadrant 2 (41) and quadrant 8 (28), however, were not differentially expressed, but their corresponding genes were differentially expressed. The proteins in quadrants 1 (63) and 9 (3) had the opposite expression patterns from their transcripts, while the proteins in quadrants 3 (141) and 7 (1480) showed the same expression patterns as their transcripts. This indicated that the expression of proteins in quadrants 3 and 7 was regulated at the transcriptional level. However, for proteins in quadrants 1, 2, 4, 6, 8, and 9, there were regulatory events at the post-transcriptional or translational levels (e.g., miRNAs regulate the translation of target genes and inhibit the expression of proteins) [29], and therefore, the gene expression cannot fully represent the abundance of these proteins.

The correlation analysis between the transcriptome and proteome data in GO function and KEGG pathway was performed to compare the gene function and metabolic pathway in terms of their similarities and differences between the two groups. GO analysis of the DEGs/DAPs upregulated in the fast oil accumulation stages of *P. volubilis* seeds (Table S11) revealed that the vacuole (GO:0005773) and lipid particle (GO:0005811) were enriched in the CC category, suggesting that oil body formation and development took place in Pv-2. Remarkably, based on GO enrichment analysis, the GO terms, including enzyme inhibitor activity (GO:0004857) and catalytic activity (GO:0003824) were enriched in the MF category; and the negative regulation of catalytic activity (GO:0043086), regulation of hydrolase activity (GO:0051336), negative regulation of hydrolase activity (GO:0051346), negative regulation of peptidase activity (GO:0010466), regulation of peptidase activity (GO:0052547), regulation of proteolysis (GO:0030162), and negative regulation of proteolysis (GO:0045861) in the BP category were all significantly enriched. This indicated that the activity of proteins decreased during seed ripening. Particularly, 5 ricin-like DEGs/DAPs including Unigene0000207, Unigene0000208, Unigene0000209, Unigene0012388, and Unigene0037089 were involved in hydrolase activity (GO:0016787), and one RIP (Unigene0012386) was also involved in hydrolase activity, acting on glycosyl bonds (GO:0016798).

KEGG pathway enrichment analysis showed that DEGs/DAPs were mainly involved in metabolism pathways (ko01100, 749 DEGs/963 DAPs), followed by biosynthesis of secondary metabolites (ko01110, 435 DEGs/515 DAPs), and ribosome (ko03010, 185 DEGs/213 DAPs) (Figure S7). In quadrant 3, 141 unigenes showed a positive relationship between mRNA enrichment and protein abundances. The upregulated DEG/DAP pairs were mainly enriched in metabolic pathways (ko01100), biosynthesis of secondary metabolites (ko01110), and phenylpropanoid biosynthesis (ko00940). Among the top 20 pathways in quadrant 3, five pathways related to lipid metabolism were enriched, and the fatty acid degradation pathway (ko00071) and glycerolipid metabolism pathway (ko00561) were also significantly enriched (Figure 3c). In quadrant 7, in Pv-2, 1480 downregulated DEGs/DAPs were involved in significantly enriched pathways of (Q-value < 0.05) ribosome (ko03010), spliceosome (ko03040), RNA transport (ko03013), and DNA replication (ko03030) (Figure 3d), which were growth-related and developmental pathways, and key enzymes were significantly downregulated in Pv-2. Plant-pathogen interaction (ko04626) and phosphatidylinositol signaling system (ko04070) were enriched in 51 and 20 DAPs, respectively, in quadrant 7.

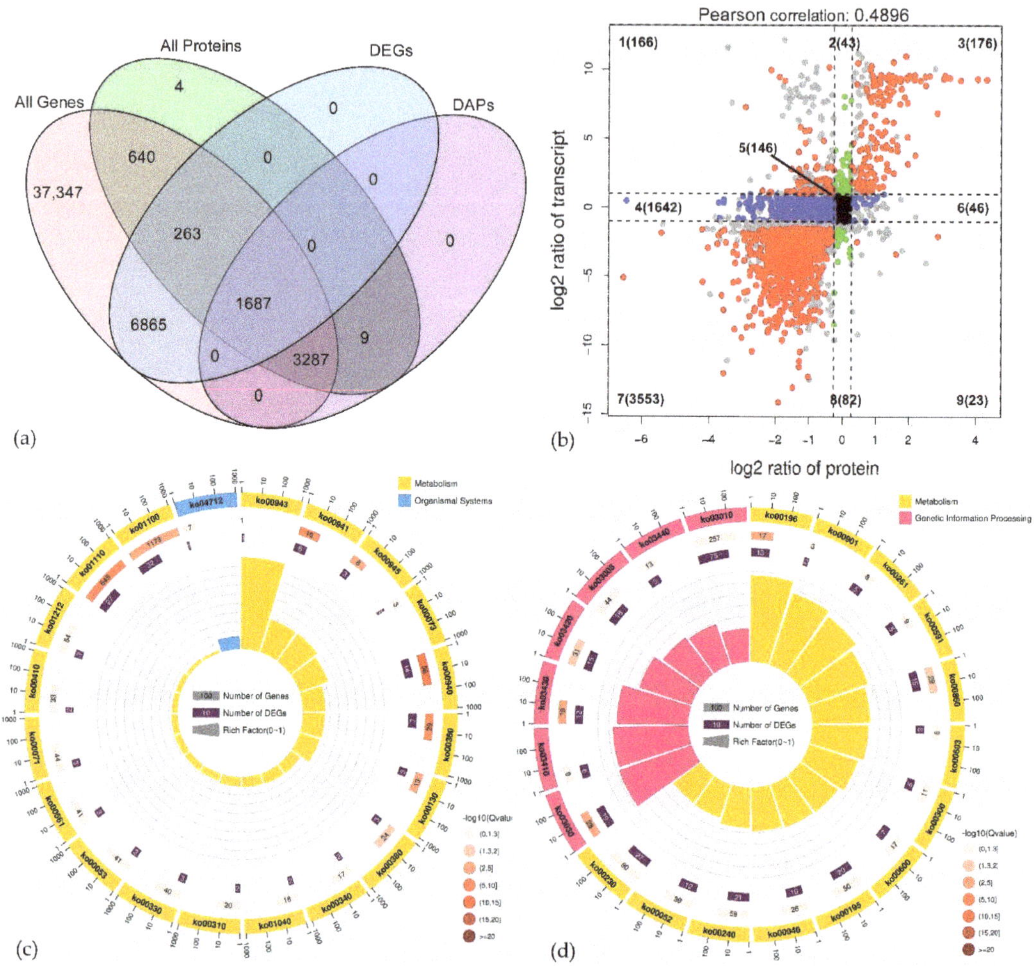

Figure 3. (**a**) Venn diagram of DEGs in all genes and DAPs in all proteins; the description of what is contained in the first panel; (**b**) scatter plot of nine-quadrant associate analyses of transcripts and proteins in Pv-1 and Pv-2. Each dot represents a gene/protein. The dashed line on the abscissa represents the FC threshold of DEGs (FC $\geq$ 2.0), the dashed line on the ordinate represents the FC threshold of DAPs (FC $\geq$ 1.2), the genes/proteins outside the threshold line are DEGs/DAPs, and the genes/proteins inside the threshold line are not DEGs/DAPs. Numbers 1–9 represent each quadrant, and the number of points in each quadrant showed in parentheses. Quadrants 1, 2, and 4 indicate that the protein abundance was lower than the RNA expression. In 3 and 7, the RNAs correspond with the related proteins. Quadrant 5 represents that the proteins and transcripts were commonly expressed with no difference. Quadrants 6, 8, and 9 indicate that the protein abundance was higher than the RNA expression. (If the FC is reached and the *p* value is not reached, it will be shown as the gray point). Top 20 of KEGG pathway enrichment of DEGs/DAPs in quadrant 3 (c) and quadrant 7 (**d**). The first circle: numbers outside of the circle are the coordinate rulers of the gene number, different colors represent different KEGG A classes; the second circle: the number and *Q*-value of the KEGG pathway in the background gene. The more genes, the longer the bars, and the smaller the *Q*-value, the redder the color; the *Q*-value is the *p*-value corrected by the FDR method. The third circle: a bar chart of the proportion of up and down-regulated genes, dark purple represents the proportion of up-regulated genes, and light purple represents the proportion of down-regulated genes; the specific values are shown below. Fourth circle: RichFactor value of each KEGG pathway.

2.6. Analysis of Physicochemical Properties of RIPs

Analysis of spatial protein structure is of great significance for understanding the function and implementation of proteins, the interaction between biological macromolecules, and the development of medicine and pharmacy [30].

Based on the transcriptomic data, 21 transcripts from RIP-seq experiments were screened, but only 8 proteins were identified based on the proteomic data. Using the DNA sequences of transcripts, we translated the encoded proteins of 21 RIP genes, and their bioinformatics analysis was performed (Table S12). The analysis of ExPASyParam showed that the length of different identified transcripts was variable, ranging from 67 to 559 amino acids. The protein encoding Unigene0004179 had the largest number of amino acids (559), followed by Unigene0052941 (552), while the proteins encoding Unigene0012387, Unigene0012388, and Unigene0000208 had the lowest number of amino acids. This may be due to the presence of variable shear strength. The theoretical isoelectric points (PI) of 21 RIPs ranged from 4.43 to 9.78, with the highest PI for the unigene0000786 and the lowest PI for unigene0030485. According to the PI values of these 21 RIPs, RIPs in *P. volubilis* had slightly higher contents of proteins, which was attributable to the PI values of 12 RIPs that were less than 7. The instability indices (II) of 21 RIPs in *P. volubilis* ranged from 11.11 to 40.28 and were all classified as stable proteins except unigene0018529 (II = 40.28). Based on aliphatic indices of 21 RIPs, the number of aliphatic amino acids in unigene0005315 was the highest (123.13), followed by unigene0012388 and unigene0018530, while the number of aliphatic amino acids in unigene0043472 protein was the lowest (64.64). Among them, 13 RIPs were hydrophilic, but 8 were hydrophobic. This indicates that there were some differences in physicochemical properties of different RIPs in *P. volubilis*.

The Batch CD-search tool, Pfam, and SMART Online software programs were used to search the protein conservative domains. The results showed that 10 transcripts contained the conservative domain ricin with $(QxW)_n$ motif (cl23784, IPR001574), while 9 transcripts contained the RIP superfamily domain with $Yx_nYx_nExxRx_nW$ motif (cl08249, IPR017989), and 4 transcripts contained the ricin B-like lectin domain with $(QxW)_3$ motif (cl40832, IPR040249). The motif of $Yx_nYx_nExxRx_nW$ plays a role in stabilizing the active center of the enzyme, and the QxW domain has been found to be associated with diverse functions such as enzyme activity, inhibition of toxicity, and signal transduction [31]. Among 21 RIPs, 15 contained the transcripts encoding the signal peptide. The modification of proteins through glycosylation not only affects the biological activity, spatial conception, and localization and transport of proteins but also plays an important role in specific biological processes such as cell communication, molecular recognition, signal transduction, etc., [32,33]. In the current study, the NetOGlyc software was used to analyze the glycosylation sites. There were 1~8 O-glycosylation sites in 12 RIPs. Unigene0018947 had 8 O-glycosylation sites, whereas Unigene0000209 had 6 O-glycosylation sites. Protein phosphorylation refers to the process of transferring the phosphate group from an ATP to amino acid residues (serine, threonine, and tyrosine) within substrate proteins catalyzed by protein kinases or binding GTP under the action of signals (usually GTP replaces GDP) [34]. The analysis using the NetPhos software showed that all 21 RIPs had phosphorylation sites (8~94), of which 94 were on the Unigene0052941, while 70 were on Unigene0004179. Consistent with the results of Rezaei-Moshaei et al. [35], the number of potential Ser and Ther phosphorylation sites were much higher than Tyr phosphorylation sites (Table S12). Furin serves as an endoproteinase in eukaryotes. It recognizes the specific amino acid sequence and after two self-cleavages in the endoplasmic reticulum in the Golgi body, it cleaves many crucial precursors of peptides and proteins in the secretory pathway, facilitating the bioactivity of precursors [33]. Based on the prediction of the presence and location of furin cleavage sites by the ProP software, there was only one furin cleavage site in four RIPs. The results of the prediction of the regions of transmembrane proteins based on the HMM method showed that 5 transcripts in RIPs contained a transmembrane helix, and transmembrane proteins were expressed as i12–34o, i7–24o, o4–26i, i5–22o, and i5–27o. The prediction of protein subcellular localization using WoLF PSORT showed that three transcripts encoding a ricin-like protein, Unigene0004179 encoding ricin B-like lectin, and Unigene0018529 encoding type I RIP might be distributed to extracellular vesicles; two transcripts encoding type I RIP might be located in the cytoplasm; Unigene0005315 encoding ricin-like, three transcripts encoding ricin B-like lectin, and two transcripts encoding type I RIP might be distributed along cytoskeletal filaments; four transcripts encoding a ricin-like protein and four transcripts encoding type I RIP might be found in the chloroplast.

According to the results of the maximum likelihood estimation of phylogenetic tree construction, the RIPs in *P. volubilis* and other plants were divided into three subfamilies (Figure 4), similar to the RIPs in *R. communis*. Subfamily I was mainly composed of ricin B-like lectin proteins, while subfamily II contained the ricin-like proteins, and subfamily III consisted of type I RIPs. The differentiation of type I RIP (subfamily III) occurred early; however, the ricin B-like lectin (subfamily I) and the ricin-like protein (subfamily II) were differentiated relatively recently, with the differentiation of ricin B-like lectin occurring a little later than that of the ricin-like protein. Therefore, it was inferred that ricin B-like lectin that belongs to type II RIP could possibly have evolved from type I RIP. This inference was consistent with Li and Liu [31].

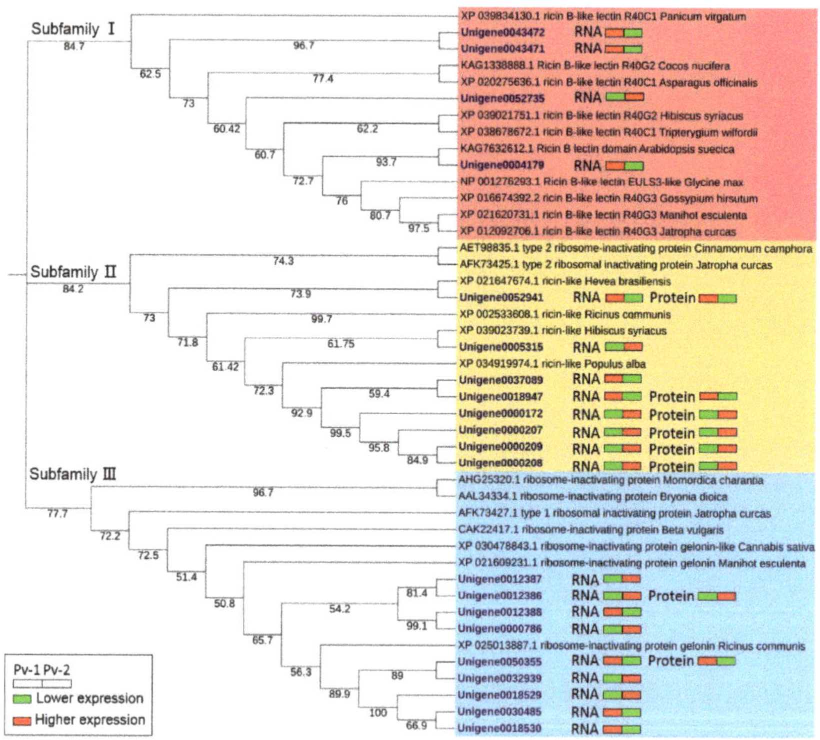

Figure 4. ML phylogenetic tree of 21 RIPs in *P. volubilis* and 23 RIPs in other plants with 1000 bootstrap replicates. The number below the branches indicates bootstrap percentages. The icons right to each RIP ID showed the expression patterns of each transcript/protein through the two seed developmental stages, with the green color indicating lower expression/abundance, while the red color indicates higher expression/abundance.

Based on the relative expressions of 21 RIP genes and the abundance of 8 RIPs in seeds at two different developmental stages from transcriptome and proteome data, 11 RIP genes were up-regulated at Pv-2 (the fast oil accumulation stage), whereas 10 RIP genes were up-regulated at Pv-1 (the initial stage). The trend of the accumulation of eight proteins was exactly consistent with the expression patterns of their corresponding genes. Subfamily II consisted of type 2 RIPs and ricin-like proteins, with the latter having the probability of being RTA. Six proteins in P. volubilis in the subfamily II were DAPs, four of which (Unigene0000172, Unigene0000207, Unigene0000208, and Unigene0000209) had the highest abundance at Pv-2. Additionally, the type I RIP (Unigene0012386) was also a DAP and abundantly accumulated at Pv-2.

2.7. The Validation of RT-qPCR Assay and Expression Patterns of RIP Genes in P. volubilis Seeds

To verify the accuracy of transcriptome data, 10 mRNAs with higher expression levels (greater than 50 in at least one sample) and relatively high read counts (usually greater than 20) in both groups were selected for RT-qPCR analysis (Figure 5a–j). The results showed that the relative expression trends of 10 genes by RT-qPCR were consistent with the transcriptome sequencing results. Meanwhile, a highly significant correlation ($R^2 = 0.95$) was found between RT-qPCR and RNA-seq data for these 10 RIP genes (Figure S8), which indicated that the transcriptome data were highly reliable.

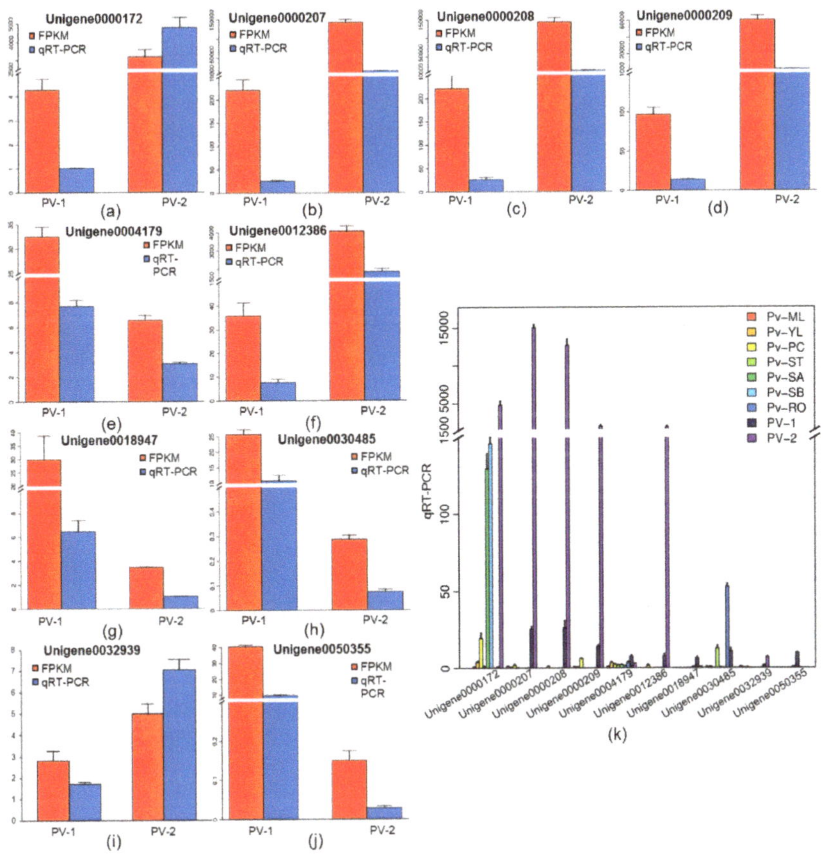

Figure 5. (**a–j**) Verification of the consistency of the RNA-Seq data using quantitative real-time PCR (qRT-PCR). (**k**) The expression patterns of the description of ten RIP genes in nine organs of *P. volubilis*. Pv-ML, mature leaf; Pv-YL, young leaf; Pv-PC, pericarp; Pv-ST, stem; Pv-SA, stem apex; Pv-SB, stem bark; Pv-RO, root young leaves. Error bars ± SD from 3 biologicals. The data were subjected to Student's *t*-test.

To further identify the RIP genes in *P. volubilis* seeds, the expression levels of 10 genes with higher RPKM values (RPKM $\geq$ 10) were determined using RT-qPCR in nine organs in *P. volubilis*. Although 10 RIP genes were all highly expressed in *P. volubilis* seeds, individual RIP genes were differentially expressed in different organs (Figure 5k). Among these 10 genes, Unigene0000172 (the transcript of ricin-like protein) had the highest expression level in Pv-2 (the fast oil accumulation stage), which was 32.73 times and 36.83 times higher than in SB and SA, respectively. Similarly, the expression levels of three transcripts (Unigene0000207, Unigene0000208, and Unigene0000209) encoding ricin-like proteins and two transcripts (Unigene0012386 and Unigene0032939) encoding type I RIP also had the highest expression levels in Pv-2, around 586.51, 482.62, 140.69, 242.85, and 4.16 times higher than in Pv-1, respectively. Furthermore, Unigene0004179 encoding ricin B-like lectin and Unigene0050355 encoding type I RIP had the highest expression levels at Pv-1. Unigene00004179 specifically had a very low expression level in all organs, and Unigene0030485 encoding type I RIP was expressed higher at Pv-RO, indicating that most ricin genes were prominently expressed in mature seeds of *P. volubilis*.

2.8. Transcriptional Regulatory Network

Gene function is often closely related to the transcriptional regulation [36,37]. We analyzed 21 transcripts encoding RIPs and found that only 4 unigenes (Unigene0004179, Unigene0043471, Unigene0043472, and Unigene0052735) encoding ricin B-like proteins had paired-end relationships with the other 29 genes (Figure 6). The "combined score" indicated the strength of data support, and 29 unigenes were screened out when the threshold was set to 0.4. Among these 29 unigenes, 5 unigenes (Unigene0012529, Unigene0012530, Unigene0012531, Unigene0012532, and Unigene0050249) encoding calcium-dependent protein kinases (CPKs), 4 unigenes (Unigene0006177, Unigene0026569, Unigene0026570, and Unigene0038545) encoding long-chain acyl-CoA synthetase 8 (LACS8), and 4 genes encoding ascorbate peroxidase 3 (APX3) represented potential targets for 3 ricin B-like genes (Unigene0043471, Unigene0043472, and Unigene0052735). Notably, 4 ricin B-like genes had low expression levels during seed development, and thus, they could not be regarded as potential targets.

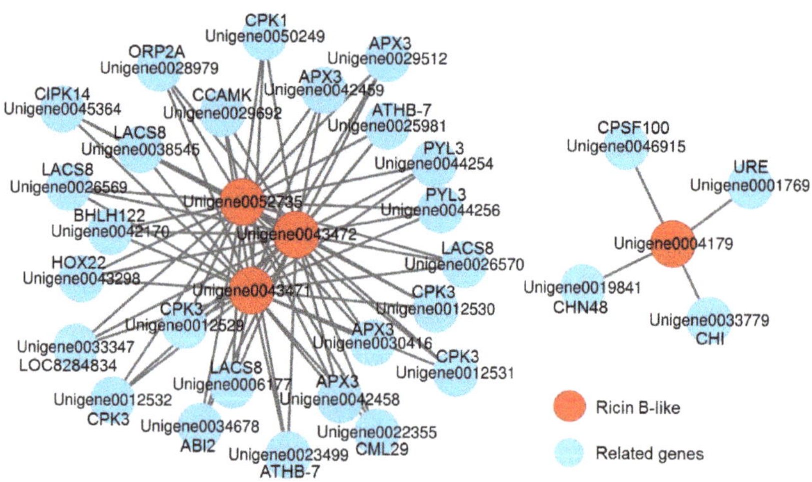

Figure 6. Transcriptional regulatory network of RIP genes in *P. volubilis* seeds.

3. Discussion

P. volubilis is a highly promising crop, primarily due to the excellent nutritional composition of its seeds [13]. In the current study, a combination of transcriptomic and proteomic approaches was used to analyze *P. volubilis* seeds at two developmental stages (the initial stage and the fast oil accumulation stage). The association analysis for two-omics data was then performed. We obtained a total of 227.61 million 150 bp paired-end reads from two libraries (an average of 37.94 million), which were higher in number than those obtained by Wang et al. [21], who reported 52.6 million 90 bp paired-end reads from two libraries, with equal division of reads between the two (26.3 million for each). Transcriptome analysis at two developmental stages of *P. volubilis* seeds revealed the presence of a total of 53,224 unigenes, and the expression of 8815 unigenes differed, with at least a two-fold change between the two libraries. Compared with the transcriptome of five developmental stages of seeds [20], the gene number (53,224) obtained in this study was higher than 44,797 in the transcriptome of five developmental stages, and the N50 (1834), average length (1019), total assembled bases (54.25 M) were rather fewer than other data (2299, 1345, 60.67 M, respectively) in the transcriptome of five developmental stages. This may be associated with the upgrade of data filtering. From the annotation of GO, KOG, and KEGG, the transcriptome of five developmental stages of *P. volubilis* seeds [20] was essentially in agreement with the results of this study. A total of 6026 proteins and 4983 DAPs were identified by proteomic analysis. It is a very common protein post-translational modification (PTM) in organisms that plays an important role in the process of signal transduction at the single-cell level [38], and the association analysis revealed that transcriptome abundance and protein levels were only moderately correlated. This may be due to the effects of translation regulation, PTM, and the rate of protein transcription [37]. As the next step, we can consider other post-transcriptional regulatory binding sites as the important regulatory targets for the development of *P. volubilis* seeds.

The publication of the caster bean genome revealed the presence of 28 genes in the ricin gene family [39]. In our study, 21 transcripts and 8 proteins were identified in the transcriptome of *P. volubilis* seeds. The length of the different transcripts identified was variable, and some smaller genes (Unigene0012387 and Unigene0012388) could be nonfunctional or pseudogenes; furthermore, start and stop codons could be predicted, making it difficult to determine whether the genes were functional or not. Bioinformatics methods were employed to analyze the nucleotide and amino acids homology between sequences of 21 RIP genes, their possible protein domains, and the topology of transmembrane proteins. The possible protein domains were ricin, ricin-B-lectin, or RIP domains. There were phosphorylation sites in all 21 RIPs and more Ser and Ther phosphorylation sites than Tyr phosphorylation sites. It has been shown that the functional effects of protein phosphorylation are site-dependent, and phosphorylation occurs at a specific site [35]. Among the 21 RIPs, 15 RIPs were signal peptides, and O-glycosylation sites were found in 11 RIPs. Four RIPs were found to have furin cleavage sites, and five had a transmembrane domain. A total of 21 RIPs were distributed in the chloroplast and cytoskeleton and extracellular and cytoplasmic regions in the seeds of *P. volubilis*. Among them, five RIPs with the highest levels of expression and highest abundance levels in Pv-2 were most likely located in extracellular regions or chloroplast. The phylogenetic tree of the sequences of 21 RIPs in *P. volubilis* and other RIPs in other plants revealed that the RIPs in *P. volubilis* were divided into three subfamilies. The clustering results were consistent with the previous study of Loss-Morais [22] and provided new evidence for the hypothesis that type II RIPs evolved from type I RIPs. Of these three subfamilies, the differentiation of subfamily I (ricin B-like lectin) occurred late, which could be due to the fact that the ricin B chain is a product of gene duplication [40,41]. Similar to the RIPs in *R. communis*, the number of type II RIPs was the largest, with the highest abundance and expression levels. It is worth noting that the four transcripts encoding ricin-like proteins and Unigene0012386 encoding type I RIPs had the highest expression levels in Pv-2, but the four transcripts encoding ricin B-like lectin protein had relatively low expression levels during seed development of *P. volubilis*. Type II RIPs were highly toxic on RTB, which promotes the entry of RTA into host cells and inhibits protein synthesis [42]. The RTB itself was non-toxic and it could recognize terminal galactose residues and facilitate the interaction between type II RIPs and the cell membrane [43]. According to proteomic data from *P. volubilis*, no RTB proteins were identified, and based on transcriptomic data, 4 transcripts encoding ricin B-like lectin had very low expression levels. Therefore, it can be inferred that the RIPs in *P. volubilis* would not be a serious threat to public safety unless the factors (e.g., biotic or abiotic stress, plant hormones) [22] affect their expression levels in the plants.

The expression patterns of 10 RIP genes in different organs of *P. volubilis* were assessed by RT-qPCR. They were all highly expressed at two developmental stages of *P. volubilis* seeds, among which, six were expressed at the highest levels in seeds at the fast oil accumulation stage. At the initial stage, however, 3 RIP genes were expressed at higher levels. The results of our study corroborated the previous findings obtained by Llediad et al. [44] and Loss-Morais et al. [22] who reported that the expressions of members of the RIP gene family were often tissue-specific and developmental stage-specific. In addition to being specifically and highly expressed in seeds, Unigene0000172 also had higher expression levels in stem and stem apex. Additionally, Unigene0030485 encoding type I RIPs had the highest expression level in roots. Type I RIPs are widely distributed in most organs of many plants, and the contents of these genes vary greatly in different organs of different plants. The toxicity of type I RIPs is generally low, but they have certain antiviral properties since the viral infection in plants promotes the passage of the RIP into the cells to inhibit the replication of the virus. Furthermore, plant RIPs play an important role in defense against various environmental stresses and can activate plant defense systems [45]. There have been many research findings that revealed that the expression of RIPs could be activated by some factors, such as phytohormones, viral infection, development, senescence, and environmental stress [46–48]. The expression of Unigene0030485 was higher in the roots of *P. volubilis*; however, it remains to be determined whether it is related to antiviral activity or various environmental stresses.

In addition, the results of co-expression analysis and protein interaction network analysis showed that there has little correlation between oil accumulation and ricin accumulation. Transcriptional regulatory network analysis showed that only 4 ricin B-like genes with low expression levels could be associated with the other 29 genes, and 4 *LACS8* transcripts represented potential targets for 3 RIP genes. LACSs play vital roles in lipid biosynthesis and fatty acid metabolism in plants [49]. However, Zhao et al. [50] reported that disruption and overexpression of *LACS8* did not affect the seed fatty acid content in *Arabidopsis*. These findings also provide evidence that the oil obtained from *P. volubilis* would be safe for human consumption.

4. Materials and Methods

4.1. Sample Collection

The 3-year-old *P. volubilis* trees introduced from Peru were cultivated at the South China Experimental Nursery (21°30′ N, 111°38′ E, 90 asl), Guangdong, China, under natural conditions. In the study, the strain number of V3, which has excellent oil quality and high yield, was selected as the research material. Based on the dynamic changes of fatty acid accumulation in developing seed [23], we analyzed the transcriptome and proteome data from samples at two developmental stages (the initial stage of seed development, Pv-1, and the fast oil accumulation stage, Pv-2), which were found to be consistent with those reported by Wang et al. [21]. The developing seeds did not start to accumulate large amounts of fatty acids, especially the α-linolenic acid, in Pv-1 and mostly in Pv-2. The mature female flowers were tagged and hand pollinated. Flowers were collected 10 DAP at Pv-1 and 100 DAP at Pv-2. In addition to the seeds (Pv-1 and Pv-2), mature leaves (Pv-ML), unfold young leaves (Pv-YL), pericap (Pv-PC), stem (Pv-ST); stem apex (Pv-SA); stem bark (Pv-SB) and root (Pv-RO) were sampled to analyze the expression patterns of target genes. Each sample was taken from three different plants, and for each stage, three biological replicates were considered.

4.2. Transcriptome Analysis

A transcriptome profile was examined at two stages of seeds using RNA sequencing (RNA-seq). Total RNA was extracted using the TRIzol reagent kit (Invitrogen, Carlsbad, CA, USA). The quality of RNA was assessed on an Agilent 2100 Bioanalyzer (Agilent Technologies, Palo Alto, CA, USA) and double-checked using RNase-free agarose gel electrophoresis. The samples with the RNA integrity number (RIN) higher than 8.0 determined by the Agilent 2100 Bioanalyzer were used for the construction of sequencing libraries. The steps involved the enrichment of mRNA by Oligo (dT) beads, RNA fragmentation, synthesis of the second-strand cDNA, size selection, and PCR amplification. The two libraries were then sequenced using Illumina HiSeq™ 4000 by Gene Denovo Biotechnology Co., (Guangzhou, China).

To get high-quality clean reads, the obtained 150 bp paired-end reads were further filtered for quality by FASTP (version 0.18.0, Shenzhen, China) [51]. The quality filtering was performed by removing the reads containing adapters, low-quality reads containing more than 50% of low-quality (Q-value $\leq$ 20) bases and reads containing more than 10% of unknown nucleotides. Due to the absence of a reference genome, de novo transcriptome assembly was carried out using the short-read assembly program Trinity (version 2.8.6, Cambridge, MA, USA) [52]. The transcriptome integrity was assessed using Benchmarking Universal Single-Copy Orthologs (BUSCO, version 3.0.2, Geneva, Switzerland) [53] that incorporated 1440 single-copy orthologous genes as the embryophyte dataset.

The unigenes were annotated by performing BLASTx searches (http://www.ncbi.nlm.nih.gov/BLAST/, accessed on 29 August 2021) with an E-value threshold of $\leq 1 \times 10^{-5}$ against the non-redundant protein database (Nr) in NCBI (http://www.ncbi.nlm.nih.gov, accessed on 29 August 2021), the SWISS-PROT protein sequence database (http://www.expasy.ch/sprot, accessed on 29 August 2021), the Gene Ontology (GO) database (http://geneontology.org/, accessed on 29 August 2021), the Kyoto Encyclopedia of Genes and Genomes (KEGG) database (http://www.genome.jp/kegg, accessed on 29 August 2021), and the COG/KOG database (http://www.ncbi.nlm.nih.gov/COG, accessed on 29 August 2021). Further analysis of unigene differential expression was carried out between two libraries using DESeq2 [54] and edgeR [55] software programs. The differentially expressed genes (DEGs) were identified based on criteria set as the FDR $\leq$ 0.05 and the fold change (FC) $\geq$ 2. Gene ontology (GO) and KEGG pathway enrichment analyses of DEGs were conducted by the hypergeometric test.

4.3. Proteomics Analysis

Total protein extraction was determined in three biological replicates of each sample using the cold acetone method. The quality of proteins was evaluated with SDS-PAGE, and the BCA Protein Assay Kit was used to determine the protein concentrations in the supernatant. Around 100 μg of protein for each sample was transferred to a new microcentrifuge tube, and the final volume was adjusted to 100 μL with 8 mol/L urea. Thereafter, 2 μL of 0.5 mol/L TCEP was added and the sample was incubated at 37 °C for 1 h, followed by the addition of 4 μL of 1 mol/L iodoacetamide into samples and incubation for 40 min at room temperature in the dark. Five volumes of −20 °C pre-chilled acetone were added to precipitate the protein extracts at −20 °C overnight. Around 1 mL pre-chilled 90% acetone aqueous solution was used to wash protein precipitate twice, followed by re-dissolving in 100 μL 100 mmol/L TEAB. To digest the proteins at 37 °C overnight, sequencing-grade modified trypsin (Promega, Madison, WI, USA) was added at the ratio of 1: 50 (an enzyme (wt): protein (wt)). C18 ZipTip was used for desalting the peptide mixture, which was then quantified by Pierce™ Quantitative Colorimetric Peptide Assay (23,275) and lyophilized by SpeedVac.

The resultant peptides were labeled by the iTRAQ-8PlexIsobaric Mass Tag Labeling Kit (Thermo Fisher Scientific, Waltham, MA, USA), pooled, and lyophilized in a vacuum concentrator. Subsequently, the peptides were redissolved and separated at high pH. Twelve separated fractions were collected from each sample and identified after drying.

Each collected peptide fraction was re-dissolved and analyzed by online nanospray LC-MS/MS on an Orbitrap Fusion Lumos coupled to the EASY-nLC 1200 system (ThermoFisher Scientific, Waltham, MA, USA). A 2 kV electrospray voltage was set on the inlet of the mass spectrometer. The mass spectrometer was operated in the data-dependent acquisition (DDA) mode and could automatically be switched between MS/MS and MS mode.

DIA raw data were processed and analyzed with Spectronaut X (Biognosys AG, Zurich, Switzerland) under default settings. The retention time prediction type was set to dynamic iRT. Data extraction was determined by Spectronaut X based on the extensive mass calibration. The Spectronaut Pulsar X was used to dynamically determine the ideal extraction window based on iRT calibration and gradient stability. The *Q*-value (FDR) cutoff at the precursor and protein levels was set to 1%. Decoy generation was set to mutate, which was similar to scrambled but only applied a random number of AA position swamps (min = 2, max = length/2). All screened precursors which were passed through filters were used for quantification. The major group quantities were calculated using the average top 3 filtered peptides, which passed the 1% *Q*-value cutoff. After Student's *t*-test, the Benjamini and Hochberg method was applied, and DAPs were filtered with the fold change (FC) > 1.2 and FDR < 0.05.

The GO, KEGG, and COG/KOG databases were used to annotate proteins and predict their functions. The significantly enriched GO terms and KEGG pathways of DAPs with *Q*-value $\leq$ 0.05 were identified. MASCOT [56] was used to analyze the DAPs with the FC > 1.2 and $p < 0.05$ at two different developmental stages of *P. volubilis* seeds.

4.4. mRNA and Protein Association Analysis

A qualitative correlation was established between the genes that were regulated in a similar direction by RNA and protein expression [57], and then the Venn diagram was plotted [58]. For the analysis of the correlation between transcriptome and proteome data, changes in protein levels complementary to the changes in the corresponding transcripts were investigated, and the correlations between DEGs and DAPs were depicted by drawing a nine-quadrant map in R (version 3.5.1, Boston, MA, USA). The correlation analysis of the GO annotation and KEGG pathway was performed between the transcriptome and proteome data, and the similarities and differences between the gene function and metabolic pathway in two groups of data were compared.

4.5. Bioinformatics Analysis of RIP Genes

Based on the NR and SwissProt annotation results, 21 transcripts of RIPs were screened. The Open Reading Frame Finder (ORF Finder) online program at NCBI (https://www.ncbi.nlm.nih.gov/orffinder, accessed on 2 March 2022) was used to determine the ORFs of these transcripts. The Conserved Domain Database (CDD) (https://www.ncbi.nlm.nih.gov/Structure/cdd/wrpsb.cgi, accessed on 2 March 2022) was used to identify conserved domains in protein sequences [59]. The ProtScale server (http://www.web.expasy.org/protscale, accessed on 2 March 2022) was used for the analysis of basic physicochemical properties of amino acids, including the number of amino acids, theoretical isoelectric point (pI), instability index (II), aliphatic index, and grand average of hydropathicity (GRAVY) [60], and the TMHMM tool (https://services.healthtech.dtu.dk/service.php?TMHMM-2.0, accessed on 2 March 2022) was used to analyze the transmembrane protein structure. A total of 23 amino acid sequences of other plant species were selected for phylogenetic tree analysis. The IQ-TREE 2.2.0 software [61] was used to construct the phylogenetic tree using the maximum likelihood (ML) method with 1000 bootstrap replicates. All amino acid sequences were aligned using the MUSCLE method.

4.6. Quantitative Reverse Transcription PCR (RT-qPCR)

The relative expression of the 10 selected genes was characterized using total RNA extracted at two developmental stages of *P. volubilis* seeds. Real-time quantitative polymerase chain reaction (RT-qPCR) was performed with all RNA samples using gene-specific oligonucleotide primers (Table S13), HiScript II Q RT SuperMix (Vazyme, Nanjing, China) with iQ SYBR Green Supermix (TaKaRa Bio, Beijing, China), and the Thermo Scientific PikoReal 96 Real-Time PCR System (Thermo Fisher, Waltham, MA, USA). All obtained values were normalized to the *Actin* gene (Unigene0042747) and then standardized to control conditions. Error bars represent the standard error of the mean (SEM), and the Student's *t*-test, a statistical significance test, was used to compare the means of groups, with $p < 0.05$ considered to be statistically significant. After performing RT-qPCR, *Ct* values were obtained using the ABI StepOnePlus™ software. The $2^{-\Delta\Delta Ct}$ method [62] was used to calculate the expression level of mRNA and FC values. Three biological replicates and three technical replicates were considered for each treatment in all experiments.

5. Conclusions

Numerous studies claimed that the plant RIPs have been connected to defense by the antifungal [63], antibacterial [64], antiviral [65], and anti-pest agent activities [43,66]. So, RIP has the potential to be widely used as biopesticides, and it has been extensively studied for its insecticidal activity. Due to its potential to inhibit protein biosynthesis, RIP can be used as a good target gene in plant genetic engineering for disease resistance. However, the related functions of RIP in *P. volubilis* need to be further studied. The screened transcripts in this study, including four transcripts (Unigene0000172, Unigene0000207, Unigene0000208, and Unigene0000209) encoding ricin-like proteins and Unigene0012386 encoding type I RIPs were highly expressed in Pv-2, and thus, they will be the focus of further research.

Supplementary Materials: The following supporting information can be downloaded at: https://www.mdpi.com/article/10.3390/ijms23179562/s1.

Author Contributions: Conceptualization, G.L.; methodology, G.L., Z.W. and Y.P.; software, G.L. and X.S.; formal analysis, G.L.; investigation, L.G.; writing—original draft preparation, G.L.; writing—review and editing, G.L. All authors have read and agreed to the published version of the manuscript.

Funding: This research was funded by the GuangDong Basic and Applied Basic Research Foundation (Project Number: 2021A1515011021 and 2019A1515110168).

Institutional Review Board Statement: Not applicable.

Informed Consent Statement: Not applicable.

Data Availability Statement: The raw data of sequenced transcriptome have been deposited to the Sequence Read Archive (SRA) at NCBI with the accession number of SRP387519 (http://www.ncbi.nlm.nih.gov/sra, accessed on 22 July 2022). The mass spectrometry data have been submitted into the iProX with the accession number of IPX0004771000 (http://www.ncbi.nlm.nih.gov/sra, accessed on 25 July 2022).

Conflicts of Interest: The authors declare no conflict of interest.

References

1. Fabbrini, M.S.; Katayama, M.; Nakase, I.; Vago, R. Plant ribosome-inactivating proteins: Progesses, challenges and biotechnological applications (and a few digressions). *Toxins* **2017**, *9*, 314. [CrossRef] [PubMed]
2. Noguchi, K.; Obuki, M.; Sumi, H.; Klussmann, M.; Morimoto, K.; Nakai, S.; Hashimoto, T.; Fujiwara, D.; Fujii, I.; Yuba, E.; et al. Macropinocytosis-Inducible Extracellular Vesicles Modified with Antimicrobial Protein CAP18-Derived Cell-Penetrating Peptides for Efficient Intracellular Delivery. *Mol. Pharm.* **2021**, *18*, 3290–3301. [CrossRef] [PubMed]
3. Yu, H.; Li, S.; Xu, N.; Liu, W. Ricin toxin and its neutralizing antibodies: A review. *Toxicon* **2022**, *214*, 47–53. [CrossRef] [PubMed]
4. Delgado, M.L.O.; Avril, A.; Prigent, J.; Dano, J.; Rouaix, A.; Worbs, S.; Dorner, B.G.; Rougeaux, C.; Becher, F.; Fenaille, F.; et al. Ricin Antibodies' Neutralizing Capacity against Different Ricin Isoforms and Cultivars. *Toxins* **2021**, *13*, 100. [CrossRef] [PubMed]
5. Van Damme, E.J.M.; Peumans, W.J.; Barre, A.; Rougé, P. Plant Lectins: A Composite of Several Distinct Families of Structurally and Evolutionary Related Proteins with Diverse Biological Roles. *Crit. Rev. Plant Sci.* **1998**, *17*, 575–692. [CrossRef]
6. Xu, N.; Dong, M.; Yang, Y.; Wang, Y.; Chang, Y.; Wan, J.; Zhu, W.; Wang, J.; Liu, W. Integrative transcriptomics, proteomics, and metabolomics data analysis exploring the injury mechanism of ricin on human lung epithelial cells. *Toxicol. Vitr.* **2019**, *60*, 160–172. [CrossRef]
7. Argent, R.H.; Roberts, L.M.; Wales, R.; Robertus, J.D.; Lord, J.M. Introduction of a disulfide bond into ricin A chain decreases the cytotoxicity of the ricin holotoxin. *J. Biol. Chem.* **1994**, *269*, 26705–26710. [CrossRef]

8. Liang, L.; Xia, J.; Liu, C.; Liu, S. Highly toxic type II ribosome-inactivating proteins ricin and abrin and their detection methods: A review. *Chin. J. Chromatogr.* **2021**, *39*, 260–270. [CrossRef]
9. Sandvig, K.; Kavaliauskiene, S.; Skotland, T. The Protein Toxins Ricin and Shiga Toxin as Tools to Explore Cellular Mechanisms of Internalization and Intracellular Transport. *Toxins* **2021**, *13*, 377. [CrossRef]
10. Zhang, Y.-H.; Wang, Y.; Yusufali, A.H.; Ashby, F.; Zhang, D.; Yin, Z.-F.; Aslanidi, G.V.; Srivastava, A.; Ling, C.-Q.; Ling, C. Cytotoxic genes from traditional Chinese medicine inhibit tumor growth both in vitro and in vivo. *J. Integr. Med.* **2014**, *12*, 483–494. [CrossRef]
11. Tyagi, N.; Tyagi, M.; Pachauri, M.; Ghosh, P.C. Potential therapeutic applications of plant toxin-ricin in cancer: Challenges and advances. *Tumor Biol.* **2015**, *36*, 8239–8246. [CrossRef] [PubMed]
12. Meneguelli de Souza, L.C.; Carvalho, L.P.; Araújo, J.S.; Melo, E.J.T.; Machado, O.L.T. Cell toxicity by ricin and elucidation of mechanism of Ricin inactivation. *Int. J. Biol. Macromol.* **2018**, *113*, 821–828. [CrossRef] [PubMed]
13. Goyal, A.; Tanwar, B.; Sihag, M.K.; Sharma, V. Sacha inchi (*Plukenetia volubilis* L.): An emerging source of nutrients, omega-3 fatty acid and phytochemicals. *Food Chem.* **2022**, *373*, 131459. [CrossRef]
14. Kodahl, N. Sacha inchi (*Plukenetia volubilis* L.)—From lost crop of the Incas to part of the solution to global challenges? *Planta* **2020**, *251*, 80. [CrossRef] [PubMed]
15. Gutiérrez, L.-F.; Quiñones-Segura, Y.; Reinoso, Z.S.; Díaz, D.L.; Abril, J.I. Physicochemical properties of oils extracted from γ-irradiated Sacha Inchi (*Plukenetia volubilis* L.) seeds. *Food Chem.* **2017**, *237*, 581–587. [CrossRef] [PubMed]
16. Sierra, D.M.C.; Rave, L.J.G.; Soto, J.A. Biological Activity of Sacha Inchi (*Plukenetia volubilis* Linneo) and Potential Uses in Human Health: A Review. *Food Technol. Biotechnol.* **2021**, *59*, 253–266. [CrossRef]
17. Chirinos, R.; Pedreschi, R.; Domínguez, G.; Campos, D. Comparison of the physico-chemical and phytochemical characteristics of the oil of two Plukenetia species. *Food Chem.* **2015**, *173*, 1203–1206. [CrossRef]
18. EFSA (European Food Safety Authority). Technical report on the notification of roasted seeds from *Plukenetia volubilis* L. as a traditional food from a third country pursuant to Article 14 of Regulation (EU) 2015/2283. *EFSA Support. Publ.* **2020**, *17*, 1817E.
19. Srichamnong, W.; Ting, P.; Pitchakarn, P.; Nuchuchua, O.; Temviriyanukul, P. Safety assessment of *Plukenetia volubilis* (Inca peanut) seeds, leaves, and their products. *Food Sci. Nutr.* **2018**, *6*, 962–969. [CrossRef]
20. Liu, G.; Wu, Z.; Peng, Y.; Shang, X.; Xie, Y.; Arnold, R.J. Transcriptome analyses reveals the dynamic nature of oil accumulation during seed development of *Plukenetia volubilis* L. *Sci. Rep.* **2020**, *10*, 20467. [CrossRef]
21. Wang, X.; Xu, R.; Wang, R.; Liu, A. Transcriptome analysis of Sacha Inchi (*Plukenetia volubilis* L.) seeds at two developmental stages. *BMC Genom.* **2012**, *13*, 716. [CrossRef] [PubMed]
22. Loss-Morais, G.; Turchetto-Zolet, A.C.; Etges, M.; Cagliari, A.; Körbes, A.P.; Maraschin, F.D.S.; Margis-Pinheiro, M.; Margis, R. Analysis of castor bean ribosome-inactivating proteins and their gene expression during seed development. *Genet. Mol. Biol.* **2013**, *36*, 74–86. [CrossRef] [PubMed]
23. Liu, G.; Chen, H.; Wu, Z.; Peng, Y.; Xie, Y. Analyses of seed development of *Plukenetia volubilis* by joint metabolomics and transcriptomics approaches. *Sci. Silvae Sin.* **2019**, *55*, 169–179.
24. Zhang, D.; Zhang, H.; Hu, Z.; Chu, S.; Yu, K.; Lv, L.; Yang, Y.; Zhang, X.; Chen, X.; Kan, G.; et al. Artificial selection on GmOLEO1 contributes to the increase in seed oil during soybean domestication. *PLoS Genet.* **2019**, *15*, e1008267. [CrossRef]
25. Zou, Z.; Zhao, Y.; Zhang, L. Genomic insights into lineage-specific evolution of the oleosin family in Euphorbiaceae. *BMC Genom.* **2022**, *23*, 178. [CrossRef]
26. Faraji, S.; Chari, G.; Najafi-Zarrini, H. Phosphatidylinositol pathway-associated genes adjust the rice growth and stress signaling: A global assay of the 5PTase family in the *Oryza sativa* genome. *Plant Gene* **2020**, *23*, 100244. [CrossRef]
27. Sun, T.; Zhang, J.; Zhang, Q.; Li, X.; Li, M.; Yang, Y.; Zhou, J.; Wei, Q.; Zhou, B. Exogenous application of acetic acid enhances drought tolerance by influencing the MAPK signaling pathway induced by ABA and JA in apple plants. *Tree Physiol.* **2022**, *22*, tpac034. [CrossRef]
28. Tang, N.; Liu, W.; Zhang, W.; Tang, D. Integrative analysis of transcriptomic and proteomic changes related to male sterility in Tagetes erecta. *Physiol. Mol. Biol. Plants* **2020**, *26*, 2061–2074. [CrossRef]
29. Oliveto, S.; Mancino, M.; Manfrini, N.; Biffo, S. Role of microRNAs in translation regulation and cancer. *World J. Biol. Chem.* **2017**, *8*, 45–56. [CrossRef]
30. Song, B.; Luo, X.; Luo, X.; Liu, Y.; Niu, Z.; Zeng, X. Learning spatial structures of proteins improves protein–protein interaction prediction. *Brief. Bioinform.* **2022**, *23*, bbab558. [CrossRef]
31. Li, X.; Liu, W. Structure, function and distribution of ribosomal inactivated proteins. *Chin. J. Cell Biol* **1997**, *19*, 69–75.
32. Strasser, R. Plant protein glycosylation. *Glycobiology* **2016**, *26*, 926–939. [CrossRef] [PubMed]
33. Wang, T.; Zhao, J.; Yang, A. Furin: An endoprotease involved in processing of a wide variety of precursor proteins. *J. Med. Mol. Biol* **2006**, *3*, 4.
34. Ardito, F.; Giuliani, M.; Perrone, D.; Troiano, G.; Lo Muzio, L. The crucial role of protein phosphorylation in cell signaling and its use as targeted therapy. *Int. J. Mol. Med.* **2017**, *40*, 271–280. [CrossRef] [PubMed]
35. Rezaei-Moshaei, M.; Bandehagh, A.; Dehestani, A.; Pakdin-Parizi, A.; Golkar, M. Molecular cloning and in-depth bioinformatics analysis of type II ribosome-inactivating protein isolated from Sambucus ebulus. *Saudi J. Biol. Sci.* **2020**, *27*, 1609–1623. [CrossRef]
36. Peng, L.; Li, E.-M.; Xu, L.-Y. From start to end: Phase separation and transcriptional regulation. *Biochim. Biophys. Acta Gene Regul. Mech.* **2020**, *1863*, 194641. [CrossRef]

37. Xiao, Y.; Yi, F.; Ling, J.; Wang, Z.; Zhao, K.; Lu, N.; Qu, G.; Kong, L.; Ma, W.; Wang, J. Transcriptomics and Proteomics Reveal the Cellulose and Pectin Metabolic Processes in the Tension Wood (Non-G-Layer) of Catalpa bungei. *Int. J. Mol. Sci.* **2020**, *21*, 1686. [CrossRef]
38. Li, Y.; Liu, Y.; Huang, X.; Ren, J. Analysis of protein phosphorylation combining capillary electrophoresis with ATP analog labeling technique. *Electrophoresis* **2022**, *43*, 548–558. [CrossRef]
39. Chan, A.P.; Crabtree, J.; Zhao, Q.; Lorenzi, H.; Orvis, J.; Puiu, D.; Melake-Berhan, A.; Jones, K.M.; Redman, J.; Chen, G.; et al. Draft genome sequence of the oilseed species Ricinus communis. *Nat. Biotechnol.* **2010**, *28*, 951–956. [CrossRef]
40. Wales, R.; Richardson, P.T.; Roberts, L.M.; Lord, J.M. Recombinant ricin B chain fragments containing a single galactose binding site retain lectin activity. *Arch. Biochem. Biophys.* **1992**, *294*, 291–296. [CrossRef]
41. Villafranca, J.; Robertus, J. Ricin B chain is a product of gene duplication. *J. Biol. Chem.* **1981**, *256*, 554–556. [CrossRef]
42. Lord, J.M.; Roberts, L.M.; Robertus, J.D. Ricin: Structure, mode of action, and some current applications. *FASEB J.* **1994**, *8*, 201–208. [CrossRef] [PubMed]
43. Stirpe, F. Ribosome-inactivating proteins: From toxins to useful proteins. *Toxicon* **2013**, *67*, 12–16. [CrossRef] [PubMed]
44. Lledías, F.; Gutiérrez, J.; Martínez-Hernández, A.; García-Mendoza, A.; Sosa, E.; Hernández-Bermúdez, F.; Dinkova, T.D.; Reyes, S.; Cassab, G.I.; Nieto-Sotelo, J. Mayahuelin, a Type I Ribosome Inactivating Protein: Characterization, Evolution, and Utilization in Phylogenetic Analyses of Agave. *Front. Plant Sci.* **2020**, *11*, 573. [CrossRef] [PubMed]
45. Zhu, F.; Zhou, Y.-K.; Ji, Z.-L.; Chen, X.-R. The Plant Ribosome-Inactivating Proteins Play Important Roles in Defense against Pathogens and Insect Pest Attacks. *Front. Plant Sci.* **2018**, *9*, 146. [CrossRef] [PubMed]
46. Wang, S.; Zhang, H.; Zheng, Y.; Li, Z.; Xiang, F.; Ding, Y.; Xiang, J. Environmental factors and phytohormones enhancing expression of α-momorcharin gene in Momordica charantia. *Biologia* **2016**, *71*, 155–160. [CrossRef]
47. Qian, Q.; Huang, L.; Yi, R.; Wang, S.; Ding, Y. Enhanced resistance to blast fungus in rice (*Oryza sativa* L.) by expressing the ribosome-inactivating protein alpha-momorcharin. *Plant Sci.* **2014**, *217*, 1–7. [CrossRef]
48. Zhu, F.; Yuan, S.; Zhang, Z.-W.; Qian, K.; Feng, J.-G.; Yang, Y.-Z. Pokeweed antiviral protein (PAP) increases plant systemic resistance to Tobacco mosaic virus infection in Nicotiana benthamiana. *Eur. J. Plant Pathol.* **2016**, *146*, 541–549. [CrossRef]
49. Ayaz, A.; Saqib, S.; Huang, H.; Zaman, W.; Lü, S.; Zhao, H. Genome-wide comparative analysis of long-chain acyl-CoA synthetases (LACSs) gene family: A focus on identification, evolution and expression profiling related to lipid synthesis. *Plant Physiol. Biochem.* **2021**, *161*, 1–11. [CrossRef]
50. Zhao, L.; Katavic, V.; Li, F.; Haughn, G.W.; Kunst, L. Insertional mutant analysis reveals that long-chain acyl-CoA synthetase 1 (LACS1), but not LACS8, functionally overlaps with LACS9 in Arabidopsis seed oil biosynthesis. *Plant J.* **2010**, *64*, 1048–1058. [CrossRef]
51. Chen, S.; Zhou, Y.; Chen, Y.; Gu, J. Fastp: An ultra-fast all-in-one FASTQ preprocessor. *Bioinformatics* **2018**, *34*, i884–i890. [CrossRef] [PubMed]
52. Grabherr, M.G.; Haas, B.J.; Yassour, M.; Levin, J.Z.; Thompson, D.A.; Amit, I.; Adiconis, X.; Fan, L.; Raychowdhury, R.; Zeng, Q.D.; et al. Full-length transcriptome assembly from RNA-Seq data without a reference genome. *Nat. Biotechnol.* **2011**, *29*, 644–652. [CrossRef] [PubMed]
53. Simão, F.A.; Waterhouse, R.M.; Ioannidis, P.; Kriventseva, E.V.; Zdobnov, E.M. BUSCO: Assessing genome assembly and annotation completeness with single-copy orthologs. *Bioinformatics* **2015**, *31*, 3210–3212. [CrossRef]
54. Love, M.I.; Huber, W.; Anders, S. Moderated estimation of fold change and dispersion for RNA-seq data with DESeq2. *Genome Biol.* **2014**, *15*, 550. [CrossRef]
55. Robinson, M.D.; McCarthy, D.J.; Smyth, G.K. EdgeR: A Bioconductor package for differential expression analysis of digital gene expression data. *Bioinformatics* **2010**, *26*, 139–140. [CrossRef] [PubMed]
56. Perkins, D.N.; Pappin, D.J.; Creasy, D.M.; Cottrell, J.S. Probability-based protein identification by searching sequence databases using mass spectrometry data. *Electrophor. Int. J* **1999**, *20*, 3551–3567. [CrossRef]
57. Meierhofer, D.; Weidner, C.; Sauer, S. Integrative Analysis of Transcriptomics, Proteomics, and Metabolomics Data of White Adipose and Liver Tissue of High-Fat Diet and Rosiglitazone-Treated Insulin-Resistant Mice Identified Pathway Alterations and Molecular Hubs. *J. Proteome Res.* **2014**, *13*, 5592–5602. [CrossRef]
58. Dou, T.; Wang, J.; Liu, Y.; Jia, J.; Zhou, L.; Liu, G.; Li, X.; Han, M.; Lin, J.; Huang, F.; et al. A Combined Transcriptomic and Proteomic Approach to Reveal the Effect of Mogroside V on OVA-Induced Pulmonary Inflammation in Mice. *Front. Immunol.* **2022**, *13*, 5592–5602. [CrossRef]
59. Lu, S.; Wang, J.; Chitsaz, F.; Derbyshire, M.K.; Geer, R.C.; Gonzales, N.R.; Gwadz, M.; Hurwitz, D.I.; Marchler, G.H.; Song, J.S.; et al. CDD/SPARCLE: The conserved domain database in 2020. *Nucleic Acids Res.* **2020**, *48*, D265–D268. [CrossRef]
60. Nie, L.; Xu, Z.; Wu, L.; Chen, X.; Cui, Y.; Wang, Y.; Song, J.; Yao, H. Genome-wide identification of protein phosphatase 2C family members in Glycyrrhiza uralensis Fisch. and their response to abscisic acid and polyethylene glycol stress. *J. Taibah Univ. Sci.* **2021**, *15*, 1260–1268. [CrossRef]
61. Minh, B.Q.; Schmidt, H.A.; Chernomor, O.; Schrempf, D.; Woodhams, M.D.; Von Haeseler, A.; Lanfear, R. IQ-TREE 2: New Models and Efficient Methods for Phylogenetic Inference in the Genomic Era. *Mol. Biol. Evol.* **2020**, *37*, 1530–1534. [CrossRef] [PubMed]
62. Livak, K.J.; Schmittgen, T.D. Analysis of relative gene expression data using real-time quantitative PCR and the $2^{-\Delta\Delta CT}$ method. *Methods* **2001**, *25*, 402–408. [CrossRef]

63. Zhu, F.; Zhang, P.; Meng, Y.-F.; Xu, F.; Zhang, D.-W.; Cheng, J.; Lin, H.-H.; Xi, D.-H. Alpha-momorcharin, a RIP produced by bitter melon, enhances defense response in tobacco plants against diverse plant viruses and shows antifungal activity in vitro. *Planta* **2013**, *237*, 77–88. [CrossRef] [PubMed]
64. Ajji, P.K.; Walder, K.; Puri, M. Functional Analysis of a Type-I Ribosome Inactivating Protein Balsamin from Momordica balsamina with Anti-Microbial and DNase Activity. *Plant Foods Hum. Nutr.* **2016**, *71*, 265–271. [CrossRef] [PubMed]
65. Sipahioğlu, H.M.; Kaya, I.; Usta, M.; Ünal, M.; Özcan, D.; Dilmen, M.Ö.; Güller, A.; Pallas, V. Pokeweed (*Phytolacca americana* L.) antiviral protein inhibits Zucchini yellow mosaic virus infection in a dose-dependent manner in squash plants. *Turk. J. Agric. For.* **2017**, *41*, 256–262. [CrossRef]
66. Akkouh, O.; Ng, T.B.; Cheung, R.C.F.; Wong, J.H.; Pan, W.; Ng, C.C.W.; Sha, O.; Shaw, P.C.; Chan, W.Y. Biological activities of ribosome-inactivating proteins and their possible applications as antimicrobial, anticancer, and anti-pest agents and in neuroscience research. *Appl. Microbiol. Biotechnol.* **2015**, *99*, 9847–9863. [CrossRef]

International Journal of
Molecular Sciences

MDPI

Article

Genome-Wide Identification of DOF Gene Family and the Mechanism Dissection of *SbDof21* Regulating Starch Biosynthesis in Sorghum

Qianlin Xiao *, Tingting Liu, Min Ling, Qiannan Ma, Wan Cao, Fangyu Xing, Tianhui Huang, Yingyi Zhang, Hong Duan and Zhizhai Liu *

College of Agronomy and Biotechnology, Southwest University, Chongqing 400716, China
* Correspondence: xiaoql1853@swu.edu.cn (Q.X.); liu003@swu.edu.cn (Z.L.)

Abstract: Starch is one of the main utilization products of sorghum (*Sorghum bicolor* L.), the fifth largest cereal crop in the world. Up to now, the regulation mechanism of starch biosynthesis is rarely documented in sorghum. In the present study, we identified 30 genes encoding the C2-C2 zinc finger domain (DOF), with one to three exons in the sorghum genome. The DOF proteins of sorghum were divided into two types according to the results of sequence alignment and evolutionary analysis. Based on gene expressions and co-expression analysis, we identified a regulatory factor, *SbDof21*, that was located on chromosome 5. *SbDof21* contained two exons, encoding a 36.122 kD protein composed of 340 amino acids. *SbDof21* co-expressed with 15 genes involved in the sorghum starch biosynthesis pathway, and the Pearson correlation coefficients (PCCs) with 11 genes were greater than 0.9. The results of qRT-PCR assays indicated that *SbDof21* is highly expressed in sorghum grains, exhibiting low relative expression levels in the tissues of roots, stems and leaves. SbDOF21 presented as a typical DOF transcription factor (TF) that was localized to the nucleus and possessed transcriptional activation activity. Amino acids at positions 182–231 of SbDOF21 formed an important structure in its activation domain. The results of EMSA showed that SbDOF21 could bind to four tandem repeats of P-Box (TGTAAAG) motifs in vitro, such as its homologous proteins of ZmDOF36, OsPBF and TaPBF. Meanwhile, we also discovered that SbDOF21 could bind and transactivate *SbGBSSI*, a key gene in sorghum amylose biosynthesis. Collectively, the results of the present study suggest that SbDOF21 acts as an important regulator in sorghum starch biosynthesis, exhibiting potential values for the improvement of starch contents in sorghum.

Keywords: sorghum (*Sorghum bicolor* L.); *SbDof21*; starch biosynthesis; transcriptional regulation

Citation: Xiao, Q.; Liu, T.; Ling, M.; Ma, Q.; Cao, W.; Xing, F.; Huang, T.; Zhang, Y.; Duan, H.; Liu, Z. Genome-Wide Identification of DOF Gene Family and the Mechanism Dissection of *SbDof21* Regulating Starch Biosynthesis in Sorghum. *Int. J. Mol. Sci.* **2022**, 23, 12152. https://doi.org/10.3390/ijms232012152

Academic Editors: Andrés J. Cortés and Hai Du

Received: 30 August 2022
Accepted: 4 October 2022
Published: 12 October 2022

1. Introduction

Starch, one of the most important carbohydrates in the world, is the main component of human food and an important raw material for industrial applications [1,2]. Human beings have currently achieved the artificial synthesis of starch from carbon dioxide [3], while the current main way to obtain starch for mankind still depends on plant photosynthesis.

In plants, starch biosynthesis occurs in the plastids and undergoes a series of complex and coordinated biochemical reactions catalyzed by multiple enzymes [1,4]. Among these enzymes, ADP-glucose pyrophosphorylase (AGPase) catalyzes the biosynthesis of adenosine diphosphate-glucose (ADPG) to provide direct precursors for starch biosynthesis, presenting as a key speed-limiting step during starch biosynthesis [5]. Starch synthase (SS), including granule-bound starch synthase (GBSS) and soluble starch synthesis (SSS), is mainly responsible for the extension of sugar chains [1,6]. GBSS functions in the synthesis of amylase, and its mutation can lead to waxy (amylose-free) grains in wheat [7,8]. SSS catalyzes linear chain elongation by adding glucose units provided by ADPG to the non-reducing end of the acceptor chains, which is important for the synthesis of amylopectin [9,10]. The starch-branching enzyme (SBE) cuts α-1,4-glucan chains and transfers the segments to glucosyl residue in the C6 position, which is important for catalyzing the formation of the branch linkage [11]. The starch-debranching enzyme (DBE) is involved in amylopectin synthesis for hydrolyzing the wrong branch linkages and also participates in the formation of starch granules and the degradation of starch [12–14]. Besides AGPase, SS, SBE and DBE, another enzyme of starch phosphorylase (SP) is reported to have a function in starch biosynthesis [15–17]. Meanwhile, all those functional enzymes involved in starch synthesis contain isozymes that are encoded by multiple genes and exhibit tissue-specific expression patterns, constituting a complex mechanism of starch biosynthesis in plants [1,4,6].

In addition, the complicated regulatory networks also play key roles in starch biosynthesis. For example, AGPase is regulated by allosteric effectors, 3-phosphoglycerate and Pi [5]; ISA1 and ISA2 can form heterodimers in rice [18] and *Arabidopsis* [19]; SS and SBE tend to function in the form of complexes [20–22]; and the phosphorylation of functional enzymes is also one of the important regulation patterns for starch biosynthesis [23,24]. Meanwhile, the gene-encoding enzymes related to starch biosynthesis are induced by signal molecules and are directly regulated by transcription factors (TFs) at the transcriptional level, which already becomes an important regulation mode of starch biosynthesis in plants [4,25]. For example, HvSUSIBA2 in barley can regulate the transcription of *Iso1* by binding to the sugar response element within the promoter [26]; OsbZIP58 can directly bind to the promoter regions of *OsAGPL3, Wx, OsSSIIa, SBE1, OsBEIIb* and *ISA2* and regulate their expression [27]. RSR1, identified by gene co-expression analysis, is a negative regulatory TF for starch biosynthesis in rice [28]; OsNAC20/26—highly expressed in rice grains—are also involved in the transcriptional regulation of starch biosynthesis [29]. In maize, TFs of ZmbZIP91 [30], ZmMYB14 [31], ZmNAC126 [32], ZmNAC128, ZmNAC130 [33], O2 and PBF [34] are also reported to regulate the transcription of maize starch biosynthesis-related genes, thereby affecting starch contents in kernels. However, the biosynthesis and regulation mechanisms of starch are rarely documented in sorghum grain.

DNA binding with the C2-C2 zinc finger domain (DOF) family, a group of plant-specific TFs, contains a conserved region of 50 amino acids with a C2-C2 finger domain at the N-terminal that recognizes the cis-element containing (A/T) AAAG core sequence [35–37] and prolamin box (P-Box) [38,39]. DOF TFs have been reported to regulate multiple aspects of plant development. In *Arabidopsis*, OBF binding protein 1 (OBP1) can regulate the transcription of *CycD3;3* and regulate the plant cell cycle [40]. *AtDof2.1*, a jasmonic acid (JA)-inducible gene, can significantly reduce the promotion of leaf senescence [41]. In rice, OsDOF12 and OsDOF4 have been reported to participate in the regulation of flowering [42], while OsDOF12 and OsDOF18 are related to the regulation of nitrogen absorption in rice roots [43]. Meanwhile, an endosperm-specific gene in maize, *ZmDof3*, regulates starch accumulation and aleurone development [44]; ZmDOF36 can directly regulate the transcription of *ZmAGPS1a, ZmAGPL1, ZmGBSSI, ZmSSIIa, ZmISA1* and *ZmISA3*, playing important roles in starch biosynthesis in maize kernels [45].

DOF TFs exhibit diverse and important functions in different plants, including the cereals of rice, maize and wheat, while few documents are focused on the DOF family of sorghum. Although Kushwaha and colleagues have preliminarily identified the DOF TFs in sorghum [46], the functional profiling of DOF TFs, especially their regulations in starch biosynthesis, still remains less dissection in sorghum. Here, we identified 30 proteins containing conserved DOF domains from the entire sorghum genome through sequence characterizing. Phylogenetic analysis was performed on the amino acids, and we divided all sorghum DOF TFs into two groups, i.e., Group A and B. *SbDof21*, a highly expressed DOF gene in grains, was selected for further dissection of transcriptional regulation to the starch biosynthesis genes in sorghum grains.

2. Results

2.1. Identification of Sorghum DOF Proteins

A total of 30 DOF proteins were finally obtained from the sorghum genome via BLASTP queries. To better distinguish the corresponding sorghum Dof genes (*SbDofs*), we temporarily named those coding genes from *SbDof1* to *SbDof30* according to their order on the corresponding chromosomes (Table S1). Meanwhile, *SbDof5* and *SbDof30* are different from those reported by Kushwaha et al. [46], showing up as newly identified Dof genes in sorghum. Among these *SbDofs*, eight are located on chromosome 1 (Chr1), six on Chr3, three on each of Chr2, Chr4, Chr8 and Chr9, two on Chr6, while only one on both Chr5 and Chr7 (Table S1, Figure 1a). The number of amino acids encoded by 30 *SbDofs* ranged from 168 aa (*SbDof15*) to 560 aa (*SbDof13*), with the molecular weight and isoelectric points (pI) ranging from 17.25 kDa (SbDOF15) to 61.40 kDa (SbDOF13) and 4.75 (SbDOF17) to 10.27 (SbDOF13), correspondingly (Table S1). Subcellular localization analysis of all sorghum DOF proteins (SbDOFs) demonstrated that SbDOF5 located on chloroplast thylakoid membrane, SbDOF10 and SbDOF16 on the mitochondrion, SbDOF12, SbDOF15 and SbDOF26 on extracellular space, SbDOF23 on chloroplast, while the remaining 23 SbDOFs all located in the nucleus (Table S1).

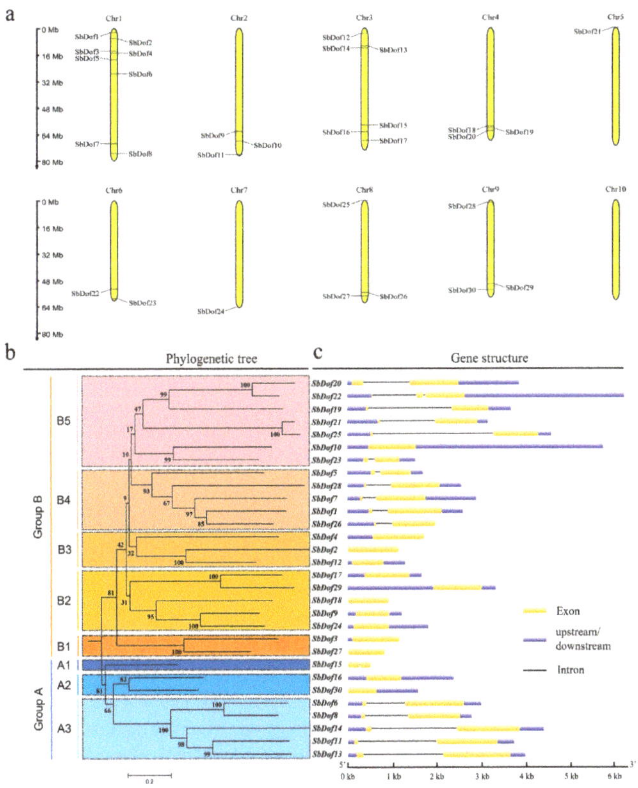

Figure 1. Characterizing 30 identified *SbDofs* across the sorghum genome: (**a**) Chromosomal localization of 30 identified *SbDofs*, the chromosome size is indicated by the left side rulers; (**b**) Phylogenetic analysis of 30 *SbDofs*, A and B refer to two groups of *SbDofs*, while A1 to A3 and B1 to B5 to the corresponding subgroups of Group A and B, figures on each clade present the percentage resulted from the bootstrap analysis; (**c**) Gene structures of 30 *SbDofs* in sorghum.

2.2. Phylogeny and Sequence Characteristics of Dofs in Sorghum

Neighbor-joining (NJ) phylogenetic trees were constructed to reveal the evolutionary relationships among the members of *SbDof*s. Thirty identified *SbDof*s were divided into two groups: Group A and Group B (Figure 1b), similar to the phylogenetic results of Maximum likelihood (ML) (Figure S1). Groups A and B were further divided into three (A1 to A3) and five (B1 to B5) subgroups, respectively (Figure 1b). *SbDof21* and *SbDof25* from subgroup B5 exhibited a relatively closer relationship and formed an independent clade in B5 (Figure 1b). Similar trends were also observed among the pairs of *SbDof20/22* in B5, *SbDof2/12* in B3, *SbDof17/29* in B2, *SbDof3/27* in B1 and *SbDof6/8* in A3 (Figure 1b).

The results of gene structure analysis showed that all *SbDof*s contained only one or two exons (CDSs) (Figure 1c). Among the divided groups and subgroups, *SbDof*s from subgroups of B1, B2 and B3 exhibited only one exon, while all *SbDof*s of B4 contained two exons, *SbDof10* in B5 has only one exon, *SbDof22* has three exons, and others possess two exons (Figure 1c). All *SbDof*s in A1 and A3 in Group A contain two exons, while those in A2 exhibit a single-exon structure (Figure 1c).

To reveal the sequence features of SbDOFs, we performed domain and motif analysis. The results showed that all the SbDOFs contain conserved zf-Dof domain at a position closer to their N-terminals. Moreover, SbDOF13 contains another PTZ00121 superfamily domain at the N-terminal close to the zf-Dof domain with an unknown function currently (Figure 2a). The amino acid (aa) sequence of the conserved zf-DOF domain was further analyzed by sequence alignment. Twenty-nine identified SbDOFs all contained two C2 domains, while SbDOF15 possessed only one C at the region of the first C2 (Figure 2b). Even though, SbDOF15 still formed its conservative zf-Dof domain.

A total of 15 conserved motifs were detected in the 30 SbDOFs, i.e., Motif1 to 15 (Figure 2c). The multilevel consensus sequence and logos of conserved motifs are shown in Table S2 and Figure S2. Among these motifs, Motif1 was conserved and existed in all SbDOFs, while the other 14 motifs (Motif2 to Motif15) were only partially detected among specific SbDOFs (Figure 2c). For example, Motif2/4/6 were common to proteins of SbDOF6/8/11/13/14, Motif2/6 co-existed at the C-terminal of these proteins and Motif4 was located at the N-terminal near Motif1 (Figure 2c). Motif3 was observed at the N-terminus of seven SbDOFs, while Motif5/7 were only detected in both SbDOF21 and SbDOF25 (Figure 2c).

2.3. SbDof21 Highly Expressed in Sorghum Grains and Co-Expressed with Maize Starch Biosynthesis-Related Genes

Based on previous studies, we dissected the expression patterns of *SbDof*s among different sorghum tissues [47]. The results showed that the expression patterns of *SbDof*s were divided into two major types (Type i and ii). Additionally, 14 genes were involved in Type i; these genes were specifically expressed in grains or other tissues, among which both *SbDof21* and *SbDof25* exhibited high expression levels in grains (Figure 3a). However, the 16 genes in type ii were basically expressed in all tissues and formed a non-specific expression group (Figure 3a).

In our previous study, genes related to sorghum starch biosynthesis can also be divided into two categories based on their expression patterns. Type I covered 15 genes that were almost all highly expressed in sorghum grains, while Type II contained the rest of the 12 genes that exhibited relatively low expression levels in all tissues [47]. Co-expression analysis based on the expression data of RNA-sequencing (RNA-Seq) revealed that *SbDof21* exhibited a similar expression pattern to sorghum starch biosynthesis-related genes, and its Pearson correlation coefficient (PCC) with 15 sorghum starch biosynthesis-related genes were greater than 0.5 (Figure 3b). The PCC between *SbDof21* and 11 sorghum starch biosynthesis-related genes (*SbAGPS1, SbAGPS2, SbAGPLS1, SbGBSSI, SbSSIIa, SbISA1, SbPUL, SbSBEI, SbSBEIIb, SbPHOL, SbBt1*) were even greater than 0.9 (Figure 3b).

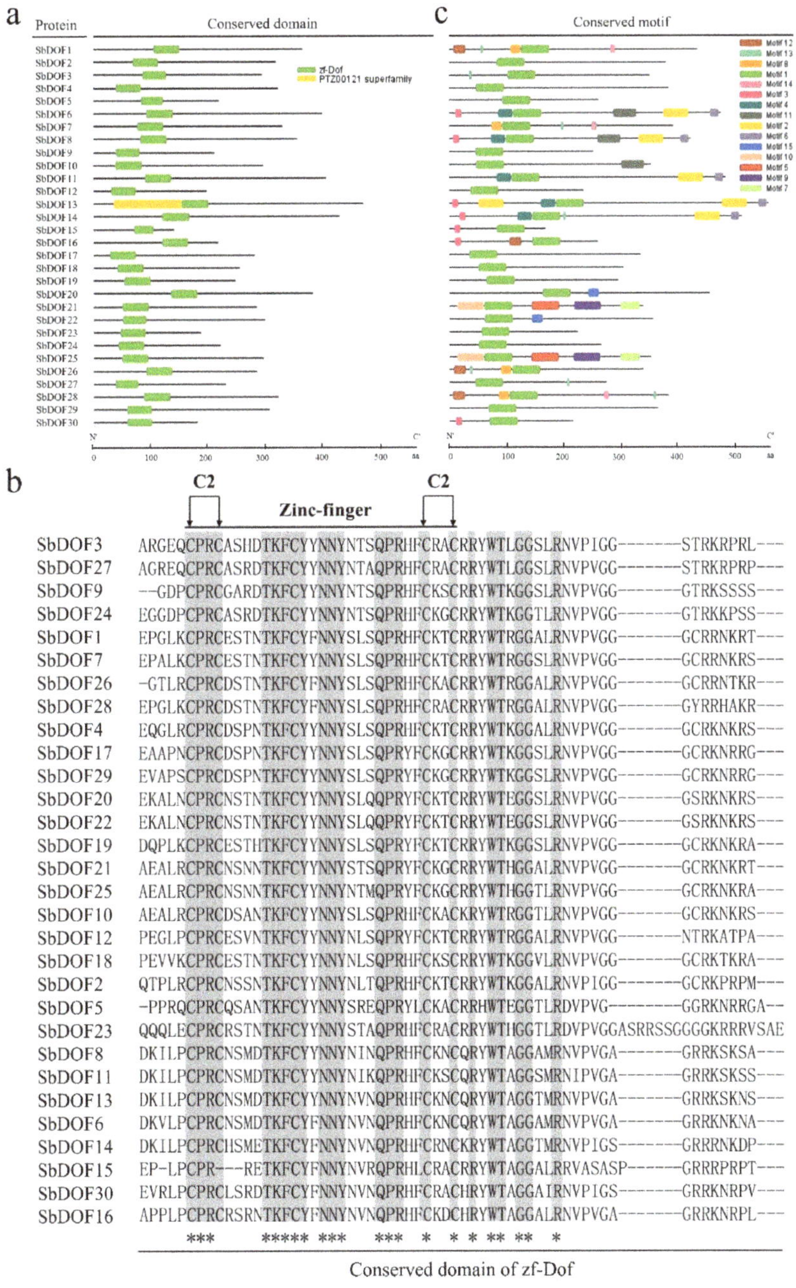

Figure 2. Sequence analysis of 30 DOF proteins in sorghum. (**a**) Conserved domain of 30 SbDOFs. The green box indicates the zf-DOF domain, and the lines represent amino acid sequences. (**b**) Sequence alignment of conserved domains of 30 SbDOFs. Gray labelled letters indicate the same amino acid sites in all proteins, * refers to the conserved amino acids among different SbDOFs. (**c**) Conserved motifs of zf-DOF among 30 identified DOF proteins in sorghum. Different colored boxes represent different motifs, and black lines represent amino acid sequences.

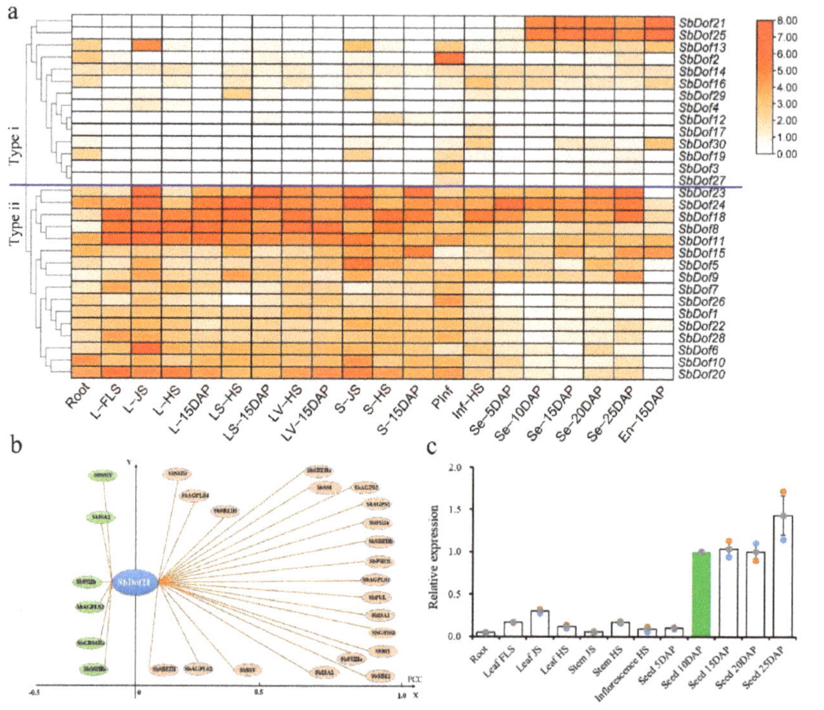

Figure 3. Expression profiling of *SbDof*s. (**a**) Expression data of 30 *SbDof*s in different tissues. (**b**) Co-expression analysis of *SbDof21* and sorghum starch biosynthesis-related genes. The vertical distance from a gene to the Y axis indicates the size of the Pearson correlation coefficient (PCC). (**c**) The expression pattern of *SbDof21* revealed via qRT-PCR. The green bar represents the relative expression level of Seed_10DAP that is standard to 1.

We further investigated the expression pattern of *SbDof21* in multiple tissues of sorghum cultivar BTx623 through qRT-PCR. The results showed that the transcripts of *SbDof21* could be detected among all 12 tissues, while the relative expression levels of *SbDof21* were apparently in four grain-related tissues of Seed_10/15/20/25 DAP rather than the other eight tissues, even the grain tissue of Seed_5DAP (Figure 3c).

2.4. Cloning and Sequence Analysis of SbDOF21

The coding sequence (CDS) of *SbDof21* contained 1023 bp and encoded a 36.122 kDa protein that consisted of 340 amino acids (Figure 4a). The protein sequence contains one zf-DOF domain at the N terminus, indicating that the cloned *SbDof21* was definitely a TF belonging to the DOF family (Figure 4b). Additionally, the results showed that another DOF gene, i.e., *SbDof25*, exhibited as the homolog of *SbDof21* in the sorghum genome, with sequence similarity of their cDNA > 71%. Meanwhile, *SbDof25* is also highly expressed in sorghum grains (Figure 3a).

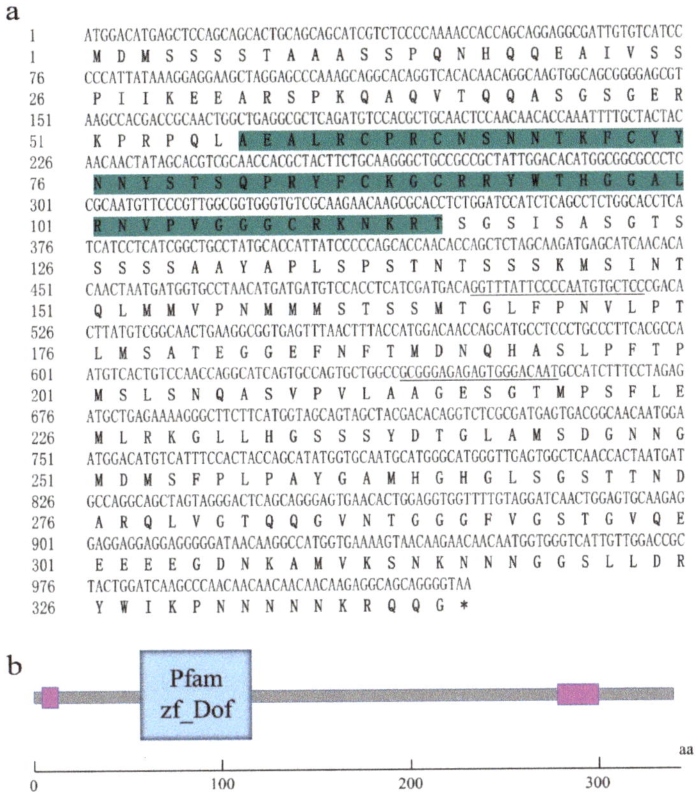

Figure 4. Coding Sequence (CDS) analysis of *SbDof21*. (**a**) The cloned CDS of *SbDof21* in sorghum cultivar (BTx623) and the corresponding coding products of amino acids. The blue background refers to the amino acids that formed the zf-DOF domain. The underlined DNA sequence refers to the primer position of qRT-PCR. (**b**) The predicted structure of the conserved domain of SbDOF21 by Smart. Pink boxes present low complexity regions.

2.5. SbDOF21 Was a Nuclear Localization Protein with Activation Activity

The transactivation activity of SbDOF21 was detected via a GAL4-based Y2H system. The experiment vector pGBKT7-SbDOF21, positive control pGBKT7-ZmMYB14 [31], and negative control pGBKT7 were transformed into yeast strain AH109 and screened on SD/-Trp plates. The transactivation activity was determined in the colonies cultivated on the SD/-Trp-His-Ura plates containing *X-α-gal*. The yeast contains the pGBKT7-SbDOF21 and pGBKT7-ZmMYB14 that could degrade the *X-α-gal* and turn blue, which demonstrated that SbDOF21 exhibited self-activating trans-activity (Figure 5a). The truncated SbDOF21 protein was prepared by removing the N-terminal or C-terminal residues, and its self-activation activity was detected. The results showed that 49 amino acids between the positions of 182 and 231 of SbDOF21 were important components of the activation domain, and the N-terminal and C-terminal sequences of SbDOF21 acted as inhibitors of its self-activation activity (Figure 5b).

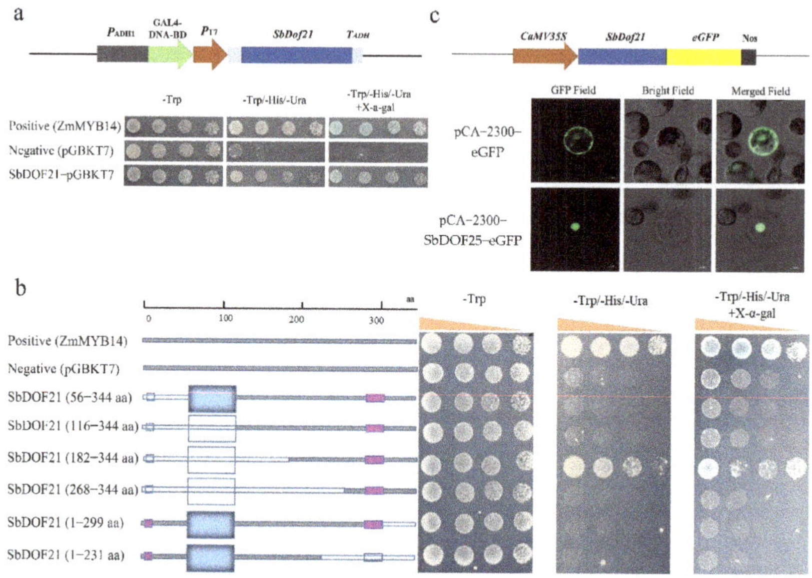

Figure 5. Functional characterization of SbDOF21. (**a**) Trans-activity assays of SbDOF21 in yeast strain AH109. ZmMYB14 with transactivation activity was used as a positive control [31]. Empty vector pGBKT7 was used as the negative control. (**b**) Transcriptional activation site identification of SbDOF21. The gray box referred to the amino acids, while the white box represented the excised sequence, pink box presented low complexity regions. Rectangular shapes refer to the zf-DOF domain. (**c**) The SbDOF21-eGFP fusion protein was driven by the 35S promoter and transiently expressed in the protoplasts of maize leaves. The eGFP driven by the 35S promoter transformed into the protoplasts of maize and was used as a control. GFP field, bright field and merged field indicate the state of fluorescent protein under three different channels.

Nuclear localization signals decide the nuclear localization of the transcription factors. In the prediction of the protein location, SbDOF21 was located in the nucleus (Table S1). We further verified the subcellular location of SbDOF21 via maize leaf protoplasts. The enhanced green fluorescent protein (eGFP) signals driven by the 35S promoter could be detected in the protoplast cytoplasm, nucleus and cell membrane, while the eGFP signals of fusion proteins with SbDOF21 were only detected in the nucleus (Figure 5c). These results indicated that SbDOF21 was localized in the nucleus.

2.6. Binding Site Detection of SbDOF21 in the Promoter of Sorghum Starch Synthesis-Related Genes

Previous studies reported that PBF protein, DOF TF, could bind to P-box (TGTAAAG) and regulate the biosynthesis of grain protein and starch [44,48]. The results of sequence alignment and phylogenetic analysis showed that SbDOF21 and ZmDOF36 [45], ZmPBF1 [34], OsPBF [39] and TaPBF [38] exhibited homologous protein relationships (Figure 6a), and similar trends were observed within the phylogenetic results of Maximum likelihood (ML) (Figure S3).

In order to further reveal whether there are DOF protein binding sites in the promoter regions of sorghum starch biosynthesis-related genes, we intercepted 2000 bp sequences upstream of the transcription start sites (TSS) of 12 genes highly expressed in grains for binding motif analysis. The results turned out that different numbers of P-Box (TGTAAAG), (T/A)AAAG, GTAAAG and TGTAAAG motifs were detected with the 2000 bp TSS regions of 12 genes (Figure 6b). Among them, the 2000 bp TSS regions of *SbGBSSI* and *SbSBEIIb* contain seven related motifs, *SbBt1* contains nine and other sequences contain more than ten related motifs. These motifs make it possible for SbDOF21 to bind to the promoters and regulate the corresponding genes during starch biosynthesis in sorghum grains.

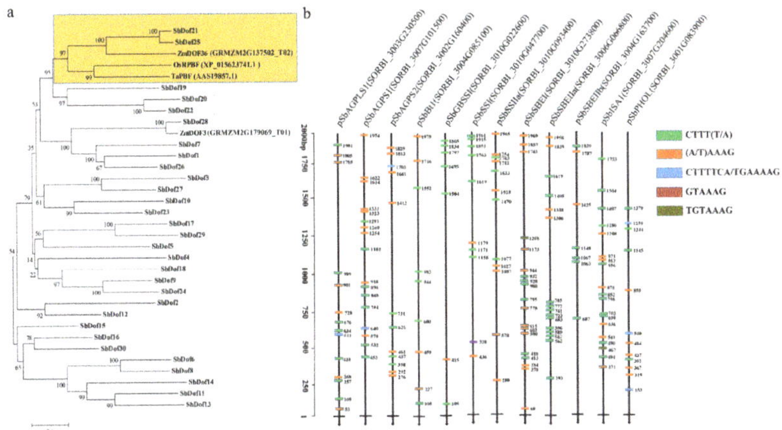

Figure 6. The orthologs analysis of SbDOF21 and P_Box and DOF protein binding sites were identified in the 2000 bp DNA sequence of sorghum starch synthesis-related genes. (**a**) Phylogenetic analysis of 30 SbDOFs, ZmDOF36 (GRMZM2G137502_T02), ZmDOF3 (GRMZM2G179069_T01), OsPBF (XP_015623741.1) and TaPBF (AAS19857.1). "Zm" means "*Zea mays* L.", "Os" is the abbreviation of "*Oryza sativa* L." and "Ta" represents "*Triticum aestivum* L." (**b**) The P_Box and DOF protein binding sites were identified in the 2000 bp DNA sequence of 12 sorghum grains highly expressed starch synthesis genes from their TSS upstream. Different colored boxes represent different motifs, arrow lines represent the sequence position upstream of TSS and numbers represent the position of motifs in the sequence.

2.7. SbDOF21 Could Bind to the P-Box (TGTAAAG) Repeats In Vitro and Transactivate the Promoter of SbGBSSI

To identify whether SbDOF21 can bind with P-Box, a DNA fragment containing four tandem TGTAAAG repeats was used for the EMSA assay, and the recombinant protein of SbDOF21 was induced and purified from *E. coli*. The free probes were detected in three lanes, while shift bands were only detected in the right two lanes with the addition of SbDOF21 (Figure 7a). The addition of a 20-fold unlabeled probe can reduce the blocking band to a certain extent (Figure 7a). These results demonstrated that SbDOF21 could bind to the P-box in vitro.

We further cloned the promoter of *SbGBSSI* (1902 bp) with biological activity. Co-transformed the pUbi-Gus, pUbi-SbDof21 and pGreenII0800-pSbGBSSI-Luc into the maize leaf protoplast and detected the activities of Renilla (REN) luciferase and luciferase (LUC). The ratio of LUC/REN served as the standard to detect whether the SbDOF21 can activate or inhibit the activity of the promoter of *SbGBSSI*. The results showed that co-transforming SbDOF21 significantly increases the biological activity of the promoter of *SbGBSSI* ($p < 0.01$, Figure 7b). Moreover, SbDOF21 also presents similar transactivation to both *SbSSII* and *SbSBEIIb* ($p < 0.05$, Figure 7c). The results of the transient expression assays also suggest that SbDOF21 acts as a transcriptional repressor to the starch biosynthesis-related genes of *SbAGPLS1*, *SbAPGS2*, *SbAGPLS4* and *SbSBEI* but exhibits no transcriptional regulations to *SbSSI*, *SbISAII3* and *PHOL* (Figure 7c).

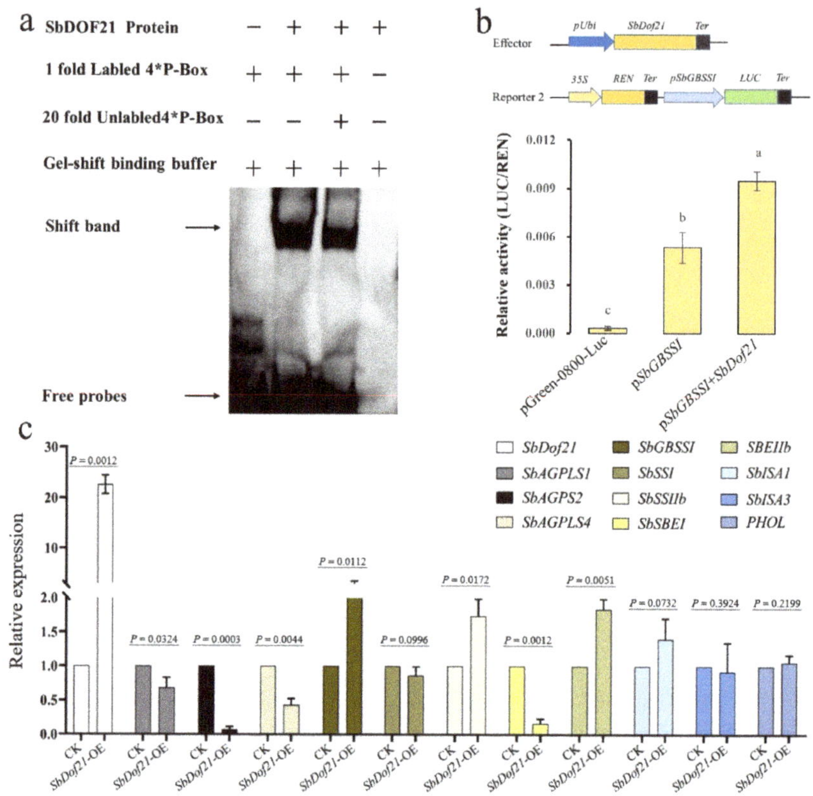

Figure 7. Study on the regulation mechanism of SbDOF21. (**a**) The binding results of SbDOF21 recombinant protein with the P-Box repeats in vitro. "+" represents the added composition and amount, and "-" means the corresponding component not added. (**b**) *SbDof21* affected the activities of *SbGBSSI* in transient expression assays. "a, b, c" are the results of the significance test at the level of $p < 0.01$. (**c**) Transient expression assays based on qRT-PCR of the functions of SbDOF21 to 11 starch biosynthesis-related genes in sorghum grains.

3. Discussion

Sorghum bicolor L. is an important C4 crop with high photosynthetic efficiency, which can be used to produce grain, feed, sugar/syrup and lignocellulosic biomass production for bioenergy [49]. Sorghum has the advantages of drought resistance, salt and alkali resistance and barren resistance, resulting in its extensive planting area, and has become the fifth cereal crop in the world [49,50]. Sorghum starch is also a carbohydrate used by human beings, which has important value in the feed industry and the brewing industry. However, there are relatively few studies on the molecular mechanism of sorghum starch synthesis [47]. Campbell and colleagues reported that the biosynthesis of sorghum starch depends on a series of functional enzymes, such as AGPase, SS, SBE and DBE [51]. These enzymes are also documented in maize, rice and other crops, and their functions have been studied [1,6,52]. The biosynthesis of starch is highly conserved in plants, and the research of other crops also provides a direction for the research of sorghum.

Previous studies document that there are different patterns in regulating the starch biosynthesis in crops, including allosteric effectors [5], protein interaction [18,20,22], protein phosphorylation [23,24], and transcriptional regulation [4,25], among which transcriptional regulation is popular and important for starch biosynthesis in plants. A large number of TFs have been reported to be involved in the transcriptional regulation of starch biosynthesis.

For example, OsbZIP58 [27], OsNAC20/26 [29] and OsRSR1 [28] in rice; ZmbZIP91 [30], ZmMYB14 [31], ZmNAC126 [32], ZmNAC128/130 [33], O2 and PBF [34] in maize; HvSUSIBA2 in barley [26]; and TaNAC019 in wheat [53]. However, the transcription regulation in sorghum starch biosynthesis is rarely documented. In the present study, we reported that *SbDof21* is a grain highly expressed gene and is involved in the transcriptional regulation of starch biosynthesis in sorghum grains, which is consistent with other reported genes involved in starch biosynthesis [4].

DOF TFs family is plant-specific, and all DOF TFs possess conserved sequence features and played important roles in diverse directions of plant development, even in starch biosynthesis in maize [40–43,45,54]. In the present study, a total of 30 Dof genes, including two new members of *SbDof5* and *SbDof30*, were identified across the entire sorghum genome (Figure 1). The gene structure of *SbDof*s usually contains one to three exons (Figure 1c), similar to those reported in *Arabidopsis* and rice [35,46]. The exon number affects the diversity and the post-transcriptional processes of genes [55]. On the other hand, genes with low exon numbers are quickly induced and expressed in response to stimuli [55,56]. Conservative motif analysis found that there was a conservative motif, i.e., Motif1, contained by all SbDOFs, which is the motif that constitutes all the conservative domains of the sorghum DOF protein.

The sequence characteristics of DOF proteins are the basis of their biological function, and the diverse functions of DOF proteins have been reported in previous studies [40,42,43,48]. Currently, DOFs are believed to play key roles in the storage of protein biosynthesis and carbohydrate metabolism of endosperm and seeds of gramineous crops [34,38,39,44,45,57]. The results of the present study revealed that the DOF member of SbDOF21 tends to function similarly in starch biosynthesis in sorghum grains. We constructed the evolutionary tree of DOF proteins reported in rice [39], maize [44,45], wheat [38] and all DOF proteins in sorghum (Figure 6a). The results further confirmed their evolutionary conservatism and proved that the conserved function of SbDOF21 was consistent with the results of the present study that SbDOF21 was involved in the regulation of grain starch biosynthesis. We also found that there are motifs of P-box or (T/A) AAAG in the upstream sequence of starch biosynthesis-related genes in sorghum (Figure 6b), and SbDOF21 can directly bind to this tandem repeated P-box in vitro (Figure 7a).

Furthermore, SbDOF21 exhibited transactivation to the promoter of *SbGBSSI* in maize leaf protoplast (Figure 7b), which indicates that SbDOF21 potentially possesses the transcriptional regulating activity to the expression of *SbGBSSI*, similar to its ortholog of *ZmDof36* during 10DAP to 20DAP in developing maize kernels [45]. Moreover, *SbDof21* also exhibits diverse potential functions to the other starch biosynthesis-related genes in sorghum grains. For example, *SbDof21* can transactivate the expression of *SbSSII* and *SbSBEIIb* rather than *SbGBSSI* and can suppress the expression of *SbAGPLS1/3*, *SbSGPS2* and *SbSBE1* while presenting no significant functions to some other genes (Figure 7c). In the transient expression assays, Wu and colleagues also observed the transactivation of ZmDOF36 to *ZmISA1* [45], but in the present study, no significant activation or suppression was detected for SbDOF21 to the expression of both *SbISA1* ($p = 0.0732$) and *SbISA3* ($p = 0.3924$, Figure 7c). ISA is one of the isoenzymes of DBE that functions in the synthesis of amylopectin [12–14]. The different activation patterns between SbDOF21 and ZmDOF36 to *ISA*s might suggest the specific transcriptional regulations of SbDOFs to starch biosynthesis in sorghum grains. In conclusion, the results of the present study suggest that SbDOF21 is a candidate vital regulator of starch biosynthesis in sorghum grain.

4. Materials and Methods

4.1. Plant Materials

BTx623, provided by Rice and Sorghum Institute, Sichuan Academy of Agricultural Sciences (Luzhou, Sichuan, China), was grown under natural conditions on the college farm (College of Agronomy and Biotechnology, Southwest University, Chongqing, China). The tissues/samples of roots, stems, leaves, inflorescence and seeds were collected at different development stages, including jointing, flowering and maturity. All collected fresh tissues/samples were immediately immersed in liquid nitrogen and then stored at −80 °C for gene cloning and expression analysis. Three biological replicates were prepared for each sample.

4.2. Identification of Sorghum DOF Proteins

To identify all putative DOF proteins across the sorghum genome, we performed BLASTP (E-values < 1.0) queries via the NCBIv3 sorghum genome sequence of the Gramene database (http://www.gramene.org/, 26 March 2022) and NCBI (https://www.ncbi.nlm.nih.gov/, 20 December 2019), and the documented DOF protein sequences from *Arabidopsis*, rice and sorghum were used as the references [35,46]. ClustalW1.83 was used for multiple candidate sequence alignment to remove redundant and incomplete proteins. The conserved domain database (https://www.ncbi.nlm.nih.gov/cdd, 26 March 2020) was applied to confirm each protein containing the conserved region of 50 amino acids with a C2-C2 finger domain. Phylogenetic trees were constructed by using MEGA 5.10 via the neighbor-joining (NJ) method combined with a passion model in contrast to the maximum likelihood method phylogeny reconstruction results, and the bootstrap replicates were set to 1000 [58]. The length (no. of amino acids), pI and Mw of the sorghum candidate DOF proteins were summarized via Expasy (http://web.expasy.org/compute_pi/, 13 April 2021).

4.3. Sequence Characterizing of Sorghum DOF Proteins

MEME (http://meme.nbcr.net/meme/cgi-bin/meme.cgi, Version 5.4.1, 28 August 2022) was used to investigate the conserved motifs of sorghum DOF proteins. The minimum motif width, maximum motif width, and no. of different motifs were specified as 6, 50 and 20, respectively [46]. Gene Structure Display Server (GSDS2.0, http://gsds.cbi.pku.edu.cn/, 15 April 2015) was used to define the gene structure [59]. The intron distribution pattern and intron/exon boundaries of the candidate genes were obtained, and the displayed results were aligned with the 5′ terminal. Putative sorghum DOF protein nuclear localization signals (NLSs) were predicted via PSORT II (http://psort.ims.u-tokyo.ac.jp/, 28 August 2022).

4.4. Cloning and Expression Analysis of SbDof21

Total RNAs were isolated from different tissues of BTx623 through RNA Extraction Kit (Tiangen, Beijing, China) according to the manufacturer's instructions. The quality and quantity of total RNA were verified via Nano Drop 1000 spectrophotometer and agarose gel. The PrimeScript^{TM} RT reagent kit with a gDNA Eraser (TaKaRa, Dalian, China) was used to obtain the first-strand cDNA and used for gene cloning and expression analysis. KOD enzymes (Toyobo, Osaka, Japan) with high fidelity were used to clone *SbDof21*. The amplified *SbDof21* products were constructed into the pMD-19T vector (TaKaRa, Dalian, China) and further verified by sequencing.

Quantitative real-time PCR (qRT-PCR) was performed via the Bio-Rad CFX96 real-time system in a total reaction volume of 10 μL Hieff qPCR SYBR Green Master Mix (Yeasen, Shanghai, China). The sorghum *eukaryotic translation initiation factor 4α* (*SbEif4α*, *SORBI_3004G039400*) was used as the internal control. All experiments were conducted four times, with three samples taken at each developmental stage. The relative transcription levels were calculated via the $2^{-\Delta\Delta CT}$ method. The primers used for real-time PCR of *SbDof21* are named SbDof21QF (forward primer) and SbDof21QR (reverse primer) (Table S2).

The expression pattern analysis of all sorghum DOF protein-coding genes is based on the RNA-sequencing (RNA-Seq) results of different tissues in our previous work [47], and TBtools is used for the heatmap drawing [60]. The co-expression analysis of *SbDof21* and sorghum starch biosynthesis-related genes was performed according to the Fragments Per Kilobase of exon model per Million mapped fragments (FPKM) of RNA-Seq data, and Pearson correlation coefficients (PCCs) were calculated with Microsoft-Excel 2016.

4.5. Functional Profiling of SbDof21 Gene

GAL4 two-Hybrid Yeast system was applied to study the self-activity of the transcription factor. The *SbDof21* and fragments deletion of *SbDof21* were sub-cloned into the pGBKT7 vector by using the sense primer with *Nde* I and the anti-sense primer with *BamH* I (Table S2). The completed carrier pGBKT7-SbDOF21 was transformed into a yeast strain, AH109, to detect the activation of the transcription factor. SD/-Trp plates were used for positive screening, and the transformants were grown for three days under dark conditions of 28 °C. The monoclones were picked into 2 mL microtubes with liquid culture for propagation. All the monoclones, including positive and negative controls, were screened on SD/-Trp-His-Ura plates with *X-α-gal* cultivated under dark conditions of 28 °C for three days to test the transcription activation.

The sub-cellular localization of SbDOF21 was analyzed by the transient expression of a fusion construct containing eGFP in the protoplasts of maize leaves. The extraction of protoplasts from maize leaves depends on the lysate liquid system constructed by 0.5 M mannitol, 1.5% cellulose, 0.5% macerozyre-R10, 10 mM EMS, 10 mM $CaCl_2$ and 0.1% BSA. The constructed vector was transformed into protoplast through the PEG-Ca^{2+} method, and the transformed protoplasts were cultured in the dark for 16 h for fluorescence detection. The subcellular localization of eGFP and fusion proteins was detected under blue excitation light at 488 nm by a fluorescence microscope LSM 800 with Airyscan (Zeiss, Jena, Germany).

4.6. Over-Expression of Recombinant SbDOF21 Protein in E. coli

SbDof21 was sub-cloned into pET32a (Takara, Dalian, China) vector for the prokaryotic expression. *BamH* I and *Hind* III were the restriction sites for vector construction. The primers are listed in Table S2. Transetta (DE3) (Transgen Biotech, Beijing, China) was used as the host cell for prokaryotic expression. When the OD_{600} value of the propagation bacterial was 0.6, the isopropyl β-d-1-Thiogalactopyranoside (IPTG) was added for induction, and the final concentration of the inducer IPTG was 0.5 mM. The strains were continued to be cultured at 16 °C and 120 rpm overnight. The strains were collected by centrifugation and were broken discontinuously by ultrasound under 120 W for 10 min. The purification of the recombinant protein was according to the instruction of the Ni-Agarose His label Kit (CWBIO, Beijing, China).

4.7. Electrophoretic Mobility Shift Assay (EMSA) Assay

The SbDOF21-His recombinant protein was purified from the prokaryotic expression system. The single-stranded oligonucleotides 5′-TGTAAAGTGTAAAGTGTAAAGTGTAAAG-3′ and reverse complementary sequence 5′-CTTTACACTTTACACTTTACACTTTACA-3′ contained 4 P-Box (TGTAAAG) sequences to detect whether SbDOF21 could bind to the P-Box. The 5′ Biotin marker is directly added by the company (Sangon Biotech, Shanghai, China) in the primer synthesis. The conjugation reaction, competitive reaction and detection of electrophoretic mobility shift assay (EMSA) were performed according to the instruction of the Chemiluminescent EMSA Kit (Beyotime Biotechnology, Shanghai, China).

4.8. Dual-Luciferase Assay in Maize Leaf Protoplast

The pUbi-SbDof21:pGreenII0800-pSbGBSSI-Luc (1:2) was the experimental group, and pUbi-Gus:pGreenII0800-pSbGBSSI-Luc (1:2) was set as the control group. All the constructs were transformed into the maize protoplast through the PEG-Ca^{2+} method. Luciferase (LUC) and Renilla (REN) luciferase activities were measured via the Dual Luciferase Assay Kit (Promega, Madison, WI, USA) and analyzed via GloMax_2020 (Thermo Fisher Scientific, Waltham, MA, USA). LUC/REN ratio was calculated to measure the relationship between the experimental and control groups. Six independent experiments were performed, and each independent experiment consisted of three replicates. The difference with SbDof21 on the promoter activity of *SbGBSSI* was tested by *t*-test.

4.9. Sorghum Leaf Protoplast Transformation

Cellulase R-10 (Yakult, Japan) and Macerozyme R-10 (Yakult, Japan) were used to obtain sorghum leaf protoplasts. The enzymolysis liquid consisted of 1.5% Cellulase R-10, 0.5% Macerozyme R-10, 0.5 M mannitol, 10 mM 2-Morpholinoethanesulphonic acid monohydrate (MES monohydrate), 10 mM $CaCl_2$ and 0.1% Bovine serum albumin (BSA). The cut sorghum leaves are immersed in the enzymolysis liquid and cracked at 25 °C at 50 rpm for 4 h. W5 buffer containing 2 mM MES, 154 mM NaCl, 125 mM $CaCl_2$ and 5 mM KCl was used for sorghum leaf protoplast washing. MMG buffer containing 15 mM $MgCl_2$, 0.4 M mannitol and 4 mM $CaCl_2$ was used for sorghum leaf protoplast suspension and transformation. The transformation solution of protoplast was composed of 40% PEG4000, 0.8 M mannitol and 1 M $CaCl_2$. The transformed protoplasts were cultured in a dark environment at 25 °C for 24 h. The cultured protoplasts were used for RNA extraction and gene expression detection. The qRT-PCR possessed three technical replications, and the p-value was calculated by t-test.

5. Conclusions

In this study, 30 genes encoding the C2-C2 zinc finger domain (DOF) were identified in the sorghum genome, including two genes of *SbDof21* and *SbDof25* that are highly expressed in sorghum grains. The co-expression analysis of *SbDof21* and starch biosynthesis-related genes confirmed *SbDof21* as a candidate regulator during starch biosynthesis. SbDOF21 is a typical DOF transcription factor that is localized to the nucleus and possesses transcriptional activation activity. Amino acids at positions 182–231 of SbDOF21 formed an important activation domain. P-Box and other DOF protein binding sites are generally observed within the sequences of 2000 bp upstream of the translation start site of sorghum starch biosynthesis-related genes. SbDOF21 can bind to four tandem repeat P-Box (TGTAAAG) sequences in vitro and regulate the transcription of these genes. Meanwhile, we found that SbDOF21 could transactivate *SbGBSSI*, a key gene in sorghum amylose biosynthesis. In summary, the results of the present study indicated that SbDOF21 acts as an important regulator in sorghum starch biosynthesis and exhibits potential values for the improvement of starch contents in sorghum.

Supplementary Materials: The following supporting information can be downloaded at: https://www.mdpi.com/article/10.3390/ijms232012152/s1.

Author Contributions: Q.X., T.L., M.L., Q.M., W.C., F.X., T.H., Y.Z. and H.D. performed the experiments, Q.X. and T.L. analyzed data; Q.X. and Z.L. provided ideas, designed the research and edited the manuscript. All authors have read and agreed to the published version of the manuscript.

Funding: This work was supported by the National Natural Science Foundation of China (32001607), Fundamental Research Funds for the Central Universities of Southwest University (S202210635326).

Institutional Review Board Statement: Not applicable.

Informed Consent Statement: Not applicable.

Data Availability Statement: Data are contained within the article or Supplementary Materials.

Conflicts of Interest: The authors declare that they have no conflicts of interest to report regarding the current study.

References

1. Stitt, M.; Zeeman, S.C. Starch turnover: Pathways, regulation and role in growth. *Curr. Opin. Plant Biol.* **2012**, *15*, 282–292. [CrossRef]
2. Bahaji, A.; Li, J.; Sánchez-López, Á.M.; Baroja-Fernández, E.; Muñoz, F.J.; Ovecka, M.; Almagro, G.; Montero, M.; Ezquer, I.; Etxeberria, E. Starch biosynthesis, its regulation and biotechnological approaches to improve crop yields. *Biotechnol. Adv.* **2014**, *32*, 87–106. [CrossRef]
3. Cai, T.; Sun, H.; Qiao, J.; Zhu, L.; Zhang, F.; Zhang, J.; Tang, Z.; Wei, X.; Yang, J.; Yuan, Q.; et al. Cell-free chemoenzymatic starch synthesis from carbon dioxide. *Science* **2022**, *373*, 1523–1527. [CrossRef]

4. Huang, L.; Tan, H.; Zhang, C.; Li, Q.; Liu, Q. Starch biosynthesis in cereal endosperms: An updated review over the last decade. *Plant Commun.* **2021**, *2*, 100237. [CrossRef]
5. Stark, D.M.; Timmerman, K.P.; Barry, G.F.; Preiss, J.; Kishore, G.M. Regulation of the amount of starch in plant tissues by ADP Glucose pyrophosphorylase. *Science* **1992**, *258*, 287–292. [CrossRef]
6. Jeon, J.S.; Ryoo, N.; Hahn, T.R.; Walia, H.; Nakamura, Y. Starch biosynthesis in cereal endosperm. *Plant Physiol. Biochem.* **2010**, *48*, 383–392. [CrossRef]
7. Delrue, B.; Fontaine, T.; Routier, F.; Decq, A.; Wieruszeski, J.; Van Den Koornhuyse, N.; Maddelein, M.; Fournet, B.; Ball, S. Waxy Chlamydomonas reinhardtii: Monocellular algal mutants defective in amylose biosynthesis and granule-bound starch synthase activity accumulate a structurally modified amylopectin. *J. Bacteriol.* **1992**, *174*, 3612–3620. [CrossRef]
8. Nakamura, T.; Yamamori, M.; Hirano, H.; Hidaka, S.; Nagamine, T. Production of Waxy (Amylose-Free) Wheats. *Mol. Gen Genet.* **1995**, *248*, 253–259. [CrossRef]
9. Ball, S.G.; Morell, M.K. From bacterial glycogen to starch: Understanding the biogenesis of the plant starch granule. *Annu. Rev. Plant Biol.* **2003**, *54*, 207–233. [CrossRef]
10. Keeling, P.L.; Myers, A.M. Biochemistry and genetics of starch synthesis. *Annu. Rev. Food Sci. Technol.* **2010**, *1*, 271–303. [CrossRef]
11. Zeeman, S.C.; Kossmann, J.; Smith, A.M. Starch: Its metabolism, evolution, and biotechnological modification in plants. *Annu. Rev. Plant Biol.* **2010**, *61*, 209–234. [CrossRef] [PubMed]
12. Myers, A.M.; Morell, M.K.; James, M.G.; Ball, S.G. Recent progress toward understanding biosynthesis of the amylopectin crystal. *Plant Physiol.* **2000**, *122*, 989–997. [CrossRef]
13. Burton, R.A.; Jenner, H.; Carrangis, L.; Fahy, B.; Fincher, G.B.; Hylton, C.; Laurie, D.A.; Parker, M.; Waite, D.; Van Wegen, S.; et al. Starch granule initiation and growth are altered in barley mutants that lack isoamylase activity. *Plant J.* **2002**, *31*, 97–112. [CrossRef]
14. Zeeman, S.C.; Smith, S.M.; Smith, A.M. The diurnal metabolism of leaf starch. *Biochem. J.* **2007**, *401*, 13–28. [CrossRef] [PubMed]
15. Zeeman, S.C.; Thorneycroft, D.; Schupp, N.; Chapple, A.; Weck, M.; Dunstan, H.; Haldimann, P.; Bechtold, N.; Smith, A.M.; Smith, S.M. Plastidial alpha-glucan phosphorylase is not required for starch degradation in Arabidopsis leaves but has a role in the tolerance of abiotic stress. *Plant Physiol.* **2004**, *135*, 849–858. [CrossRef] [PubMed]
16. Tickle, P.; Burrell, M.M.; Coates, S.A.; Emes, M.J.; Tetlow, I.J.; Bowsher, C.G. Characterization of plastidial starch phosphorylase in *Triticum aestivum* L. endosperm. *J. Plant Physiol.* **2009**, *166*, 1465–1478. [CrossRef] [PubMed]
17. Nakamura, Y.; Ono, M.; Utsumi, C.; Steup, M. Functional interaction between plastidial starch phosphorylase and starch branching enzymes from rice during the synthesis of branched maltodextrins. *Plant Cell Physiol.* **2012**, *53*, 869–878. [CrossRef] [PubMed]
18. Utsumi, Y.; Nakamura, Y. Structural and enzymatic characterization of the isoamylase1 homo-oligomer and the isoamylase1–isoamylase2 hetero-oligomer from rice endosperm. *Planta* **2006**, *225*, 75–87. [CrossRef]
19. Delatte, T.; Trevisan, M.; Parker, M.L.; Zeeman, S.C. Arabidopsis mutants Atisa1 and Atisa2 have identical phenotypes and lack the same multimeric isoamylase, which influences the branch point distribution of amylopectin during starch synthesis. *Plant J.* **2005**, *41*, 815–830. [CrossRef]
20. Hennen-Bierwagen, T.A.; Liu, F.; Marsh, R.S.; Kim, S.; Gan, Q.; Tetlow, I.J.; Emes, M.J.; James, M.G.; Myers, A.M. Starch biosynthetic enzymes from developing maize endosperm associate in multisubunit complexes. *Plant Physiol.* **2008**, *146*, 1892–1908. [CrossRef]
21. Tetlow, I.J.; Beisel, K.G.; Cameron, S.; Makhmoudova, A.; Liu, F.; Bresolin, N.S.; Wait, R.; Morell, M.K.; Emes, M.J. Analysis of protein complexes in wheat amyloplasts reveals functional interactions among starch biosynthetic enzymes. *Plant Physiol.* **2008**, *146*, 1878–1891. [CrossRef] [PubMed]
22. Hennen-Bierwagen, T.A.; Lin, Q.; Grimaud, F.; Planchot, V.; Keeling, P.L.; James, M.G.; Myers, A.M. Proteins from multiple metabolic pathways associate with starch biosynthetic enzymes in high molecular weight complexes: A model for regulation of carbon allocation in maize amyloplasts. *Plant Physiol.* **2009**, *149*, 1541–1559. [CrossRef]
23. Tetlow, I.J.; Wait, R.; Lu, Z.; Akkasaeng, R.; Bowsher, C.G.; Esposito, S.; Kosarhashemi, B.; Morell, M.K.; Emes, M.J. Protein phosphorylation in amyloplasts regulates starch branching enzyme activity and protein-protein interactions. *Plant Cell* **2004**, *16*, 694–708. [CrossRef]
24. Georgelis, N.; Shaw, J.R.; Hannah, L.C. Phylogenetic analysis of ADP-glucose pyrophosphorylase subunits reveals a role of subunit interfaces in the allosteric properties of the enzyme. *Plant Physiol.* **2009**, *151*, 67–77. [CrossRef] [PubMed]
25. Li, R.; Tan, Y.; Zhang, H. Regulators of Starch Biosynthesis in Cereal Crops. *Molecules* **2021**, *26*, 7092. [CrossRef] [PubMed]
26. Sun, C.; Palmqvist, S.; Olsson, H.; Borén, M.; Ahlandsberg, S.; Jansson, C. A novel WRKY transcription factor, SUSIBA2, participates in sugar signaling in barley by binding to the sugar-responsive elements of the iso1 promoter. *Plant Cell* **2003**, *15*, 2076–2092. [CrossRef] [PubMed]
27. Wang, J.; Xu, H.; Zhu, Y.; Liu, Q.; Cai, X. OsbZIP58, a basic leucine zipper transcription factor, regulates starch biosynthesis in rice endosperm. *J. Exp. Bot.* **2013**, *64*, 3453–3466. [CrossRef]
28. Fu, F.; Xue, H. Coexpression analysis identifies Rice Starch Regulator1, a rice AP2/EREBP family transcription factor, as a novel rice starch biosynthesis regulator. *Plant Physiol.* **2010**, *154*, 927–938. [CrossRef]
29. Wang, J.; Chen, Z.; Zhang, Q.; Meng, S.; Wei, C. The NAC transcription factors OsNAC20 and OsNAC26 regulate starch and storage protein synthesis. *Plant Physiol.* **2020**, *184*, 1775–1791. [CrossRef]

30. Chen, J.; Yi, Q.; Cao, Y.; Wei, B.; Zheng, L.; Xiao, Q.; Xie, Y.; Gu, Y.; Li, Y.; Huang, H.; et al. ZmbZIP91 regulates expression of starch synthesis-related genes by binding to ACTCAT elements in their promoters. *J. Exp. Bot.* **2016**, *67*, 1327–1338. [CrossRef]
31. Xiao, Q.; Wang, Y.; Du, J.; Li, H.; Wei, B.; Wang, Y.; Li, Y.; Yu, G.; Liu, H.; Zhang, J.; et al. ZmMYB14 is an important transcription factor involved in the regulation of the activity of the ZmBT1 promoter in starch biosynthesis in maize. *FEBS J.* **2017**, *284*, 3079–3099. [CrossRef]
32. Xiao, Q.; Wang, Y.; Li, H.; Zhang, C.; Wei, B.; Wang, Y.; Huang, H.; Li, Y.; Yu, G.; Liu, H.; et al. Transcription factor ZmNAC126 plays an important role in transcriptional regulation of maize starch synthesis-related genes. *Crop J.* **2021**, *9*, 192–203. [CrossRef]
33. Zhang, Z.; Dong, J.; Ji, C.; Wu, Y.; Messing, J. NAC-type transcription factors regulate accumulation of starch and protein in maize seeds. *Proc. Natl. Acad. Sci. USA* **2019**, *116*, 11223–11228. [CrossRef]
34. Zhang, Z.; Zheng, X.; Yang, J.; Messing, J.; Wu, Y. Maize endosperm-specific transcription factors O2 and PBF network the regulation of protein and starch synthesis. *Proc. Natl. Acad. Sci. USA* **2016**, *113*, 10842–10847. [CrossRef]
35. Yanagisawa, S. A novel DNA-binding domain that may form a single zinc finger motif. *Nucleic Acids Res.* **1995**, *23*, 3403–3410. [CrossRef]
36. Yanagisawa, S. Involvement of Maize Dof Zinc Finger Proteins in Tissue-Specific and Light-Regulated Gene Expression. *Plant Cell* **1998**, *10*, 75–89. [CrossRef]
37. Lijavetzky, D.; Carbonero, P.; Vicente-Carbajosa, J. Genome-wide comparative phylogenetic analysis of the rice and Arabidopsis Dof gene families. *BMC Evol. Biol.* **2003**, *3*, 17. [CrossRef] [PubMed]
38. Yamamoto, M.P.; Onodera, Y.; Touno, S.M.; Takaiwa, F. Synergism between RPBF Dof and RISBZ1 bZIP Activators in the Regulation of Rice Seed Expression Genes. *Plant Physiol.* **2006**, *141*, 1694–1707. [CrossRef]
39. Dong, G.; Ni, Z.; Yao, Y.; Nie, X.; Sun, Q. Wheat Dof transcription factor WPBF interacts with TaQM and activates transcription of an alpha-gliadin gene during wheat seed development. *Plant Mol. Biol.* **2007**, *63*, 73–84. [CrossRef] [PubMed]
40. Skirycz, A.; Radziejwoski, A.; Busch, W.; Hannah, M.A.; Czeszejko, J.; Kwaśniewski, M.; Zanor, M.I.; Lohmann, J.U.; Veylder, L.D.; Witt, I. The DOF transcription factor OBP1 is involved in cell cycle regulation in Arabidopsis thaliana. *Plant J.* **2008**, *56*, 779–792. [CrossRef] [PubMed]
41. Zhuo, M.; Sakuraba, Y.; Yanagisawa, S. A Jasmonate-activated MYC2-Dof2.1-MYC2 Transcriptional Loop Promotes Leaf Senescence in Arabidopsis. *Plant Cell* **2019**, *32*, 242–262. [CrossRef]
42. Qi, W.; Xue, L.; Yin, D.; Hua, Y.; Qi, X.; Zhao, X.; Li, X.; Zhu, L.; Li, S.; Li, D. Constitutive expression of OsDof4, encoding a C2-C2 zinc finger transcription factor, confesses its distinct flowering effects under long- and short-day photoperiods in rice (*Oryza sativa* L.). *BMC Plant Biol.* **2017**, *17*, 166.
43. Wu, Y.; Yang, W.; Wei, J.; Yoon, H.; An, G. Transcription Factor OsDOF18 Controls Ammonium Uptake by Inducing Ammonium Transporters in Rice Roots. *Mol. Cells* **2017**, *40*, 178–185.
44. Qi, X.; Li, S.; Zhu, Y.; Zhao, Q.; Zhu, D.; Yu, J. ZmDof3, a maize endosperm-specific Dof protein gene, regulates starch accumulation and aleurone development in maize endosperm. *Plant Mol. Biol.* **2016**, *93*, 7–20. [CrossRef]
45. Wu, J.; Chen, L.; Chen, M.; Zhou, W.; Dong, Q.; Jiang, H.; Cheng, B. The DOF-Domain Transcription Factor ZmDOF36 Positively Regulates Starch Synthesis in Transgenic Maize. *Front Plant Sci.* **2019**, *10*, 465. [CrossRef]
46. Kushwaha, H.; Gupta, S.; Singh, V.K.; Rastogi, S.; Yadav, D. Genome wide identification of Dof transcription factor gene family in sorghum and its comparative phylogenetic analysis with rice and Arabidopsis. *Mol Biol. Rep.* **2011**, *38*, 5037–5053. [CrossRef]
47. Xiao, Q.; Huang, T.; Cao, W.; Ma, K.; Liu, T.; Xing, F.; Ma, Q.; Duan, H.; Ling, M.; Ni, X.; et al. Profiling of transcriptional regulators associated with starch biosynthesis in sorghum (*Bicolor sorghum* L.). *Front. Plant Sci.* **2022**, *13*, 999747. [CrossRef]
48. Noguero, M.; Atif, R.M.; Ochatt, S.; Thompson, R.D. The role of the DNA-binding One Zinc Finger (DOF) transcription factor family in plants. *Plant Sci.* **2013**, *209*, 32–45. [CrossRef]
49. Mccormick, R.F.; Truong, S.K.; Sreedasyam, A.; Jenkins, J.; Shu, S.; Sims, D.; Kennedy, M.; Amirebrahimi, M.; Weers, B.D.; Mckinley, B. The Sorghum bicolor reference genome: Improved assembly, gene annotations, a transcriptome atlas, and signatures of genome organization. *Plant J.* **2018**, *93*, 338–354. [CrossRef]
50. Corredor, D.Y.; Salazar, J.M.; Hohn, K.L.; Bean, S.; Bean, B.; Wang, D. Evaluation and characterization of forage Sorghum as feedstock for fermentable sugar production. *Appl. Biochem. Biotechnol.* **2009**, *158*, 164–179. [CrossRef] [PubMed]
51. Campbell, B.C.; Gilding, E.K.; Mace, E.S.; Tai, S.; Tao, Y.; Prentis, P.J.; Thomelin, P.; Jordan, D.R.; Godwin, I.D. Domestication and the storage starch biosynthesis pathway: Signatures of selection from a whole sorghum genome sequencing strategy. *Plant Biotechnol. J.* **2016**, *14*, 2240–2253. [CrossRef] [PubMed]
52. Zhang, H.; Dong, S.; Gao, R.Q.; Sun, Q. Starch accumulation and enzyme activities associated with starch synthesis in maize kernels. *J. Integr. Agr.* **2007**, *6*, 808–815. [CrossRef]
53. Yanagisawa, S. The Dof family of plant transcription factors. *Trends Plant Sci.* **2002**, *7*, 555–560. [CrossRef]
54. Gao, Y.; An, K.; Guo, W.; Chen, Y.; Zhang, R.; Zhang, X.; Chang, S.; Vincenzo, R.; Jin, F.; Cao, X.; et al. The endosperm-specific transcription factor TaNAC019 regulates glutenin and starch accumulation and its elite allele improves wheat grain quality. *Plant Cell* **2021**, *33*, 603–622. [CrossRef] [PubMed]
55. Koralewski, T.E.; Krutovsky, K.V. Evolution of Exon-Intron Structure and Alternative Splicing. *PLoS ONE* **2011**, *6*, e18055. [CrossRef]
56. Heidari, P.; Puresmaeli, F.; Mora-Poblete, F. Genome-Wide Identification and Molecular Evolution of the Magnesium Transporter (MGT) Gene Family in Citrullus lanatus and Cucumis sativus. *Agronomy* **2022**, *12*, 2253. [CrossRef]

57. Marzábal, P.; Gas, E.; Fontanet, P.; Vicente-Carbajosa, J.; Torrent, M.; Ludevid, M.D. The maize Dof protein PBF activates transcription of γ-zein during maize seed development. *Plant Mol. Biol.* **2008**, *67*, 441–454. [CrossRef] [PubMed]
58. Tamura, K.; Peterson, D.; Peterson, N.; Stecher, G.; Nei, M.; Kumar, S. MEGA5: Molecular Evolutionary Genetics Analysis Using Maximum Likelihood, Evolutionary Distance, and Maximum Parsimony Methods. *Mol. Biol. Evol.* **2011**, *28*, 2731–2739. [CrossRef] [PubMed]
59. Guo, A.Y.; Zhu, Q.H.; Xin, C. GSDS:a gene structure display server. *Hereditas* **2007**, *29*, 1023–1026. [PubMed]
60. Chen, C.; Chen, H.; Zhang, Y.; Thomas, H.R.; Frank, M.H.; He, Y.; Xia, R. TBtools-an integrative toolkit developed for interactive analyses of big biological data. *Mol. Plant.* **2020**, *13*, 1194–1202. [CrossRef] [PubMed]

International Journal of *Molecular Sciences*

MDPI

Article

The Q^{c5} Allele Increases Wheat Bread-Making Quality by Regulating SPA and SPR

Zhenru Guo [1,2,†], Qing Chen [2,†], Jing Zhu [2], Yan Wang [2], Yang Li [2], Qingcheng Li [2], Kan Zhao [2], Yue Li [2], Rui Tang [2], Xiaoli Shi [2], Kenan Tan [2], Li Kong [2], Yunfeng Jiang [2], Qiantao Jiang [2], Jirui Wang [2], Guoyue Chen [2], Yuming Wei [2], Youliang Zheng [2,*] and Pengfei Qi [2,*]

[1] State Key Laboratory of Crop Gene Exploration and Utilization in Southwest China, Sichuan Agricultural University, Chengdu 611130, China; guozhenru@stu.sicau.edu.cn

[2] Triticeae Research Institute, Sichuan Agricultural University, Chengdu 611130, China; qingchen83@sicau.edu.cn (Q.C.); zhujing@stu.sicau.edu.cn (J.Z.); wangyan6@stu.sicau.edu.cn (Y.W.); liyang1@stu.sicau.edu.cn (Y.L.); liqingcheng@stu.sicau.edu.cn (Q.L.); zhaokan6137@163.com (K.Z.); 15397768229@163.com (Y.L.); 15691269656@163.com (R.T.); 18790196562@163.com (X.S.); 2020312061@stu.sicau.edu.cn (K.T.); kongli@sicau.edu.cn (L.K.); jiangyunfeng@sicau.edu.cn (Y.J.); qiantaojiang@sicau.edu.cn (Q.J.); jirui.wang@gmail.com (J.W.); gychen@sicau.edu.cn (G.C.); ymwei@sicau.edu.cn (Y.W.)

* Correspondence: ylzheng@sicau.edu.cn (Y.Z.); pengfeiqi@hotmail.com or pengfeiqi@sicau.edu.cn (P.Q.); Tel.: +86-28-82650337 (P.Q.)

† These authors contributed equally to this work.

Abstract: Common wheat (*Triticum aestivum* L.) is an important food crop with a unique processing quality. The *Q* gene positively regulates the processing quality of wheat, but the underlying mechanism remains unclear. Here, a new *Q* allele (Q^{c5}) responsible for compact spikes and good bread performance was identified. Compared with the *Q* allele widely distributed in modern common wheat cultivars, Q^{c5} had a missense mutation outside the miRNA172-binding site. This missense mutation led to a more compact messenger RNA (mRNA) secondary structure around the miRNA172-binding region, resulting in increased Q^{c5} expression during the spike development stage and a consequent increase in spike density. Furthermore, this missense mutation weakened the physical interaction between Q^{c5} and storage protein activator (SPA) in seeds and suppressed the expression of *storage protein repressor* (*SPR*). These changes increased the grain protein content and improved the bread-making quality of wheat. In conclusion, a missense mutation increases *Q* expression because of the resulting highly folded mRNA secondary structure around the miRNA172-binding site. Furthermore, this mutation improves the bread-making quality of wheat by repressing the expression of *SPR* and influencing the physical interaction between Q and SPA. These findings provide new insights into the miRNA172-directed regulation of gene expression, with implications for wheat breeding.

Keywords: breeding; seed storage proteins; mRNA secondary structure; microRNA; processing quality

Citation: Guo, Z.; Chen, Q.; Zhu, J.; Wang, Y.; Li, Y.; Li, Q.; Zhao, K.; Li, Y.; Tang, R.; Shi, X.; et al. The Q^{c5} Allele Increases Wheat Bread-Making Quality by Regulating SPA and SPR. *Int. J. Mol. Sci.* **2022**, *23*, 7581. https://doi.org/10.3390/ijms23147581

Academic Editors: Andrés J. Cortés and Hai Du

Received: 4 June 2022
Accepted: 6 July 2022
Published: 8 July 2022

Publisher's Note: MDPI stays neutral with regard to jurisdictional claims in published maps and institutional affiliations.

1. Introduction

Common wheat (*Triticum aestivum*; 2n = 6x = 42, AABBDD) originated about 8000 years ago in the Fertile Crescent of the Middle East [1,2] and has since spread worldwide to become one of the most important food crops. Common wheat resulted from the hybridization between the domesticated tetraploid wheat *Triticum turgidum* (2n = 4x = 28, AABB) and the diploid *Aegilops tauschii* (2n = 2x = 14, DD) [2]. The Food and Agriculture Organization of the United Nations estimates that wheat provides approximately 20% of the calories and 22% of the proteins consumed by humans worldwide [3].

The unique processing quality of wheat is due to its seed storage proteins (SSPs), including gliadins and glutenins (i.e., gluten) [4]. Gliadins, which are monomeric compounds, can be classified as α- (α/β-), γ-, and ω-gliadins [5]. Glutenins consist of high molecular weight glutenin subunits (HMW-GS) and low molecular weight glutenin subunits

(LMW-GS) [6], which form large macropolymers during the grain desiccation stage [7]. Glutenins strengthen wheat dough by conferring elasticity, whereas gliadins affect dough viscosity by conferring extensibility [8]. The SSP content (grain protein content; GPC) and composition are critical factors influencing the processing quality of wheat, and thus affect wheat price. Therefore, elucidating the mechanisms mediating SSP synthesis is necessary.

The synthesis of SSPs during the grain filling stage is primarily regulated by a network of *cis*-elements and their interacting transcription factors (TFs) [9–11]. Previous studies revealed that the accumulation of SSPs is controlled by several TF families, namely basic leucine zipper (bZIP), DNA binding with one finger (DOF), myeloblastosis (MYB), B3 DNA-binding domain transcription factor (B3), and no apical meristem (NAM)/ATAF subfamily/cup-shaped cotyledon (CUC) family transcription factor (NAC). The bipartite endosperm box consists of a GCN4-like motif (GLM; 5′-ATGAG/CTCAT-3′) and a prolamin box (P-box; 5′-TGTAAAG-3′). Earlier research indicated that GLM is recognized by bZIP TFs, such as BLZ1 and BLZ2 in barley [12,13] or storage protein activator (SPA) in wheat [14], whereas P-box is a binding site for DOF TFs, including prolamin box-binding factor (PBF) and scutellum and aleurone-expressed DOF (SAD) [12,15,16]. The RY repeat (5′-CATGCATG-3′), which is also a conserved motif in *SSP* gene promoters, is recognized by the B3 protein FUSCA3 [17–19]. Moreover, TaGAMyb regulates glutenin gene expression by directly binding to a 5′-C/TAACAAA/C-3′-like motif in the promoter of HMW-GS genes and by recruiting GCN5 to modulate histone acetylation [20]. Additionally, the MYB TFs TuODORANT1 and TaODORANT1 are suppressors of SSP synthesis [21]. An NAC family TF (i.e., storage protein repressor; SPR) binds to the *cis*-element 5′-CANNTG-3′ to suppress the accumulation of SSPs [22]. Another endosperm-specific NAC TF (TaNAC019) directly activates the expression of HMW-GS genes by binding to their promoters and by interacting with TaGAMyb [23]. Opaque-2, which is an endosperm-specific bZIP family member in maize, binds to the promoter of the gene encoding the 22 kD zein protein to activate transcription [24]. In wheat, SPA is an opaque-2-like protein, which activates the transcription of LMW-GS genes by binding to two GLMs [14]; furthermore, SPA also enhances the expression of HMW-GS genes by binding to two GLMs and G-box in their promoters [25,26]. Moreover, SPA can activate the expression of ω-gliadin genes [27]. In addition to binding to DNA, TFs may interact with each other (i.e., protein–protein interactions) to form relatively large TF complexes that regulate the expression of *SSP* genes [28]. Furthermore, the SPA heterodimerizing protein (SHP) represses HMW-GS and LMW-GS gene promoter activities [26].

MicroRNAs (miRNAs) comprise approximately 22 nucleotides and inhibit gene expression mainly by eliminating DNA, cleaving mRNA, and repressing translation [29]. In plants, miRNA172 targets the transcripts of *APETALA2* (AP_2) [30,31] and a few AP_2*-like* genes [32–34]. In wheat, the miRNA172–AP_2*-like* system is crucial for regulating spike morphology [35,36].

The *Q* gene encodes an AP_2 TF that regulates spike morphology along with miRNA172. This gene is located on chromosome 5AL and contains two AP_2-binding domains as well as a miRNA172-binding site in the 10th exon [37,38]. Additionally, *Q* was critical for wheat domestication [37] and de-domestication [39]. A previous study demonstrated that four overexpressed *Q* alleles (Q^{c1}–Q^{c4}) with point mutations in their miRNA172-binding site are associated with increases in spike density, GPC, and wheat bread-making quality [40]. However, the underlying molecular mechanisms remain unclear.

In this study, another wheat mutant with increased spike density and improved bread-making quality was identified. We demonstrated that a missense mutation flanking the miRNA172-binding site of the *Q* gene (Q^{c5} allele) is responsible for the mutant phenotype. How Q^{c5} increases the spike density and enhances the bread-making quality of wheat was also investigated.

2. Results

2.1. Phenotype and Segregation Analyses

Compared with wild type (WT) wheat, the common wheat mutant *R-Cp5-1*, which was identified in a "Roblin" mutant population, was shorter in plant height and had a higher spike density (Figure 1A,B,E,F). There was no significant difference between *R-Cp5-1* and WT plants regarding spikelet number per spike, grain number per spike, and tiller number (Figure 1G–I). Using the compact spike as the target trait, the 160 F_2 plants were segregated in the expected 3:1 ratio ($\chi^2 = 0.533$; $p = 0.47$), suggesting that the mutant phenotype was controlled by a single dominant locus (*Cp5*).

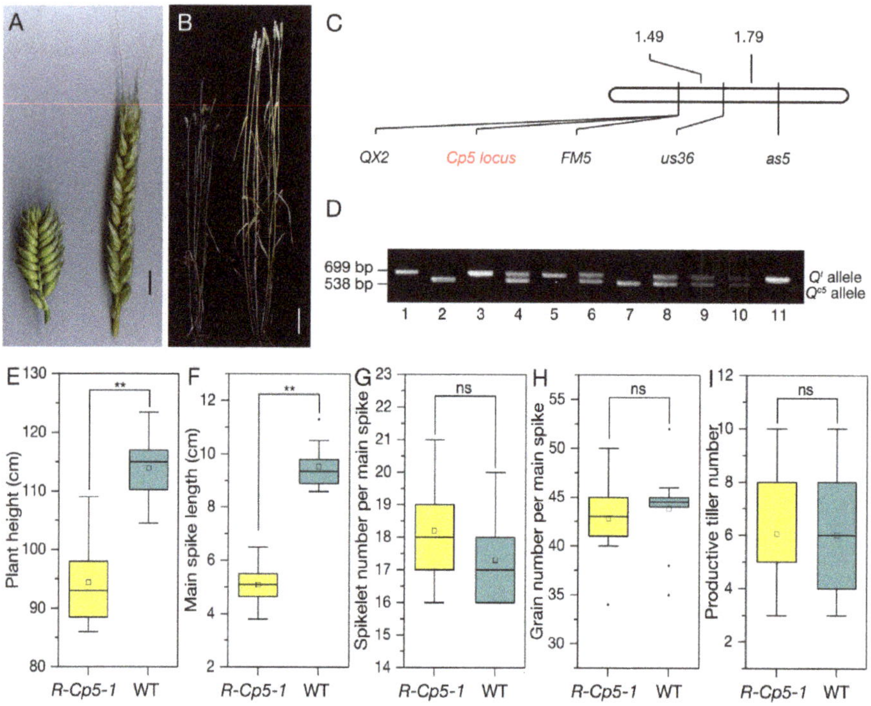

Figure 1. The mutant phenotype is controlled by the *Cp5* locus. (**A**) Spikes of *R-Cp5-1* (left) and the wild-type (WT) control (right). Scale bar, 1 cm. (**B**) *R-Cp5-1* (left) and WT (right) plants. Scale bar, 10 cm. (**C**) Segregation analysis of the *Cp5* locus. (**D**) Polymorphic molecular marker between Q^t and Q^{c5}, which co-segregated with the compact spike trait. Lane 1: QZ96; Lane 2: *R-Cp5-1*; Lanes 3–11: representative F_3 individuals from *R-Cp5-1*/QZ96. (**E–I**) Comparison between *R-Cp5-1* and WT in terms of the plant height, spike length, spikelet number per spike, grain number per spike, and tiller number ($n = 20$; **, $p \leq 0.01$; ns, not significant).

The *R-Cp5-1* mutant is phenotypically similar to the *S-Cp1-1* mutant, which carries the overexpressed Q^{c1} allele with a point mutation in the miRNA172-binding site [40]. Therefore, we cloned the *Q* alleles in *R-Cp5-1* and "Roblin". Unexpectedly, there were no variations in the miRNA172-binding region between the *R-Cp5-1* and "Roblin" *Q* sequences (Figure S1). The available molecular markers around *Q* [39] were used to determine whether the *R-Cp5-1* mutant phenotype is associated with *Q*. According to the linkage map, the markers *FM5* and *QX2* were closely linked with *Cp5* in the F_2 population (Figure 1C). Additionally, *QX2* co-segregated with *Cp5* in the F_3 population consisting of 1101 individuals (Figure 1D). Notably, *QX2* is an intragenic molecular marker of *Q* [39].

These results implied that *Q* is most likely responsible for the phenotype associated with the *Cp5* locus.

2.2. Gene Cloning and Expression Analysis

The full-length cDNA and genomic DNA sequences of *Q* in *R-Cp5-1* (GenBank No. MW419115) and "Roblin" (KX620763.1) were cloned and sequenced; then, sequence alignment revealed a missense mutation (C–T; Figures 2A,B and S1) flanking the miRNA172-binding region. The new *Q* allele was named Q^{c5}, which was added to the four reported Q^c alleles (Q^{c1}–Q^{c4}; Figure S2) [40].

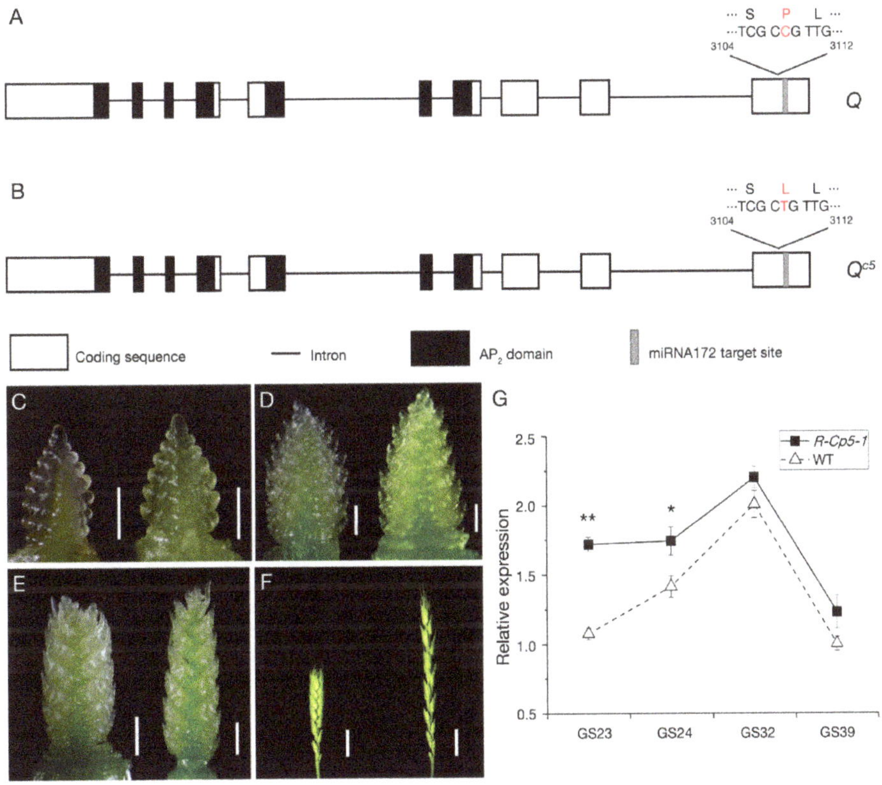

Figure 2. One missense mutation flanking the miRNA172-binding site increased the spike density and the Q^{c5} expression level. (**A**) and (**B**) Schematic diagrams of the *Q* and Q^{c5} alleles. A missense mutation (C in "Roblin" and T in *R-Cp5-1*) changes the 408th residue from proline (P) to leucine (L). (**C–F**) *R-Cp5-1* (left) and WT (right) spikes at GS23 (**C**), GS24 (**D**), GS32 (**E**), and GS39 (**F**). Scale bars, 0.05 cm in (**C**,**D**), 0.1 cm in (**E**), and 1 cm in (**F**). (**G**) Comparison of the *Q* and Q^{c5} transcription levels in spikes at GS23, GS24, GS32, and GS39. Data are presented as the mean ± standard deviation ($n = 3$; *, $p \leq 0.05$; **, $p \leq 0.01$).

A qRT-PCR analysis was performed to assess whether this missense mutation increases the *Q* expression level. The Q^{c5} and *Q* allele expression patterns in the mutant and the WT control, respectively, were similar, but Q^{c5} was consistently more highly expressed (significantly higher at Growth Stage 23 (GS23) and GS24 [41]) than the *Q* allele in spikes at four growth stages (Figure 2C–G).

To clarify how the missense mutation enhances Q^{c5} expression, the *Q* and Q^{c5} mRNA secondary structures were compared. Unlike the corresponding region in *Q*, the miRNA172-binding region of Q^{c5} was highly folded (Figure S3), which inhibited the binding of miRNA172.

2.3. Protein Structure Analysis

The secondary and tertiary (i.e., 3D) structures of the Q and Q^{c5} proteins were predicted to assess the biological significance of the amino acid substitution. The predicted structures indicated that compared with Q, Q^{c5} had more amino acid residues that can form an α-helix in the region around the changed residue (Figure S4).

2.4. Processing Quality Analysis

To determine whether Q^{c5} affects the wheat processing quality, the quality-related parameters of *R-Cp5-1* and WT were compared (Figure 3D–N). The GPC, wet gluten content, and Zeleny sedimentation value were higher for *R-Cp5-1* than for WT (Figure 3G–I). Additionally, the average bread volume was 20% greater for *R-Cp5-1* than for WT (Figure 3D–F). The glasshouse experiment also showed that GPC was significantly higher for *R-Cp5-1* than for WT (Figure S5). The RP-HPLC analysis indicated that the Glu/Gli and HMW/LMW ratios as well as the total HMW-GS, total LMW-GS, and total gliadin contents were greater for *R-Cp5-1* than for WT (Figures 3J–N and S6). In contrast to its positive effect on processing quality, Q^{c5} negatively affected grain size (Figure 3A,B) and thousand kernel weight (TKW) (Figure 3C).

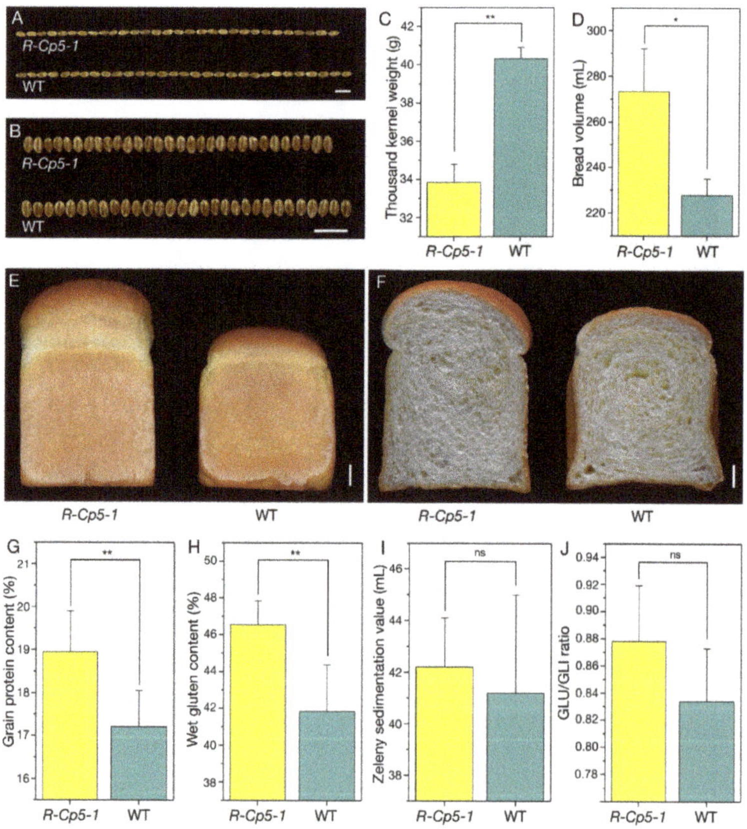

Figure 3. *Cont.*

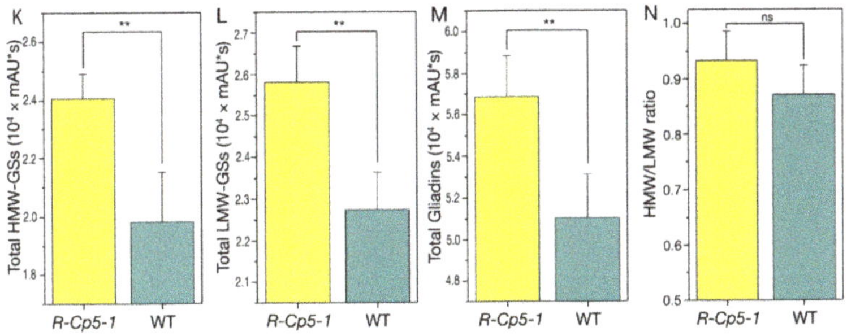

Figure 3. Q^{c5} results in relatively small seeds and increased processing quality. Length (**A**) and width (**B**) of 30 seeds from *R-Cp5-1* (upper) and WT (lower). Scale bar, 1 cm. (**C**) Thousand kernel weight of *R-Cp5-1* and WT. (**D–F**) Comparison of bread made from *R-Cp5-1* and WT wheat. Scale bar, 1 cm. (**G–I**) Comparison of the grain protein content (**G**), wet gluten content (**H**), and Zeleny sedimentation value (**I**) between *R-Cp5-1* and WT. (**J–N**) The Glu/Gli ratio (**J**), total HMW-GS (**K**), total LMW-GS (**L**), total gliadins (**M**), and HMW/LMW ratio (**N**) were greater for *R-Cp5-1* than for WT. Data are presented as the mean ± standard deviation (n = 4; *, $p \leq 0.05$; **, $p \leq 0.01$).

2.5. Mechanisms Underlying the Effect of Q^{c5} on Processing Quality

The starch granule (SG) and protein body (PB) morphological characteristics in the mature seeds of *R-Cp5-1* and WT were examined using a scanning electron microscope. Compared with the WT samples, the PBs and SGs in three representative areas of *R-Cp5-1* were more tightly arranged, with fewer and smaller gaps. Furthermore, the SGs in WT were relatively exposed, whereas the SGs in *R-Cp5-1* were more tightly packed with PBs (Figure S7).

In the subcellular localization experiment involving wheat mesophyll protoplasts, Q^{c5} and Q were distributed in the nucleus (Figure 4A). However, Q cannot directly activate the transcription of *SSP* genes (Figure S8). Therefore, we speculated that Q indirectly regulates the expression of *SSP* genes by interacting with other TFs. Hence, a yeast two-hybrid assay was performed to identify the TFs encoded in a cDNA library that can interact with Q. Among the six TFs that were obtained (Table S1), SPA encoded by TraesCS1B02G343500 was the most likely interacting partner of Q. Previous research demonstrated that SPA directly regulates the expression of *SSP* genes [14,25,26]. Accordingly, SPA was selected for further research and was verified to interact with Q. The Q^{c5} expression level was slightly, but not significantly, higher than the *Q* expression level in developing seeds, but there was no correlation between the *SPA* and *Q* expression patterns (Figure 5A,B), suggesting that *Q* does not regulate *SPA* expression. Notably, the interaction between Q^{c5} and SPA was weaker than that between Q and SPA (Figure 4B).

Because SPA is reportedly an activator of LMW-GS gene expression [14], dual-luciferase reporter assays were performed to determine whether Q and SPA can regulate *LMW-1D1* expression (Figure 4C–E). As expected, SPA significantly increased the LUC activity, whereas Q and Q^{c5} had no effect. However, compared with the effect of SPA alone, the co-infiltration of Q/Q^{c5} and SPA resulted in decreased LUC activity. The extent of this decrease was positively associated with the abundance of Q (i.e., dosage effect) (Figure 4D). The inhibitory effect of Q^{c5} was weaker than that of Q (Figure 4E) because of the weaker physical interaction between Q^{c5} and SPA than between Q and SPA (Figure 4B). Overall, the activation of LMW-GS gene expression by SPA can be inhibited by Q, but a mutation from Q to Q^{c5} weakens this inhibition.

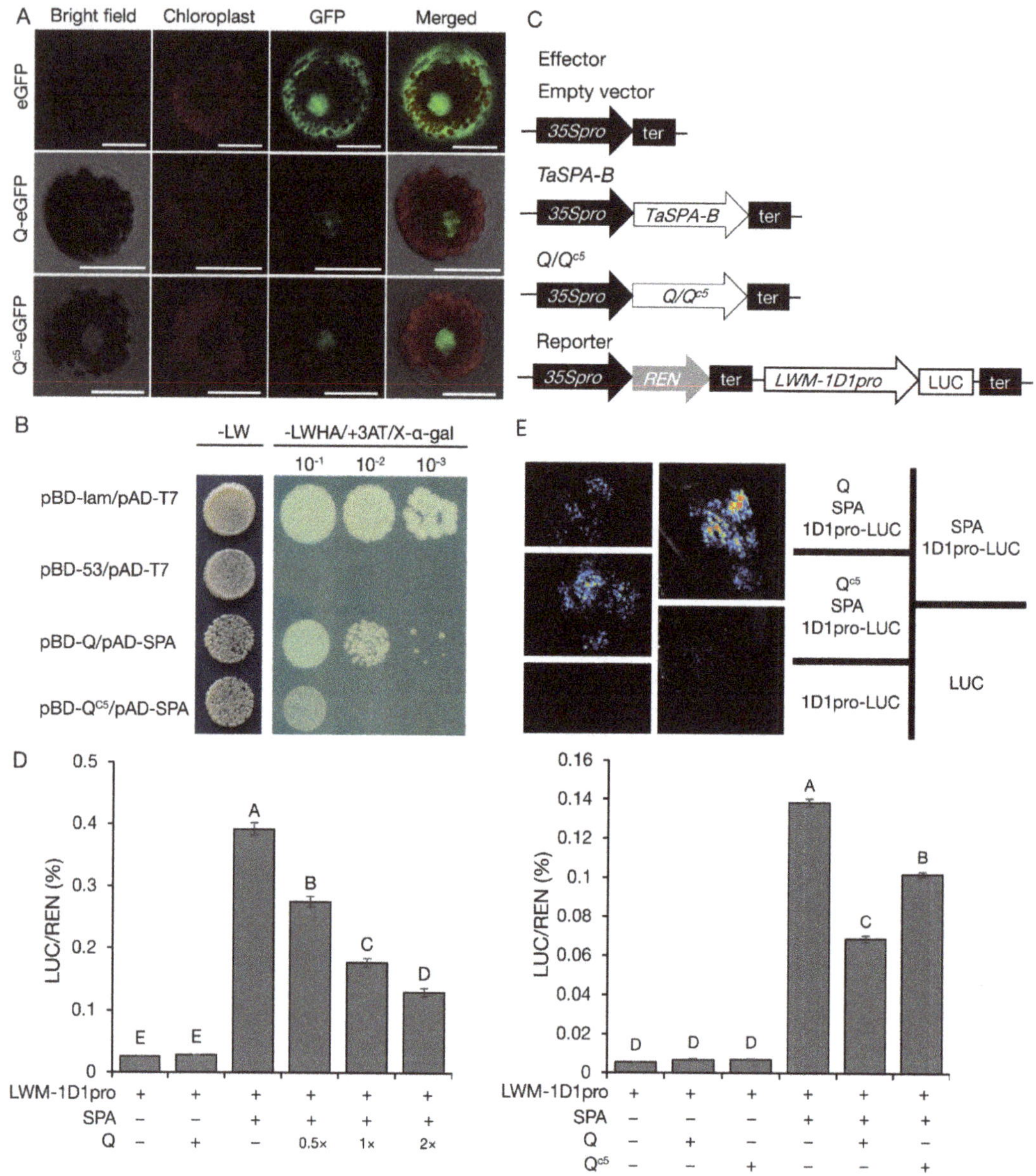

Figure 4. Q physically interacts with SPA to regulate the expression of *SSP* genes. (**A**) Q and Q^{c5} are localized to the nucleus of wheat mesophyll protoplasts. (**B**) Q^{c5} interacts with SPA more weakly than Q in yeast two-hybrid assays. Y2HGold yeast cells transformed with the labeled constructs were assayed for growth on SD medium lacking LW and for LacZ activities on SD-LWHA medium supplemented with X-α-gal. pAD, GAL4 activation domain; pBD, GAL4 DNA-binding domain; 4 mM

3AT was added to the SD-LWHA medium to repress self-activation. The pGBKT7 and pGADT7 empty vectors were used as the negative control, whereas the interaction between p53 and Lam served as the positive control. (**C–E**) Ability of Q/Q^{c5} and SPA to activate the LMW-GS gene promoter revealed by dual-luciferase transcriptional activity assays. (**C**) Schematic diagrams of the effector and reporter plasmids. *35S:SPA*, *35S:Q*, and *35S:Q^{c5}* were constructed as effectors. The promoter of the LMW-GS gene *LMW-1D1* was fused to *LUC* to generate the reporter *LMW-1D1pro:LUC*. LUC, firefly luciferase; 35Spro, CaMV 35S promoter; REN, *Renilla reniformis* luciferase; ter, terminator. (**D**) Q inhibits the ability of SPA to activate the *LMW-1D1* promoter. (**E**) Q^{c5} + SPA activates the promoter more than Q + SPA. Representative leaves are shown in the upper panel, whereas the relative luciferase activities are presented in the lower panel. LUC/REN indicates the LUC-to-REN activity ratio. Data are presented as the mean ± standard deviation ($n = 3$). Capital letters above each column indicate significant differences at $p \leq 0.01$. + and − respectively indicate the presence and absence of the indicated effector/reporter constructs, whereas × indicates the Q concentration.

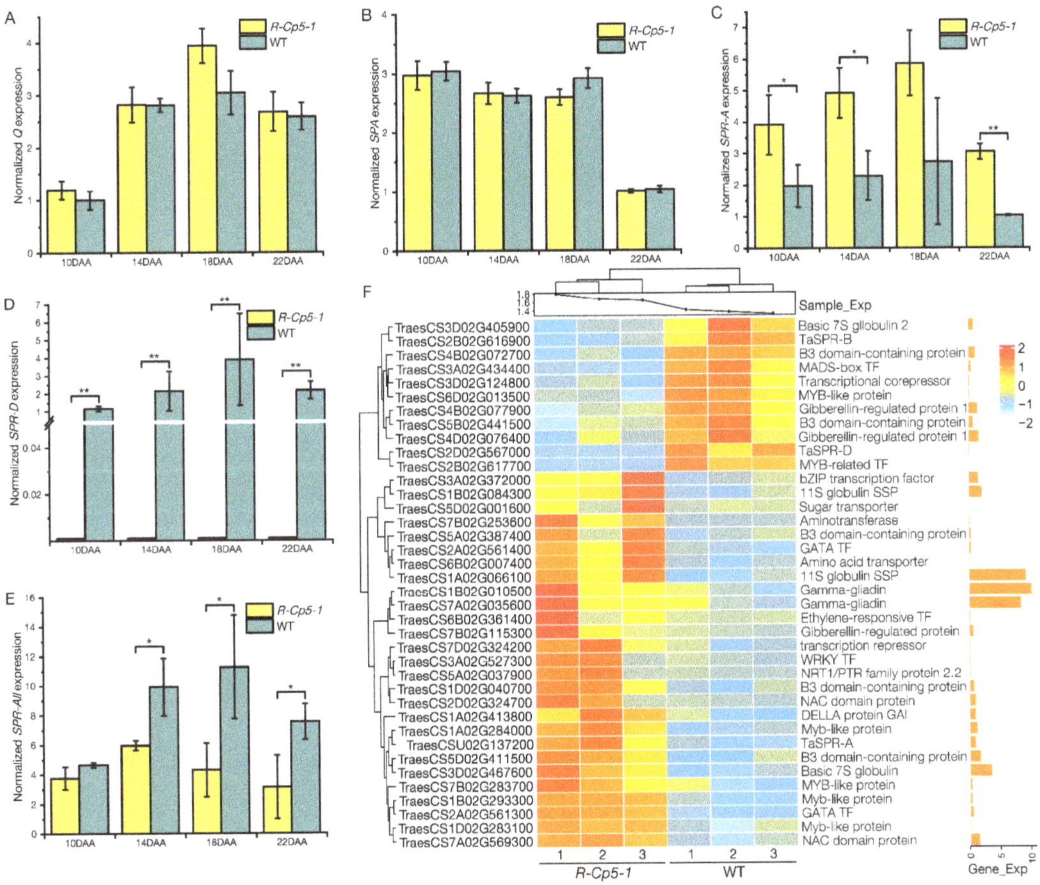

Figure 5. Relative expression of *Q* (**A**), *SPA* (**B**), *SPR-A* (**C**), *SPR-D* (**D**), and *SPR-All* (**E**) in developing seeds. Data are presented as the mean ± standard deviation ($n = 3$; *, $p \leq 0.05$; **, $p \leq 0.01$). (**F**) Heat map presenting the mean normalized expression of selected DEGs. The scale bar indicates the mean normalized expression level. Sample_Exp is the mean expression of all genes in every sample; Gene_Exp is the mean expression of each gene. The histograms reflect the relative expression levels. All DEGs are listed in Tables S5 and S6.

2.6. Effect of Q^{c5} on the Seed Transcriptome

To further clarify how Q^{c5} improves the bread-making quality of wheat, the transcriptomes of immature seeds at 14 days after anthesis (DAA) were compared between *R-Cp5-1* and WT (Table S2). A total of 597 DEGs were identified, including 391 up-regulated and 206 down-regulated genes in *R-Cp5-1* (Tables S3 and S4; Figure S9). The GO enrichment analysis revealed 31 significantly enriched terms associated with Q^{c5} (Table S5; Figure S10). A total of 90 KEGG pathways were assigned to 108 up-regulated and 17 down-regulated genes (Table S6; Figure S10).

We searched for TFs that reportedly regulate *SSP* gene expression. Only *SPR* expression was significantly altered (Table 1), suggesting that *Q* regulates the expression of *SSP* genes by modulating *SPR* expression. The homologous alleles of *SPR* in the three sub-genomes were differentially regulated. Specifically, *SPR-B* (TraesCS2B02G616900) and *SPR-D* (TraesCS2D02G567000) expression levels were down-regulated in *R-Cp5-1*, which was in contrast to the up-regulated expression of *SPR-A* (TraesCSU02G137200) (Table 1; Figure 5C–F). Additionally, *SPR-D* was unexpressed in *R-Cp5-1* and *SPR-B* was expressed at very low levels in WT and *R-Cp5-1* (Table S4). The qRT-PCR analysis of *SPR* expression indicated that the total expression of all *SPR* alleles (*SPR-All*) decreased significantly in *R-Cp5-1* developing seeds (Figure 5C–E).

Table 1. Expression of the genes encoding transcription factors that regulate *SSP* gene expression.

Gene ID	Log_2FC	Gene Name
TraesCSU02G137200	1.42	*SPR-A* [22]
TraesCS2B02G616900	−3.89	*SPR-B* [22]
TraesCS2D02G567000	−8.22	*SPR-D* [22]
TraesCS3A02G077900	−0.31	*TaNAC019-A* [23]
TraesCS3B02G092800	−0.0037	*TaNAC019-B* [23]
TraesCS3D02G078500	0.996	*TaNAC019-D* [23]
TraesCS5A02G440400	0.32	*TaSHP* [26]
TraesCS5B02G444100	0.46	*TaSHP* [26]
TraesCS5D02G447500	0.34	*TaSHP* [26]
TraesCS1B02G343500	0.13	*SPA-B* [14,24]
TraesCS1D02G332200	0.35	*SPA-D* [14,24]
TraesCS3A02G336500	−0.61	*TaGAMYB-A* [20]
TraesCS3B02G367500	0.58	*TaGAMYB-B* [20]
TraesCS3D02G329400	0.49	*TaGAMYB-D* [20]
TraesCS3A02G249100	−0.07	*TaFUSCA3* [19]
TraesCS3B02G278000	−0.38	*TaFUSCA3* [19]
TraesCS3D02G249100	0.07	*TaFUSCA3* [19]
TraesCS5A02G155900	0.47	*TaPBF*
TraesCS5B02G154100	0.49	*TaPBF*
TraesCS5D02G161000	0.14	*TaPBF*
TraesCS7A02G205100	−0.23	*TaODORANT1* [21]
TraesCS7B02G112400	0.12	*TaODORANT1* [21]
TraesCS7D02G208000	−0.04	*TaODORANT1* [21]

3. Discussion

Wheat domestication was a long process that was closely related to the development of human societies and the establishment of civilizations [42]. Additionally, *Q* is arguably the most important domestication gene in cultivated wheat. The *Q* allele was the result of a spontaneous mutation in the miRNA172-binding region of the *q* allele in wild wheat, which increased the *Q* expression level [37]. The *Q* allele confers the free-threshing characteristic of wheat and influences several domestication-related traits, including rachis fragility, glume toughness, spike architecture, flowering time, and plant height [37,38,43–47]. The Q^{c1} allele originated from a point mutation within the miRNA172-binding region of the *Q* allele, which decreases the miRNA172-directed cleavage of Q^{c1} transcripts, thereby leading to increased expression levels and enhanced wheat bread-making quality [40]. Therefore, point mutations in the miRNA172-binding site can increase the *Q* transcription level and influence multiple traits, including processing quality.

Processing quality is a major trait of interest among wheat breeders, but relatively few genetic loci with significant effects on processing quality and useful for wheat breeding have been reported. Joppa [48] detected a QTL explaining 66% of the variation in GPC. The identified gene in this QTL encodes a NAC TF that is associated with GPC increases of approximately 14 g kg^{-1}, but it also accelerates senescence [49]. Gao [23] identified an elite *TaNAC019* allele related to bread-making quality. Compared with the effects of the *Q* allele common among modern wheat cultivars, we demonstrated that Q^{c1} is correlated with a GPC increase of about 60 g kg^{-1} and a loaf volume increase of 37%. In this study, we analyzed the Q^{c5} allele to confirm *Q* affects GPC (Figure 3G; Figure S5) and loaf volume (Figure 3D–F). Unfortunately, Q^{c1} is inappropriate for breeding because it leads to decreased grain yields, extreme dwarfism, and the production of compact spikes [40]. A previous study identified Q^{c1}-*N8* as a new allele derived from Q^{c1}, with a missense mutation in the sequence encoding the second AP_2 domain [50]. This new allele reverses the unfavorable agronomic characteristics resulting from Q^{c1} while also positively affecting GPC and grain yields [50]. Wheat is an important protein source for humans [3]. Therefore, the influence of *Q* on GPC suggests that *Q* should be targeted by wheat breeding programs.

The Q^{c5} allele improves the bread-making quality of wheat by enhancing the accumulation of SSPs and by increasing the Glu/Gli ratio (Figure 3J). The unique properties of wheat depend on its SSPs [4]. An increase in the SSP contents in mutant grains inevitably leads to the production of more PBs, which combine to form a matrix protein that binds to SGs during the seed dehydration stage (Figure S7).

The value of *Q* is associated with the number of point mutations in its miRNA172-binding site. The transition from *q* to *Q* (one point mutation in the miRNA172-binding site) during wheat domestication had a profound effect on the grain yield and shape [51]. The stepwise mutation from the *Q* allele to the Q^{c1}-*N8* allele (another point mutation in the miRNA172-binding site) further increased the grain yield by increasing TKW and the grain number per spike [50]. In plants, a high sequence complementarity between a specific miRNA and its target mRNA is required for the miRNA–target interaction [52,53]. The cleavage efficiency of miRNAs can be modified by changes in their binding regions within target genes [52,54]. Several studies have focused on the interactions between miRNA172 and *AP2-like* genes in grass species [55–58]. In addition to sequence complementarity, another factor influencing miRNA regulatory effects is the mRNA secondary structure [59]. In this study, Q^{c5}, which has a point mutation outside of the miRNA172-binding site, was more highly expressed than *Q* because the miRNA172-binding region is more highly folded in Q^{c5} than in *Q* (Figure S3). This finding may form the basis of a new method for further increasing the *Q* expression level.

The C-terminal region of Q protein is related to grain size. Both Q^{c5} and Q^{c1} are transcribed at a higher level than the *Q* allele (Figure 2G) [40]. However, increased grain size is associated with Q^{c1} (Figure S11), but not Q^{c5} (Figure 3A,B). The missense mutation that changes a proline in Q to a leucine in Q^{c5} is outside the AASSGF box. Moreover, the amino acid change alters the protein structure, resulting in decreased protein activity. Proline is detrimental to the formation of α-helices [60,61], whereas leucine tends to be included in helical structures. Therefore, replacing the proline (amino acid 408) at the end of the α-helix in Q with leucine extends the α-helix in Q^{c5} by two residues. Specifically, the α-helix ends at amino acid residues 409 and 411 in Q and Q^{c5}, respectively (Figure S4B–E). The missense mutations in the first four Q^{c} alleles affect the conserved AASSGF box of Q (Figure S12), resulting in minor changes to the secondary structure of the AASSGF domain in the four Q^{c} proteins (Figures S13 and S14). Moreover, there were no differences in the 3D structures of Q and Q^{c}. Therefore, the opposite effects of Q^{c5} and Q^{c1} on grain size may be due to the diversity in their protein structures.

The results of this study have clarified the mechanism underlying the effect of Q^{c5} on wheat processing quality (Figure 6). Specifically, Q^{c5} enhances *SSP* gene expression in two ways. First, SPA can activate the transcription of glutenin genes [14], but Q has an inhibitory effect on SPA (Figure 4E). The missense mutation in Q^{c5} weakens the physical interaction between the encoded protein and SPA, leading to a weaker inhibitory effect on SPA and then increased *LMW-GS* gene expression (Figure 4D–E). Second, the missense mutation in Q^{c5} is related to the down-regulated expression of *SPR*. However, our ChIP-seq analysis did not detect the enrichment of Q in the *SPR* region (unpublished data). Accordingly, how *Q* regulates *SPR* expression remains to be elucidated.

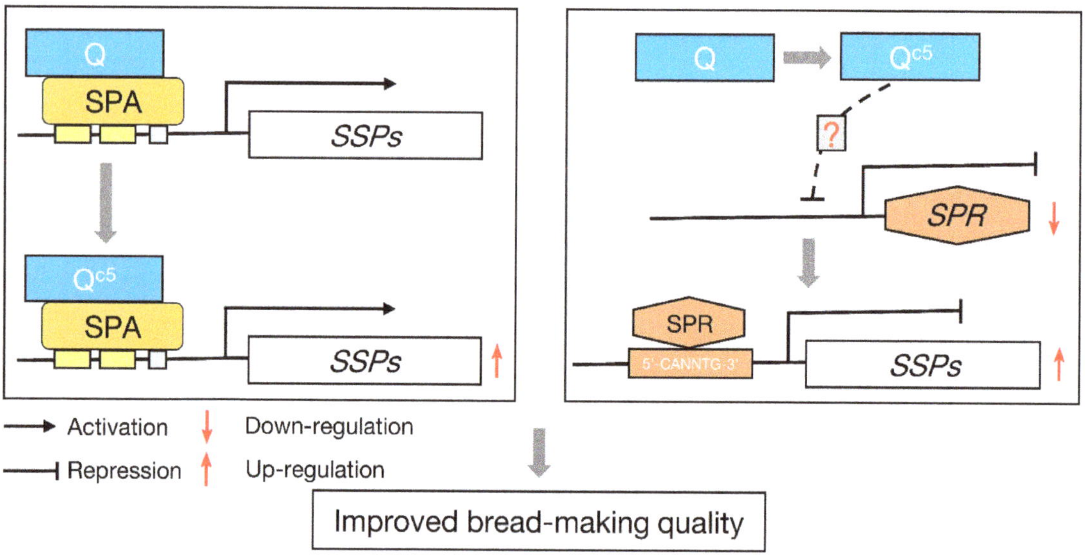

Figure 6. Schematic diagram of the regulatory effect of Q^{c5} on *SSP* gene expression in wheat seeds. The activation of *SSP* expression by SPA is inhibited by Q, but the missense mutation from *Q* to Q^{c5} weakens this inhibition, resulting in more SSP accumulation. Moreover, the missense mutation in Q^{c5} down-regulates *SPR* expression; since *SPR* functions as a repressor of *SSP* genes, the suppression of *SPR* results in up-regulated *SSP* expression levels. Taken together, the up-regulated SSPs expression results in improved bread-making quality. Solid and dotted arrows indicate direct and indirect regulation, respectively. The question mark indicates the regulatory effect of *Q* on *SPR* expression remains to be elucidated.

4. Materials and Methods

4.1. Plant Materials and Growth Conditions

Common wheat cultivar "Roblin" was treated with 0.6% ethyl methanesulfonate (Sigma-Aldrich, St Louis, MO, USA). A mutant (*R-Cp5-1*) with increased spike density and its corresponding wild-type (WT) control were isolated from a single M_7 heterozygous plant. A total of 160 F_2 individuals derived from the *R-Cp5-1* × QZ96 hybridization were used for the chi square (χ^2) test and segregation analysis. An F_3 population comprising 1101 individuals was generated from heterozygous F_2 plants (*R-Cp5-1* × QZ96) and then used for the segregation analysis. Hexaploid wheat line QZ96 has speltoid-like spikes and contains the Q^t allele (i.e., *Q* allele with a transposon insertion [39]). Plants were grown in a plot with a row spacing of 20 cm × 10 cm at the experimental farm of Sichuan Agricultural University, Chengdu, China. A nitrogen: phosphorous: potassium (15:15:15) compound fertilizer was applied (750 kg per hectare) before sowing.

A field experiment was conducted according to a randomized block design with five replicates during the 2017–2018 wheat growing seasons to compare the agronomic traits and processing parameters of *R-Cp5-1* and WT. Each replicate was planted in a

2 m × 4 m plot with a row spacing of 20 cm × 5 cm. At GS80 [41], 20 plants per line were randomly selected for an analysis of agronomic traits, including plant height (cm), main spike length (cm), spikelet number per main spike, grain number per main spike, and productive tiller number. The wheat materials were grown as described above.

One glasshouse (16 h day (23 °C)/8 h night (18 °C) cycle) experiment was conducted in 2019 to analyze gene expression in grains. Plants were grown in 2 L pots (four *R-Cp5-1* plants and four WT plants per pot) filled with a loam-based substrate (Pindstrup, Ryomgaard, Denmark). The pots were arranged according to a randomized complete block design with three replicates. Plants were watered as needed and fertilized before planting with a nitrogen: phosphorous: potassium (15:15:15) compound fertilizer.

Nicotiana benthamiana plants were grown in a glasshouse at 22 °C with a 16 h day/8 h night cycle. When plants had six leaves, the youngest leaves longer than 1 cm were infiltrated with *Agrobacterium tumefaciens* and maintained in the glasshouse for the duration of the dual-luciferase reporter assay.

4.2. DNA Extraction and Segregation Analysis

Genomic DNA was extracted from the young leaves of 160 F_2 individuals and 1101 F_3 individuals derived from the *R-Cp5-1* × QZ96 hybridization using cetyltrimethylammonium bromide [62]. The subsequent PCR amplifications were performed using the following program: 94 °C for 4 min; 35 cycles of 94 °C for 30 s, 55 °C for 30 s, and 72 °C for 30 s; 72 °C for 10 min. The PCR products were separated on 8% polyacrylamide gels or 1% agarose gels. Compact/normal spike was selected as the target trait for the segregation analysis. The spike morphology of all individuals was investigated at GS80.

Four molecular markers [39] that were polymorphic between the parents (*R-Cp5-1* and QZ96) were used for the segregation analysis; then, a linkage map was constructed using the MAP function of IciMapping 4.1 (Institute of Crop Science and CIMMYT China, CAAS, Beijing, China).

4.3. RNA Extraction and Gene Expression Analysis

Developing spikes were collected from the *R-Cp5-1* and WT plants grown for the field experiment at GS23, GS24, GS32, and GS39. Three biological replicates were collected, with at least 10 spikes per replicate. Developing grains were collected from the *R-Cp5-1* and WT plants grown for the glasshouse experiment at 10, 14, 18, and 22 DAA. Three biological replicates were collected per time-point. The collected samples were immediately frozen in liquid nitrogen and then stored at −80 °C. The plant samples were ground to a fine powder in liquid nitrogen. Total RNA was extracted from the ground material using the Plant RNA Extraction kit (Biofit, Chengdu, China) and then was quantized using the NanoDrop 8000 spectrophotometer (Thermo Scientific, Waltham, MA, USA). First-strand cDNA was synthesized from 1 μg total RNA using the PrimeScript™ II 1st Strand cDNA Synthesis Kit (Takara, Dalian, China).

All RNA samples extracted from spikes and seeds were used for qRT-PCR analysis by using the ChamQ™ Universal SYBR® qPCR Master Mix (Vazyme, Nanjing, China). All experiments were performed according to the manufacturer's instructions. Housekeeping genes, namely the genes encoding the scaffold-associated region DNA-binding protein (NCBI UniGene: *Ta.14126*) and methionine aminopeptidase 1 (*Ta.7894*) [63], were used for normalizing gene expression data. The qRT-PCR primers are listed in Table S7.

The RNA extracted from seeds at 14 DAA was used for a transcriptome analysis. Specifically, the RNA was sequenced using the Illumina NovaSeq platform at Tcuni Biotech (Chengdu, China). The RNA quantity and integrity were determined using the NanoDrop 8000 spectrophotometer (Thermo Scientific, Waltham, MA, USA) and the Agilent 4200 TapeStation (Agilent Technologies, Santa Clara, CA, USA). The RNA-seq libraries were constructed using the VAHTS® Stranded mRNA-seq Library Prep Kit for Illumina. The libraries were amplified by PCR and quantified by qRT-PCR. The default parameters of Kallisto (v0.42.4) [64] were used for quantifying transcript abundances. The edgeR (v3.12.1) program was used to detect differentially expressed genes (DEGs) between *R-Cp5-1* and "Roblin" according to the following criteria: adjusted $p < 0.05$ and an absolute fold-change > 2.0 [65]. The GOseq (v1.22.0) program was used for a Gene Ontology (GO) enrichment analysis of the DEGs [66] on the basis of the Wallenius non-central hyper-geometric distribution, which can adjust for the gene length bias among DEGs. A Kyoto Encyclopedia of Genes and Genomes (KEGG) pathway enrichment analysis of the DEGs was completed using KOBAS (v2.0) [67]. The transcripts per kilobase of exon model per million mapped reads (TPM) value was used for calculating gene expression levels.

4.4. Gene Cloning

The cDNA and genomic DNA sequences of candidate genes in *R-Cp5-1* and "Roblin" were amplified by PCR. The *HMW-Bx7, HMW-Dx5, HMW-Dy10*, and *LMW-1D1* promoters in "Roblin" were also amplified. Phanta Max Super-Fidelity DNA Polymerase (Vazyme) was used for the PCR amplifications, which were conducted in a 50 µL mixture comprising genomic DNA or cDNA, 100 µM each dNTP, 4 pmol each primer, 1 U DNA polymerase, and 25 µL 2× buffer (with 4 mM Mg^{2+}). The PCR amplifications were completed in the Mastercycler® nexus programmable thermal cycler (Eppendorf, Wesseling-Berzdorf, Germany) using the following program: 95 °C for 3 min; 35 cycles of 95 °C for 15 s, 60 °C for 15 s, and 72 °C for 1–2 min; 72 °C for 10 min. The PCR products were separated on a 1.5% agarose gel. The expected fragments were purified and inserted into the pMD19-T vector (Takara). Positive colonies were verified by Sanger sequencing (Sangon, Chengdu, China). The cloning and sequencing experiments were repeated at least three times. The primers used are listed in Table S7.

4.5. Structure Prediction

The secondary structures of the mRNA sequences of *q* (GenBank No. AY702957.1), *Q* (KX620763.1), Q^{c5} (MW419115), and another four overexpressed *Q* alleles (i.e., Q^{c1}–Q^{c4}, KX620765, KX620766, KX620767, and KX620768) [40] were predicted using the Vienna RNA fold server (http://rna.tbi.univie.ac.at/; accessed on 1 March 2018) [68,69]. The secondary structures around the miRNA172-binding site were clipped and compared. The secondary structures of the deduced peptides encoding by *q*, *Q*, Q^{c5}, and Q^{c1}–Q^{c4} were predicted using the default parameters of Antheprot 6.9.3 [70].

The three-dimensional (3D) structures of *q*, *Q*, Q^{c5}, and Q^{c1}–Q^{c4} were predicted using I-TASSER (http://zhanglab.ccmb.med.umich.edu/I-TASSER/; accessed on 10 August 2018) [71–73] as described by Zhao [74]. The full-length *Q* and Q^{c5} structures were compared using TM-align (https://zhanggroup.org/TM-align/; accessed on 27 December 2018) as described by Zhang [75]. After simulating the substitution of the 408th residue (from proline in *Q* to leucine in Q^{c5}), the resulting sequences were aligned using STRUM (https://zhanggroup.org/STRUM/; accessed on 1 January 2019) [76]. The *q*/*Q* and *Q*/Q^{c1}–Q^{c4} protein structures were also compared using TM-align. Moreover, the 3D protein structures were visualized using UCSF Chimera (version 1.13.1; http://www.rbvi.ucsf.edu/chimera/; accessed on 1 January 2019) as described by Pettersen [77].

4.6. Processing Quality Analysis

All plants grown in the experimental plot were harvested, threshed, and sun-dried at approximately 35 °C to a constant weight. Grain samples were stored at room temperature for 2 months before milling. Two grain portions (500 kernels) were randomly selected and weighed to calculate the TKW (the difference in kernel weight between the two portions was less than 5%). The grain moisture content was measured using the MJ33 moisture analyzer (Mettler Toledo, Switzerland) and adjusted to 16.5% before milling. Each grain sample (1 kg) was milled using the Chopin® CD1 Mill (Chopin Technologies, Villeneuve-la-Garenne, France). The GPC (dry weight), Zeleny sedimentation value, and wet gluten content were determined as described by Wang [78].

A baking test was performed according to the AACC method 10.09-01 [40], with some modifications. Briefly, the standard rapid-mix-test with 35 g flour (14% moisture content) was used. The test was performed using three biological replicates, with two loaves of bread per flour sample. The bread volume was determined using the BVM6630 volume meter (Perten, Hägersten, Sweden).

4.7. Reversed-Phase High-Performance Liquid Chromatography (RP-HPLC)

An RP-HPLC analysis was performed to quantify the gliadin, HMW-GS, and LMW-GS contents and calculate the glutenins-to-gliadins (Glu/Gli) and HMW-GS-to-LMW-GS (HMW/LMW) ratios as described by Dupont [79] and Zheng [80], with some modifications. The total HMW-GS, LMW-GS, and gliadin contents were estimated by integrating the relevant RP-HPLC peaks in the chromatograms. The Glu/Gli ratio was calculated according to the relative contents.

4.8. Subcellular Localization

The full *Q* and Q^{c5} coding sequences (CDSs) were cloned into the pCAMBIA2300-eGFP (enhanced green fluorescent protein) vector. The primers used are listed in Table S7. The resulting vectors were inserted into wheat mesophyll protoplasts. After a 16 h incubation, the transfected protoplasts were examined using the STELLARIS STED/EM CPD300 confocal microscope (Leica, Germany). Wheat mesophyll protoplasts were prepared and transfected as described by Brandt [81].

4.9. Yeast One- and Two-Hybrid Analyses

To screen for Q-interacting proteins, equal amounts of RNA extracted from grain samples at 5, 10, 15, 20, 25, and 30 DAA were combined for the construction of an AD-prey yeast library (pGADT7-Rec) (Oebiotech, Shanghai, China). The *Q* CDS was inserted into the pGBKT7 vector (Q-BD). Subsequently, Q-BD and the AD-prey library were used for an examination of protein–protein interactions according to the Matchmaker Gold Yeast Two-Hybrid Library screening user manual (Clontech, Saint-Germain-en-Laye, France).

To compare the physical interaction between Q/Q^{c5} and SPA (GenBank No. Y09013.1), the *SPA* CDS was cloned from the cDNA of 14 DAA grains and then incorporated into the prey vector pGADT7 (*AD-SPA*). The resulting vector and the *Q-BD* or Q^{c5}*-BD* bait vector were used for the co-transformation of Y2H Gold yeast cells. After preparing gradient dilutions, the two yeast strains containing *Q-BD* + *AD-SPA* and Q^{c5}*-BD* + *AD-SPA* were cultured on double (SD/−Leu/−Try) and quadruple (SD/−Leu/−Try/−His/−Ade + X-α-Gal) dropout SD media (Clontech) in plates incubated at 28 °C for 3 days.

To determine whether Q can bind to the promoters of *SSP* genes, yeast one-hybrid assays were performed according to the Matchmaker Gold Yeast One-Hybrid Library screening user manual (Clontech). The HMW-GS (Bx7, Dx5, and Dy10) and LMW-GS (1D1) gene promoters were used as the bait sequences. The prey sequence (*Q*) was cloned into the pGADT7-Rec vector and then inserted into the bait yeast strains. To examine the potential interaction between Q and the *SSP* gene promoters, the bait yeast strains were grown on medium lacking uracil but supplemented with different aureobasidin A concentrations. The yeast cells were subsequently added to medium lacking leucine and the resulting colonies were photographed. The primers used are listed in Table S7.

4.10. Dual-Luciferase Reporter Assay

A dual-luciferase reporter assay was performed as described by Song [82], with some modifications. The *SPA*, *Q* allele, and Q^{c5} allele CDSs were inserted into the pGreenII 62-SK vector [83] under the control of the cauliflower mosaic virus 35S promoter (*35S:SPA*, *35S:Q*, and *35S:*Q^{c5}). The *LMW-1D1* promoter was incorporated into the pGreenII 0800-*LUC* (firefly luciferase) vector, which contains the *35S:REN* (*Renilla reniformis* luciferase) cassette [83], to produce the reporter vector *LMW-1D1pro:LUC*. The primers used are listed in Table S7. The constructs were separately inserted into *A. tumefaciens* strain GV3101 cells (Weidi Biotechnology, Shanghai, China) along with the helper plasmid pSoup-19. The transformed *A. tumefaciens* cells were used in the following assay. Briefly, individual colonies carrying the above constructs were grown in YMB (yeast mannitol medium) liquid medium containing appropriate antibiotics, harvested, and resuspended in infiltration solution (10 mM $MgCl_2$, 10 mM 4-morpholineethanesulfonic acid hydrate, and 150 mM acetosyringone) for an OD_{600} of 0.2. The cell suspensions were mixed (1:1 volume or a specific ratio) and injected into *N. benthamiana* leaves. The leaves co-infiltrated with *LMW-1D1pro:LUC* and the empty pGreenII 62-SK vector served as controls. All infiltrations were repeated at least three times, with at least three leaves per infiltration.

Leaf samples were collected using a hole punch at 48–60 h post-infiltration and then the LUC and REN activities were measured using the Dual Luciferase Reporter Assay Kit (Vazyme) and the GloMax® 96 Microplate Luminometer (Promega, Madison, WI, USA) as previously described [83]. The LUC-to-REN activity (LUC/REN) ratio was used to estimate the transactivation by the effectors. Images of fluorescence were captured using the ChemiDoc™ Touch Imaging System (Bio-Rad, Hercules, CA, USA). Three biological replicates were examined per experiment.

4.11. Statistical Analysis

All data were compiled using Excel 2016 (Microsoft 2016, Redmond, WA, USA) and analyzed using the Data Processing System software (version 17.10) [84]. A one-tailed Student's *t*-test was used for comparing two groups. For multiple comparisons, the Student's *t*-test and an analysis of variance were performed (least significant difference). Data were recorded as the mean ± standard deviation.

5. Conclusions

The present study verified the positive effect of *Q* on GPC and the bread-making quality of wheat on the basis of an analysis of a new *Q* allele (Q^{c5}). A missense mutation outside of the Q^{c5} miRNA172-binding site leads to increased expression because of the associated highly folded mRNA secondary structure around the miRNA172-binding site, ultimately leading to an increase in spike density. Additionally, the missense mutation in Q^{c5} increases GPC and the bread-making quality of wheat by repressing the expression of *SPR* and modulating the interaction between Q and SPA.

Supplementary Materials: The following supporting information can be downloaded at: https://www.mdpi.com/article/10.3390/ijms23147581/s1. Reference [85] is cited in the supplementary figures.

Author Contributions: Investigation, Z.G., J.Z., Y.W. (Yan Wang), Y.L. (Yang Li), Q.L., K.Z., Y.L. (Yue Li), R.T., X.S., K.T. and L.K.; Data curation and Formal analysis, Z.G. and Q.C.; Writing—original draft, Z.G.; Methodology, Z.G., Y.Z. and P.Q.; Resources, Y.J., Q.J., J.W., G.C. and Y.W. (Yuming Wei); Writing—review and editing and funding acquisition, Y.Z. and P.Q. All authors have read and agreed to the published version of the manuscript.

Funding: This study was supported by the National Natural Science Foundation of China (grant numbers 32072054, 31971939 and 31901961), Science and Technology Department of Sichuan Province (2019YFH0066 and 2020YFH0150).

Institutional Review Board Statement: Not applicable.

Informed Consent Statement: Not applicable.

Data Availability Statement: All data supporting the findings of this study are available within this article and the Supplementary Materials published online.

Conflicts of Interest: The authors declare no conflict of interest.

References

1. Nesbitt, M.; Samuel, D. From the staple crop to extinction? The archaeology and history of hulled wheats. In *Hulled Wheats*; Padulosi, S., Hammer, K., Heller, J., Eds.; International Plant Genetic Resources Institute: Rome, Italy, 1996; Volume 4, pp. 41–100.
2. Huang, S.; Sirikhachornkit, A.; Su, X.; Faris, J.; Gill, B.; Haselkorn, R.; Gornicki, P. Genes encoding plastid acetyl-CoA carboxylase and 3-phosphoglycerate kinase of the *Triticum/Aegilops* complex and the evolutionary history of polyploid wheat. *Proc. Natl. Acad. Sci. USA* **2002**, *99*, 8133–8138. [CrossRef]
3. FAOSTAT. FAO Database. Food and Agriculture Organization of the United Nations. Available online: http://www.fao.org/faostat/en/#data (accessed on 26 July 2021).
4. Shewry, P.R. Wheat. *J. Exp. Bot.* **2009**, *60*, 1537–1553. [CrossRef] [PubMed]
5. Qi, P.F.; Wei, Y.M.; Yue, Y.W.; Yan, Z.H.; Zheng, Y.L. Biochemical and molecular characterization of gliadins. *Mol. Biol.* **2006**, *40*, 713–723. [CrossRef]
6. Payne, P.I. Genetics of wheat storage proteins and the effect of allelic variation on bread-making quality. *Annu. Rev. Physiol.* **1987**, *38*, 141–153. [CrossRef]
7. Don, C.; Mann, G.; Bekes, F.; Hamer, R.J. HMW-GS affect the properties of glutenin particles in GMP and thus flour quality. *J. Cereal Sci.* **2006**, *44*, 127–136. [CrossRef]
8. Tian, J.; Deng, Z.; Zhang, K.; Yu, H.; Jiang, X.; Li, C. Genetic detection of main quality traits in wheat. In *Genetic Analyses of Wheat and Molecular Marker-Assisted Breeding*; Science Press: Beijing, China, 2015; Volume 5, pp. 177–350.
9. Rubio-Somoza, I.; Martinez, M.; Abraham, Z.; Diaz, I.; Carbonero, P. Ternary complex formation between HvMYBS3 and other factors involved in transcriptional control in barley seeds. *Plant J.* **2006**, *47*, 269–281. [CrossRef]
10. Verdier, J.; Thompson, R.D. Transcriptional regulation of storage protein synthesis during dicotyledon seed filling. *Plant Cell Physiol.* **2008**, *49*, 1263–1271. [CrossRef]
11. Xi, D.M.; Zheng, C.C. Transcriptional regulation of seed storage protein genes in *Arabidopsis* and cereals. *Seed Sci. Res.* **2011**, *21*, 247–254. [CrossRef]
12. Vicente-Carbajosa, J.; Moose, S.P.; Parsons, R.L.; Schmidt, R.J. A maize zinc-finger protein binds the prolamin box in zein gene promoters and interacts with the basic leucine zipper transcriptional activator Opaque2. *Proc. Natl. Acad. Sci. USA* **1997**, *94*, 7685–7690. [CrossRef]
13. Oñate, L.; Vicente-Carbajosa, J.; Lara, P.; Diaz, I.; Carbonero, P. Barley BLZ2, a seed-specific bZIP protein that interacts with BLZ1 in vivo and activates transcription from the GCN4-like motif of B-hordein promoters in barley endosperm. *J. Biol. Chem.* **1999**, *274*, 9175–9182. [CrossRef]
14. Albani, D.; Hammond-Kosack, M.C.; Smith, C.; Conlan, S.; Colot, V.; Holdsworth, M.; Bevan, M.W. The wheat transcriptional activator SPA: A seed-specific bZIP protein that recognizes the GCN4-like motif in the bifactorial endosperm box of prolamin genes. *Plant Cell* **1997**, *9*, 171–184. [CrossRef] [PubMed]
15. Mena, M.; Vicente-Carbajosa, J.; Schmidt, R.J.; Carbonero, P. An endosperm-specific DOF protein from barley, highly conserved in wheat, binds to and activates transcription from the prolamin-box of a native B-hordein promoter in barley endosperm. *Plant J.* **1998**, *16*, 53–62. [CrossRef] [PubMed]
16. Diaz, I.; Martinez, M.; Isabel-LaMoneda, I.; Rubio-Somoza, I.; Carbonero, P. The DOF protein, SAD, interacts with GAMYB in plant nuclei and activates transcription of endosperm-specific genes during barley seed development. *Plant J.* **2005**, *42*, 652–662. [CrossRef]
17. Bäumlein, H.; Nagy, I.; Villarroel, R.; Inzé, D.; Wobus, U. Cis-analysis of a seed protein gene promoter: The conservative RY repeat CATGCATG within the legumin box is essential for tissue-specific expression of a legumin gene. *Plant J.* **1992**, *2*, 233–239. [CrossRef] [PubMed]
18. Moreno-Risueno, M.A.; González, N.; Díaz, I.; Parcy, F.; Carbonero, P.; Vicente-Carbajosa, J. FUSCA3 from barley unveils a common transcriptional regulation of seed-specific genes between cereals and Arabidopsis. *Plant J.* **2008**, *53*, 882–894. [CrossRef]
19. Sun, F.; Liu, X.; Wie, Q.; Liu, J.; Yang, T.; Jia, L.; Wang, Y.; Yang, G.; He, G. Functional characterization of TaFUSCA3, a B3-superfamily transcription factor gene in the wheat. *Front. Plant Sci.* **2017**, *8*, 1133. [CrossRef]
20. Guo, W.; Yang, H.; Liu, Y.; Gao, Y.; Ni, Z.; Peng, H.; Xin, M.; Hu, Z.; Sun, Q.; Yao, Y. The wheat transcription factor TaGAMyb recruits histone acetyltransferase and activates the expression of a high-molecular-weight glutenin subunit gene. *Plant J.* **2015**, *84*, 347–359. [CrossRef]
21. Luo, G.; Shen, L.; Song, Y.; Yu, K.; Ji, J.; Zhang, C.; Yang, W.; Li, X.; Sun, J.; Zhan, K.; et al. The MYB family transcription factor *TuODORANT1* from *Triticum urartu* and the homolog *TaODORANT1* from *Triticum aestivum* inhibit seed storage protein synthesis in wheat. *Plant Biotechnol. J.* **2021**, *19*, 1863–1877. [CrossRef]
22. Shen, L.; Luo, G.; Song, Y.; Xu, J.; Ji, J.; Zhang, C.; Gregová, E.; Yang, W.; Li, X.; Sun, J.; et al. A novel NAC family transcription factor *SPR* suppresses seed storage protein synthesis in wheat. *Plant Biotechnol. J.* **2021**, *19*, 992–1007. [CrossRef]
23. Gao, Y.; An, K.; Guo, W.; Chen, Y.; Zhang, R.; Zhang, X.; Chang, S.; Rossi, V.; Jin, F.; Cao, X.; et al. The endosperm-specific transcription factor TaNAC019 regulates glutenin and starch accumulation and its elite allele improves wheat grain quality. *Plant Cell* **2021**, *33*, 603–622. [CrossRef]
24. Schmidt, R.; Burr, F.A.; Aukerman, M.J.; Burr, B. Maize regulatory gene opaque-2 encodes a protein with a "leucine-zipper" motif binds to zein DNA. *Proc. Natl. Acad. Sci. USA* **1990**, *87*, 46–50. [CrossRef] [PubMed]
25. Ravel, C.; Fiquet, S.; Boudet, J.; Dardevet, M.; Vincent, J.; Merlino, M.; Michard, R.; Martre, P. Conserved cis-regulatory modules in promoters of genes encoding wheat high-molecular-weight glutenin subunits. *Front. Plant Sci.* **2014**, *5*, 621. [CrossRef] [PubMed]

26. Boudet, J.; Merlino, M.; Plessis, A.; Gaudin, J.C.; Dardevet, M.; Perrochon, S.; Alvarez, D.; Risacher, T.; Martre, P.; Ravel, C. The bZIP transcription factor SPA Heterodimerizing Protein represses glutenin synthesis in *Triticum aestivum*. *Plant J.* **2019**, *97*, 858–871. [CrossRef]
27. Guo, D.; Hou, Q.; Zhang, R.; Lou, H.; Li, Y.; Zhang, Y.; You, M.; Xie, C.; Liang, R.; Li, B. Over-expressing *TaSPA-B* reduces prolamin and starch accumulation in wheat (*Triticum aestivum* L.) grains. *Int. J. Mol. Sci.* **2020**, *21*, 3257. [CrossRef] [PubMed]
28. Rubio-Somoza, I.; Martinez, M.; Diaz, I.; Carbonero, P. HvMCB1, a R1MYB transcription factor from barley with antagonistic regulatory functions during seed development and germination. *Plant J.* **2006**, *45*, 17–30. [CrossRef]
29. Mallory, A.C.; Vaucheret, H. Functions of microRNAs and related small RNAs in plants. *Nat. Genet.* **2006**, *38*, S31–S36. [CrossRef]
30. Park, W.; Li, J.; Song, R.; Messing, J.; Chen, X. Carpel factory, a Dicer homolog, and HEN1, a novel protein, act in microRNA metabolism in *Arabidopsis thaliana*. *Curr. Biol.* **2002**, *12*, 1484–1495. [CrossRef]
31. Chen, X. A microRNA as a translational repressor of *APETALA2* in *Arabidopsis* flower development. *Science* **2004**, *303*, 2022–2025. [CrossRef]
32. Aukerman, M.J.; Sakai, H. Regulation of flowering time and floral organ identity by a microRNA and its *APETALA2*-like target genes. *Plant Cell* **2003**, *15*, 2730–2741. [CrossRef]
33. Schmid, M.; Uhlenhaut, N.H.; Godard, F.; Demar, M.; Bressan, R.; Weigel, D.; Lohmann, J.U. Dissection of floral induction pathways using global expression analysis. *Development* **2003**, *130*, 6001–6012. [CrossRef]
34. Schwab, R.; Palatnik, J.F.; Riester, M.; Schommer, C.; Schmid, M.; Weigel, D. Specific effects of microRNAs on the plant transcriptome. *Dev. Cell* **2005**, *8*, 517–527. [CrossRef] [PubMed]
35. Debernardi, J.M.; Lin, H.Q.; Chuck, G.; Faris, J.D.; Dubcovsky, J. MicroRNA172 plays a crucial role in wheat spike morphogenesis and grain threshability. *Development* **2017**, *144*, 1966–1975. [CrossRef] [PubMed]
36. Debernardi, J.M.; Greenwood, J.R.; Finnegan, E.J.; Jernstedt, J.; Dubcovsky, J. *APETALA 2*-like genes *AP2L2* and *Q* specify lemma identity and axillary floral meristem development in wheat. *Plant J.* **2020**, *101*, 171–187. [CrossRef] [PubMed]
37. Simons, K.J.; Fellers, J.P.; Trick, H.N.; Zhang, C.; Tai, Y.S.; Gill, B.S.; Faris, J.D. Molecular characterization of the major wheat domestication gene *Q*. *Genetics* **2006**, *172*, 547–555. [CrossRef] [PubMed]
38. Zhang, Z.C.; Belcram, H.; Gornicki, P.; Charles, M.; Just, J.; Huneau, C.; Magdelenat, G.; Couloux, A.; Samain, S.; Gill, B.S.; et al. Duplication and partitioning in evolution and function of homoeologous *Q* loci governing domestication characters in polyploid wheat. *Proc. Natl. Acad. Sci. USA* **2011**, *108*, 18737–18742. [CrossRef]
39. Jiang, Y.F.; Chen, Q.; Wang, Y.; Guo, Z.R.; Xu, B.J.; Zhu, J.; Zhang, Y.Z.; Gong, X.; Luo, C.H.; Wu, W.; et al. Re-acquisition of the brittle rachis trait via a transposon insertion in domestication gene *Q* during wheat de-domestication. *New Phytol.* **2019**, *224*, 961–973. [CrossRef]
40. Xu, B.J.; Chen, Q.; Zheng, T.; Jiang, Y.F.; Qiao, Y.Y.; Guo, Z.R.; Cao, Y.L.; Wang, Y.; Zhang, Y.Z.; Zong, L.J.; et al. An overexpressed *Q* allele leads to increased spike density and improved processing quality in common wheat (*Triticum aestivum*). *G3 Genes Genom. Genet.* **2018**, *8*, 771–778. [CrossRef]
41. Zadoks, J.C.; Chang, T.T.; Konzak, C.F. A decimal code for the growth stages of cereals. *Weed Res.* **1974**, *14*, 415–421. [CrossRef]
42. Eckardt, N.A. Evolution of domesticated bread wheat. *Plant Cell* **2010**, *22*, 993. [CrossRef]
43. Muramatsu, M. Dosage effect of the spelta gene *q* of hexaploid wheat. *Genetics* **1963**, *48*, 469–482. [CrossRef]
44. Muramatsu, M. The vulgare super gene, *Q*: Its universality in durum wheat and its phenotypic effects in tetraploid and hexaploid wheats. *Can. J. Genet. Cytol.* **1986**, *28*, 30–41. [CrossRef]
45. Endo, T.R.; Gill, B.S. The deletion stocks of common wheat. *J. Hered.* **1996**, *87*, 295–307. [CrossRef]
46. Förster, S.; Schumann, E.; Baumann, M.; Weber, W.E.; Pillen, K. Copy number variation of chromosome 5A and its association with *Q* gene expression, morphological aberrations, and agronomic performance of winter wheat cultivars. *Theor. Appl. Genet.* **2013**, *126*, 3049–3063. [CrossRef]
47. Faris, J.D.; Zhang, Q.; Chao, S.; Zhang, Z.; Xu, S. Analysis of agronomic and domestication traits in a durum × cultivated emmer wheat population using a high-density single nucleotide polymorphism-based linkage map. *Theor. Appl. Genet.* **2014**, *127*, 2333–2348. [CrossRef] [PubMed]
48. Joppa, L.R.; Du, C.; Hart, G.E.; Hareland, G.A. Mapping gene (s) for grain protein in tetraploid wheat (*Triticum turgidum* L.) using a population of recombinant inbred chromosome lines. *Crop Sci.* **1997**, *37*, 1586–1589. [CrossRef]
49. Uauy, C.; Distelfeld, A.; Fahima, T.; Blechl, A.; Dubcovsky, J. A NAC gene regulating senescence improves grain protein, zinc, and iron content in wheat. *Science* **2006**, *314*, 1298–1301. [CrossRef]
50. Chen, Q.; Guo, Z.R.; Shi, X.L.; Wei, M.Q.; Fan, Y.Z.; Zhu, J.; Zheng, T.; Wang, Y.; Kong, L.; Deng, M.; et al. Increasing the Grain Yield and Grain Protein Content of Common Wheat (*Triticum aestivum*) by Introducing Missense Mutations in the *Q* Gene. PREPRINT (Version 1) Available at Research Square. Available online: https://www.researchsquare.com/article/rs-899571/v1 (accessed on 18 October 2021).
51. Xie, Q.; Li, N.; Yang, Y.; Lv, Y.; Yao, H.; Wei, R.; Sparkes, D.L.; Ma, Z. Pleiotropic effects of the wheat domestication gene *Q* on yield and grain morphology. *Planta* **2018**, *247*, 1089–1098. [CrossRef]
52. Mallory, A.C.; Reinhart, B.J.; Jones-Rhoades, M.W.; Tang, G.; Zamore, P.D.; Barton, M.K.; Bartel, D.P. MicroRNA control of *PHABULOSA* in leaf development: Importance of pairing to the microRNA 5′ region. *EMBO J.* **2004**, *23*, 3356–3364. [CrossRef]

53. German, M.A.; Pillay, M.; Jeong, D.H.; Hetawal, A.; Luo, S.; Janardhanan, P.; Kannan, V.; Rymarquis, L.A.; Nobuta, K.; German, R.; et al. Global identification of microRNA-target RNA pairs by parallel analysis of RNA ends. *Nat. Biotechnol.* **2008**, *26*, 941–946. [CrossRef]
54. Liu, Q.; Wang, F.; Axtell, M.J. Analysis of complementarity requirements for plant microRNA targeting using a *Nicotiana benthamiana* quantitative transient assay. *Plant Cell* **2014**, *26*, 741–753. [CrossRef]
55. Chuck, G.; Meeley, R.; Irish, E.; Sakai, H.; Hake, S. The maize *tasselseed4* microRNA controls sex determination and meristem cell fate by targeting Tasselseed6/indeterminate spikelet1. *Nat. Genet.* **2007**, *39*, 1517–1521. [CrossRef] [PubMed]
56. Zhu, Q.H.; Upadhyaya, N.M.; Gubler, F.; Helliwell, C.A. Over-expression of miR172 causes loss of spikelet determinacy and floral organ abnormalities in rice (*Oryza sativa*). *BMC Plant Biol.* **2009**, *9*, 149. [CrossRef] [PubMed]
57. Nair, S.K.; Wang, N.; Turuspekov, Y.; Pourkheirandish, M.; Sinsuwongwat, S.; Chen, G.; Sameri, M.; Tagiri, A.; Honda, I.; Watanabe, Y.; et al. Cleistogamous flowering in barley arises from the suppression of microRNA-guided *HvAP2* mRNA cleavage. *Proc. Natl. Acad. Sci. USA* **2010**, *107*, 490–495. [CrossRef] [PubMed]
58. Houston, K.; McKim, S.M.; Comadran, J.; Bonar, N.; Druka, I.; Uzrek, N.; Cirillo, E.; Guzy-Wrobelska, J.; Collins, N.C.; Halpin, C.; et al. Variation in the interaction between alleles of *HvAPETALA2* and microRNA172 determines the density of grains on the barley inflorescence. *Proc. Natl. Acad. Sci. USA* **2013**, *110*, 16675–16680. [CrossRef]
59. Zheng, Z.; Reichel, M.; Deveson, I.; Wong, G.; Li, J.; Millar, A.A. Target RNA secondary structure is a major determinant of miR159 efficacy. *Plant Physiol.* **2017**, *174*, 1764–1778. [CrossRef]
60. Chou, P.Y.; Fasman, G.D. Conformational parameters for amino acids in helical, beta-sheet, and random coil regions calculated from proteins. *Biochemistry* **1974**, *13*, 211–222. [CrossRef]
61. Barlow, D.J.; Thornton, J.M. Helix geometry in proteins. *J. Mol. Biol.* **1988**, *201*, 601–619. [CrossRef]
62. Doyle, J.J.; Doyle, J.L.; Doyle, J.A.; Doyle, F.J. A rapid DNA isolation procedure for small quantities of fresh leaf tissue. *Phytochem. Bull.* **1987**, *19*, 11–15.
63. Long, X.Y.; Wang, J.R.; Ouellet, T.; Rocheleau, H.; Wei, Y.M.; Pu, Z.E.; Jiang, Q.T.; Lan, X.J.; Zheng, Y.L. Genome-wide identification and evaluation of novel internal control genes for Q-PCR based transcript normalization in wheat. *Plant Mol. Biol.* **2010**, *74*, 307–311. [CrossRef]
64. Bray, N.L.; Pimentel, H.; Melsted, P.; Pachter, L. Near-optimal probabilistic RNA-seq quantification. *Nat. Biotechnol.* **2016**, *34*, 525–527. [CrossRef]
65. Robinson, M.D.; McCarthy, D.J.; Smyth, G.K. edgeR: A Bioconductor package for differential expression analysis of digital gene expression data. *Bioinformatics* **2010**, *26*, 139–140. [CrossRef] [PubMed]
66. Young, M.D.; Wakefield, M.J.; Smyth, G.K.; Oshlack, A. Gene ontology analysis for RNA-seq: Accounting for selection bias. *Genome Biol.* **2010**, *11*, R14. [CrossRef]
67. Xie, C.; Mao, X.; Huang, J.; Ding, Y.; Wu, J.; Dong, S.; Kong, L.; Gao, G.; Li, C.Y.; Wei, L. KOBAS 2.0: A web server for annotation and identification of enriched pathways and diseases. *Nucleic Acids Res.* **2011**, *39*, W316–W322. [CrossRef]
68. Gruber, A.R.; Lorenz, R.; Bernhart, S.H.; Neuböck, R.; Hofacker, I.L. The vienna RNA websuite. *Nucleic Acids Res.* **2008**, *36*, W70–W74. [CrossRef] [PubMed]
69. Lorenz, R.; Bernhart, S.H.; Höner zu Siederdissen, C.; Tafer, H.; Flamm, C.; Stadler, P.F.; Hofacker, I.L. ViennaRNA Package 2.0. *Algorithms Mol. Biol.* **2011**, *6*, 26. [CrossRef] [PubMed]
70. Deleage, G.; Combet, C.; Blanchet, C.; Geourjon, C. ANTHEPROT: An integrated protein sequence analysis software with client/server capabilities. *Comput. Biol. Med.* **2001**, *31*, 259–267. [CrossRef]
71. Roy, A.; Kucukural, A.; Zhang, Y. I-TASSER: A unified platform for automated protein structure and function prediction. *Nat. Protoc.* **2010**, *5*, 725–738. [CrossRef]
72. Yang, J.; Yan, R.; Roy, A.; Xu, D.; Poisson, J.; Zhang, Y. The I-TASSER Suite: Protein structure and function prediction. *Nat. Methods* **2015**, *12*, 7–8. [CrossRef]
73. Yang, J.; Zhang, Y. I-TASSER server: New development for protein structure and function predictions. *Nucleic Acids Res.* **2015**, *43*, W174–W181. [CrossRef]
74. Zhao, K.; Xiao, J.; Liu, Y.; Chen, S.; Yuan, C.; Cao, A.; You, M.F.; Yang, D.; An, S.; Wang, H.; et al. *Rht23* (*5Dq′*) likely encodes a *Q* homeologue with pleiotropic effects on plant height and spike compactness. *Theor. Appl. Genet.* **2018**, *131*, 1825–1834. [CrossRef]
75. Zhang, Y.; Skolnick, J. TM-align: A protein structure alignment algorithm based on TM-score. *Nucleic Acids Res.* **2005**, *33*, 2302–2309. [CrossRef] [PubMed]
76. Quan, L.; Lv, Q.; Zhang, Y. STRUM: Structure-based stability change prediction upon single-point mutation. *Bioinformatics* **2016**, *32*, 2936–2946. [CrossRef] [PubMed]
77. Pettersen, E.F.; Goddard, T.D.; Huang, C.C.; Couch, G.S.; Greenblatt, D.M.; Meng, E.C.; Ferrin, T.E. UCSF Chimera—A visualization system for exploratory research and analysis. *J. Comput. Chem.* **2004**, *25*, 1605–1612. [CrossRef] [PubMed]
78. Wang, Y.; Chen, Q.; Li, Y.; Guo, Z.R.; Liu, C.H.; Wan, Y.F.; Hawkesford, M.; Zhu, J.; Wu, W.; Wei, M.Q.; et al. Post-translational cleavage of HMW-GS Dy10 allele improves the cookie-making quality in common wheat (*Triticum aestivum*). *Mol. Breed.* **2021**, *41*, 49. [CrossRef]
79. Dupont, F.M.; Chan, R.; Lopez, R.; Vensel, W.H. Sequential extraction and quantitative recovery of gliadins, glutenins, and other proteins from small samples of wheat flour. *J. Agr. Food Chem.* **2005**, *53*, 1575–1584. [CrossRef]

80. Zheng, T.; Qi, P.F.; Cao, Y.L.; Han, Y.N.; Ma, H.L.; Guo, Z.R.; Wang, Y.; Qiao, Y.Y.; Hua, S.Y.; Yu, H.Y.; et al. Mechanisms of wheat (*Triticum aestivum*) grain storage proteins in response to nitrogen application and its impacts on processing quality. *Sci. Rep.* **2018**, *8*, 11928. [CrossRef]
81. Brandt, K.M.; Gunn, H.; Moretti, N.; Zemetra, R.S. A streamlined protocol for wheat (*Triticum aestivum*) protoplast isolation and transformation with CRISPR-Cas ribonucleoprotein complexes. *Front. Plant Sci.* **2020**, *11*, 769. [CrossRef]
82. Song, Y.; Li, G.; Nowak, J.; Zhang, X.; Xu, D.; Yang, X.; Huang, G.; Liang, W.; Yang, L.; Wang, C.; et al. The rice actin-binding protein RMD regulates light-dependent shoot gravitropism. *Plant Physiol.* **2019**, *181*, 630–644. [CrossRef]
83. Hellens, R.P.; Allan, A.C.; Friel, E.N.; Bolitho, K.; Grafton, K.; Templeton, M.D.; Karunairetnam, S.; Gleave, A.P.; Laing, W.A. Transient expression vectors for functional genomics, quantification of promoter activity and RNA silencing in plants. *Plant Methods* **2005**, *1*, 13. [CrossRef]
84. Tang, Q.Y.; Zhang, C.X. Data Processing System (DPS) software with experimental design, statistical analysis and data mining developed for use in entomological research. *Insect Sci.* **2013**, *20*, 254–260. [CrossRef]
85. Gil-Humanes, J.; Pistón, F.; Martín, A.; Barro, F. Comparative genomic analysis and expression of the *APETALA2*-like genes from barley, wheat, and barley-wheat amphiploids. *BMC Plant Biol.* **2009**, *9*, 66. [CrossRef] [PubMed]

International Journal of
Molecular Sciences

MDPI

Article

HuNAC20 and *HuNAC25*, Two Novel NAC Genes from Pitaya, Confer Cold Tolerance in Transgenic *Arabidopsis*

Xinglong Hu, Fangfang Xie, Wenwei Liang, Yinhao Liang, Zhike Zhang, Jietang Zhao, Guibing Hu and Yonghua Qin *

Guangdong Provincial Key Laboratory of Postharvest Science of Fruits and Vegetables, Key Laboratory of Biology and Genetic Improvement of Horticultural Crops (South China), Ministry of Agriculture and Rural Affairs, College of Horticulture, South China Agricultural University, Guangzhou 510642, China; 15238328269@163.com (X.H.); xiefangfang202012@163.com (F.X.); Venwyliang@163.com (W.L.); liangyh131@163.com (Y.L.); poloky2@163.com (Z.Z.); zhaojietang@gmail.com (J.Z.); guibing@scau.edu.cn (G.H.)

* Correspondence: qinyh@scau.edu.cn

Abstract: NAC transcription factors are one of the largest families of transcriptional regulators in plants, and members of the gene family play vital roles in regulating plant growth and development processes including biotic/abiotic stress responses. However, little information is available about the NAC family in pitaya. In this study, we conducted a genome-wide analysis and a total of 64 NACs (named *HuNAC1-HuNAC64*) were identified in pitaya (*Hylocereus*). These genes were grouped into fifteen subgroups with diversities in gene proportions, exon–intron structures, and conserved motifs. Genome mapping analysis revealed that *HuNAC* genes were unevenly scattered on all eleven chromosomes. Synteny analysis indicated that the segmental duplication events played key roles in the expansion of the pitaya NAC gene family. Expression levels of these *HuNAC* genes were analyzed under cold treatments using qRT-PCR. Four *HuNAC* genes, i.e., *HuNAC7*, *HuNAC20*, *HuNAC25*, and *HuNAC30*, were highly induced by cold stress. HuNAC7, HuNAC20, HuNAC25, and HuNAC30 were localized exclusively in the nucleus. HuNAC20, HuNAC25, and HuNAC30 were transcriptional activators while HuNAC7 was a transcriptional repressor. Overexpression of *HuNAC20* and *HuNAC25* in *Arabidopsis thaliana* significantly enhanced tolerance to cold stress through decreasing ion leakage, malondialdehyde (MDA), and H_2O_2 and O_2^- accumulation, accompanied by upregulating the expression of cold-responsive genes (*AtRD29A*, *AtCOR15A*, *AtCOR47*, and *AtKIN1*). This study presents comprehensive information on the understanding of the NAC gene family and provides candidate genes to breed new pitaya cultivars with tolerance to cold conditions through genetic transformation.

Keywords: pitaya; genome-wide analysis; NAC gene family; cold stress; genetic transformation

Citation: Hu, X.; Xie, F.; Liang, W.; Liang, Y.; Zhang, Z.; Zhao, J.; Hu, G.; Qin, Y. *HuNAC20* and *HuNAC25*, Two Novel NAC Genes from Pitaya, Confer Cold Tolerance in Transgenic *Arabidopsis*. *Int. J. Mol. Sci.* **2022**, *23*, 2189. https://doi.org/10.3390/ijms23042189

Academic Editors: Andrés J. Cortés and Hai Du

Received: 14 January 2022
Accepted: 11 February 2022
Published: 16 February 2022

1. Introduction

Transcription factors (TFs) and *cis*-elements function in the promoter region of different stress-related genes, and thereby alter their expression in response to the stress tolerance [1,2]. Many plant TFs, including MYB, bHLH, AP2, MYC, WRKY, and NAC, have been identified and play diverse functions in various biological processes [3]. Among those TFs, NACs are one of the largest families and play important roles in diverse developmental processes in plants [4,5]. The name of the NAC gene family was derived from the three earliest characterized proteins with a particular domain (NAC domain) from petunia NAM (no apical meristem), Arabidopsis ATAF1/2, and CUC2 (cup-shaped cotyledon) [6,7].

Protein sequences of this family reveal that a typical NAC TF has a highly conserved NAC domain with about 160 amino acid residues at the N-terminal region while the C-terminal region is highly diversified in length and sequence, which is considered the transcriptional activation domain [8]. The NAC domain is further divided into five subdomains (A–E) that represent motifs for both DNA-binding and protein–protein interactions [9].

"A" sub-domain functions in the dimerization of the TF, "B", and "E" have distinctive functions of proteins while "C" and "D" are positively charged and allow the TF to bind to the DNA [10–12].

The NAC family widely exists in various kinds of plants involving diverse biological processes, including flower formation [13], lateral root formation [14], leaf senescence [15], hormone signaling [16], cell division [17], secondary wall synthesis [18], and fruit growth and ripening [19,20]. NAC TFs have attracted a lot of attention as regulators in plant tolerance during biotic and abiotic stresses such as high salinity, drought, temperature, and pathogen defense [21–24]. Therefore, the NAC TF family is of great importance for plants to resist harsh environmental conditions. Among these adverse external stimuli, cold stress has become a major environmental factor limiting plant growth and productivity throughout the world [25]. Previous studies have showed that some plant NAC TFs are involved in plant cold stress. Overexpression of *MbNAC25* in Arabidopsis could improve the tolerance to cold and salinity stresses via enhanced scavenging capability of reactive oxygen species (ROS) [26]. Transgenic Arabidopsis plants overexpressing the *GmNAC20* gene enhanced tolerance to salt and freezing stresses. *GmNAC20* may regulate cold stress tolerance through activation of the DREB/CBF–COR pathway [14]. The plasma membrane-associated transcription factor, *ANAC062*, is an important regulator in the cold tolerance signal pathway [27]. Overexpressing *LINAC2* in *Arabidopsis thaliana* could enhance tolerance to cold stress and activated the expression of many cold-responsive genes [28].

Pitaya, a tropical and subtropical plant belonging to *Hylocereus* or *Selenicereus* in the family Cactaceae, is famous for its high nutritional, economic, and medicinal values [29]. Moreover, pitaya can adapt to a wide ecological range such adverse environmental conditions as drought, heat, and poor soil due to it being a kind of succulent plant [30]. Genome-wide surveys of the NAC family have been identified in various plant species, such as Arabidopsis [8], rice [8], maize [31], pepper [32], pear [33], apple [34], and wheat [35]. However, comprehensive analysis of the NAC family in pitaya has not been reported yet. With the completion of the chromosome-level genome sequencing of pitaya [36], it provides a great opportunity to systematically study the NAC family at the genome-wide level. In this study, the NAC family in the pitaya genome were identified and their phylogeny, genomic structures, conserved motifs, chromosomal locations, synteny, and expression levels were analyzed. Transgenic arabidopsis plants harboring *HuNAC20* and *HuNAC25* were studied under cold treatment. The aim of the present study is to identify and validate the candidate *HuNAC* genes involved in cold–stress response that can be used for transgenic breeding to enhance the cold tolerance in pitaya.

2. Results

2.1. Genome-Wide Identification of NAC Family Genes

To identify pitaya NAC TF encoding genes, all proteins were annotated from the *H. undatus* genome [36]. In total, 64 NAC genes were identified and named *HuNAC1* to *HuNAC64* according to their chromosomal position. Each of these HuNAC proteins contained an NAM domain (PF02365.15), a specific conserved domain of NAC TF protein family (Supplementary File S1). These genes encoded predicted proteins ranging from 154 to 684 AA (amino acids) with isoelectric point (PI) values ranging from 5.33 to 9.19 and molecular weights from 17.95 to 75.87 KDa (Table S1). Subcellular location of these genes was predicted using an online tool from Molecular Bioinformatics Center (http://cello.life.nctu.edu.tw/) (Accessed on 13 January 2022). The subcellular localization and the protein sequences of the 64 HuNACs were listed in Supplementary File S2. Among the 64 NAC proteins, five (*HuNAC16*, *HuNAC17*, *HuNAC21*, *HuNAC52*, and *HuNAC55*) were predicted to be located in the cytoplasmic; three (*HuNAC3*, *HuNAC34*, and *HuNAC48*) were chloroplast; two (*HuNAC4* and *HuNAC*10) were mitochondrial; and the rest were localized in the nucleus.

2.2. Phylogenetic Analyses of NAC Family Members

To explore the evolutionary relationships among *HuNAC* genes, a Maximum Likelihood (ML) phylogenetic tree was constructed according to NAC protein sequences from *H. undatus* and *A. thaliana* (Figure 1). Based on the ANAC classification and NAC domain alignments of *HuNACs*, all members of the HuNACs and ANACs were divided into two major groups: Groups A and B. The tree was divided into fifteen subgroups in Group A (A1-15) according to similarities in NAC domain structures. The number of *HuNAC* genes in each subgroup varied greatly. Subgroups A14 and A13 had only one and two genes, respectively. Subgroup A1 constituted the largest clades with 12 pitaya NAC members, followed by Subgroup A3 with 10 sequences. However, no members were detected in Group B in pitaya, suggesting that they might have been lost in these subfamilies.

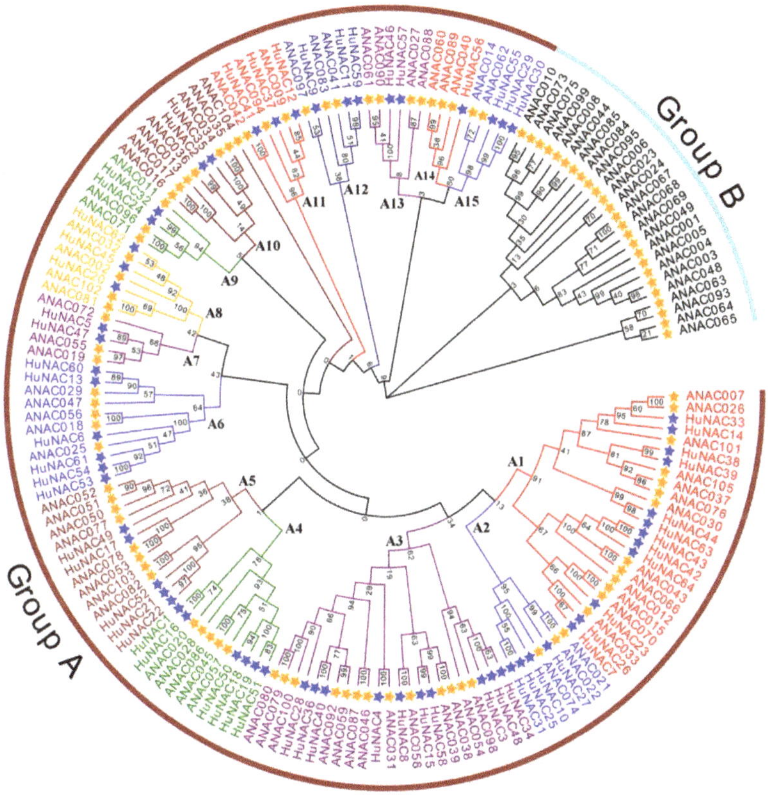

Figure 1. Phylogenetic tree analyses of NAC transcription factor family in pitaya (64 genes) and *A. thaliana* (105 genes). The full-length sequences of the NAC proteins were aligned using ClustalW, and the phylogenetic tree was constructed using the maximum-likelihood method in the MEGAX software. The Bootstrap value was 1000 replicates. HuNACs were indicated by blue stars. All members from both species were designated as Group A (A1-A15) and Group B. Each of the subfamily is indicated in a specific color.

2.3. Gene Structure and Conserved Motif Analyses of HuNACs

To better understand the similarity and diversity of the *HuNAC* genes, the exon/intron distributions of *HuNAC* and conserved motifs were analyzed. As shown in Figure 2, the 64 NAC genes were divided into 15 subcases. Genes within the same subcase had a similar exon/intron structure in terms of intron number and exon length. For example, among the 12 members in subgroup A3, 11 had two introns, while six genes in subgroup A4 had 2–5 introns.

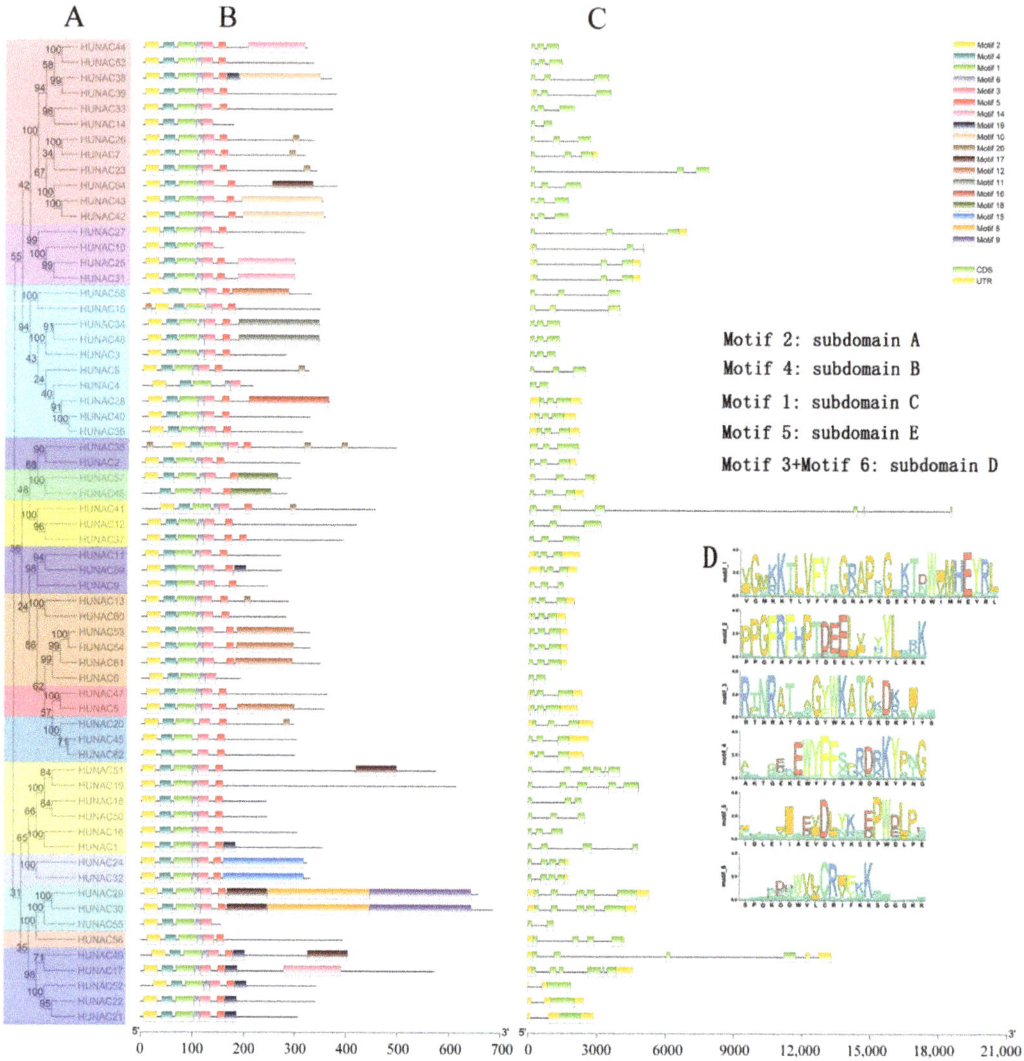

Figure 2. Phylogenetic relationship, gene structure, and conserved motif analyses of 64 HuNAC genes. (**A**) Phylogenetic tree. The phylogenetic tree was constructed using the neighbor-joining method through MEGAX software. The bootstrap analysis was conducted with 1000 replicates. (**B**) protein motif. Schematic diagrams of possible conserved motifs in HuNAC proteins. MEME tool was used to find out the conserved motifs; (**C**) gene structure. The green bars indicate the exons, and the black lines indicate the introns. Yellow bars indicate the UTR region; (**D**) the sequences of key motifs (motif 1, motif 2, motif 3, motif 4, motif 5 and motif 6).

The conserved motifs of NAC family proteins in pitaya were investigated using MEME online software. Based on this program, 20 distinct motifs were identified (Figure S1). According to frequencies of occurrence, motifs 1, 4, and 3 were the three most frequently presented motifs which were observed in all HuNAC proteins, and most of the conserved motifs were located at the N-terminus. Among the 20 motifs, motif 2, motif 4, motif 3, and motif 5 had the subdomains A, B, D, and E, respectively, and motifs 1 and 6 contained subdomain C (Figure 2D). These results are consistent with the finding that a connection existed between subfamilies and motifs [37].

2.4. Chromosomal Localization and Synteny Analyses of HuNACs

Genome chromosomal location analyses revealed that the 64 *HuNAC* genes were unevenly scattered on all 11 chromosomes (Figure 3). Chromosomes 1, 3, 4, and 8 contained 12, 8, 8, and 8 *HuNAC* genes, respectively. Chromosomes 6, 10, and 11 had only three *HuNAC* genes. The average numbers of *HuNAC* genes on the chromosome were approximately 6.0.

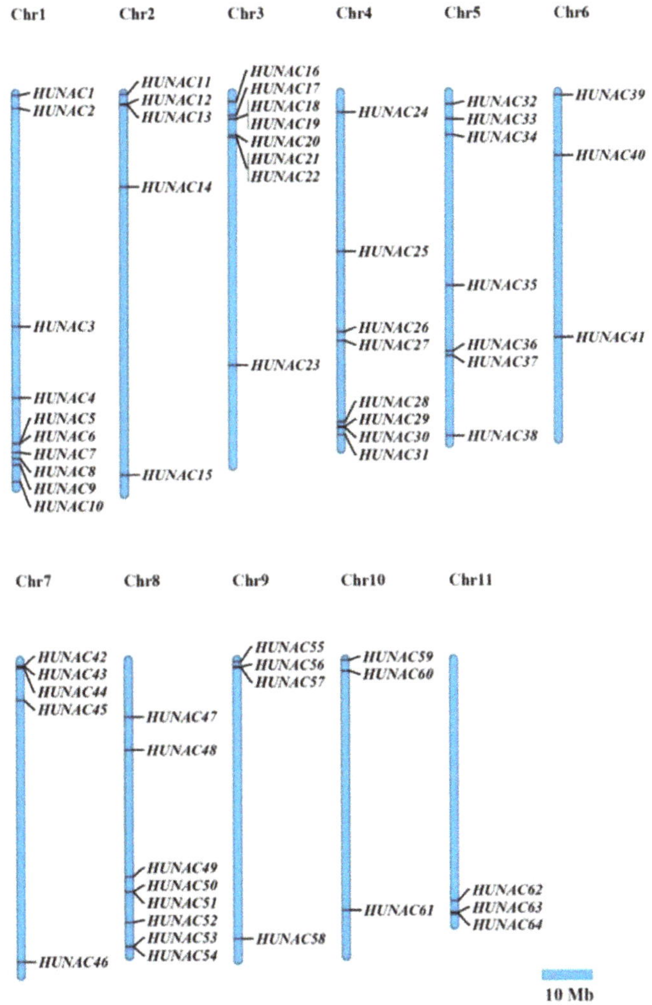

Figure 3. Schematic diagrams of the chromosomal location of the *HuNAC* genes. Eleven chromosomes with varying lengths are shown on the megabases (Mb) scale on the left, and the chromosome number is shown on top of each chromosome.

The phenomenon of gene duplication has been recognized to occur throughout plant evolution, and plays an important role in the expansion of the large gene families in plant [38]. Gene duplication events were investigated to clarify the expansion patterns of the NAC family in pitaya. Twenty segmental duplicated events with 33 *HuNAC* genes were identified in the pitaya genome (Figure 4; Supplementary File S3). The high-identity segmental duplication events in pitaya suggested that this duplication type likely plays a crucial role in the expansion of the pitaya NAC gene family.

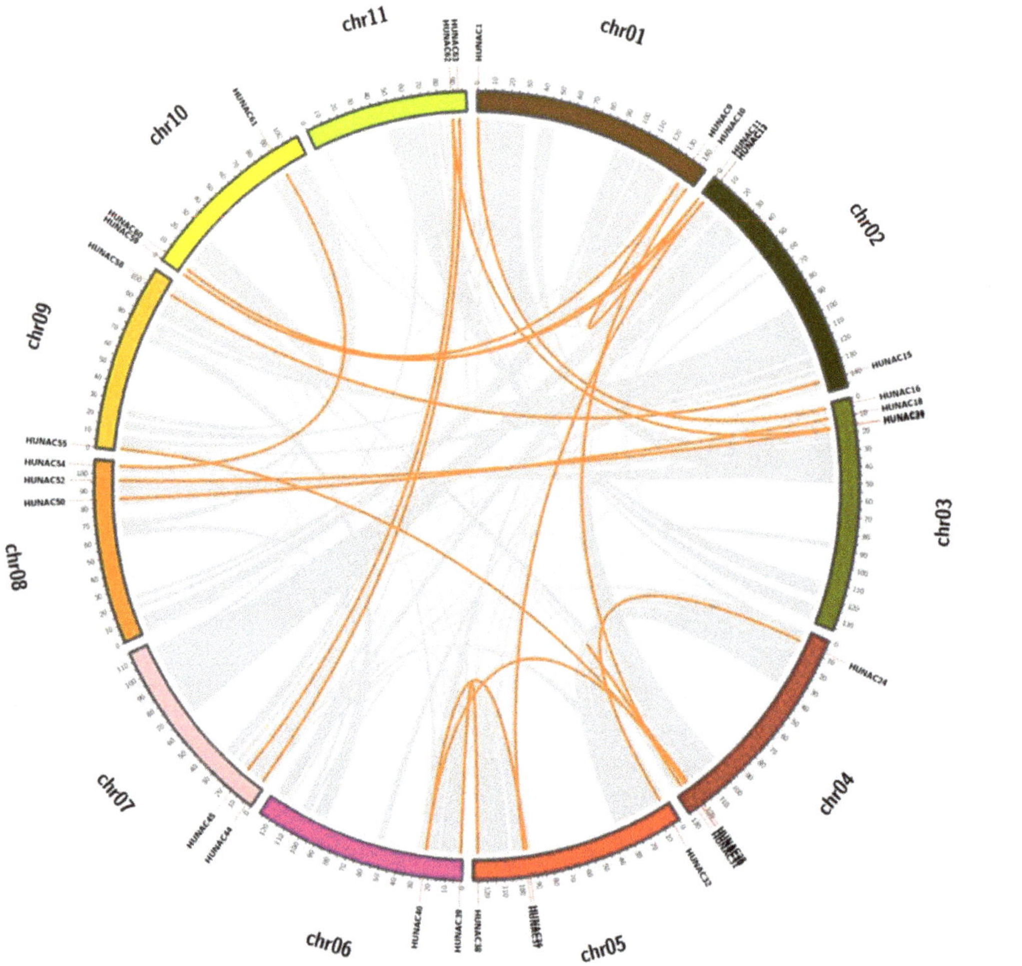

Figure 4. Schematic representations of interchromosomal relationships of the *HuNAC* genes. Gray lines represent all synteny blocks in the pitaya genome, and the orange lines indicate duplicated NAC gene pairs. The chromosome number is indicated at the top of each chromosome.

2.5. Expression Analyses of HuNAC Genes under Cold Treatment

In plants, many NAC genes are involved in the response to abiotic stresses. Therefore, the expression patterns and putative functions of all 64 *HuNAC* genes were analyzed under cold conditions, and 49 *HuNAC* genes were induced under cold treatment (Figure S2). As shown in Figure 5, compared with normal temperature, higher expression levels of *HuNAC7*, *HuNAC20*, *HuNAC25*, and *HuNAC30* were detected at a low temperature. The expression of *HuNAC7*, *HuNAC20*, *HuNAC25*, and *HuNAC30* showed a trend of gradual increase. *HuNAC7*, *HuNAC25*, and *HuNAC30* were strongly induced at 24 h after cold treatment, while expression levels of *HuNAC20* significantly increased at 36 h after cold treatment. Expressions of *HuNAC7*, *HuNAC20*, *HuNAC25*, and *HuNAC30* reached their maximum levels at 60 h after cold treatment.

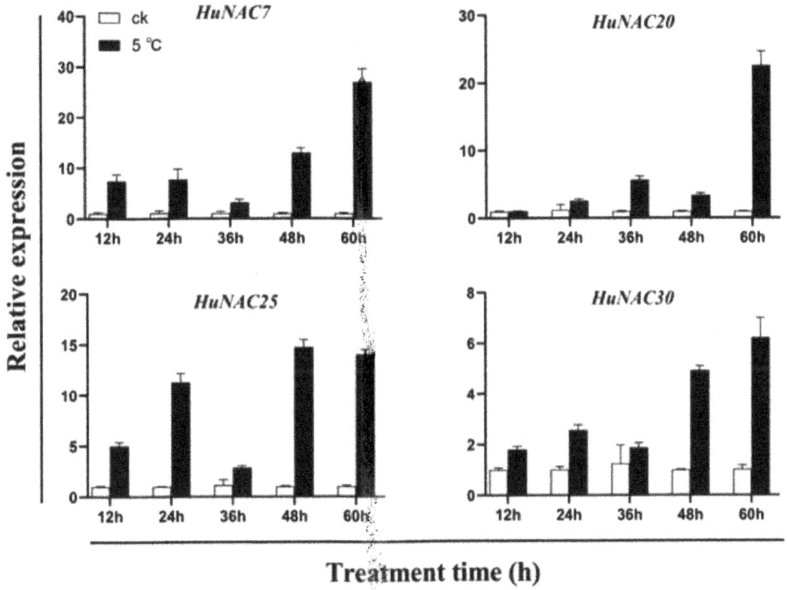

Figure 5. Expression analyses of *HuNAC7*, *HuNAC20*, *HuNAC25*, and *HuNAC30* under cold stress. Three biological replicates were used and bars represent the relative expression of different genes under cold stress.

2.6. Subcellular Localization and Transcriptional Activation Analyses of HuNACs

To analyze the subcellular localization of *HuNAC7*, *HuNAC20*, *HuNAC25*, and *HuNAC30*, their full-length coding sequences were fused with the GFP to construct 35S-HuNACs-GFP vectors, and transiently expressed in leaves of *N. benthamiana*. As shown in Figure 6, the fluorescence of HuNACs-GFP were predominately observed in the nucleus of epidermal cells, while the GFP signal of positive control was detected around the cytoplasm and the nucleus. Those results suggested that the *HuNAC7*, *HuNAC20*, *HuNAC25*, and *HuNAC30*, like the other reported NAC proteins [39], encoded nuclear proteins. Full-length coding regions of *HuNAC7*, *HuNAC20*, *HuNAC25*, and *HuNAC30* were fused to the GAL4BD to generate pGBKT7-HuNACs fusion plasmids to study the transcriptional activation abilities of HuNACs. Like the positive control (pGBKT7-53+pGADT7-T), yeast cells expressing the HuNAC20, HuNAC25, and HuNAC30 grew well on SD/-Trp-His-Ade and showed α-galactosidase activity, indicating that HuNAC20, HuNAC25, and HuNAC30 had trans-activation ability in yeast cells; however, yeast cells expressing the HuNAC7 could not grow on SD/-Trp-His-Ade (pGBKT7) (Figure 7A). Trans-activation of HuNACs were further verified in leaves of *N. benthamiana* using the dual-luciferase reporter system. Compared with the negative control, the expression of the positive control (BD-62SK-VP16), HuNAC20, HuNAC25, and HuNAC30 resulted in a higher value of the LUC/REN ratio. However, HuNAC7 significantly repressed the expression of the LUC reporter in comparison to the negative control (pBD) and the ratio of LUC/REN of HuNAC7 was 0.3-fold compared with the pBD (Figure 7C). These results demonstrated that HuNAC20, HuNAC25, and HuNAC30 are transcriptional activators while HuNAC7 is a transcriptional repressor.

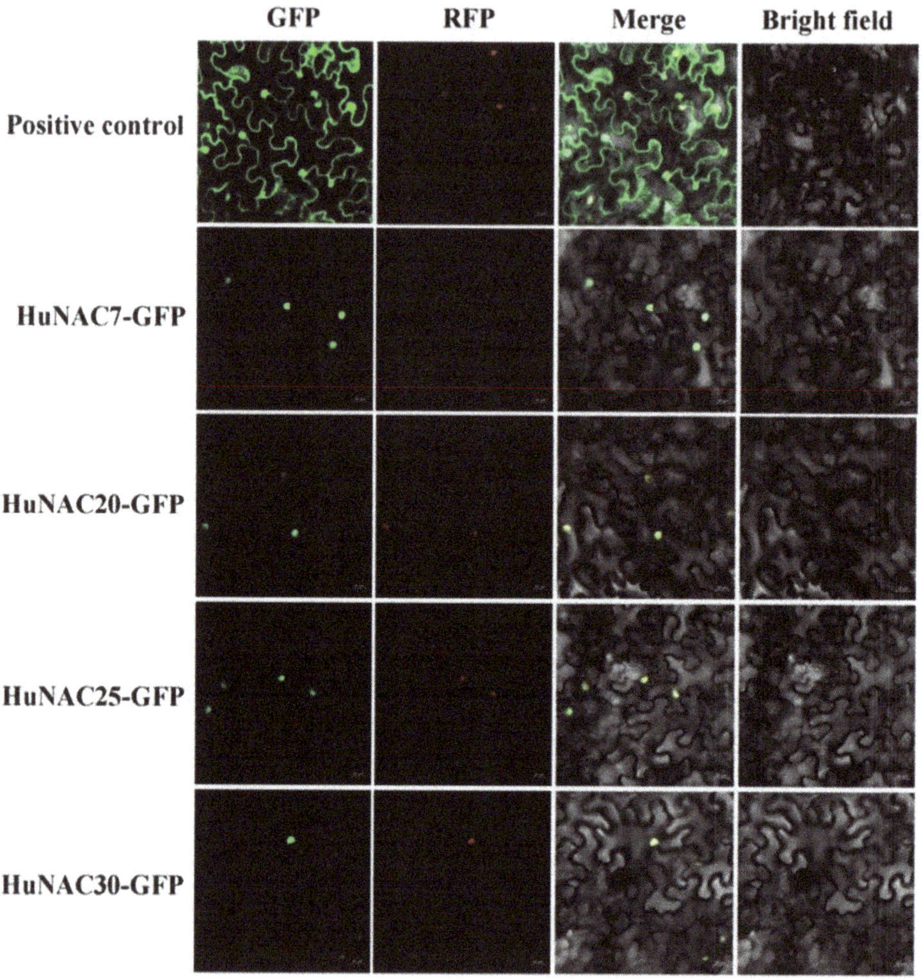

Figure 6. Subcellular localization of HuNACs in *Nicotiana benthamiana* leaves. The fusion protein and GFP-positive control were transiently expressed in *N. benthamiana* leaves by *Agrobacterium tumefaciens* strain GV3101 (pSoup-p19), respectively. Bars = 20 μm.

2.7. Phylogenetic and Sequence Analyses of HuNAC20 and HuNAC25

HuNAC20 and *HuNAC25* were cloned based on the sequences of the pitaya genome [36]. The full-length coding DNA sequences (CDSs) of *HuNAC20* and *HuNAC25* were both 891 bp, encoding 296 amino acids with molecular weights of 33.74 and 33.65 kDa and pIs of 7.69 and 6.26, respectively. Homologous analyses of HuNAC20, HuNAC25, and 18 NAC protein sequences from different plant species were aligned with DNAMAN software. The similarity between HuNAC20 and HuNAC25 was 17.7% (Figure S3a). Both HuNAC20 and HuNAC25 had a conserved subdomains (A–E) in the N-terminal region and a diverse activation domain in C-terminal, respectively. HuNAC20 and HuNAC25 both contained a conserved nuclear localization signal (NLS) (Figure S3b,c). HuNAC20 shared 78.2% similarity with CqNAC2 (XP_021772603.1) from *Chenopodium quinoa* and 77.6% similarity with BvATAF2 (QGZ00533.1) from *Beta vulgaris*. HuNAC25 had 67.2% similarity with BvNAC (XP_010679326.1) from *Beta vulgaris* and 64.0% similarity with MeNAC (XP_021630142.1) from *Manihot esculenta*.

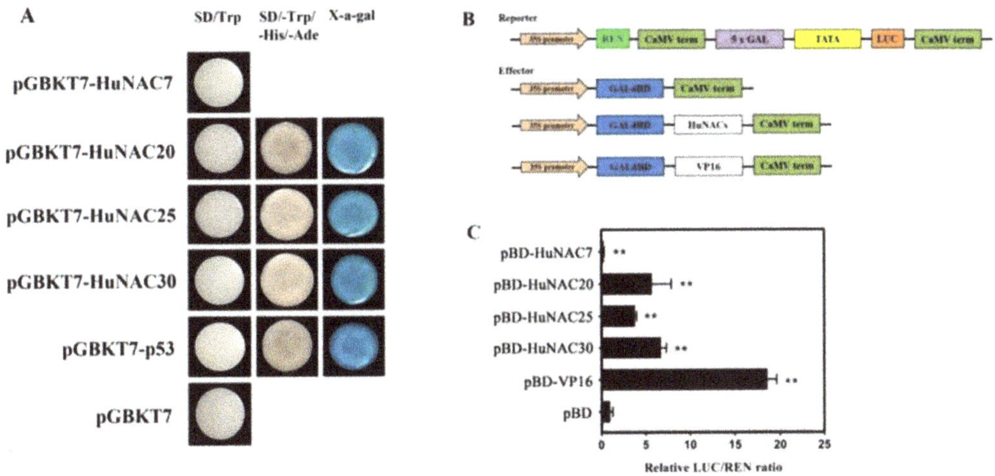

Figure 7. Transcriptional activation analyses of HuNACs. (**A**) transcriptional activation of HuNAC in yeast cells. pGBKT7 and pGBKT7-53 + pGADT7-T were used as negative and positive controls, respectively. Transcription activation was monitored according to growth status of yeast cells and an α-Gal assay; (**B**) diagrams of the reporter and effector vectors; (**C**) transcriptional activation of HuNACs in *N. benthamiana* leaves. The trans-activation ability of HuNACs is indicated by the ratio of LUC to REN. The LUC/REN ratio of the empty pBD vector (negative control) was used as a calibrator (set as 1). pBD-VP16 was used as a positive control. The asterisk indicates a significant difference at the 1% level compared to the pBD.

2.8. Overexpression of HuNAC20 and HuNAC25 in Arabidopsis Enhanced Tolerance to Freezing Stress

To further explore the function of *HuNAC20* and *HuNAC25*, we generated transgenic *A. thaliana* plants constitutively expressing *HuNAC20* and *HuNAC25* driven by the 35S promoter. Homozygous T_3 lines were obtained on the basis of 3:1 segregation for kanamycin resistance phenotype. Expression levels of the target genes in the homozygous transgenic lines were analyzed by RT-qPCR. Two independent T_3 transgenic lines with a relatively high expression of *HuNAC20* (NAC20-L1 and NAC20-L3) and *HuNAC25* (NAC25-L2 and NAC25-L3) were selected for cold tolerance experiments.

At normal conditions, no significant difference in morphology was observed between transgenic lines and WT plants. After cold acclimation (48 h at 4 °C) and recovery for 6 d at 22 °C, transgenic plants displayed better performance than wild-type under freezing treatment (−6 °C for 6 h) (Figures 8A and 9A). Most WT plants died with a survival rate at around 8.0%. The survival rates of the transgenic lines (90.0% for NAC20-L1, 88.0% for NAC20-L3, 82.0% for NAC25-L2 and 78.0% for NAC25-L3) were significantly higher than that of WT plants (8.0%) (Figures 8C and 9C). These results indicated that overexpression of *HuNAC20* and *HuNAC25* in *A. thaliana* enhances tolerance to cold stress.

2.9. Overexpression of HuNAC20 and HuNAC25 Affected Ion Leakage, MDA Contents, H_2O_2, and O_2^- Accumulation under Cold Stress

Ion leakage, MDA content, H_2O_2, and O_2^- accumulation are commonly used to assess stress resistance capacity during abiotic stresses. Under normal conditions (22 °C), no significant differences in ion leakage, MDA, H_2O_2, and O_2^- contents were detected between the transgenic lines and WT plants. Ion leakage, MDA, H_2O_2, and O_2^- contents gradually increased in both transgenic lines and WT plants after cold treatment. The levels of ion leakage in the transgenic lines were significantly lower than those in WT plants during cold (4 °C for 48 h) and freezing treatments (−6 °C for 6 h). After cold treatment for 24 h and 48 h, MDA, H_2O_2, and O_2^- contents in the transgenic lines were significantly lower than those in WT plants (Figure 10). These results suggested that overexpression of *HuNAC20* and *HuNAC25* enhances *A. thaliana* tolerance to cold stress by altering the ion leakage, MDA, H_2O_2, and O_2^- accumulation.

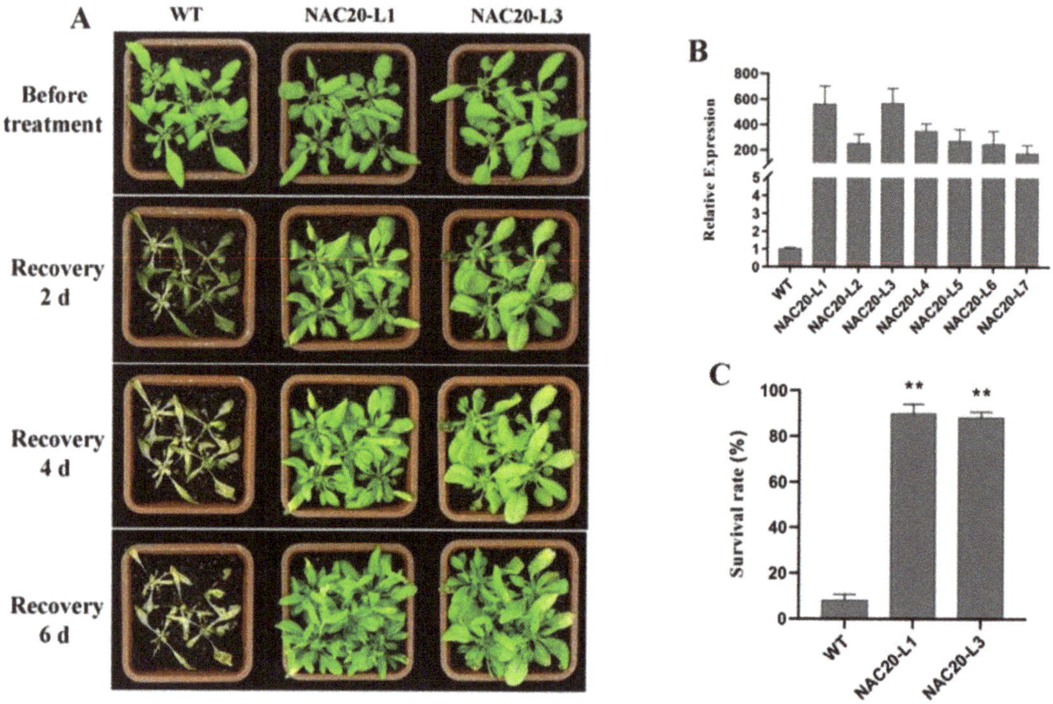

Figure 8. Overexpression of *HuNAC20* in *Arabidopsis* enhanced freezing tolerance. (**A**) Performance and (**C**) survival rates of WT and *HuNAC20* transgenic plants after freezing; (**B**) expression levels of *HuNAC20* in WT and transgenic plants. *AtACTIN2* was used as an internal control. Twenty seedlings per transgenic line were used in each freezing treatment. Data represent average values from three biological replicates (±S.D.). Asterisks indicate significant differences (** $p < 0.01$) between the transgenic lines and WT plants.

2.10. Overexpression of HuNAC20 and HuNAC25 Activated the Expression of Cold-Responsive Genes under Cold Stress

The transcript levels of cold-responsive genes including *AtRD29A*, *AtCOR15A*, *AtCOR47*, and *AtKIN1* were analyzed in *HuNAC20* and *HuNAC25* transgenic lines and WT plants under cold treatment. The expressions of these genes in transgenic lines and WT plants were relatively low under normal conditions (22 °C). Compared with WT plants, higher transcripts of *AtRD29A*, *AtCOR15A*, *AtCOR47*, and *AtKIN1* were observed in *HuNAC20* and *HuNAC25* transgenic plants after 24 h under cold treatment (Figure 11). These results indicated that *HuNAC20* and *HuNAC25* could activate expression levels of *AtRD29A*, *AtCOR15A*, *AtCOR47*, and *AtKIN1* responsible for stronger tolerance to cold stress in transgenic *A. thaliana* lines.

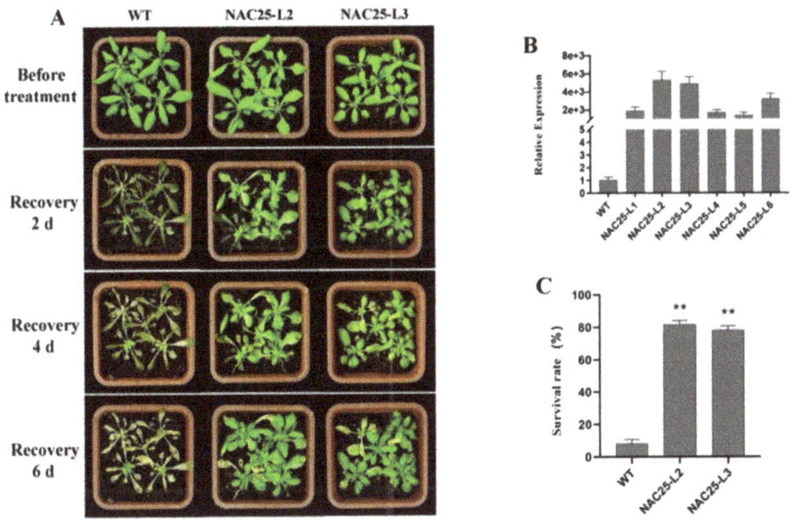

Figure 9. Overexpression of *HuNAC25* in *Arabidopsis* enhanced freezing tolerance. (**A**) performance and (**C**) survival rates of WT and *HuNAC25* transgenic plants after freezing; (**B**) expression levels of *HuNAC25* in WT and transgenic plants. *AtACTIN2* was used as an internal control. Twenty seedlings per transgenic line were used in each freezing treatment. Data represent average values from three biological replicates (±S.D.). Asterisks indicate significant differences (** $p < 0.01$) between the transgenic lines and WT plants.

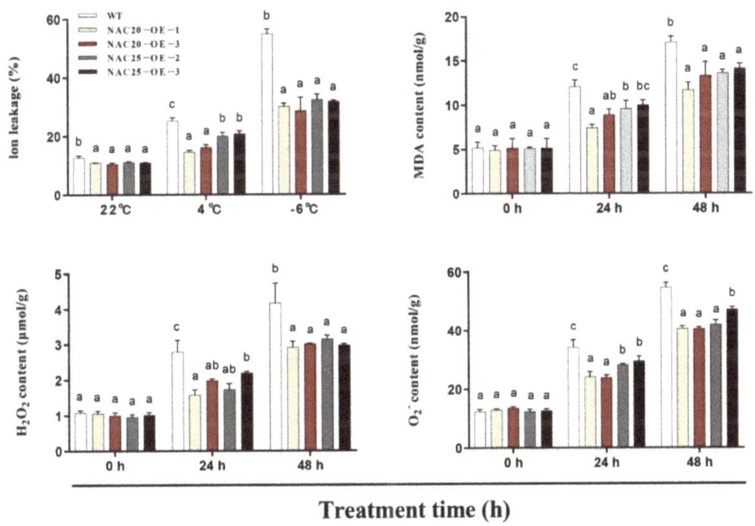

Figure 10. Changes of ion leakage, MDA content, H_2O_2, and O_2^- of WT and transgenic lines under cold stress. Ion leakage were determined in WT and transgenic lines under cold stress (4 °C for 48 h) and freezing (−6 °C for 6 h) treatments. MDA, H_2O_2, and O_2^- accumulation were determined in WT and transgenic lines under cold stress (4 °C for 48 h). Data represent average values from three biological replicates (±S.D.). The different letters above bars indicate significant differences at the $p < 0.05$ level according to Duncan's multiple comparison tests.

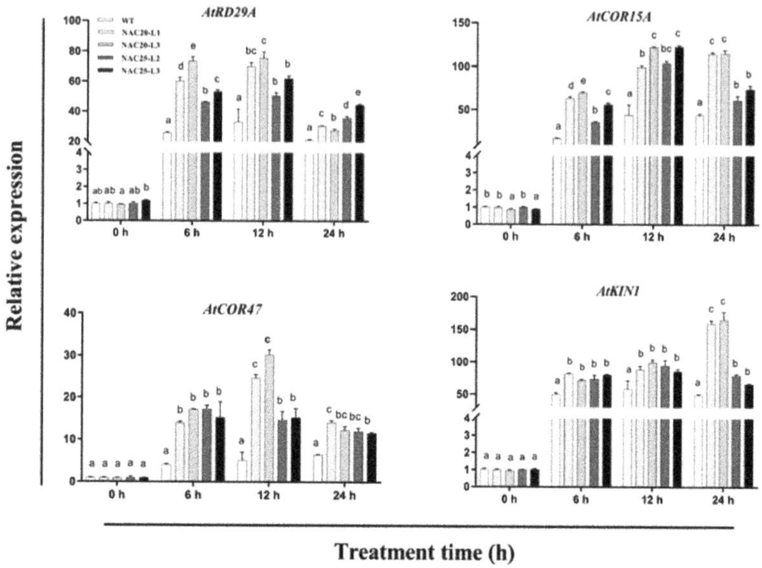

Figure 11. Expression analyses of *AtRD29A*, *AtCOR15A*, *AtCOR47*, and *AtKIN1* in WT and transgenic lines under cold stress. Data represented average values from three biological replicates (±S.D.). The different letters above bars indicated significant differences at the $p < 0.05$ level according to Duncan's multiple comparison tests.

3. Discussion

NAC is one of the largest families of TFs unique to plants and plays important roles in defense against harsh environmental conditions [40,41]. NAC TFs also play crucial roles in plant physiological processes [42]. With the development of whole genome sequencing technology, great progress has been made in identification of NAC genes from Arabidopsis [8], rice [8], cucumber [24], pepper [32], and apple [34]. However, little information is available about the NAC family in pitaya. In the present study, a total of 64 NAC genes distributed in 11 chromosomes were identified from pitaya (Figure 3). Great differences in the size, sequence, physical, and chemical properties of the proteins encoded by *HuNACs* were detected, which was consistent with NACs from the other plant species [22,24,33,38]. The phylogenetic analyses showed that these *HuNACs* were classified into 15 subgroups with *ANACs* in the Group A, and the subgroup A1 had the most *HuNACs* (19%), followed by the A3 (16%), whereas the A14 subgroup had the fewest genes (2%) (Figure 1). However, Group B had no member of *HuNACs* similar to the results from quinoa [43]. Most *HuNACs* had three exons and two introns (Figure 2), indicating that the genetic structural diversity of NAC in pitaya, which was similar to that of the other species such as kiwifruit [44], *Musa Acuminata* [45], cucumber [24], and white pear [22]. High segmental duplication events resulted in the expansion of the NAC gene family in *H. undatus* (Figure 4), which was consistent with the evolutionary pattern in pineapple [38]. Therefore, we speculated that these pitaya-specific *NAC* genes may have special roles in pitaya.

The growth and development of pitaya are frequently affected by abiotic stresses such as drought, high salinity, or extreme temperatures [46,47]. Cold stress is one of the most important influencing factors restricting pitaya production in South China. Previous studies have suggested that NAC TFs can be induced by cold stress. Overexpression of *SlNAC35* enhanced chilling tolerance of transgenic tomato [48]. Compared with non-cold control fruits, the expression level of *MaNAC1* in the fruits directly stored at 7 °C was significantly increased [49]. *VvNAC17*, a novel NAC TF, was expressed in various tissues under cold treatment and enhanced freezing tolerance in transgenic *Arabidopsis* [50]. In this study, expressions of *HuNAC7*, *HuNAC20*, *HuNAC25*, and *HuNAC30* were strongly induced under cold-stress (Figure 5), indicating that they may be involved in plant responses to cold stress. Subcellular localization is important to elucidate protein function. In our study, HuNAC7, HuNAC20, HuNAC25, and HuNAC30 were nuclear proteins (Figure 6), which is consistent with the other reported NAC proteins [16,39,45].

NAC proteins are involved in plant responses to cold stress. Overexpressing *LlNAC2* in *A. thaliana* enhanced tolerance to cold stress. Compared with WT plants, the transgenic plants had lower electrolyte leakage and higher levels of soluble sugars [28]. Overexpression of *GmNAC20* confers tolerance of freezing stress in transgenic plants by regulating downstream stress-responsive genes such as *cor6.6*, *cor78*, and *cor15A* [14]. Overexpressing *MbNAC25* in *Arabidopsis* increased cold tolerance via enhanced scavenging capability of reactive oxygen species (ROS) under low-temperature stress (4 °C) [51]. In the present study, two novel NAC TFs, i.e., *HuNAC20* and *HuNAC25*, were cloned from pitaya. *HuNAC20* and *HuNAC25* were nuclear-localized NAC-type DNA-binding proteins. Overexpression of *HuNAC20* and *HuNAC25* in *Arabidopsis* enhanced cold tolerance. Ion leakage, MDA contents, and H_2O_2 and O_2^- accumulation are commonly used to assess the severity of membrane lipid damage under stress conditions [52]. In our study, overexpression of *HuNAC20* and *HuNAC25* in *Arabidopsis* lines resulted in lower contents of ion leakage, MDA, and H_2O_2 and O_2^- under cold stress. Compared with WT, higher expression levels of the cold-responsive genes (*AtRD29A*, *AtCOR15A*, *AtCOR47*, and *AtKIN1*) were detected in the transgenic plants under cold stress (Figure 11). These results indicated that overexpression of *HuNAC20* and *HuNAC25* in *A. thaliana* could activate expressions of *AtRD29A*, *AtCOR15A*, *AtCOR47*, and *AtKIN1* under cold stress (Figure 12). The results are consistent with those NAC TFs from the other plants species such as *Lilium lancifolium* [28], *Solanum lycopersicum* [48,53], and *Malus baccata* [51]. Putative binding *cis*-elements of NAC proteins have been identified to localize in the promoter sequences of these genes [54]. However, the mechanism of interaction between *HuNAC20* and *HuNAC25* and the cold-responsive genes is still not clear yet, and further study is necessary to elucidate it.

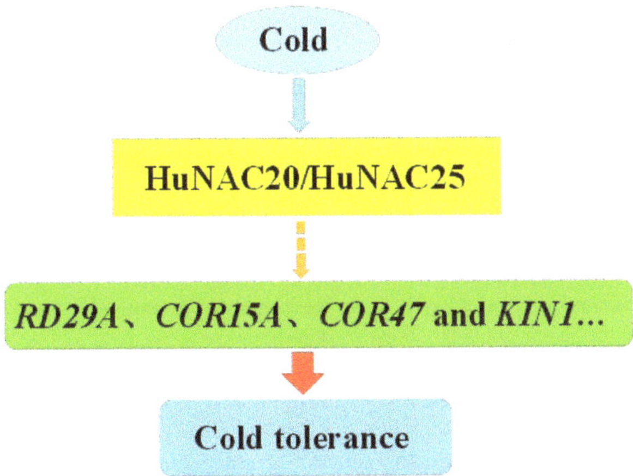

Figure 12. The proposed model of HuNAC20 and HuNAC25-mediated cold stress-responsive signaling. Cold stress induces the expression of *HuNAC20* and *HuNAC25*. *HuNAC20* and *HuNAC25* regulate the expressions of cold-responsive genes by binding to the *cis*-elements of cold-responsive proteins in their promoters, and then modulate plant tolerance to cold stress.

4. Conclusions

In summary, this study provides the first report on identification and characterization of the *NAC* gene family based on the genome-wide analyses of the *H. undatus* genome. A total of 64 *HuNAC* were identified and divided into fifteen subfamilies. The 64 *HuNAC* were unevenly scattered on all 11 chromosomes. *HuNAC7*, *HuNAC20*, *HuNAC25*, and *HuNAC30* may be involved in cold stress tolerance according to their expression patterns. *HuNAC7*, *HuNAC20*, *HuNAC25*, and *HuNAC30* were nucleus proteins. *HuNAC20*, *HuNAC25*, and *HuNAC30* had transcriptional activity while *HuNAC7* acted as a transcriptional repressor. Overexpressing *HuNAC20* and *HuNAC25* in *Arabidopsis* enhanced tolerance of cold stress by altering the expression of cold-responsive genes in transgenic plants. The results of the present study provide valuable information for a better understanding of NAC TFs involved in cold–stress response in pitaya.

5. Materials and Methods

5.1. Identification of the Pitaya NAC Gene

The Hidden Markov Model (HMM) profiles of the NAM domain PF02365.15 were downloaded from the Pfam database (https://pfam.xfam.org/) (Accessed on 13 January 2022). HMM was used to search NAM (PF02365.15) domains from the pitaya genome with values (e-value) cut-off at 1.0 [36]. The integrity of the NAM domain was determined using the online program SMART with an e-value < 0.1. The length, molecular weight, and isoelectric point of each NAC protein were predicted by the online ExPasy program (https://web.expasy.org/protparam/) (accessed on 13 January 2022).

5.2. Phylogenetic Analyses of the Pitaya NAC Gene

To investigate the phylogenetic relationship of the NAC gene families in Arabidopsis and *H. undatus*, the Arabidopsis NAC protein sequences were downloaded from the Arabidopsis Information Resource (https://www.arabidopsis.org/) (accessed on 13 January 2022). All NAC TFs were aligned using the MUSCLE set at default parameters [55]. The Maximum Likelihood (ML) phylogenetic tree was constructed using the MEGA7 program [56]. Bootstrapping was performed with 1000 replications. The tree was exhibited by the EVOLVIEW online tool (https://www.evolgenius.info/evolview/) (accessed on 13 January 2022).

5.3. Analyses of Gene Structure and Conserved Motifs in the HuNAC Family

The gene structure of each *HuNAC* gene was drawn using the TBtools software [57]. The MEME (http://meme-suite.org/tools/meme) (accessed on 13 January 2022) was used to identify the unknown conserved motifs with the following parameters: site distribution: zero or one occurrence (of a contributing motif site) per sequence, maximum number of motifs: 25, and optimum motif width $\geq$6 and $\leq$200.

5.4. Chromosomal Locations and Synteny Analyses of HuNAC Genes

The chromosome location of each *HuNAC* gene was obtained from the pitaya genome [36]. These data were then integrated and plotted using Mapchart software [58]. The Multiple Collinearity Scan toolkit (MCscanX) was applied to analyze the duplication pattern for each HuNAC followed the operation manual [59].

5.5. Plant Material

Stems (6–8 cm in height) in vitro from pitaya cultivar 'Hongguan No. 1' (*Hylocereus monacanthus*) on rooting medium (MS containing 0.5 mM IBA) for 15 d were used as materials [60]. The pitaya plants were cultured in a climate chamber at the South China Agricultural University (Guangzhou, China) under a temperature range of 23–25 °C and a 16 h light (50 μmol m^{-2} s^{-1}) and 8 h dark. Plantlets were placed in a climate cabinet at 5 °C to analyze the expression patterns of the NAC family. The stems from treatments and control were collected respectively at 12, 24, 36, 48, and 60 h after treatment, immediately frozen in liquid nitrogen and stored at −80 °C until future analysis.

Arabidopsis thaliana Columbia-0 (Col-0) was used for genetic transformation of *HuNAC20* and *HuNAC25*. Transgenic lines and WT plants were grown in 9 cm × 9 cm plastic pots containing a 1:1 mixture of sterile peat soil and vermiculite with normal management in a growth chamber [24 ± 1 °C, 16 h light (50 μmol m^{-2} s^{-1}) and 8 h dark and 65% relative humidity]. Seeds of *Nicotiana benthamiana* were planted and cultured under the same conditions.

5.6. Analyses of Ion Leakage, MDA Content, H_2O_2 and O_2^-

Three-week-old seedlings of transgenic *Arabidopsis* lines and WT were treated at −6 °C for 6 h after cold acclimation at 4 °C for 48 h, followed by 4 °C in the dark for 12 h. The plants were transferred to normal conditions (22 °C) for recovery for 6 d, and then the survival rates were counted. Photos were taken before freezing and after recovery. The rosette leaves were collected for analyses of ion leakage [61], malondialdehyde (MDA) content, and H_2O_2 and O_2^- accumulation [62,63]. For chilling treatment, the seedlings were put in a 4 °C incubator, and leaves were collected respectively at 0, 6, 12, and 24 h for expression analyses of cold-responsive genes (*AtRD29A*, *AtCOR15A*, *AtCOR47*, and *AtKIN1*) [64], and *AtACTIN2* (AT1G13320) were used as internal control [61]. Specific primers are listed in Supplementary File S4.

5.7. Gene Cloning and Expression Analyses

Total RNA was isolated using the EASYspin Plus Complex Plant RNA Kit (RN53) (Aidlab Biotechnology, Beijing) according to the manufacturer's protocol. Single-stranded cDNA was synthesized using the PrimeScript™ RT Reagent Kit with gDNA Eraser (TaKaRa, Shiga, Japan). The RT-qPCR primers were designed by BatchPrimer3 (https://probes.pw.usda.gov/cgi-bin/batchprimer3/batchprimer3.cgi) (accessed on 13 January 2022), and the *Actin(1)* reference gene was used as the internal control [65]. RT-qPCR was performed in an CFX384-Real-Time system (C1000 Touch Thermal Cycler, Bio-Rad, CA, USA) using the RealUniversal Color PreMix (SYBR Green) (TIANGEN, Beijing, China). Specific primers are in Supplementary File S4. Each experiment was repeated in triplicate using independent RNA samples. Relative gene expression levels were calculated using the $2^{-\Delta\Delta C}T$ method [66].

The full-length coding sequences of HuNACs were cloned using I-5TM2× High-Fidelity Master Mix (MCLAB, San Francisco, CA, USA) with specific primers (Supplementary File S5).

5.8. Subcellular Localization Analyses

The full lengths of HuNAC7/20/25/30 without a stop codon were subcloned into the pGreen-35S-GFP vector to fuse with the gene sequence of green fluorescent protein (GFP) (primers are listed in Supplementary File S5). Then, the pGreen- HuNAC7/20/25/30-35S-GFP and the control pGreen-35S-GFP vector were transferred into the *Agrobacterium tumefaciens* strain GV3101 (pSoup-p19), and injected into the abaxial side of 4- to 6-week-old *N. benthamiana* leaves. After 48 h of infiltration, infected leaf tissues were collected for analyses. The GFP signal was captured under a fluorescence microscope (ZEISS LCM-800, Oberkochen, Germany). All assays were repeated three times.

5.9. Transcriptional Activation Analyses in Yeast Cells

The pGBKT7-HuNACs, pGBKT7-p53 and pGBKT7 empty plasmids were transferred into the Y2HGold yeast strain independently using the lithium acetate method (PT1172-1, Clontech) (primers are listed in Supplementary File S5). The transformed yeast cells were cultured on SD/-Trp and SD/-Trp-His-Ade medium. The growth status of yeast cells and the activity of X-α-galactosidase (X-α-Gal) were observed after incubation with 20 mg/mL X-α-Gal for 10–30 min.

5.10. Dual-Luciferase Reporter Assays in N. benthamiana Leaves

For transcriptional activity analyses of HuNACs in *N. benthamiana* leaves, coding sequences of *HuNAC7/20/25/30* were cloned into the 35S promoter-driven pBD vectors to fuse with the yeast GAL4 DNA-binding domain (GAL4BD) as an effector (pBD-HuNACs). The double-reporter vector contained a firefly luciferase (LUC) driven by five copies of the GAL4-binding elements (5×GAL4) and minimal TATA region of CaMV 35S. Renilla luciferase (REN) in the same vector driven by the CaMV 35S promoter was used for normalization. The primers are listed in Supplementary File S5.

5.11. *Arabidopsis thaliana Transformation and Phenotypic Analyses*

The open read frame (ORF) of *HuNAC20* and *HuNAC25* were subcloned into the pPZP6K90 vector under the control of the 35S promoter and introduced into *Agrobacterium tumefaciens* strain GV3101. The recombinant vectors were transformed into Arabidopsis Col-0 using the floral dip method. Positive plants were screened on MS medium containing 100 mg L^{-1} kanamycin and identified by PCR detection. The expression levels of *HuNAC20* and *HuNAC25* in the homozygous T_3 transgenic lines were analyzed by RT-qPCR, and the primers used were listed in Supplementary File S5. Two homozygous T_3 transgenic lines were used for cold tolerance experiments and photographed with a digital camera (G16, Canon, City, Japan).

Supplementary Materials: The following supporting information can be downloaded at: https://www.mdpi.com/article/10.3390/ijms23042189/s1.

Author Contributions: Conceived and designed the experiments, X.H. and Y.Q.; Performed the experiments, X.H., F.X., W.L. and Y.L.; Analyzed the data: X.H., F.X. and Y.Q.; Contributed reagents/materials/analysis tools: Z.Z., J.Z. and G.H.; Wrote and revised the paper: X.H. and Y.Q. All authors have read and agreed to the published version of the manuscript.

Funding: This work was supported by the Key Realm R&D Program of Guangdong Province (2018B020202011), the Science and Technology Program of Guangzhou (202002020060), the Science and Technology Program of Zhanjiang (2019A01003), and the Key Science and Technology Planning Project of Guangzhou (201904020015).

Institutional Review Board Statement: Not applicable.

Informed Consent Statement: Not applicable.

Data Availability Statement: Data are contained within the article and Supplementary Materials.

Conflicts of Interest: The authors declare no conflict of interest.

Abbreviations

TF	Transcription factor
RT-qPCR	Reverse transcription quantitative real-time polymerase chain reaction
AA	Amino acid
PI	Isoelectric point
kDa	Kilodaltons
MDA	Malondialdehyde
ROS	Reactive oxygen species
NAM	No apical meristem
GFP	Green fluorescent protein
LUC	Luciferase
REN	Renilla
ORF	Open read frame
CDS	Coding sequence
WT	Wild type
COR	Cold regulated
IBA	Indole-3-butyric acid
NCBI	National Center for Biotechnology Information

References

1. Singh, D.; Laxmi, A. Transcriptional regulation of drought response: A tortuous network of transcriptional factors. *Front. Plant Sci.* **2015**, *6*, 895. [CrossRef] [PubMed]
2. Samo, N.; Wang, X.C.; Imran, M.; Bux, H.; Ahmed, S.; Hu, Y.G. Nac Vs: Abiotic stresses, current understanding and perspective, with special reference to the crops of Poaceae family. *Pak. J. Bot.* **2019**, *51*, 2037–2045. [CrossRef]
3. Yamasaki, K.; Kigawa, T.; Inoue, M.; Watanabe, S.; Tateno, M.; Seki, M.; Shinozaki, K.; Yokoyama, S. Structures and evolutionary origins of plant-specific transcription factor DNA-binding domains. *Plant Physiol. Biochem.* **2008**, *46*, 394–401. [CrossRef] [PubMed]
4. Puranik, S.; Sahu, P.P.; Srivastava, P.S.; Prasad, M. NAC proteins: Regulation and role in stress tolerance. *Trends Plant Sci.* **2012**, *17*, 369–381. [CrossRef]
5. Singh, S.; Koyama, H.; Bhati, K.K.; Alok, A. The biotechnological importance of the plant-specific NAC transcription factor family in crop improvement. *J. Plant Res.* **2021**, *134*, 475–495. [CrossRef]
6. Tran, L.S.P.; Nakashima, K.; Sakuma, Y.; Simpson, S.D.; Fujita, Y.; Maruyama, K.; Fujita, M.; Seki, M.; Shinozaki, K.; Yamaguchi-Shinozaki, K. Isolation and functional analysis of *Arabidopsis* stress-inducible NAC transcription factors that bind to a drought-responsive cis-element in the early responsive to dehydration stress 1 promoter. *Plant Cell* **2004**, *16*, 2481–2498. [CrossRef]
7. Olsen, A.N.; Ernst, H.A.; Lo Leggio, L.; Skriver, K. NAC transcription factors: Structurally distinct, functionally diverse. *Trends Plant Sci.* **2005**, *10*, 79–87. [CrossRef]
8. Ooka, H.; Satoh, K.; Doi, K.; Nagata, T.; Otomo, Y.; Murakami, K.; Matsubara, K.; Osato, N.; Kawai, J.; Carninci, P.; et al. Comprehensive analysis of NAC family genes in *Oryza sativa* and *Arabidopsis thaliana*. *DNA Res.* **2003**, *10*, 239–247. [CrossRef]
9. Jensen, M.K.; Kjaersgaard, T.; Nielsen, M.M.; Galberg, P.; Petersen, K.; O'Shea, C.; Skriver, K. The *Arabidopsis thaliana* NAC transcription factor family: Structure-function relationships and determinants of *ANAC019* stress signalling. *Biochem. J.* **2010**, *426*, 183–196. [CrossRef]
10. Chen, Q.F.; Wang, Q.; Xiong, L.Z.; Lou, Z.Y. A structural view of the conserved domain of rice stress-responsive *NAC1*. *Protein Cell* **2011**, *2*, 55–63. [CrossRef]
11. Delessert, C.; Kazan, K.; Wilson, I.W.; Van Der Straeten, D.; Manners, J.; Dennis, E.S.; Dolferus, R. The transcription factor *ATAF2* represses the expression of pathogenesis-related genes in *Arabidopsis*. *Plant J.* **2005**, *43*, 745–757. [CrossRef]
12. Puranik, S.; Bahadur, R.P.; Srivastava, P.S.; Prasad, M. Molecular cloning and characterization of a membrane associated NAC family gene, *SiNAC* from Foxtail Millet [*Setaria italica* (L.) P. Beauv.]. *Mol. Biotechnol.* **2011**, *49*, 138–150. [CrossRef]
13. Guo, S.; Dai, S.; Singh, P.K.; Wang, H.; Wang, Y.; Tan, J.L.H.; Wee, W.; Ito, T. A membrane-bound NAC-like transcription factor *OsNTL5* represses the flowering in *Oryza sativa*. *Front. Plant Sci.* **2018**, *9*, 555. [CrossRef]
14. Hao, Y.J.; Wei, W.; Song, Q.X.; Chen, H.W.; Zhang, Y.Q.; Wang, F.; Zou, H.F.; Lei, G.; Tian, A.G.; Zhang, W.K.; et al. Soybean NAC transcription factors promote abiotic stress tolerance and lateral root formation in transgenic plants. *Plant J.* **2011**, *68*, 302–313. [CrossRef]
15. Kim, H.J.; Nam, H.G.; Lim, P.O. Regulatory network of NAC transcription factors in leaf senescence. *Curr. Opin. Plant Biol.* **2016**, *33*, 48–56. [CrossRef]
16. He, X.J.; Mu, R.L.; Cao, W.H.; Zhang, Z.G.; Zhang, J.S.; Chen, S.Y. *AtNAC2*, a transcription factor downstream of ethylene and auxin signaling pathways, is involved in salt stress response and lateral root development. *Plant J.* **2005**, *44*, 903–916. [CrossRef]
17. Kim, Y.S.; Kim, S.G.; Park, J.E.; Park, H.Y.; Lim, M.H.; Chua, N.H.; Park, C.M. A membrane-bound NAC transcription factor regulates cell division in *Arabidopsis*. *Plant Cell* **2006**, *18*, 3132–3144. [CrossRef]
18. Zhong, R.Q.; Demura, T.; Ye, Z.H. *SND1*, a NAC domain transcription factor, is a key regulator of secondary wall synthesis in fibers of *Arabidopsis*. *Plant Cell* **2006**, *18*, 3158–3170. [CrossRef]
19. Zhou, H.; Kui, L.W.; Wang, H.L.; Gu, C.; Dare, A.P.; Espley, R.V.; He, H.P.; Allan, A.C.; Han, Y.P. Molecular genetics of blood-fleshed peach reveals activation of anthocyanin biosynthesis by NAC transcription factors. *Plant J.* **2015**, *82*, 105–121. [CrossRef]
20. Giovannoni, J.J. Genetic regulation of fruit development and ripening. *Plant Cell* **2004**, *16*, S170–S180. [CrossRef]
21. Dudhate, A.; Shinde, H.; Yu, P.; Tsugama, D.; Gupta, S.K.; Liu, S.; Takano, T. Comprehensive analysis of NAC transcription factor family uncovers drought and salinity stress response in pearl millet (*Pennisetum glaucum*). *BMC Genom.* **2021**, *22*, 70. [CrossRef] [PubMed]
22. Gong, X.; Zhao, L.Y.; Song, X.F.; Lin, Z.K.; Gu, B.J.; Yan, J.X.; Zhang, S.L.; Tao, S.T.; Huang, X.S. Genome-wide analyses and expression patterns under abiotic stress of NAC transcription factors in white pear (*Pyrus bretschneideri*). *BMC Plant Biol.* **2019**, *19*, 161. [CrossRef]
23. Wu, Q.; Bai, X.; Zhao, W.; Shi, X.D.; Xiang, D.B.; Wan, Y.; Wu, X.Y.; Sun, Y.X.; Zhao, J.L.; Peng, L.X.; et al. Investigation into the underlying regulatory mechanisms shaping inflorescence architecture in *Chenopodium quinoa*. *BMC Genom.* **2019**, *20*, 658. [CrossRef] [PubMed]
24. Liu, X.W.; Wang, T.; Bartholomew, E.; Black, K.; Dong, M.M.; Zhang, Y.Q.; Yang, S.; Cai, Y.L.; Xue, S.D.; Weng, Y.Q.; et al. Comprehensive analysis of NAC transcription factors and their expression during fruit spine development in cucumber (*Cucumis sativus* L.). *Hortic. Res.* **2018**, *5*, 31. [CrossRef] [PubMed]

25. Dong, H.; Chen, Q.; Dai, Y.; Hu, W.; Zhang, S.; Huang, X. Genome-wide identification of *PbrbHLH* family genes, and expression analysis in response to drought and cold stresses in pear (*Pyrus bretschneideri*). *BMC Plant Biol.* **2021**, *21*, 86. [CrossRef] [PubMed]
26. Han, D.G.; Du, M.; Zhou, Z.Y.; Wang, S.; Li, T.M.; Han, J.X.; Xu, T.L.; Yang, G.H. An NAC transcription factor gene from *Malus baccata, MbNAC29*, increases cold and high salinity tolerance in *Arabidopsis*. *In Vitro Cell. Dev. Biol.-Plant* **2020**, *56*, 588–599. [CrossRef]
27. Yang, Z.T.; Lu, S.J.; Wang, M.J.; Bi, D.L.; Sun, L.; Zhou, S.F.; Song, Z.T.; Liu, J.X. A plasma membrane-tethered transcription factor, *NAC062/ANAC062/NTL6*, mediates the unfolded protein response in *Arabidopsis*. *Plant J.* **2014**, *79*, 1033–1043. [CrossRef] [PubMed]
28. Yong, Y.B.; Zhang, Y.; Lyu, Y.M. A stress-responsive NAC transcription factor from tiger lily (*LlNAC2)* interacts with *LlDREB1* and *LlZHFD4* and enhances various abiotic stress tolerance in *Arabidopsis*. *Int. J. Mol. Sci.* **2019**, *20*, 3225. [CrossRef]
29. Obregón La Rosa, A.; Lozano Zanelly, G.A. Nutritional and bioactive compounds of three fruits from the peruvian highlands and jungle as a potential source of nutrients for human consumption. *Cienc. Tecnol. Agropecu.* **2021**, *22*, e1835.
30. Ibrahim, S.R.M.; Mohamed, G.A.; Khedr, A.I.M.; Zayed, M.F.; El-Kholy, A.A.S. Genus *Hylocereus*: Beneficial phytochemicals, nutritional importance, and biological relevance-A review. *J. Food Biochem.* **2018**, *42*, e12491. [CrossRef]
31. Peng, X.J.; Zhao, Y.; Li, X.M.; Wu, M.; Chai, W.B.; Sheng, L.; Wang, Y.; Dong, Q.; Jiang, H.Y.; Cheng, B.J. Genomewide identification, classification and analysis of NAC type gene family in maize. *J. Genet.* **2015**, *94*, 377–390. [CrossRef]
32. Diao, W.P.; Snyder, J.C.; Wang, S.B.; Liu, J.B.; Pan, B.G.; Guo, G.J.; Ge, W.; Dawood, M. Genome-wide analyses of the NAC transcription factor gene family in Pepper (*Capsicum annuum* L.): Chromosome location, phylogeny, structure, expression patterns, cis-elements in the promoter, and interaction network. *Int. J. Mol. Sci.* **2018**, *19*, 1028. [CrossRef]
33. Ahmad, M.; Yan, X.H.; Li, J.Z.; Yang, Q.S.; Jamil, W.; Teng, Y.W.; Bai, S.L. Genome-wide identification and predicted functional analyses of NAC transcription factors in Asian pears. *BMC Plant Biol.* **2018**, *18*, 214. [CrossRef]
34. Su, H.Y.; Zhang, S.Z.; Yuan, X.W.; Chen, C.T.; Wang, X.F.; Hao, Y.J. Genome-wide analysis and identification of stress-responsive genes of the *NAM-ATAF1,2-CUC2* transcription factor family in apple. *Plant Physiol. Biochem.* **2013**, *71*, 11–21. [CrossRef]
35. Borrill, P.; Harrington, S.A.; Uauy, C. Genome-wide sequence and expression analysis of the NAC transcription factor family in polyploid wheat. *G3-Genes Genom. Genet.* **2017**, *7*, 3019–3029. [CrossRef]
36. Chen, J.Y.; Xie, F.F.; Cui, Y.Z.; Chen, C.B.; Lu, W.J.; Hu, X.D.; Hua, Q.Z.; Zhao, J.; Wu, Z.J.; Gao, D.; et al. A chromosome-scale genome sequence of pitaya (*Hylocereus undatus*) provides novel insights into the genome evolution and regulation of betalain biosynthesis. *Hortic. Res.* **2021**, *8*, 164. [CrossRef]
37. Li, P.X.; Peng, Z.Y.; Xu, P.L.; Tang, G.Y.; Ma, C.L.; Zhu, J.Q.; Shan, L.; Wan, S.B. Genome-wide identification of NAC transcription factors and their functional prediction of abiotic stress response in peanut. *Front. Genet.* **2021**, *12*, 240. [CrossRef]
38. Liu, C.Y.; Xie, T.; Chen, C.J.; Luan, A.P.; Long, J.M.; Li, C.H.; Ding, Y.Q.; He, Y.H. Genome-wide organization and expression profiling of the R2R3-MYB transcription factor family in pineapple (*Ananas comosus*). *BMC Genom.* **2017**, *18*, 503. [CrossRef]
39. Fan, Z.Q.; Tan, X.L.; Chen, J.W.; Liu, Z.L.; Kuang, J.F.; Lu, W.J.; Shan, W.; Chen, J.Y. BrNAC055, a novel transcriptional activator, regulates leaf senescence in Chinese flowering cabbage by modulating reactive oxygen species production and chlorophyll degradation. *J. Agric. Food Chem.* **2018**, *66*, 9399–9408. [CrossRef]
40. Nuruzzaman, M.; Sharoni, A.M.; Kikuchi, S. Roles of NAC transcription factors in the regulation of biotic and abiotic stress responses in plants. *Front. Microbiol.* **2013**, *4*, 248. [CrossRef]
41. Tweneboah, S.; Oh, S.K. Biological roles of NAC transcription factors in the regulation of biotic and abiotic stress responses in solanaceous crops. *J. Plant Biotechnol.* **2017**, *44*, 1–11. [CrossRef]
42. Shao, H.B.; Wang, H.Y.; Tang, X.L. NAC transcription factors in plant multiple abiotic stress responses: Progress and prospects. *Front. Plant Sci.* **2015**, *6*, 902. [CrossRef]
43. Li, F.; Guo, X.H.; Liu, J.X.; Zhou, F.; Liu, W.Y.; Wu, J.; Zhang, H.L.; Cao, H.F.; Su, H.Z.; Wen, R.Y. Genome-wide identification, characterization, and expression analysis of the NAC transcription factor in *Chenopodium quinoa*. *Genes* **2019**, *10*, 500. [CrossRef]
44. Jia, D.F.; Jiang, Z.Q.; Fu, H.H.; Chen, L.; Liao, G.L.; He, Y.Q.; Huang, C.H.; Xu, X.B. Genome-wide identification and comprehensive analysis of NAC family genes involved in fruit development in kiwifruit (*Actinidia*). *BMC Plant Biol.* **2021**, *21*, 44. [CrossRef]
45. Li, B.; Fan, R.Y.; Yang, Q.S.; Hu, C.H.; Sheng, O.; Deng, G.M.; Dong, T.; Li, C.Y.; Peng, X.X.; Bi, F.C.; et al. Genome-wide identification and characterization of the NAC transcription factor family in *Musa Acuminata* and expression analysis during fruit ripening. *Int. J. Mol. Sci.* **2020**, *21*, 634. [CrossRef]
46. Nie, Q.; Qiao, G.; Peng, L.; Wen, X.P. Transcriptional activation of long terminal repeat retrotransposon sequences in the genome of pitaya under abiotic stress. *Plant Physiol. Biochem.* **2019**, *135*, 460–468. [CrossRef]
47. Li, A.L.; Wen, Z.; Yang, K.; Wen, X.P. Conserved miR396b-GRF regulation is involved in abiotic stress responses in pitaya (*Hylocereus polyrhizus*). *Int. J. Mol. Sci.* **2019**, *20*, 2501. [CrossRef]
48. Wang, G.D.; Liu, Q.; Shang, X.T.; Chen, C.; Xu, N.; Guan, J.; Meng, Q.W. Overexpression of transcription factor *SlNAC35* enhances the chilling tolerance of transgenic tomato. *Biol. Plantarum.* **2018**, *62*, 479–488. [CrossRef]
49. Shan, W.; Kuang, J.F.; Lu, W.J.; Chen, J.Y. Banana fruit NAC transcription factor *MaNAC1* is a direct target of *MaICE1* and involved in cold stress through interacting with *MaCBF1*. *Plant Cell Environ.* **2014**, *37*, 2116–2127. [CrossRef]
50. Ju, Y.L.; Yue, X.F.; Min, Z.; Wang, X.H.; Fang, Y.L.; Zhang, J.X. *VvNAC17*, a novel stress-responsive grapevine (*Vitis vinifera* L.) NAC transcription factor, increases sensitivity to abscisic acid and enhances salinity, freezing, and drought tolerance in transgenic *Arabidopsis*. *Plant Physiol. Biochem.* **2020**, *146*, 98–111. [CrossRef]

51. Han, D.G.; Du, M.; Zhou, Z.Y.; Wang, S.; Li, T.M.; Han, J.X.; Xu, T.L.; Yang, G.H. Overexpression of a *Malus baccata* NAC transcription factor gene *MbNAC25* increases cold and salinity tolerance in *Arabidopsis*. *Int. J. Mol. Sci.* **2020**, *21*, 1198. [CrossRef] [PubMed]
52. Hu, L.X.; Li, H.Y.; Pang, H.C.; Fu, J.M. Responses of antioxidant gene, protein and enzymes to salinity stress in two genotypes of perennial ryegrass (*Lolium perenne*) differing in salt tolerance. *J. Plant Physiol.* **2012**, *169*, 146–156. [CrossRef] [PubMed]
53. Li, X.D.; Zhuang, K.Y.; Liu, Z.M.; Yang, D.Y.; Ma, N.N.; Meng, Q.W. Overexpression of a novel NAC-type tomato transcription factor, *SlNAM1*, enhances the chilling stress tolerance of transgenic tobacco. *J. Plant Physiol.* **2016**, *204*, 54–65. [CrossRef] [PubMed]
54. Jiang, G.M.; Jiang, X.Q.; Lu, P.T.; Liu, J.T.; Gao, J.P.; Zhang, C.Q. The Rose (*Rosa hybrida*) NAC transcription factor 3 gene, *RhNAC3*, involved in ABA signaling pathway both in Rose and *Arabidopsis*. *PLoS ONE* **2014**, *9*, e109415. [CrossRef]
55. Edgar, R.C. MUSCLE: A multiple sequence alignment method with reduced time and space complexity. *BMC Bioinform.* **2004**, *5*, 113. [CrossRef]
56. Kumar, S.; Stecher, G.; Tamura, K. MEGA7: Molecular evolutionary genetics analysis version 7.0 for bigger datasets. *Mol. Biol. Evol.* **2016**, *33*, 1870–1874. [CrossRef]
57. Chen, C.J.; Chen, H.; Zhang, Y.; Thomas, H.R.; Frank, M.H.; He, Y.H.; Xia, R. TBtools: An integrative toolkit developed for interactive analyses of big biological data. *Mol. Plant.* **2020**, *13*, 1194–1202. [CrossRef]
58. Voorrips, R.E. MapChart: Software for the graphical presentation of linkage maps and QTLs. *J. Hered.* **2002**, *93*, 77–78. [CrossRef]
59. Wang, Y.P.; Tang, H.B.; Debarry, J.D.; Tan, X.; Li, J.P.; Wang, X.Y.; Lee, T.H.; Jin, H.Z.; Marler, B.; Guo, H.; et al. MCScanX: A toolkit for detection and evolutionary analysis of gene synteny and collinearity. *Nucleic Acids Res.* **2012**, *40*, e49. [CrossRef]
60. Nie, Q.; Gao, G.L.; Fan, Q.J.; Qiao, G.; Wen, X.P.; Liu, T.; Peng, Z.J.; Cai, Y.Q. Isolation and characterization of a catalase gene "*HuCAT3*" from pitaya (*Hylocereus undatus*) and its expression under abiotic stress. *Gene* **2015**, *563*, 63–71. [CrossRef]
61. Ding, Y.L.; Li, H.; Zhang, X.Y.; Xie, Q.; Gong, Z.Z.; Yang, S.H. *OST1* kinase modulates freezing tolerance by enhancing *ICE1* stability in *Arabidopsis*. *Dev. Cell* **2015**, *32*, 278–289. [CrossRef]
62. Guo, F.Q.; Crawford, N.M. *Arabidopsis* nitric oxide synthase1 is targeted to mitochondria and protects against oxidative damage and dark-induced senescence. *Plant Cell* **2005**, *17*, 3436–3450. [CrossRef]
63. Shi, J.; Fu, X.Z.; Peng, T.; Huang, X.S.; Fan, Q.J.; Liu, J.H. Spermine pretreatment confers dehydration tolerance of citrus in vitro plants via modulation of antioxidative capacity and stomatal response. *Tree Physiol.* **2010**, *30*, 914–922. [CrossRef]
64. Gilmour, S.J.; Zarka, D.G.; Stockinger, E.J.; Salazar, M.P.; Houghton, J.M.; Thomashow, M.F. Low temperature regulation of the *Arabidopsis CBF* family of *AP2* transcriptional activators as an early step in cold-induced *COR* gene expression. *Plant J.* **1998**, *16*, 433–442. [CrossRef]
65. Chen, C.B.; Wu, J.Y.; Hua, Q.Z.; Tel-Zur, N.; Xie, F.F.; Zhang, Z.K.; Chen, J.Y.; Zhang, R.; Hu, G.B.; Zhao, J.T.; et al. Identification of reliable reference genes for quantitative real-time PCR normalization in pitaya. *Plant Methods* **2019**, *15*, 70. [CrossRef] [PubMed]
66. Livak, K.J.; Schmittgen, T.D. Analysis of relative gene expression data using real-time quantitative PCR and the $2^{-\Delta\Delta C_T}$ method. *Methods* **2001**, *25*, 402–408. [CrossRef]

International Journal of
Molecular Sciences

MDPI

Article

A WRKY Protein, MfWRKY40, of Resurrection Plant *Myrothamnus flabellifolia* Plays a Positive Role in Regulating Tolerance to Drought and Salinity Stresses of *Arabidopsis*

Zhuo Huang [1,*], Jiatong Wang [1], Yuan Li [1], Li Song [1], Duo'er Chen [1], Ling Liu [1] and Cai-Zhong Jiang [2,3]

1 College of Landscape Architecture, Sichuan Agricultural University, Chengdu 611130, China; wangjiatong@stu.sicau.edu.cn (J.W.); yuanlisci@outlook.com (Y.L.); imsongli@stu.sicau.edu.cn (L.S.); 2020210028@stu.sicau.edu.cn (D.C.); myliuling1994@163.com (L.L.)

2 Department of Plant Sciences, University of California Davis, Davis, CA 95616, USA; caizhong.jiang@usda.gov

3 Crops Pathology and Genetics Research Unit, United States Department of Agriculture, Agricultural Research Service, Davis, CA 95616, USA

* Correspondence: huangzhuo@sicau.edu.cn; Tel.: +86-134-3893-4187

Abstract: WRKY transcription factors (TFs), one of the largest transcription factor families in plants, play an important role in abiotic stress responses. The resurrection plant, *Myrothamnus flabellifolia*, has a strong tolerance to dehydration, but only a few WRKY proteins related to abiotic stress response have been identified and functionally characterized in *M. flabellifolia*. In this study, we identified an early dehydration-induced gene, *MfWRKY40*, of *M. flabellifolia*. The deduced MfWRKY40 protein has a conserved WRKY motif but lacks a typical zinc finger motif in the WRKY domain and is localized in the nucleus. To investigate its potential roles in abiotic stresses, we overexpressed *MfWRKY40* in *Arabidopsis* and found that transgenic lines exhibited better tolerance to both drought and salt stresses. Further detailed analysis indicated that MfWRKY40 promoted primary root length elongation and reduced water loss rate and stomata aperture (width/length) under stress, which may provide *Arabidopsis* the better water uptake and retention abilities. MfWRKY40 also facilitated osmotic adjustment under drought and salt stresses by accumulating more osmolytes, such as proline, soluble sugar, and soluble protein. Additionally, the antioxidation ability of transgenic lines was also significantly enhanced, represented by higher chlorophyll content, less malondialdehyde and reactive oxygen species accumulations, as well as higher antioxidation enzyme activities. All these results indicated that MfWRKY40 might positively regulate tolerance to drought and salinity stresses. Further investigation on the relationship of the missing zinc finger motif of MfWRKY40 and its regulatory role is necessary to obtain a better understanding of the mechanism underlying the excellent drought tolerance of *M. flabellifolia*.

Keywords: *Myrothamnus flabellifolia*; drought tolerance; salinity tolerance; WRKY; zinc finger

Citation: Huang, Z.; Wang, J.; Li, Y.; Song, L.; Chen, D.; Liu, L.; Jiang, C.-Z. A WRKY Protein, MfWRKY40, of Resurrection Plant *Myrothamnus flabellifolia* Plays a Positive Role in Regulating Tolerance to Drought and Salinity Stresses of *Arabidopsis*. *Int. J. Mol. Sci.* **2022**, *23*, 8145. https://doi.org/10.3390/ijms23158145

Academic Editors: Andrés J. Cortés and Hai Du

Received: 28 June 2022
Accepted: 21 July 2022
Published: 24 July 2022

Publisher's Note: MDPI stays neutral with regard to jurisdictional claims in published maps and institutional affiliations.

1. Introduction

Throughout the life cycle, plants are subjected to various environmental stresses, such as drought, high salinity and extreme temperatures. Water deficit and salt stress are among the two major forms of abiotic stress that have a serious impact on plant growth and development. To adapt to such unfavorable environmental conditions, plants have evolved complex mechanisms to respond to these abiotic stresses at phenotypic (height, leaf size, root growth, etc.), physiological and biochemical (metabolism, enzymes, hormones, etc.), and molecular (protein expression, gene expression, etc.) levels [1–3].

Transcription factors are specialized proteins that bind to DNA-regulatory sequences, usually localized in the 5′-upstream region of the target genes, to activate or suppress gene transcription. Numerous studies indicated that transcription factor families, such as bZIP, bHLH, WRKY, MYB, and AP2/ERF, are involved in the plants' responses to abiotic stresses [4–8]. WRKY, one of the largest transcription factor families in plants, is important for plant growth, development, and resistance to external stresses, especially drought and salt [9–11]. For example, overexpressing *TaWRKY46* in *Arabidopsis* increased drought tolerance in abscisic acid (ABA)-dependent and ABA-independent ways [12]. Transgenic rice plants overexpressing *OsWRKY8* also exhibited improved drought tolerance [13], while cotton *GhWRKY33* can reduce the drought tolerance of transgenic *Arabidopsis* plants [14]. Moreover, overexpression of cotton *GhWRKY34* and *GhWRKY39-1* in *Arabidopsis* and tobacco can significantly improve the salt tolerance [15,16]. Through regulatory functions of the SOS (salt overly sensitive) pathway, overexpression of the *AtWRKY15* and *AtWRKY46* genes in *Arabidopsis* can improve salt tolerance [17,18].

WRKY proteins are unique to plants, with nearly 100 members in *Arabidopsis*. All WRKY proteins are named by the conserved WRKY domain comprised of two prominent structures: a short peptide WRKYGQK (WRKY motif) and the C2H2 or C2HC type zinc finger motif, which is generally highly conserved among family members. In addition, these proteins can specifically bind to the W box motif, which can differentially regulate the expression of various target genes [5]. WRKY proteins can be divided into two categories according to the number of WRKY domains, those with two WRKY domains belonging to class I and most WRKY proteins with only one WRKY domain belonging to class II, both of which have a ($C\text{-}X_{4-5}\text{-}C\text{-}X_{22-23}\text{-}H\text{-}X_1\text{-}H$) zinc finger structure. There is a class of WRKY proteins containing a zinc finger structure ($C\text{-}X_7\text{-}C\text{-}X_{23}\text{-}H\text{-}X_1\text{-}C$) which is different from the first two classes and separately grouped into the class III [5].

WRKY40 together with WRKY18 and WRKY60 have only one WRKY domain and belong to class II WRKY TF. In *Arabidopsis*, they are pathogen-induced and can interact directly with themselves and with each other. They have a partially redundant negative effect on the salicylic acid (SA)-mediated defense but a positive role in the Jasmonic acid (JA)-mediated defense [19]. Additionally, they all respond to ABA and abiotic stress. WRKY18 and WRKY60 could enhance plant sensitivity to ABA, salt, and osmotic stress, whereas WRKY40, on the other hand, antagonizes their effects [20]. Che et al. [21] reported that overexpression of *AtWRKY40* enhanced drought stress responses, presumably by interfering with the reactive oxygen species (ROS)-scavenging pathway and osmolyte accumulation process.

Research has shown that WRKY40 is integral to the growth and development of plants, as well as their resistance to external stresses. Jiang et al. [22] found that *NbWRKY40* might regulate tomato mosaic virus resistance by modulating the expression of SA, leading to the accumulation of callose (a sugar around each sieve pore) in the neck of plasmodesmata, thus interfering with virus movement. Karim et al. [23] determined that *PtrWRKY40* plays a negative role in the resistance of poplar to semi-vegetative fungi; however, it has a positive impact on the resistance of *Arabidopsis* to necrotic vegetative fungi. According to Lin et al. [24]., *PbWRKY40* could play a role in salt tolerance and organic acid accumulation in part by regulating the expression of *PbVHA-B1* Moreover, wheat *TaWRKY40* could positively regulate transcription of *TaGAPC1* to enhance tolerance to drought stress [25]. Studies on kumquat showed that *FcWRKY40* participates in the ABA-signaling pathway and acts as a positive regulator of salt tolerance by regulating genes involved in ion homeostasis and proline biosynthesis [26]. *ZmWRKY40* is induced by drought, high salinity, high temperature, and ABA. Transgenic *Arabidopsis* plants with overexpression of *ZmWRKY40* are more drought tolerant [27]. These results indicated that WRKY40 may positively regulate plant tolerance to abiotic stress.

Myrothamnus flabellifolia is the only woody resurrection plant found so far. Various aspects of drought tolerance of *M. flabellifolia* have been extensively studied, including its morphological structure [28–30], physiological and biochemical characteristics [31–33], as well as its physiological and metabolic pathways [34] during rehydration, but research on its drought tolerance mechanism is limited. Ma et al. [35] examined the transcriptome of *M. flabellifolia* and found that approximatel 295 TFs are responsive to dehydration.

MfWRKY40 was evidently upregulated in the early dehydration stage, suggesting that it may participate in response to a water deficit [35]. In this study, we cloned the *MfWRKY40* and overexpressed it in *Arabidopsis* to investigate its potential roles in tolerance to drought and salt.

2. Results

2.1. Cloning and Sequence Analysis of MfWRKY40

The cDNA sequence of the coding region of *MfWRKY40* (unigene ID: comp48667_c1__seq16) was cloned from *M. flabellifolia* by PCR amplification. The obtained nucleotide sequence is 441 bp in length and possesses an open-reading frame (ORF) encoding 146 amino acids. The calculated isoelectric point of MfWRKY40 is 6.15, and the deduced molecular weight is 16.8 kDa. A blastp search against the *Arabidopsis* genome data indicated that it showed the highest degree of homology to *AtWRKY40* of *Arabidopsis*. Multiple alignment of the deduced amino acids of *MfWRKY40* and several highly homologous WRKY proteins showed that all of them have a conservative "WRKYGQK" domain, indicating that it belongs to WRKY TF (Figure 1a). However, MfWRKY40 lacks the zinc finger structure and is significantly shorter than other WRKYs. Phylogenetic analysis indicated that MfWRKY40 was most homologous to a WRKY protein of *Bruguiera gymnorhiza* (Figure 1b).

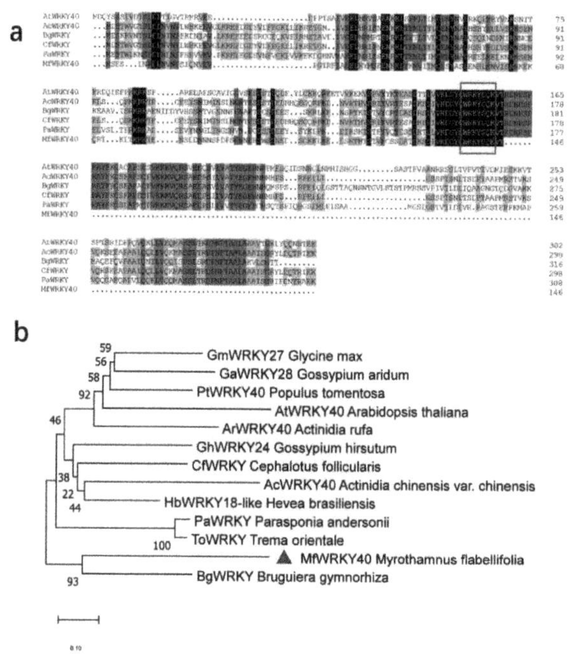

Figure 1. Multiple sequence alignment (**a**) and phylogenetic analysis (**b**) of MfWRKY40 and several highly homologous WRKY proteins. Black and gray shade in (**a**) show identical and similar amino acids, respectively. The WRKY motif was indicated by red box. (**b**) Phylogenetic reconstruction using the neighbor-joining method. The accession numbers for the sequences used were as follows: AtWRKY40 (AT1G80840) of *Arabidopsis thaliana;* ArWRKY40 (GFZ11870.1) of *Actinidia rufa;* PtWRKY40 (AZQ19202.1) of *Populus tomentosa;* GaWRKY28 (AIY62465.1) of *Gossypium aridum;* GmWRKY28 (ABC26917.1) of *Glycine max;* AcWRKY40 (PSS01081.1) of *Actinidia chinensis* var. *chinensis;* HbWRKY18-like (XP_021659791.1) of *Hevea brasiliensis;* GhWRKY24 (AGV75937.1) of *Gossypium hirsutum;* CfWRKY (GAV83654.1) of *Cephalotus follicularis;* PaWRKY (PON74566.1) of *Parasponia andersonii;* ToWRKY (PON99223.1) of *Trema orientale;* and BgWRKY (BAG15874.1) of *Bruguiera gymnorhiza.*

2.2. *MfWRKY40 Is Localized in the Nucleus*

We transiently expressed 35S::MfWRKY40-YFP in tobacco leaf epidermal cells. Confocal microscope observation showed that fluorescence of 35S::YFP could be detected in the whole cell, while strong fluorescence appeared specifically in the nucleus of 35S::MfWRKY40-YFP transformed cells (Figure 2). These results proved that MfWRKY40 is located in the nucleus, although no sequence for nuclear localization signal is predicted.

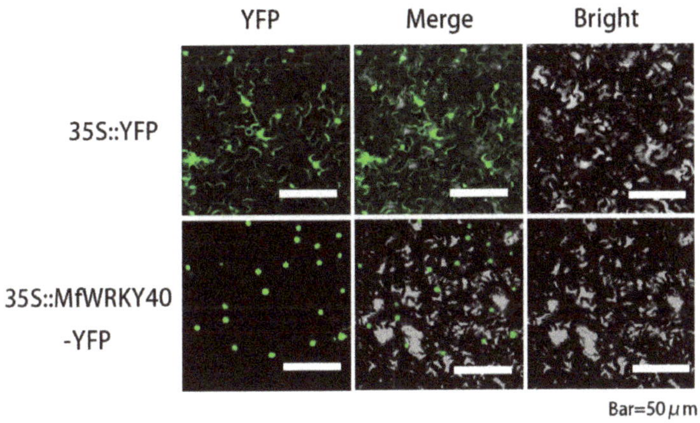

Figure 2. Subcellular localization of MfWRKY40. YFP, yellow fluorescent protein.

2.3. *Overexpressing MfWRKY40 Increased Drought and Salt Tolerance*

To analyze the potential role under abiotic stress, we introduced *MfWRKY40* into *Arabidopsis* by *Agrobacterium tumefaciens* mediated genetic transformation driven by the CaMV 35S promoter. T_1 transgenic *Arabidopsis* lines expressing *MfWRKY40* gene were acquired from kanamycin-resistance screening, and two homozygous positive T_3 transgenic lines identified by PCR were randomly selected for further analysis.

To investigate whether *MfWRKY40* is associated with abiotic stress tolerance, WT and transgenic lines were treated by drought and salinity stresses at the seedling and adult stages. For seedling treatment, the seedlings of wild type (WT) and transgenic lines (OE) were planted on solid medium containing different concentrations of mannitol and NaCl. The WT and OE lines showed similar primary root length and lateral root number at normal condition. Under the treatments of mannitol with concentrations from 200 mM to 300 mM, no significant change on lateral root number was found. The root lengths of the WT and OE lines were gradually decreased, but those of line A and line C were 1.22 and 1.29, 1.19 and 1.29, 1.75 and 1.67 folds of those of WT, respectively (Figure 3a,b). Similar to mannitol treatment, two OE lines, A and C, exhibited significantly longer primary roots, which were 1.34 and 1.31, 1.31 and 1.21, and 1.50 and 1.28 folds of those WT under 50 mM, 75 mM, and 100 mM NaCl treatments, respectively (Figure 3c,d).

For tolerance evaluation at the adult stage, four-week-old OE lines and WT plants were treated by natural drought (stopping watering) and 300 mM NaCl solution. They were morphologically similar before treatment. At the early stage of drought treatment (five days after stopping watering), the leaves of WT and OE lines began to wither. Seventeen days after stopping watering, the leaf withering became more serious, but the two OE lines were less dehydrated. After 19 days, although almost all plants were severely dehydrated, the two OE lines, especially the line A, were less withered. Three days after rewatering, the WT almost died, whereas the two OE lines were significantly recovered (Figure 4a).

The leaf withering was detected at about day five of the salt treatment. From five to ten days of salt treatment, more withered leaves were found on both the WT and OE lines, but the latter, especially the line A, had more green leaves. At day 15 of the salt treatment, most leaves of the WT and OE lines were severely wilted, but more leaf parts of the line

A were still green (Figure 4b). The above results showed that *Arabidopsis* overexpressing *MfWRKY40* exhibited better tolerance to drought and salt stresses.

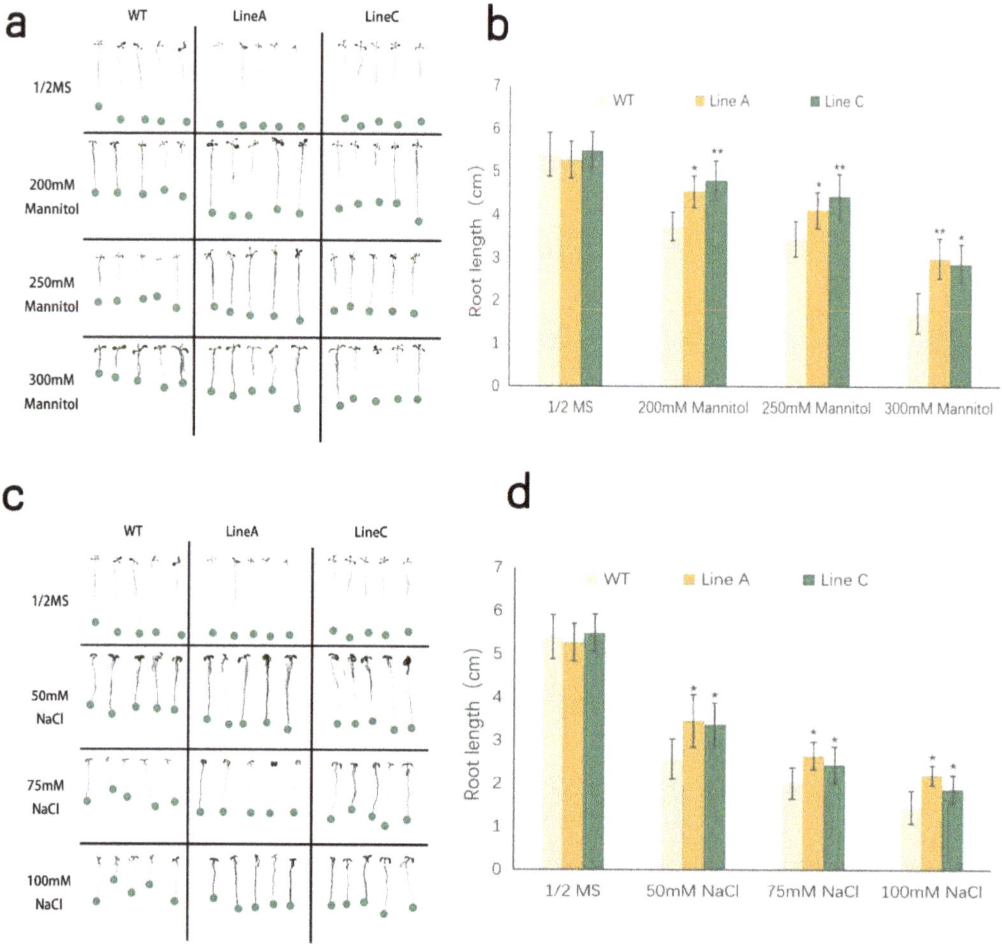

Figure 3. Analysis of drought and salinity tolerance at seedling stage. (**a**,**b**) indicated morphology of transgenic and WT seedlings grown for nine days on 1/2 MS medium with varying concentrations of mannitol and NaCl. (**c**,**d**) indicated root length of corresponding plants under different treatments. Data are presented as mean and SD values (error bar) of three independent experiments. Asterisks indicate significant difference (* $p < 0.05$, ** $p < 0.01$, by independent sample t-test) comparing to WT.

2.4. Measurement of Tolerance-Related Physiological and Biochemical Parameters

Drought and salinity will induce the reduction of the chlorophyll contents in plant leaves, which may trigger the inactivation of photosynthesis and leaf chlorosis. Under drought and salt treatments, the chlorophyll contents of the WT and OE lines were decreased compared to that under the normal condition, but lines A and C showed significantly higher chlorophyll content, which were 1.56 and 1.33 folds and 1.39 and 1.33folds of those of WT under the drought and salt treatments, respectively (Figure 4c).

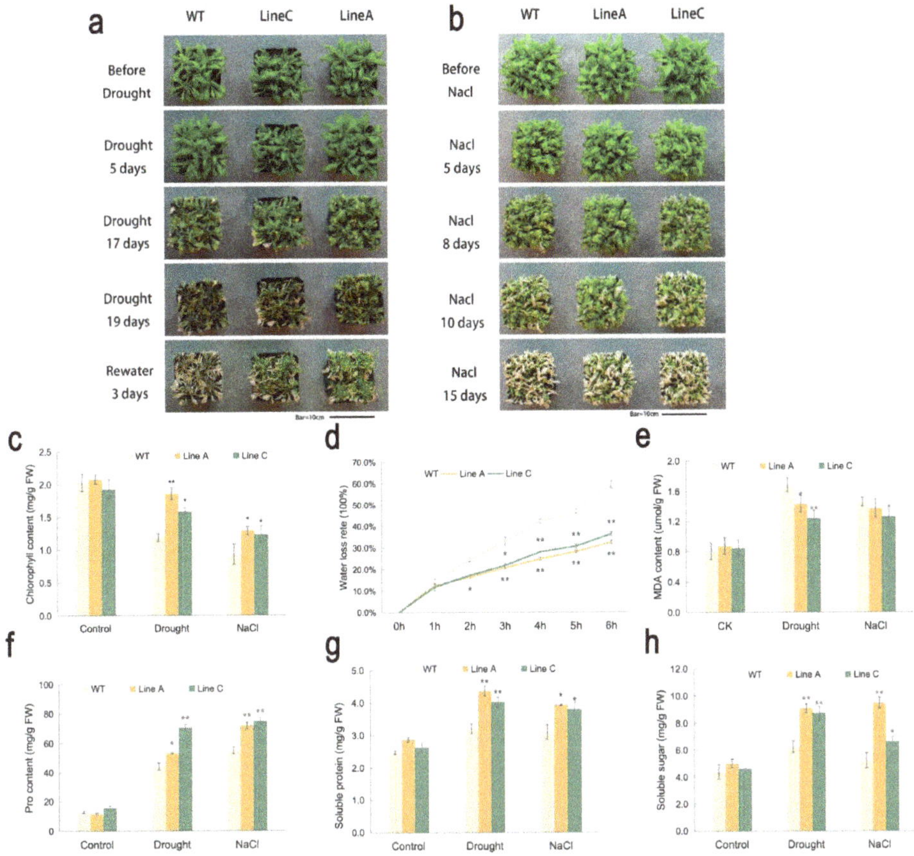

Figure 4. Analysis of drought and salinity tolerance at the adult stage. (**a**,**b**) showed the change of growth status of transgenic and WT plants during the progress of drought and salinity treatments. (**c**–**h**) showed measurements of tolerance-related physiological indexes. Data are presented as mean and SD values of three independent experiments. Asterisks indicated significant difference (* $p < 0.05$, ** $p < 0.01$, by independent sample *t*-test) comparing to WT.

We measured the dynamic water loss rate (WLR) of the detached leaves during the dehydration progress from 0 h to 6 h. The WLR of the WT and OE lines were almost the same before 1 h. From 2 h to 6 h, the average WLR of line A and line C were 13.82–34.89%, which were significantly lower than the 24.30–59.30% of WT (Figure 4d). Malondialdehyde (MDA) is an important indicator of membrane lipid peroxidation and can lead to severe cell membrane damage. Both drought and salt stresses induced MDA accumulation in the WT and OE lines. Under drought treatment, lines A and C accumulated less (86% and 74% of that of WT) MDA than WT. Under salt stress, the MDA content of WT was significantly higher than the line C (1.16 folds), while it was at a similar level with line A (1.06 folds) (Figure 4e). These results indicated that the WT plants suffered more serious membrane damage than did the OE lines (Figure 4e).

Osmotic adjustment mediated by generating osmolytes is an important feature allowing plants to adapt in osmotic stress conditions [36]. We compared the contents of three osmolytes, proline, soluble protein, and soluble sugar, between the WT and OE lines. The results showed that under the drought and salt treatments, lines A and C accumulated more proline (1.20 and 1.59 folds under drought and 1.31 and 1.36 folds under salt), soluble protein (1.36 and 1.25 folds under drought and 1.27 and 1.23 folds under salt) and soluble sugar (1.45 and 1.44 folds under drought and 1.81 and 1.27 folds under salt) than did the WT (Figure 4f–h). Additionally, line A showed a lower proline content but higher soluble protein and soluble sugar content compared to line C.

2.5. Effect of *MfWRKY40* Overexpression on Antioxidant Metabolism

Drought and salt stresses usually cause excessive accumulation of ROS, which will lead to oxidative stress on plant cells. To evaluate the ROS levels under drought and salt treatments, we used 3,3′-diaminobenzidine (DAB) and nitroblue tetrazole (NBT) for histochemical staining. The results showed that under the normal condition, both the WT and OE lines were in similar staining degrees, which were stained in very light brown or only very few parts of the leaves could be stained in blue. These results indicated that the WT and OE lines accumulated similar low contents of H_2O_2 and O_2^- under the normal conditions. However, after being treated by drought and salt stresses, the whole leaves of the WT were stained in dark brown or blue by DAB and NBT, respectively, while lines A and C were stained in a visible lighter brown by DAB, and significantly less leaf parts were stained in blue by NBT, respectively (Figure 5a,b). Additionally, we quantified the contents of H_2O_2 and O_2^- and found that line A and line C exhibited similar contents of H_2O_2 and O_2^- with WT under the normal condition, while significantly lower contents were exhibited under the drought and salt stresses (Figure 5c, d), which were 58% and 56%, 54.94% and 67.35%, 70.74% and 78.11%, and 67.21% and 72.31% of WT, respectively (Figure 5c,d). These results were consistent with those detected by histochemical staining, suggesting that the OE lines overexpressing *MfWRKY40* suffered less oxidative stress.

The peroxidase (POD) and superoxide dismutase (SOD) were the key enzymes involved in ROS scavenging. We determined the activities of the two enzymes and found that under the drought and salt treatments, the activities of POD and SOD of the WT and transgenic plants were both significantly increased. Under drought, the POD and SOD activities of line A and line C were 1.14 and 1.28 folds and 2.33 and 2.18 folds of those of the WT (Figure 5e,f), whereas they were 1.29 and 1.50 folds and 1.51 and 1.96 folds under salt treatment, respectively. These results showed that the lower ROS accumulation in the OE lines was due to, at least partly, the promoted ROS-scavenging system.

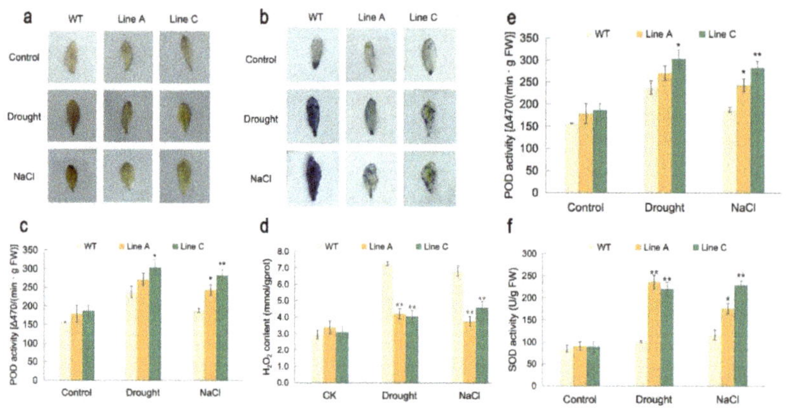

Figure 5. Analysis of ROS accumulation and activities of key antioxidant enzymes under drought and salt treatments. (**a**,**b**) showed the analysis of H_2O_2 and O_2^- accumulation by using histochemical staining with DAB and NBT, respectively. (**c**,**d**) showed the content of hydrogen peroxide (H_2O_2) and superoxide anion content (O_2^-), respectively. (**e**,**f**) indicated activities of peroxidase (POD) and superoxide dismutase (SOD) in the leaves of transgenic and WT plants, respectively. Data are presented as mean and SD values of three independent experiments. Asterisks indicated significant difference (* $p < 0.05$, ** $p < 0.01$, by independent sample *t*-test) compared to WT.

2.6. MfWRKY40 Promoted Stomatal Closure

We evaluated the stomatal closures of leaves under 300 mM mannitol treatment. Under normal conditions, most of the stomata in the WT and OE lines were open (Figure 6a), and there were no significant differences in the stomatal apertures (width/length ratio) between the OE and WT plants. Mannitol treatment reduced the stomatal apertures of the two OE lines to 0.5 and 0.52, which were significantly lower than the 0.63 of the WT (Figure 6b). These results suggest that MfWRKY40 promotes mannitol-induced stomatal closure.

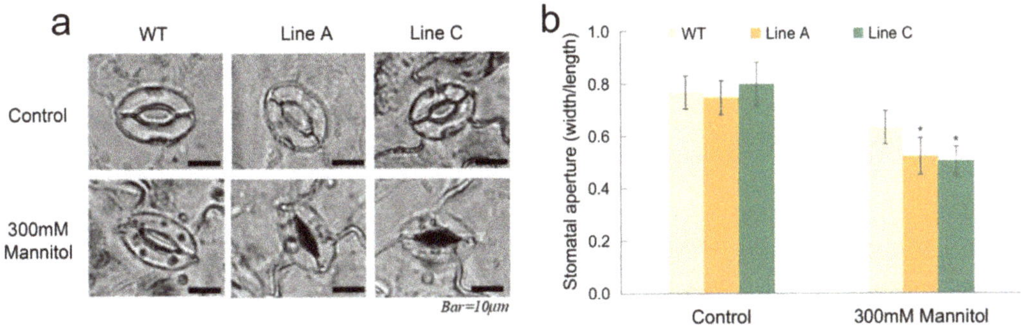

Figure 6. Measurements of stomatal aperture. (**a**) Microscopy observation of the stomatal aperture of OE and WT plants treated by 300 mM mannitol. (**b**) Measurement of the stomatal aperture with or without mannitol treatment. Data are presented as mean and SD values of three independent experiments. Asterisks indicated significant difference (* $p < 0.05$, by independent sample *t*-test) comparing to WT.

3. Discussion

Up to now, many WRKY TFs have been isolated from different plant species [24,25,37], and are involved in abiotic stress responses. In this study, *MfWRKY40*, which was upregulated at the early dehydration stage [35], was isolated from *M. flabellifolia*. It contains a highly conserved WRKY motif and has the highest homology with a WRKY protein of *Bruguiera gymnorhiza*. The nuclear localization of MfWRKY40 suggests that MfWRKY40 may function as a transcription activator or repressor. It is worth mentioning that there is no zinc finger structure in the WRKY domain of MfWRKY40, which is usually considered as a functional structure binding to the W-box in the promoter region of the target genes with the DNA sequence of (C/T)TGAC(C/T) [38]. We also search the transcriptome dataset of *M. flabellifolia* and failed to find a more intact transcript related to *MfWRKY40* (comp48667_c1__seq16). Therefore, it is unlikely to be derived from alternative splicing.

Previous studies reported that there may be a fourth class of WRKY proteins in which the members contain incomplete WRKY domains. For example, five WRKY proteins found in rice do not contain zinc finger motifs. Whether the five WRKY proteins have regulatory functions have not been studied [39]. The complex and different roles of WRKY40 in response to abiotic stress were also reported [20,22,25–27]. In the present study, our results showed that the MfWRKY40 without the zinc finger structure may function as a regulator for drought and salinity tolerance. However, whether this role is related to absence of the zinc finger is unclear.

According to our results, *MfWRKY40* OE plants showed better growth under both drought and salt treatments, indicating that MfWRKY40 plays a positive role in drought and salt responses in *Arabidopsis*. The plant organs, such as the leaves and roots, orchestrate defense mechanisms (internal or external) in response to abiotic stress [40–42]. Roots are the first organs exposed to drought and salt stresses because these stresses are inseparable from the soil medium [43–47]. Our results showed that the introduction of *MfWRKY40* into *Arabidopsis* enhanced primary root length under both stress treatments, which may benefit the plant with better water uptaking and transportation under adverse conditions.

Additionally, the lower transpiration rate, thick cuticle, and small stomatal aperture enhance the water retention ability and drought tolerance in plants [48]. In our study, the OE lines had smaller stomatal aperture and significantly lower water loss rates. These combined results suggested that MfWRKY40 also enhanced water uptake and retention capacity under stress.

Osmotic adjustment is well known to have an important role in plant adaptation to dehydration via the maintenance of turgor pressure, relative water content, stomatal conductance, and specific cellular functions [49]. The accumulation of compatible solutes in plants is thought to benefit stressed cells, either by acting as cytoplasmic osmolytes to facilitate water uptake and retention or by protecting macromolecules and their structures from stress-induced damage [49]. The common osmotic protective agents in plants include proline, soluble sugar, and soluble protein, etc. All are essential substances for the primary metabolism of plants. Several studies showed that these substances are helpful in plant tolerance when exposed to drought and salinity stresses. Among them, proline affects the genes responsible for stress tolerance and reduces the harmful effects caused by drought stress. It also restores stress injuries by stabilizing the antioxidants system though osmotic adjustments and by decreasing ROS effects [50]. For instance, proline is interconnected with the signaling pathways of the total soluble sugars and the quenching of ROS and thus, promotes drought tolerance in Arabidopsis [51]. Our research shows that under drought or salt stress, the *MfWRKY40* OE lines had higher contents of proline, soluble protein, and soluble sugar than the WT, suggesting that MfWRKY40 may contribute to regulating the accumulation of osmotic substances.

In addition, stresses can increase ROS production and lead to oxidative stress [52]. However, plant cells are well equipped with antioxidants and scavenging enzymes to maintain their levels. By histochemical staining and ROS quantification, we found significantly lower ROS accumulation in OE lines, and the promoted SOD and POD activities were also detected. These results apparently indicated that MfWRKY40 enhanced the ROS-scavenging system. This is similar to the report of *AtWRKY40* overexpression [20].

In conclusion, our results in this study showed that overexpression of *MfWRKY40* of *M. flabellifolia* can enhance the drought and salt tolerance of *Arabidopsis* by directly or indirectly promoting primary root elongation and enhancing water uptake and retention capacity, osmotic adjustment, and antioxidation system. Therefore, it may be involved in positive responses to abiotic stress. However, as MfWRKY40 protein lacks a complete typical WRKY domain, its regulatory mechanisms in abiotic stress responses need to be further investigated.

4. Materials and Methods

4.1. Plant Materials and Growth Conditions

M. flabellifolia used for gene cloning was provided by the Department of Plant Science, University of California, Davis. *Arabidopsis* ecotype *Columbia* (WT) is conserved by our lab. For stress treatment at the seedling stage, the seeds of *Arabidopsis* were sterilized with 50% bleach solution for 5 min and washed with sterile water three to five times, 1 min each time. Then the seeds were evenly planted on square petri dishes (10 × 10 cm) (15 seeds per dish) containing solid 1/2 strength Murashige and Skoog (MS) medium. After 2 days of vernalization in the dark and at 4°C, the medium was placed in a incubator under conditions of 60% relative humidity and long day (16 h light/8 h dark) treatment at 24 °C/22 °C with light intensity of approximately 100 μM photons m^{-2} s^{-1}. For stress treatment at the adult stage, WT and transgenic seeds germinated on 1/2 MS medium were planted in pots (8.5 cm deep) with square mouths (10 × 10 cm), containing the same amounts of the substrate (peat soil: vermiculite = 1:1), were grown for four weeks under the same conditions, as mentioned for seedling treatment.

4.2. Cloning and Bioinformatic Analysis of MfWRKY40

The total RNA of *M. flabellifolia* was extracted with Plant RNA Kit (Omega). Subsequently, RNA from *M. flabellifolia* was reversely transcribed using Uscript II All in One First-strand cDNA Synthesis Super Mix (InnovGene, Chengdu, China) to obtain the first strand cDNA.

According to the sequence of *MfWRKY40* (uniGene ID: comp48667_c1__seq16) [35], a pair of gene specific primers were designed by using Primer Premier 6 software to isolate the complete coding sequence from cDNA of *M. flabellifolia*. The primer sequences were as follows: forward primer: 5′-TCCCCCGGGATGTCCGAGTCCCTGAACTT-3′ (*Sma* I cut site is underlined) and reverse primer: 5′-GACTAGTTCAGGTGACCTTCTGACCAT-3′ (*Spe* I cut site is underlined). To facilitate the subsequent vector construction, the cleavage sites of two restriction endonucleases, *Sma* I and *Spe* I, were added on the 5′ end of the primers. The primers were synthesized by Qingke Biotechnology (Chengdu, China). Using the cDNA of *M. flabellifolia* as the template, the target gene was amplified by PCR employing Phantamax Super-Fidelity DNA Polymerase (Vazyme Biotech Co., Nanjing, China). The purified PCR product with expected size was cloned into a pEasy-T1 simple vector (Transgenic Biotechnology, Beijing, China) and transformed into *E. coli* strain, DH5 α. The positive clones identified by PCR were sequenced by Qingke Biotechnology Co. (Beijing, China).

The ORF Finder (https://www.ncbi.nlm.nih.gov/orffinder/, accessed on 15 July 2022) was used to find the open reading frame (ORF) of the obtained nucleotide sequence. NCBI blastp search (https://blast.ncbi.nlm.Nih.gov/Blast.cgi, accessed on 15 July 2022) was used to find proteins with high degrees of homology with MfWRKY40. The multiple sequence alignment of these homologous proteins was performed using DNAMAN 9 software. The phylogenetic tree was constructed based on the neighbor joining method using MEGA 7.0 software.

4.3. Subcellular Localization of MfWRKY40

A pair of primers was designed to amplify *MfWRKY40* without stop codon and was supplemented with recognition sites of restriction endonucleases *Hind* III and *BamH* I and sequences homologous to the terminals of the pHB-YFP vector containing gene encoding yellow fluorescent protein. The primer sequences were as follows: forward primer: 5′-*ACCAGTCTCTCTCTC*AAGCTTATGTCCGAGTCCCTGAACTT-3′ (*Hind* III site is underlined, homologous arm sequence is italic); reverse primer: 5′-*GCTCACCATACTAGT*GGATCC GGTGACCTTCTGACCAT-3′. (*BamH* I site is underlined; homologous arm sequence is italic). The amplified fragment was cloned into the vector by the homologous recombination method using REIII One-step Cloning Mix (Innovagene biotechnology Ltd., Changsha, China). The resulting construct *35S::MfWRKY40-YFP* and control construct (*35S::YFP*) were respectively transformed into *Agrobacterium tumefaciens* GV3101 by the freeze–thaw method and injected into four-week-old tobacco (*Nicotiana benthamiana*) cells. After 16 h dark treatment and two days of normal cultivation, the leaves of the transformed tobacco were cut and placed under a laser confocal scanning microscope (Nikon, Tokyo, Japan) to observe the expression of YFP in tobacco cells.

4.4. Overexpression Vector Construction and Screening of Transgenic Lines

To generate the *35S::MfWRKY40* lines, the coding sequence of *MfWRKY40* was ligated into a binary vector, *pGSA1403*, and transferred into *A. tumefaciens* LBA4404. Subsequently, the transgenic *Arabidopsis* was produced by the floral-dip transformation method [53]. The T_0 seeds were collected and screened on 1/2 MS medium supplied with kanamycin (50 μg/mL). The continuous selfing and screening with kanamycin and PCR detection were performed. The resulting T_3 homozygous positive lines were randomly selected for further analysis.

4.5. Drought and Salinity Treatments

For drought and salt treatments at the seedling, sterilized seeds of WT and two T_3 transgenic lines were germinated and vertically grown in square petri dishes containing 1/2 MS medium containing varying concentrations of mannitol (0 mM, 200 mM, 250 mM and 300 mM) and NaCl (0 mM, 50 mM, 75 mM and 100 mM). The growing conditions were mentioned above. After nine days, the primary root lengths of each line were measured. A total of 15 seedlings from each line were measured, and three biological repeated experiments were conducted.

The four-week-old plants of the WT and transgenic lines were used to explore the tolerance of mature plants to drought and salt stresses. The plants grown in pots were saturated with water before treatment. For drought stress, the watering was held for 19 days and then rewatered for three days. For salt treatment, plants were irrigated by 300 mM NaCl solution twice at three-day intervals. The whole treatment was continued for 15 days. The plants were observed and photographed regularly. Each treatment was performed in three biological repeats.

4.6. Determination of Water Loss Rate

The WT and transgenic plants were conventionally cultivated for four weeks. The 0.5 g rosette leaves were cut from similar positions of the WT plants and two T_3 transgenic lines and weighed immediately. Then the leaves were placed on filter paper at room temperature (~24 °C, 60% relative air humidity) and weighed at set time points (0, 1, 2, 3, 4, 5, and 6 h). The water loss rate representing percentage of water loss at each time point were obtained according to the fresh weights of the leaves before dehydration. The experiment was repeated three times.

4.7. Physiological Index Determination

Chlorophyll was directly extracted with 95% ethanol and quantified as previously reported [54]. The acidic ninhydrin colorimetric method was used for the determination of proline content [55]. Soluble sugar and soluble protein were quantitatively determined with Plant Soluble Sugar Content Detection Kit (Nanjing Jiancheng, Nanjing, China) and Total Protein Quantitative Determination Kit (Nanjing Jiancheng, Nanjing, China), respectively. The histochemical staining using 3,3′-diaminobenzidine (DAB) and nitroblue tetrazole (NBT) was used for visualization of the accumulation of hydrogen peroxide (H_2O_2) and superoxide anion radical (O_2^-) in the leaves, respectively [56]. A hydrogen peroxide assay kit and superoxide anion assay kit (Nanjing Jiancheng, Nanjing, China) were used to measure the contents of H_2O_2 and O_2^-, respectively. The SOD activity was determined by the diazo-blue tetrazole photo-reduction method [57]. The POD activity was determined by the guaiacol determination method [58]. The malondialdehyde (MDA) content of each strain was determined with the thiobarbituric acid method [59]. Three replicates were executed for all these experiments.

4.8. Stomatal Aperture Analysis under Drought Stress

The rosette leaves of four-week-old plants of the WT and transgenic lines were placed in the MES–KCl solution (50 mM KCl, 0.1 mM $CaCl_2$ and 10 mM MES, pH = 6.15) and exposed to light for 2.5 h. The treated leaves were then transferred to MES–KCl solutions containing 0 and 300 mM mannitol for 2 h of exposure to light, respectively. The stomata on the lower epidermis of the leaves were immediately observed and photographed by an optical microscopy (DP80, Olympus, Japan). The width–length ratios of at least 50 stomatal cells was measured. All experiments were repeated three times.

4.9. Statistical Analysis

Data were analyzed by Student's *t*-test using SPSS v. 23.0 and shown as the mean ± standard deviation (SD) of three replicates. The significance of difference was expressed as * ($p < 0.05$) or ** ($p < 0.01$).

Author Contributions: Conceptualization, Z.H.; methodology, Z.H. and C.-Z.J.; validation, Z.H., Y.L., J.W., L.S., and L.L.; data curation, Z.H., Y.L. and D.C.; writing—original draft preparation, Z.H., J.W., L.S. and D.C.; project administration, Z.H.; funding acquisition, Z.H. All authors have read and agreed to the published version of the manuscript.

Funding: This study is supported by Sichuan Science and Technology Program (No. 2022YFH0066) funded by Science and Technology Department of Sichuan Province, China, and Shuangzhi Plan funded by Sichuan Agricultural University.

Institutional Review Board Statement: Not applicable.

Data Availability Statement: Not applicable.

Acknowledgments: We thank for Chao Mao, College of Horticulture, China Agricultural University, for his kind help in the *M. flabellifolia* cultivation and for his suggestions on the experiments. Shuang-Cheng Li of Sichuan Agricultural University is appreciated for his kind help on the experiments.

Conflicts of Interest: The authors declare no conflict of interest. The funders had no role in the design of the study; in the collection, analyses, or interpretation of data; in the writing of the manuscript; or in the decision to publish the results.

References

1. Wang, K.; Chen, F.; Huang, W. Research advance on drought stress response mechanism in plants. *J. Agric. Sci. Technol.* **2019**, *21*, 19–25.
2. Chen, L.; He, D. Research advance on drought and salt resistant genes in transgenic plants. *Genom. Appl. Biol.* **2010**, *29*, 542–549.
3. Fathi, A.; Tari, D.B. Effect of drought stress and its mechanism in plants. *Int. J. Life Sci.* **2016**, *10*, 1–6. [CrossRef]
4. Singh, K.B.; Foley, R.C.; Oñate-Sánchez, L. Transcription factors in plant defense and stress responses. *Curr. Opin. Plant Biol.* **2002**, *5*, 430–436. [CrossRef]
5. Eulgem, T.; Rushton, P.J.; Robatzek, S.; Somssich, I.E. The WRKY superfamily of plant transcription factors. *Trends Plant Sci.* **2000**, *5*, 199–206. [CrossRef]
6. Stracke, R.; Werber, M.; Weisshaar, B. The R2R3-MYB gene family in *Arabidopsis thaliana*. *Curr. Opin. Plant Biol.* **2001**, *4*, 447–456. [CrossRef]
7. Mizoi, J.; Shinozaki, K.; Yamaguchi-Shinozaki, K. AP2/ERF family transcription factors in plant abiotic stress responses. *Biochim. Et Biophys. Acta (BBA)-Gene Regul. Mech.* **2012**, *1819*, 86–96. [CrossRef]
8. Soltabayeva, A.; Ongaltay, A.; Omondi, J.O.; Srivastava, S. Morphological, physiological and molecular markers for salt-stressed plants. *Plants* **2021**, *10*, 243. [CrossRef]
9. Chen, L.; Song, Y.; Li, S.; Zhang, L.; Zou, C.; Yu, D. The role of WRKY transcription factors in plant abiotic stresses. *Biochim. Biophys. Acta (BBA)-Gene Regul. Mech.* **2012**, *1819*, 120–128. [CrossRef]
10. Eulgem, T.; Somssich, I.E. Networks of WRKY transcription factors in defense signaling. *Curr. Opin. Plant Biol.* **2007**, *10*, 366–371. [CrossRef]
11. Ülker, B.; Somssich, I.E. WRKY transcription factors: From DNA binding towards biological function. *Curr. Opin. Plant Biol.* **2004**, *7*, 491–498. [CrossRef] [PubMed]
12. Li, X.; Tang, Y.; Zhou, C.; Zhang, L.; Lv, J. A wheat WRKY transcription factor TaWRKY46 enhances tolerance to osmotic stress in transgenic Arabidopsis plants. *Int. J. Mol. Sci.* **2020**, *21*, 1321. [CrossRef] [PubMed]
13. Rushton, D.L.; Tripathi, P.; Rabara, R.C.; Lin, J.; Ringler, P.; Boken, A.K.; Langum, T.J.; Smidt, L.; Boomsma, D.D.; Emme, N.J. WRKY transcription factors: Key components in abscisic acid signalling. *Plant Biotechnol. J.* **2012**, *10*, 2–11. [CrossRef] [PubMed]
14. Wang, N.-N.; Xu, S.-W.; Sun, Y.-L.; Liu, D.; Zhou, L.; Li, Y.; Li, X.-B. The cotton WRKY transcription factor (GhWRKY33) reduces transgenic Arabidopsis resistance to drought stress. *Sci. Rep.* **2019**, *9*, 724. [CrossRef] [PubMed]
15. Zhou, L.; Wang, N.-N.; Gong, S.-Y.; Lu, R.; Li, Y.; Li, X.-B. Overexpression of a cotton (*Gossypium hirsutum*) WRKY gene, GhWRKY34, in Arabidopsis enhances salt-tolerance of the transgenic plants. *Plant Physiol. Biochem.* **2015**, *96*, 311–320. [CrossRef] [PubMed]
16. Shi, W.; Hao, L.; Li, J.; Liu, D.; Guo, X.; Li, H. The Gossypium hirsutum WRKY gene GhWRKY39-1 promotes pathogen infection defense responses and mediates salt stress tolerance in transgenic Nicotiana benthamiana. *Plant Cell Rep.* **2014**, *33*, 483–498. [CrossRef] [PubMed]
17. Vanderauwera, S.; Vandenbroucke, K.; Inzé, A.; Van De Cotte, B.; Mühlenbock, P.; De Rycke, R.; Naouar, N.; Van Gaever, T.; Van Montagu, M.C.; Van Breusegem, F. AtWRKY15 perturbation abolishes the mitochondrial stress response that steers osmotic stress tolerance in Arabidopsis. *Proc. Natl. Acad. Sci. USA* **2012**, *109*, 20113–20118. [CrossRef]
18. Ding, Z.J.; Yan, J.Y.; Li, C.X.; Li, G.X.; Wu, Y.R.; Zheng, S.J. Transcription factor WRKY 46 modulates the development of Arabidopsis lateral roots in osmotic/salt stress conditions via regulation of ABA signaling and auxin homeostasis. *Plant J.* **2015**, *84*, 56–69. [CrossRef]

19. Xu, X.; Chen, C.; Fan, B.; Chen, Z. Physical and functional interactions between pathogen-induced Arabidopsis WRKY18, WRKY40, and WRKY60 transcription factors. *Plant Cell* **2006**, *18*, 1310–1326. [CrossRef]
20. Chen, H.; Lai, Z.; Shi, J.; Xiao, Y.; Chen, Z.; Xu, X. Roles of Arabidopsis WRKY18, WRKY40 and WRKY60 transcription factors in plant responses to abscisic acid and abiotic stress. *BMC Plant Biol.* **2010**, *10*, 281. [CrossRef]
21. Che, Y.; Sun, Y.; Lu, S.; Zhao, F.; Hou, L.; Liu, X. AtWRKY40 functions in drought stress response in Arabidopsis thaliana. *Plant Physiol. J* **2018**, *54*, 456–464. (In Chinese)
22. Jiang, Y.; Zheng, W.; Li, J.; Liu, P.; Zhong, K.; Jin, P.; Xu, M.; Yang, J.; Chen, J. NbWRKY40 positively regulates the response of Nicotiana benthamiana to tomato mosaic virus via salicylic acid signaling. *Front. Plant Sci.* **2021**, *11*, 603518. [CrossRef] [PubMed]
23. Karim, A.; Jiang, Y.; Guo, L.; Ling, Z.; Ye, S.; Duan, Y.; Li, C.; Luo, K. Isolation and characterization of a subgroup IIa WRKY transcription factor PtrWRKY40 from Populus trichocarpa. *Tree Physiol.* **2015**, *35*, 1129–1139. [CrossRef] [PubMed]
24. Lin, L.; Yuan, K.; Huang, Y.; Dong, H.; Qiao, Q.; Xing, C.; Huang, X.; Zhang, S. A WRKY transcription factor PbWRKY40 from Pyrus betulaefolia functions positively in salt tolerance and modulating organic acid accumulation by regulating PbVHA-B1 expression. *Environ. Exp. Bot.* **2022**, *196*, 104782. [CrossRef]
25. Zhang, L.; Xu, Z.; Ji, H.; Zhou, Y.; Yang, S. TaWRKY40 transcription factor positively regulate the expression of TaGAPC1 to enhance drought tolerance. *BMC Genom.* **2019**, *20*, 795. [CrossRef] [PubMed]
26. Dai, W.; Wang, M.; Gong, X.; Liu, J.H. The transcription factor Fc WRKY 40 of Fortunella crassifolia functions positively in salt tolerance through modulation of ion homeostasis and proline biosynthesis by directly regulating SOS 2 and P5 CS 1 homologs. *New Phytol.* **2018**, *219*, 972–989. [CrossRef]
27. Wang, C.-T.; Ru, J.-N.; Liu, Y.-W.; Yang, J.-F.; Li, M.; Xu, Z.-S.; Fu, J.-D. The maize WRKY transcription factor ZmWRKY40 confers drought resistance in transgenic Arabidopsis. *Int. J. Mol. Sci.* **2018**, *19*, 2580. [CrossRef]
28. Korte, N.; Porembski, S. A morpho-anatomical characterisation of Myrothamnus moschatus (Myrothamnaceae) under the aspect of desiccation tolerance. *Plant Biol.* **2012**, *14*, 537–541. [CrossRef]
29. Wagner, H.-J.; Schneider, H.; Mimietz, S.; Wistuba, N.; Rokitta, M.; Krohne, G.; Haase, A.; Zimmermann, U. Xylem conduits of a resurrection plant contain a unique lipid lining and refill following a distinct pattern after desiccation. *New Phytol.* **2000**, *148*, 239–255. [CrossRef]
30. Carlquist, S. Wood anatomy of Myrothamnus flabellifolia (Myrothamnaceae) and the problem of multiperforate perforation plates. *J. Arnold Arbor.* **1976**, *57*, 119–126. [CrossRef]
31. Vieweg, G.H.; Ziegler, H. Zur Physiologie von Myrothamnus flabellifolia 1. *Ber. Der Dtsch. Bot. Ges.* **1969**, *82*, 29–36.
32. Toldi, O.; Tuba, Z.; Scott, P. Vegetative desiccation tolerance: Is it a goldmine for bioengineering crops? *Plant Sci.* **2009**, *176*, 187–199. [CrossRef]
33. Scheibe, R.; Beck, E. Drought, desiccation, and oxidative stress. In *Plant Desiccation Tolerance*; Springer: Berlin/Heidelberg, Germany, 2011; pp. 209–231.
34. Zhu, J.; Liu, L.; Zhu, L. The Relationship Between Light-induced Changes in Transpiration Pull and Rehydration Times in Desiccated *Myrothamnus flabellifolia*. *Chin. Bull. Bot.* **2013**, *48*, 423.
35. Ma, C.; Wang, H.; Macnish, A.J.; Estrada-Melo, A.C.; Lin, J.; Chang, Y.; Reid, M.S.; Jiang, C.-Z. Transcriptomic analysis reveals numerous diverse protein kinases and transcription factors involved in desiccation tolerance in the resurrection plant *Myrothamnus flabellifolia*. *Hortic. Res.* **2015**, *10*, 5546. [CrossRef] [PubMed]
36. Bai, X.; Dai, L.; Sun, H.; Chen, M.; Sun, Y. Effects of moderate soil salinity on osmotic adjustment and energy strategy in soybean under drought stress. *Plant Physiol. Biochem.* **2019**, *139*, 307–313. [CrossRef] [PubMed]
37. Cui, X.; Yan, Q.; Gan, S.; Xue, D.; Wang, H.; Xing, H.; Zhao, J.; Guo, N. GmWRKY40, a member of the WRKY transcription factor genes identified from Glycine max L., enhanced the resistance to Phytophthora sojae. *BMC Plant Biol.* **2019**, *19*, 598. [CrossRef]
38. Liu, B.; Ouyang, Z.; Zhang, Y.; Li, X.; Hong, Y.; Huang, L.; Liu, S.; Zhang, H.; Li, D.; Song, F. Tomato NAC transcription factor SlSRN1 positively regulates defense response against biotic stress but negatively regulates abiotic stress response. *PLoS ONE* **2014**, *9*, e102067. [CrossRef]
39. Xie, Z.; Zhang, Z.L.; Zou, X.; Huang, J.; Ruas, P.; Thompson, D.; Shen, Q.J. Annotations and functional analyses of the rice WRKY gene superfamily reveal positive and negative regulators of abscisic acid signaling in aleurone cells. *Plant Physiol.* **2005**, *137*, 176–189. [CrossRef]
40. Bielach, A.; Hrtyan, M.; Tognetti, V.B. Plants under Stress: Involvement of Auxin and Cytokinin. *Int. J. Mol. Sci.* **2017**, *18*, 1427. [CrossRef]
41. de Zelicourt, A.; Synek, L.; Saad, M.M.; Alzubaidy, H.; Jalal, R.; Xie, Y.; Andres-Barrao, C.; Rolli, E.; Guerard, F.; Mariappan, K.G.; et al. Ethylene induced plant stress tolerance by Enterobacter sp. SA187 is mediated by 2-keto-4-methylthiobutyric acid production. *PLoS Genet.* **2018**, *14*, e1007273. [CrossRef]
42. Nadarajah, K.; Kumar, I.S. Drought Response in Rice: The miRNA Story. *Int. J. Mol. Sci.* **2019**, *20*, 3766. [CrossRef] [PubMed]
43. Kim, Y.H.; Hwang, S.J.; Waqas, M.; Khan, A.L.; Lee, J.H.; Lee, J.D.; Nguyen, H.T.; Lee, I.J. Comparative analysis of endogenous hormones level in two soybean (*Glycine max* L.) lines differing in waterlogging tolerance. *Front. Plant Sci.* **2015**, *6*, 714. [CrossRef] [PubMed]
44. Kim, Y.; Seo, C.W.; Khan, A.L.; Mun, B.G.; Shahzad, R.; Ko, J.W.; Yun, B.W.; Park, S.K.; Lee, I.J. Exo-ethylene application mitigates waterlogging stress in soybean (*Glycine max* L.). *BMC Plant Biol.* **2018**, *18*, 254. [CrossRef]

45. Koevoets, I.T.; Venema, J.H.; Elzenga, J.T.; Testerink, C. Roots Withstanding their Environment: Exploiting Root System Architecture Responses to Abiotic Stress to Improve Crop Tolerance. *Front. Plant Sci.* **2016**, *7*, 1335. [CrossRef] [PubMed]
46. Zhu, Y.X.; Gong, H.J.; Yin, J.L. Role of Silicon in Mediating Salt Tolerance in Plants: A Review. *Plants* **2019**, *8*, 147. [CrossRef] [PubMed]
47. Kaashyap, M.; Ford, R.; Kudapa, H.; Jain, M.; Edwards, D.; Varshney, R.; Mantri, N. Differential Regulation of Genes Involved in Root Morphogenesis and Cell Wall Modification is Associated with Salinity Tolerance in Chickpea. *Sci. Rep.* **2018**, *8*, 4855. [CrossRef]
48. Ullah, A.; Sun, H.; Hakim; Yang, X.; Zhang, X. A novel cotton WRKY gene, GhWRKY6-like, improves salt tolerance by activating the ABA signaling pathway and scavenging of reactive oxygen species. *Physiol. Plant.* **2018**, *162*, 439–454. [CrossRef]
49. Martinez, J.P.; Lutts, S.; Schanck, A.; Bajji, M.; Kinet, J.M. Is osmotic adjustment required for water stress resistance in the Mediterranean shrub Atriplex halimus L? *J. Plant Physiol.* **2004**, *161*, 1041–1051. [CrossRef]
50. Moustakas, M.; Sperdouli, I.; Kouna, T.; Antonopoulou, C.I.; Therios, I. Exogenous proline induces soluble sugar accumulation and alleviates drought stress effects on photosystem II functioning of Arabidopsis thaliana leaves. *Plant Growth Regul.* **2011**, *65*, 315–325. [CrossRef]
51. Ali, Q.; Anwar, F.; Ashraf, M.; Saari, N.; Perveen, R. Ameliorating effects of exogenously applied proline on seed composition, seed oil quality and oil antioxidant activity of maize (*Zea mays* L.) under drought stress. *Int. J. Mol. Sci.* **2013**, *14*, 818–835. [CrossRef]
52. Jaspers, P.; Kangasjarvi, J. Reactive oxygen species in abiotic stress signaling. *Physiol. Plant.* **2010**, *138*, 405–413. [CrossRef] [PubMed]
53. Clough, S.J.; Bent, A.F. Floral dip: A simplified method for Agrobacterium-mediated transformation of Arabidopsis thaliana. *Plant J.* **1998**, *16*, 735–743. [CrossRef] [PubMed]
54. Palta, J.P. Leaf chlorophyll content. *Remote Sens. Rev.* **1990**, *5*, 207–213. [CrossRef]
55. Bates, L.S.; Waldren, R.P.; Teare, I.D. Rapid determination of free proline for water-stress studies. *Plant Soil* **1973**, *39*, 205–207. [CrossRef]
56. Fryer, M.J.; Oxborough, K.; Mullineaux, P.M.; Baker, N.R. Imaging of photo-oxidative stress responses in leaves. *J. Exp. Bot.* **2002**, *53*, 1249–1254.
57. Giannopolitis, C.N.; Ries, S.K. Superoxide Dismutases. *Plant Physiol.* **1977**, *59*, 309–314. [CrossRef]
58. Zheng, X.; Tian, S.; Meng, X.; Li, B. Physiological and biochemical responses in peach fruit to oxalic acid treatment during storage at room temperature. *Food Chem.* **2007**, *104*, 156–162. [CrossRef]
59. Du, Z.; Bramlage, W.J. Modified thiobarbituric acid assay for measuring lipid oxidation in sugar-rich plant tissue extracts. *J. Agric. Food Chem.* **1992**, *40*, 1566–1570. [CrossRef]

International Journal of
Molecular Sciences

MDPI

Article

Effect of Overexpression of *γ-Tocopherol Methyltransferase* on α-Tocopherol and Fatty Acid Accumulation and Tolerance to Salt Stress during Seed Germination in *Brassica napus* L.

Yuan Guo †, Dong Li †, Tiantian Liu, Meifang Liao, Yuxin Li, Weitang Zhang, Zijin Liu and Mingxun Chen *

State Key Laboratory of Crop Stress Biology for Arid Areas, National Yangling Agricultural Biotechnology & Breeding Center, Shaanxi Key Laboratory of Crop Heterosis and College of Agronomy, Northwest A&F University, Yangling 712100, China
* Correspondence: cmx786@nwafu.edu.cn
† These authors contributed equally to this work.

Abstract: Rapeseed (*Brassica napus* L.) is an important oil crop and a major source of tocopherols, also known as vitamin E, in human nutrition. Enhancing the quality and composition of fatty acids (FAs) and tocopherols in seeds has long been a target for rapeseed breeding. The gene *γ-Tocopherol methyltransferase* (*γ-TMT*) encodes an enzyme catalysing the conversion of γ-tocopherol to α-tocopherol, which has the highest biological activity. However, the genetic basis of *γ-TMT* in *B. napus* seeds remains unclear. In the present study, *BnaC02.TMT.a*, one paralogue of *Brassica napus γ-TMT*, was isolated from the *B. napus* cultivar "Zhongshuang11" by nested PCR, and two homozygous transgenic overexpression lines were further characterised. Our results demonstrated that the overexpression of *BnaC02.TMT.a* mediated an increase in the α- and total tocopherol content in transgenic *B. napus* seeds. Interestingly, the FA composition was also altered in the transgenic plants; a reduction in the levels of oleic acid and an increase in the levels of linoleic acid and linolenic acid were observed. Consistently, *BnaC02.TMT.a* promoted the expression of *BnFAD2* and *BnFAD3*, which are involved in the biosynthesis of polyunsaturated fatty acids during seed development. In addition, *BnaC02.TMT.a* enhanced the tolerance to salt stress by scavenging reactive oxygen species (ROS) during seed germination in *B. napus*. Our results suggest that *BnaC02.TMT.a* could affect the tocopherol content and FA composition and play a positive role in regulating the rapeseed response to salt stress by modulating the ROS scavenging system. This study broadens our understanding of the function of the *Bnγ-TMT* gene and provides a novel strategy for genetic engineering in rapeseed breeding.

Keywords: *γ-TMT*; α-tocopherol; fatty acids; seed germination; salt stress

Citation: Guo, Y.; Li, D.; Liu, T.; Liao, M.; Li, Y.; Zhang, W.; Liu, Z.; Chen, M. Effect of Overexpression of *γ-Tocopherol Methyltransferase* on α-Tocopherol and Fatty Acid Accumulation and Tolerance to Salt Stress during Seed Germination in *Brassica napus* L.. *Int. J. Mol. Sci.* **2022**, 23, 15933. https://doi.org/10.3390/ijms232415933

Academic Editors: Andrés J. Cortés, Hai Du and Juan Manuel Ruiz Lozano

Received: 28 September 2022
Accepted: 12 December 2022
Published: 14 December 2022

1. Introduction

Rapeseed (*Brassica napus* L., AACC, 2n = 38) is one of the most important sources of edible oils worldwide. It is an allopolyploid species originating from the natural hybridisation between two ancestorial species, namely, *B. rapa* (AA, 2n = 20) and *B. oleracea* (CC, 2n = 18) [1]. The nutritional characteristics of rapeseed oil are mainly determined by its fatty acid (FA) composition and vitamin E content. The major FAs present in *B. napus* seeds (canola type) are oleic acid (C18:1), linoleic acid (C18:2), and linolenic acid (C18:3), with concentrations of 61%, 21%, and 11%, respectively [2]. Among these, C18:2 and C18:3 are polyunsaturated FAs (PUFAs) that cannot be biosynthesised in the human body. These PUFAs play an important role in the protection of eyesight and the prevention of various diseases, such as cancer and cardiovascular and inflammatory diseases [3–5]. Vitamin E, which is present in cold-pressed oil, is an essential dietary nutritional element for all mammals, and its deficiency leads to neurological disorders, ataxia, and even death [6,7]. Increasing the quantities of FAs and vitamin E has long garnered interest in rapeseed breeding [8]. Therefore, the characterisation of key genes regulating the biosynthesis of FAs and/or vitamin E in *B. napus* will be helpful in accurately designing new germplasms or varieties of crops with high PUFA and vitamin E contents using molecular biology techniques, thereby improving human health.

Vitamin E is an essential lipid-soluble antioxidant comprising tocopherols and tocotrienols. Tocopherols are mostly present in the seeds of common oil crops and the green-coloured components of higher-order plants, whereas tocotrienols are mainly found in cereal kernels and certain tropical fruits [9]. According to the number and position of the methyl substituents on the aromatic ring, tocopherols are classified into four forms: α, β, γ, and δ [10]. Among them, α-tocopherol is beneficial to human health, as it has the highest efficacy [11].

In plants, tocopherol biosynthesis occurs in plastids through a combination of two main pathways (Figure 1). The shikimate pathway leads to the synthesis of an aromatic ring from homogentisic acid (HGA) and *p*-hydroxyphenylpyruvic acid (HPP) by the enzyme HPP dioxygenase (PDS1/HPPD) in the cytoplasm [12,13]. The methylerythritol phosphate (MEP) pathway leads to the generation of a polyprenyl side chain from phytyl diphosphate (PDP) in plastids [14,15]. An additional pathway for PDP production was discovered from the phytol recycling pathway [16,17]. In tocopherol biosynthesis, the two substrates, HGA and PDP, are fused together to produce the first intermediate, 2-methyl-6-phytyl-1,4-benzoquinol (MBPQ). This reaction is mediated by the enzyme HGA phytyl transferase (VTE2/HPT). MPBQ is a substrate of either tocopherol cyclase (VTE1/TC) or MPBQ methyltransferase (VTE3/MPBQ-MT). MPBQ-MT catalyses the formation of 2,3-dimethyl-6-phytyl-1,4-benzoquinol (DMPBQ) from MPBQ. TC converts MPBQ and DMPBQ to δ- and γ-tocopherols, respectively. Finally, the last key enzyme, γ-tocopherol methyltransferase (VTE4/γ-TMT), catalyses the conversion of γ- to α-tocopherol and δ- to β-tocopherol, thereby determining the type of tocopherol [18,19]. The role of the *γ-TMT* gene has been functionally characterised in *Arabidopsis thaliana*. The loss of *A. thaliana Atγ-TMT* function results in reduced levels of α-tocopherol; however, an increased γ-tocopherol content in leaves is observed, and its overexpression leads to the approximate complete conversion of γ- to α-tocopherol and an increase in vitamin E activity in *A. thaliana* seeds [20,21]. Many useful crops with a high α-tocopherol content can be developed by introducing this gene. Transgenic soybean (*Glycine max* L.) [22], lettuce (*Latuca sativa* L.) [23,24], rice (*Oryza sativa* L.) [25], and perilla (*Perilla frutescens* L.) [26] with high levels of α-tocopherol have been developed by overexpressing the *γ-TMT* gene from the model plant *A. thaliana* (*Atγ-TMT*). Recently, transgenic soybean [27] and rapeseed (*B. juncea* L. AABB, 2n = 36) [28] with increased α-tocopherol content were developed by expressing the perilla *γ-TMT* gene. Generally, tocopherols in most common oilseeds contain minor amounts of α-tocopherol and relatively higher levels of γ-tocopherol [20,29]. Therefore, improving the γ- to α-tocopherol conversion by elevating the expression level of the *γ-TMT* gene in *B. napus* seeds would be a promising strategy.

The major tocopherol functions are related to antioxidant properties because of their ability to interact with polyunsaturated acyl groups and protect membrane lipids (especially PUFAs) from oxidative damage by scavenging lipid peroxy radicals and quenching reactive oxygen species (ROS) in photosystem II. Furthermore, they also prevent damage due to the fact of oxidation during membrane lipid peroxidation [10,30]. Tocopherols donate a hydrogen atom to lipid-free radicals, thereby neutralising the radical and terminating the autocatalytic chain reaction of lipid peroxidation and protecting cell membranes [31,32]. In plants, both tocopherol and FA biosynthesis occur in the plastid, and the tocopherol functions as an antioxidant for stabilising PUFAs against lipid oxidation in vegetable oils [32]. The balance between vitamin E and PUFA content mainly determines the susceptibility to lipid peroxidation and storage stability of rapeseed oil. However, there is a paucity of studies examining the correlation between tocopherol and FA components in rapeseed oils. A positive correlation has been demonstrated between linoleic acid (18:2) and α-tocopherol, linolenic acid (18:3) and γ-tocopherol from the vegetable oils of 14 different plant species [33]. In addition, upon studying a collection of maize hybrids, a relationship between FAs, particularly PUFAs, and tocochromanol content was observed [34]. Furthermore, the content of tocopherol and FA components in seeds from three *Brassica* oil crop species was analysed, and a significantly positive correlation was observed between α-tocopherol and the sum of C18:1 and C18:2, whereas no association was observed between the γ-tocopherol and FA components [35].

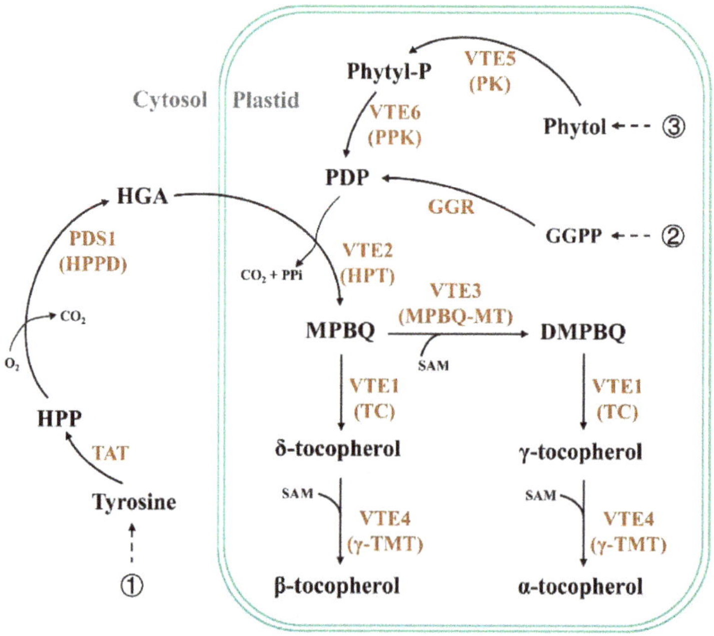

Figure 1. Simplified schematic diagram showing the tocopherols biosynthetic pathway in plants: ① shikimic pathway; ② methylerythritol phosphate pathway; ③ phytol recycling pathway. TAT: tyrosine aminotransferase; HPP: *p*-hydroxyphenylpyruvic acid; PDS1 (HPPD): HPP dioxygenase; HGA: homogentisic acid; VTE5 (PK): phytol kinase; Phytyl-P: phytyl phosphate; VTE6 (PPK): phytyl phosphate kinase; GGPP: geranylgeranyl diphosphate; GGR: geranylgeranyl reductase; PDP: phytyl diphosphate; VTE2 (HPT): HGA phytyl transferase; MPBQ: 2-methyl-6-phytyl-1,4-benzoquinol; VTE3 (MPBQ-MT): MPBQ methyltransferase; SAM: S-adenosylmethionine; DMBPQ: 2,3-dimethyl-6-phytyl-1,4-benzoquinol; VTE1 (TC): tocopherol cyclase; VTE4 (γ-TMT): γ-tocopherol methyltransferase.

In addition to their antioxidant function, recent studies have suggested additional roles for tocopherols in diverse biological processes, including germination [36], cell signalling [37], and biotic [38] and abiotic stresses, such as those related to salt [39–44], low temperature [45,46], drought [41,47], and heavy metals [40,43,48,49]. The RNAi-mediated silencing of *γ-TMT* leads to a deficiency in α-tocopherol and elevated susceptibility to salt stress, whereas the overexpression of *γ-TMT* increases the salt and heavy metal tolerance in transgenic tobacco [39,43]. In addition, the increased sensitivity to salt stress in vitamin E-deficient mutants of *A. thaliana* has been observed. Tocopherols play a key role in salt stress tolerance not only by reducing the extent of oxidative stress but also by improving ion homeostasis and the hormonal balance of leaves [42]. Furthermore, the overexpression of the *γ-TMT* gene in transgenic *B. juncea* plants enhances the tolerance to abiotic stresses, such as salt, heavy metals, and osmoticums [40].

In the present study, *BnaC02.TMT.a*, an *Atγ-TMT* orthologue in *B. napus*, was isolated using nested PCR and functionally characterised. To explore the biological functions of *BnaC02.TMT.a* in rapeseed, we measured the contents of tocopherols and FAs and conducted a germination assay on sodium chloride (NaCl) stress medium in transgenic overexpression plants. We showed that the overexpression of *BnaC02.TMT.a* increased the α- and total tocopherol content, altered the proportion of FA composition, and promoted the germination rate after NaCl treatment. Under conditions involving salt-related stress, the accumulation of ROS is an important factor affecting the inhibition of seed germination, and hydrogen peroxide (H_2O_2) and superoxide anion (O_2^-) are the main components of ROS [50,51]. To obtain a deeper insight into the mechanism underlying increased salt tolerance in transgenic plants, we analysed ROS accumulation, which was estimated using diaminobenzidine (DAB) and nitroblue tetrazolium (NBT) staining and measured the H_2O_2 and O_2^- content. Our results suggest that *BnaC02.TMT.a* prevents oxidative damage by ROS scavenging during germination in *B. napus* seeds under conditions involving salt stress. This study revealed that the overexpression of *BnaC02.TMT.a* could affect tocopherols, FAs, and the response to salt stress, which will be useful for engineering rapeseed with improved vitamin E content, FA composition, and salt tolerance.

2. Results

2.1. Sequence Analysis of Bnγ-TMT Paralogs

Upon performing the protein BLAST using *A. thaliana* Atγ-TMT in *Brassica napus* pan-genome information resource (BnPIR), four Bnγ-TMT paralogs were predicted in the genome of the *B. napus* cultivar "Zhongshuang11 (ZS11)": BnaC02T0331100ZS, BnaC02T0197500ZS, BnaA02T0247300ZS, and BnaA02T0154300ZS, which were located on the chromosomes C02 and A02 and designated *BnaC02.TMT.a*, *BnaC02.TMT.b*, *BnaA02.TMT.a*, and *BnaA02.TMT.b*, respectively. The alignment of the amino acid sequences of the four Bnγ-TMT paralogs revealed high identities ranging from 96.5% to 98.8% (Figure S1). The identity of the protein sequence between the four Bnγ-TMT proteins and Atγ-TMT was at least 88.4%, whereas BnaC02.TMT.a showed a higher identity with Atγ-TMT, reaching 89.3% (Figure S1). Moreover, all Bnγ-TMT paralogs were predicted to possess four highly conserved S-adenosylmethionine (SAM)-binding domains, similar to Atγ-TMT (Figure 2). This implies that all four paralogs of Bnγ-TMT may be functionally conservative. Phylogenetic analysis was performed to investigate the evolutionary relationships between the Bnγ-TMT and 11 γ-TMT proteins in *A. thaliana* and six other oil crops. As shown in Figure 3, BnaC02.TMT.b was most closely related to Atγ-TMT, whereas BnaC02.TMT.a was closely related to BnaA02.TMT.a. Compared with γ-TMT in other species, all four paralogs were closely related to Atγ-TMT.

2.2. Analysis of the Expression Pattern of Bnγ-TMT

To verify the function of *Bnγ-TMT*, quantitative reverse transcription PCR (RT-qPCR) was used to explore the spatiotemporal expression pattern of all *Bnγ-TMT* paralogs in various tissues of the *B. napus* cultivar "ZS11". As illustrated in Figure 4, *Bnγ-TMT* was widely expressed in various tissues, with higher expression levels in the flowers, leaves, and developing seeds (35 days after pollination (DAP)) than in the roots and stems. Notably, during seed development, the *Bnγ-TMT* expression remained relatively low at the embryogenesis stage from 10 to 15 DAP, decreased slightly from 20 to 26 DAP, and progressively increased at 29 DAP and 32 DAP, reaching a maximum at 35 DAP (Figure 4).

2.3. BnaC02.TMT.a Increased α-Tocopherol Content in B. napus Seeds

To elucidate the role of *Bnγ-TMT*, we introduced the overexpression construct *35S:Bna C02.TMT.a-6 haemagglutinin* (*6HA*) into the *B. napus* cultivar "ZS11" plants (Figure 5A). Two representative independent T_2 homozygous lines (*OE#3* and *OE#5*) were confirmed by the PCR amplification of *BnaC02.TMT.a* using the specific primers 35S_P/*BnTMT*_R (Figure 5B). Furthermore, the expression level of the *Bnγ-TMT* gene was confirmed in these transgenic plants by RT-qPCR. The expression of *Bnγ-TMT* was significantly higher in these transgenic plants than in wild-type ZS11 (Figure 5C).

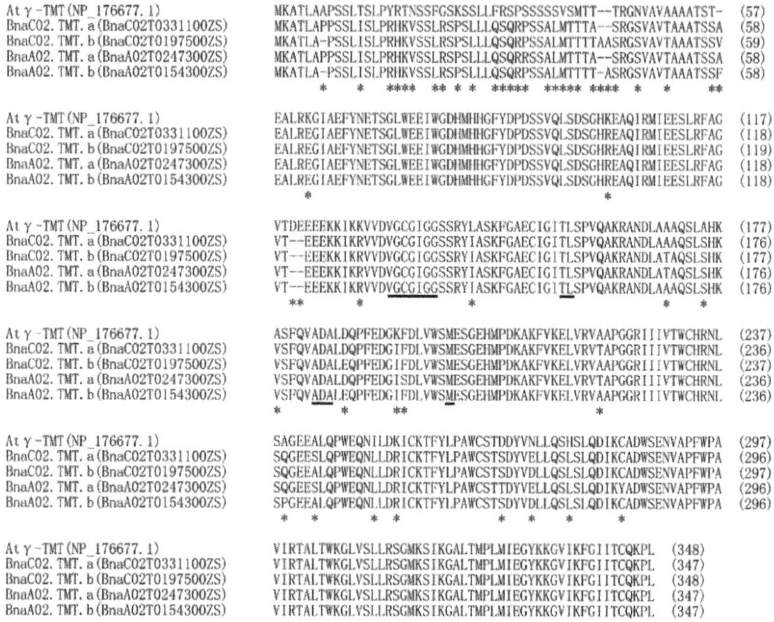

Figure 2. Alignment of amino acid sequences of γ-TMT derived from *Arabidopsis thaliana* and *Brassica napus*. The alignment was conducted using the MUSCLE program (http://www.ebi.ac.uk/Tools/msa/muscle/, accessed on 8 September 2022), and the different amino acids are indicated by asterisks. S-adenosylmethionine binding sites are underlined.

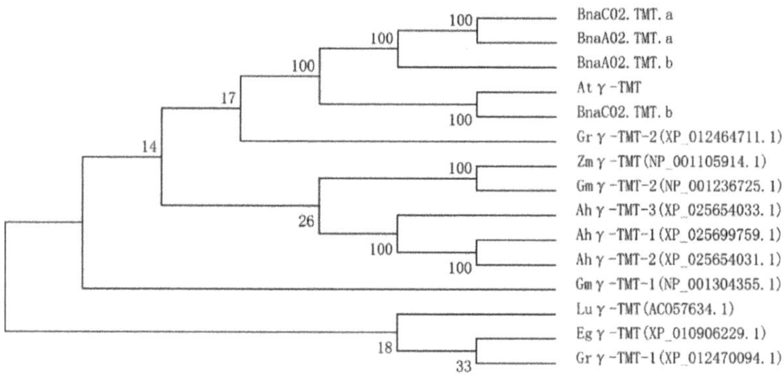

Figure 3. Phylogenetic tree analysis of γ-TMT proteins derived from *A. thaliana* and seven oil crops. A neighbour-joining tree (Jones–Taylor–Thornton model) was generated by MEGA7. A bootstrap analysis with 1000 replicates was performed to assess the statistical reliability of the tree topology. The accession numbers corresponding to the species names are listed as follows: *A. thaliana*, Atγ-TMT (NP_176677.1); *B. napus*, BnaC02.TMT.a (BnaC02T0331100ZS), BnaC02.TMT.b (BnaC02T0197500ZS), BnaA02.TMT.a (BnaA02T0247300ZS), and BnaA02.TMT.b (BnaA02T0154300ZS); *Zea mays*, Zmγ-TMT (NP_001105914.1); *Elaeis guineensis*, Egγ-TMT (XP_010906229.1); *Linum usitatissimum*, Luγ-TMT (ACO57634.1); *Glycine max*, Gmγ-TMT-1 (NP_001304355.1) and Gmγ-TMT-2 (NP_001236725.1); *Gossypium raimondii*, Grγ-TMT-1 (XP_012470094.1) and Grγ-TMT-2 (XP_012464711.1); *Arachis hypogaea*, Ahγ-TMT-1 (XP_025699759.1), Ahγ-TMT-2 (XP_025654031.1), and Ahγ-TMT-3 (XP_025654033.1).

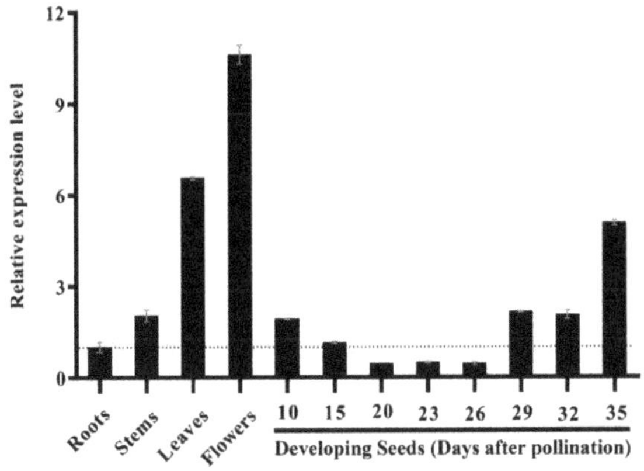

Figure 4. RT-qPCR analysis of the *Bnγ-TMT* expression in various tissues and developing seeds at different developmental stages in the *B. napus* cultivar "ZS11". The RT-qPCR results were normalised against *BnGAPDH* expression as an internal control, and the expression level of *Bnγ-TMT* in roots was set to 1. The values indicate the means of three replicates of dilutions of cDNA obtained from three independent RNA extractions. The error bars denote standard deviations.

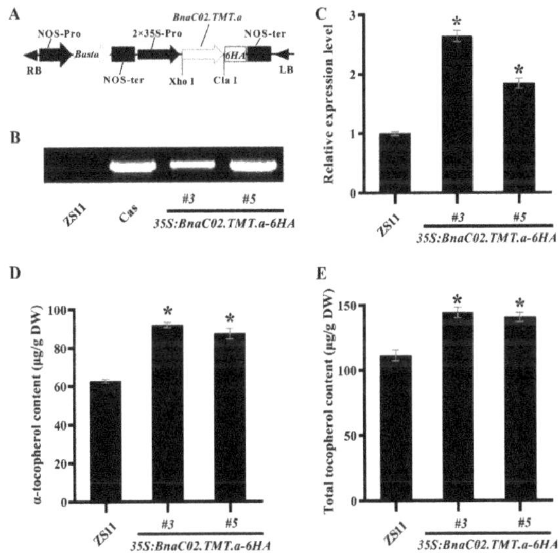

Figure 5. Characterisation of *35S:BnaC02.TMT.a-6HA* lines. (**A**) Schematic diagram of the constitutive expression cassette of the *35S:BnaC02.TMT.a-6HA* gene in the binary vector pGreen 2 × 35S used for plant transformation. RB, right border; LB, left border; NOS-pro, nopaline synthase promoter; NOS-ter, nopaline synthase terminator; Basta, glyphosate; 35S-pro, CaMV 35S promoter. (**B**) PCR-based genotyping of the *35S:BnaC02.TMT.a-6HA* transgenic plants using specific primers for the 35S_P/*BnTMT*_R. Cas, cassette. (**C**) Expression analysis of *Bnγ-TMT* in the *35S:BnaC02.TMT.a-6HA* transgenic plants using RT-qPCR. The RT-qPCR results were normalised against the expression of *BnGAPDH*, which was used as an internal control. The values are the means ± SD (n = 3). (**D**,**E**) The levels of α-tocopherol and total tocopherol in the mature seeds of the transgenic lines and control plants (ZS11). The asterisks denote statistically significant differences between the wild-type ZS11 and the *35S:BnaC02.TMT.a-6HA* transgenic plants (Student's *t*-test, $p \leq 0.05$). The error bars denote the standard deviation.

To investigate the function of *BnaC02.TMT.a* in mediating tocopherol accumulation, we measured the levels of total tocopherol and α-, β-, γ-, and δ-tocopherol in the mature seeds of wild-type ZS11 and the transgenic lines *OE#3* and *OE#5*. The levels of α-tocopherol and total tocopherol were much higher in both of the transgenic lines than those in the wild-type seeds (Figure 5D,E), whereas no apparent differences were observed in the levels of γ- and δ-tocopherol between the wild-type and transgenic plants (Figure S2). In addition, no β-tocopherol was detected in either the control or transgenic plants.

2.4. *BnaC02.TMT.a Promoted PUFA Biosynthesis in B. napus Seeds*

To elucidate the biological functions of *BnaC02.TMT.a* in seed FAs, we measured the quantities of the major FA compositions and total FAs of mature seeds between the wild-type ZS11 plants and the *BnaC02.TMT.a* homozygous transgenic lines *OE#3* and *OE#5*. In the mature seeds of the transgenic lines, a significant decrease was observed in the C18:1 (oleic acid) content along with a significant increase in the amount of C18:2 (linoleic acid) and C18:3 (linolenic acid) (Figure 6A), suggesting that α-tocopherol might affect the conversion of oleic acid to linoleic acid and linoleic acid to linolenic acid. However, no significant differences were observed with respect to the total FA content (Figure S3D) and several seed morphological traits, such as the seed coat colour (Figure S3A), seed size (Figure S3B), and dry weight (Figure S3C), between the mature seeds of the wild-type and transgenic plants.

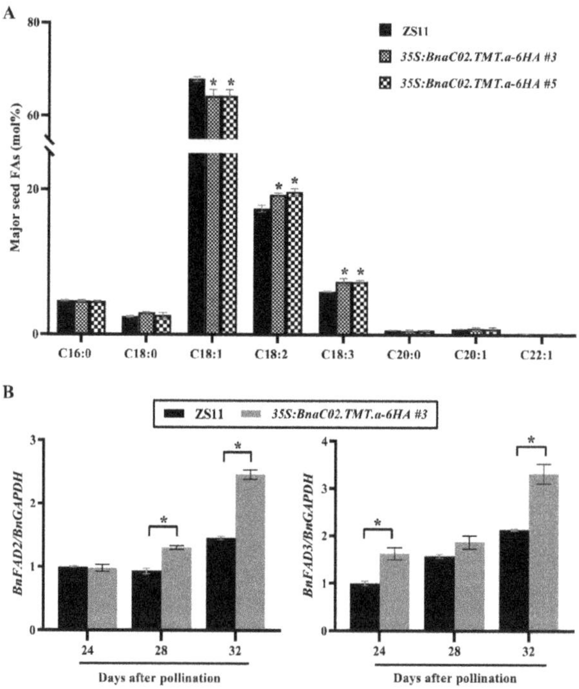

Figure 6. Effect of *BnaC02.TMT.a* on FA metabolism in seeds. (**A**) Comparison of the FA compositions of major seeds between the wild-type ZS11 and *35S:BnaC02.TMT.a-6HA* plants. The asterisks indicate significant differences in the total FA levels in the seeds (two-tailed paired Student's *t*-test, $p \leq 0.05$). The error bars denote the standard deviation. (**B**) Comparison of the expression of genes contributing to the FA modification in the developing seeds of the ZS11 and *BnaC02.TMT.a-6HA* plants. The asterisks indicate significant differences between the wild-type ZS11 and *35S:BnaC02.TMT.a-6HA* plants (two-tailed paired Student's *t*-test, $p \leq 0.05$). The error bars denote the standard deviation.

Three critical stages of seed oil accumulation (24, 28, and 32 DAP) were carefully selected to compare the expression profiles of *BnFAD2* and *BnFAD3* between the transgenic line *OE#3* and the wild-type control. As expected, the expression of *BnFAD2* was significantly higher at the seed maturation stage from 28 and 32 DAP in the transgenic seeds than in the wild-type seeds (Figure 6B), and the expression of *BnFAD3* was consistently upregulated in the developing seeds of the *OE#3* transgenic plants at 24 and 32 DAP (Figure 6B).

2.5. Functional Characterisation of BnaC02.TMT.a in the Seed Germination under Conditions Involving Salt Stress in B. napus

The effects of *BnaC02.TMT.a* on seed germination under conditions involving salt stress in the wild-type and transgenic lines *OE#3* and *OE#5* were examined in sterile redistilled water with or without 180 mM NaCl. As illustrated in Figures 7 and 8, *OE#3*, *OE#5*, and the control ZS11 showed similar seed germination rates in the medium without NaCl treatment. Notably, under stress conditions involving 180 mM NaCl, the differences in germination between the wild-type and transgenic seeds were apparent (Figures 7 and 8). The seed germination was completely inhibited during the first 40 h after sowing (HAS) and from 40 to 192 HAS, which was significantly higher in the transgenic line than in the controls (Figure 8).

Figure 7. Phenotypes of the seed germination at 96 h after sowing (HAS) in the wild-type ZS11 and *35S:BnaC02.TMT.a-6HA* plants under control conditions and after treatment with 180 mM NaCl. The emergence of a visible radicle was used as a morphological marker for seed germination. Bar = 1 cm.

Under control conditions, the ROS levels were similar between the wild-type and transgenic lines *OE#3* and *OE#5*. In contrast, upon treatment with 180 mM NaCl, a lower DAB and NBT staining intensity, especially in cotyledons, was observed with the transgenic plants, suggesting that the ROS levels were significantly suppressed in the transgenic lines *OE#3* and *OE#5* compared to those in the wild-type ZS11 (Figure 9A). Paralleling the observations described in Figure 9A, under control conditions, no significant differences were observed in the H_2O_2 and O_2^- contents between the wild-type and transgenic plants. Compared to the controls, the H_2O_2 and O_2^- contents were substantially decreased in both of the transgenic plants exposed to 180 mM NaCl (Figure 9B). Therefore, our results revealed that the overexpression of *BnaC02.TMT.a* alleviated the oxidative damage to the seeds under conditions involving salt stress by ROS scavenging, thereby promoting seed germination under such conditions.

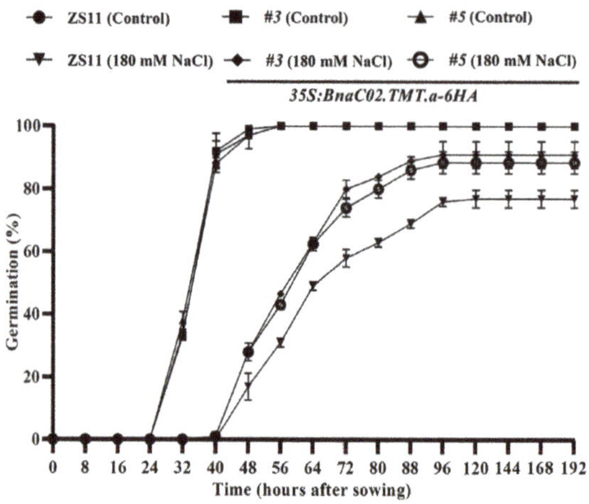

Figure 8. Time course analysis of seed germination in the wild-type ZS11 and *35S:BnaC02.TMT.a-6HA* plants under control conditions and after treatment with 180 mM NaCl. Seed germination was scored every eight hours, and the emergence of a visible radicle was used as a morphological marker. The data are expressed as the mean ± SD from three independent experiments evaluating 50 individual seeds.

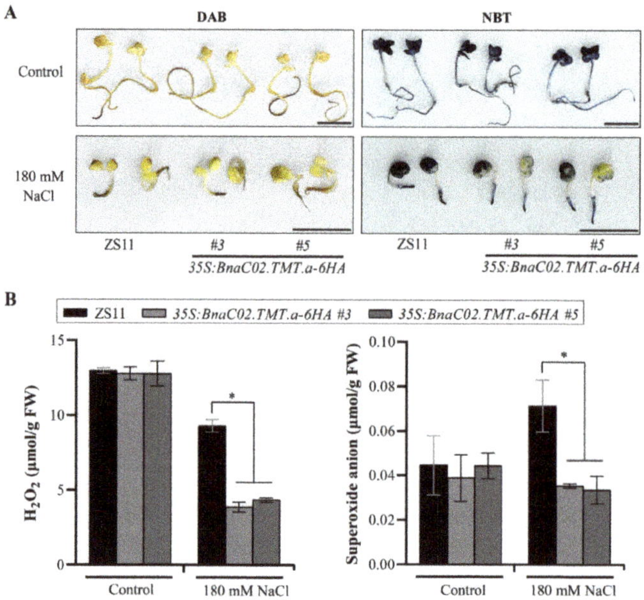

Figure 9. Effect of *BnaC02.TMT.a* on ROS accumulation. (**A**) diaminobenzidine (DAB, Left) and nitroblue tetrazolium (NBT, right) staining of ROS in the wild-type ZS11 and *35S:BnaC02.TMT.a-6HA* plants treated with 180 mM NaCl at 96 h after sowing (HAS). Bar = 1 cm; (**B**) measurements of H_2O_2 (Left) and O_2^- (right) in the wild-type ZS11 and *35S:BnaC02.TMT.a-6HA* plants in response to treatment with 180 mM NaCl at 96 HAS. The asterisks indicate significant differences between the wild-type ZS11 and *35S:BnaC02.TMT.a-6HA* plants under control conditions or after 180 mM NaCl treatment, respectively (two-tailed paired Student's *t*-test, $p \leq 0.05$). The error bars denote the standard deviation. FW, fresh weight.

Our results suggest that *Bnγ-TMT* plays a positive role in regulating the response of *B. napus* to salt stress, which might involve complex mechanisms, such as the ROS scavenging system.

3. Discussion

Rapeseed is one of the most important oil supply crops and contains a significant amount of tocopherols, which are beneficial for human health. Enhancing the quality and composition of FAs and tocopherols in seeds has long been a target for rapeseed breeding. In this study, we showed that *BnaC02.TMT.a* promoted the accumulation of α- and total tocopherol, altered the FA composition in seeds, and enhanced the tolerance to salt stress during seed germination in *B. napus*.

B. napus is an allotetraploid species formed by the natural hybridisation between *B. rapa* and *B. oleracea* [1]. A large number of chromosome duplications, rearrangements, and deletions occurred during the evolutionary processes, resulting in an average of 2–8 paralogs of each *A. thaliana* locus in the *B. napus* genome [52]. Consistently, four paralogs to Atγ-TMT were found in the *B. napus* cultivar "ZS11" genome (Figure 2). It was shown the overexpression of *Atγ-TMT* increased the α-tocopherol concentration and vitamin E activity with unaltered total tocopherol quantity in seeds, and the *Atγ-TMT* mutation exhibited an absent α-tocopherol but high levels of γ-tocopherol accumulation in the leaves of *A. thaliana* [20,21]. We found that the overexpression of *BnaC02.TMT.a* promoted α-tocopherol accumulation in *B. napus* seeds (Figure 5D). BnaC02.TMT.a shared a high amino acid identity and a close evolutionary relationship with Atγ-TMT (Figures 2 and 3). In addition, four SAM binding domains involved in methyl transfer reactions [53] were highly conserved between BnaC02.TMT.a and Atγ-TMT (Figure 2). These results suggest that BnaC02.TMT.a may exhibit a conserved role in regulating the accumulation of α-tocopherol in *A. thaliana*. In rapeseed oil, the major tocopherol form is γ-tocopherol (65%), followed by α-tocopherol (35%). Furthermore, a small proportion of δ-tocopherol (<1%), and β-tocopherol is absent [54]. *BnaA.VTE4.a1* (one *γ-TMT* paralogue from *B. napus* variety "Express") is able to increase the α-tocopherol content because of the shift from γ- to α-tocopherol, and the total tocopherol level is not altered in transgenic *A. thaliana* seeds [55]. Interestingly, we found that the overexpression of *BnaC02.TMT*.a resulted in higher levels of α-tocopherol (Figure 5D); however, no obvious changes were observed in γ-tocopherol content in *B. napus* seeds (Figure S2A), which may explain the higher levels of total tocopherol (Figure 5E). The distinct roles of *BnaA.VTE4.a1* and *BnaC02.TMT.a* in γ- and total tocopherol production might be attributed to their amino acid differences and functional divergence (Figure S4). Notably, the functional characterisation of *BnaA.VTE4.a1* was performed in a heterologous transgenic system, which may not truly reflect its function in *B. napus*. A possible reason for the unaltered γ-tocopherol content is that γ-tocopherol could be replenished by other mechanisms in the *B. napus* transgenic lines.

The levels of α-tocopherol are positively correlated with linoleic acid and PUFAs in vegetable oils [33,34]. Upon examining a collection of *B. napus* lines with low erucic acid, a significant positive correlation was found between the α-/γ-tocopherol ratio and C18:2 and PUFAs [35]. Accompanied by an increased α-tocopherol level, the FA composition was altered; that is, decreased levels of oleic acid (C18:1) and increased levels of linoleic acid (C18:2) and linolenic acid (C18:3) were observed in *35S:BnaC02.TMT.a-6HA* plants (Figure 6A). *FAD2* [56,57] and *FAD3* [58,59] are essential for the formation of C18:2 and C18:3, respectively, and higher transcript levels of *BnFAD2* and *BnFAD3* (Figure 6B) may be helpful for the accumulation of C18:2 and C18:3 in *35S:BnaC02.TMT.a-6HA* seeds. As previously reported, the association between tocopherol and FAs in oils is mostly attributed to the biochemical function of tocopherol in protecting against lipid oxidation, rendering it crucial for PUFA stability [33,34]. Furthermore, biochemical and genetic evidence indicates that tocopherols have a more limited role in photoprotection than previously assumed [45,46]. In the absence of tocopherols, the conversion of 18:2 to 18:3 in endoplasmic reticulum (ER) lipids is reduced under low temperatures, suggesting some plausible scenarios for the direct or indirect interaction of tocopherols during ER PUFA metabolism. FAD2 and FAD3 are ER-localised FA desaturases. Together with the findings indicating that *BnaC02.TMT.a* enhanced the expression of *FAD2* and *FAD3* in *35S:BnaC02.TMT.a-6HA* developing seeds (Figure 6B), we hypothesised that in addition to their antioxidant activity, the function of α-tocopherol in regulating FA formation may be attributed to the interaction between α-tocopherol and ER FA metabolism.

The role of tocopherols in abiotic stresses, such as salt [39–44], low temperature [45,46], drought [41,47], and heavy metals [40,43,48,49], has been observed. These results, together with our observations of *35S:BnaC02.TMT.a-6HA* plants under conditions involving salt stress (Figures 7–9), suggest that tocopherols play a crucial role in the alleviation of salt stress. Stresses lead to excessive ROS levels causing oxidative damage [60,61]. H_2O_2 and O_2^- are the most common forms of ROS generated during photosynthesis [50]. In this study, the overexpression of *BnaC02.TMT*.a promoted seed germination (Figures 7 and 8) and reduced the production of H_2O_2 and O_2^- in *B. napus* under conditions involving salt stress (Figure 9), indicating that *BnaC02.TMT.a* alleviated the negative effects of NaCl toxicity by scavenging ROS. This was reminiscent of the expression of *γ-TMT* in chloroplasts, accelerating the conversion of γ-tocopherol to α-tocopherol, which contributes to a decrease in susceptibility to salt [43]. Tocopherol-deficient *Arabidopsis* mutants were particularly sensitive to salt stress, highlighting the role of α-tocopherol in maintaining cellular Na^+/K^+ homeostasis and hormonal balance [42,47]. The overexpression of *Msγ-TMT* in alfalfa (*Medicago sativa* L.) can directly or indirectly affect the biosynthesis of multiple phytohormones that may play a comprehensive role in drought stress tolerance [47]. FAs or FA derivatives function as signalling molecules or hormones, carbon and energy storage materials, and surface layers that protect plants from environmental stresses [62–64]. In addition, C18:3 serves as a precursor for the biosynthesis of jasmonic acid, which is involved in the response to various stresses [65,66]. Overall, considering the changes in FA composition, especially with an increase in the proportion of C18:3 (Figure 6A), other possible reasons for the better seed germination performance of the *35S:BnaC02.TMT.a-6HA* plants in the present study might, in part, be attributed to signalling cascades connecting tocopherol levels with FAs, hormonal responses, ion homeostasis, and all aspects that warrant further investigation.

Plant growth and crop production are adversely affected by environmental stresses in nature. By uncovering the genes that underlie the tolerance adaptive trait, natural variation has the potential to be introgressed into elite cultivars [67]. Modern analytical approaches, such as genomic prediction, machine learning, and multi-trait gene editing, as well as genome–environment associations, were applied to speed up the identification and deployment of genotypic sources for climate change adaptation [67,68]. Indeed, genotypes from advanced interspecific congruity backcross exhibit promising responses to extreme conditions (i.e., heat and drought), which offers novel perspectives to breed traits of interest in the face of multiple climate changes [69]. Our study revealed that *BnaC02.TMT.a* acts as a positive regulator of tocopherol biosynthesis and salt stress response, providing a beneficial candidate for improving tocopherol activity and salt stress resistance in rapeseed breeding. Tocopherol-associated traits could be employed in traditional breeding for rapeseed improvement. Furthermore, advanced breeding strategies established in recent years have provided new avenues for harnessing tocopherols biosynthesis to improve the adaptation to stressful environments.

4. Materials and Methods

4.1. Plant Material and Growth Conditions

The *B. napus* cultivar "Zhongshuang11 (ZS11)" was used in this study. It was obtained through a material transfer agreement from the Oil Crops Research Institute of the Chinese Academy of Agricultural Sciences, Wuhan, China, and selfed for at least 10 generations. The rapeseed seeds were placed in a glass dish with three layers of filter paper soaked with sterile redistilled water and imbibed at 4 °C for one day in the dark. After imbibition, the seeds were transferred to 11 × 11 cm pots and grown in a greenhouse at 22 °C with a long day duration (LD, 16 h light/8 h dark) for six weeks. Subsequently, the plants were vernalised for four weeks at 4 °C under LD conditions in a cold chamber. After vernalisation, the plants were returned to the initial greenhouse conditions for ten weeks until harvest.

4.2. Protein Sequence and Phylogenetic Analysis

The protein sequences of γ-TMT were obtained from the National Center for Biotechnology Information (NCBI) database and *B. napus* pan-genome information resource (BnPIR) database (http://cbi.hzau.edu.cn/bnapus/index.php, accessed on 6 September 2022). The protein sequence alignment was performed using MUSCLE (http://www.ebi.ac.uk/Tools/msa/muscle/, accessed on 8 September 2022). The S-adenosyl methionine-binding domain was predicted using the Conserved Domain Search program (http://www.ncbi.nlm.nih.gov/Structure/cdd/wrpsb.cgi, accessed on 8 September 2022). The phylogenetic analysis was performed using MEGA7 to confirm the evolutionary relationships between the γ-TMT protein sequences. A bootstrap analysis with 1000 replicates was performed to assess the statistical reliability of the tree topology.

4.3. Gene Cloning and Plasmid Construction

Owing to a high sequence similarity among the four paralogs of *Bnγ-TMT*, nested PCR was used for cloning *BnaC02.TMT.a*. Two pairs of PCR primers (specific and nested primers) were designed. The specific primers were designed based on the alignment of four Bnγ-TMT paralogs, and the nested primers were designed based on the full-length coding sequence (CDS) of *BnaC02.TMT.a* (BnaC02T0331100ZS) without a stop codon derived from the BnPIR database. To obtain full-length *BnaC02.TMT.a* cDNA, the total RNA was extracted from the developing seeds of the *B. napus* cultivar "ZS11" using the SteadyPure Plant RNA Extraction Kit (Accurate Biology, Changsha, China) and was subjected to reverse transcription to obtain first-strand cDNA according to the manufacturer's instructions (TransGen, Beijing, China).

The full-length BnaC02.TMT.a CDS was amplified by nested PCR using specific and nested primers. The first pair of the PCR primers amplified a specific fragment using the primers BnaC02.TMT.a_Fc and BnaC02.TMT.a_Rc. The second pair of primers, called nested primers, was used to clone the full-length CDS of BnaC02.TMT.a. The sequence was ligated into the binary vector pGreen-35S-6HA under the control of the CaMV35S (35S) promoter, which had been digested by Xho1 and Cla1 [70]. The primers used in this study are listed in Supplementary Table S1.

4.4. Generation of B. napus Transgenic Plants

To obtain the transgenic *B. napus* plants, the hypocotyl segments of the *B. napus* cultivar "ZS11" were infected with the plasmid *35S:BnaC02.TMT.a-6HA* by *Agrobacterium tumefaciens* strain GV3101 [71]. Transgenic T_0 plants were selected on MS agar medium containing 10 μg/mL phosphinothricin, and a successful transformation was confirmed by DNA genotyping according to the PCR amplification performed using gene-specific primers. The transformants were selected by genotyping according to PCR amplification, and homozygous seeds from the T_2 generation were used for subsequent experiments after cultivation under the same aforementioned conditions.

4.5. Morphological Observation of Mature Seeds

The T_3 plants of the *BnaC02.TMT.a* homozygous transgenic lines (*OE#3* and *OE#5*) and wild-type ZS11 controls were grown in a greenhouse under the aforementioned conditions. Siliques were harvested from the middle and basal parts of the main inflorescence. Subsequently, mature seeds were randomly selected from the different lines and observed using a stereomicroscope (Olympus SZ 61, Tokyo, Japan).

4.6. Measurement of the FA Content

The T_3 plants of the *BnaC02.TMT.a* homozygous transgenic lines (*OE#3* and *OE#5*) and wild-type ZS11 controls were grown in a greenhouse under the aforementioned conditions. For the FA measurement, seeds were harvested from the middle and basal parts of the main stem of eight individual plants from each line. The FAs were extracted and analysed as previously described [72,73], with minor modifications. First, the seeds were ground in a mortar, and approximately 8 mg of seed powder was heated at 80 °C in a methanol solution containing 2.5% (*v*/*v*) H_2SO_4 for 2 h, after which the total FAs were converted to FA methyl-esters. After cooling to room temperature, the FA methyl-esters were extracted with 1 mL hexane and 2 mL 0.9% (*w*/*v*) NaCl, and the organic phase was transferred to autoinjector vials. The gas chromatography (GC-2010 Plus, Shimadzu, Japan) analysis was performed using methyl heptadecanoate as an internal standard. The concentration of each FA species was normalised to that of the internal control.

4.7. HPLC Analysis of the Tocopherol Content

The tocopherol analysis was performed by Suzhou Michy Biology (http://www.michybio.com, accessed on 8 July 2021). The dried seeds from eight individual plants (transgenic lines and wild-type ZS11) were sent to the company. Briefly, 30–80 mg of mature seeds were ground, and the powder was mixed with 1 mL of ethyl alcohol extraction buffer containing 1% butylated hydroxytoluene (BHT) and 1 mL of 100 g/L potassium hydroxide solution. After incubation for 40 min (85 °C) and cooling to room temperature, 2 mL of petroleum ether was added. After mixing and centrifugation, the lower organic layer was recovered, dried under nitrogen, and resuspended in 0.4 mL methanol for the HPLC analysis.

The tocopherol content of the organic extract was determined using an Agilent 1100 HPLC system. A total of 10 μL of the sample was chromatographically analysed using a Compass C18 column (250 × 4.6 mm length, 5 μm particle size) and a solvent system consisting of methyl alcohol as the mobile phase with a flow rate of 1 mL/min. The column temperature was 20 °C, and the sample was scanned at 294 nm. The identification and quantification of tocopherols were performed by comparing the retention time and peak area to tocopherol standards (Supplementary Table S2).

4.8. ROS Staining

As previously described [74], the seeds collected at 96 h after sowing (HAS) were incubated in the dark at 37 °C. The seeds were saturated with 0.1% (*w*/*v*) diaminobenzidine (DAB, Sigma-Aldrich, St. Louis, MO, USA) in 10 mM 2-(*N*-morpholino) ethanesulphonic acid (MES, pH 6.5) for 8 h for the H_2O_2 staining and with 0.1% (*w*/*v*) nitroblue tetrazolium (NBT, Sigma-Aldrich, St. Louis, MO, USA) in 50 mM potassium phosphate buffer (pH 6.4) for 10 min for superoxide staining. Finally, the seeds were destained with 95% (*v*/*v*) ethanol.

4.9. ROS Quantification

The H_2O_2 content was measured using the titanium–peroxide complex method, as described previously, with minor modifications [74–76]. A total of 0.1 g (fresh weight) of the seeds collected at 96 HAS were ground in liquid nitrogen and extracted with 1 mL of cooled acetone. After the vortex mixing, the samples were centrifuged for 10 min at 8000× g at 4 °C. The supernatant was then added to a solution comprising 0.1 mL of 5% (*w*/*v*) titanium sulphate and 0.2 mL of ammonia. The mixture was centrifuged for 10 min at 8000× g. After removing the supernatant, the precipitate was dissolved in 1 mL of 2 M H_2SO_4. The absorbance was immediately measured at 415 nm. The H_2O_2 content was calculated using a standard curve generated using known concentrations of H_2O_2.

The O_2^- content was quantified as previously described, with minor modifications [74,76]. First, 0.1 g (fresh weight) of the seeds at 96 HAS was ground in liquid nitrogen and added to 1 mL of K-phosphate buffer (65 mM, pH 7.8). The homogenates were centrifuged at 12,000× *g* for 20 min (4 °C). A total of 1 mL 0.1 M hydroxylamine was added to the supernatant, and the mixture was incubated at 25 °C for 20 min. Then, 1 mL of 7 mM α-naphthalenamine and 1 mL of 58 mM 4-aminobenzenesulfonic acid were added, and the samples were centrifuged for 20 min at 12,000× *g* at 4°C. Trichloromethane (1 mL) was added to eliminate the pigments, and the mixture was centrifuged at 10,000× *g* for 5 min. The absorbance of the supernatant was measured at 530 nm. The O_2^- content was determined using a standard curve generated from known concentrations of $NaNO_2$.

4.10. Germination Test

For the seed germination assay, 50 plump seeds from each treatment group were carefully picked and washed with sterile redistilled water thrice, evenly placed on a glass dish with three layers of filter paper soaked in the same volume of 180 mM NaCl solution or sterile redistilled water as the control, and imbibed at 4 °C for one day in the dark. After imbibition, the seeds were transferred to a growth chamber for germination. Every other day, the germinating seeds were transferred to a new glass dish with three layers of filter paper soaked in the same volume of 180 mM NaCl solution or sterile redistilled water until the seed germination experiment was completed. Three independent biological replicates were obtained for the control and salt treatments of the transgenic lines (*OE#3* and *OE#5*) and ZS11. Seed germination was defined as the complete emergence of the radicle through the seed coat [77]. The number of germinated seeds was calculated every 8 h until the completion of the experiment.

4.11. Gene Expression Analysis

The roots, stems, flowers, and developing seeds for the spatial and temporal expression analysis were harvested from at least eight individual plants grown in different pots. The total RNA was isolated using the SteadyPure Plant RNA Extraction Kit (Accurate Biology, Changsha, China) and reverse-transcribed using EasyScript One-Step gDNA Removal and cDNA Synthesis SuperMix (TransGen, Beijing, China) according to the manufacturer's instructions. Quantitative reverse transcription RT-PCR (RT-qPCR) was performed in 96-well blocks using SYBR Green Master Mix (Cofitt, Hong Kong, China) with three biological replicates and three technical replicates. Considering the high sequence similarity among the paralogs of *Bnγ-TMT*, the primers were designed for the conserved regions within groups of paralogs to detect joint gene expression. The relative expression levels were calculated using a modified double delta method [78]. The housekeeping gene encoding glyceraldehyde-3-phosphate dehydrogenase (BnGAPDH), a key enzyme in the process of glycolysis and gluconeogenesis, was considered as a reference, which is widely founds in various organisms and is regarded as an internal control [79]. The primers used for RT-qPCR are listed in Supplementary Table S1.

4.12. Statistical Analysis

A completely randomised design was used in this study. All data were classified using Win-Excel and reported as the mean ± standard deviation (SD). The data were analysed by one-way analysis of variance (ANOVA) using SPSS software (version 17.0, SPSS Inc., Chicago, IL, USA). Significant differences were determined by Student's *t*-test ($p \leq 0.05$).

5. Conclusions

In summary, we discovered new functions for *Bnγ-TMT* in the regulation of tocopherol and FA biosynthesis and salt tolerance in rapeseed. Since it is an allopolyploid species, four paralogs of *Bnγ-TMT* were found in the *B. napus* genome. The protein sequence and phylogenetic analyses suggested that all four Bnγ-TMT paralogs might exhibit functions similar to those shown by Atγ-TMT for determining tocopherol accumulation. Together with the expression data, these results further suggest that *Bnγ-TMT* might regulate tocopherol accumulation mainly during the maturation stage in *B. napus* seeds. Overexpression of *BnaC02.TMT.a* promotes seed α- and total tocopherol accumulation during seed maturation. Additionally, the fatty acid composition was altered. Our findings suggest that

BnaC02.TMT.a promoted the conversion of oleic acid to linoleic acid and subsequently linoleic acid to linolenic acid by upregulating the expression of *BnFAD2* and *BnFAD3* in rapeseed. Furthermore, *BnaC02.TMT.a* plays a crucial role in mediating the response to salt stress during seed germination, which involves complex mechanisms, including ROS scavenging capability, signalling cascades connecting tocopherol with FAs, hormonal responses, and ion homeostasis. This study broadens our understanding of the function of the *Bnγ-TMT* gene and provides a novel strategy for genetic engineering to improve rapeseed breeding.

Supplementary Materials: The following supporting information can be downloaded at: https://www.mdpi.com/article/10.3390/ijms232415933/s1.

Author Contributions: M.C. conceived and designed the experiments; D.L. and Y.G. conducted the experiments and analysed the data; T.L., M.L., Y.L. and W.Z. conducted parts of the experiments; Y.G. and D.L. wrote the draft of the manuscript; Z.L. and M.C. revised the manuscript. All authors have read and agreed to the published version of the manuscript.

Funding: This work was financially supported by a grant from the Yang Ling Seed Industry Innovation Center (Grant no. K3031122024), the National Natural Science Foundation of China (Grant no. 31801393), and the National Key Research and Development Program of China (Grant no. 2022YFD1200400).

Institutional Review Board Statement: Not applicable.

Informed Consent Statement: Not applicable.

Data Availability Statement: All data included in this study are available upon reasonable request by contact with the corresponding author.

Conflicts of Interest: The authors declare no conflict of interest.

Abbreviations

A. thaliana	*Arabidopsis thaliana*
ANOVA	analysis of variance
BHT	butylated hydroxy toluene
BnPIR	*Brassica napus* pan-genome information resource
B. juncea	*Brassica juncea*
B. napus	*Brassica napus*
B. oleracea	*Brassica oleracea*
B. rapa	*Brassica rapa*
CDS	coding sequence
DAB	diaminobenzidine
DAP	days after pollination
DMBPQ	2,3-dimethyl-6-phytyl-1,4-benzoquinol
ER	endoplasmic reticulum
FAs	fatty acids
GAPDH	glyceraldehyde-3-phosphate dehydrogenase
GGPP	geranylgeranyl diphosphate
GGR	geranylgeranyl reductase
HA	haemagglutinin
HAS	h after sowing
HGA	homogentisic acid
H_2O_2	hydrogen peroxide
HPP	*p*-hydroxyphenylpyruvic acid
HPPD	HPP dioxygenase
HPT	HGA phytyl transferase
LD	long day
MEP	methylerythritol phosphate
MES	2-(*N*-morpholino) ethanesulfonic acid

MBPQ	2-methyl-6-phytyl-1,4-benzoquinol
MPBQ-MT	MPBQ methyltransferase
NaCl	sodium chloride
NBT	nitroblue tetrazolium
NCBI	National Center for Biotechnology Information
O_2^-	superoxide anion
PDP	phytyl diphosphate
Phytyl-P	phytyl phosphate
PK	phytol kinase
PPK	phytyl phosphate kinase
PUFAs	polyunsaturated FAs
ROS	reactive oxygen species
γ-TMT	γ-tocopherol methyltransferase
RT-qPCR	quantitative reverse transcription PCR
SAM	S-adenosylmethionine
SD	standard deviation
TAT	tyrosine aminotransferase
TC	tocopherol cyclase
ZS11	Zhongshuang11

References

1. Chalhoub, B.; Denoeud, F.; Liu, S.; Parkin, I.A.P.; Tang, H.; Wang, X.; Chiquet, J.; Belcram, H.; Tong, C.; Samans, B.; et al. Early allopolyploid evolution in the post-Neolithic *Brassica napus* oilseed genome. *Science* **2014**, *345*, 950–953. [CrossRef]
2. Fu, T.D. Variety improvement in rapeseed. *Crop. Res.* **2007**, *3*, 159–162. (In Chinese)
3. Nestel, P.; Clifton, P.; Colquhoun, D.; Noakes, M.; Mori, T.A.; Sullivan, D.; Thomas, B. Indications for omega-3 long chain polyunsaturated fatty acid in the prevention and treatment of cardiovascular disease. *Heart Lung Circ.* **2015**, *24*, 769–779. [CrossRef]
4. Zhao, J.V.; Schooling, C.M. Role of linoleic acid in autoimmune disorders: A Mendelian randomisation study. *Ann. Rheum. Dis.* **2019**, *78*, 711. [CrossRef]
5. Djuricic, I.; Calder, P.C. Beneficial outcomes of omega-6 and omega-3 polyunsaturated fatty acids on human health: An update for 2021. *Nutrients* **2021**, *13*, 2421. [CrossRef]
6. Weinstein, S.J.; Wright, M.E.; Lawson, K.A.; Snyder, K.; Mannisto, S.; Taylor, P.R.; Virtamo, J.; Albanes, D. Serum and dietary vitamin E in relation to prostate cancer risk. *Cancer Epidemiol. Biomark. Prev.* **2007**, *16*, 1253–1259. [CrossRef]
7. Euch-Fayache, G.E.; Bouhlal, Y.; Amouri, R.; Feki, M.; Hentati, F. Molecular, clinical and peripheral neuropathy study of Tunisian patients with ataxia with vitamin E deficiency. *Brain* **2014**, *137*, 402–410. [CrossRef]
8. Hunter, S.; Cahoon, E. Enhancing vitamin E in oilseeds: Unraveling tocopherol and tocotrienol biosynthesis. *Lipids* **2007**, *42*, 97–108. [CrossRef]
9. Horvath, G.; Wessjohann, L.; Bigirimana, J.; Jansen, M.; Guisez, Y.; Caubergs, R.; Horemans, N. Differential distribution of tocopherols and tocotrienols in photosynthetic and non-photosynthetic tissues. *Phytochemistry* **2006**, *67*, 1185–1195. [CrossRef]
10. Munné-Bosch, S.; Alegre, L. The function of tocopherols and tocotrienols in plants. *Crit. Rev. Plant Sci.* **2002**, *21*, 31–57. [CrossRef]
11. DellaPenna, D.; Last, R.L. Progress in the dissection and manipulation of plant vitamin E biosynthesis. *Physiol. Plant.* **2006**, *126*, 356–368. [CrossRef]
12. Garcia, I.; Rodgers, M.; Lenne, C.; Rolland, A.; Sailland, A.; Matringe, M. Subcellular localization and purification of a *p*-hydroxyphenylpyruvate dioxygenase from cultured carrot cells and characterization of the corresponding cDNA. *Biochem. J.* **1997**, *325*, 761–769. [CrossRef] [PubMed]
13. Riewe, D.; Koohi, M.; Lisec, J.; Pfeiffer, M.; Lippmann, R.; Schmeichel, J.; Willmitzer, L.; Altmann, T. A tyrosine aminotransferase involved in tocopherol synthesis in *Arabidopsis*. *Plant J.* **2012**, *71*, 850–859. [CrossRef] [PubMed]
14. Keller, Y.; Bouvier, F.; D'Harlingue, A.; Camara, B. Metabolic compartmentation of plastid prenyllipid biosynthesis evidence for the involvement of a multifunctional geranylgeranyl reductase. *Eur. J. Biochem.* **1998**, *251*, 413–417. [CrossRef] [PubMed]
15. Kimura, E.; Abe, T.; Murata, K.; Kimura, T.; Otoki, Y.; Yoshida, T.; Miyazawa, T.; Nakagawa, K. Identification of OsGGR2, a second geranylgeranyl reductase involved in α-tocopherol synthesis in rice. *Sci. Rep.* **2018**, *8*, 1870. [CrossRef] [PubMed]
16. Valentin, H.E.; Lincoln, K.; Moshiri, F.; Jensen, P.K.; Qi, Q.; Venkatesh, T.V.; Karunanandaa, B.; Baszis, S.R.; Norris, S.R.; Savidge, B.; et al. The *Arabidopsis vitamin E pathway gene5-1* mutant reveals a critical role for phytol kinase in seed tocopherol biosynthesis. *Plant Cell* **2006**, *18*, 212–224. [CrossRef] [PubMed]
17. Dorp, K.V.; Hölzl, G.; Plohmann, C.; Eisenhut, M.; Abraham, M.; Weber, A.P.M.; Hanson, A.D.; Dörmann, P. Remobilization of phytol from chlorophyll degradation is essential for tocopherol synthesis and growth of *Arabidopsis*. *Plant Cell* **2015**, *27*, 2846–2859.
18. Mène-Saffrané, L.; DellaPenna, D. Biosynthesis, regulation and functions of tocochromanols in plants. *Plant Physiol. Bioch.* **2010**, *48*, 301–309. [CrossRef] [PubMed]

19. Fritsche, S.; Wang, X.; Jung, C. Recent advances in our understanding of tocopherol biosynthesis in plants: An overview of key genes, functions, and breeding of vitamin E improved crops. *Antioxidants* **2017**, *6*, 99. [CrossRef] [PubMed]
20. Shintani, D.; DellaPenna, D. Elevating the vitamin E content of plants through metabolic engineering. *Science* **1998**, *282*, 2098–2100. [CrossRef]
21. Bergmüller, E.; Porfirova, S.; Dörmann, P. Characterization of an *Arabidopsis* mutant deficient in γ-tocopherol methyltransferase. *Plant Mol. Biol.* **2003**, *52*, 1181–1190. [CrossRef]
22. Kim, Y.J.; Seo, H.Y.; Park, T.I.; Baek, S.H.; Shin, W.C.; Kim, H.S.; Kim, J.G.; Choi, Y.E.; Yun, S.J. Enhanced biosynthesis of α-tocopherol in transgenic soybean by introducing γ-TMT gene. *Plant Biotechnol. J.* **2005**, *7*, 203–209.
23. Kim, M.J.; Baek, S.H.; Yoo, N.H.; Yun, S.J. Transformation of *Arabidopsis* gamma-tocopherol methyltransferase into lettuce (*Lactuca sativa* L.). *Korean J. Plant Tiss. Cult.* **2000**, *27*, 435–439.
24. Cho, E.A.; Lee, C.A.; Kim, Y.S.; Baek, S.H.; de Los, R.B.; Yun, S.J. Expression of γ-tocopherol methyltransferase transgene improves tocopherol composition in lettuce (*Latuca sativa* L.). *Mol. Cell* **2005**, *19*, 16–22.
25. Zhang, G.Y.; Liu, R.R.; Xu, G.; Zhang, P.; Li, Y.; Tang, K.X.; Liang, G.H.; Liu, Q.Q. Increased α-tocotrienol content in seeds of transgenic rice overexpressing *Arabidopsis* γ-tocopherol methyltransferase. *Transgenic Res.* **2013**, *22*, 89–99. [CrossRef]
26. Ghimire, B.K.; Seong, E.S.; Lee, C.O.; Lim, J.D.; Lee, J.G.; Yoo, J.H.; Chung, I.M.; Kim, N.Y.; Yu, C.Y. Enhancement of α-tocopherol content in transgenic *Perilla frutescens* containing the *γ-TMT* gene. *Afr. J. Biotechnol.* **2011**, *10*, 2430–2439.
27. Tavva, V.S.; Kim, Y.; Kagan, I.A.; Dinkins, R.D.; Kim, K.; Collins, G.B. Increased α-tocopherol content in soybean seed overexpressing the *Perilla frutescens γ-tocopherol methyltransferase* gene. *Plant Cell Rep.* **2007**, *26*, 61–70. [CrossRef]
28. Yusuf, M.A.; Sarin, N.B. Antioxidant value addition in human diets: Genetic transformation of *Brassica juncea* with *γ-TMT* gene for increased α-tocopherol content. *Transgenic Res.* **2007**, *16*, 109–113. [CrossRef]
29. Grusak, M.A.; DellaPenna, D. Improving the nutrient composition of plants to enhance human nutrition and health. *Annu. Rev. Plant Physiol. Plant Mol. Biol.* **1999**, *50*, 133–161. [CrossRef]
30. Krieger-Liszkay, A.; Fufezan, C.; Trebst, A. Singlet oxygen production in photosystem II and related protection mechanism. *Photosynth. Res.* **2008**, *98*, 551–564. [CrossRef]
31. Liebler, D.C. The role of metabolism in the antioxidant function of vitamin E. *Crit. Rev. Toxicol.* **1993**, *23*, 147–169. [CrossRef] [PubMed]
32. Kamal-Eldin, A.; Appelqvist, L.A. The chemistry and antioxidant properties of tocopherols and tocotrienols. *Lipids* **1996**, *31*, 671–701. [CrossRef] [PubMed]
33. Kamal-Eldin, A.; Andersson, R. A multivariate study of the correlation between tocopherol content and fatty acid composition in vegetable oils. *J. Am. Oil Chem. Soc.* **1997**, *74*, 375–380. [CrossRef]
34. Goffman, F.D.; Böhme, T. Relationship between fatty acid profile and vitamin E content in maize hybrids (*Zea mays* L.). *J. Agric. Food Chem.* **2001**, *49*, 4990–4994. [CrossRef]
35. Li, Y.; Hussain, N.; Zhang, L.; Chen, X.; Ali, E.; Jiang, L. Correlations between tocopherol and fatty acid components in germplasm collections of *Brassica oilseeds*. *J. Agri. Food Chem.* **2013**, *61*, 34–40. [CrossRef]
36. Sattler, S.E.; Gilliland, L.U.; Magallanes-Lundback, M.; Pollard, M.; DellaPenna, D. Vitamin E is essential for seed longevity and for preventing lipid peroxidation during germination. *Plant Cell* **2004**, *16*, 1419–1432. [CrossRef]
37. Munné-Bosch, S.; Weiler, E.W.; Alegre, L.; Müller, M.; Düchting, P.; Falk, J. α-Tocopherol may influence cellular signaling by modulating jasmonic acid levels in plants. *Planta* **2007**, *225*, 681–691. [CrossRef]
38. Stahl, E.; Hartmann, M.; Scholten, N.; Zeier, J. A role for tocopherol biosynthesis in *Arabidopsis* basal immunity to bacterial infection. *Plant Physiol.* **2019**, *181*, 1008–1028. [CrossRef]
39. Abbasi, A.R.; Hajirezaei, M.; Hofius, D.; Sonnewald, U.; Voll, L.M. Specific roles of α- and γ-tocopherol in abiotic stress responses of transgenic tobacco. *Plant Physiol.* **2007**, *143*, 1720–1738. [CrossRef]
40. Yusuf, M.A.; Kumar, D.; Rajwanshi, R.; Strasser, R.J.; Tsimilli-Michael, M.; Govindjee; Sarin, N.B. Overexpression of γ-tocopherol methyl transferase gene in transgenic *Brassica juncea* plants alleviates abiotic stress: Physiological and chlorophyll a fluorescence measurements. *Biochim. Biophys. Acta Bioenerg.* **2010**, *1797*, 1428–1438. [CrossRef]
41. Cela, J.; Chang, C.; Munné-Bosch, S. Accumulation of γ- rather than α-tocopherol alters ethylene signaling gene expression in the *vte4* mutant of *Arabidopsis thaliana*. *Plant Cell Physiol.* **2011**, *52*, 1389–1400. [CrossRef] [PubMed]
42. Ellouzi, H.; Hamed, K.B.; Cela, J.; Müller, M.; Abdelly, C.; Munné-Bosch, S. Increased sensitivity to salt stress in tocopherol-deficient *Arabidopsis* mutants growing in a hydroponic system. *Plant Signal Behav.* **2013**, *8*, e23136. [CrossRef]
43. Jin, S.; Daniell, H. Expression of γ-tocopherol methyltransferase in chloroplasts results in massive proliferation of the inner envelope membrane and decreases susceptibility to salt and metal-induced oxidative stresses by reducing reactive oxygen species. *Plant Biotechnol. J.* **2014**, *12*, 1274–1285. [CrossRef] [PubMed]
44. Ma, J.; Qiu, D.; Xu, L.; Cui, M.; Gao, H.; Pang, Y.; Qin, Y.; Wang, X. Overexpression of alfalfa *γ-tocopherol methyltransferase* (*γ-TMT*) gene increases salt susceptibility of transgenic *Arabidopsis* in seed germination. *Environ. Exp. Bot.* **2020**, *180*, 104264. [CrossRef]
45. Maeda, H.; Song, W.; Sage, T.L.; DellaPenna, D. Tocopherols play a crucial role in low-temperature adaptation and phloem loading in *Arabidopsis*. *Plant Cell* **2006**, *18*, 2710–2732. [CrossRef]
46. Maeda, H.; Sage, T.L.; Isaac, G.; Welti, R.; DellaPenna, D. Tocopherols modulate extraplastidic polyunsaturated fatty acid metabolism in *Arabidopsis* at low temperature. *Plant Cell* **2008**, *20*, 452–470. [CrossRef]

47. Ma, J.; Qiu, D.; Gao, H.; Wen, H.; Wu, Y.; Pang, Y.; Wang, X.; Qin, Y. Over-expression of a γ-tocopherol methyltransferase gene in vitamin E pathway confers PEG-simulated drought tolerance in alfalfa. *BMC Plant Biol.* **2020**, *20*, 226. [CrossRef]
48. Collin, V.C.; Eymery, F.; Genty, B.; Rey, P.; Havaux, M. Vitamin E is essential for the tolerance of *Arabidopsis thaliana* to metal-induced oxidative stress. *Plant Cell Environ.* **2008**, *31*, 244–257. [CrossRef]
49. Zhu, Q.; Zhang, J.; Yu, H.; Li, L.; Chen, X.; Jiang, M.; Tan, M. Maize Cd-tolerant ZmVTE4 encoding γ-tocopherol-methyl-transferase alleviated Cd-toxicity through its product α-tocopherol. *Environ. Exp. Bot.* **2019**, *158*, 171–179. [CrossRef]
50. Mittler, R. Oxidative stress, antioxidants and stress tolerance. *Trends Plant Sci.* **2002**, *7*, 405–410. [CrossRef]
51. Miller, G.; Suzuki, N.; Ciftci-Yilmaz, S.; Mittler, R. Reactive oxygen species homeostasis and signalling during drought and salinity stresses. *Plant Cell Environ.* **2010**, *33*, 453–467. [CrossRef]
52. Schranz, M.E.; Lysak, M.A.; Mitchell-olds, T. The ABC's of comparative genomics in the Brassicaceae: Building blocks of crucifer genomes. *Trends Plant Sci.* **2006**, *11*, 535–542. [CrossRef]
53. Grillo, M.A.; Colombatto, S. S-adenosylmethionine and its products. *Amino Acids* **2008**, *34*, 187–193. [CrossRef] [PubMed]
54. Goffman, F.D.; Velasco, L.; Becker, H.C. Tocopherols accumulation in developing seeds and pods of rapeseed (*Brassica napus* L.). *Lipid/Fett* **1999**, *101*, 400–403. [CrossRef]
55. Endrigkeit, J.; Wang, X.; Cai, D.; Zhang, C.; Long, Y.; Meng, J.; Jung, C. Genetic mapping, cloning, and functional characterization of the *BnaX.VTE4* gene encoding a γ-tocopherol methyltransferase from oilseed rape. *Theor. Appl. Genet.* **2009**, *119*, 567–575. [CrossRef] [PubMed]
56. Okuley, J.; Lightner, J.; Feldmann, K.; Yadav, N.; Lark, E.; Browse, J. *Arabidopsis FAD2* gene encodes the enzyme that is essential for polyunsaturated lipid synthesis. *Plant Cell* **1994**, *6*, 147–158.
57. Lee, K.R.; Lee, Y.; Kim, E.H.; Lee, S.B.; Roh, K.H.; Kim, J.B.; Kang, H.C.; Kim, H.U. Functional identification of oleate 12-desaturase and ω-3 fatty acid desaturase genes from *Perilla frutescens* var. *Frutescens*. *Plant Cell Rep.* **2016**, *35*, 2523–2537. [CrossRef]
58. Shah, S.; Xin, Z.G.; Browse, J. Overexpression of the *FAD3* desaturase gene in a mutant of *Arabidopsis*. *Plant Physiol.* **1997**, *114*, 1533–1539. [CrossRef]
59. Peng, Z.; Ruan, J.; Tian, H.; Shan, L.; Meng, J.; Guo, F.; Zhang, Z.; Ding, H.; Wan, S.; Li, X. The family of peanut fatty acid desaturase genes and a functional analysis of four ω-3 *AhFAD3* members. *Plant Mol. Biol. Rep.* **2020**, *38*, 209–221. [CrossRef]
60. Choudhury, F.K.; Rivero, R.M.; Blumwald, E.; Mittler, R. Reactive oxygen species, abiotic stress and stress combination. *Plant J.* **2017**, *90*, 856–867. [CrossRef]
61. Miller, G.; Shulaev, V.; Mittler, R. Reactive oxygen signaling and abiotic stress. *Physiol. Plant.* **2008**, *133*, 481–489. [CrossRef] [PubMed]
62. Ohlrogge, J.B.; Jaworski, J.G. Regulation of fatty acid synthesis. *Annu. Rev. Plant Physiol. Plant Mol. Biol.* **1997**, *48*, 109–136. [CrossRef] [PubMed]
63. Hong, J.K.; Choi, H.W.; Hwang, I.S.; Kim, D.S.; Kim, N.H.; Choi, D.S.; Kim, Y.J.; Hwang, B.K. Function of a novel GDSL-type pepper lipase gene, *CaGLIP1*, in disease susceptibility and abiotic stress tolerance. *Planta* **2008**, *227*, 539–558. [CrossRef] [PubMed]
64. Mu, J.; Tan, H.; Zheng, Q.; Fu, F.; Liang, Y.; Zhang, J.; Yang, X.; Wang, T.; Chong, K.; Wang, X.J.; et al. *LEAFY COTYLEDON1* is a key regulator of fatty acid biosynthesis in *Arabidopsis*. *Plant Physiol.* **2008**, *148*, 1042–1054. [CrossRef] [PubMed]
65. Mueller, M.J. Enzymes involved in jasmonic acid biosynthesis. *Physiol. Plant.* **1997**, *100*, 653–663. [CrossRef]
66. Wang, J.; Song, L.; Gong, X.; Xu, J.; Li, M. Functions of jasmonic acid in plant regulation and response to abiotic stress. *Int. J. Mol. Sci.* **2020**, *21*, 1446. [CrossRef] [PubMed]
67. Cortés, A.J.; López-Hernández, F. Harnessing crop wild diversity for climate change adaptation. *Genes* **2021**, *12*, 783. [CrossRef] [PubMed]
68. Cortés, A.J.; López-Hernández, F.; Blair, M.W. Genome-environment associations, an innovative tool for studying heritable evolutionary adaptation in orphan crops and wild relatives. *Front. Genet.* **2022**, *13*, 910386. [CrossRef]
69. Burbano-Erazo, E.; León-Pacheco, R.I.; Cordero-Cordero, C.C.; López-Hernández, F.; Cortés, A.J.; Tofiño-Rivera, A.P. Multi-environment yield components in advanced common bean (*Phaseolus vulgaris* L.) × Tepary bean (*P. acutifolius* A. Gray) interspecific lines for heat and drought tolerance. *Agronomy* **2021**, *11*, 1978. [CrossRef]
70. Liu, C.; Chen, H.; Er, H.L.; Soo, H.M.; Kumar, P.P.; Han, J.; Liou, Y.C.; Yu, H. Direct interaction of *AGL24* and *SOC1* integrates flowering signals in *Arabidopsis*. *Development* **2008**, *135*, 1481–1491. [CrossRef]
71. Cardoza, V.; Stewart, C.N. Increased *Agrobacterium*-mediated transformation and rooting efficiencies in canola (*Brassica napus* L.) from hypocotyl segment explants. *Plant Cell Rep.* **2003**, *21*, 599–604. [CrossRef] [PubMed]
72. Poirier, Y.; Ventre, G.; Caldelari, D. Increased flow of fatty acids toward β-oxidation in developing seeds of *Arabidopsis* deficient in diacylglycerol acyltransferase activity or synthesizing medium-chain-length fatty acids. *Plant Physiol.* **1999**, *121*, 1359–1366. [CrossRef] [PubMed]
73. Chen, M.; Wang, Z.; Zhu, Y.; Li, Z.; Hussain, N.; Xuan, L.; Guo, W.; Zhang, G.; Jiang, L. The effect of transparent *TESTA2* on seed fatty acid biosynthesis and tolerance to environmental stresses during young seedling establishment in *Arabidopsis*. *Plant Physiol.* **2012**, *160*, 1023–1036. [CrossRef] [PubMed]
74. Yu, Y.; Wang, J.; Li, S.; Kakan, X.; Zhou, Y.; Miao, Y.; Wang, F.; Qin, H.; Huang, R. Ascorbic acid integrates the antagonistic modulation of ethylene and abscisic acid in the accumulation of reactive oxygen species. *Plant Physiol.* **2019**, *179*, 1861–1875. [CrossRef]

75. Kakan, X.; Yu, Y.; Li, S.; Li, X.; Huang, R.; Wang, J. Ascorbic acid modulation by ABI4 transcriptional repression of *VTC2* in the salt tolerance of *Arabidopsis*. *BMC Plant Biol.* **2021**, *21*, 112. [CrossRef] [PubMed]
76. Luo, X.; Dai, Y.; Zheng, C.; Yang, Y.; Chen, W.; Wang, Q.; Chandrasekaran, U.; Du, J.; Liu, W.; Shu, K. The ABI4-RbohD/VTC2 regulatory module promotes Reactive Oxygen Species (ROS) accumulation to decrease seed germination under salinity stress. *New Phytol.* **2021**, *229*, 950–962. [CrossRef] [PubMed]
77. Assmann, S.M. G protein signaling in the regulation of *Arabidopsis* seed germination. *Sci. STKE* **2005**, *308*, cm11. [CrossRef]
78. Pfaffl, M.W.; Horgan, G.W.; Dempfle, L. Relative expression software tool (REST©) for group-wise comparison and statistical analysis of relative expression results in real-time PCR. *Nucleic Acids Res.* **2002**, *30*, e36. [CrossRef]
79. Küpper, H.; Seib, L.O.; Sivaguru, M.; Hoekenga, O.A.; Kochian, L.V. A method for cellular localization of gene expression via quantitative in situ hybridization in plants. *Plant J.* **2007**, *50*, 159–175. [CrossRef]

International Journal of
Molecular Sciences

MDPI

Article

Silencing *GmBIR1* in Soybean Results in Activated Defense Responses

Dan-Dan Liu [1,†], Hu-Jiao Lan [1,†], Hashimi Said Masoud [1], Mei-Yan Ye [1], Xian-Yong Dai [1], Chen-Li Zhong [1], Sheng-Nan Tian [1] and Jian-Zhong Liu [1,2,*]

1 Institute of Plant Genetics and Developmental Biology, College of Chemistry and Life Sciences, Zhejiang Normal University, Jinhua 321004, China; dandanliuxy@163.com (D.-D.L.); 18659351751@163.com (H.-J.L.); s.masoud.hashimi@gmail.com (H.S.M.); a19557861320@163.com (M.-Y.Y.); q1178261391@163.com (X.-Y.D.); 18170141671@163.com (C.-L.Z.); xiaotian12260606@163.com (S.-N.T.)

2 Zhejiang Provincial Key Laboratory of Biotechnology on Specialty Economic Plants, Zhejiang Normal University, Jinhua 321004, China

* Correspondence: jzliu@zjnu.cn; Tel.: +86-579-8228-8053; Fax: +86-579-8228-2269

† These authors contributed to this work equally.

Abstract: Receptor-like kinases (RLKs) are a large group of pattern recognition receptors (PRRs) and play a critical role in recognizing pathogens, transducing defense signals, and mediating the activation of immune defense responses. Although extensively studied in the model plant Arabidopsis, studies of RLKs in crops, including soybean, are limited. When a *BAK1-interacting receptor-like kinase* (*BIR1*) homolog (referred to as *GmBIR1* hereafter) was silenced by the BPMV (*Bean pod mottle virus*)-induced gene silencing (BPMV-VIGS), it resulted in phenotypes that were reminiscent of constitutively activated defense responses, including a significantly stunted stature with observable cell death on the leaves of the silenced plants. In addition, both SA and H_2O_2 were over-accumulated in the leaves of the *GmBIR1*-silenced plants. Consistent with this autoimmune phenotype, *GmBIR1*-silenced plants exhibited significantly enhanced resistance to both *Pseudomonas syringae pv. glycinea* (*Psg*) and *Soybean mosaic virus* (SMV), two different types of pathogens, compared to the vector control plants. Together, our results indicated that *Gm*BIR1 is a negative regulator of immunity in soybean and the function of BIR1 homologs is conserved in different plant species.

Keywords: *Glycine max*; virus-induced gene silencing (VIGS); receptor-like kinases; immune responses; BIR1; salicylic acid

Citation: Liu, D.-D.; Lan, H.-J.; Masoud, H.S.; Ye, M.-Y.; Dai, X.-Y.; Zhong, C.-L.; Tian, S.-N.; Liu, J.-Z. Silencing *GmBIR1* in Soybean Results in Activated Defense Responses. *Int. J. Mol. Sci.* **2022**, *23*, 7450. https://doi.org/10.3390/ijms23137450

Academic Editors: Andrés J. Cortés and Hai Du

Received: 7 June 2022
Accepted: 30 June 2022
Published: 5 July 2022

1. Introduction

Pathogen-associated molecular patterns (PAMPs) are conserved microbial components that can be perceived by plasma membrane-localized pattern recognition receptors (PRRs) and trigger immunity [1]. The *PRRs* constitute the largest gene family (>600 genes) in Arabidopsis [2]. The PRRs include two types of protein families: the receptor-like kinases (RLKs) and receptor-like proteins (RLPs). While a typical RLK contains an extracellular leucine-rich repeat (LRR) domain, a transmembrane domain (TM), and a cytoplasmic kinase domain, an RLP contains only LRR and TM domains but lacks a cytoplasmic kinase domain. The LRR domain is responsible for perception of PAMPs or ligands, and the cytoplasmic kinase domain is involved in intracellular signal transduction [3]. The lack of a kinase domain in a RLP suggests that an RLP needs to pair with RLKs or another cytoplasmic kinase to transduce signals [3].

Some RLKs have been reported to sense PAMPs either from bacteria or fungi and transduce defense signals [4]. The best studied examples are FLAGELLIN SENSING2 (FLS2) and EF-TU RECEPTOR (EFR). FLS2 recognizes the 22 amino acid peptide flg22 from bacterial flagellin, while EFR recognizes an 18 amino acid peptide, elf18, derived from bacterial translation elongation factor EF-TU [5,6]. BAK1 was originally identified as a

BRI1 interacting protein, and it forms a protein complex with BRI1 during perception of Brassinosteroids (BRs) [7,8]. It is now clear that BAK1 serves as a common co-receptor for different RLKs or even RLPs that sense ligands ranging from BRs to PAMPs. Upon ligand or PAMP recognition, BAK1 is recruited to form a complex with RLKs such as FLS2, EFR and CERK1 [9–11]. The ligand-induced RLKs–BAK1 complex formation initiates dynamic trans-phosphorylation between the receptor, co-receptor, and receptor-like cytoplasmic kinases (RLCKs) such as BOTRYTIS-INDUCED KINASE1 (BIK1) [12,13], thus leading to the activation of downstream signaling.

Using isotope-coded affinity tag reagents and mass spectrometry, Gao et al. [14] identified BAK1 as BIR1-interacting protein on the plasma membrane (PM), which was subsequently confirmed by BiFC and Co-IP assays [14]. BIR1 belongs to the LRRX group of RLKs, with 620 amino acids and 5 LRRs. Compared to RLKs such as FLS2 or BAK1, BIR1 is a smaller protein with a significantly reduced number of LRR repeats [14]. Loss function of *BIR1* in Arabidopsis leads to extensive cell death, induction of *PR* genes, enhanced accumulation of SA and H_2O_2 and impairment in the activation of MPK4 by flagellin [14]. Genetic studies show that the *bir1* mutant phenotype is dependent on BAK1, EDS1, and PAD4 and loss-of-function mutations in these genes can partially suppress the autoimmune phenotype of *bir1* mutant, indicating that over-accumulation of SA is partially responsible for the autoimmune phenotype [14,15]. The kinase activity of BAK1 is required for its function in activation of defense responses in *bir1-1*.

SOBIR1 (suppressor of BIR1) was identified as a suppressor of *bir1-1* autoimmune phenotypes in a suppressor screening. Loss of function in SOBIR1 strongly suppresses cell death in *bir1-1*. Combining the *sobir1* and *pad4-1* mutations leads to suppression of all mutant phenotypes of *bir1-1*, suggesting that SOBIR1 functions in parallel with PAD4. SOBIR1 encodes an RLK with four extracellular LRRs and a cytoplasmic kinase domain [14]. BAK1 interacts with SOBIR1 only when BIR1 is absent [15–17]. In the absence of BIR1, inhibition of BAK1 by BIR1 is released and BAK1 forms an active receptor complex with SOBIR1, leading to cell death and defense responses [15]. In addition, BAK1 associates with the BIR1 and the copain-like proteins BONZAI1 (BON1)–BON3, which are required to suppress autoimmunity [18].

BIR2 is an inactive pseudokinase and a substrate of BAK1. It forms a constitutive complex with BAK1 to prevent it from interacting with FLS2. The kinase activity of BAK1 is required for BIR2-BAK1 interaction [19]. BIR2 Is released from BAK1 after PAMP treatment and BIR2 controls BAK1-FLS2 complex formation in a ligand-dependent manner [19]. *bir2* mutant plants exhibit enhanced pathogen-associated molecular pattern (PAMP) responses but do not have the dwarf morphology like *bir1-1*. Unlike BIR2, BIR1 is an active kinase, and kinase activity is at least partially required for its function. flg22 treatment leads to dissociation of BIR2 from BAK1 and over-expression of BIR2 attenuates PAMP-triggered responses [19].

Although extensively studied in the model plant Arabidopsis, the investigations of RLKs in crop plants are limited. Using the BPMV-VIGS system, we have recently identified multiple positive and negative regulators of defense responses in soybean [20–25]. Here, using the same strategy, we identified *Gm*BIR1 as a negative regulator of defense responses in soybean. Silencing *GmBIR1* led to spontaneous cell death and constitutively activated defense responses. These autoimmune responses are highly correlated with the over-accumulation of both SA and H_2O_2. Consistent with the autoimmune phenotypes, the *GmBIR1*-silenced plants displayed a significantly increased resistance to both *Soybean mosaic virus* (SMV) and *Pseudomonas syringae pv. glycinea* (*Psg*). Taken together, our results indicate that *Gm*BIR1 plays a negative regulatory role in soybean immunity, and the function of BIR1 homologs is conserved across plant species.

2. Results

2.1. Silencing Glyma.18g246400 Results in Stunted Stature and Spontaneous Cell Death in Soybean

Through BPMV-VIGS silencing approach, we successfully identified both positive and negative regulators of defense responses in soybean [20–25]. Using the same strategy, we found that when *Glyma*.18g246400 was silenced, a significantly stunted stature (Figure 1A) and spontaneous cell death on the systemic leaves were observed for the silenced plants (Figure 1B). The cell death was further confirmed by Tryplan blue staining, which specifically stains the dead cells. The massive blue patches were observed along the lager veins on the systemic leaves of the silenced plants but not on that of vector control plants (Compare Figure 1C,D). RT-PCR analysis confirmed that the transcript level of *Glyma*.18g246400 was significantly reduced in the silenced plants (Figure 1E). Stunted stature and spontaneous cell death are the signature characteristics of the constitutively activated defense responses. Consistent with this, a significantly induced expression of the *PR1* gene was observed in the silenced plants (Figure 1F). These results indicated that *Glyma*.18g246400 encodes a negative regulator of cell death and defense responses. Blast search with the full-length cDNA of *Glyma*.18g246400 revealed that there is a duplicated gene in the soybean genome, *Glyma*.09g246600, which shares 95% identity with *Glyma*.18g246400. It has been reported that VIGS can simultaneously silence genes with a homology greater than 85% at the nucleotide level [26,27]. We confirmed that the silenced phenotype observed for the *Glyma*.18g246400-silenced plants was actually a result of co-silencing both *Glyma*.18g246400 and *Glyma*.09g246600 (Figure 1G). Both *Glyma*.18g246400 and *Glyma*.09g246600 share the highest homology with Arabidopsis *BIR1* (65%), we thus referred to these two genes as *GmBIR1a* and *GmBIR1b*, respectively. Because the *bir1* mutant in Arabidopsis displays a constitutive activated defense phenotype (Gao et al., 2009), our results indicate that the function of BIR1 homologs is conserved in different plants species.

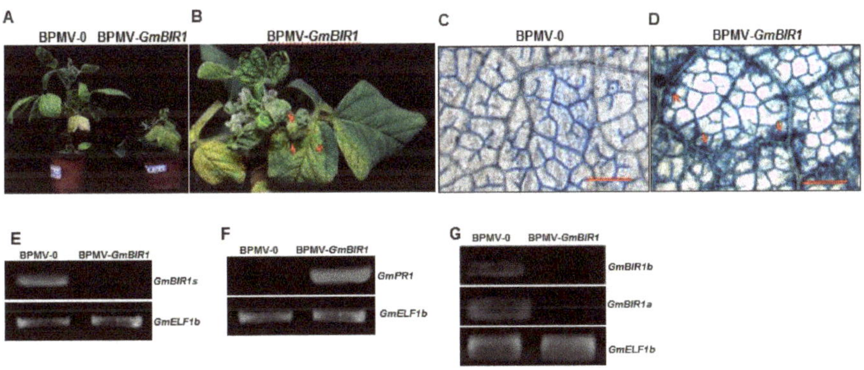

Figure 1. Silencing *GmBIR1* causes both local and systemic hypersensitive response (HR) cell death in soybean. (**A**) The phenotypes of vector control (BPMV-0) and *GmBIR1*-silenced (BPMV-*GmBIR1*) plants at 15 days post infection (dpi); (**B**) Massive cell death was observed on the systemic leaves of *GmBIR1*-silenced plants at 25 dpi. Red arrows point to the dead leaves; (**C**,**D**) Trypan Blue-stained systemic leaves from vector control (**C**) and *GmBIR1*-silenced plants (**D**). Red arrows point to the dead regions. Bar = 500 μm; (**E**) RT-PCR result indicates that the *GmBIR1* transcript level was reduced in *GmBIR1*-silenced plants relative to the vector control plants; (**F**) RT-PCR result indicates that the *GmPR1* transcript level was increased in *GmBIR1*-silenced plants relative to the vector control plants; (**G**) RT-PCR result indicates that the *GmBIR1a and GmBIR1b* transcript level were both reduced in *GmBIR1*-silenced plants relative to the vector control plants. Each experiment was repeated at least twice. Each experiment consisted of 3 replicates and at least 3 plants were used in each replicate.

2.2. Both SA and H_2O_2 Are Over-Accumulated in the GmBIR1-Silenced Plants

SA is a key hormone that positively regulates defense responses against biotrophic pathogens and is over-accumulated in the plants with activated defense responses [20,28]. To examine whether SA over-accumulation occurred in *GmBIR1*-silenced plants, both free SA and bound SA were quantified, and the levels of free SA and bound SA were 2.46- and 4.1-fold higher in the *GmBIR1*-silenced plants than in BPMV-0 control plants, respectively (Figure 2A,B). The over-accumulated level of SA suggests that SA is involved in the constitutively activated defense responses observed in *GmBIR1*-silenced plants.

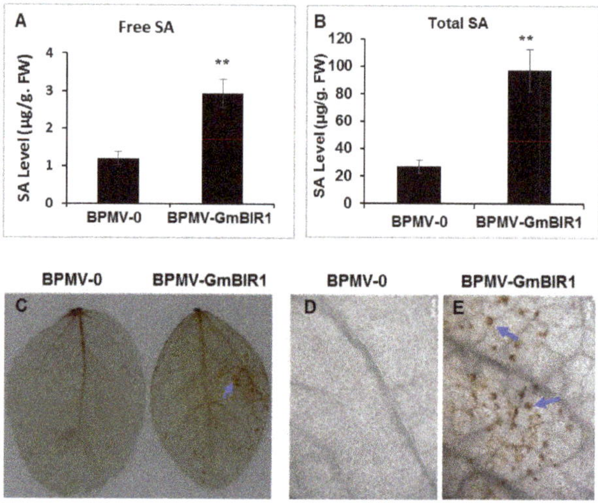

Figure 2. Elevated accumulation of SA and H_2O_2 in *GmBIR1*-silenced plants. (**A**,**B**) Free SA (**A**) and Total SA (**B**) levels were quantified in *GmBIR1*-silenced and BPMV-0 empty vector control plants at 20 days post BPMV inoculation. Error bars represent SD for three independent samples. Asterisks indicate significant differences from the control (**, $p < 0.01$, Student's *t* test). FW, fresh weight. (**C**) Presence of H_2O_2 in soybean leaves visualized by staining with DAB. (**D**,**E**) The images taken under the stereomicroscope from part of the regions shown in C, Bar = 500 μm. Oxidized DAB formed a reddish-brown deposit. (Examples of these deposits are indicated by the blue arrows).

H_2O_2 is a potent trigger of cell death [29,30] and H_2O_2 is over-accumulated in the plants with activated defense responses [20,23]. To examine if H_2O_2 is over-accumulated in the *GmBIR1*-silenced plants, the leaves collected from both vector control plants and the *GmBIR1*-silenced plants were stained with 3,3-diaminobenzidine (DAB) [31,32]. More intense brown spots were observed on the leaves of the *GmBIR1*-silenced plants compared to the BPMV-0 control plants (Figure 2C), indicating that the cell death observed in the *GmBIR1*-silenced plants was correlated with increased level of H_2O_2 accumulation.

2.3. Silencing GmBIR1 Enhances Resistance against SMV and Pseudomonas syringae pv. glycinea (Psg), R4 Strain

Activated defense responses are usually correlated with enhanced disease resistance [20–25]. To examine whether disease resistance is enhanced in *GmBIR1*-silenced plants, SMV strain N tagged with the GUS marker protein (SMV-N-GUS) [33] was inoculated via biolistic bombardment onto three individual leaves collected from both the vector control plants and the *GmBIR1*-silenced plants. Both the size of the GUS foci and the intensity of GUS staining reflect viral infectivity. At 5 days post inoculation (dpi), the SMV-N-GUS infection was visualized by GUS staining. The GUS infection foci were easily seen on the vector control leaves (Figure 3A, left panels; Figure 3B). However, the GUS spots on the *GmBIR1*-silenced plants were much smaller and only visible under a dissecting microscopy (Figure 3C). The average diameter of the GUS foci formed on the vector control leaves was four-fold of those formed on the *GmBIR1*-silenced leaves (Figure 3D). This result indicates that silencing *GmBIR1* leads to increased resistance against SMV.

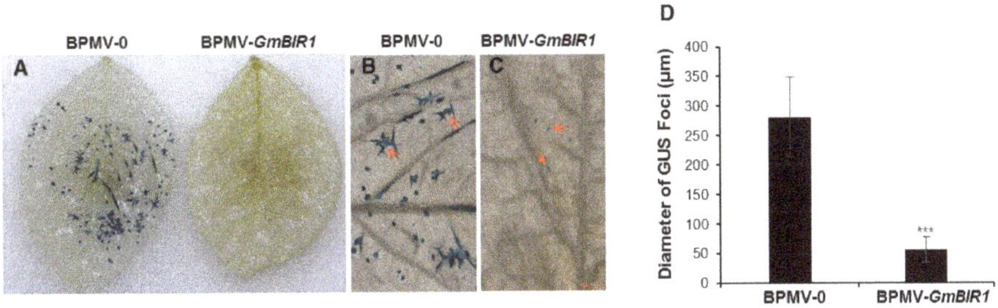

Figure 3. Silencing *GmBIR1* enhances the resistance of soybean plants to SMV. At 18 dpi with BPMV-0 or BPMV-*GmBIR1*, SMV-N-GUS was biolistically delivered onto the detached leaves of silenced and non-silenced plants. At 5 dpi with SMV-N-GUS, the replication and movement of SMV-N-GUS in the biolistically inoculated leaves was detected by GUS staining. The GUS foci were counted and measured. (**A**) The distribution of SMV-N-GUS foci on the leaves of BPMV-0 and BPMV-*GmBIR1* plants. (**B**,**C**) Infection foci of SMV-N-GUS on the leaves of BPMV-0 and BPMV-*GmBIR1* plants. Bar = 2000 μm. (**D**) Comparison of the diameters of SMV-N-GUS foci on the leaves of BPMV-0 and BPMV-*GmBIR1* plants. Error bars represent SD of the diameters of at least 30 GUS foci measured on each of three independent leaves (at least 90 foci). Asterisks indicate a significant difference from the control (***, $p < 0.001$, Student's *t* test). This experiment was repeated 3 times with similar results.

We reported that silencing *GmFLS2* compromises soybean resistance against *Pseudomonas syringae pv. glycinea* (*Psg*) R4 [24]. To test the effect of *GmBIR1* silencing on *Psg* infection, we inoculated the BPMV-0 control plants and the *GmBIR1*-silenced plants with *Psg*. At 4- and 6-dpi, the colony forming unit of *Psg* on the vector control leaves was significantly higher than on that of the *GmBIR1*-silenced leaves (Figure 4). Together, these results clearly show that silencing *GmBIR1* enhances resistance of soybean plants against two totally different types of pathogens, SMV and *Psg*.

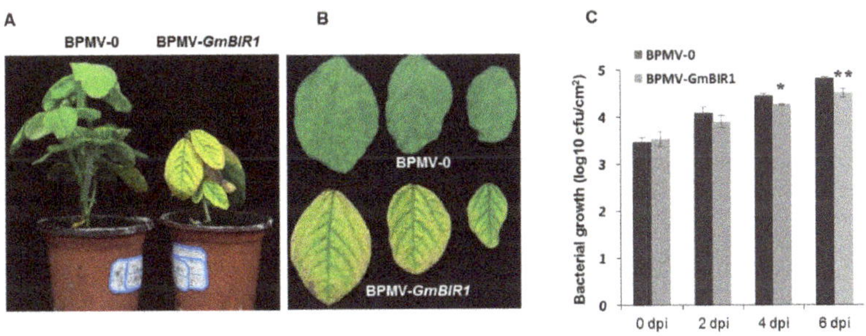

Figure 4. Silencing *GmBIR1* enhanced the resistance of soybean plants to *Pseudomonas syringae pv. glycinea* (*Psg*) R4. (**A**) Comparison of the vector control plants and the *GmBIR1*-silenced plants 4 days after spraying *Psg* suspension; (**B**) Comparison of the leaves from the vector control plants and the *GmBIR1*-silenced plants 4 days after spraying *Psg*; (**C**) Comparison of the colony forming unit (cfu)between the vector control plants and the *GmBIR1*-silenced plants after different days of *Psg* infection. * and ** represent $p < 0.05$ and 0.01, respectively (Student's *t*-test). This experiment was repeated 2 times with similar results.

2.4. *The Activated Defense Responses in GmBIR1-Silenced Plants Are Independent of GmMPK3/6 Activation*

Activation of MAPK signaling pathway is one of the signature hallmarks of activated defense responses [34]. To test whether the activated MAPK signaling pathway contributes to the activated defense responses and enhanced disease resistance observed in *GmBIR1*-silenced plants (Figures 1–4), the activation of *Gm*MPK3 and *Gm*MPK6 in response to flg22 was examined both in vector control plants and *GmBIR1*-silenced plants using the Phospho-p44/42 MAP Erk1/2 antibody that recognizes phosphorylation of Arabidopsis and soybean MPK3/4/6 at the sites corresponding to threonine (T) 202/tyrosine (Y) 204 of human (*Homo sapiens*) p44 MAPK [23,24,35]. Contrary to our expectation, activation of both *Gm*MPK3 and *Gm*MPK6 by flg22 was significantly reduced in *GmBIR1*-silenced plants compared to vector control plants (Figure 4), indicating that the activated defense responses observed in the *GmBIR1*-silenced plants are independent of *Gm*MPK3/6 activation.

2.5. *GmBIR1 Is Localized at the Plasma Membrane (PM) and in the Nucleus*

RLKs normally function at the PM. To investigate the subcellular location of *Gm*BIR1, the full-length *GmBIR1a* was fused to the C-terminus of GFP and transiently co-expressed in *N. benthamiana* leaves with a PM marker, *At*CBL1-mCherry, via agroinfiltration. As shown in Figures 5 and 6, GFP-*Gm*BIR1 was not only co-localized with *At*CBL1-mCherry but also presented in the nucleus, indicating that *Gm*BIR1 has dual subcellular localizations at the PM and in the nucleus.

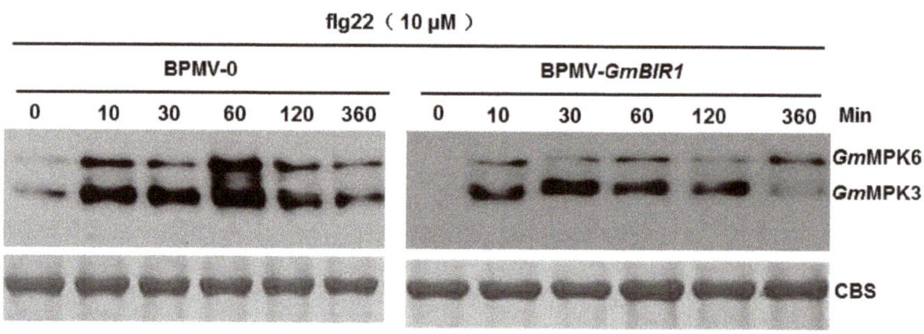

Figure 5. Silencing *GmBIR1* reduces the activation of *Gm*MPK3/*Gm*MPK6 in response to flg22 treatment. Leaf discs from indicated soybean plants were incubated on moist filter paper for 24 h to allow recovery from wounding before being treated with 10 μM flg22 or diluted DMSO. The kinase activities were detected by Western blotting using Phosph-p44/p42 MAP Erk1/2 antibody. Coomassie Blue–stained gel (CBS) used as loading controls. The experiment was repeated 3 times with similar results.

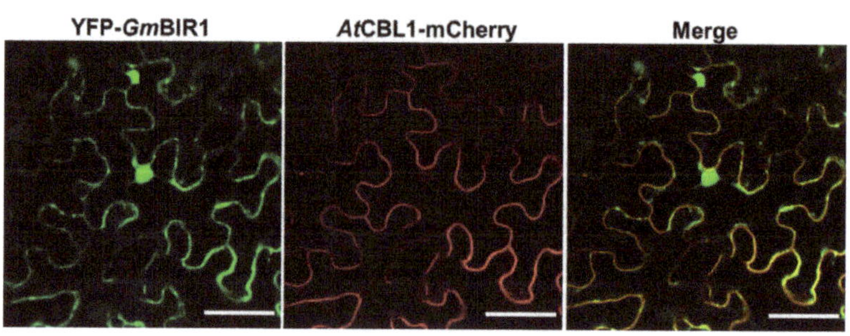

Figure 6. The subcelluar localization of GFP-*Gm*BIR1. The GFP-*Gm*BIR1 localizes both at plasma membrane (PM) and in the nucleus. The PM-localized protein *At*CBL1-mCherry was used as a control. Bar = 50 μm. This experiment was repeated 3 times with similar results.

3. Discussion

3.1. The Function of BIR1 Homologs Is Conserved from Different Plant Species

It has been reported that the extensive cell death, induction of *PR* genes, enhanced accumulation of SA and H_2O_2, impaired activation of MPK4, and enhanced resistance to different pathogens were observed for Arabidopsis *bir1-1* mutant [14,15,36–38]. Similarly, spontaneous cell death, constitutively activated autoimmune responses (including induced *PR* gene expression), accumulated levels of both SA and H_2O_2, and enhanced disease resistance to both viral and bacterial pathogens were also observed for the *GmBIR1*-silenced soybean plants (Figures 1–3). The phenotypic reminiscence between *bir1-1* mutant and the *GmBIR1*-silenced plants strongly indicate that the functions of BIR1 homologs are highly conserved across plant species in negatively regulating cell death and immunity. Given that cell death occurs in *bir1-1* mutants even under sterile conditions, BIR1 may not function in perception of PAPMs (pathogen-associated molecular patterns).

3.2. The Constitutively Activated Cell Death and Defense Responses in GmBIR1-Silenced Plants Is A Result of Co-Silencing of Both GmBIR1a and GmBIR1b

Soybean is a paleopolyploid and 75% of genes in its genome contain at least two copies [39]. Consistent with this, there are two *GmBIR1* genes, *GmBIR1a* and *GmBIR1b*, in the soybean genome. The open reading frame sequences of *GmBIR1a* and *GmBIR1b* share 95% identity at the nucleotide level. The DNA fragment used for *GmBIR1* silencing contains four stretches of sequences with 100% identity between *GmBIR1a* and *GmBIR1b* that could potentially generate siRNAs of >20 nt in length (Supplementary Figure S1). Therefore, both *GmBIR1a* and *GmBIR1b* were silenced using the same construct (Figure 1G). The autoimmune phenotype of the *GmBIR1*-silenced plants (Figure 1B) actually is a result of co-silencing of both *GmBIR1a* and *GmBIR1b*. We have previously shown that multiple copies of *GmMPK4*, *GmMPK6*, *GmMEKK1*, *GmFLS2*, and *GmATG2* genes could be successfully co-silenced using a single BPMV-VIGS construct [20,22–25]. Given that soybean is polyploidy and is recalcitrant to transformation, VIGS is currently one of the most efficient approaches to interrogate gene functions in soybean [26,27].

3.3. Does the Autoimmunity Observed in GmBIR1-Silenced Plants Result from Activated PTI or Activated ETI?

Activation of either PTI or ETI results in autoimmune responses, although the autoimmune responses resulting from ETI are much stronger and last longer than those resulting from PTI [1]. Structure analysis indicates that Arabidopsis BIR1 and BAK1 directly interact with each other through their ecto-LRR domains [36]. Under no-elicited conditions, $BIR1^{LRR}$–$BAK1^{LRR}$ interaction sequesters BAK1 and prevents formation of a signal competent complex with FLS2. In the presence of flg22, flg22-bound $FLS2^{LRR}$ outcompetes $BIR1^{LRR}$ for binding to $BAK1^{LRR}$ from the $BIR1^{LRR}$–$BAK1^{LRR}$ complex and forms a signaling competent flg22–$FLS2^{LRR}$–$BAK1^{LRR}$ complex [36] Based on this structural study, BIR1 functions to keep BAK1-mediated cell death signaling under tight control, thus preventing undesired autoimmunity, and therefore, it is possible that the autoimmunity in *bir1-1* or *GmBIR1*-silenced plants is a result of activation of BAK1-mediated PTI. Our *Psg* infection assay supported this possibility (Figure 4). However, inconsistent with this possibility, the flg22-induced activation of *Gm*MPK3 and *Gm*MPK6, which is a hallmark event of activated PTI (Meng et al., 2013), was significantly reduced in *GmBIR1*-silenced plants relative to vector control plants (Figure 4), and the activation of MPK3 and MPK6 in the Arabidopsis *bir1-1* mutant is unchanged compared to WT Col-0 [14]. Therefore, it is still unclear whether the autoimmune responses observed in *bir1-1* or *GmBIR1*-silenced plants are solely due to activated PTI, reflecting the complexity of the involvement of BIR1 in plant immunity.

Several lines of evidence indicate that autoimmune responses in *bir1-1* or the *GmBIR1*-silenced plants could result from misactivation of R proteins. Firstly, loss function of *BIR1* in Arabidopsis leads to the impairment in the activation of MPK4 by flagellin [14].

The autoimmune phenotype of *bir1-1* could be the consequence of the impaired MPK4 activation given that Arabidopsis MEKK1-MKK1/2-MPK4 signaling module negatively regulates the activation of the R proteins SUMM2 and SMN1/RPS6 [40–42]. Misactivation of SUMM2 and SMN1/RPS6 result from mutations in MEKK1-MKK1/2-MPK4 pathway leads to autoimmune responses [40,42]. However, *Gm*MPK4 activity cannot be induced either by flg22 or SA, regardless of concentration and duration of treatments [23]. Therefore, it is unclear whether *Gm*MPK4 plays a similar role as its counterpart in Arabidopsis. As the *Gm*MPK4 activation is also not observed in the vector control plants (Figure 4 and [23]), it is unlikely that the *GmBIR1*-silenced phenotypes are related to *Gm*MPK4 function. Secondly, in *Arabidopsis thaliana*, the copain-like protein BONZAI1 (BON1) interacts physically with and phosphorylates BIR1 and *bir1-1* mutant is partially suppressed by loss of function mutation in the resistance gene *SNC1* [18]. Thirdly, the autoimmune phenotype of *bir1-1* can be partially suppressed by elevated temperatures [18], which is a characteristic feature of R protein-mediated responses [43,44]. Lastly, the autoimmune phenotype of *bir1-1* is rescued by the loss of function of EDS1 and PAD4 [14], two key players in TIR type of R protein activation [45]. These results strongly suggest that the autoimmune phenotype of the *bir1-1* mutant is a result of misactivated R protein(s). Therefore, it is possible that BIR1 might participate in both PTI and ETI through interacting with different proteins, and the autoimmune responses observed in *bir1-1* and *GmBIR1*-silenced plants could be combined effects of activated PTI and ETI. The facts that RLKs/RLPs are required for NLR-mediated immunity [46] and RLKs/RLPs and NLRs mutually potentiate each other [47] support the idea that BIR1 may participate in both PTI and ETI.

3.4. Is the Enhanced Resistance of GmBIR1-Silenced plants to SMV SA-Dependent?

The enhanced resistance of *GmBIR1*-silenced plants against *Psg* could be attributed to the increased accumulation of both SA and H_2O_2 (Figure 2) and is thus SA-dependent. The enhanced SMV resistance in *GmBIR1*-silenced plants may not be solely due to increased accumulation of both SA and H_2O_2. In Arabidopsis, Guzman-Benito et al. [38] reported that the enhanced resistance of *bir1-1* to *Tobacco rattle virus* (TRV) is independent of cell death or SA defense priming. The expression of Arabidopsis *BIR1* is regulated both at TGS and PTGS level by RNA silencing pathway [38]. A threshold level of *BIR1* expression must be properly maintained and an expression beyond the threshold level (either too high or too low) triggers autoimmunity. Whether the enhanced resistance of *GmBIR1*-silenced plants to SMV is uncoupled from the cell death needs to be further investigated.

4. Materials and Methods

4.1. Plant Materials

Seeds of soybean (Glycine max 'Williams 82') used in this study were harvested from greenhouse-grown plants previously indexed for the absence of BPMV and SMV [48,49]. Soybean plants were grown in a growth chamber at 22 degrees with a photoperiod of 16 h.

4.2. Flg22 Peptides

The flg22 peptide was synthesized by HuaBio (Hangzhou, China).

4.3. BPMV-Mediated VIGS

BPMV strains, BPMV VIGS constructs, and inoculation of soybean seedlings with DNA-based BPMV constructs via biolistic particle bombardments using a Biolistic PDS-1000/He system (Bio-Rad Laboratories, Hercules, CA, USA) have been described previously [48]. The orthologs of ArabidopsisBIR1 in the soybean genome were identified by BLASTN in phytozome databases. The primers used for construction of the BPMV-*GmBIR1* (Glyma.18g246400.1) are:

GmBIR1-F: 5′-aagGGATCCCTCACTGGTCAAATTCCTGCTA-3′
GmBIR1-R: 5′-ttgGGTACCCAGGGTCCTCTTCCTTCTTCCTA-3′

The underlined sequences are *Bam*H I and *Kpn* I restriction sites, respectively, which were used for directional cloning of the PCR fragment into the BPMV VIGS (IA-D35). The letter C in bold indicates an extra nucleotide in reverse primers needed to maintain the correct open reading frame with BPMV genome.

4.4. RNA Isolation and RT-PCR

RNA isolation and RT-PCR were performed as described elsewhere [50]. The RNA samples extracted were treated with DNaseI according to the manufacturer's instructions (Invitrogen, Carlsbad, CA, USA). Primers used for RT-PCR in this study are:

GmELF1b-F: 5′- ACCGAAGAGGGCATCAAATCCC -3′
GmELF1b-R: 5′- CTCAACTGTCAAGCGTTCCTC -3′
GmBIR1-full-F: 5′- ATAGGAAGAAGGAAGAGGACC -3′
GmBIR1-full-R: 5′- TTTAGGAGTAGCCACCAAAGT -3′
GmBIR1a-F: 5′- GACTTGCTCAATATCCCAGGATAAGT-3′
GmBIR1a-R: 5′- TGTTGAGATTCCTGTAATCTCTTAACCAT -3′
GmBIR1b-F: 5′- TACTTCCTCAATATCCCAGGTGG -3′
GmBIR1b-R: 5′- TATTGAGATTCCTGCAATCTCTTGAC -3′.

The *GmBIR1a*-full primers were used for amplifying the full-length cDNA and the *GmBIR1a* and *GmBIR1b* primers are used for determining the specific transcript levels of the *GmBIR1a* and *GmBIR1b*, respectively.

4.5. Trypan Blue Staining

Cell death was detected specifically by trypan blue staining solution as described by [18]. At 18 dpi with BPMV vector only (BPMV-0) or BPMV-*GmBIR1* constructs, second fully expanded soybean trifoliolate leaves counting from the top were detached. The detached leaves were placed in trypan blue staining solution (10 mL glycerol, 10 mL lactic acid, 10 mL water, 10 mg trypan blue powder and 10 g phenol) and boiled for 1–1.5 min. The stained leaves were cleared using tap water and then dipped in the chloral hydrate solution in the shaker overnight to remove the non-specific stains. Images were captured using a stereo microscope (Olympus SZH10, Center Valley, PA, USA).

4.6. SA Quantification

SA was quantified using an Agilent 1260 HPLC system (Agilent Technologies) with a diode array detector and a fluorescence detector and a column as described previously [51]. The column was a 4.6-by-75-mm Zorbax SB-C18 with a mobile phase comprising sodium acetate (0.2 m, pH 5.5) and methanol with a series of concentration gradients. The initial methanol gradient was maintained at 3% (v/v) for 12 min, linearly increased to 7% (v/v) at 12.5 min and maintained until 38 min at a flow rate of 0.8 mL min−1 throughout the process. After 1 min, the initial reaction conditions were restored and the system needed to be equilibrated for 7 min before the next injection. Total SA from the sample treated with β-Glucosidase and free SA were detected with a 296 nm excitation wavelength and 410 nm emission wavelength. The concentration was calculated by determining the HPLC peak area according to a standard curve.

4.7. H_2O_2 Detection by DAB Staining

H_2O_2 was detected by an endogenous peroxidase-dependent in situ histochemical staining procedure using DAB (Sigma-Aldrich, St. Louis, MO, USA) [32]. Detached leaves were placed in a solution containing 1 mg mL−1 DAB (pH 5.5) for 2 h. The leaves were cleared by boiling in 96% (v/v) ethanol for 10 min. H_2O_2 production was visualized as a reddish-brown precipitate on the cleared leaves [32].

4.8. Inoculation of Pseudomonas Syringae pv. Glycinea (Psg) onto the Leaves of Soybean Plants

Psg growth assay was performed as described by [24]. *Psg* was cultivated at 28 degrees until OD = 1.3. The bacterial culture was centrifuged at 3000 rpm for 10 min and the pellet was re-suspended in double-distilled water with an OD = 1. The vector control or *GmBIR1*-silenced plants were thoroughly and evenly sprayed with the re-suspended bacterial solution on both upsides and downsides of the leaves. The sprayed plants were immediately covered with transparent plastic bags for at least 24 h.

4.9. SMV-N-GUS Inoculation, GUS Staining, and GUS Foci Measurements

At 18 dpi with BPMV vector only (BPMV-0) or BPMV-*GmBIR1* constructs, second fully expanded soybean trifoliolate leaves counting from the top were detached and biolistically inoculated with SMV-N-GUS [33,48]. Following SMV-N-GUS inoculation, the detached leaves were put into petri dishes with moist filter papers and kept on a lighted growth shelf for 5 days before GUS staining. GUS staining was performed as described by [52]. Photographs of the leaves with GUS foci were taken using a stereo microscope (Olympus SZH10, Center Valley, PA, USA). The numbers of GUS foci were counted, and the diameters of GUS foci were measured using Soft Image System analysis (IA Package; Olympus).

4.10. Western Blot Analysis for Detecting Phosphorylated MPKs

Protein was extracted from soybean leaf tissues using extraction buffer (50 mM Tris-MES pH 8.0, 0.5 M sucrose, 1 mM $MgCl_2$, 10 mM EDTA, 5 mM DTT) with protease inhibitor cocktail S8830 (Sigma-Aldrich, St. Louise, MO, USA) added as described by [23]. The extract was centrifuged at 12,000 rpm at 4 °C for 30 min and the supernatant was collected. For immunoblotting, the extracted proteins were separated by SDS-PAGE (10% acrylamide gel) and transferred to PVDF membrane (Millipore, Billerica, MA, USA) by semi-dry electro-transfer (Bio-Rad, Hercules, CA, USA). The membrane was blocked in 1× TBST buffer containing 5% milk powder for 2 h. After washing, the membrane was further incubated with anti–Phospho-p44/p42 MAPK (anti-pTEpY) diluted at 1:3000 (Cell Signaling Technology, Danvers, MA, USA), followed by incubation with secondary antibody diluted at 1:7500. The bands were visualized by incubating with chemiluminescent HRP substrate (Millipore, Billerica, MA, USA).

4.11. Subcellular Localization of GmBIR1

The *GmBIR1* (Glyma.18g246400.1) open reading frame was amplified by RT-PCR from total RNA extracted from cv Williams 82 soybean plants using the following primers:

*Gm*BIR1-PGDG-F-HinDIII: 5′-**AAGCTT**ATGAAGATGTTTATGGGTGGC-3′
*Gm*BIR1-PGDG-R-BamHI: 5′-**GGATCC**TCAATCATGTCCCTCTCGAG-3′

The letters in bold represent the cleavage sites for different restriction enzymes.

The RT-PCR products were firstly double digested with HindIII + BamHI and subsequently cloned into the PGDG vector [53] to generate the GFP-*Gm*BIR1 fusion construct. These fusion constructs and the *AtCBL1*-mCherry construct (membrane marker) were coinfiltrated into *N. benthamiana* leaves as described by [50]. Images were captured with a confocal laser-scanning microscope (Leica TCS SP5 AOBS, Wetzlar, Germany).

Supplementary Materials: The following supporting information can be downloaded at: https://www.mdpi.com/article/10.3390/ijms23137450/s1.

Author Contributions: J.-Z.L. designed the experiments and wrote the original draft; D.-D.L., H.-J.L., H.S.M., M.-Y.Y., X.-Y.D., C.-L.Z. and S.-N.T. performed the experiments; J.-Z.L. and D.-D.L. prepared figures. All authors have read and agreed to the published version of the manuscript.

Funding: This research was funded by the National Natural Science Foundation of China (grant numbers 32170761 and 31571423 to J.Z.L.).

Informed Consent Statement: No applicable.

Data Availability Statement: No applicable.

Acknowledgments: We thank Steven A. Whitham and John Hill for providing the BPMV-VIGS system, *Pseudomonas syringae pv. glycinea* R4 strain and SMV-N-GUS construct.

Conflicts of Interest: The authors declare no conflict of interest.

References

1. Jones, J.D.; Dangl, J.L. The plant immune system. *Nature* **2006**, *444*, 323–329. [CrossRef] [PubMed]
2. Shiu, S.H.; Bleecker, A.B. Receptor-like kinases from Arabidopsis form a monophyletic gene family related to animal receptor kinases. *Proc. Natl. Acad. Sci. USA* **2001**, *98*, 10763–10768. [CrossRef] [PubMed]
3. Jamieson, P.A.; Shan, L.; He, P. Plant cell surface molecular cypher: Receptor-like proteins and their roles in immunity and development. *Plant Sci.* **2018**, *274*, 242–251. [CrossRef] [PubMed]
4. Boller, T.; Felix, G. A renaissance of elicitors: Perception of microbe-associated molecular patterns and danger signals by pattern-recognition receptors. *Annu. Rev. Plant Biol.* **2009**, *60*, 379–406. [CrossRef] [PubMed]
5. Gómez-Gómez, L.; Boller, T. FLS2: An LRR Receptor–like Kinase Involved in the Perception of the Bacterial Elicitor Flagellin in Arabidopsis. *Mol. Cell* **2009**, *5*, 1003–1011. [CrossRef]
6. Zipfel, C.; Kunze, G.; Chinchilla, D.; Caniard, A.; Jones, J.D.; Boller, T.; Felix, G. Perception of the Bacterial PAMP EF-Tu by the Receptor EFR Restricts Agrobacterium-Mediated Transformation. *Cell* **2006**, *125*, 749–760. [CrossRef]
7. Li, J.; Wen, J.; Lease, K.A.; Doke, J.T.; Tax, F.E.; Walker, J.C. BAK1, an Arabidopsis LRR Receptor-like Protein Kinase, Interacts with BRI1 and Modulates Brassinosteroid Signaling. *Cell* **2002**, *110*, 213–222. [CrossRef]
8. Nam, K.H.; Li, J. BRI1/BAK1, a Receptor Kinase Pair Mediating Brassinosteroid Signaling. *Cell* **2002**, *110*, 203–212. [CrossRef]
9. Chinchilla, D.; Zipfel, C.; Robatzek, S.; Kemmerling, B.; Nürnberger, T.; Jones, J.D.G.; Felix, G.; Boller, T. A flagellin-induced complex of the receptor FLS2 and BAK1 initiates plant defence. *Nature* **2007**, *448*, 497–500. [CrossRef]
10. Heese, A.; Hann, D.R.; Gimenez-Ibanez, S.; Jones, A.M.; He, K.; Li, J.; Schroeder, J.I.; Peck, S.C.; Rathjen, J.P. The receptor-like kinase SERK3/BAK1 is a central regulator of innate immunity in plants. *Proc. Natl. Acad. Sci. USA* **2007**, *104*, 12217–12222. [CrossRef]
11. Roux, M.; Schwessinger, B.; Albrecht, C.; Chinchilla, D.; Jones, A.; Holton, N.; Malinovsky, F.G.; Tör, M.; de Vries, S.; Zipfel, C. The Arabidopsis leucine-rich repeat receptor-like kinase BAK1/SERK3 and BKK1/SERK4 are required for the innate immunity to hemibiotrophic and biotrophic pathogens. *Plant Cell* **2011**, *23*, 2440–2455. [CrossRef] [PubMed]
12. Macho, A.P.; Schwessinger, B.; Ntoukakis, V.; Brutus, A.; Segonzac, C.; Roy, S.; Kadota, Y.; Oh, M.H.; Sklenar, J.; Derbyshire, P.; et al. A bacterial tyrosine phosphatase inhibits plant pattern recognition receptor activation. *Science* **2014**, *343*, 1509–1512. [CrossRef] [PubMed]
13. Couto, D.; Zipfel, C. Regulation of pattern recognition receptor signalling in plants. *Nat. Rev. Immunol.* **2016**, *16*, 537–552. [CrossRef] [PubMed]
14. Gao, M.; Wang, X.; Wang, D.; Xu, F.; Ding, X.; Zhang, Z.; Bi, D.; Cheng, Y.T.; Chen, S.; Li, X.; et al. Regulation of cell death and innate immunity by two receptor-like kinases in *Arabidopsis*. *Cell Host Microbe* **2009**, *6*, 34–44. [CrossRef] [PubMed]
15. Liu, Y.; Huang, X.; Li, M.; He, P.; Zhang, Y. Loss-of-function of Arabidopsis receptor-like kinase BIR1 activates cell death and defense responses mediated by BAK1 and SOBIR1. *New Phytol.* **2016**, *212*, 637–645. [CrossRef] [PubMed]
16. Liebrand, T.W.; van den Berg, G.C.; Zhang, Z.; Smit, P.; Cordewener, J.H.; America, A.H.; Sklenar, J.; Jones, A.M.; Tameling, W.I.; Robatzek, S.; et al. Receptor-like kinase SOBIR1/EVR interacts with receptor-like proteins in plant immunity against fungal infection. *Proc. Natl. Acad. Sci. USA* **2013**, *110*, 10010–10015. [CrossRef]
17. Liebrand, T.W.; van den Burg, H.A.; Joosten, M.H. Two for all: Receptor-associated kinases SOBIR1 and BAK1. *Trends Plant Sci.* **2014**, *19*, 123–132. [CrossRef]
18. Wang, Z.; Meng, P.; Zhang, X.; Ren, D.; Yang, S. BON1 interacts with the protein kinases BIR1 and BAK1 in modulation of temperature-dependent plant growth and cell death in Arabidopsis. *Plant J.* **2011**, *67*, 1081–1093. [CrossRef]
19. Halter, T.; Imkampe, J.; Mazzotta, S.; Wierzba, M.; Postel, S.; Bücherl, C.; Kiefer, C.; Stahl, M.; Chinchilla, D.; Wang, X.; et al. The leucine-rich repeat receptor kinase BIR2 is a negative regulator of BAK1 in plant immunity. *Curr. Biol.* **2014**, *24*, 134–143. [CrossRef]
20. Liu, J.Z.; Horstman, H.D.; Braun, E.; Graham, M.A.; Zhang, C.; Navarre, D.; Qiu, W.L.; Lee, Y.; Nettleton, D.; Hill, J.H.; et al. Soybean homologs of MPK4 negatively regulate defense responses and positively regulate growth and development. *Plant Physiol.* **2011**, *157*, 1363–1378. [CrossRef]
21. Liu, J.Z.; Whitham, S.A. Overexpresson of a soybean nuclear-localized type III DnaJ domain-containing HSP40 reveals it roles in cell death and disease resistance. *Plant J.* **2013**, *74*, 110–121. [CrossRef] [PubMed]
22. Liu, J.Z.; Braun, E.; Qiu, W.L.; Shi, Y.F.; Marcelino-Guimarães, F.C.; Navarre, D.; Hill, J.H.; Whitham, S.A. Positive and negative roles for soybean MPK6 in regulating defense responses. *Mol. Plant Microbe Interact.* **2014**, *27*, 824–834. [CrossRef] [PubMed]
23. Xu, H.; Zhang, C.; Li, Z.C.; Wan, Z.R.; Jiang, X.X.; Tian, S.N.; Braun, E.; Mei, Y.; Qiu, W.L.; Li, S.; et al. The MAPK kinase kinase GmMEKK1 regulates cell death and defense responses. *Plant Physiol.* **2018**, *178*, 907–922. [CrossRef] [PubMed]
24. Tian, S.N.; Liu, D.D.; Zhong, C.L.; Xu, H.Y.; Yang, S.; Fang, Y.; Ran, J.; Liu, J.Z. Silencing *GmFLS2* enhances the susceptibility of soybean to bacterial pathogen through attenuating the activation of *Gm*MAPK signaling pathway. *Plant Sci.* **2019**, *292*, 110386. [CrossRef]

25. Hashimi, S.M.; Wu, N.N.; Ran, J.; Liu, J.Z. Silencing Autophagy-Related Gene 2 (ATG2) Results in Accelerated Senescence and Enhanced Immunity in Soybean. *Int. J. Mol. Sci.* **2021**, *22*, 11749. [CrossRef]
26. Liu, J.Z.; Graham, M.A.; Pedley, K.F.; Whitham, S.A. Gaining insight into soybean defense responses using functional genomics approaches. *Brief. Funct. Genom.* **2015**, *14*, 283–290. [CrossRef]
27. Liu, J.Z.; Fang, Y.; Pang, H. The current status of the soybean-soybean mosaic virus (SMV) pathosystem. *Front. Microbiol.* **2016**, *7*, 1906. [CrossRef]
28. Petersen, M.; Brodersen, P.; Naested, H.; Andreasson, E.; Lindhart, U.; Johansen, B.; Nielsen, H.B.; Lacy, M.; Austin, M.J.; Parker, J.E.; et al. Arabidopsis map kinase 4 negatively regulates systemic acquired resistance. *Cell* **2000**, *103*, 1111–1120. [CrossRef]
29. Lamb, C.; Dixon, R.A. The oxidative burst in plant disease resistance. *Annu. Rev. Plant Physiol. Plant Mol. Biol.* **1997**, *48*, 251–275. [CrossRef]
30. Delledonne, M.; Xia, Y.; Dixon, R.A.; Lamb, C. Nitric oxide functions as a signal in plant disease resistance. *Nature* **1998**, *394*, 585–588. [CrossRef]
31. Thordal-Christensen, H.; Zhang, Z.G.; Wei, Y.D.; Collinge, D.B. Subcellular localization of H_2O_2 in plants: H_2O_2 accumulation in papillae and hypersensitive response during the barley-powdery mildew interaction. *Plant J.* **1997**, *11*, 1187–1194. [CrossRef]
32. Ren, D.; Yang, H.; Zhang, S. Cell death mediated by MAPK is associated with hydrogen peroxide production in Arabidopsis. *J. Biol. Chem.* **2002**, *277*, 559–565. [CrossRef] [PubMed]
33. Wang, L.; Eggenberger, A.; Hill, J.; Bogdanove, A.J. *Pseudomonas syringae* effector avrB confers soybean cultivar-specific avirulence on Soybean mosaic virus adapted for transgene expression but effector avrPto does not. *Mol. Plant Microbe Interact.* **2006**, *19*, 304–312. [CrossRef] [PubMed]
34. Meng, X.; Zhang, S. MAPK cascades in plant disease resistance signaling. *Annu. Rev. Phytopathol.* **2013**, *51*, 245–266. [CrossRef]
35. Zhao, C.; Nie, H.; Shen, Q.; Zhang, S.; Lukowitz, W.; Tang, D. EDR1 physically interacts with MKK4/MKK5 and negatively regulates a MAP kinase cascade to modulate plant innate immunity. *PLoS Genet.* **2014**, *10*, e1004389. [CrossRef]
36. Ma, C.; Liu, Y.; Bai, B.; Han, Z.; Tang, J.; Zhang, H.; Yaghmaiean, H.; Zhang, Y.; Chai, J. Structural basis for BIR1-mediated negative regulation of plant immunity. *Cell Res.* **2017**, *27*, 1521–1524. [CrossRef]
37. Wierzba, M.P.; Tax, F.E. An allelic series of bak1 mutations differentially alter bir1 cell death, immune response, growth, and root development phenotypes in *Arabidopsis thaliana*. *Genetics* **2016**, *202*, 689–702. [CrossRef]
38. Guzmán-Benito, I.; Donaire, L.; Amorim-Silva, V.; Vallarino, J.G.; Esteban, A.; Wierzbicki, A.T.; Ruiz-Ferrer, V.; Llave, C. The immune repressor BIR1 contributes to antiviral defense and undergoes transcriptional and post-transcriptional regulation during viral infections. *New Phytol.* **2019**, *224*, 421–438. [CrossRef]
39. Schmutz, J.; Cannon, S.B.; Schlueter, J.; Ma, J.; Mitros, T.; Nelson, W.; Hyten, D.L.; Song, Q.; Thelen, J.J.; Cheng, J. Genome sequence of the palaeopolyploid soybean. *Nature* **2010**, *463*, 178–183. [CrossRef]
40. Zhang, Z.; Wu, Y.; Gao, M.; Zhang, J.; Kong, Q.; Liu, Y.; Ba, H.; Zhou, J.; Zhang, Y. Disruption of PAMP-induced MAP kinase cascade by a *Pseudomonas syringae* effector activates plant immunity mediated by the NB-LRR protein SUMM2. *Cell Host Microbe* **2012**, *11*, 253–263. [CrossRef]
41. Kong, Q.; Qu, N.; Gao, M.; Zhang, Z.; Ding, X.; Yang, F.; Li, Y.; Dong, O.X.; Chen, S.; Li, X.; et al. The MEKK1-MKK1/MKK2-MPK4 kinase cascade negatively regulates immunity mediated by a mitogen-activated protein kinase kinase kinase in Arabidopsis. *Plant Cell* **2012**, *24*, 2225–2236. [CrossRef] [PubMed]
42. Takagi, M.; Hamano, K.; Takagi, H.; Morimoto, T.; Akimitsu, K.; Terauchi, R.; Shirasu, K.; Ichimura, K. Disruption of the MAMP-Induced MEKK1-MKK1/MKK2-MPK4 Pathway Activates the TNL Immune Receptor SMN1/RPS6. *Plant Cell Physiol.* **2019**, *60*, 778–787. [CrossRef] [PubMed]
43. Whitham, S.; Dinesh-Kumar, S.P.; Choi, D.; Hehl, R.; Corr, C.; Baker, B. The product of the tobacco mosaic virus resistance gene N: Similarity to toll and the interleukin-1 receptor. *Cell* **1994**, *78*, 1101–1115. [CrossRef]
44. Yang, S.; Hua, J. A haplotype-specific Resistance gene regulated by BONZAI1 mediates temperature-dependent growth control in Arabidopsis. *Plant Cell* **2004**, *16*, 1060–1071. [CrossRef]
45. Wiermer, M.; Feys, B.J.; Parker, J.E. Plant immunity: EDS1 regulatory node. *Curr. Opin. Plant Biol.* **2005**, *8*, 383–389. [CrossRef] [PubMed]
46. Yuan, M.; Jiang, Z.; Bi, G.; Nomura, K.; Liu, M.; Wang, Y.; Cai, B.; Zhou, J.M.; He, S.Y.; Xin, X.F. Pattern-recognition receptors are required for NLR-mediated plant immunity. *Nature* **2021**, *592*, 105–109. [CrossRef]
47. Ngou, B.P.M.; Ahn, H.K.; Ding, P.; Jones, J.D.G. Mutual potentiation of plant immunity by cell-surface and intracellular receptors. *Nature* **2021**, *592*, 110–115. [CrossRef]
48. Zhang, C.; Yang, C.; Whitham, S.A.; Hill, J.H. Development and use of an efficient DNA-based viral gene silencing vector for soybean. *Mol. Plant Microbe Interact.* **2009**, *22*, 123–131. [CrossRef]
49. Zhang, C.; Bradshaw, J.D.; Whitham, S.A.; Hill, J.H. The development of an efficient multipurpose bean pod mottle virus viral vector set for foreign gene expression and RNA silencing. *Plant Physiol.* **2010**, *153*, 52–65. [CrossRef]
50. Liu, J.Z.; Blancaflor, E.B.; Nelson, R.S. The tobacco mosaic virus 126-kilodalton protein, a constituent of the virus replication complex, alone or within the complex aligns with and traffics along microfilaments. *Plant Physiol.* **2005**, *138*, 1853–1865. [CrossRef]
51. Zhang, K.; Halitschke, R.; Yin, C.; Liu, C.J.; Gan, S.S. Salicylic acid 3-hydroxylase regulates Arabidopsis leaf longevity by mediating salicylic acid catabolism. *Proc. Natl. Acad. Sci. USA* **2013**, *110*, 14807–14812. [CrossRef] [PubMed]

52. Jefferson, R.A.; Kavanagh, T.A.; Bevan, M.W. GUS fusions: Beta-glucuronidase as a sensitive and versatile gene fusion marker in higher plants. *EMBO J.* **1987**, *6*, 3901–3907. [CrossRef] [PubMed]
53. Goodin, M.; Dietzgen, R.G.; Schichnes, D.; Ruzin, S.; Jackson, A.O. pGD vectors: Versatile tools for the expression of green and red fluorescent protein fusions in Agroinfiltrated plant leaves. *Plant J.* **2002**, *31*, 375–383. [CrossRef] [PubMed]

International Journal of *Molecular Sciences*

MDPI

Article

Increasing the Grain Yield and Grain Protein Content of Common Wheat (*Triticum aestivum*) by Introducing Missense Mutations in the *Q* Gene

Qing Chen [1,2], Zhenru Guo [2], Xiaoli Shi [2], Meiqiao Wei [2], Yazhen Fan [2], Jing Zhu [2], Ting Zheng [2], Yan Wang [2], Li Kong [2], Mei Deng [2], Xinyou Cao [3], Jirui Wang [1,2], Yuming Wei [1,2], Qiantao Jiang [2], Yunfeng Jiang [2], Guoyue Chen [1,2], Youliang Zheng [2] and Pengfei Qi [2,*]

1 State Key Laboratory of Crop Gene Exploration and Utilization in Southwest China, Chengdu 611130, China
2 Triticeae Research Institute, Sichuan Agricultural University, Chengdu 611130, China
3 Crop Research Institute, Shandong Academy of Agricultural Sciences, Jinan 250100, China
* Correspondence: pengfeiqi@hotmail.com; Tel.: +86-28-82650337; Fax: +86-28-82650350

Abstract: Grain yield (GY) and grain protein content (GPC) are important traits for wheat breeding and production; however, they are usually negatively correlated. The *Q* gene is the most important domestication gene in cultivated wheat because it influences many traits, including GY and GPC. Allelic variations in the *Q* gene may positively affect both GY and GPC. Accordingly, we characterized two new *Q* alleles (Q^{s1} and Q^{c1}-*N8*) obtained through ethyl methanesulfonate-induced mutagenesis. Compared with the wild-type *Q* allele, Q^{s1} contains a missense mutation in the sequence encoding the first AP_2 domain, whereas Q^{c1}-*N8* has two missense mutations: one in the sequence encoding the second AP_2 domain and the other in the microRNA172-binding site. The Q^{s1} allele did not significantly affect GPC or other processing quality parameters, but it adversely affected GY by decreasing the thousand kernel weight and grain number per spike. In contrast, Q^{c1}-*N8* positively affected GPC and GY by increasing the thousand kernel weight and grain number per spike. Thus, we generated novel germplasm relevant for wheat breeding. A specific molecular marker was developed to facilitate the use of the Q^{c1}-*N8* allele in breeding. Furthermore, our findings provide useful new information for enhancing cereal crops via non-transgenic approaches.

Keywords: wheat quality; agronomic trait; mutation; breeding

Citation: Chen, Q.; Guo, Z.; Shi, X.; Wei, M.; Fan, Y.; Zhu, J.; Zheng, T.; Wang, Y.; Kong, L.; Deng, M.; et al. Increasing the Grain Yield and Grain Protein Content of Common Wheat (*Triticum aestivum*) by Introducing Missense Mutations in the *Q* Gene. *Int. J. Mol. Sci.* **2022**, *23*, 10772. https://doi.org/10.3390/ijms231810772

Academic Editors: Andrés J. Cortés and Hai Du

Received: 28 August 2022
Accepted: 13 September 2022
Published: 15 September 2022

1. Introduction

Common wheat (*Triticum aestivum*) is a major food crop that serves as the primary protein source in the human diet. Wheat provides approximately 18% of the calories and 20% of the proteins consumed by humans worldwide [1]. Therefore, grain yield (GY) and grain protein content (GPC) are critical traits to be considered for wheat breeding and production. Because of increased demand driven by population growth and improvements in living conditions, there is an urgent need for wheat varieties with increased GY and GPC.

Nitrogen applications during wheat production are vital for increasing GY and GPC [2–4]. In order to produce wheat with a high GY and GPC, farmers tend to apply large amounts of nitrogen fertilizer to wheat fields, which increases cultivation costs and environmental pollution. Breeding to increase the wheat GY and GPC remains a considerable challenge because of the confirmed negative relationship between the two parameters [5–8].

Wheat flour has unique processing properties that enable it to be used to make diverse end-products. The end-use quality of wheat is significantly influenced by GPC. The unique processing quality of wheat flour depends on the seed storage proteins, especially gliadins and glutenins, which account for 60–80% of the total GPC [9,10]. Gliadins are monomeric compounds that contribute to dough extensibility [11], whereas glutenins, which are polymeric compounds linked by intermolecular disulfide bonds, affect dough elasticity. Glutenins consist of high and low molecular weight glutenin subunits [12,13].

Because the *Q* gene (TraesCS5A02G473800) influences many important traits, including GY, GPC, grain threshability, grain size, spike morphology, rachis fragility, plant height, and flowering time, it plays a major role in wheat domestication, de-domestication and breeding [14–18]. This gene is located on the long arm of chromosome 5A and encodes a member of the APETALA2 (AP_2) transcription factor family [15,19]. The AP_2 transcription factors have diverse functions affecting plant development [20]. The *Q* allele originated from a spontaneous mutation to the microRNA172-binding region of the *q* allele [15]. Similarly, the introduction of another point mutation in the microRNA172-binding site of the *Q* allele resulted in the Q^{c1} allele [17]. The *q*, *Q*, and Q^{c1} transcription levels are correlated with the number of point mutations in the microRNA172-binding site [15,17]. Compared with the effects of the *Q* allele, Q^{c1} increases GPC by approximately 60 g kg^{-1}, reflecting the value of Q^{c1} for wheat breeding. However, Q^{c1} decreases the longitudinal cell size of rachises, resulting in compact spikes and decreases in GY [17]. Missense mutations in the *Q* sequence encoding the AP_2 domain can lead to decreased spike density [15,21]. These challenges underline the need to identify or generate new *Q* alleles that positively affect GY and GPC.

In this study, we characterized two new *Q* alleles, namely Q^{s1} and Q^{c1}-*N8*, which have a single missense mutation in the sequences encoding the first and second AP_2 domains, respectively. They were obtained via the chemical treatment of common wheat lines carrying the *Q* and Q^{c1} alleles, respectively. The effects of Q^{s1} and Q^{c1}-*N8* on the wheat GY and GPC were investigated.

2. Results

2.1. Phenotype of the Mutant ss1 Carrying the Q^{s1} Allele

In order to elevate the effect of missense mutations in the sequence encoding the AP_2 domain of the Q protein, a mutant carrying the Q^{s1} allele (*ss1*; sparse spike 1) was isolated from the M_2 population of the common wheat cultivar 'Shumai482' (Figure 1). The *ss1* plants produced a speltoid-like spike (Figure 2a). In contrast to the *Q* allele (GenBank No. KX580301.2), Q^{s1} has a missense mutation (GenBank No. OK041024) in the sequence encoding the first AP_2 domain (Figure 3 and Figure S1). Compared with the WT control (Figure 1), *ss1* plants were taller (Figure 2b, Table 1) and had a longer main spike (Figure 2a, Table 1) but a lower spike density (Table 1).

Regarding the examined yield-related traits, the spikelet number per the main spike, grain number per the main spike, thousand kernel weight (Table 1), and grain width (Figure 2d,e) were lower for *ss1* than for the WT control. In contrast, the grain length (Figure 2c,f) was greater for *ss1* than for the WT control. Notably, there was no significant difference in the productive tiller number between the *ss1* and WT plants (Table 1). Therefore, GY was significantly lower ($p < 0.01$) for *ss1* (0.46 kg m^{-2}) than for the WT control (0.58 kg m^{-2}) in the 2018–2019 growing season. Under our experimental conditions, the Q^{s1} allele decreased GY by 20.6% by decreasing the thousand kernel weight and the grain number per spike.

An analysis of the processing quality traits revealed a lack of a significant difference between the *ss1* and WT plants in terms of GPC, wet gluten content, gluten index, Zeleny sedimentation value, water absorption, development time, stability time, and loaf volume (Table 2; Figure 4).

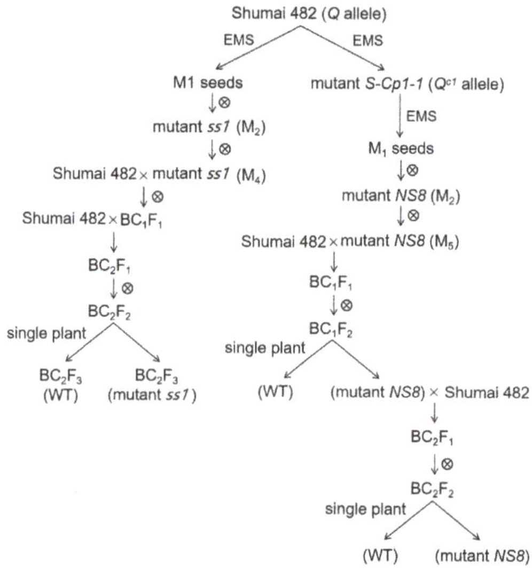

Figure 1. Outline of the generation of the mutants *ss1* and *NS8* in the *T. aestivum* cv. 'Shumai 482' genetic background.

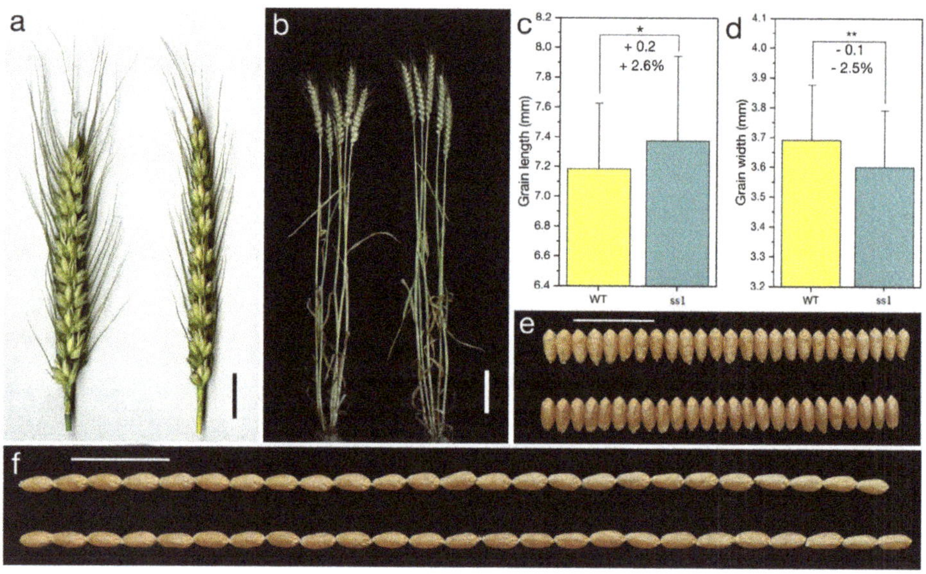

Figure 2. Phenotype of the mutant *ss1*. (**a**) Spikes of the wild-type (WT) (**left**) and *ss1* (**right**) plants. Scale bar, 2 cm. (**b**) WT (**left**) and *ss1* (**right**) plants. Scale bar, 10 cm. Comparisons of grain length (**c**) and grain width (**d**). **, $p < 0.01$; *, $p < 0.05$. Data are presented as the mean ± standard deviation. '+' and '−' indicate more and less than the WT control, respectively. '+0.2' in panel (**c**) indicates the *ss1* grain was 0.2 mm longer than the WT grain (on average). '+ 2.6%' in panel (**c**) indicates Q^{s1} increased the grain length by 2.6% (on average). (**e**) Kernel width of the WT (**upper**) and *ss1* (**lower**) samples. Scale bar, 2 cm. (**f**) Kernel length of the WT (**upper**) and *ss1* (**lower**) samples. Scale bar, 2 cm.

Figure 3. Alignment of the *Q* (GenBank No. KX580301.2), Q^{s1} (OK041024), Q^{c1} (KX580302.2), and Q^{c1}-*N8* (OK041023) open reading frames. The red arrows indicate the two specific missense mutations in the Q^{s1} and Q^{c1}-*N8* alleles. The microRNA172-binding site is boxed.

Table 1. Comparison of the agronomic traits of the mutant *ss1* and the wild-type (WT) control.

Traits	Growing Season	*ss1*	WT	E	G	E × G
Plant height (cm)	2018–2019	83.41 ± 4.05 **	79.44 ± 3.09	205.087 **	24.459 **	0.534
	2019–2020	71.82 ± 2.30 **	67.11 ± 2.00			
Spike length (cm)	2018–2019	14.65 ± 0.68 *	14.06 ± 0.75	123.331 **	15.408 **	2.680
	2019–2020	12.88 ± 0.80 **	11.87 ± 0.46			
Spikelet number per main spike	2018–2019	21.00 ± 1.15 **	22.94 ± 1.44	15.038 **	24.858 **	0.110
	2019–2020	19.93 ± 0.70 **	21.43 ± 1.45			
Spike density	2018–2019	1.43 ± 0.07 **	1.63 ± 0.09	45.434 **	84.944 **	13.78
	2019–2020	1.55 ± 0.10 **	1.80 ± 0.08			
Grain number per main spike	2018–2019	58.81 ± 5.83 *	65.13 ± 8.23	43.376 **	8.860 **	0.044
	2019–2020	50.43 ± 4.14 *	54.64 ± 4.25			
Thousand kernel weight (g)	2018–2019	44.08 ± 1.71 *	46.13 ± 0.43	50.427 **	13.780 **	0.216
	2019–2020	48.76 ± 1.25 *	51.90 ± 1.67			
Tiller number	2018–2019	5.30 ± 1.03	5.25 ± 1.33	27.478 **	0.730	1.207
	2019–2020	3.85 ± 0.74	4.25 ± 0.79			

**, $p < 0.01$; *, $p < 0.05$; E, environment; G, genotype; E × G, interaction between the environment and genotype. Data are presented as the mean ± standard deviation.

Table 2. Comparison of the processing quality parameters of the mutant *ss1* and the wild-type (WT) control.

Trails	Growing Season	*ss1*	WT	E	G	E × G
Grain protein content (%; dry weight)	2018–2019 2019–2020	13.12 ± 0.43 14.53 ± 0.60	13.45 ± 0.36 15.16 ± 0.76	30.313 **	2.562	0.323
Zeleny sedimentation value (mL)	2018–2019 2019–2020	32.65 ± 3.64 21.54 ± 2.51	27.73 ± 4.20 23.88 ± 1.98	29.114 **	0.301	4.315
Wet gluten content (%)	2018–2019 2019–2020	24.72 ± 1.60 27.21 ± 1.49	24.98 ± 3.06 29.66 ± 2.05	14.645 **	0.570	2.996
Gluten index (%)	2018–2019 2019–2020	92.87 ± 2.59 63.37 ± 9.02	87.32 ± 8.94 65.04 ± 6.69	65.599 **	0.004	0.315
Water absorption (%)	2018–2019 2019–2020	50.36 ± 0.71 58.74 ± 1.12	50.88 ± 1.67 58.28 ± 0.45	428.875 **	0.783	0.077
Development time (s)	2018–2019 2019–2020	78.00 ± 13.22 61.50 ± 10.71	77.67 ± 9.14 68.40 ± 10.46	7.440 *	0.365	2.875
Stability time (s)	2018–2019 2019–2020	260.86 ± 121.40 193.20 ± 56.34	203.83 ± 66.50 230.40 ± 54.54	0.394	0.125	1.560

**, $p < 0.01$; *, $p < 0.05$; E, environment; G, genotype; E × G, interaction between the environment and genotype. Data are presented as the mean ± standard deviation.

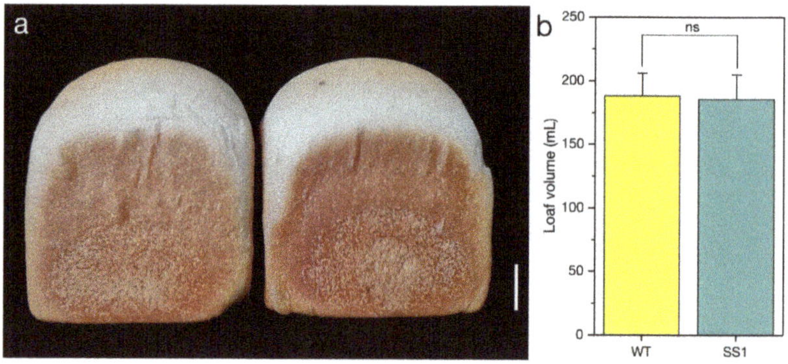

Figure 4. Q^{s1} has no effect on the bread loaf volume. (**a**) Comparison of the intact *ss1* (**left**) and wild-type (WT) (**right**) loaves. Scale bar, 2 cm. (**b**) Comparison of the *ss1* and WT loaf volumes. ns, not significant.

2.2. Characterization of the Mutant NS8 Carrying the Q^{c1}-N8 Allele

Because a missense mutation in the sequence encoding the AP_2 domain can decrease the spike density without altering processing quality parameters, the normal-spike mutant *NS8* containing the Q^{c1}*-N8* allele was isolated from the M_2 population of the mutant *S-Cp*1-1. The *NS8* plants had a normal spike, which was similar to the 'Shumai482′ spike (Figures 5a and S2a). Compared with the *Q* allele sequence, Q^{c1}*-N8* contains two missense mutations (GenBank No. OK041023), with one in the sequence encoding the second AP_2 domain and the other in the microRNA172-binding region (Figures 3 and S1). When compared with the WT control (plants with only the *Q* allele; Figure 1), the mutant *NS8* plants were shorter (Figures 5b, 6b, and S2b,j) but had a similar spike length (Figures 5a, 6d, and S2a,l) and spike density (Figures 6f and S2n).

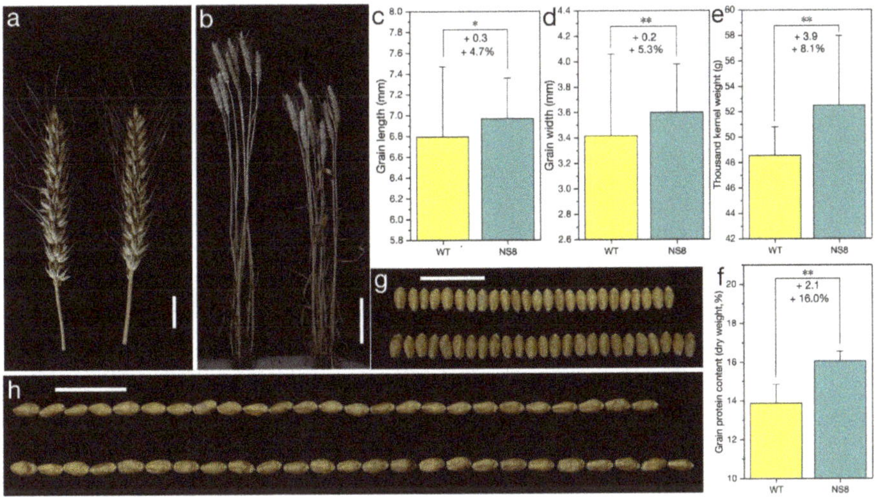

Figure 5. Phenotype of the mutant *NS8* during the 2020–2021 growing season. (**a**) Spikes of the wild-type (WT) (**left**) and *NS8* (**right**) plants. Scale bar, 2 cm. (**b**) WT (**left**) and *NS8* (**right**) plants. Scale bar, 10 cm. Comparisons of the grain length (**c**), grain width (**d**), thousand kernel weight (**e**), and grain protein contents (**f**) of the WT and *NS8* samples. **, $p < 0.01$; *, $p < 0.05$. Data are presented as the mean ± standard deviation. '+' indicates more than the WT control. '+0.3' in panel (**c**) indicates the *NS8* grain was 0.3 mm longer than the WT grain (on average). '+4.7%' in panel (**c**) indicates that Q^{c1}-*N8* increased the grain length by 4.7% (on average). (**g**) Kernel width of the WT (**upper**) and *NS8* (**lower**) samples. Scale bar, 2 cm. (**h**) Kernel length of the WT (**upper**) and *NS8* (**lower**) samples. Scale bar, 2 cm.

Of the yield-related traits, the grain number per spike (Figure 6a and Figure S2i), thousand kernel weight (Figure 5e and Figure S2e), grain width (Figure 5d,g and Figure S2d,g), and grain length (Figure 5c,h and Figure S2c,h), were greater for the mutant *NS8* than for the WT control. However, there were no significant differences in the spikelet number per the main spike (Figure 6e and Figure S2m) and productive tiller number (Figure 6c and Figure S2k) between the *NS8* and WT plants. As expected, the Q^{c1}-*N8* allele positively affected GPC (Figure 5f and Figure S2f) and GY by increasing the thousand kernel weight and grain number per spike.

In order to facilitate the use of the Q^{c1}-*N8* allele in breeding, the missense mutation in the sequence encoding the second AP_2 domain was converted to a CAPS marker (Figure 7). The CAPS primer pair (N8-CAPS-F1 + N8-CAPS-R1; Table 3) amplified a 300 bp fragment from the genomic DNA of *NS8* and WT. The 300 bp fragment of WT could be digested into 210 bp and 90 bp bands. In contrast, the amplicons of *NS8* remained undigested.

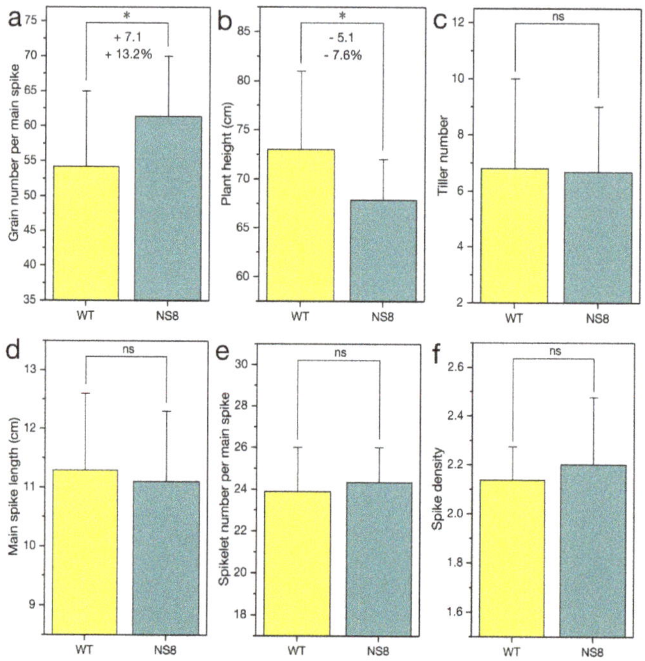

Figure 6. Effect of Q^{c1}-*N8* on agronomic traits during the 2020–2021 growing season. The subfigures (**a**–**f**) are the comparisons of the grain number per main spike, plant height, tiller number, main spike length, spikelet number per main spike, and spike density of the WT and *NS8* samples, respectively. The data in the bar graphs are presented as the mean ± standard deviation. *, $p < 0.05$; ns, not significant. '+' and '−' indicate more and less than the wild-type (WT) control, respectively. '+7.1' in panel (**a**) indicates *NS8* plants had 7.1 more grains per main spike than the WT plants (on average). '+13.2%' in panel (**a**) indicates that Q^{c1}-*N8* increased the grain number per main spike by 13.2% (on average).

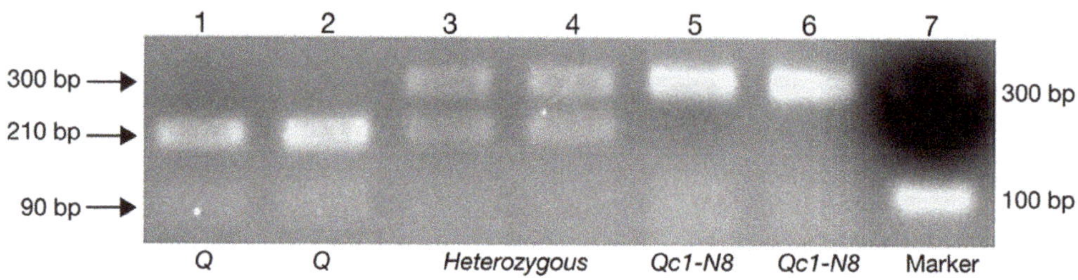

Figure 7. Gel separation of digested and undigested PCR products amplified by using the primer pair N8-CAPS-F1 + N8-CAPS-R1. Lanes 1 and 2 are the digested results of the *Q* allele. Lanes 3 and 4 are the digested results of the Q/Q^{c1}-*N8* alleles. Lanes 5 and 6 are the undigested results of the Q^{c1}-*N8* allele. Lane 7 is the DNA marker.

Table 3. Primers used in this study.

Primers Name	Sequences (5′-3′)	Reference	Objective
AP2startF	ATGGTGCTGGATCTCAATGTGGAGTCGCCGGCGGA	[15]	Cloning of the genomic DNA sequence of *Q* gene
AP2.8R	CGCGGCCAAATCGGGGCAAAGGAATTCAAACGA	[15]	
AP2.2-1F	ATCTTAGCTGTATGGGCTCGTG	This study	
AP2.2-1R	TCAACGGAGATAGGGGTGTG	This study	
AP2.2-2F	AGGCTCCACATAAGTATATGATCGAGTC	This study	
AP2.2-2R	CTTAATTTCAGGAACGAACTTGTCG	This study	
AP2.16F	CTGCTTGGTGCGCTGCTCCACCAGCTTACTGAAA	[15]	
AP45.1R	CAGAAGGCCCAACGGTTAACGCAACAATGGC	[15]	
Q-mRNA-F2-123	TCGGAGATGGTGCTGGAT	This study	Cloning the full open reading frame of *Q* gene
Q-mRNA-R1-1479	GCCAGCTTCAGTTGTCCG	This study	
QF7	GACCAGCCAGTAGTGTCACC	This study	Genotyping of Q^{c1}-*N8* allele
QR7	TCTTGCAGTTCCATCCGTCC	This study	
N8-CAPS-F1	ACTAGAGTGAGTGAGCAAAGATT	This study	CAPS marker for Q^{c1}-*N8* allele
N8-CAPS-R1	AGCAACTGTTAGGCTCCACATAA	This study	

3. Discussion

Because GY and GPC determine the profitability of wheat production, they are the primary traits upon which wheat breeders and growers focus. More specifically, GY is critical for ensuring food security, especially in many developing countries, while GPC is a crucial index for assessing the nutritional and unique processing quality of wheat [22]. Increasing both GY and GPC is important to improving the availability of high-quality food and, by extension, the living standards of humans. As an essential macronutrient for plants, nitrogen is crucial for establishing a balance between GY and GPC during wheat production [3,4,23]. Accordingly, the most frequent cultivation practice used by farmers to increase GY and GPC is the application of nitrogen fertilizers [4,24], but this leads to increased cultivation costs, decreased nitrogen use efficiency, and increased environmental pollution. Moreover, after incremental additions of nitrogen fertilizer, GPC reaches a maximum and then remains constant [24]. In this study, we created a new *Q* allele (i.e., Q^{c1}-*N8*), which breaks the negative relationship between GY and GPC and can synchronously increase the wheat GY and GPC. Thus, Q^{c1}-*N8* may be useful for breeding more profitable wheat varieties.

'Shumai482′ is an elite commercial wheat cultivar with a relatively high GY and GPC. The Q^{c1}-*N8* allele in the 'Shumai482′ genetic background can increase both GY and GPC. Notably, Q^{c1}-*N8* decreased the plant height (Figure 5b, Figure 6b and Figure S2b,j), thereby enhancing lodging resistance. Lodging is still a major factor limiting global wheat production, especially in regions with heavy rain and strong winds, because it leads to serious decreases in GY. We are currently assessing the breeding value of Q^{c1}-*N8* in multiple environments, in diverse genetic backgrounds, and in field plot experiments.

As important parts of the Q protein, the AP_2 domains are critical for DNA binding and for physical interaction with other proteins. Modifying the amino acids in the AP_2 domains of the Q protein may reverse the unfavorable agronomic traits of the mutant *S-Cp*1-1 carrying the Q^{c1} allele [17] by affecting the expression of downstream genes and the interaction between Q and other proteins. For example, the transcription factor TaLAX1 physically interacts with Q to antagonistically regulate grain threshability and spike morphology [25]. However, most downstream genes of Q for regulating agronomic traits remain unknown. The mutation in the Q^{s1} allele results in a single amino acid change in the first AP_2 domain (Figure 3 and Figure S1), which negatively affects the thousand kernel weight (Table 1). The overexpressed Q^{c1} allele has a missense mutation in the mircoRNA172-binding site (Figure 3 and Figure S1) that increases the thousand kernel weight [17]. It is likely that at least some missense mutations in the sequences encoding the AP_2 domains and those in the microRNA172-binding site have the opposite effect on the thousand kernel weight. The Q^{c1}-*N8* allele has two missense mutations, with one in the sequence encoding the AP_2 domain and the other in the microRNA172-binding site (Figure 3 and Figure S1); this allele is associated with an increase in the thousand kernel weight (Figure 5e and Figure S2e). Therefore, the opposite effects of the two-point mutations in the Q^{c1}-*N8* allele are relatively well balanced to increase the thousand kernel weight. However, further increases in GY require the creation of new alleles with mutation(s) beyond the AP_2 domain-encoding sequences; four previously reported Q^c alleles (i.e., Q^{c1}–Q^{c4}) [17] might be useful for generating new alleles.

Common wheat is a hexaploid species (AABBDD; 2n = 6x = 42) that contains three homologous genomes (i.e., A, B, and D). An earlier study revealed the dosage effect of the *Q* gene in wheat [26]. To date, only the *Q* gene copy in the A genome has been optimized. To further improve GY and GPC, the stepwise optimization of the *Q* copies in the B and D genomes of common wheat and related species is ongoing.

Durum wheat (*Triticum turgidum* ssp. *durum*) is a tetraploid species (AABB; 2n = 4x = 28) and is the main and preferred raw material for pasta production [27]. The GPC is a determining factor influencing durum wheat quality, and grains with a high GPC tend to produce good cooking quality pasta [27–29]. The Q^{c1}-*N8* allele, which is located in the A genome, may be useful for durum wheat breeding.

In addition to the *Q* gene, many other plant genes include microRNA-binding sites, including some genes encoding a conserved AP_2 domain. The directed evolution of these genes via the introduction of point mutations in their microRNA-binding sites and other domain-encoding sequences may be an efficient and effective way to ensure global food security.

Increases in GY and GPC are also required for other major cereal crops, such as rice, maize, barley, sorghum, and foxtail millet, all of which carry *Q* gene orthologs and homologs [20,30–34]. Moreover, these orthologous and homologous genes seem to have conserved functions among cereals [31,33,34]. Similar to the allele development in this study (i.e., *Q* allele to Q^{c1} and then to Q^{c1}-*N8*), elite alleles for the *Q* gene orthologs and homologs can be created by the stepwise optimization of their expression (e.g., by introducing point mutations in the mircoRNA172-binding site or in other elements) and by enhancement of the activities of the encoded proteins (e.g., by introducing point mutations in the sequences encoding the AP_2 domains or other domains), affecting specific downstream gene(s) and interacting protein(s). As they are applied in breeding programs involving non-transgenic methods, these attributes illustrate the utility of creating a set of elite alleles of the *Q* gene orthologs and homologs to increase the GY and GPC of cereal crops.

4. Materials and Methods

4.1. Plant Materials and Growth Conditions

The seeds of common wheat cultivar 'Shumai482′ (*Q* allele) and its compact-spike mutant *S-Cp*1-1 (Q^{c1} allele) [17] were treated with 0.8% and 0.4% ethyl methanesulfonate (Catalog number: M0880-100G; Sigma-Aldrich, St Louis, MO, USA), respectively. Seeds from the leading spikes of the M_1 plants were harvested and sown to generate the M_2 population. The mutant *ss1* (sparse spike 1) was obtained from the M_2 population of 'Shumai482′. The mutant *NS8* (normal spike 8) was isolated from the M_2 population of *S-Cp*1-1. The *Q* genes of *ss1* (Q^{s1} allele) and *NS8* (Q^{c1}-*N8* allele) were sequenced.

The mutants were backcrossed with 'Shumai482′ to assess the effects of Q^{s1} and Q^{c1}-*N8* on agronomic traits and processing quality parameters. Ten BC_2F_3 homozygous lines (five with the *Q* allele and five with the Q^{s1} allele) and 10 BC_2F_4 homozygous lines (five with the *Q* allele and five with the Q^{s1} allele) (Figure 1) were grown at the experimental farm of Sichuan Agricultural University in Wenjiang (30°43′16″ N, 103°52′15″ E) during the 2018–2019 and 2019–2020 wheat growing seasons, respectively. Field trials were performed using a randomized block design. Each line was cultivated in a 2 m × 3 m area, with a row spacing of 20 cm × 5 cm. The BC_1F_2 and BC_2F_2 plants carrying *Q* or Q^{c1}-*N8* (Figure 1) were grown with a row spacing of 20 cm × 10 cm in Wenjiang during the 2020–2021 and

2021–2022 growing seasons, respectively. A nitrogen:phosphorous:potassium (15:15:15) compound fertilizer was applied before sowing (450 kg per hectare).

At the GS87 growth stage [35], agronomic traits, including plant height (cm), spike length (cm), spikelet number per spike, grain number per spike, and productive tiller number, were recorded. Spike density was calculated as the ratio of the spike length to the spikelet number per spike. For the BC_2F_3 and BC_2F_4 homozygous lines with the *Q* or Q^{s1} allele, 20 representative plants of each line were examined. For the homozygous BC_1F_2 and BC_2F_2 plants carrying the *Q* or Q^{c1}-*N8* allele, the agronomic traits of 20–30 plants were also evaluated.

After harvesting samples and drying them under the sun at approximately 35 °C to a constant weight, the thousand kernel weight (g), grain length (mm), and grain width (mm) were determined. For each BC_2F_3 and BC_2F_4 homozygous line with the *Q* or Q^{s1} allele, the thousand kernel weight was measured by randomly selecting 1000 seeds. For the BC_1F_2 and BC_2F_2 plants carrying the *Q* or Q^{c1}-*N8* allele, the thousand kernel weight was measured on the basis of 200 randomly selected mature seeds. In order to measure the grain length and width, 100 randomly selected seeds were scanned using the Epson Eu-88 A3 Transparency Unit (Seiko Epson, Nagano, Japan). The resulting images were analyzed using the WinSEEDLE Analysis System (Regent Instruments, QC, Canada).

4.2. Gene Cloning

Young leaves collected from individual plants at the GS13 growth stage [35] were ground to a fine powder in liquid nitrogen. Genomic DNA and total RNA were extracted from the ground materials using Plant DNA/RNA extraction kits, respectively (Catalog numbers: DN32-100 and RN33050; Biofit, Chengdu, China). First-strand cDNA was synthesized using the Prime Script™ 1st Strand cDNA Synthesis Kit (Catalog number: 6110A; Takara, Dalian, China). All kits were used as recommended by the manufacturers.

The *Q* cDNA and genomic DNA sequences of the mutants *ss1* and *NS8* were cloned and sequenced. The PCR amplifications were completed in a 50 μL volume consisting of genomic DNA or cDNA, 200 μM dNTPs, 10 μM each primer, 1 U Phanta Max Super-Fidelity DNA Polymerase (Catalog number: P505-d1/d2/d3; Vazyme, Nanjing, China), and 25 μL 2× supplied buffer (with Mg^{2+}). The PCR was performed using the Mastercycler Pro thermal cycler (Eppendorf, Hamburg, Germany) with the following program: 95 °C for 5 min; 35 cycles of 95 °C for 45 s, 60–68 °C for 30 s, and 72 °C for 2 min; 10 min at 72 °C. The PCR products were separated on a 1.5% agarose gel (Catalog number: 5260; Takara). The target fragments were purified using the FastPure Gel DNA Extraction Mini Kit (Catalog number: DC301-01; Vazyme) and then inserted into the pCE2 TA/Blunt-Zero vector using the 5 min TA/Blunt-Zero Cloning Kit (Catalog number: C602-01; Vazyme). Positive colonies were sequenced by Sangon Biotech (Chengdu, China). The cloning and sequencing experiments were repeated at least three times. Sequences were analyzed using DNAMAN (version 8) (Lynnon Biosoft, San Ramon, CA, USA). The primers used are listed in Table 3.

4.3. Genotyping for Q^{c1}-*N8*

Genomic DNA extracted from individual plants in the BC_1F_2 and BC_2F_2 population of the mutant *NS8* was used as the PCR template. The QF7 + QR7 primer pair (Table 3) flanking the microRNA172-binding site was used. The PCR amplifications were performed as described above. The PCR products were sequenced to determine the presence/absence of Q^{c1}-*N8*.

4.4. Development of CAPS Marker for Q^{c1}-*N8*

The point mutation in the sequence encoding the second AP_2 domain of the Q^{c1}-*N8* allele was converted to the CAPS (Cleaved Amplified Polymorphic Sequence) marker by using DNAMAN (version 8). This CAPS marker (N8-CAPS-F1 + N8-CAPS-R1; Table 3) was tested in the BC_1F_2 and BC_2F_2 populations of the mutant *NS8*. The PCR amplifications were performed as described above. The PCR products (about 5 μg) were digested with five units of the restriction enzyme *Bbv*I (Catalog number: R0173S; New England Biolabs, Ipswich, MA, USA) along with given buffer at 37 °C for 120 min. The digested fragments were separated on a 1.5% agarose gel.

4.5. Processing Quality Analysis

Mature grains were dried under the sun, cleaned, and stored at room temperature for 2 months. The GPC (dry weight) was measured as described by [36]. In order to assess the effect of Q^{s1} on processing quality, the moisture content of the *ss1* (Figure 1) and wild-type (WT) grains were adjusted to 16.5% before the samples were milled using the CD1 Laboratory Mill (CHOPIN Technologies, Villeneuve-la-Garenne Cedex, France). The Zeleny sedimentation value, wet gluten content, gluten index, and dough rheological properties were determined as previously described [36]. A farinograph (Brabender GmbH & Co., Duisburg, Germany) was used to determine the rheological properties.

A baking test was performed according to a slightly modified version of AACC method 10.09-01 [37]. Specifically, a standard rapid mix test involving 50 g flour (14% moisture content) was conducted. There were two loaves of bread per flour sample. The loaf volume was determined using the BVM6630 volume meter (Pertern, Stockholm, Sweden), as described by the manufacturer.

4.6. Statistical Analysis

All data were calculated using Excel 2010 (Microsoft, Redmond, WA, USA). The significance of the differences in the mean values for the agronomic traits and processing quality parameters between the WT and mutant samples was determined according to Student's *t*-test implemented in the Data Processing System (DPS) software (version 18.10) (Zhejiang University, Hangzhou, China) [38]. The DPS software was also used to perform an analysis of variance.

5. Conclusions

In this study, we characterized two new *Q* alleles (Q^{s1} and Q^{c1}-*N8*) and demonstrated that the negative correlation between GY and GPC of wheat could be broken by Q^{c1}-*N8* via a non-transgenic approach. Q^{c1}-*N8* synchronously increases the wheat GY and GPC, and a specific molecular marker was developed to facilitate its use in breeding. It is possible to further increase both GY and GPC by progressively optimizing the *Q* gene in the three genomes of wheat.

Supplementary Materials: The following supporting information can be downloaded at: https://www.mdpi.com/article/10.3390/ijms231810772/s1.

Author Contributions: Conceptualization, P.Q.; data curation, P.Q., Q.C., Z.G., M.W., L.K., X.C., Y.W. (Yuming Wei), Y.J. and G.C.; formal analysis, P.Q., Q.C. and X.S.; methodology, P.Q., Q.C., Z.G., X.S., M.W., J.W., Q.J., Y.J. and Y.Z.; investigation, P.Q., Q.C., Z.G., X.S, M.W., Y.F., J.Z., T.Z., Y.W. (Yan Wang) and L.K.; resources, M.D., X.C., J.W., Y.W. (Yuming Wei), Q.J., G.C. and Y.Z.; validation, Q.C. and L.K.; visualization, P.Q., Q.C., Z.G., X.S, M.W. and Y.F.; project administration, P.Q.; writing—original draft preparation, P.Q. and Q.C.; writing—review and editing, P.Q., Q.C., Z.G. and X.S.; funding acquisition, P.Q., Q.C., Y.W. (Yuming Wei) and Y.Z. All authors have read and agreed to the published version of the manuscript.

Funding: This study was supported by the National Natural Science Foundation of China (grant numbers 32072054, 31971939, and 31901961), the Science and Technology Department of Sichuan Province (2019YFH0066, and 2020YFH0150). The wheat mutants are available upon request for wheat breeding.

Institutional Review Board Statement: Not applicable.

Informed Consent Statement: Not applicable.

Data Availability Statement: All data supporting the findings of this study are available within this article and the supplementary material published online.

Conflicts of Interest: The authors declare no conflict of interest.

References

1. FAOSTAT. FAO Database. Food and Agriculture Organization of the United Nations. Available online: http://www.fao.org/faostat/en/#data (accessed on 26 July 2021).
2. Kichey, T.; Heumez, E.; Pocholle, D.; Pageau, K.; Vanacker, H.; Dubois, F.; Le Gouis, J.; Hirel, B. Combined agronomic and physiological aspects of nitrogen management in wheat highlight a central role for glutamine synthetase. *New Phytol.* **2006**, *169*, 265–278. [CrossRef] [PubMed]
3. Zhang, Y.; Zhang, Y.; Liu, N.; Su, D.; Xue, Q.; Stewart, B.A.; Wang, Z. Effect of source-sink manipulation on accumulation of micronutrients and protein in wheat grains. *J. Plant Nutr. Soil Sci.* **2012**, *175*, 622–629. [CrossRef]
4. Zheng, T.; Qi, P.F.; Cao, Y.L.; Han, Y.N.; Ma, H.L.; Guo, Z.R.; Wang, Y.; Qiao, Y.Y.; Hua, S.Y.; Yu, H.Y.; et al. Mechanisms of wheat (*Triticum aestivum*) grain storage proteins in response to nitrogen applications and its impacts on processing quality. *Sci. Rep.* **2018**, *8*, 11928. [CrossRef] [PubMed]
5. Laidig, F.; Piepho, H.P.; Dirk Rentel, D.; Thomas Drobek, T.; Meyer, U.; Huesken, A. Breeding progress, environmental variation and correlation of winter wheat yield and quality traits in German official variety trials and on-farm during 1983–2014. *Theor. Appl. Genet.* **2017**, *130*, 223–245. [CrossRef]
6. Mirosavljevic, M.; Momcilovic, V.; Zivancev, D.; Acin, V.; Jockovic, B.; Mikic, S.; Takac, V.; Dencic, S. Genetic improvement of grain yield and bread-making quality of winter wheat over the past 90 years under the Pannonian Plain conditions. *Euphytica* **2020**, *216*, 184. [CrossRef]
7. Subira, J.; Peña, R.J.; Álvaro, F.; Ammar, K.; Ramdani, A.; Royo, C. Breeding progress in the pasta-making quality of durum wheat cultivars released in Italy and Spain during the 20th Century. *Crop Pasture Sci.* **2014**, *65*, 16–26. [CrossRef]
8. Tabbita, F.; Pearce, S.; Barneix, A.J. Breeding for increased grain protein and micronutrient content in wheat: Ten years of the *GPC-B1* gene. *J. Cereal Sci.* **2017**, *73*, 183–191. [CrossRef]
9. Rasheed, A.; Xia, X.; Yan, Y.; Appels, R.; Mahmood, T.; He, Z. Wheat seed storage proteins: Advances in molecular genetics, diversity and breeding applications. *J. Cereal Sci.* **2014**, *60*, 11–24. [CrossRef]
10. Shewry, P.R. Wheat. *J. Exp. Bot.* **2009**, *60*, 1537–1553. [CrossRef]
11. Qi, P.F.; Wei, Y.; Yue, Y.W.; Yan, Z.H.; Zheng, Y.L. Biochemical and molecular characterization of gliadins. *Mol. Biol.* **2006**, *40*, 713–723. [CrossRef]
12. Payne, P.I. Genetics of wheat storage proteins and the effect of allelic variation on bread-making quality. *Ann. Rev. Plant Physiol.* **1987**, *38*, 141–153. [CrossRef]
13. Wieser, H. Chemistry of gluten proteins. *Food Microbiol.* **2007**, *24*, 115–119. [CrossRef] [PubMed]
14. Jiang, Y.F.; Chen, Q.; Wang, Y.; Guo, Z.R.; Xu, B.J.; Zhu, J.; Zhang, Y.Z.; Gong, X.; Luo, C.H.; Wu, W.; et al. Re-acquisition of the brittle rachis trait via a transposon insertion in domestication gene *Q* during wheat de-domestication. *New Phytol.* **2019**, *224*, 961–973. [CrossRef] [PubMed]
15. Simons, K.J.; Fellers, J.P.; Trick, H.N.; Zhang, Z.; Tai, Y.; Gill, B.S.; Faris, J.D. Molecular characterization of the major wheat domestication gene *Q*. *Genet.* **2006**, *172*, 547–555. [CrossRef]
16. Xie, Q.; Li, N.; Yang, Y.; Lv, L.; Yao, H.; Wei, R.; Sparkes, D.L.; Ma, Z.Q. Pleiotropic effects of the wheat domestication gene *Q* on yield and grain morphology. *Planta* **2018**, *247*, 1089–1098. [CrossRef] [PubMed]
17. Xu, B.J.; Chen, Q.; Zheng, T.; Jiang, Y.F.; Qiao, Y.Y.; Guo, Z.R.; Cao, Y.L.; Wan, G.Y.; Zhang, Y.Z.; Zong, L.J.; et al. An overexpressed *Q* allele leads to increased spike density and improved processing quality. *G3 Genes Genom. Genet.* **2018**, *8*, 771–778. [CrossRef]
18. Guo, Z.; Chen, Q.; Zhu, J.; Wang, Y.; Li, Y.; Li, Q.; Zhao, K.; Li, Y.; Tang, R.; Shi, X.; et al. The Q^{c5} allele increases wheat bread-making quality by regulating SPA and SPR. *Int. J. Mol. Sci.* **2022**, *23*, 7581. [CrossRef]
19. Endo, T.R.; Gill, B.S. The deletion stocks of common wheat. *J. Hered.* **1996**, *87*, 295–307. [CrossRef]
20. Solomon, C.S.; Drea, S. Besides and beyond flowering: Other roles of *EuAP2* genes in plant development. *Genes* **2019**, *10*, 994. [CrossRef]
21. Greenwood, J.R.; Finnegan, E.J.; Watanabe, N.; Trevaskis, B.; Swain, S.M. New alleles of the wheat domestication gene *Q* reveal multiple roles in growth and reproductive development. *Development* **2017**, *144*, 1959–1965. [CrossRef]
22. Weegels, P.L.; Hamer, R.J.; Schofield, J.D. Functional properties of wheat glutenin. *J. Cereal Sci.* **1996**, *23*, 1–18. [CrossRef]
23. Masoni, A.; Ercoli, L.; Mariotti, M.; Arduini, I. Post-anthesis accumulation and remobilization of dry matter, nitrogen and phosphorus in durum wheat as affected by soil type. *Eur. J. Agron.* **2007**, *26*, 179–186. [CrossRef]
24. Barnex, A.J. Physiology and biochemistry of source-regulated protein accumulation in the wheat grain. *J. Plant Physiol.* **2007**, *164*, 581–590. [CrossRef]
25. He, G.; Zhang, Y.; Liu, P.; Jing, Y.; Zhang, L.; Zhu, Y.; Kong, X.; Zhao, H.; Zhou, Y.; Sun, J. The transcription factor TaLAX1 interacts with Q to antagonistically regulate grain threshability and spike morphogenesis in bread wheat. *New Phytol.* **2021**, *230*, 988–1002. [CrossRef] [PubMed]
26. Muramatsu, M. Dosage effect of the spelta gene *q* of hexaploid wheat. *Genetics* **1963**, *48*, 469–482. [CrossRef] [PubMed]

27. Sissons, M. Role of durum wheat composition on the quality of pasta and bread. *Food* **2008**, *2*, 75–90.
28. Porceddu, E.; Turchetta, T.; Masci, S.; D'Ovidio, R.; Lafiandra, D.; Kasarda, D.D.; Impiglia, A.; Nachit, M.M. Variation in endosperm protein composition and technological quality properties in durum wheat. *Euphytica* **1998**, *100*, 197–205. [CrossRef]
29. Walsh, D.E.; Gilles, K.A. The influence of protein composition on spaghetti quality. *Cereal Chem.* **1971**, *48*, 544–553.
30. Chuck, G.; Meeley, R.; Hake, S. Floral meristem initiation and meristem cell fate are regulated by the maize AP_2 genes *ids1* and *sid1*. *Development* **2008**, *135*, 3013–3019. [CrossRef]
31. Dong, Z.; Alexander, M.; Chuck, G. Understanding grass domestication through maize mutants. *Trends Genet.* **2019**, *35*, 118–128. [CrossRef]
32. Gil-Humanes, J.; Piston, F.; Martin, A.; Barro, F. Comparative genomic analysis and expression of the APETALA2-like genes from barley, wheat, and barley-wheat amphiploids. *BMC Plant Biol.* **2009**, *9*, 66. [CrossRef] [PubMed]
33. Lee, D.; An, G. Two AP2 family genes, *Supernumerary Bract* (*SNB*) and *Osindeterminate Spikelet 1* (*OsIDS1*), synergistically control inflorescence architecture and floral meristem establishment in rice. *Plant J.* **2012**, *69*, 445–461. [CrossRef] [PubMed]
34. Wang, J.; Lin, Z.; Zhang, X.; Liu, H.; Zhou, L.; Zhong, S.; Li, Y.; Zhu, C.; Lin, Z. *krn1*, a major quantitative trait locus for kernel row number in maize. *New Phytol.* **2019**, *223*, 1634–1646. [CrossRef] [PubMed]
35. Zadoks, J.C.; Chang, T.T.; Konzak, C.F. A decimal code for the growth stages of cereals. *Weed Res.* **1974**, *14*, 415–421. [CrossRef]
36. Wang, Y.; Chen, Q.; Li, Y.; Guo, Z.R.; Liu, C.H.; Wan, Y.F.; Hawkesford, M.; Zhu, J.; Wu, W.; Wei, M.Q.; et al. Post-translational cleavage of HMW-GS *Dy10* allele improves the cookie-making quality in common wheat (*Triticum aestivum*). *Mol. Breed.* **2021**, *41*, 49. [CrossRef]
37. AACC International. *Basic Straight-Dough Bread-Baking Method—Long Fermentation (Method 10-09.01)*; Cereals & Grains Association: St. Paul, MN, USA, 2010; Available online: http://methods.aaccnet.org/ (accessed on 8 February 2014).
38. Tang, Q.Y.; Zhang, C.X. Data Processing System (DPS) software with experimental design, statistical analysis and data mining developed for use in entomological research. *Insect Sci.* **2013**, *20*, 254–260. [CrossRef]

International Journal of
Molecular Sciences

MDPI

Article

Sequencing and Genomic Analysis of Sorghum DNA Introgression Variant Line R21 and Recipient Rice Jin Hui 1 Revealed Repetitive Element Variation

Ting Zhang [1,2,†], Xiaodong Li [1,2,†], Zijun Zhao [1,2], Renhong Wu [1,2], Zhenglin Yang [1,2] and Guanghua He [1,2,*]

1 College of Agronomy and Biotechnology, Southwest University, Chongqing 400716, China
2 Academy of Agricultural Sciences, Southwest University, Chongqing 400716, China
* Correspondence: heghswu@163.com; Tel.: +86-13-88-321-9316
† These authors contributed equally to this work.

Abstract: Transferring the genome of distant species to crops is an efficient way to create new germplasms. However, the molecular mechanisms involved are unclear. In this study, a new rice restorer line R21 with heat tolerance was created by introgressing the genomic DNA of sorghum into the recipient restorer line Jin Hui 1. Assembly of rice R21 and Jin Hui 1 genomes was performed using PacBio sequencing technology. Comparative genome analysis and coverage statistics showed that the repetitive sequence atr0026 was a candidate introgression fragment of sorghum DNA. Sequence similarity analysis revealed that atr0026 was distributed at different copy numbers on the telomeric position of chromosomes 9 or 10 in R21, Jin Hui 1, and several rice varieties, indicating that the repetitive sequence from sorghum was highly conserved in rice. The repeat annotation in Gramineae indicated that ribosomal DNA loci that existed in atr0026 may be cause a rearrangement of chromosomes 9 and 10 of the R21 genome, resulting in a copy number variation at the 5′ end of it. Our study lays the foundation for further elucidation of the molecular mechanisms underlying the heat tolerance of sorghum DNA introgression variant line R21, which is of great significance for guiding crop genetic breeding.

Keywords: rice (*Oryza sativa L.*); sorghum (*Sorghum bicolor* (L.) Moench); distant species; genome; heat tolerance; new germplasm

Citation: Zhang, T.; Li, X.; Zhao, Z.; Wu, R.; Yang, Z.; He, G. Sequencing and Genomic Analysis of Sorghum DNA Introgression Variant Line R21 and Recipient Rice Jin Hui 1 Revealed Repetitive Element Variation. *Int. J. Mol. Sci.* **2022**, *23*, 11864. https://doi.org/10.3390/ijms231911864

Academic Editors: Andrés J. Cortés and Hai Du

Received: 26 August 2022
Accepted: 30 September 2022
Published: 6 October 2022

1. Introduction

Rice (*Oryza sativa L.*) is a major food crop, feeding more than half of the world's population [1]. As the global population continues to grow, the demand for rice is also increasing. In recent years, global warming has intensified, and high temperatures severely constrain rice yield, posing a severe threat to world food production and security [2,3]. Therefore, the creation of heat-tolerant rice germplasm resources has become a major scientific problem that needs to be solved urgently at present.

Distant cross is one of the most common methods of conventional breeding to create new germplasm. Transferring the genomes of distant species to varieties of recipient plants can generate a wealth of variant material, thus achieving a comprehensive improvement of agronomic traits in the recipient [4–6]. Currently, molecular breeding for the direct introduction of genomic DNA from distantly related species into different rice recipients has been widely used [7]. YeWei B is a variant line obtained by introgressing the genomic DNA of wild rice of *O. minuta* into the hybrid rice maintainer line V20B, which showed a narrowed and reduced leaf angle of the top three leaves and improved seed setting rate and seed quality [8]. Genome sequencing of YeWei B showed that 21 of the 28 selected genes associated with differential traits had variants, including 208 single nucleotide polymorphisms (SNPs) and 98 insertion-deletion (InDel) variants [7]. Ma et al. introduced the maize genomic DNA into recipient rice cells by distant molecule hybridization and obtained 18 rice variant line materials with significantly altered phenotypes such as plant height, purplish stems and leaf sheaths, and aerial roots. A total of 308 base mutations were found in these variant lines, and more than half of them were transitions [9]. *ERV1* is a variant line with significantly improved yield-related traits, which is obtained by introgressing genomic DNA of *Oryza eichingeri* into the recipient rice *RH78* [10]. Liu et al. used the pollen-tube pathway to introgress *Fargesia spathacea* Franch DNA into the rice. The obtained variant line T1 had a reduced plant height and increased 1000-grain weight, and T2 had an increased number of effective panicles and grains per panicle but a reduced panicle length [11]. Thus, transferring genomes of distance species has been widely used in the molecular improvement of crops, and it has created a large number of new germplasms and varieties of crops with significantly improved traits [8,12–14]. However, the molecular mechanism associated with this process is still unclear.

In this study, the genomic DNA of the high photosynthetic efficiency dryland crop sorghum (*Sorghum bicolor* (L.) Moench) was introgressed into the rice recipient restorer line Jin Hui 1 to create a new rice restorer line R21, which showed heat tolerance at the flowering stage [15,16]. We used PacBio sequencing technology for the de novo assembly of genomes of R21 and Jin Hui 1, performed comparative analysis between genomes with *Japonica*, and identified abundant genomic differences. The whole-genome coverage statistics showed that the potential introgression fragment of Sorghum DNA was located within 45.4–45.6 Mb of chromosome 5 of Sorghum. Further comparative analysis demonstrated that a 5.8 kb sequence among 200 kb intervals can be screened as a candidate introgression fragment due to its highest homology with the R21 genome. The fragment annotated as a repeat sequence (atr0026) in sorghum was distributed at different copy numbers on the telomeric position of chromosomes 9 or 10 in R21, Jin Hui 1, and several rice varieties, indicating that this repeat sequence from sorghum is highly conserved in rice. The repeat annotation from Gramineae indicated that the repeat atr0026 on sorghum contains a conservative 18s ribosomal DNA loci, and its introgression may be cause the genome rearrangement of chromosomes 9 and 10 of the R21, resulting in the presence of CNV at its 5′ end, which may be responsible for the heat tolerance of R21. These results lay the foundation for an in-depth analysis of the molecular mechanism related to germplasm innovation through exogenous gene introgression, which is of great theoretical and practical significance for guiding crop genetic breeding.

2. Results

2.1. Comparative Analysis of Genomes of R21 and Jin Hui 1

The rice genome IRGSP-1.0 (Oryza_sativa—Ensembl Genomes 53) was used as the reference genome. The assembled genomes of R21 and Jin Hui 1 were aligned against reference genome respectively by Minimap2, and presence/ absence variations (PAVs) were identified by Assemblytics software. Based on the alignment results, we re-corrected the chromosome numbering of the assembled genomes of R21 and Jin Hui 1 firstly (Figure 1A,B) (Supplemental Table S1), and analyzed the PAV variations, respectively (Figure 1C,D). The comparative analysis between genomes showed abundant genomic variations, among which 11,880 insertions, 9893 deletions, 83 tandem expansions, 15 tandem contractions, 2159 repeat expansions and 2126 repeat contractions were found in R21; 11,953 insertions, 10,010 deletions, 78 tandem expansions, 17 tandem contractions, 2189 repeat expansions and 2138 repeat contractions were found in Jin Hui 1 (Figure 1C,D).

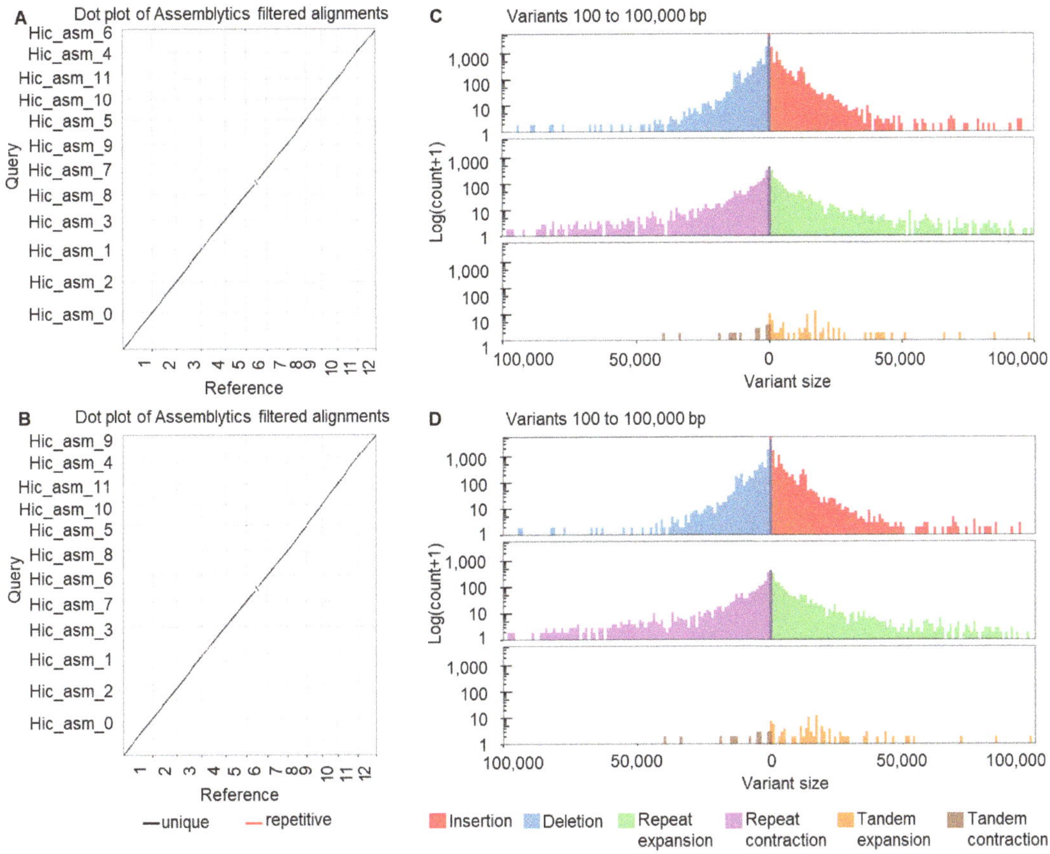

Figure 1. Comparative genome analysis of R21 and Jin Hui 1. (**A**,**B**) Comparative analysis of assembled genomes with rice reference genome IRGSP-1.0 of R21 (**A**) and Jin Hui 1 (**B**); (**C**,**D**) Variation analysis of assembled genomes of R21 (**C**) and Jin Hui 1 (**D**).

2.2. Identification of Insertion/Homologous Fragments in Sorghum

Based on the PacBio sequencing data of R21 and Jin Hui 1, we mapped HiFi reads to the sorghum genome and found that the mapping rate of R21 was 20.4%, which was 0.4% higher than that of Jin Hui 1 (20%) (Figure 2A). After that, we calculated the depth of coverage of HiFi reads on the sorghum genome and the depth of difference, and we identified a total of three candidate intervals (top 1%) on sorghum chromosome 5, in which the depth of difference in coverage of HiFi reads on 45.4–45.5 Mb and 45.5–45.6 Mb reached 170 (Figure 2B). Since the two intervals were adjacent to each other, we combined them into one interval of 200 kb for the subsequent analysis (Figure 2B).

A

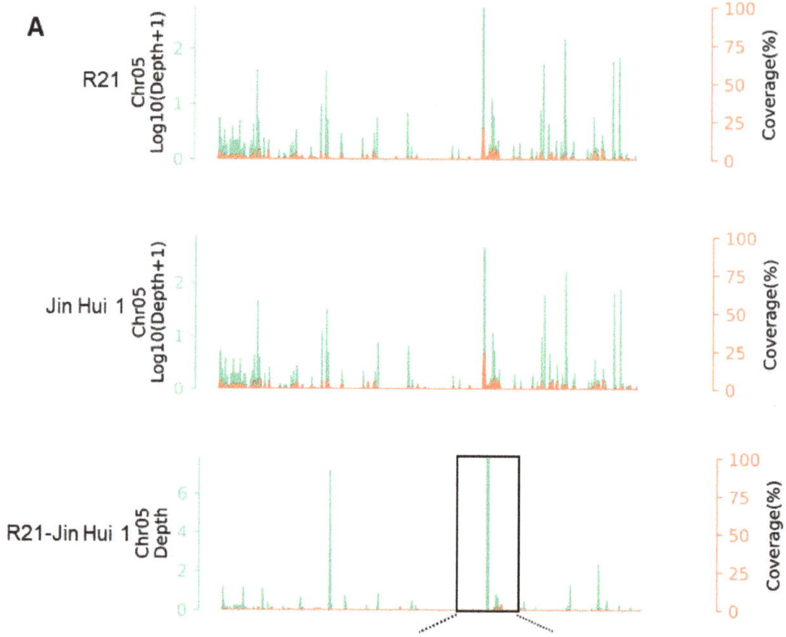

B

startpos	endpos	coverage	meandepth
45,400,000	45,500,000	−3.776	173.196
45,500,000	45,600,000	−2.3569	226.956

Figure 2. Identification of insertions/homologous fragments in sorghum. (**A**) Distribution of depth of coverage of chromosome 5 between R21 and Jin Hui 1; (**B**) Region of significant difference in depth of coverage between R21 and Jin Hui 1 (top 1%). Green represents depth, Orange represents coverage.

2.3. *Collinearity Analysis of Insertion/Homologous Fragments of Sorghum with R21 and Jin Hui 1 Genomes*

We extracted 200 kb DNA sequences from the sorghum genome on the 45.4–45.6 Mb interval on chromosome 5 and used BLAST to align to the assembled R21 and Jin Hui 1 genomes, respectively. Based on the alignment results, we found that this candidate sequence of sorghum differed significantly in the number of matches on chromosomes 9 and 10 on both R21 and Jin Hui 1 genomes (Figure 3A). Second, despite the difference in the number of matches, homologous sequences were indeed found on both R21 and Jin Hui 1 genomes, suggesting that this candidate sequence of sorghum itself is likely to be highly homologous to rice (Supplemental Figures S1 and S2). Finally, by analyzing the alignment information (including length, identity, and coverage) within all match regions, we hypothesized that the length of this introgressed sequence in sorghum is approximately 5 kb, mainly affecting the telomeric region of R21 genome chromosomes 9 or 10 (Figure 3B). In summary, the introgression of the sorghum homologous sequence resulted in a difference in coverage depth between R21 and Jin Hui 1; the size of the sorghum homologous introgressed sequence was about 5.8 kb.

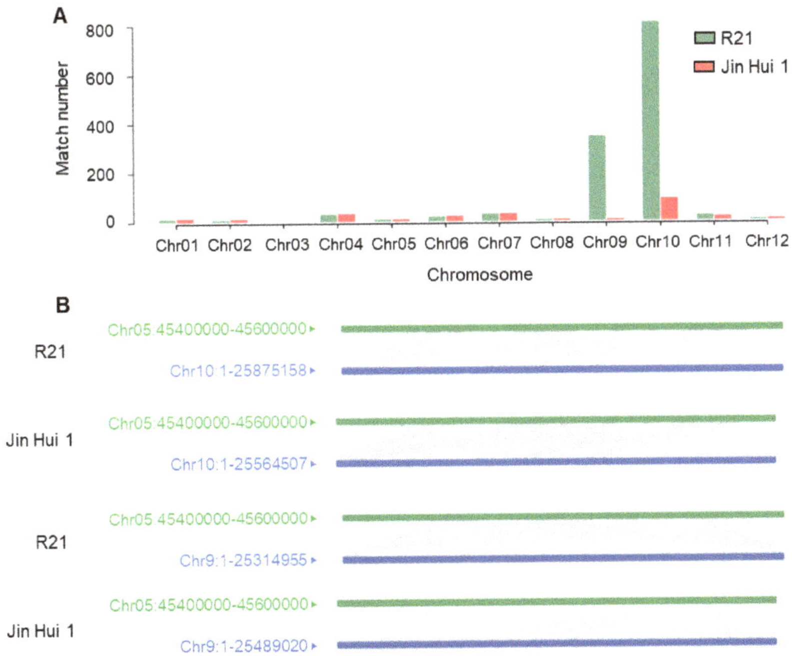

Figure 3. Collinearity analysis of sorghum insertions/homologous fragments with R21 and Jin Hui 1 genomes. (**A**) The alignment results of sorghum candidate analysis sequences blast against two rice genomes; (**B**) The collinearity results of sorghum candidate analysis sequences with chromosomes 9 and 10 of R21 and Jin Hui 1 genomes. Green labels represent the 45.4–45.6 Mb interval of chromosome 5 in sorghum, blue labels represent the full length of chromosomes 9 and 10 in rice.

2.4. Comparative Analysis of Sorghum Insertion/Homologous Fragment, R21, Jin Hui 1 and Japonica Genomes

We extracted a 5.8 kb fragment from the candidate region of sorghum (Chr05: 45.4–45.6 Mb), which has the highest similarity with the rice genome as the candidate insertion sequence, and then compared with R21, Jin Hui 1 and IRGSP of the Japonica genome by BLAST. First of all, according to the results of sorghum genome repeat sequence annotation, we found that the 5.8 kb insertion sequence is a repeat sequence atr0026, and it was repeatedly annotated in the Chr05: 45.4–45.6 Mb interval (Figure 4A,B). Secondly, based on the previously calculated reads coverage rate of 1 kb in Chr05: 45.4–45.6 Mb of sorghum, we found that the region with reads coverage is almost consistent with the position distribution of repeat atr0026 (Figure 4A,B). To confirm the presence of the repeat atr0026 on sorghum in other species, we used the BLAST function (with parameters e-value 1×10^{-5} and word length 11) on RGAP (http://rice.uga.edu/, accessed on 20 September 2022) [17] to match the sequence to the Gramineae Repeat Database. According to the results, there is a repeat sequence belonging to ribosomal DNA in wheat (*Triticum aestivum*), rice (*Oryza sativa*), maize (*Zea mays*), and Foxtail Barley (*Hordeum jubatum*), which can be matched to the front end of the repeat atr0026. Among them, there is another repeat sequence in rice and wheat that can be matched to the back end of the repeat atr0026. This result indicated that the repeat atr0026 on sorghum contains a conservative 18s ribosomal DNA loci, which exists in Gramineae species (Supplemental Table S2 and Supplemental Figure S3).

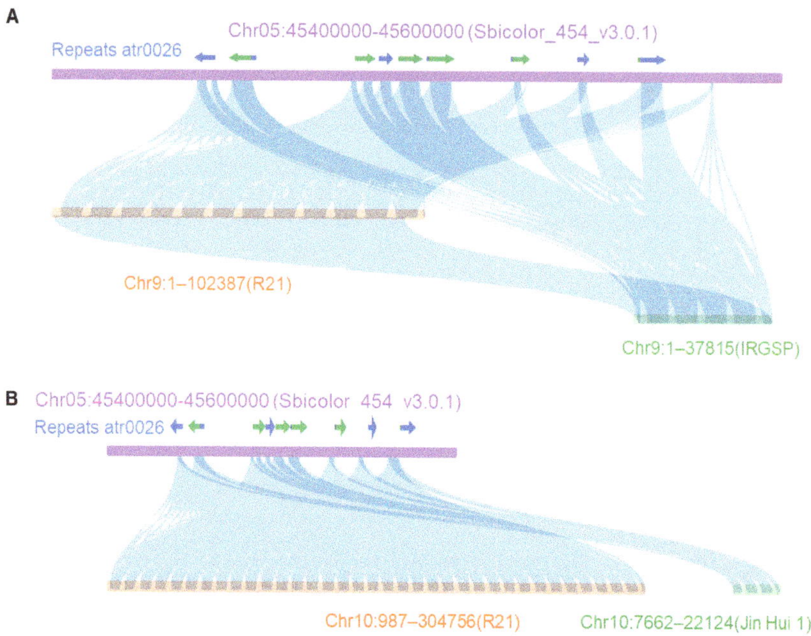

Figure 4. Collinearity analysis of sorghum insertion sequences. (**A**) Collinearity analysis between Sorghum insertion sequence and chromosome 9 of rice (from top to bottom: sorghum Chr05: 45.4–45.6 Mb fragment, repeat annotation atr0026, chromosome 9 segment of R21 genome, chromosome 9 segment of IRGSP genome); (**B**) Collinearity analysis between sorghum insertion sequence and chromosome 10 of rice (from top to bottom: sorghum Chr05: 45.4–45.6 Mb fragment, repeat annotation atr0026, chromosome 10 segment of R21 genome, partial fragment of chromosome 10 in Jin Hui 1 genome). Consistency between sequences was greater than 90%. Arrows represent repeats atr0026, and blue and green are used to distinguish the atr0026 overlapped.

Finally, based on the results of sequence alignment, we found that the 5.8 kb repeat sequence atr0026 in sorghum has multiple high homology alignments on the chromosome 9 of R21 and IRGSP genomes (sequence consistency is greater than 90%), and the high homology alignment frequency (12 times) on the Chr9 chromosome of the R21 genome is much higher than that on the Chr9 chromosome of the IRGSP genome (4 times). However, no homologous region was found on chromosome 9 of the Jin Hui 1 genome (Figure 4A). In addition, in the reference genome of Japonica, there are two candidate genes, *LOC_Os09g00999* and *LOC_Os09g01000*, with unknown function in the region corresponding to this repeat sequence. At the same time, the sequence also has several high homology alignments on Chr10 chromosomes of R21 and Jin Hui 1 genomes, and the number of high homology alignments on chromosome 10 of R21 genome (35 times) is much higher than that of the Jin Hui 1 genome (4 times). However, no homologous region was found on chromosome 10 of the IRGSP genome (Figure 4B). In summary, the introgression of the sorghum sequence may be cause of a repetitive region occurring on chromosome 9 and 10 in R21, resulting in CNV (copy number variation) at the 5′ end, up to 47 times.

2.5. Collinearity Analysis between Sorghum Candidate Sequences and Rice Pan Genome

To confirm whether the 5.8 kb repetitive sequence of sorghum is homologously conserved in all current rice varieties, we also compared the sequence with 33 rice genomes on the rice pan genome website (http://ricerc.sicau.edu.cn/, accessed on 20 September 2022) by BLAST [18]. Based on the comparison results, we found that this 5.8 kb repeat sequence atr0026 on the sorghum genome is distributed on the genomes of different rice materials, where it is mainly found on chromosome 9 or 10 (Figure 5). According to statistics, this sequence repeats 110 and 154 times on chromosomes 9 and 10 in 27 rice varieties, respectively, and there are no homologous sequences in six rice varieties. Among the 27 rice varieties, N22 is a strong heat-tolerant rice variety, which originated from India with high temperature and a high humidity environment [19]. It is worth noting that in N22, the candidate sequence of sorghum appeared 34 times on chromosome 9 (Figure 5).

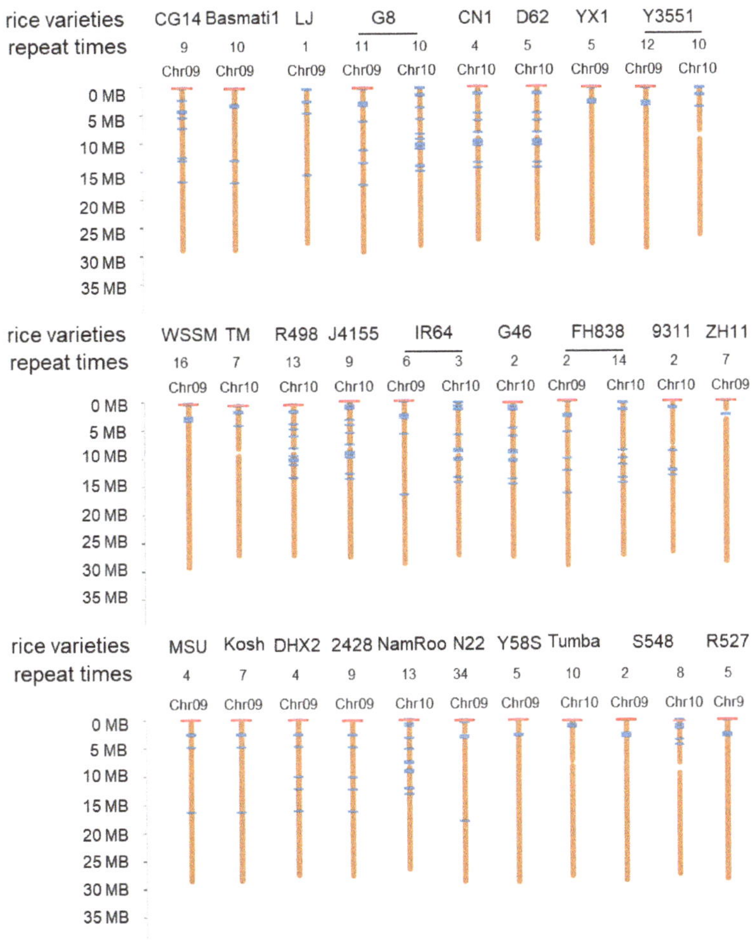

Figure 5. Homology alignment of 5.8 kb sequence from sorghum with 33 rice genomes on chromosomes (over 90% sequence consistency).

3. Discussion

So far, many examples have shown that the introgression of favorable exogenous genes in rice plays a vital role in improving the yield, quality, and resistance, which is an effective tool for the innovation of rice germplasm resources [20–23]. Among the distant hybrids of rice, crosses with sorghum have been more intensively and maturely studied. Sorghum is a C4 plant with high photosynthetic efficiency, fast growth rate, good photosynthetic products, high adaptable, strong heat tolerance, and stable and high yield [24]. Some excellent characteristics of sorghum are transferred to rice through hybridization, and new rice varieties with high yield and high quality can be bred. *SbSGL* encodes a DUF1645 superfamily protein in sorghum, and its transfer to rice significantly increased the size and number of glume cells and post-fertilization filling rate, which in turn led to a highly significant increase in grain length and 1000-grain weight [20]. Qin et al. introgressed sorghum cDNA into the rice and more than 92% of the 965 transgenic lines obtained different phenotypic changes, including longer leaves and increased effective spike number [21]. In this study, R21 was a rice variant line obtained by introgressing sorghum DNA into the rice restorer line Jin Hui 1, and the content proline, SOD and POD activities, and MDA of R21 were significantly increased after being subjected to high temperature and drought stress, indicating that R21 has excellent heat tolerance and drought tolerance [15]. Therefore, exogenous gene introgression can significantly improve rice traits and is an effective way of creating new germplasm resources and crop varieties.

The key to hybrid rice breeding is to produce stress-resistant, high-quality, and high combining ability sterile and restorer lines with excellent agronomic traits, which is an essential requirement for assembling widely adapted, highly dominant rice hybrid combinations. R21 is an excellent restorer parental material, which has excellent characteristics, such as strong heat and drought resistance, compact plant shape, stiff stalks, strong resistance to falling, upright leaves, neat panicles, long spikes and high seed-setting rate [15,16]. Many breeders use R21 as the parent material to cultivate hybrid rice varieties. Based on data from the Chinese Rice Data Center [25], seven varieties bred by using R21 passed the examination 13 times. The regional and production test results showed that the average yield of the seven varieties out of the the 13 examined increased by 5.33% and 8.88%, and the highest yield increases were 10.26% and 16.26%, which were higher than the varieties approved for the same period. Among them, Fu You 1, bred from the combination of male sterile line II-32A and R21, is the first national approved rice variety selected by an exogenous DNA introgression technique in China. Fu You 1 has passed the national approval of the upper, middle and lower reaches of the Yangtze River, the Wuling Mountains, as well as the approval of Hainan and Vietnam. The average yield increase in Fu You 1 in the regional test and production test is higher than that of the varieties in the same period. From 2003 to 2021, there were 1181 hybrid Indica rice varieties that were examined and approved by the Ministry of Agriculture in the Yangtze River basin, and only six varieties were certified in three zones simultaneously. Fu You 1 is not only one of them but also ranks first in the average yield increase in regional and production trials (8.35% on average).

Genome sequencing is the most direct and effective method for analyzing the genomic variation. DNA sequencing technology has undergone three technological revolutions [26]. In this study, we found that after the introgression of Sorghum DNA, there are obvious differences between the R21 and Jin Hui 1 genomes. Repeat atr0026 on sorghum contains a conservative 18s ribosomal DNA loci, which exists in Gramineae species. It has been proved that ribosomal DNA is a kind of common characteristic in Oryza and even in Gramineae, but because of the positional variability or transpositional nature of ribosomal DNA, the position of ribosomal DNA is different in different species, and this difference may be determined in the process of evolution [27]. Therefore, we speculate that since atr0026 on sorghum contains ribosomal DNA loci, when it introgressed into the rice genome, it affected the location and number of rDNA of R21 on chromosome 9 and chromosome 10, which in turn led to the fragment highly homologous to atr0026 recurring many times on rice. The repetitive sequence atr0026 is distributed in different degrees in all published rice pan genomes, with 34 occurrences on chromosome 10 of the strongly heat-tolerant variety N22 and up to 47 occurrences on chromosomes 9 and 10 of R21, further suggesting that the number of duplications of the repeat atr0026 is closely associated with the formation of traits in R21.

4. Materials and Methods

4.1. Plant Materials

The two plant materials used in this study were the japonica restorer line Jin Hui 1 and the variant line R21. R21 was obtained by using Jin Hui 1 as the acceptor and sorghum

DNA as the donor. The sorghum DNA is injected into the young panicle of Jin Hui 1 by the pollen-tube pathway at the meiotic stage. R21 has excellent heat tolerance. After five years and 13 generations of directional selection and comparative identification of the variant line, the traits can be inherited stably [16].

4.2. Rice Genome Sequencing and Assembly

In this study, PacBio sequencing was completed in Novogene Bioinformatics Technology Co., Ltd. High molecular weight DNA was first extracted from the seedlings of R21 and Jin Hui 1. A SMRT bell library was constructed as previously described [28] and sequenced on the PacBio Sequel II platform (Pacific Biosciences). The PacBio HiFi reads obtained by sequencing were assembled de novo using Hifiasm software [29]. The integrity of the assembled genome was assessed using BUSCO [30] and GEGMA software [31].

4.3. Comparative Analysis and Variant Identification of Rice Genome

The two rice assembled genomes of R21 and Jin Hui 1 were compared with the rice genome IRGSP-1.0 (Oryza_sativa—Ensembl Genomes 53) using Minimap2 software [32], and PAV variants were identified by Assemblytics software [33].

4.4. Screening and Collinearity Analysis of Homologous Sequences in Sorghum

The HiFi reads obtained by PacBio third-generation sequencing were stored in BAM format files. Therefore, we first converted the format of files from BAM to FASTQ by SAMtools software [34], and then, the HiFi reads of the two rice materials were aligned with the sorghum reference genome (https://phytozome-next.jgi.doe.gov/, accessed on 20 September 2022) using Minimap2. After obtaining the comparison results, we used SAMtools to count the coverage and sequencing depth within each 100 kb sequence of the sorghum genome, and we calculated the difference in the coverage depth of the two rice materials within the same sequence.

According to the coverage depth difference region, the coverage depth and coverage rate of each 1 kb sequence were further calculated. Then, the DNA sequence corresponding to this interval was extracted from the sorghum genome as a candidate analysis sequence, which was aligned with the assembled R21 and Jin Hui 1 and the rice pan-genome, respectively, using BLAST to confirm the collinearity of the sequences [35].

4.5. Repeat Annotation in Gramineae

The DNA sequence of repeat was submitted to RGAP (http://rice.uga.edu/, accessed on 20 September 2022) for identifying and classifying this repeat in the Gramineae Repeat Database [36] using the BLAST function with an e-value 1×10^{-5} and word length of 11.

5. Conclusions

In this study, a new rice restorer line R21 with heat tolerance was created by introgressing the genomic DNA of sorghum into the recipient restorer line Jin Hui 1. We used PacBio sequencing technology for the de novo assembly of genomes of R21 and Jin Hui 1. The candidate introgression fragment of sorghum DNA was a repetitive sequence (atr0026), which was distributed at different copy numbers on the telomeric position of chromosomes 9 or 10 in R21, Jin Hui 1, and several rice varieties, indicating that the repetitive sequence from sorghum is highly conserved in rice. The repeat annotation in Gramineaes indicated that ribosomal DNA loci existing in atr0026 may be cause a rearrangement of chromosomes 9 and 10 of the R21 genome, resulting in a CNV at the 5′ end of it. Our study lays the foundation for further elucidation of the molecular mechanisms underlying the heat tolerance of sorghum DNA introgression variant line R21, which is of great significance for guiding crop genetic breeding.

Supplementary Materials: The following supporting information can be downloaded at: https://www.mdpi.com/article/10.3390/ijms231911864/s1.

Author Contributions: G.H. contributed to the conception of the study. G.H., T.Z. and X.L. drafted the manuscript. X.L., Z.Z. and R.W. contributed to the data analyses. T.Z. and Z.Y. conceived and designed the experiments. G.H. and T.Z. revised the manuscript. All authors have read and agreed to the published version of the manuscript.

Funding: This work was supported by the Foundation for Innovative Research Groups of the National Natural Science Foundation of Chongqing (cstc2021jcyj-cxttX0004), Chongqing outstanding scientist foundation (cstc2022ycjh-bgzxm0073), and Fundamental Research Funds for the Central Universities (SWU-KT22041).

Institutional Review Board Statement: Not applicable.

Informed Consent Statement: Not applicable.

Data Availability Statement: The raw sequence data have been deposited in the Genome Sequence Archive (GSA) database in BIG Data Center (https://ngdc.cncb.ac.cn/, accessed on 30 September 2023), under accession number PRJCA011922. All other data are available from the corresponding author upon reasonable request.

Conflicts of Interest: The authors declare no conflict of interest.

References

1. Guo, Y.; Fu, Y.; Hao, F.; Zhang, X.; Wu, W.; Jin, X.; Robin Bryant, C.; Senthilnath, J. Integrated phenology and climate in rice yields prediction using machine learning methods. *Ecol. Indic.* **2021**, *120*, 106935. [CrossRef]
2. Liu, Y.L.; Liu, X.Y.; Wang, X.; Gao, K.; Qi, W.W.; Ren, H.M.; Hu, H.R.; Sun, D.Y.; Bai, J.T.; Zheng, S.Z. Heterologous expression of heat stress-responsive AtPLC9 confers heat tolerance in transgenic rice. *BMC Plant Biol.* **2020**, *20*, 514. [CrossRef] [PubMed]
3. Xu, J.M.; Henry, A.; Sreenivasulu, N. Rice yield formation under high day and night temperatures-A prerequisite to ensure future food security. *Plant Cell Environ.* **2020**, *43*, 1595–1608. [CrossRef] [PubMed]
4. Kopecky, D.; Martin, A.; Smykal, P. Interspecific hybridization and plant breeding: From historical retrospective through work of Mendel to current crops. *Czech J. Genet. Plant* **2022**, *58*, 113–126. [CrossRef]
5. Gallois, P.; Lindsey, K.; Malone, R.; Kreis, M.; Jones, M. Gene rescue in plants by direct gene transfer of total genomic DNA into protoplasts. *Nucleic Acids Res.* **1992**, *20*, 3977–3982. [CrossRef]
6. Peng, Y.; Hu, Y.Y.; Mao, B.G.; Xiang, H.T.; Shao, Y.; Pan, Y.L.; Sheng, X.B.; Li, Y.K.; Ni, X.M.; Xia, Y.M.; et al. Genetic analysis for rice grain quality traits in the YVB stable variant line using RAD-seq. *Mol. Genet. Genom.* **2016**, *291*, 297–307. [CrossRef]
7. Sun, Y.D.; Hu, Y.Y.; Peng, Y.; Mao, B.G.; Wang, J.J.; Zhao, B.R. Phenotypic Difference and Gene Variation in Agronomic Traits between a Rice Variant with Introduced Exogenous Genomic DNA and Its Receptor. *Hybrid Rice* **2013**, *28*, 62–67.
8. Zhao, B.R.; Xing, Q.H.; Xia, H.A.; Yang, H.H.; Jin, D.M.; Liu, X.; Wang, S.W.; Wang, B.; Yuan, L.P. DNA polymorphism among Yewei B, V20B, and *Oryza minuta* J. S. Presl. ex C. B. Presl. *J. Integr. Plant Biol.* **2005**, *47*, 1485–1492. [CrossRef]
9. Ma, Y.F. Specific DNA Fragments Analysis of Eighteen Rice Variants Transformed with Maize DNA. Master's Thesis, Henan Normal University, Xinxiang, China, 2014.
10. Hu, Y.Y.; Mao, B.G.; Xia, Y.M.; Peng, Y.; Zhang, D.; Tang, L.; Shao, Y.; Li, Y.K.; Zhao, B.R. Spike-Stalk Injection Method Causes Extensive Phenotypic and Genotypic Variations for Rice Germplasm. *Front. Plant Sci.* **2020**, *11*, 575373. [CrossRef]
11. Liu, H.; Liang, F.M.; Zhang, D.J.; Xu, J.Y.; Li, Z.X. Genetic background, agronomic traits and yield of rice after introgression of *Fargesia spathacea* Franch. *J. South. Agric.* **2018**, *49*, 619–627.
12. Haque, M.A.; Rafii, M.Y.; Yusoff, M.M.; Ali, N.S.; Yusuff, O.; Datta, D.R.; Anisuzzaman, M.; Ikbal, M.F. Advanced Breeding Strategies and Future Perspectives of Salinity Tolerance in Rice. *Agronomy* **2021**, *11*, 1631. [CrossRef]
13. Galvez, L.C.; Banerjee, J.; Pinar, H.; Mitra, A. Engineered plant virus resistance. *Plant Sci.* **2014**, *228*, 11–25. [CrossRef]
14. Khush, G.S.; Brar, D.S. Alien introgression in rice. *Nucl. India* **2017**, *60*, 251–261. [CrossRef]
15. Wei, X.; Li, Y.F.; Xiang, J.; Ling, Y.H.; Yang, Z.L.; He, G.H.; Zhao, F.M. Effect of Drought Stress on Physiological Characters of Rice Variant Line R21 by Introgression Sorghum DNA. *Southwest China J. Agric. Sci.* **2015**, *28*, 930–934.
16. Su, D.Z.; Shi, S.Z.; Xiao, C.H.; Du, F.Y.; Liang, S.Y.; Sun, X.H.; Wang, Q. Utilization of Restorer R21 and Jiahui 825 in Rice Breeding. *China Rice* **2015**, *21*, 80–82.
17. Kawahara, Y.; de la Bastide, M.; Hamilton, J.P.; Kanamori, H.; McCombie, W.R.; Ouyang, S.; Schwartz, D.C.; Tanaka, T.; Wu, J.Z.; Zhou, S.G.; et al. Improvement of the *Oryza sativa* Nipponbare reference genome using next generation sequence and optical map data. *Rice* **2013**, *6*, 4. [CrossRef]
18. Qin, P.; Lu, H.W.; Du, H.L.; Wang, H.; Chen, W.L.; Chen, Z.; He, Q.; Ou, S.J.; Zhang, H.Y.; Li, X.Z.; et al. Pan-genome analysis of 33 genetically diverse rice accessions reveals hidden genomic variations. *Cell* **2021**, *184*, 3542. [CrossRef]
19. Rang, Z.W.; Zhou, Q.M. Physiological responses of heat-tolerant variety "Nagina 22" to high temperature stress in rice (*Oryza sativa* L.). *J. Plant Genet. Resour.* **2012**, *13*, 1045–1049.
20. Zhang, B.; Zhang, X.; Xu, G.Y.; Li, M.J.; Cui, Y.C.; Yin, X.M.; Yu, Y.; Xia, X.J.; Wang, M.L. Expression of sorghum gene *SbSGL* enhances grain length and weight in rice. *Mol. Breed.* **2018**, *38*, 40. [CrossRef]

21. Qin, K.Z.; Qiu, P.; Wen, J.Y.; Zhu, Y.G.; Li, N.W.; Li, S.Q. High throughput transformation of a Sorghum cDNA library for rice improvement. *Plant Cell Tiss. Org.* **2016**, *125*, 471–478. [CrossRef]
22. Prasetiyono, J.; Fatimah; Trijatmiko, K.R.; Sustiprijatno; Ma, S.; Nafisah; Supriyanta. Improvement of Inpari 30 and Situ Bagendit rice varieties for tolerance to drought through spike-stalk injection method. *IOP Conf. Ser. Earth Environ. Sci.* **2020**, *482*, 12029. [CrossRef]
23. Septiningsih, E.M.; Hidayatun, N.; Sanchez, D.L.; Nugraha, Y.; Carandang, J.; Pamplona, A.M.; Collard, B.; Ismail, A.M.; Mackill, D.J. Accelerating the development of new submergence tolerant rice varieties: The case of Ciherang-Sub1 and PSB Rc18-Sub1. *Euphytica* **2015**, *202*, 259–268. [CrossRef]
24. Fu, J.; Xu, Z.G.; Zou, W.X. On breeding rice through wide hybridization between rice (*Oryza sativa*) and sorghum (*Sorghum bicolor*). *J. Hunan Agric. Univ.* **1997**, *23*, 14–19.
25. China Rice Data Center. Available online: https://ricedata.cn/ (accessed on 20 September 2022).
26. Schadt, E.E.; Turner, S.; Kasarskis, A. A window into third-generation sequencing. *Hum. Mol. Genet.* **2010**, *19*, R227–R240. [CrossRef] [PubMed]
27. Shishido, R.; Sano, Y.; Fukui, K. Ribosomal DNAs: An exception to the conservation of gene order in rice genomes. *Mol. Gen. Genet.* **2000**, *263*, 586–591. [CrossRef]
28. Pendleton, M.; Sebra, R.; Pang, A.; Ummat, A.; Franzen, O.; Rausch, T.; Stutz, A.M.; Stedman, W.; Anantharaman, T.; Hastie, A.; et al. Assembly and diploid architecture of an individual human genome via single-molecule technologies. *Nat. Methods* **2015**, *12*, 780–786. [CrossRef] [PubMed]
29. Cheng, H.Y.; Concepcion, G.T.; Feng, X.W.; Zhang, H.W.; Li, H. Haplotype-resolved de novo assembly using phased assembly graphs with hifiasm. *Nat. Methods* **2021**, *18*, 170. [CrossRef]
30. Simão, F.A.; Waterhouse, R.M.; Loannidis, P.; Kriventseva, E.; Zdobnov, E.M. BUSCO: Assessing genome assembly and annotation completeness with single-copy orthologs. *Bioinformatics* **2015**, *31*, 3210–3212. [CrossRef]
31. Parra, G.; Bradnam, K.; Korf, I. CEGMA: A pipeline to accurately annotate core genes in eukaryotic genomes. *Bioinformatics* **2007**, *23*, 1061–1067. [CrossRef]
32. Li, H. Minimap2: Pairwise alignment for nucleotide sequences. *Bioinformatics* **2018**, *34*, 3094–3100. [CrossRef]
33. Nattestad, M.; Schatz, M.C. Assemblytics: A web analytics tool for the detection of variants from an assembly. *Bioinformatics* **2016**, *32*, 3021–3023. [CrossRef]
34. Danecek, P.; Bonfield, J.K.; Liddle, J.; Marshall, J.; Ohan, V.; Pollard, M.O.; Whitwham, A.; Keane, T.; McCarthy, S.A.; Davies, R.M.; et al. Twelve years of SAMtools and BCFtools. *GigaScience* **2021**, *10*, b8. [CrossRef]
35. Altschul, S.F.; Gish, W.; Miller, W.; Myers, E.W.; Lipman, D.J. Basic local alignment search tool. *J. Mol. Biol.* **1990**, *215*, 403–410. [CrossRef]
36. Ouyang, S.; Buell, C.R. The TIGR Plant Repeat Databases: A collective resource for the identification of repetitive sequences in plants. *Nucleic Acids Res.* **2004**, *32*, D360–D363. [CrossRef]

MDPI AG
Grosspeteranlage 5
4052 Basel
Switzerland
Tel.: +41 61 683 77 34

International Journal of Molecular Sciences Editorial Office
E-mail: ijms@mdpi.com
www.mdpi.com/journal/ijms

www.ingramcontent.com/pod-product-compliance
Lightning Source LLC
Chambersburg PA
CBHW061458220926
51850CB00004B/121
* 9 7 8 3 7 2 5 8 1 4 9 8 5 *